T0313880

History of Exercise Physiology

Charles M. Tipton, PhD
University of Arizona
EDITOR

Human Kinetics

Library of Congress Cataloging-in-Publication Data

History of exercise physiology / Charles M. Tipton, editor.
p. ; cm.
Includes bibliographical references and index.
I. Tipton, Charles M., 1927- editor of compilation.
[DNLM: 1. Exercise--physiology. 2. Physiology--history. 3. History of Medicine. 4. Physical Exertion--physiology. QT 11.1]
QP321
612'.044--dc23

2013038987

ISBN-10: 0-7360-8369-3
ISBN-13: 978-0-7360-8369-0

Acquisitions Editor: Myles Schrag; **Developmental Editor:** Amanda S. Ewing; **Assistant Editors:** Casey A. Gentis and Anne E. Mrozek; **Copyeditor:** Amanda M. Eastin-Allen; **Indexer:** Susan Danzi Hernandez; **Permissions Manager:** Dalene Reeder; **Cover Designer:** Keri Evans; **Photograph (cover):** Reproduced, with permission, from A. Krogh and J. Lindhard, 1920, *Biochemical Journal* 14: 290-363. © the Biochemical Society. **Photo Production Manager:** Jason Allen; **Art Manager:** Kelly Hendren; **Associate Art Manager:** Alan L. Wilborn; **Printer:** Sheridan Books

The image on the cover is the cycle ergometer and Jacquet-type respiration chamber used by August Krogh and Johannes Lindhard in 1920 to measure respiratory gas exchange using an open-circuit approach. For more information on their study, see chapter 17.

Printed in the United States of America 10 9 8 7 6 5 4 3 2 1

The paper in this book is certified under a sustainable forestry program.

Human Kinetics
Website: www.HumanKinetics.com

United States: Human Kinetics
P.O. Box 5076
Champaign, IL 61825-5076
800-747-4457
e-mail: humank@hkusa.com

Canada: Human Kinetics
475 Devonshire Road Unit 100
Windsor, ON N8Y 2L5
800-465-7301 (in Canada only)
e-mail: info@hkcanada.com

Europe: Human Kinetics
107 Bradford Road
Stanningley
Leeds LS28 6AT, United Kingdom
+44 (0) 113 255 5665
e-mail: hk@hkeurope.com

Australia: Human Kinetics
57A Price Avenue
Lower Mitcham, South Australia 5062
08 8372 0999
e-mail: info@hkaustralia.com

New Zealand: Human Kinetics
P.O. Box 80
Torrens Park, South Australia 5062
0800 222 062
e-mail: info@hknewzealand.com

E4847

Contents

Preface

Exercise physiology is a dynamic and exciting scientific discipline for the 21st century. However, the history of exercise physiology is incomplete and fragmentary despite concerted efforts by distinguished authors (1-5, 8) and me (6, 7). This situation exists not because publishers lack interest in the subject but rather because they have concerns about the profitability of publishing a text on the subject. Hence, Dr. Rainer Martens and Human Kinetics are commended for publishing a comprehensive and complete exercise physiology text that can serve as an important reference for students, instructors, investigators, historians, librarians, and interested individuals.

The information presented in each chapter of this text is by one or more nationally and internationally recognized authors. Part I begins with the antiquity and evolution of exercise physiology from two river civilizations and those from ancient Greece and Rome. It acquaints the reader with early laboratories in Denmark and Germany and at Harvard University as well as with the contemporary Copenhagen Muscle Research Centre. Part I also emphasizes the emergence of exercise physiology in the United States from physical education classrooms and laboratories into a scientific discipline. It also contains a chapter that describes two developments in the United States that solidified exercise physiology as a scientific discipline as well as a chapter that details the relationship between exercise physiology and the new, unknown world of genomics.

Part II also features chapters by nationally and internationally known investigators. Part II, which picks up after the discipline of exercise physiology has been established, emphasizes the acute and chronic effect of exercise on multiple physiological bodily systems. Although some information in this part is repetitious, this practice is not detrimental to the acquisition of physiological knowledge or understanding. Most chapters in this part contain a table that highlights significant research milestones (findings) since 1910.

The authors of this text were selected based on their scientific knowledge and expertise. The same approach was followed when selecting the reviewers: Each chapter was evaluated by me and a recognized external expert. Individuals who graciously agreed to evaluate manuscripts are Robert B. Armstrong, David R. Bassett Jr., Jack W. Berryman, Susan A. Bloomfield, Katarina T. Borer, G. Edgar Folk Jr., Phillip Gardiner, L. Bruce Gladden, Laurie Hoffman-Goetz, Edward T. Howley, Timothy L. Musch, Stephen M. Roth, Michael N. Sawka, Bengt Saltin, Xiaocai Shi, Charles E. Wade, and Hugh G. Welch.

The authors of these chapters did not contribute simply for the modest honorarium or the limited acclaim from their professional societies. Rather, they contributed because they believe it is essential for physiologists to understand the past before approaching the future.

Finally, acknowledgement is extended to Myles Schrag of Human Kinetics, who served as acquisitions editor. His patience, guidance, and wisdom were essential in bringing this text to completion and are deeply appreciated.

Charles Tipton

References

1. Berryman JW. Ancient and early influences. In: Tipton CM, ed. *Exercise Physiology: People and Ideas*. New York: Oxford University Press, 2003, 1-38.
2. Brooks GB, Fahey TD, White TP, Baldwin KM. *Exercise Physiology*. 3rd ed. Mountain View, CA: 2000.
3. McArdle WD, Katch FI, Katch VL. *Exercise Physiology*. 5th ed. Philadelphia: Lippincott Williams & Wilkins, 2001.
4. Powers SK, Howley ET. *Exercise Physiology*. 5th ed. Boston: McGraw-Hill, 2004.
5. Robergs RA, Roberts SO. *Exercise Physiology*. St. Louis: Mosby, 1997.
6. Tipton CM. Historical perspective: Origin to recognition. In: Tipton CM, ed. *ACSM's Advanced Exercise Physiology*. Philadelphia: Lippincott Williams & Wilkins, 2006, 11-38.
7. Tipton CM. Part II: A contemporary historical perspective. In: Massengale JD, Swanson RA, eds. *The History of Exercise and Sport Science*. Champaign, IL: Human Kinetics, 1997.
8. Wilmore JH, Costill DL, Kenney WL. *Physiology of Sport and Exercise*. 4th ed. Champaign, IL: Human Kinetics, 2008.

Contributors

P.-O. Åstrand, MD
Department of Physiology
Karolinska Institute
Stockholm, Sweden

Kenneth M. Baldwin, PhD
Department of Physiology and Biophysics
School of Medicine
University of California
Irvine, California, USA

Katarina T. Borer, PhD
School of Kinesiology
University of Michigan
Ann Arbor, Michigan, USA

Claude Bouchard, PhD
Human Genomics Laboratory
Pennington Biomedical Research Center
Baton Rouge, Louisiana, USA

George A. Brooks, PhD
Exercise Physiology Laboratory
Department of Integrative Biology
University of California
Berkeley, California, USA

Andrew R. Coggan, PhD
Cardiovascular Imaging Laboratory
Mallinckrodt Institute of Radiology
Washington University School of Medicine
St. Louis, Missouri, USA

V. Reggie Edgerton, PhD
Department of Integrative Biology and Physiology
Brain Research Institute
University of California, Los Angeles
Los Angeles, California, USA

Peter A. Farrell, PhD
Department of Kinesiology
East Carolina University
Greenville, North Carolina, USA

G. Edgar Folk Jr., PhD
Department of Molecular Physiology and Biophysics
University of Iowa
Iowa City, Iowa, USA

Henrik Galbo, MD, DMS
Department of Infectious Diseases and Rheumatology
Rigshospitalet
University of Copenhagen
Copenhagen, Denmark

Phillip F. Gardiner, PhD
Spinal Cord Research Center
Health, Leisure, and Human Performance Research Institute
University of Manitoba
Winnipeg, Canada

Grant C. Goulet, PhD
Assistant Research Scientist
School of Kinesiology
University of Michigan
Ann Arbor, Michigan, USA

Fadia Haddad, PhD
Department of Physiology and Biophysics
School of Medicine
University of California
Irvine, California, USA

Ylva Hellsten, DMSc
Department of Nutrition, Exercise and Sports
University of Copenhagen
Copenhagen, Denmark

Wildor Hollmann, MD
Institute for Cardiology and Sports Medicine
Cologne, Germany

Michael Kjaer, MD, DMSc
Institute of Sports Medicine, Bispebjerg Hospital
Faculty of Health and Medical Sciences, University of Copenhagen
Copenhagen, Denmark

G. Patrick Lambert, PhD
Department of Exercise Science
Creighton University
Omaha, Nebraska, USA

M. Harold Laughlin, PhD
Biomedical Sciences
Medical Pharmacology and Physiology
Dalton Cardiovascular Research Center
University of Missouri
Columbia, Missouri, USA

Benjamin D. Levine, MD
Department of Internal Medicine, Cardiology
University of Texas Southwestern Medical Center
Dallas, Texas, USA
and
The Institute For Exercise and Environmental Medicine
Presbyterian Hospital
Dallas, Texas, USA

Robert M. Malina, PhD
Department of Kinesiology and Health Education
University of Texas at Austin
Austin, Texas, USA

Sarah L. Manske, PhD
McCaig Institute for Bone and Joint Health
University of Calgary
Calgary, Alberta, Canada

Jere H. Mitchell, MD
Department of Internal Medicine, Cardiology
University of Texas Southwestern Medical Center
Dallas, Texas, USA

Pope Moseley, MD
Department of Internal Medicine
University of New Mexico
Albuquerque, New Mexico, USA

P. Darrell Neufer, PhD
East Carolina Diabetes and Obesity Institute
Departments of Physiology and Kinesiology
East Carolina University
Greenville, North Carolina, USA

Jaume Padilla, PhD
Biomedical Sciences
University of Missouri
Columbia, Missouri, USA

Jacques R. Poortmans, PhD
Professor Emeritus
Faculty of Motor Sciences
Free University of Brussels
Brussels, Belgium

Peter B. Raven, PhD
Department of Integrative Physiology
North Texas Health Science Center
Fort Worth, Texas, USA

Bruno Roseguini, MS
Biomedical Sciences
University of Missouri
Columbia, Missouri, USA

Suzanne Schneider, PhD
Department of Health, Exercise and Sports Sciences
University of New Mexico
Albuquerque, New Mexico, USA

Roy J. Shephard, MD, PhD, DPE, LLD (Hon. Caus.)
Faculty of Kinesiology & Physical Education
University of Toronto
Toronto, Canada

Grant H. Simmons, PhD
Biomedical Sciences
University of Missouri
Columbia, Missouri, USA

Peter G. Snell, PhD
Department of Internal Medicine, Cardiology
University of Texas Southwestern Medical Center
Dallas, Texas, USA

Susan A. Ward, MA, DPhil
Human Bio-Energetics Research Centre
Crickhowell, United Kingdom

Brian J. Whipp, PhD
Human Bio-Energetics Research Centre
Crickhowell, United Kingdom

Edward J. Zambraski, PhD
Division Chief
U.S. Army Research Institute of Environmental Medicine
Natick, Massachusetts, USA

Ronald F. Zernicke, PhD, DSc
School of Kinesiology and Department of Orthopaedic Surgery
University of Michigan
Ann Arbor, Michigan, USA

PART I

Antiquity, Early Laboratories, and Entering the 21st Century

CHAPTER 1

Antiquity to the Early Years of the 20th Century

Charles M. Tipton, PhD

Introduction

For humans, the history of exercise began with the evolutionary transformation from a quadrupedal to an upright bipedal posture that continued with the migration of *Homo sapiens* from the plains of Africa to the banks of the Tigris, Euphrates, Nile, and Yellow Rivers and to the establishment of the river civilizations (99, 106). During this time, humans transitioned from being hunter–gatherers to becoming members of an agrarian society. A constant factor during these eons of time was the ravages of disease and its consequences. Emerging from these circumstances was a concern for health and a primitive template for the physiology of disease.

Influences From Select River Civilizations

According to Baas, a medical historian, antiquity ends with the death of Galen but can begin at any earlier time depending on the era of the event (1). For this presentation, the term *antiquity* relates to the time period between 3000 BCE and 200 CE and pertains to river and adjacent civilizations that include individuals from Assyria, Babylonia, China, Egypt, Mesopotamia, and Persia (figure 1.1). Although their contributions to architecture, art, literature, philosophy, and science were disparate, all civilizations sought benevolence from supernatural entities as protection against devils, evil spirits, demons, curses, or anger of the gods, which most believed caused disease and health disorders. Appeasement of the gods was also facilitated by prayers, incantations, ceremonies, trephination of the skull, and kneading the body in the direction of the feet in order for the spirit to escape (52, 159, 174).

Indus River Valley Civilizations

The existence of an Indus River Valley civilization was unknown until the archeological excavations of Sir John Marshall during the 1920s (107). Carbon dating results from Mohenjo-Daro indicated the existence of an Indus River Valley civilization dating from 3300 BCE that was contemporary to the one in Mesopotamia (161). The excavations revealed an advanced culture with major concern for sanitation and public health as demonstrated by a public water and sewage-drainage system, massive public-bathing buildings, freshwater tanks, soakage pits for sewage, and private baths and lavatories (29, 107). Skeletal remains found during the excavation revealed a wide range of diseases and disorders, including metal poisoning, arteriosclerosis, osteomyelitis, cancer, and infectious diseases. Trephined skulls were also found, suggesting that epilepsy and related brain disorders were present (29, 107).

During the millennia that followed, the Hindu culture was established and the sacred texts containing the cultural and religious beliefs, medical concepts, and health practices of India were written. The oldest, the Rig Veda (1500 BCE), contained 1,028 hymns that sought the protection of the gods against disease. Such protection included the elimination of devils and demons, removal of curses and evil spirits of ancestors or enemies, and the preservation of the three humors of the body (83, 159). The authors and origin of the humoral (dosas) theory are unknown, but the theory was known to the medical community between 1500 and 800 BCE (94). It was identified as the tridosa doctrine, and its primary intent was to explain the phenomena of disease and health and of life and death (136). Inherent in the theory is that the elements of water, fire, air, earth, and ether were responsible for the formation of the human body. Originally interacting with the body were three nutrient substances, known as dosas (humors), that were derived from the wind, sun, and moon: vayu (or vata), pitta, and kapha (94, 136). These

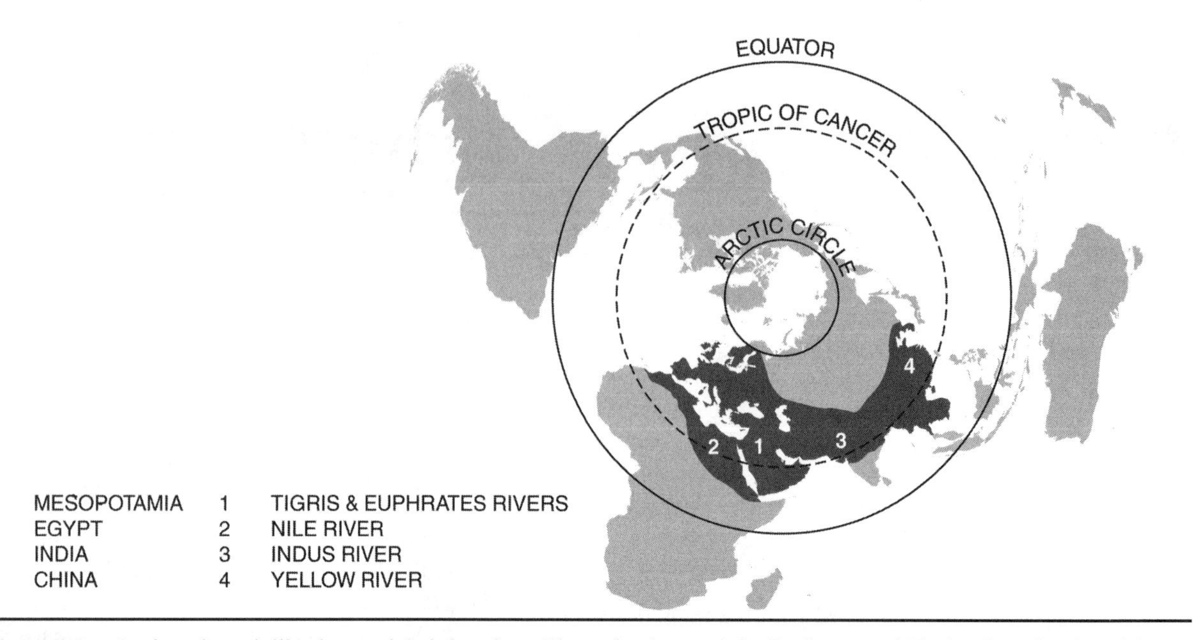

Figure 1.1 Ancient river civilizations and their locations. The region between the Euphrates and Tigris Rivers has been identified with the Garden of Eden. The emergence and development of the various civilizations was not systematic or sequential. However, not all river civilizations made significant contributions to the history of exercise physiology.

Reprinted, by permission, from G. Majno, 1991, *The healing hand* (Cambridge, MA: Harvard University). Redrawn, by permission, from Bishop, 1939, p. 46, with permission from the American Oriental Society.

dosas were associated with air, bile, and phlegm, respectively. According to the tridosa doctrine, these dosas controlled and regulated all bodily functions and had to be in equilibrium for health to be present. Because changes in dosa concentrations did not occur spontaneously, it was essential not to displace them because displacement meant illness and disease. Besides known diseases, conditions that could displace dosas were climatic diseases, select foods, poisons, fatigue, psychic changes, and heavy exercise (94, 136). Medical historians agree that in ancient India a physician named Susruta existed who gained prominence for his knowledge and writings on surgery, urology, ophthalmology, and obstetrics (161) (figure 1.2). Susruta defined exercise as a sense of weariness of the body that should be taken every day and was the first physician to prescribe exercise for health reasons (161). He advocated exercising (walking, running, jumping, swimming, diving, participating in games, and wrestling) to maintain equilibrium among the dosas and prevent the kapha dosa from being increased. Regular moderate exercise (to the point of labored breathing) was promoted to produce training effects (table 1.1), whereas heavy exercise was opposed because it altered the equilibrium among the dosas resulting in consumption, cachexia, asthma, vomiting, and fever. Obesity, a disease of the vayu dosa, was caused by a sedentary lifestyle whereas diabetes, a disorder of the urinary tract, was alleviated by exercise, dietary changes, and, apparently, riding an elephant (161).

Around 200 BCE, the physician named Caraka (Charka) became prominent for his teaching and writings in medicine. He also proclaimed that disease was the result of a disturbance of the equilibrium be-

Figure 1.2 Susruta of India (ca. 600 BCE) was the first physician to advocate moderate exercise to restore equilibrium among the dosas (humors), maintain health, and minimize the problems of diabetes and obesity. The author considers Susruta to be the father of exercise physiology.

Reprinted from C.M. Tipton, 2008, "Susruta of India: An unrecognized contributor to the history of exercise physiology," *Journal of Applied Physiology* 104: 1553-1556. Used with courtesy from American Physiological Society.

Table 1.1 Influence of Chronic Exercise, According to Susruta of India (600 BCE)

Promoted the growth of limbs and enhanced muscle development
Increased the strength, endurance, tone, and stoutness of muscles
Reduced body corpulence
Improved the digestive process
Delayed the occurrence of muscular fatigue
Increased perspiration during exercise
Improved bodily appearances
Improved mental alertness, retentive memory, and keen intelligence
Provided resistance against disease and physical decay

Additional information on the effects of training can be found in references 159 and 161.

tween the dosas and that health occurred when equilibrium had been established. Physical activity was advocated to prevent an increase in the kapha humor and as a cure for diabetes. Caraka was against excessive exercise because it caused dyspnea, heart disease, gastrointestinal disorders, and, potentially, death (147, 148).

Yellow River Civilization

Contrary to the impression created by some medical historians, the Yellow River civilization was contemporary to those located on the banks of the Nile, Tigris, and Euphrates Rivers. The health of the ancient Chinese was determined by whether the matrix of demons, devils, and spirits had gained possession of the body. If successful, disease occurred. With time, there was one devil for every disease. Cures or recoveries were attributed to the gods (2).

The authenticity of records is uncertain and controversial. The Shang dynasty, which existed between 1800 and 800 BCE, was the first dynasty to maintain useful records (167). Many of the listed illnesses were regarded as being the result of evil wind spirits and other diseases mentioned within the records were considered to be the consequences of curses from ancestors or punishments for previous sins (167). Recovery necessitated prayers, incantations, astrology, and, frequently, breathing exercises. Gordon (52) and returning missionaries (25) cite 2500 BCE for the practice of breathing exercises; however, Unschuld has effectively challenged this date (167). The Chou dynasty (1050-256 BCE) is associated with the introduction of the yin–yang doctrine, which states that the human body is composed of three parts yin and three parts yang. Yang was associated with health and life, whereas yin was identified with disease. Yang and yin were each associated with select diseases and demonstrated mutual affinities and antagonism toward each other. Equilibrium had to exist before harmony or health could be present (126, 169). Stretching and deep-breathing exercises were regarded as yang exercises that promoted health because they stimulated energy flow and eliminated the bad air. A yin exercise was any activity that was performed at a high or strenuous intensity. Such exercises were disruptive to the equilibrium between the principles and were considered to be unhealthy (78).

During the Eastern Han dynasty (25-250 CE), Hua T'O (Hua Tuo) became recognized as a great surgeon (figure 1.3). He endorsed exercise for its yang effect and followed the movement patterns of deer, tigers, bears, monkeys, and birds (78, 180). His movement patterns have become identified with the Five Animal Frolics and with Daoyin healing exercises. Hua T'O advocated moderate exercise for his followers because it expelled the bad air, enhanced circulation, strengthened legs, improved appetite and digestion, and prevented sickness and old age (78, 180).

Influences From Civilizations in Greece (3000-200 BCE)

Archeologists have shown that two civilizations and one classical period were associated with ancient Greece. They were the Minoans (3000-1100 BCE), the Mycenaeans (1350-1050 BCE), and the Greeks living during the fourth and fifth centuries, respectively (13).

Figure 1.3 Hua T'O of China (ca. 25-220 CE) was a physician who endorsed exercise because its yang effect promoted health.

Reprinted from C. Wong and W. Lien-The, 1936, *History of Chinese medicine,* 2nd ed. (Shanghai: National Quarantine Service).

The two tribes of importance were those that migrated to what became the city-states of Sparta and Rome. Although the dialects and social structure of these city-states differed, the inhabitants of each accepted disease and illness as punishment from the gods and believed that healing and recovery required benevolence from other gods (150).

Influences From Sparta

The Dorians who settled in the city of Sparta established a governance that was markedly different from what the Ionians developed when they settled Athens. The Dorians had an oligarchic form of government that suppressed sedition, required obedience to the state, and expected citizens to be prepared for war at all times. Disease and illness were in the realm of the gods, and health was defined as being fit for combat. Unhealthy and ill babies were taken to the Apothetae to die. Males were expected to be warriors and females were expected to be fit mothers of warriors. Strenuous and combative exercises and athletics were encouraged and focused on making Spartans better warriors. Festival games that required athletes to endure pain and perform feats of endurance were featured events. There was little public concern if the participants suffered injuries or death. Only in Sparta did females compete against males, especially in wrestling. The Spartans made meaningful contributions to the importance of physical training, the value of athletics, and the development and maintenance of physical fitness, especially for the purposes of war. However, their contributions to exercise physiology were minimal (77, 118).

Select Influences of Greek Philosophers and Physicians Before Hippocrates

Thales (639-544 BCE), who founded a school at Miletus, believed that water was the basic element in plant and animal life and the source of earth and air (7). Anaximander (611-547 BCE) believed that humans and other living creatures had their beginnings in water and that the universe existed because of a balance between opposing forces (87). Anaximenes (610-545 BCE) held the view that air was divine, capable of stimulating blood and the heart, and responsible for substance, motion, and life (170). Pythagoras (570-490 BCE) founded a school that emphasized philosophy and science. He expounded a complex philosophy concerning man, God, and the universe in which the universe consisted of the elements of fire, earth, water, and air (similar to the tridosa doctrine mentioned previously) and possessed the qualities of moisture, dryness, heat, and cold. He did not believe that gods were the causes of disease but rather that disease and illness were the result of dissolute behavior and lack of harmony (equilibrium) among the elements, qualities, and tendencies of the body. Health represented a state of harmony and could be enhanced by improving dietary habits, taking long walks, and participating in activities such as running, wrestling, discus throwing, and boxing (170). Alcmaeon (ca. 500 BCE), a graduate of the Pythagorean School, believed that health represented a harmonious equilibrium among the qualities of wet, dry, hot, bitter, sweet, and so on and that disease occurred when any one quality attained supremacy. Factors or conditions that could provoke diseases included an excess of either heat or cold, a lack or excess of food, physical fatigue, phlegm from the head, black and yellow bile from the blood, substances from bone marrow or water, or personal hardships. To elicit cures, it was necessary to know what elements or qualities were responsible (52, 135).

Empedocles (504-443 BCE), a disciple of Pythagoras, stated that all matter contained elements of water, earth, fire, and ether [a view similar to what the Hindu and Chinese believed (159)] and that transformation was possible without origination or destruction (96, p. 102). These changes were as follows (48, p. 77):

hot + dry = fire	cold + dry = earth
hot + wet = air	cold + wet = water
hot + wet = blood	cold + wet = phlegm
hot + dry = yellow bile	cold + wet = black bile

Inherent in this theory was that disease would occur with any disturbance of the relationships between these elements and that health was present when these elements were in equilibrium (126).

Hippocrates, the Humoral Doctrine, and His Contributions to the Exercise Physiology and Exercise Prescription

Hippocrates (460-370 BCE) (figure 1.4) is associated with at least 76 texts related to Greek medicine that are collectively known as the *Corpus Hippocraticum*. Because some of the information is contradictory, redundant, and confusing, scholars cannot ascertain which information is authentic. Besides his contributions to exercise physiology and exercise prescription, few historians doubt that he was responsible for separating medicine from philosophy, isolating medicine from religion and magic, and the emergence of rational medicine and that he was the father of scientific medicine (48, 52). He also has been regarded by many historians as the father of the humoral doctrine (figure 1.5). However, this distinction ignores the contributions of Pythagoras, Empedocles, and Alcmaeon.

Hippocrates accepted the concept that the human body was a composite of the four elements and that four qualities had to be in perfect equilibrium to prevent disease and maintain health. Specifically, diseases could emerge when an imbalance existed among yellow bile, black bile, phlegm, or blood, which had been categorized as humors (123). Changes in seasons and other climatic factors could also elicit diseases (66, 67). These collective relationships can be seen in figure 1.5. However, not all of Hippocrates' texts are consistent with the identification of the humors; different combinations were mentioned in *Diseases IV, Ancient Medicine* and *Affection I.*

Hippocrates believed that nature (physis) could cure disease and that health could be acquired by establishing equilibrium between the respective humors. This in turn provided a rationale for the treatment and management of disease. He felt that exercise was necessary in the treatment and management of humoral diseases and wrote, "Collected humor may grow warm, become thin, and purge itself away" (68, p. 363). Concerning the need for proper food and exercise in maintaining a healthy state, he stated, "Food alone will not keep a man well, he must also take exercise. For food and exercise, while possessing opposite qualities, yet work together to produce health" (68, pp. 227-229). He also believed that exercise would aid digestion. Exercise to Hippocrates meant walking, running, wrestling, swinging the arms, push-ups, shadow boxing, and ball punching at a moderate intensity (98). Exercise required warming up, and excessive exercise was to be avoided because illness, disease, or even death could result. Inactivity and idleness were potential causes for disease, and on these topics he proclaimed, "Those due

Figure 1.4 Hippocrates of Cos (ca. 460-370 BCE), a physician, is acknowledged as the father of medicine and the major advocate of the humoral theory. He believed that exercise was essential for maintaining equilibrium among the humors and was the first to provide a detailed exercise prescription for health.

Reprinted from C. Singer, 1922, *Greek biology and Greek medicine* (Oxford: Oxford University Press).

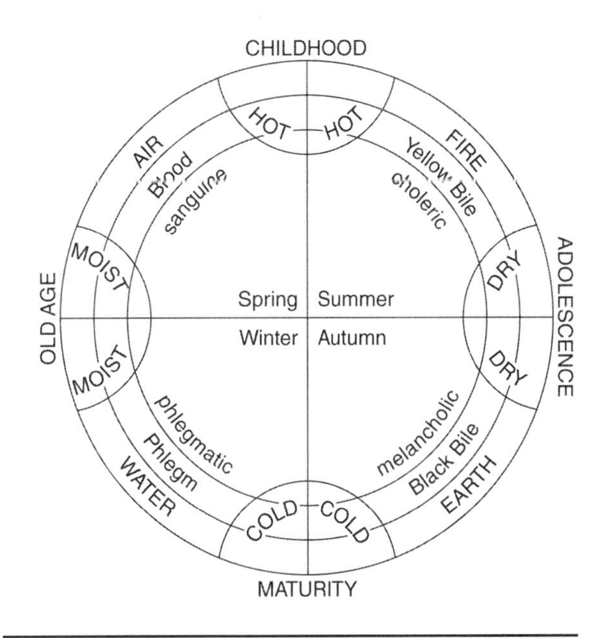

Figure 1.5 Berryman diagram of the various components and stages of the humoral theory.

Reprinted by permission of Jack Berryman, Ph.D. of the University of Washington in Seattle, Washington.

to exercise are cured by rest, and those due to idleness are cured by exercise" (69, p. 25).

Although Susruta advocated exercise for disease conditions, Hippocrates was the first to prescribe the amount of exercise to perform. For instance, if consumption—regardless of whether it was caused by phlegm, bile, or blood in the spinal marrow—had been diagnosed, the treatment regimen would include the prescribed exercise along with select food and liquids, medication, vapor baths, and sleep. As detailed in *Disease III and Internal Afflictions* (70), a patient with consumption should during the first month walk 20 stades (3.70 km) the first day and subsequently increase the distance by 5 stades (0.93 km)/d until a total distance of 100 stades (18.5 km)/d had been achieved. This daily distance was maintained until 30 d had elapsed. Because treatment was expected to require a year, Hippocrates' exercise prescription included implicit instructions on how the exercise should progress during that time (70).

Role and Influence of Physicians, Trainers, and Gymnasiums

Music and gymnastics were the core of Greek education. The Greeks believed music trained the mind while gymnastics trained the body (47). Education was voluntary but favored the wealthy. It began at age 7 yr and continued until the individual was 14 yr, when it ended for the poorer class. The Greek male devoted portions of each day to the palaestra (a building that served as a wrestling school and that had rooms for undressing, bathing, and viewing athletes) and in the gymnasium (an area where exercise was performed). The gymnasium contained a running track, a palaestra, and areas where athletes could exercise and train. Education included lectures from mathematicians, rhetoricians, and grammarians and daily practice in the palaestra and gymnasiums until students were 17 or 18 yr old (47). The palaestra was a public institution. When not required for school purposes it was used by older pupils and was the site of several dialogues by Plato.

Gymnasiums, which were dedicated to Apollo, were established in both Sparta and Athens. However, they achieved prominence and importance during the eras of Hippocrates and Plato (38). According to Park, physicians were associated with gymnasiums and had duties that included preparing medical prescriptions, performing venesections and massages, reducing dislocations, and advising trainers (gymnasts) on the execution of gymnastics exercises (128). Such exercises included running for speed and endurance, wrestling, boxing, jumping, throwing the javelin or discus, dancing, walking for long distances, and playing games with balls (38). Also associated with gymnasiums were coaches (paidotribes) who interacted with the athletes. As a group, gymnasium physicians promoted exercise over conventional medical treatments to restore health (125).

By the time of Hippocrates, many athletes had professional trainers whose primary role was to improve skill acquisition and dietary practices (47). Trainers began to receive recognition in the era of Hippocrates, but their prominence came when Galen's influence was supreme. There is evidence that ancient trainers advised their athletes to use repetitions to acquire skill and endurance, overload to increase strength (e.g., bending wrought-iron rods; lifting heavier animals, stones, or boxes; wearing heavier halters during jumping exercises; and throwing heavier javelins), chasing animals to enhance speed, and running in the sand for long distances to improve performance in aerobic activities (47, 59, 117, 181). Insight into the practices of trainers was revealed by the writings of Lucius Flavius Philostratus II (170-244 CE), a Sophist philosopher who lived in Greece before moving to Rome. He was of the opinion that training could be more meaningful and vigorous if the trainers had more knowledge about individual differences and diet while advising rest, sunbaths, and using weights when jumping. He argued against the rigid tetrad training system in which the first day of training was devoted to preparation, the second to being intensely engaged to fatigue, the third to recreation, and the fourth to skill attainment and moderate exertion (181). However, Hippocrates, Plato, and Galen had minimum respect for trainers because the trainers lacked current scientific information.

Influences From the Roman Republic and Empire (900 BCE-200 CE)

Archeological investigations indicate that ancient tribes settled the hills of Rome around 900 BCE. Like other ancient tribes, these tribes believed that spirits existed in animate and inanimate objects, which had become the objects of their worship. With time, their worship was directed toward gods and goddesses. They believed that disease was a result of divine displeasure and used prayers and ceremonies to seek benevolence and forgiveness (144). They accepted the gods and goddesses of Greece with Greek medicine being introduced to Rome in 219 BCE. The first physician, Archagathus of

Peloponesia, was labeled a butcher and was encouraged to leave. However, he did prescribe exercise for patients with dropsy and consumption (38).

Influences From the Roman Army and Gladiators

The transformation of the Roman Republic into the Roman Empire (510 BCE to 161-180 CE) was due, in part, to the organization and effectiveness of the Roman Army (173). According to Vegetius (168, p. 65), Roman military leaders considered that daily exercise in arms was more conducive to health than did physicians. The army expected its recruits to be healthy and adhere to a lifestyle that included reasonable and proper exercise, sleep and rest, plus a well-regulated diet (173). Once in the army, recruits were expected to be able to march 32 km in 5 h following a regular cadence and 38.4 km in 4 h at a faster pace. Drills were performed with swords, wickerworth shields, and spears that were much heavier than normal equipment in order to strengthen the arms (172, 173). In addition, recruits were instructed to perform drills at the same intensity as in combat and leaders were ordered to conduct long marches to fatigue the stragglers. Scipio, in the 210 AD war with Spain, adopted the tetrad system of training for his warriors (86).

The nadir of gladiator contests occurred during the reign of Augustus (27 BCE-14 AD), when there were an estimated 10,000 combatants (88). At first there were private schools for the preparation of gladiators, but in 100 CE gladiator preparation became the responsibility of the state. Schools were staffed with physicians, trainers, masseurs, bandage specialists, and tailors. Trainers had the responsibility of improving the gladiators' physical fitness for combat readiness. Consequently, they conducted repetitive and exhaustive drills at maximum intensity and used shields, swords, and spears that were heavier than required. Their most famous physician was Claudius Galenus, or Galen of Pergamon (125).

Influences of Claudius Galenus

According to West, the beliefs of Galen of Pergamon (129-210 AD) dominated medical and health practices for more than 1,400 yr (175) (figure 1.6). Galen accepted the element theory of Pythagoras and Alcmaeon and the humoral doctrine credited to Hippocrates. He regarded health as an unimpaired capacity to function that occurred because of good mixtures and proportions of the elements and their qualities (45) and believed that disease or illness occurred because of bad mixtures and their proportions (58). An excess or deficiency in the qualities—especially heat and cold or any of the four humors—was regarded as pathological disease or as a disease state (123).

Conditions responsible for health or disease were classified as natural (kata physin, or healthy), nonnatural (ou kata physin, or hygenic), or contrary (para physin, or diseased) (45). Included with the nonnatural conditions was exercise (12). Consistent with the Indian tridosa and the Greek humoral doctrine, health required that a normal balance be maintained among the body constituents. Exercise was important for achieving health, and lack or excess of exercise could cause illness (although Galen failed to quantify the relationship). For movement or gymnastics to be exercise it had to be sufficiently vigorous to cause an increase in breathing (45). *Work* and *exercise* were equivalent terms. Shadow boxing, leaping, discus throwing, ball activities, and rope climbing were regarded as exercise whereas digging, rowing, plowing, riding, fighting, walking, or running could be either work or exercise (45). Exercise was classified as slow, swift, atonic, vigorous, gentle, or violent (45). Ball exercises and running were examples of swift exercise; digging, climbing a rope, and lifting heavy weights were considered to be vigorous exercise; and throwing a discus and continuous jumping were classified as violent exercise (45). Galen advocated moderation, and his favorite exercises involved the small ball (46).

Galen and his contemporaries (Philostratus and Areteaus) recommend gymnastics and ball exercises for individuals recovering from illnesses or for patients with

Figure 1.6 Claudius Galenus, or Galen of Pergamon (129-200 CE), was a physician in Rome whose views on medicine, physiology, and exercise influenced philosophers, physicians, educators, and coaches for more than a thousand yr.

Reprinted from J.B. West, 1996, *People and ideas: Respiratory physiology* (Bethesda, MD: American Physiology Society), 141. Used with permission of American Physiology Society.

arthritis, dropsy, gout, depression, tuberculosis, vertigo, or epilepsy (45, 46, 100, 109). He reported that acute exercise would increase pulse rate, elevate ventilation and body warmth (temperature), promote seating, ready the metabolism, enhance digestion, improve the diffusion and acidity of fluids, clean out fine pores, enhance the evacuation of excrements, balance humors, and reduce the concentration so phlegm, yellow bile, and black bile in the blood (45, 46). Chronic exercise or training was associated with a reduced flesh (a thinner body), improved muscle strength, improved mass and tone (hard muscles), elevated blood volume, and achievement of the unqualified good condition associated with Milo the wrestler, Hercules, or Achilles (45, 46).

Galen had little respect for athletes because they lacked awareness of the natural good of their soul (intelligence) and because of their daily practices. He wrote:

> Because they are always occupied in the business of amassing flesh and blood, their souls are as if it was extinguished in a heap of mire, unable to contemplate anything clearly, mindless as beasts without reason. (46, p. 47)

To Galen, an athletic state was not a natural condition and had no practical value. Moreover, he believed that training cultivated disease rather than health. Because athletes were usually in peak physical condition, it was impossible for them to improve. Hence, their fate was deterioration (46).

From Galen to the 16th Century

This time period included the spread of Christianity, the establishment of the religion of Islam, the darkness of the Middle Ages, and the excitement of the Renaissance (1400-1700) as well as the rise, universal acceptance, and fall of the physiological and medical influences of Galen. The demise began with Paracelsus (1493-1541), who in 1527 informed medical students and the general public that he no longer followed the teachings of Galen or believe that changes in humors were responsible for the occurrence of diseases. In addition, he threw texts by both Galen and Avicenna into a bonfire (156). The demise heighted when Vesalius (1514-1564) made it known that he had rejected the anatomical concepts of Galen (111).

Influence of Avicenna

The global influence of Galen, a prolific author, was facilitated by the Arabic and Latin translations of his extensive publications. However, his influence was initially higher in the Islamic world because his concepts were accepted by a Persian physician named Abu-Ali al-Husayn Ibn-Sina, or Avicenna (980-1037), who included them in his more than 100 books that were distributed throughout Arabic communities. Avicenna's most renowned work was *Canon of Medicine* (53), which Osler identified as the most famous medical textbook ever written (126).

Avicenna incorporated Galen's physiological concepts pertaining to humors, elements, qualities, members, facilities, operations, and spirits into a complex system of principal members (heart, brain, liver) with each having a controlling influence on the functions of other organs. Inherent in each separate system were virtues (actions), operations (functions), and faculties (select characteristics of organs). There were three virtues (vital, animal, natural) associated with spiritus (vital spirit) for distribution throughout the body. The most relevant was the vital virtue, which was found in the heart and blood and distributed by the arteries. It was the source of innate heat, responsible for heart rate and respiration, and contained the spiritus that Hippocrates had labeled *pneuma* (53).

Avicenna promoted the exercise concepts of Galen and advocated moderate exercise as beneficial (a necessary factor with rest) for living because it balanced the body by expelling residues and impurities (93, p. 24). He also noted that moderate exercise (walking) would repel the bad humors. To Avicenna, exercise effects were dependent on the degree (intensity) and amount (frequency and duration) of exercise performed as well as the amount of rest taken. He cautioned his readers that excessive exercise could adversely affect the innate heat and lead to a state akin to death (53). A focal point was the production and consequences of innate heat, which he felt was responsible for the large and strong pulse observed during exercise. In addition, Avicenna emphasized that exercise was associated with elevated body temperatures (would make the body very hot), that perspiration occurred with physical exhaustion, and that exercise should stop when sweating ceased because an abundant sweat was a symptom of moist illnesses. He recommended chronic exercise for weak and undeveloped limbs and promoted breathing exercises for individuals with respiratory weakness (53). Individuals in geographical areas that were not receptive to or influenced by the writings of Avicenna followed the nonnatural concepts of Galen on matters that pertained to the effects of exercise.

During the second half of the 16th century (1553), a Spanish physician named Cristobal Mendez (1500-1561) published in Latin *Book of Bodily Exercise* (115). When translated, the text consisted of 82 pages containing 40 distinctive chapters that had 4 treatises: exercise and its benefits, the division of exercise, common exercises and which ones are the best, and the time convenient for exercise and its value. The contents were essentially Greek and Roman concepts concerning exercise and emphasized their importance for health. To be beneficial, exercise must be moderate, performed frequently, enjoyable, and continuous (intermittent exercise failed to consume and dissipate humors, causing them to leave by pores opened by the heat of movement), and be associated with a shortness of breath (caused by the increased heat in the heart and the need for more air via increased ventilation). The text extolled the health virtues of exercise and the need to avoid overexercising with several chapters devoted to the heat produced by exercise and its consequences. While noting that movement per se would increase body heat, he mentioned that the movements of exercise cause the blood, spirits, and bile to rub with the parts causing subtlety and lightness (115, p. 19). Although the text has historical significance, it had limited circulation and a minimal influence on the medical profession.

Select Features of the 17th Century

This era has been characterized as the age of the scientific revolution because experimentation replaced speculation, explanations became mechanistic, mathematics was introduced into the language of science, and measurements became a component of medicine. Jan Baptista Van Helmont (1580-1664) did not accept the concept that innate heat was responsible for life and advocated that fermentation in the left ventricle was the reason for animal heat. Moreover, Franciscus Sylvius (1614-1672) rejected the humoral theory of Hippocrates and believed that health was the result of the interactions between bodily acids and bases and their neutralization (104). Thus, in physiology and especially in exercise physiology, the 17th century was an era during which only the nonnatural concepts of Galen were accepted and followed (12). At the end of the century, few physicians advocated strenuous or violent exercise because they felt, as did Hippocrates and Galen, that it was unhealthy (158).

Cardiovascular System

William Harvey's (1578-1657) epic discovery of a continuous circulation in 1628 (62, 141) was the most important scientific event of the century (104). Harvey challenged Galen's concept of circulation and established that cardiac output and the distribution of blood to the periphery was dependent on the strength of the pulse, which meant that an increase in heart rate had to occur during exercise (62). Harvey appears to be the first to suggest that physical exercise (training) is associated with cardiac enlargement (i.e., a thicker, more powerful and muscular heart) (62).

William Lower (1631-1693) of Oxford University was a strong advocate of Harvey's concepts, especially the one pertaining to the heart as muscle. [Others who reported the heart to be a muscle were Leonardo da Vinci (1452-1519) and Niels Stensen (1638-1686)]. Unlike Hippocrates, Lower felt that the heart was not the source of heat production in the body and believed that the movement of the heart was caused by the inflow of spirits through the nerves (103). He also stated that during violent exercise (maximal intensity) the movements of the heart were accelerated in proportion to the amount of blood that was "driven and poured in great abundance into its ventricles as a result of movements of muscles" (103, p. 122). In addition, he stated that exercise increased blood flow to the brain and enhanced health.

Respiratory System

During this century, von Helmont reported that air was a composite of different gases whereas Robert Boyle (1627-1691) demonstrated that air was necessary for life and established the foundation for the gas laws (158). In 1674, John Mayow (1641 1679) of Oxford University published *Medico-Physical Works*, which clarified the results of Borelli on the mechanics of breathing by emphasizing that the lungs passively followed the movements of the thorax. Mayow was emphatic that respiration had no role in cooling the blood or promoting blood flow (108). In addition, he posited that breathing facilitated the contact between air and blood and enabled the transfer of niter particles to the blood, resulting in a reaction between the sulfurous and combustible particles, which elevated body temperature and changed the color of blood. When performed, violent exercise increased the frequency of breathing and produced an intense heat due to the effervescence created by the presence and number of nitroaerial particles (44, 108, 140).

Neuromuscular Systems

Giovanni Borelli (1608-1679) of Italy believed that a spirituous juice (*succus nerveus*) was carried by nerves to the brain and skeletal muscles and that the inflation and contraction of muscles to create sensations and limb movements because of its effervescent reaction with blood causing an inflation and contraction of muscles (43, 140). Stensen, a physician in Denmark, reported that skeletal muscles contracted in a geometric manner that decreased their length and increased their width (140). Near the end of the century, Johann Bernoulli (1667-1748) of Switzerland published his dissertation related to the mechanics and movement of muscles in which he described muscles as small machines (10). He employed differential calculus to explain muscle functions and described the contraction as follows:

> When the mind wishes that a limb of the body moves, some agitation of animal spirits occurs in the brain so that, by twitching the origin of some nerve, they shake the spirituous juice contained inside over its whole length and, because of the irritation of the origin of the nerve, the last droplet of nervous juice is driven out by the slight vibration of the other orifice. (10, p. 139)

To Bernoulli, muscular strength was related to the availability of the spirituous juice being released and tiredness (muscular fatigue) was related to the quantity of consumed spirits. On this matter he stated, "I also assume that carrying same weight at the same height during the same time consumes the same quantity of spirits" (10, p. 135). William Crome (1663-1684), an English physician, associated violent exercise with painful and stiff skeletal muscles that were influenced by inadequate blood flow, profuse sweating, and the removal of spirits. In fact, he felt that the presence of sweat on skeletal muscles was responsible for both the swelling and contraction of skeletal muscles (178, p. 122).

Temperature Regulation

As mentioned, Mayow believed that the elevated body temperature produced by muscle contractions was the result of the increased effervescence produced by nitro-aerial particles interacting during violent exercise (108). Another result of violent movements was the production of sweat (108). Sweat glands were first identified in 1664 by Steno, who reported that fluid (insensible perspiration and sweat) passed out of them (137). During the 17th century, there was a concerted effort by scientists and physicians to develop accurate instruments for securing objective measurements of bodily functions. Around 1617, Galileo Galilei added a scale to his thermoscope, making it a clinical thermometer (140). His student Santorio Santorio (1561-1636) of Padua, Italy, developed a clinical thermometer as well as a sensitive human-weighing scale (chair) to quantify insensible perspiration (138) and weight changes in humans that occur while sitting, sleeping, and eating and after urination, defecation, and exercise. Thus, Santorio became the first investigator to conduct metabolic balance studies (140, 158).

According to Santorio, if individuals with sound bodies did not perspire, this condition could be corrected by exercise (143, aphorism 34). He explained the origin of sweat as "the fluid evacuated by violent exercise . . . that it originated from uncoted [unheated] juices" (143, aphorism 1) and stated that to maintain a youthful face individuals should avoid excessive sweating in the heat. He also believed that violent exercise would reduce body weight, advance the aging process, and promote an early death (143, aphorism 19).

Finally, in 1664, Nicolas Steno (1638-1686) discovered the existence of sweat glands (*fonts sudores*), a finding that the ancient Greeks had postulated almost two millennia earlier (140).

Select Features of the 18th Century

The 18th century is known as the era of enlightenment because of the introduction of various contrasting philosophies and because of the efforts by several physicians to systematically arrange and organize physiological knowledge for the benefit of medicine (104, 158).

Respiratory and Metabolic Systems

The latter half of the 18th century is associated with the discoveries of respiratory gases. Joseph Black (1728-1799) rediscovered carbon dioxide, Carl W. Scheele (1742-1786) of Sweden reported the existence of oxygen ("fiery air"), and Joseph Priestly discovered oxygen but named it dephlogisticated air (140, 175). However, the laboratory and animal calorimetry studies of the brilliant French chemist Antoine Lavoisier (1743-1794) demonstrated that respirable air (he called it oxygine) entered the lung and left it as a chalky aeriform acid (carbonic acid due to CO_2 reacting with

Figure 1.7 Eighteenth century wooden sketch of the first experiment that measured oxygen consumption during exercise. Marie-Anne Lavoisier is depicted recording data. Seguin was the subject performing foot treadle exercise to produce work, and an unidentified physician was observing the health status of Seguin. The location of Lavoisier during the experiment is unknown.

Reprinted from D. McKie, 1972, *Antoine Lavoisier, scientist, economist, social reformer* (New York: Henry Schuman), 53.

Table 1.2 Oxygen Consumption Results From the 1789 Work Experiment of Lavoisier and Seguin

Experimental condition	Cubic lb/h of work	L/h	L/min	kg/min of work performed
Fasting and resting at 26 °C	1,210	24.002	0.400	0.000 (because no work was done)
Fasting and resting at 12 °C	1,344	26.660	0.444	0.000 (because no work was done)
Resting after food consumption	1,800-1,900	37.689	0.628	0.000 (because no work was done)
Work after fasting	3,200	63.477	1.058	1,469

Oxygen consumption was recorded in pounces; 1,000 pounces represented 18.8363 L/h. Work was performed using a foot treadle that lifted 7.343 kg to a height of 613 pieds; a pied represents 0.325 m. In a 15 min time period, a distance of approximately 200 m was accomplished. The food consumed was assumed to be Seguin's breakfast.

Data from Benedict and Cathart 1913, p. 6; Flint 1875, p. 138; Seguin and Lavoisier 1789, pp. 566-584.

H_2O) in almost equal volumes. In fact, he found that for every 100 g of oxygen absorbed in respiration, only 81 parts reappeared as carbonic acid gas (110). He also knew that oxygine reacted with blood and would change the color of blood to red (140, 175). From his experiments, Lavoisier concluded that respiration was a slow combustion process that occurred in the lungs between carbon and hydrogen, that it required oxygine, and that it resulted in the formation of heat and water. To determine the effect of rest, food consumption, and exercise on respiration he conducted experiments with his chemist colleague Seguin as the subject, his wife Marie-Anne Lavoisier as the recorder, and a physician to listen to heart sounds and record heart rates (146, 160) (figure 1.7). Seguin performed work on a foot treadle while sitting, while fasting, and after eating breakfast (table 1.2). Although they were wrong about the combustion process being confined to the lungs and used approximations for their results, these collective experiments were the first in humans to demonstrate the transformation of energy with special reference to the influences of external temperatures, digestion, and exercise (160). During this era, authors promoted the concept that exercise enhanced digestion by aiding the descent and evacuation of bowel contents (158).

Cardiovascular System

Although Harvey's discovery of the circulation of blood continued to be emphasized during the early decades of the century, John Foyer (1649-1734), an Oxford University physician, provided meaningful insights with his invention of an accurate pulse watch in 1707 (42). Foyer used pulse rates to classify individuals into various humoral states and to identify select disorders and diseases. For instance, individuals with pulse rates between 75 and 80 beats/min were classified as being hot in the first degree and were identified with a choleric disposition. This disposition was affected by hot air, seasons, baths, medicines, exercise, and retained excrements. Individuals with resting pulse rates between 60 and 65 beats/min were classified as being cold in the second degree and were associated with a melancholic disposition. According to Foyer, this specific classification involved the spleen, dystrophic tumors, and various states of cachexia (42). He considered a normal pulse rate to be between 70 and 75 beats/min with changes being the result of increased blood temperature and elevations in spirits. Exercise was known to increase pulse rates because of its effect on humoral rarefactions. Interestingly, a pulse rate of 140 beats/min was the highest value recorded. Moreover, if rates became higher, death could occur. Foyer reported a rate of 112 while walking and values of 76 and 90 before and after riding a horse, respectively (42).

Pulse rates were also reported by the Irish physician Bryan Robinson (1680-1746), who recorded a value of 64 beats/min for individuals in the recumbent position. The value changed to 68 beats/min when sitting, to 78 beats/min when standing, and 100 beats/min after walking 4 miles (6.4 km) in 1 h. Rates between 140 and 150 beats/min were observed when individuals ran as hard as possible (139).

Stephen Hales (1677-1761), an English clergyman of Teddingon who was labeled by Rowell as the father of hemodynamics (141), measured the blood pressure of animals after exercise and reported that exercise was responsible for a brisk circulation, an increased number of systoles of the heart, and an improved blood flow to the dilated and agitated lungs, stomach, and guts (intestines) (55, p. 88). Robinson reported that muscle blood flow was related to the force created by contracting muscles (139), whereas Lavoisier and Seguin made the observation that the quantity of oxygine consumed was proportional to the number of pulsations multiplied by the number of inspirations (74).

Neuromuscular Systems

Friedrich Hoffmann (1660-1742), a German systematist who influenced physiological thinking during this century, advocated that motion was essential for life and that all vital actions could be explained on a mechanical basis. However, particles were required. He believed that muscle contraction occurred because of a nervous fluid ("nervous ether") that originated in the blood and was distributed to the nerves. When this fluid came into contact with muscles, they dilated and initiated contraction (73, 140).

The Swiss scholar Albrect von Haller (1708-1777) was a student of Herman Boerhaave, professor at the University of Gottingen, and the author of the influential *First Lines of Physiology* (57). Like his mentor, von Haller, he emphasized that nervous tissue exhibited sensibility (sensed pain) whereas skeletal and cardiac tissues represented irritability. Consequently, when skeletal muscles were repeatedly stimulated "they will raife a weight egual to, or much greater than, that of the whole human body itfelf" (57, p. 238) and "grow ftrong with exercife while their 'brawny' parts become thicker" (57, p. 243; in this quote, "f" is pronounced as a "s" and "g" is pronounced as a "q").

The Irish physician Bryan Robinson felt that the forces that caused limb movements were controlled by the will and acted on nerves. To him, nerves were the principal instruments of sensation and motion (139, p. 91). He advocated moderate exercise to increase muscle strength and size and acknowledged that laboring (trained) individuals had larger and stronger muscles than did sedentary individuals. During this time, James Keill (1673-1719), an English physician, made the observation that muscle strength was related to the number of fibers present (85). Around 1760, John Theophilus Desaguliers (1683-1744), an English priest, curator, and admirer of strongmen, helped to develop a dynamometer that was used to accurately measure the muscular strength of different individuals (56, 131). In a text pertaining to the benefits of therapeutic exercise, the Paris physician Joseph Clement Tissot (1747-1826) indicated that motion (exercise) would increase muscle size and strength and acknowledged that laboring (trained) individuals had larger and stronger muscles than did sedentary individuals (165).

Temperature Regulation

During this time, Adair Crawford (1748-1795) used the specific heat values formulated by Black and reported that the quantity of heat produced by an animal consuming oxygen was similar to the results achieved when the same amount of oxygen was utilized by the combustion of charcoal or wax. He concluded the heat was derived from the transformation of oxygen into fixed air (CO_2) or water (110)—a conclusion similar to the one formulated by Lavoisier (74). Bryan Robinson stated that exercise would give warmth to the skin if a fever was present (139), whereas Tissot was of the opinion that exercise would increase sweating because it removed the sour salts and related parts from the blood (165). In addition, Robinson conducted experiments and reported that individuals who traveled 2 miles (3.2 km) in 30 min would produce 8 to 9 oz of sweat. He also studied the relationship between sweating and urine production and indicated that the volume relationship between sweating and urine production was approximately 6:1 at the start of exercise and 16:1 near the end of exercise (139).

Others who made contributions were Lavoisier, who noted that insensible perspiration functioned to control the production of body heat (74); Black, who determined the latent heat of evaporation (110); and Franklin and Blagden, who reported that a cooling effect would occur from the evaporation of sweat (110).

Investigations of the 19th and Early 20th Centuries (Until 1915) That Enhanced the Recognition of Exercise Physiology

Central, Peripheral, and Autonomic Nervous Systems

In the last decades of the 19th century, German investigators were dominating European physiology and promoting the concept that the central nervous system was responsible for locomotion. However, the most important study pertaining to exercise physiology was conducted by August Krogh (1862-1949) and Johannes Lindhard (1870-1949) who observed exercise responses in less than 1 s, which they attributed to neural mechanisms associated with an "irradiation of impulses from the motor cortex rather than a reflex from the muscles" (90, p. 122). Little did they realize that this observation became the foundation for the concept of the central command.

During 1888, Julius Geppert (1856-1937) and Nathan Zuntz (1847-1920) conducted hyperpnea experiments that advanced the concept that during exercise neural and autonomic systems released into the circulation substances that stimulated directly the respiratory region of the brain and neural receptors in exercising muscle to activate the respiratory system (51). Later, Jons Erik Johansson (1862-1935) demonstrated with a rabbit that passive leg exercise would increase its heart rate, which suggested that the cardiovascular centers in the brain had been activated (82). In 1895, Heinrich Ewald Hering (1886-1948) of Prague exercised rabbits and indicated that an accelerated heart rate was the result of increased neural activity of the accelerator nerve coupled with a decrease in the activity of the vagus nerve (64). Hunt came to the same conclusion 4 yr later (80).

In 1904, Wilbur Bowen of Michigan, United States, measured the latency periods of exercising subjects on a bicycle ergometer after the initial heart rate cycle and suggested that the increase occurred because of a decrease in the restraining influences of the inhibitory center (vagal influences) (16). A decade later, Gasser and Meek investigated exercised dogs that had been vagal sectioned and adrenalectomized and that had received a muscarinic blocking agent (atropine). They concluded that the inhibition of vagal impulses or vagal withdrawal was the most economical means by which an increase in heart rate could occur (50).

In 1866, Walter Holbrook Gaskell (1847-1914) concluded from his anatomic studies that almost every tissue was innervated by nerve fibers (autonomic) that had opposite characteristics and that these differences could be explained by chemical changes (49). He also emphasized that the term *involuntary nervous system* should be used to describe the functions of what is identified as the sympathetic and parasympathetic nervous systems (134). The adrenal gland studies of Walter Cannon (1871-1945), conducted between 1911 and 1914, highlighted the importance of the sympathetic nervous system in an exercise response and the significance of glucose and adrenaline (epinephrine) availability in delaying the occurrence of muscular fatigue. (These studies are discussed in "Gastrointestinal, Endocrine, and Immune Systems.")

Muscular System

The availability of dynamometers during the 1860s promoted the measurement of strength and the development of norms for soldiers in the U.S. Army and for college students at Amherst College and the YMCA International College at Springfield College (109). During the last decade of the 19th century, Angelo Mosso (1846-1910) in Italy developed the first ergograph capable of recording work performance and evidence of the existence of muscular fatigue (121). Although Mosso supported the concept of central fatigue (171), he had strong evidence for the presence of peripheral fatigue, which incorporated the fatiguing events of neuromuscular transmission and muscular contraction (121).

Morpurgo at the University of Siena in Italy reported in 1897 the histological results of a study of two dogs that performed 2 mo of wheel running. The cross-sectional area of the sartorius muscles of the dogs increased 54% with no changes occurring in the number of fibers, nuclei, or spindles. He concluded that the hypertrophy observed was the result of an increase in sarcoplasm (120).

Theodore Hough at Massachusetts Institute of Technology in the United States had subjects exercise on an ergometer. The subjects subsequently complained of muscle soreness, which Hough attributed to diffusible waste products and torn tissues (76). In an earlier fatigue study, Hough concluded that nerve cells fatigued more rapidly than did muscle cells (75).

The association between muscular performance and lactic acid concentration began in 1910 when Ryffel mentioned that running 12 laps in 2 min 45 s caused blood lactic acid levels to increase from 12.5 mg/100 ml of blood to 71 mg/100 ml of blood and that urine values were elevated from 4 mg/h to 362 mg/h after 17 min of recovery (142). Chapters by Coggan and Brooks later in this text discuss the significance of blood and tissue changes in lactic acid concentrations.

Although Heidenhain and Fick experimented with isolated muscle preparations, it was A.V. Hill who developed and perfected an in vitro muscle preparation that enabled him to quantify the work performed and the mechanical efficiency of the responses (71). After 1915, Fenn in his laboratory measured the heat produced and related it to the work performed whereas Hill reported on the force–velocity relationships during muscle contraction, which has become a milestone in muscle physiology research (110).

Cardiovascular System

According to McDowall (113), in 1830 Francoise Magendie (1783-1855) raised the issue of whether the muscle contraction (muscle pump) is important in returning fluid in the legs to the heart. In the later decades of the 19th century, animals were extensively utilized to elucidate cardiovascular responses. In 1864, Etienne Marey (1830-1904) measured an increase in the blood pressure of an exercising horse with his sphygmograph (140). In 1881, Zadek recorded an increase in arterial pressures of humans during exercise (182). Eleven years later Grebner and Grunbaum reported that the blood pressure increase with exercise was between 50 and 60 mmHg (65). In 1898, Leonard Hill in England utilized a Hill-Barnard sphygmomanometer to measure the arterial blood pressure of a subject who ran 400 yards (365.8 m). Before exercise, a pressure between 120 and 130 mmHg was recorded (72); after exercise, a pressure between 110 and 115 mmHg was noted. Hill concluded that "arterial pressure becomes depressed below the normal resting pressure after severe muscular work" (72, p. xxvii). A similar finding was reported by Cook and Pembry in 1913 with trained and nontrained subjects (32).

McCurdy at Springfield College was among the first to measure blood pressure during dynamic and resistive exercise. He had students perform maximum back and leg lifts and reported that the mean values had increased from 111 mmHg to 180 mmHg, which he attributed to an increase in intra-abdominal and intrapulmonary pressures (112).

In 1894, the maximum heart rate with exercise was identified as being between 160 and 170 beats/min (30); in 1913, it was 180 (32). However, it was a decade or more before values in excess of 200 beats/min were recorded. In 1908, Pembrey and Todd had a trained and an untrained subject run up and down the same stairs for 30 s. Heart rate and blood pressure were measured before and after exercise. The experiment was repeated on 15 different d. From their findings they concluded that a trained subject had a markedly higher heart rate and faster recovery response than did a nontrained subject (133). Five years later, Cook and Pembrey conducted a similar study that included half-mile runs with more trained and nontrained subjects. Again they concluded that the heart rate response of trained subjects had a greater response range and a more rapid recovery (32). Cotton summarized the European results pertaining to the bradycardia of training and reported that numerous observations between active and inactive subjects or sportsmen and nonsportsman suggested that trained populations had lower rates (33). Cook and Pembrey also concluded that trained subjects had lower heart rates than did nontrained individuals (32).

During 1887 in France, J.B. Auguste Chauveau (1837-1917) and Kaufman measured the blood flow to the levator muscle of the lower lip in a horse and found that chewing increased flow fivefold (28). Before 1900, Nathan Zuntz (1847-1920) of Germany developed a treadmill suitable for horses and with Hagerman used the direct Fick to measure cardiac output during exercise. Not surprisingly, they reported that exercise caused an increase in cardiac output (141, 183). By 1914, Starling's law of the heart was announced (3). Lindhard used nitrous oxide to measure the cardiac output of exercising male subjects and recorded values that were fivefold higher than resting that had plateaued at 28.6 L/min. However, one trained subject had a peak value of 30.6 L/min. (102). Lindhard also compared the cardiac outputs and stroke volumes of a trained and a nontrained subject at similar heart rates (148 and 150 beats/min) and reported cardiac outputs of 22.6 and 10.7 L/min, and stroke volumes of 153 and 71 ml/beat, respectively (102). In 1915, Boothby measured the cardiac output of one subject during submaximal work and concluded that the flow increased proportionally with oxygen consumption, which paralleled the increase in ventilation (14).

Between 1884 and 1912, numerous studies involving wild and domesticated rabbits, dogs, and birds supported the concept that animals that were active, exercised systematically, and ran or flew long distances had larger and heavier hearts than those that were inactive (155). During 1914, Nicolai and Zuntz were the first to measure changes in cardiac dimensions during exercise using X-rays of subjects (n = 4) who walked on a treadmill at a grade of 24% at a speed 80 m/min. Measurements taken before, during, and after exercise revealed that the size of the shadow increased during exercise and decreased below normal after exercise. It was concluded that the heart dilated during exercise and constricted once exercise had ceased (122). In 1915, Williamson carefully considered the role of respiration in measuring heart dimensions of subjects who ran up and down stairs. He reported that exercise reduced dimensions; however, he was criticized for this observation because the results were not obtained during the run (177).

In 1899, the Harvard University Athletic Committee directed Eugene Darling, MD, to investigate the physiological effects of rowing to determine whether its effects on the heart were detrimental to the health of varsity crew members (34). His investigations were concerned with the size of the heart as determined by percussion, the occurrence of abnormal sounds, and the character of the pulse. He determined that the heart was enlarged but not to the point of being pathological (34). Two years later he assessed the effect of training on members of the crew and football teams and reported that cardiac hypertrophy was present in both groups but that no ill effects could reasonably be attributed to training (35).

Respiratory System

The penal system of England during the middle of the 19th century required prisoners to walk on massive treadwheels for punitive and health reasons (24). They were scheduled to exercise on the treadwheels 3 d/wk for a duration of 15 min until 4 h/d of exercise had been completed (109). This practice attracted the attention of Edward Smith (1819-1874), who was a physician, respiratory physiologist, social reformer, and an advocate of public health (figure 1.8). Smith also was concerned about the health and well-being of the prisoners (24). Thus, between the years 1856 and 1859, he conducted seminal respiratory studies on the effects of exercise on the prisoners and himself.

With the aid of a mask, a gasometer for respiratory volumes, an absorption chamber containing potassium hydrate, and a dehumidifier chamber, he reported mean values for inspired volumes (liters) and respiratory and heart rates for swimming, rowing, walking at 4 miles (6.4 km)/h, and walking at 3 miles (4.8 km)/h while carrying a 50.9 kg load. Walking on the treadwheel required 43 steps to travel 8.73 m/min. Inspired volumes ranged from 23.1 to 39.1 L/min, respiratory rates ranged from 20 to 30 breaths/min, and heart rate values ranged from 114 to 189 beats/min (153). He also measured the carbonic acid formed during exercise and reported that walking at 2 miles (3.2 km)/h produced 1.15 g (18 grains), walking at 3 miles (4.8 km)/h yielded 1.66 g (26 grains), and walking on the treadwheel produced 3.06 g (48 grains) (154). His research established a foundation for subsequent respiratory and metabolic studies and effectively demonstrated that the production of CO_2 was linearly related to the intensity of exercise (158). Smith also measured the carbon dioxide produced while riding a horse, and climbing stairs (165).

In 1885, Miescher-Rusch promoted the concept that the increase in tension of carbonic acid in the blood stimulated the activation of the respiratory center (116). Three years later, a hyperpnea investigation by Geppert and Zuntz led to the hypothesis that either the brain or a blood metabolite from contracting muscle activated the

Figure 1.8 Edward Smith of England who measured the respiratory responses of prison subjects walking on treadwheels.

Reprinted from W.D. McArdle, F.I. Katch, and V.L. Katch, 2001, *Exercise physiology: Energy, nutrition and human performance,* 5th ed. (Philadelphia: Lippincott Williams & Wilkins).

respiratory center during exercise (51). After 1910, Zuntz favored the concept that a blood-borne substance was responsible for the hyperpnea of exercise (54). However, it was the experiments of Krogh and Lindhard during 1913 that conclusively demonstrated that the onset of exercise increased the excitability of the respiratory center, which they believed was the result of circulating hydrogen ions (90).

Oxygen Consumption: Metabolic Systems and Their Substrates

Zuntz (figure 1.9) appears to be the first to directly measure oxygen consumption during exercise, first in animals (notably horses) and then in humans (54). A massive respiration chamber was required for horses whereas a Zuntz-Geppert respiratory apparatus was required for humans (54, 140). Although the human methodology was initially criticized by contemporaries, Zuntz and colleagues were able to secure oxygen-consumption values of individuals who were marching, climbing, cycling, or swimming as well as individuals who were playing the piano or typing (54). As presented by Bainbridge (3), Lindhard's data reported in 1915 demonstrated a linear relationship between O_2 consumption (~0.6-3.2 L/min) and work performed (~200-1,500 $kg \cdot m^{-1} \cdot min^{-1}$) as well as a linear relationship between pulmonary ventilation (~0.6-68.0 L/min) and oxygen consumption (102). The highest $\dot{V}O_2$ max value reported by Lindhard was 3.20 L/min for a subject who weighed 75 kg and had a cardiac output of 35.0 L/min.

Figure 1.9 Nathan Zuntz (1847-1920) was a distinguished German investigator who was nominated for the Nobel Prize three times. He was considered to be the first person to accurately measure oxygen consumption in animals and humans during exercise. Here, he is shown wearing a respiratory apparatus unit that contained a dry gas meter to measure oxygen consumption.

Reprinted, by permission, from H.C Gunga, 2009, *Nathan Zuntz* (Amsterdam: American Physiological Society), 127, 147. Used with courtesy from American Physiological Society.

Using questionable measuring procedures, Chauveau and Kaufman reported that chewing by the active muscles of the horse increased oxygen consumption 18-fold when expressed as cubic centimeters of oxygen utilized per gram of muscle per minute (28).

In 1842, Carl Liebig proclaimed that nitrogenous food sources were necessary to restore the equilibrium between waste and supply and were directly proportional to the amount of tissue being utilized (101). From their extensive studies with a calorimeter, von Pettehofer and von Voit indicated that Liebig was incorrect concerning whether proteins were the substrate of choice for muscular activity (110). However, it was the studies and apparatus of Regnault and Reiset that established the concept of the respiratory quotient (RQ) and its value in ascertaining the substrate being oxidized (110). In 1891, Katzenstein in Zunt's laboratory reported an RQ value of 0.80 for resting, walking, and climbing (84). Before the turn of the century, Chauveau of France published RQ values of 0.70 for resting and 0.84, 0.87, 0.97, and 0.87 for climbing stairs; these values changed to 0.84 after 60 min of rest (27). Zuntz and Chauveau disagreed on the interpretation of their data; however, both agreed that proteins were not the primary substrate during exercise and that fats and carbohydrate were involved (5, 91).

The most careful and important metabolic studies concerning substrate utilization conducted during the early decades of the 20th century were performed between 1909 and 1913 by Benedict and Cathcart (9). They reported that exercise increased RQ regardless of whether a diet was high or low in carbohydrate. When diets were high in carbohydrate, the exercise RQ increased from 0.85 to 0.90 before decreasing to 0.78. On the other hand, when diets were low in carbohydrate, the RQ changed from 0.79 to 0.82 before decreasing to 0.75. Their main conclusion was that carbohydrate was the primary substrate for muscular work (9).

Hemopoietic, Body Fluid, Renal, and Temperature Regulation Systems

In 1894, J. Mitchell conducted an early study on whether acute exercise would increase the number of red blood cells and the concentration of hemoglobin in blood. Mitchell, who had subjects run a moderate distance before participating in a massage experiment, found increases in both measures (119). In 1901, Zuntz and Schumburg recorded the red cell counts of five individuals who performed repeated marches between 18 and 25 km that required approximately 7 h to complete while carrying packs weighing between 22 and 31 kg. They found an average increase of 9% in the number of red blood cells (184). At the University of Pennsylvania in the United States, Hawk conducted 50 experiments with athletic students and team members who walked, sprinted, or ran distances ranging from 50 yd (45.7 m) to 3.5 miles (5.6 km); cycled 4 miles (6.4 km); broad jumped 6 times; swam 184 ft (56.1 m); and participated in water polo games that lasted between 3 and 15 min. Increases were reported in all events, and mean increases ranged from 10% to 23% (63). The increased numbers were explained by formation of new corpuscles, copious sweating and urine loss, fluids shifting from the blood to the muscles, elevation of blood pressure, and the shift of dormant or storage cells into the circulation. Of the explanations, Hawk favored the shift of dormant cells (63).

A related study conducted in 1915 by Schneider and Havens involved several athletes who ran from 0.8 mile to 2 miles (1.3-3.2 km), ran up several flights of stairs, or performed 15 min on a cycle ergometer (145). Like previous investigators, they found that exercise resulted in increases that ranged from 3.2% to 22.8% depending on the severity of the exercise. Hemoglobin also increased between 3.5% and 10.9%, but no observations indicated that erythrocytes had been destroyed (145). They also studied the effect of training on three subjects. Two subjects showed an increased number of erythrocytes (3.6%) and elevated hemoglobin values (9.4%).

Because methodology was not available until 1919 to measure fluid compartments, volumes, shifts, or any changes had to be assessed by weight loss. However, as early as 1866, Archibald Maclaren, an ardent sportsman, advocated drinking fluids to replace the loss caused by sweating (105). Hunt advanced the same argument some 46 yr later (79).

Until the 19th century, health status was assessed by the examination of urine turbidity, sediments, smell, taste, and color (104). Austin Flint Jr. was one of the first physicians to collect urine from subjects during exercise. He measured urinary water, nitrogenous products, and the presence of sulfur, phosphates, and chloride in a subject who walked 100 miles (160.9 km) in less than 22 h. However, he failed to standardize the diet and record the types of fluids that were ingested (40).

In 1878, Justus von Leube in Germany reported the presence of albumin in the urine of soldiers who had participated in a strenuous march. He regarded the findings to be a normal result of the physiological process of filtration through the pores of the glomerular membrane (97). During 1906, Blades and colleagues collected urine from soldiers who had marched 100 km and found that 40% of the soldiers had elevated protein levels. In fact, their urine profile of proteins, red blood cells, and casts was similar to that found with the disease condition known as acute parenchymatous nephritis, which suggested that prolonged exercise was harmful to the individual. (4). In 1912, Barauch published urinalysis results of Olympic runners who participated in the marathon race. All runners showed evidence for albumin and many had casts, red blood cells, and acetone bodies, although there was no evidence for glucose being filtered. Not surprisingly, this evidence contributed to the controversy as to whether heavy, prolonged exercise was pathological in nature (6).

The foundations of temperature regulation were enhanced in the middle of the 19th century by the contributions of James Joule (1818-1889) and Julius Robert von Mayer (1814-1878). Using a horse and a paddle-wheel apparatus capable of determining the mechanical equivalent of heat, Joule provided the means to quantify the energy produced by the heat of exercise (15). von Mayer formulated the concepts responsible for the law of the conservation of energy—better recognized as "energy can neither be created or destroyed," or the essence of the first law of thermodynamics (19). In 1891, Max Rubner experimentally validated the law using direct and indirect calorimetry procedures (110).

Near the end of century, Pembry and Nicol measured skin (five sites) and rectal temperatures of a subject during two walking experiments. When air temperatures were between 18.5 °C and 19.0 °C, mean increases for the skin sites ranged from 0.1 °C to 2.2 °C and rectal temperature was elevated by 1.3 °C (132). These same investigators compared temperature readings from oral and rectal sites and concluded that oral measurements were unreliable and a poor indicator of rectal temperatures. When rectal temperatures were recorded during exercise, they ranged from 38.3 °C to 40.0 °C (132). During the 1904 Olympics, rectal temperatures were measured after the marathon. The highest value recorded was 38.0 °C, and increases of 1.15 °C to 1.90 °C were recorded for the event (3).

Gastrointestinal, Endocrine, and Immune Systems

Physician Walter Beaumont (1785-1853) in 1833 reported his observations of a subject who had experienced a gunshot wound that left a gastric fistula (8). Beaumont conducted 70 experiments involving his patient's stomach that also included exercise and concluded that moderate exercise enhances secretions and fluids in the stomach and promotes digestion whereas heavy and fatiguing exercise impairs the process (8). Near the turn of the century, Salvioli of Italy reported that food in the stomach of dogs left faster when the dogs exercised than when they were idle, and Spirig of Switzerland claimed, after pumping test food out of the stomach of male subjects, that inactivity decreased stomach motility (155).

In the Harvard studies conducted in 1899, Darling studied the physiological effects of crew training as it pertained to gastrointestinal function and to the possibility of diarrhea. Darling reported that infrequent incidences of indigestion and diarrhea were more a nutritional issue than a physiological issue (34). Cannon's research at Harvard on gastric hunger contractions led A.J. Carlson (1875-1956) to include exercise in his investigation on gastrointestinal function. Carlson observed that standing or walking on the treadmill had little or no influence on gastrointestinal function whereas running had an inhibitory effect. When gastric fistulas were performed on dogs, he found that the effects were similar to what had been reported for humans (23).

Although many interesting and exciting endocrine studies emerged after 1850, few were concerned with the effects of exercise. In 1849, Thomas Addison (1793-1860) proposed that an association existed between the adrenal gland (cortex) and the incidence of muscular fatigue (166). Seven years later, Brown-Sequard (1817-1894) suggested that the cortex would detoxify the metabolites of muscular activity (17). Jacobj of Germany demonstrated in 1892 that electrical stimulation of the splanchnic nerves to the adrenal glands caused the release of a substance that increased the amplitude of contractile tissue (81). Three years later, Oliver and Schafer discovered that an extract from the adrenal gland had the same effect on contractile tissue as stimulation of the splanchnic nerves (124). Walter Cannon used this information to plan the experiment of de la Paz, which demonstrated that the blood of cats contained epinephrine when the cats heard the sound of barking dogs (21). With Nice, Cannon investigated in animals the role of epinephrine in delaying the onset of muscular fatigue (22). One year later he proposed the concept of flight or fight, which for many decades was used to explain the endocrine response to exercise. On the importance of the adrenal gland in an emergency situation, Cannon wrote:

> The organism with which the aid of increased adrenal secretion can best muster its energies can best call forth sugar to supply the laboring muscles, can best lessen fatigue, and can best send blood to the parts essential in run or flight for life, is most likely to survive. (20, p. 372)

Immune System

In ancient Rome, the term *immunity* referred to a protection against infectious disease (104). As the field of immunology evolved, minimal attention was devoted to the effects of exercise. When the topic did require notice, it concerned infections and the role of leukocytes in the process. In 1890, Charrin and Roger reported that rats fatigued by heavy exercise had reduced resistance to infection following subcutaneous inoculation with anthrax bacilli (26). In the 1901 marching study conducted by Zuntz and Schumburg, in which hemopoietic measurements were made, they reported elevations in neutrophils and lymphocytes (184). Three years later, Hawk summarized the influences of 13 athletic events on white cell counts; namely leukocytes were increased in all by a mean percentage of 57% (63). The resting leukocyte values of the athletic subjects were 11% higher (8,000/ml) than those of the nonathletic subjects. According to Schneider and Havens, athletes had increased leukocyte counts that ranged from 3% to 23% after running events. When they conducted differential white cell counts, the polymorphonecular counts increased by 9% to 45% whereas the mononuclear counts were elevated by 14% to 55%. These changes were attributed to the redistribution of cells and fluids during exercise, and no attempt was made to relate them to an infectious process (145).

19th Century and Early 20th Century Authors, Physicians, and Educators Who Helped Establish a Discipline With Courses and Laboratories in Exercise Physiology

Background

Medical historians Lyons and Pertrucelli (104) described the 19th century as the beginning of modern medicine. Rothschuh (140), the physiology historian, labeled the 19th century as the beginning whereas Berryman (11), the exercise science historian, identified this as the era when physicians began to recognize the health benefits of exercise. The author considered the 19th century, especially the latter half, as the beginning of exercise physiology because the discipline was identified by Byford (18) and because formal courses were introduced in select institutions and taught by physicians.

Contributions by Authors and Designated Authorities

In 1807, John Sinclair published a text that discussed the concepts of training for humans and animals that incorporated the views of Herodicus, Asclepiades, Galen, Sir Francis Bacon, Bryan Robinson, and the trainers of his era (151). According to the text, training reduced fat tissue, increased muscle mass, hardened bones, enhanced perspiration, improved the wind (lungs), lengthened breath-holding times, and hastened recovery times (151). Sinclair felt that training would reduce the incidence and duration of fatigue in horses and delay the process of being worn out. In addition, he believed that the core of the various training programs consisted of purging, puking, sweating, dieting, and exercising (151).

In 1835, Robley Dunglison (1798-1869), a faculty member in the hygiene department at the University of Maryland in the United States, was acknowledged by Berryman for writing the first text on preventive medicine for medical students (12). He advocated exercising for one's health and felt that the lack of exercise was responsible for a loss of function of the nervous, circulatory, muscular, digestive, secretory, and excretory systems, which could lead to select diseases and bodily disorders (36). Dunglison emphasized traveling exercises, which meant running long distances in order to improve digestive and mental functions. He was against violent exercise because it caused less oxygen to be inspired and more CO_2 to be produced, resulting in suffocation; because it caused aneurysms and hemorrhaging; and because it caused hernias, dislocations, and sprains. On the other hand, he favored moderate exercise because it promoted blood flow, enhanced the actions of the heart, increased muscle firmness and bulk, and reduced fat around the muscles.

In 1836, Andrew Combe (1787-1847), an Edinburgh physician, published a physiology textbook concerning the achievement of mental and physical education (31). He emphasized the health benefits of exercise and its influences on the various systems of the body, including the central nervous system. He advocated moderate exercise and scheduled rest periods. He felt that exercise was responsible for an increase in muscle power that was mediated by the central nervous system. Lack of exercise was associated with muscle weakness and diseases of the lungs (31).

In 1842, the brilliant chemist Justus Liebig (1803-1873) of Munich published *Animal Chemistry*, in which he championed proteins as the substrate for muscular activity. However, he made this statement without experimental evidence (101). Not surprisingly, it had a profound effect in promoting metabolic studies and in creating controversy.

William H. Byford (1817-1890), a physician and professor at Rush Medical School in Chicago, was the first to incorporate the physiology of exercise into a text that was published in 1855 (18). Byford, who had previously conducted thermal, circulatory, respiratory, and secretory experiments with exercising animals and humans, was an advocate of the health benefits of exercise and was concerned by the indifference the medical profession exhibited toward the topic. His article was an attempt to inform and educate physicians on the benefits of exercise and to encourage them to conduct research in this area (18). He defined exercise as "the voluntary discharge of any or all the animal functions as intellect, sensation, locomotion, and voice" and characterized exercise as an example of "vascular excitement, increased heat, redness of the surface, and augmented secretion and excretion" (18, p. 33). However, he provided no insight into modality, intensity, frequency, or duration of exercise. Byford attributed the healthful benefits of exercise to include enhanced growth and development, improved digestion, increased blood flow and its distribution to organs and tissues via capillary exchange, and to increased secretions from the skin, liver, and kidney (18).

In 1863, Charles Westhall, a renowned English runner, published a text on training that was noteworthy because the information it contained pertaining to exercise prescription, exercise specificity, and overtraining was similar to current views on these topics (176).

The most influential author in the latter half of the 19th century was Austin Flint Jr. (1836-1915) (figure 1.10), who studied with Dalton in Vermont in the United States and Bernard in France before being appointed as a professor of physiology and physiological anatomy at Bellevue Hospital in New York (109, 140). Early in his career, Flint advocated in his texts the benefits of exercise and was careful to cite the contributions of others (39). In 1870, he became involved in the controversy concerning whether proteins served as a substrate for muscle during exercise. If so, increased nitrogen would be found in urine during exercise. Therefore, he studied the urinary nitrogen excretion of an individual who walked 318 miles (511.7 km) in 5 d and found that 154 parts of nitrogen were excreted for every 100 parts ingested. He concluded that violent exercise caused muscle breakdown and resulted in an in-

Table 1.3 Select Conclusions by Austin Flint Concerning Exercising Muscle

Conclusion number	Conclusion
V	Experiments show that excessive and prolonged muscular exercise may increase the waste or wear of certain constituents of the body to such a degree that this wear is not repaired by food. Under these conditions there is an increased discharge of nitrogen particularly in the urine.
VII	By systemic exercise of the general muscular system of particular muscles, with proper intervals of repose for repair and growth, muscles may be developed in size, hardness, power, and endurance. The only reasonable theory that can be offered in explanation of the process is the following: While exercise increases the activity of dissimulation of the muscle substance a necessary accompaniment of this is an increased activity of the circulation of the muscles, for the purpose of removing the products of their physiological wear. This increased activity of the circulation is attended with an increased activity of the nutritive processes, provided the supply of nutriment be sufficient, also, that the exercise be succeeded by proper periods of rest. It is in this way only that we can comprehend the process of development of muscles by training; the conditions in training being exercise, rest following the exercise, and appropriate alimentation, the food furnishing nitrogenized matter to supply the waste of the nitrogenized parts of the tissues.
VIII	All that is known with regard to nutrition and disassimilation of muscles during ordinary or extraordinary work teaches that such work is always attended with destruction of muscular substance, which may not be completely repaired by food, according to the amount of work performed and the quantity and kind of alimentation.

Adapted from Flint 1878, pp. 94-96.

Figure 1.10 The physician Austin Flint Jr. (1836-1915) of Bellevue Hospital Medical College in New York. In 1875 he authored *A Textbook of Human Physiology*, which was adopted by numerous medical colleges in the United States. His chapters on nearly all the systems of the body discussed the physiological changes that occurred with exercise. It is plausible that this text was among the first to scientifically discuss the effects of exercise.

Reprinted from W.D. McArdle, F.I. Katch and V.L. Katch, 2001, *Exercise physiology: Energy, nutrition and human performance*, 5th ed. (Philadelphia: Lippincott Williams & Wilkins).

creased loss of proteins (40). In 1875, he published the first of several editions of *A Textbook of Human Physiology*, which was adopted by numerous medical schools throughout the United States. The text discussed topics in exercise physiology pertaining to circulation, respiration, metabolism, temperature regulation, muscle function and fatigue, bone strength, and longevity and included experimental data from Chauveau, Bryan Robinson, Lavoisier and Seguin, and Edward Smith (39). Three years later he published a text pertaining to the source of muscular power that contained the experimental results of Liebig, Lehmann, Fick and Wislicenus, Parkes, Pavy, and his data with Weston (41). Select conclusions from this text are found in table 1.3.

During 1886, Edward Hartwell of Johns Hopkins University published two manuscripts pertaining to the physiology of exercise, which was the topic of a speech presented to the American Association for the Advancement of Physical Education (60, 61). This topic was selected because the subject matter was frequently misstated and its effects were often overlooked. He felt that exercise was essential to achieving a healthy state and that failure to exercise could be associated with an incomplete oxidation of food, the accumulation of effete products, disordered digestion, an enfeebled

Table 1.4 Select Results From the 1893 Text of Kolb Concerning Berliner-Ruder Club Members

Maximum effort of rowers was associated with heart rates in excess of 230 beats/min, and radial artery pressures were 185 mmHg. After several months of training, heart rates were reduced by 16 beats/min and blood pressure was reduced by 20 mmHg.
From sphygmographic results, Kolb concluded that maximal exertion from rowing increased both the rate and work of the heart, altered cardiac dilation, elevated blood flow velocity, and produced muscle and cardiac hypertrophy. However, for the trained rower, cardiac hypertrophy was not pathological in nature.
When rowing, respiratory rates increased from 12 frequencies/min to more than 60/min. When rates approached 100/min, muscle failure occurred. After training, rowers had lower resting and exercise frequencies.
He attributed the dyspnea associated with rowing to lack of oxygen and increased carbon dioxide levels acting on the respiratory center.
Vital capacity measurement after rowing exhibited reductions, which was attributed to an increased volume of blood in the lungs.
Kolb served as a subject for many racing experiments, including those that measured the percentage of CO_2 expired during and after rowing. It increased from 4.3% to 6.0% during the race and continued to increase to 9.0% after the event ended.
Rectal temperatures recorded during rowing were 104 °F, which caused no apparent harm to the individual. However, it was important for rowers to be hydrated during training and competition. In a chapter on nervous insufficiency, Kolb mentioned the problems of overtraining and depression.

Adapted from Kolb 1893.

nervous system, flabby muscles, impaired secretions, onset of ill health, and occurrence of diseases (60, p. 301). He estimated that an individual should walk 8 to 9 miles (14.5 km) daily to achieve a healthy state.

Hartwell also believed that exercise enhanced the development of both muscle and nerve cells and that both cell types would decrease with disuse. He felt that inheritance was affected by exercise but provided no evidence for his statements (61). In 1890, *The Physiology of Bodily Exercise* by the French physician Fernand Lagrange was translated into English (95) and distributed to instructors teaching exercise physiology. The text covered such subjects as muscular work, fatigue, habituation to work, and the effects of exercise on the brain. Some suggest that the Lagrange text was the first to be published in the discipline of exercise physiology. However, McArdle and colleagues (109) challenged this idea because the text included incomplete information, inadequate documentation, and unscholarly presentation. It is of interest that George Wells Fitz of Harvard University also expressed this same concern soon after publication of the book (109). The author agrees with the assessments by Fitz and McArdle and colleagues concerning the scholarly nature of the text and the value of the information included; however, all must acknowledge that it was among the first to be published.

In 1893, *Physiology of Sport* by the German physician and acknowledged sportsman George Kolb was translated and published in English (89). The book is noteworthy because it contained experimental results from members of the Berliner-Ruder Club who were predominately rowers and because it was one of the earliest texts on exercise physiology. Salient results from the club members are listed in table 1.4.

R. Tait McKinzie, director of physical education at the University of Pennsylvania, is remembered more for his sculptures of athletes performing sporting events than for his 1909 text *Exercise in Education and Medicine* (114). The text was written for students and practitioners of physical training, teachers of youths, and students and practitioners of medicine in order to provide a perspective on the role of exercise in the education and treatment of disease and abnormal conditions. He defined exercise as follows:

> The term *exercise* as used here comprises all movement, voluntary or passive, including manipulations by the hand of the operator or by a machine, designed to act on the muscles, the blood vessels, the

> nervous system, the skin, and the abdominal organs. (114, p. 9)

The text contains a limited amount of information pertaining to exercise physiology but extensively details systems of physical training and physical education programs in municipalities, schools, universities, foreign countries, and so on. Approximately 50% of the text was devoted to exercise and medicine associated with diseases and ailments.

McKinzie considered active exercise to be an effort that could be violent in nature, involve extensive muscle groups, be associated with hypertrophy and muscle damage, and would lead to fatigue. Moreover, exercise of effort provided no time for the scavengers to act and was unable to develop constitutional vigor to the same degree as endurance exercise. Endurance exercise was associated with less effort (reduced intensity), longer durations (>1 h), and elimination of poisonous waste matter before fatigue occurred. Endurance exercises were advocated because they powerfully affected the heart, lungs, and muscular and nervous systems. Passive exercise, which essentially was massage and manipulation, was advocated for conditions of fatigue because it had a beneficial effect on muscles, ligaments, and the circulatory, respiratory, and nervous systems.

In his text, McKinzie included the findings of Hough (muscle soreness; 76), Mosso (fatigue; 121), Bowen (effects of cycling on heart rate; 16), McCurdy (blood pressure and lifting weights; 112), and Hawk (red blood cells; 63). He expressed concern about exercise causing severe fatigue that could influence the hearts of growing adolescents due to the relationship between heart volume and arterial diameter. He substantiated his concern by citing published data stating that heart volume increased 12-fold during adolescent growth whereas arterial diameter changed only 3-fold. McKinzie was worried about the presence of albumin in urine and the elevation in rectal temperature with heavy exercise as rectal temperature could reach 104 °F and be associated with a constant fever. Surprisingly, no mention was made of the contributions of Robinson, Seguin and Lavoisier, Hartwell, Flint Jr., or Kolb (114).

Introduction of Exercise Physiology into the Classroom and Laboratory

Although Byford promoted the concept of exercise physiology, his article (18) had no impact in spurring educational or medical institutions to offer classes or related degrees in this area and had minimal influence on establishing a discipline. However, several decades after the United States Civil War, interest increased in gymnastics, outdoor recreation, competitive athletics, physical training, and hygiene that became implemented by select physical education departments that prepared students for careers in schools, athletic clubs, gymnasiums, or agencies such as the YMCA. Intrinsic to health and hygiene courses was dissemination of information pertaining to the physiology of the circulatory, respiratory, muscular, digestive, and excretory systems (128). To teach hygiene and physical-training courses and to supervise gymnasium activities, nearly all institutions employed physicians, such as Dudley Allen Sargent and George Wells Fitz of Harvard University; Luther Halsey Gulik and James H. McCurdy of the YMCA College in Springfield, Massachusetts; Delphine Hanna of Oberlin College; and Thomas Denison Woods at Stanford University (92, 129, 158, 164).

Collectively, these physicians taught some aspects of exercise physiology to their physical education students, and it is likely that Dudley A. Sargent of Harvard University was the first to do so because the Lawrence Scientific School began offering summer classes for teachers in 1847. It is unclear who was the first to schedule a laboratory, but it has been cited that George Wells Fitz of Harvard University scheduled laboratory experiments for students (164). Fitz is renowned as an educator for developing a 4 yr degree program that emphasized anatomy, physiology, and physical training for individuals who envisioned a career in directing gymnasium activities. In addition, Fitz taught a course in exercise physiology theory during the fourth year of the Harvard curriculum (129, 158). Students who did not wish to supervise gymnasium activities were eligible to enter the second year of the curriculum for medical students. The program graduated nine students before closing in 1899 due to financial, philosophical, and political problems (129).

Approximately six to seven decades later, universities in the United States established graduate programs in physical education that included and emphasized exercise physiology, thereby recapturing the scientific rigor that was lost at Harvard (157).

Closing Remarks

This chapter contains extensive historical data pertaining to exercise physiology that spans more than 2,000 years. However, the most important individuals and their contributions are listed in table 1.5.

Table 1.5 Select Historical Milestones in Exercise Physiology (600 BCE-1915 CE)

Year	Investigator	Historical milestone	References
600 BCE	Susruta	Indian physician who advocated moderate exercise to promote an equilibrium state between the dosas (humors) for health reasons and to prevent disease. He was the first physician to advocate moderate exercise to minimize the health consequences of diabetes and obesity. He was also the first to record the effects of training.	136, 161
460-370 BCE	Hippocrates	Grecian physician who advocated moderate exercise to achieve an equilibrium state between the various humors in order to promote health and to minimize the effects of health disorders and diseases. He was the first physician to provide a written exercise prescription for a patient with a disease (consumption).	70, 159
129-210	Galen	Roman physician whose intelligence and dominating personality coupled with a prolific publication record pertaining to medicine, physiology, exercise physiology, and the role of exercise in the practice of medicine left a legacy that lasted more than 1,000 yr.	12, 45, 46, 159
980-1037	Avicenna	Persian physician who published in Arabic *Canon of Medicine*, which made Galen's perspective on medicine and exercise physiology available to the Arabic world and Europe and contained new information on the thermal response during exercise.	53, 93
1553	Mendez	A Spanish physician who published (in Latin) the first textbook devoted to the effects of exercise on the body. The text contained the perspective of Galen.	115
1628	Harvey	An English physician who discovered the nature of circulation of blood throughout the body, observed heart rates increase with exercise, and reported strenuous exercise by animals would increase heart weight.	62
1636	Santorio	An Italian professor who invented a sensitive balance chair capable of measuring insensible perspiration and changes in body weight that were used in metabolic studies concerning food consumption and exercise. He also developed instruments to measure hear rates and body temperature.	143
1734	Robinson	An Irish physician whose textbook (second edition) contained experimental data pertaining to the effects of exercise and training on the circulatory and thermoregulatory systems.	139
1789	Lavoisier and Seguin	The brilliant French chemist and his assistant conducted the first experimental study on the metabolic transformations that occurred during exercise. It was also the first time that oxygen consumption had been measured during exercise.	146, 158

(continued)

Table 1.5 *(continued)*

Year	Investigator	Historical milestone	References
1833	Beaumont	A U.S. Army surgeon who reported observations on the gastric responses to exercise using a patient with a healed gastric fistula. Light to moderate exercise enhanced gastric secretions and the digestive process whereas maximal and exhaustive exercise had an inhibitory effect.	8
1856-1859	Smith	An English physician who conducted extensive respiratory investigations that included prison inmates walking on treadwheels. A notable finding was that a significant relationship existed between CO_2 production and intensity of exercise.	153, 154
1875	Flint Jr.	A physician and author of the first physiological textbook used in American medical schools that contained experimental data and scientific explanations for the physiological effects of exercise.	39
1888	Geppert and Zuntz	German investigations whose studies on the hyperpnea of exercise were responsible for the concept that an exercise stimulus from either the brain or active muscles was activating the respiratory center.	51
1890	Charrin and Roger	French investigators who inoculated rats with subcutaneous injections of anthrax bacilli and made the immunological observation that heavy exercise diminished their resistance to infection.	26
1891	Mosso	An Italian physician and physiologist who developed the first ergograph. Published the text *La Fatica.* Included were his results that characterized muscular fatigue and demonstrated chronic exercise would increase muscle strength and endurance while delaying the onset of fatigue.	121
1892	Wells Fitz	A Harvard University physician and professor who established a rigorous undergraduate program in the department of anatomy, physiology, and physical training in the Lawrence Scientific School that prepared students for a leadership role in gymnasiums or for entrance into the second year of medical school. Students were required to enroll in physiology of exercise theory and laboratory courses.	92, 109, 129, 164
1893	Kolb	Author of a translated German text on the physiology of sport that contained physiological responses of rowers to heavy and maximal exercise.	89
1896	Chauveau and Zuntz	A French veterinarian and a German physician who used the respiratory quotient (RQ) to determine substrate utilization during resting and when climbing stairs. Because their results were markedly different, a metabolic controversy prevailed between them during the first decade of the 20th century.	27, 84

1898	Hill and Nicol	First investigator to report the existence of postexercise hypotension.	72
1898	Pembry and Nicol	English investigators who measured oral temperatures during exercise and found them to be unreliable for research purposes. When rectal temperatures were measured, the highest value recorded was 40.0 °C, or 104 °F.	158
1902	Hough	A U.S. investigator who used an ergograph to produce muscular fatigue in human subjects. He reported the presence of muscle soreness, which was attributed to diffusible waste products acting on nerve endings or to the possible tearing of muscle tissue.	75
1904	Bowen	A University of Michigan investigator who observed that subjects exercising on a bicycle ergometer exhibited a latency after the initial cardiac cycle, which he suggested was the result of a loss of latency from a restraining influence (e.g., vagal influence).	16
1913	Krogh and Lindhard	Copenhagen investigators who conducted an exercise cardiorespiratory study and reported that the rapid physiological responses indicated an "irradiation of impulses from the motor cortex rather than a reflex from the muscles." Scholars such as Rowell and Mitchell attribute this observation as the beginning of the central command concept.	91
1914	Cannon	A Harvard University physician who investigated the emergency functions of the adrenal gland and indicated that they were responsible for the flight or fight response. The terminology subsequently changed to *fight or flight response*, which was used extensively by exercise physiologists for many decades to explain the role of the sympathetic nervous system during exercise.	20
1915	Lindhard	A Copenhagen investigator who perfected the nitrous oxide method to measure cardiac output during maximal exercise. He reported human cardiac output results approaching 30.0 L/min and oxygen consumption values reaching 3.2 L/min.	102

References

1. Baas JH. *Outlines of the History of Medicine and the American Profession.* Vol 1. Huntington, NY: Krieger, 1971.
2. Bahita SL. *A History of Medicine with Special Reference to the Orient*. New Delhi: Office of the Medical Council of India, 1977.
3. Bainbridge FA. *The Physiology of Muscular Exercise.* London: Longman, 1919.
4. Baldes, Heishelmeim, Metzer. Untersuchungen uber den einfluss grosser koperanstrengungen auf zirkulationapparat, nieren und nervensystem. *Muenschen Med Wschr.* 53: 1865-1866, 1906.
5. Barnard RJ, Holloszy JO. The metabolic systems: Aerobic metabolism and substrate utilization in exercising skeletal muscle. In: Tipton CM, ed. *Exercise Physiology: People and Ideas.* New York: Oxford University Press, 2003, 292-321.
6. Barauch JH. Physiological and pathological effects of severe exertion (the marathon race). *Am Phys Ed Rev.* 16: 1-11, 144-150, 200-205, 262-268, 325-334, 1912.
7. Barnes J. *The Presocratic Philosophers.* Vol 1. London: Routledge, 1979.
8. Beaumont. W. *Experiments and Observations on the Gastric Juice and the Physiology of Digestion.* Facsimile of the original edition of 1833. New York: Dover Publications, 1959.
9. Benedict FG, Cathcart EF. *Muscular Work.* Washington, DC: Carnegie Institute of Washington, 1913.
10. Bernoulli J. *Dissertations on the Mechanics of Effervescence and Fermentation and on the Mechanics of the Movement of the Muscles by Johann Bernoulli.* Philadelphia: American Philosophical Society, 1997.
11. Berryman JW. Ancient and early influences. In: Tipton CM, ed. *Exercise Physiology: People and Ideas.* New York: Oxford University Press, 2003, 1-38.
12. Berryman, JW. Exercise and the medical tradition from Hippocrates through Antebellum America: A review essay. In: Berryman JW, Park RJ, eds. *Sport and Exercise Sciences: Essays in the History of*

Sport Medicine. Urbana, IL: University of Illinois Press, 1992, 1-57.

13. Biers WR. *The Archeology of Greece: An Introduction.* Ithaca, NY: Cornell University Press, 1980.
14. Boothby W. A determination of the circulation rate in man at rest and at work. *Am J Physiol.* 37: 383-417, 1915.
15. Bottomley JT. James Prescott Joule. *Nature.* 26: 617-620, 1882.
16. Bowen WP. Changes in heart-rate, blood pressure, and duration of systole resulting from bicycling. *Am J Physiol.* 11: 59-77, 1904.
17. Brown-Sequard CE. Recherches experimentales sur la physiologie et la pathologie des capsules surrenals. *C R Acad Sci.* 43: 422-425, 1856.
18. Byford WH. On the physiology of exercise. *Am J Med Sci.* 30: 32-42, 1855.
19. Caneva KL. *Robert Mayer and the Conservation of Energy.* Princeton, NJ: Princeton University Press, 1993.
20. Cannon WB. The emergency function of the adrenal medulla in pain and the major emotions. *Am J Physiol.* 33: 356-372, 1914.
21. Cannon WB, de La Paz D. Emotional stimulation of adrenal secretion. *Am J Physiol.* 28: 64-70, 1911.
22. Cannon WB, Nice LB. The effect of adrenal secretion on muscular fatigue. *Am J Physiol.* 32: 44-60, 1913.
23. Carlson AJ. *The Control of Hunger in Health and Disease.* Chicago: The University of Chicago Press, 1916.
24. Chapman CB. Edward Smith (1818-1874) physiologist, human ecologist, reformer. *J History Med Allied Sci.* 22: 1-26, 1967.
25. Chancerel PG. *Historique de les Gymnastique Medicale.* These pour le doctorat en medecine, faculte de medecine de Paris, 1864.
26. Charrin AR, Roger G-H. A l'etude experimentale du surmenage; son influence du l'infecction. *Arch Physiol Normal Pathol.* 2: 273-283, 1890.
27. Chauveau A. Source et nature du potentiel directment utilise dans le travail musculaire, d'apres les echanges respiratoires, chez l'homme en etat d'abstinence. *C R Acad Sci (Paris).* 122:1163-1221, 1896.
28. Chauveau A., Kaufman M. Experiences pour la determination du coefficient de l'activite nutritive et respiratoirs des muscles en repos et en travail. *C R Acad Sci (Paris).* 104:126, 1887.
29. Chowdhury AKR, Chawdhury KR III. *Man, Malady and Medicine.* Calcutta, India: Das Gupta & Co., 1988.
30. Christ H. The influence of muscular work on heart action [translated from German]. *Deutsch Arch Clin Med.* 53: 102-140, 1894.
31. Combe A. *The Principles of Physiology Applied to the Preservation of Health, and to the Improvement of Physical and Mental Education.* New York: Harper, 1836.
32. Cook R, Pembrey MS. Observations on the effects of muscular exercise on man. *J Physiol.* 45: 429-446, 1913.
33. Cotton FS. The relation of athletic status to pulse rate, in men and women. *J Physiol.* 76: 39-51, 1932.
34. Darling E. The effects of training: A study of the Harvard University Crew. *Boston Med Surg J.* 141: 229-233, 1899.
35. Darling E. The effects of training: Second paper. *Boston Med Surg J.* 144: 550-559, 1901.
36. Dunglison R. *On the Influence of Atmosphere and Locality; Change of Air and Climate; Seasons; Food; Clothing; Bathing; Exercise; Sleep; Corporeal and Intellectual Pursuits, etc. etc. on Human Health; Constituting Elements of Hygiene.* Philadelphia: Carey, Lea, & Blanchard, 1835.
37. Edelstein L. *Ancient Medicine: Selected Papers of Ludwick Edelstein.* Baltimore: Johns Hopkins Press, 1967.
38. Elliot JS. *Outlines of Greek and Roman Medicine.* Boston: Milford House, 1971.
39. Flint A. *A Textbook of Physiology.* New York: Appleton and Company, 1875.
40. Flint A Jr. *On the Physiological Effects of Severe and Protracted Muscular Exercise: With Special Reference to its Influence Upon the Secretion of Nitrogen.* New York: Appleton-Century-Crofts, 1871.
41. Flint A Jr. *On The Source of Muscle Power.* New York: Appleton, 1878.
42. Foyer SJ. *The Physician's Pulse-Watch, or, An Essay to Explain the Old Art of Feeling the Pulse, and to Improve it by the Help of a Pulse Watch.* London: Sam Smith and Benj. Walford, 1707.
43. Foster M. *Lectures on the History of Physiology During the Sixteenth, Seventeenth, and Eighteenth Centuries.* London: Dover Publications, 1970.
44. Frank RG Jr. *Harvey and the Oxford Physiologists.* Berkeley, CA: University of California Press, 1980.
45. Galen. *Galen's Hygiene* (Des sanitate tuenda) [translated by Green RM]. Springfield, IL: Thomas, 1951.
46. Galen. *Galen: Selected Works* [translated by Singer PN]. New York: Oxford University Press, 1997.
47. Gardiner EN. *Athletics of the Ancient World.* Oxford: Oxford University Press, 1955.
48. Garrison FH. *An Introduction to the History of Medicine.* 2nd ed. Philadelphia: Saunders, 1917, 1-16, 66-103.
49. Gaskell WH. On the structure, distribution and function of the nerves which innervate the visceral and vascular systems. *J Physiol.* 7: 1-80, 1886.

50. Gasser HS, Meek WJ. A study of the mechanisms by which muscular exercise produces acceleration of the heart. *Am J Physiol.* 34: 48-71, 1914.
51. Geppert J, Zuntz N. Uber die Regulation der Atmung. *Arch Ges Physiol.* 42: 189-244, 1888.
52. Gordon BL. *Medicine Through Antiquity.* Philadelphia: F.A. Davis Company, 1949.
53. Grunner OC. *A Treatise on the Canon of Medicine of Avicenna, Incorporating a Translation of the First Book.* New York: Augustus M. Kelley, 1970.
54. Gunga H-C. *Nathan Zuntz.* Amsterdam: American Physiological Society, 2009.
55. Hales S. *Statical Essays: Containing Haemastaticks.* New York: Hafner, 1964.
56. Hall JR. John Theophilus Desaguliers, 1663-1774. In: Gillispie CC, ed. *Dictionary of Scientific Biography.* New York: Scribner & Sons, 1971, 43-46.
57. Haller A. *First Lines of Physiology, the Source of Science.* Vol 1. New York: Johnson, 1966.
58. Hankinson RJ. *Galen on Antecedent Causes.* Cambridge: Cambridge University Press, 1998.
59. Harris HA. *Greek Athletes and Athletics.* London: Hutuchinson, 1964.
60. Hartwell EM. On the physiology of exercise (Part I). *Boston Med Surg J.* 116: 297-302, 1887.
61. Hartwell EM. On the physiology of exercise (part II). *Boston Med Surg J.* 116. 321-324, 1887.
62. Harvey W. *Exercitation Anatomica De Motu Cordis et Sanguinis Animalibus.* 3rd ed. Springfield, IL: Charles C Thomas, 1941.
63. Hawk PB. On the morphological changes in the blood after muscular exercise. *Am J Physiol.* 10: 384-400, 1904.
64. Herring HE. Uber die Beziehung der extracardialen Herznerven zur steigerung Herzchlagzahl dei Muskelthatkeit. *Pflug Arch Ges Physiol.* 40: 429-492, 1895.
65. Herxheimer H. *The Principles of Medicine in Sport* [translated by Tuttle WW, Knowlton CC]. Berlin: Leipzig, 1933, 13-18.
66. Hippocrates. *Hippocrates* [English translation by Jones WHS]. Vol I. Cambridge, MA: Harvard University Press, 1923.
67. Hippocrates. *Hippocrates* [English translation by Jones WHS]. Vol II. Cambridge, MA: Harvard University Press, 1923.
68. Hippocrates. *Hippocrates* [English translation by Jones WHS]. Vol IV. Cambridge, MA: Harvard University Press, 1923.
69. Hippocrates. *Hippocrates* [English translation by Jones WHS]. Vol V. Cambridge, MA: Harvard University Press, 1923.
70. Hippocrates. *Hippocrates* [English translation by Potter P]. Vol VI. Cambridge, MA: Harvard University Press, 1923.
71. Hill AV. The absolute mechanical efficiency of the contraction of an isolated muscle. *J Physiol.* 46: 435-469, 1913.
72. Hill L. Arterial pressure in man while sleeping, resting, working, bathing. *J Physiol.* 22: xxvi-xxx, 1898.
73. Hoffmann FH. *Fundamental Medicinae* [translation and introduction by King LS]. New York: American Elsevier, 1971.
74. Holmes FL. *Lavoisier and the Chemistry of Life: An Exploration of Scientific Creativity.* Madison, WI: University of Wisconsin Press, 1995.
75. Hough T. Ergographic studies in muscular soreness. *Am J Physiol.* 5: 240-265, 1901.
76. Hough T. Ergographic studies in muscular soreness. *Am J Physiol.* 7: 76-92, 1902.
77. Hooker JT. *Ancient Spartans.* London: Dent, 1980.
78. Huard P, Wong M. *Chinese Medicine.* New York: McGraw-Hill, 1964.
79. Hunt EH. The regulation of temperature in extremes of dry heat. *J Hygiene.* 12: 478-488, 1912.
80. Hunt R. Direct and reflex acceleration of the mammalian heart with some observations on the relations of the inhibitory and accelerator nerves. *Am J Physiol.* 2: 395-470, 1899.
81. Jacobj C. Beitrage zur physiologischen und pharmakologischen Kenntniss der Darmbewegungen mit besonder Berucksichtigung der Beziehung der Nebenniere zu denselben. *Arch Exp Pathol Pharmak.* 29: 171-211, 1892.
82. Johansson JE. Uber die Einwirkung der Muskelthatigkeit auf die Athmung und die Herzthatkeit. *Skan Arch Physiol.* 5: 20-66, 1893.
83. Kaegi A. *The Rigveda* [translated by Arrowsmith R]. Boston: Ginn and Company, 1902.
84. Katzenstein G. Ueber die Einwirkung der Muskelthatigkeit auf den Stoffverbrauch des Menschen. *Pflug Arch Ges Physiol.* 49: 330-404, 1891.
85. Keill J. *An Account of Animal Secretion, the Quantity of Blood in the Human Body, and Muscular Motion.* London: George Strahan, 1708.
86. Keppie L. *The Making of the Roman Army.* London: B.T. Batsford, 1984.
87. Kirk GS, Raven JE, Schofield M. *The Presocratic Philosphers.* 2nd ed. Cambridge, Great Britain: Cambridge University Press, 1983.
88. Kohne E, Ewigleben E. *Gladiators and Casears.* Berkeley, CA: University of California Press, 2000.
89. Kolb G. *Physiology of Sport.* 2nd ed. London: Krohne & Sesemann, 1893.
90. Krogh A, Lindhard J. The regulation of respiration and circulation during the initial stages of muscular work. *J Physiol.* 47: 112-136, 1913.
91. Krogh A, Lindhard J. The relative value of fat and carbohydrate as sources of muscular energy. *Biochem J.* 14: 290-363, 1920.

92. Kroll W. *Perspectives in Physical Education.* New York: Academic Press, 1971.
93. Krueger HC. *Avicenna's Poem on Medicine.* Springfield, MO: Charles C Thomas, 1963.
94. Kutambiah P. *Ancient Indian Medicine—Orient.* Madras: Longmans, 1962.
95. Lagrange F. *Physiology of Bodily Exercise.* New York: D. Appleton and Company, 1890.
96. Lambridis H. *Empedocles, A Philosophical Investigation.* University, AL: University of Alabama Press, 1976.
97. Leube, von, W. Ueber die ausscheidung von eiweiss im harn ges gesunden menschen. *Virchow Archiv Pathol Anat Physiol Klin Med.* 72: 145-157, 1878.
98. Levine EB. *Hippocrates.* New York: Twayne, 1971.
99. Lewin R. *Human Evolution.* 5th ed. Malden, MA: Blackwell, 2005.
100. Licht S. *Therapeutic Exercise.* 2nd ed. New Haven: Elizabeth Light, 1965.
101. Liebig von, JF. *Animal Chemistry, or Organic Chemistry in its Application to Physiology and Pathology.* London: Taylor & Walton, 1842.
102. Lindhard J. Uber das Minutenvolumen des Herzen bei Ruhe und bei Muskelslarbeit. *Pflug Arch.* 161: 233-383, 1915.
103. Lower RA. *Treatise on the Heart on the Movement and Colour of the Blood and on the Passage Of the Chyle in the Blood.* In: RT Gunther, ed. *Early Science in Oxford.* Printed for the subscribers, 1932.
104. Lyons S, Petrucelli RJ. *Medicine: An Illustrated History.* New York: Abrams, 1994.
105. Maclaren A. *Training in Theory and Practice.* London: Macmillan, 1866.
106. Majno G. *The Healing Hand: Man and Wound in the Ancient World.* Cambridge, MA: Harvard University Press, 1975.
107. Marshall J. *Mohenjo-Daro and the Indus Civilization.* Vol I. Delhi, India: Indological Book House, 1973.
108. Mayow J. *Medicao-Physical Works, Being a Translation of Tractatus Quinque Medico Physici.* London: Simpkin, Marshall, Hamilton, Kent, & Co., 1907.
109. McArdle WD, Katch FI, Katch VL. *Exercise Physiology: Energy, Nutrition and Human Performance.* 5th ed. Philadelphia: Lippincott Williams & Wilkins, 2001.
110. McCollum EV. *The History of Nutrition.* Boston: Houghton, Mifflin, Co., 1957, 119-122.
111. McComas AJ. The neuromuscular system. In: Tipton CM, ed. *Exercise Physiology: People and Ideas.* New York: Oxford University Press, 2003, 39-97.
112. McCurdy JH. The effect of maximum muscular effort on blood pressure. *Am J Physiol.* 5: 95-103, 1901.
113. McDowall RJS. *The Control of the Circulation of the Blood.* London: Longsman & Green, 1938.
114. McKinzie RT. *Exercise in Education and Medicine.* Philadelphia: Saunders, 1909.
115. Mendez C. *Book of Bodily Exercise.* Published in 1553. Baltimore: Waverly Press, 1960.
116. Miescher-Rusch F. Bemerkungen zur Lehre von den Athembewegungen. *Arch f Anat u Physiol.* (Abstract) p. 355, 1885.
117. Miller SG. *Arete, Greek Sports From Ancient Sources.* Berkeley, CA: University of California Press, 1991.
118. Mitchell H. *Sparta.* Westport, CT: Greenwood, 1985.
119. Mitchell JK. The effect of massage on the number and hemoglobin value of red blood cells. *Am J Med Sci.* 107: 502-515, 1894.
120. Morpurgo B. Uber Activitats-Hypertrophie der wirkurllichen Muskeln. *Virchowsn Arch.* 150: 522-544, 1897.
121. Mosso A. *Fatigue* [translated by Drummond M, Drummond WB]. Putnam, 1904.
122. Nicolai GF, Zuntz N. Fullung und Entleerung des Herzens bei Ruhe und Arbeit. *Berliner Klin Wochenschr.* 51: 821-824, 1914.
123. Nutton V. Medicine in the Greek World, 800-500 B.C. In: Conrad LI, Neven M, Nutton V, Wear A, eds. *The Western Medical Tradition.* Cambridge, England: Cambridge University Press, 1995, 11-38.
124. Oliver G, Schafer EA. The physiological effects of extracts of the suprarenal capsules. *J Physiol.* 18: 230-276, 1895.
125. Olivia V. *Sports and Games in the Ancient World.* London: Orbis Publishing Company, 1984.
126. Osler W. *The Evolution of Modern Medicine.* New York: Armo Press, 1922.
127. Park R. *An Epitome on the History of Medicine.* Philadelphia: F.A. Davis, 1897.
128. Park RJ. Athletes and their training in Britain and America, 1800-1914. In: Berryman JW, Park RJ, eds. *Sport and Exercise Science.* Urbana, IL: University of Illinois Press, 1992.
129. Park RJ. The rise and demise of Harvard's B.S. program in anatomy, physiology and physical training: A case of conflicts of interest and scarce resources. *Res Quart Exerc Sport.* 63: 246-260, 1992.
130. Patterson SW, Piper H, Starling EH. The regulation of the heart beat. *J Physiol.* 48: 465-513, 1914.
131. Pearn J. Two early dynamometers: An historical account of the earliest measurements to study human muscular strength. *J Neurobiol Soc.* 37: 127-134, 1978.
132. Pembry MS, Nicol BA. Observations upon the deep and surface temperature of the human body. *J Physiol.* 23: 386-406, 1898.

133. Pembry MS, Todd AH. The influence of exercise upon the pulse and blood pressure. *J Physiol* . 37: lxvi-lxvii, 1908.
134. Pick J. *The Autonomic Nervous System.* Philadelphia: Lippincott, 1970.
135. Prioreschi P. *A History of Medicine*. Vol. 2. 2nd ed. Omaha, NE: Horatius Press, 1994.
136. Ray P, Gupa H, Roy M. *Susruta Samhita.* New Delhi, India: Indian National Science Academy, 1980.
137. Renbourn ET. The history of sweat and sweat rashes from the earliest times to the end of the 18th century. *J Hist Med Allied Sci.* 14: 202-1959.
138. Renbourn ET. The natural history of insensible perspiration: A forgotten doctrine of health and disease. *Med Hist.* 4: 135-152, 1960.
139. Robinson BA. *A Treatise of Animal Oeconomy.* 2nd ed. Dublin: George Ewing and Edwin Smith, 1734.
140. Rothschuh KE. *History of Physiology.* Huntington, NY: Robert E Krieger Publishing Company, 1973.
141. Rowell LB. The cardiovascular system. In: Tipton CM, ed. *Exercise Physiology: People and Ideas.* New York: Oxford University Press, 2003, 98-137.
142. Ryfell JH. Experiments in lactic acid formation in man. *J Physiol.* 39: xxix-xxxii, 1910.
143. Santorio S. *Medicina Statica, or, Rules of Health in Eight Sections of Aphorisms.* London: John Starkey, 1676.
144. Scarborough J. *Roman Medicine.* Ithaca, NY: Cornell University Press, 1969.
145. Schneider EC, Havens LC. Changes in the blood after muscular activity and during training. *Am J Physiol.* 36: 239-259, 1915.
146. Seguin A, Lavoisier A. Premier memoire sur las respiration des animaux. *Mem Acad R Sci.* 566-584, 1789.
147. Sharma PS, Dash VB. *Agnivesa's Caraka Samhita.* Vol II. Varanasi, India: Chowkhamba Sanskrit Series Office, 1976.
148. Sharma PS, Dash VB. *Agnivesa's Caraka Samhita.* Vol III. Varanasi, India: Chowkhamba Sanskrit Series Office, 1977.
149. Siegel RE. Galen on 'natural', 'non-natural' and 'counternatural' conditions of health and disease In: *Galen on Psychology, Psychopathology, and Function and Diseases of the Nervous System.* Basel: Karger, 1973, 227-230.
150. Sigerist HE. *A History of Medicine.* Vol II. New York: Oxford University Press, 1961.
151. Sinclair J. *The Code of Health and Longevity; or, A Concise View of the Principles Calculated for the Preservation of Health and the Attainment of Long Life.* Edinburgh: Arch. Constable & Co., 1807.
152. Singer C. *Greek Biology and Greek Medicine.* Oxford: Oxford University Press, 1922.
153. Smith E. Inquires into the quantity of air inspired throughout the day and night and under the influence of exercise, food, medicine, temperature, etc. *Proc Royal Soc.* 8: 451-454, 1857.
154. Smith E. Experimental inquires into the chemical and other phenomena of respiration and their modifications by various physical agencies. *Phil Trans.* 149: 681-714, 1859.
155. Steinhaus AH. Chronic effects of exercise. *Physiol Rev.* 13: 103-145, 1933.
156. Tan SY, Yeow ME. Medicine in Stamps Paracelsus (1493-1541): The man who dared. *Singapore Med J.* 44: 5-7, 2003.
157. Tipton CM. Exercise physiology, part II. In: Massengal JD, Swanson RA, eds. *The History of Exercise and Sport Science.* Champaign, IL: Human Kinetics, 1997, 396-438.
158. Tipton CM. Historical perspective: Origin to recognition. In: Tipton CM, ed. *ACSM'S Advanced Exercise Physiology*. Philadelphia: Lippincott Williams & Wilkins, 2006, 11-38.
159. Tipton CM. Historical perspective: The antiquity of exercise, exercise physiology and the exercise prescription for health. In: Simopoulos AP, ed. *World Review of Nutrition and Dietetics.* Vol 98. Basel: Karger, 2008, 198-245.
160. Tipton CM. Milestone of discovery. In: Tipton CM, ed. *ACSM's Advanced Exercise Physiology.* Philadelphia: Lippincott Williams & Wilkins, 2006, 34.
161. Tipton CM. Susruta of India: An unrecognized contributor to the history of exercise physiology. *J Appl Physiol.* 104: 1553-1556, 2008.
162. Tipton CM. The autonomic nervous system. In: Tipton CM, ed. *Exercise Physiology: People and Ideas.* New York: Oxford University Press, 2003, 188-236.
163. Tipton CM. The language of exercise. In: Tipton CM, ed. *ACSM's Advanced Exercise Physiology.* Philadelphia: Lippincott Williams & Wilkins, 2006, 3-10.
164. Tipton CM, Berryman JA, Kroll W. George Wells Fitz of Harvard: A forgotten advocate of exercise physiology. *FASEB J.* (Abstract) p. 53, 1997.
165. Tissot J-C. *Gymnastique Medicinale et Chirurgicale* [translated by Light E, Light S]. New Haven, CT: Elizabeth Light, 1964.
166. Turner CD. *General Endocrinology.* 4th ed. Philadelphia: W.B. Saunders, 1966.
167. Unschuld PU. History of Chinese medicine. In: Kiple K, ed. *Cambridge World History of Human Disease*. Cambridge, Great Britain: Cambridge University Press, 1993.
168. *Vegetius: Epitome of Military Science* [translated with notes and introduction by Milner NP]. Liverpool: Liverpool University Press, 1993.

169. Veith I. *The Yellow Emperor's Classic of Internal Medicine*. Berkeley, CA: University of California Press, 2002.
170. Vogel CJDE. *Pythagoras and Early Pythagoreanism.* Assen, The Netherlands: Royal Van Gorcum & Co., 1966.
171. Waller A. The sense of effort: An objective study. *Brain.* 14: 179-249, 1891.
172. Watson GR. *The Roman Soldier.* Bristol: Thames & Hudson, 1969.
173. Webster G. *The Roman Imperial Army.* 3rd ed. Norman, OK: University of Oklahoma Press, 1998.
174. Wellcome Historical Medical Museum. *Prehistoric Man in Health and Sickness.* London: Oxford University Press, 1951.
175. West JB. Pulmonary blood flow and gas exchange. In: West JB, ed. *Respiratory Physiology: People and Ideas*. New York: Oxford University Press, 2003, 1-38.
176. Westhall C. *The Modern Method of Training for Running, Walking, Rowing and Boxing, Including Hints on Exercise, Diet, Clothing, and Advise to Trainers.* 7th ed. London: Ward, Lock, & Tyler, 1863.
177. Williamson CS. The effects of exercise on the normal and pathological heart: Based on the study of one hundred cases. *Am J Med Sci.*149: 492-503, 1915.
178. Wilson LG. William Croone's theory of muscular contraction. *Notes and Records, Royal Soc Lon.* 16: 158-178, 1961.
179. Withington ET. *Medical History from the Earliest Times: A Popular History of the Healing Art.* London: Scientific Press, 1894.
180. Wong C, Lien-Teh W. *History of Chinese Medicine.* 2nd ed. Shanghai: National Quarantine Service, 1936.
181. Woody T. Philostratus: Concerning gymnastics. *Res Quart.* 7: 3-26, 1936.
182. Zadek L. Die Messung des Blutdruks am Menschen mittelsdes Basch'schen apparatus. *Zeitschr f Klin Med.* 2: 509, 1881.
183. Zuntz N, Hagermann O. Untersuchungen uber den Stoffwechsel des Pferdes bei Ruhe und Arbeit. *Landwirtschaft Jahr* 27 Erganbd 3: 1-338, 1898.
184. Zuntz N, Schumburg W. *Studien zu einer Physiologie des Marches.* Berlin, Hirschwald, 1901.

CHAPTER 2

Influence of Scandinavian Scientists in Exercise Physiology

P.-O. Åstrand, MD

This article, based on an invited lecture presented at the 1988 Annual Meeting of the American College of Sports Medicine, originally appeared in the following publication: Åstrand P-O. "Influence of Scandinavian scientists in exercise physiology." *Scand J Med Sci Sports* 1: 3-9, 1991. It is reprinted, by permission, from P-O. Åstrand, 1991, "Influence of Scandinavian scientists in exercise physiology," *Scandinavian Journal of Medicine & Science in Sports* 1: 3-9 (with minor edits for style).

By 1890 Christian Bohr had interpreted the O_2 and CO_2 transfer between alveolar air and capillary blood as an energy-demanding secretion of the gases. August Krogh (1874-1949) was a brilliant student in biology. Interested from childhood in insects, he studied plant and animal physiology at the university. Krogh was fascinated by Bohr's lectures and had a chance to work with him (1897). The first joint effort was a study of gas exchange over the skin and lung of the frog. The results were published in 1898 with Bohr as author. Krogh said that an acknowledgement was good enough and that he did not like to publish in German! For evident reasons he became involved in Bohr's studies of gas exchange in the lungs. When studying the frog's respiration, Krogh found that the available equipment used to analyze gas tensions in air and fluids was not accurate enough. He was a genius and very imaginative in inventing and constructing instruments and equipment (including those that measured gas tension). His microtonometer made it possible to describe the Bohr effect. It was gradually improved and allowed him to follow the gas tension of tiny air bubbles in fluid (such as blood) with a volume of 0.03 mm^3. In 1905 Krogh taught a course in physiological chemistry for medical students. He was not happy because there were too many students for laboratory experiments to be efficient and because one of the students, Marie Jorgensen (1874-1943), was a woman. However, he was very impressed by her talents. In 1905 they married and in 1907 she received her medical degree. Krogh was skeptical about the accuracy of the measurements behind Bohr's gas secretion theory and in 1906 August and Marie Krogh performed experiments that convinced them that gas exchange in the lungs took place exclusively by diffusion. The results were not published, however, until 1910. Krogh was still in Bohr's laboratory. Bohr had been kind and supportive of the young Krogh, but Bohr was still convinced that his secretion theory was correct. He refused to discuss the matter and the situation in the laboratory became tense. Krogh considered scientific controversy to be an important and stimulating way to approach the truth. However, he did not want to publish the controversial data until he and his wife had waterproof scientific support for the diffusion theory. They named the seven papers published in 1910 "the seven small devils" (1). They spent the summer in Greenland studying the diet and metabolism of Eskimos (figure 2.1). This experience strengthened their interest in measuring metabolic rate during exercise under well-controlled conditions. When back in Copenhagen, August Krogh was appointed lecturer in physiology, but it was not until 1910 that he could move into his own department (in English: the Laboratory of Zoophysiology, Copenhagen University). The Krogh family had their home in the same building. A physician from Hasselbalch's laboratory, Johannes Lindhard (1870-1947), became August Krogh's close coworker. He received a position in the department of theory of gymnastics at the University of Copenhagen in 1909 but had no laboratory facilities. Krogh therefore offered him space. Together they began to study the regulation of respiration and circulation both at rest and during and after exercise. Their first joint publication was "Measurements of the Blood Flow Through the Lungs of Man" (2).

No. 1—20 on	J. L.	age 41	weight 62·5 kg
No. 21—36 on	A. K.	age 37	weight 59·5 kg
No. 37—41 on	Mrs. M. K.	age 37	weight 65·0 kg
No. 42—43 on	H. P.	age 14	weight 53·5 kg

Figure 2.1 Original presentation of the subjects (2).

Reprinted, by permission, from P-O. Åstrand, 1991, "Influence of Scandinavian scientists in exercise physiology," *Scandinavian Journal of Medicine & Science in Sports* 1: 3-9.

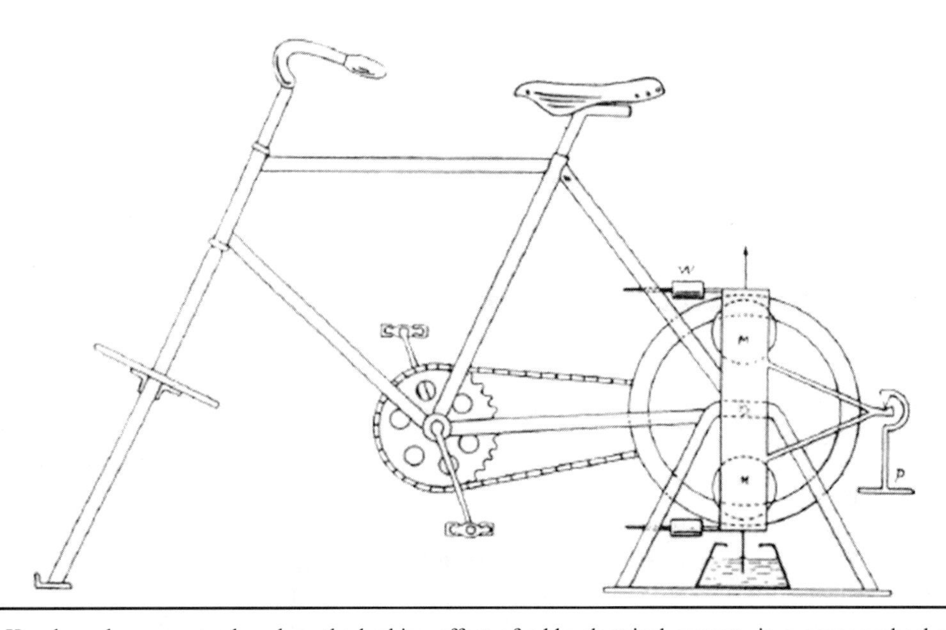

Figure 2.2 Krogh cycle ergometer based on the braking effect of eddy electrical currents in a copper wheel rotating in a magnetic field (M). The current was automatically adjusted so that the braking force could balance the weight on the stand P.

Reprinted, by permission, from P-O. Åstrand, 1991, "Influence of Scandinavian scientists in exercise physiology," *Scandinavian Journal of Medicine & Science in Sports* 1: 3-9.

August and Marie Krogh, Johannes Lindhard, and a 14-yr-old boy were all subjects. Cardiac output was measured using a nitrous oxide method; the highest value reported was 21.6 L/min. In 1913 August Krogh published a description of his new cycle ergometer (figure 2.2) and spirometer. Krogh and Lindhard studied the initial respiratory and circulatory responses to exercise, launching the hypothesis of central commands (figure 2.3). They studied the redistribution of blood flow in the body during exercise. Marie Krogh defended her doctoral thesis in 1914. It was published in 1915 as "The Diffusion of Gases Through the Lungs of Man" (3) and was based mainly on studies on August, Marie, and Marie's patients with different respiratory diseases. The CO method she used to determine the diffusing capacity of the lungs at rest and during exercise became of great importance in clinical medicine after a 35 yr lag.

In 1920 August Krogh was awarded the Nobel Prize in physiology or medicine for his discovery of the capillary motor regulating mechanism. The three papers supporting the prize were published in 1919. In those days the time between publication and the Nobel Prize was not always so long. From summaries of these papers:

> In striated muscles the capillaries are arranged with such regularity along the muscle fibers that each capillary can be taken to supply a definite cylinder of tissue, the average cross-section of which can be determined by counting the capillaries in a known area of the transverse section. A formula is given which allows the calculation of the oxygen pressure head which is necessary and sufficient to supply the muscle with oxygen from the capil-

> laries. The necessary oxygen pressure head deduced from the total number of capillaries is in all cases extremely low. (4) Microscopic observations, chiefly made by reflected light, are recorded to show that in the resting muscles of frogs and guinea pigs most of the capillaries are in a state of contractions and closed to the passage of blood. By tetanic stimulation of the muscles or by gentle massage a large number of capillaries are likewise opened. The number of open capillaries per mm- of the muscular cross section has been counted in such preparations of resting and working muscles. The average diameter of the open capillaries in resting muscles is much less than the dimensions of the red corpuscles which become greatly deformed during their passage. In working muscles the capillaries are somewhat wider. (5)

Repeatedly, new studies that apply modern techniques are published. In general, they confirm Krogh's findings published in 1919.

Krogh was very broad in his research activities and was a true representative for comparative physiologists.

THE REGULATION OF RESPIRATION AND CIRCULATION DURING THE INITIAL STAGES OF MUSCULAR WORK. BY A. KROGH AND J. LINDHARD.

(From the Laboratory of Zoophysiology, University of Copenhagen.)

[*Reprinted from the Journal of Physiology, Vol. XLVII. Nos. 1 & 2, October 17, 1913.*]

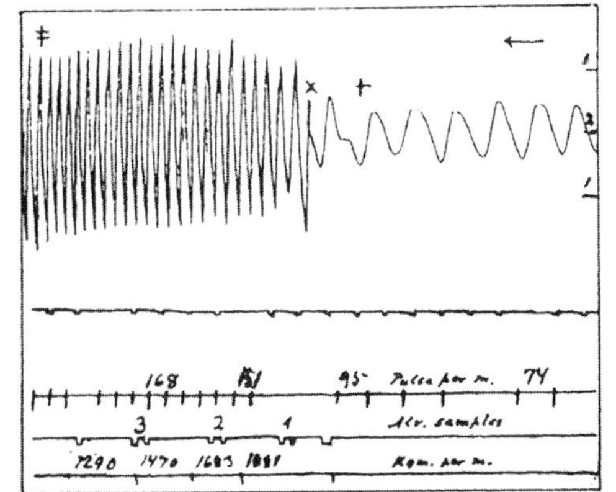

Fig. 1. J. L. Transition from rest to work. Exp. X. Scale in liters. Time in 1/10 minutes. + Ready, × Begin, ∗ Stop.

Figure 2.3 A recording of the tidal air on a spirometer (constructed by Krogh) at rest and at the beginning of exercise.

Reprinted, by permission, from P-O. Åstrand, 1991, "Influence of Scandinavian scientists in exercise physiology," *Scandinavian Journal of Medicine & Science in Sports* 1: 3-9.

He interspaced his mammalian research with studies of gas exchange in invertebrates and lower vertebrates. He studied the exchange of CO_2 between air and sea water, the composition of the atmosphere, the physiology of the blue whale, and the development of the eggs of fishes. The following is a summary of a paper illustrating his scientific skill:

> The tracheal system in the hind legs of a grasshopper can be rapidly and extensively ventilated by the respiratory movements of the insect. About 20% of the air contained can be renewed by one breath. The oxygen percentage in the air of the hind legs is very high during rest (16%) but after exhausting muscular exertions it becomes very low (5%). (6)

As mentioned, Krogh was a genius in inventing devices. In his workshop at the laboratory such devices were built for sale, which gave essential economic support for the research.

In 1910 Krogh was invited to London by John Scott Haldane, who supported the hypothesis that the lungs were glands that secreted oxygen, at least during stressful situations such as exercise and at high altitude. At the lecture, Ernest Henry Starling supported Krogh's diffusion theory. He also visited Claude Gordon Douglas in Oxford. The friendship with his British colleagues lasted over the years. Another good friend from the old days was Frances G. Benedict of the Nutrition Laboratory at the Carnegie Institute of Washington in Boston.

Krogh and Lindhard, with backgrounds in animal physiology and medicine, respectively, created a stimulating environment with a prolific range of research areas. Krogh was not interested in sport, but exercise was an important tool for attacking physiological problems. Lindhard taught students in physical education and, for evident reasons, was interested in studying various sports. Göran Liljestrand from the Karolinska Institute in Stockholm worked at the laboratory from 1918 to 1920. Together with Lindhard he measured cardiac output and oxygen uptake in swimmers, free or tethered. Rowers were also studied. In 1918 Liljestrand measured oxygen uptake during cross-country skiing. Researchers from all over the world visited the laboratory for various periods of time.

The Three Musketeers

In the 1930s, many basic studies were conducted at the laboratory. Frequent authors were Erling Asmussen,

Erik Hohwü-Christensen, and Marius Nielsen, who published individually, together, or in other combinations. August Krogh named them the Three Musketeers. Christensen was born (1904) and raised in a part of Denmark that was dominated by Germany. In 1924 he began his studies in biology, chemistry, philosophy, and geography. Three years later he entered the physical education department in the university. An important part of his education took place in Lindhard's laboratory. Together they planned projects for his thesis. He studied cardiac output with a modified Grollman acetylene method, studied body temperature and blood sugar during heavy exercise, and compared arm versus leg exercise and the effects of training (8). One could say that the five papers published in 1931, and defended with Krogh and Lindhard as opponents, would qualify today for five PhD theses.

As mentioned, Marie and August Krogh studied diet and metabolism in Eskimos. In 1920 August Krogh and Lindhard published a study on the relative value of fat and carbohydrate as sources of muscular energy. They could determine a respiratory quotient (RQ) with an accuracy of 0.005. Chauveau had proposed that carbohydrate was the main substrate for muscular exercise and that fat could not be combusted directly. Later this issue also became a controversy between Lindhard and Archibald Vivian Hill. In correspondence, Lindhard never wrote but he thought, "Hill, that idiot who cannot even determine RQ!" In 1939 Christensen and Hansen published five articles that clearly showed that the proportion of fat and carbohydrate as energy-yielding substrates depends on work rate, duration of heavy exercise, diet during the 3 d before a standard power output, and physical condition—state of training (9). They also illustrated that hypoglycemia could induce fatigue and that glucose intake 3 h before exercise could dramatically induce fatigue in the early stage of exercise because of hypoglycemia.

The literature often refers to their articles, but many authors likely have not studied the original papers, published in German. A common quotation is that Christensen and Hansen "only" determined the effect on endurance time during exercise preceded by various diets. Figure 2.4 is based on their data. By carefully measuring oxygen uptake and RQ, they were able to calculate the total energy output from the metabolism of both fat and carbohydrate.

When Jonas Bergström in 1963 reintroduced the needle for muscle biopsies, originally designed by Duchenne, a new field in human skeletal muscle histochemistry was opened. It was quite logical that at an early stage Christensen's and Hansen's protocols, published in 1939, were repeated at Christensen's department in Stockholm. The old data were confirmed, now with direct determination of changes in muscle glycogen concentrations (10).

Krogh had been interested in the effects of temperature on respiration in fishes, frogs, and insects but also in temperature regulation in humans. When Christensen began at the laboratory, physiologists were debating whether the increase in body temperature during exercise was a failure of the capacity to eliminate pro-

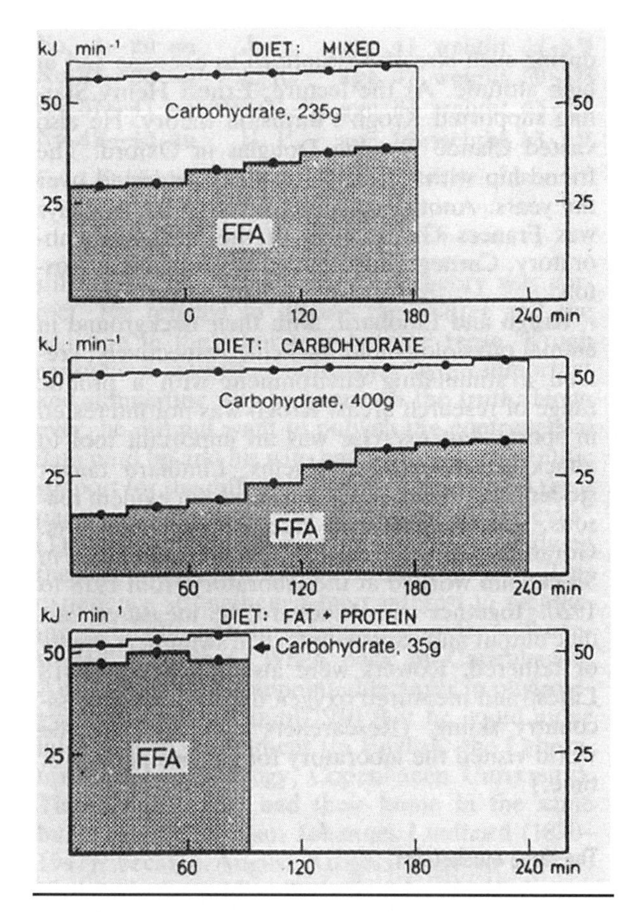

Figure 2.4 Increase in free fatty acid metabolism in prolonged exercise. One well-trained subject exercised on a cycle ergometer at 183 W after consuming a mixed diet and then at 176 W after consuming a carbohydrate-rich diet for 3 d. In another experiment, exercising at 176 W was preceded by a 3 d period of eating fat and protein and excluding carbohydrate from the diet. The subject exercised until exhausted. The total energy output was calculated from the measured oxygen uptake and respiratory quotient (RQ) during 15 min periods; the energy yield from carbohydrate and free fatty acids was estimated from the RQ values. The calculated total carbohydrate consumption (g) is presented. Note how exercise time and the diet affect the choice of substrate. At a given rate of exercise, the endurance time varied from 93 to 240 min depending on the diet. The subject's maximal oxygen uptake was not determined (9).

Reprinted, by permission, from P-O. Åstrand, 1991, "Influence of Scandinavian scientists in exercise physiology," *Scandinavian Journal of Medicine & Science in Sports* 1: 3-9; Data from Christensen and Hansen 1939.

duced heat or a local phenomenon when rectal temperature was measured during cycling or running. Or was it regulated? Christensen found that rectal temperature increased to a steady state and was related to work rate. It also increased in arm exercise. One of Christensen's subjects in many studies was Marius Nielsen (born 1903), a graduate in biology who took over the studies on temperature regulation. In a climatic chamber with a Krogh cycle ergometer on a scale that could register weight changes as small as 2 g, good control of radiation, convection, sweat, and water evaporation was achieved. The body's heat production was calculated from oxygen uptake and mechanical efficiency. Nielsen found that rectal temperature was regulated according to power output and within a wide range independent of the environmental temperature (11). For me, it was self-evident to include illustrations from Nielsen's study in a modern textbook of physiology (12, pp. 595-598). Studies related to temperature control continued. Bodil Nielsen, Erling Asmussen's daughter, was one of Marius' subjects in 1956 and later became a student. There have been many publications by Nielsen and Nielsen (these Nielsens were not married) over the years, and Bodil Nielsen continues the temperature line at the August Krogh Institute in Copenhagen.

Marius Nielsen's thesis, published 2 yr before the temperature monograph, summarized careful studies on the respiratory response to various combinations of exercise at different work rates, hypoxia, and metabolic and respiratory acidosis under strictly controlled conditions. It is a classic paper in respiratory physiology (13).

In 1930 Erling Asmussen (born 1907) graduated in zoology and botany. After that he attended both Krogh's and Lindhard's lectures. He met Christensen, who knew that Lindhard was looking for an assistant. Asmussen said that he would try, and because he did not ask for a salary Lindhard accepted. The first task was to isolate single fibers from musculus semitendinosus in frogs for studies of electrical events and mechanisms behind the contraction of the fiber. He received a modest salary in 1933 and defended his thesis on the mechanical reaction of the skeletal muscle fiber with Krogh and Lindhard as very satisfied opponents. In 1932 Lindhard invited Fritz Buchtal (born 1907), who felt very unsafe in Germany, to come to the department. Buchtal was an expert in electrical recordings from muscles and nerves and gradually took over the neuromuscular activities of the department. One result of the joint efforts was the DISA electromyograph. During World War II both Buchtal and August Krogh went to Sweden and Krogh demonstrated too openly against the Germans. After the war Buchtal got a new laboratory in clinical neurophysiology.

Asmussen gradually began to work with Christensen and Nielsen, and the Three Musketeers were very productive. Space limitations, however, permit only a sketchy summary. They studied the regulation of respiration and circulation at rest and during exercise, including the question of how peripheral and centrally induced impulses contribute to regulation. They studied the circulatory response to different body positions, with the subjects lying in a horizontal position on a tilting table alternately tilted 60° up or down. In the head-down position the blood volume in the legs was reduced, and with inflated blood pressure cuffs on the thighs the legs were still empty, even in the head-up position. Exercise could therefore be performed with variations of the central blood volume. References to many of the classical studies conducted during this very productive time can be found in a review by Asmussen (14). It also reflects many of the activities at the Laboratory for the Theory of Gymnastics at University of Copenhagen during the 1940s and 1950s. (From 1927 the laboratory was located at the Rockefeller Institute and from 1970 it was located at the August Krogh Institute.)

Later in his career Asmussen returned to his interest in skeletal muscles and was then interested in human muscles in situ. He studied factors behind maximal strength; eccentric, isometric, and concentric activation; strength; and performance in growing individuals.

Ole Bang was a physician at a laboratory for medicine and physiology, and his teacher was Ejnar Lundsgaard. Bang collaborated in his PhD studies with Christensen and his group. He made extensive and careful studies on lactate concentrations during and after exercise of different intensities and duration (13). In my opinion, many recent studies on lactate response to exercise merely confirm Bang's data (figure 2.5)! Since Lundsgaard's name was mentioned: At that time Meyerhof, Nobel Prize winner for 1922 (together with A.V. Hill) for his discovery of the fixed relationship between the consumption of oxygen and the metabolism of lactic acid in muscle, stated as follows: "In anaerobically exercising, normal muscle, phosphagen [at that time PCr] is not at all broken down but only transformed into a labile form." In 1931 Lundsgaard published data showing that a muscle poisoned by monoiodoacetic acid, which completely inhibits lactate production, could contract for some time before it stopped in a state of rigor. Phosphagen was broken down—and Meyerhof found that he was right.

Ove Böje followed traditional lines and studied oxygen diffusion in the lungs at rest and during exercise and blood sugar concentration during exercise after the consumption of different diets (13).

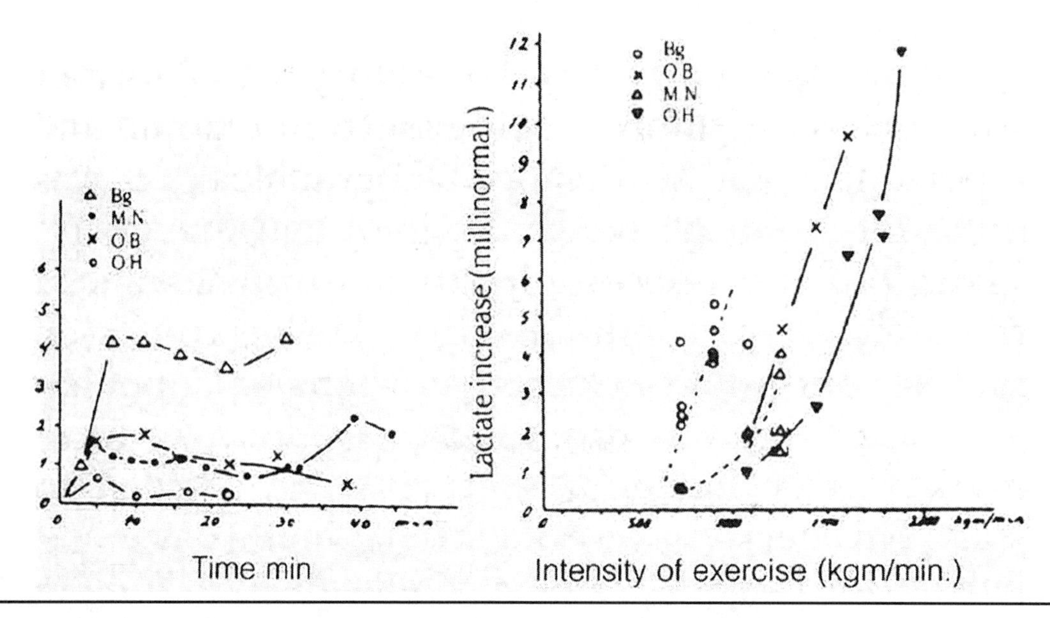

Figure 2.5 Original graphs from Bang's thesis on blood lactate concentrations during prolonged exercise (left) and during an incremental test (right) in four subjects of different training status.

Reprinted, by permission, from P-O. Åstrand, 1991, "Influence of Scandinavian scientists in exercise physiology," *Scandinavian Journal of Medicine & Science in Sports* 1: 3-9; Data from Christensen and Hansen 1939.

The Three Musketeers Abroad

August Krogh frequently made trips abroad, including to Boston to visit one of the leading institutes in exercise physiology of which L.J. Henderson (known for the Hasselbach-Henderson formula) was head. (After 1927, D. Bruce Dill was head.) It was the Harvard Fatigue Laboratory. Krogh recommended that his young colleagues widen their views by visiting this laboratory. With a Rockefeller fellowship, Christensen joined the staff in 1935. He participated in a famous expedition to the Andes in Chile together with Dill, H.T. Edwards, and W.H. Forbes. They measured pulmonary ventilation, oxygen uptake, heart rate, and blood gases at different altitudes up to 5,340 m at rest and during exercise on a cycle ergometer. They noticed that oxygen uptake, heart rate, and pulmonary ventilation reduced to standard temperature and that pressure (dry) was constant at a given submaximal work rate (15). In 1938 Marius Nielsen visited the Harvard Fatigue Laboratory, and in 1939 Asmussen went over. Together, they participated in field studies on sharecroppers working in the cotton fields in hot, humid Mississippi in the United States. Nielsen also had a sojourn at the John B. Pierce Foundation Laboratory in New Haven, Connecticut with C.-E.A. Winslow and A.P. Gagge as stars.

Asmussen participated in an expedition to Mt. Evans in Colorado (4,300 m). In fact, high-altitude physiology had a long tradition in Copenhagen, even if the highest hill in Denmark is only 201 m! Responses to high altitude provide fascinating physiology. In 1913 Krogh and Lindhard obtained a low-pressure chamber and Lindhard spent prolonged times there. He wrote most of his thesis during a 25 d stay at a simulated altitude of 60.7 kPa (455 mmHg)! In the 1930s hypoxia experiments included tests of air force pilots.

One of the Musketeers Goes to Sweden

In 1941 Christensen was elected as the first professor at a new department in physiology at the College of Physical Education (Gymnastik- och Idrottshögskolan) in Stockholm. It was a dramatic change for him. He came from a stimulating and rich scientific environment to nothing in Stockholm. It was definitely a challenge. Not until 1944 could he move into a new laboratory with two former students as assistants, two refugees from Estonia, and a person for the workshop. Geographically it was quite far from the medical school and university. Therefore, the opportunity to perform advanced research was very limited at first. He has confessed that he seriously considered returning to Copenhagen, where he was promised a position. Because Denmark was occupied by Germany, he decided to stay. This decision was of extreme importance and affected the development, in Sweden, of human biology, including exercise in various environments (such as the workplace), and in leisure.

Jöns Johansson (1862-1938) was professor of physiology at the Karolinska Institute, a medical school, in 1901. He had close contact with Krogh and sent his foremost student, Göran Liljestrand (1886-1968), to work with Krogh and Lindhard. When back in Sweden he became professor of pharmacology, where he continued his strong engagement in physiological research in the field of respiration and circulation but now worked on animals. For many years he was chief editor of *Skand Arch Physiol* (first volume issued in 1889; in 1940 the name was changed to *Acta Physiologica Scandinavica*). From 1918 to 1960 he also served as secretary of the Nobel Committee for Physiology or Medicine. He was a strong advocate behind the choice of Christensen for the position in Stockholm.

Another Swede should be mentioned: Elis S. Henschen, professor of medicine. He was the first to recognize the athlete's heart by carefully performed percission. He concluded that "skiing causes an enlargement of the heart and that this enlarged heart can perform more work than the normal heart. There is, therefore, a physiological enlargement of the heart due to athletic activity: the athlete's heart." He also stated, "Big hearts win races!" (16).

I was a student at the college from 1944 to 1946 and found Christensen's lectures to be very stimulating. He advised me to read Lundsgaard's excellent textbook in physiology for medical students. I apparently behaved well and was offered a position as assistant at his department. I accepted happily. (In those days a salary of US $64/mo was enough!) He recommended following up on Sid Robinson's thesis on physical fitness, published in 1938 (17), and said that we should include girls and women. Altogether 227 subjects were studied (age range: 4-33 yr). The dissertation was slightly delayed because in 1947 I began studying at the medical school. My opponent in the 1952 event was Torgny Sjöstrand (18). In Copenhagen he participated in some of the studies on the circulatory response to tilting. He is the father of clinical physiology as a specialty in Sweden, and many representatives from this discipline have made important contributions in exercise physiology.

What about Norway and Finland? Finland is not part of Scandinavia. This summary only discusses the scientific activities up to about 1950. I should also mention Per Scholander, born in Oslo in 1905, because he invented the ingenious gas analyzer so familiar to exercise physiologists. This analyzer made it possible to analyze, with a chemical method, smaller gas samples with higher concentrations of oxygen than with the Haldane apparatus. His area of interest included physiology of diving, climatic adaptations of arctic and tropical animals and plants, and the osmosis mechanism.

Conclusion

I have mentioned that August Krogh did not like to publish in German. Before World War II, most of the other publications were in German. This can explain why many papers from this era are neglected!

August Krogh and Johannes Lindhard worked together for many years but never became August and Johannes when they talked to each other. Lindhard retired in 1935 and Krogh in 1945, but both were scientifically active for the rest of their lives. Krogh was not interested in sport but he was very active physically. He loved walking, rowing, canoeing, and sailing. He brought a kayak home from Greenland. His daughter Bodil emphasized that he was a good father and husband. The family often enjoyed long weekend trips.

At our department we enjoy tremendously seeing Hohwü-Christensen almost every day. (Note: Hohwü-Christensen has passed away since the publication of the original article.)

How should one recruit young, upcoming scientists? At a lecture Krogh gave to the American Academy of Arts and Sciences in Boston on May 10, 1939, someone asked, "You actually advocate a system which would discourage students from going into physiological research?" Krogh replied, "So much better. Students should be discouraged from choosing a career involving research, and we would welcome only those students whose urge is strong enough to overcome discourages and difficulties. But then we must take [good] care of them!"

Acknowledgments

I appreciate very much invaluable information from conversations with Erik Hohwü-Christensen and written material by Erling Asmussen and Bodil Schmidt-Nielsen (now Chagnon). Bengt Saltin suggested definite improvements in my manuscript. I thank Margareta Fästh for typing the manuscript, Styrbjörn Bergelt for preparing the illustrations, and Andrew Cresswell for checking the English.

References

1. Schmidt-Nielsen B. August and Marie Krogh and respiratory physiology. *J Appl Physiol.* 1984; 57: 293-303.
2. Krogh A, Lindhard J. Measurements of the blood flow through the lungs of man. *Skand Arch Physiol.* 1912; 27: 103-125.
3. Krogh M. The diffusion of gases through the lungs of man. *J Physiol.* 1915; 99: 271-300.
4. Krogh A. The number and distribution of capillaries in muscles with calculations of the oxygen pressure head necessary for supplying the tissue. *J Physiol.* 1919; 52: 409-415.
5. Krogh A. The supply of oxygen to the tissues and the regulation of the capillary circulation. *J Physiol.* 1919; 52: 457-474.
6. Krogh A. On the composition of the air in the tracheal system of some insects. *Skand Arch Physiol.* 1913; 29: 29-36.
7. Krogh A, Lindhard J. The relative value of fat and carbohydrate as sources of muscular energy. *Biochem J.* 1920; 14: 290-363.
8. Christensen EH. Beiträge zur Physiologie schwerer Köperlicher Arbeit. 5 Mitteilungen. *Arbeitsphysiologie.* 1931; 128-202, 453-502.
9. Christensen EH, Hansen O. I, II, III, IV, V. *Skand Arch Physiol.* 1939; 81: 137-189.
10. Bergström J, Hermansen L, Hultman E, Saltin B. Diet, muscle glycogen, and physical performance. *Acta Physiol Scand.* 1967; 71: 140-150.
11. Nielsen M. Die Regulation der Körpertemperatur bei Muskelarbeit. *Scand Arch Physiol.* 1938; 79: 193-230.
12. Åstrand P-O, Rodahl K. *Textbook of Work Physiology.* New York: McGraw-Hill; 1986.
13. Bang O, Böje O, Nielsen M. Contributions to the physiology of severe muscular work. *Skand Arch Physiol.* 1936; 74(Suppl. 10): 1-208.
14. Asmussen E. Muscular exercise. In: Fenn WO, Rahn H, eds. *Handbook of Physiology—Respiration II.* Washington, DC; 1965: 939-978.
15. Christensen EH. Der Kreislauf in grossen Höhen. *Skand Arch Physiol.* 1937; 76: 75-100.
16. Henschen ES. Skidlöpning och skidtäfling. In: *Reports from Uppsala University (Annual Report, 1897)*; 1898: 1-69.
17. Robinson S. Experimental studies of physical fitness in relation to age. *Arbeitsphysiologie.* 1938; 10: 251-327.
18. Åstrand P-O. *Experimental Studies of Physical Working Capacity in Relation to Sex and Age.* Copenhagen: Munks-Gaard; 1952: 1-171.

CHAPTER 3

Contributions From the Harvard Fatigue Laboratory

Charles M. Tipton, PhD

G. Edgar Folk Jr., PhD

Introduction

Despite its brief history (1927-1947), no physiology laboratory in America is more revered than the Harvard Fatigue Laboratory. Described as "the first laboratory for the comprehensive study of man" (5), it was perhaps more influential and effective in promoting scientific and collaborative research in exercise physiology (76). This chapter discusses the laboratory's contributions to the study of the acute and chronic effects of exercise and to the effects of altitude and temperature on the exercise response. This chapter includes information from the laboratory's extensive published record as well as from the perspective of the last surviving member of its faculty and staff (figure 3.1) (41).

Harvard University (1920-1927)

In 1920, physiology at Harvard University was represented by four departments: physiology, comparative physiology, applied physiology, and physical chemistry. These departments were collectively known as the laboratories of physiology (53). Three years later, the department of industrial hygiene was transferred from the medical school to the recently established school of public health, facilitated by the efforts of Roger Lee (53). Subsequently, the department of industrial hygiene was renamed the department of applied physiology. At this time, Walter B. Cannon (1871-1945) was chairman of the department of physiology, David Edsall was dean of both the medical school and the school of public health, Lawrence Joseph Henderson (1888-1984) was director of the department of physical chemistry (figure 3.2), Wallace B. Donham was dean of the business school, and A. Laurence Lowell was president of the university (53). Also at this time, Dr. Arlie V. Bock was establishing a laboratory in Massachusetts

Figure 3.1 *(a)* G. Edgar Folk (1914-) of the University of Iowa in the United States who became a staff member of the Harvard Fatigue Laboratory in 1943. *(b)* A 1990 photograph of Professor Folk while he was bird watching on his farm in Iowa City, Iowa. Folk remained at the Harvard Fatigue Laboratory until its closure in 1947. He subsequently spent six yr at Bowdoin College in Maine conducting research and teaching biology to undergraduate students. In 1953, he accepted an appointment in the department of physiology at the University of Iowa. Professor Folk continues to attend APS meetings, write manuscripts, and enjoy life as an elder statesman for physiology.

(a) Courtesy of G. Edgar Folk, Jr. *(b)* From the collection of Charles M. Tipton.

General Hospital after being a physician in World War I, a Moseley Traveling Fellow in Sir Joseph Barcroft's laboratory for two years learning exercise protocols and blood-equilibration techniques, and after being a participant in Barcroft's high-altitude expedition in Peru (23) (figure 3.2).

David Bruce Dill (1891-1986) (figure 3.3) received his PhD in chemistry in 1925 from Stanford University and accepted a 2 yr National Research Council fellowship to work in the laboratory of L.J. Henderson to study the physical chemistry of proteins. However, after his arrival at Harvard, he was assigned to a Massachusetts General Hospital laboratory, where the senior staff physician was Dr. John H. Talbott. There he was reassigned to study with Dr. Bock the physiochemical properties of blood (46). When the fellowship was terminated, Dill was appointed as assistant professor of biochemistry in the school of public health, a position he retained until 1936. In addition, from 1927 to 1947, he held a professorship in the department of industrial hygiene at Harvard University (72). Although during his tenure at Harvard Henderson remained as director of the laboratory, it was Dill's leadership, organizational ability, and scientific insights that made the laboratory famous throughout the world. Between 1947 and 1961, Dill served as director of research for the U.S. Army Chemical Research and Development Laboratory. During this interval, he became president of the American Physiological Society (1950-1951) and later (1960-1961) was elected president of the American College of Sports Medicine (45, 74).

Figure 3.2 Harvard University faculty members who became the foundation for the Harvard Fatigue Laboratory. This 1926 photograph shows Lawrence Joseph Henderson (seated on the bench with the pipe), who was appointed director of the Harvard Fatigue Laboratory in 1927, Arlie V. Bock (seated next to Henderson), John S. Lawrence (standing on the left), Lewis Hurxthal (standing in the center), and David Bruce Dill (standing on the right).

Reprinted from D.B. Dill and V. Arlie, 1981, "Block-physiologist," *The Physiologist* 24: 11-13. With permission of American Physiological Society.

Henderson and the Establishment of the Harvard Fatigue Laboratory

Henderson, a recipient of an MD from Harvard in 1902, was professor of biological chemistry from 1919 until 1934 and the Abbot and James Lawrence professor of chemistry from 1934 until his death in 1942 (18) (figure 3.2). Besides chemistry and biology, Henderson had broad interests in sociology, psychology, anthropology, and the philosophy of Vilfredo Pareto, an Italian engineer and socialist (18, 46). He and Elton Mayo (a professor of industrial hygiene who, like Henderson, believed that workers should be studied in the workplace)

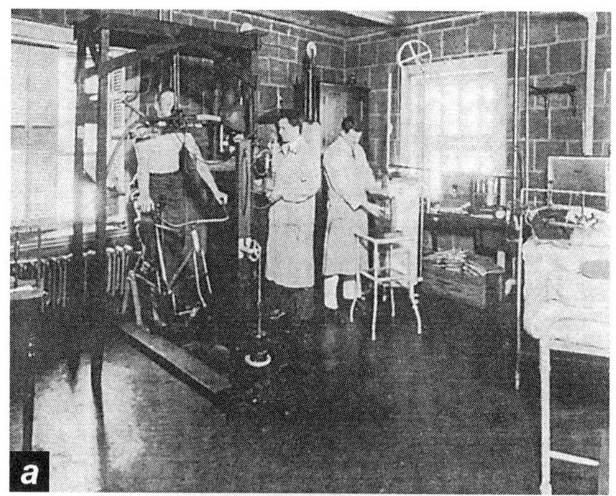

Figure 3.3 *(a)* A 1926 photograph of David Bruce Dill (standing near the subject on the bicycle ergometer) in the laboratory of Dr. Bock. Standing near the gasometer is Lewis Hurxthal. *(b)* A 1938 photograph of David Bruce Dill after he had assumed leadership responsibilities from Henderson to direct the research activities of the Harvard Fatigue Laboratory. As a result of his leadership and accomplishments, the legacy of the laboratory belongs to him.

(a) Reprinted from D.B. Dill and V. Arlie, 1981, "Block-physiologist," The Physiologist, 24: 11-13. With permission of American Physiological Society. *(b)* From the collection of Charles M. Tipton.

developed the concept of establishing a laboratory to conduct research on industrial hazards (46). Such a laboratory would study the "group psychology, the social problems, and physiology of fatigue of normal man . . . not only as individual factors in determining physical and mental health, but more especially to determine their interrelatedness and the effect upon work" (46, p. 20).

With leadership from Henderson, the support of advisory and planning committees that included most of the previously mentioned deans and professors, the endorsement of President Lowell, and funding from the Laura Spelman Rockefeller Memorial and the Rockefeller Foundation, the Harvard Fatigue Laboratory was established in 1927 with Lawrence Joseph Henderson as its official director (46). According to Chapman, the term *fatigue* was selected because all parties believed it was important. However, because they could not agree on a definition, it did not force research activities into a specific departmental shape (19, p. 19). Its home was in the basement of Harvard Business School.

As noted, Dill was designated (although never officially appointed) to organize and direct the research program of the laboratory and quickly assumed the duties and responsibilities of Henderson. Hence, his curriculum vitae in the Mandeville Special Collections Library at the University of California in the United States lists him as the informal director of the laboratory between 1927 and 1946 (72). The Horvaths identified the senior members of the laboratory from 1927 to before World War II as Henderson, Dill, Bock, and Talbott (figure 3.2) (46).

Contributions to Undergraduate and Graduate Student Education

Although the education of undergraduate and graduate students was not a purpose for establishing the laboratory, the laboratory did provide opportunities for undergraduate and graduate students to be introduced to research and become involved with projects that all pertained to physiology and sometimes to exercise physiology. Although the laboratory offered no courses or degrees, it offered opportunities for students to conduct senior theses under the supervision of select faculty members. Twelve undergraduates were involved in activities of the laboratory. Henry Taylor became a renowned exercise physiologist at the University of Minnesota in the United States (figure 3.4). Richard Riley was recognized as an outstanding respiratory physiologist at Johns Hopkins University, and John Pappenheimer and Clifford Barger received acclaim as physiologists on the faculty of Harvard University (46). Sid Robinson and Steven Horvath (figures 3.5 and 3.6) conducted research for their PhD dissertations in the laboratory but received their degrees in biological sciences. Dill served as chairman of Robinson's committee but was unable to do so for Horvath because Horvath had married Dill's daughter (73). G. Edgar Folk Jr. received an MA from Harvard University in 1937 and served as a research associate in the laboratory from 1943 to 1947 (figure 3.1). In 1947, he received a PhD in the biological sciences with John Welch as his advisor (74). Additionally, Pappenheimer, Robinson, and Horvath served along with 13 other staff members as tutors for students in biochemistry (46).

Contributions to Collaborative, University, National, and International Research and Postdoctoral Training

The Harvard Fatigue Laboratory is remembered as a unique laboratory because during its era it served as a mecca for collaborative research by established or beginning investigators and fellows from Harvard University, American universities, and foreign universities or institutes. According to the Horvaths, 41 investigators were from Harvard or other American universities and 35 fellows arrived from 15 foreign countries. Affiliated investigators or fellows who subsequently became prominent in areas directly or indirectly related to exercise physiology were Adolph, Asmussen, Brouha, Chapman, Christensen, Cotton, Margaria,

Figure 3.4 Henry Longstreet Taylor (1912-1993) was a Harvard University undergraduate student during the early years of the Harvard Fatigue Laboratory. Later he was associated with Ancel Keys at the University of Minnesota and became renowned for his knowledge, investigations, and mentoring in exercise physiology.

From the collection of Charles M. Tipton.

Metheney, Missuro, Nielsen, and Scholander (46). Although August Krogh, a Nobel laureate in 1920 (56), has been cited by many authors as being a fellow or a collaborative investigator associated with the laboratory, this listing is an error. According to conversations between C.M. Tipton and Krogh's daughter, Bodil Schmidt-Nielsen, Krogh was a visitor but never a collaborator.

Figure 3.5 Sid Robinson (1902-1982) of Indiana University in the United States was Dill's first PhD student. His thesis (64), which pertained to maximal oxygen consumption and aging, has become a classic that continues to be cited.

From the collection of Charles M. Tipton.

Figure 3.6 Steven Horvath (1911-2007) of the University of California at Santa Barbara was the second PhD student mentored by Dill at the Harvard Fatigue Laboratory. Because he was married to Dill's daughter, Horvath had to select another professor to serve as chairman (73). Steve and Elizabeth Horvath were responsible for the most comprehensive text concerning the history of the Harvard Fatigue Laboratory (46).

From the collection of Charles M. Tipton

Although published evidence is lacking, it is the belief of the authors that the impressive, successful, and productive collaborative practices of the Harvard Fatigue Laboratory for investigators and fellows were responsible for the post-World War II establishment of governmental fellowships for postdoctoral training.

Contributions to the Literature of Exercise Physiology

Perspective

By 1927, the existing body of knowledge was sufficient to establish the discipline of exercise physiology. Because of the activity of the Harvard Fatigue Laboratory and the subsequent leadership of Dill, after 1927 exercise physiology began to be accepted more as a component of physiology and less as a component of physical education (73). Contrary to the impression created by early exercise physiology textbooks, this specific discipline was only one of several areas investigated by faculty and staff. Of the 316 peer-reviewed publications listed by McFarland and colleagues that were attributed to the Harvard Fatigue Laboratory (56), 17% pertained to exercise physiology. Although the Horvaths' chapter titled "Exercise Physiology—A Study in Homeokinesis" is informative, it lacks specificity. Hence, this section specifies the areas and select citations that relate to acute and chronic exercise. Between 1928 and 1932, investigators affiliated with the Harvard Fatigue Laboratory published 7 manuscripts with the title "Studies in Muscular Activity," which were included in the third edition of the 1933 text *The Physiology of Muscular Exercise* (6).

Central Nervous and Autonomic Nervous Systems

Surprisingly, no publications dealt with the central nervous system during exercise, although the subject was effectively covered in *The Physiology of Muscular Exercise*. However, the Harvard Fatigue Laboratory did investigate the effects of sympathectomy on the exercise responses of dogs. In 1935, Saman of Belgium reported that select sympathectomy procedures increased the maximum capacity for work, a finding that differed from the 1929 results of Campos, Cannon, and colleagues (17). Between 1936 and 1939, Lucien Brouha (from Belgium), Cannon, and Dill studied the influence of cardiac sympathectomy using surgical procedures that completely eliminated the cardioaccelerator fibers that leave via the vagus nerve and enter the vagosympathetic trunk. Their investigations indicated that dogs could perform heavy exercise, but there was no evidence for an augmented work response. Cardiac acceleration occurred but it was not the result of epinephrine, elevated temperature, or sympathin (norepinephrine). Rather, it was the result of augmented vagal inhibition and increased vagal cardioacceleration (14). In addition to heart rate and work performed, measurements were made of blood lactate and rectal temperatures. When comparisons were made at the same workload, the results were similar to those recorded in normal dogs. However, when bilateral vagotomy was performed on sympathectomized animals, both parameters exhibited marked increases, cardiac acceleration decreased, respiratory responses were impaired, and the amount of work performed was reduced (16).

Respiratory System

In early studies (9, 71) authors served as the subjects, the bicycle ergometer or treadmill were the exercise modalities, and measurements pertained to respiratory rates, pulmonary ventilation, oxygen consumption, and arterial and venous PO_2 and PCO_2 results. Frequently, Clarence DeMar, a marathon runner, was a subject. Upper-limit respiratory values were 25 to 37 beats/min for rates, 50 to 90 L/min for V_E, and 40 to 70 mmHg for venous CO_2 levels. Linear relationships existed between metabolic rate and ventilation. However, a curvilinear relationship prevailed between blood flow and ventilation (9).

In a study conducted by Dill and colleagues, 10 nontrained subjects were given the task of running 20 min on a treadmill at a speed of 9.3 km/h to determine the variability of cardiorespiratory and hematological responses (34). They reported that ventilation would reach a steady state in 3 to 4 min if lactic acid was not accumulating rapidly and that there was no relationship between a decrease in pH and an increase in V_E (34). To gain insight into the nature of the stimuli (neural reflex or chemical) that increased the ventilator response to exercise, they curtailed circulation to the arms and legs in separate light- and heavy-exercise experiments (3). They concluded that the chemical stimulus was stronger than the neural stimuli for increasing ventilation but were unable to eliminate the influence of practice on the response.

Cardiovascular System

Interestingly, the first publication attributed to the Harvard Fatigue Laboratory pertained to the use and perfection of the Haldane method for the measurement of blood flow (8). During ergometric exercise by four subjects, oxygen consumption values ranged from 1.75 to 2.5 L/min, heart rates ranged from 108 to 177 beats/min, and cardiac output results were between 14.8 and 25.0 L/min. Stroke volumes ranged from 90 to 186 ml/beat, and a linear relationship existed between oxygen consumption and stroke volume until oxygen consumption reached 1 L/min after which stroke volume began to plateau. Linear relationships also were reported for blood flow and oxygen consumption and for heart rate and oxygen consumption (8). With exercise, systolic blood pressure increased from 55 to 120 mmHg with diastolic pressures exhibiting increases that seldom exceeded 10 mmHg (8).

Blood as a Physiochemical System

In 1925, Dill was recruited to Harvard as a National Research Council fellow to be with Henderson, who had a strong interest in blood as a physiochemical system. In fact, Henderson published an article in 1921 in *Journal of Biological Chemistry* using the same terminology (46).Of the publications attributed to the Harvard Fatigue Laboratory, 16% pertained to blood (45) with only a few being concerned with the effects of acute and chronic exercise. The 1927 article from Henderson's laboratory is a classic in that it contains resting and exercise (7 times the resting metabolic rate) results pertaining to oxygen capacity (volume percentages), oxygen and carbon dioxide dissociation curves, and profiles for serum, cells, and whole blood values for water, bicarbonate, proteins, chloride, pH, hydrogen ions, carbonic acid, and total CO_2 (7). A related article that included the measurement of lactic acid at sea level and at 6,214 m led to the conclusion that the concentration of lactic acid was maximal when oxygen consumption was also at its maximal level (31). Later, this conclusion became important for the study of the relationship between lactic acid and the oxygen debt.

Recognizing that work cannot continue when blood lactic acid concentrations begin to exceed 10 mEq/L, Dennig, Talbott, Edwards, and Dill investigated the effect of acidosis and alkalosis on the capacity for work. Because of the complications of exercising "sick" individuals, they studied this triad relationship by providing ammonium chloride or sodium bicarbonate before exercising (22). They found that the buffering capacity of blood proteins was decreased by approximately 11% when the carbonic acid buffering capacity was reduced by 50%. Consequently, the ability to neutralize the presence of lactic acid and to increase an oxygen debt is markedly reduced when exercising in an acidotic state. Hence, the ability to accumulate an oxygen debt will be increased if an alkalotic state existed before exercising (22).

Keys and Taylor had subjects run to exhaustion and reported that blood oxygen capacity increased by 10%, serum proteins increased by 17%, new red blood cells entered the circulation, and serum colloid osmotic pressure was decreased (51). Other Harvard Fatigue Laboratory studies indicated that heavy exercise was associated with a 5% to 10% increase in red blood cells (24), a 200% to 300% increase in leukocytes (37), and a femoral vein pH value of 6.78, which is among the lowest ever recorded for an exercising human (24).

In the first decade of the existence of the Harvard Fatigue Laboratory, blood glucose levels were frequently measured during exercise (32, 38, 39) with the intent of learning the relationship between blood glucose concentrations and diet, intensity of exercise, and the respiratory quotient (RQ). Edwards, Margaria, and Dill concluded that a small positive correlation existed between blood sugar level and RQ but the "blood sugar concentration is not the governor" (38, p. 209).

Oxygen Consumption and Metabolic Systems

With the leadership of Bock, the Harvard Fatigue Laboratory quickly developed, perfected, and utilized methodology for measuring the oxygen consumption of exercising subjects (10, 46). Inherent in their methodology was the practice of prescribing exercise for determining oxygen consumption in units of basal or resting metabolic rates (RMR) units. They soon found that during steady-state conditions nontrained subjects in general could seldom exceed 10 times their RMR values whereas trained subjects were able to achieve values 20 or more times higher (46). Dill's approach was to prescribe exercise in accordance with RMR. Light to moderate work was listed as 1 to 3 times RMR, heavy work was listed as 3 to 8 times RMR, and maximal or severe work was always higher than 8 times RMR (46). Measurements pertinent to oxygen uptake, oxygen debt, and substrate utilization were secured during steady-state conditions when subjects ran to exhaustion on a treadmill at 15.7 km/h, and RQ values were used to determine substrate utilization (38). From their respiratory data they felt that carbohydrates were the primary source of energy for muscle contraction and that glycogen was the substrate for moderate work. They also indicated that the low quotients were associated with light workloads and the utilization of fat as substrate, that trained subjects such as DeMar would experience different quotients at specific exercise intensities, and that severe or maximal exercise would be associated with the highest quotients (10, 46).

In 1933, Rodolfo Margaria from Italy investigated along with Edwards and Dill the mechanisms responsible for the oxygen debt (55) using the concept of A.V. Hill (44), the lactacid and alactacid terminology of Lundsgaard (54) for nonoxidative energy systems in muscle, and the exercise lactic acid data from Owles (62), Dill and colleagues (31), and Clapham (55). They reported that the fast component, or alactacid mechanism, was independent of any lactic acid formation (later attributed to restoration of myoglobin and venous O_2 stores and the resynthesis of adenosine triphosphate) and reached a limit of 2.5 L in their subjects. The slow component, or the lactacid mechanism, indicated that oxygen consumption was attributed to the oxidation of lactic acid, was a function of time, had a velocity constant of 0.02, "would come into play only when there was reason to believe that the work is carried on in anaerobic metabolism" (55, p. 714), and would reach an upper limit of 5.0 L in their subjects (figure 3.7). Later studies demonstrated that the removal of lactic acid after strenuous exercise was enhanced by continuous movement when compared with resting conditions (46, 60) and that strenuous exercise or competitive activity could elevate resting oxygen consumption by as much as 25% 15 h later (40). Although the oxygen debt explanation of Margaria, Edwards, and Dill prevailed for more than 50 yr, careful and extensive research near the end of the 20th century demonstrated that lactic acid was not the cause for the oxygen debt (11, 12).

Hormones and Their Influences

In 1932, the effects of epinephrine on substrate utilization was unknown. Therefore Dill, Edwards, and de Meio injected intramuscularly (IM) a 0.1% solution of epinephrine into subjects who were exercising at 7 times their resting metabolic rate. They found that epinephrine had no influence on protein metabolism, was associated with a decrease in the excretion of acetone bodies, and had a marked effect on carbohydrate metabolism as determined by RQ measurements (30). Courtice, Douglas, and Priestly challenged this interpretation because they felt that an increase in lactic acid was responsible (2). Thus, Asmussen, Wilson, and Dill (2) conducted an experiment using the exercise protocol of Courtice, Douglas, and Priestly and a design that included carbohydrate diets, IM injections of insulin, and IM injections of epinephrine. They found that the following occurred with exercise alone (2):

1. Epinephrine was associated with increased RQ values and elevated blood glucose and lactate values.
2. Insulin was associated with increased RQ values, decreased blood glucose, and increased blood lactate concentrations.
3. Combined epinephrine and insulin IM injections were associated with elevated RQ values,

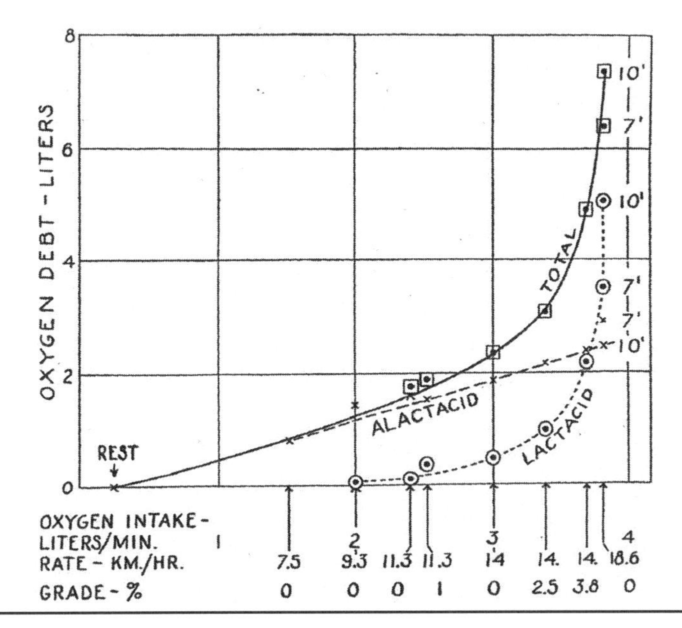

Figure 3.7 Alactacid and lactacid oxygen debts as a function of the metabolic rate in the subject, Clapham

Reprinted from R. Margaria, H.T. Edwards, D.B. Dill, 1933, "The possible mechanisms of contracting and paying the oxygen debt and the role of lactic acid in muscular contraction," *American Journal of Physiology* 106: 689-715. With permission of American Physiology Society.

minimal changes in blood glucose concentrations, and higher blood lactate levels.

4. Providing a high carbohydrate diet increased RQ and blood glucose concentrations.

Additionally, they found that neither insulin nor epinephrine had any influence in altering blood acetone levels. They concluded that the elevated RQ values found from the injections of both hormones had increased the metabolism of carbohydrate (2).

Missiuro and colleagues investigated the influences of adrenal cortical extracts on three subjects for 3 to 5 d with respect to their effect on resting and exercise heart rates, oxygen consumption, blood lactic acid concentration, erythrocyte and leukocyte levels, and blood pressures during recovery after exercise (58). The extract had no striking influence on any resting parameters except the neutrophil counts of two subjects. As for the exercise conditions, the extract appeared to improve the efficiency of light walking and the blood pressure recovery profiles (58).

Renal Function During Exercise

In 1931, Edwards, Richards, and Dill reported that proteinuria was present after a football game with the increase being associated with the number of minutes played (39). Six years later, Edwards and colleagues investigated changes in urine flow and renal blood urea clearance in exercising subjects and in football players. In addition, they assessed the effects of a subcutaneous injection (1 mg) of epinephrine before the activity began (36). The resting measures were regarded as low and ranged from 78% to 125% of normal whereas during exercise (4.2-7.4 times the RMR) the values ranged from 39% to 117% representing an inverse relationship with the intensity of exercise. With football players, clearance values were roughly inversely proportional to playing time. Urine flow decreased with the intensity of exercise whereas the injection of adrenaline appeared to reduce the resting clearance values. They explained the low clearance result with exercise as occurring because of a reduction in renal blood flow. However, they had no measurements for verification purposes (36).

Fatigue

Inspection of the bibliographic listings by McFarland and colleagues (56) or by the Horvaths (46) indicated that there was a limited number of publications on fatigue. Although Chapman stated that the term *fatigue* was chosen in the naming of the Harvard Fatigue Laboratory in order to not force their research activities into a departmental mode, it nevertheless provided a convenient explanation for the absence of specific studies. Insights on the topic come from a 1943 article by Forbes, who stated that a major difficulty in studying fatigue was the problem of measurement (42). He wrote:

> The first two or three years (of the Laboratory) were spent in essentially establishing the normal values for everything

> we could measure in blood, urine, and sweat and in studying normal circulation and respiration. Having done this we were in a position to see what was changed in industrial workers by a long day at their jobs. Nothing that we could measure was changed significantly even though the man was tired—a result expected I think by Professors Henderson and Dill but not so clearly foreseen by the younger men. (42, p. 156-157)

Thus, it is understandable that when fatigue was measured it related more to exercise and environmental conditions than to workers in industry.

In 1932, Talbott prepared a public radio message on *The Effects of Fatigue* for the Committee of Public Education in the Massachusetts Medical Society (69). He stated that there were two types of fatigue: physical and functional. Physical fatigue was identified with physical exertion whereas functional fatigue was mental in nature. Physical fatigue was associated with elevated oxygen consumption and increased respiratory and heart rates. Also increased were red and white blood cells and blood levels of lactic acid, which had spilled over from the skeletal muscles. One problem of the increased blood lactic acid level was the lowering of carbon dioxide combining power. However, Talbott was reluctant to ascribe physical fatigue to a single factor and noted that multiple factors were responsible for the subjective sensation of physical fatigue. He did indicate that hormones were involved in the delay of fatigue sensations and cited adrenaline (epinephrine) and cortin (adrenal cortical extracts that contain a mixture of hormones, notably corticosone), although he was uncertain about the nature of their specific influences. Besides hormones, he felt that bodily constitution and the speed of chemical reactions were important considerations in explaining the differences between the fatigue of a marathon runner and an individual diagnosed with the fatigue associated with neurasthenia. Like Henderson, he felt that any discussion of fatigue should consider sociological factors (69).

In 1933 at a personnel conference concerned with fatigue and work efficiency, Dill spoke on the nature of fatigue (27). He indicated that there were two categories of fatigue: one in which large organic changes occur and another in which no such changes occur. The former requires physiological measurement whereas the latter necessitates "the tools of the psychologists."

Because organic changes occur only when the capacity for physical work is taxed, Dill discussed the factors that affected the capacity for work. Included were the supply of oxygen to the active tissues, the magnitude of the anaerobic reserve, the increase in lactic acid, the active mass involved (arms versus legs), the substrate available for the work response as well as its rate of depletion, the effectiveness of the heat-dissipative mechanisms, and the ability to be mechanically efficient and coordinated when performing work. He also noted that training and acclimatization would improve work capacity. Physical fatigue could be described as a variety of phenomena. However, he felt that none would affect the normal factory worker because the output of physical energy by the worker was but a small percentage of the worker's total capacity. However, if physical fatigue did occur in factory workers, he recommended evaluating whether the individual was suited for the job, whether training should be undertaken, and whether better control of environmental conditions should be considered (27).

Exercise and Temperature Regulation

A study was conducted by Dill and colleagues in the Canal Zone. In this study, five subjects exercised on a bicycle ergometer at an established workload in environmental conditions of 34 °C and 50% humidity, which would be exhaustive in 37 to 60 min. They compared the subjects' responses with those exercising in 12 °C (29). Exercising in the heat was associated with a linear increase in rectal temperature that peaked at 40.1 °C, an upward sloping heart rate curve that peaked at 180 beats/min, peak oxygen consumptions of 2.220 L/min with 53 L/min for ventilation and 20 L for cardiac output, systolic blood pressure of 189 mmHg, and lactic acid concentrations of 5.4 mEq/L. When exercising at 12 °C, no subjects exhibited exhaustion, and rectal temperature and heart rate (peak reduction of 55 beats/min) were consistently and markedly lower with blood lactic acid, with cardiac output, RQ, and mechanical efficiency being elevated. Systolic blood pressures of group members were increased. The authors associated exhaustion in the heat with the impaired ability to dissipate heat, limited distribution of blood to tissues, and to the exhaustion of cardiac muscle by lactic acid (29).

During 1932, the Harvard Fatigue Laboratory staff (n = 10) conducted field studies in Boulder City, Nevada, in the United States using workers building the Hoover Dam. One year earlier, 15 deaths and numerous health problems had occurred as a result of working in peak temperatures of 43.3 °C and sleeping in temperatures plateauing near 37.8 °C (5, 70). After securing baseline data, Talbott and colleagues studied heart rates plus blood and urine constituents of 15 male workers and 7 staff members after an 8 h working day for the summer months (70). They concluded that body weight decreased during the early days of exposure, fluid intake increased as temperature increased, urine volume exhibited minimal changes, urine-specific gravity increased before returning to baseline values, nitrogen excretion was reduced with exposure to heat, and sodium chloride in sweat increased during the early days of heat exposure and decreased thereafter. Staff members exhibited a decrease of 0.3% in blood oxygen capacity, which was not significant, nor were other blood measurements obtained from both groups (70).

Results from other studies were instrumental in changing behaviors and practices pertaining to the frequent consumption of fluids, the addition of salt to diets and drinking solutions, and to the recognition and prevention of heat cramps. Additionally, Dill and colleagues conducted case studies on differences between humans and dogs in the dissipation of heat after desert walks. Their main conclusion was that the dog was superior to man in dissipating heat in the laboratory whereas the dog was inferior to man in the hot, arid desert (28).

In the summer of 1937, the Harvard Fatigue Laboratory staff returned to Boulder City for the Harvard desert expedition, where Dill and colleagues (33) extended the studies that were initiated earlier, especially those pertaining to sweat and its composition. Sweat rate and composition were measured in 6 subjects during the winter months in Boston. The subjects walked on the treadmill with a metabolic rate that was approximately 5 times RMR in environmental conditions of 43 °C and 10% relative humidity. When the researchers retested the same subjects in Boulder City, they found that the concentration of chloride in sweat was approximately twice as high when collected in Boston compared to Boulder City but that the sweating rate was the same. Moreover, after arriving in Boulder City, it required from 1 to 10 d for the chloride concentration in sweat to exhibit signs of acclimatization. They concluded that sweat became more dilute with adaptation to high temperatures and that its inorganic components became more concentrated as the nitrogen excretion became reduced. These researchers also concluded that the inability to prevent the loss of sodium chloride was associated with the incidence of heat cramps (33). Concepts in perspiration that evolved from the Boulder City experiments were incorporated into the following conclusions published by Dill in his 1938 text (26, pp. 47-48):

1. In males, skin areas that are responsible for heat dissipation are abundantly provided with sweat glands and are stimulated via the nervous system when other means of heat elimination are inadequate.
2. Sweat varies in composition depending on its rate of production, the degree of acclimatization, and individual variation.
3. Sweat composition varied in accordance to external temperature, physical activity, and the degree of exposure to the sun.
4. Sweat glands excrete lactic acid; however, it is of minimal value to the economy of muscular exercise.
5. Adaptation to high temperature requires an increased capacity to produce sweat, greater sensitivity of the temperature regulatory system, and economy of salt.

Beginning in 1939, the Harvard Fatigue Laboratory, led by Robinson, conducted studies in the Mississippi Delta (Benoit) as well as in laboratories in Boston and in Bloomington, Indiana (all in the United States). Caucasian and African American subjects were obtained from staff, student, sharecropper, and servant populations (46, 65). The indoor-exercise tests were conducted on a treadmill at 5.6 km/h at a grade of 8.6% that was equivalent to 7 times the basal metabolic rate and continued for 2 h or less. The environmental testing conditions were between 15 °C and 25 °C in Boston and Bloomington and between 28 °C and 33 °C in Mississippi, with a relative humidity of 45% in the northern locations and 80% in the Delta. In addition, 5 Caucasian staff members and 5 African American sharecroppers performed a 2 h outdoor walk at an approximate speed of 6.7 km/h. The energy requirement of the outdoor walk was similar to that recorded for the indoor walk.

From their extensive results the researchers concluded the following (65):

1. When compared with their sweating rates in the environmental conditions in Boston, the staff members from the Harvard Fatigue Laboratory increased their sweating rates by 200% after 10 d of treadmill walking in the environmental conditions in Mississippi and by 250% after 6 wk of walking in the environmental conditions in Boston, Massachusetts.
2. The changes in heart rates closely parallel the changes in rectal temperature.
3. The African American sharecroppers had the lowest body temperatures of any group when tested at any site. They also had lower temperatures when compared with northern African American college students.
4. Caucasian sharecroppers had higher rates of sweating than African American sharecroppers

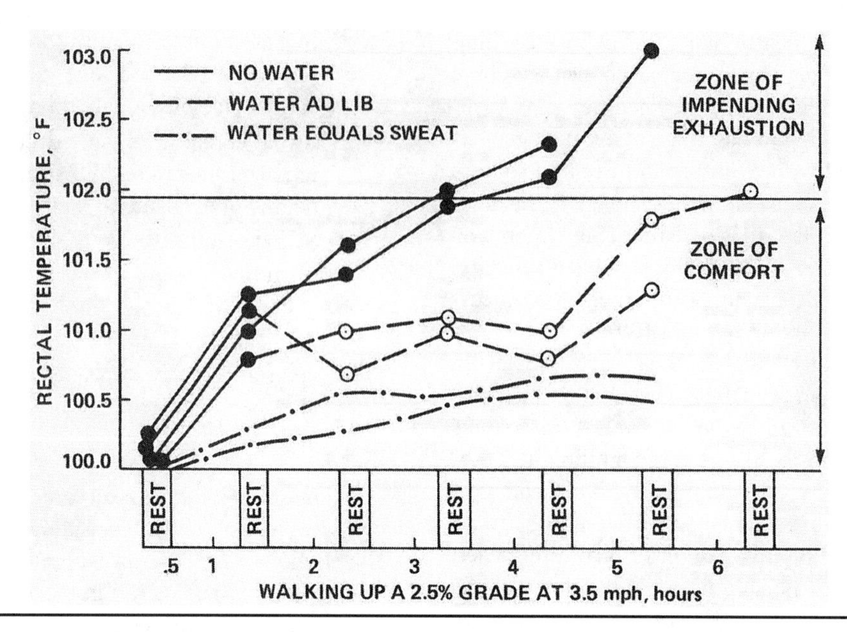

Figure 3.8 Influence of water consumption on rectal temperature and their relationships with walking performance in the heat.

Reprinted from G.C. Pitts, R.E. Johnson, and F.C. Consolazio. "Work in the heat as affected by intake of water, salt and glucose," *American Journal of Physiology* 142: 253-276, 1944. With permission of American Physiology Society.

because their walking was less mechanically efficient.

5. African American sharecroppers were superior to all groups in temperature regulation.

Robinson and colleagues investigated the process of acclimation in five subjects who walked on a treadmill at a speed of 5.6 km/h on a grade of 5.6% or 4% until exhausted in conditions of 40 °C and 23% relative humidity (67). The walks were continued for 10 and 23 d and the durations were between 60 and 90 min/walk. Signs of acclimatization appeared approximately 7 d later and became apparent for the group after 18 d. Group heart rates were 178 beats/min at the beginning of the study and 150 beats/min at the end of the study. Skin and rectal temperatures were 36.9 °C and 39.9 °C before the walks began and 35.7 °C and 39.1 °C, respectively, after acclimatization had occurred. Sweat production (kg/h) was 1.41 at the start and 1.76 at the end. They reported that 80% of the changes recorded were evident within the first 7 d and that minimal changes in metabolic rate had occurred (67).

Using these results to determine an acclimatization state, Pitts, Johnson, and Consolazio investigated performance in the heat as affected by water, salt, and glucose intake (63). Six acclimated subjects walked on treadmills at 5.6 km/h on grades ranging from 2.5% to 4.0% in conditions of 35 °C to 37.8 °C with either 30% or 80% relative humidity for periods ranging from 1 to 6 h. They reported that the subjects' best performance occurred when the water lost from sweating was replaced by water being consumed on an hourly basis. Failure to follow such a schedule led to serious physiological problems (figure 3.8). On the other hand, replacement of salt on an hourly basis had no meaningful advantage. The same appeared true for the replenishment of glucose (63).

Moreira and colleagues conducted a related experiment in a hot environment in which water or adrenal extracts were given to determine whether they provided any benefit for working in moist heat for more than 180 min. The results revealed that providing adrenal extracts to subjects working in the heat had no physiological benefit or advantage (59).

Exercise and High Altitude

The texts of the Horvaths (46) and Dill (26) extensively discussed and documented the Harvard Fatigue Laboratory's involvement with high-altitude research. Interesting and detailed information was provided concerning experiments conducted in Leadville, Colorado (3,038 m), as well as the measurements obtained at 2,810, 3,660, 4,700, 5,340, and 6,140 m in Chile and at heights associated with trans-Andean and pan-American flights to remote parts of the world. These texts also mentioned the oxygen affinity and secretion dispute between Barcroft and Haldane and the perspectives of various investigators about this dispute. However, the overwhelming majority of the information pertained to respiratory and blood measurements obtained from stationary or nonexercised subjects.

In 1931, Dill and colleagues measured select physiological responses of staff members (n = 3 or 4) at rest and while performing maximal work for 20 min on a bicycle ergometer at Boston (sea level) and at Leadville (7,740 m). Compared with results found at sea level, mechanical efficiency was similar, oxygen consumption was reduced by approximately 16%, oxygen saturation was lower by 4%, pH was more alkaline by 0.03 units, and approximately 20% less lactic acid was present. Heart rate was maximal at an O_2 consumption of 2.6 L in Boston and 2.1 L in Leadville. V_E exhibited extensive variability but exhibited an approximate 8% decline at Leadville (31).

Five years later (1936), Edwards reported the lactic acid results obtained from staff members and residents from Quilcha, Chile, and from Leadville, Colorado. Staff members and residents performed submaximal and maximal work on a bicycle ergometer at four heights ranging from 2,180 m to 5,340 m (35).

After 6 wk of acclimatization, there was no evidence of an increase in lactic acid even at the highest altitude. In most of the subjects an increase in work intensity with elevation was associated with an increase in lactic acid that was less than what was recorded at sea level. Edwards was surprised that at higher grades of work anoxemia did not elicit a greater-than-normal increase in lactic acid, postulated that training or acclimatization could be responsible for this observation, and suggested that it could be a protective mechanism of the body (35). He also suggested that the proposed protective mechanism may be the result of an inadequate oxygen supply to the essential muscles (e.g., the heart or the diaphragm).

As a foreign fellow of the Harvard Fatigue Laboratory, Asmussen conducted with Consolazio a circulatory case study on Mt. Evans in Colorado (4,300 m) that lasted 15 d (1). They observed increased resting cardiac outputs and decreased resting blood volumes after the first 3 to 5 d that never completely returned to sea-level values. Exercise was performed on a bicycle ergometer at workloads of 330 and 490 $kg \cdot m^{-1} \cdot min^{-1}$. Peak heart rates (~130-138 beats/min) and cardiac outputs (~20-22 L/min) occurred within the first 4 d, after which they gradually decreased toward baseline values. These changes were associated with increased activity of the chemosensitive reflexes and an increased blood volume and red cell count (1).

In an address to the American Heart Association in 1941, Dill commented on his research finding at high altitude. He emphasized that reductions in work capacity would occur, that acclimatization would not be completed in a 3 wk period, that the concentration of lactic acid after heavy exercise at 13,545 ft was approximately 50% lower than reported for comparable exercise at sea level, and that the maximal heart rate became decreased as the altitude increased (25).

Exercise Training and Physical Fitness

The first training study of the Harvard Fatigue Laboratory appears to have been published by Talbott and colleagues in 1929 (71). Talbott was the subject, and within 1 mo he was able to increase his peak oxygen uptake by 30% and his steady-state O_2 consumption by 17% while decreasing his nitrogen excretion by 6%. He attributed the nitrogen changes to an increase in muscle hypertrophy (71). Later, Bock compared the trained state of DeMar, the foremost marathon runner of his time, with the status of select subjects used in laboratory studies. He reported that DeMar's resting heart rate was 38% slower, exercise heart rate with hard work was 33% slower, exercise stroke volume was 30 ml/beat or 17% higher, and systolic blood pressure during work of 165 mmHg was similar to others who had performed 50% less work. Bock mentioned that DeMar had minimal changes in blood lactic acid concentrations when consuming 3 L of oxygen whereas his nontrained counterparts had approximately 100 mg/100 ml of blood when consuming 2 L/min of oxygen (4).

In 1937, Robinson and colleagues (66) published "New Records in Human Power" in which they compared the physiological responses of five renowned distance runners (Lash, Cunningham, Romani, Venzke, and Fenske) with responses from nontrained subjects who had also been tested in the Harvard Fatigue Laboratory. The researchers were particularly impressed with the ability of Lash to utilize 4.96, 5.08, and 5.1 L of oxygen in the last 3 min of an 18.1 km/h run on a 4% grade while producing 47.5 mg/100 ml of lactic acid, or approximately one half of what would be expected in nontrained subjects. When compared with the nontrained runners, the elite runners had lower exercise heart rates with submaximal intensities, higher heart rates with maximum intensities, exhibited faster heart rate recovery values after exercise, elevated a higher metabolic rate by 150%, and demonstrated higher ventilation volumes (66).

Using concepts published by Robert E. Johnson (47, 48) (figure 3.9) and Edward C. Schneider (68), Brouha and colleagues (13, 15, 43) developed, perfected, and tested a step test that became known as the Harvard Step Test. While the test was being developed, Johnson and colleagues summarized the known data relative to fitness (table 3.1).

The Harvard Step Test used recovery heart rates to measure the physical ability of males to perform hard muscular work. According to Brouha, the test was a measure of the ability of the cardiovascular system "to adapt itself to hard work and recover from what it has done" (13, p. 86). The test consisted of stepping up and down on a bench (51 cm high) at a cadence of 30 times/min for 5 min. Recovery heart rates were recorded for 30 s after 1, 2, and 3 min had elapsed after stopping. The Physical Fitness Index (PFI) was obtained as follows (13, 43):

PFI = duration of exercise (s) × 100 / 2 × sum of pulse counts in recovery

By 1943, the test had been administered to 660 male students at Andover Academy (43) and 2,200 male students at Harvard University. It was also recommended for the selection of Army combat officers (15). The test was also used during World War II to classify convalescent servicemen for assignment to reconditioning programs and for the return to active duty (50).

Physical Fitness and Aging

For his PhD thesis Sid Robinson assessed the ability of 93 nonathletic males ranging in age from 6 to 91 yr to perform moderate and maximal work on a treadmill (64). Moderate work (~7-8 times the basal metabolic rate) consisted of running on a treadmill for 15 min at a grade of 8.6% and a speed of 5.6 km/h. Maximal, or exhaustive, work was established as the condition required to produce exhaustion within 5 min when running on a treadmill set at a grade of 8.6%. Although Robinson used the term *physical fitness* to describe his study, no index was developed or used to classify the fitness levels of the subjects. He was the first to report the relationship between age and maximal oxygen consumption for normal males between 6 and 78 yr of age (figure 3.10). Relevant findings pertaining to exercise physiology are listed in table 3.2.

Figure 3.9 Robert E. Johnson (1911-2002) of the University of Illinois in the United States. Johnson, who was a Rhodes Scholar with a PhD in biochemistry, obtained an MD from Harvard while on the staff of the Harvard Fatigue Laboratory. After closure of the laboratory, he became chairman of the department of physiology at the University of Illinois in Urbana in the United States.

From the collection of Charles M. Tipton

Exercise and Gender

In 1942, Metheny, Bouha, Johnson, and Forbes reported the results of a comparative study on the exercise performance of young males and females between 19 and 27 yr of age (57). The study included 17 female graduate students who were selected for their good health status and 30 male students from Harvard University who were chosen randomly from a pool of 250 male students. Of these males, 10 were regarded as good, 10 were considered average, and 10 were listed as poor as collectively determined from their scores on the Recovery and Work Indexes (47, 48).

Two treadmill tests were performed. The first was the moderate work test used by Robinson, which consisted of walking at 5.6 km/h on an 8.6% grade for 15 min. The second consisted of running on a treadmill at 11.2 km/h with an 8.6% grade until exhausted. The results from the moderate work test indicated the following:

1. There were no marked differences between males and females with regard to oxygen consumption, pulmonary ventilation, RQ, or blood glucose concentrations.
2. At similar oxygen consumptions ($ml \cdot kg^{-1} \cdot min^{-1}$), females had a more rapid increase in heart rates and demonstrated higher values.
3. The heart rate recovery rates for both populations were approximately the same.

However, the responses between the two groups performing exhaustive or maximal exercise were markedly different; these differences are summarized in table 3.3. The authors then compared the 8 best females with the 10 poorest males and reported the following:

1. The run times to exhaustion were similar.
2. Ventilation, when expressed on a body-weight basis, was similar for both groups.
3. Oxygen consumption ($ml \cdot kg^{-1} \cdot min^{-1}$) was 41.6 for the females and 48.7 for the males (~15%).

Table 3.1 Comparison of Select Physiological Responses Between Physically Fit and Unfit Males of Similar Body Weights to the Same Rate of Physical Work

Measurement	Fit subject	Unfit subject
Comparison: Light work that both groups can maintain at steady state conditions		
Oxygen consumption	Lower	Higher
HR during work	Lower	Higher
Stroke volume during work	Larger	Smaller
BP during work	Lower	Higher
Blood lactate during work	Lower	Higher
Recovery of HR to resting baseline	Faster	Slower
Recovery of BP to resting baseline	Faster	Slower
Comparison: Exhausting (maximal) work that neither group can maintain at steady state conditions		
Maximal oxygen consumption	Higher	Lower
Maximal HR during work	Lower	Higher
Recovery of BP to resting baseline	Faster	Slower
Recovery of HR to resting baseline	Faster	Slower

BP = blood pressure; HR = heart rate.
Adapted from Johnson, Brouha, and Darling 1942, p. 502; Horvath and Horvath 1973, p. 109.

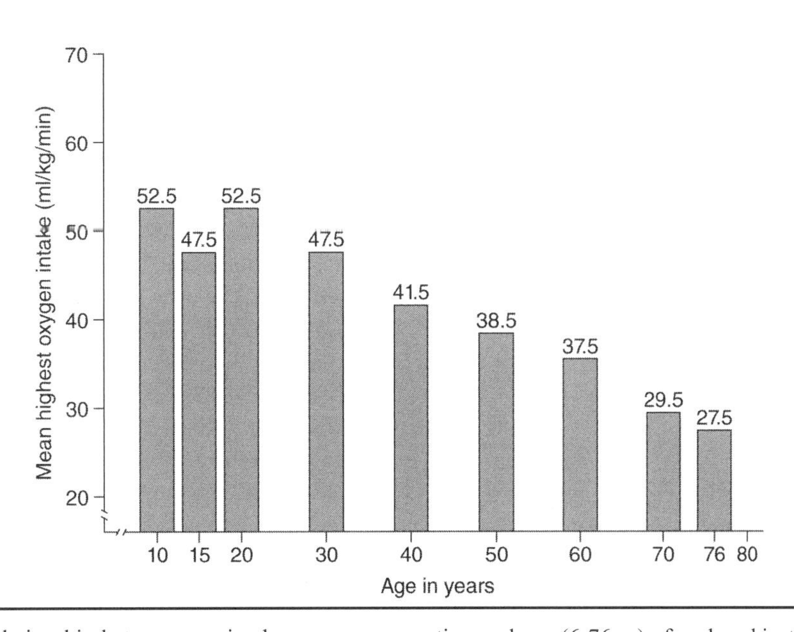

Figure 3.10 The relationship between maximal oxygen consumption and age (6-76 yr) of male subjects. Units above each bar represent mean values for that specific group.
Adapted from Robinson 1938.

Table 3.2 Select Results From the Robinson Physical Fitness Study

Males between the ages of 10 and 22 yr had the highest maximal heart rates. After age 22 yr, maximal heart rate linearly declined with age.
The mechanical efficiency when exercising at 5.6 km/h on an 8.6% grade was similar for all subjects who were less than 80 yr of age.
The RQ of performing moderate work exhibited a gradual increase with advancing age and indicated a greater percentage of carbohydrate utilization for work.
Blood lactic acid concentrations measured during moderate work demonstrated a gradual elevation with advancing age.
Respiratory rates and ventilation volumes expressed on a body-weight basis with moderate work were the highest in the 10 yr age group and declined and plateaued in the 20 and 30 yr age groups, respectively.
The highest oxygen consumption results recorded during maximal or exhaustive work were in males between 10 and 20 yr of age. Subjects in the 17 yr age group had the highest absolute (3.61 L/min) and relative (52.8 $ml \cdot kg^{-1} \cdot min^{-1}$) values. Age group relative percentages compared to the 20 yr group were progressively lower with aging ranging from 9.5% for the 30 yr old group to 48% for the 75 yr group.
The highest respiration rate with exhaustive work was found in the 6 yr age group; rates progressively declined as age increased. Ventilation volumes were highest in males between 18 and 35 yr of age. The 35 yr age group had a mean value of 122.4 L/min.
Lactic acid concentrations with exhaustive work ranged from 25 mg/100 ml of blood for the 6 yr age group to 97.1 mg/100 ml of blood for the 35 yr age group. After age 35, blood lactic acid concentrations progressively declined to 18.1 mg/100 ml for subjects who were 75 yr old.

Adapted from Robinson 1938.

4. Maximum heart rates were identical for both groups (194 beats/min).
5. Blood lactate concentrations were almost identical for both groups (112 and 113 mg/100 ml of blood for males and females, respectively).
6. Systolic blood pressure recorded in females was 9% lower than that reported for males.

The authors concluded that these selected females equaled in every respect the 10 poorest males when performing strenuous, exhausting exercise (57).

Nutrition and Performance

Various exhaustive tests for evaluating and classifying fitness were incorporated into studies that focused on nutrition. In 1942, Wald, Brouha, and Johnson investigated the role of vitamin A deficiency in fitness and fatigue. Subjects walked on a treadmill at a speed of 5.6 km/h and a grade of 8.6% for 15 min (moderate work), ran on a treadmill at a speed of 11.2 km/h and a grade of 8.6% for 5 min (exhaustive work), or performed the Harvard Step Test (75). After 1 mo on a diet that included high levels of vitamin A, subjects consumed a diet that included low levels of vitamin A for 12 and 18 wk before being evaluated by the various tests. Measurements pertained to oxygen consumption, pulmonary ventilation, heart rates, blood lactate concentrations, and RQ values. The results clearly demonstrated that a vitamin A deficiency for the time period investigated had no significant effect on the physiological parameters investigated or on recorded fitness scores (75).

Johnson and colleagues (49) investigated in 10 male subjects the effects of a vitamin B complex deficiency on work performance [chopping, sawing, and splitting logs; constructing breakwaters; and walking between 7 and 20 miles (11.3 and 32.2 km)] daily. They evaluated the subjects' work performance using fitness scores as determined by the Harvard Step Test (13) and assorted treadmill tests that were similar to those conducted by Wald and colleagues (75). Symptoms of depression, fatigue, loss of appetite, and irritability appeared within 6 d, and the Fitness Index decreased by 60% after 3 wk. In 2 subjects who performed treadmill tests, ventilation and oxygen consumption results exhibited reductions of 8% and 26%, respectively. The authors concluded that the fitness required for hard muscular work could not be maintained without daily intake of the vitamin B complex (49).

To test the 1904 concept that restricting protein in the diet to 50 to 60 g/d would improve vigor for hard

Table 3.3 Select Results From Males and Females Performing Maximal or Exhaustive Work

Measurement	Female means	Male means
Subjects (no.)	17	30
Run time (s)	108	216
Ventilation (ml·kg−1·min−1)	975	1,112
Oxygen consumption (ml·kg−1·min−1)	40.9	51.3
Blood lactate (mg/100 ml of blood)	112	119
Blood glucose (mg/100 ml of blood)	156	144
Maximum heart rate (beats/min)	197	194
Systolic blood pressure (mmHg)	163	181

Adapted from Metheny et al. 1942.

muscular work (20), Darling and colleagues conducted a study with 24 subjects living in a Civilian Public Service Camp. One group consumed between 95 and 113 g/d of protein, a second group consumed between 157 and 192 g/d of protein, and the third group consumed approximately 53 g/d of protein (21). The duration of the test period was 8 wk. The work responsibilities of the subjects consisted of kitchen, forestry, farming, road repair, trail, and office activities. The mean caloric consumption was 3,300 kcal/d (range: 2,400-5,000 kcal/d). In addition, the subjects performed a Pack Test, which was similar to the Harvard Step Test except the subject wore a pack that contained a weight that was equal to 33% of his body weight. The duration of the test was 5 min. Recovery heart rates were summed to determine a fitness score.

At the end of the study, the average fitness score increased by 15 points in the normal group, 1 point in the group consuming the high-protein diet, and 6 points in the group consuming the restricted-protein diet. In addition, the authors could find no evidence that the health of the subjects had been impaired or that the subjects had experienced any change in the ability to perform work tasks. Hence, they concluded that a 2 mo diet of 50 to 55 g/d of protein was neither beneficial nor deleterious for physical vigor or efficiency (21).

Select Aspects of the Harvard Fatigue Laboratory During World War II

The advent of World War II brought marked changes pertaining to the leadership of the Harvard Fatigue Laboratory as well as the mission of the laboratory and its staff. Besides fitness and nutrition, its mission included studying the energetic costs of military tasks in extreme heat or cold as well as laboratory and field conditions and evaluating specially designed cold-weather clothing, as tested in tents and sleeping bags located in a climatic chamber. Electrically heated clothing for use by high-altitude pilots was also tested in the climatic chambers before being evaluated under field circumstances. Specific details on the wartime activities of the laboratory can be found in reference 41.

Beginning in 1941, Dill followed the exodus of Consolazio, Graybiel, Knehr, Pecora, Pitts, Talbott, and Wilson from the laboratory and accepted an assignment as a commissioned officer at the Aeromedical Laboratory at Wright Field in Ohio in the United States (41, 46). Later, he was assigned to the Quartermaster Corps, located in the nation's capital (41). Forbes was appointed acting director, and his responsibilities increased with the untimely death of Henderson in 1942. Dill remained in contact with the Harvard Fatigue Laboratory and participated in select decisions pertaining to laboratory projects (41). Key scientists who joined the staff of the laboratory during War World II were Sid Robinson and G. Edgar Folk Jr. (46).

Closure, Dissolution, and Dispersion

The official date for the closure of the Harvard Fatigue Laboratory is 1947, although the process began between 1942 and 1943 when the laboratory changed its name to the Laboratory for Industrial Hygiene (46) with the intent, after wartime, to focus on industrial-hygiene issues. However, the laboratory continued to be known as and referred to as the Harvard Fatigue Laboratory throughout World War II and up to 4 wk before its closure. Dill returned in 1945 and continued his duties for 2 yr before assuming a position as director of medical research for the U.S. Army Chemical Research and Development Laboratory (45). Subsequently, he became the director of the Laboratory of Patho-Environmental Physiology at the University of Nevada at Las Vegas in the United States (72). With the end of World War II and the change in research focus of the Harvard Fatigue Laboratory, staff members followed the example of Dill and accepted appointments with other institutions or organizations in order to establish physiological research laboratories. Individuals who contributed to the body of knowledge pertaining to exercise physiology were Ancel Keys and Henry Taylor at the University of Minnesota, Sid Robinson at Indiana University, Robert Johnson at the University of Illinois, and Steven Horvath at the University of California at Santa Barbara. G. Edgar Folk Jr. left the laboratory to establish an environmental physiology laboratory at the University of Iowa.

Chapman was correct in labeling the Harvard Fatigue Laboratory the "magnificent anomaly" (19) because its contributions to exercise physiology have been enhanced, rather than diminished, with time.

References

1. Asmussen E, Consolazio F. The circulation in rest and work on Mount Evans (4300 m). *Am J Physiol.* 132: 555-563, 1941.
2. Asmussen E, Wilson JW, Dill DB. Hormonal influences on carbohydrate metabolism during work. *Am J Physiol.* 130: 600-607, 1940.
3. Barman JM, Moreira MF, Consolazio F. The effective stimulus for increased pulmonary ventilation during muscular exertion. *J Clin Invest.* 22: 53-56, 1943.
4. Bock AV. On some aspects of the physiology of muscular exercise. *New Engl J Med.* 13: 638-642, 1929.
5. Bock AV, Dill DB. A resume of some physiological reactions to high external temperature. *New Engl J Med.* 209: 442-444, 1931.
6. Bock AV, Dill DB. *The Physiology of Muscular Activity.* 3rd ed. London: Longmans Green, 1931.
7. Bock AV, Dill DB, Hurxthal LM, Lawrence JS, Coolidge TC, Dailey ME, Henderson LJ. Blood as physiochemical system. V. The composition and respiratory exchanges of normal human blood during work. *J Physiol.* 65: 749-766, 1927.
8. Bock AV, Dill DB, Talbott JH. Studies in muscular activity. Determination of the circulation of blood in man at work. *J Physiol.* 66: 121-132, 1928.
9. Bock AV, Van Caulaert C, Dill DB, Folling A, Hurxthal LM. Studies in muscular activity. III. Dynamical changes occurring in man at work. *J Physiol.* 66: 136-161, 1928.
10. Bock AV, Van Caulaert C, Dill DB, Folling A, Hurxthal LM. Studies in muscular studies. IV. The "steady state" and the respiratory quotient during work. *J Physiol.* 66: 162-174, 1928.
11. Brooks GA, Fahey TA, White TP, Baldwin KM. *Exercise Physiology.* 3rd ed. Mountain View, CA: Mayfield Publishing Co., 2000, 197-223.
12. Brooks GA, Gladen LB. The metabolic systems: Anaerobic metabolism (glycolytic and phosphagen). In: Tipton CM, ed. *Exercise Physiology: People and Ideas.* New York: Oxford University Press, 2003, 332-360.
13. Brouha LC. The Step Test: A simple method of measuring physical fitness for hard muscular work in young men. *Res Quart* 14: 31-36, 1943.
14. Brouha L, Cannon WB, Dill DB. The heart rate of the sympathectomized dog in rest and exercise. *J Physiol.* 37: 345-359, 1936.
15. Brouha LC, Health CW, Graybiel A. The Step Test: A simple method of measuring physical fitness for hard muscular work in adult men. *Rev Can Biol.* 2: 86-91, 1943.
16. Brouha L, Nowak SJG, Dill DB. The role of the vagus in the cardio-accelerator action of muscular exercise and emotion in sympathectomized dogs. *J Physiol.* 95: 454-463, 1939.
17. Campos FA de M, Cannon WB, Lundin H, Walker TT. Some conditions affecting the capacity for prolonged muscular work. *Am J Physiol.* 87: 680-701, 1929.
18. Cannon, WB. Biographical memoir of Lawrence Joseph Henderson, 1878-1942. *National Academy of Science Biographical Memoirs.* 23: 31-58, 1934.
19. Chapman CB. The long reach of Harvard's Fatigue Laboratory. *Persp Biol Med.* 34: 17-33, 1990.
20. Chittdenden RH. *Phsiological Economy in Nutrition.* New York: Frederick A. Stokes and Co., 1904.

21. Darling RC, Johnson RE, Pitts GC, Consolazio FC, Robinson PF. Effects of variations in dietary protein on physical well being of men doing manual work. *J Nutr.* 28: 273-281, 1944.
22. Dennig H, Talbott JH, Edwards HT, Dill DB. Effect of acidosis and alkalosis upon capacity for work. *J Clin Invest.* 9: 601-613, 1931.
23. Dill DB, Arlie V. Bock, pioneer in sports medicine, December 30, 1888-August 11, 1984. *Med Sci Sports Exerc.* 17: 401-404, 1985.
24. Dill DB. Blood changes in exercise. *Lancet.* 56: 313-315, 1935.
25. Dill DB. Effects of physical strain and high altitudes on the heart and circulation. *Am J Physiol.* 23: 443-454, 1942.
26. Dill DB. *Life, Heat, and Altitude.* Cambridge, MA: Harvard University Press, 1938.
27. Dill DB. The nature of fatigue. *Personnel.* 9: 113-117, 1933.
28. Dill DB, Bock AV, Edwards HT. Mechanisms for dissipating heat in man and dog. *Am J Physiol.* 104: 36-43, 1933.
29. Dill DB, Edwards HT, Bauer PS, Levenson EJ. Physical performance in relation to external temperature. *Arbeitsphysiol.* 4: 408-418, 1931.
30. Dill DB, Edwards HT, de Meio RH. Effects of adrenalin injection in moderate work. *Am J Physiol.* 111: 9-20, 1935.
31. Dill DB, Edwards HT, Folling A, Oberg SA, Pappenheimer Jr. AM, Talbott JH. Adaptations of the organisms to changes in oxygen pressure. *J Physiol.* 71: 47-63, 1931.
32. Dill DB, Edwards HT, Mead S. Blood sugar regulation in exercise. *Am J Physiol.* 111: 21-30, 1934.
33. Dill DB, Hall FG, Edwards HT. Changes in composition of sweat during acclimatization to heat. *Am J Physiol.* 123: 412-419, 1938.
34. Dill DB, Talbott JH, Edwards HT. Studies in muscular activity. VI. Responses of several individuals to a fixed task. *J Physiol.* 69: 267-305, 1930.
35. Edwards HT. Lactic acid in rest and work at high altitude. *Am J Physiol.* 116: 367-375, 1936.
36. Edwards HT, Cohen MI, Dill DB. Renal function in exercise. *Arbeitsphysiolog.* 9: 610-619, 1937.
37. Edwards HT, Woods WB. A study of leucocytosis in exercise. *Arbeitsphysiol.* 6: 73-83, 1932.
38. Edwards HT, Margaria R, Dill DB. Metabolic rate, blood sugar and the utilization of carbohydrate. *Am J Physiol.* 108: 203-209, 1934.
39. Edwards HT, Richards HT, Dill DB. Blood sugar, urine sugar and urine in exercise. *Am J Physiol.* 98: 352-356, 1931.
40. Edwards HT, Thorndike A, Dill DB. The energy requirement in strenuous muscular exercise. *N Engl J Med.* 213: 532-535, 1935.
41. Folk GE. The Harvard Fatigue Laboratory: Contributions to World War II. *Adv Physiol Ed.* 34: 119-127, 2010.
42. Forbes WH. Problems arising in the study of fatigue. *Psychosom Med.* 5: 155-157, 1943.
43. Gallagher JR, Brouha LC. A simple method of testing the physical fitness of boys. *Res Quart.* 14: 23-30, 1943.
44. Hill AV, Lupton H. Muscular exercise, lactic acid, and the supply and utilization of oxygen. *Q J Med.* 16: 135-171, 1923.
45. Horvath SM, Horvath EC. David Bruce Dill, President 1950-1951, American Physiological Society. *Physiologist.* 22: 1-2, 1979.
46. Horvath SM, Horvath EC. *The Harvard Fatigue Laboratory: Its History and Contributions.* Englewood Cliffs, NJ: Prentice Hall, 1973.
47. Johnson RE, Brouha L. Pulse rate, blood lactate, and duration of effort in relation to ability to perform strenuous exercise. *Rev Can Biol.* 1: 171-178, 1942.
48. Johnson RE, Brouha L, Darling RC. A test of physical fitness for strenuous exertion. *Rev Can Biol.* 1: 491-503, 1942.
49. Johnson RE, Darling RC, Forbes WH, Brouha L, Egana E, Graybiel A. The effects of a diet in part of the vitamin B complex upon men doing manual labor. *J Nutr.* 24: 585-595, 1942.
50. Karpovich PV, Starr M, Weiss R. Physical fitness tests for convalescents. *JAMA.* 126: 873-877, 1944.
51. Keys A, Taylor H. The behavior of the plasma colloids in recovery from brief severe work and the question as to the permeability of the capillaries to proteins *J Biol Chem.* 109: 56-67, 1935.
52. Knehr CA, Johnson RE, Brouha L. Pulse rate, blood lactate, and duration of effort in relation to ability to perform strenuous exercise. *Rev Can Biol.* 1: 171-178, 1942.
53. Laszlo AC. Physiology of the future: Institutional styles at Columbia and Harvard. In: Geison GL, ed. *Physiology in the American Context,* 1850-1940. Bethesda, MD: American Physiological Society, 1987, 67-96.
54. Lundsgaard, E. Untersuchungen über Muskel-Kontraktion ohne Milchsaurebildung. *Biochem Zeit.* 217: 162-177, 1930.
55. Margaria R, Edwards HT, Dill DB. The possible mechanisms of contracting and paying the oxygen debt and the role of lactic acid in muscular contraction. *Am J Physiol.* 106: 689-715, 1933.
56. McFarland R, Russell H, Loring L. *The Fatigue Laboratory, Harvard University, Bibliography 1921-1960.* Cambridge, MA: Harvard University Medical Library, 1-38.
57. Metheny E, Brouha L, Johnson RE, Forbes WH. Some physiologic responses of women and men to

moderate and strenuous exercise: A comparative study. *Am J Physiol.* 137: 318-326, 1942.

58. Missiuro V, Dill DB, Edwards HT. The effects of adrenal cortical extract in rest and work. *Am J Physiol.* 121: 549-554, 1938.
59. Moreira M, Johnson RE, Forbes AP, Consolazio F. Adrenal cortex and work in the heat. *Am J Physiol.* 126: 169-176, 1945.
60. Newman EV, Dill DB, Edwards HT, Webster FA. The rate of lactic acid removal in exercise. *Am J Physiol.* 118: 457-461, 1937
61. Nobel Prizes. Available: http://nobelprize.org/nobelprizes/medicine/laureates/.
62. Owles WH. Alterations in the lactic acid content of the blood as a result of light exercise and associated changes in the CO_2-combing power of the blood and in the alveolar CO_2 pressure. *J Physiol.* 69: 214-237, 1930.
63. Pitts GC, Johnson RE, Consolazio F. Work in the heat as affected by intake of water, salt and glucose. *Am J Physiol.* 142: 253-276, 1944.
64. Robinson S. Experimental studies of physical fitness in relation to age. *Arbeitsphysiol.* 10: 251-323, 1938.
65. Robinson S, Dill DB, Wilson JW, Nielsen M. Adaptations of white men and Negroes to prolong work in humid heat. *Am J Trop Med.* 21: 261-287, 1941.
66. Robinson S, Edwards HT, Dill DB. New records in human power. *Science.* 2208: 409-410, 1937.
67. Robinson S, Turrell ES, Belding HS, Horvath SM. Rapid acclimatization to work in hot climate. *Am J Physiol.* 140: 168-176, 1943.
68. Schneider EC. *Physiology of Muscular Activity.* 2nd ed. Philadelphia: W.B. Saunders, 1940.
69. Talbott JH. The effects of fatigue. *New Engl J Med.* 208: 658-659, 1933.
70. Talbott JH, Edwards HT, Dill DB, Drastich L. Physiological response to high environmental temperature. *Am J Trop Med.* 13: 381-397, 1933.
71. Talbott JH, Folling JA, Henderson LJ, Dill DB, Edwards HT, Johnson RE, Berggren L. Studies in muscular activity. V. Changes and adaptations in running. *J Physiol.* 66: 445-463, 1929.
72. The Register of David Bruce Dill/Harvard Fatigue Laboratory Reprints, 1924-1985, MSS 0517. Mandeville Special Collections Library, Geisel Library, University of California, San Diego. Available: http://Orpheus.ucsd.edu/speccoll/testing/html/msso517a.html.
73. Tipton CM. Contemporary exercise physiology: Fifty years after the closure of the Harvard Fatigue Laboratory. *Exerc Sport Sci Rev.* 26: 315-339, 1998.
74. Tipton CM. Living history: G. Edgar Folk, Jr. *Adv Physiol Ed.* 32: 111-117, 2008.
75. Wald G, Brouha L, Johnson RE. Experimental human vitamin A deficiency and the ability to perform muscular exercise. *Am J Physiol.* 137: 551-556, 1942.
76. Wilmore JH, Costill DL, Kenney WL. *Physiology of Sport and Exercise.* Champaign, IL: Human Kinetics, 2007.

CHAPTER 4

Contributions From German Laboratories

Wildor Hollmann, MD

Introduction

The designation *exercise physiology* does not exist in Germany as an autonomous special field. Rather, exercise physiology is integrated in the generic special area of sports medicine (table 4.1). This area was first defined in 1958 by Wildor Hollmann in connection with the founding of the Institute for Cardiovascular Research and Sports Medicine in Cologne, Germany. The formulation: "Sports medicine is the theoretical and practical aspect of medicine examining the influence of exercise, training, and sports, as well as lack of exercise, on healthy and diseased persons of all ages to apply the results for prevention, therapy, and rehabilitation as well as for the athlete himself" (134). This definition was taken over by the World Federation of Sports Medicine (FIMS) in Tokyo in 1977.

In the framework of this definition, German sports medicine placed prevention, exercise therapy, and rehabilitation in the foreground. This presumes knowledge in exercise physiology and the capability to carry out the corresponding performance diagnostics.

The editors of this book wanted to reference exercise physiology only and not describe aspects of preventive and rehabilitative medicine. Therefore, results of German sport medicine investigations regarding clinical aspects are not considered here.

Exercise Physiology Research in Germany 1700 to 1910

In 1719, Friedrich Hoffmann (1660-1742), a physician from Halle, obtained his doctoral degree with experimental investigations about the effects of physical exercise on the human cardiovascular system and on digestion. In 1796, the physician Christoph Wilhelm Hufeland (1762-1836) published a textbook about the significance of different types of physical exercise for human health and life expectancy (66).

In the 19th century, the physiologist Emil Dubois-Reymond (1818-1896) studied the role of gymnastics in the human body (21, 22). The physician and chemist Max von Pettenkofer (1818-1901) detected the existence of creatine in the skeletal muscle and described a strengthening of the circulatory system in connection with physical training (111).

The first ergometer in the present-day sense of the word was developed by the German physician Speck in 1883. When using this cranked ergometer, the test person was in the standing position (figure 4.1). The crank friction could be modified by pulling up a screw. Speck determined the resistance with weights hung on the crank. The number of revolutions could be determined by noting the number of turns of twine wound up around the crank axle. The exhaled air was collected in a double spirometer and the postexperimental air composition was determined, thus enabling conclusions to be drawn about the individual performance reaction (2, 125).

In 1887 the Viennese physician Friedrich Gaertner (1847-1917) presented a mechanically braked ergometer, based on Speck's equipment that could measure the work performed in kilogram-meter. This device, called the ergostat, later went into standard production (112).

The first heyday of exercise physiology began with the veterinarian Nathan Zuntz (1847-1920) in Berlin (figure 4.2), who developed the first motor-driven treadmill in 1889 (figure 4.3). His work is described in chapter 1 (142).

Of substantial significance in the application of aspects of exercise physiology in medicine were Adolf Theophil Ferdinand Hueppe (1852-1938) and Ferdinand August Schmidt (1852-1929). In 1899 Hueppe wrote a fundamental work titled *A Textbook of Hygiene* that covered the whole of the exercise physiology discipline of those days. This was followed by the work

Table 4.1 Milestones of Select Findings from 1880 to 2010 in Germany

Year	Investigator	Milestone	Reference
1883	Speck	Development of the first ergometer	125
1889	Zuntz and Schumburg	Development of the first treadmill	142
1895	Zuntz and Schumburg	Development of a respiratory apparatus and transportable gas measuring gauges	142
1906	Kuelbs	First experimental demonstration of the enlargement of the heart and other internal organs in connection with physical aerobic training	77
1929	Knipping	Development of spiroergometry	74
1954	Hollmann	Introduction of the bicycle ergometer for routine clinical examinations	44
1955	Hollmann	Development of a microphone-based device for automatically measuring blood pressure during ergometric work	44
1959	Hollmann	First description of the aerobic–anaerobic transition by means of the combined determination of the minute volume of ventilation and of the lactic acid level in arterial blood	45
1966	Hollmann et al., Hollmann and Venrath	Introduction of hypoxia exercise training in the laboratory	52, 53
1973	Hollmann and Liesen	Introduction of hyperoxia exercise training in the laboratory	56
1976	Hollmann and Hettinger	Development of the tint electrical and computerized bicycle ergometer	57
1976	Hollmann and Hettinger	Development of the 4 mmol/L lactate threshold	56
1987	Herholz et al.	First publication of the regional blood supply in the human brain during ergometric exercise	38
1991	Herzog et al.	First publication of the glucose consumption in different sections of the human brain after bicycle ergometric work	40

titled *Hygiene of Physical Exercises*, published in 1911. In 1893 Schmidt published the book *Physical Exercise According to the Exercise Value*, which deals with an overview of suitable physical exercises for different ages (122, 123).

From 1904 to 1906 Kuelbs examined the influence of physical training on the internal organs, particularly the heart. He took two dogs from the same litter and trained one of them 5 times/wk for about 2 h on a treadmill, which he had taken over from Zuntz; the other dog had a normal, everyday life. After 1 yr both animals were killed and all internal organs were measured and examined. Although both dogs had more or less the same body weight, the heart of the dog trained on the treadmill was approximately 33% larger than that of the untrained dog. Similar differences were also manifest in the weights of the liver, kidneys, adrenal glands, spleen, and lungs (77).

Over the subsequent decades Arthur Mallwitz (1880-1968) became the most important promoter for German, and later international, sports medicine. He obtained his doctor's degree at Halle University (Saale) in 1908 with what is presumed to be the first ever dissertation in sports medicine. In his dissertation, titled *Maximum Performances With Special Consideration of Sports Done at the Olympic Games*, Mallwitz considered, above all, the investigations of Hueppe and Schmidt as well as those of Zuntz and his school. In

1910 a comprehensive book about exercise physiology in connection with numerous kinds of sports was edited by Weissbein (141).

Exercise Physiology Research in Germany 1911 to 1933

German Empire Committee for Scientific Research in Sport and Physical Exercise

The First Congress for Scientific Research in Sport and Physical Exercise took place in Oberhof, Germany, from September 20 to 23, 1912. The congress was conducted by the professor for internal medicine of the University Clinic Berlin Charité, Friedrich Kraus (1858-1936), and was chaired by Schmidt and Hueppe. Approximately 70 doctors participated (93). On September 21, 1912, the German Empire Committee for Scientific Research in Sport and Physical Exercise—the first national medical sport association worldwide—was founded under the chairmanship of Kraus, who was elected president.

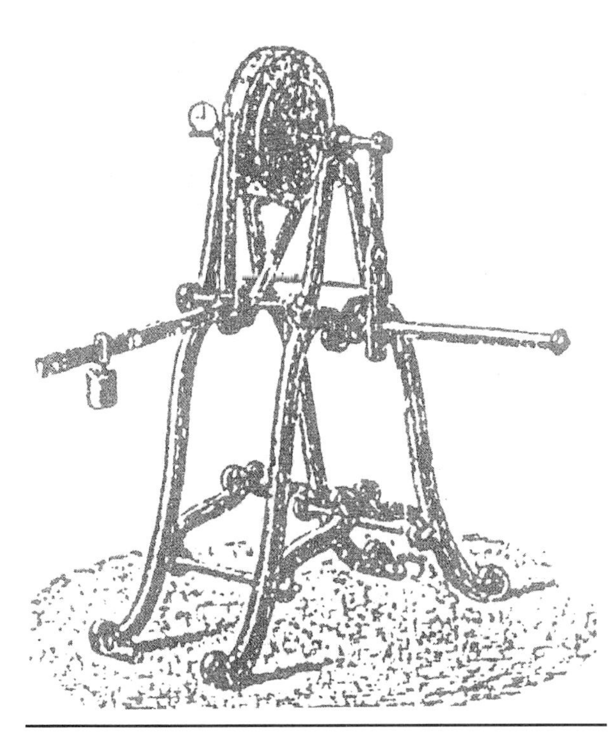

Figure 4.1 The first ergometer was developed by Speck in 1883. Gaertner produced it in series.

Hollmann W, Strueder HK, Predel HG, Tagarakis CVM. Spiroergometrie. Stuttgart, Schattauer, 2006.

Period From 1913 to 1924

In 1913, Arthur Mallwitz was appointed the first full-time sport physician worldwide by the Prussian State Gymnastics Institute. Mallwitz then took up his office under the official designation of sport physician (94).

At the same time the physician and chemist Otto Meyerhof (1884-1951) recognized the regulating relationship between respiration rate and the suppression of glycolysis in muscles. He demonstrated that the consumption of one molecule of oxygen prevents the formation of two molecules of lactate. Meyerhof's findings supported the assumption that the enzymes fanning lactate also function aerobically but that the lactate formed is resynthesized to carbohydrate at the expense of the energy provided by respiration (98).

Hugo Wilhelm Knipping (1895-1984), physician and internist at Hamburg University, developed a gas metabolism apparatus for examination of humans at rest in 1924 and constructed a dynamo cranked ergometer in 1928. In 1929 he combined an enlarged gas metabolism apparatus for exercise examinations with the crank ergometer and designated the method as spiroergometry (73, 74). It was the genesis of precise and simple clinical exercise diagnostics. Engineering, however, was not able to construct the equipment to satisfy all technical requirements until 1949 (75, 76).

German Medical Association for the Promotion of Physical Exercise (1924-1933)

Because of the political confusion and economic difficulties after the finish of World War I in 1918, approximately 6 yr elapsed before research in sports medicine

Figure 4.2 Nathan Zuntz (1847-1920).

Reprinted, by permission, from W. Hollmann and K. Tittel, 2008, *History of German sports medicine* (Gera: Druckhaus Gera), 15.

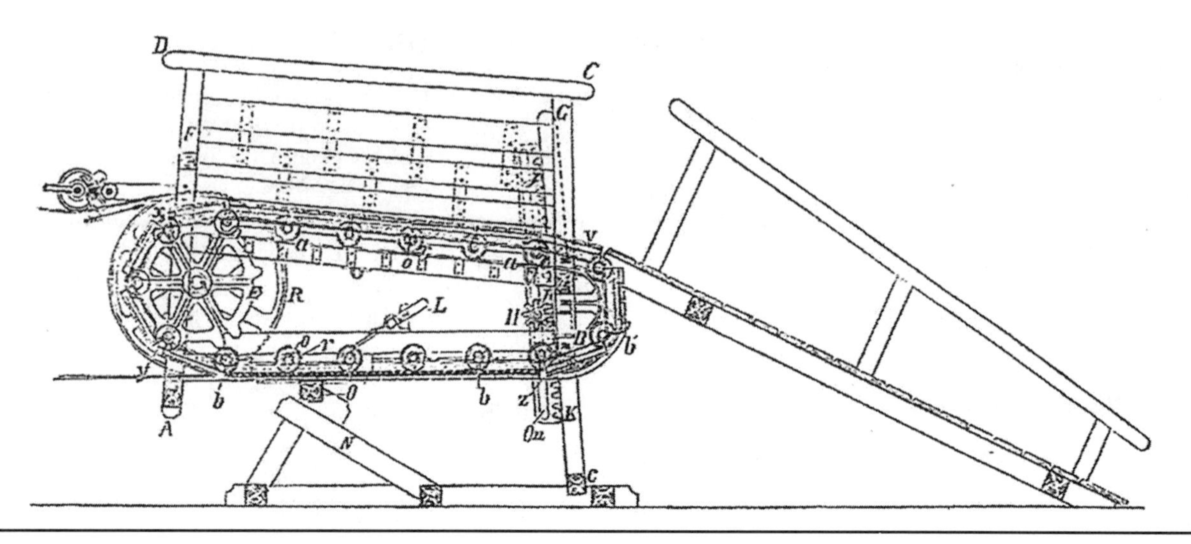

Figure 4.3 The world's first motorized treadmill, developed by Zuntz and Lehmann in 1889 in Berlin. This apparatus was shown in the international hygiene exhibition in Dresden in 1911.

Hollmann W, Strueder HK, Predel HG, Tagarakis CVM. Spiroergometry. Stuttgart, Schattauer, 2006.

and thus exercise physiology was renewed in 1924. The founding of the German Physician Association for the Promotion of Physical Exercise in 1924 was the impetus for this.

In 1925 Max Rubner (1854-1932), director of the Robert Koch Institute in Berlin and an internationally recognized metabolism researcher, ascertained that the skeletal musculature in adults amounts to 43% of the total body mass and that the total musculoskeletal system inclusive of bones, heart, and lungs amounts to 61%. The greatest proportion of the musculature was allotted to the legs (56%) and the upper extremities (28%). The head and torso muscles contributed to 16% of the total body mass (120).

Approximately 3,200 kcal could meet the daily energy requirements for physical exercise for a 70 kg person whereas 2,600 calories was sufficient for simple office work. The metabolism of a farm worker during harvesting would require 4,300 kcal. The peak values of the daily calorie consumption for lumberjacks would amount to approximately 6,000 kcal for a 70 kg person. Long-distance racing cyclists, however, could reach a peak daily metabolic rate of 11,000 kcal. Rubner believed that the limiting factor for the uptake of calories in the human body was the intestinal system. A repeat of such great exertion over several days would inevitably lead to loss of body weight (119).

As early as 1925, Herbert Herxheimer (1894-1985) and others precisely described the psychic and physical effects of overtraining, including a decrease in maximal oxygen uptake, reduced appetite, tendency to sweat, shivering, jerky reflexes, and pronounced respiratory arrhythmia with a clear preponderance of the parasympathetic part. Obstipation and painful stomach spasms also could be observed. Herxheimer published all his sports medicine knowledge in his famous book in 1933.

Herxheimer also described the connection between heart size and distance running. He found an increase in the transverse diameters of the heart in the following ascending order: boxing, swimming, middle-distance running, long-distance running, marathon running, and long-distance skiing. Particularly large hearts would work with a large stroke volume. This would result in the corresponding bradycardia at rest. The blood pressure of endurance athletes would be low (39).

Ludwig Aschoff (1866-1942) in Freiburg ascertained that there is one physiological work hypertrophy of the heart muscle as an adaptation to increased muscular activity. The enlargement remains within moderate limits. Sport does not cause a pathological hypertrophy. If a conspicuous general hypertrophy of the heart exists, it is a matter of pathological circumstances (e.g., cardiac valve ailments, hypertony or glomerular-tubular cirrhosis of the kidney). It is impossible for sport to cause a fatality in an individual with a healthy heart. A muscular cardiac insufficiency is to be treated by taking it easy. Puberty is a dangerous period regarding possible functional damage of the heart through sport.

Another significant exercise physiologist of the 1920s was Richard Herbst (1893-1949) in Köenigsberg East Prussia. He conducted numerous experimental investigations on the behavior of maximum oxygen absorption in humans of different ages and in different training conditions. He published his findings in 1928, 4 yr after the first description of maximum oxygen uptake by Hill. According to Herbst (1928), endurance-trained subjects had a higher oxygen intake than did untrained subjects. Also, after reaching the maximum oxygen uptake value, ventilation could be increased still further. Herbst used the Douglas bag with the subsequent respiratory gas analysis method for runners and cyclists. Further, he examined running distances between 100 m and marathon using the Douglas bag and gas analysis. The values obtained at that time (e.g., marathon with an energy consumption of 3,050 kcal, 10 m sprint with an energy consumption of 50 kcal) agree with measurements made nowadays. Sustained running loads of more than 3 min were determined from the magnitude of the maximal oxygen uptake. Thus, these parameters were a measure of performance. The lung ventilation volume would still increase further after reaching the maximum oxygen absorption. Cardiac output is represented as the most important factor limiting physical performance (37).

Exercise Physiology Research in Germany 1933 to 1950

Because of the political dictatorship from 1933 to 1945, exercise physiology and sports medicine was assessed solely from the point of view of military training. This substantially impeded progress in fundamental research.

Exercise Physiology Research in Germany Since 1950

After the end of World War II the German cities lay in ruins. The universities and all large laboratories were destroyed, including those for sports medicine. Thus, the reconstruction had to begin from scratch. The German Sport Physicians Association was reinstated on October 14, 1950, 1 yr after the Federal Republic of Germany had been founded. A consortium of sport physicians was grounded in the German Democratic Republic (East Germany) in 1953, which became the Medical Scientific Association for Sports Medicine in 1956. The German Democratic Republic had been founded as an independent state in 1950 (63).

Development of Spiroergometry

As mentioned, Knipping and his teacher Ludolph Brauer (1865-1951) developed spiroergometry in Hamburg as early as 1929. This was the birth of clinical exercise diagnostics. The previous methods for assessing the capability of the heart, circulation, respiration, and metabolism such as the Master Step Test (95) allowed only spot checks for determining the gas metabolic values; a physiological determination was needed for further processing (15, 76).

For two decades the technology was not able to meet all natural scientific requirements in spiroergometry. The first equipment that satisfied the required physical conditions was manufactured in 1949. At this time Knipping was the director of the Medical University Clinic in Cologne. At his instigation, the new equipment was installed at the German Sport University Cologne in order to determine cardiopulmonary and metabolic mean values at different submaximal loads for every age (44).

The examinations of our laboratory started in connection with the new apparatus in 1949. Mean values of oxygen uptake and ventilation were determined in healthy males and females from childhood to old age and were related to different steps of working intensity using Knipping's crank ergometer. In the following years we extended the program to measure oxygen uptake, CO_2 expiration, ventilatory equivalent, heart rate, oxygen pulse, and later (from 1955) arterial blood pressure during a stepwise increasing of wattage to the maximum capacity of the subject. An improvement of the apparatus allowed measurements of maximum oxygen uptake of 6,000 ml/min. Consequently, it was possible to secure maximum performance measurements from elite athletes (50).

In 1954, Hollmann replaced the Knipping crank ergometer with a bicycle ergometer for clinical examinations. That was the beginning of routine bicycle ergometer examinations for clinical purposes (44).

In 1955, the Hamburg physicist Franz Nicolai (1910-1978) successfully developed the first telemetry equipment for electrocardiograph and heart rate worldwide. Thus, the corresponding values of athletes could be recorded over many kilometers (44).

In 1955, Harald Mellerowicz (1919-1996) started in Berlin with ergometric examinations of athletes. Because gas metabolism values could not be registered, he determined heart rate with a photoelectric cell on the earlobe and correlated the heart frequency to the watt rate. Later, he performed many investigations using spiroergometry about the effects of physical training. He used twins for this purpose because the twins had nearly identical genetic prerequisites (96-98).

In the earlier German Democratic Republic, which was founded in 1950 at the end of World War II by the Soviet Union as a soviet satellite state, the Medical University Clinic in Leipzig developed into a center for sport physiological examinations. The director of this clinic was the internationally recognized gerontologist Max Buerger (1885-1966). His senior physicians Josef Noecker (1919-1989) and Volkmar Boehlau

(1917-1988) conducted metabolic examinations of ergometer work as well as examinations of trace elements in skeletal muscle when active and after training (11, 106, 107).

Erich Arthur (E.A.) Mueller (1900-1972) from the Max Planck Institute for work physiology studied heart rate behavior during variably dosed and varyingly long continuous bicycle ergometer work (101). His main focus was the continuous endurance capability of a human, initially in an industrial environment but later in the field of sport. In order to determine the degree of fatigue experienced during physical exercise, he defined the recovery pulse sum as the sum of all heartbeats after the end of the work in relation to the resting heart frequency (102). The recovery pulse sum becomes larger when the loading intensity above the long-term performance limit becomes higher and when the loading is applied longer. E.A. Mueller also defined the long-term performance limit as the highest loading that is possible without causing an increase in the recovery pulse sum, independent of the duration of loading. The oxygen delivery corresponded to the oxygen requirement. Thus, early signs of fatigue did not occur when the pulse rate was 40 beats/min higher than the rest value. This corresponded to a total pulse rate of 100 to 120 beats/min in healthy people in their 30s and 40s (103).

Edgar Atzler (1887-1937) and Otto Graf (1894-1965) coined the phrase *lohnende Pausen* ("worthwhile breaks") to describe the recovery phrases during work intervals (4, 32). One could accelerate the recovery process and thus reduce the degree of fatigue during the interval work by cooling the skin (108). If leg work was done using a foot ergometer in a water bath at 40° C, the pulse rate increased linearly with the ascending working period and the recovery took a very long time. At a water temperature of 30° C, the pulse rate increased somewhat less and the recovery was quicker. If the water temperature was only 20° C, the pulse rate adjusted to a steady-state value at constant loading intensity. Finally, if the same amount of work was done at a water temperature of 16° C, the pulse rate per minute during the working period was lower and the recovery was completed after a few minutes. The cause for this behavior is an increase in the muscle blood circulation that occurs after a reduction of the skin blood circulation because of the lower temperature.

Research Results Obtained Using Spiroergometry

In 1953 Knipping mandated that every patient of the Medical University Clinic should undergo a spiroergometric examination if the patient exhibited no contraindications. The goal of this order was to detect very early a heart muscle insufficiency (working insufficiency of the heart) so that a very early treatment with digitalis could start. Thus, a massive databank was built up with time. Knipping called the maximum oxygen uptake *vita maxima*, in contrast to *vita minima* (basal metabolic rate). Together with coworkers he determined that 1,500 ml/min of oxygen was the minimum value for a trouble-free life. A value below 900 ml/min was seen as contraindicative with regard to serious surgical operations (76).

The first international publication on mean values of maximum oxygen uptake for crank ergometer work while standing with respiration of 100 (Vol%) O_2 in healthy males and females between 10 and 80 yr of age appeared in 1955 (136).

The examination results from the German Sport University in Cologne were combined with those of the Medical University Clinic in Cologne for the period from 1949 to 1963. The findings of 2,834 clinically healthy persons from 8 to 80 yr of age were included. Females reached the highest oxygen uptake by 15 to 16 yr of age (1,700 ml/min) and males reached the highest uptake at 19 yr of age (3,300 ml/min). The values began to decline for both men and women after 30 yr of age; the decline was slower in women than in men. Men and women who participated in endurance sports had significantly higher oxygen uptake values in their 80s than did untrained persons (50). All clinical examinations were conducted on the bicycle ergometer. Paul D. White, personal physician of U.S. President Eisenhower, was a guest in the Medical University Clinic in Cologne when he witnessed the changeover to the bicycle ergometer. This prompted him to propagate the use of the bicycle ergometer for clinical purposes in the United States.

In 1955 the first equipment for measuring blood pressure during ergometric examinations was developed at the Cologne Medical University Clinic by Hollmann and Sander, an engineer. For the first time a frequency-adjusted microphone was placed at the elbow joint and the Korotkow noise was acoustically and optically recorded. The measurement could be switched via three diaphragm pumps that were arranged in series (Hollmann and the engineer, Sander). The Elag Company in Cologne produced the first serial equipment in 1958. It was in the world exhibition for medical electronics in New York in 1960 and was the only one of its kind (48).

Development of the Aerobic–Anaerobic Transition 1955 to 1959

The spiroergometric examinations of patients at the Medical University Clinic in Cologne were carried out using the maximum oxygen uptake per minute as described for the first time by Archibald Hill (1886-1977) and colleagues in 1924. This procedure had two disadvantages: Patients with heart or lung defects were exposed to risk, and the assessment of maximum performance levels depended on the motivation of the person examined. Another way to measure the individual physical performance capacity was therefore sought in connection with increasing ergometer work. At first our laboratory investigated the level of pyruvate acid during increasing exercise but obtained no useful results. Then we determined the lactic acid in venous blood but also obtained no constructive results. We then measured the lactic acid in the arterial blood and observed in every third minute of increasing ergometer work an ascent parallel to the increase in the minute volume of ventilation. In order to reach the exact point of the aerobic–anaerobic transition we constructed the tangent of the minute volume of ventilation curve as well as that of the arterial lactic acid level and then drew the vertical from the point of contact of the respective curves to the abscissa. The watt values of respective oxygen uptake per minute were noted at this point and designated as the point of optimum efficiency of ventilation, or the largest oxygen uptake per minute that could be reached before an increase in the curve of arterial lactate and minute volume of ventilation (identical with the smallest ventilatory equivalent). From 1959 onward, clinic patients and top athletes were examined using this procedure, which was independent of individual motivation, demanded no maximum loading, and had great reliability (46, 47, 49). Five years later, Wasserman and McIlroy referred to the point of optimum efficiency of ventilation as the aerobic–anaerobic threshold. Later this term became the accepted international expression.

Physical Exercise Training

In the second half of the 1950s our laboratory addressed the question of which quality, quantity, and intensity of physical activity and number of repeats per week were necessary to counter health disabilities such as cardiac infarction. We examined the training schedules and determined the physical capability of top athletes and examined the effects of a minimal exercise-training program. This program consisted of training 3 times/wk for 10 min (or 1 or 2 times/wk for 30 min) at 70% of the individual's maximal oxygen uptake on a bicycle ergometer. The result was surprising: For a completely untrained person, such a minimal training program resulted in an unexpected significant increase in cardiopulmonary capacity in connection with peripheral and central hemodynamic and metabolic adaptations (51). We concluded that such a program could reduce the risk of a heart attack. Numerous examinations in later years confirmed these assumptions, particularly the epidemiological studies, which were conducted still later (10, 13, 14, 109, 139).

In Freiburg, Herbert Reindell (1908-1990) and colleagues (Kirchhoff, Koenig, and Roskamm) examined the health-related significance of athletic hearts (113, 114). With the help of a special X-ray method, the researchers determined the normal heart sizes in healthy males and females to be 700 to 800 ml and 400 to 550 ml, respectively. Because the heart size of a healthy person is dependent on body weight, the ratio of heart volume (kg) to body weight was determined. The normal values for untrained males and females were 11.7 ± 0.19 and 9.7 ± 0.15, respectively. In contrast, elite athletes engaged in endurance sports reached values of 18 to 20 (114). In 1964 our laboratory observed in a professional racing cyclist who was a triple world champion a heart size of 1,700 ml, a body weight of 80 kg, and a heart volume:body weight ratio of 21.2 (double the normal value of an untrained person). Physical inactivity (absolute bed rest) caused a decrease in heart volume of 10% in only 9 d (51).

In 1959, Ingeborg Bausenwein (1920-2008) scheduled a congress called "The Woman in Sport" to help disprove old beliefs about the harmfulness of strength training on the uterus and on childbearing ability in women. The congress also addressed the possibilities and limits of menstrual deferment using hormones and the effect on performance capabilities in women (5).

Altitude Training

In 1963, the International Olympic Committee decided that the 1968 Olympic Games would take place in Mexico City. This decision determined the direction of exercise physiology research in Germany in the 1960s. The stadium in Mexico City is located at an altitude of 2,240 m. Therefore, the percentage of oxygen in the inhaled air is reduced to 16 Vol% O_2. The question was how the physical performance capacity of athletes changed under these conditions.

A low oxygen content of 16 Vol% affected endurance performance within 2 min. Thus, an athlete's performance in an 800 m distance event that can be completed in approximately 1 min 44 s was not adversely affected, whereas performance in the 1,500 m distance event, which has an average completion time of 3 min 20 s, was markedly changed. These adverse effects increase up to a distance of 10,000 m, when performance capacity was reduced by about 6%, even when the athlete is extensively acclimatized to the altitude. In the marathon, greater altitude-related reductions in performance were not observed because of the slower running velocity (52, 97, 116).

In other studies it could be observed that cardiopulmonary performance at sea level increased after altitude training (89). A rowing ergometer test was conducted near the beginning (d 4 and 5) and end (d 21) of a course of training at an altitude of 1,840 m as well as 3 and 4 d after leaving this altitude. The test subjects were elite rowers. The test comprised 8 min of submaximal loading near the endurance limit and 6 min of near maximum loading by which a rowing competition was simulated. The performance was determined from the braking force, given by the braking weight and the number of revolutions in a chosen period.

In the first days of the stay at altitude, maximum oxygen consumption was reduced by 10% to 15%. The performance, heart rate, and minute volume of ventilation turned out higher. A somewhat similar decline of the 4 mmol/L threshold occurred. In the course of the 3 wk of altitude training, performance and maximum oxygen uptake increased by 5.2% and 4.9%, respectively, but did not reach the values for conditions at sea level (82). Compared with the values before altitude training, maximum performance and maximum oxygen uptake increased and ventilatory equivalent decreased on returning to sea level. The same volume of cardiac output per minute resulted in a higher oxygen transport capacity.

Increasing altitude acts as an increase in the loading intensity. Beyond that, however, the total hemoglobin content and the hemoconcentration enlarge and thereby the viscosity of the blood. The higher minute volume of ventilation cause an alkalosis of the blood, which is counteracted by an increased expulsion of bicarbonate buffers by the kidneys. After the German National Swimming Team stayed at 2,600 m for 3 wk and conducted 4 h of swim training daily, the pH value decreased more steeply with increasing arterial lactic acid content than it did under conditions at sea level. The reduced plasma volumes for enlarged erythrocyte volumes contained a better CO_2 buffer but exhibited a lower lactate bounding (90).

The relationship between hematocrit and hemoglobin content at medium altitudes in practice was characterized only through plasma loss (83). To what extent the production of erythrocytes in connection with the enhanced erythropoietin production contributes to this depends on the altitude. The relationship between HCO_3 and lactate exhibited a steeper decline at altitude. After the first week of a stay at medium altitude this effect was further amplified by bicarbonate loss and the decrease in plasma volume. At given loading levels at altitude, the pCO_2 in arterial blood decreased. Even at that time it was ascertained, in view of an increased formation of erythrocytes, that one must pay attention to adequate iron intake (90).

Important risk factors for overtraining at altitude are increases in urea above a value of 8.5 mmol and increases in creatine phosphokinase above about 400 to 450 mmol/L (90). As a rule, urea should not be augmented during altitude training. This can cause a catabolic situation that, subsequent to altitude training, can have a negative effect on performance at lower altitudes.

After training for 4 h/d over a period of 4 wk at an altitude of 2,050 m, the German National Rowing Team (10 subjects) exhibited an improvement in endurance performance of 60 to 80 W at sea level. When examining the German National Canoe Team (12 athletes), we observed an increase in oxygen uptake of 12.5% at a pulse rate of 170 beats/min (82). During a spiroergometric examination, the oxygen saturation in the femoral venous blood of the active musculature decreased clearly below the reference value, which was recorded before the altitude training began. The blood sugar levels were reduced and the lactic acid level was lower with the corresponding increase in pH value. The collective findings could be seen as an improvement in oxygen utilization in the active muscles with a corresponding shift to the right of the anaerobic threshold (58).

Cardiac output and the arteriovenous oxygen difference are the deciding performance-limiting factors at sea level for loading during general aerobic conditions. However, this does not apply to the conditions for physical exertion at medium or higher altitudes. Henceforth, the diffusion capacity of the lungs becomes the most important performance-limiting factor; oxygen partial pressure and hemoglobin saturation with oxygen are reduced when the blood flows through the capillaries of the lung. Under heavy loading at altitude, the period of erythrocyte and pulmonary alveoli contact is too brief to achieve complete oxygen saturation of the hemoglobin (82).

Dieter Boening (1939–) and colleagues in Hannover and later in Berlin (1972) described at the end of altitude training a steeper slope in the CO_2 binding curve with the effect of enhanced buffering against carbonic acid (44). The half-saturation pressure P50 for standard conditions in plasma (pH 7.4) exhibited the expected increase in magnitude, which was markedly higher after 1 wk and then decreased to below the initial value during the first day after returning to sea level. The changes ran parallel to the diphosphoglycerate concentration only in part. The hematocrit value increased by 7% whereas the hemoglobin concentration increased by only 2%. This change was apparently caused by a swelling of the erythrocytes, which leads to a reduction in cellular hemoglobin concentration. The diphosphoglycerate continuously increased in magnitude before falling to below the initial values at rest after the test person descended to sea level for extended periods. No significant changes in plasma electrolyte sodium, potassium, and calcium could be determined. The thyroid gland hormones, particularly T3, significantly increased in magnitude; significant increases in the aldosterone concentration before and after the stay at altitude were not observed. Among other things, the thyroid gland was activated by cold conditions (13).

Hypoxia Training in the Laboratory

In 1963 our laboratory examined the influence of endurance training conducted 5 times/wk, each session 30 to 45 min in duration, under hypoxia conditions (12 Vol % O_2 in the inspired air) in the laboratory. Before starting these tests, the subjects were already in good condition from endurance training. Significant performance improvements, however, could be observed when subjects were compared with a trained control group under normoxia conditions. A significant (5%) increase in maximum oxygen uptake and a reduction in the pulse rate to submaximal loading steps resulted. The lactate level during submaximal loading intensities decreased just as significantly in the subjects as in the control group with a corresponding increase in blood pH value (52, 82).

In 1966, the German Democratic Republic built the largest hypoxic physical training center in the world, named Kienbaum. Many Olympic gold medalists and world champions trained in this center (129).

A reduced partial pressure of oxygen in the inspired air resulted in a 15% to 20% increase in the systolic pressure of the arteria pulmonalis. Thus, the right heart works against increased resistance, which could be a training effect for this region of the heart. The lactate level decreased at a given submaximal loading level and subsequent to hypoxia training (82).

Hyperoxia Training in the Laboratory

In 1967 our laboratory investigated the effects of hyperoxia training in the laboratory. To do this we chose the closed spirograph system with exercise on the treadmill, whereby 100 Vol% O_2 was inhaled. The subjects trained 5 times/wk for 45 min/session at 70% of their maximum oxygen uptake. After 3 mo of training we observed an increase in the performance capability, which resulted from this form of training. When compared with a control group, performance values in the trained subjects exceeded the normal values by 18%. It was known at that time that the acute maximum oxygen uptake during respiration with oxygen can be increased, on average, by 10% (44). The augmentation of the maximal oxygen uptake, a decrease in the minute volume of ventilation, and a lower heart frequency at a given submaximal loading level are typical training improved the predispositions to sustain muscle work for a longer than was possible before this type of training.

In order to examine specific effects of local muscle training under hyperoxia, 12 healthy sport students were divided into two equal groups. The first group performed one-leg training with 100 Vol% O_2 respiration while seated on a bicycle ergometer, and the second group conducted the same training but with 21 Vol% O_2 respiration. The training period lasted for 8 wk. The training format consisted of one-leg training 5 times/wk for 1 h/session with a loading intensity of approximately 70% of maximum oxygen uptake under both hyperoxia and normoxia.

Maximum oxygen uptake before training—subject to normoxia as well as hyperoxia for one-leg and two-leg exercise—was approximately the same magnitude in both test groups. In the one-leg training group, maximum oxygen uptake increased by 9% after training under normal atmospheric conditions and increased 27% under hyperoxia conditions. In the two-leg training group, maximum oxygen uptake increased by 18% in the hyperoxia group under normoxic conditions and 12% in the normoxia-trained group. Maximum oxygen uptake increased 23% in the one-leg training group after training with 100 Vol% O_2 (59).

The maximum oxygen uptake of the group with untrained control legs increased by 8% in both the hyperoxia- and normoxia-trained group. For given loading levels, especially with the higher loading ranges, a reduction in oxygen uptake for both the trained and untrained legs was determined. This result applied to both training groups. At a 150 W loading level the reduction in oxygen consumption was approximately 16%.

When subjects trained in the submaximal working range with both legs or with one leg, the minute volume of ventilation exhibited a highly significant reduction after training. The greatest difference occurred when the one leg group, trained under hyperoxia conditions, performed an exercise loading of 150 W; the minute volume of ventilation was 28% lower. A difference of only 4% was observed during an examination under oxygen respiration.

After training no difference could be determined in the pretraining exercising values for the control leg in the hyperoxia-trained group. At a loading level of 150 W, lactate content of the vena femoralis was reduced 28% in the hyperoxia-trained group and 25% in the normoxia-trained group; these reductions were highly significant. The aerobic–anaerobic threshold in the hyperoxia-trained group shifted from approximately 100 W to 150 W. The corresponding values in the normoxia-trained group were approximately 100 W before training and 125 W after. Thus, the prerequisites for long-lasting submaximal endurance work under hyperoxia training were significantly enhanced (59).

The untrained control leg in the hyperoxia-trained group exhibited no significant differences before and after training. The lactate content, however, was reduced significantly in the control leg of the normoxia-trained group. This finding corresponds well with the similarly reduced heart frequency and the diminished ventilatory minute volume for given submaximal loading levels in this group.

The increase in maximum oxygen uptake after hyperoxia training was explained as follows. In healthy persons and under normal conditions, the oxygen saturation of hemoglobin is already at an optimum and cannot be increased under respiration with oxygen. The fivefold increase in the oxygen partial pressure in the tidal air, however, allows additional oxygen to be physically dissolved in the plasma. Although the normal amount of oxygen amounts to 0.3 ml/100 ml of plasma, a four- to fivefold increase of this amount can be assumed because of the linear relationship between the physically dissolved oxygen in the plasma and the partial pressure of the oxygen. These differences have little effect on the test person in repose with a cardiac output of 51 beats/min; however, the opposite is true in the limiting range of 20 to 40 L/min (59).

Further Examinations of Exercise Performance

In the 1950s and early 1960s, Mueller and Hettinger stated that no strength increase occurs when the intensity of isometric muscle work is less than 15% to 20%. Moreover, 20% to 30% of the maximum isometric load is insufficient for an augmentation of strength. If the intensity of the muscle strain is reduced by the presence of a plaster cast, an 8 d immobilization results in a reduction of maximum isometric strength of about 20%. In 14 d a loss of approximately of 28% would be expected (41).

If the isometric loading intensity of a muscle is higher than 30% of maximum static strength, this initially results in an increase in muscle strength. Already a single stimulus of 10 s duration causes an increase in maximum isometric strength in completely untrained persons. The maximum possible training stimulus is reached at a contraction intensity of about 70% of maximum force.

According to Hettinger, the training stimulus is completely effective only if the duration of the training stimulus is approximately 20% to 30% of the maximum time until exhaustion. Isometric muscle contractions with 70% of maximal strength need a minimal contraction time of 3 s to induce training effects (41).

After reviewing the data of Mueller and Hettinger and the reviews concerning their results, Lorenz and colleagues concluded that only select principles were valid for training purposes (88). Specifically, Lorenz and colleagues findings were obtained from untrained males in their 30s. The higher the initial standard of performance, the greater the muscular force that must be applied. At the same time, the closer the test person is to his or her individual limit of performance, the lesser the training success. To avoid training at low, ineffective levels, elite athletes should always work at their highest possible loading intensities.

In contrast to static forms of training, dynamic forms of training that utilize concentric and eccentric muscular contractions have the advantage of including the forces inherent in movement and coordination that are relevant for the specific type of sport.

In scientific studies on muscle strength, the equipment used must meet the physical and biological requirements of the subjects to be tested. In 1960, Hettinger designed the Hettinger strength-measurement chair for this purpose (41). In 1968, our laboratory developed strength-training equipment that was driven by a motor and an eccentric connecting rod that allowed us to conduct user-defined static concentric and eccentric strength training (57).

In 1972, two employees from Chrysler in the United States visited us at the Institute for Research in Cardiology and Sports Medicine at the German Sport University in Cologne. They asked whether it was possible to design a new type of ergometer that had no external similarities to a bicycle, was controlled electronically, and was completely computerized. Chrysler paid the development costs. In 1974, we developed a new type of bicycle ergometer (64). Chrysler had previously transferred the rights to the German company Keiper in Kaiserslautern. The new equipment, named Dynavit, became the forerunner of all later bicycle ergometers of this type (64).

Age and Performance Capacity

The physical performance behavior in persons from childhood to very old age was one of the research focal areas in Cologne from 1950. In a representative number of subjects in Germany our laboratory ascertained that maximum oxygen uptake reached a maximum at about 16 yr of age in females and 19 to 20 yr of age in males.

These values remained more or less constant until 30 yr of age, when they decreased in untrained persons. During the third decade of life the sex-related differences are about 30%; these values decline with increasing age to minimal differences at 80 yr of age. Healthy persons who engaged in aerobic endurance training after 50 yr of age exhibited significantly higher performance values that continued after 90 yr of age (50). The determination of relative maximum oxygen uptake gave comparable results. The maximum minute volume of ventilation peaked as early as 14 yr of age in females and 17 yr of age in males (50).

The introduction of echocardiographic examinations in the 1960s enabled the determination of normal values of the diameter of the cardiac septum, cardiac walls, and the left and right cardiac ventricles. Physical inactivity reduced the ventricular dimensions whereas endurance training increased them (117). In this context it was ascertained that accelerated juveniles with a larger heart have a higher maximum oxygen uptake per kilogram of body weight than do normal or less advanced juveniles.

In the 1970s, our laboratory conducted sport medical examinations on a population of boys and girls aged 6 to 8 yr who had begun intense endurance swim training at the elite level. Significant functional and morphologic differences became apparent as early as 2 yr after training. Echocardiographic examinations showed significant enlargement of the end diastolic diameter of the heart and an increase in the end diastolic wall diameter and the septum. These results confirmed that endurance training before puberty caused cardiac enlargement. Examinations were continued over a period of 20 yr and demonstrated that the training-induced cardiac enlargement ceased in both male and female swimmers when the maximal physiological growth limit had been obtained. No pathological heart enlargement or pathological symptoms were reported (118).

Clinically healthy individuals between 55 and 80 yr of age who have been untrained for years can structurally and functionally benefit from a training program because of adaptations in the cardiopulmonary system, musculoskeletal system, and connective tissue. The findings of our laboratory collected over several decades demonstrated that training can compensate for age-related losses of function by as much as 30% to 40% (83, 133).

4 mmo1/L Lactate Threshold

After first describing the aerobic–anaerobic transition in connection with the registration of minute volume of ventilation and arterial lactate level in 1959, our laboratory reported the existence of a 4 mmol/L lactate threshold (89). This original finding pertained to a lactate steady state that has proven to be effective in prescribing and monitoring training programs for elite athletes (91).

In normal athletic subjects a maximum exercise period of approximately 3 min is associated with a blood lactate concentration between 14 and 16 mmol/L in the first 1 to 3 min of the recovery period. Elite middle-distance runners exhibit blood lactate concentrations between 20 and 27 mmol/L during the recovery period (89, 91, 92).

For a maximum exercise period of 3 min in highly trained athletes, the degree of cellular acidosis in active musculature is a vital performance-limiting factor. Such a performance presupposes a maximum rate of glycolysis and intracellular pH values of 6.2 (85, 91). This is why phosphofructokinase and other glycolytic enzymes are reduced with the corresponding decrease in glycolysis rate.

In the relationship between maximum lactate concentration in the postexercise phase, the x-axis represents running velocity performance and the y-axis represents the maximum level of postexercise lactate. Mader and colleagues plotted the lactate concentration maxima for two 600 m test runs in such a lactate-velocity diagram (91). The position of the respective curves of the maximum lactate concentrations as a function of the velocity characterizes the individual performance capabilities of each test person. The lower the performance capacity, the earlier the formation of lactate begins during an unchanged running velocity. We described this procedure as the "two velocities test." It has proven very useful over time in the field of performance diagnostics for athletes. A computer program could be developed that allows a precise differentiation between factors that limit physical performance capacity (92).

Muscle Bioptic Studies

Other studies in the 1970s used the results of biopsies of the vastus lateralis to assess qualitative and quantitative differences associated with training programs. For this purpose, our institute preferred to use one-leg training; the other leg was the control leg. This method has the advantage of using identical genotypes. In the 1970s, our laboratory found that endurance-trained subjects exhibited a capillary density that was 28% greater than that of normal individuals. A comparison of the number of capillaries per muscle fiber resulted in a 41% higher value. No significant differences in the skeletal muscle capillaries could be recognized when athletes not participating in endurance sports were compared with nonathlete subjects (123). The following were observed in trained subjects:

- increased number and size of mitochondria,
- increased activity of the aerobic enzymes,
- increased amount of intramuscular glycogen, and
- increased myoglobin.

In the 1970s, Pette and colleagues from the University of Constance in Germany developed the potential for electrically influencing the composition of muscle fibers. Using electrical simulations in animal tests, the researchers were able to transform fast muscle fibers into slow muscle fibers and vice versa (110).

Other Areas of Exercise Physiology

Heinz Liesen (1941–) and Bertin Dufaux (1943–) described enlarged glycoprotein levels in the blood after acute aerobic exercise (85). Very long and intensive endurance exercise caused an activation of the complementary system and a reduction of proteinase inhibitors. In particular, the acute phase reactants (e.g., ceruloplasmin, transferrin, haptoglobin) and the complimentary factors C1-inhibitor and C3-activator increased for several days after the end of exercise (23, 28, 84, 85).

Wilfried Kindermann (1940–) in Saarbruecken and colleagues investigated the aerobic and anaerobic metabolism and the athlete's heart in different sports in connection with determination of glucagon, insulin, and other hormones (69, 70).

Using radioactive isotopes, the influence of different forms of exercise and physical training on blood flow through the lungs, liver, heart, spleen, and kidneys was determined. In the liver, spleen, and kidneys the blood supply was reduced by 10% to 30% independent of the loading intensity. The upper lobes of the lung had a reduced blood supply in the resting condition. However, the blood supply in this region of the lung increased with increased ergometer work (16).

Kurt Tittel (1920–) worked in Leipzig in connection with the "Deutsche Hochschule fuer Koerperkultur." His examinations with Wutscherk created new knowledge about the connections between human body dimensions and specific performance capacity in different sports (134).

Georg Neumann (1949–) was one of the leading exercise physiologists in the former GDR. He enlarged the knowledge of training methods in track and field as well as in bicycling competitions (104, 105).

Manfred Lehmann (1944-2001) investigated catecholamines, receptors, and hormones and their importance for maximal physical performance capacity. His theories about the causes of overtraining in connection with reactions of the hypothalamus found international acceptance (78-81).

Juergen Steinacker (1955–) and his group (Liu and colleagues) at the Medical University Clinic in Ulm investigated heat shock proteins in connection with exercise and physical training. During exercise a heat shock protein increased to stabilize cellular proteins in order to prevent denaturation. Steinacker also examined its role as chaperone. Its task is to pleat deoxyribonucleic acid correctly (126, 127).

Aloys Berg (1946) in Freiburg studied the influence of nutrition on physical performance capacity. He first described the effect of enzyme yeast cells on the immune system of top athletes with reduction of the stress reaction (7-9).

The Muenster research group of Klaus Voelker (1948–) and colleagues studied physiological factors (i.e., energy consumption, blood pressure, and hemodynamic reactions), for instance in connection with climbing stairs. It was ascertained that males in their 30s with average capabilities require a loading of 180 steps/d in order to achieve unchanged capacity. An increase in the number of steps represented training effects similar to those for running. Atrial natriuretic peptide levels increased with the intensity of work in both men and women (137, 138).

Human Brain and Physical Activity 1985 to 2009

Before the 1980s there was no reliable method of examination for studying the influence of exercise and training on the human brain. In 1985, our laboratory together with the Nuclear Research Center in Juelich and the Max Planck Institute for Brain Research in Cologne began to undertake joint studies. During bicycle ergometer work of 25 W, the cerebral blood flow in various regions of the brain increased by 30%, in connection with a loading of 100 W until 40% (38) (figure 4.4). Using static loading we could not observe significant changes in cerebral blood flow. The cause for the different responses could be associated with proprioceptors in various tendons. We hypothesized that a simultaneously increased production of nitric oxide and potassium as well as other chemical substances could be responsible for rapidly transporting neurotransmitters and other substances to peripheral destinations.

Kenny de Meirleir (1949–) and colleagues performed investigations pertaining to the influence of neurotransmitters on physical performance capacity.

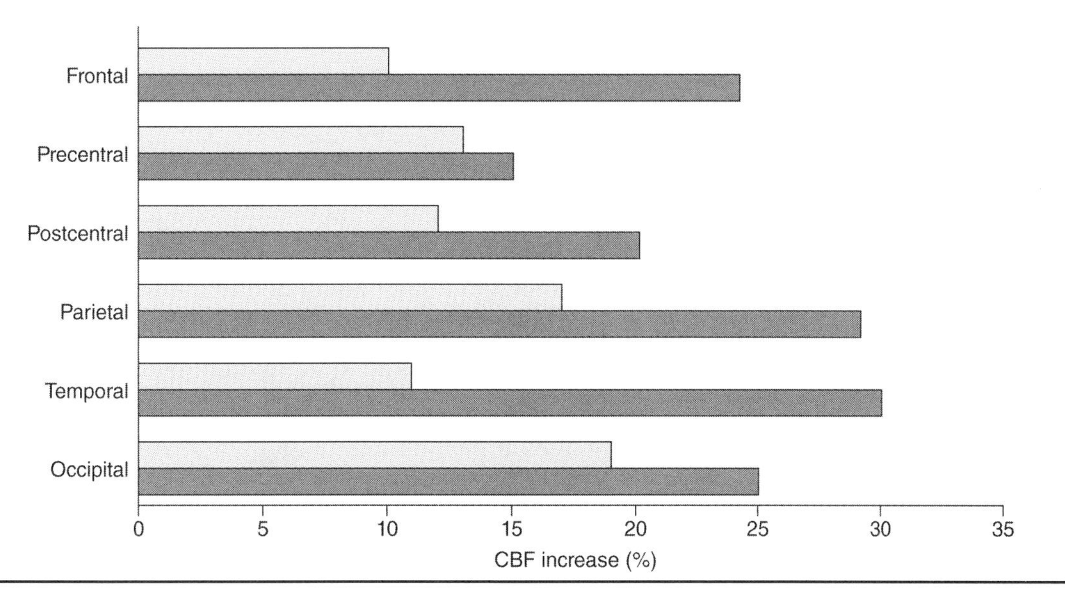

Figure 4.4 Increase in the regional blood supply in the brain during a work intensity of 25 W (white) and 100 W (dark). Subjects (6 healthy males in their 30s) performed exercise on a bicycle ergometer for 10 min.

With kind permission from Springer Science+Business Media: *Journal of Neurology*, "Regional cerebral blood flow in man at rest and during exercise," 234, 1987, pgs. 9-13. K. Herholz et al., figure 1.

The dopamine agonist pergolide increased the neurotransmitter effect of dopamine. At increasing physical loading the systolic blood pressure, heart rate, and levels of noradrenaline, dopamine, and lactate were reduced (19). The increase in prolactin, which generally occurs during physical exercise, was suppressed whereas luteinizing hormone increased. The known augmentation of adrenocorticotropic hormone in the blood did not occur, nor was there a marked increase in blood levels of growth hormone. During heavy bicycle ergometer exercise, an antagonist to serotonin (ketanserine) was associated with a reduction in systolic blood pressure at an unchanging maximum oxygen uptake (20).

To determine whether endorphins or other substances elicit a euphoric mood, our laboratory initiated a study in which an artificial dental crown was connected to the pulpa (tooth pulp) of volunteers (1). Immediately after the exercise period the subjects received electrical shocks. Directly after the end of the work the electrical shock threshold increased by 26%; using a placebo, it increased by 31%. When the subjects received the drug naloxone (an opioid inhibitor), the pain threshold after work was significantly lowered by 30%. Thus, the significance of opiates in the brain for altering mood after long-lasting aerobic exercise was proven. It could have been nature's way of endowing humans with a larger capacity for endurance performance when hunting or fleeing danger in prehistoric times.

Studies on cerebral metabolism have shown that 30 min of ergometer exercise at 60% $\dot{V}O_2max$ decreases cerebral blood glucose consumption in the fore and middle regions of the brain and increases glucose consumption in the occipital region of the brain (40) (figure 4.5). Rojas Vega and colleagues observed an increase in brain-derived neurotrophic factor in the blood during ergometer exercise (115).

Summary

In Germany, exercise physiology is a part of sports medicine. The first experimental investigations about the effects of physical exercise on the human cardiovascular system and digestion were carried out in the 18th century. The 19th century was characterized by the development of the first cranked ergometer and the first motorized treadmill. At the beginning of the 20th century, numerous physiological studies examined the effects of different sports on healthy individuals. In 1912, the first national sport medicinal association worldwide was founded. In the following two decades, research focused on the chemical processes in the skeletal muscle at rest and during physical work. The development of spiroergometry in 1929 was a crucial step in the assessment and evaluation of physical capacity in sports and clinics.

After World War II, the technical part of spiroergometry fulfilled all the scientific criteria and became fundamental in the assessment of physiological performance. In 1954 the crank ergometer was replaced by a bicycle ergometer. This was followed by the devel-

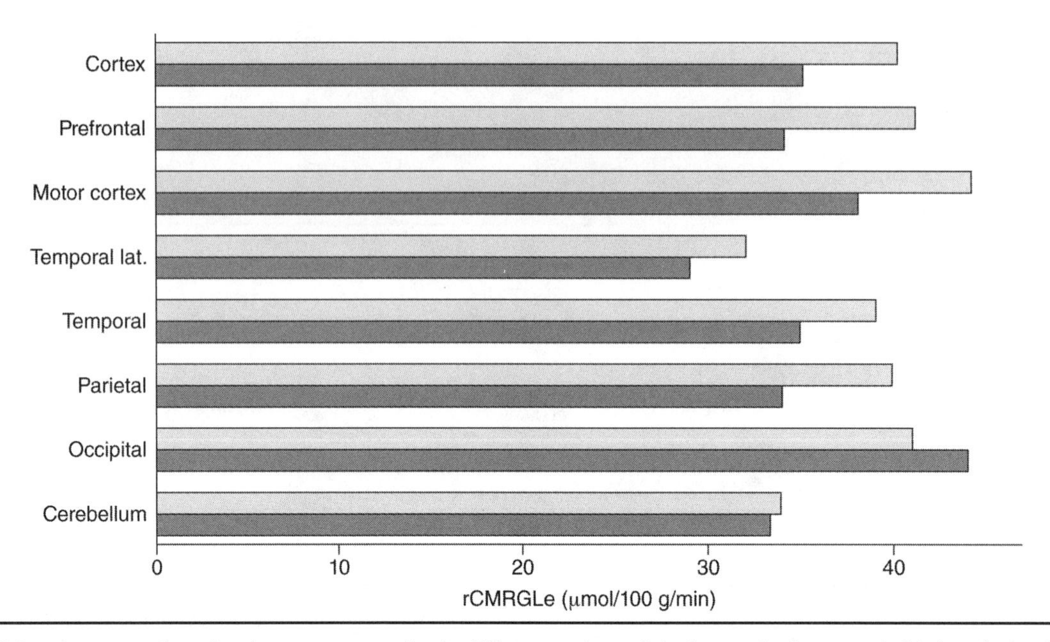

Figure 4.5 Average values for glucose consumption in different regions of the human brain at rest (white) and after 30 min of ergometer exercise at 60% of $\dot{V}O_2max$ (dark).

Adapted from H. Herzog, C. Unger, T. Kuwert, H.G. Fischer, D. Scholz, W. Hollmann, L.E. Feinendegen, 1991, "Physical exercise does not increase cerebral metabolic rate of glucose utilization," *Vth International Symposium on Cerebral Blood Flow and Metabolism* By permission of W. Hollmann.

opment of special devices that measure blood pressure during actual exercise. In 1959, the aerobic–anaerobic transition was described for the first time by means of combined measurement of the minute volume of ventilation and lactate level. Training under hypoxic and hyperoxic conditions in the laboratory was introduced in 1963 and 1967, respectively. At that time numerous studies were performed under middle-altitude conditions as well. The effects of acute and chronic physical exercise on microbiological processes were of interest in the last three decades of the 20th century. Another area of research dealt with the effects of physical training during childhood and youth as well as the role of regular exercise in the elderly.

Examinations of the influence of ergometer exercise on hemodynamic and metabolic adaptations in the human brain started in 1985. Other investigations referred to the influence of physical exercise and training on neurotransmitters and hormones. Today the main area of study, besides brain research, is the influence of exercise and training on molecular and cellular processes.

One of the most impressive detections was the extensive plasticity of the human brain. Physical exercise is a stimulus for the formation of synapses and spines as well as for angiogenesis and neurogenesis (18, 29, 30). Physical exercise results in

- angiogenesis,
- neurogenesis,
- hypertrophy of synapses,
- spine augmentation, and
- growth of neurites and dendrites (18, 19, 29, 30, 64).

Heiko K. Strueder (1965–) and colleagues observed in a double-blind experiment a significant decrease in endurance performance when serotonin reuptake was inhibited by the presence of paroxetine (130, 131).

Metabolic changes in the human brain during physical exercise play a key role in the plasticity of the brain processes similar to alterations in endogenous peptides, amino acid transport through the blood–brain barrier, and influences from neurotransmitters (129-131). Exercise also induces gene expression, thereby affecting brain plasticity. Physical activity has a positive influence on cognition and age-related processes. In older persons an unchanged cognitive capacity is associated with activation of brain regions that are larger than those found in young persons. In our laboratory this phenomenon could not be observed in marathon runners between 60 and 70 yr of age (121). Therefore, physical exercise has a protective effect on the cardiovascular system as well as on the brain and cognition.

References

1. Arentz T, De Meirleir K, Hollmann W. The role of the endogenous opioide peptides during bicycle ergometer work [in German]. *Dtsch Z Sportmed.* 37: 210-219, 1986.
2. Arnold A. History of sports medicine and tasks of sports medicine [in German]. In: Arnold A, ed. *Textbook of Sports Medicine* [in German]. Leipzig: Barth, 1956, 1960.
3. Aschoff L. *Pathological and anatomical basis of heart insufficiency* [in German]. Jena: Fischer, 1906.
4. Atzler E. Occupational physiology [in German]. *Ergebnisse der Arbeitsphysiologie.* 40: 325-341, 1938.
5. Bausenwein I, Damm F, Hillmer-Vogel U. 40 years of female sports medicine in Germany. In: Tittel K, Arndt KH, Hollmann W, eds. *Sports Medicine Yesterday—Today—Tomorrow* [in German]. Leipzig: Barth, 1993.
6. Berchtold NC, Kesslak JP, Pike CJ, Adlard PA, Cotman CW. Estrogen and exercise interact to regulate brain-derived neurotrophic factor mRNA and protein expression in the hippocampus. *Eur J Neurosci.* 14: 1992-2002, 2001.
7. Berg A, Keul J. Women in the performance sports. Maximum permissible load and capabilities of adaptation. In: Heck H, Hollmann W, Liesen H, Rost R, eds. *Sport, Performance, and Health* [in German]. Cologne: Deutscher Aerzteverlag, 1983.
8. Berg A, Koenig D, Schlachter H, Keul J. The quality of different fatty acids and their metabolic importance for peripheral regulations in athletes [in German]. *Dtsch Z Sportmed.* 44(special edition): 445-450, 1993.
9. Berg A, Hamm M. *Enzyme yeast cells in competitive sports* [in German]. Heidelberg: Haug-Verlag, 2006.
10. Blair SN. Physical activity, physical fitness, and health. In: The Club of Cologne, ed. *Joint Meeting Cologne of WHO and FINIS 1994: Health Promotion and Physical Activity.* 11-36, 1994.
11. Boehlau V. *Examination of the physical performance capacity* [in German]. Leipzig: Barth, 1955.
12. Boening D, Schweigart U, Tibes U, Hemmer P. In vivo and in vitro investigations of the oxygen dissociation curve of blood of trained and untrained subjects during exercise. *Pfluegers Arch* 332: 78-88, 1972.
13. Bouchard CW, Hollmann W, Venrath H, Herkenrath G, Schluessel H. *Minimal amount of physical training for the prevention of cardiovascular diseases. ATI World Congress for Sports Medicine.* Cologne Berlin: Deutscher Aerzteverlag, 1966.
14. Bouchard C. Physical activity, fitness and health. In: The Club of Cologne, ed. *Joint Meeting Cologne of WHO and FIMS 1994: Health Promotion and Physical Activity.* 37-49, 1994.
15. Brauer L, Knipping HW. The so-called spirographic deficit and the arterial blood gas analysis [in German]. *Med Klin.* 44: 1429-1435,1949.
16. Buskies W. *Blood Volumina in Lungs, Heart, Liver, Kidney and Brain During Ergometer Exercise in Connection With Air and Oxygen Respiration* [in German]. Dissertation. Cologne: German Sport University, 1986.
17. Candia V, Elbert I, Altenmueller E, Rau H, Schaefer T, Taub E. Constrained-induced movement therapy for focal hand dystonia in musicians. *Lancet.* 353: 42, 1999.
18. Cotman CW, Berchtold NC. Exercise: A behavioral intervention to enhance brain health and plasticity. *Trends Neurosci.* 25: 295-301, 2002.
19. De Meirleir KJ, Gerlo F, Hollmann W, Vanhaelst L. Cardiovascular effects of pergolide mesylate during dynamic exercise. *Brit J Clin Pharmacol.* 23: 633P-634P, 1987.
20. De Meirleir K, Smitz J, Vansteirteghem A, Hollmann W. Serotonine antagonism during exercise in man. *Acta Cardiol.* 42: 360-361, 1987.
21. Du Bois-Reymond E. *Parallel Bars and the So Called Rational Gymnastic* [in German]. Berlin: 1862.
22. Du Bois-Reymond E. *Specific Physiology of the Muscle and Movement* [in German]. Berlin: Hirschwald, 1903.
23. Dufaux B, Order U, Hollmann W. Serum-C-reactive protein and immune complex after exercise and training [in German]. *Dtsch Z Sportmed.* 14: 252-257, 1982.
24. Dufaux B, Hoeftken K, Hollmann W. Acute phase protein and immune complexes during several days of severe physical exercise. In: Knuttgen HG, Vogel IA, Poortmans J, eds. *Biochemistry of Exercise.* Champaign, IL: Human Kinetics, 1983.
25. Dufaux B, Order U, Geyer H, Hollmann W. C-reactive protein serum concentrations in well-trained athletes. *Int Sports Med.* 5: 102-108, 1984.
26. Dufaux B, Order U. Complement activation after prolonged exercise. *Clin Chem Acta.* 179: 45-51, 1989.
27. Dufaux B, Order U, Liesen H. Effect of a short maximal physical exercise on coagulation, fibrinolysis, and complement system. *Int J Sports Med.* 12(Suppl. 1): 38-46, 1991.
28. Dufaux B, Kothe A, Heine O, Prinz U, Rost R. Effects of physical conditioning on hemostasis and some new aspects on mechanisms of exercise and associated fibrinolysis. *Eur J Clin Chem.* A: 207-212, 1995.
29. Elbert T, Pantev C, Wienbruch C, Rockstroh B, Taub E. Increased cortical representation of the

fingers of the left hand in string players. *Science.* 270: 305-307, 1995.

30. Eriksson PS, Perfilieva E, Bjork-Eriksson T, Alborn AM, Nordborg C, Peterson DA, Gage FH. Neurogenesis in the adult human hippocampus. *Nat Med.* 4: 1313-1317, 1998.
31. Gaertner F (1887) cited in Prinz JP. Bicycle ergometer 1897-1992 [in German]. In: Tittel K, Arndt KH, Hollmann W. *Sports Medicine—Yesterday, Today, Tomorrow* [in German]. Leipzig Berlin Heidelberg: Barth, 1993.
32. Graf O. The question of the working and rest time during assembly-line production [in German]. *Arbeitsphysiol.* 11: 503-510, 1943.
33. Graf O. Exercise duration and rest arrangement in assembly-line production [in German]. *Arbeitsphysiol.* 12: 348-360, 1943.
34. Hartung M, Venrath H, Hollmann W, Isselhard W, Jaenckner D. *The Regulation of Ventilation During Physical Work* [in German]. Cologne-Opladen: Westdeutscher Verlag, 1966.
35. Heck H. *Lactate in the Performance Diagnostic* [in German]. Schorndorf: Hofmann, 1990.
36. Henschen SE. *Ski and Ski Competition* [in German]. Jena: Mitteilungen Medizinische Klinik Upsala, 1899.
37. Herbst R. The gas metabolism for the determination of the physical performance capacity [in German]. *Dtsch Arch Kith Med.* 162: 33-42, 1928.
38. Herholz K, Buskles W, Rist M, Pawlik G, Hollmann W, Heiss WD. Regional cerebral blood flow in man at rest and during exercise. *Neural.* 234: 9-13, 1987.
39. Herxheimer H. *Outlines of the Sports Medicine for Physicians and Students* [in German]. Leipzig: Thieme, 1933.
40. Herzog H, Unger C, Kuwert T, Fischer HG, Scholz D, Hollmann W, Feinendegen LE. *Physical Exercise Does Not Increase Cerebral Metabolic Rate of Glucose Utilization.* Vth International Symposium on Cerebral Blood Flow and Metabolism, Miami, Florida, United States, 1991.
41. Hettinger T. *Isometric Muscle Training* [in German]. Stuttgart: Thieme, 1968.
42. Hill AV, Long CHN, Lupton H. Muscular exercise, lactic acid and the supply and utilisation of oxygen. *Proc Roy Sac B.* 96: 438-444, 1924.
43. Hoffmann F. *Effects of Physical Exercise on Cardiovascular System and Digestion* [in German]. Halle: University, 1719.
44. Hollmann W. *The Influence of Physical Exercise on Heart and Circulation* [in German]. Darmstadt: Steinkopff, 1959.
45. Hollmann, W. The maximal O_2 update in older persons [in German]. German Sports University, Cologne, 1959.
46. Hollmann W. *The Relationship Between pH, Lactate Acid, Potassium in the Arterial and Venous Blood, the Ventilation, POW and Pulse Frequency During Increasing Spiro-Ergoinetric Work in Endurance Trained and Untrained Persons.* Pan-American Congress for Sports Medicine, Chicago, Illinois, United States, 1959.
47. Hollmann W. The pan-American congress of sports medicine in Chicago (September 1st + 2nd, 1959) [in German]. *Der Sportarzt.* 10: 300-301, 1959.
48. Hollmann W, Venrath H, Valentin H, Spellerberg B. The arterial blood pressure in humans during dosed physical exercise [in German]. *Z Kreisl Forschg.* 48: 162-168, 1959.
49. Hollmann W. A submaximal determination of the physical performance capacity [in German]. *Fortschr Med.* 79: 439-445, 1961.
50. Hollmann W. *Maximal and submaximal performance capacity of the athlete* [in German]. Muenchen: Barth, 1963.
51. Hollmann W. *Physical Training for the Prevention of Cardiovascular Diseases* [in German]. Stuttgart: Hippokrates, 1965.
52. Hollmann W, Venrath H, Herkenrath G. *Physical Training Under Hypoxic Conditions* [in German]. Cologne-Berlin: Deutscher Aerzteverlag, 1966.
53. Hollmann W, Venrath H. The cardio pulmonary system and the strength of the skeletal muscle during exercise in connection with different oxygen contents in the air [in German]. *Schweiz Z Sportmed.* 14: 27-33, 1966.
54. Hollmann W, Gruenewald B, Bouchard C. The reduction of the cardiopulmonary capacity during aging and obviations by minimal programs of physical training [in German]. *Arbeitsmed, Sozialmed und Arbeitshygiene.* 3: 88-96, 1967.
55. Hollmann W, Bouchard C. The relationships between chronological and biological age in relation to spiroergometric parameters, heart volume and skeletal muscle strength in boy 8-18 years of age [in German]. *Z Kreishnifforschung.* 59: 160-176, 1970.
56. Hollmann W, Liesen H. The influence of hypoxia and hyperoxia training in a laboratory on the cardiopulmonary capacity. In: Keul J, ed. *Limiting Factors of Physical Performance.* Stuttgart: Thieme, 1973.
57. Hollmann W, Hettinger T. *Sports Medicine—Fundamental Principles of Exercise and Training* [in German]. Stuttgart: Schattauer, 1976.
58. Hollmann W, Mader A, Liesen H. The influence of middle altitude and physical training on metabolic and hemodynamic factors. In: Deetjen P, Humpeler E, eds. *Medical Aspects of Altitude* [in German]. Stuttgart: Thieme, 1981.
59. Hollmann W, Liesen H, Egberts E, Heck H, Philippi H, Mader A, et al. The influence of a one leg bicycle ergometer training on spiroergometrical parameters and the arterial lactate level in connection with

normoxy and hyperoxy [in German]. *Dtsch Z Sportmeci.* 35: 81-91, 134-141, 1984.

60. Hollmann H, Schuerch P, Heck H, Liesen H, Mader A, Rost R, et al. Cardiopulmonary reactions and aerobic–anaerobic threshold during different forms of exercise [in German]. *Dtsch Z Sportmed.* 38: 144-156, 1987.
61. Hollmann W, Fischer HG, De Meirleir K, Herzog H, Herholz K, Feinendegen LE. The brain: Regional cerebral blood flow, metabolism and psyche during ergometer exercise. In: Bouchard C, Shephard RJ, Stephens T, eds. *Physical activity, fitness, and health. International proceedings and consensus statement.* Champaign, IL: Human Kinetics, 1994.
62. Hollmann W, Strueder HK, Predel HG, Tagarakis CVM. *Spiroergometry* [in German]. Stuttgart: Schattauer, 2006.
63. Hollmann W, Tittel K. *History of German Sports Medicine* [in German]. Gera: Druckhaus Gera, 2008.
64. Hollmann W, Strueder HK. *Sports Medicine: Textbook for Physical Activity, Training and Preventive Medicine* [in German]. Stuttgart: Schattauer, 2009.
65. Hueppe F. *Textbook of Hygiene* [in German]. Berlin: Hirschwald, 1899.
66. Hufeland CW. *Macrobiotic* [in German]. Berlin: 1842.
67. Jeschke D, Heitkamp HCh, Locher R, Schneider D, Simon M, Zintl W. Aerobic capacity and anaerobic threshold during different exercise methods on the bicycle ergometer. In: Mellerowicz H, Franz W, eds. *Standardization, Calibration and Methods in Ergometry* [in German]. Erlangen: Perimed, 1983.
68. Keul J, Haralambie G. Metabolism during physical exercise. In: Hollmann W, ed. *Central Topics of Sports Medicine* [in German]. Berlin-Heidelberg: Springer, 1986.
69. Kindermann W, Keul J, Huber G. *Anaerobic Provision of Energy in Top Athletism* [in German]. Schorndorf: Hofmann, 1977.
70. Kindermann W, Schnabel A, Schmitt WM, Biro G, Cassens J, Weber E. Catecholamines, growth hormone, cortisol, insulin, and sex hormones in anaerobic and aerobic exercise. *Eur J Appl Physiol.* 49: 389-397, 1982.
71. Klaus EJ, Noack HD. *Woman and Sport* [in German]. Stuttgart: Thieme, 1961.
72. Klaus EJ. Sports medicine: Development, aims and tasks [in German]. *Hippokrates.* 35: 111-113, 1964.
73. Knipping HW. A simple apparatus for the exact measurement of gas metabolism in the medical clinic [in German]. *Klin Wschr.* 5: 553-555, 1926.
74. Knipping HW. The investigation of the economy of muscle work in healthy and sick persons [in German]. *Z Ges Exp Med.* 66: 517-523, 1929.
75. Knipping HW, Moncrieff A. The ventilatory equivalence for oxygen. *Quart J Med.* 1: 17-22, 1932.
76. Knipping HW, Bolt W, Valentin H, Venrath H. *Investigation and Assessment of the Cardiological Patient* [in German]. Stuttgart: Enke, 1955.
77. Kuelbs F. Experimental investigation about the heart muscle [in German] *Schmiedeberger's Arch Pathol Pharmakol.* 55: 288-298, 1906.
78. Lehmann M, Dickhuth HH, Schmid P, Porzig H, Keul J. Plasma catecholamines, beta-adrenergic receptors, and isoproterenol sensitivity in endurance trained and non-endurance trained volunteers. *Eur J Appl Physiol.* 52: 362-368, 1984.
79. Lehmann M, Keul J. Free catecholamines, heart rates, lactate levels, and oxygen uptake in competition weight lifters, cyclists, and untrained controlled subjects. *Int J Sports Med.* 7: 18-24, 1986.
80. Lehmann M, Fosters C, Keul J. Overtraining in endurance athletes. A brief review. *Med Sci Sports Exerc.* 25: 854-860, 1993.
81. Lehmann M, Petersen KG, Khalaf AN, Kerp L, Keul J. Influence of a 6 weeks endurance training on the hypophysis function in men. In: Tittel K, Arndt KH, Hollmann W, eds. *Sports Medicine—Yesterday, Today, Tomorrow* [in German]. Leipzig Berlin Heidelberg: Barth, 1993.
82. Liesen H, Hollmann W. The effects of a two week altitude training on the physical performance capacity at sea level, measured by spiroergometry and metabolic parameters [in German]. *Sportarzt Sportmed.* 23: 157-163, 1972.
83. Liesen H, Heikkinen E, Suominen H, Michel D. The effects of a 12 week endurance training on the performance capacity and the skeletal muscle metabolism in untrained man of the 6th and 7th decade of life [in German]. *Sportarzt Sportmed.* 2: 26-34, 1975.
84. Liesen H, Dufaux B, Hollmann W. Endurance training and proteinase inhibitors: Importance for the regulation of the plasma volume in rest and during physical exercise [in German]. *Dtsch Z Sportmed.* 2: 37-43, 1978.
85. Liesen H, Hollmann W. *Endurance Sports and Metabolism* [in German]. Schorndorf: Hofmann, 1981.
86. Liu Y, Steinacker JM, Dehnert C, Menhold E, Baur S, Lormes W, Lehmann M. Effect of "living high, training low" on the cardiac functions at sea level. *Int J Sports Med.* 19: 380-384, 1998.
87. Liu Y, Mayr S, Opitz-Gress A, Zeller C, Lormes W, Baur S, Lehmann M, Steinacker JM. Human skeletal muscle HSP (70) response to training in highly trained rowers. *Appi Physiol.* 86: 101-106, 1999.
88. Lorenz R, Hausdorf J, Reschke D, Tusker F. Effects of isometric maximal muscle contractions with different pauses and repetition number. In: Carl K, Quade K, Stehle T, eds. *Strength Training in the*

Sport Scientific Research. Cologne: Sport & Buch Strauß, 1995.

89. Mader A, Liesen H, Heck H, Philippi H, Rost R, Schuerch P, Hollmann W. The determination of the sports specific endurance performance in the laboratory [in German]. *Sportarzt Sportmed.* 4: 80-86; 5: 109-15, 1976.
90. Mader A, Madsen O, Hollmann W. The lactazide delivery of energy in training and competition [in German]. *Leistungssport.* 10: 263-275, 408-421, 1980.
91. Mader A, Heck M. A theory of the metabolic origin of "anaerobic threshold." *Int J Sports Med.* 7(Suppl. 1): 45-65, 1986.
92. Mader A. Glycolysis and oxidative phosphorylation as a function of cytosolic phosphorylation state and power output of the muscle cell. *Eur J Appl Physiol.* 88: 317-338, 2003.
93. Mallwitz A. *Congress of Sports Physicians in Oberhof 1912* [in German]. Muenchen: Lehmann, 1925.
94. Mallwitz A. *25 Years of Sports Medical Research* [in German]. Leipzig: Thieme, 1936.
95. Master AM, Oppenheimer ET. A simple exercise tolerance test for circulatory efficiency with standard tables for normal individuals. *Am J Med Sci.* 177: 229-243, 1929.
96. Mellerowicz H, Nowacki PE. Comparison of respiration and hemodynamic functions during identical physical work in a standing, seating and lying position [in German]. *Z Kreislaufforschg.* 50: 1002-1010, 1961.
97. Mellerowicz H, Moller W. *Biological and Medical Principles of Training* [in German]. Berlin: Springer, 1972.
98. Mellerowicz H. *Ergometry: Textbook of the Physical Performance Measurement* [in German]. Muenchen: Urban & Schwarzenberg, 1975.
99. Meyerhof O. *Metabolism in the Human Cell* [in German]. Goettingen: Vandenhoeck & Ruprecht, 1913.
100. Meyerhof O. *The Chemical Mechanisms in the Skeletal Muscle and the Context With Performance Capacity and the Emergence of Heat* [in German]. Berlin: Springer, 1930.
101. Mueller EA. The relationship between maximum of work and energy consumption during static and dynamic work [in German]. *Arbeitsphysiol.* 62-68, 1937.
102. Mueller EA. A performance pulse index as measurement of the physical performance capacity [in German]. *Arbeitsphysiol.* 14: 271-278, 1950.
103. Mueller EA. *The Physical Tiredness. Hdb Ges Arbeitsmed, Bd 1, Arbeitsphysiologie* [in German]. Berlin: Urban & Schwarzenberg, 1961.
104. Neumann G, Pfuetzner A, Hottenrott K. *Endurance Training* [in German]. Aachen: Meyer & Meyer, 1993.
105. Neumann G. Physiological basis of bicycling [in German]. *Dtsch Z Sportmed.* 51: 169-175, 2000.
106. Noecker J. *Textbook of the Biology of Physical Exercises* [in German]. Berlin: Sportverlag, 1953.
107. Nowacki PE. *75 Years of Sports Medicine at the University of Gießen* [in German]. Gießen: 1994.
108. Nukada A, Mueller EA. Skin temperature and physical performance capacity in extremities during dynamic exercise [in German]. *Int Z Physiol.* 16: 61-68, 1955.
109. Paffenbarger RS, Wing AL, Hyde RT. Physical activity as an index of heart attack risk in college alumni. *Am J Epidemiol.* 108: 161-175, 1978.
110. Pette D. The adaptive potential of the skeletal muscle [in German]. *Dtsch Z Sportmed.* 50: 262-271, 1999.
111. Pettenkofer M, Volt C. Investigations about the metabolism in normal men [in German]. *Z Biol.* 2: 459-573, 1866.
112. Prinz JP. In the beginning was the oxygen: 200th birthday of spirometry [in German]. *Dtsch Z Sportmed.* 41: 392-402, 1990.
113. Reindell H, Kirchhoff W. Combined examinations of the cardiopulmonary system during work. First information: Investigations in normal persons and top athletes [in German]. *Dtsch Med Wschr.* 81: 592-598, 1956.
114. Reindell H, Koenig K, Roskamm H. *Functional Diagnostic of the Healthy and Diseased Heart* [in German]. Stuttgart: Thieme, 1967.
115. Rojas Vega S, Strueder HK, Hollmann W. Plasma prolactin concentration increases after hypercapnia acidosis. *Horm Metab Res.* 35: 598-601, 2003.
116. Roskamm H, Samek L, Weidemann H, Reindell H. *Capability and Altitude* [in German]. Ludwigshafen: Knoll, 1968.
117. Rost R. *Echocardiography of the Athlete's Heart* [in German]. Schorndorf: Hofmann, 1982.
118. Rost R, Gerhardus H, Schmidt K. Effects of a performance training in swimming on the cardiopulmonary system in childhood and 20 years later [in German]. *Med Welt.* 36: 65-71, 1985.
119. Roux W. *Collected Texts about the Development of Organism. Vol. 1: Functional Adaptation* [in German]. Leipzig: Engelmann, 1895.
120. Rubner M. Nutrition in sports [in German]. In: Kohlrausch W, ed. *Congress of Sports Physicians 1925.* Jena: Fischer, 1925.
121. Schmidt D, Krause BJ, Herzog LL, Strueder HK, Klose C, Wouters E, Hollmann W, Mueller-Gaertner HW. Age-dependent changes in activation patterns during encoding and retrieval of visually presented

word-pair associates. *Neuroimage.* 9: 908-917, 1999.

122. Schmidt FA. *Physical Education According to the Exercise Value* [in German]. Leipzig: Voigtlaender, 1893.
123. Schmidt FA. *Physiology of Physical Education* [in German]. Leipzig: Voigtlaender, 1914.
124. Schoen FA, Hollmann W, Leresen H, Waterloh E. *Electron Microscopically Findings in M. Vastus Lateralis of Untrained and Endurance Trained Persons as Well as the Relationship to the Maximal Oxygen Uptake and Lactate Production* [in German]. Bad Nauheim: Deutscher Sportaerztekongress, 1978.
125. Speck M. The influence of muscle work on the respiration [in German]. *Dtsch Arch Klin Med.* 45: 461-466, 1889.
126. Steinacker JM, Bard SA, eds. *The Physiology and Pathophysiology of Exercise Tolerance.* New York: Plenum Press, 1996.
127. Steinacker TM, Wang L, Lorme W, Reissnecker S, Liu Y. Structural adaptations to training of the skeletal muscle [in German]. *Dtsch Z Sportmed.* 53(12): 354-362, 2002.
128. Strauzenberg SE, Clausnitzer C. The importance of the steroidynamic in order to judge fatigue after physical work [in German]. *Theorie Praxis Koerperkult.* 21: 133-134, 1972.
129. Strauzenberg SE, Guertler H, Hannemann D, Tittel K, eds. *Sports Medicine: Basis of Sports Medical Care.* Leipzig: Barth, 1990.
130. Strueder HK, Hollmann W, Donike M, Platen P, Weber K. Effect of O_2 availability on neuroendocrine variables at rest and during exercise: O_2 breathing increases plasma prolactin. *Eur J Appl Physiol.* 74: 443-449, 1996.
131. Strueder HK, Hollmann W, Platen P, Rost R, Weicker H, Weber K. Hypothalamic-pituitary-adrenal and-gonadal axis function after exercise in sedentary and endurance trained elderly males. *Eur J Appl Physiol.* 77: 285-293, 1998.
132. Strueder HK, Hollmann W, Platen P, Rost R, Weicker H, Kirchoff, Weber K. Neuroendocrine system and mental function in sedentary and endurance trained elderly males. *Int J Sports Med.* 20: 159-166, 1999.
133. Suominen H, Heikkinen H, Heikkinen E, Liesen E, Michel D, Hollmann W. Effects of 8 weeks' endurance training on skeletal muscle metabolism in 56-70 year old sedentary men. *Eur J Appl Physiol.* 37: 173-180, 1977.
134. Tittel K, Wutscherk H. *Anthrometry of Sports.* Vol 6 [in German]. Leipzig: Barth, 1972.
135. Tittel K, Wesseling J. *75 Years FIMS: A Historical Review in Words and Pictures.* Munich-Leipzig: Mumm Soeffge, 2005.
136. Valentin H, Venrath H, Mallinckrodt V, Guerakar M. The maximal oxygen uptake in different classes of age [in German]. *Z Altersforschg.* 9: 291-299, 1955.
137. Voelker K. *The Blood Pressure During Static Step Test and During 24-h Profile. The Relationship to the Blood Pressure Reaction in Bicycle Ergometer Tests as Well as the ANP-Behaviour During Different Exercise Conditions* [in German]. Cologne: German Sport University, 1990.
138. Voelker K, Rosskopf P, Feuerbach U, Krestin M, Prinz U, Hollmann W. ANP levels during different kinds of physical activity in men and women [in German]. *Dtsch Z Sportmed.* 43: 450-458, 1992.
139. Vuori I. Physical activity and health: Finish experience in the 1990s. In: The Club of Cologne, eds. *Joint Meeting Cologne of WHO and FIMS 1994: Health Promotion and Physical Activity.* 189-200, 1994.
140. Wasserman K, McIlroy MB. Detecting the threshold of anaerobic metabolism in cardiac patients during exercise. *Am J Cardiol.* 14: 844-852, 1964.
141. Weissbein S. *Hygiene of Sports* [in German]. Leipzig: Grenthlein, 1910.
142. Zuntz N, Schumburg W. *Studies to a Physiology of March* [in German]. Berlin: Hirschwald, 1901.

CHAPTER 5

PhD Specialization and Incorporating Molecular Biology Into Exercise Biology and Physiology Research

P. Darrell Neufer, PhD

Charles M. Tipton, PhD

Introduction

Near the end of the 19th century, exercise physiology evolved from the classrooms and laboratories of physical education. Teaching responsibilities were initially provided by individuals with medical degrees (39). During ensuing decades, departments of physical education became established components of higher education and provided leadership for exercise physiology instruction and research. This was necessitated in part because departments of physiology exhibited minimal interest in including exercise physiology as a subdiscipline of physiology (similar to that afforded to cellular, comparative, gastrointestinal, or neural physiology) or in offering courses on the subjects to undergraduate or graduate students (38). By 1960, exercise physiology was entrenched as a responsibility of departments of physical education. Not surprisingly, many courses in biomechanics, motor learning, exercise psychology, and therapeutic exercise were being offered on a regular basis.

With the exception of some courses in anatomy and adaptive physical education, very few departments offered basic science courses (38). However, in 1963, former president of Harvard University J.B. Connat published the results of a 2 yr study of graduate programs that prepared teachers in the United States. The report was a scathing indictment of higher education teaching and training programs; the graduate programs in physical education were designated as an example of the system at its worst (18). Included in the report was the recommendation that programs in physical education incorporate more rigorous and demanding classes in the mathematical, chemical, biological, and physical sciences for graduate students (18).

Although startling to numerous educators, the report was not surprising to many departmental chairs in the Big Ten Conference, who had collectively decided in the late 1950s to drastically revamp their graduate programs in physical education (38). In fact, Louis E. Alley, chair at the University of Iowa in the United States, had previously lamented:

> Much of the research in physical education is concerned with trivial problems and is of poor quality. In many instances, it appears that the researcher does not know enough about his problem to realize he knows little about it. (19)

PhD Specialization: Effect and Implementation

During the early 1960s and continuing for more than two decades, graduate programs in physical education at institutions in the Big Ten and Pacific Coast Conferences and at Pennsylvania State University in the United States had a marked effect on the emergence and acceptance of exercise physiology as an academic discipline (38). Graduate programs with an emphasis on exercise physiology, such as that at the University of

Iowa, changed their entrance requirements so that students were required to take courses in biology or zoology, chemistry, mathematics, and physics before being permitted to take graduate courses in biochemistry, human or medical physiology, pharmacology, zoology, and exercise physiology. Thus, during the early years of specialization, graduate students with an interest in exercise physiology were enrolled in numerous undergraduate courses.

The implementation of specialization programs in exercise physiology was not an instantaneous or uniform process. However, programs initiated by the two conferences and Pennsylvania State University beginning in 1963 served as effective models for others to follow. Moreover, one can state with some degree of certainty that the majority of active and productive investigators in exercise physiology in the United States during the last decade of the 20th century and the first decade of the 21st century were either first- or second-generation descendants of the specialization era.

Limitations of the Specialization Era

Associated with the specialization era was the transformation of many departments of physical education into departments of exercise science, kinesiology, movement science, or combinations thereof. Although the transformation enhanced the scientific image, acceptance, and identity of the areas of specialization (e.g., exercise physiology, biomechanics, motor learning, exercise psychology, anatomy, and combinations thereof), it did not change the nature of the specialization.

Although specialization in exercise physiology required students to be cognizant of the principles of biochemistry and versed in biochemical methodology, it did not prepare them to be at the forefront of biochemical advances or technology. This situation, coupled with the fact that the specialization movement did not include exercise biochemistry as a designated area of emphasis, meant that very few investigators in the exercise science research community were sufficiently competent in the principles and methodology of biochemistry to meet the challenges created by the advent of molecular biology.

According to Blair (5), a prominent and productive exercise scientist, the 1970s were classified as the decade of biochemistry whereas the 1980s was regarded as the beginning of exercise molecular biology. We concur with the sequence but not with the details and contend that the biochemistry era extended from 1971 to 1987 (table 5.1) and overlapped with the exercise molecular biology era that began with the study by Williams and colleagues in 1986 (42).

Frank W. Booth and the Void in Molecular Biology in the Exercise Sciences

Introduction and the Holloszy Experience

In 2010 at the annual meeting of the American College of Sports Medicine (ACSM), Frank W. Booth, professor of biomedical sciences at the University of Missouri at Columbia in the United States (figure 5.1), was presented the prestigious ACSM Honor Award for his extraordinary achievements in sports medicine and exercise science. The award included the statement, "Professor Booth is commended for his 'single-handed advocacy efforts' to bring exercise science research into the realm of molecular biology" (8).

In 1970, Booth was a first-generation graduate of the specialization program at the University of Iowa with an emphasis in exercise physiology (38). After a 2 yr postdoctoral experience at the school of aerospace medicine, he secured a second postdoctoral appointment in the department of preventive medicine at Washington University in St. Louis, Missouri. John O. Holloszy (figure 5.2) was his supervisor from 1973 to 1975. Holloszy, after graduating with a medical degree from Washington University, served for 2 yr as a postdoctoral fellow in the biochemistry laboratory of Carl Cori (21), who was a Nobel laureate in 1947. Holloszy's biochemical and physiological research endeavors were recognized with the 1987 Honor Award by ACSM, the 1994 Honor Award by the Biochemistry of Exercise Research Group, and the 2000 Olympic Prize in Sport Science (21).

In Holloszy's laboratory, Booth enhanced his biochemistry acumen and became actively involved in studies concerned with rates of protein synthesis and degradation for mitochondrial enzymes and cytochrome c protein in response to dynamic or resistance animal-exercise programs. Two phenotypes produced indicated that the regulation of protein expression differed between the two exercise programs (20). Booth and Holloszy also discovered that the half-time for achieving a steady-state increase in cytochrome c concentration with endurance training was similar to the half-time for decreasing cytochrome c once training stopped (6-8 d), indicating that dynamic exercise training produced an

Table 5.1 Select Biochemical Procedures, Methodologies, and Technologies That Contributed to the Evolution of Exercise Molecular Biology

Date	Contribution	Reference
1971	Kenner and Aboderin describe a new fluorescent probe for protein and nucleoprotein conformation.	23
1972	Paul Berg performs the research that lays the foundation for recombinant DNA technology.*	31
1973	Stanley Cohen and Herbert Boyer use recombinant DNA technology to create a genetically modified organism. In 1974 a voluntary moratorium is initiated; this moratorium is lifted after the National Institutes of Health established strict guidelines. Also in 1974, Rudolf Jaenisch creates the first transgenic mouse.	40
1974	David Baltimore, Sol Spiegelman, and Philip Leder independently develop molecular probes for specific messenger RNAs.	25
1975	Edwin Southern develops the blot method for the detection of specific DNA sequences in DNA samples. The method becomes known as the Southern blot.	34
1975	Walter Gilbert and Frederick Sanger develop and perfect methods for determining and sequencing DNA.*	28, 38
1977	James Alwine, David Kemp, and George Stark develop the blotting method for the detection of specific RNAs from total RNA samples. The method becomes known as the Northern blot.	1
1983	James Mullis develops the polymerase chain reaction technique for amplifying single fragments of DNA.*	22, 32
1987	Kulesh and colleagues describe DNA microarray technology that can be used for gene expression levels, for detecting single nucleotide polymorphisms, or for resequencing mutant genomes.	22, 26
1995	Schena and colleagues miniaturized microarrays for gene expression profiling are perfected.	33, 34

*Denotes that a result was acknowledged with a Nobel Prize for either chemistry or physiology and medicine.
DNA = deoxyribonucleic acid; RNA = ribonucleic acid.

increase in the rate of synthesis rather than a reduction in the rate of degradation (11). However, these findings contrasted with the findings of a much longer and different cytochrome c half-life (32 d sedentary, 42 d trained) previously reported by the Holloszy laboratory (35). Booth was designated to resolve the controversy and focused his attention on the methods and assumptions used to assess cytochrome c based on the turnover of radiolabeled heme. When turnover rates of cytochrome c were measured by labeling with ^{14}C-leucine or with delta-^{14}C-aminolevulinate (a heme precursor), the turnover rates were much slower than the estimates obtained by directly measuring changes in cytochrome c protein expression. This indicated that significant recycling or reutilization of the labels accounted for the artificially longer turnover rates. Booth and Holloszy, therefore, became the first investigators to demonstrate that endurance-exercise training markedly increased mitochondrial biogenesis (11).

Booth and His Contributions at Houston

In 1975, Booth accepted an appointment in the department of physiology at the University of Texas Medical School at Houston in the United States, where he remained until 1999. In the early 1980s, Booth and his graduate student Peter Watson investigated the changes in actin synthesis and α-actin messenger ribonucleic acid (mRNA) content in animal models of decreased (immobilization) contractile activity (41). In order to measure α-actin mRNA in the skeletal muscles

Figure 5.1 Frank W. Booth, PhD (1943–), is currently professor in the departments of biomedical sciences, medical pharmacology and physiology, and nutrition and exercise physiology at the University of Missouri. By advocacy and example, he facilitated the incorporation of molecular biology principles and methodology into the research projects of the exercise science community.
Courtesy of Frank W. Booth.

of rats, they needed to learn molecular biology methodology including how to isolate total RNA from tissues, Northern blotting, preparing and radiolabeling complementary deoxyribonucleic acid (DNA) probes, and nuclear hybridization (table 5.1).

Figure 5.2 John O. Holloszy, MD (1933–), is currently professor in the department of internal medicine at Washington University in St. Louis, Missouri. After postdoctoral training in biochemistry, he became renowned for his biochemical investigations of adaptations to chronic exercise. He served as supervisor of Frank W. Booth's postdoctoral training and as a collaborator on studies during the biochemistry era (see table 5.2).
Courtesy of John O. Holloszy.

They sought and received advice and assistance from colleagues with expertise in molecular biology, including Joseph Stein, Uri Nudel, and Susan Berget of nearby Rice University (8). By comparing the change in skeletal α-actin mRNA (which was significantly decreased after 168 h but not after 6 or 72 h of immobilization) with the change in protein synthesis (which was significantly decreased after 6 and 72 h), they concluded that a change in the content of skeletal α-actin mRNA did not contribute significantly to the rapid onset of the reduction in actin synthesis rate during the initial hours of immobilization (40). Booth was of the opinion that the translation of skeletal α-actin must have been altered to explain the early decline in actin protein synthesis with immobilization, and that a change at the pretranslation level must have occurred to account for the later decrease in skeletal α-actin-specific mRNA. This study (41) was the first to report the combined determinations of both protein synthesis rates and mRNA for the same gene (skeletal muscle α-actin) in a model for reduced contractile activity (8).

In 1986, Sandy Williams (figure 5.3) and colleagues published the first molecular biology study on the effects of increased muscle contractile activity (electrical stimulation) on the adaptive properties of mRNA (42). This paper by Williams, a physician with expertise in biochemistry and molecular biology, had a profound influence on Booth and motivated him to incorporate molecular biology approaches into his own laboratory research. He set his sights on identifying the mechanism responsible for the increase in actin expression that occurs during recovery from immobilization. In 1987, Booth along with Morrison and Muller compared changes in α-actin mRNA and α-actin protein synthesis in the hind-limb muscles of rats recovering from 7 d of immobilization (30). The α-actin mRNA concentration decreased to 53% of the control values by the last day of immobilization and then increased during the first and second recovery days in parallel with the increase in the actin protein synthesis rate, thus implying regulation initiated at the level of transcription. Further increases in actin synthesis, however, appeared to shift regulation to the level of translational because actin protein synthesis increased more than α-actin mRNA from the second to fourth recovery days (30). Two years later, Morrison and colleagues compared the effects of reduced daily running (7%-14% of the duration associated with 24 h of chronic electrical stimulation of skeletal muscles) on mitochondrial protein mRNA levels (29). Female rats ran 100 min/d for 2 wk and exhibited a training effect with elevations in citrate synthase activity (31%-40%) and cytochrome c mRNA

Table 5.2 Select Contributions of Frank W. Booth to Exercise Molecular Biology

Contribution	Reference
Booth authored or coauthored publications describing molecular biology methodology. He emphasized its importance in exercise science research, encouraged investigators to learn and use techniques, and advocated acceptance from the exercise science community.	3, 4, 7, 9, 13
Booth and Holloszy were the first to show an estimated increase in protein synthesis rate (subsequently labeled mitochondrial biogenesis) for cytochrome c (a mitochondrial protein) in a skeletal muscle with dynamic exercise training.	11, 12
Booth and colleagues were the first to determine the responses of α-actin mRNA, actin protein, and cytochrome c mRNA and cytochrome c protein synthesis in skeletal muscles of an animal model and to deduce changes in pretranslational and posttranslational mechanisms.	30
Thomason and Booth were the first to induce ectopic expression of an exogenous gene in skeletal muscle using a viral vector. This event concomitantly occurred with the 1990 publication of "Direct Gene Transfer Into Mouse Muscle In Vivo" in *Science* (volume 247, pp. 1465-1468) by the laboratory group of J.A. Wolff.	36
Booth and his laboratory were the first in an exercise model to perform promoter deletion analysis in skeletal muscle of a living animal and to successfully identify an exercise-responsive element.	16
Carson, Schwartz, and Booth were the first in an exercise model to identify two transcription binding factors on a gene promoter and their combinatorial regulation of mRNA transcription in skeletal muscle of a living animal.	15
Booth and his laboratory were the first to publish results on the usage of microarrays to determine global gene expression in different types of skeletal muscle fibers.	14

mRNA = messenger ribonucleic acid.

levels (17%-56%). α-Actin mRNA increased approximately 60% in the fast-twitch muscles of the hind limbs but not in the slow-twitch muscles, which unexpectedly demonstrated that the synthesis rate of a specific protein could increase without a concurrent elevation in protein concentration (29) and that α-actin mRNA can increase when no significant increase in muscle mass occurs (27).

The 1986 study by Williams and colleagues identified a void in the exercise science community: the lack of competent and qualified investigators to conduct meaningful studies in molecular biology. Thus, in 1988 and 1989, Booth alone and with Babij submitted manuscripts to professional journals representing the exercise and sport sciences, sports medicine, and physiology with the intent to educate, encourage, and challenge readers to master and use molecular biology methodologies to investigate the mechanistic challenges of the future (2, 3, 7, 9). The manuscripts undoubtedly influenced some exercise science departments to modify their laboratory experiences so that future students could become proficient in the theory and practice of molecular biology technology.

In 1998, Booth along with colleagues from two departments at the University of Texas Medical School at Houston published an article that encouraged readers to use transgenic animals in exercise research and indicated the various uses of molecular biology methodologies in clinical situations in exercise physiology (13). Although it is virtually impossible to quantify Booth's effect on the exercise science community, in 2002 investigators from two laboratories that were not associated with the specialization movement published important molecular biology manuscripts: Allen, Harrison, and Leinwand from the University of Colorado at Boulder in the United States proposed three molecular approaches for evaluating genes involved in an exercise response that had the potential for providing more effective exercise regimens for humans, and Esser and colleagues from the University of Illinois at Chicago, using microarrays (table 5.1) performed on total RNA or polysomal RNA, reported that translational mechanisms contributed to acute gene expression after high-resistance contractions and that both the response to resistance exercise and the growth response during progression of the cell cycle share a common profile (17).

Select Studies By Booth and His Laboratory During the Early Years (1988-2001) of Exercise Molecular Biology

In 1988, Wong and Booth reported that chronic concentric resistance training (electrical stimulation for 16 wk) with progressively increased weights produced muscle enlargement whereas no changes occurred when weights were not included (43). To determine the role of muscle protein synthesis rates in the process, in 1990 they undertook a study of high- and low-resistance chronic weight training using electrical stimulation of rat gastrocnemius muscles (44). They found that a duration as short as 8 min increased protein synthesis by approximately 50% whereas high-frequency contractions did not increase gastrocnemius mass or α-actin mRNA levels. These findings indicated that exercise protein synthesis rates are regulated by the number of repetitions (192 vs. 24) and not by the degree of resistance imposed on the muscle. The lack of change in mRNA suggested that translational and posttranslational mechanisms (including protein degradation) are likely responsible for this model of stimulated concentric exercise (44).

In the study that employed chronic concentric exercise, rat tibialis anterior (TA) muscles exhibited increases in mass whereas the gastrocnemius (GAST) muscles did not. This suggested that the regulation of expression was different between the two muscles (44). Because the resistance animal model used by Booth's laboratory resulted in the simultaneous concentric contraction of the rat GAST muscles and eccentric contraction of the TA in same leg, they focused on the eccentric contractions of the TA and made comparisons with changes in GAST muscles. From their extensive results, they concluded that the acute increase in the protein synthesis rate of both the eccentric resistance-exercised TA and the concentric resistance-exercised GAST occurred primarily through translational and posttranslational mechanisms (44). In addition, the increased muscle mass of the TA muscles with 10 wk chronic training (when compared with the GAST muscles) was due in part to prolonged acute increases in protein synthesis rates. It is interesting to note that a lower number of contractile repetitions by TA muscles were required to stimulate protein synthesis pathways. The researchers felt that their results supported the concept that muscle protein adaptations to resistance training are related to the number of repetitions, the relative resistance experienced by the muscle, or the type of contractile work (concentric or eccentric) being performed (44).

Figure 5.3 R. Sanders (Sandy) Williams, MD (1948–), is currently president of J. David Gladstone Institute in San Francisco. His expertise in molecular biology, especially molecular cardiology, contributed to his membership in the National Academy of Sciences. He is considered by the authors to be responsible for conducting the first molecular biology study relevant to the exercise science community.
Courtesy of R. Sanders Williams.

In 1990, Booth and his laboratory became interested in the process of transferring a foreign gene into skeletal muscles in order to determine how exercise, or lack thereof, would alter the transcriptional control of genes normally targeted by changes in contractile activity in skeletal muscle (8). Undeterred by his lack of knowledge and experience in this new area, he followed the same approach he used in mastering the hurdles associated with learning molecular biology: seeking knowledge of and familiarity with the published literature on the topic, seeking advice from individuals and colleagues with experience in gene transfer, selecting techniques that allow visualization of foreign proteins, obtaining a replication-deficient retroviral vector and mastering the technical information associated with its use, and conducting preliminary experiments with talented postdoctoral trainees such as Donald Thomason. When Thomason injected the replication-incompetent retrovirus containing the foreign bacterial β-galactosidase into skeletal muscles of mature rats, the gene was expressed and a technique was established with muscle-specific promoter-gene constructs to investigate the physiological regulation of skeletal muscle gene expression in intact adult mammals (36).

In a 1990 article, Booth and Thomason (36) mentioned that little was known about the regulatory sequences of genes that are activated to cause an accumulation of contractile tissue in muscles of animals. Although they initiated research on the actin promoter in 1992, the findings were not published until 1995, reflecting the effort and time required in forging a new area of research (16). To conduct the study, they selected 3-wk-old chickens and selected the anterior latissimus dorsi (ALD) muscle to determine the effect of stretch overload. ALD muscles were injected with plasmids containing various lengths of the chicken skeletal α-actin promoters (from −2,090 to −77 base pairs upstream of the transcription start site), which were generously provided by Robert J. Schwartz, professor and colleague in the department of molecular and cell biology at the University of Texas Medical School at Houston (8, 16). The chicken skeletal α-actin promoters were cloned into a plasmid vector upstream of the gene encoding for firefly luciferase reporter gene, providing a unique and easily detected readout of transcriptional activity originating from the α-actin promoter. Weights added to the left wings of the chicks (the right wings served as the controls) increased the mass of the ALD muscles by 110% at the end of the 6 d experiment. Luciferase activity from the four actin promoter constructs (−2,090, −424, −202, and −99) were 127%, 179%, 134%, and 378% higher, respectively, in the stretched ALD muscles than in their contralateral controls. Interestingly, luciferase reporter gene activity generated from the −77 deletion construct was not different between the stretched and control muscles. Booth, Carson, and colleagues cautiously interpreted these results to mean that the location of the chicken skeletal α-actin promoter region that responded to mechanical stretching was located in the gene sequence region between −99 and −77 (16). It had been reported that the serum response element 1 (SRE-1) in the chicken skeletal α-actin promoter is located between −91 and −81 base pairs (6).

In 1996, Carson, Schwartz, and Booth published their findings on the α-actin regulatory elements and transcription factors that were responsible for conferring stretch-overload responsiveness during enlargement of chicken ALD muscles (16). Again, molecular biology colleagues contributed to Booth's research ventures: Robert Schwartz provided the minimal promoter constructs that contained sequence mutations in putatative DNA binding sites for specific transcription factors. They were injected into ALD muscles, which subsequently demonstrated that SRE-1 and transcriptional enhancer factor 1 (TEF-1) were sufficient to confer stretch-overload responsiveness. Moreover, when either the SRE-1 or TEF-1 response elements were mutated, both basal and stretch-induced activation of the α-actin promoter was lost. In addition, the serum response factor and the SRE-1 binding complex exhibited faster migration in mobility shift assays from ALD nuclear extracts stretched on d 3 and 6 when compared with control results, providing evidence of posttranslational modifications to serum response factor or TEF-1 (15). The authors concluded that both SRE-1 and TEF-1 binding sites are sufficient for stretch-induced overload but, in the context of the α-actin promoter, a combinatorial regulation including both elements is necessary.

With Zhen Yan, Booth returned in 1996 to investigating the mechanism by which cytochrome c mRNA was enhanced by increased contractile activity in skeletal muscles. They electrically stimulated the TA muscles for various durations and found that cytochrome c protein and mRNA concentrations began to increase after 9 d. This was associated with a concomitant decline in RNA–protein interaction in the 3′ untranslated region of the cytochrome c mRNA, a region associated with regulating the stability of mRNAs. They subsequently found that a component in the 150,000 g pellet fraction of the homogenate from stimulated muscle was capable of inhibiting the RNA–protein interactions in control TA muscles. They concluded that an inhibitory factor was responsible, in continuously stimulated muscles, for decreasing RNA–protein interactions in the 3′ untranslated region of cytochrome c mRNA (46).

In 1998, Yan and Booth investigated the mechanisms associated with the vast differences in cytochrome c promoter and mRNA levels between slow-twitch oxidative muscles (soleus) and fast-twitch glycolytic muscles (vastus lateralis) (45). Richard Scarpulla of the department of cellular and molecular biology at Northwestern University in the United States generously provided the plasmids that contained varying lengths of the rat somatic cytochrome c gene that were used in the DNA injection experiments. Surprisingly, Yan and Booth found no differences between slow- and fast-twitch muscles in the activities of the −726, −631, −489, −326, −215, and −149 cytochrome c promoter constructs, suggesting that additional elements outside of the −726 to +610 region of the cytochrome c gene may be required to increase cytochrome c mRNA concentration. They also suggested that changes in posttranslational control could be a contributing factor (8, 45).

Molecular biology investigations changed in 1995 with the introduction of miniaturized complementary DNA microarrays that enabled investigators to purchase DNA or "gene chips" to analyze hundreds and eventually thousands of mRNAs simultaneously (34). Where previously researchers were limited to 1 to 2 mRNA analyses per experiment, by 2001 it was possible to interrogate thousands of mRNAs in a single sample (8). During that same year, Booth, Campbell, and colleagues used miniaturized microarray technology to investigate mRNA differences between white quadriceps femoris muscle tissue and red soleus muscle tissue from mice (14). They identified 49 mRNA sequences that were effectively differentiated between the two muscle types. Specifically, the red soleus muscle had higher mRNA levels of myoglobin, CD36 fatty acid transport protein, malate dehydrogenase, muscle LIM protein, peroxisome proliferator-activated receptor-α, osteoblast factor 2, and cardiac isoforms of myosin heavy chains, myosin light chains, and lactate dehydrogenase than did white quadriceps femoris muscle tissue. On the other hand, the quadriceps tissue was associated with higher mRNA values for calmodulin, calcineurin catalytic subunit, parvalbumin, myosin heavy chain IIb, select glycolytic pathways steps, and CCAAT-enhancer binding protein-δ (14). In the study by Campbell and colleagues, Booth was able to use the microarray facilities of Eric P. Hoffman at the Research Center for Genetic Medicine Research at Children's National Medical Center in Washington, D.C. as a result of his interactions in the molecular biology community (8).

Since the publication of the paper by Campbell and colleagues (14), approximately 1,250 articles on mRNA and exercise have been listed by PubMed, indicating the tremendous interest in—and explosive nature of—the topic in the exercise and molecular biology scientific community. The significance of these studies and their scientific directions are currently unclear and await review by future scholars.

Closure and a Perspective

Unlike the specialization era, which was a collective outcome of educators with resources from several universities (38), the incorporation of molecular biology principles and methodologies into exercise science research—notably physiology—was predominately a singular effort by Frank W. Booth, whose vision, example, dedication, persistent educational efforts, scientific presentations, research publications, and external funding facilitated the acceptance and adoption of molecular biology into the laboratories and classrooms of exercise science departments.

Although Booth's vision has been achieved, his perspective has yet to be realized, namely the necessity of gene expression profiling studies that include large sample sizes and polymorphism analysis to allow for comparison between, for example, high and low responders to dynamic training (37). Booth believed that such investigations will provide insight into the mechanisms by which low responders to exercise can become high responders, whether low responders for one health outcome may be high responders for a different health outcome, and whether exercise prescriptions can be individualized based on genetic profiling, nutritional habits, and pharmacological interactions (8).

References

1. Alawine JC, Kemp DJ, Stark GR. Method of detection of specific RNAs in argose gels by transfer to diazbenzloxymethyl-paper and hybridization with DNA probes. *Proc Natl Acad Sci.* 74: 5350-5354, 1977.
2. Allen DL, Harrison BC, Leinwand LA. Molecular and genetic approaches to studying exercise performance and adaptation. *Exerc Sport Sci Rev.* 30: 99-105, 2002.
3. Babij P, Booth FW. Biochemistry of exercise: Advances in molecular biology relevant to the adaptation to muscle to exercise. *Sports Med.* 5:137-143, 1988.
4. Babij P, Booth FW. Sculpturing new muscle phenotypes. *News Physiol Sci.* 3: 100-102, 1988.
5. Blair SN. Sports medicine and exercise science in the 21st century. *Sports Med Bull.* 34: 8, 1999.
6. Bonini JA, Hofman C. A rapid, accurate, nonradioactive method for quantifying RNA on agarose gels. *Biotechniques.* 11: 708-709, 1991.
7. Booth FW. Application of molecular biology in exercise physiology. *Exerc Sport Sci Rev.* 17: 1-27, 1989.
8. Booth FW. Exercise biology and molecular biology. Unpublished personal communications forwarded to Charles M. Tipton, July 24, 2010.
9. Booth FW. Perspectives on molecular and cellular exercise physiology. *J Appl Physiol.* 65: 146-171, 1988.
10. Booth FW, Baldwin KM. Muscle plasticity:energy demand and supply processes. In: *The Handbook of Physiology. Exercise: Regulation and Integration of Multiple Systems.* Ed. Rowell LB and Shepherd JT., American Physiological Society, Bethesda, MD 1075-1123, 1996.
11. Booth FW, Holloszy JO. Cytochrome c turnover in rat skeletal muscles. *J Biol Chem.* 252: 416-419, 1977.
12. Booth FW, Holloszy JO. Effect of thyroid hormone administration on synthesis and degradation of cytochrome c in rat liver. *Arch Biochem Biophys.* 167:674-677, 1975.
13. Booth FW, Tseng BS, Fluck M, Carson JA. Molecular and cellular adaptation of muscle in response to physical training. *Acta Physiol Scand.* 162: 343-350, 1998.
14. Campbell WG, Gordon SE, Carlson CJ, Pattison JS, Hamilton MT, Booth FW. Differential global gene expression in red and white skeletal muscle. *Am J Physiol.* 280: C763-C768, 2001.

15. Carson JA, Schwartz RJ, Booth FW. SRF and TEF-1 control of chicken skeletal alpha-actin gene during slow-muscle hypertrophy. *Am J Physiol.* 270: C1624-C1633, 1996.
16. Carson JA, Yan Z, Booth FW, Coleman ME, Schwartz RJ, Stump CS. Regulation of skeletal alpha-actin promoter in young chickens during hypertrophy caused by stretch overload. *Am J Physiol.* 268: C918-C924, 1995.
17. Chen YW, Nader GA, Baar KR, Fedele MJ, Hoffman EP, Esser KA. Response of rat muscle to acute resistance exercise defined by transcriptional and translational profiling. *J Physiol.* 545: 27-41, 2002.
18. Connat JB. *The Education of American Teachers.* New York: McGraw-Hill, 1963.
19. Forker BE. The Alley contribution and legacy. In: Tipton CM, Hay HG, eds. *Specialization in Physical Education: The Alley Legacy.* Iowa City: Department of Physical Education—Men; 1984, 7.
20. Holloszy JO, Booth FW. Biochemical adaptations to endurance exercise in muscles. *Ann Rev Physiol.* 38: 273-291, 1976.
21. John O. Holloszy. Available: http://drtc.im.wustl.edu/members/files/holloszy_john_biosketch.pdf.
22. Kary Mullis Nobel Lecture, December 8, 1994. Available: http://nobelprize.org/nobel_prizes/chemistry/laureates/1993/mullis-lecture.html.
23. Kenner RA, Aboderin AA. New fluorescent probe and nucleoprotein conformation. Binding of 7-(p-methoxybenzylamino)-4-nitrobenzoxadiazole to bovine trypsinogen and bacterial ribosomes. *Biochemistry.* 24: 4433-4440, 1971.
24. Knudsen S. *Guide to Analysis of DNA Microarray Data.* 2nd ed. Hoboken, NJ: Wiley-Liss, 2002
25. Kolata GA. Molecular probes: A new way to study gene expression. *Science.* 183: 63-64, 1974.
26. Kulesh DA, Clive DR, Zaarlenga DA, Greene JJ. Identification of interferon-modulated proliferation-related cDNA sequences. *Proc Natl Acad Sci.* 84: 8453-8457, 1987.
27. Laurent GJ, Sparrow MP, Millward DJ. Turnover of muscle protein in the fowl. Changes in rates of protein synthesis and breakdown during hypertrophy of the anterior and posterior latissimus dorsi muscles. *Biochem J.* 176: 407-417, 1978.
28. Maxam AM, Gilbert W. A new method for sequencing DNA. *Proc Natl Acad Sci.* 74: 560-564, 1977.
29. Morrison PR, Biggs RB, Booth FW. Daily running for 2 wk and mRNAs for cytochrome c and alpha-actin in rat skeletal muscle. *Am J Physiol.* 257: C936-C939, 1989.
30. Morrison PR, Muller GW, Booth FW. Actin synthesis rate and mRNA level increase during early recovery of atrophied muscle. *Am J Physiol.* 253: C205-C209, 1987.
31. Nobel Laureate Paul Berg. Available: http://news.Stanford.edu/news/october3/berg-103.html.
32. Polymerase chain reaction. Wikipedia, the free encyclopedia (http://en.wikipedia.org/wiki/Polymerase_chain_reaction).
33. Sanger F, Nicklen S, Coulson AR. DNA sequencing with chain-terminating inhibitors. *Proc Natl Acad Sci.* 74: 5463-5467, 1977.
34. Schena M, Shalon D, Davis RW, Brown PO. Quantitative monitoring of gene expression patters with complimentary DNA microarray. *Science.* 270: 467-470, 1995.
35. Terjung RL, Winder WW, Baldwin KM, Holloszy JO. Effect of exercise on the turnover of cytochrome c in skeletal muscle. *J Biol Chem.* 248: 7404-7406, 1973.
36. Thomason DB, Booth FW. Stable incorporation of a bacterial gene into adult rat skeletal muscle in vivo. *Am J Physiol.* 258: C578-C581, 1990.
37. Timmons JA, Knudsen S, Rankinen T, Koch LG, Sarzynski M, Jensen T, Keller P, et al. Using molecular classification to predict gains in maximal aerobic capacity following endurance exercise training in humans. *J Appl Physiol.* 108: 1487-1496, 2010.
38. Tipton CM. Exercise physiology, part II: Contemporary historical perspective. In: Massengale JD, Swanson RA, eds. *The History of Exercise and Sport Science.* Champaign, IL: Human Kinetics; 1997, 396-438.
39. Tipton CM. Antiquity to the early years of the 20th century. In: Tipton CM, ed. *History of Exercise Physiology.* Champaign, IL: Human Kinetics; 2014.
40. Transgenic history. Available: www.transgenic-mouse.com/transgenesis-history.php.
41. Watson PA, Stein JP, Booth FW. Changes in actin synthesis and alpha-actin mRNA content in rat muscle during immobilization. *Am J Physiol.* 247: C39-C44, 1994.
42. Williams RS, Salmons S, Newsholme EA, Kaufman RE, Mellor J. Regulation of nuclear and mitochondrial gene expression by contractile activity in skeletal muscle. *J Biol Chem.* 2661: 376-380, 1986.
43. Wong TS, Booth FW. Protein metabolism in rat gastrocnemius muscle after stimulated chronic concentric exercise. *J Appl Physiol.* 69: 1709-1717, 1990.
44. Wong TS, Booth FW. Protein metabolism in rat tibialis anterior muscle after stimulated chronic eccentric exercise. *J Appl Physiol.* 69: 1718-1724, 1990.
45. Yan Z, Booth FW. Cytochrome c promoter activity in soleus and white vastus lateralis muscles in rats. *J Appl Physiol.* 85: 973-978, 1998.
46. Yan Z, Salmons S, Dang YI, Hamilton MT, Booth FW. Increased contractile activity decreases RNA-protein interaction in the 3'-UTR of cytochrome c mRNA. *Am J Physiol.* 271: C1157-C1166, 1996.

CHAPTER 6

Contributions From Copenhagen Muscle Research Center

Peter B. Raven, PhD

Michael Kjaer, MD, DMSc

Ylva Hellsten, DMSc

Historical Background

The Scandinavian School of Exercise Physiology began in Copenhagen at the time when Peter Ludvig Panum (1820-1885), an infectious-disease epidemiologist, returned from Kiel, Germany, to be professor of physiology at the University of Copenhagen and director of a laboratory investigating respiration, digestion, and metabolism (3). Christian Bohr (1855-1911), Panum's most successful student, became the successor after Panum's death in 1885. Because of his early training with Carl Ludwig in Leipzig, Germany, Bohr developed a lifelong interest in the transport of oxygen and carbon dioxide in the blood. He described the effects of carbon dioxide on the oxyhemoglobin dissociation curve (i.e., the Bohr effect), an effect that is beneficial for the unloading of oxygen from hemoglobin in the exercising muscle. In addition to being the father of Niels Bohr, the great Danish nuclear physicist, Christian Bohr's fame in the physiology world was enhanced by many of his students, the two most famous being August Krogh (1874-1949), who won the Nobel Prize in 1920 for identifying the phenomenon of capillary recruitment in muscle during exercise, and Karl Albert Hasselbalch (1874-1962), known for defining the logarithmic Henderson-Hasselbalch equation and defining metabolic acidosis by standardizing the measurement of pH of the blood. Of note is that the Henderson of the collaboration with Hasselbalch is L.J. Henderson, the founding director of the Harvard Fatigue Laboratory in 1927. (See chapter 3 for more on the contributions of the Harvard Fatigue Laboratory.)

After serving as an assistant in Hasselbalch's laboratory, Johannes Lindhard (1870-1947) left in 1909 to become a family physician. He joined the University of Copenhagen and because of space limitations was invited by Krogh to collaborate and share his laboratory space. The collaboration between Krogh and Lindhard, which is considered the birth of human integrative physiology or exercise physiology in Copenhagen, further defined specific areas of investigation, now referred to as the neural control of circulation and respiration. For a more detailed history, see the proceedings of the 2 d symposium held in Copenhagen in May 2009 celebrating the centennial of Krogh and Lindhard's collaboration and their effect on exercise physiology (8). In addition, further historical information recording the effect on exercise physiology of the Danish scientists, especially Krogh himself and the PhD graduates mentored by Johannes Lindhard—Erik Hohwü-Christensen (1904-1996), Erling Asmussen (1907-1991), and Marius Nielsen (1903-2000), euphemistically known as the Three Musketeers—can be found in chapter 2 (3). The coupling of the physiologist Krogh and the medical doctor Lindhard was a fruitful symbiosis between natural science and medical science and resulted in many investigations into the interactions between clinical conditions and the natural scientists. Krogh also received a substantial grant from the League of Nations to investigate the beneficial effects of exercise on the human body and to evaluate the risks of hard physical labor, thus further establishing a link between the art and science of medicine and the natural sciences in Denmark. This tradition of investigating the interaction between medicine and the natural sciences resulted in the development of integrative and invasive human research in exercise physiology in Denmark.

For the next 20 yr the Laboratory of Gymnastics at the University of Copenhagen continued as the Danish hub of exercise physiology research. During this time, the exercise physiology program at the Karolinska Institute in Stockholm, Sweden, was rapidly emerging as another center of investigation. The development of another Scandinavian center of excellence for exercise physiology occurred because Christensen accepted a position as professor of the physiology of exercise and hygiene (Kroppsövningarnas Fysiologi och Hygien) at the College of Physical Education associated with the Karolinska Institute in Stockholm in 1941 in neutral Sweden, which was not occupied by Germany. In 1944, Christensen recruited Per-Olof Åstrand, a student in physical education, to work with him as a student assistant. On graduation in 1946, Åstrand entered the medical school at the Karolinska Institute and defended his thesis in physiology in 1952. Later in the 1950s, Bengt Saltin, a medical student at the Karolinska Institute, joined the group around Christensen, where von Döbeln and Irma Ryhming were also active. During her tenure with Christensen, Irma Rhyming married Per-Olof Åstrand and subsequently published her research using her married name of Åstrand. Saltin completed his MD degree in 1961 and finished his PhD in physiology in 1964.

Additional examples of the ongoing interactions between the natural sciences and the activities of the medical faculty occurred at the Panum Institute, where in the 1970s Henrik Galbo, MD, established a metabolic-oriented research group. In addition, Jan Praetorius Clausen, MD, at the Frederiksberg Hospital in the 1970s and Niels Henry Secher, MD, DMSc, in the 1980s established cardiovascular research groups at the National Hospital of Denmark (Rigshospitalet).

In 1964, Carlton Chapman (1915-2000), chief of cardiology and founder and director of the Pauline and Adolph Weinberger Laboratory for Cardiopulmonary Research at the University of Texas Southwestern Medical Center at Dallas in the United States from 1953 to 1966, visited Christensen and Åstrand at the Karolinska Institute in 1966. During this visit he recruited Bengt Saltin and C. Gunnar Blomqvist to join him and Jere H. Mitchell to participate in one of the first National Institutes of Health program projects awarded, titled "Response and Adaptation to Exercise." Two outcomes of the collaboration of Saltin, Mitchell, and Blomqvist that were very influential on exercise physiology are the continued, career-long collaboration and friendship that exists between Saltin and Mitchell and their research in Chapman's planned Dallas Bed Rest and Training Study (8). This study identified the benefits of habitual physical activity in determining one's cardiovascular capacity and demonstrated the cardiovascular deterioration that results from prolonged (21 d) bed rest. The findings of this study provided a scientific rationale for initiating the practice of early ambulation postsurgery, especially after myocardial infarction, and the justification of developing exercise rehabilitation therapy after traumatic and cardiovascular injuries and disease. During his MD and PhD training from 1960 to 1964, Saltin was involved in publishing 15 peer-reviewed manuscripts reporting on both groups' research that investigated cardiorespiratory work capacity during aerobic and anaerobic work with and without heat stress.

In 1967, Bengt Saltin returned to the Karolinska Institute as an assistant professor in physiology and was promoted to associate professor of applied physiology at the Karolinska in 1968. During this time he turned his focus to investigating the effects of exercise and exercise training on skeletal muscle. During his PhD training Saltin became interested in Bergström's adaptation of the muscle biopsy needle, developed by Duchenne (of muscular dystrophy fame), for analyzing muscle fiber type using histochemical analyses. Indeed, at the time of his return to the Karolinska, even though he was well established as a cardiovascular physiologist, he made a conscious decision to focus his research primarily on skeletal muscle and pursued the identification of the molecular mechanisms involved in the utilization of the delivered oxygen in both exercise-trained and nontrained muscles. Evidence of this focus is his many publications between 1966 and 1973 in collaboration with Armstrong, Bergström, Gollnick, Graham, Hermansen, Hultman, Karlsson, Knuttgen, and Pernow. Saltin acknowledges that Gollnick played a major role in his involvement in muscle biochemistry research and the effects of exercise training on skeletal muscle enzymes. Gollnick and Saltin collaborated throughout the years until Gollnick's untimely death in June 1991 and contributed to the research collaboration axis between the United States and Denmark.

Despite Saltin's stated decision to focus on skeletal muscle, he maintained a broad spectrum of interests in exercise physiology. He began his career—and remains to this day—intrigued by all integrative physiology questions that are unanswered, whether at the level of organ systems or molecular physiology. For example, between 1966 and 1973 he published 79 peer-reviewed publications. In addition to the individuals previously identified, his collaborating investigators included the thermal regulation group of Gagge, Hardy, Nadel, and Stolwijk at the Pierce Foundation of Yale and members of the cardiovascular groups interested in the role of exercise in health and disease and altitude, including Bevegård, Blomqvist, Ekblom, Grimby, Hartley, Mitchell, Rowell, and Secher.

The University of Copenhagen memorialized August Krogh (1874-1949) by designating a new building the August Krogh Institute in 1970. Erling Asmussen and Marius Nielsen were named codirectors of the institute. By this time, Asmussen had turned his research focus back to skeletal muscle. Because of ill health, Nielsen retired and Bengt Saltin was recruited to take over the professorship in human physiology at the August Krogh Institute in 1973. Before relocating his research program, Saltin sought the advice of Christensen, who encouraged him to accept the position. He stayed at the August Krogh Institute as professor until 1990. During this time Saltin was involved in the publication of another 112 peer-reviewed publications with collaborators at the University of Copenhagen and other universities in Scandinavia, Europe, Canada, and the United States. The bulk of these publications were focused on the effects of exercise-training protocols and dietary substrates on skeletal muscle metabolism, muscle fiber types, and specific skeletal muscle metabolic or hormonal diseases. In the 1980s, Saltin began to address the question of the limitations of oxygen delivery to active muscle and asked the question "What is the maximal oxygen delivery to a maximally active skeletal muscle?" This question led Saltin and colleagues to develop a one-leg knee-extension exercise protocol on a modified Krogh cycle ergometer in order to isolate the quadriceps muscle (1). In addition, the measurement of maximal skeletal muscle blood flow during the one-leg knee-extension exercise required the adaptation of the thermal dilution technique of measuring cardiac output to measure femoral vein blood flow (2). Saltin was joined by Rowell from the United States. The data from these studies identified that the maximal skeletal muscle blood flow (250 $ml \cdot kg^{-1} \cdot min^{-1}$) and oxygen uptake (350 ml of $O_2 \cdot kg^{-1} \cdot min^{-1}$) were far in excess of what had been previously reported (19).

Founding of the Copenhagen Muscle Research Center

It was fortunate for Denmark, and the University of Copenhagen in particular, when the Danish National Research Foundation (DNRF) was formed. The DNRF is an independent foundation and granting agency with the purpose of advancing Danish fundamental research in all scientific disciplines. A board was created, and Professor Peder Olesen Larsen was appointed as chairman of the board and acting director of the foundation. Larsen was a professor of chemistry and played a central role in Denmark's university politics and research administration, both before and during his time as the chair and director of DNRF (1991-1998). His vision of establishing centers of scientific excellence along with his political flair and personal interactions enabled him to coordinate a number of supporting groups in successfully funding the DNRF. Larsen and the other directors of the DNRF envisioned that each of the centers would require a central figure as a strong leader; this has remained a key demand of the DNRF. The DNRF required that each of the funded research centers be directed by an individual recognized as a strong pillar of science. This person would, because of their expertise, be granted the right to lead both the research and appropriation management.

It was of great concern to the human integrative physiologists of Denmark when Saltin accepted an invited appointment to a professorship at the Institute for Physiology and Pharmacology at the Karolinska Institute in Stockholm, Sweden, in 1990. However, Larsen was familiar with the history of Danish research in human integrative physiology and, as director of the DNRF, he had a great amount of influence in the creation of the Copenhagen Muscle Research Center (CMRC). Because of the DNRF leadership requirement, the funding of the CMRC was tied to the goal of persuading Saltin to submit an application for establishing the CMRC and to return from the Karolinska Institute to be the director of the CMRC (figure 6.1).

CMRC (1994-2003)

At its inception, Saltin recruited a number of senior researchers to be part of the CMRC (1994-1998). From the faculty of health sciences at the Panum Institute came Henrik Galbo, at the time an adjunct professor in exercise medicine at the Physiological Institute (figure 6.2), and Bjørn Quistorff (figure 6.3), a professor at the Biochemical Institute of the University of Copenhagen. From the faculty of science at the August Krogh Institute came Erik A. Richter (figure 6.2), an associate professor in the department of human physiology, and Carsten Juel (figure 6.4), an associate professor from the department of biology. From the faculty of health sciences came two researchers from the University Hospital: senior physician Niels H. Secher (figure 6.5) from the department of anesthesiology and consultant senior physician Bente Klarlund Pedersen (figure 6.6) from the department of infectious diseases.

In connection with the 5 yr renewal of funding for the CMRC (1999-2003), an additional nine senior investigators became part of the senior researchers at the

CMRC. The CMRC researchers continued to work at three locations: the August Krogh Institute, the Panum Institute, and the Rigshospitalet (University Hospital). The CMRC researchers housed at the August Krogh Institute are Erik A. Richter, Carsten Juel, Bente Kiens, Jens Bangsbo, and Ylva Hellsten. Erik A. Richter was promoted to full professor and Jens Bangsbo received a tenured position. The group consisting of Bengt Saltin, Niels H. Secher, Susanne Vissing, and Bente Klarlund Pedersen continued their research at the Rigshospitalet. At the Panum Institute, Henrik Galbo, now full professor, Michael Kjær, and Thorkil Ploug continued to be part of the CMRC. Bjørn Quistorff at the Panum

Figure 6.1 Bengt Saltin, MD, PhD, DMSc, and professor of human physiology, who was selected to be the original director of the Copenhagen Muscle Research Center.

Courtesy of Inge Holm, Administrator of the Centre of Inflammation and Metabolism in Copenhagen.

Figure 6.2 Henrik Galbo, MD, DMSc (right), professor of internal medicine and rheumatology, and Erik A. Richter, MD, DMSc (left), professor of human physiology and exercise physiology. Both were recruited by Saltin to be original members of the Copenhagen Muscle Research Center.

Courtesy of Inge Holm, Administrator of the Centre of Inflammation and Metabolism in Copenhagen.

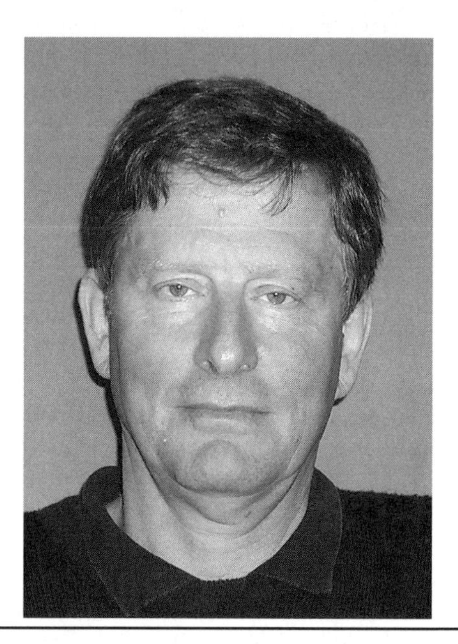

Figure 6.3 Bjørn Quistorff, MD, DSc, professor of biochemistry, who was recruited by Saltin to be an original member of the Copenhagen Muscle Research Center.

Courtesy of Inge Holm, Administrator of the Centre of Inflammation and Metabolism in Copenhagen.

Figure 6.4 Carsten Juel, PhD, DSc, professor of muscle physiology, who was recruited by Saltin to be an original member of the Copenhagen Muscle Research Center.

Courtesy of Inge Holm, Administrator of the Centre of Inflammation and Metabolism in Copenhagen.

NMR center no longer formally worked in the CMRC, although he continued to collaborate with the CMRC investigators. Bengt Saltin continued as director and was supported by an advisory group comprising Henrik Galbo from the Panum Institute, Erik A. Richter from the August Krogh Institute, and Henning Langberg from the PhD graduate student and postdoctoral fellows association. At that time in European institutions, appointment to professor was usually limited to one per department or program; only in rare situations of achievement does a department have two professors.

Based on the concept that exercise physiology is the *sine qua non* of integrative physiology, the investigators of the CMRC modified and used techniques developed in animals to address many basic physiological and pathophysiological mechanisms in healthy, disabled, and diseased individuals. However, when the questions addressed required more in-depth investigation, in vitro animal, cell culture, and cellular and molecular biologic techniques were employed.

Figure 6.5 Niels H. Secher, MD, PhD, professor of anesthesiology, who was recruited by Saltin to be an original member of the Copenhagen Muscle Research Center.

Courtesy of Inge Holm, Administrator of the Centre of Inflammation and Metabolism in Copenhagen.

Figure 6.6 Bente Klarlund Pedersen, MD, DMSc, professor of integrative medicine, who was recruited by Saltin to be an original member of the Copenhagen Muscle Research Center.

Courtesy of Inge Holm, Administrator of the Centre of Inflammation and Metabolism in Copenhagen.

In the CMRC, distinct groups of senior investigators focus their investigations primarily on one or more of the following areas of emphasis:

- skeletal muscle—growth and differentiation;
- skeletal muscle—metabolism of carbohydrate and lipids;
- skeletal muscle—nucleotides, electrolytes, and pH regulation;
- skeletal muscle—immune response (cytokines); and
- cardiovascular regulation, oxygen uptake, and mitochondrial regulation.

Although the main focus of the CMRC was integrated human physiology, the subjects that CMRC researchers explored during the period of direct DNRF funding included the whole field of physiology with the focus on basic mechanisms. Investigations were performed on all experimental levels, from in vitro subcellular molecular structures to in vivo healthy or diseased animal and human models. Subsequently, research plans that were established after the DNRF funding period began continued to involve interaction between studies in humans and other species and studies on less integrated systems (e.g., isolated organs, cell cultures, and organelles such as plasma membrane, vesicles, mitochondria, and nuclei). The studies on lower levels of organization were not expected to predict events in intact humans, but they did offer more precise, controlled experimental conditions and interpretations on the molecular level. Furthermore, the studies regularly gave rise to new ideas and the development of new techniques that, in turn, were applied in studies on humans. The principal investigators at the CMRC realized that their approach may appear complex, but a close collaboration developed between CMRC scientists with mutual interests in solving intact-human mechanisms and other investigators working on less integrated systems. The discussions created a stimulating dynamic milieu that continues to have exceptional potential for identifying translational mechanisms.

The proposed investigations required the simultaneous evaluation of the regulation and interplay of several major physiological functions (e.g., cardiovascular, ventilatory, endocrine, and substrate delivery systems). This naturally resulted in the collaboration between CMRC scientists with different specific research groups in Copenhagen and Denmark and in other countries. Positron emission tomography makes possible the visualization of the brain nuclei involved in the regulation of exercise responses. For studies of ventilation and circulation (cardiac output, arterial blood pressure, and regional blood flow), all relevant invasive and noninvasive techniques were available. In particular, CMRC investigators developed and validated Doppler ultrasound techniques for continuous recording of blood flow in the brain and legs and routinely recorded muscle sympathetic nerve activity using the well-established technique of microneurography (22, 23).

Accomplishments (1994-2003)

Skeletal Muscle: Growth and Differentiation

Using whole-muscle and single-fiber preparations from conditioned muscle (e.g., sprint and endurance training, functional electrical stimulation) and deconditioned muscle (e.g., spinal cord injured, after space flight, bed rest, congestive heart failure) of humans, investigators identified that the muscle plasticity identified in animal models exists in humans. This muscle plasticity was found to be related to transformation of myosin heavy chain (MHC) isoforms. Importantly, the work also identified that regardless of muscle group, muscle fibers expressing a specific MHC had similar velocity. In addition, the investigators adapted molecular techniques of messenger ribonucleic acid (mRNA) isolation to human muscle. They identified the mRNA for the MHC isoforms and reported a marked mismatch between the mRNA content and its concomitant protein content or vice versa. This indicated that when the muscle was used differently (i.e., during training or detraining), the fibers underwent a period of transition.

Metabolism: Carbohydrate and Lipids; Nucleotides, Electrolytes, and pH regulation; and Immune Response

Although the one-leg knee-extension exercise protocol and the concomitant measurement of quadriceps muscle blood flow using thermal dilution or Doppler ultrasound technology were developed before the establishment of the CMRC, these protocols and measurement techniques were and remain fundamental to the CMRC's investigations of exercising skeletal muscle. In addition to the exercise and measurement protocols, in 1998 the investigators of the CMRC developed innovative uses of microdialysis technology in order to investigate chemical changes in the interstitial fluid of the exercising muscle. Initially, the CMRC investigators validated the interstitial measurements of potassium ions using thallium isotope 201 and identified that interstitial potassium concentrations increased markedly during exercise (17). Subsequently, the microdialysis technique and the one-leg knee-extension exercise protocol were used to identify changes in interstitial glucose and lactate concentrations during exercise (12) and interstitial pH and ammonia concentrations in healthy subjects and patients with metabolic myopathies (26).

During the first 3 yr of operation, CMRC investigators developed modifications of the microdialysis technique for use in muscle and subcutaneous fat tissue. This technique was used to evaluate metabolic events and signals stimulating afferent nerve activity and local vasodilatation in the muscle. Subsequently, the microdialysis technique was refined to allow measurement of more substrates and potential signaling molecules than was possible in the 1990s. In 1997, mass spectrometer facilities were added to the CMRC's broad spectrum of physiological and biochemical assessment techniques. This allowed the investigators to use stable isotopes without health hazards and measure the kinetics of several substrates in blood and muscle simultaneously. Nuclear magnetic resonance spectrometry makes possible the noninvasive measurement of important metabolic variables (e.g., glycogen, pH, ADP, and ATP). By combining it with nuclear magnetic resonance imaging, it was possible to locate metabolic events in muscle more precisely. However, the muscle biopsy technique remains a central tool for sampling the skeletal muscle tissue and enabling the application of the newest biochemical, molecular biology, and histochemical techniques. In addition, healthy men and women, liver-transplant patients without sympathetic innervations, diabetic patients, adrenalectomized patients, paraplegic patients, or subjects with epidural anesthesia are recruited to address specific, relevant, integrative questions. Depending on the specific questions being addressed, the experimental techniques employed included one or more of the following: specific kinase inhibitors, insulin and hyperglycaemic clamps, stable isotopes, muscle biopsy, microdialysis, and pharmacologic blockade. The major outcomes are demonstrated in figure 6.7 and are summarized in the following sections.

Muscle

1. Muscle substrates are in part recruited from the bloodstream and include long-chain fatty acids bound to albumin or hydrolyzed from circu-

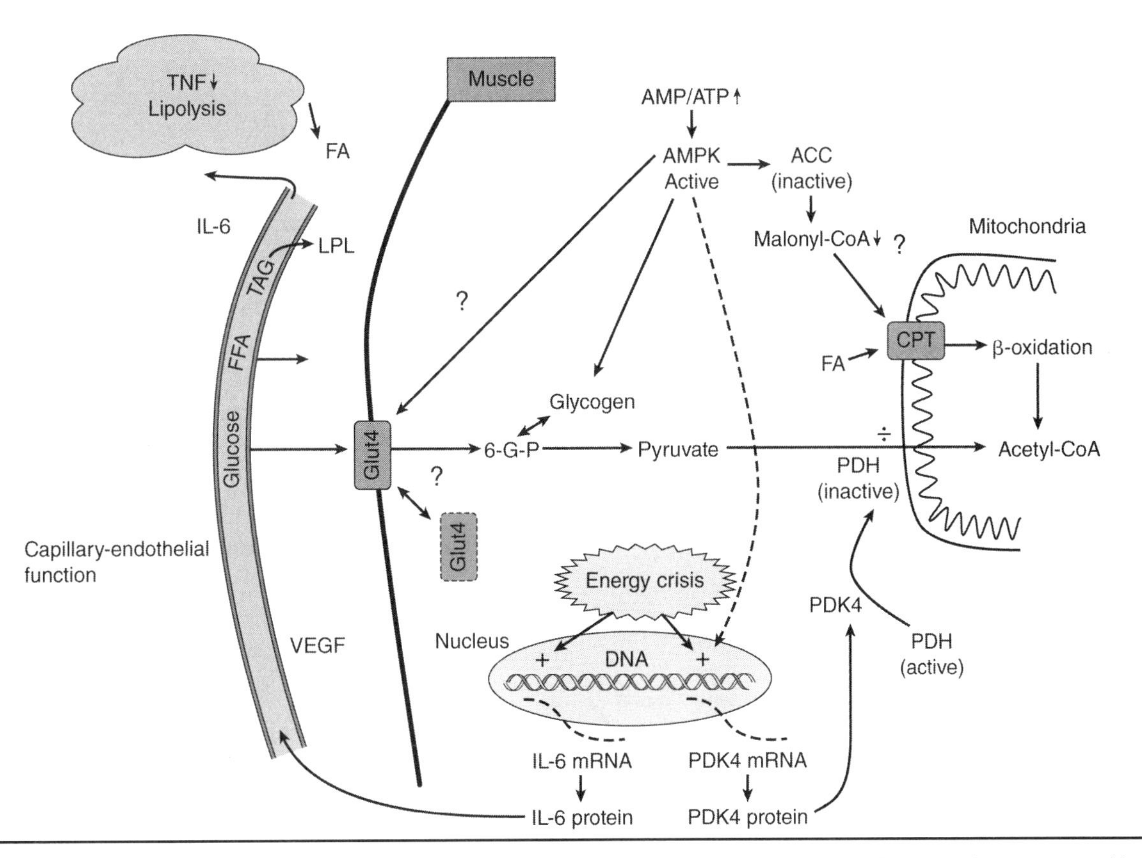

Figure 6.7 Current interpretations of the cellular events in the energy metabolism of human skeletal muscles as proposed by investigators at the Copenhagen Muscle Research Center.

Reprinted from Professor Bengt Saltin, *Annual report to the Danish National Research Foundation.*

lating triacylglycerol (catalyzed by lipoprotein) and glucose that has been transported across the sarcolemma by the glucose transporter GLUT4. A critical step in controlling energy metabolism appears to be the transport of fatty acids and pyruvate across the mitochondrial membrane.

2. In rat skeletal muscle a major regulatory reaction is the AMP activation of protein kinase, which brings about a reduction in malonyl-CoA via phosphorylation of acetyl-CoA carboxylase. This reaction enables FA to be bound to CPT, which is a prerequisite for FA to be transported into the mitochondria. AMP activation of protein kinase also changes glucose metabolism from anabolic to a catabolic mode, possibly by activation of glucose transport via recruitment of glucose transporters (GLUT4) to the plasma membrane and deactivation of glycogen synthase, all in the course of restoring energy balance.
3. The pyruvate dehydrogenase kinase gene is activated by muscle contraction, which appears to contribute to the deactivation of pyruvate dehydrogenase inhibiting the decarboxylation of pyruvate and the formation of acetyl-CoA. This process is enhanced when muscle glycogen stores are low (energy crisis).
4. In prolonged exercise, the interleukin (IL)-6 gene is activated by muscle contractions and the IL-6 protein is produced and released into the blood stream for possible action in the fat pad, thus enhancing lipolysis and inhibiting production of tumor necrosis factor. When muscle glycogen stores become low, activation of the pyruvate dehydrogenase kinase gene is more rapid and pronounced.
5. Initial work suggests that there is a link between muscle contraction-induced nitric oxide (NO) production and angiogenesis, possibly by an action via the vascular endothelial growth factor.

Cardiovascular System

In the area of blood flow regulation, the CMRC investigators began by measuring ATP, ADP, AMP, and ade-

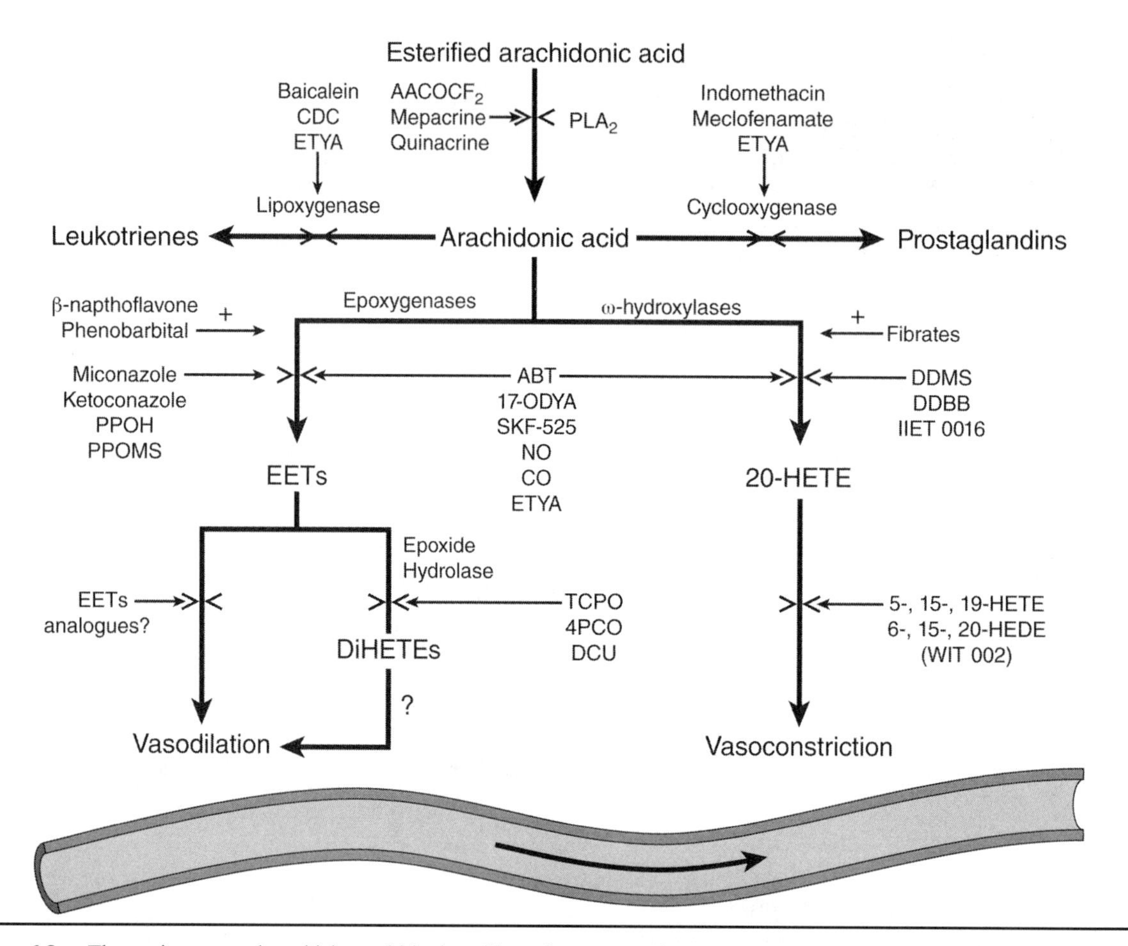

Figure 6.8 The various steps by which arachidonic acid can be converted to compounds regulating vasodilatation and vasoconstriction. Research at the Copenhagen Muscle Research Center has shown that using either Indomethacin to inhibit cyclooxygenase or Sulphenazole to block the formation of EETs combined with an NOS blockade markedly affects skeletal muscle blood flow during exercise in humans.

Reprinted from Professor Bengt Saltin, *Annual report to the Danish National Research Foundation.*

nosine in the interstitial fluid of the muscle using microdialysis at different work rates and found a close relationship between adenosine concentration and flow rate. Subsequently, it was identified that both adenosine and bradykinin increase in the interstitial fluid of the muscle during contraction. In order to study redundancy between vasoactive compounds, subjects were exercised with microdialysis probes in the leg during a control situation and during blockade by L-NAME. None of the measured vasodilators were enhanced in response to NO synthase blockade; hence, the redundancy concept was not initially verified. However, with the Doppler ultrasound technique the first detailed description of rate of increase and magnitude of elevation in human blood flow identified a very quick first phase (T½: <1 s) and a quick second phase (T½: <10 s). The first phase is caused purely by mechanical factors whereas the second phase was initially thought to be linked with shear stress-induced NOS activation. Furthermore, inhibiting the NOS activation with either L-NMMA or L-NAME resulted in a reduction in resting blood flow to one half or less but had no effect on blood flow during any phase of exercise. However, another set of investigations into the initial absence of redundancy mechanisms used double blockade of NOS (L-NAME or L-NMMA) and either sulfaphenazole to block cytochrome P450 (CYP2C) or indomethacin to block cyclooxygenase. These investigations demonstrated a significant reduction in muscle hyperemia, indicating the presence of redundant vasodilator mechanisms (figure 6.8).

An additional link between muscle metabolic demands and microcirculatory flow is the fact that muscle cells produce adenosine, which is a potent vasodilator in human skeletal muscle. However, a final answer to the question of redundancy awaits the development of a selective adenosine receptor blocker for human use.

Table 6.1 Copenhagen Muscle Research Center Conferences

Conferences	Month and year	Organizers	Publication
First conference: "Exercise and the Circulation in Health and Disease," Royal Danish Academy of Sciences and Letters, Copenhagen	November 1995	B. Saltin and N. Secher	Conference lectures resulted in a graduate-level textbook: Saltin B, Boushel R, Secher NH, Mitchell JH, eds. *Exercise and Circulation in Health and Disease.* Champaign, IL: Human Kinetics, 1999.
Second conference: "Muscle Metabolism: Regulation, Exercise, and Diabetes," Royal Danish Academy of Sciences and Letters, Copenhagen	October 1997	E.A. Richter, B. Kiens, H. Galbo, and B. Saltin	Conference lectures resulted in a book: Richter EA, Kiens B, Galbo H, Saltin B, eds. *Skeletal Muscle Metabolism in Exercise and Diabetes. Advances in Experimental Medicine and Biology.* Vol. 441. New York: Plenum Press, 1998.
Third conference: *Acta Physiologica Scandinavica* Symposium "Skeletal Muscle Oxygen Availability and Utilization," Rungstedgaard	August 1999	B. Saltin, J. Bangsbo, Y. Hellsten, and R. Boushel	Proceedings published in *Acta Physiologica Scandinavica*, Vol. 168, 2000.

During exercise, sympathetically mediated vasoconstriction of local blood flow in the skeletal muscle has to be overcome during exercise to allow for the blood flow to meet the metabolic demands of the muscle. Thus, the muscle may produce a metabolite, which partially inhibits the action of norepinephrine (functional sympatholysis) in the exercising muscle in direct relation to exercise intensity. Strong evidence has been provided that NO may have this effect, possibly acting via the adenosine triphosphate-sensitive K^+ channel on the smooth muscle cells. In exercise that intensely engages a dominant fraction of the muscle mass, the pump capacity of the heart may not be sufficient to provide all active muscles with optimal blood flow. In such situations, muscle sympathetic nerve activity increases in direct relation to work intensity such that sympathetically mediated vasoconstriction balances the locally induced vasodilatation. It is important to note that the term *functional sympatholysis* does not imply complete absence of sympathetically mediated vasoconstrictor control of the vasculature in the exercising muscle. It does imply that a proportional reduction in the degree of vasoconstriction in the exercising muscle is related to the exercise intensity (15). Arterial and cardiopulmonary baroreflexes appear to have a crucial role in regulating central sympathetic outflow. Their operating point is reset in direct relation to the intensity of exercise as a function of central command, exercise pressor reflex activity, and central blood volume (18).

In addition to the CMRC's publications, another measure of productivity over the years has been the effect of CMRC-organized scientific meetings and research education. Table 6.1 lists international and national conferences. Furthermore, based on a philosophy that the performance of research investigation is a process, the faculty of the August Krogh Institute and the CMRC decided to provide advanced PhD theoretical and practical technique courses (see table 6.2). The first course was scheduled in December 1995 in Copenhagen. Because of the success of the CMRC's intensive PhD course in the summer of 1996 (see table 6.2), a number of formalized courses were established for postdoctoral fellows and PhD students. These courses were named "The Baltic Summer School" (see table 6.3). These summer school programs became collaborations between three universities: University of Copenhagen, Denmark; University of Lund, Sweden; and University of Kiel, Germany. The summer courses alternated among the three universities with support from the European Union and NorFa Corp. Bengt Saltin was the director, and the secretariat of the school was based at the CMRC and funded by Hovedstadens Sygehusfællesskab.

These conferences, schools, and symposia provided many students the opportunity to interact with the leaders in the field without being intimidated by a perception that the faculty were unapproachable. In addition, the social times students spent with faculty and

Table 6.2 Specialized PhD Courses

Course name	Instructors and students	Schedule
Methods in the Study of Substrate Transport in Muscle and Metabolism in Man	CMRC faculty and Danish students only.	2 wk in December 1995
Control Mechanisms for Human Movement and Muscle Energetics	CMRC faculty and 7 invited faculty from outside Denmark. Attended by 49 students (27 Scandinavians, 14 from the rest of Europe, and 8 from North America).	1 wk in June 1996
Molecular Muscle Biology	Faculty: 2 from the Panum Institute, 3 from the United States, and 1 from Germany. Attended by 8 Danish students and 7 from outside Denmark.	2 wk in January 1999
Oxygen Utilization in Skeletal Muscle	Faculty: 2 from CMRC and 4 invited faculty from the United States.	1 wk in August 2000
Tracer Methodology	CMRC faculty.	1 wk in October 2002
Integrative Human Cardiovascular Control	Directed by Mikael Sander and Niels Secher. Faculty: CMRC and invited international faculty. Attended by 25-30 Danish and international PhD students and postdoctoral fellows.	1 wk in May each year since 2002

CMRC = Copenhagen Muscle Research Center.

peers enabled collaboration, recruitment of students for fellowships, and interaction between national and international laboratories.

By its final year of DNRF funding (2003), the CMRC had grown to involve more than 250 people, including faculty, postdoctoral fellows, and graduate students along with administrative, clinical, and technical support staff. During the period of direct DNRF funding (1994-2003) the research generated 42 PhD theses, 514 peer-reviewed publications, and 143 book chapters and symposia proceedings.

End of an Era

At the CMRC closing conference in 2003, Bengt Saltin wrote in the report summing its 10 yr era, "Indeed, the CMRC has become the meeting place for young and more senior human physiologists from all over the world."

He further wrote:

> The list of international guests that have visited CMRC during these ten years is long, and the professional levels span from medical students, undergraduate and PhD students, and postdoctoral and clinical fellows to established senior researchers and guest professors from a long row of countries. All of whom had stays ranging from days, weeks, and months. The postgraduate research training, the large number of international meetings, and the establishment of specialized summer schools have contributed significantly not only to Danish scientists' development but has to a large extent also attracted foreigners resulting in Copenhagen being recognized as the center for dissemination of human integrative physiological science at the very highest level.

In 2002 the evaluation team noted that ". . . no other training/research program in the world has created such an impact on integrated physiology of exercise in the last ten years."

The CMRC is known far outside the Danish borders and is still today one of the strongest brands in integrated physiology and physical activity. At the start of the CMRC the ultimate goal was, and remains even now, "to understand how the skeletal muscle functions and the role of exercise (muscle contraction) in the development of optimum health." Figure 6.9 shows a schematic representation of this goal.

Table 6.3 Baltic Summer School

Theme	Venue and date	Organizers and instructors	Students
Current trends in immunology and signal transduction	University of Kiel, Germany, September 2000	CMRC organizer and lecturer: B.K. Pedersen	Theoretical course was attended by 60 students from many international countries. 20 students continued with laboratory courses; 3 of these students worked at a laboratory course arranged by B.K. Pedersen at the CMRC.
Cardiovascular systems in health and disease	University of Lund, Sweden, May 2001	CMRC organizers and lecturers: B. Saltin and M. Sander; CMRC lecturer: C. Juel	Theoretical course was attended by 60 students from many international countries. 20 students continued with laboratory courses; 4 of these students worked in 2 courses arranged by Y. Hellsten, J. Bangsbo, N.H. Secher, and B. Saltin at the CMRC.
Signaling in muscle metabolism	CMRC, University of Copenhagen, Denmark, August 2002	CMRC organizers and lecturers: E.A. Richter, B. Kiens, B. Stallknecht, H. Galbo, and H. Pilegaard; CMRC lecturers: F. Dela, G. Van Hall, M. Kjær, B.K. Pedersen, T. Ploug, and B. Saltin	70 students were chosen from more than 130 international applicants. In addition, 23 applicants were selected to participate in the subsequent 8 laboratory courses arranged by the CMRC faculty.

CMRC = Copenhagen Muscle Research Center.

New Programmatic Outcomes

From 1994 to 2003 the CMRC provided physical and personnel resources for clinical projects that were not directly related to the key research issues of the CMRC.

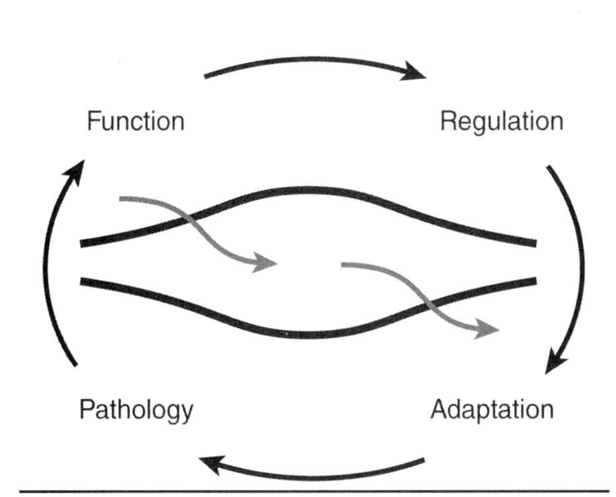

Figure 6.9 The goal of the Copenhagen Muscle Research Center.

Reprinted from Professor Bengt Saltin, *Annual report to the Danish National Research Foundation.*

This was especially true for the work by five CMRC members (M. Kjær, B.K. Pedersen, N.H. Secher, J. Vissing, and G. Van Hall) but also to some extent for the research of others. The five CMRC researchers have developed exercise physiology areas of interest that have a specific effect on exercise physiology independent of the CMRC.

Niels H. Secher: Department of Anesthesia, Rigshospitalet

Before entering medical school at the University of Copenhagen, Niels Secher won gold medals as the North American champion in 1967 and the world champion in 1970 for single scull rowing on the waters above Niagara Falls. He received his MD in 1975 and a DMSc in physiology in 1984 and completed a specialty in anesthesia in 1984. During his anesthesia fellowship in the 1980s he joined forces with Jere Mitchell to investigate the role of central command and the exercise pressor reflex in cardiovascular regulation during exercise. In these studies they used curare to weaken the exercising muscle. In one study they had subjects attempt to perform the same amount of work (11) and in the other study they had the subjects perform arm or leg exercise while the afferent nerve traffic from the muscles was blocked with local anesthesia (6). Roles for both central command (the central nervous system) and neural influence from the exercising muscles were demonstrated. As demonstrated with epidural anesthesia (10, 21) and in paraplegic patients (5), the exercise pressor reflex was crucial for the blood pressure response to dynamic exercise.

During his clinical practice of anesthesia Secher became intrigued with the phenomenon of syncope. In collaboration with Johannes Van Lieshout from the Amsterdam Medical Center in the Netherlands, he delved into the regulation of cerebral blood flow (CBF) during conditions of central hypovolemia (25). In addition, because of his history as an elite athlete, the realization that the brain played an important role in exercise performance drew his attention to whether the brain plays a role in the phenomenon of central or peripheral fatigue during maximal performance. Through the years his research group has reported that CBF increases during dynamic exercise but not during static exercise, indicating that changes in CBF reflect the neuronal integration of movement. Subsequently, Secher's group has confirmed that cerebral oxygenation is a balance between the cerebral metabolic rate for oxygen and CBF. Paradoxically, CBF decreases during maximal exercise (in response to a lowering of the arterial carbon dioxide tension with the exponential increase in ventilation) despite an increase in the cerebral metabolic rate for oxygen. In addition, they have demonstrated that cerebral metabolism during exercise is both aerobic and anaerobic (as evidenced by a release of lactate) and that lactate concentrations in the jugular vein may be elevated sixfold during exhaustive exercise in hypoxia. Similarly, they found that the brain takes up and metabolizes some of the lactate produced by the working muscles associated with plasma adrenaline (epinephrine) concentrations. They concluded that these observations indicate that brain oxygenation and its glycogen content are used during recruitment of motorneurons and that, when the glycogen content is exhausted, central fatigue ensues. These effects on brain metabolism are exacerbated by exercising in the heat (20).

In keeping with the Danish tradition of international collaboration, Secher has continued his collaborations with a variety of established and young investigators in and outside of Denmark. He especially encourages young investigators who bring new ideas on blood pressure, central blood volume, and CBF regulation during orthostasis, exercise, and heat stress. These include Crandall, Fadel, Ide, Lundby, Ogoh, Pawelczyk, Pott, Rasmussen, Seifert, Smith, Volianitis, and Williamson, to name several.

Bente Klarlund Pedersen: Center of Inflammation and Metabolism

Because of the reduction in guaranteed support in the CMRC and her career-long interest in infectious diseases and the skeletal muscle as an endocrine organ, Bente Klarlund Pedersen successfully applied to the DNRF for funding of a Center of Inflammation and Metabolism (CIM). The CIM was funded in 2005 and Pedersen was appointed director. The CIM has just recently been awarded its second 5 yr period of funding.

In the early 1990s Pedersen initiated the very first studies using exercise as a tool for studying the immune system and opened a new research field called exercise immunology (16). In 1993 the International Society of Exercise and Immunology was formed. While looking for a mechanistic explanation for exercise-induced immune changes, Pedersen identified an exercise factor being released from the contracting skeletal muscle as the myokine IL-6. The study of the IL-6 myokine from its subcellular production in skeletal muscle to its secretion and how it affected other organs led her to conclude that skeletal muscle was indeed an endocrine organ. These findings resulted in the development of another new field of exercise physiology research: muscle as an endocrine organ (16). Her research with IL-6 questioned a well-established concept and has introduced a new and controversial perspective suggesting that IL-6 is not just a proinflammatory cytokine that causes detrimental effects but rather that the muscle-derived IL-6 possesses important metabolic properties and has anti-inflammatory effects.

Pedersen is internationally recognized as a leader in the field of inflammation and metabolism when it comes to conducting mechanistic in vivo studies in human beings. In these studies she has been instrumental in combining cytokine infusion; tracer techniques; insulin clamps; pharmacological modulation of lipid levels; exercise and inactivity interventions; and measurements of fluxes over muscle, adipose tissue, and the brain, including biopsies from muscle and adipose tissue, establishment of primary cultures from biopsies, and the use of classic molecular biology methods. Her overall research strategy is based on a highly multidisciplinary research concept—from man to molecule—that requires collaboration across disciplines, institutions, and borders. Her research environment attracts many high-profile international researchers who benefit from the human studies and share their expertise with regard to cutting-edge molecular biology. Her commitment to the CIM and its research focus, training of PhD graduates and postdoctoral fellows, and research collaborations with Danish and international colleagues is representative of the history of the Danish School of Exercise Physiology and the CMRC.

Michael Kjaer: Institute of Sports Medicine

In addition to being appointed chief physician and obtaining a professorship in sports medicine at Copenhagen University Hospital at Bispebjerg, Kjaer established a research group to focus on changes and adaptation in the connective tissue in relation to training and overuse of muscle and tendon. The focus of the work of the Institute of Sports Medicine is to understand sports overuse injury in tendon. Early findings demonstrate that the connective tissue provides a much more dynamic protein turnover than previously thought. Overuse is not just an issue of mechanical failure but rather is due to a mismatch between adaptation and tissue loading. The group explores the mechanisms underlying the regeneration potential of tendon and the connective tissue of the muscle. Initial findings suggest that connective tissue cells possess a major capacity to repair but are inhibited by changes in tissue circulating factors during both chronic injury and aging. In addition, investigators at the Institute of Sports Medicine have found that aging skeletal muscle has a reduced repair capacity, which appears to be due to age-induced inhibition of the skeletal muscle stem cells. Research from the Institute of Sports Medicine has identified a closer interplay between muscular contraction and connective tissue than earlier thought. These findings have resulted in the development of guidelines for individuals for training skeletal muscle at an optimal physical stress without getting injured.

John Vissing: Neurology and Metabolic Diseases Rigshospitalet

John Vissing joined his wife Suzanne Vissing as a postdoctoral fellow with Ron Victor and Jere Mitchell at the University of Texas Southwestern Medical Center at Dallas in 1985. During his fellowship he developed an international collaboration with Ron Haller, a neurologist at the University of Texas Southwestern Medical Center, to examine the sympathetic activity and circulatory control of patients with McArdle's disease during dynamic exercise. This collaboration continues today and has resulted in more than 20 yr of clinical and basic science investigations using patients with point deletions of glycolytic or mitochondrial enzymes (7). Their collaboration is recognized worldwide and results in consultations for patients from many countries. Vissing is a full professor in neurology at the Rigshospitalet and is part of the faculty of the CMRC. He heads an established muscle group and recruits patients with specific point deletion myopathies, including mitochondrial dysfunction, for studies that compare them with healthy experimental subjects at various levels of organization (i.e., from genes to whole-body function). His research examines substrate (carbohydrate and lipids) mobilization and utilization, glucose transport, muscle pH and its regulation, and effects of training.

Gerrit van Hall: Stable Isotope Laboratory Rigshospitalet

Gerrit van Hall was recruited to the CMRC from the University of Maastricht in the Netherlands. He established the Stable Isotope Laboratory (SIL) and collaborates with all the CMRC-funded investigators involved in studies of metabolic function during exercise and environmental stressors in healthy and diseased populations. He was appointed professor, a joint position between the Panum Institute and the Rigshospitalet, in April 2010. The SIL facility, under his leadership, encompasses a multitude of mass spectroscopy equipment, has provided new avenues of investigation for the muscle group and the CIM, and has greatly expanded the effect of the investigations. He has recently developed a special interest in lactate kinetics (24).

A New Beginning: CMRC 2003 to Present

In 2003, the last year of stable and generous funding from the DNRF, the future of the CMRC was a concern to not only the Danish investigators but to the many international collaborators and collaborating institutions worldwide. Indeed, the mantra of DNRF was for the CMRC to secure incorporation into the host institutions. Fortunately for the CMRC the Copenhagen Hospital Cooperation and the University of Copenhagen established a financial base to enable a 3 yr period of continuation.

From 2003 to 2008, a steering group for the CMRC consisting of Drs. Dela, Galbo, Richter, Kiens, Kjaer, and Pedersen was established with financial input from Copenhagen University Hospitals, the University of Copenhagen, and the faculties of natural science and health sciences at the university. Saltin was required to relinquish his administrative position because of age, but the Rigshospitalet continue to provide him space and salary. In 2008, direct financial support from the universities stopped and only support from the hospital continued, thus reducing the role of the CMRC steering group. In addition, a Center of Excellence at the faculty of health sciences, led by Pedersen since 1995, formed a new research school combining natural sciences and health sciences faculties with the Academy of Muscle Biology, Exercise, and Health Research (AMBEHR). AMBEHR is led by Kjaer. These combined structures encourage continuous contact between former CMRC researchers.

The CMRC was created to develop a collaborating research network based on laboratories at the Copenhagen Hospital Cooperation (i.e., the Rigshospitalet and the Bispebjerg Hospitalet). In 2008 the group at the Rigshospitalet, consisting of Saltin, Secher, and Pedersen, continued in the CMRC and both Secher and Pedersen, director of the CIM, became professors. In addition, Vissing was appointed as professor in neurology and specialized in the treatment of patients with muscle metabolic disorders. Galbo, Kjær, and Dela are professors at the Rigshospitalet and Bispebjerg Hospital. All of the senior members of the CMRC have become internationally recognized in their fields of expertise.

International Effect

The CMRC is the only research group in the world that encompasses both cardiovascular and metabolic research, ranging from studies on isolated genes to integrated studies in intact healthy or diseased human beings. Thus, a direct comparison between the CMRC and a similar research group nationally or internationally is not feasible. However, it is generally accepted that the CMRC stands at the forefront of exercise physiology research and continues to have a major effect internationally. At the 2002 Congress of the International Union of Physiological Sciences in New Zealand, six CMRC members were invited as keynote lecturers (some gave two lectures); the presenter of the prestigious August Krogh lecture also came from the CMRC. Likewise, several CMRC members have given keynote lectures at various annual meetings of national physiological societies in Europe, including multiple lectures at the Physiological Society meetings in the United Kingdom. Recently, a CMRC member was selected as the 2002 Federation of European Biological Societies lecturer. CMRC members have contributed one handbook chapter, two well-cited reviews in *Physiological Reviews*, and several topical reviews in *Journal of Physiology* and other physiological or biochemical journals.

During the first half of the 20th century, skeletal muscle metabolism and integrative physiology attracted the attention of major groups of investigators in Denmark, most of Europe (especially Oxford University and King's College, University of London in the United Kingdom), and North America (Harvard Fatigue Laboratory). After a period of reduced research on skeletal muscle, this tissue is again in focus. Researchers now realize that the functional status of skeletal muscle plays a major role in the development of diseases such as type II diabetes and cardiovascular disorders, which may be the stimulus for the renewed interest. Moreover, as a larger fraction of the world's population grows older, maintenance of optimal muscle metabolism and function is critical for the individual and society. Although the CMRC's research was never focused solely on diabetes, cardiovascular diseases, or aging muscles, the outcomes of the research have contributed, and continue to contribute, significantly to the basic understanding of these areas. The CMRC is unique in its historical perspective and translational approach to its investigations with respect to experience and access to techniques. These qualities enable the CMRC investigators to conduct comprehensive and penetrating studies of relevant problems and remain recognized as the leaders in human integrative physiology.

A major difference between the CMRC and many of the exercise physiology research institutes, laboratories, or programs established in the United States (with the exceptions of the Institute of Exercise and Environmental Medicine in Dallas, Texas; Penn State Hershey Heart and Vascular Institute in Hershey, Pennsylvania; and the department of anesthesia at the Mayo Clinic in Rochester, Minnesota) is that the senior investigators are practicing physicians who view exercise physiology research as a component of medical research.

As mentioned previously, the authors believe there are no research groups with which the CMRC can be directly compared. Thus, in a self-evaluation the CMRC administration compared their publications with others in similar fields of research in two categories: metabolic research and cardiovascular research. In the cardiovascular field they chose four researchers who in 2002 were conducting research at different universities in the United States: Clifford at the University of Wisconsin in Milwaukee in the United States, who used dogs and humans; Laughlin at the University of Missouri in Columbia in the Unites States, who used rats and pigs; Joyner at the Mayo Clinic, who used healthy humans and patients; and Victor, then at the University of Texas Southwestern Medical Center and now at Cedars Sinai in Los Angeles, who used rats, healthy humans, and patients. Each of these researchers had experience in collaborating with the CMRC investigators in Copenhagen. Collectively, these four researchers and their groups had twice the number of cardiovascular investigators as did the CMRC. However, the CMRC matched the combined number of publications published in journals with similarly high impact factors, resulting in a similar or higher total impact (number of publication times impact factors). In regard to the CMRC's metabolic research program, the comparison was made with three groups. The first group was selected because it formed a conglomerate of researchers from four universities in Canada and had a broad investigative approach that was similar to that of the CMRC. Their group of investigative personnel was 50% larger than the CMRC's group of metabolic investigators. The other three groups (Holloszy in St. Louis, and Wallberg-Henriksson's group and Zierath's group at the Karolinska Institute, Stockholm) specifically focused on insulin resistance. Each of these groups had one third the number of personnel of the CMRC metabolic group. When the comparison was made with the Canadian researchers, the CMRC was equal in the number of publications but published in journals with higher impact factors. When compared with the two groups from St. Louis and the Karolinska Institute working primarily on insulin resistance, the CMRC was equal in both quantity and quality of publication.

Because the impact factor of a particular journal carries a great deal of weight with European funding agencies and in judging the quality of research in European academic institutions, CMRC investigators are encouraged to present their physiological results in *Journal of Physiology (London)*. This journal has the highest impact factor (~4.5) among the nonspecialized physiology journals. In 2002 the University of Copenhagen ranked second in the world in regard to number of publications in *Journal of Physiology* over the last 6 yr. It was ranked at the same level as Oxford University and ahead of Cambridge University.

Significance of the CMRC to Exercise Science

In the initial years, CMRC researchers actively published 30 to 40 publications/yr. From 2001 to 2003 they published 80 to 100 articles/yr—and the number has been steadily increasing ever since. During the DNRF-supported period from 1994 to 2003, approximately 500 original papers were published. In the research areas of metabolic and cardiovascular exercise, comparing the number of times CMRC articles have been cited with the number of times papers by other leading research groups around the world have been cited indicates that the CMRC has a strong standing internationally. During the first 10 yr of CMRC's existence, 30 senior researchers (not counting guest professors) from abroad visited the CMRC for extended periods (>1 mo) and 15 foreigners participated in PhD or postdoctoral fellow training courses at the CMRC. A number of these international colleagues achieved tenured faculty positions at universities outside of Denmark and yet continue their collaborations with the CMRC and its adjunct programs. These include Alonzo-Gonzalez, Boushel, Calbet, Clifford, Crandall, Graham, Harridge, MacLean, Pawelczyk, and Williamson, to name several.

For Danish exercise scientists, the CMRC has led to extensively more collaboration between several groups and the chance for younger researchers to immerse themselves in muscle research. At the national level, the exercise science field was afforded more respect and had a more central role in the national psyche compared with other research fields. Most importantly, it was internationally recognized that the CMRC set the standards for exercise science research. That is, if one was part of the elite in the CMRC research group in Denmark, then one was highly rated internationally. The discoveries at the CMRC were instantly compared and cited by international groups, whereby the results were spread much faster and the quality of the research was evaluated. This constant competition has contributed considerably to the drive for CMRC to be recognized as the leading center of human integrative physiology research in the world.

At the termination of the DNRF funding, the CMRC was evaluated and the international panel (consisting of two American and two Australian scientists) observed with disbelief the lack of financial opportunities available to this internationally unique center. It was highlighted that in the United States it would be possible to continue supporting such a center financially and was pointed out that 75 yr earlier the Americans had failed to financially support the leading American exercise research group, the Harvard Fatigue Laboratory, which had catastrophic consequences for the United States' role in future research in the field of exercise science. It was further noted that it takes a long time to establish a research tradition in a given field but no time at all to break the line if its financial umbilical cord is cut.

Acknowledgments

The primary author acknowledges his coauthors for their important contributions to the recognition of the historical accomplishments of the CMRC. The history of its accomplishments was gleaned from the CMRC's annual reports to the DNRF and was provided to the primary author by Bengt Saltin and Inge Holm. The CMRC, and Bengt Saltin in particular, wish to recognize Inge Holm for her conscientious commitment to providing administrative support in ensuring the productivity and success of the CMRC. Finally, the specific quotes attributed to Saltin and the external reviewers and figures 6.1 through 6.3 are available in Saltin's final report to the DNRF and on the CMRC website of the University of Copenhagen.

References

1. Andersen P, Adams RP, Sjogaard G, Thorboe A, Saltin B. Dynamic knee extension as model for study of isolated exercising muscle in humans. *J Appl Physiol* 59: 1647-1653, 1985.
2. Andersen P, Saltin B. Maximal perfusion of skeletal muscle in man. *J Physiol* 366: 233-249, 1985.
3. Astrand P-O. Influence of Scandinavian scientists in exercise physiology. *Scand J Med Sci Sports* 1: 3-9, 1991.
4. Brassard P, Seifert T, Wissenberg M, Jensen PM, Hansen CK, Secher NH. Phenylephrine decreases frontal lobe oxygenation at rest but not during moderately intense exercise. *J Appl Physiol* 108: 1472-1478, 2010.
5. Dela F, Mohr T, Jensen CM, Haahr HL, Secher NH, Biering-Sorensen F, Kjaer M. Cardiovascular control during exercise: Insights from spinal cord-injured humans. *Circulation* 107: 2127-2133, 2003.
6. Friedman DB, Johnson JM, Mitchell JH, Secher NH. Neural control of the forearm cutaneous vasoconstrictor response to dynamic exercise. *J Appl Physiol* 71: 1892-1896, 1991.
7. Haller RG, Vissing J. Drilling for energy in mitochondrial disease. *Arch Neurol* 66: 931-932, 2009.
8. Hellsten Y, Saltin B. The legacy of the Copenhagen School: In the footsteps of Lindhard and Krogh. *Acta Physiol (Oxf)* 199: 347-348, 2010.
9. Hoffner L, Nielsen JJ, Langberg H, Hellsten Y. Exercise but not prostanoids enhance levels of vascular endothelial growth factor and other proliferative agents in human skeletal muscle interstitium. *J Physiol* 550: 217-225, 2003.
10. Ide K, Boushel R, Sorensen HM, Fernandes A, Cai Y, Pott F, Secher NH. Middle cerebral artery blood velocity during exercise with beta-1 adrenergic and unilateral stellate ganglion blockade in humans. *Acta Physiol Scand* 170: 33-38, 2000.
11. Leonard B, Mitchell JH, Mizuno M, Rube N, Saltin B, Secher NH. Partial neuromuscular blockade and cardiovascular responses to static exercise in man. *J Physiol* 359: 365-379, 1985.
12. MacLean DA, Bangsbo J, Saltin B. Muscle interstitial glucose and lactate levels during dynamic exercise in humans determined by microdialysis. *J Appl Physiol* 87: 1483-1490, 1999.
13. Mortensen SP, Gonzalez-Alonso J, Nielsen JJ, Saltin B, Hellsten Y. Muscle interstitial ATP and norepinephrine concentrations in the human leg during exercise and ATP infusion. *J Appl Physiol* 107: 1757-1762, 2009.
14. Mortensen SP, Nyberg M, Thaning P, Saltin B, Hellsten Y. Adenosine contributes to blood flow regulation in the exercising human leg by increasing prostaglandin and nitric oxide formation. *Hypertension* 53: 993-999, 2009.
15. Norton KH, Boushel R, Strange S, Saltin B, Raven PB. Resetting of the carotid arterial baroreflex during dynamic exercise in humans. *J Appl Physiol* 87: 332-338, 1999.
16. Pedersen BK, Febbraio MA. Muscle as an endocrine organ: Focus on muscle-derived interleukin-6. *Physiol Rev* 88: 1379-1406, 2008.
17. Rapela CE, Green HD, Denison Jr. AB. Baroreceptor reflexes and autorregulation of cerebral blood flow in the dog. *Circ Res* 21: 559-568, 1967.
18. Raven PB, Fadel PJ, Ogoh S. Arterial baroreflex resetting during exercise: A current perspective. *Exp Physiol* 91: 37-49, 2006.
19. Rowell LB, Saltin B, Kiens B, Christensen NJ. Is peak quadriceps blood flow in humans even higher during exercise with hypoxemia? *Am J Physiol* 251: H1038-H1044, 1986.
20. Secher NH, Seifert T, Van Lieshout JJ. Cerebral blood flow and metabolism during exercise: Implications for fatigue. *J Appl Physiol* 104: 306-314, 2008.
21. Strange S, Secher NH, Pawelczyk JA, Karpakka J, Christensen NJ, Mitchell JH, Saltin B. Neural control of cardiovascular responses and of ventilation during dynamic exercise in man. *J Physiol* 470: 693-704, 1993.
22. Sundlof G, Wallin BG. Human muscle nerve sympathetic activity at rest. Relationship to blood pressure and age. *J Physiol* 274: 621-637, 1978.
23. Vallbo AB, Hagbarth KE, Torebjork HE, Wallin BG. Somatosensory, proprioceptive, and sympathetic activity in human peripheral nerves. *Physiol Rev* 59: 919-957, 1979.
24. van Hall G. Lactate kinetics in human tissues at rest and during exercise. *Acta Physiol (Oxf)* 199: 499-508, 2010.
25. Van Lieshout JJ, Wieling W, Karemaker JM, Secher NH. Syncope, cerebral perfusion, and oxygenation. *J Appl Physiol* 94: 833-848, 2003.
26. Vissing J, MacLean DA, Vissing SF, Sander M, Saltin B, Haller RG. The exercise metaboreflex is maintained in the absence of muscle acidosis: Insights from muscle microdialysis in humans with McArdle's disease. *J Physiol* 537: 641-649, 2001.

CHAPTER 7

Genomics, Genetics, and Exercise Biology

Claude Bouchard, PhD

Robert M. Malina, PhD

Introduction

This chapter reviews major advances in our understanding of the genomics and genetics of traits relevant to exercise biology and highlights several of the scientists who made significant contributions. Emphasis is on developments over the past 50 yr or so, beginning with the early 1960s. Several reports of interest were published before the 1960s but they were few and far between. For instance, Francis Galton, using twins and pedigree records, wrote extensively in the 19th century on the inheritance of several traits that are still of interest to physical performance (65). German investigators were also interested in the issue early in the 20th century (229, 230).

The relative contributions of genetic and environmental factors to phenotypic variation in a variety of behavioral and biological characteristics have long been a topic of interest. The seemingly simple dichotomy was often expressed as "nature and nurture" in the past. The phrase was introduced by Francis Galton in the 1870s (65). Nature and nurture, more often expressed as "nature versus nurture," influenced research in developmental and educational psychology (e.g., 84) and motor skills for many years before it began to influence research on traits of interest to exercise physiology (100). Indeed, the acquisition of competence in movement skills during infancy and early childhood has long been a focus of those interested in the genetics of development. Many studies were driven by the question "What is more important in motor development: nature or nurture, or biology or environment?" The question was later extended to measures of motor performance and then later to muscular strength and physiological fitness.

Two classes of studies are reviewed. The first relates to genetic epidemiology paradigms in which the focus was on the magnitude of the genetic variance for a relevant trait and characteristics of inheritance patterns. The studies relied heavily on statistical genetics methodologies but required much larger sample sizes than were commonly used in exercise genetics. These studies dominated the landscape of exercise genetics research from the 1960s to the early 1990s. The second class of research focused on the identification of genes and DNA sequence variants contributing to human variation, including performance. Such studies have dominated the past two decades of exercise genomics. Although exercise genetics research on genes and sequence variants began with a flurry of underpowered studies on candidate genes, the field has moved on briskly in recent years and new and more powerful approaches have been incorporated.

It is essential to initially define genomics and genetics in the context of the present chapter. A genome refers to the complement of all genes of an organism. The term *human genomics* has evolved to signify the science that investigates the physical features and properties of the human genome. In the broad sense, it also includes the study of transcripts produced from genomic templates. If genomics focuses on the characteristics of the genome, genetics is the science of inheritance, that is, the transmission of traits across generations. In the simplest situation, when the observed trait is determined by a single gene with two alleles, the pattern of inheritance is defined as Mendelian and can be specified as recessive, dominant, or codominant. Examples of such traits can be found in blood serology and in tissue antigenic properties. However, when multiple genes with variable penetrance contribute to variation in a trait, it becomes much more challenging to define the mode of inheritance. Phenotypes of interest to exercise biologists and physiologists tend to be of the latter category and are referred to as quantitative and polygenic phenotypes.

Comprehensive reviews of the exercise genetics literature have appeared over the past 40 yr and have contributed to the growing interest in the subject matter over the years. Among these reviews, references 24 to 26, 45, and 123 have undoubtedly been the most influential. To this brief list one must add the series developed by a group of exercise geneticists with a focus on the status of the gene map for performance and fitness traits that was published yearly from 2001 to 2009 in *Medicine and Science in Sports and Exercise* (31, 164, 173-176, 244) as well as the more recent attempt by the same group to emphasize the most significant findings in exercise genomics and genetics (177). Finally, the latest and most comprehensive review is the volume edited by Bouchard and Hoffman and published in 2011 under the aegis of the International Olympic Committee (20).

In writing this historical chapter we have relied heavily on research papers published in English. This is not likely to be a major limitation today because most scientists publish their best science in that language. However, it was not as common in the 1950s through the 1970s for scientists from non-English-speaking countries to publish their results in what is commonly recognized today as the language of science. We have made an effort to track down papers published in German, Russian, Spanish, Polish, Japanese, French, and Czech but recognize that we have been only partly successful. In this regard, Robert M. Malina has been a leader in the effort to pull together published genetic epidemiology data from a number of countries and cultures (figure 7.1).

Genetic Epidemiology Studies

This section reviews early studies that provided a foundation for contemporary efforts aimed at defining the genomics and genetics of complex exercise biology traits and their adaptation to exercise training. These studies were essentially observational, were performed on pairs of twins or members of nuclear families, and focused on motor fitness and developmental issues. A number of observational studies performed in rodents were also informative, particularly with regard to the level of spontaneous physical activity.

Observational Twin Studies

Motor Development

Early studies compared the attainment of two developmental milestones, sitting and walking alone, in monozygotic (MZ) and dizygotic (DZ) twins (14, 230) but later compared scales of motor development (43, 64, 241). Data for other movement patterns in twins were largely limited to case studies of single pairs (69, 70, 136, 210). The data yielded mixed results. MZ twins were more similar than DZ twins in the timing of walking but not in the timing of independent sitting behavior. On the other hand, within-pair correlations for

Figure 7.1 Robert M. Malina (1937–), PhD, FACSM, FAAAS, professor emeritus in the department of kinesiology and health education at the University of Texas at Austin and research professor in the department of kinesiology at Tarleton State University, Stephenville, Texas, both in the United States. Malina earned doctoral degrees in physical education (University of Wisconsin, 1963) and biological anthropology (University of Pennsylvania, 1968) and honorary degrees from the Katholieke Universiteit te Leuven, Belgium (1989), the University School of Physical Education in Cracow, Poland (2001), the University School of Physical Education in Wrocław, Poland (2006), and the University of Coimbra, Portugal (2008). His primary area of interest is the biological growth and maturation of children and adolescents, with a focus on motor development and performance, youth sport and young athletes, and the potential influence(s) of physical activity and training for sport. Another interest is the growth and nutritional status of populations in Southern Mexico, including an ongoing project in a rural and an urban community in Oaxaca that began in 1968. He is a foreign member of the Polish Academy of Sciences and the recipient of the Honor Award of the North American Society for Pediatric Exercise Medicine (2002), the Franz Boas Award of the Human Biology Association (2006), the Clark Hetherington Award of the AAKPE (2007), Distinguished Scholar Award of the North American Society for the Psychology of Sport and Physical Activity (2009), and the Honor Award of the ACSM (2013).

Courtesy of Robert M. Malina.

MZ and DZ twins on a motor-development scale did not differ. However, compared with singletons, twins tend to show lower levels of achievement on motor scales.

Motor Performance and Muscular Strength

Test batteries used to assess performance or fitness of children commonly included items assessing a variety of movement skills and muscular strength. One of the earliest studies (139) compared 4-yr-old MZ and DZ twins on several jumping and throwing tasks. Performances of MZ twins were more similar than those of DZ twins. Three studies from Japan in the 1940s and 1950s (90, 96, 140) and several studies from Eastern Europe in the 1970s [Poland (208, 209), former Czechoslovakia (109, 110, 112), former East Germany (236), and the former Soviet Union (200, 250)] compared the strength and motor performances of school-age MZ and DZ twins. Others focused on static strength (226) and muscular force (104) in adolescent twins. Overall, relative differences were smaller within MZ pairs. However, controlling for leisure-time physical activity in a study of late adolescent twins had a negligible effect on the intraclass correlation for MZ pairs but reduced the correlation for DZ pairs (56). The latter results highlighted the potential importance of controlling for habitual physical activity in attempting to quantify genotypic contributions to muscle strength and perhaps motor performance.

In contrast to performance-related skills such as jumping, running, and throwing, several studies from Japan (90, 96), Eastern Europe (109, 208), New Zealand (239, 240), and the United States (225, 227) considered similarities among twins in balance, manual dexterity, fine motor skill, speed of limb movement, hopping, and stunt-based tests.

The studies of Maciej Skład (208, 209) (figure 7.2) in Poland added a novel dimension to twin studies of performance by extending observations to the biomechanical structure of a 60 m sprint in twins aged 11 to 15 yr. Stride length, tempo, and various trunk and limb angles were more similar in MZ twins than in DZ twins. Similar observations were noted for a sprint in Japanese twins aged 6 to 9 yr (73, 74). The study of Polish twins aged 11 to 15 yr suggested a greater genotypic contribution to the kinematic structure of the run in males than in females. By inference, it is possible that the sprint performances of adolescent girls may be more amenable to environmental influences. Corresponding observations on the kinematics of throwing and swimming in Japanese twins aged 6 to 9 yr showed similar intrapair differences in MZ and DZ twins (73, 74).

More recently, the group at the Catholic University of Leuven, Belgium, under the guidance of Gaston Beunen, has considered several motor performance and strength phenotypes in adolescent twins, both cross-sectionally and longitudinally (10, 11) (figure 7.3). Using structural equation modeling (in contrast to traditional approaches using intrapair correlations), the models permitted estimates of additive genetic (A) and specific (E) and common (C) environmental sources of phenotypic variance. The AE model best fit the majority of performance tasks in 10-yr-old Flemish twins. Estimated genetic effects were greater for static strength (arm pull), power (vertical jump), muscular strength and endurance (leg lifts, flexed-arm hang), and flexibility (sit-and-reach) compared with balance (flamingo stand), running speed and agility (shuttle run), and speed of arm movement (plate tapping). Of interest, the ACE model provided a better fit for running speed and agility and for flexibility (122). Other analyses in the

Figure 7.2 Maciej S. Skład (1931-2010), PhD. Skład earned doctoral degrees in physical education (Physical Education University, Warsaw, 1962) and anthropology (University of Wroclaw, 1973) and served as provost or vice provost of the Physical Education University in Warsaw, director of the Sport Institute, and head of the department of anthropology. He spent 1 yr of postdoctoral study at the Institute of Medical Genetics and Study on Twins in Rome in 1974. The main topics of his research included longitudinal surveys of twins, especially on the influence of genetic and environmental factors on motor development, motor learning, and physiologic parameters of children and youths; the role of intensive sport exercises on processes of physical development, stages of puberty, and body composition in young athletes; and, in the last years of his research activities, social inequalities in physical and motor development of children from rural eastern regions of Poland. Skład was the author of more than 60 research publications, 20 other articles, and 2 books and was a member of the Polish Anthropology Association and of the Committee of Anthropology and Committee of Physical Education of the Polish Academy of Sciences.

Courtesy of Józef Pilsudski, University of Physical Education in Warsaw.

Figure 7.3 Gaston Beunen (1945–2011), PhD, FACSM, was part of the department of biomedical kinesiology, faculty of kinesiology and rehabilitation sciences, Katholieke Universiteit te Leuven. Beunen graduated in 1967 from the Institute of Physical Education, Catholic University of Leuven, Belgium, and obtained his PhD in 1973. Subsequently, he was appointed in the faculty of kinesiology and physical education and became full professor in 1985. Based on the Leuven Growth Study of Belgian Boys, he investigated the associations between biological maturation, somatic growth, and physical performance. In addition, he was interested in individual developmental patterns in physical performance during the adolescent period and how somatic growth, body composition, and physical performance track over time. Subsequently, he directed the Leuven Longitudinal Study on Lifestyle, Fitness, and Health and became interested in the adult health outcomes of adolescent fitness, physical activity, and body composition. He initiated the Leuven Longitudinal Twin Study (which looked at heritability of somatic growth patterns, biological maturation, body composition, physical performance, and physical activity) and the Leuven Genes for Muscular Strength project (which became a resource for the identification of genetic markers for muscle mass, strength, and function). Finally, from 2002 to 2006 he was the coordinator of the Flemish Sport, Physical Activity, and Health Centre, which conducted surveys on physical activity, physical performance, and health indicators as well as several clinical trials in different age groups.
Photo courtesy of Gaston Beunen.

Leuven study focused on the vertical jump and arm-pull strength followed longitudinally during the interval of the adolescent spurt in height. Additive genetic factors were the primary sources of variation in the jump in both sexes, whereas unique environmental factors contributed to sex differences in the stability of arm-pull strength (157, 158). The results also suggested that genetic and environmental influences may vary with age per se and with the timing and tempo of the adolescent growth spurt. The Leuven group also considered maximal dynamic strength (isokinetic dynamometry) in twins (9). Heritabilities varied between eccentric and concentric elbow flexor strength and with contraction velocities in young adult male twins (221).

Structural equation modeling was also used in a study of Japanese twins aged 10 to 15 yr (151). Grip strength, standing long jump, and the sit-and-reach were best fit with an AE model, and agility (side step) was best fit with an ACE model. In contrast, both sit-ups and running speed were best fit by a CE model (i.e., no genetic effect was found).

Variable results were noted in Japanese female adolescent twins selected for similar environmental backgrounds and levels of habitual physical activity (129). The latter may suggest a potentially important role for homogeneity of environmental background among adolescents. Studies of age effects on the heritability of dynamic strength have been extended to male twins aged 35 to 69 yr (186) and postmenopausal female twins (4).

Physiological Fitness

Studies of maximal aerobic power among twins were pioneered by Vassilis Klissouras (figure 7.4) and colleagues in the early 1970s (99-101, 104, 165). Results of these and subsequent twin studies (22, 23, 56, 58, 85, 103, 121, 212) varied considerably, giving heritabilities ranging from near 0% to almost 90%. Interestingly, controlling for leisure-time physical activity in late-adolescent twins resulted in just as much variability within MZ twin pairs as within DZ twin pairs (56), which emphasized the need to control for habitual physical activity in estimates of heritability or to ensure that all subjects are sedentary or that no differences in activity level exist between the two sets of twins. A 2009 review of twin studies provided evidence that estimates of heritability based on structural equation modeling of maximal aerobic power phenotypes were heterogeneous but more consistent than those made using other quantitative methods (159).

Corresponding information on anaerobic performances in twins is limited. Although tasks and analytical strategies vary, results indicate significant genetic effects. For example, a heritability of 0.74 for maximal power developed in 5 s in a Wingate test in young adult male twins (36) and heritabilities of 0.97 and 0.85 for peak power developed in a Wingate test in female preadolescent and adolescent twins, respectively, were reported (129). Of particular interest for anaerobic performance is the proportion of various fiber types. Initial observations in small samples of twins indicated surprisingly high heritabilities (0.99 in males, 0.93 in females) for the proportion of Type I fibers in the vastus lateralis muscle (105). Subsequent research conducted with sets of MZ and DZ twins, nontwin brothers and sisters, and parent and offspring dyads allowing for the quantification of sampling and technical variation, resulted in an estimate of the genetic variance that

Figure 7.4 Vassilis Klissouras (1937–), PhD, FACSM, emeritus professor of ergophysiology, department of sport medicine and biology of physical activity, National and Kapodistrian University of Athens, Greece. Klissouras was nurtured in a culture rich with legends. The story of Diagoras, who on his glorious dying day was carried on the shoulders of his sons around the stadium of Olympia, made a lasting impression on him. Diagoras in his youth was an Olympic champion and all of his sons later on won first place in the Olympic Games. Klissouras wondered if Diagoras' athletic prowess had been passed on in some way to his offspring. Klissouras completed his doctoral studies at Springfield College in the early 1960s under the supervision of Peter V. Karpovich. Later, as a young professor at McGill University, he began using the twin model to explore the genetic basis of human adaptive variation. In the late 1960s, he tested identical and fraternal twin boys from Montreal and obtained an incredibly high heritability (93%) for $\dot{V}O_2max$. These results were published in 1971 in a landmark paper in *Journal of Applied Physiology*. The current focus of his research at the University of Athens is the heritability of the plasticity and excitability of the brain and epigenetic influences on performance.

Photo courtesy of Vassilis Klissouras.

reached 45% for the proportion of Type I muscle fibers in the vastus lateralis (26, 201).

Observational Family Studies

Motor Performance and Muscular Strength

Father–son similarities were noted in tests of running and jumping when both were of college age (42). Similarities in running, jumping, and throwing performances have been noted in American (126) and rural Mexican (125) siblings of school age, and similarities in the vertical jump were observed in Polish siblings aged 3 to 42 yr (213, 215).

Reaction time and movement time of siblings were measured by Galton's laboratory in the 1880s and subsequently reported in the 1980s (91). Similarities of performances were considered between Czech parents and teenage sons in three fine motor tasks (111) and between rural (242, 243) and urban (214) Polish parents and offspring spanning a broader age range in several fine motor and balance tasks. Familial resemblance in several psychomotor tasks was also considered among biological and nonbiological relatives in the Quebec Family Study (163) and in a community of Mennonites (53). Estimates of transmissibility of variance across generations varied by task and were generally low in both analyses.

Studies of familial similarities in measures of muscular strength, most often grip strength, are more extensive [e.g., England (91), the former Czechoslovakia (111), Poland (213, 215, 242, 243), the United States (126, 144), and Mexico (125)]. These studies were limited to siblings and parents and offspring. Correlations between parents and offspring were more variable than those between siblings. The Quebec Family Study (162, 163) and Canada Fitness Survey (161) considered several indicators of muscular strength and endurance in a variety of biologically and nonbiologically related family members. Except in MZ twins, results highlighted the potential role of environment and lifestyle shared by family members. The heritability coefficients were generally low.

Physiological Fitness

The first papers on exercise and physiological indicators were published in the 1970s. Parent–child similarities in the heart rate response to a step test were evaluated in the Tecumseh Community Health Study (143, 144). Although parent–children correlations were significant, parental age and fatness influenced the correlations. Similar relationships were noted between parents and children and between siblings for estimated submaximal power output in the Canada Fitness Survey (161).

Resemblances among different kinds of biological and nonbiological relatives in maximal (22, 114, 118) and submaximal (23, 162, 163) indicators of aerobic fitness were considered in the Quebec Family Study (19). Familial similarities were more pronounced in first-degree relatives, and variation remained considerably greater between families than within families when smoking, fatness, habitual physical activity, and economic status were statistically controlled.

Even though the HERITAGE Family Study was designed to investigate the contributions of genetic factors to the ability to respond to exercise training, it also considered cardiorespiratory fitness indicators and hemodynamic variables during submaximal exercise in parents and adult children in a sedentary state (21). After controlling for age, sex, and body mass, the variance in

$\dot{V}O_2max$ was almost three times greater between families than within families (18). Estimated maximal heritability of $\dot{V}O_2max$ in sedentary adults was 0.51. Familial aggregation, indicative of a significant genetic component, was also apparent for several submaximal exercise phenotypes of cardiorespiratory fitness (18, 160) and for submaximal exercise heart rate, blood pressure, stroke volume, and cardiac output per minute in the untrained state (3, 13).

Corresponding familial data for anaerobic performance phenotypes are very limited. The findings were summarized in a paper published more than a decade ago, and little research has been reported since on the topic (202). Based on a cohort of MZ and DZ twins plus biological and adopted siblings, Simoneau reported a genetic component on the order of 50% for maximal power output per unit of body mass as measured in a 10s maximal work output test on a modified cycle ergometer (202, 204).

Observational Rodent Studies

A number of observational studies performed on rodent strains are relevant to exercise traits. For instance, a comparison of inbred strains of mice revealed large differences among strains for the amount of spontaneous activity assessed by the number of revolutions of a running wheel. A fivefold range was observed between the least active (A2G and DBA/2) strains compared with the most active (C57BL/Lac and BALB/C) strains (61). The results suggested a significant role for genetic differences in the level of voluntary activity. These early observations were reinforced by subsequent studies. In a recent report on this issue, Lightfoot and colleagues compared spontaneous wheel-running activity across 41 inbred strains of mice (115). There was a 27-fold range in the distance run, a 24-fold range in duration of activity, and a 3-fold range in speed of running among strains at the extreme of the distributions for the three traits.

Other rodent studies provided support for the concept of a significant genetic component to endurance performance. In one study from the laboratory of Steven Britton and Lauren Koch, 6 untrained rats of each sex from 11 inbred strains were tested for maximal running capacity on a treadmill (7). COP and MNS rats were the lowest performers whereas DA and PVG rats were the better runners based on duration of the run, distance run, and vertical work performed. There was a 2.5-fold difference between the COP and DA strains. Heritability of endurance performance was estimated at 50% in these untrained rodents. Of interest, the findings from the observational studies of rodents are quite concordant with those obtained from studies of human families and twins.

Experimental Genetic Studies

Experimental Twin Studies

Experimental studies conducted with pairs of twins have been of three types. One design focused on MZ twins in which one twin was given a specific practice or training protocol while the other served as a control with no special practice or training (cotwin design); a second approach relied on pairs of MZ twins in which both members of each pair were exposed to an experimental intervention (27); and in a third design both MZ and DZ twins were exposed to an intervention protocol.

Motor Development

Well before twins were used in experimental studies of exercise physiology traits, several early studies used the cotwin protocol in evaluating motor development of preschool children (69, 82, 136, 139). The trained twins showed some improvements in motor proficiency; however, neuromuscular maturation, which was viewed as genotypic, was seemingly more important.

Motor Learning

Five studies of MZ and DZ twins aged 8 to 18 yr, spanning about 50 yr across time, considered the influence of specific practice protocols on mechanical ability (32, 237), several fine motor skills (130, 137, 207), and balance (239). Results suggested that rates of learning motor skills were more similar in MZ twins than in DZ twins, but estimates of genetic contributions varied from task to task and during the time course of learning (i.e., over practice trials or training sessions). Among adult MZ and DZ twins reared apart, heritability of rotary-pursuit performance increased with practice (62).

Response to Strength Training

The role of genetic factors in responses to strength training has been addressed in studies from Quebec and Leuven. Responses of young adult male MZ twin pairs to an isokinetic knee extension–flexion training protocol (218) and young adult male MZ and DZ twins to a resistance-training protocol for the elbow flexors (219) indicated little evidence for a genotype–strength training interaction effect. There also was no clear genotype–training interaction for the responses of several metabolic enzyme maximal activities from biopsies of the vastus lateralis muscles obtained before and after 10 wk of isokinetic training (creatine kinase, hexokinase, malate dehydrogenase, and 3-hydroxyacyl CoA dehydrogenase) in twin pairs. However, there was a significant genotype–training interaction effect for a marker enzyme of the tricarboxylic acid cycle, as shown by the significant intraclass correlation for the response of oxoglutarate dehydrogenase activity (218).

Response to Endurance Training

The group of Vassilis Klissouras used the cotwin protocol to evaluate the influence of endurance training on maximal aerobic power in youths. In a first study (235), trained 10- and 16-yr-old twins improved more than their nontrained brothers did whereas trained and nontrained 13-yr-old twins did not differ. A second study (44) followed 9 sets of male MZ twins aged 11 to 14 yr. Both the trained and nontrained twins improved in $\dot{V}O_2$max, but the difference was not significant. The cotwin design is excellent for the quantification of the effect size of an intervention but is not well suited for testing the hypothesis that a genotype–interaction effect occurs in response to a treatment protocol. The results of these two studies indicate that it is more challenging to induce a substantial training effect by endurance exercise in teenagers than in adults, a phenomenon that has been repeatedly evidenced in exercise-training studies performed in growing children and adolescents. Moreover, although the ages include the age range of the adolescent growth spurt for most boys, the data did not permit evaluation of maturity variation in the timing of the growth spurts in height and maximal aerobic power. $\dot{V}O_2$max has its own growth spurt, and its peak occurs, on average, close to that for height (124).

The issue of interaction between genotype and training in the response of maximal oxygen uptake to standardized training was addressed for the first time in several studies of young adult MZ twins performed in the laboratory of Claude Bouchard in Quebec (see 19 for a review). A paper published in 1984 noted that the gains in $\dot{V}O_2$max in response to 20 wk of endurance training were more similar within MZ pairs than between MZ pairs; that is, variance between genotypes (twin pairs) was considerably greater than within genotypes (169). These observations were subsequently confirmed in an experiment designed to verify whether the early findings could be replicated in a different sample of MZ twins (80). A genetically determined ability to improve $\dot{V}O_2$max was also observed in response to a high-intensity intermittent-training program and to a moderate-intensity exercise program combined with a clamping of energy intake for the 93 d of the exercise protocol (19, 30, 203).

Response to High-Intensity Training

Only one study has been reported on the effects of the interaction of genotype and high-intensity training on predominantly anaerobic performance. The response of short-term anaerobic performance (10 s power output) to a high-intensity intermittent training was minimally affected by genotype, whereas the gains in long-term anaerobic performance (90 s power output) were largely determined by genetic factors in young adult male MZ twins (203). The changes induced by high-intensity training on the activities of creatine kinase, hexokinase, lactate dehydrogenase, malate dehydrogenase, and oxoglutarate dehydrogenase in the vastus lateralis muscle also showed moderate to high intrapair correlations in the same study.

Experimental Family Studies

Another design that merits further study uses nuclear families as the experimental unit. Only one study has exposed samples of families (both white and black ancestry) to a standardized exercise-training protocol (21). The study, known as the HERITAGE Family Study, was a large undertaking that involved 5 laboratories and enrolled more than 500 whites, aged 17 to 65 yr, from 100 nuclear families and more than 250 blacks from as many nuclear families. The study was conceived by Claude Bouchard in the late 1980s and has been funded by the National Institutes of Health since 1992. Bouchard (at the time at Laval University, Quebec City) was the overall principal investigator for the study, and four colleagues served as principal investigators at the other four sites: D.C. Rao at Washington University in St. Louis, Arthur Leon at the University of Minnesota, James Skinner at Indiana University, and Jack Wilmore at the University of Texas at Austin, all in the United States (figure 7.5). Two scientists provided leadership as project directors of the study (Jacques Gagnon for 7 yr and Tuomo Rankinen for 12 yr). Since 2008, Rankinen has also served as one of the principal investigators of the study, together with Bouchard and Rao. The HERITAGE Family Study has generated approximately 160 publications to date, more than 100 of which address exercise genetics questions.

Among the array of findings from the HERITAGE Family Study, the most relevant to this section is the evidence for a significant genetic component to the changes induced by 20 wk of supervised exercise for $\dot{V}O_2$max; a maximal heritability of 0.47 was found (15). Significant genotype–exercise training interaction effects were also reported for changes in submaximal heart rate, stroke volume, and cardiac output at 50 W and at 60% of maximal aerobic power (2, 3). Similar findings were reported for changes in vastus lateralis

Figure 7.5 Principal investigators of the HERITAGE Family Study. (*a*) Claude Bouchard (1939–), PhD, FACSM, FASN, FAHA, FAAAS, John W. Barton Sr. chair in genetics and nutrition, Human Genomics Laboratory, Pennington Biomedical Research Center, Baton Rouge, Louisiana. Bouchard, who received his PhD from the University of Texas at Austin, has authored and coauthored more than 1,000 publications. He was the recipient of the Honor Award from the Canadian Association of Sport Sciences in 1988 and the Honor Award from ACSM in 2002. He was awarded honoris causa doctorates from the Katholieke Universiteit Leuven in 1998, the University of South Carolina in 2009, the University of Guelph and Brock University in 2011, and the University of Ottawa in 2012. Before coming to Pennington Biomedical Research Center, he was on the faculty at Laval University, Quebec City, for more than 30 yr. He is a fellow of ACSM, the American Society of Nutrition, the American Heart Association, and the American Association for the Advancement of Science. (*b*) Art Leon (1931–), MD, MS, Henry L. Taylor professor in exercise science and health enhancement, director of the Laboratory of Physiological Hygiene (LPH) and Exercise Science, school of kinesiology, University of Minnesota. Leon, a cardiologist, is the former director of applied physiology and nutrition in the original LPH and later the division of epidemiology, school of public health at the University of Minnesota. He has investigated the effect of exercise on the heart and coronary circulation in animal models, the association of physical activity with CHD and all-cause mortality in the MRFIT, and the effects of supervised exercise training on blood lipids, glucose–insulin dynamics, body composition, claudication, and other risk factors. His publications include more than 300 peer-reviewed papers, 3 books, and 58 book chapters. He has received research grants as principal investigator and coinvestigator from the National Institutes of Health (NIH), the American Heart Association (AHA), and pharmaceutical companies. He received the Wm. G. Anderson Award from AAKPE, Citation Award from ACSM, Leading Cardiologist of the World from International Association of Cardiologists, and Horse Collar Knight from University of Kuopio (Finland), department of physiology. He is also a Colonel MC USAR (retired). (*c*) D.C. Rao (1946–), PhD, professor of biostatistics, genetics, psychiatry, and mathematics and director of the division of biostatistics in the school of medicine at Washington University in St. Louis. Rao's primary research interest is genetic epidemiology of common complex diseases, with particular emphasis on cardiovascular disease and metabolic diseases and their risk factors, notably hypertension. His research has been supported by multiple NIH grants. He has published more than 550 research articles and edited 8 books. He was president of the International Genetic Epidemiology Society (IGES) in 1996. He received the IGES Leadership Award in 1997 and the Champion of Public Health Award from Tulane University in 2005. (*d*) James S. Skinner (1936–), PhD, professor emeritus, Indiana University. Skinner was president of the ACSM (1979-1980) and vice president of the International Council of Sports Science and Physical Education (1994-2000). He was cochair of the 2010 and 2011 Exercise is Medicine World Congress and chair of the International Advisory Council. He has been involved in research grants totaling more than $50 million and has published more than 280 research articles and 5 books. He has investigated relationships between exercise, training, and health for 45 yr and has lectured about these relationships in English, French, German, and Spanish in 58 countries. (*e*) Jack Wilmore (1938–), PhD, professor emeritus, University of Texas at Austin. Wilmore retired as a distinguished professor of health and kinesiology at Texas A&M University. He was an endowed professor and chair of kinesiology and health education at the University of Texas at Austin and professor and chair of exercise and sport sciences at the University of Arizona. He has investigated the role of exercise in the prevention and treatment of obesity and CHD as well as mechanisms of alterations in physiological function with training and detraining. His publications include more than 320 peer-reviewed research articles, 53 review chapters, and 15 books. He has received research grants from NIH, NASA, USAF, and pharmaceutical companies. He received the Honor Award from ACSM in 2006 and the Hetherington Award from AAKPE in 2010.

(*a*) Photo courtesy of Claude Bouchard. (*b*) Photo courtesy of Art Leon. (*c*) Photo courtesy of D.C. Rao. (*d*) Photo courtesy of James S. Skinner. (*e*) Photo courtesy of Jack Wilmore.

muscle enzyme activities associated with glycolytic and oxidative pathways in response to the endurance-training program (172, 181). Overall, results of the HERITAGE Family Study provide substantial evidence for the presence of family lines in response to systematic endurance training when compliance with the requirements of the program is not an issue. Although based on parents and adult offspring, it is probable that a similar genotype–training interaction effect exists among children and adolescents.

Experimental Rodent Studies

Experimental studies based on rodent models have been at the origin of multiple advances in exercise biology.

This is also the case for exercise genetics. One approach was pioneered by the team of Steven Britton and Lauren Koch, now at the University of Michigan in the United States (figure 7.6). Selective breeding for endurance running in untrained rats was undertaken for 15 successive generations. At generation 15, the $\dot{V}O_2max$ per kilogram of body mass was 49% higher in the high-capacity runners compared with the low-capacity runners (97). Overall, the observations on rats selected for endurance capacity indicated that the divergence between the high and low lines requires cardiac, hemodynamic, pulmonary, skeletal muscle, endocrine, and body mass adaptations. In an earlier report, they showed that about 40% of the variation in running performance between the high and low lines of untrained rats was determined by genetic differences (102).

Figure 7.6 Steven Britton (1948–), PhD, professor, left, and Lauren Koch (1964–), PhD, associate professor, right, of the department of anesthesiology at the University of Michigan Medical School. Koch earned her PhD in physiology from the Medical College of Ohio in 1991 and Britton obtained his PhD in physiology from Texas Tech University School of Medicine in 1978. In 1996, Koch and Britton, then at the Medical College of Ohio, started a large-scale selective-breeding program to develop strains of rats that contrast for intrinsic (i.e., untrained) aerobic treadmill-running capacity. They hypothesized that artificial selection of rats for low and high aerobic-exercise capacity would yield models that contrast sharply and account for the divide between health and disease. If true, this would support the notion that impaired exercise oxygen metabolism is a common feature that mechanistically underlies complex diseases. More than 15 yr later, after 30 generations of selection, the low-capacity runners and high-capacity runners differ in aerobic running capacity by more than 500%. The low-capacity runners score poorly on cardiovascular risks and features of the metabolic syndrome, including higher blood pressure, insulin, random glucose, fasting glucose, free fatty acids, visceral fat, and triglycerides. The high-capacity runners fare much better for health indicators such as maximal oxygen consumption, heart function, nitric oxide formation in blood vessels, economy of oxygen use, and abundance of proteins required for mitochondrial function in skeletal muscle. This approach represents a shift from the focus on individual genes and pathways to highly interconnected organs, tissues, and molecular networks. These rats are maintained as an international collaborative resource that currently includes contributing scientists from 30 institutions in 11 countries.

Photo courtesy of University of Michigan Medical School.

These results in rats are compatible with those reported recently in mice. Animals were bred for high voluntary wheel-running activity. Selected lines performed about threefold more revolutions per day compared with unselected control lines (95, 138). In contrast to the study paradigm of Britton and Koch in which the selection of rats was based on an imposed endurance run, the mice selection in these studies pioneered by the laboratories of Garland and Pomp was dependent on voluntary wheel running. Interestingly, $\dot{V}O_2max$ adjusted for body mass differed markedly between lines selectively bred for high voluntary wheel running and unselected control mice. For instance, after generation 36 of selective breeding, the high wheel-running mice achieved a $\dot{V}O_2max$ 24% higher than that of control mice under normoxic conditions (179, 180). Comparable differences were also observed under hypoxic and hyperoxic conditions.

Even though the observational and experimental genetic epidemiology studies from the 1960s to more recent times may seem foreign to exercise genomics and genetics, they are not. They provide us with an appropriate foundation for the justification to move forward to the next phase and to request the millions of dollars in research funding that are necessary to adequately define the contribution of genomics and genetics to human variation in exercise behavior, health-related fitness, performance-related fitness, and gene–exercise interactions on disease risk factors, morbidities, and premature death.

Genomics Studies

Several research designs and technologies have been used over the past two decades in the effort to identify genomic regions, genes, and sequence variants associated with exercise behavior and exercise biology traits. This section reviews advances made over time and key laboratories and investigators involved in these studies. Candidate genes, genome-wide linkage scans, genome-wide association (GWA) studies, and contributions from combinations of transcriptomics and genomics are reviewed. Efforts to identify genetic differences between elite athletes and sedentary controls are also briefly summarized.

Candidate Genes

A candidate gene is a gene that has a theoretical relationship with the behavioral, physiological, or metabolic system regulating the trait of interest. Candidate genes for exercise biology traits have been defined on the basis of advances in exercise physiology studies and animal models and on purely theoretical grounds. Transgenic (overexpression of a gene or genes), knockdown (reduced expression level), and knockout (ablation of a gene or genes) mice have generated a number of candidate genes that were subsequently investigated for potential involvement in human variation. The expression level of a gene has also been used as a candidate phenotype against which DNA variants can be tested for associations. The complete transcriptome, which is the full set of ribonucleic acid (RNA) transcripts in a given cell or tissue, has been used to define new panels of candidate genes for further genomics studies (e.g., 72, 224).

Candidate genes were very prominent in the early phase of exercise genomics. Many of the earlier studies were quite simple in design and were launched primarily because of the availability of whole blood stored in freezers. They were generally case-control or cross-sectional cohort studies with unrelated subjects. In the case of continuous traits (e.g., $\dot{V}O_2max$), the association with a candidate gene or marker is tested by comparing mean trait values across genotypes or between carriers and noncarriers of a specific allele. In the case-control design, testing for a relation between a trait and a candidate gene marker is based on the comparison of allele and genotype frequencies between two informative groups of subjects, one with the phenotype of interest (e.g., elite power and strength athletes—the "cases") and the other without the phenotype of interest (the "controls").

Special Case of Athletes Versus Controls

Luigi Gedda from the Gregor Mendel Institute of Medical Genetics and Twin Research at the University of Rome was perhaps the pioneer in the effort to understand the importance of genetic differences between athletes and nonathletes (67, 68) (figure 7.7). The initial efforts were based on surveys of Italian twins, specifically on 351 pairs of twins who were engaged in competitive sport. The results of the surveys indicated familial aggregation among athletes and the potential importance of zygosity in concordance of sport selection; participation in markedly different sports occurred in 85% of DZ twins in contrast to only 6% of MZ twins (68). The familial aggregation of athletes was also noted by others, such as Grebe (75, 76) of Germany, Linc and Fleischmann (117) of Czechoslovakia, and Jokl (92, 93), who at the time was in the United States. The work of Gedda and his longtime collaborator, Paolo Parisi, was the most systematic in this regard and focused on genealogical and twin methods. The research of Parisi continues today with the Italian Registry of Twin Athletes (154, 155). A higher prevalence of MZ twins among swimmers has been one of their findings (154). In a large sample of adult British female twins, athlete status (nonathlete vs. local-, county-, or national-level athlete) had a heritability of 0.66 (50).

The first documented attempt to identify differences in genetic markers for performance-related phenotypes dates to the late 1960s. A group of geneticists, taking advantage of the 1968 Olympic Games in Mexico, in-

Figure 7.7 Luigi Gedda (1902-2000), MD, Mendel Institute of Medical Genetics and Twin Research in Rome, right, and Paolo Parisi (1940–), PhD, professor of biology and rector, Italian University Sport and Movement, University of Rome "Foro Italico," left. Gedda was the first to confirm, through classic pedigree studies in the mid-1950s, the common knowledge that elite athletes often cluster in families and to then attempt to quantify the role of genetic factors using the twin concordance model. He studied a large sample of MZ and DZ twin pairs and found considerably higher concordance rates in the former than in the latter with respect to various aspects of sport participation, such as kind of sport practiced and performance level. In later decades, Gedda's longtime associate Paolo Parisi conducted epidemiological and cell genetics research, also through the aid of a special registry of Italian twin athletes, in order to explore the genetic basis of athletic performance. Of particular interest to him is the relation of physical activity to health, aging, and chronic disease. One major line of research of his institute focuses on the role of exercise-induced oxidative stress in genetic regulation processes.

Photo courtesy of Paolo Parisi.

vestigated common blood genetic markers in an attempt to discriminate between Olympic athletes and controls (45). Interestingly, phenylthiocarbamide nontasters were underrepresented among the athletes. The effort to document genetic differences between elite athletes and sedentary controls was continued on the occasion of the 1976 Olympic Games in Montreal (37, 41). A slightly higher frequency of the A1 allele of the ABO blood group was observed in endurance athletes participating in the 1976 Summer Olympic Games than in reference populations (37). However, the other plasma and red blood cell markers (ABO, MNSs, Rhesus, Duffy, Kell, P, LDH, MDH, phosphoglucomutase, adenosine deaminase, adenylate kinase, esterase D, haptoglobin, transferrin, hemoglobin, glucose-6-phosphate dehydrogenase, and acid phosphatase) did not differ between athletes and controls (37, 41, 45). These early attempts were all based on polymorphisms in red blood cell antigens and enzymes. Later, skeletal muscle gene product variants were screened (17, 128) and selected enzyme markers were investigated for their putative effects on a variety of skeletal muscle and cardiorespiratory endurance indicators. Genetic variants of skeletal muscle creatine kinase (*CKM*) and adenylate kinase 1 (*AK1*) were screened for using isoelectric focusing in 295 subjects (16). A variant form of both enzymes was identified with an allele frequency of 1% (*CKM*) and 3.5% (*AK1*). There was no difference in $\dot{V}O_2max$ between the carriers of the variant alleles and matched controls homozygous for the nonvariant allele (16). None of these studies yielded reliably significant genomic predictors of performance or fitness.

Genomic differences between athletes and untrained controls began to be investigated in a more systematic fashion with the launch of the GENATHLETE study in the laboratory of Claude Bouchard at Laval University in 1993. The main collaborators on GENATHLETE included Louis Perusse, Marcel Boulay, and the late Jean-Aime Simoneau (all from Laval University), Rainer Rauramaa (Kuopio, Finland), and particularly Bernd Wolfahrt (from Germany). The study focused on DNA sequence differences between a panel that has now attained more than 300 elite endurance athletes with a $\dot{V}O_2max$ no lower than 75 $ml \cdot kg^{-1} \cdot min^{-1}$ and more than 300 sedentary controls with a $\dot{V}O_2max$ no higher than 50 $ml \cdot kg^{-1} \cdot min^{-1}$. The participants are all white males from Canada, Germany, Finland, and the United States. Single nucleotide polymorphisms (SNPs) in several genes have been investigated to date, but none have provided strong evidence for differences in allele and genotype frequencies between athletes and controls (16, 178, 184, 185, 245-247). However, a common variant and haplotype in the hypoxia inducible factor 1, alpha subunit (*HIF1A*) gene were found to be more prevalent in athletes compared with controls in a 2010 report based on the GENATHLETE cohort (54), an observation that was defined in an accompanying editorial as an important milestone (194).

Another effort to delineate the differences between athletes and nonathletes is under way in Spain under the leadership of Jonatan Ruiz and Alejandro Lucia. They have used samples of Spanish Caucasian endurance athletes (about 100), power athletes (about 50), and nonathlete controls. A number of differences at candidate genes between athletes and controls, as well as between endurance and power athletes, have been reported (34, 71, 190). Taking advantage of the concept of a total genotype score developed by Williams and Folland (238), Ruiz and colleagues in 2009 attempted to predict world-class endurance athlete status based on the frequency of variant alleles at 7 candidate genes genotyped in 43 Spanish endurance athletes and 123 controls (191).

Yannis Pitsiladis from the faculty of biomedical and life sciences at the University of Glasgow has taken advantage of striking performance differences at the elite level in athletes of various ethnic backgrounds to investigate the potential role of variation in nuclear and mitochondrial DNA (figure 7.8). For instance, distance runners from Kenya and Ethiopia and sprinters from Jamaica won 25% of all track and field medals at the 2008 Olympic Games in Beijing (166). The effort to impute ethnic differences in sport performance to genetic differences is clearly challenging, particularly because only about 10% of human genetic variation can be found between the major ethnic groups. Pitsiladis' laboratory has explored differences in sequence variants in mitochondrial DNA and in Y chromosome markers between Ethiopian athletes, Ethiopian controls, Kenyan athletes, and Kenyan controls. Mitochondrial DNA haplogroups found in Kenya were different from those found in Ethiopia (197). Differences in haplogroup distributions were observed between athletes and controls in Kenya but not in Ethiopia. The full significance of this observation needs to be established. Analyses of Y chromosome haplogroups in endurance athletes from Ethiopia revealed differences between them and samples from the more general Ethiopian population (145). Population stratification was ruled out as the explanation for such differences. One important conclusion reached by Pitsiladis and colleagues is that the clusters of Kenyan and Ethiopian elite runners are distinct from each other from a genetic point of view (166).

Two genes have received considerable attention in the exercise genomics literature: angiotensin I-converting enzyme (*ACE*) and actinin alpha 3 (*ACTN3*).

Both have been strong favorites of exercise scientists over the past decade or so.

ACE Gene

The potential role of *ACE* sequence differences in human physical performance was first investigated in the 1990s by Hugh Montgomery from the University College London (figure 7.9). He and his coworkers reported in 1997 that 10 wk of physical training in British army recruits induced greater increases in left ventricular mass and septal and posterior wall thickness in the *ACE* D/D homozygotes than in the I allele carriers (141). A few years later, the same group of investigators confirmed the finding by reporting that the training-induced increase in left ventricular mass in another cohort of army recruits was 2.7 times greater in the D/D genotype compared with the I/I homozygotes (147). The I allele was associated with higher muscular endurance gains after 10 wk of physical training in British army recruits (142).

The story began in 1990 when it was reported that a 287 bp I/D polymorphism in intron 16 of the *ACE* gene was associated with plasma ACE activity (183). ACE activity was highest in D/D homozygotes and lowest in

Figure 7.8 Yannis P. Pitsiladis (1967–), MMedSci, PhD, FACSM, integrative and systems biology research theme, faculty of biomedical and life sciences, University of Glasgow, Scotland. Pitsiladis is a reader in exercise physiology at the Institute of Biomedical and Life Sciences at the University of Glasgow and founding member (and previous director) of the International Centre for East African Running Science, set up to investigate the physiological, genetic, psychosocial, and economic determinants of the phenomenal success of East African distance runners in international athletics. Recent projects include the study of West African sprinters (including elite sprinters from Jamaica and the United States) and the study of world-class swimmers (e.g., why there are very few black swimmers). He is a visiting professor in medical physiology at Moi University (Eldoret, Kenya), Addis Ababa University (Addis Ababa, Ethiopia), and University of Technology (Kingston, Jamaica). Pitsiladis has a particular research interest in mitochondrial DNA and Y chromosome markers and their potential role in ethnic differences in performance.

Photo courtesy of Yannis P. Pitsiladis.

Figure 7.9 Hugh Montgomery (1962–), MB, BS, BSc, FRCP, MD, FRGS, director of University College London Institute for Human Health and Performance; consultant intensivist at Whittington Hospital; and professor of intensive care medicine at University College London. Montgomery obtained his BSc degree in neuropharmacology and in circulatory and respiratory physiology in 1984 before graduating with a medical degree in 1987. He has since trained and accredited in general internal medicine, cardiology, and intensive care medicine. He is now professor of intensive care medicine at University College London, where he also directs the Institute for Human Health and Performance. In the early 1990s, he began using gene–environment interaction as a means to explore human physiology. Working with army recruits, he first identified a role for the angiotensin I-converting enzyme (*ACE*) I/D polymorphism in the regulation of human cardiac growth. This work was extended to broader measures of performance, including high-altitude mountaineering aptitude. Mechanistically, a role in the regulation of metabolic efficiency was suggested, leading to phase IV studies of *ACE* inhibition in the treatment of cancer cachexia. Studies of downstream gene variants (e.g., those of the bradykinin B2 receptor gene) have offered further mechanistic insight. Montgomery has a special interest in high-altitude performance: He was science lead for the Caudwell Xtreme Everest Expedition in 2007 and has identified genes under selection pressure in high-altitude populations.

Photo courtesy of Hugh Montgomery.

I/I homozygotes. More than 50 reports have subsequently dealt with the potential role of the *ACE* I/D genotype in some aspects of fitness or performance. Overall, the results are heterogeneous and often contradictory. For instance, in postmenopausal women, I/I homozygotes had a higher $\dot{V}O_2max$ than D/D homozygotes (79), whereas the opposite was observed in Chinese males (251). In the HERITAGE Family Study, no associations were found between the *ACE* I/D polymorphism and maximal and submaximal exercise $\dot{V}O_2$ and power output phenotypes in healthy, sedentary blacks and whites (41). This literature was recently reviewed by Montgomery and colleagues (206). In the aggregate, it appears that the *ACE* I/D polymorphism may contribute to human variation in skeletal and cardiac muscle growth and functional properties as well as in adaptation to hypoxia. However, its contribution to variation in human physical performance remains a matter of debate.

ACTN3 Gene

Kathryn North and colleagues from the Children's Hospital at Westmead in Australia discovered a common variant in the gene for actinin-3, a structural actin binding protein found in skeletal muscle fast-twitch fibers (figure 7.10). The polymorphism (R577X) replaces an arginine residue by a premature stop codon at position 577 and results in ACTN3 deficiency. About 16% of the world population is homozygous for the XX null genotype; the frequency of the null allele is highest in Caucasians and Asians and lowest in Africans (8). The first report on *ACTN3* genotype and performance was from North's laboratory in 2003. It indicated that the frequency of the stop codon mutation was lower in sprinters than in controls and endurance athletes (248). Several studies have concluded that the stop codon variant (X577X) was lower in sprint and strength athletes than in nonathletes (55, 57, 153, 189, 192, 248), but other reports have been negative (149, 198, 249). Interestingly, a number of studies based on direct measures of muscle strength and power have also been published and the results are contradictory; that is, the X allele is negatively associated with these muscle phenotypes in some studies whereas the converse is found in others (39, 51, 66, 132, 133, 146, 150, 228). North has reviewed the biology of the *ACTN3* polymorphism in recent publications (8, 120). In the aggregate, the data suggest that there is probably a role of the X allele in skeletal muscle function and muscle performance, but the data remain inconclusive at present. More studies with appropriate power and designs, as well as more extensive functional studies, are warranted.

Other Candidate Genes

A series of studies has dealt with the contributions of other candidate genes to human variation in muscular strength and power. These genes were identified because of their biological relevance to muscle contractile and sarcomeric proteins, myogenesis and muscle regeneration, muscle mass atrophy and sarcopenia, hormonal regulation of muscle mass, energy transfer to sustain muscle contraction, and other properties. The evidence from candidate gene studies of muscle strength and power was recently reviewed by Thomis (220).

An example of such genes is the myostatin (*MSTN*) gene. Mutations in *MSTN* leading to inactive or defective gene product have been shown to double or

Figure 7.10 Kathryn North (1960–), MD, FRACP, Director, Murdoch Childrens Research Institute, and David Danks Professor of Child Health Research at University of Melbourne, Australia. In the process of studying genes implicated in muscle disease, North and colleagues at the Children's Hospital at Westmead discovered a common variant in the actinin alpha 3 (*ACTN3*) gene. *ACTN3* encodes a structural protein found in fast-twitch skeletal muscle fibers. The *ACTN3* gene variant results in complete deficiency of actinin 3 in almost 20% of the general population. North showed that ACTN3 deficiency is rare in sprint athletes, suggesting that this protein plays a crucial role in the function of fast-twitch muscle fibers. North and her team developed a strain of mice engineered to be completely deficient in ACTN3 and showed that the muscle of these knockout mice displays an increase in oxidative metabolism. This metabolic shift could explain why ACTN3 deficiency is detrimental to sprint activities that require fast or glycolytic metabolism. Her analysis of deoxyribonucleic acid (DNA) samples from individuals from around the world suggests that the ACTN3 deficiency provided some benefit to the ancestors of modern Europeans and Asians after their migration out of Africa, resulting in its increase in frequency due to natural selection. The benefit provided by ACTN3 deficiency may have allowed them to adapt to the more hostile environments of Eurasia. North's group plans to use mouse and human studies to determine the effect of ACTN3 deficiency on muscle aging, response to exercise, and the progression of inherited muscle diseases such as the muscular dystrophies.

Photo courtesy of Kathryn North.

even triple skeletal muscle mass in mice, cows, and sheep. Functional *MSTN* mutations are very rare in humans. However, a boy appeared extraordinarily muscular at birth and at 4.5 yr of age was described as muscular and very strong (196). It turned out that he was a homozygote for a G/A mutation located five nucleotides downstream of exon 1, which abolishes a normal splice donor site and activates a cryptic splice site further downstream in intron 1, resulting in a truncated, inactive myostatin peptide. A number of polymorphisms in *MSTN* have been investigated for their associations with muscle hypertrophy or muscle strength and power (40, 60, 107, 199, 231), but the overall findings have been inconsistent.

Another class of candidate genes of great interest for muscular strength and power is that of the insulin-like growth factors and related binding proteins. These molecules play important roles in muscle growth and repair and in muscle response to exercise. Polymorphisms in insulin-like growth factors 1 (*IGF1*) and 2 (*IGF2*) have been rather consistently associated with muscle strength and its response to training, as well as other relevant endophenotypes (52, 81, 108, 193, 195).

The laboratories of James Hagberg and Stephen Roth in the department of kinesiology at the University of Maryland in the United States have pioneered a number of candidate gene research avenues since the late 1990s (figure 7.11). Their efforts focused on sequence variants in *ACE*; nitric oxide synthase 3 (*NOS3*); angiotensinogen (*AGT*); *ACTN3*; angiotensin II receptor, type 1 (*AGTR1*); tumor necrosis factor (*TNF*); nuclear factor of kappa light polypeptide gene enhancer in B-cells 1 (*NFKB1*); activin A receptor, type IIB (*ACVR2B*); adenosine monophosphate deaminase 1 (*AMPD1*); vascular endothelial growth factor A (*VEGFA*); vitamin D receptor (*VDR*); and *MSTN*, among others (see 59, 78, 79, 81, 134, 156, 167). Hagberg has a strong interest in exercise blood pressure and other hemodynamic phenotypes and through most of his career has consistently focused on the role of exercise in health-related outcomes and therapeutic applications, with an emphasis on the importance of genetic differences at key genes. Roth, whose primary interest is in muscle size and strength, has contributed a primer on exercise genetics (187) and has a longstanding interest in the role of genetic variation in talent selection for athletic performance. Both have trained a good number of graduate students in exercise genetics and have been longtime contributors to the fitness and performance gene map series that was published in *Medicine and Science in Sports and Exercise* until 2009.

When the last version of the fitness and performance gene map was published based on the publications available by the end of 2007, there were 214 autosomal, 7 X chromosome, and 18 mitochondrial gene and other loci entries (31). Most of the entries were related to specific candidate genes and were supported by more than 350 peer-reviewed papers. With the exception of the *ACE* and *ACTN3* common variants, few of the genes had been replicated in multiple studies. Moreover, after almost 25 yr of candidate gene studies in exercise biology, this line of research continues to be plagued by small sample sizes and lack of appropriate statistical power (31, 177). Studies with small sample sizes occasionally report positive findings, especially if they include several traits and perform multiple statistical tests for association. Fortunately, these positive findings are seldom replicated and usually fall into oblivion.

Figure 7.11 Stephen M. Roth (1973–), PhD, FACSM, associate professor, left, and James M. Hagberg (1950–), PhD, FACSM, professor, right, department of kinesiology, school of public health, University of Maryland. Hagberg completed his graduate studies at the University of Wisconsin. He has been on the faculty of a number of universities and at the University of Maryland for nearly 20 yr, where he is a distinguished scholar-teacher. He has been funded by the National Institute of Aging for genetics-based investigations and has published numerous papers identifying genetic markers that associate with the degree to which endurance-exercise training improves cardiovascular disease risk factors in older men and women. Stephen M. Roth completed graduate studies in exercise physiology at the University of Maryland in 2000 and postdoctoral training in human genetics at the University of Pittsburgh. In 2003 he was recruited back to the University of Maryland as director of the Functional Genomics Laboratory. He has been funded by the National Institutes of Health to perform a variety of genetics investigations. In addition to several research publications, Roth is author of a textbook titled *Genetics Primer for Exercise Science and Health*. The department of kinesiology at the University of Maryland was one of the first to develop a research focus in the area of exercise genomics. As a group, they have published more than 60 peer-reviewed publications in the areas of exercise genomics.

Photo courtesy of Stephen M. Roth and James M. Hagberg.

Genomic Linkage Scan Studies

A genetic locus contributing to a complex multifactorial trait is referred to as a quantitative trait locus (QTL). In animal models and in human studies, QTLs are defined as positions on chromosomes. In rodent models, QTLs are typically identified on the basis of crosses between informative strains, based on the cosegregation of a chromosome segment with the trait of interest. This approach works best for loci with large effect sizes.

In humans, the statistical linkage test uses either a regression-based method or a variance components modeling approach. Briefly, in the regression method, the phenotypic resemblance of siblings is modeled as the mean-corrected cross-product of the siblings' trait values (152). The phenotypic covariance of the sibling pairs is modeled as a function of allele sharing or "identical by descent." In the variance components linkage methods, the trait variance is decomposed into additive effects of a trait locus, a residual familial background, and a residual nonfamilial component (1, 168). The linkage testing is performed using the likelihood ratio test contrasting a null hypothesis model of no linkage with an alternative hypothesis model in which the variance due to the trait locus is estimated.

A difference between linkage and association studies is that association targets alleles or genotypes at a specific gene or genetic marker whereas linkage aims to identify a specific chromosomal region. Thus, linkage analysis is used to identify chromosomal regions that harbor gene(s) affecting quantitative traits (hence QTLs), even if there is no a priori knowledge of the gene(s). In humans, linkage studies require family or pedigree data; the basic observation unit is a pair of relatives (usually siblings), not an individual subject.

Although the approach has had some success with disease traits, the application to complex, multifactorial phenotypes has been more laborious. One of the reasons seems to be that QTLs for such traits have generally small effect sizes, and linkage analysis does not seem to have sufficient sensitivity to detect them. Unfortunately, exercise biology traits belong to this category.

QTLs for Physical Activity Level

Several linkage studies focusing on physical activity level have been reported to date. The mouse crosses have yielded a number of significant and suggestive QTLs with little overlap among the studies.

An intercross of two inbred strains—one with low endurance capacity in the untrained state (COP strain) and the other with a high running capacity (DA strain) —was performed, and 224 segregating animals were obtained (233). A genome-wide linkage scan performed on these animals indicated a significant QTL on chromosome 16 with a maximum logarithm of odds (LOD) score of 4.0. In a follow-up study, chromosome 16 of DA rats (high running capacity) was transferred into the genetic background of COP rats (low running capacity). The congenic COP rats had significantly greater aerobic running capacity than the wild-type COP rats, indicating allelic variation on chromosome 16 that affects baseline endurance capacity, as observed in sedentary animals (234).

Timothy Lightfoot has made some interesting observations regarding the genetic basis of voluntary and spontaneous exercise using rodent models (figure 7.12). In one experiment, he and his colleagues intercrossed inbred mouse strains displaying high and low maximal exercise endurance capacity (116). A genome-wide linkage scan detected a significant QTL on the X chromosome with a maximum LOD score of 2.26 at 57.9 cM, suggesting that a gene encoded in that region influences aerobic capacity. Epistatic interactions among QTLs were also explored (113), and epistatic QTLs ac-

Figure 7.12 J. Timothy Lightfoot (1960–), PhD, FACSM, Omar Smith endowed professor and director of the Sydney and J.L. Huffines Institute for Sports Medicine and Human Performance at Texas A&M University. Lightfoot received his PhD from the University of Tennessee, completed a research consultantship with NASA in the biomedical laboratory at Kennedy Space Center, and then completed a 3 yr National Institutes of Health postdoctoral research fellowship in the division of physiology at Johns Hopkins University. Lightfoot is fascinated with why some people choose to be very active and some choose to be inactive, especially given that it is well known that physical activity prevents and ameliorates many diseases and chronic conditions. As such, his research is centered on discovering and understanding the genetic mechanisms that control daily physical activity. He used inbred strains of mice and cross-breeding experiments to identify genomic regions and polymorphisms associated with indicators of spontaneous voluntary activity. He was on the faculty at the University of North Carolina, Charlotte, before moving to Texas A&M University.

Photo courtesy of Timothy Lightfoot.

counted for a large fraction of the variance in physical activity traits.

More recently, Lightfoot and colleagues reported on a major effort to define the genetic map of the spontaneous level of activity in mice. Voluntary wheel-running activity was measured at about 9 wk in 41 inbred strains of mice (115). Of the 41 strains, 38 could be used for haplotype association mapping with more than 8 million SNPs as derived from the Perlegen Sciences database (63). Twelve QTLs were identified—three observed in males and females, five observed only in males, and four observed only in females. Of importance in this study, female mice ran on average 24% farther, 13% longer, and 11% faster on a daily basis than males. Three QTLs were identified for the distance run daily, and they remained significant under most analytical scenarios used. Surprisingly, no QTLs for duration of daily activity or for running speed that were significant in both sexes could be identified.

Thoroughbred horses have been selected for exceptional racing performance over the past 300 to 400 yr. It has been estimated that as much as 95% of the paternal lineage comes from a single founder stallion and that 10 founder mares contributed about 72% of the maternal lineage. The selection process over many generations has enhanced many traits that favor a high running performance, such as a large lung volume and high hemoglobin concentration, cardiac output, skeletal muscle mass, mitochondrial density, oxidative enzyme activity, and glycogen storage capacity. The average $\dot{V}O_2max$ in Thoroughbreds is greater than 200 $ml \cdot kg^{-1} \cdot min^{-1}$. Gu and colleagues performed a genome-wide scan to identify genomic regions linked to performance in Thoroughbreds based on 394 microsatellite markers genotyped in 112 Thoroughbreds and 52 non-Thoroughbred horses (77). Although this approach did not have sufficient sensitivity to identify specific genes, statistical significance was found for 17 genomic regions.

Four reports have dealt with genome-wide linkage scans for physical activity traits in human cohorts. In the Quebec Family Study, the strongest evidence of linkage was observed on chromosome 2p22-p16 for a measure of physical inactivity (205). Suggestive linkages were also found on 13q22 and 7p11 for activity and inactivity traits. Genome-wide linkage scans for participation in competitive sport in 700 female DZ twins (1,946 markers) and for exercise participation in Dutch sibling pairs (361 markers) indicated suggestive QTLs on chromosomes 3q22-q24, 4q31-q34, and 19p13.3 (49, 50), and a genome-wide linkage scan with a 10 cM microsatellite panel in the Viva La Familia Study revealed QTLs on chromosome 18q12.2-q21.1 for sedentary and light activities, with LOD scores of 4.07 and 2.79, respectively (35).

QTLs for Muscular Strength and Power

A genome-wide linkage scan for maximal walking speed and knee extensor performance was undertaken in 94 pairs of elderly DZ female twins from Finland (223). The strongest evidence of linkage was detected on chromosome 8q24 for leg-extensor power and on 15q13.3 for isometric knee-extensor strength. In the Leuven Genes for Muscular Strength Study, maximum isometric and torque-length knee-strength characteristics were measured in 283 male siblings aged 17 to 36 yr from 105 families (46). A genome-wide linkage scan using a panel of 6,008 SNPs provided evidence of linkage for 6 phenotypes on 13 genomic regions. The strongest QTL (LOD = 4.09) was detected on chromosome 14q24.3 for knee torque-length flexion. Additional QTLs for the same trait were found on 9q21.32 and 10q26 (47). Other linkage regions included chromosomes 1q21.3, 2p23.3, 6p25.2, and 18q11.2 for knee torque-length extension, chromosomes 2q14.4 and 15q23 for knee slope extension, chromosomes 4p14 and 18q23 for knee ratio extension, chromosome 2p24.2 for knee torque flexion, and chromosome 7p12.3 for knee ratio flexion (46, 47). There were no common QTLs between the Finnish and Belgian studies.

QTLs for Cardiorespiratory Fitness Phenotypes

In rats, a genome-wide linkage scan was performed using an intercross of two inbred strains, one with low running capacity (COP) and one with high running capacity (DA). A genome-wide linkage scan found a significant QTL on chromosome 16 at 30 cM with a maximum LOD score of 4.0. Suggestive QTLs were also detected on chromosomes 16 (62 cM; LOD = 2.9) and 3 (~4 cM; LOD = 2.2) (233). In a follow-up study, chromosome 16 of DA rats was introgressed into the genetic background of COP rats. The DA chromosome 16 congenic COP rats had significantly greater aerobic running capacity than the COP rats (234). Lightfoot and colleagues intercrossed inbred mouse strains displaying high and low maximal exercise endurance capacity (116). A genome-wide linkage scan detected a significant QTL on the X chromosome with a maximum LOD score of 2.26 at 57.9 cM.

In the HERITAGE Family Study, more than 500 microsatellite markers were used to identify genomic regions linked to $\dot{V}O_2max$ and maximal power output in sedentary subjects (28, 182). Suggestive evidence of linkage with $\dot{V}O_2max$ in the untrained state was detected on chromosomes 7q32, 7q36, and 11p15, and

QTLs for maximal power output were found on chromosomes 10q23, 13q33, and 18q11. In the same cohort, genome-wide linkage analysis was also used to identify genes for exercise-training response phenotypes.

After the laboratory of Claude Bouchard moved from Laval University to the Pennington Biomedical Research Center, he and Tuomo Rankinen united in the Human Genomics Laboratory to investigate the genetics of the ability to respond to regular exercise. Rankinen has led several genetic linkage and association projects over the past decade (figure 7.13). QTLs for training-induced changes in submaximal exercise (50 W) stroke volume (SV50) and heart rate (HR50) were found on chromosomes 10p11 and 2q33.3-q34, respectively (170, 211). The SV50 response QTL on 10p11 was narrowed down to a 7 Mb region using dense microsatellite mapping. Genes in the region were tested for associations by genotyping a dense panel of SNPs in the gene loci. The strongest associations were found with SNPs in the kinesin family member 5B (*KIF5B*) gene locus (5). The SNP that showed the strongest association with ΔSV50 was found to modify the *KIF5B* promoter activity. Further, inhibition and overexpression studies showed that changes in *KIF5B* expression level altered mitochondrial localization and biogenesis. *KIF5B* inhibition led to diminished biogenesis and perinuclear accumulation of mitochondria whereas overexpression enhanced mitochondrial biogenesis.

The QTL for HR50 response to training on chromosome 2q33.3-q34 was localized in a 10 Mb region, and the area was fine-mapped with a dense panel of almost 1,500 SNPs (171). The strongest evidence of association was detected with two SNPs located in the 5′-region of the cyclic adenosine monophosphate responsive element binding protein 1 (*CREB1*) gene, and the associations remained significant after accounting for multiple testing. The most significant SNP (rs2253206) explained almost 5% of the variance in training-induced changes in HR50, and the common allele homozygotes and heterozygotes had, respectively, about 57% and 20% greater decrease in HR50 than the minor allele homozygotes did. The same SNP, which is located about 2.6 kb upstream of the first exon of *CREB1*, was also shown to modify promoter activity in vitro; the A-allele, which was associated with a blunted HR50 response, showed significantly greater promoter activity in the C2C12 cell model than the G-allele did.

GWA Studies

The development of high-throughput SNP genotyping methods has markedly increased the ability to capture a large fraction of variation in the genome of large numbers of individuals at a reasonable cost. It is estimated that there are about 10 million SNPs with a minor allele frequency of 5% and more in the human genome. Common variants in close physical proximity are often transmitted together across generations as a unit or a block; therefore recombination among them is absent or infrequent. Therefore, it is possible to tag a set of SNPs by genotyping only the most representative SNP, which is commonly referred to as the tagSNP. As a result, genotyping 500,000 to 1 million SNPs has been considered sufficient to capture most of the common SNPs of the human genome. However, in the

Figure 7.13 Tuomo Rankinen (1967–), PhD, FACSM, associate professor, Human Genomics Laboratory, Pennington Biomedical Research Center, Baton Rouge, Louisiana. He received his MSc and PhD in clinical nutrition and exercise medicine from the University of Kuopio in Finland and completed a postdoctoral fellowship in exercise genetics at Université Laval in Sainte-Foy, Quebec, where he began his longstanding collaboration with Claude Bouchard. His research has focused on the health effects of regular physical activity and diet, with special emphasis on the genetic and molecular basis of human variation in responsiveness to regular exercise. Another area of interest is the effects of gene–physical activity interaction and gene–obesity interaction on risk factors for cardiovascular disease and type 2 diabetes, especially blood pressure and other hemodynamic phenotypes. He has published his findings in more than 200 peer-reviewed scientific papers, including several on the role of genes in blood pressure and cardiorespiratory fitness. He was the recipient of the New Investigator Award from the ACSM in 2001. Rankinen was the principal investigator of the HYPGENE project and is currently a principal investigator of the HERITAGE Family Study.

Photo courtesy of Tuomo Rankinen.

application of these guiding principles it is critical to take into account the important differences in allele frequencies and linkage disequilibrium patterns that have been observed among the major ethnic groups.

The ability to assay hundreds of thousands of DNA sequence variants in a single experiment has made GWA studies a reality. The first report based on this technology was published in 2005 (98), and research has since advanced at a very rapid pace. There are currently more than 600 published GWA studies covering 150 diseases and complex traits in the peer-reviewed literature. About 800 SNPs have been associated with a trait at the genome-wide significance level ($P < 5 \times 10^{-8}$) (127). The catalog of these studies and their findings, which is updated regularly, can be accessed at www.genome.gov/gwastudies.

In addition to SNPs and low-frequency or rare single base mutations, another source of human genomic polymorphisms comes in the form of copy number variants. The true value of copy number variants for physiological and metabolic traits continues to be a matter of debate. Overall, GWA studies have become an essential tool in the effort to identify SNPs contributing to human variation in complex biological and behavioral traits. However, few of the significant SNPs uncovered and replicated to date have been resolved at the gene or transcript level.

The first GWA study of interest to this chapter dealt with physical activity level. It was based on two cohorts: 1,644 unrelated individuals from the Netherlands Twin Register and 978 subjects living in Omaha, Nebraska, in the United States (48). Leisure-time physical activity level was quantified using questionnaires, and MET-hours were calculated based on the type, frequency, and duration of reported activities (excluding occupational work, commuting-related activities, and household chores, including gardening). Exercisers were defined as subjects who reported at least 4 MET-hours/wk. The exerciser versus nonexerciser classification was used as the primary phenotype for the GWA analyses. The final genotype data set included 1.6 million measured or imputed SNPs.

None of the 1.6 million SNPs reached the commonly used threshold of genome-wide significance ($P = 5 \times 10^{-8}$). However, SNPs in three genomic regions showed P-values less than 1×10^{-5}. The strongest associations were observed on chromosome 10q23.2 at the 3'-phosphoadenosine 5'-phosphosulfate synthase 2 (*PAPSS2*) gene locus. The odds ratio for being an exerciser was 1.32 ($P = 3.81 \times 10^{-6}$) for the common T-allele of SNP rs10887741. Mechanisms by which *PAPSS2* could affect exercise participation are unknown. The associations with previously reported physical activity candidate genes and physical activity linkage regions were explored as well, but they generated little evidence in support of significant associations.

The GWA approach is not restricted by a priori hypotheses, as is the case in candidate gene studies. Moreover, a GWA study covers almost the entire genome uniformly and at a much more dense level than microsatellite-based linkage scans. It has greater power to detect small gene effects compared with linkage scans. A critical feature of any genetics study is replication. This implies that the findings of an individual study must be replicated in other large cohorts with a similar phenotype and study design. If the associations are replicated and replicated again, the case for the contribution of a DNA sequence variant to the trait of interest becomes considerably stronger. Exercise biologists are just beginning to take advantage of this new technology, but we are likely to see a growing incorporation of GWA studies in exercise genomics.

Transgenic and Knockout Mice Studies

Overexpressing, knocking down, or knocking out a gene (or an allelic version of a gene) in skeletal muscle, cardiac muscle, or other tissues of mice are powerful strategies for illuminating the biology of exercise or exercise deprivation and for documenting the effects of genetic differences. These technologies have been used in numerous studies to address questions of relevance to exercise biologists. Detailed review of this dynamic field of research is beyond the scope of this chapter. Instead, focus will be on one example to illustrate the usefulness of transgenic mouse models.

In one experiment, peroxisome proliferator-activated receptor delta (*Ppard*) was overexpressed in skeletal muscle of mice (232). The overexpression of an activated form of PPARD in skeletal muscles increased the formation of Type I muscle fibers and improved running capacity in sedentary mice. Indeed, the transgenic mice were able to run as much as twice the distance that wild-type littermates could run. It is as if the mice had a higher level of fitness in the untrained state. Moreover, they were resistant to obesity induced by a high-fat diet and less prone to glucose intolerance. These observations have to be evaluated in view of the fact that endurance training as such promotes an increase in PPARD protein content in skeletal muscle of mice (119). Because the results suggest in the aggregate that PPARD is a molecular regulator of muscle fiber type distribution and potentially of endurance training-induced changes in muscle metabolism, it would be useful to undertake true genetics studies in order to determine whether allelic variations in *Ppard* contribute to the response to dynamic exercise.

Combining Transcriptomics and Genomics

The use of microarray technology to study the expression profile of a large number of transcripts has been a common tool of exercise molecular biology studies for about a decade. For instance, one report dealt with cardiac expression levels of 11,904 transcripts in middle-aged and old male mice derived from sedentary and spontaneously physically active breeding lines (33). They concluded that regular exercise may delay several aging-related changes in cardiac gene expression. In another study, the effects of unloading and low-intensity activity on the gene expression pattern of the soleus muscle were investigated in rats (12). The differentially regulated transcripts were related to genes involved in protein synthesis and degradation, transcription factors, and glucose metabolism.

Microarray technologies were rapidly extended to human experiments. For instance, the effects of a 9 wk strength-training program on vastus lateralis gene expression profile were studied in 20 sedentary subjects (188). The strength-training program consisted of performing unilateral knee-extension exercises of the dominant leg 3 times/wk. A total of 69 genes showed greater than 1.7-fold difference in expression levels from pre- to posttraining when all subjects were pooled. Fourteen of the genes were identified in all age-by-sex subgroups; 12 showed decreased expression levels after the training program and 2 showed increased expression levels. Another example comes from a study of the effects of resistance exercise on specific skeletal muscle transcripts of myogenic factors associated with myofiber hypertrophy (106, 131).

Many contributions in transcriptomics and proteomics applied to exercise biology came from the laboratory of Eric Hoffman at the Children's National Medical Center in Washington, D.C. (figure 7.14). Hoffman had earlier played a key role in the identification of dystrophin as the deficient gene in Duchesne muscular dystrophy. He and M.J. Hubal from his laboratory partnered with a number of exercise physiologists (P.D. Thompson, P.M. Clarkson [deceased], L.S. Pescatello, and others) to investigate skeletal muscle gene expression and SNPs as determinants of skeletal muscle strength in the untrained state and in response to resistance-exercise training in the FAMuSS project (39, 89, 222) as well as skeletal muscle traits in men and women exhibiting the metabolic syndrome in the STRRIDE study (with W.E. Kraus from Duke University School of Medicine as the principal investigator) (83). As an example of integration between transcript abundance, protein content and localization, and DNA differences, Hoffman and colleagues undertook a series of studies focused on muscle-damage mechanisms induced by eccentric exercise (38, 87, 88). Inflammatory factors were shown to be upregulated after eccentric exercise, and three genes—chemokine ligand 2 (*CCL2*); ZFP36 ring finger protein (*ZFP36*); and CCAAT/enhancer binding protein, delta (*CEBPD*)—exhibited upregulation with the first exercise bout and further upregulation with the second exercise session. Additional data supported a role for *CCL2* in muscle repair as opposed to muscle damage (86, 88).

Figure 7.14 Eric Hoffman (1958–), PhD, chairman of the department of integrative systems biology at George Washington University School of Medicine and Health Sciences and director of the Research Center for Genetic Medicine at Children's National Medical Center, Washington, D.C. Hoffman obtained his PhD from Johns Hopkins University and completed postdoctoral training at Harvard Medical School. He came to the exercise biology field from muscular dystrophy, where he had identified the protein defect in its common form, which turned out to be dystrophin in Duchenne muscular dystrophy. Working with muscle biopsies from dystrophy patients made his lab adept at messenger RNA profiling applications to muscle. Later, he teamed up with noted exercise biologists Priscilla Clarkson and Paul Thompson to study the genetics and genomics of exercise in normal volunteers. He then led a study of the genetics of muscle strength, size, and response to resistance training in a large sample of college-age volunteers; the study is known as FAMuSS. He has published about 300 peer-reviewed papers and recently coedited with Claude Bouchard a volume titled *Genetic and Molecular Aspects of Sports Performance* in the International Olympic Committee Medical Commission Publication series. Hoffman runs a state-of-the-art genetics research laboratory with expertise in genomics, transcriptomics, and proteomics. His expertise and interest in systems biology led to his recent appointment as chairman of the department of integrative systems biology at George Washington University.

Photo courtesy of Eric Hoffman.

The gene expression profiling studies were subsequently taken one step further to generate new candidate genes to be tested for the role of allelic variation. The first of such studies in the field of exercise biology was based on a subsample of the HERITAGE Family Study cohort of whites. It focused on transcripts associated with the exercise-training response of insulin sensitivity as derived from frequent-sampling intravenous glucose tolerance tests and the Bergman minimal model (216). Total RNA was extracted before and after the exercise program from vastus lateralis muscle samples of 8 subjects who were high responders and 8 age-, sex-, and BMI-matched insulin sensitivity nonresponders out of 78 subjects with muscle biopsies. RNA samples were pooled within each responder group, labeled with fluorescent dyes, and hybridized onto in situ-generated microarrays containing 18,861 transcripts. A total of 47 transcripts were differentially expressed (at least a difference of +40% or −40%) between both groups at baseline, whereas another 361 transcripts showed differential expression posttraining. Five genes—v-ski sarcoma viral oncogene homolog (*SKI*); four and a half LIM domains 1 (*FHL1*); titin (*TTN*); pyruvate dehydrogenase kinase, isozyme 4 (*PDK4*); and C-terminal binding protein 1 (*CTBP1*)—that exhibited at least a 50% difference in expression between high responders and nonresponders either at baseline or posttraining were selected for validation experiments. Quantitative real-time polymerase chain reaction confirmed the microarray-based expression patterns for all except *CTBP1*. Three SNPs were genotyped in the *FHL1* gene encoded on Xq26 and tested for associations with exercise training-induced changes in insulin metabolism phenotypes (217). SNP rs9018 was associated with disposition index (P = .016) and glucose disappearance index (P = .008) changes in white females, and a suggestive association existed with fasting insulin training response in white males (P = .04). Another SNP (rs2180062) was associated with fasting insulin (P = .012), insulin sensitivity (P = .046), disposition index (P = .006), and glucose disappearance index (P = .03) training responses in white males (217). It appears that *FHL1* is the first gene (encoded on X) that allows researchers to distinguish between those who respond favorably to regular exercise and those who exhibit an adverse response pattern in terms of insulin and glucose homeostasis indicators.

Recently, the strategy of combining transcriptomics and genomics was taken a step further. James A. Timmons and colleagues used a combination of global skeletal muscle gene expression profiling and DNA sequence variants to identify genes associated with $\dot{V}O_2max$ training response (224). RNA expression profiling of pretraining skeletal muscle samples identified a panel of 29 transcripts that were strongly associated with $\dot{V}O_2max$ training response in a study of 24 sedentary males trained for 8 wk and in a replication training study. Next, tagSNPs of the 29 predictor transcripts were genotyped in the HERITAGE Family Study. A multivariable regression analysis using the predictor gene SNPs and a set of SNPs from positional cloning and candidate gene studies of the HERITAGE Family Study identified a set of 11 SNPs that explained 23% of the variance in $\dot{V}O_2max$ training response. Of these 11 SNPs, 7 were from the RNA predictor transcript set and 4 were from the HERITAGE QTLs and candidate genes projects. In the editorial published along with the paper of Timmons and colleagues in *Journal of Applied Physiology*, Marcas Bamman noted that the research was truly an outstanding contribution that will inform and inspire future directions (6).

Timmons and colleagues have also recently reported on a training-responsive transcriptome (TRT) and the key regulatory molecules that coordinate this complex network of transcripts (94). Runt-related transcription factor 1 (RUNX1), SRY-box 9 (SOX9), and paired box 3 (PAX3) transcription factor binding sites are overrepresented in the TRT. At least 100 of the 800 TRT transcripts were differentially regulated between low and high responders to aerobic training. Proangiogenic and tissue-development networks were among the strongest candidates to serve as regulators of adaptation to endurance training. Furthermore, several high-responder transcripts exhibited DNA sequence variants in the HERITAGE Family Study cohort of whites and those associated with the $\dot{V}O_2max$ response to exercise training.

Even though no genetic publication has yet been generated, the Training Interventions and Genetics of Exercise Response (TIGER) project, which started in 1993, is designed to investigate candidate genes and gene expression patterns as predictors of success in achieving long-term changes in body composition and exercise adherence (see NCT01127919 at clinicaltrials.gov). Subjects are recruited from the student population of the University of Houston, Texas, and the goal is to enroll a total of 35,000 students. The principal investigator of TIGER is Molly Bray, who is currently at the University of Texas at Austin in the United States.

The small body of data available to date strongly suggests that understanding the profile of acute exercise- or training-related changes in gene expression, its time-course dependence, and age-related variation can provide panels of new and exciting candidates for

Table 7.1 Milestone Studies in Exercise Genetics or Genomics Research

Year	Investigator(s)	Contribution	Reference
1971	Klissouras	Heritability of $\dot{V}O_2max$ in twins	100
1972	Skład	Twin studies of strength and performance	209
1984	Prudhomme et al.	Genotype–training interaction in $\dot{V}O_2max$ gains	169
1986	Simoneau et al.	Genotype–training interactions on anaerobic capacity and muscle metabolism gains	203
1999	Bouchard et al.	Familial aggregation of $\dot{V}O_2max$ response to training	15
2000	Bouchard et al.	Genome-wide linkage scan for $\dot{V}O_2max$ and its response to training	28
2002	Roth et al.	Gene expression profiling and muscle strength response to resistance training	188
2009	De Moor et al.	Genome-wide association for sport participation	48
2010	Timmons et al.	Transcriptomic and genomic predictors of $\dot{V}O_2max$ response to training	224
2011	Bouchard et al.	Genomic predictors of the maximal O_2 uptake response to standardized exercise training programs	252

genetics studies of questions central to exercise physiology.

Future Directions

This chapter on the historical landmarks of the field of exercise genetics and exercise genomics is not intended to be comprehensive. The focus is on research design, technologies, and major findings that are of particular interest to exercise physiologists. Historically, it all began with twin, family, and pedigree observational studies. The focus expanded from genetic epidemiology studies to DNA sequence variation and the role it plays in exercise behavior, cardiorespiratory fitness in the untrained state, muscular strength, cardiovascular and metabolic adaptation to acute exercise, and responsiveness to regular endurance or resistance exercise. This chapter reviewed evidence from animal models and human studies, with an emphasis on candidate genes, genome-wide linkage results, GWA findings, expression arrays, and combinations of these approaches. Table 7.1 lists nine key studies and their research questions. Each of these papers represented the first publication in which a new research paradigm or new technology was applied to exercise genomics or genetics.

We end this journey by providing a few comments on issues that need to be addressed in order for the field of exercise genetics and genomics to move forward. A common weakness of the early genomics studies on human exercise was that they were based on small sample sizes and thus were grossly underpowered. Because most of the sequence variants seem to have small effect sizes, it is particularly critical to have ample statistical power to have a serious probability of identifying them. This has been recognized as a major issue by all those working in this field of research. One can now identify many laboratories that are actively engaged in exercise genomics, which bodes well and should favor the establishment of large cohorts of individuals phenotyped for appropriate exercise physiology-related traits through collaborative efforts from multiple laboratories. The greatest challenge will be to develop adequate resources for exercise genomics studies of the response to regular exercise because compliance with the exercise regimen needs to be of the highest quality for this type of research to be successful (29).

Currently, a hotly debated topic is how much of the genetic architecture of common complex traits is attributable to common and rare DNA sequence variants. The common variant hypothesis states that a trait is affected by several common DNA variants, each with minor effect size, whereas the rare variant scenario proposes that a large number of rare, but relatively high-impact, variants contribute to the trait variance. The fact that common SNPs found to be associated with common diseases in the GWA studies do not fully explain the trait heritability has renewed the interest in the rare variant hypothesis. The next-generation sequencing techniques allow for an adequate exploration of the rare variant hypothesis. In addition, copy number variants encompassing deletions, insertions, and duplications that range in size from a few thousand to several millions base pairs are thought to contribute to common human traits. Moving forward, exercise genomics needs to incorporate them all.

Human exercise genomics will undoubtedly generate new gene targets, of which further validation in human studies will be challenging. Validating a new gene target and defining the contributions of specific alleles require in silico studies; cell-based investigations; informative strains of rodents; transgenic, knockdown, or knockout mice; selective breeding for the level of expression of the targeted gene; and other appropriate tools depending on the gene and pathways involved. Here the rich experience of exercise physiologists who have devoted decades of research to in vitro studies and animal experimentation is an invaluable resource. Moreover, there is an urgent need for transcriptomics, proteomics, and metabolomics to be incorporated into exercise genomics research, with the goal of illuminating genetic associations and understanding the mechanisms driving human heterogeneity.

One area that is missing from this chapter is the epigenomic events that may contribute to human heterogeneity in the biological response to exercise, particularly their interaction with human variation in DNA sequences. The topic was not covered because there is, as of yet, little to report. But we expect that this will become a rapidly growing research area in exercise biology. This is exemplified by two recent papers that have addressed relevant questions. One focused on the effects of an acute exercise bout on skeletal muscle histone deacetylases and repression of transcription (135). The second dealt with the effects of regular exercise on methylation at the *ASC* gene locus in young and older subjects based on peripheral blood DNA extracts (148). The official name of the *ASC* gene is *PYCARD*, an adaptor protein comprising two protein–protein interaction domains, an N-terminal PYRIN-PAAD-DAPIN domain, and a C-terminal caspase-recruitment domain. These two papers raise a number of issues that will undoubtedly be the topics of much research and numerous debates over the next decade.

The next generation of advances in the guidelines on exercise in preventive and therapeutic medicine will have to be grounded in the principles of personalized medicine. Success will be measured to a large extent by our ability to identify the favorable and adverse responders for given physiological traits to given exercise regimens. This challenge cannot be met without more and better exercise genomics and genetics research.

References

1. Almasy L, Blangero J. Contemporary model-free methods for linkage analysis. *Adv Genet* 60: 175-193, 2008.
2. An P, Perusse L, Rankinen T, Borecki IB, Gagnon J, Leon AS, Skinner JS, Wilmore JH, Bouchard C, Rao DC. Familial aggregation of exercise heart rate and blood pressure in response to 20 weeks of endurance training: The HERITAGE family study. *Int J Sports Med* 24: 57-62, 2003.
3. An P, Rice T, Gagnon J, Leon AS, Skinner JS, Bouchard C, Rao DC, Wilmore JH. Familial aggregation of stroke volume and cardiac output during submaximal exercise: The HERITAGE Family Study. *Int J Sports Med* 21: 566-572, 2000.
4. Arden NK, Spector TD. Genetic influences on muscle strength, lean body mass, and bone mineral density: A twin study. *J Bone Miner Res* 12: 2076-2081, 1997.
5. Argyropoulos G, Stutz AM, Ilnytska O, Rice T, Teran-Garcia M, Rao DC, Bouchard C, Rankinen T. KIF5B gene sequence variation and response of cardiac stroke volume to regular exercise. *Physiol Genom* 36: 79-88, 2009.
6. Bamman MM. Does your (genetic) alphabet soup spell "runner"? *J Appl Physiol* 108: 1452-1453, 2010.
7. Barbato JC, Koch LG, Darvish A, Cicila GT, Metting PJ, Britton SL. Spectrum of aerobic endurance running performance in eleven inbred strains of rats. *J Appl Physiol* 85: 530-536, 1998.
8. Berman Y, North KN. A gene for speed: The emerging role of alpha-actinin-3 in muscle metabolism. *Physiology (Bethesda)* 25: 250-259, 2010.
9. Beunen G, Thomis M. Gene powered? Where to go from heritability (h2) in muscle strength and power? *Exerc Sport Sci Rev* 32: 148-154, 2004.
10. Beunen G, Thomis M. Genetics of strength and power characteristics in children and adolescents. *Pediatr Exerc Sci* 15: 128-138, 2003.
11. Beunen GP, Peeters MW, Maes HH, Loos RJF, Claessens AL, Derom C, Vlietinck R, Thomis MA. The Leuven Longitudinal Twin Study (LLTS): Major findings. *Twin Res Hum Genet* 10: 15-18, 2007.
12. Bey L, Akunuri N, Zhao P, Hoffman EP, Hamilton DG, Hamilton MT. Patterns of global gene expression in rat skeletal muscle during unloading and low-intensity ambulatory activity. *Physiol Genom* 13: 157-167, 2003.
13. Bielen EC, Fagard RH, Amery AK. Inheritance of acute cardiac changes during bicycle exercise: An echocardiographic study in twins. *Med Sci Sports Exerc* 23: 1254-1259, 1991.
14. Bossik LJ. K voprosu o roli nasledsvennosti i sredi v fiziologii i patologii detskova vovrasta. *Trudy Medytsinko-Biologicheskova Instituta* 3: 33-56, 1934. (as cited by Sklad, 2008.)

15. Bouchard C, An P, Rice T, Skinner JS, Wilmore JH, Gagnon J, Perusse L, Leon AS, Rao DC. Familial aggregation of V̇O2max response to exercise training: Results from the HERITAGE Family Study. *J Appl Physiol* 87: 1003-1008, 1999.
16. Bouchard C, Chagnon M, Thibault MC, Boulay MR, Marcotte M, Cote C, Simoneau JA. Muscle genetic variants and relationship with performance and trainability. *Med Sci Sports Exerc* 21: 71-77, 1989.
17. Bouchard C, Chagnon M, Thibault MC, Boulay MR, Marcotte M, Simoneau JA. Absence of charge variants in human skeletal muscle enzymes of the glycolytic pathway. *Hum Genet* 78: 100, 1988.
18. Bouchard C, Daw EW, Rice T, Perusse L, Gagnon J, Province MA, Leon AS, Rao DC, Skinner JS, Wilmore JH. Familial resemblance for V̇O2max in the sedentary state: The HERITAGE family study. *Med Sci Sports Exerc* 30: 252-258, 1998.
19. Bouchard C, Dionne FT, Simoneau JA, Boulay MR. Genetics of aerobic and anaerobic performances. *Exerc Sport Sci Rev* 20: 27-58, 1992.
20. Bouchard C, Hoffman EP, eds. *Genetic and Molecular Aspects of Sports Performance*. West Sussex, UK: International Olympic Committee/Blackwell, 2011.
21. Bouchard C, Leon AS, Rao DC, Skinner JS, Wilmore JH, Gagnon J. The HERITAGE family study. Aims, design, and measurement protocol. *Med Sci Sports Exerc* 27: 721-729, 1995.
22. Bouchard C, Lesage R, Lortie G, Simoneau JA, Hamel P, Boulay MR, Perusse L, Theriault G, Leblanc C. Aerobic performance in brothers, dizygotic and monozygotic twins. *Med Sci Sports Exerc* 18: 639-646, 1986.
23. Bouchard C, Lortie G, Simoneau JA, Leblanc C, Theriault G, Tremblay A. Submaximal power output in adopted and biological siblings. *Ann Hum Biol* 11: 303-309, 1984.
24. Bouchard C, Malina RM. Genetics for the sport scientist: Selected methodological considerations. *Exerc Sport Sci Rev* 11: 275-305, 1983.
25. Bouchard C, Malina RM. Genetics of physiological fitness and motor performance. *Exerc Sport Sci Rev* 11: 306-339, 1983.
26. Bouchard C, Malina RM, Perusse L. *Genetics of Fitness and Physical Performance*. Champaign, IL: Human Kinetics, 1997.
27. Bouchard C, Perusse L, Leblanc C. Using MZ twins in experimental research to test for the presence of a genotype-environment interaction effect. *Acta Genet Med Gemel* 39: 85-89, 1990.
28. Bouchard C, Rankinen T, Chagnon YC, Rice T, Perusse L, Gagnon J, Borecki I, An P, Leon AS, Skinner JS, Wilmore JH, Province M, Rao DC. Genomic scan for maximal oxygen uptake and its response to training in the HERITAGE Family Study. *J Appl Physiol* 88: 551-559, 2000.
29. Bouchard C, Rankinen T, Timmons JA. Genomics and genetics in the biology of adaptation to exercise. *Compr Physiol* 1: 1603-1648, 2011.
30. Bouchard C, Tremblay A, Despres JP, Theriault G, Nadeau A, Lupien PJ, Moorjani S, Prudhomme D, Fournier G. The response to exercise with constant energy intake in identical twins. *Obes Res* 2: 400-410, 1994.
31. Bray MS, Hagberg JM, Perusse L, Rankinen T, Roth SM, Wolfarth B, Bouchard C. The human gene map for performance and health-related fitness phenotypes: The 2006-2007 update. *Med Sci Sports Exerc* 41: 35-73, 2009.
32. Brody D. Twin resemblance in mechanical ability, with reference to the effects of practice on performance. *Child Dev* 8: 207-216, 1937.
33. Bronikowski AM, Carter PA, Morgan TJ, Garland T Jr., Ung N, Pugh TD, Weindruch R, Prolla TA. Lifelong voluntary exercise in the mouse prevents age-related alterations in gene expression in the heart. *Physiol Genom* 12: 129-138, 2003.
34. Buxens A, Ruiz JR, Arteta D, Artieda M, Santiago C, Gonzalez-Freire M, Martinez A, Tejedor D, Lao JI, Gomez-Gallego F, Lucia A. Can we predict top-level sports performance in power vs endurance events? A genetic approach. *Scand J Med Sci Sports* 21: 570-579, 2011.
35. Cai G, Cole SA, Butte N, Bacino C, Diego V, Tan K, Goring HH, O'Rahilly S, Farooqi IS, Comuzzie AG. A quantitative trait locus on chromosome 18q for physical activity and dietary intake in Hispanic children. *Obesity* 14: 1596-1604, 2006.
36. Calvo M, Rodas G, Vallejo M, Estruch A, Arcas A, Javierre C, Viscor G, Ventura JL. Heritability of explosive power and anaerobic capacity in humans. *Eur J Appl Physiol* 86: 218-225, 2002.
37. Chagnon YC, Allard C, Bouchard C. Red blood cell genetic variation in Olympic endurance athletes. *J Sports Sci* 2: 121-129, 1984.
38. Chen YW, Hubal MJ, Hoffman EP, Thompson PD, Clarkson PM. Molecular responses of human muscle to eccentric exercise. *J Appl Physiol* 95: 2485-2494, 2003.
39. Clarkson PM, Devaney JM, Gordish-Dressman H, Thompson PD, Hubal MJ, Urso M, Price TB, Angelopoulos TJ, Gordon PM, Moyna NM, Pescatello LS, Visich PS, Zoeller RF, Seip RL, Hoffman EP. ACTN3 genotype is associated with increases in muscle strength in response to resistance training in women. *J Appl Physiol* 99: 154-163, 2005.
40. Corsi AM, Ferrucci L, Gozzini A, Tanini A, Brandi ML. Myostatin polymorphisms and age-related sarcopenia in the Italian population. *J Am Geriatr Soc* 50: 1463, 2002.
41. Couture L, Chagnon M, Allard C, Bouchard C. More on red blood cell genetic variation in Olympic athletes. *Can J Appl Sport Sci* 11: 16-18, 1986.
42. Cratty BJ. A comparison of fathers and sons in physical ability. *Res Quart* 31: 12-15, 1960.

43. Dales RJ. Motor and language development of twins during the first three years. *J Genet Psychol* 114: 263-271, 1969.
44. Danis A, Kyriazis Y, Klissouras V. The effect of training in male prepubertal and pubertal monozygotic twins. *Eur J Appl Physiol* 89: 309-318, 2003.
45. de Garay AL, Levine L, Carter JEL, Program of Genetics and Human Biology. *Genetic and Anthropological Studies of Olympic Athletes*. New York: Academic Press, 1974.
46. De Mars G, Windelinckx A, Huygens W, Peeters MW, Beunen GP, Aerssens J, Vlietinck R, Thomis MA. Genome-wide linkage scan for contraction velocity characteristics of knee musculature in the Leuven Genes for Muscular Strength Study. *Physiol Genom* 35: 36-44, 2008.
47. De Mars G, Windelinckx A, Huygens W, Peeters MW, Beunen GP, Aerssens J, Vlietinck R, Thomis MA. Genome-wide linkage scan for maximum and length-dependent knee muscle strength in young men: Significant evidence for linkage at chromosome 14q24.3. *J Med Genet* 45: 275-283, 2008.
48. De Moor MH, Liu YJ, Boomsma DI, Li J, Hamilton JJ, Hottenga JJ, Levy S, Liu XG, Pei YF, Posthuma D, Recker RR, Sullivan PF, Wang L, Willemsen G, Yan H, De Geus EJC, Deng HW. Genome-wide association study of exercise behavior in Dutch and American adults. *Med Sci Sports Exerc* 41: 1887-1895, 2009.
49. De Moor MH, Posthuma D, Hottenga JJ, Willemsen G, Boomsma DI, De Geus EJ. Genome-wide linkage scan for exercise participation in Dutch sibling pairs. *Eur J Hum Genet* 15: 1252-1259, 2007.
50. De Moor MH, Spector TD, Cherkas LF, Falchi M, Hottenga JJ, Boomsma DI, De Geus EJ. Genome-wide linkage scan for athlete status in 700 British female DZ twin pairs. *Twin Res Hum Genet* 10: 812-820, 2007.
51. Delmonico MJ, Kostek MC, Doldo NA, Hand BD, Walsh S, Conway JM, Carignan CR, Roth SM, Hurley BF. Alpha-actinin-3 (ACTN3) R577X polymorphism influences knee extensor peak power response to strength training in older men and women. *J Gerontol A Biol Sci Med Sci* 62: 206-212, 2007.
52. Devaney JM, Hoffman EP, Gordish-Dressman H, Kearns A, Zambraski E, Clarkson PM. IGF-II gene region polymorphisms related to exertional muscle damage. *J Appl Physiol* 102: 1815-1823, 2007.
53. Devor EJ, Crawford MH. A commingling analysis of quantitative neuromuscular performance in a Kansas Mennonite community. *Am J Phys Anthropol* 63: 29-37, 1984.
54. Doring F, Onur S, Fischer A, Boulay MR, Perusse L, Rankinen T, Rauramaa R, Wolfarth B, Bouchard C. A common haplotype and the Pro582Ser polymorphism of the hypoxia-inducible factor-1alpha (HIF1A) gene in elite endurance athletes. *J Appl Physiol* 108: 1497-1500, 2010.
55. Druzhevskaya AM, Ahmetov II, Astratenkova IV, Rogozkin VA. Association of the ACTN3 R577X polymorphism with power athlete status in Russians. *Eur J Appl Physiol* 103: 631-634, 2008.
56. Engstrom LM, Fischbein S. Physical capacity in twins. *Acta Genet Med Gemellol (Roma)* 26: 159-165, 1977.
57. Eynon N, Duarte JA, Oliveira J, Sagiv M, Yamin C, Meckel Y, Sagiv M, Goldhammer E. *ACTN3* R577X polymorphism and Israeli top-level athletes. *Int J Sports Med* 30: 695-698, 2009.
58. Fagard R, Bielen E, Amery A. Heritability of aerobic power and anaerobic energy generation during exercise. *J Appl Physiol* 70: 357-362, 1991.
59. Fenty-Stewart N, Park JY, Roth SM, Hagberg JM, Basu S, Ferrell RE, Brown MD. Independent and combined influence of AGTR1 variants and aerobic exercise on oxidative stress in hypertensives. *Blood Press* 18: 204-212, 2009.
60. Ferrell RE, Conte V, Lawrence EC, Roth SM, Hagberg JM, Hurley BF. Frequent sequence variation in the human myostatin (GDF8) gene as a marker for analysis of muscle-related phenotypes. *Genomics* 62: 203-207, 1999.
61. Festing MF. Wheel activity in 26 strains of mouse. *Lab Anim* 11: 257-258, 1977.
62. Fox PW, Hershberger SL, Bouchard TJ Jr. Genetic and environmental contributions to the acquisition of a motor skill. *Nature* 384: 356-358, 1996.
63. Frazer KA, Eskin E, Kang HM, Bogue MA, Hinds DA, Beilharz EJ, Gupta RV, Montgomery J, Morenzoni MM, Nilsen GB, Pethiyagoda CL, Stuve LL, Johnson FM, Daly MJ, Wade CM, Cox DR. A sequence-based variation map of 8.27 million SNPs in inbred mouse strains. *Nature* 448: 1050-1053, 2007.
64. Freedman DG, Keller B. Inheritance of behavior in infants. *Science* 140: 196-198, 1963.
65. Galton F. The history of twins as a criterion of the relative powers of nature and nurture. *Royal Anthropol Inst Great Brit Ireland J* 6: 391-406, 1876.
66. Gaskill SE, Rice T, Bouchard C, Gagnon J, Rao DC, Skinner JS, Wilmore JH, Leon AS. Familial resemblance in ventilatory threshold: The HERITAGE Family Study. *Med Sci Sports Exerc* 33: 1832-1840, 2001.
67. Gedda L. Genetic evaluation of athletes [in Italian]. *Acta Gerontol (Milano)* 4: 249-260, 1955.
68. Gedda L. Sports and genetics. A study on twins (351 pairs). *Acta Genet Med Gemellol (Roma)* 9: 387-406, 1960.
69. Gesell A, Thompson H. Learning and growth in identical infant twins: An experimental study by the method of co-twin control. *Genet Psychol Monogr* 6: 1-124, 1929.
70. Gifford S, Murawski BJ, Brazelton TB, Young GC. Differences in individual development within a pair of identical twins. *Int J Psychoanal* 47: 261-268, 1966.
71. Gomez-Gallego F, Ruiz JR, Buxens A, Artieda M, Arteta D, Santiago C, Rodriguez-Romo G, Lao JI, Lucia A. The -786 T/C polymorphism of the NOS3

gene is associated with elite performance in power sports. *Eur J Appl Physiol* 107: 565-569, 2009.

72. Goring HH, Curran JE, Johnson MP, Dyer TD, Charlesworth J, Cole SA, Jowett JB, Abraham LJ, Rainwater DL, Comuzzie AG, Mahaney MC, Almasy L, MacCluer JW, Kissebah AH, Collier GR, Moses EK, Blangero J. Discovery of expression QTLs using large-scale transcriptional profiling in human lymphocytes. *Nat Genet* 39: 1208-1216, 2007.
73. Goya T, Amano Y, Hoshikawa T, Matsui H. Longitudinal study on selected sports performance related with the physical growth and development of twins. In: *XIIIth International Congress of Biomechanics, Book of Abstracts*. Perth: University of Western Australia, 1991, 139-141.
74. Goya T, Amano Y, Hoshikawa T, Matsui H. Longitudinal study on the variation and development of selected sports performance in twins: Case study for one pair of female monozygous (MZ) and dizygous (DZ) twins. *Sports Sci* 14: 151-168, 1993.
75. Grebe H. Families of athletes. *Acta Gerontol (Milano)* 5: 318-326, 1956.
76. Grebe H. Sport activities of twins. *Acta Gerontol (Milano)* 4: 275-296, 1955.
77. Gu J, Orr N, Park SD, Katz LM, Sulimova G, MacHugh DE, Hill EW. A genome scan for positive selection in thoroughbred horses. *PLoS ONE* 4: e5767, 2009.
78. Hagberg JM. Genes, exercise, and cardiovascular phenotypes. In: Bouchard C, Hoffman EP, eds. *Genetic and Molecular Aspects of Sports Performance*. West Sussex, UK: International Olympic Committee/Blackwell, 2011, 249-261.
79. Hagberg JM, Ferrell RE, McCole SD, Wilund KR, Moore GE. VO2max is associated with *ACE* genotype in postmenopausal women. *J Appl Physiol* 85: 1842-1846, 1998.
80. Hamel P, Simoneau JA, Lortie G, Boulay MR, Bouchard C. Heredity and muscle adaptation to endurance training. *Med Sci Sports Exerc* 18: 690-696, 1986.
81. Hand BD, McCole SD, Brown MD, Park JJ, Ferrell RE, Huberty A, Douglass LW, Hagberg JM. NOS3 gene polymorphisms and exercise hemodynamics in postmenopausal women. *Int J Sports Med* 27: 951-958, 2006.
82. Hilgard JR. The effect of early and delayed practice on memory and motor performance studied by the method of co-twin control. *Genet Psychol Monogr* 14: 493-567, 1933.
83. Hittel DS, Kraus WE, Tanner CJ, Houmard JA, Hoffman EP. Exercise training increases electron and substrate shuttling proteins in muscle of overweight men and women with the metabolic syndrome. *J Appl Physiol* 98: 168-179, 2005.
84. Holzinger KJ. The relative effect of nature and nurture influences on twin differences. *J Educ Psychol* 20: 241-248, 1929.
85. Howald H. Ultrstructure and biochemical function of skeletal muscle in twins. *Ann Hum Biol* 3: 455-462, 1976.
86. Hubal M, Wang Z, Hoffman EP. Systems biology through time series data—A strength of muscle remodeling. In: Bouchard C, Hoffman EP, eds. *Genetic and Molecular Aspects of Sports Performance*. West Sussex, UK: International Olympic Committee/Blackwell, 2011, 319-329.
87. Hubal MJ, Chen TC, Thompson PD, Clarkson PM. Inflammatory gene changes associated with the repeated-bout effect. *Am J Physiol Regul Integr Comp Physiol* 294: R1628-R1637, 2008.
88. Hubal MJ, Devaney JM, Hoffman EP, Zambraski EJ, Gordish-Dressman H, Kearns AK, Larkin JS, Adham K, Patel RR, Clarkson PM. CCL2 and CCR2 polymorphisms are associated with markers of exercise-induced skeletal muscle damage. *J Appl Physiol* 108: 1651-1658, 2010.
89. Hubal MJ, Gordish-Dressman H, Thompson PD, Price TB, Hoffman EP, Angelopoulos TJ, Gordon PM, Moyna NM, Pescatello LS, Visich PS, Zoeller RF, Seip RL, Clarkson PM. Variability in muscle size and strength gain after unilateral resistance training. *Med Sci Sports Exerc* 37: 964-972, 2005.
90. Ishidoya Y. Sportfähhigeit de Zwillinge. *Acta Genet Med Gemellol (Roma)* 6: 321-326, 1957.
91. Johnson RC, McClearn GE, Yuen S, Nagoshi CT, Ahern FM, Cole RE. Galton's data a century later. *Am Psychol* 40: 875-892, 1985.
92. Jokl E. The contribution of twin research to the physiology of exercise. *Acta Genet Med Gemellol (Milano)* 5: 115-122, 1956.
93. Jokl E, Jokl P, International Council of Sport and Physical Education Research Committee. *The Physiological Basis of Athletic Records*. Springfield, IL: Charles C. Thomas, 1968.
94. Keller P, Vollaard NB, Gustafsson T, Gallagher IJ, Sundberg CJ, Rankinen T, Britton SL, Bouchard C, Koch LG, Timmons JA. A transcriptional map of the impact of endurance exercise training on skeletal muscle phenotype. *J Appl Physiol* 110: 46-59, 2011.
95. Kelly SA, Nehrenberg DL, Peirce JL, Hua K, Steffy BM, Wiltshire T, Pardo-Manuel de Villena F, Garland T Jr., Pomp D. Genetic architecture of voluntary exercise in an advanced intercross line of mice. *Physiol Genom* 42: 190-200, 2010.
96. Kimura K. The study on physical ability of children and youths: On twins in Osaka City [in Japanese]. *Jinrui-Gaku Zasashi (Anthropological Society of Nippon)* 64: 172-196, 1956.
97. Kirkton SD, Howlett RA, Gonzalez NC, Giuliano PG, Britton SL, Koch LG, Wagner HE, Wagner PD. Continued artificial selection for running endurance in rats is associated with improved lung function. *J Appl Physiol* 106: 1810-1818, 2009.
98. Klein RJ, Zeiss C, Chew EY, Tsai JY, Sackler RS, Haynes C, Henning AK, SanGiovanni JP, Mane SM, Mayne ST, Bracken MB, Ferris FL, Ott J, Barn-

stable C, Hoh J. Complement factor H polymorphism in age-related macular degeneration. *Science* 308: 385-389, 2005.

99. Klissouras V. Genetic limit of functional adaptability. *Int Z Angew Physiol* 30: 85-94, 1972.
100. Klissouras V. Heritability of adaptive variation. *J Appl Physiol* 31: 338-344, 1971.
101. Klissouras V, Pirnay F, Petit JM. Adaptation to maximal effort: Genetics and age. *J Appl Physiol* 35: 288-293, 1973.
102. Koch LG, Meredith TA, Fraker TD, Metting PJ, Britton SL. Heritability of treadmill running endurance in rats. *Am J Physiol* 275: R1455-R1460, 1998.
103. Komi PV, Karlsson J. Physical performance, skeletal muscle enzyme activities, and fibers types of monozygous and dizygous twins of both sexes. *Acta Physiol Scand* 462(Suppl.): 1-28, 1979.
104. Komi PV, Klissouras V, Karvinen E. Genetic variation in neuromuscular performance. *Int Z Angew Physiol* 31: 289-304, 1973.
105. Komi PV, Viitasalo JHT, Havu M, Thorstensson A, Sjodin B, Karlsson J. Skeletal muscle fibers and muscle enzyme activities in monozygous and dizygous twins of both sexes. *Acta Physiol Scand* 100: 385-392, 1977.
106. Kosek DJ, Kim JS, Petrella JK, Cross JM, Bamman MM. Efficacy of 3 days/wk resistance training on myofiber hypertrophy and myogenic mechanisms in young vs. older adults. *J Appl Physiol* 101: 531-544, 2006.
107. Kostek MA, Angelopoulos TJ, Clarkson PM, Gordon PM, Moyna NM, Visich PS, Zoeller RF, Price TB, Seip RL, Thompson PD, Devaney JM, Gordish-Dressman H, Hoffman EP, Pescatello LS. Myostatin and follistatin polymorphisms interact with muscle phenotypes and ethnicity. *Med Sci Sports Exerc* 41: 1063-1071, 2009.
108. Kostek MC, Delmonico MJ, Reichel JB, Roth SM, Douglass L, Ferrell RE, Hurley BF. Muscle strength response to strength training is influenced by insulin-like growth factor 1 genotype in older adults. *J Appl Physiol* 98: 2147-2154, 2005.
109. Kovar R. Contemporary findings on genetic conditions of human motor activities [in Czech]. *Acta Universitatis Carolinae Gymnica*: 69-76, 1974.
110. Kovar R. Genetic analysis of motor performance. *J Sports Med Phys Fitness* 16: 205-208, 1976.
111. Kovar R. Investigation on similarity between parents and their descendants in some motor manifestations [in Czech]. *Teor Praxe Těl Vých* 19: 93-98, 1981.
112. Kovar R. Motor performance in twins. *Acta Genet Med Gemellol (Roma)* 24: 194, 1975.
113. Leamy LJ, Pomp D, Lightfoot JT. An epistatic genetic basis for physical activity traits in mice. *J Hered* 99: 639-646, 2008.
114. Lesage R, Simoneau JA, Jobin J, Leblanc J, Bouchard C. Familial resemblance in maximal heart rate, blood lactate and aerobic power. *Hum Hered* 35: 182-189, 1985.
115. Lightfoot JT, Leamy L, Pomp D, Turner MJ, Fodor AA, Knab A, Bowen RS, Ferguson D, Moore-Harrison T, Hamilton A. Strain screen and haplotype association mapping of wheel running in inbred mouse strains. *J Appl Physiol* 109: 623-634, 2010.
116. Lightfoot JT, Turner MJ, Knab AK, Jedlicka AE, Oshimura T, Marzec J, Gladwell W, Leamy LJ, Kleeberger SR. Quantitative trait loci associated with maximal exercise endurance in mice. *J Appl Physiol* 103: 105-110, 2007.
117. Linc R, Fleischmann J. Sport, talent and genetics. In: Novotný VV, ed. *Anthropological Congress Dedicated to Aleš Hrdlicka; Proceedings*. Prague: Academia, 1971, 267-270.
118. Lortie G, Bouchard C, Leblanc C, Tremblay A, Simoneau JA, Theriault G, Savoie JP. Familial similarity in aerobic power. *Hum Biol* 54: 801-812, 1982.
119. Luquet S, Lopez-Soriano J, Holst D, Fredenrich A, Melki J, Rassoulzadegan M, Grimaldi PA. Peroxisome proliferator-activated receptor delta controls muscle development and oxidative capability. *FASEB J* 17: 2299-2301, 2003.
120. MacArthur DG, North KN. The *ACTN3* gene and human performance. In: Bouchard C, Hoffman EP, eds. *Genetic and Molecular Aspects of Sports Performance*. West Sussex, UK: International Olympic Committee/Blackwell, 2011, 204-214.
121. Maes H, Beunen G, Vlietinck R, Lefevre J, van den Bossche C, Claessens A, Derom R, Lysens R, Renson R, Simons J, Vanden Eynde B. Heritability of health- and performance-related fitness: Data from the Leuven Longitudinal Twin Study. In: Duquet W, Day JAP, eds. *Kinanthropometry IV*. 1st ed. London: E & F Spon, 1993, 140-149.
122. Maes HH, Beunen GP, Vlietinck RF, Neale MC, Thomis M, Vanden Eynde B, Lysens R, Simons J, Derom C, Derom R. Inheritance of physical fitness in 10-yr-old twins and their parents. *Med Sci Sports Exerc* 28: 1479-1491, 1996.
123. Malina RM, Bouchard C, editors. *Sport and Human Genetics*. Champaign, IL: Human Kinetics, 1986.
124. Malina RM, Bouchard C, Bar-Or O. *Growth, Maturation, and Physical Activity*. Champaign, IL: Human Kinetics, 2004.
125. Malina RM, Little BB, Buschang PH. Sibling similarities in the strength and motor performance of undernourished school children. *Hum Biol* 58: 945-953, 1986.
126. Malina RM, Mueller WH. Genetic and environmental influence on the strength and motor performance of Philadelphia school children. *Hum Biol* 53: 163-179, 1981.
127. Manolio TA. Genomewide association studies and assessment of the risk of disease. *N Engl J Med* 363: 166-176, 2010.

128. Marcotte M, Chagnon M, Cote C, Thibault MC, Boulay MR, Bouchard C. Lack of genetic polymorphism in human skeletal muscle enzymes of the tricarboxylic acid cycle. *Hum Genet* 77: 200, 1987.
129. Maridaki M. Heritability of neuromuscular performance and anaerobic power in preadolescent and adolescent girls. *J Sports Med Phys Fitness* 46: 540-547, 2006.
130. Marisi DQ. Genetic and extragenetic variance in motor performance. *Acta Genet Med Gemellol (Roma)* 26: 197-204, 1977.
131. Mayhew DL, Kim JS, Cross JM, Ferrando AA, Bamman MM. Translational signaling responses preceding resistance training-mediated myofiber hypertrophy in young and old humans. *J Appl Physiol* 107: 1655-1662, 2009.
132. McCauley T, Mastana SS, Folland JP. ACE I/D and ACTN3 R/X polymorphisms and muscle function and muscularity of older Caucasian men. *Eur J Appl Physiol* 109: 269-277, 2010.
133. McCauley T, Mastana SS, Hossack J, Macdonald M, Folland JP. Human angiotensin-converting enzyme I/D and alpha-actinin 3 R577X genotypes and muscle functional and contractile properties. *Exp Physiol* 94: 81-89, 2009.
134. McCole SD, Brown MD, Moore GE, Ferrell RE, Wilund KR, Huberty A, Douglass LW, Hagberg JM. Angiotensinogen M235T polymorphism associates with exercise hemodynamics in postmenopausal women. *Physiol Genom* 10: 63-69, 2002.
135. McGee SL, Fairlie E, Garnham AP, Hargreaves M. Exercise-induced histone modifications in human skeletal muscle. *J Physiol* 587: 5951-5958, 2009.
136. McGraw MB. *Growth, A Study of Johnny and Jimmy*. New York: Appleton-Century, 1935.
137. McNemar Q. Twin resemblance in motor skills, and the effect of practice thereon. *Pedagog Semin J Gen* 42: 70-99, 1933.
138. Meek TH, Lonquich BP, Hannon RM, Garland T Jr. Endurance capacity of mice selectively bred for high voluntary wheel running. *J Exp Biol* 212: 2908-2917, 2009.
139. Mirenva AN. Psychomotor education and the general development of preschool children: Experiments with twin controls. *Pedagog Semin J Gen* 46: 433-454, 1935.
140. Mizuno T. Similarity of physique, muscular strength and motor ability of identical twins. *Bull Fac Educ Tokyo U* 1: 190-191, 1956.
141. Montgomery HE, Clarkson P, Dollery CM, Prasad K, Losi MA, Hemingway H, Statters D, Jubb M, Girvain M, Varnava A, World M, Deanfield J, Talmud P, McEwan JR, McKenna WJ, Humphries S. Association of angiotensin-converting enzyme gene I/D polymorphism with change in left ventricular mass in response to physical training. *Circulation* 96: 741-747, 1997.
142. Montgomery HE, Marshall R, Hemingway H, Myerson S, Clarkson P, Dollery C, Hayward M, Holliman DE, Jubb M, World M, Thomas EL, Brynes AE, Saeed N, Barnard M, Bell JD, Prasad K, Rayson M, Talmud PJ, Humphries SE. Human gene for physical performance. *Nature* 393: 221-222, 1998.
143. Montoye HJ, Gayle R. Familial relationships in maximal oxygen uptake. *Hum Biol* 50: 241-249, 1978.
144. Montoye HJ, Metzner HL, Keller JB. Familial aggregation of strength and heart rate response to exercise. *Hum Biol* 47: 17-36, 1975.
145. Moran CN, Scott RA, Adams SM, Warrington SJ, Jobling MA, Wilson RH, Goodwin WH, Georgiades E, Wolde B, Pitsiladis YP. Y chromosome haplogroups of elite Ethiopian endurance runners. *Hum Genet* 115: 492-497, 2004.
146. Moran CN, Yang N, Bailey ME, Tsiokanos A, Jamurtas A, MacArthur DG, North K, Pitsiladis YP, Wilson RH. Association analysis of the ACTN3 R577X polymorphism and complex quantitative body composition and performance phenotypes in adolescent Greeks. *Eur J Hum Genet* 15: 88-93, 2007.
147. Myerson SG, Montgomery HE, Whittingham M, Jubb M, World MJ, Humphries SE, Pennell DJ. Left ventricular hypertrophy with exercise and ACE gene insertion/deletion polymorphism: A randomized controlled trial with losartan. *Circulation* 103: 226-230, 2001.
148. Nakajima K, Takeoka M, Mori M, Hashimoto S, Sakurai A, Nose H, Higuchi K, Itano N, Shiohara M, Oh T, Taniguchi S. Exercise effects on methylation of ASC gene. *Int J Sports Med* 31: 671-675, 2010.
149. Niemi AK, Majamaa K. Mitochondrial DNA and ACTN3 genotypes in Finnish elite endurance and sprint athletes. *Eur J Hum Genet* 13: 965-969, 2005.
150. Norman B, Esbjornsson M, Rundqvist H, Osterlund T, von Walden F, Tesch PA. Strength, power, fiber types, and mRNA expression in trained men and women with different ACTN3 R577X genotypes. *J Appl Physiol* 106: 959-965, 2009.
151. Okuda E, Horii D, Kano T. Genetic and environmental effects on physical fitness and motor performance. *Int J Sport Health Sci* 3: 1-9, 2005.
152. Palmer LJ, Jacobs KB, Elston RC. Haseman and Elston revisited: The effects of ascertainment and residual familial correlations on power to detect linkage. *Genet Epidemiol* 19: 456-460, 2000.
153. Papadimitriou ID, Papadopoulos C, Kouvatsi A, Triantaphyllidis C. The *ACTN3* gene in elite Greek track and field athletes. *Int J Sports Med* 29: 352-355, 2008.
154. Parisi P, Casini B, Di Salvo V, Pigozzi F, Pittaluga M, Priazi G, Klissouras V. The Registry of Italian Twins (RITA): Background, design, and procedures, and twin daya analysis on sport participation—An application to twin swimmers. *Eur J Sport Sci* 1: 1-12, 2001.

155. Parisi P, Pigozzi F, Klissouras V. Genetic limits of sport performance: A study of twin athletes. *Bull Int Council Sport Sci Phys Educ* 26, 1999.

156. Park JY, Farrance IK, Fenty NM, Hagberg JM, Roth SM, Mosser DM, Wang MQ, Jo H, Okazaki T, Brant SR, Brown MD. NFKB1 promoter variation implicates shear-induced NOS3 gene expression and endothelial function in prehypertensives and stage I hypertensives. *Am J Physiol Heart Circ Physiol* 293: H2320-H2327, 2007.

157. Peeters MW, Thomis MA, Maes HH, Beunen GP, Loos RJ, Claessens AL, Vlietinck R. Genetic and environmental determination of tracking in static strength during adolescence. *J Appl Physiol* 99: 1317-1326, 2005.

158. Peeters MW, Thomis MA, Maes HH, Loos RJ, Claessens AL, Vlietinck R, Beunen GP. Genetic and environmental causes of tracking in explosive strength during adolescence. *Behav Genet* 35: 551-563, 2005.

159. Peeters MW, Thomis MAI, Beunen G, Malina RM. Genetics and sports: An overview of the pre-molecular era. In: Collins M, ed. *Genetics and Sports*. Basel: Karger, 2009, 28-42.

160. Perusse L, Gagnon J, Province MA, Rao DC, Wilmore JH, Leon AS, Bouchard C, Skinner JS. Familial aggregation of submaximal aerobic performance in the HERITAGE Family Study. *Med Sci Sports Exerc* 33: 597-604, 2001.

161. Perusse L, Leblanc C, Bouchard C. Inter-generation transmission of physical fitness in the Canadian population. *Can J Sport Sci* 13: 8-14, 1988.

162. Perusse L, Leblanc C, Tremblay A, Allard C, Theriault G, Landry F, Talbot J, Bouchard C. Familial aggregation in physical fitness, coronary heart disease risk factors, and pulmonary function measurements. *Prev Med* 16: 607-615, 1987.

163. Perusse L, Lortie G, Leblanc C, Tremblay A, Theriault G, Bouchard C. Genetic and environmental sources of variation in physical fitness. *Ann Hum Biol* 14: 425-434, 1987.

164. Perusse L, Rankinen T, Rauramaa R, Rivera MA, Wolfarth B, Bouchard C. The human gene map for performance and health-related fitness phenotypes: The 2002 update. *Med Sci Sports Exerc* 35: 1248-1264, 2003.

165. Pirnay F, Klissouras V, Deroanne R, Petit J-M. Aptitude physique des jumeaux. *Med du Sport* 49: 29-33, 1975.

166. Pitsiladis YP. Ethnic differences in sports performance. In: Bouchard C, Hoffman EP, eds. *Genetic and Molecular Aspects of Sports Performance*. West Sussex, UK: International Olympic Committee/Blackwell, 2011, 121-132.

167. Prior SJ, Hagberg JM, Paton CM, Douglass LW, Brown MD, McLenithan JC, Roth SM. DNA sequence variation in the promoter region of the VEGF gene impacts VEGF gene expression and maximal oxygen consumption. *Am J Physiol Heart Circ Physiol* 290: H1848-H1855, 2006.

168. Province MA, Rice TK, Borecki IB, Gu C, Kraja A, Rao DC. Multivariate and multilocus variance components method, based on structural relationships to assess quantitative trait linkage via SEGPATH. *Genet Epidemiol* 24: 128-138, 2003.

169. Prudhomme D, Bouchard C, Leblanc C, Landry F, Fontaine E. Sensitivity of maximal aerobic power to training is genotype-dependent. *Med Sci Sport Exer* 16: 489-493, 1984.

170. Rankinen T, An P, Perusse L, Rice T, Chagnon YC, Gagnon J, Leon AS, Skinner JS, Wilmore JH, Rao DC, Bouchard C. Genome-wide linkage scan for exercise stroke volume and cardiac output in the HERITAGE Family Study. *Physiol Genom* 10: 57-62, 2002.

171. Rankinen T, Argyropoulos G, Rice T, Rao DC, Bouchard C. CREB1 is a strong genetic predictor of the variation in exercise heart rate response to regular exercise: The HERITAGE Family Study. *Circ Cardiovasc Genet* 3: 294-299, 2010.

172. Rankinen T, Bouchard C, Rao DC. Familial resemblance for muscle phenotypes: The HERITAGE Family Study. *Med Sci Sport Exer* 37: 2017, 2005.

173. Rankinen T, Bray MS, Hagberg JM, Perusse L, Roth SM, Wolfarth B, Bouchard C. The human gene map for performance and health-related fitness phenotypes: The 2005 update. *Med Sci Sports Exerc* 38: 1863-1888, 2006.

174. Rankinen T, Perusse L, Rauramaa R, Rivera MA, Wolfarth B, Bouchard C. The human gene map for performance and health-related fitness phenotypes. *Med Sci Sports Exerc* 33: 855-867, 2001.

175. Rankinen T, Perusse L, Rauramaa R, Rivera MA, Wolfarth B, Bouchard C. The human gene map for performance and health-related fitness phenotypes: The 2001 update. *Med Sci Sports Exerc* 34: 1219-1233, 2002.

176. Rankinen T, Perusse L, Rauramaa R, Rivera MA, Wolfarth B, Bouchard C. The human gene map for performance and health-related fitness phenotypes: The 2003 update. *Med Sci Sports Exerc* 36: 1451-1469, 2004.

177. Rankinen T, Roth SM, Bray MS, Loos R, Perusse L, Wolfarth B, Hagberg JM, Bouchard C. Advances in exercise, fitness, and performance genomics. *Med Sci Sports Exerc* 42: 835-846, 2010.

178. Rankinen T, Wolfarth B, Simoneau JA, Maier-Lenz D, Rauramaa R, Rivera MA, Boulay MR, Chagnon YC, Perusse L, Keul J, Bouchard C. No association between the angiotensin-converting enzyme ID polymorphism and elite endurance athlete status. *J Appl Physiol* 88: 1571-1575, 2000.

179. Rezende EL, Garland T Jr., Chappell MA, Malisch JL, Gomes FR. Maximum aerobic performance in lines of Mus selected for high wheel-running activity: Effects of selection, oxygen availability and the mini-muscle phenotype. *J Exp Biol* 209: 115-127, 2006.

180. Rezende EL, Gomes FR, Malisch JL, Chappell MA, Garland T Jr. Maximal oxygen consumption in re-

lation to subordinate traits in lines of house mice selectively bred for high voluntary wheel running. *J Appl Physiol* 101: 477-485, 2006.

181. Rico-Sanz J, Rankinen T, Joanisse DR, Leon AS, Skinner JS, Wilmore JH, Rao DC, Bouchard C. Familial resemblance for muscle phenotypes in the HERITAGE Family Study. *Med Sci Sports Exerc* 35: 1360-1366, 2003.
182. Rico-Sanz J, Rankinen T, Rice T, Leon AS, Skinner JS, Wilmore JH, Rao DC, Bouchard C. Quantitative trait loci for maximal exercise capacity phenotypes and their responses to training in the HERITAGE Family Study. *Physiol Genom* 16: 256-260, 2004.
183. Rigat B, Hubert C, Alhenc-Gelas F, Cambien F, Corvol P, Soubrier F. An insertion/deletion polymorphism in the angiotensin I-converting enzyme gene accounting for half the variance of serum enzyme levels. *J Clin Invest* 86: 1343-1346, 1990.
184. Rivera MA, Dionne FT, Wolfarth B, Chagnon M, Simoneau JA, Perusse L, Boulay MR, Gagnon J, Song TM, Keul J, Bouchard C. Muscle-specific creatine kinase gene polymorphisms in elite endurance athletes and sedentary controls. *Med Sci Sports Exerc* 29: 1444-1447, 1997.
185. Rivera MA, Wolfarth B, Dionne FT, Chagnon M, Simoneau JA, Boulay MR, Song TM, Perusse L, Gagnon J, Leon AS, Rao DC, Skinner JS, Wilmore JH, Keul J, Bouchard C. Three mitochondrial DNA restriction polymorphisms in elite endurance athletes and sedentary controls. *Med Sci Sports Exerc* 30: 687-690, 1998.
186. Ropponen A, Levalahti E, Videman T, Kaprio J, Battie MC. The role of genetics and environment in lifting force and isometric trunk extensor endurance. *Phys Ther* 84: 608-621, 2004.
187. Roth SM. *Genetics Primer for Exercise Science and Health*. Champaign, IL: Human Kinetics, 2007.
188. Roth SM, Ferrell RE, Peters DG, Metter EJ, Hurley BF, Rogers MA. Influence of age, sex, and strength training on human muscle gene expression determined by microarray. *Physiol Genom* 10: 181-190, 2002.
189. Roth SM, Walsh S, Liu D, Metter EJ, Ferrucci L, Hurley BF. The ACTN3 R577X nonsense allele is under-represented in elite-level strength athletes. *Eur J Hum Genet* 16: 391-394, 2008.
190. Ruiz JR, Arteta D, Buxens A, Artieda M, Gomez-Gallego F, Santiago C, Yvert T, Moran M, Lucia A. Can we identify a power-oriented polygenic profile? *J Appl Physiol* 108: 561-566, 2010.
191. Ruiz JR, Gomez-Gallego F, Santiago C, Gonzalez-Freire M, Verde Z, Foster C, Lucia A. Is there an optimum endurance polygenic profile? *J Physiol* 587: 1527-1534, 2009.
192. Santiago C, Gonzalez-Freire M, Serratosa L, Morate FJ, Meyer T, Gomez-Gallego F, Lucia A. ACTN3 genotype in professional soccer players. *Br J Sports Med* 42: 71-73, 2008.
193. Sayer AA, Syddall H, O'Dell SD, Chen XH, Briggs PJ, Briggs R, Day IN, Cooper C. Polymorphism of the IGF2 gene, birth weight and grip strength in adult men. *Age Ageing* 31: 468-470, 2002.
194. Schoenfelder M. Genetics-based performance talent research: Polymorphisms as predictors of endurance performance. *J Appl Physiol* 108: 1454-1455, 2010.
195. Schrager MA, Roth SM, Ferrell RE, Metter EJ, Russek-Cohen E, Lynch NA, Lindle RS, Hurley BF. Insulin-like growth factor-2 genotype, fat-free mass, and muscle performance across the adult life span. *J Appl Physiol* 97: 2176-2183, 2004.
196. Schuelke M, Wagner KR, Stolz LE, Hubner C, Riebel T, Komen W, Braun T, Tobin JF, Lee SJ. Myostatin mutation associated with gross muscle hypertrophy in a child. *N Engl J Med* 350: 2682-2688, 2004.
197. Scott RA, Fuku N, Onywera VO, Boit M, Wilson RH, Tanaka M, Goodwin WH, Pitsiladis YP. Mitochondrial haplogroups associated with elite Kenyan athlete status. *Med Sci Sports Exerc* 41: 123-128, 2009.
198. Scott RA, Irving R, Irwin L, Morrison E, Charlton V, Austin K, Tladi D, Deason M, Headley SA, Kolkhorst FW, Yang N, North K, Pitsiladis YP. ACTN3 and ACE genotypes in elite Jamaican and US sprinters. *Med Sci Sports Exerc* 42: 107-112, 2010.
199. Seibert MJ, Xue QL, Fried LP, Walston JD. Polymorphic variation in the human myostatin (GDF-8) gene and association with strength measures in the Women's Health and Aging Study II cohort. *J Am Geriatr Soc* 49: 1093-1096, 2001.
200. Shvarts VB. Twin investigations and physical exercise: To the 100th anniversary of twin method. *Teoriya i Praktika Fizicheskoi Cultury* 39: 19-21, 1976.
201. Simoneau J-A, Bouchard C. Genetic determination of fiber type proportion in human skeletal muscle. *FASEB J* 9: 1091-1095, 1995.
202. Simoneau JA, Bouchard C. The effects of genetic variation on anaerobic performance. In: van Praagh E, ed. *Pediatric Anaerobic Performance*. Champaign, IL: Human Kinetics, 1998, 5-21.
203. Simoneau JA, Lortie G, Boulay MR, Marcotte M, Thibault MC, Bouchard C. Inheritance of human skeletal muscle and anaerobic capacity adaptation to high-intensity intermittent training. *Int J Sports Med* 7: 167-171, 1986.
204. Simoneau JA, Lortie G, Leblanc C, Bouchard C. Anaerobic alactacid work capacity in adopted and biological siblings. In: Malina RM, Bouchard C, eds. *Sport and Human Genetics*. Champaign, IL: Human Kinetics, 1986, 165-171.
205. Simonen RL, Rankinen T, Perusse L, Rice T, Rao DC, Chagnon Y, Bouchard C. Genome-wide linkage scan for physical activity levels in the Quebec Family study. *Med Sci Sports Exerc* 35: 1355-1359, 2003.

206. Skipworth JRA, Puthucheary ZA, Rawal J, Montgomery HE. The *ACE* gene and performance. In: Bouchard C, Hoffman EP, eds. *Genetic and Molecular Aspects of Sports Performance*. West Sussex, UK: International Olympic Committee/ Blackwell, 2011, 195-203.
207. Skład M. The genetic determination of the rate of learning of motor skills. *Studies Phys Anthropol* 1, 1975.
208. Skład M. Physical and motor development of twins [in Polish]. *Materiały i Prace Antropologyczne* 85: 3-102, 1973.
209. Skład M. Similarity of movement in twins. *Wychowanie Fizycznie i Sport* 16: 119-141, 1972.
210. Smith NW. Twin studies and heritability. *Hum Dev* 19: 65-68, 1976.
211. Spielmann N, Leon AS, Rao DC, Rice T, Skinner JS, Rankinen T, Bouchard C. Genome-wide linkage scan for submaximal exercise heart rate in the HERITAGE Family Study. *Am J Physiol Heart Circ Physiol* 293: H3366-H3371, 2007.
212. Sundet JM, Magnus P, Tambs K. The heritability of maximal aerobic power: A study of Norwegian twins. *Scand J Med Sci Sports* 4: 181-185, 1994.
213. Szopa J. Familial studies on genetic determination of some manifestations of muscular strength in man. *Genet Pol* 23: 65-79, 1982.
214. Szopa J. Genetic control of fundamental psychomotor properties in man. *Genet Pol* 27: 137-150, 1986.
215. Szopa J. Variability and genetic determination of some aspects of muscular strength in man: Results of family studies. *Materialy i Prace Antropologiczne* 103: 131-154, 1983.
216. Teran-Garcia M, Rankinen T, Koza RA, Rao DC, Bouchard C. Endurance training-induced changes in insulin sensitivity and gene expression. *Am J Physiol Endocrinol Metab* 288: E1168-E1178, 2005.
217. Teran-Garcia M, Rankinen T, Rice T, Leon AS, Rao DC, Skinner JS, Bouchard C. Variations in the four and a half LIM domains 1 gene (FHL1) are associated with fasting insulin and insulin sensitivity responses to regular exercise. *Diabetologia* 50: 1858-1866, 2007.
218. Thibault MC, Simoneau JA, Cote C, Boulay MR, Lagasse P, Marcotte M, Bouchard C. Inheritance of human muscle enzyme adaptation to isokinetic strength training. *Hum Hered* 36: 341-347, 1986.
219. Thomis MA, Beunen GP, Maes HH, Blimkie CJ, Van Leemputte M, Claessens AL, Marchal G, Willems E, Vlietinck RF. Strength training: Importance of genetic factors. *Med Sci Sports Exerc* 30: 724-731, 1998.
220. Thomis MAI. Genes and strength and power phenotypes. In: Bouchard C, Hoffman EP, eds. *Genetic and Molecular Aspects of Sports Performance*. West Sussex, UK: International Olympic Committee/ Blackwell, 2011, 159-176.
221. Thomis MAI, Beunen GP, Van Leemputte M, Maes HH, Blimkie CJ, Claessens AL, Marchal G, WIllems E, Vlietinck RF. Inheritance of static and dynamic arm strength and some of its determinants. *Acta Physiol Scand* 163: 59-71, 1998.
222. Thompson PD, Moyna N, Seip R, Price T, Clarkson P, Angelopoulos T, Gordon P, Pescatello L, Visich P, Zoeller R, Devaney JM, Gordish H, Bilbie S, Hoffman EP. Functional polymorphisms associated with human muscle size and strength. *Med Sci Sports Exerc* 36: 1132-1139, 2004.
223. Tiainen KM, Perola M, Kovanen VM, Sipila S, Tuononen KA, Rikalainen K, Kauppinen MA, Widen EI, Kaprio J, Rantanen T, Kujala UM. Genetics of maximal walking speed and skeletal muscle characteristics in older women. *Twin Res Hum Genet* 11: 321-334, 2008.
224. Timmons JA, Knudsen S, Rankinen T, Koch LG, Sarzynski M, Jensen T, Keller P, Scheele C, Vollaard NB, Nielsen S, Akerstrom T, MacDougald OA, Jansson E, Greenhaff PL, Tarnopolsky MA, van Loon LJ, Pedersen BK, Sundberg CJ, Wahlestedt C, Britton SL, Bouchard C. Using molecular classification to predict gains in maximal aerobic capacity following endurance exercise training in humans. *J Appl Physiol* 108: 1487-1496, 2010.
225. Vandenberg SG. The hereditary abilities study: Hereditary components in a psychological test battery. *Am J Hum Genet* 14: 220-237, 1962.
226. Venerando A, Milani-Comparetti M. Twin studies in sport and physical performance. *Acta Genet Med Gemellol (Roma)* 19: 80-82, 1970.
227. Vickers VS, Poyntz L, Baum MP. The Brace scale used with young children. *Res Q* 13, 1942.
228. Vincent B, De Bock K, Ramaekers M, Van den Eede E, Van Leemputte M, Hespel P, Thomis MA. ACTN3 (R577X) genotype is associated with fiber type distribution. *Physiol Genom* 32: 58-63, 2007.
229. von Verschuer O. Twin research from the time of Francis Galton to the present day. *Proc R Soc Lond B Biol Sci* 128: 62-81, 1939.
230. von Verschuer O, Kinkelin WM, Zipperlin V. II. Die vererbungsbiologische Zwillingsforschung, Ihre biologischen Grundlagen, Studien an 102 eineiigen und 45 gleichgeschlechtlichen zweieiigen Zwilingsund an 2 Drillingspaaren. *Ergebnisse Inneren Medizin Kinderheilkund* 31: 35-120, 1927.
231. Walsh S, Metter EJ, Ferrucci L, Roth SM. Activintype II receptor B (ACVR2B) and follistatin haplotype associations with muscle mass and strength in humans. *J Appl Physiol* 102: 2142-2148, 2007.
232. Wang YX, Zhang CL, Yu RT, Cho HK, Nelson MC, Bayuga-Ocampo CR, Ham J, Kang H, Evans RM. Regulation of muscle fiber type and running endurance by *PPAR* delta. *Plos Biol* 2: 1532-1539, 2004.
233. Ways JA, Cicila GT, Garrett MR, Koch LG. A genome scan for loci associated with aerobic running capacity in rats. *Genomics* 80: 13-20, 2002.

234. Ways JA, Smith BM, Barbato JC, Ramdath RS, Pettee KM, DeRaedt SJ, Allison DC, Koch LG, Lee SJ, Cicila GT. Congenic strains confirm aerobic running capacity quantitative trait loci on rat chromosome 16 and identify possible intermediate phenotypes. *Physiol Genom* 29: 91-97, 2007.
235. Weber G, Kartodihardjo W, Klissouras V. Growth and physical training with reference to heredity. *J Appl Physiol* 40: 211-215, 1976.
236. Weiss V. The heritabilities of tests of physical fitness, calculated from the performances of ten years old twin pairs [translated from German]. *Arztl Jugendkd* 68: 167-172, 1977.
237. Wilde GJ. An experimental study of mutual behaviour imitation and person perception in MZ and DZ twins. Implications for an experimental-psychometric analysis of heritability coefficients. *Acta Genet Med Gemellol (Roma)* 19: 273-279, 1970.
238. Williams AG, Folland JP. Similarity of polygenic profiles limits the potential for elite human physical performance. *J Physiol* 586: 113-121, 2008.
239. Williams LRT, Gross JB. Heritability of motor skill. *Acta Genet Med Gemellol (Roma)* 29: 127-136, 1980.
240. Williams LRT, Hearfield V. Heritability of a gross motor balance task. *Res Q* 44: 109-112, 1973.
241. Wilson RS, Harpring EB. Mental and motor development in infant twins. *Dev Psychol* 7: 277-287, 1972.
242. Wolanski N. Heredity and psychomotor traits in man. In: Malina RM, Bouchard C, eds. *Sport and Human Genetics*. Champaign, IL: Human Kinetics, 1986, 123-129.
243. Wolanski N, Kasprzak E. Similarity in some physiological, biochemical and psychomotor traits between parents and 2-45 years old offspring. *Studies Human Ecol* 3: 85-131, 1979.
244. Wolfarth B, Bray MS, Hagberg JM, Perusse L, Rauramaa R, Rivera MA, Roth SM, Rankinen T, Bouchard C. The human gene map for performance and health-related fitness phenotypes: The 2004 update. *Med Sci Sports Exerc* 37: 881-903, 2005.
245. Wolfarth B, Rankinen T, Muhlbauer S, Ducke M, Rauramaa R, Boulay MR, Perusse L, Bouchard C. Endothelial nitric oxide synthase gene polymorphism and elite endurance athlete status: The Genathlete study. *Scand J Med Sci Spor* 18: 485-490, 2008.
246. Wolfarth B, Rankinen T, Muhlbauer S, Scherr J, Boulay MR, Perusse L, Rauramaa R, Bouchard C. Association between a beta2-adrenergic receptor polymorphism and elite endurance performance. *Metabolism* 56: 1649-1651, 2007.
247. Wolfarth B, Rivera MA, Oppert JM, Boulay MR, Dionne FT, Chagnon M, Gagnon J, Chagnon Y, Perusse L, Keul J, Bouchard C. A polymorphism in the alpha2a-adrenoceptor gene and endurance athlete status. *Med Sci Sports Exerc* 32: 1709-1712, 2000.
248. Yang N, MacArthur DG, Gulbin JP, Hahn AG, Beggs AH, Easteal S, North K. ACTN3 genotype is associated with human elite athletic performance. *Am J Hum Genet* 73: 627-631, 2003.
249. Yang N, MacArthur DG, Wolde B, Onywera VO, Boit MK, Lau SY, Wilson RH, Scott RA, Pitsiladis YP, North K. The ACTN3 R577X polymorphism in East and West African athletes. *Med Sci Sports Exerc* 39: 1985-1988, 2007.
250. Zatsiorsky VM, Sergienko LP. The influence of hereditary and environmental factors on the development of motor performance in humans: Materials of twin investigations. *Teoriya i Praktika Fizicheskoi Cultury* 38: 22-29, 1975.
251. Zhao B, Moochhala SM, Tham S, Lu J, Chia M, Byrne C, Hu Q, Lee LK. Relationship between angiotensin-converting enzyme ID polymorphism and VO2max of Chinese males. *Life Sci* 73: 2625-2630, 2003.
252. Bouchard C, Sarzynski MA, Rice TK, Kraus WE, Church TS, Sung YJ, Rao DC, Rankinen T. Genomic predictors of the maximal O2 uptake response to standardized exercise training programs. J Appl Physiol. 110:1160-1170, 2011. PMID: 21183627. PMCID: PMC3098655.

PART II

A Century of Discoveries (1910-2010)

CHAPTER 8

The Sensorimotor Nervous System

Phillip F. Gardiner, PhD

V. Reggie Edgerton, PhD

Abstract

Major developments in our knowledge of the sensorimotor system in the past 100 yr have significantly affected our understanding of exercise. The development in the mid-20th century of techniques for recording intramuscular and surface electromyography in humans, as well as for recording from single neurons in reduced animal preparations, has led to increased understanding of how motor units are used during various activities and the effects of fatigue and training on neuromuscular recruitment patterns. Possible fatigue mechanisms in the sensorimotor system that have been investigated, primarily during the latter half of the 20th century, include failure of the neuromuscular junction (NMJ), decreased motoneuronal excitability, and decrease in net excitation of motoneurons due to changes in excitatory and inhibitory influences from periphery or supraspinal centers. Examining the role of supraspinal influences in the fatigue process and the effects of training at this level have been made possible by the development of techniques such as transcranial magnetic stimulation and transcranial electrical stimulation. Proposed sensorimotor mechanisms by which exercise training may promote enhanced performance include improved function of the NMJ, altered excitability of spinal neurons and reflex pathways, modifications in spinal cord synaptology, and changes in supraspinal control. Scientists have also begun to appreciate the role of regular exercise and physical activity in promoting neural health, including enhanced cognitive function throughout the life span. For example, regular exercise can promote brain angiogenesis, neurogenesis, learning and memory, and increased production of neurotrophins and their associated proteins. Exercise studies conducted during the past 30 yr using animals and humans with spinal cord injuries have demonstrated the remarkable plasticity of the spinal cord locomotor circuitry. This information may prove vital to the development of rehabilitative strategies for promoting recovery after a range of neuromotor deficits.

Introduction

The study of neural components of exercise physiology is a relatively recent endeavor compared with the study of other systems discussed in this text. Neuroscience in general is a relatively new area of investigation. A paper on giant-squid axons published by Hodgkin and Huxley (109) in 1939 was the first published intracellular recording of action potential—the basic unit of currency of the nervous system. Recordings from mammalian neurons and muscle fibers in situ using glass microelectrodes were first made in the 1950s, at which time important concepts such as the chemical basis of synaptic transmission, its relationship to excitation and inhibition, and its quantal nature were established (25, 77). The discovery of nerve growth factor, which led to the discovery of other neural growth factors now recognized as instrumental in neural plasticity of various types, was announced by Rita Levi-Montalcini (1909–), Stanley Cohen (1922–), and Viktor Hamburger (1900-2001) in the 1950s (43). Electromyographic equipment was not commercially available to researchers until the early 1950s (139). The Society for Neuroscience, which brings together one of the largest groups of scientists in the world (the annual meeting currently attracts in excess of 30,000 delegates), was founded in 1971. The first annual meeting, held in Washington, D.C., attracted 1,396 delegates. In 1990, George H.W. Bush signed a proclamation declaring the 1990s the "decade of the brain" with the purpose of enhancing public awareness of the benefits to be derived from brain research. All of these developments (table

8.1), so vital to the development of our basic understanding of the functioning of the nervous system, are among the most crucial in advancing the field of exercise neuroscience, and exercise science in general, to where we are today. In this discussion of the historical aspects of the role of the nervous system in exercise, therefore, most of the seminal research cited is from the period following the key developments of the mid-20th century.

Neuromuscular Recruitment During Exercise

The discovery of the heterogeneity of muscle fibers, and how they support different types of voluntary movements, was a major contribution to understanding the nervous system and exercise. Denny-Brown (1901-1981) in 1929 confirmed that white (gastrocnemius) and red (soleus) muscles in the cat had different functional properties and hypothesized that slow muscles would be ideal for postural activities (48). In following years, it was recognized that both red and pale skeletal muscles comprised different types of fibers—at least most of the time. (Cat and guinea pig soleus muscles, which exclusively comprise slow-twitch fibers, are exceptions.) During the 1960s information regarding the properties and significance of muscle phenotypes began to accumulate rapidly. There was considerable interest from the neural perspective because it was known that muscle phenotypic changes could be associated with a number of human muscular diseases. In addition, the general concept existed that many if not most muscle diseases had some neural origin. A series of classical papers published in the 1960s demonstrated that if the nerves to a slow muscle and a fast muscle were severed and the fast nerve was sutured to the slow muscle and vice versa, the slow muscle would become faster and the fast muscle would become slower (28). The question about the mechanism of this neural influence on protein expression (and therefore the physiological properties of skeletal muscle) remains open, but the papers clearly showed the interdependence of the motor neuron and the muscle fibers.

At this time the concept was fairly well ingrained that there were slow and fast muscles and that slow and fast muscles were red and white, respectively. At the same time multiple observations demonstrated that there were many exceptions to this generalization, but the significance of these exceptions was generally dismissed. As the histochemical techniques became more established (largely due to the studies of W. King Engel and colleagues, then at the National Institutes of Health) (75), it became possible to begin to examine—at least qualitatively—the relationship between the biochemical properties of single muscle fibers associated with the metabolic properties and contractile properties with the assumption that the contractile properties were related to the histochemical technique referred to as myosin adenosine triphosphatase (ATPase). This was a reasonable assumption given that a close relationship had been reported between the speed at which the myosin could hydrolyze adenosine triphosphate and the speed of muscle shortening in a wide range of animals that have muscles with a wide difference in velocity of shortening. However, the histochemical technique was considered suspect by some simply because it did not match the presumed speed of muscle and its color (i.e., red for slow or white for fast). There was also some concern because the histochemical myosin ATPase staining reaction conducted under well-controlled conditions resulted in either dark or light staining intensity when observed microscopically. The explanation for this clear dichotomy is related to the pH sensitivity of the specific myosin phenotypes, then known only as slow or fast. Further subcategories of the myosin stains were developed. These subdivisions were eventually related to further subdivisions of the different myosin phenotypes identified with quantitative biochemical techniques, including molecular differences in myosin. At that time there was still a limitation of performing quantitative biochemical techniques on single muscle fibers. The myosin ATPase histochemical procedure, and the many permutations and adaptations that have been applied to this technique, formed the basis for a significant volume of subsequent research concerning exercise and fiber-type recruitment and adaptation—at least, until the use of antibodies directed toward specific myosin heavy chain isoforms. These antibodies, which arrived in the 1980s, allowed one to demonstrate the coexistence of several myosin heavy chain types in single fibers and how this related to exercise training. As the histochemical techniques were being used more commonly, another approach to muscle phenotypes was being directly pursued using neurophysiological techniques that permitted one to functionally isolate single alpha motor axons or penetrate single motor neurons with a microelectrode, thus allowing one to measure conduction velocity and other neurophysiological properties of the motor neurons as well as the isometric contractile properties of single motor units when the motor neuron was activated. This approach reflected the net properties of all muscle fibers that are simultaneously activated when the motor neuron or axon is activated. By functionally isolating single axons, Elwood Henneman (1915-1995) and colleagues (220) were able to

Table 8.1 Milestones in the History of Exercise and the Sensorimotor Nervous System

Year	Milestone	Reference
1911	Brown demonstrates that the spinal cord, with no supraspinal or afferent input, is capable of generating locomotor-like patterns of muscle activation.	26
1929	Lord Adrian uses a concentric needle electrode to demonstrate recruitment and rate coding of motor units during voluntary movement.	2
1939	Hodgkin and Huxley publish the first recording of an action potential from a giant-squid axon.	109
Mid-1950s	Cohen, Levi-Montalcini, and Hamburger discover nerve growth factor.	43
Mid-1950s	Demonstration that motoneurons innervating fast and slow muscles in the cat possess different afterhyperpolarization durations.	66
1961	Ikai and Steinhaus demonstrate that maximal contractions can be increased through hypnosis, startling response, and drugs.	114
1960s	Development of muscle histochemical techniques for distinguishing different fiber types.	74
1960s	Cross-reinnervation studies of Buller and colleagues demonstrate the importance of the motor nerve in determining muscle properties.	28
1960s	Kernell describes the phenomenon of late adaptation of motoneurons.	126
1960s	Henneman and colleagues propose the size principle.	106
Late 1960s	Burke and colleagues describe functional properties of different muscle unit types, and of the motoneurons innervating them, in the cat gastrocnemius.	29
1969	Introduction of the concept of muscle wisdom to explain the decline in motor unit firing rate during fatigue.	147
1971	First annual meeting of the Society for Neuroscience held in Washington, D.C.	
Mid-1970s	First use of glycogen-depletion techniques to demonstrate recruitment of human muscle fibers during exercise.	92
1973	Stein and colleagues demonstrate orderly recruitment in the human adductor pollicis using spike-triggered averaging.	155
1975	Demonstration that resistance training enhances motor unit synchronization and alters short-and long-loop reflex responses.	153
1979	Moritani and de Vries present evidence from surface electromyography of a neural component to resistance training.	160
1980	Schwindt and Crill describe persistent inward currents in motoneurons in vitro.	183
1982	Edgerton and colleagues begin to investigate the effects of training in spinalized cats.	121
Mid-1980s	Bigland-Ritchie and collaborators propose a fatigue reflex based on their experiments.	17
Mid-1980s	Sale and colleagues show an effect of resistance training on reflex potentiation.	179
Mid-1980s	Wolpaw and collaborators demonstrate that monkeys can be trained to alter the amplitude of the monosynaptic reflex response.	216
1988	Publication of *Neuromechanical Basis of Kinesiology* by R. Enoka.	76

(continued)

Table 8.1 *(continued)*

Year	Milestone	Reference
1990	United States President George H.W. Bush declares the 1990s the "decade of the brain."	
1994	Satellite symposium of the Society for Neuroscience, titled "Neural and Neuromuscular Aspects of Muscular Fatigue," held in Miami. Proceedings, titled "Fatigue: Neural and Muscular Mechanisms," published in 1995.	80
1995	Cotman and colleagues show that exercise causes increased expression of brain-derived neurotrophic factor in the rat hippocampus.	161
1996	Gandevia and colleagues demonstrate a possible supraspinal source of fatigue using transcranial magnetic stimulation.	82
1997	Gage and colleagues show that exercise increases neurogenesis in the rat hippocampus.	125
1998	Gorassini and collaborators present evidence of plateau potentials in human motoneurons.	97
2002-2003	The laboratory of Gardiner shows that the biophysical properties of motoneurons change in response to endurance training.	11
2004	Neurobiology of exercise workshop held in Baton Rouge, Louisiana.	

study the relationship between the physiological properties of the motor neuron (axon size and conduction velocity) and its response to a wide range of synaptic inputs as well as the contractile properties of each motor unit. Based on a series of studies, they formulated what is now commonly known as the size principle, discussed in more detail later on. In one study they compared the histochemical properties of the muscles with the physiological properties of the axon and recognized that there were significant exceptions to the concept that red and white muscles were analogous to slow and fast, respectively. A clear example of the puzzlement of the relationship between red and white muscles and their speeds was demonstrated by Peter and colleagues (168), who noted that the hexokinase activity generally associated with the glycolytic potential of a muscle fiber was actually higher in red muscle than in white muscle. This was a surprise because it was assumed at the time that white (glycolytic) muscle had high glycolytic capacity and that red muscle had very low glycolytic capacity. Edgerton and Simpson (71) explained this apparent contradiction in a 1969 paper that pointed out that the glycolytic potential of the muscle fiber rather than the redness (and thus the vascularity and oxidative potential of the muscle) was directly linked to the speed of the muscle. It was shown that the oxidative potential of a muscle fiber with a high glycolytic potential could range from very low to the highest level among a population of muscle fibers in a given muscle regardless of its speed (i.e., red muscle could be slow or fast but white muscle was consistently fast). It was also noted that there was a very tight linkage between fibers with the highest myosin ATPase activity and the highest glycolytic potential. Finally, it was pointed out that fibers that had an intermediate level of oxidative potential (thus called intermediate fibers) had low myosin ATPase and the lowest glycolytic potential and that the percentage of fibers with this histochemical profile matched the percentage of motor units in the muscle that were demonstrated to be slow or fast physiologically. In the subsequent 5 yr, a series of papers verified this initial interpretation based on histochemistry using quantitative biochemical techniques.

A more direct verification of this interpretation was demonstrated by Burke and colleagues at the motor unit level (33). By isolating single motor units and determining the contractile properties of a population of motor units in the medial gastrocnemius of the cat, they found that the contractile properties matched the histochemical profiles as predicted (see table 3 in Edgerton and Simpson [71]) and subsequently verified at the muscle level by quantitative biochemical techniques. One additional property defined by Burke and colleagues that was not studied by Henneman and colleagues was the fatigability of the motor unit. Again, as predicted, motor units that were fast and had high oxidative capacity were highly resistant to fatigue, as were the slow high-oxidative fibers, as demonstrated when the physiological and biochemical properties of whole muscles with a different proportion of phenotypes were examined biochemically and physiologically. These concepts were summarized in a review by Burke and Edgerton in the first edition (1975) of *Exercise and Sport Science Reviews*, which widely disseminated this information to the rapidly growing exercise physiology and neuroscience communities (31).

While improvements in histochemical techniques were occurring and being applied to human and animal muscles, Elwood Henneman (figure 8.1*a*) and his colleagues at Harvard University were finding that motor units in the cat were recruited in a stereotypic pattern; that is, there was order in the sequence of recruitment of motor units as force demands increased. Results published from the Henneman laboratory in 1965 (see figure 8.2) (106,107) provided evidence that formed the basis for the size principle of motor unit recruitment. In their experiments, they found that the recruitment of motoneurons of the hind-limb muscles of decerebrate cats in order to stretch proceeded in the order the of smallest, slowest conducting axons (which innervated small, slow-contracting muscle fibers) to the largest, most rapidly conducting axons (which innervated fast muscle fibers). They also suggested that the size of the axon, and therefore of the motoneuron cell body, was the most important determinant of this recruitment order in a given motoneuron pool because recruitment order did not change when a variety of excitatory and inhibitory stimuli—from ipsilateral and contralateral sources, alone or in combination—were superimposed on the stretch stimulus. Although the size principle is attributed to and verified experimentally by Henneman and his colleagues, Denny-Brown and Pennybacker, in their observations of recruitment of the human biceps brachii, published in 1938 (49), made reference to the possibility of the existence of an order of motor unit recruitment extending from small force to large force. In addition, in a 1929 paper Denny-Brown noted that motor units from red extensors were recruited before those in pale extensors during the stretch reflex in the cat. However, little was known of how this preference for recruitment was related to the actual size of motor units, particularly across multiple motor pools (muscles) (47).

The first definitive published work relating a property of motoneurons to the contractile properties of the innervated muscle fibers was that of Bessou and colleagues, published in 1963 in French. In this work, axon conduction velocities were found to covary with muscle contractile speed (15). Robert Burke (1934–) (see figure 8.1*b*) at the National Institutes of Health confirmed in his experimental research from the late 1960s and early 1970s that the properties of muscle fibers and their innervating motoneurons were indeed linked. Slow muscle fibers were innervated by small, excitable motoneurons, and motoneuronal excitability gradually decreased as one moved up the recruitment ladder in a continuum toward faster, larger motor units. Burke's work also showed, as one might expect from Eccles' original work on the afterpotentials of motoneurons innervating fast and slow muscles (66), that fast and slow motor units in the same muscle had afterpotentials of a different time course. In addition, Burke demonstrated that densities and types of synaptic contacts onto motoneurons were significant in determining recruitment order (29, 32). Daniel Kernell (127) presented evidence that small motoneurons were not merely scaled-down versions of large motoneurons but that they possessed membranes that had higher relative resistivity, thus rendering small motoneurons that much

Figure 8.1 *(a)* Elwood Henneman (1915-1996), who proposed the size principle of motor units while a professor in the faculty of medicine at Harvard University. *(b)* Robert Burke (1934–) of the National Institutes of Health, who provided the first detailed description of the properties of different motor unit types.

(a) From the collection of Charles M. Tipton. *(b)* Photo courtesy of Robert E. Burke.

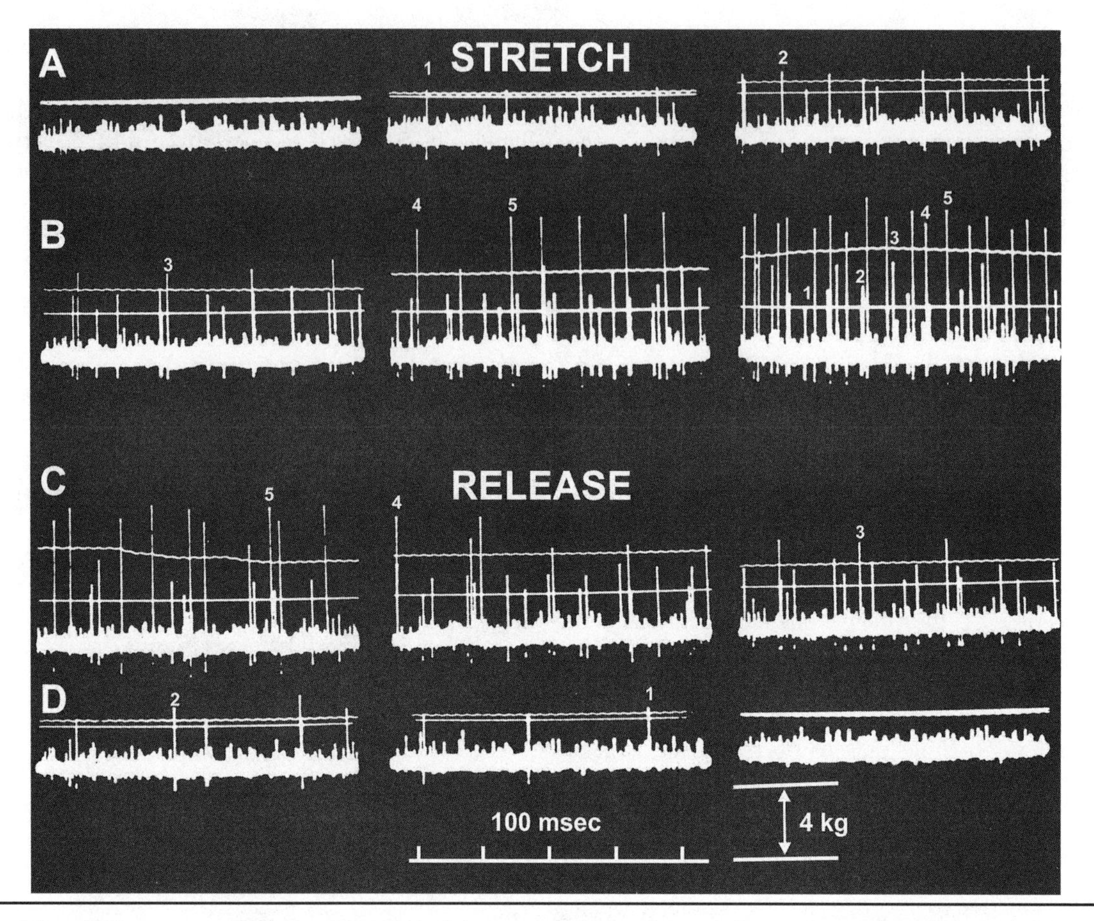

Figure 8.2 A Henneman experiment demonstrated the systematic recruitment of motor axons according to size during a muscle stretch in the cat. Five motoneurons are recruited and derecruited in order of spike amplitude (numbered 1 to 5).

Reprinted from E. Henneman, 1965, "Functional significance of cell size in spinal motoneurons," *Journal of Neurophysiology* 28: 560-580. With permission of American Physiological Society.

more excitable. All of this information provides a reasonably understandable blueprint for how motor units might be recruited during exercises with differing force demands.

But was there a way to demonstrate orderly recruitment in an intact animal or human model? Two major research foci addressed this issue: the demonstration of muscle fiber use during exercise using the glycogen-depletion technique and the recording of single motor unit action potentials using indwelling and, later, surface electromyography (EMG) electrodes.

As more was learned about the relationships between motor neurons and the muscle fibers that they innervate, interest and speculation regarding how these different types of motor units were used in normal movements increased. One of the first direct demonstrations of this, by Edgerton and colleagues in 1970, histochemically identified the phenotype of the muscle fibers that had apparent depletion of the phosphorylase enzyme immediately after a brief bout of exercise on a treadmill (70). It was demonstrated with remarkable clarity that only a few minutes of exercise on the treadmill resulted in a highly selective phenotype showing phosphorylase depletion. These were the muscle fibers with the higher oxidative capacity. These results were particularly intriguing given that it had just been reported that a selective depletion of muscle fibers with the lowest oxidative capacity occurred when a muscle nerve was stimulated electrically (137). The logical interpretation is that when the nerve is stimulated electrically there is selective activation of the largest axons, which tend to be the ones that innervate the motor units that are the most fatigable and have the lowest oxidative potential. Their low oxidative capacity makes them more susceptible to fatigue and to glycogen depletion. Although this has never been verified, the presumed relationship between apparent phosphorylase depletion and glycogen depletion in a single muscle fiber can be attributed to the observation that the phosphorylase enzyme is tightly bound to glycogen and that this enzyme becomes free floating in a tissue slice upon glycogen depletion. The phosphorylase depletion data are also consistent with the size principle; that is, running at a moderate speed requires recruitment of only the more excitable, smaller motor neurons and that the larger, faster fibers with low oxidative capacity are unlikely to be recruited to a significant level and therefore glycogen depletion would be spared. This was another example demonstrating that electrical stimulation reverses the recruitment order that occurs in normal movement. In 1968, Kugelberg and Edstrom (137) provided another possible approach to this issue. They stimulated muscles of anesthetized rats via the motor nerve and noted that the histological stain for glycogen showed decreased content in nearly all fibers. Because glycogen is slow to resynthesize, this provided a window into fiber use during activation. Gollnick and Saltin (figure 8.3) seized the opportunity to use this technique in voluntarily exercising humans. Using leg muscle biopsy material taken before and after exercise using the Bergstrom biopsy needle (14), they provided the initial histological evidence from human subjects that motor units are recruited in stereotypic patterns during exercise (92-94). Their results were consistent, for the most part, with the size principle, within the limitations that are inherent in using metabolic indexes of muscle fiber recruitment (e.g., fibers use glycogen differently depending on their metabolic economy, their initial glycogen levels, and the relative state of anaerobiosis in the muscle during exercise). Vollestad and colleagues (210, 211) continued this work by using quantitative histochemistry of glycogen in individual fibers in biopsies taken at various intervals during exercise. This allowed them to determine the time course of disappearance of glycogen as exercise progressed. Their results showed, for example, how Type I and IIA fibers are initially recruited at 75% of $\dot{V}O_2$max and how Type IIAB and IIB fibers (now referred to as IIAX and IIX, respectively) are recruited to maintain exercise pace as Type I and IIA fibers deplete their glycogen stores and become fatigued. Their results also showed a more rapid decline in fiber glycogen as exercise intensity increased. In an attempt to avoid the limitations of the glycogen-depletion techniques, Beltman and colleagues (13) showed the usefulness of measuring the change in the phosphocreatine:creatine ratio in freeze-dried segments of biopsied muscle fiber samples. Apparently, recruitment using this technique can be measured after only seven brief isometric contractions, which is not possible using the glycogen-depletion technique.

The use of indwelling and surface EMG electrodes also contributed significantly to the knowledge of how force is increased voluntarily during exercise. In 1929, Adrian and Bronk (2) were recording motor nerve activity during reflex generation in decerebrate-cat preparations using the technically challenging divided-nerve preparations, wherein nerves were dissected until only a few, or even one, axons remained for recording. They decided to try a concentric needle electrode and record the motor nerve activity from muscles. This would be technically easier to do, assuming correctly that motor axon and muscle fiber action potentials were one for one. Their electrode was basically a 36-gauge needle

Figure 8.3 *(a)* Philip Gollnick of Washington State University, Pullman, in the United States. *(b)* Bengt Saltin of the University of Copenhagen. In the 1970s, Gollnick and Saltin, using muscle glycogen-depletion patterns in histochemical sections of muscle biopsy material, demonstrated the recruitment patterns of muscle fiber types during various intensities and durations of exercise in humans.

(a) From the collection of Charles M. Tipton. *(b)* Photo courtesy of Bengt Saltin.

with an enamelled copper wire in the center. Their results from the triceps brachii demonstrated for the first time in man that increased voluntary drive resulted in both an increased firing rate of low-threshold motor units and recruitment of additional motor units as force increased:

> With a slight degree of voluntary contraction sharply isolated action currents appear at a rate which may begin as low as 6 a sec and rises gradually as the contraction develops . . . in the later stages of the record, as the contraction becomes more nearly maximal, so many fresh rhythms have appeared that it is impossible to pick out the individual series. . . . We conclude that the voluntary contraction in man is maintained, like the reflex contractions in the cat, by a series of nerve impulses which range from 5 to 50 or more a sec in each nerve fibre, and that the gradation in force is brought about by changes in the discharge frequency in each fibre and also by changes in the number of fibres in action. (2)

Using a type of electrode that is similar to that used in the Adrian and Bronk model, Milner-Brown, Stein, and Yemm in 1973 (155) used the technique of spike-triggered averaging, first developed in Henneman's laboratory (149), to demonstrate that lower threshold units in the human first dorsal interosseus muscle had slower, smaller twitches and that twitches became faster contracting and larger as contractile force increased. This constituted a fine demonstration that the ideas formulated by Burke and Henneman in the cat were in fact similar in humans. In a companion paper, Milner-Brown and colleagues (154) demonstrated how recruitment of motor units was the most important mechanism for initially increasing force and that rate coding becomes more important at higher forces. Kukulka and Clamann (138) extended these findings by illustrating that muscles differed in their profiles of recruitment versus rate coding to increase force. Their results showed that no significant recruitment of additional motor units occurred in the human adductor pollicis after 30% of maximal voluntary contraction (MVC) whereas recruitment was still evident in the biceps brachii at 88% MVC.

There have been several challenges and qualifiers to the size principle. Judith Smith and colleagues at the University of California, Los Angeles (191) demonstrated that the normal recruitment order of soleus before gastrocnemius, which adhered to the size principle, could be reversed under certain circumstances. They found that when a piece of tape was applied to a cat's hind foot, the cat would perform rapid paw shakes in an attempt to remove it, using a recruitment pattern in which the gastrocnemius predominated and the soleus was virtually silent. A significant caveat of interpreting this as an example of a violation of the size principle is that although both muscles are synergists to ankle extension, they are in fact separate muscles; the gastrocnemius is also a knee flexor. Examples of changes in the recruitment order of human motor units were reported as early as 1963 by Basmajian (8) and have been reported periodically since. Over the years hundreds of studies have challenged the size principle of motor unit recruitment. There are many exceptions to a precise and identical recruitment order for all movements; however, in most cases the robustness of this size principle has proven to be an important fundamental feature of the neural control of movement. Numerous studies have been undertaken to determine the mechanism of this recruitment order and to determine which features of the size of the motor unit are important in defining its order of recruitment. Multiple features of the motor neuron and muscle unit generally fit the size criterion. Among these many features, however, recruitment order is most clearly defined by the force that the motor unit can generate. When comparing pairs of motor units that have been isolated at the axonal level, the motor unit recruited earlier was consistently the one that generated the least force (44). For the most part, this applied even to motor units across two synergistic motor pools. Based on a rather definitive study in which all the muscle fibers of a single unit are identified by repetitive activation and glycogen depletion of muscle fibers, it is known that the maximum force that a motor unit can generate is largely a function of the number of muscle fibers of that muscle unit (24). The net size of all muscle fibers of the motor unit contributes to a small degree, as does the difference in the type of myosin. (Fibers with slow myosin generate slightly less tension per cross-sectional area than do fibers with fast myosin.) But approximately 85% of the variability that determines the maximum forced potential of a motor unit can be attributed to the number of fibers.

In the early 1980s, Loeb enunciated the idea of task groups of motor units (which implied a functional compartmentalization of the motor apparatus) based on afferent partitioning and supraspinal influence, which could explain how the same muscle functioning in two different tasks could have different motor unit recruitment profiles (144). These different profiles do not reflect a difference in recruitment order in a motor pool, however. There were several demonstrations of the task specificity of motor unit recruitment, which may or may not indicate violation of the size principle. John Desmedt (52) illustrated how the generation of ballistic contractions was accompanied by the generation of very high initial motor unit firing rates. He also demonstrated that recruitment order was consistent, with very few exceptions, during isometric, slow, and very fast contractions. Romaiguère and colleagues (170) showed that during slowly increasing and decreasing ramp contractions, the force at which motor units ceased firing was on average 25% lower than the force at which recruitment occurred on the increasing ramp. This hysteresis effect during increasing and decreasing force is the rule rather than the exception; firing frequencies that are generally lower for a given force on the down ramp probably reflect some catch-like contractile mechanism in the muscle.

Tax and colleagues (199) demonstrated another interesting phenomenon regarding motor unit recruitment. Their experiments showed that motor units in the biceps brachii had a lower force threshold during slow isotonic contractions than during isometric contractions and were recruited differently during isotonic extensions than during flexions. They found similar results during free movements and during movements imposed by an immovable torque motor, suggesting that these differences in recruitment were centrally generated. Van Bolhuis and colleagues (202) showed that motor unit firing rates did not all change the same way when changing from an isometric contraction to an anisometric (sinusoidal movements) contraction. Some units showed higher rates with anisometric contractions and others showed lower rates.

Finally, the concept of motor unit rotation has always generated significant interest in that the ability to switch among motor units during a prolonged submaximal task could help offset fatigue. Several demonstrations of motor unit rotation may have actually been expressions of the complexity of the architecture of the muscle investigated and of the task. However, since the 1990s, studies of the trapezius (214), supraspinatus (118), and several forearm and distal leg muscles (9, 10) have provided some convincing evidence for motor unit rotation (figure 8.4). These studies have all added to the understanding of how complex motor unit recruitment is during voluntary movements and of the added complexity when considering anisometric contractions that involve alterations in contractile speed, length, and task.

The realization came in the 1980s that the source of variations in recruitment thresholds among motor units could be the differences among motoneurons in the generation of persistent inward currents (figure 8.5). It has been known since the work of Schwindt and Crill in the 1980s (183) that the membranes of motoneurons possess properties of voltage-dependent, noninactivating inward currents that can generate prolonged depolarizations, termed plateau potentials. These plateau potentials can be the basis for self-sustained firing when facilitated by monoamines such as serotonin and norepinephrine. Gorassini (97) and colleagues reported in 1998 that they could alter the relative behavior of a pair of tibialis motor units during a low-force isometric contraction by vibrating the muscle tendon for 0.5 to 1.5 s. Vibration often evoked the recruitment of a second motor unit that continued firing after vibration was terminated, often for several minutes. This second unit sometimes continued firing even when the subject was asked to decrease the firing frequency of the control unit. This strategy of comparing the behavior of a test unit with that of the control unit under different circumstances in order to reveal the presence of plateau potentials has since gone through several permutations, and it is not universally accepted as evidence that it represents the activity of persistent inward currents. Nonetheless, it provides us with a mechanism for plasticity in the way motor units are recruited during voluntary effort and provides another site where adaptations with exercise training could occur (figure 8.6).

There has been some question about whether motor unit types as described in animal experiments are pertinent to human muscles. For example, Sica and McComas, using graded stimulation of the motor nerve, found no relationship between isometric twitch tension and contraction time of motor units in the extensor hallucis brevis muscle (187). Stephens and colleagues (197), however, found a better fit with first dorsal interosseus and gastrocnemius motor units; the latter experiment was performed using intramuscular microstimulation. More recently, it has been demonstrated in human long flexor and thenar units that weak units are not necessarily slow and strong units are not necessarily fast and that there are more intermediate fatigue-resistance units in human muscles than in the cat gastrocnemius (21).

Involvement of the Nervous System in Fatigue

Much research over the past century has been devoted to finding the extent to which, and the associated mechanisms by which, the nervous system limits exercise performance. Besides anecdotal evidence (e.g., the effects of hypnosis or the startle reaction on generation of strength), neural involvement in fatigue has been sug-

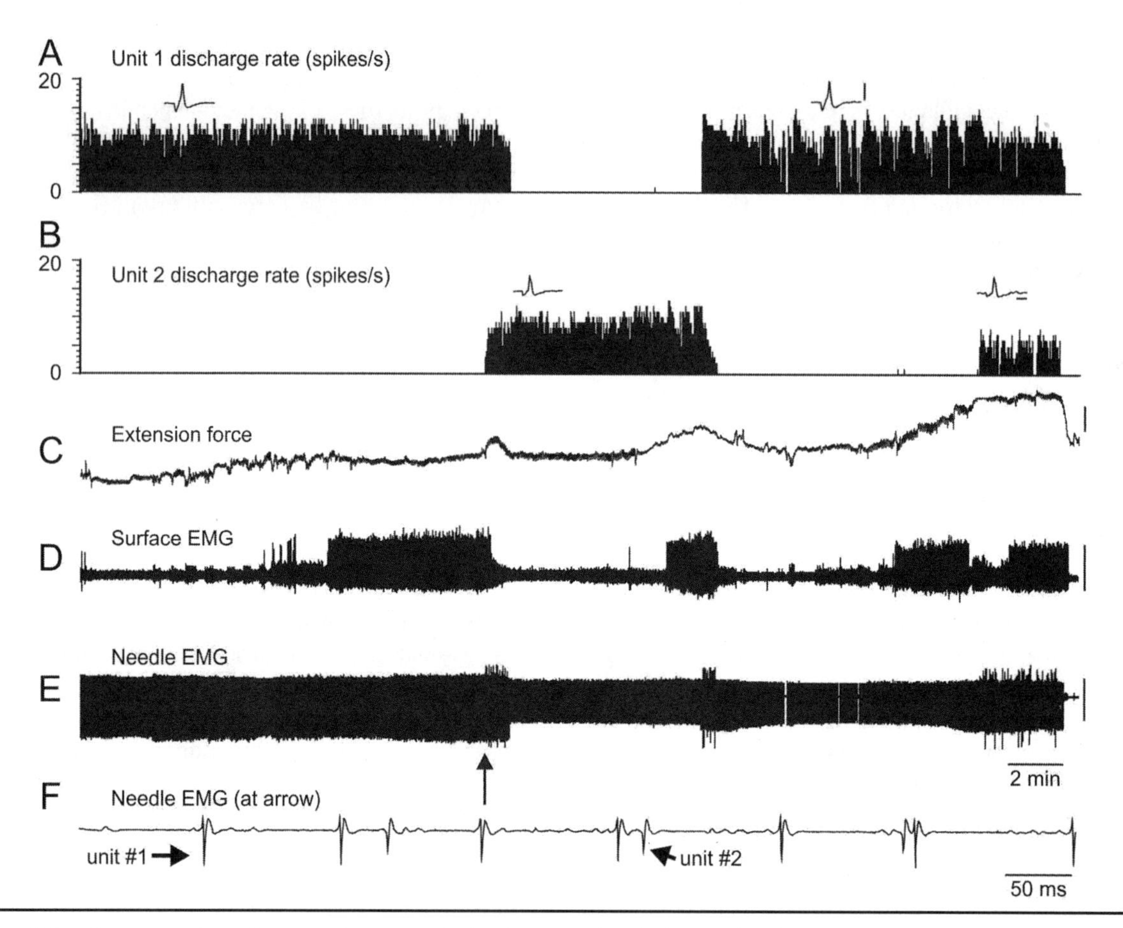

Figure 8.4 Demonstration of motor unit rotation in the flexor carpi radialis muscle of a human subject from the experiments of Bawa at Simon Fraser University. Unit 1 *(a)*, recruited at a low extension force, falls silent after unit 2 begins to discharge tonically *(b)*. Unit 1 then resumes firing as unit 2 stops. *(c)* Extension isometric force. *(d)* Unrectified surface electromyography. *(e)* Indwelling needle electrode electromyography. *(f)* Expanded time scale of part e showing activity at the onset of unit 2 discharge.

gested by several experimental results. The finding that the M-wave (the compound muscle action potential evoked by supramaximal electrical stimulation of the muscle nerve) and the voluntary EMG amplitude both decreased during MVC of the first dorsal interosseus muscle suggested NMJ failure (196). Bigland-Ritchie and colleagues later reported that MVCs were accompanied by parallel declines in surface EMG and the firing rates of individual motor units, with no change in the M-wave, suggesting that limitations occurred in the spinal cord or above (18, 20). This latter finding has since been reproduced in many studies, and the decrease in motor unit firing remains unexplained. This stimulated many researchers to search for the mechanisms of this decline in firing rate and to investigate fatigue in more prolonged types of activity (the MVC studies were restricted to <5 min). Possible neural sites of failure (that would include mechanisms other than the decrease in motoneuron firing rate) that have been investigated include the NMJ, motor axon propagation, motoneuron loss of excitability, loss of required excitation (from supraspinal centers and peripheral afferents), and inhibitory influence somewhere at or above the spinal cord.

NMJ Failure

The chance that failure at this level may be involved in fatigue is probably remote for large or force contractions that are short lasting (like those in the contraction studies by Merton and Bigland-Ritchie) and higher in activities of submaximal force maintained for relatively long periods of time. NMJ failure in human subjects is virtually impossible to ascertain, so this issue remains somewhat of a mystery. Types of failure include failure of the descending action potential to invade all terminal axon branches, failure of sufficient acetylcholine release to produce a large enough depolarization in the postsynaptic region (which can be due to reduction in acetylcholine release, reduced sensitivity of the postsynaptic membrane, or a combination of both), or a combination of both of these factors. Both of these possibilities have been demonstrated in in situ or in vitro preparations, usually using non-physiologically high rates of stimulation, and neither has been demonstrated in human studies.

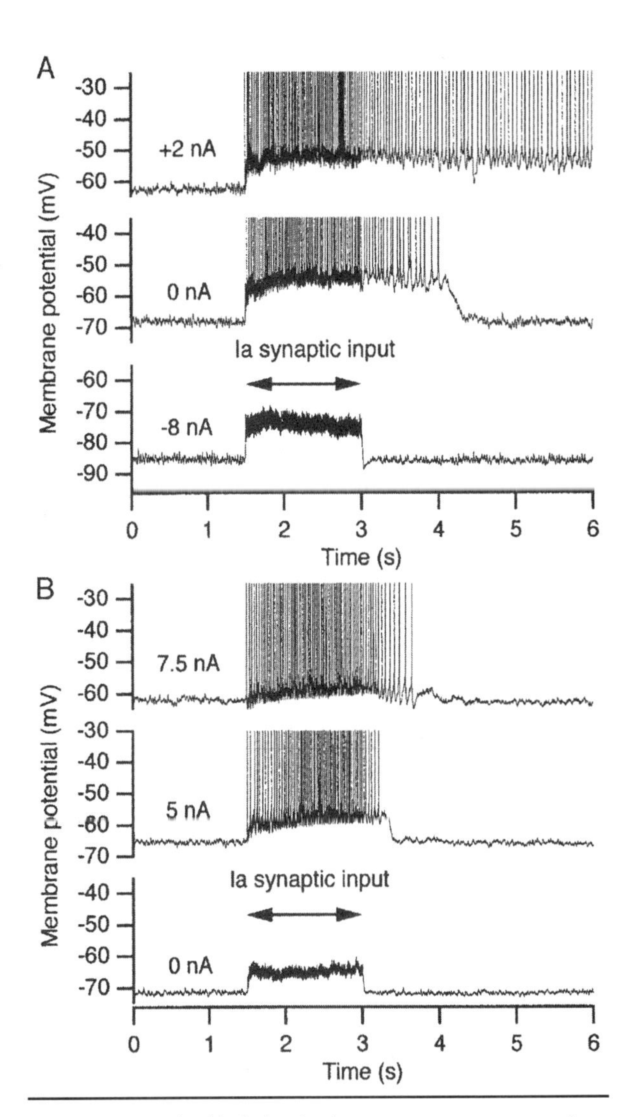

Figure 8.5 Bistable behavior in two cat motoneurons. In the top cell *(a)*, Ia synaptic input results in sustained firing after synaptic activation is turned off. Sustained firing occurs with no added somatic current injection and is enhanced with injection of a 2-nA depolarizing current. The bottom cell *(b)* shows less marked bistability under similar conditions.

Reprinted from R.H. Lee and C.J. Heckman, 1998, "Bistability in spinal motoneurons in vivo: Systematic variations in rhythmic firing patterns," *Journal of Neurophysiology* 80: 572-582. With permission of American Physiological Society.

There is ample evidence that the NMJ will fail if pushed at high enough frequencies via electrical stimulation. The decline in endplate potential amplitude with repetitive stimulation of the NMJ, causing Wedensky inhibition, had been known since the early 1940s (65). Krnjevic and Miledi showed this in 1958 but claimed that there was little evidence that depletion of neurotransmitters was involved because miniature end plate potential (mepp) characteristics were unaffected by fatigue (135). Thesleff was instrumental in demonstrating that at least part of this decrease was a desensitization of the acetylcholine receptors due to excess acetylcholine (201). Both of these latter possibilities take on added significance when one considers that they adapt to exercise training (discussed later).

Perhaps the most compelling argument to date for possible neuromuscular transmission failure during exercise is the in vitro work from the laboratory of Gary Sieck at the Mayo Clinic (136). In these experiments, neurotransmission failure in the rat diaphragm in vitro was tested by measuring force generation during stimulation of the nerve, interspersed with direct muscle stimulation, at stimulation frequencies that were within the range one would expect in vivo. They found that muscle force declined as time of stimulation proceeded, as one would expect, but the decline was significantly less for indirect muscle stimulation than for direct stimulation. In addition, the difference in force generation resulting from indirect and direct muscle stimulation increased with time, suggesting that the NMJ played an increasingly important role with time. More recently, Van Lunteren and Moyer (204) demonstrated, once again in vitro but with physiological ranges of frequency, that increasing the nerve terminal calcium conductance using 3,4-diaminopyridine decreased the difference in force between indirect and direct stimulation. This suggested that calcium conductance at the terminal may limit neurotransmission during continued activity.

Decreased Motoneuron Excitability

Marsden, Meadows, and Merton presented an abstract at the Physiological Society meetings in 1969 titled "Muscle Wisdom" (the text of that abstract is unavailable). In an abstract for the same society in 1971, they reported that motor units in the adductor pollicis decreased during an MVC from as high as 100 imp/s to about 20 imp/s in 30 s (148). The idea of muscle wisdom in which the declining firing rate was appropriate to the slowing of muscle unit contractile speed originated at this time. This decline in motoneuron firing was subsequently studied in more detail by Bigland-Ritchie and colleagues (19) and generated renewed interest in the phenomenon of motoneuronal late

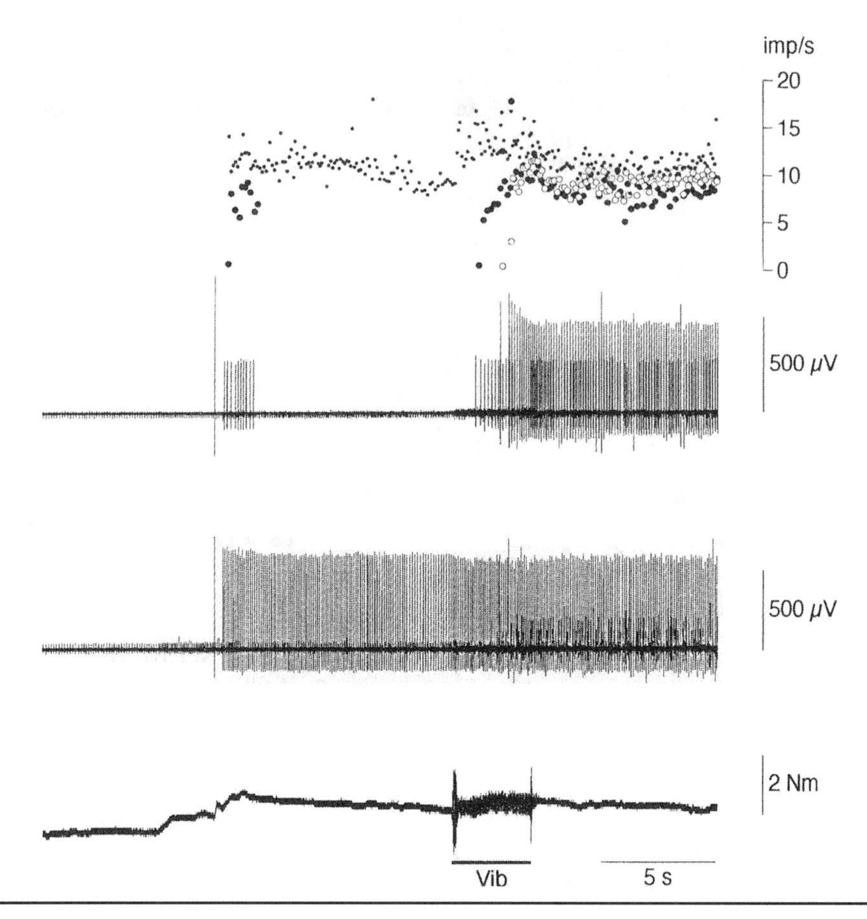

Figure 8.6 Evidence of plateau potentials in human motor units. Dorsiflexion force is shown at the bottom; vibration of the tendon (Vib) is shown. Traces of motor unit spikes show that one unit (second from bottom) maintained its firing frequency constant (except for during vibration) and that two other units (second from top) were recruited during the vibration and continued firing when vibration ceased. The top trace shows two units (larger open and filled circles) activated by vibration.

Reprinted from O. Kiehn and T. Eken, 1997, "Prolonged firing in motor units: Evidence of plateau potentials in human motoneurons?" *Journal of Neurophysiology* 78: 3061-3068. With permission of American Physiology Society.

adaptation that had been initially described by Daniel Kernell while at the Karolinska Institute in the 1960s (figure 8.7) (126, 128). This phenomenon was described as being different from accommodation and perhaps the result of summing of conductances of successive spikes that would slow the firing rate despite constant excitation. Late adaptation was clearly shown to not constitute motoneuronal fatigue because increased firing rates could be achieved during adaptation by increasing current strength. In an experiment in which Kernell and Monster measured this phenomenon in motoneurons and the simultaneous expression of muscle unit forces in the anesthetized cat, late adaptation was later analysed as a possible source of loss of muscle force during fatigue (129). Data from Kernell's laboratory in Amsterdam and Douglas Stuart's laboratory in Tucson, Arizona, in the United States (195) showed late adaptation to be more pronounced in fast motor units than in slow motor units, which seemed to be a good fit with the muscle fatigue profiles of the muscle fibers. Data also showed that late adaptation was present during intermittent and continuous excitation and that it was not the result of motoneuron damage due to microelectrode impalement because it was also present when the motoneuron was stimulated extracellularly (195). Late adaptation is much less evident in decerebrate cats during the generation of fictive locomotion compared with the nonlocomotor state (27), leaving many to doubt its pertinence in the fatigue process. It is also not a phenomenon that occurs to all motor units simultaneously. During submaximal contractions maintained to fatigue, when overall excitation to the pool is increasing, some units show decreasing firing rates at the same time that others show increasing firing rates, as shown by Garland and colleagues in 1994 (85).

The H-reflex has also been used to attempt to determine changes in basic motoneuron excitability that

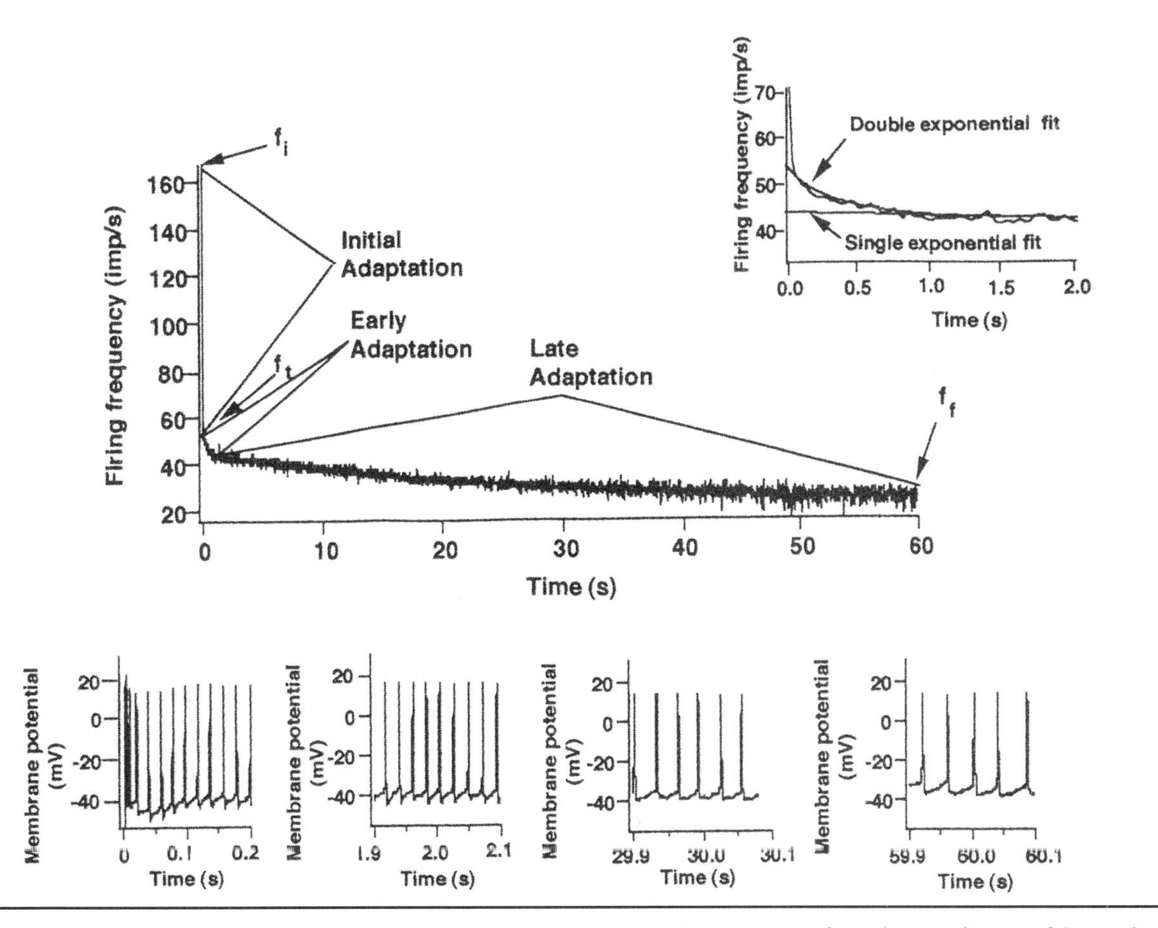

Figure 8.7 The three phases of motoneuron adaptation in rat hypoglossal motoneurons from the experiments of Sawczuk and colleagues.

Reprinted from A. Sawczuk, R.K. Powers, and M.D. Binder, 1995, "Spike frequency adaptation studied in hypoglossal motoneurons of the rat," *Journal of Neurophysiology* 73: 1799-1810. With permission of American Physiology Society.

might lead to late adaptation in human motoneurons (where late adaptation cannot be measured directly) despite knowledge that this reflex response can be influenced by many factors other than motoneuron excitability (35). The most frequent response to fatigue has been a decreased H-reflex amplitude (87), although the decrease was shown to be maintained during recovery when the muscle was kept ischemic, thereby suggesting a peripheral source for at least a portion of the altered H-reflex. Recently, evidence for a decreased excitability of motoneurons was provided by Butler and colleagues in Australia (34), who noted that the motor-evoked potential (MEP) resulting from electrical stimulation of descending axons at the cervical level had a decreased amplitude of elbow flexors at the end of an MVC. An important finding in their results was that the decreased MEP recovered regardless of whether the arm was kept ischemic during recovery, suggesting that signals from the periphery were not involved. The jury is still out about whether decreases in motoneuron excitability, as opposed to altered excitation per se, are important in fatigue.

Decreased Motoneuron Excitation and Increased Inhibition

Because muscle spindles supply excitatory influence to motoneurons, one might propose that this source of excitation might decline during fatigue, thus contributing to the decreased firing rate. It was made clear by Nelson and Hutton in 1985, and supported by later research, that group Ia and II spindle afferents do not decrease—and in fact might even increase—in dynamic sensitivity following fatigue (162). Hagbarth and colleagues in 1986 reported that blockade of gamma motoneurons innervating the tibialis anterior using small doses of prilocaine injected near the nerve trunk reduced motoneuron firing rates and force during an MVC (101). These investigators also pointed out that stimulating the spindles via tendon vibration could attenuate the decrease in EMG during maximal isometric contractions. The importance of spindle afferent activity for generation of maximal force was subsequently supported by the demonstration from the laboratory of Macefield and Gandevia in Australia (146) that spindle afferent firing rates declined during a constant-load isometric contraction of the EDL at 60% MVC, even while central drive, as evidenced by increasing EMG, was increasing. Experiments in this lab also showed how pharmacological blockade of tibialis anterior spindle afferents reduced the firing frequencies of motor units during an MVC.

Bigland-Ritchie (figure 8.8) and colleagues (16, 219) conducted ingenious experiments beginning in the 1980s that suggested that inhibitory signals emanating from fatiguing muscles might be providing at least part of the source for the declining motoneuron firing rates during fatigue. Their approach was to render the muscle ischemic during the fatiguing contraction or during recovery and measure motoneuron firing rates and voluntary forces (figure 8.9). The logic used to support this approach was that if stimulation of metabolite-sensitive afferents in fatiguing muscles (figure 8.10) was providing an inhibitory influence to the motor pool, then firing rates would surely fail to recover if the muscle was kept ischemic during recovery. Kaufman's laboratory had already provided sufficient evidence, confirmed by several afterward, that group III and IV afferents were stimulated by a number of chemical substances that increased in muscle during exercise. In fact, these afferents showed increased firing rates even at low levels of activity in walking decerebrate cats (124). Bigland-Ritchie and colleagues found that the reduced firing rates that occurred with a sustained MVC did not recover while the muscle was kept ischemic, which was consistent with the presence of a fatigue reflex. Garland, Garner, and McComas (86) confirmed these results and further reported that the phenomenon occurred when the muscle was electrically stimulated to fatigue as well. This muscle ischemia paradigm has been used since that time to suggest, for example, that fatigue of the medial gastrocnemius influences the performance of the lateral gastrocnemius, presumptively through effects of the reflex on close synergists (177); that decreased motoneuron excitability was too slow to be explained by motoneuron adaptation or reduced spindle support (64); and that flexors and extensors may differ in the effect of this fatigue reflex (34).

The 1990s saw the initiation of the use of transcranial magnetic stimulation (TMS), first described by Barker and colleagues in 1985 (7), as a tool for investigating fatigue mechanisms. The different sites stimulated by TMS and transcranial electrical stimulation (described by Merton and Morton in 1980) (152), and the interpretations of the changes in the short-latency MEPs that it evoked in muscles, allowed for more information on possible fatigue sites at the supraspinal level. In 1996, results from the laboratory of Simon Gandevia (figure 8.11) and colleagues revealed that MEP amplitude increased during an MVC of elbow flexors, as did the silent period following the MEP, which globally suggested increased intracortical inhibition or reduced voluntary motor drive (figure 8.12). Furthermore, the MEP amplitude and silent period results were unaffected by tendon vibration or ischemia administered at the point of fatigue, indicating that peripheral changes with fatigue were not the source of these changes. By showing that similar changes in MEP amplitude did not occur with electrical stimulation at

Figure 8.8 *(a)* Brenda Bigland-Ritchie, who, while at the John B. Pierce Laboratory in New Haven in the United States, provided evidence of a fatigue reflex resulting from sensory information emanating from maximally contracting muscles in humans. *(b)* Roger Enoka of the department of integrative physiology at the University of Colorado at Boulder in the United States, who has contributed significantly to the understanding of the neurobiological components of fatigue in humans and has demonstrated the interactions of aging, fatigue, and training on motor unit recruitment patterns.

(a) From the collection of Charles M. Tipton, *(b)* Courtesy of Roger Enoka.

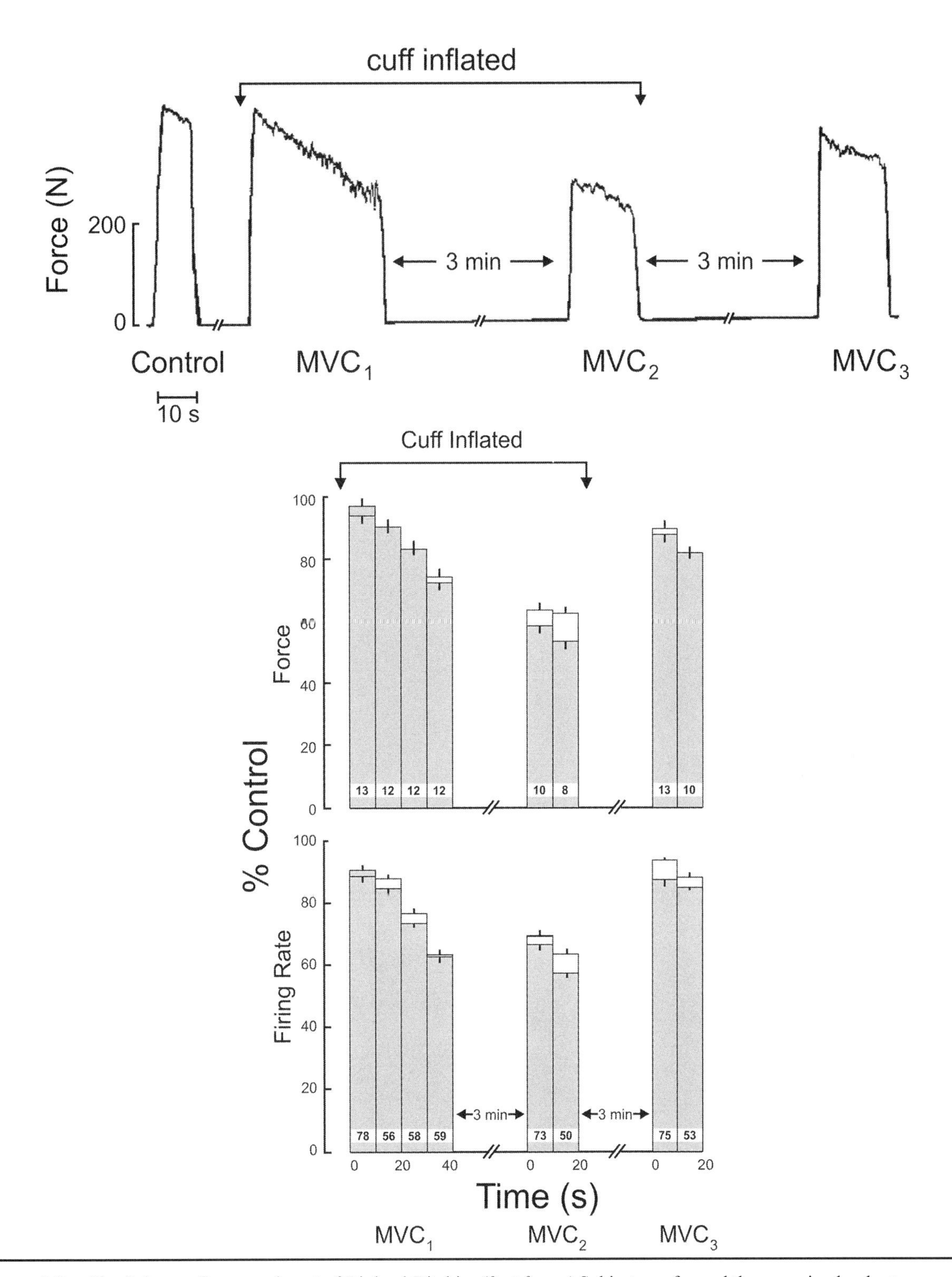

Figure 8.9 The fatigue reflex experiment of Bigland-Ritchie. *(first figure)* Subjects performed three maximal voluntary contractions of the quadriceps. A pneumatic cuff inflated around the upper thigh to block blood supply to the muscle group was applied as shown. *(second figure)* Force (top) and motor unit firing rates (bottom) were influenced by the ischemia, indicating an effect of muscle metabolic status on regulation of motoneuron firing rate.

Reprinted from J. Woods, F. Furbush, and B. Bigland-Ritchie, 1987, "Evidence for a fatigue-induced reflex inhibition of motoneuron firing rates," *Journal of Neurophysiology* 58: 125-137. With permission of American Physiology Society.

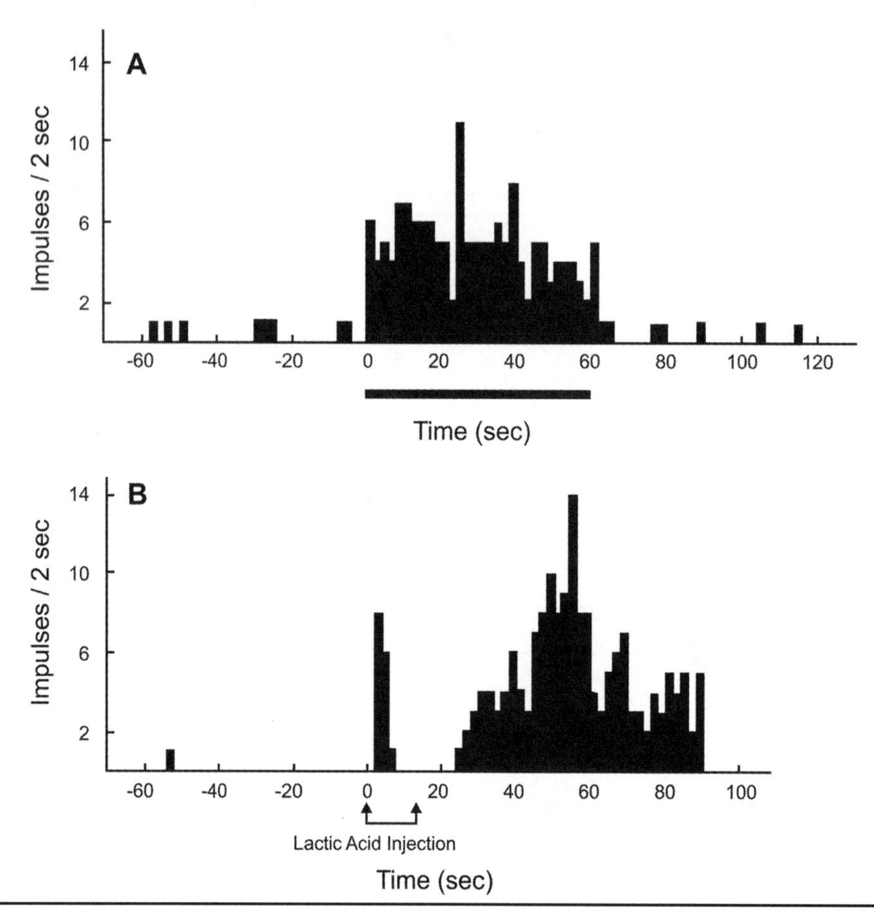

Figure 8.10 Effects of static contraction (*a*) and lactic acid injection into the femoral nerve (*b*) on discharge of group IV afferents in the triceps surae of an anesthetized cat.

Reprinted from D.M. Rotto and M.P Kaufman, 1988, "Effect of metabolic products of muscular contraction on discharge of group III and IV afferents," *Journal of Applied Physiology* 64: 2306-2313. With permission of American Physiology Society.

the cervicomedullary level, they revealed that the source of these fatigue-induced changes was supraspinal. Finally, Gandevia and colleagues showed that fatigue increased the amplitude of the twitch response evoked by TMS, suggesting that a source of excitation upstream from the motor cortex was becoming suboptimal for maximal muscle force generation (82). Although this approach is still used in the Gandevia laboratory as well as in other laboratories to investigate fatigue, using it to derive mechanisms has proved difficult, especially since the time course of changes in MEPs during exercise and recovery does not correlate well with other indexes of central fatigue, such as reduced motor unit firing rates and voluntary activation measured via the superimposed twitch technique (reviewed in reference 81). Interestingly, the fatigue-related changes described by Gandevia and colleagues do not require MVCs. Sogaard and colleagues (194) and Smith and colleagues (193) showed increased silent periods and increased TMS-evoked twitches after prolonged contractions of the biceps brachii at 15% and 5% of MVC, respectively (figure 8.13). Decreased cortical excitability following the marathon has also been reported (171).

A Neural Component of the Resistance-Training Response

Several observations of the results of strength training originally led to the belief that there was a neural component to the training response (figure 8.14). The inability to attribute increased performance during resistance-type exercise to the standard morphological parameters of muscle size was perhaps the most common factor contributing to the belief in a neural component of strength training. For example, in 1953 Hettinger and Muller (108) reported strength gains of up to 160% after 1 wk of strength training. This phenomenon has been consistently reported since that time. Another observation supporting nervous system involvement in strength training is the cross-education effect, in which the increase in strength resulting from training of one limb results in increased strength in the contralateral, untrained limb. This phenomenon was investigated in some detail by Hellebrandt and colleagues and described in papers published between 1947 and 1951 (104, 105) and is now a regularly observed phenomenon. Herbert de Vries, in the 1974 text *Physiology of Exercise for Physical Education and Athletics*, stated that "it seems likely that the cross-education is brought about by an overflow of nervous energy from neurons in the motor cortex which innervate the crossed pyramidal fibers, to a smaller number of neurons that supply the uncrossed fibers" (53). In his 2006 review of the literature on cross-education that included a meta-analysis of the literature, Timothy Carroll and colleagues in New South Wales concluded that cross-education is a real, neurally mediated phenomenon and proposed mechanisms for the cross-education effect, although to date these mechanisms remain unverified (37).

Figure 8.11 *(a)* Simon Gandevia of the University of New South Wales, whose laboratory has contributed to the knowledge concerning motor control in general and the supraspinal components of fatigue in particular. *(b)* Jacques Duchateau of the Free University of Brussels (Université Libre de Bruxelles), whose laboratory has described neurophysiological responses to training, disuse, and fatigue in humans.

(a) Photo courtesy of Simon Gandevia. *(b)* Photo courtesy of Jacques Duchateau.

Other more recent evidence has added to the notion of nervous system involvement in the strength-training response. This evidence includes an increase in the amplitude of the surface integrated electromyographic (IEMG) signal during maximal contractions of various speeds, including isometric contractions, as demonstrated by the research of Komi and colleagues (133), Moritani and de Vries (160), and Häkkinen and Komi (102). The most frequently cited research in this area is that of Moritani and de Vries, published in 1979, which, using IEMG, showed the separate contributions of muscle hypertrophy and neural factors to the strength gain seen after 6 wk of isometric training of the elbow flexors. Although comparisons of EMG signals on the same subjects separated by several weeks to several months might raise questions of reproducibility, this change in IEMG during maximal effort has been too consistent to ignore; it has been substantiated by subsequent investigators via normalization to the M-wave resulting from electrical stimulation of the motor nerve. Nonetheless, interpretation of these results must take into consideration that IEMG records electrical activity of a subpopulation of the activated muscle fibers. The origin of the increased IEMG, which supposes that muscle activation by the motor system was suboptimal in either magnitude (training results in increased recruitment or rate coding?) or pattern (altered synchronization of motor units?) before training, remains to this day unresolved. As an example, the question of whether individuals are capable of recruiting all of their motor units during an MVC, using permutations of the twitch-interpolation technique introduced by Merton in 1954 (151), remains unanswered.

There are many other, more indirect observations that have been attributed to neural effects, including demonstrations of specificity of the training response to the movement pattern during training (169), joint angle specificity (84), and velocity specificity (156), all of which have continued to be demonstrated in the literature. In 1989, Jones and colleagues (123) published a review in which these neural changes might be ex-

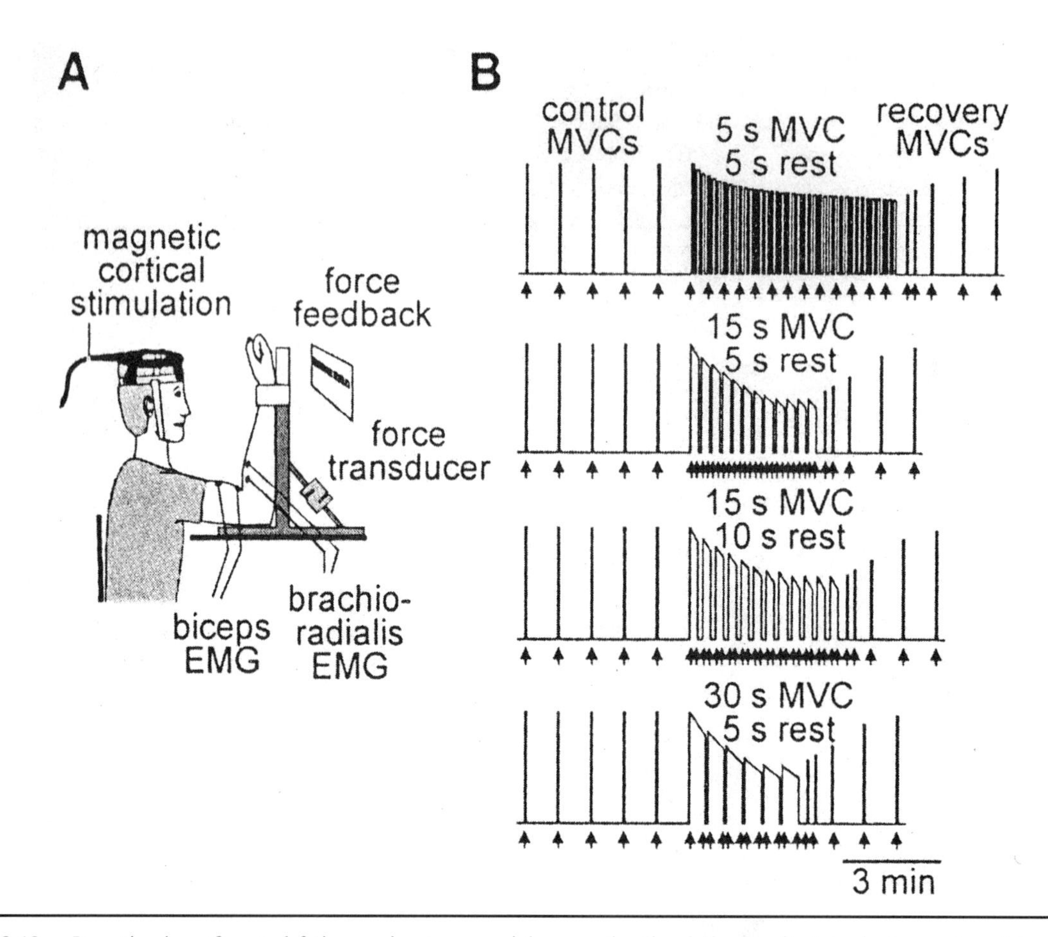

Figure 8.12 Investigation of central fatigue using transcranial magnetic stimulation. Left: A typical experimental protocol involving fatigue of the elbow flexors. Arrows represent stimuli delivered to the cortex. Right: The fatiguing protocol results in a gradual increase in the amplitude of the muscle force evoked by transcranial magnetic stimulation (top and bottom traces) while voluntary torque is decreasing (filled circles; showing force at start and end of each 15 s maximal voluntary contraction joined by a line).

Reprinted from J.L. Taylor et al., 2000, "Supraspinal fatigue during intermittent maximal voluntary contractions of the human elbow flexors," *Journal of Applied Physiology* 89: 305-313, 2000. With permission of American Physiology Society.

plained by alterations in muscle composition and architecture, arguments that have since gained some support from the experiments of Blazevich (23) on changes in muscle fascicular geometry with training. Although the extent to which these indirect measures indicate neural effects of resistance-type training remains controversial to this day, more direct evidence of the effects of training on nerves and neural circuits has been generated in the past 25 yr and is presented later in this discussion.

More objective measures of neural activation have supplied intriguing evidence that something is going on in the nervous system during resistance training. In 1975, Milner-Brown and colleagues at the University of Alberta (153) were curious about why synchronization of motor unit firing seemed more prominent on the side contralateral to neuropathies of nerve injuries as well as in individuals performing manual jobs requiring brief, powerful contractions. They conducted a study on resistance-training effects (first dorsal interosseus muscle) on motor unit synchronization as well as short- and long-loop reflex responses. They confirmed that motor unit synchronization increased as a result of resistance training and showed evidence of effects of enhanced responses of corticospinal connections. Semmler and Nordstrom (184) later confirmed the increased synchrony of motor units in strength-trained individuals using a more precise analysis of discharge patterns of pairs of motor units. They also found evidence of increased motor unit coherence, which is a function of oscillatory input to motor units that originates in cortical and subcortical areas (185). Sale and colleagues (178, 180) later substantiated the reflex results of Milner-Brown and colleagues in several other muscles, demonstrating that the phenomenon was not specific to hand muscles, and with longer periods of re-

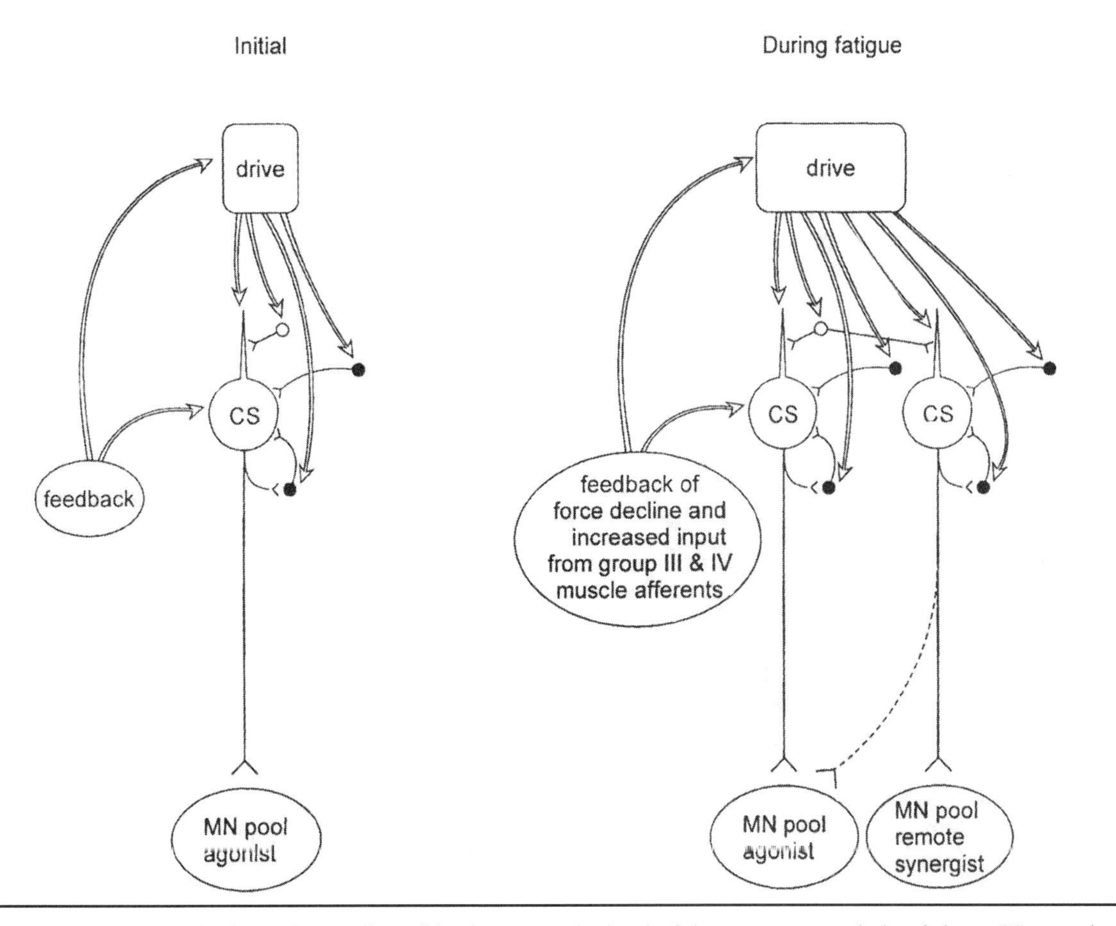

Figure 8.13 Simon Gandevia's scheme of possible changes at the level of the motor cortex during fatigue. CS = corticospinal cell; MN = motoneuron. Solid cells are inhibitory; open cells are excitatory.

Reprinted, by permission, from S.C. Gandevia, 2001, "Spinal and supraspinal factors in human muscle fatigue," *Physiological Review* 81: 1725-1789. With permission of American Physiology Society.

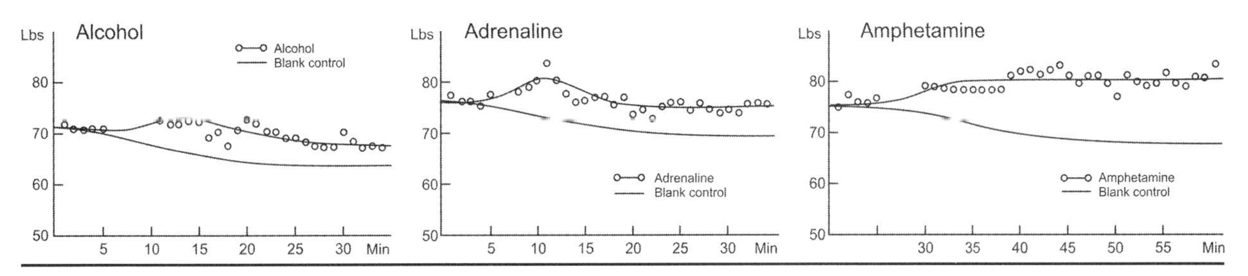

Figure 8.14 Average forearm flexor performance of 10 subjects after consumption of 15 to 20 ml of 95% ethanol (left), intramuscular intake of 0.5 ml of a 0.1% solution (center), and oral intake of 30 mg of amphetamine sulphate (right). Maximal voluntary contraction occurred once/min for 30 min. Solid lines indicate performance of the same subjects in absence of intervention.

Reprinted from M. Ikai and A. Steinhaus, 1961, "Some factors modifying the expression of human strength," *Journal of Applied Physiology* 16: 157-163. With permission of American Physiological Society.

sistance training, thus adding evidence for a central neural effect. They measured reflex potentiation during maximal contractions and found that it was generally greater after resistance training, suggesting that training increased the excitability of motoneurons during maximal effort. In 2002, Carroll and colleagues, using TMS (transsynaptic excitation of corticospinal neurons) and transcranial electric stimulation (direct depolarization of corticospinal neurons) presented evidence that spinal circuitry, and not the motor cortex, adapts to resistance training of the index finger abductors (38).

The evidence of a training effect on the recruitment patterns of single motor units has been elusive and results have been equivocal. Nonetheless, two investigators have demonstrated the clearest results concerning this issue. Roger Enoka (1949–) (see figure 8.8*b*) and colleagues at the University of Colorado in the United States demonstrated clearly that training at submaximal levels of force reduced the discharge variability of motor units in a hand muscle in elderly subjects performing slow shortening and lengthening contractions; this consequently improved manual dexterity (134). Results from the laboratory of Jacques Duchateau (1954–) (see figure 8.11*b*) in Brussels indicated that ballistic-type training of ankle dorsiflexors for 12 wk resulted in a more rapid onset of the muscle EMG and higher maximal instantaneous motor unit firing rates in the tibialis anterior during the ballistic task (figure 8.15). In addition, firing rates showed a less marked decrease and a higher incidence of doublet firing after training (203).

Responses of Single Nerve Cells to Exercise Training

Most of the evidence regarding neural adaptations to endurance training has been provided using more direct methods of measuring neural structure and function. These studies often use animal models, which were not feasible for use in experiments on the determinants of voluntary maximal efforts.

Evidence of the responsiveness of spinal neurons to increased activity dates back to the 1890s, although there was some confusion in the literature about whether the investigators were considering chronic or acute responses to increased activity. It was known rather early that neuromuscular exhaustion affected indexes of motoneuronal protein synthesis and degradation rates (60, 96, 113) and increased phosphatases, including ATPase (122, 212). It was not until 1957 that unequivocal evidence existed that permanent morphological changes occurred in motoneurons with endurance-type training. In a paper published that year, Edstrom (73) demonstrated that the nucleoli of guinea pig motoneurons were significantly elevated after daily treadmill running of at least 30 min/d for 29 d and concluded that neuronal protein synthesis was elevated. More recent research demonstrating the effects of chronic exercise on motoneurons was that of Gerchman et al. (88) at Michigan State University in the United States, who showed in a 1976 paper that alpha motoneurons from endurance-trained rats demonstrated slightly higher levels of an oxidative enzyme (malate dehydrogenase) and less stress in response to an acute exercise bout (increased acid phosphatase and decreased glucose-6-phosphate were blunted in trained motoneurons) compared with nontrained controls. Although the finding of increased resistance of motoneurons to acute stress after endurance training has been borne out in subsequent literature, the increased mitochondrial content has not. Evidence from the Edgerton (39) and Ishihara (116) laboratories in the 1980s and 1990s substantiated, using quantitative histochemistry, that small motoneurons had more succinic dehydrogenase activity than did large motoneurons and that altered neuromuscular activity had little effect. This latter finding was also consistent with the findings from Daniel Kernell's laboratory in Amsterdam in the 1980s in which motoneuronal succinic dehydrogenase remained unchanged despite chronic electrical stimu-

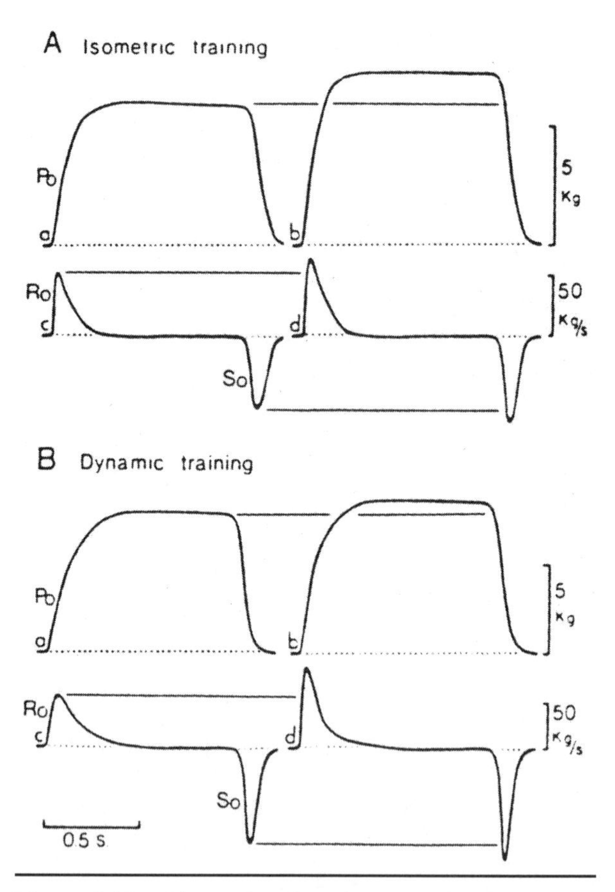

Figure 8.15 The study of Duchateau and Hainaut examined dynamic training versus isometric training. Three months of moderate isometric (*a*) or dynamic (*b*) training of the adductor pollicis resulted in different responses of peak isometric strength (top traces) and maximal rate of force development (first derivative of force; bottom traces).

Reprinted, by permission, from J. Duchateau and K. Hainaut, 1984, "Isometric or dynamic training: differential effects on mechanical properties of a human muscle," *Journal Applied of Physiology* 56: 296-301. With permission of American Physiology Society.

lation of the gastrocnemius muscle to produce dramatic biochemical and functional changes in the latter (61). Because motoneuron cell bodies are activated antidromically in this paradigm, this experiment provided further evidence that oxidative potential of motoneurons is not a limiting factor in their function.

Acquiring knowledge of the effects of training on the physiological properties of nerve cells in general, and on motoneurons in particular, required a reduced model rather than the exercising human. The investigation of electrophysiological changes in motoneurons is a relatively recent development given that recording from spinal neurons using microelectrode techniques did not take place until the 1940s and that an appropriate animal model incorporating the characteristics of trainability and recordability had to be exploited. Previous pioneering work on intracellular recording from motoneurons in cats by John Eccles (1903-1997) and colleagues in the 1950s revealed that motoneurons innervating fast and slow muscles had different action potential properties, especially different afterpotentials, most likely reflecting the differing functions of these two muscle types (66). The seminal research of Robert Burke at the National Institutes of Health in the 1960s added to this information, illustrating that various electrophysiological properties of motoneurons covary with the functional properties of the muscle fibers that they innervate, such that properties of motoneurons decreased in excitability and in duration of afterpotentials and increased in axon conduction velocity as one moved from the smallest and slowest motor unit to the largest and fastest (30). Several groups subsequently illustrated the plasticity of many of these motoneuronal properties resulting from interventions such as spinal cord transection (45), axotomy with and without reinnervation (78), and chronic stimulation of the target muscle to induce fiber-type changes (79). In the mid-1980s, Phillip Gardiner (1949–) (while on sabbatical leave) and Daniel Kernell (figure 8.16) worked together to investigate the properties of motoneurons in the rat (83), an animal model that had been used and then abandoned in the 1960s for the technically less difficult feline model. Gardiner continued to use this model in Montreal to demonstrate the changes in motoneuron biophysical properties that accompany endurance training (11, 12). These reported changes include hyperpolarization of resting membrane potential and voltage threshold (the membrane voltage at which the action potential spike begins a steep climb to the peak), increased rate of increase of the action potential, and increased amplitude of the afterhyperpolarization. Interestingly, Cleary and colleagues (42) had shown several of these changes in the motoneurons innervating the siphon in Aplysia, who underwent classical conditioning of the siphon withdrawal reflex, suggesting that motoneuron changes with endurance training are similar to classic learning responses. To date, then, evidence shows that, like several other interventions, endurance training alters fundamental properties of motoneurons. How these changes relate to changes in other neurons in the system and how the networks involved in movement are altered remain to be established.

In terms of single-cell responses to training, there is very little evidence from cells other than motoneurons. Gómez-Pinilla and colleagues at the University of California, Los Angeles have demonstrated that dorsal root

Figure 8.16 *(a)* Phillip Gardiner of the University of Manitoba in Canada with Daniel Kernell in Kernell's laboratory at the University of Amsterdam. Gardiner described the effects of increased and decreased physical activity on the metabolic and biophysical properties of rat motoneurons. Kernell provided the first detailed analyses of motoneuron late adaptation as a possible source of motor unit fatigue. *(b)* Reggie Edgerton of the department of integrative biology and physiology at the University of California, Los Angeles. Edgerton's laboratory made the first forays into the effects of training on the transected spinal cord.

(a) Photo courtesy of Phillip Gardiner. *(b)* Photo courtesy of Reggie Edgerton.

ganglion cells are also sensitive to increases in chronic activity, showing alterations in gene expression and enhanced neurite outgrowth in vitro (159).

Exercise Training and the NMJ

Because the NMJ has historically been the most studied synapse in the nervous system, it is quite surprising that the knowledge of how it adapts to exercise training is so poor. Surely part of this dearth of knowledge is attributable to the impression that the NMJ is indefatigable under conditions where firing rates are within reasonable physiological ranges. Krnjevic and Miledi (135) reported in 1958 that spontaneous discharge of miniature end-plate potentials was still evident even when their rat phrenic nerve and diaphragm muscle preparations were stimulated at frequencies much higher than one would expect in vivo, thus demonstrating that neurotransmitter was not depleted. Since that time, however, various experimental results have convinced exercise physiologists that the NMJ is indeed worth investigating as a site that adapts to increased activity. The effects of decreased usage of various forms have been studied more than the effects of increased usage. However, that unique story has not been included here because the events that occur with decreased usage may not necessarily represent a mirror image of events that occur with increased usage and in fact often muddy the waters in attempts to discern the effects of increased activity.

The morphology of the NMJ differs among the various fiber types, which may or may not be a function of their differences in activation. Padykula and Gauthier (167) in 1970 pointed out differences in several architectural features of NMJs among red, white, and intermediate muscle fiber types in the rat diaphragm (using the nomenclature based on staining for oxidative metabolism), which might support the concept of activity-related adaptations at the ultrastructural level. In 1985, Ogata and Yamasaki (164) expanded on this information using improved electron microscopic techniques, confirming most of the findings of Padykula and colleagues and adding information on differences in terminal branching (length, diameter, number). Despite continued research on this question and on the issue of adaptation to exercise training by several groups, the literature on the effects of training on morphology of the NMJ has been equivocal and only a few common changes have emerged from various studies. These include increased nerve terminal and postsynaptic membrane areas with endurance training.

Some of the first evidence for the functional adaptability of NMJs to increased activity was pharmacological and biochemical. In 1954, Jewell and Zaimis (119) showed that increased (compensatory overload via synergist tenotomy) and decreased (tenotomy) activity of the cat soleus and tibialis anterior changed the sensitivity of these muscles to the depolarizing blocking drug decamethonium, thus demonstrating a qualitative change in the properties of the NMJ with altered activity. In 1966, Lloyd Guth and colleagues at the National Institutes of Health (100) demonstrated that compensatory hypertrophy of the soleus and plantaris of up to 14 d resulted in increases in muscle cholinesterase content to values 10% (plantaris) to 25% (soleus) higher than controls. Crockett and colleagues (46) followed up on this latter observation in 1976 by showing that an increase occurred in cholinesterase in the endplate region of the vastus lateralis of endurance-trained guinea pigs. This was a significant addendum to the literature because acetylcholinesterase exists in endplate and nonendplate areas of the muscle. Fast forward to the 1990s when Gisiger and colleagues at the Université de Montréal succeeded in demonstrating that a particular form of acetylcholinesterase, the globular G4 form, was particularly sensitive, showing an increase in response to increased activity. Their hypothesis emanating from the results of several studies was that this form of the enzyme, found in soluble form in the perijunctional space (as opposed to the form attached at the NMJ, where each acetylcholinesterase unit consists of three G4 units joined to the postsynaptic membrane by a collagen tail), served as a sink to mop up excess acetylcholine before it could accumulate and cause desensitization of the acetylcholine (90, 91, 117).

Like with NMJ morphology, it is tempting to believe that physiological differences in NMJ among fiber types may be the direct consequence of varied functional demands and therefore training effects. The physiological differences between NMJ on fast and slow muscle fibers were made clear by Gertler and Robbins in 1978 when they demonstrated that endplate potentials of fast muscle fibers were larger but more prone to rundown in endplate potential amplitude in response to repetitive stimulation compared with those of slow muscle fibers (89). Thus, fast NMJs had a larger safety factor, defined as EPP/(Eap − Em), where Eap is the threshold potential for generating an action potential and Em is the membrane potential. Wood and Slater (218) estimated the safety factors of fast and slow muscle NMJs to be 5.0 and 3.5, respectively (where 1.0 represents the amplitude of the endplate potential required to generate a muscle fiber action potential). Various groups have shown that several other properties are different between fast and slow NMJs. These properties include the amount of G4 acetylcholinesterase (higher in fast; 4), density of acetylcholine receptors (higher in fast; 198), sodium channel density on the postsynaptic membrane (higher in fast; 176), and voltage dependency of activation and inactivation of the sodium current (more negative in fast; 175). Evidence for the effects of training on NMJ physiology is meager. In 1991, Dorlöchter and colleagues (62) exposed mice to voluntary exercise in wheels and measured physiological properties of NMJs of the extensor digitorum longus in vitro. Their results showed unambiguously that endplate potentials doubled and endplate potential amplitude remained above control levels when the NMJ was challenged with high-frequency stimulation (100/s for 1 s). Tests with presynaptic and postsynaptic blockers in vitro convinced them that the adaptations involved increased transmitter release. The results of Dorlöchter and colleagues were confirmed in the laboratory of Phillip Gardiner at the Université de Montréal in the hind-limb NMJs of rats subjected to endurance training and compensatory overload. The researchers used an in situ preparation in which muscle contractions were inhibited using μ-conotoxin, thus allowing the measurement of full endplate potentials (3, 51).

Consistent with the proposed increase in the area of the NMJ with endurance training, Desaulniers and colleagues demonstrated that the number of acetylcholine receptors also increases in several muscles, including the diaphragm, with endurance training (50). These results, as well as the results comparing the morphology and physiology of slow and fast NMJs, show that training does not turn fast NMJs into slow NMJs. Otherwise, endplate potential amplitudes and NMJ areas would decrease.

What are the mechanisms involved in these adaptations? Harold Atwood and colleagues at the University of Toronto, using an experimental model (141) that chronically stimulated the motoneurons innervating crayfish claw and abdominal muscles and measuring the chronic adaptations, demonstrated that phasic NMJs (high initial neurotransmitter release that dropped off quickly during moderate-frequency stimulation) could be changed to tonic NMJs (lower initial neurotransmitter release but less drop-off during stimulation) within a few days after initiation of chronic stimulation. In 1986, they found that neurotransmitter release was not necessary for these adaptations because axons that were pharmacologically isolated from the source of stimulation still showed adaptations at the NMJ (142). In 1988, they reported that subthreshold depolarization of the motoneuron, and not necessarily the production of action potentials, produced the adaptation (143). Protein synthesis was not required for the adaptation, as reported by this group in 1990 (163). Later, this group reported that neurotransmitter fatigue resistance and change in the endplate potential could be evoked separately using various combinations of stimulation conditions and therefore that these two properties were controlled by two separate mechanisms (150). Finally, calcium influx into the motoneuron cell body was shown to be required for the adaptive response (111). Whether these mechanisms relate to the mammalian system has not been established.

Synaptogenesis, Neurogenesis, Motor Map Organization, and Exercise

Although the concept that an enriched environment can evoke plasticity in several regions of the nervous system has been around for at least a century (reviewed in reference 208), the interpretation of exercise as an enriched environment that can produce similar neural changes was not evident. In the 1990s, William Greenough and colleagues at the University of Illinois in the United States, who had previously demonstrated that motor learning tasks resulted in neuronal morphological changes, conclusively showed that learning acrobatic motor tasks but not treadmill exercise increased synaptogenesis and angiogenesis in the cerebellar cortex. At the same time, however, exercise (1 h/d for 30 d) resulted in only angiogenesis in the same brain region (22, 115). The research was important in demonstrating that the presumptive learning-associated responses were not attributable solely to the repetitive use of synapses that one would see in rhythmic exercise but rather required the additional challenge of a changing environment. The laboratory of Fred Gage (figure 8.17*a*) carried this research further in demonstrating that an enriched environment increased the normal rate of development of new neurons in the hippocampus (a brain structure important for memory) of mice and that this increased neurogenesis occurred in mice that were housed in running wheels (125, 207). Interestingly, unlike what had been found previously for other brain regions, environmental enrichment and exercise—but not a water-maze learning task—increased cell proliferation in the hippocampus. The Gage laboratory subsequently demonstrated that the increased neurogenesis had functional significance in that exercise-induced enhancement of neurogenesis in the dentate gyrus of mice was associated with improved water-maze performance and improved long-term potentiation in vitro, which has been suggested as a model for certain forms of learning and memory (206). This research, conducted during the same period that the Cotman laboratory was demonstrating the effects of exercise on brain neurotrophins in rodents (referred to later), has provided a basis for the promotion of exercise as a therapeutic modality for the preservation and improvement of memory and cognition and for the consideration of the possible beneficial effects of physical activity in individuals with conditions such as Alzheimer's disease, Huntington's disease, or Parkinson's disease and even the normal aging process (205).

While the literature on cortical motor map plasticity in response to skill training, especially after trauma and as it concerns individuals possessing unusually well-developed motor skills (such as skilled musicians), is quite extensive (132), little information is available on the effects of exercise training independent of the skill component. Jeffrey Kleim and colleagues, originally at the University of Lethbridge and more recently at the University of Florida in the United States, attempted to address this difficult issue. Their research confirmed that voluntary running exercise increases motor cortical angiogenesis but not movement representations in the motor cortex (131) and that strength training has no effect on the cortical map but may in fact induce synaptogenesis and increase neuronal excitability in the spinal cord (1).

Figure 8.17 *(a)* Fred (Rusty) Gage, whose laboratory at the Salk Institute demonstrated that enriched environments and exercise enhance brain neurogenesis. *(b)* Fernando Gomez-Pinilla of the department of integrative biology and physiology at the University of California, Los Angeles, whose laboratory provided evidence of how physical activity, learning, and nutritional factors control brain neurotrophins.

(a) Photo courtesy of Fred Gage. *(b)* Photo courtesy of Fernando Gomez-Pinilla.

Training Adaptations in Reflex Pathways

Plasticity of the monosynaptic reflex was first demonstrated in 1941 by Eccles and McIntyre, who showed that the reflex responses in the cat to stimulation of dorsal roots that had been quiescent (severed extraganglionically several weeks previously) were different from those of the control (67). In the 1980s, the group of Wolpaw and colleagues at State University of New York at Albany revealed the intriguing phenomenon whereby the monosynaptic reflex could in fact be trained to increase or decrease in amplitude. In the experimental paradigm, monkeys were trained with a juice reward to increase or decrease the EMG response to a stretch perturbation of the forearm flexors. The changes became persistent with continued training in that increases or decreases in the amplitude of the stretch reflex were apparent when animals were anesthetized and for up to 3 d after removal of supraspinal influences by spinalization (217). The changes appeared to involve changes in intrinsic motoneuron properties, at least in the downregulation group (36). It was conceded that the changes were attributable to adaptations at several possible levels, including the motoneuron itself, the afferent-motoneuron synapse, and interneurons, and that the corticospinal tract was essential to generating the training effect. These investigators also demonstrated that these altered reflexes were associated with improved locomotor activity (41), suggesting (and eventually demonstrating) the potential for the use of the conditioning paradigm in cases where locomotion is compromised (40).

Several investigators subsequently verified the extrapolation of this adaptation to humans, including a demonstration of its potential usefulness for reducing hyperreflexia in stroke patients (182, 215). Whether this form of plasticity is significant in determining performance during exercise, where the reflex response itself is not the focus of training, has not been established, in part due to the complexities of the acute plasticity of reflex responses. Nonetheless, several groups have described chronic adaptations to exercise training of different types, both in cross-sectional and longitudinal studies. No firm consensus has emerged, as discussed by Paul Zehr (223).

Evolution of the Concept of Spinal Learning

Until the 1970s there was a widespread concept that the brain and particularly the spinal cord had little capacity for plasticity. Over the past several decades, voluminous data have demonstrated that in fact the opposite is true: Both are highly plastic and significant changes occur continuously throughout one's life. Some of these changes occur within milliseconds and others occur over a period of months and even years. These continuously changing properties that occur as a result of multiple biochemical and anatomical mechanisms have different time scales. A second concept that has dominated the field of learning a motor skill is that motor learning occurs only in the neural networks of the cerebellum. Of course, learning is one manifestation of plasticity, and it is generally recognized that multiple forms of learning with different time courses and different mechanisms exist. However, the concept of neural networks in spinal cord learning was generally recognized and accepted largely only in the past one or two decades. This is despite the evidence to the contrary found as early as the 1950s.

Experiments leading up to the concept of spinal learning have been reviewed previously (72). Briefly, several experiments demonstrated changes in monosynaptic and polysynaptic responses after a complete transection of the spinal cord, which would be expected. However, it was demonstrated that the magnitude of these changes could be modulated with repetitive activity (190). In a series of compelling experiments, unilateral tonic activity was induced with unilateral disruption of vestibular function. When this tonic activity was allowed to persist for approximately 30 min, this tonic activity remained even after a complete spinal transection (54). In another impressive demonstration of a learning-related phenomenon, it was demonstrated that a rat with a complete midthoracic spinal cord transection could learn to avoid a noxious stimuli applied to the foot (120). This experiment has been repeated numerous times with the appropriate controls, demonstrating that this behavior represents a clear case of one form of learning (98). In fact, the mechanisms involved in this response are known to be linked to BDNF, a neurotrophic factor also known to be important in learning among supraspinal networks such as those in the hippocampus (188).

During the same time period that these experiments were occurring, other researchers were conducting studies to determine whether more complex learning phenomena such as standing and stepping could occur in the spinal cord. In the late 1970s, the Edgerton laboratory at the University of California, Los Angeles began to examine the effects of exercise training in spinalized kittens on muscle and spinal cord properties (121). The first proposal that the spinal circuitry was capable of learning a complex motor skill was presented in a symposium at the annual meeting of the American College of Sports Medicine in 1985. Data were presented that demonstrated that a complete spinal animal could modify the trajectory of the hind limb during stepping in order to avoid a tripping stimulus applied during the swing phase of the step cycle (110). In 1983, the Edgerton group reported that recovery of locomotor capability was better in kittens that were spinalized at 2 wk versus 12 wk and that daily treadmill training had a significant positive effect on the recovery of locomotor patterns (192). Over the next two decades, numerous experiments demonstrated different forms of learning in spinal cord circuits. In 1986, it was established by the Edgerton group (145), and confirmed the following year by Serge Rossignol and colleague at the Université de Montréal (6), that daily treadmill walking involving weight bearing for several months postsurgery in an adult cat had a positive effect on hindlimb locomotor capacity.

Rossignol and colleagues demonstrated in 1998 that a daily injection of clonidine during training enhanced the recovery of locomotion during the early phase (10 d after transection). They continued their work to examine the potential beneficial effects of other agonists and antagonists of various neurotransmitter systems administered during the early phase after spinal cord transection with the goal of instituting effective training to the spinal cord as soon as possible after the injury (172).

In the late 1990s, the Edgerton group began to examine the quality of the training stimulus and how it influenced the adaptive responses. They reported, for example, that step training for 30 min/d was more effective than a similar duration of stand training in maintaining the structural and functional properties of the soleus muscle in cats (174). In addition, step training and stand training had very little transferability of positive effects, suggesting that training type was highly specific (110). Although the mechanisms behind this specificity are completely unknown, a wide range of molecular, cellular, and systems-level changes occur in the spinal locomotor circuitry with modified levels of activity (for a review, see reference 68).

In 2001, the Edgerton group reported initial experiments with spinalized rats. In these experiments, a robotic arm was used to passively track hind-limb trajectories during treadmill locomotion and to actively modulate limb trajectory by imposing forces at specific phases of the locomotor cycle. They found that the spinal cord could learn to adopt different locomotor trajectories in response to these perturbations, thus providing further evidence for the learning capacity of the spinal cord independent of supraspinal control (69).

These results on the effects of exercise training on the spinal cord below a partial or complete interruption of supraspinal influence had relevance to the rehabilitation of human patients with spinal cord injuries, and several groups have been influential in driving this field forward. The demonstration that humans with complete spinal cord lesions could demonstrate step-like EMG patterns of hind-limb muscles when partially supported on a moving treadmill (56, 59) opened up the possibility that experimental results from animals might have pertinence for rehabilitation in humans. So far, the consensus appears to be that treadmill training as a rehabilitative strategy is effective in patients with partial spinal cord lesions and some control of lower-limb function. The main players in this effort since the mid-1990s include S. Harkema, formerly of the University of California, Los Angeles and currently at the University of Kentucky in the United States (103), H. Barbeau at McGill University in Montréal (5), V. Dietz in Zurich (55), and A. Wernig from Bonn (213).

Supaspinal Control of Posture and Locomotion

Given the importance of the control of posture and locomotion in almost any exercise or sport, the evolution of the present ideas regarding the control of these important tasks is rather obvious. Years and hundreds of studies have provided an extensive body of literature that cannot be fully or even superficially acknowledged in this chapter. In this extensive amount of information, there is general agreement that the primary sources of descending control of locomotion are the cortex, reticular nucleus, vestibular nucleus, and rubrospinal tract. The motor cortex gives rise to multiple descending projections, with considerable attention given to the corticospinal tract that decussates and influences the spinal circuitry associated primarily with the contralateral limbs. It remains unclear, however, how the cortex and each of the other descending systems actually initiate control or modulate posture and locomotion. The motor cortex plays a very important role in executing more skilled movements that are less repetitive and in adjusting the basic activation patterns during locomotion in a more variable environment. For example, basic locomotor patterns can be relatively normal without corticospinal input.

A series of remarkable experiments by Shik and colleagues in Moscow in the 1960s demonstrated that stimulation of the mesencephalic locomotor region (MLR; a 1 mm long strip of cells in the nucleus cuneiformis) could elicit locomotion in a decerebrated cat (165). They demonstrated that a simple tonic stimulation of the MLR region could activate reticulospinal neurons, which, in turn, stimulated the spinal centers producing locomotion—largely circuits that are now considered to be central pattern generators. These experiments provided the basis for a new perspective on the level of automaticity that exists in the sensorimotor system. Shik and colleagues also observed that the reticulospinal neurons became more active during locomotion than when the animal is at rest. The reticulospinal neurons receive input from the MLR and from the cerebellum. The reticulospinal tract descends in the ventrolateral funiculi of the spinal cord projecting to multiple levels of the spinal cord. They demonstrated that the reticulospinal tract in cats was necessary for eliciting locomotion when stimulating the MLR.

Although the exact manner in which these neurons induce locomotion is not clear, there is some evidence that the MLR region is controlled by inhibition and that the initiation of stepping may be induced by disinhibition. When the MLR is stimulated, the decerebrated cat can generate locomotion and demonstrate awareness of spatial orientation. In animals that have a chronic (several weeks) lesion in the subthalamic locomotor region, locomotor behavior is nearly normal.

A hypothesis was presented by Orlovsky, Deliagina, and Grillner (166), who each worked in the Shik laboratory at one time, that during locomotion each limb is modulated by supraspinal input via groups of spinal neurons called controllers. It remains unclear exactly what these controllers are anatomically or how many there are, but one can imagine a number of such controllers that could interact to control individual joints. Shik and Orlovsky (186) proposed a two-level automatism control system for locomotion. One level provides nonspecific tonic input that determines the intensity of locomotion (speed and grade). It was suggested that the second level was responsible for making fine adjustments in the control of the limbs, including maintaining equilibrium. It was also suggested that this level of the control system interacts with sensory information, such as visual and proprioceptive inputs, to make fine adjustments in the locomotor pattern. It is now clear that the reticulospinal neurons and the MLR play important roles in providing input to these limb controllers.

There remain differences in opinion regarding the relationship of the neural control of posture compared with that of locomotion. One hypothesis is that the control systems are rather distinct, and an alternative hypothesis is that the neural programs for posture and locomotion are highly integrated and share rather extensively the spinal circuitry controlling these motor tasks. Considerable progress on the neural-control interactions of these tasks is being made, particularly in the laboratories of Deliagina, Orlovsky, and Grillner.

The cerebellum also plays an important role in motor control. A major function of the cerebellum is assisting in the control of limb movements by modulating supraspinal motor centers. It mediates sensory feedback from the spinal cord and modifies the motor output accordingly. The cerebellum also receives information from CPGs of the limbs to modulate the motor output. In addition, the cerebellum compares different inputs and provides a means of correcting intended movements (189).

Automaticity in Posture and Locomotion

Specific regions in the brain stem can initiate and control very complex motor behaviors, apparently with little or no necessity for conscious control, resulting in the generation of largely automatic responses. It is often assumed that the initiation of a movement, even the more automatic ones such as stepping, is triggered by a conscious event in the motor cortex. Today the question of what is consciousness remains an active point of debate. However, there seems to be a continuum of consciousness ranging from a totally simple reflex without any conscious awareness or capability of control to modulation of a task in which one is fully and continuously aware of every aspect of the movement. What is the level of consciousness when one begins to step compared with the level of automaticity during a simple monosynaptic reflex? Even the efficacy of a monosynaptic response can be modulated by conscious control in a rat, monkey, or human. A human with a low-thoracic spinal cord injury and no supraspinal control below the lesion can learn to stand and initiate steps using sensory information associated with unilaterally bearing weight and manipulating the hip position (59). This spinal stepping can be initiated consciously and voluntarily although the subject initiates the process reflexively. Thus, the subject manipulates the afferent inflow by controlling critical biomechanical and neurophysiological signals by manipulating other parts of the body. Based on these and many other observations, it becomes very difficult to formulate a concise definition of consciousness in motor control.

There is a strong element of automaticity in the neural control of most movements, both from the brain and spinal cord. For example, once the decision to walk across a room is made, very little conscious effort is required to perform the details of that task. It seems that the central nervous system is designed so that much of the intricate decision making associated with posture and locomotion occurs automatically. Supraspinal and spinal sources contribute to the pathways responsible for this automaticity. For example, a subthalamic cat, decerebrated just rostrally to the mammillary bodies, can walk in response to exteroceptive stimulation and can sometimes walk spontaneously. A cat decerebrated at the caudal border of the mammillary bodies cannot walk with or without exteroceptive stimulation. Therefore, the subthalamic locomotor region lies between these two transection levels. When the subthalamic locomotor region remains intact, the animal can walk by itself, go to a plate of food, and even kill and eat a mouse. Such a cat, however, cannot perform locomotion voluntarily.

The question of whether respiration and cardiovascular adjustments to exercise are neurally, hormonally, or metabolically driven has been raised for decades. Relative to this question, the concept of automaticity that is evident in the coordination of the neural control of locomotion is also a factor in the control of respiration and cardiovascular function. For example, not only does the stimulation of the MLR generate remarkably normal locomotion, it also leads to a modulation of cardiovascular and respiratory function that is appropriate for locomotion. When the MLR of a mesencephalic cat is stimulated and the limbs are stepping on a treadmill, researchers have observed a 20% to 35% higher arterial blood pressure, a 70% to 90% greater cardiac output, and an increase of 15% to 20% in heart rate relative to the resting conditions. In addition, pulmonary ventilation can increase threefold as a result of increased tidal volume and frequency of breathing (189). These observations demonstrate the highly integrated and automated nature of the neural system in

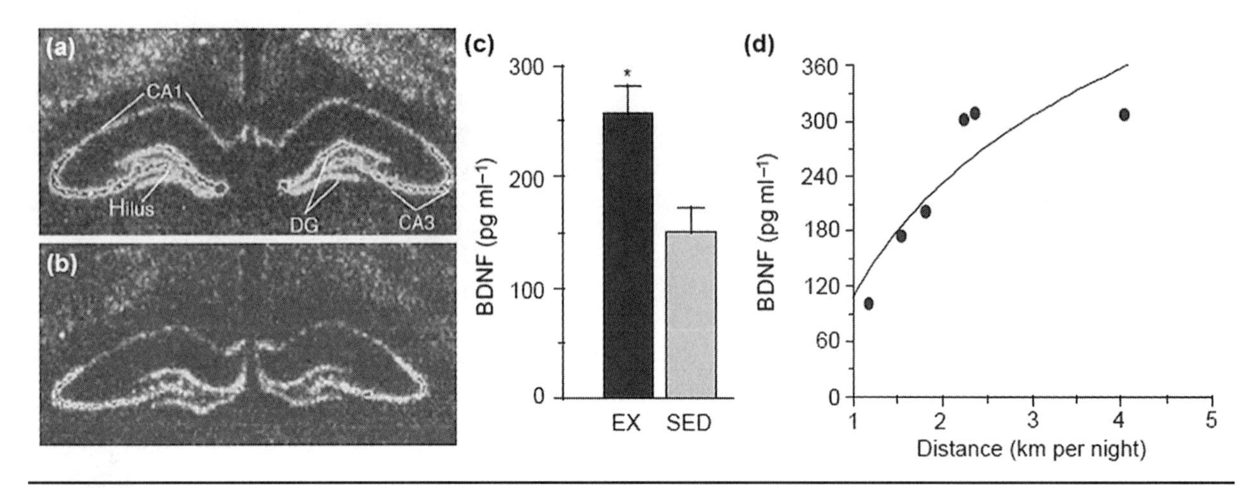

Figure 8.18 Results of the paper published by Neeper and colleagues in 1995. *(a)* Expression of brain-derived neurotrophic factor (BDNF) in rat dentate gyrus (DG), hilus, CA1 through CA3 regions, and cortex was increased in rats who performed 7 d of voluntary wheel activity compared with sedentary *(b)* rats. *(c)* Five days of wheel running increased BDNF protein in the hippocampus. *(d)* An increase in BDNF protein was related to average distance run during 14 d.

controlling locomotion with some assurance that the cardiovascular and respiratory systems will be modulated accordingly. It would be interesting, however, to perform experiments whereby the MLR was stimulated in a paralyzed preparation to exclude the involvement of mechanisms directly linked to activation of the muscle tissue and the consequential metabolic effect of muscular contractions.

Neurotrophins

A significant volume of evidence supports the positive effects of increased physical activity on cognitive function and on rehabilitation from neurotrauma. With the discovery of neurotrophins in the 1940s, possible mechanisms for the previously demonstrated changes in cognitive function and neurotransmitter systems with exercise began to take shape. A definite link between exercise and brain neurotrophins was not established, however, until the publication in 1995 of a one-page paper by Neeper and colleagues from Carl Cotman's laboratory at the University of California, Irvine (figure 8.18) (161). This paper demonstrated how rats given access to exercise wheels for up to 7 d showed increased messenger ribonucleic acid (mRNA) levels for the hippocampus and caudal one third of the neocortex after only 2 d of exercise and how these levels remained elevated for the 7 d of exercise. The authors also illustrated a dose-response relationship, with a significant positive correlation between mean distance that the rat ran per day and mRNA levels in both the hippocampus and caudal neocortex. The last sentence of the paper set the stage for future studies: "Exercise-induced upregulation of BDNF could help increase the brain's resistance to damage and degeneration through BDNF's support of neuronal growth, function, and survival." Later studies using microarray analysis of hippocampi from exercised rats revealed clusters of upregulated genes (synaptic trafficking, signal transduction pathways, and transcription regulators), all related to BDNF, suggesting a central role for the latter in brain plasticity with exercise (158).

However, it was recognized that these brain centers are not directly involved in motor function and that stronger support for the notion that neurotrophins play a role in adaptation of motor systems would be forthcoming with a demonstration of changes in neurotrophins in a motor system. Gómez-Pinilla collaborated with Reggie Edgerton at the University of California, Los Angeles to demonstrate that both the mRNA and proteins for BDNF and NT-3 were upregulated in the soleus muscle and in lumbar motoneurons after 5 d of treadmill exercise in rats (95). These adaptations were later substantiated by this group along with adaptations in other related proteins including the BDNF receptor trkB and several important modulators downstream from BDNF, including the proteins synapsin I, GAP-43, and CREB, as well as trkC, the receptor for NT-3 (221). The Gómez-Pinilla group also demonstrated effects of exercise on gene expression and neurite outgrowth of sensory neurons (159).

It was recognized that the finding of plasticity of neurotrophins and their related proteins in the exercise condition might have implications for the use of exercise as a rehabilitative tool when the nervous system is compromised. Indeed, exercise enhanced recovery from traumatic brain injury through an elevation of BDNF (99). Exercise was also shown to decrease the deficit in cognitive function in rats caused by a high-fat diet (157) and to enhance regeneration of sensory neurons (159). A general hypothesis emerged that involved BDNF as a central component (referred to by Carl Cotman as brain fertilizer) that linked exercise to neural plasticity and that implicated healthy lifestyle choices (e.g., physical activity, diet) in sculpting neural health (209). Later it was shown that elevation of insulin-like growth factor 1 during exercise interfaces with BDNF cascades to promote increased cognitive function (57). With respect to effects in the spinal cord, Gómez-Pinilla and his group demonstrated that exercise was instrumental in attenuating the allodynia that accompanies spinal cord injury, primarily via a BDNF-related mechanism (112). In addition, in collaboration with the Edgerton group, it was revealed that BDNF plays a role in the recovery of locomotor performance after spinal cord injury (222).

The new developments in the understanding of exercise responses in neurotrophin systems heralded an expert workshop in December of 2004 in Baton Rouge, Louisiana, to discuss interpretations and future directions in the area of neurobiology of exercise. A summary of discussions during this workshop was published (58). Among the recommendations emanating from this forum were the need for more information on the long-term effects of exercise on neural systems, exercise dose-response issues and neural health, the effects of exercise on various disease states, and the mechanisms by which neuroplasticity-related molecules perform their function.

References

1. Adkins D, Boychuk J, Remple M, Kleim JA. Motor training induces experience-specific patterns of plasticity across motor cortex and spinal cord. *J Appl Physiol* 101: 1776-1782, 2006.
2. Adrian ED, Bronk DW. The discharge of impulses in motor nerve fibres. Part II. The frequency of discharge in reflex and voluntary contractions. *J Physiol* 67: 119-151, 1929.
3. Argaw A, Desaulniers P, Gardiner PF. Enhanced neuromuscular transmission efficacy in overloaded rat plantaris muscle. *Muscle Nerve* 29: 97-103, 2004.
4. Bacou F, Vigneron P, Massoulié J. Acetylcholinesterase forms in fast and slow rabbit muscle. *Nature* 296: 661-664, 1982.
5. Barbeau H, Danakas M, Arsenault B. The effects of locomotor training in spinal cord injured subjects: A preliminary study. *Restor Neurol Neurosci* 5: 81-84, 1993.
6. Barbeau H, Rossignol S. Recovery of locomotion after chronic spinalization in the adult cat. *Brain Res* 412: 84-95, 1987.
7. Barker AT, Jalinous R, Freeston IL. Non-invasive magnetic stimulation of human motor cortex. *Lancet* 1: 1106-1107, 1985.
8. Basmajian JV. Control and training of individual motor units. *Science* 141: 440-441, 1963.
9. Bawa P, Murnaghan C. Motor unit rotation in a variety of human muscles. *J Neurophysiol* 102: 2265-2272, 2009.
10. Bawa P, Pang M, Olesen K, Calancie B. Rotation of motoneurons during prolonged isometric contractions in humans. *J Neurophysiol* 96: 1135-1140, 2006.
11. Beaumont E, Gardiner P. Effects of daily spontaneous running on the electrophysiological properties of hindlimb motoneurones in rats. *J Physiol* 540: 129-138, 2002.
12. Beaumont E, Gardiner PF. Endurance training alters the biophysical properties of hindlimb motoneurons in rats. *Muscle Nerve* 27: 228-236, 2003.
13. Beltman JGM, Sargeant AJ, Haan H, Van Mechelen W, De Haan A. Changes in PCr/Cr ratio in single characterized muscle fibre fragments after only a few maximal voluntary contractions in humans. *Acta Physiol Scand* 180: 187-193, 2004.
14. Bergström J. Percutaneous needle biopsy of skeletal muscle in physiological and clinical research. *Scand J Clin Lab Invest* 35: 609-616, 1975.
15. Bessou P, Emonet-Denand F, Laporte Y. Relation between the conduction rate of motor nerve fibers and the contraction time of their motor units. *C R Hebd Seances Acad Sci* 256: 5625-5627, 1963.
16. Bigland-Ritchie B, Dawson N, Johansson R, Lippold O. Reflex origin for the slowing of motoneurone firing rates in fatigue of human voluntary contractions. *J Physiol (Lond)* 379: 451-459, 1986.
17. Bigland-Ritchie B, Furbush F, Woods JJ. Fatigue of intermittent submaximal voluntary contractions: Central and peripheral factors. *J Appl Physiol* 61: 421-429, 1986.
18. Bigland-Ritchie B, Johansson R, Lippold O, Smith S, Woods J. Changes in motoneurone firing rates during sustained maximal voluntary contractions. *J Physiol (Lond)* 340: 335-346, 1983.
19. Bigland-Ritchie B, Johansson R, Lippold OCJ, Woods JJ. Contractile speed and EMG changes during fatigue of sustained maximal voluntary contractions. *J Neurophysiol* 50: 313-324, 1983.
20. Bigland-Ritchie B, Kukulka CG, Lippold OC, Woods JJ. The absence of neuromuscular transmission failure in sustained maximal voluntary contractions. *J Physiol* 330: 265-278, 1982.

21. Bigland-Ritchie BR, Fuglevand AJ, Thomas CK. Contractile properties of human motor units: Is man a cat? *Neuroscientist* 4: 240-249, 1998.
22. Black JE, Isaacs KR, Anderson BJ, Alcantara AA, Greenough WT. Learning causes synaptogenesis, whereas motor activity causes angiogenesis, in cerebellar cortex of adult rats. *Proc Natl Acad Sci USA* 87: 5568-5572, 1990.
23. Blazevich AJ. Effects of physical training and detraining, immobilisation, growth and aging on human fascicle geometry. *Sport Med* 36: 1003-1017, 2006.
24. Bodine SC, Roy RR, Eldred E, Edgerton VR. Maximal force as a function of anatomical features of motor units in the cat tibialis anterior. *J Neurophysiol* 57: 1730-1745, 1987.
25. Brock LG, Coombs JS, Eccles JC. The nature of the monosynaptic excitatory and inhibitory processes in the spinal cord. *Proc R Soc Lond B Biol Sci* 140: 170-176, 1952.
26. Brown TG. The intrinsic factors in the act of progression in the mammal. *Proc R Soc Lond* 84: 308-319, 1911.
27. Brownstone RM, Jordan LM, Kriellaars DJ, Noga BR, Shefchyk SJ. On the regulation of repetitive firing in lumbar motoneurones during fictive locomotion in the cat. *Exp Brain Res* 90: 441-455, 1992.
28. Buller AJ, Lewis DM. Further observations on mammalian cross-innervated skeletal muscle. *J Physiol* 178: 343-358, 1965.
29. Burke R. Group Ia synaptic input to fast and slow twitch motor units of cat triceps surae. *J Physiol (Lond)* 196: 605-630, 1968.
30. Burke RE. Motor unit types of cat triceps surae muscle. *J Physiol (Lond)* 193: 141-160, 1967.
31. Burke RE, Edgerton VR. Motor unit properties and selective involvement in movement. *Ex Sport Sci Rev* 3: 31-81, 1975.
32. Burke RE, Jankowska E, Bruggencate GT. A comparison of peripheral and rubrospinal synaptic input to slow and fast twitch motor units of triceps surae. *J Physiol* 207: 709-732, 1970.
33. Burke RE, Levine DN, Zajac FE, Tsairis P, Engel WK. Mammalian motor units: Physiological-histochemical correlation in three types in cat gastrocnemius. *Science* 174: 709-712, 1971.
34. Butler JE, Taylor JL, Gandevia SC. Responses of human motoneurons to corticospinal stimulation during maximal voluntary contractions and ischemia. *J Neurosci* 23: 10224-10230, 2003.
35. Capaday C. Neurophysiological methods for studies of the motor system in freely moving human subjects. *J Neurosci Methods* 74: 201-218, 1997.
36. Carp JS, Wolpaw JR. Motoneuron plasticity underlying operantly conditioned decrease in primate H-reflex. *J Neurophysiol* 72: 431-442, 1994.
37. Carroll TJ, Herbert RD, Munn J, Lee M, Gandevia SC. Contralateral effects of unilateral strength training: Evidence and possible mechanisms. *J Appl Physiol* 101: 1514-1522, 2006.
38. Carroll TJ, Riek S, Carson RG. The sites of neural adaptation induced by resistance training in humans. *J Physiol* 544: 641-652, 2002.
39. Chalmers GR, Roy RR, Edgerton VR. Adaptability of the oxidative capacity of motoneurons. *Brain Res* 570: 1-10, 1992.
40. Chen Y, Chen XY, Jakeman LB, Chen L, Stokes BT, Wolpaw JR. Operant conditioning of H-reflex can correct a locomotor abnormality after spinal cord injury in rats. *J Neurosci* 26: 12537-12543, 2006.
41. Chen Y, Chen XY, Jakeman LB, Schalk G, Stokes BT, Wolpaw JR. The interaction of a new motor skill and an old one: H-reflex conditioning and locomotion in rats. *J Neurosci* 25: 6898-6906, 2005.
42. Cleary LJ, Lee WL, Byrne JH. Cellular correlates of long-term sensitization in Aplysia. *J Neurosci* 18: 5988-5998, 1998.
43. Cohen S, Levi-Montalcini R, Hamburger V. A nerve growth-stimulating factor isolated from sarcomas 37 and 180. *Proc Natl Acad Sci USA* 40: 1014-1018, 1954.
44. Cope T, Clark B. Motor-unit recruitment in the decerebrate cat: Several unit properties are equally good predictors of order. *J Neurophysiol* 66: 1127-1138, 1991.
45. Cope TC, Nelson SG, Mendell LM. Selectivity in synaptic changes caudal to acute spinal cord transection. *Neurosci Lett* 20: 289-294, 1980.
46. Crockett JL, Edgerton VR, Max SR, Barnard RJ. The neuromuscular junction in response to endurance training. *Exp Neurol* 51: 207-215, 1976.
47. Denny-Brown D. On the nature of postural reflexes. *Proc R Soc Lond B Biol Sci* 104: 252-301, 1929.
48. Denny-Brown D. The histological features of striped muscle in relation to its functional activity. *Proc R Soc Lond B Biol Sci* 104: 371-411, 1929.
49. Denny-Brown D, Pennybacker J. Fibrillation and fasciculation in voluntary muscle. *Brain* 61: 311-334, 2010.
50. Desaulniers P, Lavoie P-A, Gardiner PF. Endurance training increases acetylcholine receptor quantity at neuromuscular junctions of adult rat skeletal muscle. *NeuroReport* 9: 3549-3552, 1998.
51. Desaulniers P, Lavoie PA, Gardiner PF. Habitual exercise enhances neuromuscular transmission efficacy of rat soleus muscle in situ. *J Appl Physiol* 90: 1041-1048, 2001.
52. Desmedt J. The size principle of motoneuron recruitment in ballistic or ramp voluntary contractions in man. In: Desmedt JE, ed. *Motor Unit Types, Re-*

cruitment and Plasticity in Health and Disease. New York: Karger, 1999, 97-136.

53. de Vries H. *Physiology of Exercise for Physical Education and Athletics*. Dubuque, IA: Brown, 1974.
54. Di Giorgio AM. Neurotropic substances and persistence of asymmetry of the extremities in spinal animals. *Boll Soc Ital Biol Sper* 36: 1151-1152, 1960.
55. Dietz V. Body weight supported gait training: From laboratory to clinical setting. *Brain Res Bull* 76: 459-463, 2008.
56. Dietz V, Colombo G, Jensen L, Baumgartner L. Locomotor capacity of spinal cord in paraplegic patients. *Ann Neurol* 37: 574-582, 1995.
57. Ding Q, Vaynman S, Akhavan M, Ying Z, Gomez-Pinilla F. Insulin-like growth factor I interfaces with brain-derived neurotrophic factor-mediated synaptic plasticity to modulate aspects of exercise-induced cognitive function. *Neuroscience* 140: 823-833, 2006.
58. Dishman RK. The new emergence of exercise neurobiology. *Scand J Med Sci Sports* 16: 379-380, 2006.
59. Dobkin BH, Harkema S, Requejo P, Edgerton VR. Modulation of locomotor-like EMG activity in subjects with complete and incomplete spinal cord injury. *J Neurol Rehab* 9: 183-190, 1995.
60. Dolley DH. The morphological changes in nerve cells resulting from over-work in relation with experimental anemia and shock. *J Med Res* 21: 95-114, 1909.
61. Donselaar Y, Kernell D, Eerbeek O. Soma size and oxidative enzyme activity in normal and chronically stimulated motoneurones of the cat's spinal cord. *Brain Res* 385: 22-29, 1986.
62. Dorlöchter M, Irintchev A, Brinkers M, Wernig A. Effects of enhanced activity on synaptic transmission in mouse extensor digitorum longus muscle. *J Physiol (Lond)* 436: 283-292, 1991.
63. Duchateau J, Hainaut K. Isometric or dynamic training: Differential effects on mechanical properties of a human muscle. *J Appl Physiol* 56: 296-301, 1984.
64. Duchateau J, Hainaut K. Behaviour of short and long latency reflexes in fatigued human muscles. *J Physiol (Lond)* 471: 787-799, 1993.
65. Eccles J, Katz B, Kuffler SW. Nature of the "end-plate potential" in curarized muscle. *J Neurophysiol* 4: 362-387, 1941.
66. Eccles JC, Eccles RM, Lundberg A. Durations of after-hyperpolarization of motoneurones supplying fast and slow muscles. *Nature* 179: 866-868, 1957.
67. Eccles JC, McIntyre AK. The effects of disuse and of activity on mammalian spinal reflexes. *J Physiol* 121: 492-516, 1953.
68. Edgerton VR, Courtine G, Gerasimenko YP, Lavrov I, Ichiyama RM, Fong AJ, Cai LL, Otoshi CK, Tillakaratne NJK, Burdick JW, Roy RR. Training locomotor networks. *Brain Res Rev* 57: 241-254, 2008.
69. Edgerton VR, De Leon RD, Harkema SJ, Hodgson JA, London N, Reinkensmeyer DJ, Roy RN, Talmadge RJ, Tillakaratne NJ, Timoszyk W, Tobin A. Retraining the injured spinal cord. *J Physiol* 533: 15-22, 2001.
70. Edgerton VR, Simpson D, Barnard RJ, Peter JB. Phosphorylase activity in acutely exercised muscle. *Nature* 225: 866-867, 1970.
71. Edgerton VR, Simpson DR. The intermediate muscle fiber of rats and guinea pigs. *J Histochem Cytochem* 17: 828-838, 1969.
72. Edgerton VR, Tillakaratne NJK, Bigbee AJ, De Leon RD, Roy RR. Plasticity of the spinal neural circuitry after injury. *Annu Rev Neurosci* 27: 145-167, 2004.
73. Edstrom J-E. Effects of increased motor activity on the dimensions and the staining properties of the neuron soma. *J Comp Neurol* 107: 295-304, 1957.
74. Engel WK. Adenosine triphosphatase of sarcoplasmic reticulum triads and sarcolemma identified histochemically. *Nature* 200: 588-589, 1963.
75. Engel WK, Cunningham GG. Rapid examination of muscle tissue. An improved trichrome method for fresh-frozen biopsy sections. *Neurology* 13: 919-923, 1963.
76. Enoka RM. *Neuromechanical Basis of Kinesiology*. Champaign, IL: Human Kinetics, 1988.
77. Fatt P, Katz B. An analysis of the end-plate potential recorded with an intracellular electrode. *J Physiol* 115: 320-370, 1951.
78. Foehring R, Sypert G, Munson J. Properties of self-reinnervated motor units of medial gastrocnemius of cat. II. Axotomized motoneurons and time course of recovery. *J Neurophysiol* 55: 947-965, 1986.
79. Foehring R, Sypert G, Munson J. Motor-unit properties following cross-reinnervation of cat lateral gastrocnemius and soleus muscles with medial gastrocnemius nerve. II. Influence of muscle on motoneurons. *J Neurophysiol* 57: 1227-1245, 1987.
80. Gandevia S, McComas A, Stuart DG, Thomas CK. *Fatigue: Neural and Muscular Mechanisms*. New York: Plenum Press, 1995.
81. Gandevia SC. Spinal and supraspinal factors in human muscle fatigue. *Physiol Rev* 81: 1725-1789, 2001.
82. Gandevia SC, Allen GM, Butler JE, Taylor JL. Supraspinal factors in human muscle fatigue: Evidence for suboptimal output from the motor cortex. *J Physiol (Lond)* 490: 529-536, 1996.
83. Gardiner P, Kernell D. The "fastness" of rat motoneurons: Time-course of afterhyperpolarization in relation to axonal conduction velocity and muscle unit contractile speed. *Pflugers Arch* 415: 762-766, 1990.

84. Gardner GW. Effect of isometric and isotonic exercise on joint motion. *Arch Phys Med Rehabil* 47: 24-30, 1966.
85. Garland SJ, Enoka RM, Serrano LP, Robinson GA. Behavior of motor units in human biceps brachii during a submaximal fatiguing contraction. *J Appl Physiol* 76: 2411-2419, 1994.
86. Garland SJ, Garner SH, McComas AJ. Reduced voluntary electromyographic activity after fatiguing stimulation of human muscle. *J Physiol* 401: 547-556, 1988.
87. Garland SJ, McComas AJ. Reflex inhibition of human soleus muscle during fatigue. *J Physiol (Lond)* 429: 17-27, 1990.
88. Gerchman LB, Edgerton VR, Carrow RE. Effects of physical training on the histochemistry and morphology of ventral motor neurons. *Exp Neurol* 49: 790-801, 1975.
89. Gertler RA, Robbins N. Differences in neuromuscular transmission in red and white muscles. *Brain Res* 142: 160-164, 1978.
90. Gisiger V, Bélisle M, Gardiner PF. Acetylcholinesterase adaptation to voluntary wheel running is proportional to the volume of activity in fast, but not slow, rat hindlimb muscles. *Eur J Neurosci* 6: 673-680, 1994.
91. Gisiger V, Sherker S, Gardiner PF. Swimming training increases the G_4 acetylcholinesterase content of both fast ankle extensors and flexors. *FEBS Lett* 278: 271-273, 1991.
92. Gollnick PD, Armstrong RB, Sembrowich WL, Shepherd RE, Saltin B. Glycogen depletion pattern in human skeletal muscle fibers after heavy exercise. *J Appl Physiol* 34: 615-618, 1973.
93. Gollnick PD, Piehl K, Karlsson E, Saltin B. Glycogen depletion patterns in human skeletal muscle fibers after varying types and intensities of exercise. In: Howald H, Poortmans J, eds. *Metabolic Adaptations to Prolonged Physical Exercise*. Basel, Switzerland: Birkhauser Verlag, 1975, 416-421.
94. Gollnick PD, Piehl K, Saubert CW, Armstrong RB, Saltin B. Diet, exercise, and glycogen changes in human muscle fibers. *J Appl Physiol* 33: 421-425, 1972.
95. Gómez-Pinilla F, Ying Z, Opazo P, Roy RR, Edgerton VR. Differential regulation by exercise of BDNF and NT-3 in rat spinal cord and skeletal muscle. *Eur J Neurosci* 13: 1078-1084, 2001.
96. Gomirato G. Quantitative evaluation of the metabolic variations in the spinal motor root cells, studied by biophysical method and following adequate stimulation (muscular fatigue); action on metabolism of vitamin B12. *J Neuropathol Exp Neurol* 13: 359-368, 1954.
97. Gorassini MA, Bennett DJ, Yang JF. Self-sustained firing of human motor units. *Neurosci Lett* 247: 13-16, 1998.
98. Grau JW, Crown ED, Ferguson AR, Washburn SN, Hook MA, Miranda RC. Instrumental learning within the spinal cord: Underlying mechanisms and implications for recovery after injury. *Behav Cogn Neurosci Rev* 5: 191-239, 2006.
99. Griesbach GS, Hovda DA, Molteni R, Wu A, Gomez-Pinilla F. Voluntary exercise following traumatic brain injury: Brain-derived neurotrophic factor upregulation and recovery of function. *Neuroscience* 125: 129-139, 2004.
100. Guth L, Brown WC, Ziemnowicz JD. Changes in cholinesterase activity of rat muscle during growth and hypertrophy. *Am J Physiol* 211: 1113-1116, 1966.
101. Hagbarth K-E, Kunesch E, Nordin M, Schmidt R, Wallin E. Gamma loop contributing to maximal voluntary contractions in man. *J Physiol (Lond)* 380: 575-591, 1986.
102. Häkkinen K, Komi PV. Electromyographic and mechanical characteristics of human skeletal muscle during fatigue under voluntary and reflex conditions. *Electroencephalog Clin Neurophysiol* 55: 436-444, 1983.
103. Harkema S, Hurley S, Patel U, Requejo P, Dobkin B, Edgerton VR. Human lumbrosacral spinal cord interprets loading during stepping. *J Neurophysiol* 77: 797-811, 1997.
104. Hellebrandt FA. Cross education: Ipsilateral and contralateral effects of unimanual training. *J Appl Physiol* 4: 136-144, 1951.
105. Hellebrandt FA, Parrish AM, Houtz SJ. The influence of unilateral exercise on the contralateral limb. *Arch Phys Med Rehabil* 28: 76-85, 1947.
106. Henneman E. Functional significance of cell size in spinal motoneurons. *J Neurophysiol* 28: 560-580, 1965.
107. Henneman E, Somjen G, Carpenter DO. Excitability and inhibitility of motoneurons of different sizes. *J Neurophysiol* 28: 599-620, 1965.
108. Hettinger T, Muller EA. Muscle capacity and muscle training. *Arbeitsphysiologie* 15: 111-126, 1953.
109. Hodgkin A, Huxley AF. Action potentials recorded from inside a nerve fiber. *Nature* 144: 710-711, 1939.
110. Hodgson JA, Roy RR, de Leon R, Dobkin B, Edgerton VR. Can the mammalian lumbar spinal cord learn a motor task. *Med Sci Sports Exerc* 26: 1491-1497, 1994.
111. Hong SJ, Lnenicka GA. Long-term changes in the neuromuscular synapses of a crayfish motoneuron produced by calcium influx. *Brain Res* 605: 121-127, 1993.
112. Hutchinson KJ, Gómez-Pinilla F, Crowe MJ, Ying Z, Basso DM. Three exercise paradigms differentially improve sensory recovery after spinal cord contusion in rats. *Brain* 127: 1403-1414, 2004.

113. Hydén H. Protein metabolism in the nerve cell during growth and function. *Acta Physiol Scand Supplementum* 6: 1-136, 1943.
114. Ikai M, Steinhaus A. Some factors modifying the expression of human strength. *J Appl Physiol* 16: 157-163, 1961.
115. Isaacs KR, Anderson BJ, Alcantara AA, Black JE, Greenough WT. Exercise and the brain: Angiogenesis in the adult rat cerebellum after vigorous physical activity and motor skill learning. *J Cereb Blood Flow Metab* 12: 110-119, 1992.
116. Ishihara A, Roy RR, Edgerton VR. Succinate dehydrogenase activity and soma size of motoneurons innervating different portions of the rat tibialis anterior. *Neuroscience* 68: 813-822, 1995.
117. Jasmin BJ, Gisiger V. Regulation by exercise of the pool of G_4 acetylcholinesterase characterizing fast muscles: Opposite effect of running training in antagonist muscles. *J Neurosci* 10: 1444-1454, 1990.
118. Jensen BR, Pilegaard M, Sjogaard G. Motor unit recruitment and rate coding in response to fatiguing shoulder abductions and subsequent recovery. *Eur J Appl Physiol Occup Physiol* 83: 190-199, 2000.
119. Jewell PA, Zaimis EJ. Changes at the neuromuscular junction of red and white muscle fibres in the cat induced by disuse atrophy and by hypertrophy. *J Physiol* 124: 429-442, 1954.
120. Jindrich DL, Joseph MS, Otoshi CK, Wei RY, Zhong H, Roy RR, Tillakaratne NJ, Edgerton VR. Spinal learning in the adult mouse using the Horridge paradigm. *J Neurosci Methods* 182: 250-254, 2009.
121. Johnson DJ, Smith LA, Eldred E, Edgerton VR. Exercise-induced changes of biochemical, histochemical, and contractile properties of muscle in cordotomized kittens. *Exp Neurol* 76: 414-427, 1982.
122. Jonek J, Konecki J, Grzybek H, Olkowski Z. Electron microscopic localization of acid phosphatase and thiamine pyrophosphatase activities in the motoneurons of exercised mice. *Histochemie* 23: 116-119, 1970.
123. Jones DA, Rutherford OM, Parker DF. Physiological changes in skeletal muscle as a result of strength training. *Q J Exp Physiol* 74: 233-256, 1989.
124. Kaufman MP, Longhurst JC, Rybicki KJ, Wallach JH, Mitchell JH. Effects of static muscular contraction on impulse activity of groups III and IV afferents in cats. *J Appl Physiol Respirat Environ Exerc Physiol* 55: 105-112, 1983.
125. Kempermann G, Kuhn H, Gage FH. More hippocampal neurons in adult mice living in an enriched environment. *Nature* 386: 493-495, 1997.
126. Kernell D. The adaptation and the relation between discharge frequency and current strength of cat lumbosacral motoneurones stimulated by long-lasting injected currents. *Acta Physiol Scand* 65: 65-73, 1965.
127. Kernell D. Input resistance, electrical excitability, and size of ventral horn cells in cat spinal cord. *Science* 152: 1637-1640, 1966.
128. Kernell D. The early phase of adaptation in repetitive impulse discharges of cat spinal motoneurones. *Brain Res* 41: 184-186, 1972.
129. Kernell D, Monster A. Motoneurone properties and motor fatigue. An intracellular study of gastrocnemius motoneurones of the cat. *Exp Brain Res* 46: 197-204, 1982.
130. Kiehn O, Eken T. Prolonged firing in motor units: Evidence of plateau potentials in human motoneurons? *J Neurophysiol* 78: 3061-3068, 1997.
131. Kleim JA, Cooper NR, VandenBerg PA. Exercise induces angiogenesis but does not alter movement representations within rat motor cortex. *Brain Res* 934: 1-6, 2002.
132. Kleim JA, Jones TA. Principles of experience-dependent neural plasticity: Implications for rehabilitation after brain damage. *J Speech Lang Hear Res* 51: S225-S239, 2008.
133. Komi PV, Viitasalo JT, Rauramaa R, Vihko V. Effect of isometric strength training of mechanical, electrical, and metabolic aspects of muscle function. *Eur J Appl Physiol Occup Physiol* 40: 45-55, 1978.
134. Kornatz KW, Christou EA, Enoka RM. Practice reduces motor unit discharge variability in a hand muscle and improves manual dexterity in old adults. *J Appl Physiol* 98: 2072-2080, 2005.
135. Krnjevic K, Miledi R. Presynaptic failure of neuromuscular propagation in rats. *J Physiol* 149: 1-22, 1959.
136. Kuei JH, Shadmehr R, Sieck GC. Relative contribution of neurotransmission failure to diaphragm fatigue. *J Appl Physiol* 68: 174-180, 1990.
137. Kugelberg E, Edstrom L. Differential histochemical effects of muscle contractions on phophorylase and glycogen in various types of fibres: Relation to fatigue. *J Neurol Neurosurg Psychiat* 31: 415-423, 1968.
138. Kukulka CG, Clamann HP. Comparison of the recruitment and discharge properties of motor units in human brachial biceps and adductor pollicis during isometric contractions. *Brain Res* 219: 45-55, 1981.
139. Ladegaard J. Story of electromyography equipment. *Muscle Nerve* S128-S133, 2002.
140. Lee RH, Heckman CJ. Bistability in spinal motoneurons in vivo: Systematic variations in rhythmic firing patterns. *J Neurophysiol* 80: 572-582, 1998.
141. Lnenicka GA, Atwood HL. Long-term facilitation and long-term adaptation at synapses of a crayfish phasic motoneuron. *J Neurobiol* 16: 97-110, 1985.

142. Lnenicka GA, Atwood HL. Impulse activity of a crayfish motoneuron regulates its neuromuscular synaptic properties. *J Neurophysiol* 61: 91-96, 1986.
143. Lnenicka GA, Atwood HL. Long-term changes in neuromuscular synapses with altered sensory input to a crayfish motoneuron. *Exp Neurol* 100: 437-447, 1988.
144. Loeb GE. Motoneurone task groups: Coping with kinematic heterogeneity. *J Exp Biol* 115: 137-146, 1985.
145. Lovely RG, Gregor RJ, Roy RR, Edgerton VR. Effects of training on the recovery of full-weight-bearing stepping in the adult spinal cat. *Experim Neurol* 92: 421-435, 1986.
146. Macefield G, Hagbarth K-E, Gorman R, Gandevia SC, Burke D. Decline in spindle support to alpha-motoneurones during sustained voluntary contractions. *J Physiol* 440: 497-512, 1991.
147. Marsden CD, Meadows JC, Merton P. "Muscular wisdom" that minimizes fatigue during prolonged effort in man: Peak rates of motoneuron discharge and slowing of discharge during fatigue. In: Desmedt JE, ed. *Motor Control Mechanisms in Health and Disease*. New York: Raven Press, 1983, 169-211.
148. Marsden CD, Meadows JC, Merton PA. Isolated single motor units in human muscle and their rate of discharge during maximal voluntary effort. *J Physiol* 217: 12P-13P, 1971.
149. Mendell LM, Henneman E. Terminals of single Ia fibers: Location, density, and distribution within a pool of 300 homonymous motoneurones. *J Neurophysiol* 34: 171-187, 1971.
150. Mercier AJ, Bradacs H, Atwood HL. Long-term adaptation of crayfish neurons depends on the frequency and number of impulses. *Brain Res* 598: 221-224, 1992.
151. Merton PA. Voluntary strength and fatigue. *J Physiol* 123: 553-564, 1954.
152. Merton PA, Morton HB. Stimulation of the cerebral cortex in the intact human subject. *Nature* 285: 227, 1980.
153. Milner-Brown HS, Stein RB, Lee RG. Synchronization of human motor units: Possible roles of exercise and supraspinal reflexes. *Electroencephalogr Clin Neurophysiol* 38: 245-254, 1975.
154. Milner-Brown HS, Stein RB, Yemm R. Changes in firing rate of human motor units during linearly changing voluntary contractions. *J Physiol* 230: 371-390, 1973.
155. Milner-Brown HS, Stein RB, Yemm R. The contractile properties of human motor units during voluntary isometric contractions. *J Physiol (Lond)* 228: 285-306, 1973.
156. Moffroid MT, Whipple RH. Specificity of speed of exercise. *Phys Ther* 50: 1692-1700, 1970.
157. Molteni R, Wu A, Vaynman S, Ying Z, Barnard RJ, Gómez-Pinilla F. Exercise reverses the harmful effects of consumption of a high-fat diet on synaptic and behavioral plasticity associated to the action of brain-derived neurotrophic factor. *Neuroscience* 123: 429-440, 2004.
158. Molteni R, Ying Z, Gómez-Pinilla F. Differential effects of acute and chronic exercise on plasticity-related genes in the rat hippocampus revealed by microarray. *Eur J Neurosci* 16: 1107-1116, 2002.
159. Molteni R, Zheng JQ, Ying Z, Gómez-Pinilla F, Twiss JL. Voluntary exercise increases axonal regeneration from sensory neurons. *Proc Natl Acad Sci USA* 101: 8473-8478, 2004.
160. Moritani T, de Vries HA. Neural factors versus hypertrophy in the time course of muscle strength gain. *Am J Phys Med* 58: 115-131, 1979.
161. Neeper SA, Gómez-Pinilla F, Choi J, Cotman C. Exercise and brain neurotrophins. *Nature* 373: 109, 1995.
162. Nelson DL, Hutton RS. Dynamic and static stretch responses in muscle spindle receptors in fatigued muscle. *Med Sci Sports Exerc* 17: 445-450, 1985.
163. Nguyen PV, Atwood HL. Expression of long-term adaptation of synaptic transmission requires a critical period of protein synthesis. *J Neurosci* 10: 1099-1109, 1990.
164. Ogata T, Yamasaki Y. The three-dimensional structure of motor endplates in different fiber types of rat intercostal muscle. *Cell Tissue Res* 241: 465-472, 1985.
165. Orlovsky GN, Severin FV, Shik ML. Locomotion induced by stimulation of the mesencephalon. *Dokl Akad Nauk SSSR* 169: 1223-1226, 1966.
166. Orlovsky GN, Deliagina T, Grillner S. *Neuronal Control of Locomotion: From Mollusc to Man.* Oxford: Oxford University Press, 1999.
167. Padykula H, Gauthier G. The ultrastructure of the neuromuscular junctions of mammalian red, white, and intermediate skeletal muscle fibers. *J Cell Biol* 46: 27-41, 1970.
168. Peter JB, Jeffress RN, Lamb DR. Exercise: Effects on hexokinase activity in red and white skeletal muscle. *Science* 160: 200-201, 1968.
169. Rasch PJ, Morehouse LE. Effect of static and dynamic exercises on muscular strength and hypertrophy. *J Appl Physiol* 11: 29-34, 1957.
170. Romaiguère P, Vedel J-P, Pagni S. Comparison of fluctuations of motor unit recruitment and de-recruitment thresholds in man. *Exp Brain Res* 95: 517-522, 1993.
171. Ross EZ, Middleton N, Shave R, George K, Nowicky A. Corticomotor excitability contributes to neuromuscular fatigue following marathon running in man. *Exp Physiol* 92: 417-426, 2007.

172. Rossignol S, Giroux N, Chau C, Marcoux J, Brustein E, Reader TA. Pharmacological aids to locomotor training after spinal injury in the cat. *J Physiol* 533: 65-74, 2001.
173. Rotto DM, Kaufman MP. Effect of metabolic products of muscular contraction on discharge of group III and IV afferents. *J Appl Physiol* 64: 2306-2313, 1988.
174. Roy RR, Talmadge RJ, Hodgson JA, Zhong H, Baldwin KM, Edgerton VR. Training effects on soleus of cats spinal cord transected (T12-13) as adults. *Muscle Nerve* 21: 63-71, 1998.
175. Ruff R, Whittlesey D. Na currents near and away from endplates on human fast and slow twitch muscle fibers. *Muscle Nerve* 16: 922-929, 1993.
176. Ruff RL. Na current density at and away from end plates on rat fast- and slow-twitch skeletal muscle fibers. *Am J Physiol Cell Physiol* 262: C229-C234, 1992.
177. Sacco P, Newberry R, McFadden L, Brown T, McComas AJ. Depression of human electromyographic activity by fatigue of a synergistic muscle. *Muscle Nerve* 20: 710-717, 1997.
178. Sale DG, Upton ARM, McComas AJ, MacDougall JD. Neuromuscular function in weight-trainers. *Exp Neurol* 82: 521-531, 1983.
179. Sale DG, MacDougall JD, Upton ARM, McComas AJ. Effect of strength training upon motoneuron excitability in man. *Med Sci Sports Exerc* 15: 57-62, 1983.
180. Sale DG, MacDougall JD, Upton ARM, McComas AJ. Effect of strength training upon motoneuron excitability in man. *Med Sci Sports Exerc* 15: 57-62, 1983.
181. Sawczuk A, Powers RK, Binder MD. Spike frequency adaptation studied in hypoglossal motoneurons of the rat. *J Neurophysiol* 73: 1799-1810, 1995.
182. Schneider C, Capaday C. Progressive adaptation of the soleus H-reflex with daily training at walking backward. *J Neurophysiol* 89: 648-656, 2003.
183. Schwindt P, Crill W. Role of a persistent inward current in motoneuron bursting during spinal seizures. *J Neurophysiol* 43: 1296-1318, 1980.
184. Semmler JG, Nordstrom MA. Motor unit discharge and force tremor in skill- and strength- trained individuals. *Exp Brain Res* 119: 27-38, 1998.
185. Semmler JG, Sale MV, Meyer FG, Nordstrom MA. Motor-unit coherence and its relation with synchrony are influenced by training. *J Neurophysiol* 92: 3320-3331, 2004.
186. Shik ML, Orlovsky GN. Neurophysiology of locomotor automatism. *Physiol Rev* 56: 465-501, 1976.
187. Sica REP, McComas AJ. Fast and slow twitch units in a human muscle. *J Neurol Neurosurg Psychiat* 34: 113-120, 1971.
188. Silva AJ. Molecular and cellular cognitive studies of the role of synaptic plasticity in memory. *J Neurobiol* 54: 224-237, 2003.
189. Sirota MG, Shik ML. Locomotion of the cat on stimulation of the mesencephalon. *Fiziol Zh SSSR Im I M Sechenova* 59: 1314-1321, 1973.
190. Skinner RD, Houle JD, Reese NB, Berry CL, Garcia-Rill E. Effects of exercise and fetal spinal cord implants on the H-reflex in chronically spinalized adult rats. *Brain Res* 729: 127-131, 1996.
191. Smith JL, Betts B, Edgerton VR, Zernicke RF. Rapid ankle extension during paw shakes: Selective recruitment of fast ankle extensors. *J Neurophysiol* 43: 612-620, 1980.
192. Smith JL, Edgerton VR, Eldred E, Zernicke RF. The chronic spinalized cat: A model for neuromuscular plasticity. In: Haber B, Perez-Polo J, Hashim G, Stella A, eds. *Neuromuscular System Regeneration*. NewYork: Alan Liss, 1983, 357-373.
193. Smith JL, Martin PG, Gandevia SC, Taylor JL. Sustained contraction at very low forces produces prominent supraspinal fatigue in human elbow flexor muscles. *J Appl Physiol* 103: 560-568, 2007.
194. Sogaard K, Gandevia SC, Todd G, Petersen NT, Taylor JL. The effect of sustained low-intensity contractions on supraspinal fatigue in human elbow flexor muscles. *J Physiol* 573: 511-523, 2006.
195. Spielmann JM, Laouris Y, Nordstrom MA, Robinson GA, Reinking RM, Stuart DG. Adaptation of cat motoneurons to sustained and intermittent extracellular activation. *J Physiol (Lond)* 464: 75-120, 1993.
196. Stephens JA, Taylor A. Fatigue of maintained voluntary muscle contraction in man. *J Physiol* 220: 1-18, 1972.
197. Stephens JA, Usherwood TP. The mechanical properties of human motor units with special reference to their fatiguability and recruitment threshold. *Brain Res* 125: 91-97, 1977.
198. Sterz R, Pagala M, Peper K. Postjunctional characteristics of the endplates in mammalian fast and slow muscles. *Pflugers Arch* 398: 48-54, 1983.
199. Tax AAM, Van der Gon JJD, Gielen CCAM, van den Tempel CMM. Differences in the activation of m. biceps brachii in the control of slow isotonic movements and isometric contractions. *Exp Brain Res* 76: 55-63, 1989.
200. Taylor JL, Allen GM, Butler JE, Gandevia SC. Supraspinal fatigue during intermittent maximal voluntary contractions of the human elbow flexors. *J Appl Physiol* 89: 305-313, 2000.
201. Thesleff S. Motor end-plate desensitization by repetitive nerve stimuli. *J Physiol (Lond)* 148: 659-664, 1959.
202. Van Bolhuis BM, Medendorp WP, Gielen CCAM. Motor unit firing behavior in human arm flexor

muscles during sinusoidal isometric contractions and movements. *Exp Brain Res* 117: 120-130, 1997.

203. Van Cutsem M, Duchateau J, Hainaut K. Changes in single motor unit behaviour contribute to the increase in contraction speed after dynamic training in humans. *J Physiol* 513: 295-305, 1998.
204. Van Lunteren E, Moyer M. Effects of DAP on diaphragm force and fatigue, including fatigue due to neurotransmission failure. *J Appl Physiol* 81: 2214-2220, 1996.
205. van Praag H. Neurogenesis and exercise: Past and future directions. *Neuromolecular Med* 10: 128-140, 2008.
206. Van Praag H, Christie BR, Sejnowski TJ, Gage FH. Running enhances neurogenesis, learning, and long-term potentiation in mice. *Proc Natl Acad Sci USA* 96: 13427-13431, 1999.
207. Van Praag H, Kempermann G, Gage FH. Running increases cell proliferation and neurogenesis in the adult mouse dentate gyrus. *Nat Neurosci* 2: 266-270, 1999.
208. Van Praag H, Kempermann G, Gage FH. Neural consequences of environmental enrichment. *Nat Rev Neurosci* 1: 191-198, 2000.
209. Vaynman S, Gomez-Pinilla F. Revenge of the "sit": How lifestyle impacts neuronal and cognitive health through molecular systems that interface energy metabolism with neuronal plasticity. *J Neurosci Res* 84: 699-715, 2006.
210. Vollestad NK, Blom PCS. Effect of varying exercise intensity on glycogen depletion in human muscle fibres. *Acta Physiol Scand* 125: 395-405, 1985.
211. Vollestad NK, Tabata I, Medbo JI. Glycogen breakdown in different human muscle fibre types during exhaustive exercise of short duration. *Acta Physiol Scand* 144: 135-141, 1992.
212. Wawrzyniak M. The activity of adenosine triphosphatase in motoneurones of the spinal cord of the guinea-pig. *Folia Histochem Cytochem* 1: 57-94, 1963.
213. Wernig A. Weight-supported treadmill vs over-ground training for walking after acute incomplete SCI. *Neurology* 67: 1900, 2006.
214. Westgaard RH, De Luca CJ. Motor unit substitution in long-duration contractions of the human trapezius muscle. *J Neurophysiol* 82: 501-504, 1999.
215. Wolf S, Levin R. Preliminary analysis on conditioning of exaggerated triceps surae stretch reflexes among stroke patients. *Proc Biofeedback Soc Am* 18: 17, 1987.
216. Wolpaw JR, Kieffer VA, Seegal RF, Braitman DJ, Sanders MG. Adaptive plasticity in the spinal stretch reflex. *Brain Res* 267: 196-200, 1983.
217. Wolpaw JR, Lee CL. Memory traces in primate spinal cord produced by operant conditioning of H-reflex. *J Neurophysiol* 61: 563-572, 1989.
218. Wood SJ, Slater CR. Safety factor at the neuromuscular junction. *Prog Neurobiol* 64: 393-429, 2001.
219. Woods J, Furbush F, Bigland-Ritchie B. Evidence for a fatigue-induced reflex inhibition of motoneuron firing rates. *J Neurophysiol* 58: 125-137, 1987.
220. Wuerker RB, McPhedran M, Henneman E. Properties of motor units in a heterogeneous pale muscle (m. gastrocnemius) of the cat. *J Neurophysiol* 28: 85-99, 1965.
221. Ying Z, Roy RR, Edgerton VR, Gómez-Pinilla F. Voluntary exercise increases neurotrophin-3 and its receptor TrkC in the spinal cord. *Brain Res* 987: 93-99, 2003.
222. Ying Z, Roy RR, Zhong H, Zdunowski S, Edgerton VR, Gomez Pinilla F. BDNF-exercise interactions in the recovery of symmetrical stepping after a cervical hemisection in rats. *Neuroscience* 155: 1070-1078, 2008.
223. Zehr EP. Training-induced adaptive plasticity in human somatosensory reflex pathways. *J Appl Physiol* 101: 1783-1794, 2006.

CHAPTER 9

The Autonomic Nervous System in Exercise: An Integrative View

Katarina T. Borer, PhD

This selective overview of the evolving knowledge about the autonomic nervous system (ANS; table 9.1) in exercise highlights the advances in experimental analytical power through time in three sections. The first section outlines the evolution of knowledge about ANS structure and function. The second describes how the understanding of ANS support of increased energy requirements during exercise grew. The final section examines the progression of thinking about the role of the ANS in energy recovery after acute exercise, adaptations to habitual exercise or training, and ANS regulation of whole-body energy balance. Table 9.2 lists the abbreviations used in this chapter.

Organization of the ANS

The ANS is often described as the reptilian brain controlling both utilization of energy for movement and energy intake and conservation. Benigne Winslow (1669-1760) identified "les grands nerfs sympathiques," nerves arising from the paravertebral sympathetic ganglion chain, as the key part controlling the first, or sympathetic (S), function (320), and John Newport Langley (1852-1925) identified the parasympathetic nervous system (PS) as the controller of the second function. Langley also recognized that the two divisions of the ANS are antagonistic yet complementary (175). In the early 20th century, Cannon's view was that the cranial PS nerves served to conserve body resources whereas the role of sacral nerves was to empty the colon and bladder and cause penile erection. He also noted that S messengers inhibited PS actions such as salivary and gastric juice secretion, appetite, meal digestion (47), and gastric contractions associated with hunger (53).

With the evolution of homeothermy in mammals and birds (34), maintaining a stable body temperature required faster and more efficient movements, and regulation of plasma glucose concentration became necessary for supplying the fuel needs of the muscle and brain. Homeothermy thus led to homeostatic regulation of oxygen, metabolism, and temperature, posited by Walter Bradford Cannon (1871-1945) (49) in the internal environment (milieu interieur, or extracellular fluid) conceptualized by Claude Bernard (1813-1878) (24). Bernard proposed that the balance between S and PS divisions preserved the stability of the internal environment (24). The ANS bears a close functional connection with the endocrine system, which is described in more detail elsewhere (31, 108). The ANS imposes a rhythmic pattern on gene expression, autonomic neuronal discharges, heart contraction, and growth cycles. In the temporal domain, ANS controls many physiological rhythms with periods lasting minutes, hours (circhoral or ultradian), a day (circadian), a month (circamensal), a season, or even a year (circannual). These ANS rhythms lower the cost of thermoregulation and homeostasis by allowing energy turnover in tissues and organs, restricting physical activities and feeding to the opportune portions of the solar day and season, and limiting metabolic energy expenditure during inopportune phases.

Autonomic Neurotransmitters

Epinephrine (E), norepinephrine (NE), and acetylcholine (Ach) mediate ANS actions. Their hormonal actions were recognized before their neurotransmitter functions. Oliver and Schäfer in 1895 demonstrated inotropic and chronotropic effects of adrenal gland extracts on isolated animal hearts (218). Walter B. Cannon at Harvard Medical School discovered that blood containing E elicited dilatation of the pupils, pallor, acceleration of the heart, sweating, release of

Table 9.1 Landmark Discoveries About the Autonomic Nervous System and Exercise

Year	Investigators	Finding	Reference
1914	Walter Bradford Cannon	Attributed autonomic manifestations that accompany physical stress to epinephrine	46
1937	M. Alam and F.H. Smirk	Provided the first demonstration of the chemoreflex, a hypertensive response to circulatory occlusion during exercise	5
1963	J.G. Devlin	Reported that exercise suppressed plasma insulin concentration during exercise	75
1973	John Wahren, Philip Felig, Rosa Hendler, and Gunvor Ahlborg	Initiated studies of substrate exchange and catecholamine and pancreatic hormone responses to exercise and recovery of exercise using arteriovenous cannulation of limb and splanchnic circulations	309
1975 and 1977	Henrik Galbo, Jens J. Holst, and Niels J. Christensen	Demonstrated exercise intensity-dependent increases in plasma catecholamine secretion at exercise intensities above 75% of $\dot{V}O_2max$, plasma insulin increases at 100% of $\dot{V}O_2max$, and differential responsiveness of epinephrine relative to norepinephrine to glucopenia during exhaustive dynamic exercise	109 and 111
1976	Bengt Saltin	Demonstrated decreased sympathetic activation of cardiorespiratory function at the same relative intensity when trained leg, but not untrained leg, was exercised	251
1976	Lennart Fagraeus and Dag Linnarsson	Demonstrated that the near-instantaneous acceleration of heart rate at the start of exercise results from vagal withdrawal rather than from beta adrenergic stimulation	93
1978	William W. Winder	Discovered that endurance training reduces secretion of epinephrine and norepinephrine at the same absolute work rate	319
1986	Eva Jansson, Paul Hjemdahl, and Lennart Kaijser	Provided evidence that epinephrine stimulates muscle glucose uptake and metabolism during exercise	141
1987	Peter Arner with Hans Wahrenberg, Peter Engfeldt, and Jan Bolinder	Demonstrated a selective increase in beta adrenergic stimulation of lipolysis after prolonged exercise in gluteal adipose tissue in vitro	310
1987	Gabrielle Savard	Quantified norepinephrine spillover during exercise in active and inactive skeletal muscles	255
1987	Allyn L. Mark with B.G. Wallin, Douglas R. Seals, and Ronald G. Victor	Used microneurography to document the role of increased muscle sympathetic nerve activity in raising systolic and diastolic pressures during chemoreflex elicitation by isometric exercise	190
1987	Charles E. Wade	Documented the vasoconstrictive role of angiotensin II in potentiating increases in systolic blood pressure without affecting renal function during exercise	307
1988 and 1990	Ronald G. Victor and S.L. Pryor	Linked chemoreflex stimulation of muscle sympathetic nerve activity to anaerobic metabolism during inadequate oxygen supply by showing its absence in patients with genetic muscle glycolytic deficiency (McArdle's disease)	223 and 301

1989	Patricia A. Deuster	Established that endurance training does not alter epinephrine and norepinephrine responses to submaximal exercise performed at the same relative intensities	74
1989	J.M. Moates, David H. Wasserman, and Alan D. Cherrington	Established the important role of epinephrine in hepatic glucose production by gluconeogenesis during dynamic exercise lasting more than 2 h	203
1990	Loring B. Rowell and D.S. O'Leary	Determined that vasoconstriction through increased sympathetic vasoconstriction is initiated only after the parasympathetic withdrawal has run its course and heart rate has increased to 100 beats/min	244
1993	Raffaello Furlan and Alberto Malliani	Documented the delayed restoration of parasympathetic tone after maximal exercise to exhaustion using spectral analysis of heart rate variability	106
1993 and 1994	Yves Papelier and Jeffrey T. Potts	Provided evidence for upward resetting of the baroreflex operating point, but not its gain, during exercise at different intensities	220 and 221
2000	Douglas Seals and Christopher Bell	Attributed reduced leg blood flow during exercise in older individuals to sympathetic compensation for obesity leading to beta adrenergic insensitivity due to chronic elevation of plasma norepinephrine	261
2003	Don D. Sheriff, Loring B. Rowell, and Allen M. Scher	Demonstrated the necessary contribution of the baroreflex in containing vascular conductance and increasing systolic pressure at the onset of, and during, dynamic exercise	266

sugar from the liver, relaxation of the bronchioles (first discovered by Kahn in 1907; 147), suppression of muscle fatigue, rapid coagulation of blood, and redistribution of blood from the viscera to the heart, lungs, central nervous system (CNS), and limbs (46, 47, 51, 52). Cannon studied the actions of circulating E using a bioassay consisting of a strip of intestinal smooth muscle at a time when no method for direct measurement of this catecholamine was available (47).

Chemical identification and naming of ANS neurotransmitters followed a separate path. Ach had been synthesized by Adolph von Baeyer (1835-1917) in 1867 (304). John Abel at Johns Hopkins University isolated the benzoyl derivative of adrenal extract and named it epinephrine (1), whereas Takamine in 1902 named its more purified form adrenaline (291), thus starting an international dichotomy in the designation of this catecholamine. Cannon advanced *sympathin* as another name for E. With Arturo Rosenblueth, he observed both the excitatory and inhibitory effects of E and incorrectly attributed them to inhibitory sympathin I and excitatory sympathin E mediator substances rather than to stimulation by E of different receptor types (52).

Catecholamines as neurotransmitters were recognized at the beginning of 20th century. Henry Hallett Dale (1875-1968) in 1914 noticed the resemblance between Ach actions and the effects of stimulation of the vagus nerve (70). Otto Loewi (1873-1961) in 1926 collected blood from the frog heart after stimulation of the vagus nerve and demonstrated deceleration of the heart rate (HR) in another frog after application of this "Vagusstoff" (184), although the involvement of the vagus in HR deceleration was postulated by Dixon as early as 1906 (79). This established Ach as a neurotransmitter in autonomic ganglia, the myoneural junction, and some CNS synapses (71). Langley in 1901 (176) and Cannon in 1921 (48) recognized that S nerve stimulation produced a substance with an action similar to that of adrenal gland extract. However, recognition of NE as a neurotransmitter was delayed until midcentury when Ulf von Euler (1905-1983) identified postganglionic S neurons as its source and the adrenergic receptors as its target (305, 306). In 1957, Julius Axelrod (1912-2004) documented the role of the enzyme phenylethanolamine N-methyltransferase (PNMT) in the biosynthesis of E from NE (15). For these discoveries, von Euler and Axelrod were awarded the Nobel Prize for physiology and medicine in 1970. Current understanding is that NE is released from postganglionic S nerve terminals and E is synthesized from NE by PNMT in S ganglion cells in

Table 9.2 Abbreviations Used in This Chapter

Ach	acetylcholine	MPOA	medial preoptic area of the hypothalamus
ANG	angiotensin	MSNA	muscle sympathetic nerve activation
ANP	atrial natriuretic peptide	NA	nucleus ambiguus (source of vagal neurons innervating the heart, medullary depressor area)
ANS	autonomic nervous system	NE	norepinephrine or noradrenalin
ARC	arcuate hypothalamic nucleus	nNOS	neuronal nitric oxide synthetase
ATP	adenosine triphosphate	NO	nitric oxide
AVP	arginine vasopressin or antidiuretic hormone	NTS	nucleus of the tractus solitarius (target of visceral afferent neurons)
BAT	brown adipose tissue	OCN	osteocalcin
CLPO	caudolateral preoptic nucleus	PNMT	phenylethanolamine N-methyltransferase
CNS	central nervous system	POA	preoptic hypothalamic area
DMH	dorsomedial hypothalamic nucleus	PS	parasympathetic
E	epinephrine or adrenalin	PVN	paraventricular
eNOS	endothelial nitric oxide synthetase	RMPO	rostromedial preoptic nucleus
FFA	free fatty acid	RVLM	rostral ventrolateral medulla (medullary pressor area)
HGP	hepatic glucose production	S	sympathetic
HR	heart rate	SAP	systolic arterial pressure
IML	intermediolateral column of spinal gray matter	SCN	suprachiasmatic nucleus (the master circadian clock)
Km	Michaelis equation constant reflecting the affinity of an enzyme for its substrate	SV	stroke volume
LH	lateral hypothalamus	VC	vascular conductance (blood flow at a given arterial pressure)
MAPK	mitogen-activated protein kinase	VIP	vasoactive intestinal peptide
MO	medulla oblongata	VMH	ventromedial hypothalamic nucleus
		WAT	white adipose tissue

the adrenal medulla. During evolution and embryonic development (the latter due to activation by cortisol of PNMT), these postganglionic S neurons lose their axon. Thus, E is released as a hormone into systemic circulation in response to stimulation by S preganglionic neurons that originate in the intermediolateral (IML) section of the gray matter in the thoracic and lumbar spinal cord (justifying the term *sympathoadrenal system*).

The concept of exclusive catecholamine signaling in the S division of the ANS and Ach signaling in the PS division was modified after the discovery that monoamines, amino acids, neuropeptides, adenosine triphosphate (ATP), and nitric oxide (NO) are colocalized with catecholamines and Ach in autonomic nerves (reviewed in 44). Categorized globally as nonadrenergic–noncholinergic neurotransmitters, they are released upon activation of autonomic nerves. ATP is found in autonomic ganglia, neuroeffector junctions, and the enteric nervous system and on sensory nerves, where it participates in pain sensing. NO synthesis is activated by cholinergic autonomic neurons and freely diffuses into smooth muscle, acts on guanylate cyclase, and causes vascular relaxation. A role for neuropeptide Y in cholinergic S nerves to sweat glands and some S nerves was also documented. Ach cotransmitters include vasoactive intestinal polypeptide (VIP), ATP, and NO, all with vasodilatory action in different vascular beds (185).

Autonomic Receptors

Alpha and beta adrenergic receptors were discovered in 1948 by Raymond Ahlquist at the Medical College of Georgia, who compared vasoactive potencies of six sympathomimetic amines on smooth muscles (4, 114). The division into β_1 and β_2 subtypes found mainly on myocardial and smooth muscle cells and adipocytes was proposed by Robert Furchgott in 1967 (104) and was followed by the discovery of β_3 adrenergic receptors (83). In 1977, alpha adrenergic receptors were divided into α_1 and α_2 subtypes by Berthelson and Pettinger (25). That same year, Langer localized α_2 receptors responsible for catecholamine reuptake on presynaptic membranes (173). By the end of the 1970s, the catecholamine signal transduction pathways were described: beta adrenoceptor coupling to adenylyl cyclase (178), inhibition of this transduction by α_2 adrenoceptor signaling and by coupling of α_1 signaling to calcium and phosphatidylinositol turnover (92). Currently, three beta adrenoceptors and three each of α_1 (α_{1A}, α_{1B}, and α_{1D}) and α_2 (α_{2A}, α_{2B}, and α_{2C}) types are recognized (22, 222). With time, a distinct pattern of distribution of ANS receptors in tissues and organs was recognized (alpha receptors predominating in the vascular smooth muscles, beta receptors predominating in the heart, skeletal muscle, and airway smooth muscle, and both found in the endocrine glands, liver, and adipose tissue) and their distinct end-organ actions coupled to different enzymes and ion channels were discovered (222). By the end of the 20th century, chromosomal gene location, molecular composition, and transmembrane structure of adrenergic receptors were elucidated (162), interactions with G regulatory proteins were identified, and cloned modifications were produced (178, 222).

Muscarinic receptors on targets of the vagus and other cranial and sacral PS nerves were characterized in the 1950s, and nicotinic receptors, the site of action of Ach in autonomic ganglia, the myoneural junction, and some CNS synapses were characterized in the 1970s (57). Five muscarinic receptors, M1 to M5, were recognized by the 1980s. Elucidation of the structural homology to adrenergic receptors and of signaling cascades of muscarinic receptors followed. M1 and M4 receptors stimulate phosphatidylinositol turnover whereas M2 and M3 receptors inhibit adenylyl cyclase (288).

Neural Substrates of ANS Reflexes

Langley's early attempt at a synthetic view of ANS function (177) was accompanied by several studies implicating the medulla oblongata (MO) in the control of circulation (Ditmar in 1873; 78), inhibition of HR (Dixon in 1906; 79), and control of blood pressure (Ranson and Billingsley in 1916; 225). Another comprehensive review of the integrative function of central ANS circuits appeared seven decades after Langley's book (183).

ANS pathways are organized at two levels of complexity: those that support reflex homeostatic regulation of the internal environment and those that reciprocally connect the brain centers that modulate autonomic reflexes with those that exert endocrine and autonomic influence over behaviors such as exercise, feeding, drinking, and sodium hunger.

Sensory input from the vasculature, heart, lungs, and gastrointestinal tract travels in PS glossopharyngeal (IXth) and vagus (Xth) nerves to the nucleus of the tractus solitarius (NTS) in the dorsomedial MO to influence autonomic reflexes (figure 9.1). These afferents include high-pressure arterial baroreceptors in the aortic arch and carotid sinus that sense changes in arterial pressure; low-pressure cardiac baroreceptors in the atrial walls and great veins at their confluence with the atria that sense changes in venous return to the heart; peripheral chemoreceptors in carotid bodies and the aortic arch that detect changes in arterial pO_2, pCO_2, and pH; and slowly adapting and rapidly adapting pulmonary stretch receptors. Additional sensory input to the NTS comes from the central chemoreceptors on the ventral surface of the MO that monitor changes in pCO_2, joint mechanoreceptors that signal limb movement through undifferentiated nerve endings, muscle receptors that transmit the information about intracellular metabolic changes via group III myelinated mechanoreceptors and group IV unmyelinated chemoreceptors, and gut afferents that travel in the sensory vagus. These visceral messages are not consciously detected. Pain and temperature receptors on the body surface, taste afferents in the facial and glossopharyngeal cranial nerves, and visual sensing of danger in the external environment are consciously perceived and trigger autonomic and endocrine reflexes as well as behaviors such as fight, flight, or feeding. Visceral afferent fibers comingle with a high proportion of somatic sensory fibers so that the pain in the heart or portions of the gut is referred to a large and diffuse body surface area.

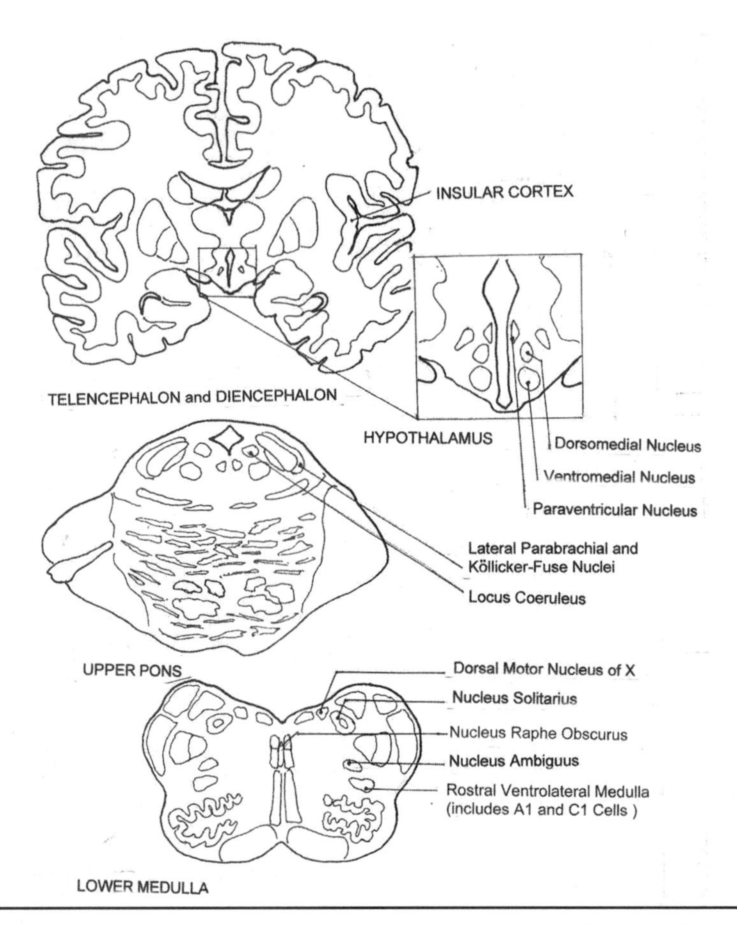

Figure 9.1 Coronal views of representative CNS regions involved in the control of autonomic function. The trunk of the trigeminal (Vth) cranial nerve is seen in the pontine section.

Drawing by S.W. Newman adapted from *Atlas of the human brain in section*, 2nd ed., Roberts, Hanaway, and Morest, Lea and Febiger, 1987; and *Neuroanatomy: An atlas of structures, sections and systems*, 6th ed., Lippincott, Williams and Wilkins, 2004.

The NTS is the target of multiple visceral modalities that partially overlap (182, 183), and its commissural part serves as a first-order integrative center. Additional integration involves reciprocal connections with higher brain centers that, in turn, have projections to the S preganglionic neurons in the IML column of thoracic and lumbar spinal gray matter as well as to preganglionic PS neurons in the dorsal motor nucleus of the vagus controlling gut organs and to the nucleus ambiguus (NA) controlling the heart, vasculature, and respiratory organs (figure 9.1).

Reflexes for Supply and Transport of Oxygen

Heart pump and pulmonary pump reflexes are closely coordinated in the MO to adjust circulatory transport of respiratory gases to their exchange sites in the lungs and tissues. This was anticipated as early as the 19th century when Traube (298) and Hering (128) described respiratory movements of the vascular system. In 1936, Anrep discovered respiratory sinus arrhythmia, whereby respiratory rhythm increases HR due to a partial withdrawal of the inhibitory vagal tone (9).

Cardiovascular and respiratory functions are jointly influenced by arterial baroreceptors, respiratory chemoreceptors, and lung stretch receptors. Peripheral chemoreceptors in carotid bodies and the aortic arch and chemosensitive neurons on the ventral surface of the medulla and in the rostral ventrolateral medulla (RVLM) trigger deeper and faster respiration in response to increases in pCO_2 and H^+ and decreases in pO_2. Peripheral chemosensitive afferents account for one third of the respiratory drive and central chemosensitive neurons account for two thirds. The chemoreflex triggers inspiration and increases HR and vasoconstriction through concurrent S activation and suppression of vagal activity. Coordination of cardiorespiratory control is facilitated by the close proximity of brain centers controlling these functions. In the dorsal pons (figure 9.1, center), the locus coeruleus and the parabrachial and Kölliker-Fuse nuclei ("pneumotaxic center" or dorsal respiratory group) modulate respiratory patterns of the ventral respiratory center that constitutes a pressor area (282). The ventral respiratory center is responsible for rhythmic activation of the heart, inspiration, and synchronous S vasomotor discharges. Among its components, inspiration is initiated by the Bötzinger nucleus and carried out by rostral and caudal respiratory nuclei and the pre-Bötzinger nucleus, and the RVLM and associated A1 and C1 adrenergic nuclei synchronize S activation of the heart and vasoconstriction with inspiration. The RVLM acts as a pacemaker that imposes a rhythmic S control over cardiovascular function. Although arterial baroreceptors normally entrain this cardiovascular rhythm, the rhythm originates in the RVLM and persists, albeit desynchronized from HR, after baroreceptor denervation. S outflow to the heart, vascular beds, pupillary muscles, and kidney vasculature is synchronized at 2 to 6 Hz.

Lung stretch and baroreceptor afferents to the NTS inhibit inspiratory activity, slow the heart, and lower blood pressure through activation of the NA, the origin of vagal depressor outflow to the cardiovascular and respiratory organs. The NA lies immediately dorsal to the ventral respiratory center (figure 9.1) and exerts a tonic depressor action over respiration and cardiovascular function that coincides with expiration. A number of other brain centers are involved in the control of cardiorespiratory reflexes. These include reciprocal connections between the medullary cardiorespiratory centers with an A5 NE cell group, the parabrachial nucleus, limbic structures including the central amygdala and basal nucleus of the stria terminalis, and the paraventricular (PVN) and lateral hypothalamic (LH) nuclei (figure 9.1).

Thermoregulatory Reflexes

The involvement of the brain in reflex regulation of body temperature was noted as early as the 19th century (13, 219). The current view is that thermoregulation is served by feed-forward or open-loop reflexes initiated by skin temperature receptors that project to thermosensitive neurons in the preoptic area (POA) of the hypothalamus and by central temperature sensors in this part of the brain. Rostromedial (RMPO) and caudolateral (CLPO) nuclei in the POA and the rostral raphe nucleus in the pons act as integrative centers for the peripheral and central temperature information that mounts the defenses against heat and cold. RMPO and CLPO areas exert tonic vasoconstrictive influence over skin circulation through raphe and RVLM projections to preganglionic neurons in the IML column of the thoracolumbar spine (292). Postganglionic S neurons from the superior cervical and middle cervical ganglia innervate the skin of the head, and those from stellate ganglia innervate the skin of the arms. Lumbar S fibers from the paravertebral chain innervate the skin of the trunk and lower limbs. Vasomotor reflexes are elicited by S noradrenergic nerves to skin arterioles, venules, and shunt anastomoses, and piloerection is elicited by S nerves to muscles attached to skin hair. S fibers to brown adipose tissue (BAT) beds elicit nonshivering thermogenesis (256), and S cholinergic efferent nerves to the eccrine sweat glands trigger the sudomotor reflex (27).

In the heat, RMPO and CLPO neurons inhibit action of the rostral raphe and S vasoconstrictor neurons. This elicits a cooling vasomotor reflex via S cholinergic nerves by increasing skin blood flow (145, 152). The identity of the cotransmitter responsible for vasodilatation in the heat is uncertain; VIP and NO are possible candidates (145). Stimulation by RMPO and CLPO of sweating is mediated by S activation of muscarinic cholinergic pathways (145) rather than S activation of alpha and beta adrenergic receptors (254).

Medial preoptic area (MPOA) neurons stimulate the dorsomedial hypothalamus (DMH) and rostral raphe nucleus to coordinate S defenses against cold conditions (205). Cold-responsive neurons in the RMPO stimulate dorsal raphe neurons and the RVLM to exert a tonic skin vasoconstriction in the cold. BAT thermogenesis is turned on by activation of the RMPO nucleus, which then inhibits a tonic influence from the CLPO to the DMH and increases S outflow to BAT (208). Prolonged exposure to cold results in BAT hypertrophy, a process that again depends on S activation and involvement of the VMH nucleus (116, 290) but is complemented and augmented by hormonal reflexes. Adipokine leptin increases S outflow to activate the expression of the thermogenic uncoupling protein UCP1 in BAT (64). It also stimulates the release of thyrotropin-releasing hormone and activates the melanocortin receptor MC4R in the brain to increase thermogenesis (215).

Besides their role in thermoregulation, vasomotor reflexes participate in blood pressure regulation when, as during intense exercise, thermoregulatory and nonthermoregulatory demands compete for limited blood volume. In addition, afferents from warm skin to the lateral parabrachial nucleus and RVLM accelerate HR in the heat (207).

Reflexes Supplying Metabolic Fuels

Coordinated cardiorespiratory reflexes match oxygen intake and delivery to support the metabolic cost of exercise. It is not surprising then that the ANS coordinates mobilization and utilization of metabolic fuels with cardiorespiratory reflexes.

Hepatic Glucose Production Reflexes

Plasma glucose concentration is regulated through closed-loop negative feedback reflexes because of the brain's dependence on this fuel and its versatility in supporting muscle energy needs. The free fatty acids (FFA), on the other hand, are mobilized in an episodic open-loop manner when the metabolism of carbohydrate is not obligatory. It has been known since 1843 that the brain is involved in the regulation of plasma glucose, when Claude Bernard observed that stimulation by puncture ("piqūre") of the floor of the fourth ventricle increased plasma glucose concentration (23). This effect was initially attributed to E (50), then to NE in 1918 (284) and to the sympathoadrenal system in 1924 (49). In 1964, the effect was definitively attributed to S nerves rather than to circulating E (275). With the early focus on the ventromedial (VMH) and LH nuclei as controllers of feeding behavior and with the early focus on the role of glucose in this control, the VMH and regions posterolateral to the VMH were electrically stimulated and found to elicit hyperglycemia mediated by splanchnic S nerves (100). VMH stimulation increased the activity of the hepatic glycogenolytic enzyme phosphorylase (269, 270) and the gluconeogenic enzyme phosphoenolpyruvate carboxykinase and reduced liver glycogen content (270). The suprachiasmatic (SCN) hypothalamic master clock also was found to exert temporal modulation of this process because glucose elevation could be elicited by electrical stimulation of VMH during only one half of the nycthemeral cycle (322) and was abolished by SCN lesions (214). Subsequently, the projections from the adrenergic cell groups A1 through A3 and C3 in the mediodorsal MO and from the RVLM to S preganglionic neurons in the IML, rather than the hypothalamus (61, 206), were found to be essential for elicitation of hepatic glucose production (HGP) (188, 234, 235).

Hyperglycemia results from reflex elicitation of HGP in response to the sensing of cellular glucose need (glucopenia). Glucosensitive and glucoresponsive neurons were described in the VMH and LH nuclei (7) as well as in the MO (234). Peripheral glucoreceptors in the liver were reported in 1969 (212). The glucosensitive afferent fibers in the hepatic branch of the vagus project to the NTS. Neural projections from the LH area to the pancreas facilitate insulin secretion, and those to the liver facilitate hepatic glycogen synthesis; both are mediated by the vagus (101, 233, 269). The counteractive ANS arm in this homeostatic feedback loop consists of the inhibition of pancreatic insulin secretion by the VMH and S nerves (100). Intracerebral administration of glucose and insulin blocks VMH activation of S outflow and its inhibition of insulin secretion (250). The counterbalancing roles of the S and PS nervous system in blood glucose regulation are clearly seen in reciprocal discharge patterns of the efferent pancreatic branch of the vagus and of S adrenal nerves in response to changes in blood glucose concentration (213).

HGP also is controlled by purely hormonal reflexes, chief of which is the counterregulation by glucagon of the hypoglycemic effects of insulin. Counterregulation is complemented by the hypothalamic and hindbrain control of S outflow (8, 100), which directly activates HGP, elicits E secretion, and facilitates hyperglycemic actions of glucagon and cortisol (236). The ANS also is involved in stimulation of glucose uptake by skeletal muscle, heart, and adipose tissue via α_1, β_1, and β_3 adrenoceptors (59, 136, 200). Such glucose uptake is non-insulin-dependent and, when elicited by E, is associated with suppression of insulin secretion (217).

Sympathoadrenal reflexes supplying metabolic fuels are discrete rather than indiscriminate. Fasting suppresses S neural stimulation of metabolism while at the same time the secretion of E is amplified (325). Similarly, cold exposure stimulates gluconeogenesis through selective activation by S nerves of α_1 adrenoceptors (271).

Reflex S Elicitation of Lipolysis

The involvement of S nerves in the control of lipolysis in both white adipose tissue (WAT) and BAT has only recently been recognized. It was assumed that the interactions between the brain and WAT were predominantly mediated by hormones despite the microscopic evidence of WAT innervation as early as the end of the 19th century (80) and recurrent reports that stimulation of the VMH elicits lipolysis (289) whereas lesions (216) and knife-cut isolation of this area (37) reduce it. Leptin (259) and several other adipokines were seen as afferent messengers, and growth hormone, cortisol, glucagon, atrial natriuretic peptide (ANP) (28, 263), leptin (103), and E (43) were viewed as principal triggers of lipolysis. New techniques for retrograde transneuronal mapping of synaptic connections have revealed afferent projections from WAT and BAT to the NTS (20). On the efferent side, PVN, DMH, and SCN hypothalamic nuclei, and outside the hypothalamus, MPOA, were known to mediate S control of lipolysis. The RVLM and catecholaminergic cell groups C1 and A5 were recently implicated in this control as well (16, 19, 326). Neural elicitation of lipolysis from WAT and BAT is episodic and controlled in an open-loop fashion. WAT lipolysis occurs during the postabsorptive period and in response to fasting, glucopenia, and the energy costs of exercise and cold exposure. All of these circumstances are accompanied by the lowering of plasma concentrations of insulin and the withdrawal of its powerful antilipolytic actions (232).

The catecholamines released from S nerve endings and the adrenal medulla trigger lipolysis from dispersed subcutaneous and intra-abdominal fat pads to mobilize the release of FFAs into circulation for use by skeletal and cardiac muscles. Lipolysis from individual WAT pads depends on the type of stimulus. In a seasonally breeding hamster, subcutaneous groin and intra-abdominal retroperitoneal fat pads respond to food deprivation, cold, and glucoprivation. Dorsal subcutaneous fat is mobilized in response to glucoprivation, and epididymal fat is mobilized in response to cold and food deprivation (40). BAT lipolysis is a thermoregulatory response to cold exposure that augments uncoupled oxidation of FFAs in the tissue (20).

Selective activation of lipolysis reflects differential expression and affinities of the antilipolytic α_2 and lipolytic β_1, β_2, and β_3 adrenoceptors in individual fat depots. Peter Arner in the late 1980s used a microdialysis technique to study regional differences in adrenergic receptor distribution and their affinities for catecholamines. Subcutaneous fat depots have 30% more α_2 than β adrenoceptors, especially in women (170, 197). In subcutaneous fat depots, lipolysis rate is lower in femoral and gluteal subcutaneous depots than in abdominal subcutaneous and omental depots (11). Fat depot activation is affected by exercise intensity, which alters the rate of S nerve discharges and tissue NE turnover and concentration. Low NE concentrations activate β_1 receptors that, along with β_2 receptors, rapidly desensitize (169), whereas higher and prolonged NE exposure resulting from higher rates of S nerve discharges activates β_3 adrenoceptors that are more resistant to desensitization (113). Difference in adrenoceptor affinities for catecholamines also accounts for selective activation of lipolysis in individual fat pads. α_2 adrenoceptors have the highest affinity for catecholamines and therefore exert tonic suppression over lipolysis at low level of S activation or higher plasma insulin concentrations. The relative order of affinity of fat cell adrenergic receptors for catecholamines is $\alpha_2 > \beta_1 \geq \beta_2 > \beta_3$ for NE and $\alpha_2 > \beta_2 > \beta_1 > \beta_3$ for E (169).

The SCN master clock controls circadian changes in NE turnover in sympathetically innervated tissues (311). This highlights the central role of the ANS in increasing carbohydrate metabolism during diurnal active periods and lipid metabolism during nocturnal periods of rest and sleep (180). Finally, β_3 stimulation of WAT produces hypotensive and vasodilatory effects that trigger baroreflex activation of the HR, demonstrating the interrelatedness of cardiorespiratory and metabolic S actions (293).

Role of the ANS in Immunomodulation

The ANS also influences immune defenses (132). Lymphoid organs are innervated by the splenic S nerve and thymic vagal efferent nerve. Their activities are influenced by stress, stimulation, or ablation of the hypothalamus. Release of NE in the spleen affects natural killer lymphocyte cytotoxicity. The thymic vagal branch may be involved in immunomodulation, whereas hepatic vagal afferents may transmit cytokine and neuropeptide signals to the NTS.

ANS Support of Increased Energy Requirements During Exercise

Exercise presents an extreme physiological challenge to cardiorespiratory and metabolic functions because it can require an up to 700% increase in cardiac output and a 1,000% increase in minute ventilation and metabolism. In addition, some forms of extreme endurance activity may increase energy expenditure 800% above daily resting energy balance. Such large increases in metabolism create chemical, thermoregulatory, and bioenergetic disturbances in the internal environment that require acute autonomic adjustments and delayed autonomic compensations.

This section addresses the role of the ANS in the autonomic control of cardiorespiratory, thermoregulatory, and metabolic functions mainly during acute exercise. It focuses on how the ANS matches increases in cardiorespiratory and circulatory functions to metabolic needs at different intensities and durations of exercise and how the ANS establishes higher operating points in its reflexes and different priorities among the homeostatic functions it controls.

Reflex Autonomic Control of Cardiorespiratory Function During Exercise

Autonomic Control of Heart Function During Exercise

Exercise-associated increases in cardiac output require increases in HR and stroke volume (SV) and adjustments in vascular resistance. Precise control of HR and respiration during exercise is essential. The almost instantaneous acceleration of HR at the start of exercise is launched by withdrawal of vagal tone (figure 9.2), an observation first reported in 1895 by Henrich Ewald

Hering in Czechoslovakia (129). A decade later, Wilbur Bowen at the University of Michigan in the United States attributed the 1 s latency in HR acceleration at the start of cycling to reduced vagal action (35). The central role of vagal withdrawal in this phenomenon was inferred in 1914 by Gasser and Meek (115) and was simulated in the 1970s by vagal resection and muscarinic blockade of vagal action with atropine (figure 9.2), both of which preempted HR acceleration at the start of exercise (93). This vagal involvement was documented for dynamic exercise in 1970s (93, 99) and for isometric exercise in 1987 (187).

Vagal withdrawal is almost complete by the time the ventilatory threshold is reached at about 60% of maximal effort and at a HR of between 100 and 130 beats/min (244, 323). At such low exercise intensities, no S nerve activity to skeletal muscle is detected (301). S activation of HR progressively increases as relative exercise intensities increase above 50% of maximal effort (209, 249) and is reflected in proportional increases in plasma NE and E concentrations (111, 124).

S nerves produce end-organ effects through NE spillover (leakage of the transmitter into extracellular fluid and into circulation) that was first measured in the body as a whole and was subsequently measured in individual organs in 1979 by Murray Esler in Australia (88, 89). NE spillover distribution at rest is 2% to 3% in the heart, 30% in the lungs, from 20% to 50% in muscle, about 25% in the kidneys, approximately 10% in the splanchnic region, and about 5% in the skin. In 1987, Gabrielle Savard at Karolinska Institute in Stockholm, Sweden, showed that NE spillover was increased in contracting skeletal muscle (255).

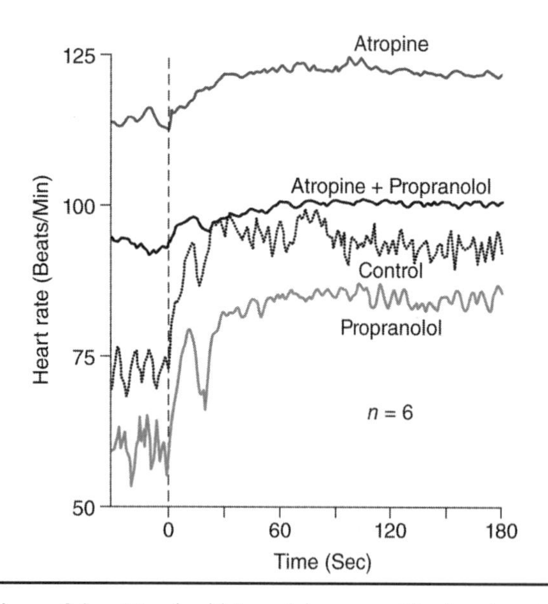

Figure 9.2 Vagal withdrawal is responsible for almost instantaneous HR acceleration at the start of exercise. The effect is atropine suppressible but not propranolol suppressible. Absence of vagal activity lowers HR variability.

Reprinted from L. Fagraeus and D. Linnarsson, 1976, "Autonomic origin of heart rate fluctuations at the onset of muscular exercise," *Journal of Applied Physiology* 40: 679- 682. With permission of American Physiology Society.

The primacy of beta adrenergic stimulation of the HR was established in 1965 when receptor blockade with propranolol was found to reduce left ventricular inotropic action during exercise by 34% and produce comparable reductions in cardiac output, coronary blood flow, and myocardial oxygen consumption (82, 87). After selective blockers for β_1 (atenolol) and β_2 (ICI 118551) adrenoceptors became available between 1970 and 1980, the predominant role of the former receptor type in ventricular contractility was demonstrated during maximal exercise in dogs (194).

The ANS frequently engages endocrine reflexes that amplify or modify some of its actions. A case in point is S elicitation of the hydromineral hormones renin, angiotensin (ANG) II, aldosterone (119), arginine vasopressin (159), and ANP, also localized in the adrenal medulla and S ganglia (137). Increased S activation and NE spillover in the kidney during exercise triggers the renin–ANG–aldosterone reflex through stimulation of beta adrenergic receptors on the granular cells in the macula densa region of the kidney tubule. Secretion of hydromineral hormones increases as exercise intensity increases (295). These hormones affect both central and peripheral S action, and all facilitate increases in blood pressure during exercise. ANG II, a powerful vasoconstrictor, contributes an about 10 mmHg increase to blood pressure elevation during maximal exercise without having any renal action (307). It causes upward resetting of the baroreflex, an effect that is independent of its hypertensive action (41) and that requires the integrity of the area postrema in the MO (195). Moreover, ANG II may play an essential role in the pressor action of the RVLM because excitatory vasomotor tone ends when the ANG I and ANG II receptors are blocked (138). Arginine vasopressin, on the other hand, affects cardiovascular function by enhancing neurotransmission in S ganglia, increasing constriction of smooth muscles, and potentiating the baroreflex in the area postrema by acting on arginine vasopressin 1 receptors (67). At higher exercise intensities, ANP increases blood pressure by acting on the anterior hypothalamus and augments ANG II hypertensive action (98).

Autonomic Control of Respiratory Function During Exercise

As was the case with HR acceleration, acceleration of breath frequency and increases in tidal volume at the start of exercise are almost instantaneous and proportional to exercise intensity (171). Exercise does not affect the discharge rate of peripheral chemoreceptors.

For instance, ventilatory frequency during exercise does not respond to hypoxia, increases in pCO_2, or vagotomy. Thus the near-instantaneous increase in breathing rate at the start of exercise appears to be a consequence of withdrawal of the tonic inhibitory vagal influence originating in the NA and cardiorespiratory bulbopontine nuclei (171).

Vasomotor Reflexes and ANS Control of Blood Pressure During Exercise

Supporting an up to 1,000% increase in metabolism during maximal exercise depends on providing the contracting muscle with a sufficient supply of oxygen and nutrients with only a 5.5 L supply of blood. The ANS supports this feat by routing blood flow to only the most actively contracting muscle mass while increasing cardiac output and systolic arterial pressure (SAP). S activation of cardiac function alone is not sufficient to adequately increase SAP during exercise in the absence of S vasoconstriction of splanchnic, renal, and cutaneous vascular beds. These circulations have high vascular conductance (VC; blood flow at a given arterial pressure) at rest because of their size, the significant proportion of cardiac output that they receive (27%, 22%, and 10% for splanchnic, renal, and cutaneous vascular beds, respectively), and high compliance (distensibility) of the arteries in these vascular beds. For instance, liver arterial compliance is as high as 20 $ml \cdot mmHg^{-1} \cdot kg^{-1}$ whereas the vasculature in the approximately 30 kg of skeletal muscle has very low compliance of 0.48 $ml \cdot mmHg^{-1} \cdot kg^{-1}$ and receives at rest only about 15% of the cardiac output (245).

As exercise starts, VC in skeletal muscle increases dramatically in large part due to vasodilatory events in the skeletal muscle vasculature (discussed later). The redistribution of blood to the contracting muscle was recognized as early as 1794 when Scottish anatomist and surgeon John Hunter stated that "blood goes where it is needed." Increased muscle blood flow with the onset of contraction was reported in 1876 by William Gaskell. The rapid increase in total VC at the onset of exercise usually produces a short-lived decline in SAP (figure 9.3, lower panel). The baroreflex corrects this blood pressure error within 20 s by triggering increased S activation of the heart (figure 9.3, central panel) and vasoconstriction in inactive vascular beds (266) (figure 9.3, top panel). If the baroreflex is inactivated by a blockade of aortic and carotid mechanoreceptors or is malfunctioning in individuals with autonomic dysfunction (26), maintenance of normal cardiac output by electrical stimulation of the heart cannot prevent a large decline in SAP (166). The recognition of the baroreflex role in blood pressure regulation dates to 1923 and Heinrich Ewald Hering (130), but it took a quarter of a century of research with different animal models to elucidate its importance in blood pressure regulation during exercise (296). By 2003, the baroreflex vasoconstriction of visceral and inactive skeletal vascular beds was found to play a central role in adjusting VC to match necessary increases in SAP.

At exercise intensities below the ventilatory threshold and HR of about 100 to 130 beats/min, no S nerve activity or vasoconstriction in contracting skeletal muscle is detected. A decline in skeletal muscle intracellular pH and related metabolic signals of inadequate oxygen supply and anaerobic metabolism normally trigger a chemoreflex or metaboreflex representing an increased muscle S nerve activation (MSNA) (301) that is absent in patients with genetic muscle phosphorylase deficiency (i.e., McArdle's disease) (223). The first description of chemoreflex should be credited to M. Alam and F.H. Smirk at Cairo University in Egypt. They demonstrated in 1937 that occlusion of blood flow in a limb during serial contractions dramatically increased blood pressure and that this hypertension was sustained for the duration of occlusion beyond the cessation of contractions (5).

Muscle chemoreflex increases S discharges to the skeletal muscle in response to chemical error signals associated with reduced muscle perfusion (244). As such, it aggravates the error signal that triggered it. However, it serves as a delayed backup system to supply additional oxygen to underperfused muscle by increasing cardiac output, SAP, and vasoconstriction in contracting skeletal muscle during intense exercise even in the absence of an operational baroreflex. S activation of vasoconstriction is not the only contributor to maintenance of SAP during exercise. Increases in SV during exercise also contribute to exercise-associated increases in cardiac output. SV increases are critically dependent on the return of venous blood to the atria. This return depends in part on the operation of muscle and respiratory pumps during exercise (245).

Blood flow to muscle (hyperemia) can increase between 400% and 500% between rest and maximal exercise, but how this happens is only partially understood. Contributions to exercising muscle hyperemia by neural, mechanical, or chemical signals either resident in the muscle and its vascular endothelium or transported in circulation have been examined (146, 299). Direct neural influences on muscle vasculature cannot account for the magnitude of vasodilation during exercise. Unlike cats and dogs, humans do not appear to have S cholinergic vasodilatory nerves, and the role of a withdrawal of S vasoconstrictor tone was discounted by normal hyperemia when S transmission was experimentally blocked (230). Ach spillover from the motoneuron end-plates also fails to account for human exercise hy-

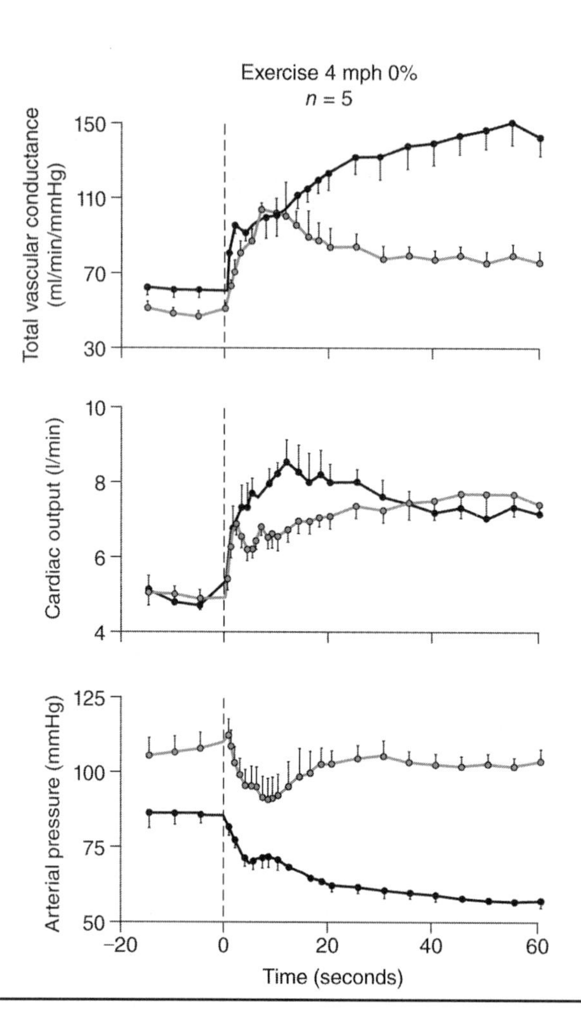

Figure 9.3 Despite controlled cardiac output produced by electrical pacing, dogs exercising with S ganglionic blockade displayed 100% increases in vascular conductance and progressive declines in systemic arterial pressure (solid circles) compared with intact dogs (open circles).

Reprinted from D.D. Sheriff, L.B. Rowell, and A.M. Scher, 2003, "Is rapid rise in vascular conductance at onset of dynamic exercise due to muscle pump?" *American Journal of Physiology* 265: H1227-H1234. With permission of American Physiological Society.

peremia, as does mechanical pressure from the muscle pump.

With the discovery of NO by Furchgott and Zavadsky in 1980 (105), the role of substances released by the muscle or its vascular endothelium have been scrutinized. NO is synthesized by the enzyme NO synthetase from the terminal nitrogen atom of L-arginine in the presence of O_2 and the cofactors nicotinamide adenine nucleotide diphosphate, flavin adenine dinucleotide, flavin mononucleotide, heme, and tetrahydrobiopterin. NO is synthesized by the neuronal NO synthetase (nNOS) in the sarcolemma. Additional vasodilatory substances released by the contracting skeletal muscle are adenosine, ATP, and K. The vascular endothelium in skeletal muscle releases vasodilatory prostaglandins and NO synthesized by endothelial NO synthetase (eNOS). Shear stress (163) as well as Ach that is present in and released by some endothelial cells (76, 202) activate eNOS. Blockers specific to eNOS (N^G-monomethyl-L-arginine), both nNOS and eNOS (N^G-nitro-L-arginine methyl ester), prostanoids (the cyclooxygenase inhibitor ketorolac), and adenosine were used to identify obligatory factors for exercise hyperemia. Neither NO or adenosine (191) nor K or prostanoids (94), alone or in combination, fulfilled that role. Vasodilatory effects of NO were found to be more pronounced at rest and during recovery from exercise than during exercise (224).

Failure to substantiate the involvement of NO in hyperemia during dynamic exercise was, in part, a consequence of the corrective baroreflex compensation for a decline in SAP resulting from NO vasodilatation (266). However, even an increase in NO hyperemia after sinoaortic denervation (253) did not match the increase in muscle VC during exercise. This increase in vasodilatation after baroreflex denervation demonstrates sympa-

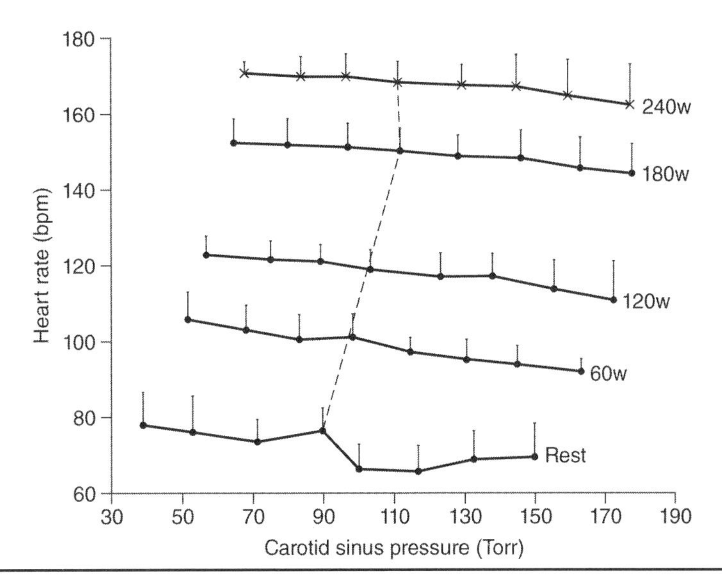

Figure 9.4 The shift of baroreflex operating point (indicated by the dashed line) to higher blood pressures with increases in exercise intensity (expressed in watts).

Reprinted from Y. Papelier et al., 1994, "Carotid baroreflex control of blood pressure and heart rate in man during dynamic exercise," *Journal of Applied Physiology* 77: 502-506. With permission of American Physiological Society.

tholysis during exercise, (vasodilatory antagonism of S vasoconstriction), a concept first introduced in the 1960s by Remensnyder (231). However the magnitude of sympatholysis is modest because NO and prostaglandin antagonize S vasoconstriction by only about 10% (77). These various local vasodilatory metabolites released by working muscle are more effective in suppressing α_2 adrenoceptors prevalent on terminal arteriolar sphincters in Type II glycolytic muscle fibers than in suppressing α_1 adrenoceptors on resistance arterioles prevalent in Type I fibers (294). Of the circulating vasodilatory factors, E acting on vascular β_2 adrenergic receptors at most increases vasodilatation during exercise by 100% (230). Other circulating factors that cause vasodilatation are partial pressures of oxygen and carbon dioxide and ATP that is released from hemoglobin as it offloads oxygen in the muscle. The dilatory capacity of these molecules is insufficient to account for the large increase in muscle VC during exercise (146).

Despite the currently unsatisfactory level of understanding of the mechanism of contracting muscle hyperemia (146), it should be added that NO mediates interactions between PS and S divisions of the ANS. In the periphery, vagus-associated nNOS activation causes vasodilatation in the kidney (297) and vagal chronotropic effects on the left ventricle (36). Therefore, a decline in central NO production during vagal withdrawal at the onset of exercise could account for the rapid vasoconstriction of renal circulation (295) and for rapid HR acceleration (figure 9.3). In the CNS, nNOS is localized in the hypothalamic and medullary centers involved in the control of cardiorespiratory function. This enzyme is found in neural terminals impinging on NA neurons (150) and on NTS neurons (90), where it tonically restrains S outflow (240) and elicits a decline in SAP (181).

The baroreflex plays an important role in adjusting S vasoconstriction to maintain SAP; however, this reflex does not prevent the significant and proportional increases in HR and in systolic blood pressure that are observed with increases in work output (97). At the onset of exercise, the operating point of the baroreflex (blood pressure that triggers compensatory changes in cardiac output) is reset to a higher blood pressure level so that prevailing blood pressure is perceived as a hypotensive error signal (figure 9.4) (220, 221). The near-instantaneous increase in HR at the onset of exercise and its dependence on an early baroreflex response to the decrease in SAP (figure 9.3) has been attributed to the operation of central neural command in these processes (135). This concept, initially proposed by Krogh and Lindhard in 1913 (168) and more fully developed by Rowell in 1993 (242), attributes simultaneous initiation of physical activity and cardiorespiratory activation to the brain central command. The brain regions subsequently implicated in central command include the mesencephalic cuneiform nucleus and lateral and posterior hypothalamic areas as motor areas in initiation of movement, the withdrawal of vagal tone from NA, and the rapid activation of medullary cardiorespiratory responses from structures such as the NTS and rostral and caudal ventrolateral medulla (139).

Reflex Autonomic Control of Thermoregulation During Exercise

Because heat is the byproduct of energy metabolism, thermal homeostasis is challenged by muscle contractions during exercise. S activation is the final common pathway for heat loss reflexes. A thermoregulatory vasomotor reflex leads to a rapid increase in cutaneous VC so that a portion of the cardiac output is now diverted from the exercising muscle to the skin (267). Increased heat load during exercise also increases the sudomotor (sweating) reflex that becomes the principal means of heat dissipation by evaporation. HR increases in the heat through reduced vagal modulation that allows increased diversion of cardiac output toward the skin (39). As core temperature approaches 38 °C, increases in skin blood flow (38) and cardiac output (243) decline suddenly. This may reflect the operation of reflex inhibition of S cholinergic nerves because it is unaffected by adrenergic blockade (152, 156).

Autonomic Control of Vasomotor Heat Dissipation

Increases in skin blood flow in response to an increasing core temperature during exercise are rapid and mainly mediated by a cotransmitter in cholinergic S nerves rather than by withdrawal of S tone (145, 152). The S outflow actively adjusts cutaneous vasoconstrictor tone to balance the competing requirements of maintaining adequate SAP during exercise and heat loss through increases in cutaneous vasodilatation (267). Increases in skin blood flow due to core temperature elevation can be prevented by either sympathectomy (238) or a chemical presynaptic blockade of cholinergic neurotransmission but not by muscarinic receptor blockade with atropine (152). The usual cotransmitters in the S nerves are ATP and neuropeptide Y, and the usual cotransmitter in the cholinergic PS nerves is VIP. However, the identity of the transmitter involved in active skin vasodilatation during exercise is at present unknown.

Autonomic Control of Sudomotor Heat Dissipation

Although the sweat gland duct was described in 1600 and the existence of the sweat gland was recognized in 1800, the role of sweating in thermoregulation was discovered in the 20th century (268). Afferent input to the anterior hypothalamus for adjustments in vasomotor and sudomotor reflexes arises in skin temperature receptors, receptors in the brain sensing increases in core temperature, and metaboreceptors activated during ischemic isometric exercise (268). However, sweating from eccrine glands appears to be initiated by central command neurons in the motor cortex because it begins within seconds of the onset of exercise, before any increases in skin and core temperature are registered. Neural sources for activation of preganglionic S neurons in the IML column of the spinal cord include the POA, the tegmentum of the pons, and the medullary raphe nuclei. Sweating is mediated to a greater extent by S activation of muscarinic cholinergic receptors (145) rather than by S action on alpha and beta adrenergic receptors (254, 264).

Reflex Autonomic Control of Fuel Mobilization and Utilization at Different Exercise Intensities and Durations

Progressive increases in work intensity lead to recruitment of different types of skeletal muscle fibers and different patterns of fuel utilization. At low exercise intensities, Type I oxidative muscle fibers are recruited and mainly metabolize circulating FFAs (240). At moderate exercise intensities, Type IIa oxidative-glycolytic fibers are recruited, and approximately one half of the energy metabolized is provided by carbohydrate of both hepatic and muscle origin. The other half is supplied by lipids, again in equal measure from adipose tissue and muscle. At high exercise intensities, glycolytic Type IIx fibers predominantly utilize muscle glycogen and, to a lesser extent, circulating glucose and FFAs, muscle triglycerides, and plasma glucose (240).

The difference in the pattern of fuel utilization is a product of different patterns of autonomic and endocrine control. At low exercise intensities, S nerves discharge at a rate of about 1 impulse/s and stimulate adipose tissue lipolysis without eliciting peripheral vasoconstriction (96, 241). Moderate-intensity exercise doubles blood flow to adipose tissue (12) and increases blood flow to muscle more than tenfold (134). Thus, low to moderate exercise intensities facilitate FFA transport from adipose tissue to skeletal muscle and fat oxidation by Type I muscle fibers. S inhibition of insulin secretion during exercise contributes powerfully to activation of lipolysis. Adipose tissue lipolysis therefore benefits from both a decrease in plasma insulin concentration and an increase in sympathoadrenal activation. A rich supply of beta adrenergic receptors facilitates lipid utilization in oxidative skeletal muscle fibers at low exercise intensities that allow lipolytic action of beta receptors to overcome the predominant antilipolysis of alpha adrenoceptors (174). Experiments using selective beta adrenergic blockade identified β_1, rather than β_2, adrenoceptors as mediators of adipose tissue lipolysis during exercise (186, 300). A study of exercise-associated lipolysis in subjects with spinal cord injury that reduced S nerve traffic to areas below the injury more than to areas above the injury showed no impairment based on the location of subcutaneous adipose tissue, thus suggesting that circulating E plays a more important role in the process than do S nerves (283).

Stimulation of Hepatic Alpha Adrenoceptors by S Nerves Promotes Hepatic Gluconeogenesis (91)

At moderate exercise intensities of between 50% and 75% of maximal effort, E secretion from the adrenal medulla augments the effects of NE spillover out of the S nerve terminals in the muscle, liver, and adipose tissue. Plasma E and NE concentrations, which are increased four- to sixfold during moderate-intensity exercise, contribute to increased glycogenolysis and carbohydrate metabolism in the liver and the muscle (141) and to increased lipolysis in adipose tissue and muscle lipid uptake (174). Type IIx glycolytic fibers are recruited at high exercise intensities. Due to their lack of sensitivity to catecholamines, their reliance on glycogen utilization is controlled by contractile activity (118).

Cellular metabolism at rest and during exercise is controlled in a redundant fashion by the ANS and by hormones (31, 108), a process studied and clarified by David Wasserman (figure 9.5), among others. Hormones control exercise metabolism distinctly so that a decline in insulin facilitates FFA mobilization and HGP (315), S neural activation—but not circulating E—facilitates hepatic glycogenolysis (55), circulating E acting on beta adrenergic receptors stimulates FFA mobilization from the adipose tissue (30) and muscle glycogenolysis (10), growth hormone shifts metabolism away from protein catabolism toward lipid mobilization and utilization and hepatic gluconeogenesis, and glucagon promotes hepatic glycogenolysis and gluconeogenesis in response to reduced glucose availability but plays a minor role in the control of lipolysis (109).

The question that exercise physiologists have grappled with over the past 60 yr is how much of exercise metabolism is under homeostatic negative-feedback control and how much is driven by central neural command in a feed-forward fashion. Studies with rats, cats, dogs, and humans indicate that during low- to moderate-intensity exercise, blood glucose regulation operates through homeostatic negative feedback between insulin and its principal counterregulatory hormones glucagon and E (312). When also of moderate duration, increases in plasma glucagon account for 60% of increases in HGP (314) and declines in plasma insulin account for 55% of increases in HGP (315). Because these two activation processes account for more than 100% of HGP, their mechanisms partially overlap.

Thus, a decline in plasma insulin and an increase in plasma glucagon concentrations can stimulate HGP and FFA mobilization without S efferent neural activation of liver enzymes (158, 277, 316), innervation of pancreatic beta cells (62), presence of the adrenal medulla (86, 158), or sensory afferent messages (157). This does not negate the usual physiological role of such neural stimulation during exercise but it does underscore that it is not essential.

Neural contribution to exercise metabolism increases with increases in exercise intensity. At low to moderate exercise intensities, S nerves inhibit insulin secretion (252). This facilitates lipolysis and HGP in part by increasing liver sensitivity to glucagon (60). Glucagon stimulates liver uptake of gluconeogenic precursors, and gluconeogenesis contributes approximately 20% to 25% of HGP (308). As exercise increases in intensity, S

Figure 9.5 David H. Wasserman (professor of molecular physiology and biophysics, Vanderbilt University Diabetes Research and Training Center, Nashville, Tennessee, United States) in collaboration with Alan D. Cherrington clarified the conditions of endurance exercise that require catecholamine involvement in support of hepatic glucose production. His contributions to this chapter are found in references 30, 62, 63, 107, 164, 165, 203, and 312 through 316.

Photo courtesy of David Wasserman.

spillover increases in skeletal muscle, the liver, and the pancreas (63). HGP increases up to fivefold largely by glycogenolysis, which contributes about 85% of glucose production (308). Beta adrenergic stimulation of the pancreas facilitates insulin secretion (252) and causes some increase in insulin concentration at high exercise intensities (111). This favors a shift in fuel metabolism from FFAs to carbohydrate. At the same time, plasma glucose concentration increases as it is less tightly regulated but appears to be under a feed-forward central S command. Other lines of evidence supporting a central feed-forward control of metabolism at high exercise intensities are initial refractoriness of HGP to supranormal increases in plasma glucose (317) and increased fuel mobilization during exercise in response to electrical stimulation of the posterior hypothalamic locomotor area (303).As was the case with differential actions of the sympathoadrenal system at rest, the metabolic actions of E and NE differ during exercise. Secretion of E and glucagon change in response to variations in plasma glucose concentration, in contrast to NE spillover from S nerve terminals (109). These effects have been studied and elucidated by Henrik Galbo (figure 9.6) and colleagues. During exhaustive exercise at 60% of relative intensity, beta adrenergic blockade accelerates decline in plasma glucose, increase in plasma glucagon, and time to exhaustion compared with exercise in a nonmedicated condition. Intravenous glucose infusion during beta adrenergic blockade increased plasma glucose, normalized plasma glucagon, and restored plasma E to near-control conditions without having an effect on plasma NE (figure 9.7).

The involvement of S nerves in coordinating a variety of physiological adjustments to exercise underscores their greater global rather than particular actions. An example of pleiotropic S actions during exercise is the involvement of carotid bodies in glucoregulation. In addition to their role in chemoreflex increases in blood pressure during isometric exercise, carotid bodies are chemosensitive to glucose availability. They enhance glucagon counterregulatory response to decreasing blood glucose concentration (164) and they transiently uncouple HGP from muscle glucose uptake at the onset of exercise through a feed-forward increase in glucose production (165).

As duration of exercise increases, plasma glucose declines in parallel with muscle and liver glycogen depletion. Carbohydrate metabolism becomes dependent on the slower gluconeogenic glucose production that cannot keep pace with the rate of glucose utilization. Beyond 120 to 150 min of moderate-intensity exercise, catecholamines assume a more dominant role in fuel mobilization and metabolism (203). Exercise in a 60 to 72 h fasted state leads to a significant decline in plasma glucose and insulin and to exaggerated increases in plasma E and glucagon (29, 110). High concentrations of circulating catecholamines during such negative energy balance produce glucose almost entirely by hepatic gluconeogenesis (29) and suppress peripheral glucose uptake by about 50% (123). This demonstrates that autonomic and hormonal controls of metabolism are as sensitive to reductions in energy availability during exercise as they are during rest, but specific pathways of fuel mobilization are selectively activated depending on the hepatic glycogen supply.

Role of the ANS in Energy Recovery After Acute Exercise, Adaptations to Habitual Exercise or Training, and Regulation of Whole-Body Energy Balance

Recovery After Acute Exercise

Acute exercise affects cardiorespiratory and circulatory adjustments and metabolism through initial vagal with-

Figure 9.6 Henrik Galbo, professor of internal medicine at the University of Copenhagen in Denmark, defined the conditions of exercise such as intensity, energy deprivation, and glucose availability that increase S activation and plasma concentrations of catecholamines. His contributions to this chapter are found in references 108 to 112, 157, 158, 174, 277, 303, and 317.

Photo courtesy of Henrik Galbo.

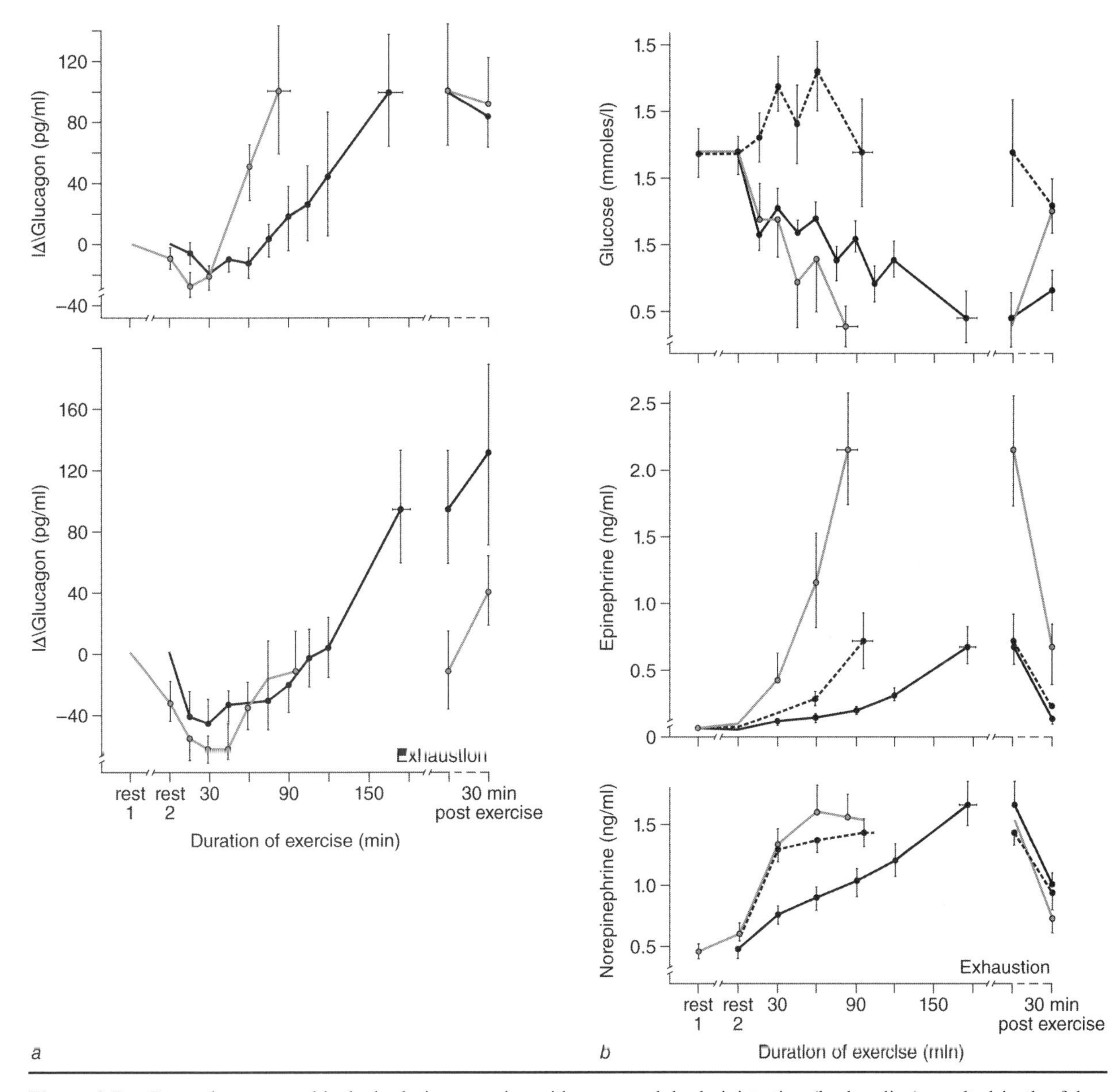

Figure 9.7 Beta adrenoceptor blockade during exercise with propranolol administration (broken line) resulted in the following: faster rate of glucagon release (stippled line) than in the absence of drug treatment (solid line; *a*, top); slower glucagon release with concurrent glucose infusion (*a*, bottom); faster decline in blood glucose than in the untreated condition or in drug with concurrent glucose infusion (*b*, top); slower E release with concurrent glucose infusion (*b*, center); and no suppression by concurrent glucose infusion of high rate of NE release (*b*, bottom).

Reprinted from H. Galbo, N.J. Christensen, and J.J. Holst, 1977, "Glucose-induced decrease in glucagon and epinephrine responses to exercise in man," *Journal of Applied Physiology* 42: 525-530. With permission of American Physiological Society.

drawal followed by increases in S outflow in proportion to the intensity of exercise. Prolonged exercise additionally produces hepatic and muscle glycogen depletion. Thus, one must examine how autonomic control of cardiorespiratory, circulatory, and metabolic functions changes when exercise ends. The use of spectral analysis of HR variability for the resolution of S and PS contributions to the control of HR (189) has facilitated studies of cardiac control during recovery from exercise. In untrained individuals, the delay in the restoration of vagal tone is proportional to the intensity and duration of exercise. After 20 min of supine exercise at intensities producing 2, 3, and 4 mM of plasma lactate concentration, vagal tone returned to baseline after 15 to 30 min (117). On the other hand, after a single bout of maximal exercise to exhaustion, the HR remained elevated 1 h after exercise although respiratory rate and SAP returned to normal. Full recovery of vagal tone was delayed for 48 h (106).

Adjustments in peripheral resistance and VC after 1 h of large-muscle exercise at 60% of maximal effort are delayed; this is another example of dissociation in the timing of recovery of individual components of autonomic function after cessation of exercise. This type and duration of exercise is associated with an approximately 2 h period of hypotension. S nerve activity and vascular resistance in the calf muscle are reduced such that vasoconstriction is lower for any increase in S activity, and the baroreflex setpoint declines because S response is reduced for any change in diastolic pressure (125).

Reliance of exercise metabolism on the breakdown of stored metabolic fuels can be viewed as a cumulative error signal in liver and muscle fuel store homeostasis. The resynthesis of metabolic fuel stores is coordinated in part by changes inherent in tissue and enzyme sensitivities to fuel depletion and in part by ANS control. A lower Km (Michaelis equation constant reflecting the affinity of an enzyme for its substrate) of muscle hexokinase compared with the Km of liver glucokinase facilitates earlier repletion of muscle than of liver glycogen after glycogen-depleting exercise (95). In the same vein, the degree of muscle glycogen depletion in different muscle fiber types leads to a proportional increase in the activation of glycogen synthase (66). Autonomic influences on the recovery of muscle and liver glycogen stores also depend on the magnitude of depletion of these stores. During acute, higher-intensity exercise of 1 h duration, liver and muscle glycogen undergo about 35% to 40% depletion. Plasma insulin concentration increases within 2 to 8 min after cessation of exercise, which accounts for the rapid decrease in HGP (309). A concurrent rapid decline in S activity restores plasma E concentration to pre-exercise levels (122). Plasma glucagon concentration remains above normal, contributes to the uptake of gluconeogenic precursors, and accounts for doubling the gluconeogenic component of HGP above the 20% baseline levels (3).

Recovery after prolonged exercise of more than 2 to 3 h in duration is marked by substantial depletion of muscle and liver glycogen stores. This leads to a persistent decrease in plasma glucose concentration and sustained elevations of plasma glucagon and catecholamine concentrations. High plasma glucagon concentration facilitates a 65% to 80% increase in gluconeogenic contribution to HGP that declines in absolute terms to the extent of hepatic glycogen depletion. Low plasma glucose also contributes to sustained elevation in plasma E, which in turn stimulates muscle glycolysis and lactate production (3). At the same time, muscle insulin sensitivity is increased (321). Largely due to this increase, muscle glucose uptake remains three to four times higher than before exercise and contributes to muscle glycogen resynthesis (3).

Additional complexities in the role of ANS during recovery from exercise occur when exercise of longer duration (90-120 min) is repeated after several hours of rest under conditions of either low or high energy availability. When such exercise is performed at both times in a postabsorptive state, plasma glucose may remain reduced over several hours (33), HGP may be reduced, and muscle uptake of glucose may be greatly elevated up to 24 h later (107). Although the nature of these changes is not fully understood, they may reflect an alteration in ANS regulation of plasma glucose consisting of downward resetting of counterregulatory glycemic thresholds and of S activation of HGP.

Adaptations to Habitual Exercise or Training

In contrast to man-made machines, human organs that support physical activities have the capacity to modify their structure and function to better accommodate the physiological requirements of exercise. These adjustments include changes in gene expression and in the synthesis of structural proteins and enzymes. They either are direct cellular effects of mechanotransduction or are mediated by endocrine and sympathoadrenal responses to altered metabolic function in tissues and organs affected by exercise. Because catecholamine release during acute exercise is proportional to the magnitude of work effort and plays a central role in the coordination of cardiorespiratory adjustments and metabolic fuel delivery to exercising muscle, there has been a continued interest in the possible contribution of catecholamines to adaptations to exercise training. The difficulty in distinguishing direct, specific effects of catecholamines in training adaptations from indirect effects generated by their acute effects on muscle contractions is exemplified by training-induced changes in heart structure.

Catecholamines could play a role in the hypertrophy of the left heart ventricle in response to endurance training. α_{1A} adrenergic activation of cardiomyocytes in tissue culture causes their hypertrophy, although the effect can be produced by mechanical stretch alone as well as by ANG II and some growth factors. Alpha effects are mediated by activation of phospholipase C and protein kinase C and cleavage of inositol biphosphate (IP-2) and further through mitogen-activated protein kinase (MAPK) signaling pathways (257). When combined with mechanical stretch, catecholamine stimulation of MAPK activity and *c-fos* gene expression is mediated by ANG II. Activation of MAPK in cardiomyocytes requires activation of both protein kinase A by beta adrenoceptors and protein kinase C by α_1 adrenoceptors. The activity of the anabolic MAPK signaling is dependent on the cross-talk between these signaling pathways (324). Thus, it is difficult to infer whether the catecholamines contribute to cardiac hypertrophy directly or indirectly through stimulation of heart contractility.

The scientific interest regarding ANS function and exercise training has revolved around three issues: training-induced changes in catecholamine release and spillover, changes in tissue responsiveness to catecholamine or changes in adrenergic and cholinergic receptor properties, and effect of any such changes on the reflexes and other functions controlled by the ANS during exercise.

Training-Induced Changes in Catecholamine Release and Spillover

One question regarding the autonomic (or hormonal) basis for training adaptations is whether a change in basal concentration of autonomic or endocrine messengers can account for these adaptations. Chronic elevations of catecholamines do not account for training adaptations because, despite the reduced response of catecholamines to an absolute exercise workload within 3 wk of high-intensity aerobic training (figure 9.8, left bars) (319), this response remains unchanged at the same relative workloads (figure 9.8, right bars) (74).

Another way to address this question is to assess whether NE spillover during exercise undergoes differential changes in individual organs of trained individuals and contributes unequally to the overall reduction in circulating concentrations of NE. The reduction in whole-body NE spillover after an endurance-training program that produces significant increases in aerobic power is reported to range between 23% and 35% (142, 286). The 24% decline in total NE spillover is associated with a 12% increase in total VC. Although the change in regional NE spillover after training has been measured infrequently, it is clear that endurance training affects it in individual organs differently. NE spillover in the heart is not affected at all by endurance training (201). On the other hand, renal NE spillover after endurance training is reduced by 41% and represents 66% of the total decline in NE spillover (201). Training has a disproportionately strong effect in suppressing S activation of the kidney because 22% of total NE spillover takes place in the kidney in untrained individuals at rest (89). The 41% decline in kidney catecholamine spillover increases renal VC by 10% and contributes 18% to the reduction in total VC (201). A functional manifestation of the adaptation of renal circulation to exercise training is absence of the usual reduction in renal blood flow and increased VC in response to an orthostatic challenge (65).

Because an absolute reduction in catecholamine release and NE spillover does not hold the clues about why catecholamine release is unchanged at the same relative intensities, catecholamine actions after training also need to be examined in terms of greater tissue responsiveness to their reduced concentrations and in terms of possible changes in adrenergic and cholinergic receptor properties.

Change in Tissue Responsiveness and Sensitivity to ANS Messengers After Training

Adaptive changes to aerobic-exercise training include increased capacity of tissue to perform a given function and increased tissue responsiveness to ANS messengers. Examples of the former include increased cardiorespiratory capacity in supplying muscle with oxygen and fuels and in muscle capacity to oxidize the fuels. Well-known structural and functional adaptations after endurance training are increases in left ventricular volume, increases in blood volume and its oxygen carrying capacity, expanded muscle capillarization and metabolic fuel stores, expanded muscle mitochondrial

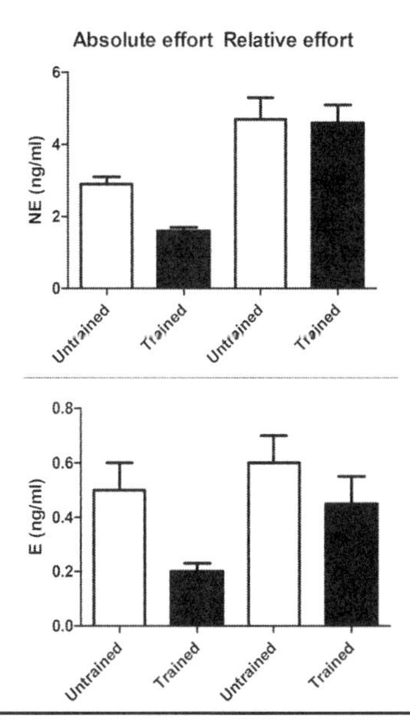

Figure 9.8 Left bars: Reduction in catecholamine response to 5 min of exercise at absolute effort of 1,500 kpm/min (kilopond meters per minute) before (open bar) and after (solid bar) 3 wk of high-intensity training. Modified from the data in figure 2 in reference 319. Right bars: No significant difference was found in catecholamine response to exercise at 95% of relative intensity in untrained (open bar) and highly trained (solid bar) individuals.

Data from Deuster et al. 1989.

mass and oxidative capacity, and much more. Adaptations to resistance-exercise training include structural and functional changes in skeletal and cardiac muscles that increase muscle strength and power. Diminished sympathoadrenal responses to the same absolute exercise workload after training may occur because of the increased relative contribution of training-induced changes in circulatory, metabolic, and contractile properties of adapted muscles; ANS messenger receptor numbers and sensitivity; and participation of central neural command. Stated differently, structural and functional changes to endurance training that increase the capacity to distribute and utilize metabolic fuels generate fewer cellular signals of reduced energy availability at any submaximal exercise intensity and are likely to elicit a correspondingly lower S response.

Supporting this interpretation are the results of the one-leg training paradigm applied by Bengt Saltin and colleagues in 1976 in which a single limb was subjected to training and acute autonomic responses to exercise of a trained limb are compared with exercise responses of the untrained limb (251). Because this study design eliminates the potential role of systemic endocrine or neural effects on exercise response, lower observed HR response and higher aerobic capacity during acute exercise by the trained leg than by the untrained leg suggest that increased muscle oxidative adaptations after training produce a smaller stimulus for elicitation of an S response.

Changes in receptor properties of ANS messengers also deserve attention. Interest in the contribution of changes in sensitivity of autonomic receptors in response to training arose in the mid-1980s. Quite early, it became obvious that training did not produce uniform change in autonomic receptors throughout the body but rather differentially elicited changes in individual organs related to the functional adaptations to training. For example, in 1996 it was reported that numbers of β_1 adrenergic receptors and their postreceptor signaling increased with endurance training in adipose tissue but decreased in the left ventricle of the heart (17, 18, 211). Information about changes in receptor properties in response to endurance training largely focused on the heart, blood vessels, and adipose tissue.

Changes in Cardiovascular Receptor Properties With Training

Training reduces blood pressure and increases vascular relaxation at rest and during exercise. Analysis of the mechanism of this change focused on the altered vascular responsiveness to autonomic stimulation after training.

Most studies examined the relaxation responses of segments of rat blood vessels that were precontracted by exposure to a vasoconstrictor such as NE. Specific adrenergic receptor agonists were used (phenylephrine and NE for α_1, clonidine for α_2, and dobutamine for β_1 receptors; isoproterenol and isoprenaline as β_1 and β_2 agonists; and procaterol for β_2 receptors), and sometimes they were combined with the selective β_1 blocker metoprolol or the nonselective beta blocker propranolol to reveal the alpha adrenergic actions of E.

Five conclusions emerged from these studies. First, training led to reduced sensitivity of the abdominal aorta (73, 280), thoracic aorta (140), and coronary arteries (239) to the vasoconstrictive actions of NE or phenylephrine and to increased sensitivity to α_2 agonists (58) and Ach (73). The changes in sensitivity were evident as leftward shifts in transmitter actions in tissue responses to increasing agonist concentrations. An early study also suggests increased responses to the vasodilatory effects of isoprenaline and decreased responses to the vasoconstrictive effects of phenylephrine (127). Second, in almost all studies these training adaptations were inherent in the endothelial layer of the blood vessel rather than its underlying smooth muscle (58, 73, 140, 239, 280). The endothelial dependence of the training effect was mediated by the sensitization of release of NO and endothelium-derived hyperpolarization factor to α_2 stimulation (58). Third, reduced sensitivity of conduit arteries to β_1 agonists developed after a threshold training duration was achieved (280), whereas opposite effects (increased sensitivity to NE) were seen after a brief exposure to exercise training (199). Fourth, actions of endothelial agents other than adrenergic or cholinergic agonists also were altered after endurance training. Sensitivity to the vasoconstrictive actions of VIP was reduced (239) after training and the vasodilatory effect of bradykinin was increased (199). Finally, changes in sensitivity to training were clearly demonstrated only for β_1 vasoconstrictors (140, 280) and for α_2 adrenoceptors responsible for catecholamine reuptake (58); other changes in tissue responsiveness could have resulted from alterations in autonomic receptor numbers. A case in point is the enhanced cholinergic responsiveness of sweat glands after exercise training, resulting from an increase in the secretory capacity of the sweat gland and muscarinic receptor numbers but not from a change in receptor affinity (318). Thus, increased capacity of a trained organism to deliver fuels and dissipate heat energy after training may account for decreased sympathoadrenal activation during submaximal exercise.

Training was found to increase cardiac responsiveness to β_1 adrenergic stimulation in terms of inotropic action (281), oxygen consumption, and glucose utilization (278). However, aside from the already-mentioned rat study (211), information about receptor numbers and sensitivity was not available. Increased ventricular volume and improved endothelial responses to blood flow require less heart activation and permit expansion of the range over which the vagus controls HR and cardiac output during exercise (262, 274). As S respiratory and circulatory responses during submaximal exercise retain their relationship to HR and cardiac output, exercise training expands the range over which both the vagus and S nerves exert their control.

Changes in Adipose Tissue Receptor Properties With Training

Most interest by far was focused on the possible training-induced changes in catecholamine action on adipose tissue because of the known catecholamine-mediated lipolysis during exercise (134) and after energy restriction (160). In humans, enhanced whole-body lipolysis after training was documented in studies of glycerol kinetics (161). Lipolysis during exercise at the same 70% relative intensity was between 30% and 40% higher than was lipolysis in the untrained state. Measures of lipolytic responsiveness to catecholamines after training were most often taken in subcutaneous abdominal tissue by in situ microdialysis (72, 204, 285) or in dissociated adipocytes in vitro (68, 294). Lipolysis from intra-abdominal fat depots after training usually was studied in rats (84).

Increased lipolytic responsiveness to β_1 adrenergic agonists after endurance training was the consistent finding in most of the studies (68, 72, 84, 204, 285). No changes in the affinity of adipocyte adrenoceptors to β_1 adrenergic agonists were found (68). On the other hand, maximal lipolytic responsiveness was increased after training at high physiological and supraphysiological concentrations of β_1 agonists (68, 283); this was consistent with the reported increases in adipocyte beta adrenergic receptor numbers (211). A similar pattern of changes to endurance training was reported for intra-abdominal fat depots in rats (84, 85). E-stimulated lipolysis and the activity of HSL as well as blood flow were increased in the retroperitoneal and mesenteric fat depots to a greater extent than in subcutaneous adipose tissue, whereas no such adaptations were observed in the soleus and extensor digitorum muscles of trained rats. In contrast to the increased responsiveness of beta adrenergic receptors, endurance training increased the affinity and thus lipolytic sensitivity to ANP (204).

The previously discussed sex difference in the number and distribution of adrenergic receptor types also affects adrenergic responsiveness to catecholamines after training. Endurance-trained women display increased lipolytic responsiveness to beta adrenergic agonists to an even greater extent than do males (69), but unlike males they also show reduced antilipolytic responsiveness to stimulation of α_2 adrenoceptors (69, 197, 237).

Lipolytic responsiveness to beta agonists after training can be influenced by prandial state at the time of testing as well as exercise-associated loss of body fat, thus making causal distinction between fat loss and training difficult. When endurance-trained swine were exposed to dietary restriction, increased in vitro responsiveness of their subcutaneous adipose tissue was associated with a reduced number of adenosine α_1 receptors. By contrast, acute overfeeding of trained animals was associated with higher numbers of α_1 receptors and less of an increase in beta adrenergic responsiveness (54). To dissociate the two actions, postexercise adipose tissue was pretreated with adenosine deaminase in some studies of its lipolytic responsivess to catecholamines in order to remove the confounding antilipolytic actions of α_1 receptors (68, 69, 237).

In almost all studies reporting increased lipolytic responsiveness to beta adrenergic agonists, exercise training caused some loss of body fat. A direct correlation was found between the magnitude of increase in beta adrenergic responsiveness and decrease in adipocyte size (197). When increased dietary intake during endurance training matched energy expenditure to prevent body weight and body fat loss, no increase in beta adrenergic lipolytic responsiveness or stimulatory effect on adipose tissue blood flow was found (133).

The interdependence between negative energy balance and fat loss on one hand, and changes in adrenergic responsiveness after endurance training on the other, strengthens the inference that changes in ANS function reflect training adaptations or bioenergetic changes in peripheral tissues. After endurance training, adipose tissue and muscle become more insulin sensitive; however, this effect appears to be linked more to negative energy balance caused by dietary restriction (14) than to fat loss (133). Fat loss, whether achieved through exercise training or dieting, leads to reduced NE spillover (286, 297). NE spillover is negatively correlated with whole-body insulin sensitivity and positively correlated with plasma leptin concentration (287). Thus, as was the case with altered autonomic control of cardiorespiratory function in response to training, changes in adipocyte size and the level of body fat stores appear to drive changes in catecholamine secretion and lipolytic responsiveness.

Changes in Muscle Receptor Properties With Training

The effects of training on muscle responses to catecholamine action are not as distinct as alterations in the responsiveness of cardiovascular and adipose tissues. Effects of endurance training were examined in different rat muscle fiber types because beta adrenoceptor abundance is known to be about three times higher in predominantly oxidative soleus muscles than in predominantly glycolytic fast-twich vastus muscles (193). Endurance training of rats was reported to increase the number of beta adrenergic receptors in oxidative Type I and Type II muscles as well as adenylate cyclase activity in all three fiber types in one study (42), but in another study no changes were found in beta receptor number (192) or in responsiveness of lipoprotein lipase (LPL) to E in an oxidative and a glycolytic muscle (85).

A possible involvement of catecholamines in protein turnover in response to endurance training is inferred from animal research. E secreted by the rat adrenal medulla and NE released from adrenergic terminals inhibit Ca^{2+}-dependent protein degradation in oxidative muscles by increasing the levels of the protease inhibitor calpastatin. The effect appears to be mediated by the production of cyclic adenosine monophosphate and protein kinase A through the catecholamine stimulation of β_2 and β_3 adrenergic receptors. Suppression of protein degradation by catecholamines is most effective under the conditions of hormonal and nutritional deficiency (210). In addition to the enzymatic action of catecholamines in controlling cardiorespiratory and thermoregulatory reflexes, promoting fuel mobilization, and suppressing protein degradation, E infusion has been shown to affect expression of a number of skeletal muscle genes that may contribute to the metabolic and antiproteolytic effects of endurance training in humans (302). Of 30 genes involved in carbohydrate metabolism affected by E infusion, several promoting glycogen synthesis were downregulated and the gene for the glycogen synthetase kinase-3 enzyme was upregulated. E infusion induced genes for glycogen phosphorylase, several enzymes facilitating glycolysis, and some apolipoproteins and downregulated several lipid transport genes. Finally, the genes for ubiquitin degradation enzymes were among the most strongly downregulated, whereas there was no effect on genes mediating protein synthesis.

Alteration by Training of Autonomic Reflexes

In view of previously discussed changes in S activation of cardiovascular targets after training, training adaptations in the baroreflex and the S component of the metaboreflex (MSNA) have received considerable attention. Changes in baroreflex sensitivity after training are usually assessed from the relationship between the duration of heart interbeat intervals (a measure of HR) and mean blood pressure in response to an orthostatic challenge or agents with vasoconstrictor (phenylephrine) or vasodilator (nitroprusside) actions. An alternative method is to measure changes in MSNA and diastolic blood pressure after pharmacological manipulations of blood pressure. In cardiac baroreflex, approximately 70% of both vagal and S components of HR change are mediated by arterial baroreceptors and 30% are mediated by cardiopulmonary baroreceptors. Postganglionic S vasoconstrictor nerve traffic or MSNA are tracked with fine electrodes inserted into peroneal, tibial, or median nerves. Microneurography from these nerves also can assess changes in sudomotor, pilomotor, and vasomotor reflexes directed at the skin.

In a study that compared the effects of dieting-induced body fat loss with the effects of a combination of dietary restriction and endurance training, both treatments produced a similar 9% body-weight loss and a 23% decline in NE spillover. This increased baroreflex sensitivity by 5.2 ms interbeat intervals/mm change in blood pressure (286). While the magnitude of the increase in baroreflex sensitivity was correlated with the magnitude of NE spillover decline, baroreflex sensitivity was attenuated during weight-loss maintenance whereas NE spillover response to the two treatments persisted. Six months of endurance training led to a 25% increase in cardiac baroreflex sensitivity tested with the α_1 agonist phenylephrine in subjects with borderline hypertension (276). Ten weeks of such training led to 41% and 45.5% increases in cardiac and MSNA measures of baroreceptor sensitivity, respectively, in response to increases in blood pressure. Corresponding measures of increased baroreceptor sensitivity to decreases in blood pressure were 50% and 102.6% (120). Endurance training also normalized cardiac and MSNA measures of baroreflex sensitivity in hypertensive subjects who were challenged with phenylephrine and nitroprusside (179). In most exercise-training studies, subjects lost body fat mass. Because body fat loss by dietary restriction also reduced NE spillover and increased baroreflex sensitivity (6, 121, 287), it is not certain to what extent endurance training rather than associated body fat loss contributed to reflex sensitization in these studies. Doug Seals (figure 9.9) has contributed to the understanding of S involvement in age-associated declines in baroreflex sensitivity and vascular compliance and the counteractive role of exercise training.

The training effects on MSNA are relevant in the control of hypertension because of the vasoconstrictive role of S innervations of the muscle vasculature. A variety of studies have analyzed systemic and local influences on acute and training-induced changes in MSNA. The best-known trigger of MSNA is activation of group III mechanosensitive afferents and group IV chemosensitive afferents by declining pH in the muscle during ischemia (301) caused by isometric or fatiguing dynamic exercise (21). This was illustrated by the high correlation between the metabolic muscle profiles and MSNA responses during isometric exercise (246). Isometric forearm flexion at the same relative intensity that engaged a higher proportion of fast-twitch glycolytic muscle fibers produced greater MSNA response than did isometric dorsiflexion by the anterior tibialis muscle, which has a lower proportion of such fibers. The lowest MSNA response was obtained by plantar flexing the soleus, a muscle with a higher proportion of oxidative fibers and capillary supply than the other two muscles. MSNA is also responsive to arterial and pulmonary baroreceptors as well as temperature receptors that are affected by postural changes and shifts in blood distribution and pressure. Significant increases in HR and core temperature during prolonged low-intensity exercise at 40% of maximal effort precede by several minutes, and are highly correlated with, increases in MSNA (247). Because exercise-associated activation of MSNA (126, 229) and NE spillover (255) occur in both exercising and resting limbs, a thermoregulatory diversion of blood to the skin when core temperature increases during prolonged exercise is the probable trigger of the baroreflex-mediated increase in S outflow to the muscle. Even central command may be involved in increases in MSNA during acute exercise, as was shown by comparing MSNA and HR responses to 16 min of exercise at different intensities. At an exercise intensity below 40% of $\dot{V}O_2max$ that elicits an HR of 100 beats/min, MSNA decreased below the resting level. As exercise intensity increased to 75% of $\dot{V}O_2max$ in three additional steps, there was a linear increase in MSNA (248). This parallels the involvement of central command in vagal withdrawal at exercise intensities below 40% of $\dot{V}O_2max$ and in S activation above this threshold (209, 244, 301, 323).

Figure 9.9 Douglas R. Seals, professor of distinction in the department of integrative physiology at the University of Colorado (Boulder, Colorado, United States) has made major contributions to the understanding of hypertension and of the role of cateholamines in the control of blood pressure and circulation during exercise. His contributions to this chapter are found in references 190, 260, 261, and 262.

Photo courtesy of Doug Seals.

Given the complex controls over MSNA, it is not surprising that the effects of training present an unclear picture (227). The clearest evidence for an effect of endurance training on MSNA is that it causes an adaptive reduction in MSNA during exercise (120, 273) but not at rest (227). The type of exercise used in training is critical in producing adaptive MSNA changes because neither resistance exercise (56) nor isometric exercise (228)—both of which would be expected to elicit metaboreflex—produce this adaptation. The length and effectiveness of training as well as the age of the subjects may influence whether a reduction of MSNA during exercise is seen (227). The final important training outcome is the specificity of MSNA change to the muscles subjected to training or to the dominant limbs habitually used in physical activity (226, 249, 273). The training-induced decline in MSNA manifests at lower exercise intensities where muscle pH does not change and the metaboreflex is unlikely to be triggered (273). Thus, MSNA training effects reflect metabolic adaptations achieved by the trained muscle.

Regulation by ANS of Whole-Body Energy Balance

Three lines of research prompt the hypothesis that the ANS is critically involved in the regulation of whole-body energy balance and adult body mass through the control of spontaneous movement and ingestive behaviors, metabolism, and temporal organization of energy intake and energy expenditure. These lines of research include the role of the ANS in the coincident onset of adult spontaneous activity and energy regulation; the functional relationship between spontaneous physical activity and body mass regulation; and the temporal organization of physical activity, feeding, and metabolism.

Role of the ANS in the Onset of Spontaneous Physical Activity and Energy Regulation in Adults

At the midpoint of the 20th century, Gordon Kennedy described the coincidence of three developmental events that characterize adult mammals: onset of high volumes of spontaneous physical activity, cessation of body mass growth, and initiation of body mass regulation (153, 154). Regulation of body mass consists of the defense of lean tissue, body fat, and bone against losses through compensatory increases in food seeking and metabolic adjustments for weight regain (2). Kennedy also demonstrated that the integrity of the medial basal hypothalamus, including the VMH and ARC nuclei, establishes the level at which body mass is regulated in part through changes in the level of physical activity (155). The weight plateau of the rat oscillates around a set point in parallel with changes in levels of spontaneous locomotor activity. Body mass, especially its adipose component, increases in parallel with drastic reduction in spontaneous locomotor activity after lesions of the medial basal hypothalamus. These findings prompted a homeostatic concept of energy regulation based on the control by the ARC and VMH hypothalamic nuclei of either increased feeding and decreased energy expenditure when body mass losses are incurred or decreased feeding and increased energy expenditure when the adiposity hormone leptin signals an accumulation of excess body fat (258). According to this view, lesions of ARC neurons interrupt the target of leptin negative feedback that disrupts regulation of body fat, the key variable of energy regulation.

This view neglected two facts: first, that energy regulation is preserved after medial basal hypothalamic lesions because lesioned animals defend their larger body mass against losses after the higher fatness level is attained (155), and second, that any changes in body mass also involve adjustments in the mass of bone and lean tissues. With each kilogram of body fat gained or lost, 16.5 g of bone mineral is gained or lost (143) and changes in body fat level are accompanied by changes in lean mass. Collectively, these findings indicate that the medial basal hypothalamus, one of the key neural sites of ANS regulation of body mass, sets the level at which the fat component of body mass is metabolically and behaviorally maintained and determines the levels of spontaneous physical activity and energy expenditure.

With the ongoing development in the 21st century of powerful research tools such as gene manipulation, selective neuronal chemical lesioning, and anterograde and retrograde neuronal tracings, the role of the ANS in energy regulation can be reinterpreted. The new research tools helped Gerard Karsenty and his research team demonstrate that S nerves regulate bone mass, muscle, and adipose tissue mass in adult mammals through the regulation of bioactivity of the bone hormone osteocalcin (OCN) (151). Although this new concept of energy regulation has omitted measurements of physical activity, it allows an integrative view of the role of the ANS in regulating adult body mass.

A key finding of the Karsenty research was that increased S tone suppresses the bioactivity of the bone hormone OCN by activating an *Esp* gene that increases γ-carboxylation of the hormone (81). Upregulation of this gene reduces osteoblast numbers and blocks increases in bone mineralization and size. The effect is dependent on beta adrenergic receptors on the osteoblasts, and when these are genetically ablated a high bone phenotype is observed that is similar to the phenotypes of animals with lesions of the medial basal hypothalamus or of genetic models that lack the ability to produce leptin protein (ob/ob mice) or its receptors (db/db mice). Most of these phenotypes are characterized by obesity, inactivity, and low levels of metabolic energy expenditure.

Another finding illuminating the role of S neural outflow in energy regulation was the demonstration that regulation of bone mass required the integrity of the VMH hypothalamus and that regulation of body fat, blood glucose, and feeding required the integrity of the ARC hypothalamic nucleus (151). Because leptin was implicated in the alterations in body fat in some of the obese phenotypes through its activation of S outflow, the neural site of its inhibitory actions was of interest. The original hypothesis that circulating leptin acts directly on the medial hypothalamus (259) proved incorrect because mice with deletions of LRb leptin receptors on pro-opiomelanocortin neurons in the medial basal hypothalamus were not obese yet they responded to intracerebroventricular leptin administration, thus indicating another neural site of leptin action (151). Instead, leptin acts on the raphe nuclei located in the midline of the rostral pons and extending to the rostral MO (figure 9.1) to inhibit the separate serotonergic projections to the ARC and VMH nuclei. The separation of leptin actions on bone and body fat was demonstrated with chemical lesioning of ARC neurons with monosodium glutamate and the chemical lesioning of the neurons in the VMH nucleus with gold thioglucose. When the two nuclei are selectively chemically lesioned, lesions confined to the ARC nucleus cause obesity and hypoactivity whereas lesions confined to the VMH nucleus produce increases in bone mass.

When body fat level and energy availability increase, plasma leptin concentration increases and blocks the serotonergic projections from the raphe nuclei to the mediobasal hypothalamus from promoting body mass gain. This leptin action facilitates S outflow through the ARC nucleus that suppresses feeding and increases lipolysis, thermogenesis, and physical activity. Leptin suppresses bone accretion by facilitating S outflow through the VMH nucleus. When the three components of body mass decline due to energy expenditure of exercise or dietary restriction, plasma leptin concentration, and S tone decrease. This permits greater PS and serotonergic activation of anabolic processes. Serotonergic suppression of S outflow through the ARC and VMH nuclei enhances OCN bioactivity, which in turn stimulates the secretion of the adipokine adiponectin and an associated increase in insulin sensitivity. Thus, leptin appears to be the link between body energy status and neural substrates responsible for the activation of S outflow, thermogenesis, and physical activity. Its actions restrain bone mass gain on one hand and anabolic PS actions that favor feeding and body mass accretion on the other.

Role of the ANS in the Functional Relationship Between Spontaneous Physical Activity and Body Mass Regulation

The interactions between the ANS and exercise are usually considered from the perspective of the effects of volitional physical activity on autonomic and hormonal effects on human physiology. However, energy status can affect one's motivation to be physically active (32). In stark contradiction to the prevailing view that spontaneous physical activity is homeostatically connected to energy regulation (259), levels of spontaneous physical activity are inversely and nonhomeostatically related to body fat levels. The extremes of this relationship range from weight losses associated with feed-forward increases in wheel running to the point of death in food-restricted rats, similar to compulsive running in anorexics, to the almost complete inactivity in the morbidly obese (32). This reciprocal relationship stems from suppression by leptin of the motivation to locomote and seek food. The neural substrate where leptin exerts this inhibitory action is the nucleus accumbens, one of the forebrain basal ganglia, and its associated limbic circuits. Body mass losses that lead to declines in plasma leptin concentration increase the incentive value of engaging in physical activity and eating food. Associated increases in insulin sensitivity favor fat resynthesis and weight regain under the influence of PS serotonergic activation. Body fat and body mass gains increase plasma leptin and block further increases in body mass through inactivation of OCN by increases in S outflow. In addition to deleterious health consequences of insulin resistance, increased S activation by plasma leptin in the overweight and obese state increases vascular vasoconstrictor tone and reduces blood flow in the lower limbs during exercise (261).

Role of the ANS in the Temporal Organization of Physical Activity, Feeding, and Metabolism

Under the direction of the SCN master clock (198), the ANS orchestrates daily rhythms of increased activity, cardiorespiratory function, blood pressure, feeding, and energy expenditure during the wakeful phase of the diurnal period and reduced energy expenditure through suppression of all of these functions during sleep (265, 279). The SCN influences ANS circuits involved in the control of energy balance through direct projections to the dorsal and ventral PVN, the source of S outflow controlling the circadian rhythm of the heart, respiration, circulation, and blood pressure (198). SCN also projects to the dorsal subparaventricular zone for control of the circadian rhythm of body temperature and to the ventral subparaventricular zone for control of the sleep–wakefulness cycle and daily rhythm of spontaneous physical activity (locomotion) (102).

DMH cell bodies also receive SCN projections and play a role in the circadian pattern of sleep–wakefulness, locomotor activity, corticosteroid secretion, and feeding. Most viscera receive SCN-dependent circadian time cues via their PS or S innervations. SCN fibers also project to the ARC, VMH, and ventral part of the LH, influencing the circadian control of food intake, metabolism, and thermogenesis.

A large number of metabolic enzymes involved in lipid, amino acid, glucose, and glycogen metabolism and in the citric acid cycle are more active during the wakeful portion of the diurnal cycle. Among those are glycogen phosphorylase, cytochrome oxidase, lactate dehydrogenase, acetyl-CoA carboxylase, malic enzyme, fatty acid synthase, and glucose-6-phosphate dehydrogenase. It is not clear whether all of these rhythms are truly endogenous or synchronized to times of feeding, physical activity, and metabolic demand. The diurnal variation in peripheral glucose uptake and HGP, on the other hand, depends on the integrity of the SCN and their projections to PVN and S outflow (45, 149). Daily rhythm in lipid metabolism appears to be linked to circadian changes in S and PS activity via SCN neural projections to the MPOA (16) and circadian variation in leptin concentration.

Direct evidence for the ANS control of circadian rhythms in processes and behaviors important for the regulation of energy is that severe disruption in these rhythms associated with lesions of the medial basal hypothalamus results in obesity and hypoactivity (148). After such lesions, feeding and metabolism are no longer subject to diurnal increases and nocturnal declines. Instead, they operate in a disorganized and continuous fashion without a connection to periods of high physical activity and thermogenic energy expenditure.

References

1. Abel JJ, Crawford AC. On the blood-pressure raising constituent of the suprarenal capsule. *Bull Johns Hopkins Hosp* 8: 151-157, 1897.
2. Adolph EF. Urges to eat and drink in rats. *Am J Physiol* 151: 110-125, 1947.
3. Ahlborg G, Felig P. Lactate and glucose exchange across the forearm, legs, and splanchnic bed during and after prolonged leg exercise. *J Clin Inves* 69: 45-54, 1982.
4. Ahlquist RP. A study on the adrenotropic receptors. Am J Physiol 153: 586-600, 1948.
5. Alam M, Smirk FH. Observation in man upon a blood pressure raising reflex arising from the voluntary muscles. *J Physiol* 89: 372-383, 1937.
6. Alvarez GE, Davy BM, Ballard TP, Beske SD, Davy KP. Weight loss increases cardiovagal baroreflex function in obese young and older men. *Am J Physiol Endocrinol Metab* 289: E665-E669, 2005.
7. Anand BK, Chhina GS, Sharma KN, Dua S, Singh B. Activity of single neurons in the hypothalamic feeding centers: Effect of glucose. *Am J Physiol* 207: 1146-1154, 1964.
8. Andrew SF, Dinh TT, Ritter S. Localized glucoprivation of hindbrain sites elicits corticosterone and glucagon secretions. *Am J Physiol* 292: R792-R798, 2007.
9. Anrep GV, Pascual W, Rössler R. Respiratory variations of the heart rate. The reflex mechanism of the sinus arrhythmia. *Proc Roy Soc London Series B* 119: 191-217, 1936.
10. Arnall DA, Marker JC, Conlee RK, Winder WW. Effect of infusing epinephrine on liver and muscle glycogenolysis during exercise in rats. *Am J Physiol* 250: E641-E649, 1986.
11. Arner P. Differences in lipolysis between human subcutaneous and omental adipose tissue *Ann Med* 27: 435-438, 1995.
12. Arner P, Kriegholm E, Engfeldt P, Bolinder J. Adrenergic regulation of lipolysis in situ at rest and during exercise. *J Clin Invest* 85: 893-898, 1990.
13. Aronsohn E, Sachs J. Die beziehungen des Gehirns zur Körperwarme and zum Fieber. *Pflügers Arch* 37: 232-300, 1885.
14. Assali AR, Ganor A, Beigel Y, Shafer Z, Hershcovici T. Insulin resistance in obesity: Body weight or energy balance. *J Endocr* 171: 293-296, 2001.
15. Axelrod J. O-Methylation of epinephrine and other catechols in vitro and in vivo. *Science* 126: 400-401, 1957.
16. Bamshad M, Aoki VT, Adkison MG, Warren WS, Bartness TJ. Central nervous system origins of the sympathetic nervous system outflow to white adipose tissue. *Am J Physiol* 275: R291-R299, 1999.
17. Barbier J, Rannou-Bekono F, Marchais J, Berthon PM, Delamarche P, Carré F. Effect of training on β1 β2 β3 adrenergic and M2 muscarinic receptors in rats heart. *Med Sci Sports Exerc* 36: 949-954, 2004.
18. Barbier J, Reland S, Ville N, Rannou-Bekono F, Wong S, Carré F. The effects of exercise training on myocardial adrenergic and muscarinic receptors. *Clin Auton Res* 16: 61-65, 2006.
19. Bartness TJ, Bamshad M. Innervation of mammalian white adipose tissue: Implications for the regulation of total body fat. *Am J Physiol* 275: R1399-R1411, 1998.
20. Bartness TJ, Vaughan CH, Song CK. Sympathetic sensory innervation of brown adipose tissue. *Intl J Obesity* 34(Suppl. 1): S36-S42, 2010.
21. Batman BA, Hardy JC, Leuenberger UA, Smith MB, Yang QX, Sinoway LI. Sympathetic nerve activity during prolonged rhythmic forearm exercise. *J Appl Physiol* 76: 1077-1081, 1994.
22. Berkowitz DE, Price DT, Bello EA, Page SO, Schwinn DA. Localization of messenger RNA for three distinct alpha 1-adrenergic receptor subtypes in human tissues: Evidence for species heterogeneity and implications for human pharmacology. *Anesthesiology* 81: 1235-1244, 1994.
23. Bernard C. Chiens rendus diabetiques. *Compt Rend Soc Biol* 1 :60, 1849.
24. Bernard C. *Leçons Sur les Phénomènes de la Vie Communs aux Animaux et aux Végétaux*. Paris: Ballière, 1878-1879.
25. Berthelson S, Pettinger WA. A functional basis for the classification of alpha-adrenergic receptors. *Life Sci* 21: 595-606, 1977.
26. Bevegård S, Jonsson B, Karlof I. Circulatory response to recumbent exercise and head-up tilting in patients with disturbed sympathetic cardiovascular control (postural hypotension). *Acta Med Scand* 172: 623-636, 1962.
27. Bini G, Hagbarth K-E, Hynninen P, Wallin BG. Thermoregulatory and rhythm-generating mechanisms governing sudomotor and vasoconstrictor outflow in human cutaneous nerves. *J Physiol London* 306: 537-552, 1980.
28. Birkenfeld AL, Boschmann M, Moro C, Adams F, Heusser K, Franke G, Berlan M, Luft FC, Lafontan M, Jordan J. Lipid mobilization with physiological atrial natriuretic peptide concentrations in humans. *J Clin Endocrinol Metab* 90: 3622-3628, 2005.
29. Björkman O, Ericsson LS. Splanchnic glucose metabolism during leg exercise in 60-hour-fasted

human subjects. *Am J Physiol* 245: E444-E448, 1983.

30. Björkman O, Miles P, Wasserman D, Lickley L, Vranic M. Regulation of glucose turnover during exercise in pancreatectomized, totally insulin-deficient dogs. Effects of beta-adrenergic blockade. *J Clin Invest* 81:1759-1767, 1988.
31. Borer KT. *Exercise Endocrinology*. Champaign, IL: Human Kinetics, 2003.
32. Borer KT. Nonhomeostatic control of human appetite and physical activity in regulation of energy balance. *Exerc Sport Sci Rev* 38: 114-121, 2010.
33. Borer KT, Wuorinen E, Lukos J, Denver J, Porges S, Burant C. Two bouts of exercise before meals, but not after meals, lower fasting blood glucose. *Med Sci Sports Exer* 41: 1606-1614, 2009.
34. Boulant JA. Hypothalamic mechanisms in themoregulation. *Fed Proc* 40: 2843-2850, 1981.
35. Bowen WP. Changes in heart-rate, blood pressure, and duration of systole resulting from bicycling. *Am J Physiol* 11: 59-77, 1904.
36. Brack KE, Patel VH, Mantravardi R, Coote JH, Ng, GA. Direct evidence of nitric oxide release from neuronal nitric oxide synthase activation in the left ventricle as a result of cervical vagus innervation. *J Physiol* 587: 3045-3054, 2009.
37. Bray GA, Sclafani A, Novin D. Obesity-inducing hypothalamic knife-cuts: Effects on lipolysis and blood insulin levels. *Am J Physiol* 243: R455-R449, 1982.
38. Brengelman GL, Johnson JM, Hermansen L, Rowell LB. Altered control of skin blood flow during exercise at high internal temperature. *J Appl Physiol* 43: 790-794, 1977.
39. Brenner IK, Thomas S, Shephard RJ. Autonomic regulation of the circulation during exercise and heat exposure: Inferences from heart rate variability. *Sports Med* 26: 85-99, 1998.
40. Brito NA, Brito MN, Bartness TJ. Differential sympathetic drive to adipose tissues after food deprivation, cold exposure or glucoprivation. *Am J Physiol* 294: R1445-R1452, 2008.
41. Brooks VL, Eli KR, Wright RM. Pressure-independent baroreflex resetting produced by chronic infusion of angiotensin II in rabbits. *Am J Physiol* 265: H1275-H1282, 1993.
42. Buckenmeyer PJ, Goldfarb AH, Partilla JS, Piñeyro MA, Dax EM. Endurance training, not acute exercise, differentially alters beta-receptors and cyclase in skeletal fiber types. *Am J Physiol* 258: E71-E77, 1990.
43. Bukowiecki L, Lupien J, Follea N, Paradis A, Richard D, LeBlanc J. Mechanism of enhanced lipolysis in adipose tissue of exercise-trained rats. *Am J Physiol* 239, E422-E429, 1980.
44. Burnstock G. Autonomic neurotransmission: 60 years since Sir Henry Dale. *Ann Rev Pharmacol Toxicol* 49: 1-30, 2009.
45. Cailotto C, La Fleur SE, Van Heijningen C, Wortel J, Kalsbeek A, Feenstra M, Pevet P, Buijs RM. The suprachiasmatic nucleus controls the daily variation of plasma glucose via the autonomic output to the liver: Are the clock genes involved? *Eur J Neurosci* 22: 2531-2540, 2005.
46. Cannon WB. The emergency function of the adrenal medulla in pain and the major emotions. *Am J Physiol* 33: 356-372, 1914.
47. Cannon WB. *Bodily Changes in Pain, Hunger, Fear and Rage*. New York: Appleton & Co., 1920.
48. Cannon WB. Studies on the conditions of activity in endocrine glands. VIII. Some effects on the denervated heart of stimulating the nerves of the liver. *Am J Physiol* 58: 353-364, 1921.
49. Cannon WB. Organization for physiological homeostasis. *Physiol Rev* 9: 399-431, 1929.
50. Cannon WB, McIver MA, Bliss SW. Studies on the conditions of activity of endocrine glands. XIII. A sympathetic and adrenal mechanism for mobilizing sugar in hypoglycemia. *Am J Physiol* 69: 46-66, 1924.
51. Cannon WB, Nice LB. The effect of adrenal secretion on muscular fatigue. *Am J Physiol* 32: 44-60, 1913.
52. Cannon WB, Rosenblueth A. Studies on conditions of activity in endocrine organs. Sympathin E and sympathin I. *Am J Physiol* 104: 557-574, 1933.
53. Cannon WB, Washburn AL. An explanation of hunger. *Am J Physiol* xxix: 441-454, 1912.
54. Carey GB. Cellular adaptations in fat tissue of exercise-trained miniature swine: Role of excess energy intake. *J Appl Physiol* 88: 881-887, 2000.
55. Carlson KI, Marker JC, Arnall DA, Terry ML, Yang HT, Lindsay LG, Bracken ME, Winder WW. Epinephrine is unessential for stimulation of liver glycogenolysis during exercise. *J Appl Physiol* 58: 544-548, 1985.
56. Carter JR, Ray CA, Downs EM, Cooke WH. Strength training reduces arterial blood pressure but not sympathetic neural activity in young normotensive subjects. *J Appl Physiol* 94 : 2212-2216, 2003.
57. Changeux J-P, Kasai M, Lee CY. Use of snake venom toxin to characterize the cholinergic receptor protein. *Proc Natl Aad Sci USA* 67: 1241-1247, 1970.
58. Chen H-I, Cheng S-Y, Jen CJ. Chronic exercise enhances vascular responses to clonidine in rats by increasing endothelial a2-adrenergic receptor affinity. *Chin J Physiol* 42: 61-66, 1999.
59. Chernogubova E, Hutchinson DS, Nedergaard J, Bengtsson T. α_1- and β_1-adrenoceptor signaling fully compensates for β_3 deficiency in brown adipocyte norepinephrine-stimulated glucose uptake. *Endocrinol* 146: 2271-2284, 2005.

60. Cherrington AD, Lacy WW, Chiasson JL. Effect of glucagon on glucose production during insulin deficiency in the dog. *J Clin Invest* 62: 664-677, 1978.
61. Chun S-J, Niijima A, Nagai S, Nagai K. Effect of bilateral lesions of the suprachiasmatic nucleus on hyperglycemia caused by 2-deoxy-D-glucose and vasoactive intestinal peptide in rats. *Brain Res* 809: 165-174, 1998.
62. Coker RH, Koyama Y, Lacy DB, Williams PE, Rhèaume N, Wasserman DH. Pancreatic innervation is not essential for exercise-induced changes in glucagon and insulin or glucose kinetics. *Am J Physiol* 277: E1122-E1129, 1999.
63. Coker RH, Krishna MG, Zinker BA, Allen EJ, Lacy DB, Wasserman DH. Sympathetic drive to liver and nonhepatic splanchnic tissue during prolonged exercise is increased in diabetes. *Metabolism* 46: 1327-1332, 1997.
64. Commins SP, Watson PM, Levin N, Beiler RJ, Gettys TW. Central leptin regulates the UCP1 and ob genes in brown and white adipose tissue via different beta adrenoceptor subtypes. *J Biol Chem* 275: 33059-33067, 2000.
65. Conboy EE, Fogelman AE, Sauder CL, Ray CA. Endurance training reduces renal vasoconstriction to orthostatic stress. *Am J Physiol Renal Physiol* 298: F279-F284, 2010.
66. Conlee RK, Hickson RC, Winder WW, Hagberg JM, Holloszy JO. Regulation of glycogen resynthesis in muscles of rats following exercise. *Am J Physiol* 235: R145-R150, 1978.
67. Cox BF, Hay M, Bishop VS. Neurons in area postrema mediate vasopressin-induced enhancement of the baroreflex. *Am J Physiol* 258: H1943-H1946, 1990.
68. Crampes F, Beauville M, Riviere D, Garrigues M. Effect of physical training in humans on the response of isolated fat cells to epinephrine. *J Appl Physiol* 61:25-29, 1986.
69. Crampes F, Riviere D, Beauville M, Marceron M, Garrigues M. Lipolytic response of adipocytes to epinephrine in sedentary and exercise-trained subjects: Sex-related differences. *Eur J Appl Physiol Occup Physiol* 59: 249-255, 1989.
70. Dale HH. The action of certain esters and ethers of choline and their relation to muscarine. J *Pharmacol* 6: 147-190, 1914.
71. Dale HH, Feldberg W, Vogt M. Release of aceylcholine at voluntary motor nerve endings. *J Physiol* 86: 353-380, 1936.
72. De Glisezinski I, Crampes F, Harant I, Berlan M, Hejnova J, Langin D, Rivière D, Stich V. Endurance training changes in lipolytic responsiveness of obese adipose tissue. *Am J Physiol* 275: E951-E956, 1998.
73. Delp MD, McAllister RM, Laughlin MH. Exercise training alters endothelium-dependent vasoreactivity of rat abdominal aorta. *J Appl Physiol* 75: 1354-1363, 1993.
74. Deuster PA, Chrousos GP, Luger A, DeBolt JE, Bernier LL, Trostmann UH, Kyle SB, Montgomery LC, Loriaux DL. Hormonal and metabolic responses of untrained, moderately trained, and highly trained men to three exercise intensities. *Metab Clin Exp* 38: 141-148, 1989.
75. Devlin JG. The effect of training and acute physical exercise on plasma insulin-like activity. *Ir J Med Sci* 453: 423-425, 1963.
76. Dietz NM, Engelke KA, Samuel TT, Fix RT, Joyner MJ. Evidence for nitric oxide-mediated sympathetic forearm vasodilatation in humans. *Physiol* 498: 531-540, 1997.
77. Dinenno FA, Joyner MJ. Combined NO and PG inhibition augments alpha-adrenergic vasoconstriction in contracting human skeletal muscle. *Am J Physiol* 287: H2576-H2584, 2004.
78. Dittmar C. Über die Lage des sogennanten Gefasscentrums in der medulla oblongata. *Ber Verh Saechs Wiss Leipzig Math Phys Kl* 25: 449-469, 1873.
79. Dixon WE. Vagus inhibition. *Br Med J* ii: 1807, 1906.
80. Dogiel AS. Die sensiblen Nervenendigungen im Herzen und in den Blutgefassen der Saugethiere. *Arch Mikrosk Anat* 52: 44-70, 1898.
81. Ducy P, Amling M, Takeda S, Priemel M, Schilling AF, Beil FT, Shen J, Vinson C, Rueger JM, Karsenty G. Leptin inhibits bone formation through a hypothalamic relay: A central control of bone mass. *Cell* 100: 197-207, 2000.
82. Ekstrom-Jodal B, Haggendal E, Malmberg R, Svedmyr N. The effect of adrenergic-receptor blockade on coronary circulation in man during work. *Acta Med Scand* 10: 245-248, 1972.
83. Emorine LJ, Marullo S, Briend-Sutren MM, Patey G, Tate K, Delavier-Klutchko C, Strosberg AD. Molecular characterization of the human beta 3-adrenergic receptor. *Science* 245: 1118-1121, 1989.
84. Enevoldsen LH, Stallknecht B, Fluckey JD, Galbo H. Effect of exercise training on in vivo lipolysis in intra-abdominal adipose tissue in rats. *Am J Physiol* 279: E585-E592, 2000.
85. Enevoldsen LH, Stallknecht B, Langfort J, Petersen LN, Holm C, Ploug T, Galbo H. The effect of exercise training on hormone-sensitive lipase in rat intra-abdominal adipose tissue and muscle. *J Physiol* 536: 871-877, 2001.
86. Ensinck JW, Walter RM, Palmer JP, Brodows RG, Campbell RG. Glucagon responses to hypoglycemia in adrenalectomized man. *Metabolism* 25: 227-232, 1976.
87. Epstein SE, Robinson SH, Kahler RL, Braunwald E. Effect of beta adrenergic blockade on the cardiac response to maximal and submaximal exercise in man. *J Clin Invest* 44: 1745-1753, 1965.
88. Esler M, Jennings G, Lambert G, Meredith I, Horne M, Eisenhofer G. Overflow of catecholamine neuro-

transmitters to the circulation: Source, fate, and functions. *Physiol Rev* 70: 963-985, 1990.
89. Esler M, Jennings G, Leonard P, Sacharias N, Burke F, Johns J, Blombery P. Contribution of individual organs to total noradrenaline release in humans. *Acta Physiol Scand* 104(Suppl. 527): 11-16, 1984.
90. Esteves FO, McWilliam PN, Batten TF. Nitric oxide producing neurones in the rat medulla oblongata that project to nucleus tractus solitarii. *J Chem Neuroanat* 20: 195-197, 2000.
91. Exton JH. Mechanisms involved in alpha-adrenergic phenomena: Role of calcium ions in actions of catecholamines in liver and other tissues. *Am J Physiol* 238: E3-E12, 1980.
92. Fain JN, Garcia-Sainz JA. Role of phosphatidylinositol turnover in alpha 1 and of adenylate cyclase inhibition in alpha 2 effects of catecholamines. *Life Sci* 26: 1183-1194, 1980.
93. Fagraeus L, Linnarsson D. Autonomic origin of heart rate fluctuations at the onset of muscular exercise. *J Appl Physiol* 40: 679-682, 1976.
94. Farouque HM, Meredith IT. Relative contribution of vasodilatory prostanoids, NO, and KATP channels to human forearm metabolic vasodilation. *Am J Physiol* 284: H2405-H2411, 2003.
95. Fell RD, McLane JA, Winder WW, Holloszy JO. Preferential resynthesis of muscle glycogen in fasting rats after exhausting exercise. *Am J Physiol* 238: R328-R332, 1980.
96. Fredholm GB, Rosell S. The effect of alpha- and beta-adrenergic blocking agents on free fatty acid release from subcutaneous adipose tissue in vivo. *Acta Pharm Toxicol* 25(Suppl. 4): 20, 1967.
97. Freund BJ, Shizuru EM, Hashiro GM, Claybaugh JB. Hormonal, electrolytic, and renal responses to exercise are intensity dependent. *J Appl Physiol* 70: 900-906, 1991.
98. Freund BJ, Wade CE, Claybaugh JR. Effects of exercise on atrial natriuretic factor: Release mechanisms and implications for fluid homeostasis. *Sports Med* 6: 364-377, 1988.
99. Freyschuss U. Elicitation of heart rate and blood pressure increase on muscle contraction. *J Appl Physiol* 28: 758-761, 1970.
100. Frohman LA, Bernardis LL. Effect of hypothalamic stimulation on plasma glucose, insulin and glucagon levels. *Am J Physiol* 221: 1596-1603, 1971.
101. Frohman LA, Ezdinli EZ, Javid R. Effect of vagotomy and vagal stimulation on insulin secretion. *Diabetes* 16: 443-448, 1967.
102. Froy O. Metabolism and circadian rhythms—Implications for obesity. *Endocr Rev* 31: 1-24, 2010.
103. Fruhbeck G, Aguado M, Martinez JA. In vitro lipolytic effect of leptin on mouse adipocytes: Evidence for a possible autocrine/paracrine role of leptin. *Biochem Biophys Res Commun* 240: 500-504, 1997.
104. Furchgott RF. The pharmacological differentiation of adrenergic receptors. *Ann NY Acad Sci* 139: 553-570, 1967.
105. Furchgott RF. The 1996 Albert Lasker Medical Research Awards. The discovery of endothelium-derived relaxing factor and its importance in the identification of nitric oxide. *J Am Med Assoc* 276: 1186-1188, 1996.
106. Furlan R, Piazza S, Dell'Orto S, Gentile E, Cerutti S, Pagani M, Malliani A. Early and late effects of exercise and athletic training on neural mechanisms controlling heart rate. *Cardiovasc Res* 27: 482-288, 1993.
107. Galassetti P, Mann S, Tate D, Neill RA, Costa F, Wasserman DH, Davis SN. Effects of antecedent prolonged exercise on subsequent counterregulatory responses to hypoglycemia. *Am J Physiol* 280: E908-E917, 2001.
108. Galbo H. *Hormonal and Metabolic Adaptations to Exercise*. New York: Thieme-Stratton, 1983.
109. Galbo H, Christensen NJ, Holst JJ. Glucose-induced decrease in glucagon and epinephrine responses to exercise in man. *J Appl Physiol* 42: 525-530, 1977.
110. Galbo H, Christensen NJ, Mikines KJ, Sonne B, Hilsted J, Hagen C, Fahrenkrug J. The effect of fasting on the hormonal response to graded exercise. *J Clin Endocrin Metab* 52: 1106-1112, 1981.
111. Galbo H, Holst JJ, Christensen NJ. Glucagon and plasma catecholamine responses to graded prolonged exercise in man. *J Appl Physiol* 38: 70-76, 1975.
112. Galbo H, Richter EA, Holst JJ, Christensen NJ. Diminished hormonal responses to exercise in trained rats. *J Appl Physiol* 43: 953-958, 1977.
113. Galitzky J, Reverte M, Portillo M, Carpene C, Lafontan M, Berlan M. Coexistence of β1, β2, and β3-adrenoceptors in dog fat cells and their differential activation by catecholamines. *Am J Physiol* 264: E402-E412, 1993.
114. Garcia-Sainz JA. Adrenaline and its receptors: One hundred years of research. *Arch Med Res* 26: 205-212, 1995.
115. Gasser HS, Meek WJ. A study of the mechanisms by which muscular exercise produces acceleration of the heart. *Am J Physiol* 34: 48-71, 1914.
116. Géloën A, Collet AJ, Bukowiecki LJ. Role of sympathetic innervation in brown adipocyte proliferation. *Am J Physiol* 263: R1176-1181, 1992.
117. Gladwell VF, Sandercock GR, Birch SL. Cardiac vagal activity following three intensities of exercise in humans. *Clin Physiol Funct Imaging* 30: 17-22, 2010.
118. Greenhaff PL, Ren JM, Söderlund K, Hultman E. Energy metabolism in single human muscle fibers during contraction without and with epinephrine infusion. *Am J Physiol* 260: E713-E718, 1991.
119. Gordon RD, Kuchel O, Liddle GW, Island DP. Role of sympathetic nervous system in regulating renin

and aldosterone production in man. *J Clin Invest* 46: 599-605, 1967.

120. Grassi G, Seravalle G, Calhoun DA, Mancia G. Physical training and baroreceptor control of sympathetic nerve activity in humans. *Hypertension* 23: 294-301, 1994.
121. Grassi G, Seravalle G, Colombo M, Bolla G, Cattaneo BM, Cavagnini F, Mancia G. Body weight reduction, sympathetic nerve traffic, and arterial baroreflex in obese normotensive humans. *Circulation* 97: 2037-2042, 1998.
122. Hagberg JM, Hickson RC, McLane JA, Ehsani AA, Winder WW. Disappearance of norepinephrine from circulation following strenuous exercise. *J Appl Physiol* 47: 1311-1314, 1979.
123. Hagenfeldt L, Wahren J. Human forearm muscle metabolism during exercise. VI. Substrate utilization in prolonged fasting. *Scand J Clin Lab Invest* 27: E422-E429, 1971.
124. Haggendal J, Hartley LH, Saltin B. Arterial noradrenaline concentration during exercise in relation to the relative work levels. *Scand J Clin Lab Invest* 26: 337-342, 1970.
125. Halliwill JR, Taylor JA, Eckberg DL. Impaired sympathetic vascular regulation in humans after acute dynamic exercise. *J Physiol* 495: 279-288, 1996.
126. Hansen J, Thomas GD, Jacobsen TN, Victor RG. Muscle metaboreflex triggers parallel sympathetic activation in exercising and resting human skeletal muscle. *Am J Physiol* 266: H2508-H2514, 1994.
127. Harri MN. Physical training under the influence of beta-blockade in rats. II. Effects on vascular reactivity. *Eur J Appl Physiol Occup Physiol* 42: 151-157, 1979.
128. Hering E. Über den Einfluss der Athmung auf den Kreislauf. 1. Über Athembewegungen des Gefässsystems. Sber Akad Wiss Wien. *Math Nat Kl II Abtl* 60: 829-856, 1869.
129. Hering HE. Über die Beziehung der extracardialen Herznerven zur Steigerung der Herzschlagzahl der Muskelthatigkeit. *Arch Gesammte Physiol* 40: 429-492, 1895.
130. Hering HE. Der Karotisdruckversuch. *Munch Med Wochenschr* 70: 1287-1290, 1923.
131. Hinoi E, Gao N, Jung DY, Yadav V, Yoshizawa T, Myers MG Jr., Chua SC Jr., Kim JK, Karsenty G. The sympathetic tone mediates leptin's inhibition of insulin secretion by modulating osteocalcin bioactivity. *J Cell Biol* 183: 1235-1242, 2008.
132. Hori T, Katafuchi T, Take S, Shimizu N, Niijima A. The autonomic nervous system as a communication channel between the brain and the immune system. *Neuroimmunomodulat* 2: 203-215, 1995.
133. Horowitz JF, Braudy RJ, Martin WH III, Klein S. Endurance exercise training does not alter lipolytic or adipose tissue blood flow sensitivity to epinephrine. *Am J Physiol* 277: E325-E331, 1999.
134. Horowitz JF, Klein S. Lipid metabolism during endurance exercise. *Am J Clin Nutr* 72(Suppl.): 558S-563S, 2000.
135. Houk JC. Control strategies in physiological systems. *FASEB* J 2: 97-107, 1988.
136. Hutchinson DS, Bengtsson T. Alpha1A-adrenoceptors activate glucose uptake in L6 muscle cells through a phospholipase C-, phosphatidylinositol-3 kinase-, and atypical protein kinase C-dependent pathway. *Endocrinol* 146: 901-912, 2005.
137. Inagaki S, Kubota Y, Kito S, Kangawa K, Matsuo H. Immunoreactive atrial natriuretic polypeptides in the adrenal medulla and sympathetic ganglia. *Regul Pept* 15: 249-260, 1986.
138. Ito S, Sved AF. Blockade of angiotensin receptors in rat rostral ventralateral medulla. *Am J Physiol* 270: R1317-R1323, 1996.
139. Iwamoto GA, Wappel SM, Fox GM, Buetow KA, Waldrop TG. Identification of diencephalic and brainstem cardiorespiratory areas activated during exercise. *Brain Res* 726: 109-122, 1996.
140. Izawa T, Morikawa M, Mizuta T, Nagasawa J, Kizaki T, Oh-ishi S, Ohno H, Komabayashi T. Decreased vascular sensitivity after acute exercise and chronic exercise training in rat thoracic aorta. *Res Commun Mol Pathol Pharmacol* 93: 331-342, 1996.
141. Jansson E, Hjemdahl P, Kaijser L. Epinephrine-induced carbohydrate metabolism during exercise in male subjects. *J Appl Physiol* 60: 1466-1470, 1986.
142. Jennings G, Nelson L, Nestel P, Esler M, Korner PI, Burton D, Bazelmans J. The effects of changes in physical activity on major cardiovascular risk factors, hemodynamics, sympathetic function and glucose utilization in man: A controlled study of 4 levels of activity. *Circulation* 73: 30-40, 1986.
143. Jensen LB, Quaade F, Sorensen OH. Bone loss accompanying voluntary weight loss in obese humans. *J Bone Miner Res* 9: 459-463, 1994.
144. Jordan D. Autonomic changes in affective behavior. In Loewy AD, Spyer KM, eds. *Central Regulation of Autonomic Functions*. New York: Oxford University Press, 1990, 349-366.
145. Joyner MJ, Halliwill JR. Sympathetic vasodilatation in human limbs. *J Physiol* 526: 471-480, 2000.
146. Joyner MJ, Wilkins BW. Exercise hyperaemia: Is anything obligatory but the hyperaemia? *J Physiol* 583: 855-860, 2007.
147. Kahn R. Zur Physiologie der Trachea. Arch Physiol 398-426, 1907.
148. Kakolewski JW, Deaux E, Christense J, Case B. Diurnal patterns in water and food intake and body weight changes in rats with hypothalamic lesions. *Am J Physiol* 221: 711-718, 1971.
149. Kalsbeek A, Ruiter M, La Fleur SE, Cailotto C, Kreier F, Buijs RM. The hypothalamic clock and its control of glucose homeostasis. *Prog Brain Res* 153: 283-307, 2006.

150. Kamendi H, Dergacheva O, Wang X, Huang ZG, Bouairi E, Gorini C, Mendelowitz D. NO differentially regulates neurotransmission to premotor cardiac vagal neurons in the nucleus ambiguus. *Hypertension* 48: 1137-1142, 2006.
151. Karsenty G, Oury F. The central regulation of bone mass, the first link between bone remodeling and energy metabolism. *J Clin Endocrinol Metab* 95: 4795-4801, 2010.
152. Kellogg DL Jr., Pérgola PE, Piest KL, Kosiba WA, Crandall CG, Grossmann M, Johnson JM. Cutaneous active vasodilation in humans in mediated by cholinergic nerve transmission. *Circulat Res* 77: 1222-1228, 1995.
153. Kennedy GC. The development with age of hypothalamic restraint upon the appetite of the rat. *J Endocrinol* 16: 9-17, 1957.
154. Kennedy GC, Mitra J. Body weight and food intake as initiating factors for puberty in the rat. *J Physiol* 166: 408-418, 1963a.
155. Kennedy GC, Mitra J. Hypothalamic control of energy balance and the reproductive cycle in the rat. *J Physiol* 166: 396-407,1963b.
156. Kenny WL, Tankersley CG, Newswanger DL, Puhl SM. A1-adrenergic blockade does not alter control of skin blood flow during exercise. *Am J Physiol* 260: H855-H861, 1991.
157. Kjaer M, Secher NH, Bach FW, Sheikh S, Galbo H. Hormonal and metabolic responses to exercise in humans: Effect of sensory nervous blockade. *Am J Physiol* 257: E95-E101, 1989.
158. Kjaer M, Engfred K, Fernandes A, Secher NH, Galbo H. Regulation of hepatic glucose production during exercise in humans: Role of sympathoadrenergic activity. *Am J Physiol* 265: E275-E283, 1993.
159. Klein LA. Beta-adrenergic (isoproterenol) regulation of antidiuretic hormone. *Invest Urol* 12, 285-290, 1975.
160. Klein S, Peters EJ, Holland OB, Wolfe RR. Effect of short- and long-term β adrenergic blockade on lipolysis during fasting in humans. *Am J Physiol* 257: E65-E73, 1989.
161. Klein S, Weber JM, Coyle EF, Wolfe RR. Effect of endurance training on glycerol kinetics during strenuous exercise in humans. *Metabolism* 45: 357-361, 1996.
162. Kobilka BK, Dixon RA, Frielle HG, Dohlman HG, Bolanowski MA, Sigal IS, et al. cDNA for the human beta 2-adrenergic receptor: A protein with multiple spanning domains and encoded by a gene whose chromosomal location is shared with that of a receptor for platelet growth factor. *Proc Natl Acad Sci USA* 84: 46-50, 1987.
163. Koller A, Kaley G. Endothelial regulation of wall shear stress and blood flow in skeletal muscle microcirculation. *Am J Physiol* 260: H862-H868, 1991.
164. Koyama Y, Coker RH, Stone EE, Lacy DB, Jabbour K, Williams PE, Wasserman DH. Evidence that carotid bodies play an important role in glucoregulation in vivo. *Diabetes* 49: 1434-1442, 2000.
165. Koyama Y, Coker RH, Denny JC, Lacy DB, Jabbour K, Williams PE, Wasserman DH. Role of carotid bodies in control of the neuroendocrine response to exercise. *Am J Physiol* 281: E742-E748, 2001.
166. Krasney JA, Levitzky MG, Koehler RC. Sinoaortic contribution to the adjustment of systemic resistance in exercising dogs. *J Appl Physiol* 36: 679-685, 1974.
167. Kriegsfeld LJ, LeSauter J, Silver R. Targeted microlesions reveal novel organization of the hamster suprachiasmatic nucleus. *J Neurosci* 24: 2449-2457, 2004.
168. Krogh A, Lindhard J. The regulation of respiration and circulation during the initial stages of muscular work. *J Physiol* 47: 112-136, 1913.
169. Lafontan M, Bousquet-Melou A, Galitzky J, Barbe P, Carpene C, Langin D, et al. Adrenergic receptors and fat cells: Differential recruitment by physiological amines and homologous regulation. *Obes Res 3*(Suppl. 4): 507S-514S, 1995.
170. Lafontan M, Dang-Tran L, Berlan M. Alpha-adrenergic antilipolytic effect of adrenaline in human fat cells of the thigh: Comparison with adrenaline responsiveness of different fat deposits. *Eur J Clin Invest* 9: 261-266, 1979.
171. Lahiri S, Mei SS, Kao FF. Vagal modulation of respiratory control during exercise. *Respir Physiol* 23: 133-146, 1975.
172. Lands AM, Arnold A, McAuliff JP, Luduena FP, Brown TG Jr. Differentiation of receptor systems activated by sympathomimetic amines. *Nature* 214: 597-598, 1967.
173. Langer SZ. Presynaptic receptors and their role in the regulation of transmitter release. *Br J Pharmacol* 60: 481-497, 1977.
174. Langfort J, Ploug T, Ihlemann J, Saldo M, Holm C, Galbo H. Expression of hormone-sensitive lipase and its regulation by adrenaline is skeletal muscle. *Biochem J* 340: 459-465, 1999.
175. Langley JN. On the union of cranial autonomic (visceral) fibres with the nerve cells of the superior cervical ganglia. *J Physiol* 23: 240-270, 1898.
176. Langley JN. Observations of the physiological action of extracts of supra-renal bodies. *J Physiol* 27: 237-256, 1901.
177. Langley JN. *The Autonomic Nervous System*. Cambridge: Heffner & Sons, 1921.
178. Lefkowitz RJ, Limbird LE, Mukherjee C, Caron MG. The beta-adrenergic receptor and adenylate cyclase. *Biochim Biophys Acta* 457: 1-39, 1976.
179. Laterza MC, de Matos LD, Trombetta IC, Braga AM, Roveda F, Alves MJ, Krieger EM, Negrão CE, Rondon MU. Exercise training restores baroreflex sensitivity in never-treated hypertensive patients. *Hypertension* 49: 1298-1306, 2007.

180. LeMagnen J, Devos M. Metabolic correlates of the meal onset in the free food intake of rats. *Physiol Behav* 5: 805-814, 1970.
181. Lewis S, Ohta H, Machado B, Bates J, Talman W. Microinjection of S-nitrosocysteine into the nucleus tractus solitarii decreases arterial pressure and heart rate via activation of soluble guanylate cyclase. *Eur J Pharmacol* 202: 135-136, 1991.
182. Loewy AD, Burton H. Nuclei of the solitary tract: Efferent projections to the lower brain stem and spinal cord of the cat. *J Comp Neurol* 181: 421-450, 1978.
183. Loewy AD, Spyer KM, eds. *Central Regulation of Autonomic Functions*. New York: Oxford University Press, 1990.
184. Loewi O, Navratil E. Über humorale Übertragbarkeit der herzenwirkung. X Mitteilung. Über das Schicksal des Vagusstoff. *Pflugers* 214: 678-688, 1926.
185. Lundberg JM. Pharmacology of cotransmission in the autonomic nervous system: Integrative aspects on amines, neuropeptides, adenosine triphosphate, amino acids and nitric oxide. *Pharmacol Rev* 48: 113-178, 1996.
186. Lundborg P, Aström H, Bengtsson C, Fellenius E, Von Schenck H, et al. Effect of beta-adrenoceptor blockade on exercise performance and metabolism. *Clin Sci* 61: 299-305, 1981.
187. Maciel BC, Gallo L Jr., Marin Neto JA, Martins LEB. Autonomic nervous control of the heart rate during isometric exercise in normal man. *Pflugers Arch* 408: 173-177, 1987.
188. Madden CJ, Stocker SD, Sved AF. Attenuation of homeostatic responses to hypotension and glucoprivation after destruction of catecholaminergic rostral ventrolateral medulla neurons. *Am J Physiol* 291: R751-R759, 2006.
189. Malliani A, Pagani M, Lombardi F, Cerutti S. Cardiovascular neural regulation explored in the frequency domain. *Circulation* 84: 482-492, 1991.
190. Mark AL, Wallin BG, Seals DR, Victor RG. Mechanism of contrasting pressor responses to static and dynamic exercise: New insights from direct intraneural recordings of sympathetic nerve activity in humans. *Trans Am Clin Climatol Assoc* 98: 98-104, 1987.
191. Martin EA, Nicholson WT, Eisenach JH, Charkoudian N, Joyner MJ. Bimodal distribution of vasodilator responsiveness to adenosine due to difference in nitric oxide contribution: Implications for exercise hyperemia. *J Appl Physiol* 101: 492-499, 2006.
192. Martin WH III, Coggan AR, Spina RJ, Saffitz JE. Effects of fiber type and training on beta-adrenoceptor density in human skeletal muscle. *Am J Physiol* 257: E736-E742, 1989.
193. Martin WH III, Murphree SS, Saffitz JE. Beta-adrenergic receptor distribution among muscle fiber types and resistance arterioles of white, red, and intermediate skeletal muscle. *Circ Res* 64: 1096-1105, 1989.
194. Mass H, Gwirtz PA. Myocardial flow and function after regional β-blockade in exercising dogs. *Med Sci Sports Exerc* 19: 443-450, 1987.
195. Matsukawa S, Reid IA. Role of area postrema in the modulation of baroreceptor control of heart rate by angiotensin II. *Circ Res* 67: 1462-1473, 1990.
196. Mauriege P, Galitzky J, Berlan M, Lafontan M. Heterogenous distribution of β and α-2 adrenoceptor bionding sites in human fat cells from various fat deposits: Functional consequences. *Eur J Clin Invest* 17: 156-165, 1987.
197. Mauriège P, Prud'Homme D, Marcotte M, Yoshioka M, Tremblay A, Després JP. Regional differences in adipose tissue metabolism between sedentary and endurance-trained women. *Am J Physiol* 273: E497-E506, 1997.
198. Maywood ES, O'Neill JS, Chesham JE, Hastings MH. Minireview: The circadian clockwork of the suprachiasmatic nuclei—Analysis of a cellular oscillator that drives endocrine rhythms. *Endocrinology* 148: 5624-5634, 2007.
199. McAllister RM, Laughlin MH. Short-term exercise training alters responses of porcine femoral and brachial arteries. *J Appl Physiol* 82: 1438-1444, 1997.
200. McConville P, Lakatta EG, Spencer RG. Greater glycogen utilization during 1- than 2-adrenergic receptor stimulation in the isolated perfused rat heart. *Am J Physiol* 293: E1828-E1835, 2007.
201. Meredith IT, Friberg P, Jennings GL, Dewar EM, Fazio VA, Lambert GW, Esler MD. Exercise training lowers resting renal but not cardiac sympathetic activity in humans. *Hypertension* 18: 575-582, 1991.
202. Milner P, Kirkpatrick KA, Ralevic V, Toothill V, Pearson J, Burnstock G. Endothelial cells cultured from human umbilical vein release ATP, substance P and acetylcholine in response to increased flow. *Proceed Roy Soc* B 241: 245-248, 1990.
203. Moates JM, Lacy DB, Goldstein RE, Cherrington AD, Wasserman DH. Metabolic role of the exercise-induced increment in epinephrine in the dog. *Am J Physiol* 255: E428-E436, 1988.
204. Moro C, Pasarica M, Elkind-Hirsch K, Redman LM. Aerobic exercise training improves atrial natriuretic peptide and catecholamine-mediated lipolysis in obese women with polycystic ovary syndrome. *J Clin Endocrinol Metab* 94: 2579-2586, 2009.
205. Morrison SF, Nakamura K, Madden CJ. Central control of thermogenesis in mammals. *Exp Physiol* 93: 773-797, 2008.
206. Müller EE, Pecile A, Cocchi D, Oligiati VR. Hyperglycemic or feeding response to glucoprivation and hypothalamic glucoreceptor. *Am J Physiol* 226: 1100-1109, 1974.

207. Nadel ER. Recent advances in temperature regulation during exercise in humans. *Fed Proc* 44: 2286-2292, 1985.
208. Nakamura K, Morrison SF. Central efferent pathways mediating skin cooling-evoked sympathetic thermogenesis in brown adipose tissue. *Am J Physiol* 292: R127-R136, 2007.
209. Nakamura Y, Yamamoto Y, Muraoka I. Autonomic control of heart rate during physical exercise and fractal dimension of heart rate variability. *J Appl Physiol* 74: 875-881, 1993.
210. Navegantes LCC, Migliorini RH, Kettelhut, IC. Adrenergic control of protein metabolism in skeletal muscle. *Cur Opin Clin Nutr Metab Care* 5: 281-286, 2002.
211. Nieto JL, Laviada ID, Guillén A, Haro A. Adenylyl cyclase system is affected differently by endurance physical training in heart and adipose tissue. *Biochem Pharmacol* 51: 1321-1329, 1996.
212. Niijima A. Afferent impulse discharges from glucoreceptors in the liver of guinea pig. Ann NY *Acad Sci* 157: 690-700, 1969.
213. Niijima A. Studies of the nervous regulatory mechanisms of blood sugar levels. *Pharm Biochem Behav* 3(Suppl. 1): 1390-1430, 1975.
214. Niijima A, Nagai K, Nagai N, Nakagawa H. Light enhances sympathetic and suppresses vagal outflows and lesions including SCN eliminate these changes in rats. *J Auton Nerv Syst* 40: 155-160, 1992.
215. Nillni EA. Regulation of the hypothalamic thyrotropin releasing hormone (TRH) neuron by neuronal and peripheral inputs. *Front Neuroendocrinol* 31: 134-156, 2010.
216. Nishizawa Y, Bray GA. Ventromedial hypothalamic lesions and the mobilization of fatty acids. *J Clin Invest* 61: 714-721, 1978.
217. Nonogaki K. New insights into sympathetic regulation of glucose and fat metabolism. *Diabetologia* 43: 533-549, 2000.
218. Oliver G, Schäfer EA. The physiological effects of extracts of the suprarenal capsules. *J. Physiol* 18: 230-276, 1895.
219. Ott I. Heat center of the brain. *J Nerv Ment Dis* 14: 152, 1877.
220. Papelier Y, Escourrou P, Gauthier JP, Rowell LB. Carotid baroreflex control of blood pressure and heart rate in man during dynamic exercise. *J Appl Physiol* 77: 502-506, 1994.
221. Potts JT, Shi SR, Raven PB. Carotid baroreflex responsiveness during dynamic exercise in humans. *Am J Physiol* 265: H1928-H1938, 1993.
222. Price DT, Lefkowitz RJ, Caron MG, Berkowitz D, Schwinn DA. Localization of messenger RNA for three distinct alpha 1-adrenergic receptor subtypes in human tissues: Implications for human adrenergic physiology. *Mol Pharmacol* 45: 171-175, 1994.
223. Pryor SL, Lewis SF, Haller RG, Bertocci LA, Victor RG. Impairment of sympathetic activation during static exercise in patients with muscle phosphorylase deficiency (McArdle's disease). *J Clin Invest* 85: 1444-1449, 1990.
224. Radegran G, Saltin B. Nitric oxide in the regulation of vasomotor tone in human skeletal muscle. *Am J Physiol* 276: H1951-H1960, 1999.
225. Ranson SW, Billingsley PR. Vasomotor reactions from stimulation of the floor of the fourth ventricle. *Am J Physiol* 41: 85-90, 1916.
226. Ray CA. Sympathetic adaptations to one-legged training. *J Appl Physiol* 86:1583-1587, 1999.
227. Ray CA, Hume KM. Sympathetic neural adaptations to exercise training in humans: Insights from microneurography. *Med Sci Sports Exerc* 30: 387-391, 1998.
228. Ray CA, Carrasco DI. Isometric handgrip training reduces arterial pressure at rest without changes in sympathetic nerve activity. *Am J Physiol* 279: H245-H249, 2000.
229. Ray CA, Mark AL. Sympathetic nerve activity to nonactive muscle of the exercising and nonexercising limb. *Med Sci Sports Exerc* 27: 183-187, 1995.
230. Reed AS, Tschakovsky ME, Minson CT, Halliwill JR, Torp RD, Nauss LA, Joyner MJ. Skeletal muscle vasodilatation during sympathoexcitation is not neurally mediated in humans. *J Physiol* 525: 253-262, 2000.
231. Remensnyder JP, Mitchell JH, Sarnoff SJ. Functional sympatholysis during muscular activity. *Circ Res* 11: 370-380, 1962.
232. Richelsen B, Pedersen SB, Moller-Pedersen T, Bak JF. Regional differences in triglyceride breakdown in human adipose tissue: Effects of catecholamines, insulin, and prostaglandin E2. *Metabolism* 40: 990-996, 1991.
233. Rinaman L, Miselis RR. The organization of vagal innervation of rat pancreas using cholera toxin-horseradish peroxidase conjugate. *J Auton Nerv Sys* 21: 109-125, 1987.
234. Ritter RC, Slusser PG, Stone S. Glucoreceptors controlling feeding and blood glucose: Location in the hindbrain. *Science* 213: 451-452, 1981.
235. Ritter S, Dinh TT, Zhang Y. Localization of hindbrain glucoreceptive sites controlling food intake and blood glucose. *Brain Res* 856: 37-47, 2000.
236. Ritter S, Watts AG, Dinh TT, Sanchez-Watts G, Pedrow C. Immunotoxin lesion of hypothalamically projecting norepienphrine and epinephrine neurons differentially affects circadian and stressor-stimulated corticosterone secretion. *Endocrinology* 144: 1357-1367, 2003.
237. Riviere D, Crampes F, Beauville M, Garrigues M. Lipolytic response of fat cells to catecholamines in sedentary and exercise-trained women. *J Appl Physiol* 66: 330-335, 1989.
238. Roddie IC. Circulation to skin and adipose tissue. In Shepherd JT, Abboud FM, eds. *Handbook of Physi-*

ology, Section 2: The Cardiovascular System. Vol III: Peripheral and Organ Blood Flow. Maryland: Am Physiol Soc USA, 1983, 285-317.

239. Rogers PJ, Miller TD, Bauer BA, Brum JM, Bove AA, Vanhoutte PM. Exercise training and responsiveness of isolated coronary arteries. *J Appl Physiol* 71: 2346-2351, 1991.
240. Romijn JA, Coyle EF, Sidossis LS, Gastaldelli A, Horowitz JF, Endert E, Wolfe RR. Regulation of endogenous fat and carbohydrate metabolism in relation to exercise intensity and duration. *Am J Physiol* 265: E380-E391, 1993.
241. Rosell S. Release of free fatty acids from subcutaneous adipose tissue in dogs following sympathetic nerve stimulation. *Acta Physiol Scand* 67: 343-351, 1966.
242. Rowell LB. *Human Cardiovascular Control.* New York: Oxford University Press, 1993.
243. Rowell LB, Marx HJ, Bruce RA, Conn RD, Kusumi F. Reductions in cardiac output, central blood volume, and stroke volume with thermal stress in normal men during exercise. *J Clin Invest* 45: 1801-1816, 1966.
244. Rowell LB, O'Leary DS. Reflex control of the circulation during exercise: Chemoreflexes and mechanoreflexes. *J Appl Physiol* 69: 407-418, 1990.
245. Rowell LB, O'Leary DS, Kellogg DL. Integration of cardiovascular control systems in dynamic exercise. In: Rowell LB, Shepherd JT, eds. *Handbook of Physiology, Section 12. Exercise: Regulation and Integration of Multiple Systems.* New York: Oxford University Press, 1996, 770-838.
246. Saito M. Differences in muscle sympathetic nerve response to isometric exercise in different muscle groups. *Eur J Appl Physiol* 70: 26-35, 1995.
247. Saito M, Sone R, Ikeda M, Mano T. Sympathetic outflow to the skeletal muscle in humans increases during prolonged light exercise. *J Appl Physiol* 82: 1237-1243, 1997.
248. Saito M, Tsukanaka A, Yanagihara D, Mano T. Muscle sympathetic nerve response to graded cycling. *J Appl. Physiol* 75: 663-667, 1993.
249. Saito M, Watanabe H, Mano T. Comparison of muscle sympathetic nerve activity during exercise in dominant and non-dominant forearm. *Eur J Appl Physiol* 66: 108-115, 1993.
250. Sakaguchi T, Bray GA. Intrahypothalamic injection of insulin decreases firing rate of sympathetic nerves. *Proc Natl Acad Sci USA* 84: 2012-2014, 1987.
251. Saltin B, Nazar K, Costill DL, Stein E, Jansson E, Essen B, Gollnick D. The nature of the training response: Peripheral and central adaptations of one-legged exercise. *Acta Physiol Scand* 96: 289-305, 1976.
252. Samols E, Weir GC. Adrenergic modulation of pancreatic A, B, and D cells. *J Clin Invest* 63: 230-238, 1979.
253. Sartori C, Lepori M, Scherrer U. Interaction between nitric oxide and sympathetic nervous system in cardiovascular control in humans. *Pharmacol Therapeut* 106: 209-220, 2005.
254. Sato K. The physiology, pharmacology, and biochemistry of the eccrine sweat gland. *Rev Physiol Biochem Pharmacol* 79: 51-131, 1977.
255. Savard G, Strange S, Kiens B, Richter EA, Christensen NJ, Saltin B. Noradrenaline spillover during exercise in active versus resting skeletal muscle in man. *Acta Physiol Scand* 131: 507-515, 1987.
256. Scarpace PJ, Matheny M. Leptin induction of UCP1 gene expression is dependent on sympathetic innervation. *Am J Physiol* 275: E259-E264, 1998.
257. Scheuer J. Catecholamines in cardiac hypertrophy. *Am J Cardiol* 83: 70H-74H, 1999.
258. Schwartz MW, Niswender KD. Adiposity signaling and biological defense against weight gain: Absence of protection or central hormone resistance? *J Clin Endocrinol Metab* 89: 5889-5897, 2004.
259. Schwartz MW, Woods SC, Porte D Jr., Seeley RJ, Baskin DG. Central nervous system control of food intake. *Nature* 404: 661-671, 2000.
260. Seals, DR. The autonomic nervous system. In Tipton CM, ed. *Advanced Exercise Physiology*. Champaign, IL: Human Kinetics, 2003, 197-245.
261. Seals DR, Bell C. Chronic sympathetic activation: Consequence and cause of age-associated obesity? *Diabetes* 53: 276-284, 2000.
262. Seals DR, Chase PB. Influence of physical training on heart rate variability and baroreflex circulatory control. *J Appl Physiol* 66: 1886-1895, 1989.
263. Sengenes C, Berlan M, De Glisezinski I, Lafontan M, Galitzky J. Natriuretic peptides: A new lipolytic pathway in human adipocytes. *FASEB J* 14: 1345-1351, 1977.
264. Shamsuddin AKM, Reddy MM, Quinton PM. Iontophoretic β-adrenergic stimulation of human sweat glands: Possible assay for cystic fibrosis transmembrane conductance regulator activity *in vivo*. *Exp Physiol* 93: 969-981, 2008.
265. Shea SA, Hilton MF, Hu K, Scheer FA. Existence of an endogenous circadian blood pressure rhythm in humans that peaks in the evening. *Circ Res* 108: 980-984, 2011.
266. Sheriff DD, Rowell LB, Scher AM. Is rapid rise in vascular conductance at onset of dynamic exercise due to muscle pump? *Am J Physiol* 265: H1227-H1234, 2003.
267. Shibasaki M, Davis SL, Cui J, Low DA, Keller DM, Durand S, Crandall CG. Neurally mediated vasoconstriction is capable of decreasing skin blood flow during orthostasis in heat-stressed human. *J Physiol* 575: 953-959, 2006a.
268. Shibasaki M, Wilson TE, Crandall CG. Neural control and mechanisms of eccrine sweating during heat stress and exercise. *J Appl Physiol* 100: 1692-1701, 2006b.

269. Shimazu T. Central nervous system regulation of liver and adipose tissue metabolism. *Diabetologia* 20: 343-356, 1981.
270. Shimazu T, Ogasawara S. Effects of hypothalamic stimulation of gluconeogenesis and glycolysis in rat liver. *Am J Physiol* 228: 1787-1793, 1975.
271. Shiota M, Tanaka T, Sugano T. Effect of norepinephrine on gluconeogenesis in perfused livers of cold-exposed rats. *Am J Physiol* 249: E281-E286, 1985.
272. Sinoway L, Shenberger J, Leaman G, Zelis R, Gray K, Baily R, Leuenberger U. Forearm training attenuates sympathetic responses to prolonged rhythmic forearm exercise. *J Appl Physiol* 81: 1778-1784, 1996.
273. Sinoway L, Rea RF, Mosher TJ, Smith MB, Mark, AL. Hydrogen ion concentration is not the sole determinant of muscle metaboreceptor responses in humans. *J Clin Invest* 89: 1875-1884, 1992.
274. Smith ML, Hudson DL, Graitzer HM, Raven PB. Exercise training bradycardia: The role of autonomic balance. *Med Sci Sports Exer* 21: 40-44, 1989.
275. Sokal JE, Sarcione EJ, Henderson AM. The relative potency of glucagon and epinephrine as hepatic glycogenolytic agents: Studies with the isolated perfused rat liver. *Endocrinol* 74: 930-938, 1964.
276. Somers VK, Conway J, Johnston J, Sleight P. Effects of endurance training on baroreflex sensitivity and blood pressure in borderline hypertension. *Lancet* 337:1363-1368, 1991.
277. Sonne B, Mikines KJ, Richter EA, Christensen NJ, Galbo H. Role of liver nerves and adrenal medulla in glucose turnover of running rats. *J Appl Physiol* 59: 1640-1646, 1985.
278. Soto PF, Herrero P, Schechtman KB, Waggoner AD, Baumstark JM, Ehsani AA, Gropler RJ. Exercise training impacts the myocardial metabolism of older individuals in a gender-specific manner. *Am J Physiol* 295: H842-H850, 2008.
279. Spengler CM, Shea SA. Endogenous circadian rhythm of pulmonary function in healthy humans. *Am J Respir Crit Care Med* 162: 1038-1046, 2000.
280. Spier SA, Laughlin MH, Delp MD. Effects of acute and chronic exercise on vasoconstrictor responsiveness of rat abdominal aorta. *J Appl Physiol* 87: 1752-1757, 1999.
281. Spina RJ, Turner MJ, Ehsani AA. Beta-adrenergic-mediated improvement in left ventricular function by exercise training in older men. *Am J Physiol* 274: H397-H404, 1998.
282. Spyer KM, Gourine AV. Chemosensory pathways in the brainstem controlling cardiorespiratory activity. *Phil Trans R Soc B* 364: 2603-2610, 2009.
283. Stallknecht B, Lorentsen J, Enevoldsen LH, Bülow J, Biering-Sorensen F, et al. Role of sympathoadrenergic system in adipose tissue metabolism during exercise in humans. *J Physiol* 536: 283-294, 2001.
284. Stewart GN, Rogoff J. The relation of the adrenals to piqūre hypoglycemia and to glycogen content of the liver. *Am J Physiol* 46: 90-116, 1918.
285. Stich V, de Glisezinski I, Galitzky J, Hejnova J, Crampes F, Rivière D, Berlan M. Endurance training increases the beta-adrenergic lipolytic response in subcutaneous adipose tissue in obese subjects. *Int J Obes Relat Metab Disord* 23: 374-381, 1999.
286. Straznicky NE, Grima MT, Eikelis N, Nestel PJ, Dawood T, Schlaich MP, Chopra R, Masuo K, Esler MD, Sari CI, Lambert GW, Lambert EA. The effects of weight loss versus weight loss maintenance on sympathetic nervous system activity and metabolic syndrome components. *J Clin Endocrinol Metab* 96: E503-E508, 2011.
287. Straznicky NE, Lambert EA, Lambert GW, Masuo K, Esler MD, Nestel PJ. Effects of dietary weight loss on sympathetic activity and cardiac risk factors associated with the metabolic syndrome. *J Clin Endocrinol Metab* 90: 5998-6005, 2005.
288. Strosberg AD. Structure function relationship of proteins belonging to the family of receptors coupled to GTP binding proteins. *Eur J Biochem* 196, 1-10, 1991.
289. Takahashi A, Shimazu T. Hypothalamic regulation of lipid metabolism in the rat: Effect of hypothalamic stimulation on lipolysis. *J Auton Nerv Sys* 4: 195-205, 1981.
290. Takahashi A, Shimazu T. Hypothalamic regulation of lipid metabolism in the rat: Effect of hypothalamic stimulation on lipogenesis. *J Auton Nerv Sys* 6: 225-235, 1982.
291. Takamine J. The isolation of the active principle of the suprarenal gland. *J Physiol* 27: xxix-xxx, 1902.
292. Tanaka M, McKinley MJ, McAllen RM. Roles of two preoptic cell groups in tonic and febrile control of rat tail sympathetic fibers. *Am J Physiol* 296: R1248-R1257, 2009.
293. Tavernier G, Galitzky J, Bousquet-Melou A, Notastruc JL, Berlan M. The positive chronotropic effect induced by BRL 37344 and CGP 12177, two beta-3 adrenergic agonists, does not involve cardiac beta adrenoceptors but baroreflex mechanisms. *J Pharm Exp Ther* 263: 1083-1090, 1992.
294. Thomas GD, Hansen J, Victor RJ. Inhibition of α_2-adrenergic vasoconstriction during contraction of glycolytic, not oxidative, rat hindlimb muscle. *Am J Physiol* 266: H920-H929, 1994.
295. Tidgren B, Hjemdahl P, Theodorsson E, Nussberger J. Renal neurohormonal and vascular responses to dynamic exercise in humans. *J Appl Physiol* 70: 2279-2286, 1991.
296. Tipton CM. The autonomic nervous system. In Tipton CM, ed. *Exercise Physiology: People and Ideas*. New York: Oxford University Press, 2003, 188-254.

297. Toda N, Okamura T. Modulation of renal blood flow and vascular tone by neuronal nitric oxide synthase-derived nitric oxide. *J Vasc Res* 48: 1-10, 2011.
298. Traube D. Über periodische Thatigskeits-Ǎusserungen des vasomotorischen und Hemmungs-Nervencentrums. *Cbl Med Wiss* 56: 881-885, 1865.
299. Tschakovsky ME, Joyner MJ. Nitric oxide and muscle blood flow in exercise. *Appl Physiol Nutr Metab* 33: 151-161, 2008.
300. Van Baak MA. Beta-adrenoceptor blockade and exercise. *An update. Sports Med* 5: 209-225, 1988.
301. Victor RG, Bertocci LA, Pryor SL, Nunnally RL. Sympathetic nerve discharge is coupled to muscle cell pH during exercise in humans. *J Clin Invest* 82: 1301-1305, 1988.
302. Viguerie N, Clement K, Barbe P, Courtine M, Benis A, Larrouy D, et.al.. In vivo epinephrine-mediated regulation of gene expression in human skeletal muscle. *J Clin Endocr Metab* 89: 2000-2014, 2004
303. Vissing J, Iwamoto GA, Rybicki KJ, Galbo H, Mitchell JH. Mobilization of glucoregulatory hormones and glucose by hypothalamic locomotor centers. *Am J Physiol* 257: E722-E728, 1989.
304. von Baeyer A. Notes from the organic laboratory of the Gewerbeacademie in Berlin. I. On the neurin. *Liebigs Ann* 142: 322-326, 1867.
305. von Euler US. Sympathin E and noradrenaline. *Science* 107: 422, 1948.
306. von Euler U, Hellner S. Excretion of noradrenaline and adrenaline in muscular work. *Acta Physiol Scand* 26: 183-191, 1952.
307. Wade CE, Ramee SR, Hunt MM, White CJ. Hormonal and renal responses to converting enzyme inhibition during maximal exercise, *J Appl Physiol* 63: 1796-1800, 1987.
308. Wahren J, Felig P, Ahlborg G, Jorfeldt L. Glucose metabolism during leg exercise in man. *J Clin Invest* 50: 2715, 1971.
309. Wahren J, Felig P, Hendler R, Ahlborg G. Glucose and amino acid metabolism during recovery after exercise. *J Appl Physiol* 34: 838-845, 1973.
310. Wahrenberg H, Engfeldt P, Bolinder J, Arner P. Acute adaptation in adrenergic control of lipolysis during physical exercise in humans. *Am J Physiol* 253: E383-E390, 1987.
311. Warren WS, Champney TH, Cassone VM. The hypothalamic suprachiasmatic nucleus regulates circadian changes in norepinephrine turnover in sympathetically innervated tissue. *Soc Neurosci Abstr* 21: 178, 1995.
312. Wasserman DH. Regulation of glucose fluxes during exercise in the postabsorptive state. *Annu Rev Physiol* 57: 191-218, 1995.
313. Wasserman DH, Cherrington AD. Hepatic fuel metabolism during muscular work: Role and regulation. *Am J Physiol* 260: E811-E824, 1989.
314. Wasserman DH, Spalding JA, Lacy DB, Colburn CA, Goldstein RE, Cherrington AD. Glucagon is a primary controller of hepatic glycogenolysis and gluconeogenesis during muscular work. *Am J Physiol* 257: E108-E117, 1989.
315. Wasserman DH, Williams PE, Lacy DB, Goldstein RE, Cherrington AD. Exercise-induced fall in insulin and hepatic carbohydrate metabolism during muscular work. *Am J Physiol* 256: E500-E509, 1989.
316. Wasserman DH, Williams PE, Lacy DB, Bracy D, Cherrington AD. Hepatic nerves are not essential to the increase in hepatic glucose production during muscular work. *Am J Physiol* 259: E195-E203, 1990.
317. Wiersma MM, Vissing J, Steffens AB, Galbo H. Effects of glucose infusion on hormone secretion and hepatic glucose production during heavy exercise. *Am J Physiol* 265: R1333-R1338, 1993.
318. Wilson TE, Monahan KD, Fogelman A, Kearney ML, Sauder CL, Ray CA. Aerobic training improves in vivo cholinergic responsiveness but not sensitivity of eccrine sweat glands. *J Invest Dermatol* 130: 2328-2330, 2010.
319. Winder WW, Hagberg JM, Hickson RC, Ehsani AA, McLane JA. Time course of sympathoadrenal adaptation to endurance exercise training in man. *J Appl Physiol* 45: 370-374, 1978.
320. Winslow JB. *Exposition Anatomique du Corp Humain*. Paris: Desprez, 1732.
321. Wojtaszewski JF, Hansen BF, Gade J, Kiens B, Markuns JF, Goodyear LJ, Richter EA. Insulin signaling and insulin sensitivity after exercise in human skeletal muscle. *Diabetes* 49: 325-331, 2000.
322. Yamamoto H, Nagai K, Nakagawa H. Role of the suprachiasmatic nucleus in glucose homeostasis. *Biomed Res* 5: 55-60, 1984.
323. Yamamoto Y, Hughson RL, Nakamura Y. Autonomic nervous system responses to exercise in relation to ventilatory threshold. *Chest* 101: 206S-210S, 1992.
324. Yamazaki T, Yazaki Y. Molecular basis of cardiac hypertrophy. *Z Kardiol* 89: 1-6, 2000.
325. Young JBV, Rosa RM, Landsberg L. Dissociation of sympathetic nervous system and adrenal medullary responses. *Am J Physiol* 247: E35-E40, 1984.
326. Youngstrom TG, Bartness TJ. Catecholaminergic innervation of white adipose tissue in the Siberian hamster. *Am J Physiol* 268: R744-R751, 1995.

CHAPTER 10

The Respiratory System

Brian J. Whipp, PhD

Susan A. Ward, MA, DPhil

Brian J. Whipp passed away before the publication of this book.

Introduction

History consists of the who, what, why, where, and how of events. Historians provide a filter regarding which events are considered and in what depth. Many more scientists have contributed to its conceptual and technical advances than is possible to include, and many of those included justify more extensive assessment, especially those relating to environmental influences and impaired systemic functioning (table 10.1). For convenience, this chapter is structured into the following categories: pulmonary mechanics, pulmonary gas exchange, ventilatory control, and the perception of breathing.

Pulmonary Mechanics

The study of pulmonary mechanics in the 20th century began auspiciously with Aron's description of his daring puncturing of the intrapleural space of a healthy subject to determine its pressure (11). Although Carl Ludwig (1816-1895) (158) had successfully accomplished this in experimental animals a half-century earlier, it had been thought inappropriate for studies on humans. As early as 1878, Luigi Luciani (1840-1919) at the University of Parma (who had previously worked with Ludwig in Leipzig) used the pressure changes in the highly compliant esophagus to estimate the intrapleural pressure (157). However, the method was not shown to be a valid and reliable index in humans until Hermanus Johannes Buytendijk's doctoral studies at the University of Groningen (40). This proved to be of significant practical import for understanding the static mechanical properties of the lungs in situ and also of the chest wall (i.e., when coupled with the thoracic recoil pressure expressed at the shuttered airway opening with the respiratory muscles relaxed) (5). The static pressure–volume relationships of the lung and chest wall under these conditions allowed the elastic components of the work of breathing to be determined. The work done in overcoming the resistances to air and tissue flow could be established from the deviation from the static pressure–volume relationships during the breathing cycle, although the work done in gas compression or chest wall distortion could not. This could be clearly characterized in the Campbell diagram, the useful graphical display of lung volume against transthoracic pressure by Moran Campbell (1925-2004) and colleagues at McMaster University in Canada (41) [which differs somewhat from that proposed by Herman Rahn (1912-1990) and colleagues (209)]. Although the practical details of these measurements were developed by numerous subsequent investigators (with an insightful review and conceptual critique by Agostoni and Mead (5) in 1964), the groundwork for understanding the pressure–volume relationships, the forces generating air flow, and the consequent demands of the work of breathing had actually been outlined as early as the 1920s by Fritz Rohrer (1888-1926) (219) and von Neergaard and Wirtz (242) in Basel. Also, the development of whole-body plethysmography by Arthur DuBois and colleagues at the University of Pennsylvania in the United States (74) and by Jere (Jeremiah) Mead (1920-2009) at the Harvard School of Public Health (176) subsequently allowed absolute thoracic gas volume and airway resistance to be determined.

Work of Breathing

The determinants of the thoracic volume change and consequent air flow generation have important implications for both the pattern and work of breathing during exercise: A given ventilation ($\dot{V}_E$) achieved with large tidal volume (V_T) and low breathing frequency (f_B) incurs a high elastic cost, whereas $\dot{V}_E$ when achieved with a low V_T and high f_B incurs a high flow-resistive cost. Rohrer (218) importantly estimated the optimum

Table 10.1 Key 20th and 21st Century Advances in the Respiratory Physiology of Exercise

Year	Investigators	Reference	Significant advance
1900	Aron	11	Described direct measurement of intrapleural pressure in a healthy human.
1905	Haldane and Priestley	103	Stressed the importance of respiratory center PCO_2 in the control of the exercise hyperpnea.
1910	Krogh	146	Ruled out O_2 secretion as a determinant of pulmonary O_2 exchange.
1913	Krogh and Lindhard	147	Cortical irradiation as a mediator of the control of the exercise hyperpnea.
1918	Liljestrand	155	Respiratory muscle O_2 consumption increases with concave-upward curvilinearity as a function of ventilation.
1921	Winterstein	267	Stressed the importance of the respiratory center $[H^+]$ in ventilatory control.
1925	Rohrer	218	Modeled breathing pattern optimization using minimum work criterion.
1927	Douglas	71	Alveolar ventilation responded in proportion to CO_2 output during a dietary-induced increase in the respiratory quotient in a resting subject.
1932	Herxheimer and Kost	112	Showed that ventilation was highly correlated with CO_2 output during exercise transients.
1938	Enghoff	83	Modified the Bohr equation for practical determination of physiological dead space.
1946	Asmussen and Nielsen	15	Peripheral chemoreceptor contribution to the proportionally greater increase in ventilation at high work rates.
1946	Lilienthal et al.	156	Reported exercise-induced arterial hypoxemia in normal subjects at high work rates.
1949	Rahn; Riley and Cournand	207, 215	Developed the concepts of mean alveolar gas and ideal alveolar gas, respectively.
1958	Hyatt et al.	125	Established flow-volume and isovolume pressure–flow relationships for understanding the determinants of air flow and flow limitation.
1959	Milic-Emili and Petit	181	Determined that the optimal frequency of breathing in exercise occurred at the value that provided the minimum work of breathing.
1960	Yamamoto and Edwards	269	Introduced an oscillations (of arterial PCO_2-H^+) hypothesis for the control of the exercise hyperpnea.
1963	Campbell and Howell	42	Proposed length–tension inappropriateness as a dyspneagenic mediator.
1963	Kao	133	Used cross-circulation studies to suggest peripheral neurogenesis as the dominant control mechanism of the exercise hyperpnea.

1966	Farhi	84	Detailed consideration of pulmonary inert-gas transfer as related to ventilation-to-perfusion relationships.
1966	Shephard	224	Estimated respiratory muscle O_2 consumption in exhaustive exercise.
1967	Grodins	99	Applied engineering control theory to modeling the exercise hyperpnea.
1967	Macklem and Mead; Pride et al.	167, 206	Developed concepts of equal pressure point and choke point for understanding mechanisms of expiratory flow limitation.
1970	Guz et al.	100	Showed that bilateral vago-glossopharyngeal nerve blockade in conscious humans could ameliorate dyspnea.
1971	Grimby et al.	97	Demonstrated expiratory flow limitation during high-intensity exercise in normal subjects with high aerobic capacity.
1972	Clark and von Euler	48	Analyzed components of the pattern of breathing in the context of central nervous system timing and drive correlates.
1972	McCloskey and Mitchell	164	Demonstrated the importance of small-diameter muscle afferents in the control of the exercise hyperpnea.
1973	Beaver et al.	29	Formulated a breath-by-breath algorithm for real-time determination and display of ventilatory and pulmonary gas exchange responses during exercise.
1974	Wagner et al.	244	Developed the multiple inert gas elimination technique for characterizing the components of pulmonary gas exchange.
1974	Wasserman et al.	253	Proposed cardiodynamic control of the exercise hyperpnea.
1975	Wasserman et al.	254	Established the importance of the carotid bodies in mediating the compensatory hyperventilation for exercise metabolic acidosis.
1977	Casaburi et al.	44	Demonstrated the close dynamic coupling of ventilation to CO_2 output using sinusoidal exercise over a range of work-rate forcing frequencies.
1977	Gledhill et al.	94	Described the ventilation-perfusion distribution to become less uniform during exercise.
1980	Eldridge and Gill-Kumar	79	Demonstrated the importance of neurogenic mediation of the non-steady-state phase of the exercise hyperpnea.
1983	Killian and Campbell	138	Advocated a primary role for inspiratory muscle force in exertional dyspnea.
1987	Kaufman and Rybicki	135	Identified mechanical and humoral stimuli for group III and IV muscle afferent activation in exercise.
1987	Poon et al.	203	Designed a servo-controlled inspiratory-assist device for the study of ventilatory control in exercise.
1989	Ward and Whipp	247	Used an isopneic method to show that the carotid chemoreceptors contribute to exercise dyspnea beyond any effect of ventilation per se.
1993	Shea et al.	223	Concluded, based on studies in patients with congenital central hypoventilation syndrome, that central chemoreceptors do not contribute to ventilatory control in moderate exercise.
1997	Hopkins et al.	119	Provided evidence for alveolar-capillary stress failure during high-intensity exercise in humans.

(continued)

Table 10.1 *(continued)*

Year	Investigators	Reference	Significant advance
2002	Babcock et al.	20	Definitively demonstrated respiratory muscle fatigue in exercise.
2004	Aliverti et al.	7	Used an optoelectronic device for remote breath-by-breath sensing of thoracic volume and lung gas-store changes during exercise.
2004	Haouzi et al.	110	Demonstrated the importance of peripheral vascular conductance or tissue pressure in the control of the exercise hyperpnea.

f_B in 1925 based on minimum-work considerations using a simple, linear, lumped-parameter model. The model predictions were consonant with those observed in normal resting subjects. The model assumptions, such as constant inspiratory flow, were subsequently refined, initially by Arthur Otis, Wallace Fenn (1893-1971), and Herman Rahn ["or any permutation you may prefer," as Leon Farhi (1923-2003) discerningly judged (85, p. 1567)] at the University of Rochester's department of physiology and vital economics, by including nonlinear resistance and sinusoidal air flow. The optimum f_B during exercise was again consistent with minimum-work considerations (192). Mead, however, found an even better prediction of exercise f_B if the minimum average respiratory muscle force was considered a major determinant, that is consistent with a minimum energy cost of the hyperpnea (175). Kieran Killian and colleagues at McMaster University later suggested the interesting possibility that minimizing inspiratory muscle force may also minimize the associated respiratory sensation (140). This was subsequently developed further by Neil Cherniack (1931-2009) at Case Western Reserve University in Cleveland, Ohio. (46).

Josef Milic-Emili (figure 10.1), who provided compelling work on this and other topics in respiratory mechanics, was stimulated by study and collaboration with investigators such as Rodolfo Margaria, Jere Mead, David Bates, and Peter Macklem. His wide-ranging work on respiratory mechanics has been consistently consequential and often definitive. For example, in 1959 he and Petit (181) determined the work rate of breathing over a wide f_B range at several whole-body work rates. The minimum work rate of breathing occurred at progressively higher frequencies as alveolar ventilation ($\dot{V}_A$) increased, but in each case it closely matched that chosen spontaneously—although Mead, in the following year (175), proposed that the shape around the minimum may be appreciably broader, which may help explain the highly variable intersubject contribution of f_B to the exercise hyperpnea.

Breathing Pattern

Among the many insights provided in the 1930s, 1940s, and 1950s—by what might be termed the Scandinavian school of exercise physiologists, including Erling Asmussen (1907-1991), Erik Howhü-Christensen (1904-1996), and Marius Nielsen (1903-2000)—was the demonstration that the V_T increase during exercise normally encroached on both inspiratory and expiratory reserve volumes and that neither total lung capacity nor residual volume changed appreciably with exercise (e.g., 13). Furthermore, V_T changes dominated the $\dot{V}_E$–V_T relationship as WR increased, to a high fraction of the subject's vital capacity.

The typically linear $\dot{V}_E$–V_T relationship was shown to evidence a positive V_T intercept (e.g., 115, 180), a region that has been termed range 1 (48), with f_B consequently increasing in a quasihyperbolic fashion (261). At higher levels, the $\dot{V}_E$ response becomes dominated (in some cases exclusively) by f_B changes (range 2). The transition between these ranges is thought to reflect mechanical factors related to the high elastic work of

Figure 10.1 Josef Milic-Emili (left) with Rodolfo Margaria (right) in Milan in 1974.

Reprinted, in modified form, from J. Mead, 1996, Mechanics of lung and chest wall. In *Respiratory physiology: People and ideas,* edited by J.B. West (Bethesda, MD: American Physiological Society), 201. With permission of American Physiological Society.

breathing at a high lung volume or a critical lung volume threshold for reflex activation of vagal pulmonary mechanoreceptors (e.g., 48, 261). In some subjects f_B can manifest such a striking further increase, with impending exercise limitation, that V_T actually decreases (range 3).

In the 1970s, considerations of the breathing pattern during exercise were extended to the inspiratory and expiratory components (e.g., 12, 172), based on the key analysis of Clark and von Euler (48). This showed, for example, that the range 1 f_B or breath duration (T_T) change is dominated by reduction of expiratory duration (T_E) and that its subsequently marked decrease in ranges 2 and 3 results from the shortening of both inspiratory duration (T_I) and T_E. This approach allowed ventilatory response patterns to be viewed more mechanistically with respect to what have been termed inspiratory drive and timing components (48). The $\dot{V}_E$ response, characterized as $V_T/T_I \cdot T_I/T_T$, yields V_T/T_I as the mean inspiratory flow (with implications for optimizing breathing pattern) and T_I/T_T as the inspiratory duty cycle (with implications for the neural control of breathing) (183). The V_T/T_I increase during exercise was shown to be qualitatively similar to that of $\dot{V}_E$, with T_I/T_T typically increasing in normal subjects from approximately 0.4 at rest to approximately 0.5 at maximum exercise (e.g., 12, 172), reflecting the greater proportional contribution of T_E to the f_B response.

Air Flow Limitation and Dynamic Hyperinflation

Robert Hyatt (figure 10.2) and colleagues at the department of physiology and biophysics and the department of internal medicine at the Mayo Clinic played a major role in advancing the understanding of pulmonary mechanical functioning during exercise by considering not only the volume–time and flow–time characteristics of the breath but also the flow–volume and pressure–flow relationships. Hyatt had been a student fellow in the laboratory of George H. Whipple (Nobel laureate in physiology or medicine, 1934) while a medical student at the University of Rochester and was subsequently a research fellow with Donald Fry at the National Heart Institute; their seminal and productive collaboration continued at the Mayo Clinic. The 1958 paper by Hyatt and colleagues (125), which can justifiably be termed a classic, provided the impetus for numerous subsequent papers to address the profile of the spontaneously achieved flow–volume relationship during exercise in the context of that attained, most commonly at rest, with maximum volitional effort. This paper showed that, even at maximum exercise in normal subjects, neither inspiratory and expiratory air flow nor the volume excursions encroached on the values established for the forced maneuver (151, 190). Coupled with the demonstration that the ratio of maximum exercise $\dot{V}_E$ to maximal voluntary ventilation at rest was only approximately 60% to 70% (97, 198) (i.e., the subjects had a significant breathing reserve) (248), the observations of Hyatt and colleagues (125) were consistent with the mechanical performance of the lung not limiting maximum exercise performance. Grimby and colleagues (97), Hesser and colleagues (113), and Klas and Dempsey (142) did demonstrate, however, that subjects exhibiting high aerobic capacity could evidence flow limitation at, and close to, maximal exercise as manifest by the profile of the spontaneous expiratory flow–volume curve impacting on its maximum-effort outer envelope.

The notion of the effort-dependent and effort-independent air flow domains—for which the brilliant insights of Macklem and Mead (167) and Pride and colleagues (206) regarding the equal pressure point or choke point concepts were explanatory—was crucial to understanding this behavior. Since the work of Hyatt and colleagues (125), it was recognized that, at high lung volumes, progressively greater expiratory efforts produced progressively greater expiratory air flow. At lower volumes, progressively greater expiratory efforts did not yield further increases in volume-specific expiratory air flow, at least above a particular minimum value (figure 10.3). That particular minimum value corresponded to the transpulmonary pressure on the isovolume pressure–flow relationship above which expiratory flow became constant: the maximum effective pressure (125) (figure 10.3). The remarkable feature of the normal exercise response at which flow limitation became apparent was that the maximum effective

Figure 10.2 Bob Hyatt (1925–) enjoying a relaxing moment with Mary Ellen Wohl (formerly chief of the pulmonary division at Boston's Children's Hospital) and Dot Mead (wife of Jere Mead) in 1960.

Photo courtesy of Robert Hyatt.

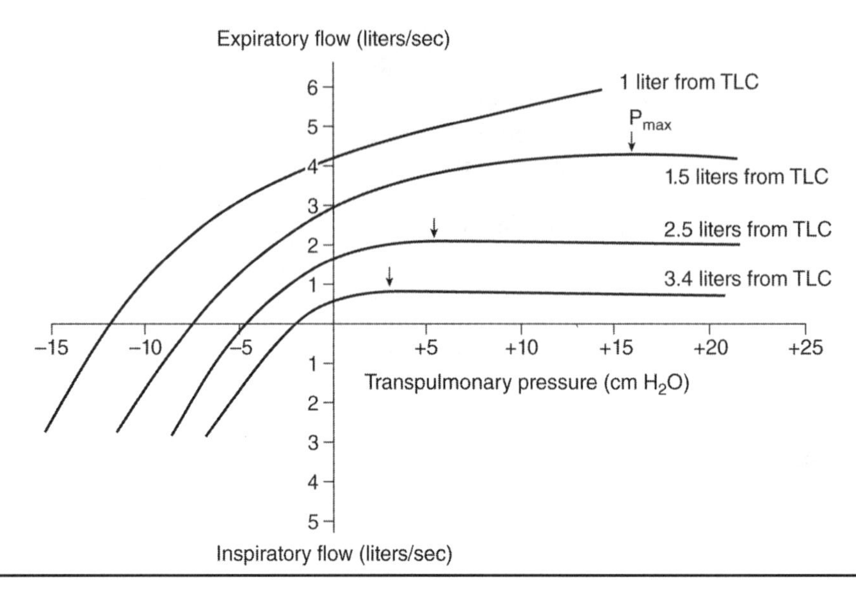

Figure 10.3 Isovolume pressure–flow curves obtained from a normal subject, each measured at a different volume, expressed as difference from total lung capacity (TLC). The transpulmonary pressures at which expiratory flow limitation was estimated to occur (P_{max}) are indicated by arrows. Note that flow limitation was not evident for the uppermost curve (1 L from TLC).

Reprinted, by permission, from S. Olafsson and R.E. Hyatt, 1969, "Ventilatory mechanics and expiratory flow limitation during exercise in normal subjects," *Journal of Clinical Investigation* 48: 564-573.

pressure was not exceeded. That is, normal subjects did not typically generate intrapleural pressures that simply increased the work of breathing, via compression of downstream airways (167, 206), but that were ineffective for increasing air flow. However, patients with chronic obstructive pulmonary disease (COPD) were subsequently shown to generate intrapleural pressures during exercise in excess of that being maximally effective (204).

Another feature of the spontaneously generated flow–volume curve during exercise, predominantly in subjects with high airway resistance or low lung recoil, was that volume-specific air flow not infrequently exceeded that on the forced maneuver at rest (98). The mechanisms proposed to contribute to this seeming anomaly included the difficulty of appropriately placing the spontaneous exercise curve with respect to the volume axis on the resting maneuver, bronchodilatation resulting from increased circulating catecholamines, the effect of gas compression dissociating changes in thoracic volume from those measured at the mouth, and the inappropriateness of the forced maneuver as a frame of reference for the spontaneous curve. Based upon Otis' conceptualization (191), Younes argued insightfully (270) that, during the forced maneuver in subjects with a wide distribution of pulmonary mechanical time constants, the fast time-constant units dominate early in expiration (i.e., at high lung volume) and that they would therefore empty at lower operating lung volumes early in the spontaneous maneuver. However, as shown as early as 1951 by Dayman (60), maximum air flow in such subjects is not achieved with maximum effort. This was clearly evident in the demonstration by Hyatt and colleagues (125) that the isovolume pressure–flow relationship exhibited negative effort dependence above the maximally effective value in such subjects (figure 10.3). Long time-constant units therefore predispose to both air flow limitation and airway closure. Consequently, as f_B increases at high WRs, end expiratory lung volume (EELV) begins to increase (e.g., 190, 270). Although this dynamic hyperinflation provides a more favorable volume for air flow generation, it predisposes to inspiratory muscle fatigue and dyspneic sensations as a result of increased lung elastance and reduced inspiratory capacity.

Interventions that reduce dynamic hyperinflation can markedly improve exercise tolerance in COPD patients, especially during constant WR tasks (e.g., 82, 189). Interestingly, Johnson and colleagues (128) demonstrated that, in otherwise normal subjects, the age-dependent reduction in lung recoil changes the normally persistent decrease of EELV (e.g., 64, 270) to a profile in which it initially decreases but then increases progressively as WR and f_B increase further, with evidence of limited expiratory air flow. The question of whether the spontaneous expiratory air flow profile during exercise is actually limited was addressed with a typically perceptive approach by Milic-Emili and colleagues (144). Rather than increasing the driving pressure for air flow by generating a more positive intrapleural pressure, they developed a device that could abruptly provide a more negative mouth pressure during exhalation. If air flow increased, the spontaneous expiration was not flow limited; if it did not increase, it was flow limited.

Oxygen Cost of Breathing

The increase in respiratory muscle power as a function of $\dot{V}_E$ during exercise was shown, notably by Margaria and Milic-Emili and their colleagues (171) among numerous others (e.g., 191), to exhibit a concave-upward curvilinearity (figure 10.4) and to increase in moderately fit subjects at near-maximum exercise to approximately 30% to 40% of that achieved on a maximal voluntary ventilation maneuver (113, 142). Respiratory muscle oxygen consumption ($\dot{V}_{RM}O_2$) was also shown to increase with a concave-upwards profile as a function of increasing $\dot{V}_E$. This was initially shown by Liljestrand in 1918 over a relatively small $\dot{V}_E$ range (155) and was subsequently shown by others over the entire range of exercise tolerance (figure 10.4), as summarized by Otis (191). Whereas $\dot{V}_{RM}O_2$ is small at resting levels of $\dot{V}_E$, it can be a significant contributor to whole-body $\dot{V}O_2$ at near-maximal levels of exercise. Shephard (224), for example, reported $\dot{V}_{RM}O_2$ to approach 0.6 L/min for a $\dot{V}_E$ in the region of 130 L/min [i.e., ~15% of the total $\dot{V}O_2$ for a subject with a maximum $\dot{V}O_2$ ($\dot{V}O_2max$) of ~4 L/min]. Similar values have been subsequently reported by Aaron and colleagues (1). However, estimating $\dot{V}_{RM}O_2$ in exercising humans is technically difficult because of its small size relative to whole-body $\dot{V}O_2$ and because the methods used to induce hyperpnea (commonly volition or altered inspired gas composition) do not provide pressure and pattern profiles that adequately reflect the actual exercise response. A further confounding influence is that if arterial PCO_2 ($PaCO_2$) is allowed to decrease in tests at rest that attempt to mimic the exercise hyperpnea, then—as shown by Karetzky and Cain in an important but unfortunately widely neglected study (134)—whole-body $\dot{V}O_2$ is increased by the alkalosis by approximately 10% per 10 mmHg reduction in $PaCO_2$. This is large compared with the actual $\dot{V}O_2$ cost of exercise ventilation and presumably accounts, in part, for the extremely wide range of published exercise $\dot{V}_{RM}O_2$ values.

There is, nonetheless, general agreement that the respiratory muscles make a progressively greater proportional contribution to whole-body $\dot{V}O_2$ as $\dot{V}_E$ increases. Is there some critical level of $\dot{V}_E$ above which further increases would notionally demand the entire increase in the body's $\dot{V}O_2$? Margaria and colleagues (171) estimated the critical level to be between 130 and 150 L/min.

Jerry Dempsey (figure 10.5), who is among the very few investigators to have led teams that made noteworthy contributions to each of the categories of respiratory investigation, addressed this issue. Dempsey had been steered into respiratory physiological research by John Rankin at the University of Wisconsin in the United States; Ellsworth Buskirk, Vladimir Fencl, and Charles Tipton are also cited as academic influences. He and his colleagues at the University of Wisconsin–Madison in the United States suggested that the critical level of $\dot{V}_E$ was not likely to be reached during exercise. They determined the ventilatory $\dot{V}O_2$ cost of

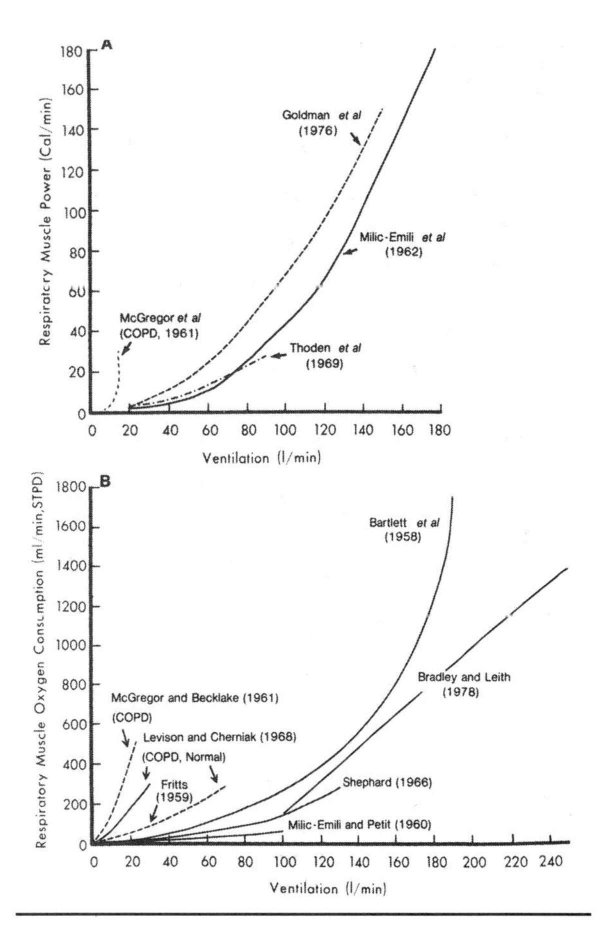

Figure 10.4 (*a*) Relationship between respiratory muscle power and ventilation in normal subjects during exercise and in patients with chronic obstructive pulmonary disease (COPD) during volitional hyperventilation. (*b*) Relationship between respiratory muscle oxygen consumption and ventilation during exercise and volitional hyperventilation in normal subjects and patients with COPD.

Reprinted from *Clinics in Chest Medicine*, Vol. 5(1), R.L. Pardy, S.N. Hussain, and P.T. Macklem, "The ventilatory pump in exercise," pgs. 35-49, copyright 1984, with permission of Elsevier.

high-intensity exercise in a meticulously designed study in which resting subjects were required to replicate the $\dot{V}_E$, breathing pattern, and EELV observed during exercise as well as the esophageal and transdiaphragmatic pressures (1). The subjects did not reach the critical level of $\dot{V}_E$, even at maximal exercise. The investigators judged by extrapolation that this critical value would be obtained only if the maximally achieved $\dot{V}_E$ were some 30% greater. The body would be expected to partition the nutrient blood flow for oxygen utilization as required by the various muscles contributing to the exercise. However, the Wisconsin group in particular provided evidence for a preferential hierarchy for blood flow ($\dot{Q}$) distribution during exercise: When the air flow demands of the respiratory muscles get sufficiently high, a proportion of their $\dot{Q}$ is provided at the expense of a compromised $\dot{Q}$ to the locomotor muscles—argued to be mediated by sympathetic reflex vasoconstriction (66, 106) (figure 10.6). In fact, when the work of breathing was reduced during high-intensity exercise by means of an inspiratory-assist device, locomotor muscle $\dot{Q}$ increased (106), the perception of limb fatigue decreased, and exercise tolerance consequently increased (107).

Respiratory Muscle Fatigue

Do the high O_2 demands of the respiratory muscles during spontaneous exercise result in their fatigue, and to what extent? The diaphragm—the dominant inspiratory muscle—was shown to be relatively resistant to fatigue, at least up to high levels of exercise, presumably as a result of its high oxidative capacity and capillarity. The conclusion, initially, was that any contribution of respiratory muscle fatigue to exercise performance was likely to be small at best (e.g., 166). However, using more refined techniques, including supramaximal bilateral phrenic nerve stimulation effects on the intrapleural and gastric pressure components of the isovolume transdiaphragmatic pressure, sustained, high-intensity, constant WR exercise was shown to induce fatigue of both diaphragm (20, 168) and abdominal muscles (235, 240)—an effect that could, to a large extent, be abolished by reducing the demands for respiratory muscle power generation via sufficient inspiratory mechanical assist (20). Interestingly, however, these effects were not evident when resting subjects volitionally reproduced the diaphragmatic work profile of the exercise (21).

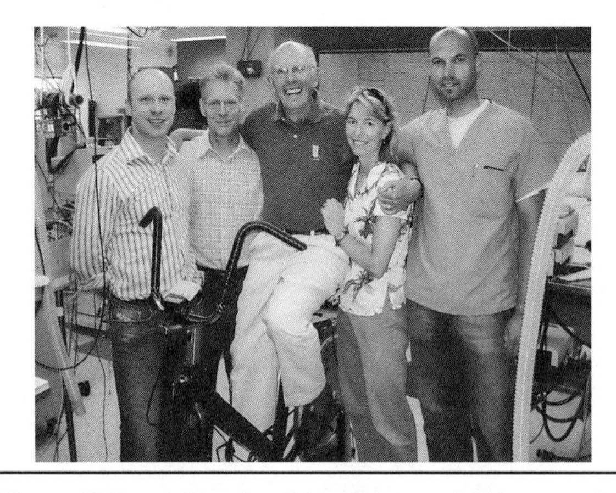

Figure 10.5 A beaming Jerry Dempsey (1938–; center) with a group of scientific collaborators in 2008 (left to right: Lee Romer, Marlowe Eldridge, Margaret Rankin, and Markus Amann) at the John Rankin Laboratory of Pulmonary Medicine at the University of Wisconsin–Madison.
Photo courtesy of Jerome A. Dempsey.

Remote Sensing of Thoracic Volume Change

The last decade of the 20th century and the first decade of the 21st have been notable for the work of Andrea Aliverti of the Politechnico di Milano and his colleagues, including the eminent Peter Macklem (1931-2011) of McGill University in Canada. They developed an optoelectronic remote-sensing technique to determine the breath-to-breath contour of thoracic volume change during exercise (7). This device uses small, strategically placed reflective discs arranged over the thorax and abdomen. The position of these discs is sensed by several television cameras linked to an automatic motion analyzer and transformed by computer algorithm into thoracic volume change. Also, when coupled with simultaneous measurement of volume and air flow changes at the mouth, this device allows the gas compression component of respiratory work and power to be determined during spontaneous breathing. However, currently, many of the mechanistic implications of such remote sensing during exercise remain to be exploited.

Pulmonary Gas Exchange

It is hard to conceive of a more significant dispute regarding pulmonary gas exchange in need of resolution than that being waged at the start of the 20th century. Was simple passive diffusion sufficient for the exchange process at the lungs, as advocated by Eduard Pflüger (1829-1910; founder of *Archiv für die Gesamte Physiologie*, the eponymous *Pflügers Archiv*) (196) in 1868 at the University of Bonn, or did the process involve active oxygen secretion, as favored by Pflüger's former pupil Carl Ludwig, a perspective consistent with

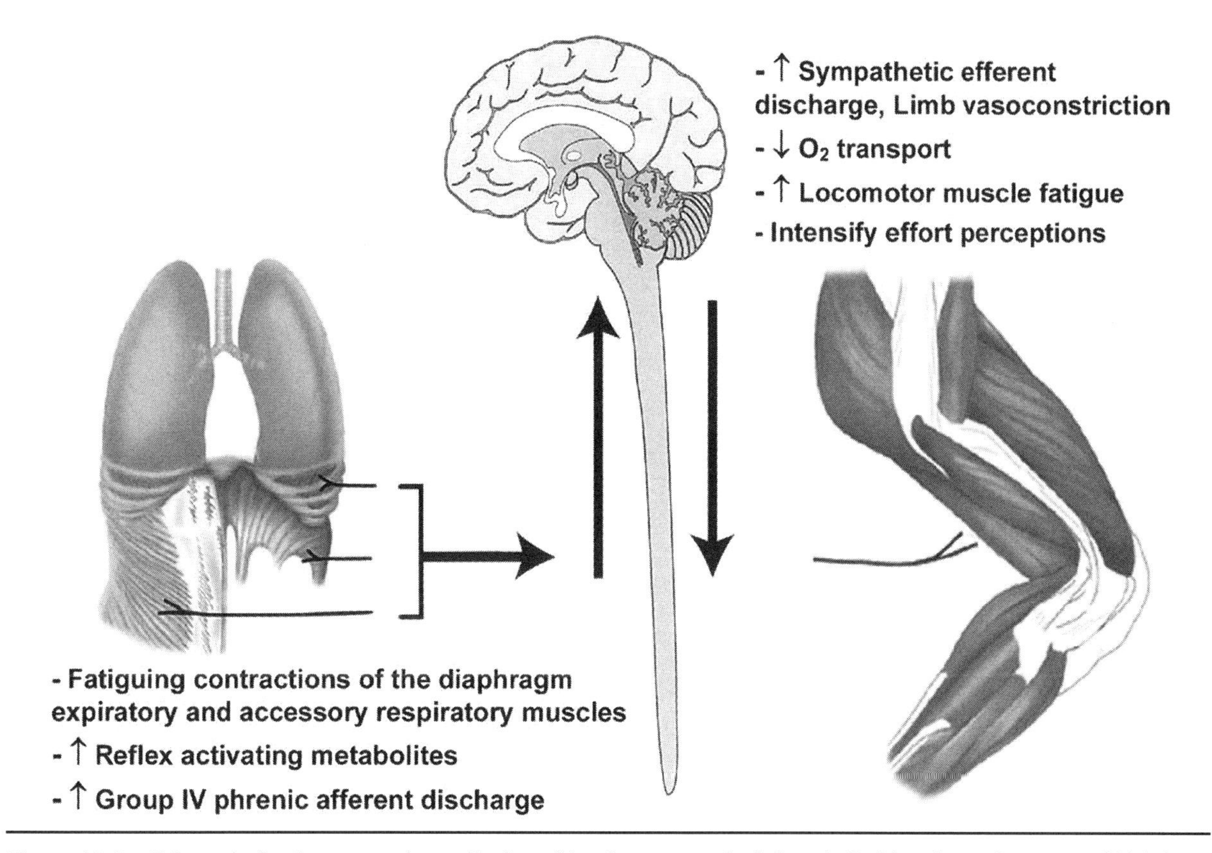

Figure 10.6 Schematic for the proposed contribution of inspiratory muscle fatigue in limiting the performance of high-intensity endurance exercise. Phrenic afferent stimulation evokes supraspinally mediated sympathetic reflex vasoconstriction in limb muscles, exacerbating limb muscle fatigue and perceptions of effort.

his longstanding interest in secretory physiology? Christian Bohr (1855-1911) at the University of Copenhagen, who had studied with Ludwig, provided seemingly strong support for oxygen secretion with evidence of arterial PO_2 (PaO_2) exceeding alveolar PO_2 (P_AO_2) (32). This was sufficient to gain the support of the eminent English physiologist John Scott Haldane (1860-1936) of Oxford's University Laboratory of Physiology (102). The uncertainty about whether PaO_2 was in fact higher than P_AO_2 [or, remarkably, even inspired PO_2 (P_IO_2)] reflected in part the technical difficulties in accurately estimating mean P_AO_2 existing at that time. By 1910, however, the issue was finally settled decisively in favor of diffusion by August Krogh (1874-1949; 1920 Nobel laureate in physiology or medicine) at the University of Copenhagen (e.g., 146), who had previously been an assistant in Bohr's laboratory. Significant contributions were made by Krogh's wife Marie.

Not only was PaO_2 not higher than P_AO_2, but concerns began to emerge that pulmonary O_2 transfer may not be an entirely efficient process and that the alveolar-to-arterial (A-a) PO_2 difference might actually be positive. Geppert and Zuntz recognized that "even if we knew the exact composition of the alveolar air, it remains to be proven that an absolute equilibrium in tension between the arterial blood and the alveolar air is established" (93, p. 195). Furthermore, influences of regional variations of ventilation to perfusion ($\dot{V}_A/\dot{Q}$) on pulmonary gas exchange were introduced by Haldane (101) and Krogh and Lindhard (149) in the early 20th century.

(A-a)PO_2 Difference

Accurate and precise quantitation of the (A-a)PO_2 difference and how it might be influenced during exercise had to await the conceptual (and virtually parallel) advances in 1949 of Hermann Rahn regarding mean alveolar air (207) and of Richard Riley (1911-2001) and André Cournand at Bellevue Hospital in New York regarding ideal alveolar air (215). The (A-a)PO_2 difference could now be considered in terms of an ideal lung compartment, wherein gas exchange was efficient, together with conceptually useful but figmentary shunt and alveolar dead space compartments. However, the constraints of the three-compartment gas exchange model, of necessity, consigned low-$\dot{V}_A/\dot{Q}$ and high-$\dot{V}_A/\dot{Q}$ units to shunt-like and dead space-like compartments, respectively. This provided a tractable means of characterizing gas exchange abnormalities, with the ideal P_AO_2 (P_ACO_2 being assigned to equal $PaCO_2$) providing a frame of reference for the directly measured PaO_2. It was only at higher WRs, however, that a widening of this (A-a)PO_2 difference could be consistently demonstrated (16, 130, 156, 250). That is, despite P_AO_2 being increased consequent to the compensatory hyperventilation associated with metabolic (largely lactic) acidosis, PaO_2 did not increase pari passu. Indeed, the work of Lilienthal, Riley, and colleagues in 1946 at the Naval School of Aviation Medicine in Pensacola in United States (156) showed that PaO_2 could actually decrease below resting levels at high WRs in some subjects; this phenomenon was later termed exercise-induced arterial hypoxemia (EIAH) (65). This was subsequently confirmed in the 1950s and 1960s by investigators such as Holmgren and Linderholm in Stockholm (117) and Doll, Keul, and colleagues in Freiburg (70). Furthermore, Shepherd (225) used computer modeling to demonstrate that a compromised pulmonary O_2 diffusing capacity (D_LO_2) could predispose to a decrease of arterial O_2 saturation at high WRs, especially in subjects with a high $\dot{V}O_2max$. It was not until the latter part of the century that these predictions were confirmed and that the profile of change of (A-a)PO_2 difference during exercise could be deconvoluted into its contributing components, notably by the seminal studies by Peter Wagner (figure 10.7), John West, and their colleagues at the University of California, San Diego (213, 244) and by Dempsey and colleagues at the University of Wisconsin (65, 94). Wagner's initial interest in the nonuniform distribution of ventilation and blood flow in the lung was fostered during his study period with John Read while still a medical student at Sydney University. His subsequent collaboration with John West at the University of California, San Diego proved to be remarkably productive: The development and application of the multiple inert gas elimination technique (MIGET) to characterize $\dot{V}_A/\dot{Q}$ distribution in health and disease (244) was judged to be primus inter pares.

The late 1940s and early 1950s also saw the emergence of two innovative and influential developments that provided the first quantitative, graphically based attempts at analyzing the (A-a)PO_2 difference: the Rahn-Fenn O_2-CO_2 diagram (208) and the Riley-Cournand four-quadrant diagram (216). By the early 1970s, the consensus was that although diffusion limitation had the potential to contribute to a widened (A-a)PO_2 difference at high WRs, right-to-left shunt (especially in combination with a reduced mixed venous PO_2) was thought to make the major contribution at lower WRs. This had been suggested by Lilienthal and colleagues (156) as early as 1946.

Ventilation Perfusion Distribution

A major role for $\dot{V}_A/\dot{Q}$ maldistribution during exercise seemed to have been ruled out when tracer-gas techniques showed that the topographic distributions of $\dot{V}_A$ and $\dot{Q}$ were each improved during exercise (35, 258) with the (A-a)PN_2 difference being effectively abolished (22). This view was comprehensively turned around with the development of MIGET by Peter Wagner, Herbert Salzman, and John West (244) in the 1970s. This was a beautifully conceived advance that evolved from the earlier insightful work of Seymour Kety at the University of Pennsylvania in Philadelphia

Figure 10.7 Peter Wagner (1944–) of the division of physiology in the department of medicine at the University of California, San Diego in 2002. Wagner is an important contributor to the understanding of the determinants and limitations of gas exchange in the lung and locomotor muscles during exercise.
Photo courtesy of Peter D. Wagner.

(137) and Leon Farhi at the State University of New York in Buffalo (84) regarding the determinants and consequences of the pulmonary exchange of inert gases. Thus, if a mixture of inert gases (dissolved in saline) with a wide range of partition coefficient (λ) is infused into the mixed-venous blood until a steady state is achieved, the excretion of each species into alveolar gas [i.e., the ratio of its arterial to mixed-venous ($\bar{v}$) partial pressure] was shown to be determined exclusively by λ and the $\dot{V}_A/\dot{Q}$ of the involved lung units. Using iterative digital computation, a functionally continuous characterization of the $\dot{V}_A/\dot{Q}$ distribution could be determined from the collective retention and excretion data—a concept that provoked intense debate. The method provided the true right-to-left shunt from those regions for which $\dot{V}_A/\dot{Q} = 0$. Furthermore, as the PaO_2 predicted from a particular $\dot{V}_A/\dot{Q}$ shunt profile and any hypoventilation (assessed by measuring $PaCO_2$) can be established, any shortfall relative to the measured PaO_2 could be ascribed to diffusion impairment.

In striking contrast to earlier findings, both the Wagner and Dempsey groups demonstrated more rather than less $\dot{V}_A/\dot{Q}$ maldistribution during exercise with the MIGET technique (figure 10.8). This constituted the major component of the widened $(A\text{-}a)PO_2$ difference, and the remainder was ascribable to shunt; essentially, none was ascribable to diffusion impairment (e.g., 94, 243). The causes of this wider $\dot{V}_A/\dot{Q}$ maldistribution during exercise remain largely unresolved, although Dempsey and Wagner, in their influential 1999 review

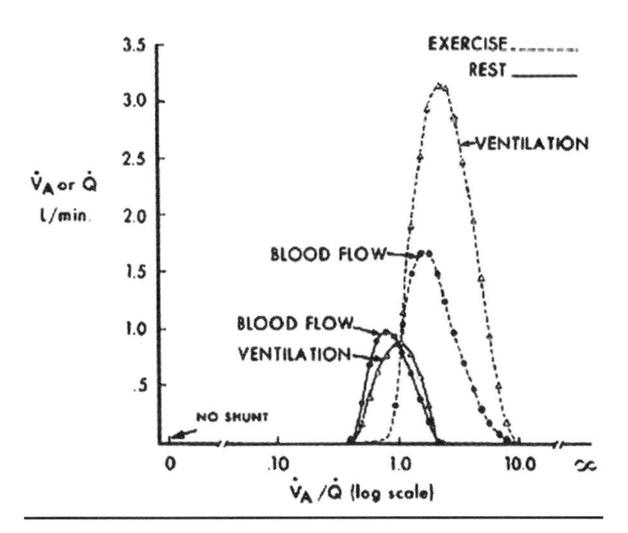

Figure 10.8 Profiles of alveolar ventilation ($\dot{V}_A$) and perfusion ($\dot{Q}$) in a normal, healthy subject at rest (——) and during moderate-intensity exercise (- - -).

(67), considered the following: altered airway or vascular tone by mediators such as histamine; excessive airway secretions induced by high airway flows; and interstitial edema, consequent to the high intravascular pressures associated with a high $\dot{Q}$, influencing regional alveolar expansion or vascular resistance.

Diffusion Limitation

Wagner and colleagues (243) also presented evidence that diffusion limitation could contribute to the $(A\text{-}a)PO_2$ difference at high WRs. This was consistent with the supposition of Dempsey and colleagues in 1984 (65) to account for EIAH demonstrable in some, but not all, highly fit endurance athletes at or near $\dot{V}O_2max$. They ascribed the diffusion component to a critical shortening of pulmonary capillary transit time. Although the PaO_2 decrement yielded only small reductions in arterial O_2 saturation, it proved to be an important contributor to exercise tolerance: When the hypoxemia was prevented by appropriately increased P_IO_2, exercise tolerance increased (205). Deserving of mention at this juncture is the perceptive analysis of Robert Johnson and colleagues in 1965 at the University of Texas Southwestern Medical School in Dallas in the United States (129) who, using a combination of modeling and direct experimentation (albeit on patients with alveolar capillary block due to diffuse interstitial infiltrates), demonstrated the importance of pulmonary capillary transit time and its distribution as a potential limiter of $\dot{V}O_2max$ through EIAH.

The conceptually elegant $D/\dot{Q} \cdot \beta$ analysis (D, $\dot{Q}$, and β being D_LO_2, pulmonary blood flow, and the slope of the O_2 dissociation curve over the region of interest, respectively) developed in the early 1980s by Johannes Piiper (1882-1973) and Peter Scheid at the University of Göttingen (200) and subsequently considered by investigators such as Michael Hughes at Imperial College in London (122), provided greater insight into such considerations. For a given initial driving pressure ($P_AO_2 - P_{\bar{v}}O_2$, where $P_{\bar{v}}O_2$ is the mixed-venous PO_2), this composite equilibrium coefficient term determines whether diffusion equilibrium will be attained. Interestingly, despite D_LO_2, $\dot{Q}$, and β each increasing with exercise, the equilibrium coefficient actually decreases, and to a greater extent in highly fit endurance athletes. This is because of the higher $\dot{Q}$ (and therefore reduced transit time) coupled with the greater peripheral O_2 extraction causing O_2 exchange to occur over a steeper region of the O_2 dissociation curve. Dempsey and Wagner emphasized that this predisposition to diffusion limitation and EIAH in such highly fit subjects would be compounded if exercise-induced interstitial edema

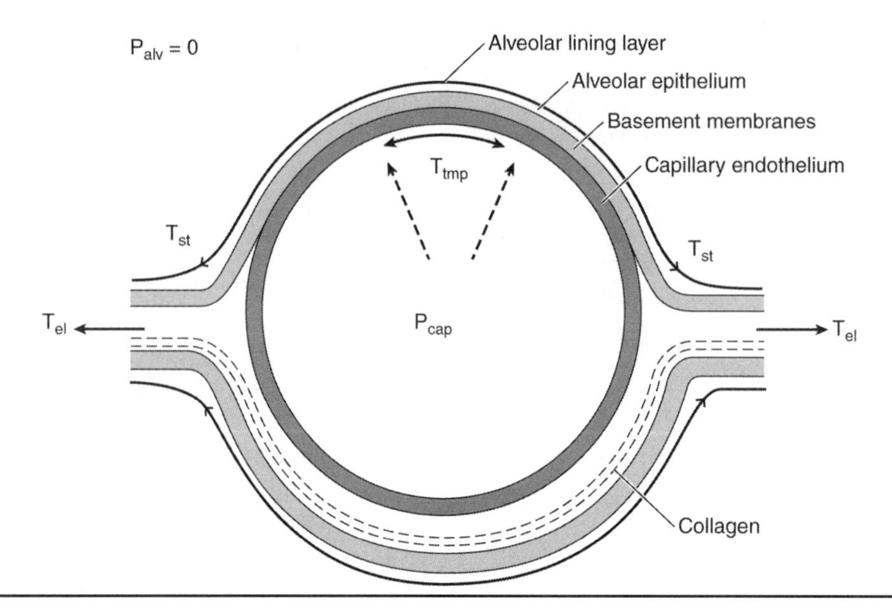

Figure 10.9 Interaction between forces acting on the pulmonary capillary-alveolar membrane. Elevated capillary transmural pressure [difference between capillary pressure (P_{cap}) and alveolar pressure (P_{alv})] increases circumferential tension in the capillary wall (T_{tmp}). Lung inflation increases longitudinal alveolar-wall tension (T_{el}). Alveolar lining layer surface tension (T_{st}) assists capillary patency.

Reprinted from J.B. West et al., 1991, "Stress failure in pulmonary capillaries," *Journal of Applied Physiology* 70: 1731-1742. With permission of American Physiological Society.

were to develop (e.g., 163, 257), thereby constraining the D_LO_2 increase (67). Any disruption of the pulmonary capillary interface (John West's stress failure concept; figure 10.9) (259) at high WRs would also be contributory (119, 257).

Physiological Dead Space

The original Bohr equation for determining the dead space of the lung (V_D) suffered in practice from the imprecision of estimating a truly mean alveolar PCO_2 ($P_{\bar{A}}CO_2$). This led to considerable debate early in the 20th century, most notably between Krogh and Haldane, about how V_D changed with exercise. On one hand, using the discrete sampling of gas from a sharp expiration at the end of a normal inspiration to estimate $P_{\bar{A}}CO_2$, Haldane and Priestley (103) concluded that V_D increased with WR. In contrast, Krogh and Lindhard (148), employing a single-breath hydrogen technique, argued for no change. Recognizing that $P_{\bar{A}}CO_2$ fluctuated throughout the respiratory cycle, Krogh and Lindhard were properly concerned that "the whole problem about the average composition of the alveolar air turns upon the correct determination of the dead space" (148, p. 437). But, of course, the reverse is equally true! The problem of how to estimate the time-average $P_{\bar{A}}CO_2$ had to await the technique of DuBois and colleagues in 1952 (75).

Enter Henrik Enghoff! In 1938, he proposed the decisive step of replacing $P_{\bar{A}}CO_2$ with the directly sampled $PaCO_2$ to determine V_D (or the "volumen inefficax") (83). This modified Bohr-Enghoff equation was embraced by basic and clinical investigators alike. By the 1960s, it was generally agreed that V_D increases during exercise consequent to the progressive lung expansion. This is a consequence of the compliance of the conducting airways being appreciably less than that of the gas exchange units, such that the dead space fraction of the breath (V_D/V_T) decreases (e.g., 130, 250). The concept of what constitutes anatomical or series dead space was critically re-examined by Ewald Weibel's scrupulous morphometric analysis of the lung, along with its role in O_2 exchange, at the University of Berne in Switzerland in the 1960s (255).

In the 1980s, Gordon Cumming at the Midhurst Medical Research Institute in the United Kingdom developed an ingenious extension of the Fowler single-breath technique (54) in which the cumulative volume of expired CO_2 (VCO_2) is determined throughout the exhalation as a function of the cumulative expired volume (V_T) (i.e., a volume-weighted integral of the expired PCO_2 profile). The resulting lag-linear VCO_2–V_T relationship yielded a V_T intercept equivalent to the series dead space with a slope of $P_{\bar{A}}CO_2$. Norman Lamarra and colleagues (150, 260) used this technique to determine the breath-by-breath profiles of the series dead space and $P_{\bar{A}}CO_2$ during a range of exercise formats. They showed that end tidal PCO_2 ($P_{ET}CO_2$) can exceed both $P_{\bar{A}}CO_2$ and $PaCO_2$ by 6 mmHg or more at high WRs.

Aware that $P_{ET}CO_2$ does not provide an adequate estimate of $P_{\bar{A}}CO_2$, especially during exercise, Norman Jones and colleagues (131) developed an empirical relationship to predict $P_{\bar{A}}CO_2$ from $P_{ET}CO_2$ and V_T in normal subjects. This provided a reasonable group-average estimate; the individual-subject estimates, however, varied by up to an unacceptably large ±4 mmHg. Similar variations were also shown in studies utilizing the time-average $P_{\bar{A}}CO_2$ provided by the DuBois intrabreath reconstruction (217), the midexpiratory P_ACO_2 value (8), and the flow-weighted $P_{\bar{A}}CO_2$ estimate (260). If $PaCO_2$ is needed, it must be measured therefore—although samples from the dorsum of the heated hand, with appropriately induced local hyperemia, have been shown to provide adequate estimation (88).

Kinetic Profiles

Although Krogh and Lindhard (147), Hill and Lupton (116), Hansen (104), Margaria and colleagues (170), and Henry and DeMoor (111) had considered inferences that could be drawn from the $\dot{V}O_2$ time course during the non-steady-state phase of exercise in normal subjects in the first half of the 20th century [as had Meakins and Long (178) in patients with cardiovascular disease], Whipp and Wasserman published a paper in 1972 (263) that used breath-to-breath determination and display. This heralded a larger-scale, and still-continuing, interest in pulmonary gas exchange kinetics during exercise. The computer-based technique (29)—a considerably more extensive version of a method adumbrated by Auchincloss and colleagues (19)—developed from a collaboration between William Beaver (a physicist-engineer at Varian Associates in Palo Alto, California) and Karlman Wasserman [initially at Stanford University and then at the University of California, Los Angeles-affiliated Harbor General Hospital in Torrance, California (now the Harbor-UCLA Medical Center), to which he moved in 1967]. This, with its subsequent fine-tuning, allowed other investigators—including Richard Hughson, Dag Linnarson, Paolo Cerretelli, Pietro di Prampero, Thomas Barstow, David Poole, Bruno Grassi, and Andy Jones and their colleagues, among many others—to productively explore both the physiological mechanisms and the performance consequences of the components of the pulmonary gas exchange response transients. Karlman Wasserman (figure 10.10) and colleagues especially pioneered its now large-scale application to clinical investigation (248). Wasserman's PhD research, under the tutelage of Hyman Mayerson at Tulane University, was on capillary permeability to macromolecules; he also gained his MD at that institution while a faculty member in its department of physiology. An internship in internal medicine at Johns Hopkins and a fellowship with Julius Comroe (1911-1984) at the Cardiovascular Research Institute at the University of California, San Francisco preceded faculty appointments at Stanford University and then, for the remainder of his career, at the Harbor General Hospital in Torrance, California (now the Harbor-UCLA Medical Center). This was notable for both its research productivity and for the number of scientists who trained in his division who went on to make significant contributions to basic and clinical investigations.

Control of Ventilation

The understanding of the humoral control of $\dot{V}_E$ at the beginning of the 20th century was heavily influenced by two sets of investigations. One set is those of the Swiss physiologist Johann Friedrich Miescher-Rusch (1844-1895), who demonstrated in humans that it took only small increases in inspired PCO_2 to provoke significant ventilatory stimulation whereas much larger P_IO_2 decrements were required before there was a ventilatory response. Miescher-Rusch famously wrote that "over the oxygen supply of the body, carbon dioxide spreads its protecting wings" (179). The second set of investigations is those of Léon Fredericq (1851-1935) of the University of Liege Institute of Physiology in Belgium (89), who used carotid–carotid arterial and

Figure 10.10 Karlman Wasserman (1927–; left) and William Beaver (right), pioneers in developing and utilizing breath-by-breath ventilatory and pulmonary gas exchange measurements to study exercise transients, photographed in 2002 with their colleague Xing-Guo Sun in Wasserman's office in the division of respiratory and critical care physiology and medicine at the Harbor-UCLA Medical Center, Torrance, California, United States.

Photo courtesy of Xing-Guo Sun and Karlman Wasserman.

jugular–jugular venous flow transpositions in dogs (nominally A and B) to show that tracheal occlusion in dog A induced hyperventilation in dog B and that $\dot{V}_E$ was reduced in dog A as a consequence. Similarly, when P_ACO_2 and $PaCO_2$ were reduced by hyperventilation of dog A, hypopnea resulted in dog B. The PCO_2 in the arterial inflow to the brain, and presumably the respiratory center, was therefore an (the?) important stimulus to breathing.

These concepts were extended by John Scott Haldane (1860-1936) in Oxford. As a result of many varied experiments on human subjects, he along with his student J.G. Priestley asserted in 1905 that "the regulation of the rate of alveolar ventilation in breathing depends, under normal conditions, exclusively on the CO_2 pressure in the respiratory centre" (103, p. 266). But how soon they would regret the use of the term *exclusively*. Among the findings demanding a correction were those from Haldane's own laboratory. A.E. Boycott in 1908 (34) and C.G. Douglas in 1909 (72) showed, respectively, that P_ACO_2 actually decreased during the hyperpnea of sufficient hypoxia or of heavy exercise. Haldane then modified his earlier assertion to state that "the hyperpnoea of muscular exertion is due solely to rise of CO_2 pressure in the centre if the work is moderate" (72, p. 440)—despite the failure to consistently demonstrate an increase in P_ACO_2 during moderate exercise.

Hans Winterstein (1879-1963) attempted to reconcile the available evidence in 1911 by proposing, while at the University of Rostock, that the ventilatory drive derived from the arterial hydrogen ion concentration ($[H^+a]$) (266). However, because this was shown to be incompatible with the alkalemia resulting from hypoxic hyperpnea, Winterstein subsequently modified his theory to shift the site of the H^+ stimulus from blood to that of the cells of the respiratory centers (267). These investigators, however, were significantly disadvantaged: Neither the function of the aortic and carotid bodies as peripheral chemoreceptors nor the function of the central chemoreceptors as distinct from the pontomedullary respiratory complex had been determined.

Ventilatory–Pulmonary Gas Exchange Coupling

The issue of whether the hyperpnea of moderate-intensity exercise was attributable, in whole or in significant part, to sensors cephalad to the lung responding to $PaCO_2$-H^+a was, in large part, resolved—as arterial blood sampling during exercise became more common [having been undertaken via radial artery puncture as early as 1923 by Barr and Himwich (28)]—by the demonstration that, in the steady state, $PaCO_2$ during dynamic exercise under laboratory conditions was typically regulated at or close to normal resting levels (e.g., 105, 250). In contrast, hypocapnic alkalemia resulted during sustained isometric exercise in humans (201) and in dynamic exercise in some experimental animals, such as the pony (194).

The challenge remained to understand the relationships between $\dot{V}_E$ and the indices of metabolic rate, both as determinants of $PaCO_2$ regulation and as clues about the control mechanisms. This was determined by C.G. Douglas (1882-1963) (71), who ingested sugar to increase his respiratory quotient. He made the crucial discovery that $\dot{V}_A$ increased in proportion to the increase in $\dot{V}CO_2$. Norman Jones subsequently showed that when the steady-state respiratory quotient at a given WR was decreased as a result of endurance training, there was a close correlation between the decrease of $\dot{V}CO_2$ and that of $\dot{V}_E$ but not between the $\dot{V}_E$ and the $\dot{V}O_2$ responses (236). Consequently, if there was a control link between $\dot{V}_E$ and metabolic rate, it seemed to be mediated, somehow, via a CO_2-linked process. But, as for any control system, clues about ventilatory control lie in the characteristics of its transient, rather than steady-state, response to a stressor. The interrelationships among $\dot{V}_E$, $\dot{V}CO_2$, and $\dot{V}O_2$ throughout an exercise transient (in this case, the recovery phase) were addressed by Herxheimer and Kost in 1928 (112). Their results were in agreement with those obtained in the steady state of moderate exercise: $\dot{V}_E$ changed linearly with $\dot{V}CO_2$ but not $\dot{V}O_2$.

These relationships were subsequently considered using a wide range of dynamic WR functions, including the impulse, ramp, and pseudorandom binary sequences and, perhaps most tellingly by Casaburi and colleagues (44), constant-amplitude sinusoidal WR variations over a wide range of forcing frequencies. The results demonstrated that as sinusoidal frequency increased, the amplitude of the oscillating $\dot{V}_E$ response decreased but remained highly and linearly correlated to that of the decreasing $\dot{V}CO_2$ oscillation. The relationship was appreciably poorer with $\dot{V}O_2$. At sufficiently high forcing frequencies ("sufficiently" depending on the response-time constants), the relationship extrapolated to—or very close to—the origin. That is, the $\dot{V}_E$ oscillation was abolished despite the central command mechanisms continuing to drive the force-generating muscle units over the WR oscillation and the postulated feedback mechanisms from the contracting muscles still, presumably, being operative. It seemed, therefore, that the ventilatory control system could not keep up with the amplitude demands of the task at high forcing frequencies. A rapid neurogenic mechanism should!

These transient dynamics were suggestive of a control link that was related not to the metabolic production CO_2 rate ($\dot{Q}CO_2$) but rather to its pulmonary exchange rate (262). That is, during the transient phase, pulmonary gas exchange is dissociated from that of muscle because the tissue capacitance for CO_2 of the blood and skeletal muscle is appreciably greater than that for O_2 (86, 199). In fact, during the on-transient of a WR step, a transient metabolic alkalosis is evident in both the force-generating units of the muscle (e.g., 220, 230) and its venous effluent (discussed in 248). This is a result of the net proton trapping associated with phosphocreatine splitting to regenerate the adenosine triphosphate used to fuel the contraction (136). As a result of this transient alkalosis, a component of the metabolically produced CO_2 is retained in the muscle, leading to the time constant of the $\dot{V}CO_2$ transient being typically at least 50% greater than that of muscle $\dot{Q}CO_2$ and of $\dot{V}O_2$. This suggested that the component of the metabolically produced CO_2 that does not reach the lungs for exchange does not influence the exercise $\dot{V}_E$ (262).

Although the wealth of evidence documenting the close correlation between $\dot{V}_E$ and $\dot{V}CO_2$ over a wide range of exercise formats was suggestive of CO_2-linked control, the absence of a consistently demonstrable stimulus in arterial blood was a major hurdle to elucidating the control mechanism(s). However, if the changes in mean arterial or cerebrospinal fluid PCO_2-H^+ could not explain the CO_2-linked mechanism, might the mediation be related to other CO_2-linked processes? Several, including mixed-venous chemoreceptor mediation (10, 214), were proposed in the 1960s and 1970s. However, the suggestion was undermined by the demonstration that CO_2-laden blood infused into the pulmonary artery or right atrium of anesthetized or awake dogs, respectively, did not stimulate $\dot{V}_E$ until it reached arterial chemoreceptor sites (51, 234).

In 1961, William Yamamoto (figure 10.11), working with McIver Edwards, then at the University of Pennsylvania, changed the focus of what might constitute an appropriate stimulus for a CO_2-coupled hyperpnea. They proposed a mechanism that was both elegant in its simplicity and far-reaching in its potential application (269): Might a signal be manifest from the characteristics of the intrabreath oscillation of $PaCO_2$ and arterial pH (pHa) despite mean $PaCO_2$ and pHa being unchanged? The characteristics of this oscillation, a consequence of the phasic ventilation of the alveoli, would be a direct function of $P_{\bar{V}}CO_2$. And because $P_{\bar{V}}CO_2$ is a function of the metabolic rate, the potential for a CO_2-coupled drive to breathe, independent of its mean arterial level, was both obvious and attractive.

In 1967 Band and Semple (23), at the University of London's Middlesex Hospital Medical School, used a rapidly responding pH electrode, which they developed, that was capable of tracking the pH consequence of PCO_2 variations. They demonstrated pHa oscillations that were consistent with those expected from the $PaCO_2$ variations, initially in the anesthetized cat and subsequently in humans during exercise (24). Although the magnitude of the pHa oscillation was not systematically increased with exercise, a close correlation was found between the maximum rate of change of the pHa oscillation downstroke and the magnitude of the hyperpneic response (52). But did chemoreceptors capable of appropriately sensing the oscillatory signals exist? This was initially accomplished by Thomas Hornbein (of "high altitude" fame) and colleagues at the University of Washington in the United States, who demonstrated oscillations in carotid chemoreceptor afferent-neural discharge with the period of the respiratory cycle (120).

Robert (Torrance (1923-1999), along with his student Andrew Black, at Oxford demonstrated that the apparent sensitivity of the pontomedullary respiratory complex to afferent signals—from the carotid bodies, for example was phasic rather than constant (31). Signals arriving during the ongoing inspiratory phase of the breath stimulated $\dot{V}_E$ whereas signals arriving during the expiratory phase of the breath provided little or no additional stimulation. Consequently, $\dot{V}_E$ would be stimulated if the nadir, or maximum rate of change, of the pHa oscillation arrived at the carotid bodies during inspiration whereas the same signal phase

Figure 10.11 William Yamamoto (center) in discussion with Fred Grodins (left) and Robert Gelfand in 1982 at the second Oxford Conference, titled "Modeling and Control of Breathing," held at the University of California Conference Center, Lake Arrowhead, California, United States.

Reprinted from J.A. Dempsey and B.J. Whipp, 2003, The respiratory system. In *Exercise physiology: People and ideas,* edited by C. Tipton (United Kingdom: Oxford University Press), 154. By permission of Oxford University Press, Inc.

shifted by 180° would have little or no effect. Dan Cunningham (1919-1996) and colleagues, also at Oxford, addressed this issue in humans. However, they were unable to demonstrate any consistent relationship between their estimate of the arrival time of the peak of the $PaCO_2$ oscillation at the carotid bodies and the magnitude of the exercise hyperpnea (232). This seemed to rule out a significant obligatory role for the oscillation mechanism, at least as mediated by phase coupling. This issue seemed to be subsequently clinched by Karlman Wasserman and colleagues, who demonstrated that subjects who had undergone bilateral carotid body resection (BCBR) evidenced a normal, steady-state $\dot{V}_E$ response during moderate exercise (161).

Another ingenious mechanism for CO_2-linked mediation of the exercise hyperpnea (the disequilibrium theory) was proposed by Crandall and Forster (50) at the University of Pennsylvania. They advocated that the pH measured in the blood sampled from a systemic artery might not be reflective of the pH that was sensed by the carotid bodies, the difference being the time that had elapsed between drawing the sample and its measurement. If, as was thought at the time, carbonic anhydrase was restricted to the interior of the erythrocyte, the unanalyzed equilibrium reaction between CO_2 and H_2O, and H^+ and bicarbonate (HCO_3^-) downstream of the lung not only would continue during transit to the chemoreceptors but also would continue further in the syringe sample. The chemoreceptor signal would be greater than that measured subsequently in the sampled blood. This notionally intriguing proposal, however, failed to meet the demands of a series of challenges, the major challenge being that an accessible source of carbonic anhydrase (capable of catalyzing the dehydration reaction) was located at the pulmonary capillary endothelial surface (e.g., 76). But the repeated failure of the mechanisms proposed to account for the apparent coupling between $\dot{V}_E$ and $\dot{V}CO_2$ under such a wide range of conditions during exercise did not, for many, dispel the allure of the closeness of the correlation between the variables.

Cardiodynamic Hyperpnea

Among the observations that were strikingly apparent regarding the ventilatory and gas exchange responses was that the widely recognized rapid onset of the exercise hyperpnea was not hyperventilatory in humans, as evidenced by $P_{ET}CO_2$ decreasing and the respiratory exchange ratio increasing—although, naturally, in some subjects this could be seen. The early-phase $\dot{V}_E$ increase at exercise onset (i.e., prior to the transit-delayed influence of increased tissue metabolic rate) was therefore proportional to the increase in pulmonary blood flow per se, as had been posited as early as 1913 by Krogh and Lindhard (147) and subsequently demonstrated by Saunders and colleagues (53). Wasserman and colleagues therefore questioned whether the early hyperpneic response might be mechanistically linked to primary changes in pulmonary blood flow (253) (i.e., a cardiodynamic hyperpnea with a proposed signaling mechanism downstream of the lung, proportional to the CO_2 flux to the lung). This hypothesis had much in common with Douglas' 1927 proposal that "the activity of the respiratory centre is proportional to the mass of CO_2 produced by the body and carried to the lungs" (71, p. 213). The CO_2 flux hypothesis was tested by several groups of investigators. Experiments involved CO_2 being loaded into or out of the systemic venous inflow to the lung in several animal species. Spirited debate followed, however, among those who reported that the ventilatory changes were isocapnic (e.g., 197, 252) and those who argued that the response was nothing more than could be accounted for by the animal's inhaled CO_2 response curve (e.g., 154, 212).

Wasserman's group continued to address the issue. They infused the beta adrenergic-blocking drug propranolol intravenously into exercising humans to decrease cardiac output ($\dot{Q}_T$) (249). The consequent decrease of $\dot{V}_E$, while consistent with cardiodynamism, was only transient as $\dot{V}_E$ subsequently increased back to the control level. The authors argued that this was a result of the CO_2 flux returning to normal as $P_{\bar{v}}CO_2$ increased. There was little or no change in P_ACO_2. They also stimulated $\dot{Q}_T$ with intravenous infusions of the beta agonist isoproterenol in cats. Their experiment confirmed a close correlation between the induced hyperpnea and the magnitude of the $\dot{Q}_T$ increase. As the hyperpnea could be abolished by acute hyperventilation despite a marked increase in $\dot{Q}_T$, the normal response was considered to be somehow related to CO_2. Fred Eldridge (figure 10.12), however, challenged this notion (80) by demonstrating with Gill-Kumar that isoproterenol-mediated hyperpnea was a result of direct stimulation of the carotid bodies. Eldridge had undertaken both undergraduate and medical studies at Stanford University and would subsequently join the faculty of its department of medicine. A fellowship in Ronald Christie's department at St. Bartholomew's Hospital in London led to an involvement in respiratory physiology and renewed his interest in the control of breathing. Recognizing that the question of respiratory control needed to be approached from a more neurophysiological perspective, he took a sabbatical in the physiology department at the University of Utah in the United States to develop this expertise. The subsequent benefits to respiratory physiological research can hardly be overstated. Much of his research was performed at the University of North Carolina in Chapel Hill in the United States, to which he moved from Stanford in 1973 and where he led an outstanding team of investigators.

Andrew Huszczuk and colleagues in the Wasserman group (124) then addressed the issue by reducing CO_2 delivery to the lung in the steady state of electrically induced exercise in the anesthetized dog. They diverted a fraction of the venous return from the inferior vena cava to the descending aorta, having cleared the appropriate amount of CO_2 via an extracorporeal gas exchanger. $\dot{V}_E$ decreased in appropriate proportion to the reduced $\dot{V}CO_2$ demands at the lung, resulting in arterial isocapnia, which was evident from the continuous display of $PaCO_2$ monitored by means of an indwelling CO_2 electrode. A different mechanism for the cardiodynamic hyperpnea was subsequently proposed by members of this group in 1982. They hypothesized that the $\dot{Q}_T$-coupled hyperpnea was mechanically mediated, being transmitted via cardiac afferents in response to what they termed right ventricular strain (132). Except for the locus of the afferent signal, this had much in common with Harrison and colleagues' previous hypothesis that an increase in central-venous and right-ventricular pressure increased respiration via a vagally mediated reflex (108). However, although several groups had demonstrated that hyperpnea could be evoked by stimulating particular afferents from the heart (e.g., 143, 239), an obligatory role for the proposed cardiodynamic mechanism seemed to be ruled out by the lack of a systematic reduction in the magnitude or rapidity of onset of the exercise hyperpnea either in humans who had undergone cardiac transplantation (25, 237) or during treadmill exercise in awake calves with implanted artificial pneumatically-driven (Jarvik-type) hearts (123).

Figure 10.12 Fred Eldridge (1924–), who made outstanding contributions to both the control and perception of breathing during exercise, in his laboratory in the department of physiology at the University of North Carolina in 1982.

Photo courtesy of Frederic L. Eldridge.

Neurohumoral Considerations

The early 1960s provided two publications that were greatly influential in stimulating interest in this control challenge. The first was the conference proceedings of the Haldane Symposium, held in Oxford in 1960 to commemorate the centenary of Haldane's birth (56). The section devoted to the hyperpnea of exercise included a detailed account of experiments by F.F. Kao (1919-1992) of the State University of New York Downstate Medical Center, Brooklyn and his colleagues (133) who had undertaken a series of cross-circulation experiments in anesthetized dogs. These experiments were so bold and technically challenging that we are unaware of them being subsequently replicated. This led Kao to conclude that "the primary exercise stimulus is, therefore, neural. The humorally borne CO_2 refills the pulmonary circulation and this prevents the hypocapnia that normally accompanies hyperventilation" (133, p. 499). Not surprisingly, robust discussion followed because this was diametrically opposed to the proposal Geppert and Zuntz referred to in their classic 1888 publication (93), from which the authors concluded that the exercise hyperpnea was controlled by some substance formed in the exercising limbs and transferred to the respiratory center.

The other influential document during this period was the chapter titled "Control of Respiration in Muscular Exercise" (63) by the outstanding French physiologist Pierre Dejours (1922-2009; figure 10.13), then of the Laboratoire de Physiologie, Faculté de Médicine in Paris, in the American Physiological Society's 1964 *Handbook of Physiology*. In this insightful, contemporary, state-of-the-art review, he concluded that the available experimental evidence supported a neurohumoral control nexus for the exercise hyperpnea. The neural component was proposed to account for the rapid $\dot{V}_E$ response at exercise onset, the immediacy of which simply seemed incompatible with a humorally mediated drive from metabolites released from the working limbs and acting on central sites of stimulation. The subsequent slower component was proposed to be mediated by supplementary humoral mechanisms (i.e., stimuli transferred from the exercising limbs to sensors cephalad to the lungs). The temporal features of these response profiles were subsequently termed phase 1 and phase 2, bringing the total response to the phase 3 steady state.

Neurogenic Mediation

The 1970s and 1980s provided challenges to the then largely ingrained notions that humoral mediation of the exercise hyperpnea needs to be slow and that neural mediation needs to be fast, at least of necessity. These challenges were spearheaded by two outstanding respiratory physiologists who, for a time in the 1960s, were contemporaries in Stanford University's department of medicine: Karlman Wasserman and Frederic Eldridge. Wasserman's proposal of a rapid cardiodynamic mechanism for the exercise hyperpnea is discussed previously. Eldridge introduced a major conceptual advance that provided a change of perspective to considerations of the exercise hyperpnea (discussed in 81). With Gill-Kumar he proposed, and provided evidence (in the cat) for, a slow neural component that supplemented any ongoing fast neural component (79) (figure 10.14). This could account, in large part, for both the steady-state and transient phases of the moderate-exercise hyperpnea being mediated via neural mechanisms. The "in large part" allows for fine tuning of the response by means of a small humoral component mediated by the carotid bodies (81).

August Krogh with his colleague Johannes Lindhard (1870-1947) had proposed in 1913 that the rapid component of the exercise hyperpnea was neither a reflex from the exercising muscles nor a direct, cortically mediated stimulation of the respiratory muscles (147). Rather, they considered it to be a consequence of the cortical somatomotor command to skeletal muscle contraction irradiating to the brainstem respiratory center, thereby eliciting neurally mediated ventilatory responses proportional to the magnitude of what was subsequently termed central command by Goodwin and colleagues (95). But a substantial body of more recent evidence demonstrated that the cerebral cortex is not the sole origin of the central component of the hyperpnea, i.e., decorticate animals are capable of spontaneous locomotion accompanied by increases in both respiratory and cardiovascular activity (69, 81).

Figure 10.13 The eminent French physiologist Pierre Dejours (left) with Brian Whipp (1937-2011) at the 1977 International Union of Physiological Sciences meeting in Paris, France. In the background, Herman Rahn and Autar Paintal can be seen in deep discussion.

Photo courtesy of Brian J. Whipp and Susan A. Ward.

Focal chemical and electrical stimulation of subcortical regions such as the hypothalamic paraventricular locomotor region and the H2 fields of Forel was demonstrated to lead to rapid hyperpnea (and associated tachycardia and arterial hypertension) in conjunction with locomotor activity (e.g., 81). Efferent neural projections were also demonstrated from these hypothalamic regions to the medullary respiratory and cardiovascular control sites, such as the nucleus of the solitary tract, the nucleus ambiguus, and the dorsal vagal nucleus (e.g., 162, 221). Furthermore, these evoked responses appear not to depend on muscular contraction and increased metabolic rate because they were shown to be essentially unaffected by muscle paralysis (e.g., 81). Several studies, including the influential fictive locomotory model of Eldridge and colleagues, reported that the normal exercise $\dot{V}_E$ response was abolished when the hypothalamic regions were lesioned (discussed in 81). The coarse resemblance between these hypothalamically induced responses and those seen during exercise in the awake animal suggested that central neurogenesis is the major and essential component of ventilatory and cardiovascular control during dynamic exercise in humans. Only a modest feedback component from humoral or mechanical sources related to muscle contraction would be required. Support for this contention was provided in humans by experiments that demonstrated that hyperventilation occurs during steady-state exercise when the degree of conscious effort required to accomplish the task was greater than normal [e.g., after partial muscle curarization (14) or by simultaneous activation of antagonistic musculature (95)]. Barcroft previously noted, "Even if the subject (on a bicycle ergometer) thought he was taking exercise and was not in reality doing so, the total ventilation increased" (27, p. 312). More recently, Thornton and colleagues (238) extended previous studies supporting this notion that had used hypnotic suggestion (57, 186). Thus, they demonstrated marked ventilatory and heart rate responses to imagined cycling exercise in hypnotized human subjects at rest and were able to document simultaneous activation of cerebral locomotor regions by means of positron emission tomography. The $\dot{V}_E$ response, however, seemed to be quite different from that of normal volitional exercise in that it was both predominantly tachypneic and consistently hyperventilatory.

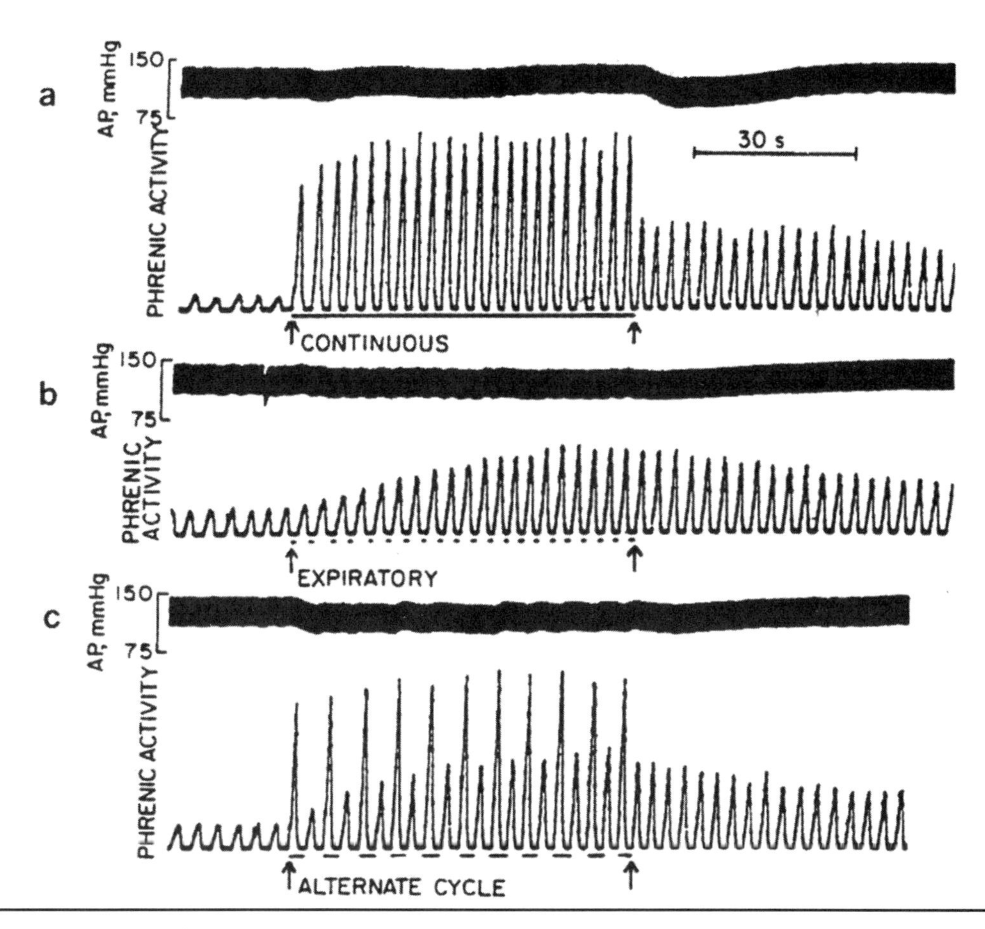

Figure 10.14 Arterial blood pressure (ΔP) and integrated phrenic nerve activity responses to various carotid sinus nerve stimulation (—) formats in a paralyzed, vagotomized, artificially ventilated cat (normal alveolar PCO_2). Upper panel: Onset of continuous stimulation elicits an immediate, rapid increase in phrenic activity (direct) followed by a slower, gradually developing component (potentiation). At offset, these responses decline in a correspondingly similar fashion. Middle panel: Onset of continuous expiratory-phase stimulation elicits only a slow potentiation component, which declines similarly at offset. Lower panel: Onset of alternate-breath stimulation elicits a direct component on stimulated breaths whereas a potentiation component develops on the nonstimulated breaths. At offset, the direct component declines rapidly (as in the upper panel) and the potentiation component declines slowly (as in the middle panel).

The primacy of a corticogenically determined the exercise hyperpnea was challenged by the 1943 case study of Asmussen and colleagues regarding a patient with tabes dorsalis (17); this favored muscle reflex mediation. Indeed, a wide range of receptor afferents and free nerve endings of the contracting skeletal muscles was also proposed to contribute to the ventilatory responses to exercise (e.g., 164, 222). Support for the concept was provided by the demonstration that the hyperpnea could be abolished by dorsal spinal-root section, attesting to the exercising-muscle origin of the response, and that differential blockade of the small-diameter group III and IV somatic projections from the exercising limbs eliminated the response (164, 184). One subpopulation of these small-diameter afferents was demonstrated to respond to local mechanical distortion and increased intramuscular pressure. A second group responded to humoral mediators such as intramuscular levels of H^+, potassium ions (K^+), bradykinin, and arachidonic acid derivatives, as exemplified by the elegant studies of Marc Kaufman and colleagues conducted at the University of California, Davis (e.g., 135). More recently, Philippe Haouzi and Andrew Huszczuk, with their colleagues at the University of Nancy in France and at the Harbor-UCLA Medical Center in California, provided evidence for a control mechanism intimately related to the peripheral microvasculature, responding to altered vascular conductance or tissue pressure in the exercising muscles themselves (e.g., 110)—shrewdly termed a "plethysmographic" mechanism by Waldrop and Iwamoto (245) in their "Point–

Counterpoint" interchange with Haouzi (109) on the matter in *Journal of Applied Physiology*.

Despite the formidable body of supporting evidence, doubts were raised regarding neurogenic mediation as an obligatory component of the exercise hyperpnea. For example, Dejours (63) showed that the magnitude of the phase 1 $\dot{V}_E$ response changed little over a wide range of imposed WRs despite the increase in the number of motor units recruited to do the work and magnitude of the central drive that commands their recruitment. Furthermore, when the WR increment is imposed from a background of light exercise (i.e., with a number of motor units already having been recruited), the phase 1 component is both smaller and more slowly developing (discussed in 262). A correspondingly slow initial hyperpnea was also demonstrated when rest-to-work transitions were imposed with the subject in the supine position (256). Why would body position per se so markedly reduce the contribution of either the central command or limb afferent neurogenic control mechanisms? Thus, although neurogenesis as the mechanism mediating the rapid component of the exercise hyperpnea seemed beyond serious challenge, the nature and functional organization of the signals themselves are by no means as clearly understood.

The results of the sinusoidal exercise studies described earlier (44) were also thought to be revealing in this regard, suggesting that the central command necessary to drive the force-generating muscle units and the force-related feedback from the muscles play a trivially small role in the exercise hyperpnea under these conditions or that these mechanisms themselves exhibit slow neural dynamics that are, somehow, closely matched to the rate of pulmonary CO_2 exchange. A further serious challenge to an obligatory role for central command and neurogenic feedback from the contracting muscles came from studies in which the subject's intrinsic ventilatory drive was dissociated from the motor task. Poon and colleagues (203) originally, then others (145), used a servo-assisted positive-pressure ventilator synchronized with the respiratory cycle to take over a proportion of the normal inspiratory flow. Were $\dot{V}_E$ to be dictated exclusively by the magnitude of the central command and muscle feedback mechanisms, the subject's intrinsic ventilatory drive would presumably remain unchanged and the overall $\dot{V}_E$ would naturally be expected to increase, thus inducing hypocapnia. But this was not the case! Rather, there was a proportional compensatory reduction in the subject's intrinsic ventilatory component.

Furthermore, when the influence of somatic limb afferents was removed or markedly reduced in normal humans who had undergone spinal-epidural anesthesia to induce muscle sensory blockade, the ventilatory (and cardiovascular) response to dynamic exercise was not appreciably impaired—at least, over the range of metabolic rates achieved and despite the absence of functioning afferent projections from the exercising limbs. Studies were also performed in subjects who had suffered complete spinal cord transection and in whom dynamic exercise was induced by direct electrical stimulation of the quadriceps muscles (3, 36, 262). If the proposed neurogenic central command and muscle feedback control mechanisms were both obligatory and major components of the exercise hyperpnea, then attenuated $\dot{V}_E$ responses would be expected. However, both the magnitude and time course of the $\dot{V}_E$ responses were reported to be essentially normal, at least with respect to the slower-than-normal $\dot{V}CO_2$ response. The $\dot{V}_E$ time course when pneumatic cuffs, placed around the thighs, were inflated to suprasystolic pressures at the end of exercise (the recovery profile was used because inflation during exercise was typically painful) was also hard to reconcile with muscle-metabolite stimulation of $\dot{V}_E$. The $\dot{V}_E$ recovery profile was shown to be markedly more rapid than normal (e.g., 126) despite the exercise-induced metabolites being trapped in the muscle tissue.

Long-Term Potentiation

Martin and Mitchell (173) put forth in 1993 a novel proposal for ventilatory control during exercise: an involvement of long-term potentiation (LTP) resulting from repeated exposure to physical activity, akin to associative motor learning. Goats were reported to consistently overbreathe during a standard treadmill task after they had been trained to exercise repeatedly at that WR with an added ventilatory stimulus, provided by an increased external dead space. Before this intervention, $PaCO_2$ during unencumbered exercise averaged only an approximately 0.5 mmHg decrease (i.e., hypocapnia was not characteristic in these goats); however, the decrease was approximately 3 mmHg postintervention. The authors concluded that the neural mechanisms controlling the exercise hyperpnea evidence plasticity resulting from the hyperpneic history, possibly via central serotonergic mediation. Such LTP of the exercise hyperpnea would be consistent with Somjen's postulate that functionally error-free physiological control systems may operate through central nervous system control that "anticipates present and future needs on the basis of past experience" (226, p. 184). Attractive though this postulate might seem, however, LTP has not been consistently demonstrable in exercising humans (e.g., 45, 187).

Respiratory Compensation for Metabolic Acidosis

WRs engendering a metabolic (predominantly lactic) acidosis evidence an additional drive to $\dot{V}_E$ that results in arterial hypocapnia. This control component, which constrains the decrease of pHa by providing respiratory compensation for the acidosis, was shown in an important study by Erling Asmussen with Marius Nielsen in 1946 (15) to involve the arterial chemoreceptors responding—they conjectured—to a work-derived substance transported from the contracting muscles. This was based on the rapid and sustained reduction of $\dot{V}_E$ at high WRs when the inspirate was switched to 100% O_2. Pierre Dejours, with Louis-Michel Leitner, addressed this issue in greater detail (152) and demonstrated, in the cat, that there was a high degree of correlation between both the magnitude and time course of the $\dot{V}_E$ response to abrupt and surreptitious hyperoxia and that of the afferent impulse traffic in the carotid sinus nerve. They concluded that the rapid decrement in $\dot{V}_E$ during exercise consequent to abrupt hyperoxia was a result of suppressing peripheral chemosensitivity. This procedure (the Dejours test), with its breath-by-breath time resolution, was subsequently applied to humans during exercise. The magnitude of the V_E decrement was considered to represent the component of the ongoing ventilatory drive attributable to the peripheral chemoreceptors (e.g., 62).

Subsequent studies have demonstrated that the magnitude of this component of the exercise hyperpnea is highly variable among human subjects but that, on average, it accounts for approximately 15% to 20% of the hyperpnea, for both moderate and heavy exercise (e.g., 262). A range of plausible peripheral chemoreceptor candidate stimuli has been proposed, including the arterial concentrations of H^+, K^+, adenosine, and catecholamines; osmolarity; and even, for more prolonged exercise, body temperature. The relative importance of these stimuli and the characteristics of their interaction remain to be determined. The aggregate response, although dominating the hyperventilatory component, appears to contribute only 15% to 20% of the entire hyperpnea, at least as evidenced by the Dejours test. However, as the subject with the largest compensatory increase in $\dot{V}_E$ will have the smallest decrease in pHa, this suggests that H^+ per se is unlikely to be the stimulus (68). Central chemoreceptors are thought not to mediate a significant component of the compensatory hyperventilation. Neither H^+ nor K^+ ions, for example, cross the blood–brain barrier, at least as reflected by bulk cerebrospinal fluid measurements. However, as demonstrated by both Leusen (153) and Bisgard and colleagues (30), hyperventilation-induced arterial hypocapnia results in an alkaline cerebrospinal fluid. This would be expected to reduce ongoing central chemoreceptor activity and also to stimulate efferent projections to the carotid chemoreceptors, which have been demonstrated by Majcherczyk and Willshaw to inhibit afferent chemosensory discharge in the cat (169). Because there is no reason to suspect that these mechanisms are not also operative in humans, the net response to the metabolic acidosis of muscular exercise is consistent with integrated peripheral and central chemoreceptor mediation.

James Duffin of the University of Toronto, however, proposed a nonchemoreceptor mechanism for this component of the exercise hyperpnea in humans (174). The development of muscle fatigue during heavy exercise would demand additional motor units to be recruited to maintain total force production (this was supported by locomotor muscle electromyographic evidence) and compensate for the reduction in force generation in acidotic muscle units. This would be expected to supplement both the central command and muscle-afferent neurogenic control components.

The availability of human subjects who had undergone BCBR, predominantly for the relief of dyspnea (initially asthmatic patients in remission and then more generally those with chronic obstructive lung disease), provided a further level of insight into factors controlling the compensatory hyperventilation. The studies of Yoshiyuki Honda (1926-2003) and colleagues at the Chiba University School of Medicine in Japan (118) and, more extensively, the studies of the Wasserman group (254, 262) demonstrated an absence of the normal hyperventilatory response in the heavy-intensity domain of both incremental (figure 10.15) and sustained, constant WR exercise; that is, the decrease in pHa was greater for a given decrease in arterial ([HCO_3^-]([HCO_3^-]a) during heavy-intensity exercise and $PaCO_2$ was actually higher than the corresponding control values (254). As the Wasserman group had shown that the impairment to air flow in the asthmatic patients was modest at most, it was argued that the carotid bodies were the dominant or exclusive mediators of the compensatory hyperventilation for the acute metabolic acidosis of exercise. Rausch and colleagues (210) added to the quantitative aspects of these findings by investigating the response to constant WR exercise designed to induce the same arterial [lactate] increase, but against a background of varied oxygenation, in normal subjects. The degree of metabolic acidemia associated with a given degree of metabolic acidosis was found to be highly dependent on peripheral chemoreceptor sensitivity. A then-surprising additional finding of this experiment was that a small compensatory hyperventilation was evident despite peripheral chemosensitivity presumably being abolished by sustained hyperoxia (80% inspired [O_2]) throughout the experiment. The mechanism of this slower compensatory component awaits elucidation.

What also emerged from studies on these subjects was that there was no demonstrable role for the aortic bodies as ventilatory chemosensors in humans; unlike the control subjects, the BCBR subjects evidenced no $\dot{V}_E$ decrease in response to the abrupt transition from euoxia or hypoxia to 100% O_2 as the inspirate, either at rest or during moderate and heavy exercise. Despite the absence of peripheral chemosensitivity in BCBR subjects, the steady-state $\dot{V}_E$ response to moderate exercise was not appreciably different during air breathing compared with either normal controls or an appropriate group of asthmatic controls. In fact, the steady-state $\dot{V}_E$ during moderate exercise was not different between air breathing and 12% O_2 (161) in BCBR subjects. Also, patients with impaired central chemosensitivity, as evidenced by an absence of CO_2 chemosensitivity (223) (congenital central hypoventilation syndromes), were shown to have effectively normal $\dot{V}_E$ responses to exercise, which regulated PCO_2 close to resting levels; the PCO_2 set point is commonly elevated in these patients (e.g., 160, 223). Similarly, the regulatory features of the exercise hyperpnea were shown not to differ from normal in subjects with chronic hyperventilatory syndrome, in whom resting pHa was normal despite $PaCO_2$ being stable at approximately 30 mmHg (127). The magnitude of the $\dot{V}_E$ response was, naturally, greater as a regulatory requirement at the reduced PCO_2 set point. The time course of the transient $\dot{V}_E$ response to the steady state, however, was markedly slowed in BCBR subjects (254, 262). Furthermore, the time course was shown also to be slowed under conditions of reduced peripheral chemosensitivity in normal subjects (hyperoxia, induced metabolic alkalosis, infused dopamine) and speeded under conditions of increased peripheral chemosensitivity (hypoxia, induced metabolic acidosis). Therefore, the carotid bodies were proposed to be important contributors to the dynamics of the exercise hyperpnea (as discussed in 262).

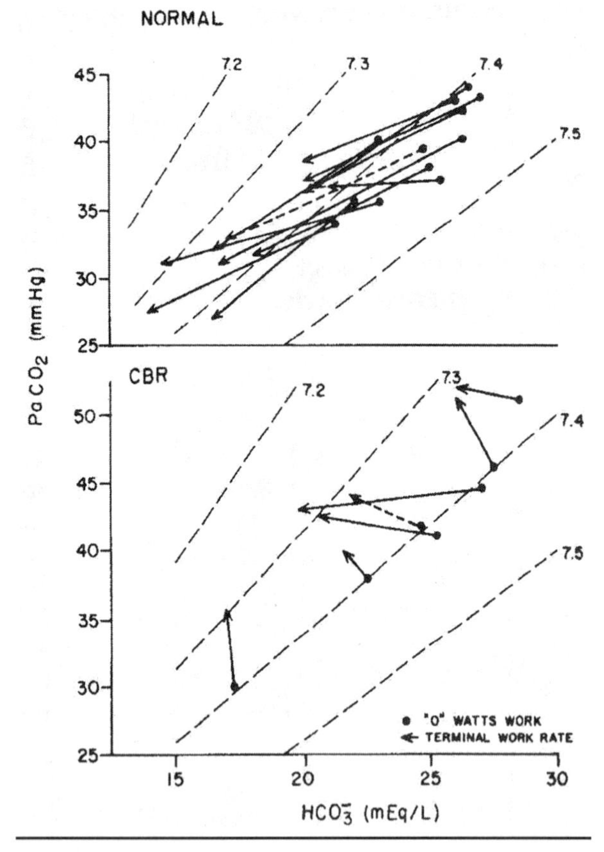

Figure 10.15 Arterial PCO_2 ($PaCO_2$) as a function of arterial bicarbonate ([HCO_3^-]) at unloaded pedaling (•) and peak work rate (←) for incremental exercise in normal subjects (upper panel) and subjects who had previously undergone bilateral carotid body resection (lower panel). Group mean responses are shown as dashed arrows. Dashed lines indicate arterial pH isopleths.

Reprinted from K. Wasserman et al., 1975, "Effect of carotid body resection on ventilatory and acid-base control during exercise," *Journal of Applied Physiology* 39: 354-358. With permission from American Physiological Society.

The advent of rapid-incremental and continuous-ramp exercise resulted in an initially surprising finding in normal subjects, the context for which is provided by the reformulation of the Henderson-Hasselbalch equation (262) in which $PaCO_2$ is replaced by its ventilatory determinants. That is,

$$pHa = pK' + \log\{([HCO_3^-]a/25.6) \cdot (\dot{V}_E/\dot{V}CO_2) \cdot (1 - V_D/V_T)\},$$

where the constant pK′ normally has a value of 6.1 and $\dot{V}_E/\dot{V}CO_2$ is the ventilatory equivalent for CO_2. That is, as expected, $\dot{V}_E$ increased disproportionately with respect to $\dot{V}O_2$ as the lactic acidosis developed. In contrast, however, the hyperventilatory compensation for the metabolic acidosis, as reflected in the increase in $\dot{V}_E/\dot{V}CO_2$, did not occur until $\dot{V}O_2$ had increased to approximately half of the remaining metabolic rate range to the maximum $\dot{V}O_2$ (e.g., 248). This WR range, over which the PCO_2 remained relatively stable, was termed the region of isocapnic buffering. The metabolic rate at which $\dot{V}_E/\dot{V}CO_2$ began to increase and hyperventilation became manifest was termed the respiratory compensation point by the Wasserman group (251) and the threshold of decompensated metabolic acidosis by Reinhard and colleagues (211). Why the hyperventilatory response to the metabolic acidosis of exercise is so delayed remains to be resolved, as the time course of the hypoxic $\dot{V}_E$ response mediated by the carotid bodies in humans is rapid (with a time constant of approximately 5-10 s) (73). However, the time course of the carotid chemoreceptor response to H^+ is appreciably slower, possibly reflecting a slow intracellular expression of the metabolic acidosis (37) or slow signal transduction via an H^+-sensitive, type I, voltage-sensitive, tandem-P-domain K^+ channel (38). The precise mechanism awaits resolution.

Modeling Considerations

It is evident that although a congeries of mechanisms capable of stimulating $\dot{V}_E$ under some circumstances during exercise in humans has been demonstrated, there is not at present a satisfying resolution to the integrative aspects of the control; that is, one that accounts for the actual dynamic features of the intensity-dependent responses. Rising to Defares' 1964 challenge that "respiratory control during exercise has long been regarded as forbidden territory for cybernetic analysis" (61, p. 671), investigators began to address the control challenge differently by considering insights that could emerge from the application of engineering control theory to the issue. Pride-of-place for this must be ceded to Fred Grodins (1915-1989; figure 10.11), the founding chairman of the department of biomedical engineering at the University of Southern California in Los Angeles. Grodins, using the benefits of analog and then digital computer developments, extended Gray's multiple factor theory (96) to considerations of the exercise hyperpnea (e.g., 99). His melding of direct experimental evidence with control-system model considerations provided a stimulus for numerous others.

Among these, Chi-Sang Poon of the division of health sciences and technology at Harvard–Massachusetts Institute of Technology, for example, proposed an optimal control model, which has the notionally exciting benefit that it dispenses with the need for an explicit exercise stimulus, in which "the controller is programmed to minimize the total cost due to chemical stimulation and the work of breathing" (202, p. 191). George Swanson, currently in the department of kinesiology at California State University at Chico, proposed a scheme motivated by optimal controller behavior in which ventilatory variations from a trial ventilation (set by the brain) would result in a consequent variation in $PaCO_2$ that would itself be a function of the metabolic $\dot{Q}CO_2$ (233). This hunting-and-seeking behavior, evident in the real breath-to-breath behavior, would be incorporated into a feed-forward feed-back control scheme that would iteratively yield the appropriate minimum-cost $\dot{V}_E$.

These and other insights from modeling the dynamic features of the exercise hyperpnea have formed an important component of the continuing, and now triennial, Oxford conferences on modeling and the control of breathing. These originated in a discussion in 1971 between Dan Cunningham, a respiratory physiologist at the University of Oxford, and Richard Herczynski (1926-2009), a mathematician and fluid dynamicist at the Polish Academy of Sciences in Warsaw, who wondered whether consideration of the control of breathing from the perspectives of their particular disciplines might be to the benefit of both sides (55). The resulting inaugural conference, held at the University Laboratory of Physiology in Oxford in 1978, was titled "Modeling of a Biological Control System: The Regulation of Breathing" (43). This conference and the proceedings of the eleven subsequent conferences provide an important historical perspective on the growth, development, and occasional demise of considerations of both the exercise hyperpnea and pulmonary gas exchange during exercise.

Exertional Dyspnea

The recognition of sensations of breathlessness or dyspnea (from the Ancient Greek δυσπνοια, meaning difficult or labored breathing) during or after exertion is likely to have been among man's earliest recognitions of physiological function. In Sophocles' *Antigone*, for example, there is reference to a messenger charged with conveying bad news to Creon, King of Thebes. The messenger, distinguishing exertional dyspnea from emotional dyspnea, states, "My lord, if I am out of breath it is not from haste. I have not been running" (227, p. 132).

Although considerations of exertional dyspnea in the early 20th century were discussed largely in the context of pathological conditions such as congestive heart failure and respiratory mechanical dysfunction, Means interestingly noted the following in 1924: "In the normal man the hyperpnoea induced by muscular work finally reaches a level at which the respiratory organs meet with some embarrassment in providing the necessary ventilation; then dyspnoea results. Dyspnoea may therefore be physiologic as well as pathologic" (177, p. 351). But even by the 1960s, formal considerations of dyspnea in the normal exercising subject were relatively limited. For example, the American Physiological Society's *Handbook of Physiology* series on respiration mentions dyspnea only in the context of unusual conditions, such as inert gas narcosis during hyperbaria. What received greater attention is how exercise might influence the breath-hold breaking point (185), although there was no explicit discussion of the relationship between breath-hold duration (T_{BH}) and dyspneic sensation per se.

As late as 1991, Lewis Adams and Abraham Guz at the University of London's Charing Cross Hospital Medical School stated, "Shortness of breath on exertion is a common experience in healthy subjects . . . . Despite its obvious importance as a sensory consequence of exercise, there has been relatively little formal research to define more carefully its sensory characteristics or to evaluate its influence on behavioural aspects of exercise" (4, p. 466). In fact, Guz and colleagues, particularly Lewis Adams and Douglas Corfield, are notable for their significant contributions to understanding exertional dyspnea in health and in disease. Major conceptual advances were also made by E. Moran Campbell, one of the pre-eminent British physician-scientists, who started his research career in the department of physiology at the University of London's Middlesex Hospital Medical School and subsequently moved to McMaster University in Canada where, as chairman of the department of medicine, he established an influential research group that included notable investigators such as Norman Jones, chief of the respiratory division, and Kieran Killian. After undergraduate medical training and study of internal medicine at the National University of Ireland, University College Galway, Killian joined the McMaster group for training in respiratory medicine. This led to a long-standing and productive collaboration that advanced the understanding of exertional and other sources of dyspnea. In particular, his collaboration with Simon Gandevia (of the Prince of Wales Hospital and University of New South Wales in Sydney, Australia) has advanced our thinking considerably with regard to the contribution of perception of respiratory load and effort in dyspneagenesis during exercise and to whether exertional dyspnea and breathlessness actually reflect different aspects of respiratory sensation.

Methods of Assessment

Until the late 1960s, assessment of exertional dyspnea focused on patient populations and was based largely on indirect instruments such as case histories, questionnaires, and indices of respiratory mechanical limitation, such as the ratio of $\dot{V}_E$max to maximal voluntary ventilation, called the dyspnea index (268). Although these indices might be objective descriptors of the predisposition to ventilatory limitation during exercise, they could not provide an a priori indication of how intensely a particular level of $\dot{V}_E$ will be perceived by a particular individual. For this, the pioneering psychophysical work of Stanley Smith Stevens of Harvard in the 1950s provided the impetus (231). This led to reliable assessment of dyspneic sensation during exercise by means of appropriately constructed rating scales. The attraction of the visual analog scale, first used to monitor dyspnea by Aitken (6), is its simple linear structure anchored to fixed extremes (i.e., 0 and 100, corresponding to minimal and maximal intensities of sensation, respectively, with the intervening portion of the scale remaining undefined). This and the Borg category scale (33) [first used for monitoring dyspnea by Burdon and colleagues in 1982 (39)] have been widely used to quantify exertional dyspnea. Each rating scale has proved convenient to use and has acceptable reproducibility (e.g., 2, 229, 264). Implicit in such closed-ended scales is that the maximum intensity experienced by a subject does not exceed the upper end of the scale. But, were this constraint not to be imposed, might subjects rate supramaximally? El-Manshawi and colleagues (77), for example, have indicated that their normal exercising subjects did rate above 10 on the modified Borg scale. These implications are yet to be formally addressed.

Because of the subjective nature of these approaches, with ratings critically depending on exactly how the subject is instructed to rate dyspnea and on the extent and nature of the subject's prior experience, some investigators have sought more objective measures during exercise (i.e., quantitative and descriptor free)—but with limited success. Ward and colleagues (246), for example, explored the extent to which the relationship between dyspnea and the reduction of T_{BH} during exercise might serve such a role. However, large reductions in T_{BH} in the low WR range were associated with only modest increases in dyspneic sensation, whereas progressively smaller decrements in T_{BH} were associated with progressively larger increases in dyspnea at higher WRs.

Mechanisms

It has been common practice to search for the dyspneageneic mechanisms during exercise among those contributing to control of the exercise hyperpnea. However, as early as the 1930s, Christie cautioned that "it must not be imagined that dyspnea and hyperpnea are synonymous" (47, p. 423). Researchers have subsequently often lost sight of this, as Julius Comroe (1911-1984), the inaugural director of the University of California, San Francisco's Cardiovascular Research Institute, emphasized in the opening statement of Howell and Campbell's influential 1965 symposium on breathlessness (121): "Many physiologists and clinicians have written articles entitled 'Dyspnea' which were in fact articles on the regulation of respiration, the causes of hyperpnea or the causes of hyperventilation—which, in fact, had nothing to do with the sensation of dyspnea" (49, p. 1). However, as discussed earlier, Cherniack questioned whether dyspneic sensation(s) might actually contribute to an optimization strategy of ventilatory control (46), such as proposed by Poon (202).

Psychophysical principles imply that afferent influences during exercise have the potential to contribute to a cortical sensory impression that is expressed as dyspnea (e.g., 138). However, merely demonstrating that a particular mechanism has neural connections with the cerebral cortex (e.g., 78) and can under certain experimental conditions (e.g., imposition of elastic or resistive loads; inspiratory occlusion) elicit or exacerbate dyspnea is not sufficient to ascribe to it an obligatory role in mediating exertional dyspnea. The prevailing view for much of the 20th century has been that respiratory mechanical factors were the major determinants of exertional dyspnea, with pride of place going to Campbell and Howell's 1963 length-tension inappropriateness hypothesis. That is, the degree of misalignment between inspiratory muscle intrafusal and extrafusal fibers, as they shortened during lung inflation, was translated into dyspneic sensation via muscle spindle action (42). With Killian's input, this hypothesis was subsequently extended to accommodate the respiratory effort required for force generation: "The role of exercise in the genesis of dyspnea is, in our opinion, quantitative. There is little unique about exercise itself. Many factors . . . contribute to dyspnea; but the central unifying feature is the force generated by the inspiratory muscle" (138, p. 474).

Campbell and Howell (42) made a clear distinction between the awareness of increased ventilation during relatively moderate exercise and a more overt unpleasant sensation at higher WRs when V_T started to encroach on the upper, and stiffer, region of the lung compliance curve, the seeds of which are implicit in the writings of the physician Sir John Floyer (1649-1734). He noted, in the parlance of the times, "If my pulse be 90, I am always pursive, but 95 makes me asthmatick" (87). How inspiratory muscle force or effort might be sensed during exercise remains far less clear. Several putative mechanisms have been considered, including central command, respiratory muscle and chest wall proprioception, and vagal pulmonary influences. Not only do these each appear dyspneagenic under some experimental situations (e.g., 4, 139, 271), but such receptor inputs also have been demonstrated to project to higher central nervous system regions (e.g., 18, 78, 91). However, although the nature and extent of their involvement during normal exercise conditions remain more conjectural, El-Manshawi and colleagues (77) have suggested that dyspnea may actually protect the respiratory muscles from fatigue, raising the possibility that factors diverting attention from respiratory sensations at high WRs in athletes may extend exercise tolerance as respiratory muscle fatigue develops.

The parallels that have been drawn between effort perception related to locomotor muscle activity and respiratory muscle activity by investigators such as Zechman and Wiley (271) and Killian and Gandevia (139) have suggested common mediation; that is, some combination of central command and respiratory muscle proprioception. With chest wall vibration (a specific muscle spindle stimulus), reduced dyspnea during exercise has been demonstrated in COPD patients despite $\dot{V}_E$ being unchanged (90). Studies on normal subjects undergoing respiratory muscle paralysis at rest, however, are not generally in support of an essential contribution from respiratory muscle proprioception (as discussed in 4). Indeed, the demonstration that dyspnea at a given WR was less marked than that at a lower WR at which $\dot{V}_E$ was matched volitionally (with isocapnia maintained) supported the notion that the level of dyspnea is not simply dictated by the level of $\dot{V}_E$ (and therefore the degree of respiratory muscle tension) but rather was more likely to involve mechanisms operating through the brain stem respiratory integrating regions.

These considerations on the sensing of effort were readily extended to include the influence of corollary discharge of motor command signals to higher brain regions involved in respiratory perception. Killian and Campbell wrote, "Although the evidence is circumstantial, a conscious awareness of the outgoing motor command may be dominant in our appreciation not only of muscular effort but also of respiratory distress" (138, p. 474). This sense of respiratory motor output was seen to be distinct from sensations directly related to changes in muscle length or tension and was attributed to a corollary discharge from brain stem respiratory neurons to the sensory cortex during automatic reflex breathing or from cortical motor centers to the sensory cortex during voluntary respiratory efforts (e.g., 78, 139).

With regard to vagal afferent mediation, Guz and colleagues, who daringly used bilateral local-anesthetic blockade of the vagus and glossopharyngeal nerves in resting subjects, demonstrated both amelioration of dyspnea and increase in T_{BH} (100). Subsequently, bupivacaine-mediated airway anesthesia was reported to reduce dyspnea during incremental exercise (265). Later statements from these investigators, however, suggest that the reported reduction was not "consistent" (4); this was reinforced by their later demonstration that dyspnea (and $\dot{V}_E$) during exercise was unaffected after heart–lung transplantation (26, 141). However, what might prove to be of significance are the juxtapulmonary capillary or J-receptors, which Paintal (e.g., 193) has shown to be responsive to increases in pulmonary–interstitial and pulmonary–capillary pressure. The possibility of pulmonary–capillary stress failure or pulmonary edema at near-maximal WRs in highly fit endurance athletes (e.g., 119, 163, 257) also raises the possibility that J-receptor stimulation may contribute to both tachypnea and dyspnea at or near maximal exercise (e.g., 193).

Interestingly, a role for respiratory chemoreceptors in dyspneagenesis during exercise was not widely considered until the latter portion of the 20th century. For example, the 1984 National Heart, Lung, and Blood Institute workshop on respiratory sensations and dyspnea stated "there is no evidence that changes in PCO_2 or PO_2 per se are consciously perceived. Rather, it appears that changes in respiratory sensation produced by increasing chemical drive are a consequence of increases in respiratory efferent activity" (9, p. 1052). Indeed, as for ventilatory control, more recent studies on congenital central hypoventilation syndrome patients support the view that the central chemoreceptors do not, in themselves, provide any unique drive to dyspnea during exercise. That is, such subjects evidence increases in both $\dot{V}_E$ and dyspnea during exercise that are comparable with those of control subjects (228).

This does not appear to be the case for the carotid chemoreceptors, however. Early clues that the carotid chemoreceptors might influence dyspnea during exercise, beyond their effects on $\dot{V}_E$, came from reports that inhalation of O_2 during exercise reduced dyspnea and improved exercise tolerance (114, 188) and, inferentially, from later breath-holding studies showing that the hypoxia-induced reduction in breath-hold duration was not manifest in subjects who had previously undergone BCBR (59). Might therefore the prolongation of T_{BH} that Guz and colleagues (100) documented after vagoglossopharyngeal nerve block have reflected interference with the transmission of carotid body afferent discharge through the glossopharyngeal nerves? In the last two decades of the 20th century, better-controlled hypoxic inhalation studies have shown rather convincingly that the influence of the carotid bodies on dyspnea during exercise is out of proportion to their effect on $\dot{V}_E$ (4, 228, 247).

Much still remains to be learned about how such putative inputs are integrated to generate a cortical sensory impression. There seems to be considerable potential in combining classical exercise paradigms with more recently emerging techniques for interrogating central nervous system regions in real time, such as positron emission tomography (e.g., 195), functional magnetic resonance imaging (e.g., 165, 241), respiratory-related evoked potentials (e.g., 58), and transcranial magnetic stimulation for identification of connections between the cerebral cortex and respiratory muscles (e.g., 92).

Dyspnea as a Factor Limiting Exercise Tolerance

While exertional dyspnea is known to be an important contributor to exercise intolerance in several pulmonary diseases, might it also be contributory in highly fit endurance athletes in whom extremely high metabolic rate and degree of metabolic acidemia can bring $\dot{V}_E$ to, or closely toward, their limits of thoracic mechanical functioning? This may be compounded, in some instances, by increased carotid chemoreceptor drive from developing arterial hypoxemia (65) as well as influences arising from respiratory muscle fatigue (e.g., 20, 168, 235). Finally, interstitial pulmonary edema associated with high pulmonary–vascular pressures (e.g., 119, 163, 257) is a further potential influence via J-reflex activation (193). However, little is actually known about exertional dyspnea in the elite athlete—except that it is typically not a self-reported perception.

References

1. Aaron EA, Seow KC, Johnson BD, Dempsey JA. Oxygen cost of exercise hyperpnea: Implications for performance. *J Appl Physiol* 72: 1818-1825, 1992.
2. Adams L, Chronos N, Lane R, Guz A. The measurement of breathlessness induced in normal subjects: Validity of two scaling techniques. *Clin Sci* 69: 7-16, 1985.
3. Adams L, Frankel H, Garlick J, Guz A, Murphy K, Semple SJG. The role of spinal cord transmission in the ventilatory response to exercise in man. *J Physiol* 355: 85-97, 1984.
4. Adams L, Guz A. Dyspnea on exertion. In: Whipp BJ, Wasserman K, eds. *Pulmonary Physiology and Pathophysiology of Exercise*. New York: Dekker, 1991, 2449-2494.
5. Agostoni E, Mead J. Statics of the respiratory system. In: Fenn WO, Rahn H, eds. *Handbook of Physiology: Respiration.* Vol I. Washington, DC: American Physiological Society, 1964, 387-409.
6. Aitken RCB. Measurements of feelings using visual analogue scales. *Proc Roy Soc Med* 62: 989-993, 1962.

7. Aliverti A, Kayser B, Macklem PT. Breath-by-breath assessment of alveolar gas stores and exchange. *J Appl Physiol* 96: 1464-1469, 2004.
8. Allen CJ, Jones NL, Killian KJ. Alveolar gas exchange during exercise: A single-breath analysis. *J Appl Physiol* 57: 1704-1709, 1984.
9. Altose M, Cherniack N, Fishman AP. Respiratory sensations and dyspnea. *J Appl Physiol* 58: 1051-1054, 1985.
10. Armstrong W, Hurt HH, Blide RW, Workman JM. The humoral regulation of breathing. *Science* 133: 1897-1906, 1961.
11. Aron E. Der intrapleurale Druck beim lebenden gesunden Menschen. *Virchow's Arch Path Anat* 160: 226-234, 1900.
12. Askanazi J, Milic-Emili J, Broell JR, Hyman AI, Kinney JM. Influence of exercise and CO_2 on breathing pattern of normal man. *J Appl Physiol* 47: 192-196, 1979.
13. Asmussen E, Christensen EH. Die Mittelkapazität der Lungen bei erhöhtem O_2-Bedarf. *Skand Arch Physiol* 82: 201-211, 1939
14. Asmussen E, Johansen SH, Jorgensen M, Nielsen M. On the nervous factors controlling respiration and circulation during exercise. Experiments with curarization. *Acta Physiol Scand* 63: 343-350, 1965.
15. Asmussen E, Nielsen M. Studies on the regulation of respiration in heavy work. *Acta Physiol Scand* 12: 171-188, 1946.
16. Asmussen E, Nielsen M. Alveolo-arterial gas exchange at rest and during work at different O_2 tensions. *Acta Physiol Scand* 50: 153-166, 1960.
17. Asmussen E, Nielsen M, Weith-Pedersen G. Cortical or reflex control of respiration during muscular work? *Acta Physiol Scand* 6: 168-175, 1943 .
18. Aubert M, Legros J. Topography of the projections of visceral sensitivity on the cerebral cortex of the cat. I. Study of the cortical projections of the cervical vagus in the cat anesthetized with nembutal. *Arch Ital Biol* 108: 423-446, 1970.
19. Auchincloss JH, Gilbert R, Baule GH. Effect of ventilation on oxygen transfer during early exercise. *J Appl Physiol* 21: 810-818, 1966.
20. Babcock MA, Pegelow DF, Harms CA, Dempsey JA. Effects of respiratory muscle unloading on exercise-induced diaphragm fatigue. *J Appl Physiol* 93: 201-206, 2002.
21. Babcock MA, Pegelow DF, McClaran SR, Suman OE, Dempsey JA. Contribution of diaphragmatic power output to exercise-induced diaphragm fatigue. *J Appl Physiol* 78: 1710-1719, 1995.
22. Bachofen H, Hobi HJ, Scherrer M. Alveolar-arterial N_2 gradients at rest and during exercise in healthy men of different ages. *J Appl Physiol* 34: 137-142, 1973.
23. Band DM, Semple SJG. Continuous measurement of blood pH with an indwelling arterial glass electrode. *J Appl Physiol* 22: 854-857, 1967.
24. Band DM, Wolff CB, Ward J, Cochrane GM, Prior J. Respiratory oscillations in arterial carbon dioxide tension as a control signal in exercise. *Nature* 283: 84-85, 1980.
25. Banner N, Guz A, Heaton R, Innes JA, Murphy K, Yacoub M. Ventilatory and circulatory responses at the onset of exercise in man following heart or heart-lung transplantation. *J Physiol* 399: 437-449, 1988.
26. Banner NR, Lloyd MH, Hamilton RD, Innes JA, Guz A, Yacoub MH. Cardiopulmonary response to dynamic exercise after heart and combined heart-lung transplantation. *Br Heart J* 61: 215-223, 1989.
27. Barcroft J. *Features in the Architecture of Physiological Function*. Cambridge: Cambridge University Press, 1934.
28. Barr DP, Himwich HE. Studies in the physiology of muscular exercise: III. Development and duration of changes in acid-base equilibrium. *J Biol Chem* 55: 539-555, 1923.
29. Beaver WL, Wasserman K, Whipp BJ. On-line computer analysis and breath-by-breath graphical display of exercise function tests. *J Appl Physiol* 34: 128-132, 1973.
30. Bisgard GE, Forster HV, Byrnes B, Stanek K, Klein J, Manohar M. Cerebrospinal fluid acid-base balance during muscular exercise. *J Appl Physiol* 45: 94-101, 1978.
31. Black AMS, Torrance RW. Respiratory oscillations in chemoreceptor discharge in control of breathing. *Respir Physiol* 13: 221-237, 1971.
32. Bohr C. Ueber die lungenathmung. *Skand Arch Physiol* 2: 236-268, 1891.
33. Borg GAV. Psychophysical bases of perceived exertion. *Med Sci Sports Exerc* 14: 377-381, 1982.
34. Boycott AE, Haldane JS. The effects of low atmospheric pressures on respiration. *J Physiol* 37: 355-377, 1908.
35. Bryan AC, Bentivoglio LG, Beerel F, MacLeish H, Zidulka A, Bates DV. Factors affecting regional distribution of ventilation and perfusion in the lung. *J Appl Physiol* 19: 395-402, 1964.
36. Brice AG, Forster HV, Pan LG, Funahashi A, Lowry TF, Murphy CL, Hoffman MD. Ventilatory and $PaCO_2$ responses to voluntary and electrically induced leg exercise. *J Appl Physiol* 64: 218-225, 1988.
37. Buckler KJ, Vaughan-Jones RD, Peers C, Lagadic-gossmann D, Nye PCG. Effects of extracellular pH, PCO_2 and HCO_3^- on intracellular pH in isolated type-I cells of the neonatal rat carotid body. *J Physiol* 444: 703-721, 1991.

38. Buckler KJ, Williams BA, Honore E. An oxygen, acid and anaesthetic sensitive TASK-like background potassium channel in rat arterial chemoreceptor cells. *J Physiol* 525: 135-142, 2000.

39. Burdon JG, Juniper EF, Killian KJ, Hargreave FE, Campbell EJ. The perception of breathlessness in asthma. *Am Rev Respir Dis* 126: 825-828, 1982.

40. Buytendijk JH. *Esophageal Pressure and Lung Elasticity*. Doctoral thesis, University of Groningen, The Netherlands, 1949.

41. Campbell EJM. *The Respiratory Muscles and the Mechanics of Breathing*. Chicago: Year Book, 1958.

42. Campbell EJ, Howell JB. The sensation of breathlessness. *Br Med Bull* 19: 36-40, 1963.

43. Carson ER, Cunningham DJC, Hercynski R, Murray-Smith DJ, Petersen ES. *Modelling of a Biological Control System: The Regulation of Breathing*. Oxford: Institute of Measurement and Control, 1978.

44. Casaburi R, Whipp BJ, Wasserman K, Beaver WL, Koyal SN. Ventilatory and gas exchange dynamics in response to sinusoidal work. *J Appl Physiol* 42: 300-311, 1977.

45. Cathcart A, Herrold N, Turner AP, Wilson J, Ward SA. Absence of long-term modulation in response to external dead-space loading during moderate exercise in humans. *Eur J Appl Physiol* 93: 411-420, 2005.

46. Cherniack NS. Respiratory sensation as a respiratory controller. In: Adams L, Guz A, eds. *Respiratory Sensation*. New York: Dekker, 1996, 213-230.

47. Christie RV. Dyspnoea: A review. *Quart J Med* 7: 421-454, 1938.

48. Clark FJ, von Euler C. On the regulation of depth and rate of breathing. *J Physiol* 222: 267-295, 1972.

49. Comroe JH. Some theories of the mechanism of dyspnoea in breathlessness. In: Howell JBL, Campbell, EJM, eds. *Breathlessness*. Oxford: Blackwell Scientific, 1965, 1-7.

50. Crandall ED, Forster RE. Rapid ion exchange across the red cell membrane. *Adv Chem Ser* 118: 65-87, 1973.

51. Cropp GJA, Comroe JH Jr. Role of mixed venous CO_2 in respiratory control. *J Appl Physiol* 16: 1029-1033, 1961.

52. Cross BA, Davey A, Guz A, Katona PG, MacLean M, Murphy K, Semple SJ, Stidwill R. The pH oscillations in arterial blood during exercise: A potential signal for the ventilatory response in the dog. *J Physiol* 329: 57-73, 1982.

53. Cummin ARC, Tyawe VI, Saunders KB. Ventilation and cardiac output during the onset of exercise and during voluntary hyperventilation in humans. *J Physiol* 370: 567-583, 1986.

54. Cumming G. Gas mixing in disease. In: Scadding JG, Cumming G, eds. *Scientific Foundations of Medicine*. London: Heinemann, 1981.

55. Cunningham DJC. Oral history project. Physiological Society audio recording, 1994.

56. Cunningham DJC, Lloyd BB, eds. *The Regulation of Human Respiration*. Philadelphia: Davis, 1963.

57. Daly WJ, Overley T. Modification of ventilatory regulation by hypnosis. *J Lab Clin Med* 68: 279-285, 1966.

58. Davenport PW, Vovk A. Cortical and subcortical central neural pathways in respiratory sensations. *Respir Physiol Neurobiol* 167: 72-86, 2009.

59. Davidson JT, Whipp BJ, Wasserman K, Koyal SN, Lugliani R. Role in the carotid bodies in the sensation of breathlessness during breath-holding. *N Engl J Med* 290: 819-822, 1974.

60. Dayman H. Mechanics of airflow in health and in emphysema. *J Clin Invest* 30: 1175-1190, 1951.

61. Defares JG. Principles of feedback control and their application to the respiratory control system. In: Fenn WO, Rahn H, eds. *Handbook of Physiology: Respiration.* Vol I. Washington, DC: American Physiological Society, 1964, 649-680.

62. Dejours P. Control of respiration by arterial chemoreceptors. *Ann NY Acad Sci* 109: 682-695, 1963.

63. Dejours P. Control of respiration in muscular exercise. In: Fenn WO, Rahn H, eds. *Handbook of Physiology: Respiration.* Vol I. Washington, DC: American Physiological Society, 1964, 631-648.

64. Dempsey JA, Adams L, Ainsworth DM, Fregosi RF, Gallagher CG, Guz A, Johnson BD, Powers SK. Airway, lung and respiratory muscle function. In: Rowell LB, Shepherd JT, eds. *Handbook of Physiology: Section 12. Exercise: Regulation and Integration of Multiple Systems*. New York: Oxford University Press, 1996, 448-514.

65. Dempsey J, Hansen P, Henderson K. Exercise-induced hypoxemia in healthy persons at sea level. *J Physiol* 355: 161-175, 1984.

66. Dempsey JA, Sheel AW, St. Croix CM, Morgan BJ. Respiratory influences on sympathetic vasomotor outflow in humans. *Respir Physiol Neurobiol* 130: 3-20, 2002.

67. Dempsey JA, Wagner PD. Exercise-induced arterial hypoxemia. *J Appl Physiol* 87: 1997-2006, 1999.

68. Dempsey JA, Whipp BJ. The respiratory system. In: Tipton C, ed. *Exercise Physiology: People and Ideas*. Oxford: Oxford University Press, 2003, 138-187.

69. DiMarco AF, Romaniuk JR, von Euler C, Yamamoto Y. Immediate changes in ventilation and respiratory pattern with onset and offset of locomotion in the cat. *J Physiol* 343: 1-16, 1983.

70. Doll E, Keul J, Brechtel A, Limon-Lason R, Reindell H. Der Einfluss korperlicher Arbeit auf die ar-

teriellen Blutgase in Freiburg und in Mexico City. *Sportarzt und Sportsmedizin* 8: 317-325, 1967.

71. Douglas CG. Co-ordination of the respiration and circulation with variations in bodily activity. *Lancet* 210: 213-218, 1927.
72. Douglas CG, Haldane JS. The regulation of normal breathing. *J Physiol* 38: 420-440, 1909.
73. Downes JJ, Lambertsen CJ. Dynamic characteristics of ventilatory depression in man on abrupt administration of O_2. *J Appl Physiol* 21: 447-453, 1966.
74. DuBois AB, Botelho SY, Comroe JH Jr. A new method for measuring airway resistance using a body plethysmograph. *J Clin Invest* 35: 327-355, 1955.
75. DuBois AB, Britt AG, Fenn WO. Alveolar CO_2 during the respiratory cycle. *J Appl Physiol* 4: 535-548, 1952.
76. Effros RM, Chang RSY, Silverman P. Carbonic anhydrase activity of the pulmonary vasculature. *Science* 199: 427-429, 1978.
77. El-Manshawi A, Killian KJ, Summers E, Jones NL. Breathlessness during exercise with and without resistive loading. *J Appl Physiol* 61: 896-905, 1986.
78. Eldridge FL, Chen Z. Respiratory sensation: A neurophysiological perspective. In: Adams L, Guz A, eds. *Respiratory Sensation*. New York: Dekker, 1996, 19-67.
79. Eldridge FL, Gill-Kumar P. Central neural respiratory drive and afterdischarge. *Respir Physiol* 40: 49-63, 1980.
80. Eldridge FL, Gill-Kumar P. Mechanisms of hyperpnea induced by isoproterenol. *Respir Physiol* 40: 349-363, 1980.
81. Eldridge FL, Waldrop TG. Neural control of breathing. In: Whipp BJ, Wasserman K, eds. *Pulmonary Physiology and Pathophysiology of Exercise*. New York: Dekker, 1991, 309-370.
82. Emtner M, Porszasz J, Burns M, Somfay A, Casaburi R. Benefits of supplemental oxygen in exercise training in nonhypoxemic chronic obstructive pulmonary disease patients. *Am J Resp Crit Care Med* 168: 1034-1042, 2003.
83. Enghoff H. Volume inefficax. Bemerkungen zur Frage des schädlichen Raumes. *Upsala Lakaref Forh* 44: 191-218, 1938.
84. Farhi LE. Ventilation-perfusion relationship and its role in alveolar gas exchange. In: Caro CG, ed. *Advances in Respiratory Physiology*. London: Arnold, 1966, 148-197.
85. Farhi LE. World War II and respiratory physiology: The view from Rochester, New York. *J Appl Physiol* 69: 1565-1570, 1990.
86. Farhi LE, Rahn H. Gas stores in the body and the unsteady state. *J Appl Physiol* 7: 472484, 1955.
87. Floyer J. *The Physician's Pulse Watch; or, an Essay to Explain the Old Art of Feeling the Pulse and to Improve It By the Help of a Pulse-Watch*. Vol I. London: S. Smith and B. Walford, 1707.
88. Forster HV, Dempsey JA, Thomson J, Vidruk R, DoPico GA. Estimation of arterial PO_2, PCO_2, pH and lactate from arterialised venous blood. *J Appl Physiol* 32: 134-137, 1972.
89. Fredericq L. Sur la cause de l'apnée. *Arch Biol (Liege, Paris)* 17: 561-576, 1901.
90. Fujie T, Tojo N, Inase N, Nara N, Homma I, Yoshizawa Y. Effect of chest wall vibration on dyspnea during exercise in chronic obstructive pulmonary disease. *Respir Physiol Neurobiol* 130: 305-316, 2002.
91. Gandevia SC, Macefield G. Projection of low threshold afferents from human intercostal muscles to the cerebral cortex. *Respir Physiol* 77: 203-214, 1989.
92. Gandevia SC, Killian K, McKenzie DK, Crawford M, Allen GM, Gorman RB, Hales JP. Respiratory sensations, cardiovascular control, kinaesthesia and transcranial stimulation during paralysis in humans. *J Physiol* 470: 85-107, 1993.
93. Geppert J, Zuntz N. Ueber die Regulation der Atmung. *Arch Ges Physiol* 42: 189-244, 1888.
94. Gledhill N, Froese AB, Dempsey JA. Ventilation to perfusion distribution during exercise in health. In: Dempsey JA, Reed CE, eds. *Muscular Exercise and the Lung*. Madison, WI: University of Wisconsin Press, 1977, 325-343.
95. Goodwin GM, McCloskey DI, Mitchell JH. Cardiovascular and respiratory responses to changes in central command during isometric exercise at constant muscle tension. *J Physiol* 226: 173-190, 1972.
96. Gray JS. The multiple factor theory and the control of ventilation. *Science* 103: 739-744, 1946.
97. Grimby G, Saltin B, Wilhelmsen L. Pulmonary flow-volume and pressure-volume relationship during submaximal and maximal exercise in young well-trained men. *Bull Physiopathol Respir* 7: 157-172, 1971.
98. Grimby G, Stiksa J. Flow-volume curves and breathing patterns during exercise in patients with obstructive lung disease. *Scand J Clin Lab Invest* 25: 303-313, 1970.
99. Grodins FS. Some simple principles and complex realities of cardiopulmonary control in exercise. *Circ Res* XX-XXI(Suppl.): I171-I178, 1967.
100. Guz A, Noble MIM, Eisele JH, Trenchard D. Experimental results of vagal blockade in cardiopulmonary disease. In: Porter R, ed. *Breathing: Hering-Breuer Centenary Symposium*. London: Churchill, 1970, 315-329.
101. Haldane JS. *Respiration*. New Haven, CT: Yale University Press, 1922.
102. Haldane JS, Lorrain Smith J. The absorption of oxygen by the lungs. *J Physiol* 22: 231-258, 1897.

103. Haldane JS, Priestley JG. The regulation of the lung-ventilation. *J Physiol* 32: 225-266, 1905.

104. Hansen E. Über dir Sauerstoffschuld bei körperlicher Arbeit. *Arbeitsphysiol* 8: 151-171, 1935.

105. Hansen JE, Stelter GP, Vogel JA. Arterial pyruvate, lactate, pH and PCO_2 during work at sea level and high altitude. *J Appl Physiol* 23: 523530, 1967.

106. Harms CA, Babcock MA, McClaran SR, Pegelow DF, Nickele GA, Nelson WB, Dempsey JA. Respiratory muscle work compromises leg blood flow during maximal exercise. *J Appl Physiol* 82: 1573-1583, 1997.

107. Harms CA, Wetter TJ, St. Croix CM, Pegelow DF, Dempsey JA. Effects of respiratory muscle work on exercise performance. *J Appl Physiol* 89: 131-138, 2000.

108. Harrison TR, Harrison WG, Marsh JP. Reflex stimulation of respiration from increase in venous pressure. *Am J Physiol* 100: 417-419, 1932.

109. Haouzi P. Counterpoint: Supraspinal locomotor centers do not contribute significantly to the hyperpnea of dynamic exercise. *J Appl Physiol* 100: 1079-1082, 2006.

110. Haouzi P, Chenuel B, Huszczuk A. Sensing vascular distension in skeletal muscle by slow conducting afferent fibers: Neurophysiological basis and implication for respiratory control. *J Appl Physiol* 96: 407-418, 2004.

111. Henry FM, DeMoor JC. Lactic and alactic oxygen consumption in moderate exercise of graded intensity. *J Appl Physiol* 8: 608-614, 1956.

112. Herxheimer H, Kost R. Das verhaltnis von sauerstoffaufnahme und kohlen-saurausscheidung zur ventilation bei harter muskelarbeit. *Z Klin Med* 108: 240-247, 1932.

113. Hesser CM, Linnarsson D, Fagraeus L. Pulmonary mechanics and work of breathing at maximal ventilation and raised air pressure. *J Appl Physiol* 50: 747-753, 1981.

114. Hewlett AW, Barnett GD, Lewis JK. The effect of breathing oxygen enriched air during exercise upon pulmonary ventilation and upon the lactic acid content of blood and urine. *J Clin Invest* 3: 317-325, 1926.

115. Hey EN, Lloyd BB, Cunningham DJC, Jukes MGM, Bolton DPG. Effects of various respiratory stimuli on the depth and frequency of breathing in man. *Respir Physiol* 1: 193-205, 1966.

116. Hill AV, Lupton H. Muscular exercise, lactic acid and the supply and utilization of oxygen. *Quart J Med* 16: 135-171, 1923.

117. Holmgren A, Linderholm H. Oxygen and carbon dioxide tensions of arterial blood during heavy and exhaustive exercise. *Acta Physiol Scand* 44: 203-215, 1958.

118. Honda Y. Respiratory and circulatory activities in carotid body-resected humans. *J Appl Physiol* 73: 1-8, 1992.

119. Hopkins SR, Schoene RB, Henderson WR, Spragg RG, Martin TR, West JB. Intense exercise impairs the integrity of the pulmonary blood-gas barrier in elite athletes. *Am J Respir Crit Care Med* 155: 1090-1094, 1997.

120. Hornbein TF, Griffo ZJ, Roos A. Quantitation of chemoreceptor activity: Interrelation of hypoxia and hypercapnia. *J Neuro Physiol* 24: 561-568, 1961.

121. Howell JBL, Campbell EJM, eds. *Breathlessness*. Oxford: Blackwell Scientific, 1965.

122. Hughes JMB. Diffusive gas exchange. In: Whipp BJ, Wasserman K, eds. *Pulmonary Physiology and Pathophysiology of Exercise*. New York: Dekker, 1991, 143-171.

123. Huszczuk A, Whipp BJ, Adams TD, Fisher AG, Crapo RO, Elliott CG, Wasserman K, Olsen DB. Ventilatory control during exercise in calves with artificial hearts. *J Appl Physiol* 68: 2604-2611, 1990.

124. Huszczuk A, Whipp BJ, Oren A, Shors EC, Pokorski M, Nery LE, Wasserman K. Ventilatory responses to partial cardiopulmonary bypass at rest and exercise in dogs. *J Appl Physiol* 61: 575-583, 1986.

125. Hyatt RE, Schilder DP, Fry DL. Relationship between maximum expiratory flow and degree of lung inflation. *J Appl Physiol* 13: 331-338, 1958.

126. Innes JA, Solarte I, Huszczuk A, Yeh E, Whipp BJ, Wasserman K. Respiration during recovery from exercise: Effects of trapping and release of femoral blood flow. *J Appl Physiol* 67: 2608-2613, 1989.

127. Jack S, Rossiter HB, Pearson MG, Ward SA, Warburton CJ, Whipp BJ. Ventilatory responses to CO_2 inhalation, hypoxia and exercise in idiopathic hyperventilation. *Am J Respir Crit Care Med* 170: 118-125, 2004.

128. Johnson BD, Badr S, Dempsey JA. Impact of the aging pulmonary system on the response to exercise. *Clin Chest Med* 15: 229-246, 1994.

129. Johnson RL Jr., Taylor HF, DeGraff AC Jr. Functional significance of a low diffusing capacity for carbon monoxide. *J Clin Invest* 44: 789-800, 1965.

130. Jones NL, McHardy GJ, Naimark A, Campbell EJ. Physiological dead space and alveolar-arterial gas pressure differences during exercise. *Clin Sci* 31: 19-29, 1966.

131. Jones NL, Robertson DG, Kane JW. Difference between end-tidal and arterial PCO_2 in exercise. *J Appl Physiol* 47: 954-960, 1979.

132. Jones PW, Huszczuk A, Wasserman K. Cardiac output as a controller of ventilation through changes in right ventricular load. *J Appl Physiol* 53: 218-244, 1982.

133. Kao FF. An experimental study of the pathways involved in exercise hyperpnea, employing cross-circulation techniques. In: Cunningham DJC, Lloyd BB, eds. *The Regulation of Human Respiration*. Philadelphia: Davis, 1963, 461-502.
134. Karetzky MS, Cain SM. Effect of carbon dioxide on oxygen uptake during hyperventilation in normal man. *J Appl Physiol* 28: 8-12, 1970.
135. Kaufman MP, Rybicki KJ. Discharge properties of group III and IV muscle afferents: Their responses to mechanical and metabolic stimuli. *Circ Res* 61(Suppl. I): 60-65, 1987.
136. Kemp G. Lactate accumulation, proton buffering and pH change in ischemically exercising muscle. *Am J Physiol Reg Int Comp Physiol* 289: R895-R901, 2005.
137. Kety SS. The theory and applications of the exchange of inert gas at the lungs and tissues. *Pharmacol Rev* 3: 1-41, 1951.
138. Killian KJ, Campbell EJM. Dyspnea and exercise. *Ann Rev Physiol* 5: 465-479, 1983.
139. Killian KJ, Gandevia SC. Sense of effort and dyspnea. In: Adams L, Guz A. *Respiratory Sensation*. New York: Dekker, 1996, 181-199.
140. Killian KJ, Summers E, Basalygo M, Campbell EJM. Effect of frequency on perceived magnitude of added loads to breathing. *J Appl Physiol* 58: 1616-1621, 1985.
141. Kimoff RJ, Cheong TH, Cosio MG, Guerraty A, Levy RD. Pulmonary denervation in humans. Effects on dyspnea and ventilatory pattern during exercise. *Am Rev Resp Dis* 142: 1034-1040, 1970.
142. Klas JV, Dempsey JA. Voluntary versus reflex regulation of maximal exercise flow: Volume loops. *Am Rev Respir Dis* 139: 150-156, 1989.
143. Kostreva DR, Zuperku EJ, Purtock RV, Coon RL, Kampine JP. Sympathetic afferent nerve activity of right heart origin. *Am J Physiol* 229: 911-915, 1975.
144. Koulouris NG, Valta P, Lavoie A, Corbeil C, Chasse M, Braidy J, Milic-Emili J. A simple method to detect expiratory flow limitation during spontaneous breathing. *Eur Respir J* 8: 306-313, 1995.
145. Krishnan B, Zintel T, McParland C, Gallagher CG. Lack of importance of respiratory muscle load in ventilatory regulation during heavy exercise in humans. *J Physiol* 490: 537-550, 1996.
146. Krogh A. On the mechanism of the gas-exchange in the lungs. *Skand Arch Physiol* 23: 248-278, 1910.
147. Krogh A, Lindhard J. The regulation of respiration and circulation during the initial stages of muscular work. *J Physiol* 47: 112-136, 1913.
148. Krogh A, Lindhard J. On the average composition of the alveolar air and its variations during the respiratory cycle. *J Physiol* 47: 431-445, 1914.
149. Krogh A, Lindhard J. The volume of the dead space in breathing and the mixing of gases in the lungs of man. *J Physiol* 51: 59-90, 1917.
150. Lamarra N, Whipp BJ, Ward SA. Physiological inferences from intra-breath measurement of pulmonary gas exchange. *Proc Ann Internat Conf IEEE Eng Med Biol Soc,* 1988, 825-826.
151. Leaver DJ, Pride NB. Flow-volume curves and expiratory pressures during exercise in patients with chronic airway obstruction. *Scand J Respir Dis* 77(Suppl.): 23-27, 1971.
152. Leitner L-M, Dejours P. The speed of response of chemoreceptors. In: Torrance RW, ed. *Arterial Chemoreceptors*. Oxford: Blackwell, 1968, 79-88.
153. Leusen I. Aspects of the acid-base balance between blood and cerebrospinal fluid. In: Brooks C, Kao FF, Lloyd BB, eds. *Cerebrospinal Fluid and the Regulation of Ventilation*. Oxford: Blackwell, 1965, 5589.
154. Lewis SM. Awake baboon's ventilatory response to venous and inhaled CO_2 loading. *J Appl Physiol* 39: 417-422, 1975.
155. Liljestrand G. Untersuchungen über die Atmungsarbeit. *Skand Arch Physiol* 35: 199-293, 1918.
156. Lilienthal JR Jr., Riley RL, Proemmel DD, Franke RE. An experimental analysis in man of the oxygen pressure gradient from alveolar air to arterial blood during rest and exercise at sea level and at altitude. *Am J Physiol* 147: 199-216, 1946.
157. Luciani L. Delle oscillazioni della pressione intratoracica e intraddominale. *Arch Sci Med* 2: 177-224, 1878.
158. Ludwig C. Beiträge zur Kenntniss des Einflusses der Respirationsbewegungen auf den Blutlauf im Aortensysteme. *Archiv für Anat Physiol und Wissenschaftliche Med* 242-302, 1847.
159. Ludwig C. Zusammenstellung der Untersuchungen über Blutgase. *Zeitschrift Kaiserlich Königlich Gesellschaft der Ärzte in Wien* 1: 145-166, 1865.
160. Lugliani R, Whipp BJ, Brinkman J, Wasserman K. Doxapram hydrochloride: A respiratory stimulant for patients with primary alveolar hypoventilation. *Chest* 76: 414-419, 1979.
161. Lugliani R, Whipp BJ, Seard C, Wasserman K. The effect of bilateral carotid body resection on ventilatory control at rest and during exercise in man. *N Engl J Med* 285: 1105-1111, 1971.
162. Luiten PGM, ter Horst GJ, Karst H, Steffens AB. The course of paraventricular hypothalamic efferents to autonomic structures in medulla and spinal cord. *Brain Res* 329: 374-378, 1985.
163. MacKechnie JK, Leary WP, Noakes TD, Kallmeyer JC, MacSearraigh ET, Olivier LR. Acute pulmonary oedema in two athletes during a 90-km running race. *S Afr Med J* 56: 261-265, 1979.

164. McCloskey DI, Mitchell JH. Reflex cardiovascular and respiratory responses originating in exercising muscle. *J Physiol* 224: 173-186, 1972.
165. McKay LC, Adams L, Frackowiak RS, Corfield DR. A bilateral cortico-bulbar network associated with breath holding in humans, determined by functional magnetic resonance imaging. *Neuroimage* 40: 1824-1832, 2008.
166. Macklem PT. Discussion of "inspiratory muscle fatigue as a factor limiting exercise" by Grassino A, Gross D, Macklem PT, Roussos C, Zagelbaum G. *Bull Eur Physiopathol Respir* 15: 111-115, 1979.
167. Macklem PT, Mead J. Resistance of central and peripheral airways measured by a retrograde catheter. *J Appl Physiol* 22: 395-440, 1967.
168. Mador MJ, Magalang UJ, Rodis A, Kufel TJ. Diaphragmatic fatigue after exercise in healthy human subjects. *Am Rev Respir Dis* 148: 1571-1575, 1993.
169. Majcherczyk S, Willshaw P. Inhibition of peripheral chemoreceptor activity during superfusion with an alkaline c.s.f. of the ventral brainstem surface of the cat. *J Physiol* 231: 26P, 1973.
170. Margaria R, Mangili F, Cuttica F, Cerretelli P. The kinetics of oxygen consumption at the onset of muscular exercise in man. *Ergonomics* 8: 49-54, 1965.
171. Margaria R, Milic-Emili G, Petit JM, Cavagna G. Mechanical work of breathing during muscular exercise. *J Appl Physiol* 15: 354-358, 1960.
172. Martin BJ, Weil JV. CO_2 and exercise tidal volume. *J Appl Physiol* 46: 322-325, 1979.
173. Martin PA, Mitchell GS. Long term modulation of the exercise ventilatory response in goats. *J Physiol* 470: 601-617, 1993.
174. Mateika JH, Duffin J. Coincidental changes in ventilation and electromyographic activity during consecutive incremental exercise tests. *Eur J Appl Physiol Occup Physiol* 68: 54-61, 1994.
175. Mead J. Control of respiratory frequency. *J Appl Physiol* 15: 325-336, 1960.
176. Mead J. Volume displacement body plethysmograph for respiratory measurement in human subjects. *J Appl Physiol* 15: 736-740, 1960.
177. Means JH. Dyspnoea. *Medicine* 3: 309-416, 1924.
178. Meakins J, Long CNH. Oxygen consumption, oxygen debt and lactic acid in circulatory failure. *J Clin Invest* 4: 273-293, 1927.
179. Miescher-Rusch F. Bemerkungen zur Lehre von den Ahembewegungen. *Arch Anat u Physiol (Leipzig)* 6: 355-380, 1885.
180. Milic-Emili G, Cajani F. La frequenza dei respiri in funzione della ventilazione durante la marcia. *Boll Soc Ital Biol Sper* 33: 825-827, 1957.
181. Milic-Emili G, Petit JM. Il lavoro meccanico della respirazione a varia frequenza respiratoria. *Arch Sci Biol (Bologna)* 43: 326-330, 1959.
182. Milic-Emili G, Petit JM, Deroanne R. Mechanical work of breathing during exercise in trained and untrained subjects. *J Appl Physiol* 17: 43-46, 1962.
183. Milic-Emili J, Whitelaw WA, Grassino AE. Measurement and testing of respiratory drive. In: Hornbein T, ed. *The Regulation of Breathing*. New York: Dekker, 1981, 675-743.
184. Mitchell JH, Schmidt RF. Cardiovascular reflex control by afferent fibers from skeletal muscle receptors. In: Shepherd JT, Abboud FM, Geiger SR. *Handbook of Physiology. The Cardiovascular System: Peripheral Circulation and Organ Blood Flow.* Sect 2, Vol III. Bethesda, MD: Am Physiol Soc, 1983, 623-658.
185. Mithoefer JC. Breath holding. In: Fenn WO, Rahn H. *Handbook of Physiology, Section 3: Respiration.* Vol 2. Washington, DC: American Physiological Society, 1965, 1011-1025.
186. Morgan WP, Raven PB, Drinkwater BL, Horvath SM. Perceptual and metabolic responsivity to standard bicycle ergometry following various hypnotic suggestions. *Int J Clin Exp Hypnosis* 21: 86-101, 1973.
187. Moosavi SH, Guz A, Adams L. Repeated exercise paired with "imperceptible" dead space loading not alter $\dot{V}_E$ of subsequent exercise in humans. *J Appl Physiol* 92: 1159-1168, 2002.
188. Nielsen M, Hansen O. Maximale korperliche Arbeit bei Atmung O_2-reicher Luft. *Skand Arch Physiol* 76: 37-59, 1937.
189. O'Donnell DE, Fluge T, Gerken F, Hamilton A, Webb K, Aguilaniu B, Make B, Magnussen H. Effects of tiotropium on lung hyperinflation, dyspnoea and exercise tolerance in COPD. *Eur Respir J* 23: 832-840, 2004.
190. Olafsson S, Hyatt RE. Ventilatory mechanics and expiratory flow limitation during exercise in normal subjects. *J Clin Invest* 48: 564-573, 1969.
191. Otis A. The work of breathing. In: Fenn WO, Rahn H, eds. *Handbook of Physiology, Respiration.* Vol I. Washington, DC: American Physiological Society, 1964, 463-476.
192. Otis AB, Fenn WO, Rahn H. Mechanics of breathing in man. *J Appl Physiol* 2: 592-607, 1950.
193. Paintal AS. Vagal sensory receptors and their reflex effects. *Physiol Rev* 53: 159-227, 1973.
194. Pan LG, Forster HV, Bisgard GE, Dorsey SM, Busch MA. Cardiodynamic variables and ventilation during treadmill exercise in ponies. *J Appl Physiol* 57: 753-759, 1984.
195. Peiffer C, Costes N, Hervé P, Garcia-Larrea L. Relief of dyspnea involves a characteristic brain activation and a specific quality of sensation. *Am J Respir Crit Care Med* 177: 440-449, 2008.
196. Pflüger E. Uber die Diffusion des Sauerstoffs, den Ort und die Gesetze der Oxydationsprozesse im tier-

ischen Organismus. *Arch Gesamte Physiol* 6: 43-190, 1892.

197. Phillipson EA, Bowes G, Townsend ER, Duffin J, Cooper JD. Role of metabolic CO_2 production in ventilatory response to steady-state exercise. *J Clin Invest* 68: 768-774, 1981.
198. Pierce AK, Luterman D, Loudermilk J, Blomqvist G, Johnson RL Jr. Exercise ventilatory patterns in normal subjects and patients with airway obstruction. *J Appl Physiol* 25: 249-254, 1968.
199. Piiper J. Physiological equilibria of the gas cavities of the body. In: Fenn WO, Rahn H, eds. *Handbook of Physiology, Respiration.* Vol 2. Washington, DC: American Physiological Society, 1965, 1205-1218.
200. Piiper J, Scheid P. Blood gas equilibration in lungs. In: West JB, ed. *Pulmonary Gas Exchange.* Vol II. New York: Academic Press, 1980, 132-161.
201. Poole DP, Ward SA, Whipp BJ. Control of blood gas and acid-base status during isometric exercise in humans. *J Physiol* 396: 365377, 1988.
202. Poon CS. Optimal control of ventilation in hypoxia, hypercapnia and exercise. In: Whipp BJ, Wiberg DM, eds. *Modelling and Control of Breathing.* New York: Elsevier, 1983, 189-196.
203. Poon CS, Ward SA, Whipp BJ. Influence of inspiratory assistance on ventilatory control during moderate exercise. *J Appl Physiol* 62: 551-560, 1987.
204. Potter WA, Olafsson S, Hyatt RE. Ventilatory mechanics and expiratory flow limitation during exercise in patients with obstructive lung disease. *J Clin Invest* 50: 910-919, 1971.
205. Powers SK, Lawler J, Dempsey JA, Dodd S, Landry G. Effects of incomplete pulmonary gas exchange on $\dot{V}O_2$max. *J Appl Physiol* 66: 2491-2495, 1989.
206. Pride NB, Permutt S, Riley RL, Bromberger-Barnea B. Determinants of maximal expiratory flow from the lungs. *J Appl Physiol* 23: 646-662, 1967.
207. Rahn H. A concept of mean alveolar air and the ventilation-blood flow relationships. *Am J Physiol* 158: 21-30, 1949.
208. Rahn H, Fenn WO. *A Graphical Analysis of the Respiratory Gas Exchange.* Washington, DC: American Physiological Society, 1955.
209. Rahn H, Otis AB, Chadwick LE, Fenn WO. The pressure-volume diagram of the thorax and lung. *Am J Physiol* 146: 161-178, 1946.
210. Rausch SM, Whipp BJ, Wasserman K, Huszczuk A. Role of the carotid bodies in the respiratory compensation for the metabolic acidosis of exercise in humans. *J Physiol* 444: 567-578, 1991.
211. Reinhard U, Mueller PH, Schmuelling RM. Determination of anaerobic threshold by the ventilation equivalent in normal individuals. *Respiration* 38: 36-42, 1979.
212. Reischl P, Gonzalez F Jr., Greco EC Jr., Fordyce WE, Grodins FS. Arterial PCO_2 response to intravenous CO_2 in awake dogs unencumbered by external breathing apparatus. *J Appl Physiol* 47: 1099-1104, 1979.
213. Rice AJ, Thornton AT, Gore CJ, Scroop GC, Greville HW, Wagner H, Wagner PD, Hopkins SR. Pulmonary gas exchange during exercise in highly trained cyclists with arterial hypoxemia. *J Appl Physiol* 87: 1802-1812, 1999.
214. Riley RL. The hyperpnea of exercise. In: Cunningham DJC, Lloyd BB, eds. *The Regulation of Human Respiration.* Oxford: Blackwell, 1963, 525-534.
215. Riley RL, Cournand A. "Ideal" alveolar air and the analysis of ventilation-perfusion relationships in the lungs. *J Appl Physiol* 1: 825-847, 1949.
216. Riley RL, Cournand A. Analysis of factors affecting the partial pressures of oxygen and carbon dioxide in gas and blood of lungs: Theory. *J Appl Physiol* 4: 77-101, 1951.
217. Robbins PA, Conway J, Cunningham DA, Khamnei S, Paterson DJ. A comparison of indirect methods for continuous estimation of arterial PCO_2 in men. *J Appl Physiol* 68: 1727-1731, 1990.
218. Rohrer F. Physiologie der Atembewegungen. In: Bethe A, von Bergmann G, Embden G, Ellinger A, eds. *Handbuch der Normalen und Pathologischen Physiologie.* Vol 2. Berlin: Springer, 1925, 70-127.
219. Rohrer F. Der Strömungswiderstand in den menschlichen Atemwegen und der Einfluss der unregelmässigen Verweigung des Bronchialsystems auf den Atmungsverlauf in verschiedenen Lungenbezirken. *Plug Arch ges Physiol* 162: 225-299, 1925.
220. Rossiter HB, Ward SA, Kowalchuk JM, Howe FA, Griffiths JR, Whipp BJ. Dynamic asymmetry of phosphocreatine concentration and O_2 uptake between the on- and off-transients of moderate- and high-intensity exercise in humans. *J Physiol* 541: 991-1002, 2002.
221. Saper CB, Loewy AD, Swanson LW, Cowan WM. Direct hypothalamo-autonomic connections. *Brain Res* 117: 305-312, 1976.
222. Senapati JM. Effect of stimulation of muscle afferents on ventilation of dogs. *J Appl Physiol* 21: 242-246, 1966.
223. Shea SA, Andres LP, Shannon DC, Banzett RB. Ventilatory responses to exercise in humans lacking ventilatory chemosensitivity. *J Physiol* 469: 623-640, 1993.
224. Shephard RJ. The oxygen cost of breathing during vigorous exercise. *Q J Exp Physiol Cogn Med Sci* 51: 336-350, 1966.
225. Shepherd RH. Effect of pulmonary diffusing capacity on exercise tolerance. *J Appl Physiol* 12: 487-488, 1958.
226. Somjen GC. The missing error signal—Regulation beyond negative feedback. *NIPS* 7: 184-185, 1992.

227. Sophocles. Antigone. In: *The Theban Plays* [translated by Watling EF]. Middlesex: Penguin, 1974, 126-162.

228. Spengler CM, Banzett RB, Systrom DM, Shannon DC, Shea SA. Respiratory sensations during heavy exercise in subjects without respiratory chemosensitivity. *Respir Physiol* 114: 65-74, 1998.

229. Stark RD Gambles SA, Lewis JA. Methods to assess breathlessness in healthy subjects: A critical evaluation and application to analyse the acute effects of diazepam and promethazine on breathlessness induced by exercise or by exposure to raised levels of carbon dioxide. *Clin Sci* 61: 429-439, 1981.

230. Steinhagen C, Hirche HJ, Nestle HW, Bovenkamp U, Hosselmann I. The interstitial pH of the working gastrocnemius muscle of the dog. *Pflügers Archiv Eur J Physiol* 367: 151-156, 1976.

231. Stevens SS. In: Stevens G, ed. *Psychophysics: Introduction to its Perceptual, Neural, and Social Prospects*. New York: Wiley Interscience, 1975.

232. Strange-Petersen E, Whipp BJ, Drysdale DB, Cunningham DJC. Carotid arterial blood gas oscillations and the phase of the respiratory cycle during exercise in man: Testing a model. *Adv Exp Med Biol* 99: 335-342, 1978.

233. Swanson GD. An optimal controller motivated variation hypothesis. In: Benchetrit G, ed. *Concepts and Formulations in the Control of Breathing*. Manchester: Manchester University Press, 1987, 143-153.

234. Sylvester JT, Whipp BJ, Wasserman K. Ventilatory control during brief infusions of CO_2-laden blood in the awake dog. *J Appl Physiol* 35: 178-186, 1973.

235. Taylor BJ, How SC, Romer LM. Exercise-induced abdominal muscle fatigue in healthy humans. *J Appl Physiol* 100: 1554-1562, 2006.

236. Taylor R, Jones NL. The reduction by training of CO_2 output during exercise. *Eur J Cardiol* 9: 53-62, 1979.

237. Theodore J, Morris AJ, Burker CM, Glanville AR, VanKessel A, Baldwin JC, Stinson EB, Shumway NE, Robin ED. Cardiopulmonary function at maximum tolerable constant work rate exercise following human heart-lung transplantation. *Chest* 92: 433-439, 1987.

238. Thornton JM, Guz A, Murphy K, Griffith AR, Pedersen DL, Kardos A, Leff A, Adams L, Casadei B, Paterson DJ. Identification of higher brain centres that may encode the cardiorespiratory response to exercise in humans. *J Physiol* 533: 823-836, 2001.

239. Uchida Y. Tachypnea after stimulation of afferent cardiac sympathetic nerve fibers. *Am J Physiol* 230: 1003-1007, 1976.

240. Verges S, Schulz C, Perret C, Spengler CM. Impaired abdominal muscle contractility after high-intensity exhaustive exercise assessed by magnetic stimulation. *Muscle Nerve* 34: 423-430, 2006.

241. von Leupoldt A, Sommer T, Kegat S, Baumann HJ, Klose H, Dahme B, Büchel C. The unpleasantness of perceived dyspnea is processed in the anterior insula and amygdala. *Am J Respir Crit Care Med* 177: 1026-1032, 2008.

242. von Neergaard K, Wirtz K. Über eine Methode zur Messung der Lungenelastizität am lebenden Menschen, insbesondere beim Emphysem. *Z Klin Med* 105: 35-50, 1927.

243. Wagner PD, Gale GE, Moon RE, Torre-Bueno JR, Stolp WB, Saltzman HA. Pulmonary gas exchange in humans exercising at sea level and simulated altitude. *J Appl Physiol* 61: 260-270, 1986.

244. Wagner PD, Saltzman HA, West JB. Measurement of continuous distributions of ventilation-perfusion ratios: Theory. *J Appl Physiol* 36: 588-599, 1974.

245. Waldrop TG, Iwamoto GA. Point: Supraspinal locomotor centers do contribute significantly to the hyperpnea of dynamic exercise. *J Appl Physiol* 100: 1077-1079, 2006.

246. Ward SA, Macias D, Whipp BJ. Is breath-hold time an objective index of exertional dyspnoea in humans? *Europ J Appl Physiol* 85: 272-279, 2001.

247. Ward SA, Whipp BJ. Effects of peripheral and central chemoreflex activation on the isopnoeic rating of breathing in exercising humans. *J Physiol* 411: 27-43, 1989.

248. Wasserman K, Hansen JE, Sue DY, Stringer WW, Sietsema KE, Sun X-G, Whipp BJ. *Principles of Exercise Testing and Interpretation*. 5th ed. Philadelphia: Lippincott Williams & Wilkins, 2012.

249. Wasserman K, Mitchell RA, Burger AJ, Casaburi R, Davis JA. Mechanism of the isoproterenol hyperpnea in the cat. *Respir Physiol* 38: 359-376, 1979.

250. Wasserman K, Van Kessel A, Burton GB. Interaction of physiological mechanisms during exercise. *J Appl Physiol* 22: 71-85, 1967.

251. Wasserman K, Whipp BJ, Casaburi R, Beaver WL, Brown HV. CO_2 flow to the lungs and ventilatory control. In: Dempsey JA, Reed, CE, eds. *Muscular Exercise and the Lung*. Madison, WI: University of Wisconsin Press, 1977, 105-135.

252. Wasserman K, Whipp BJ, Casaburi R, Huntsman D, Castagna J, Lugliani R. Regulation of arterial PCO_2 during intravenous CO_2 loading. *J Appl Physiol* 38: 651-656, 1975.

253. Wasserman K, Whipp BJ, Castagna J. Cardiodynamic hyperpnea: Hyperpnea secondary to cardiac output increase. *J Appl Physiol* 36: 457-464, 1974.

254. Wasserman K, Whipp BJ, Koyal SN, Cleary MG. Effect of carotid body resection on ventilatory and acid-base control during exercise. *J Appl Physiol* 39: 354-358, 1975.

255. Weibel ER. *Morphometry of the Human Lung*. Berlin: Spring, 1963.

256. WeilerRavell D, Cooper DM, Whipp BJ, Wasserman K. The control of breathing at the start of exercise as influenced by posture. *J Appl Physiol* 55: 1460-1466, 1983.

257. Weiler-Ravell D, Shupak A, Goldenberg I, Halpern P, Shoshani O, Hirschhorn G, Margulis A. Pulmonary oedema and haemoptysis induced by strenuous swimming. *Br Med J* 311: 361-362, 1979.

258. West JB, Dollery CT. Distribution of blood flow and ventilation-perfusion ratio in the lung measured with radioactive CO_2. *J Appl Physiol* 15: 405-410, 1960.

259. West JB, Tsukimoto K, Mathieu-Costello O, Prediletto R. Stress failure in pulmonary capillaries. *J Appl Physiol* 70: 1731-1742, 1991.

260. Whipp BJ, Lamarra N, Ward SA, Davis JA, Wasserman K. Estimating arterial PCO_2 from flow-weighted and time-averaged alveolar PCO_2 during exercise. In: Swanson GD, Grodins FS, eds. *Respiratory Control: Modelling Perspective*. New York: Plenum, 1990, 91-99.

261. Whipp BJ, Ward SA. The control of ventilation. In: Wilson A, ed. *Pulmonary Function Testing: Indications and Interpretations*. Orlando: Grune & Stratton, 1985, 201220.

262. Whipp BJ, Ward SA. The coupling of ventilation to pulmonary gas exchange during exercise. In: Whipp BJ, Wasserman K, eds. *Pulmonary Physiology and Pathophysiology of Exercise*. New York: Dekker, 1991, 271-307.

263. Whipp BJ, Wasserman K. Oxygen uptake kinetics for various intensities of constant load work. *J Appl Physiol* 33: 351-356, 1972.

264. Wilson RC, Jones PA. A comparison of the visual analogue scale and modified Borg scale for the measurement of dyspnoea during exercise. *Clin Sci* 76: 277-282, 1989.

265. Winning AJ, Hamilton RD, Shea SA, Knott C, Guz A. The effect of airway anaesthesia on the control of breathing and the sensation of breathlessness in man. *Clin Sci* 68: 215-225, 1985.

266. Winterstein H. Die Regulierung der Atmung durch das Blut. *Arch Ges Physiol* 138: 167-184, 1911.

267. Winterstein H. Die Reaktionstheorie der Atmungsregulation. *Arch Ges Physiol* 187: 293-298, 1921.

268. Wright GW, Filley GF. Pulmonary fibrosis and respiratory function. *Am J Med* 10: 642-661, 1951.

269. Yamamoto WS, Edwards MW Jr. Homeostasis of carbon dioxide during intravenous infusion of carbon dioxide. *J Appl Physiol* 15: 807-818, 1960.

270. Younes M. Determinants of thoracic excursion during exercise. In: Whipp BJ, Wasserman K, eds. *Pulmonary Physiology and Pathophysiology of Exercise*. New York: Dekker, 1991, 1-65.

271. Zechman FW Jr., Wiley RL. Afferent inputs to breathing: Respiratory sensation. In: Widdicombe JG, Cherniack N, eds. *Handbook of Physiology, Respiration (Control)*. Washington, DC: American Physiological Society, 1986, 449-474.

CHAPTER 11

The Oxygen Transport System: Maximum Oxygen Uptake

Peter G. Snell, PhD

Benjamin D. Levine, MD

Jere H. Mitchell, MD

Summary

Our understanding of the oxygen transport system arises from studies that have been performed for more than 200 yr. In the 20th century the early contributions of N. Zuntz, F.G. Benedict, A. Krogh, A.V. Hill, and others made the measurement of oxygen consumption possible and laid the foundation for the exponential increase in published studies that has occurred since the 1950s. Many of the concepts articulated in these early investigations hold true today. Hill's explanation of a physiological plateau limited by oxygen delivery for the observed leveling off of oxygen uptake at higher workloads is still relevant. The academic interest in $\dot{V}O_2$max and the effects of age, gender, level of habitual activity, and genetics led to important clinical applications. Studies have clarified the role of various organ systems involved in the transfer of oxygen from ambient air to the mitochondria in muscle, where it becomes the final acceptor of hydrogen ions in the process of generating adenosine triphosphate (ATP) from the breakdown of glucose and fat. Thus, muscular contraction activates multiple systems, including the lungs, heart, circulation, muscle, and endocrine organs. It was apparent well before the mechanisms were unraveled that exercise was beneficial and that inactivity, epitomized by bed rest and the weightlessness of space travel, was detrimental and predominated by the loss of muscle mass, bone density, exercise capacity, cardiovascular regulatory mechanisms. These findings led to changes in clinical practice and the prescription of exercise for a variety of metabolic and cardiovascular diseases. Ischemic heart attack patients, who were previously confined to bed for recovery, are now required to become physically active as soon as possible. Techniques from molecular biology have been used to study how exercise stimulates pathways involved in the transcription of hundreds of genes and the very subtle interplay between central nervous system factors and reflexes to match and optimally distribute the available cardiac output to active muscle and other central organs during exercise.

Introduction

Since the late 1950s the measurement of $\dot{V}O_2$max has assumed great importance as an integrator of the respiratory, cardiovascular, and muscular systems during the performance of muscular work in health and disease. Consequently, as shown in figure 11.1, published research on $\dot{V}O_2$max and its meaning has exponentially increased.

In part, this has been due to major refinements in measurement techniques. The use of Douglas bags for collecting expired air and the assessment of O_2 and CO_2 in expired air samples by the laborious Scholander apparatus have given way to computerized systems interfaced with pneumotachs and rapidly responding O_2 and CO_2 analyzers. Online technology allows researchers to perform breath-by-breath analysis of subjects while the subjects are actually exercising, bringing a new level of sophistication to protocols and test performance. In recent times, the use of gas exchange has allowed the use of indexes of hyperventilation such as respiratory exchange ratio ($\dot{V}CO_2/\dot{V}O_2$) and ventilatory equivalent ($V_E/\dot{V}O_2$) to help indicate when the limits of cardiovascular capacity are approached.

Early Contributions

As discussed by Mitchell (1928–) and Saltin (1933–) in their 2003 review (124), the history of the oxygen transport system in the past 100 yr begins with the work of A.V. Hill (1886-1977), who, along with colleagues Lupton and Long, was the first to systematically explore the limits of oxygen consumption during maximal exercise. In their classic papers on muscular exercise, lactic acid, and the supply and utilization of oxygen (69), they designated this limit the maximum oxygen intake, which for A.V. Hill was 4,055 ml/min at the time. They did emphasize that the maximum value for $\dot{V}O_2$ varied among individuals and they identified the main links in the oxygen transport system, which incorporated the combined and closely integrated functions of the respiratory, circulatory, and skeletal muscle systems. The concept of maximal oxygen consumption was fundamental to the use of exercise as a means of understanding basic physiological regulations. In their 1923 article, Hill and Lupton summarized data in the literature and added their own measurements of oxygen uptake obtained when running at increasing speeds in the open field, including one bout leading to exhaustion. This led to the following statement, which has had a profound influence on exercise physiology to this day:

> However much the speed be increased beyond this limit, no further increase in oxygen intake can occur: The heart, lungs, circulation, and the diffusion of oxygen to the active muscle-fibers have attained their maximal activity. (68)

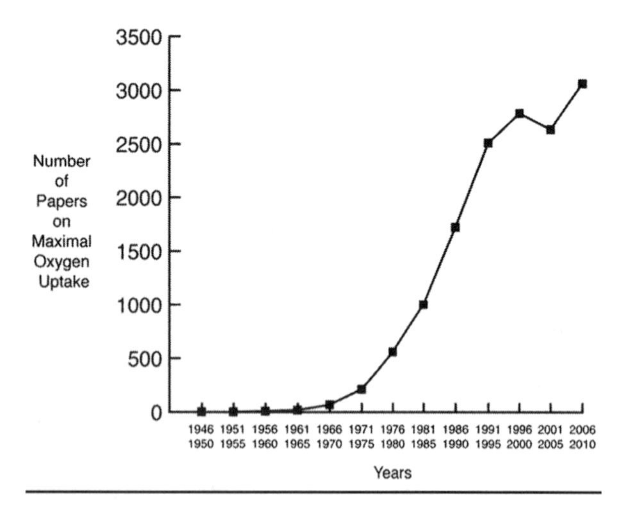

Figure 11.1 Publication history from 1946 of papers on maximal oxygen uptake in which the term *maximal oxygen uptake* or an equivalent is found in the title or as a key word in the abstract. From a search of Ovid MEDLINE, Ovid MEDLINE In-Process, and Ovid OLD MEDLINE by Helen Mayo.

Adapted from Mitchell and Saltin 2003.

Hill's research was based on the studies of several 18th- and 19th-century scientists, including Antoine Lavoisier (1743-1794) in Paris, who first described how respiration involved the combustion of carbon in oxygen, and Nathan Zuntz (1847-1920), who constructed the first treadmill in his Cologne laboratory and conducted energy expenditure studies during hypoxia and exercise in humans and dogs. Later, physiologists Francis Benedict (1870-1957), August Krogh (1874-1949), and Goran Liljestrand (1886-1960) made important contributions linking components involved in the transport of oxygen from ambient air to its final role in the oxidation of carbon substrates in exercising muscles.

Since Lavoisier's research at the end of the 18th century, the understanding of the limits of humans in performing physical activity has intrigued investigators with interest in both applied and basic science. However, before quantitative studies of gas exchange in humans could be performed, technical developments were crucial, such as the Tissot spirometer, Douglas' invention of a bag for air collection, and Haldane's absorption method for measuring CO_2 and O_2—all accomplished in the early 1900s (35, 56, 182).

Thus, some 140 yr after Lavoisier, there was a basis for Hill to measure oxygen consumption and, more importantly, to provide an understanding of the meaning of such a measurement and its maximal value. After Hill, determinations of maximal oxygen uptake became the common method for characterizing cardiorespiratory fitness in health and disease. It is still the gold standard.

Studies have also been performed to define the functional role of each link in the oxygen transport chain and to determine the regulation and adaptability of these links. Francis G. Benedict made measurements of oxygen consumption during light, moderate, and heavy exercise on a cycle ergometer. During his research training with Wilbur O. Atwater (1844-1907) at Wesleyan College in Connecticut in the United States, Benedict changed his interest from biochemistry to physiology, a transformation that led to his outstanding contributions to the understanding of energy expenditure at rest and during exercise. Atwater had studied with the

German chemists and physiologists Voigt and Rubner, and later with Zuntz, and he returned with the latest techniques for measuring respiration and metabolism. Benedict, like Atwater, received training in Europe with Voigt. Benedict developed a highly sophisticated calorimeter that made it possible to very precisely measure heat production, energy turnover, and respiratory exchange ratios (9). He found that oxygen consumption progressively increased during exercise as workloads increased and that the highest value obtained in a 32-yr-old former professional cyclist was 3,265 ml/min (49.5 $ml \cdot kg^{-1} \cdot min^{-1}$) (9, 13). However, he made no attempt to determine whether a maximal value was obtained in any of his studies. Benedict's other important contributions were the precise measurement of gross and net efficiency in bicycle exercise and the degree to which fat and carbohydrate were metabolized to provide the energy required during rest and exercise (14).

Goran Liljestrand studied at the Karolinska Institute in Stockholm with Erik Johan Johansson (1862-1938), who in 1895 had suggested a coupling between central and peripheral neural mechanisms in regulating the heart rate in response to exercise (77). Liljestrand, together with N. Stenstrom, took these investigations to the field in an impressive series of studies in humans. They determined the oxygen uptake when walking, running, swimming, rowing, and cross-country skiing (111, 112). Also, measurements of blood pressure and cardiac output were made during swimming and rowing (109, 110). They indirectly used the term *maximal oxygen intake* and demonstrated that slightly higher values were observed in cross-country skiing than in running (111). An unresolved question is whether these early measurements performed by Liljestrand, Stenstrom, and others were true maximal values in the various exercise modes.

August Krogh (1874-1949), working in the laboratory of Christian Bohr (1855-1911), studied the mechanisms by which oxygen was transported from the alveoli to the blood in the pulmonary circulation. Bohr, like most physiologists at the time (e.g., Haldane, Douglas, Starling), favored the notion that oxygen was secreted into the blood, whereas Krogh worked on the hypothesis that O_2 transport was brought about by simple diffusion. His studies published in 1910 convincingly proved that O_2 even at large oxygen intakes could be transported by diffusion (96). August Krogh did not receive the Nobel Prize for this research. Instead, he was awarded the Nobel Prize in 1920 for his equally outstanding research on the function of capillaries (95).

During the decade of 1910, Krogh's interest in using exercise as an intervention while studying cardiorespiratory regulation was initiated by Johannes Lindhard (1870-1947). Their cooperation resulted in several significant contributions over the next 15 yr. One major contribution was the detailed elaboration of the role of cortical irradiation that instantaneously affects pulmonary ventilation and heart rate at the onset of exercise. The first study, published in 1913 (98), was followed by experiments with heavy and maximal exercise that involved measurements during both exhaustive exercise and the recovery phase (97, 98). They demonstrated that oxygen uptake during recovery is far greater than the lack of oxygen uptake at the start of exercise. Moreover, in a separate study, they proved that carbohydrate as well as fat could be directly combusted by contracting skeletal muscles (99).

Determination of Maximal Oxygen Consumption

Methods of Producing Exercise

Early in the 1900s there were well-functioning exercise machines on which one could precisely control the speed and incline (treadmills) or the brake on a flywheel either electronically or mechanically. In 1954, von Dobeln (184) introduced the sinus pendulum with a known weight. By changing the angle of the axis from its vertical position via a ribbon, the pendulum precisely controlled the brake of the flywheel. Researchers were also interested in measuring oxygen uptake in several other forms of exercise. Expired air was collected in bags and analyzed for volume and O_2 and CO_2 content. Liljestrand and Stenstrom applied a similar procedure in their studies on swimming, skating, rowing, and cross-country skiing (111, 112). Together with Lindhard, Liljestrand also introduced tethered exercise in which the swimmer or the rowing boat was firmly attached to a bridge, making the subject exercise in one spot. This simplified the collection of the expired air (109, 110) and allowed for other measurements (e.g., blood pressure, cardiac output) to be made using a foreign gas method.

These early exercise devices have become more sophisticated. Tethered swimming has given way to the swimming flume in which water is circulated at electronically controlled speeds. Exercise treadmills are large enough to allow cross-country skiers to use roller skis and poles. Cycles and rowing ergometers have computer displays of power output, and some models are interfaced with a heart rate transmitter so that power output may be automatically regulated to a predetermined heart rate.

Evaluation Criteria for $\dot{V}O_2max$

In 1923, A.V. Hill (figure 11.2) and H. Lupton (68) reviewed the work of Benedict and Cathcart from 1913 (14), Liljestrand and Stenstrom's articles from 1920, and reports of others who had measured oxygen uptake in various exercises (14, 200). It was quite apparent that peak oxygen uptake was highest in cross-country skiing, slightly lower in running or bicycling, and even lower in swimming. In the running experiments by Liljestrand and Stenstrom (111), it appeared likely that $\dot{V}O_2max$ was reached in at least some subjects because little or no increase in $\dot{V}O_2$ was observed with increasing treadmill speed. Indeed, a decrease from the peak values was apparent at times. Hill and Lupton acknowledged this finding and highlighted the point that this leveling off in O_2 uptake at super-high speeds for the subject does not mean an improved running efficiency, as proposed by Liljestrand and Stenstrom (111), but rather that aerobic metabolism cannot contribute more and the anaerobic energy yield makes up for the difference in regard to energy demand at the actual speed (68). It is surprising that it took so long for the interplay between aerobic and anaerobic energy yields during heavy exercise to be defined, given that it was known in the previous century that lactic acid was formed during intense exercise (36). In contrast, understanding of ATP and phosphocreatine and their roles in providing the energy for the mechanical output of skeletal muscles had to await the work of Lundsgaard (116) and Lohmann (114) in the early 1930s. It is noteworthy that although well-designed treadmills were available, Hill preferred walking and running in the field or on the track for his experiments. To determine the velocity of the runner, he developed a sophisticated electromagnetic system that provided split times for every 25 yards. In the experiments on Hill himself, a leveling off in $\dot{V}O_2$ was observed, not as a function of increasing speed of running but with time at the highest velocity, which was 260 m/min (68). The expected $\dot{V}O_2$ for this velocity is 50 to 51 $ml \cdot kg^{-1} \cdot min^{-1}$.

Robert Herbst (1890-1962) was the first to carefully apply the leveling-off criterion to establish $\dot{V}O_2max$ in his studies in the late 1920s (63). His subjects ran in place with higher and higher step frequency. In 52 male subjects aged 19 to 24 yr, he found $\dot{V}O_2max$ values that ranged from 1,815 to 4,022 ml/min. He was also the first to express $\dot{V}O_2max$ in milliliters per kilogram per minute. The two principles of establishing $\dot{V}O_2max$—a plateau in $\dot{V}O_2$ with time at an exercise intensity leading to exhaustion within 3 to 6 min or a plateau with increasing exercise loads—should give similar results. Although research by Henry Longstreet Taylor (1912-1983) and colleagues in 1955 showed that the latter was the better method (179), a study by Åstrand and Saltin (8) showed that both methods give similar maximal values provided that a sufficient warming-up period is allowed when using Hill's approach. Recently, a study of a large number of athletes (59) clearly documented the presence of a plateau. Hawkins and colleagues showed a similar leveling of oxygen uptake using both a treadmill test with increasing workloads and a supramaximal test lasting 3 min at a treadmill speed and grade demanding 80 $ml \cdot kg^{-1} \cdot min^{-1}$ of oxygen uptake (figure 11.3).

Noakes has challenged whether Hill actually demonstrated a plateau in $\dot{V}O_2$ and thus measured a true $\dot{V}O_2max$ (130). As argued by Benjamin Levine (1956–) (103), a large part of the debate instigated by Noakes hinges on whether downstream factors, predominantly muscle motor recruitment, alone drive $\dot{V}O_2max$ or whether $\dot{V}O_2max$ has upstream limits independent of muscle motor recruitment. Indeed, Noakes articulated what he considers a new model of integrated performance physiology, which he called the central governor model (131-134). In this formulation, a neural mechanism or central governor shuts down the body by putting a brake on muscle motor recruitment at very high work rates to prevent a disturbance of homeostasis. Therefore, during an incremental exercise test, the highest $\dot{V}O_2$ achieved does not reflect a true maximal ability to transport oxygen to the tissues and use it to make ATP available for physical work because there remains reserve that subjects do not choose to

Figure 11.2 Archibald Vivian Hill (1886-1977), who was the first to describe that a point is reached during increasing levels of heavy exercise where "no further increase in oxygen intake can occur: The heart, lungs, circulation, and the diffusion of oxygen to the active muscle-fibers have attained their maximal activity" (67).

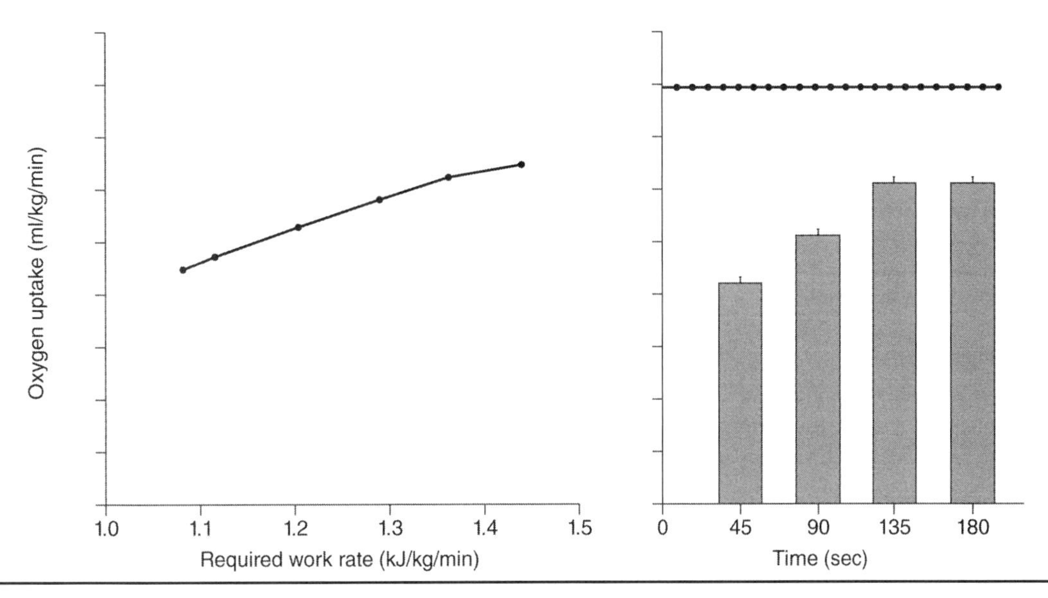

Figure 11.3 Determination of maximal oxygen uptake. Left: Treadmill test with incremented increases in workload calculated from treadmill speed and grade as kilojoules per kilogram per minute. Right: Supramaximal treadmill test at imposed workload of 80 $ml \cdot kg^{-1} \cdot min^{-1}$ of oxygen uptake.

Reprinted, by permission, from M.N. Hawkins et al., 2007, "Maximal oxygen uptake as a parametric measure of cardiorespiratory capacity," *Medicine Science in Sports and Exercise* 39: 103-107.

evoke. However, because of Krogh's and Lindhard's findings, cardiovascular scientists have appreciated that a central mechanism plays a role in the cardiovascular response to exercise (50, 98, 121, 126, 196, 197). Also, when skeletal muscle motor units are inhibited by curare, which weakens the muscle contraction, the blood pressure and heart rate responses to exercise are augmented (102, 123) as a function of increasing central command. Feedback to the brain from mechanically and metabolically sensitive skeletal muscle afferents also plays an essential role in increasing sympathetic nervous system outflow (121, 126) and in regulating the augmentation in cardiac output and the distribution of muscle blood flow—a response that is tightly regulated with little effect of age, sex, or fitness (46, 118). Indeed, when such signals are disturbed, the cardiovascular response to exercise is dramatically altered. For example, patients with muscle metabolic disorders may have 3 to 5 or, in extraordinary cases, more than 10 times the increase in cardiac output normally seen for a given increase in oxygen uptake (57, 107, 177). Countless papers on the topic of cardiovascular regulation in healthy and patient populations, both animal and human, demonstrate the intimate connection between skeletal muscle and the central nervous system. These papers were reviewed thoroughly in a recent themed issue of *Experimental Physiology* (142).

A.V. Hill appears to have formulated his conclusions from the experiments conducted on himself. More importantly, he was the one who conceived the physiological meaning of $\dot{V}O_2max$. Together with Lupton, he was the first to make a clear distinction between O_2 intake and demand. In a critical comment on the conclusion reached by Liljestrand and Stenstrom—that running economy improved at higher speeds—Hill and Lupton wrote:

> It was apparently more economical to run fast than slow! Now, the opposite is notoriously the case, and these observations of Liljestrand and Stenstrom (of which, on technical grounds, we have no criticism) obviously need an explanation. The explanation is simple: The subjects of their experiments were not in a genuine steady state at the higher speeds. In the case, e.g., of their subject N.S. (111, p. 183) it is clear that the maximum oxygen intake of about 3.3 liters per min was attained at a speed of about 186 meters per min. Hence, however fast N.S. ran above this speed he did not use more oxygen, not because he did not require it, but because he could not get it. (68)

It is very fortunate that Hill decided in 1920 to apply his research talents to the study of exercising humans. His Nobel Prize-winning studies of isolated frog skeletal muscle made him uniquely qualified to study similar problems in exercising people. In this regard, he wrote:

> As it proved, work on hard muscular exercise in man provided the same opportunity of getting accurate and reproducible results as is found in experiments on isolated muscle. Indeed, in some ways, man was the better experimental object; when trained, he can repeat the same performance again and again. And it remained a matter always of satisfaction, sometimes even of excitement, as the work evolved, to find how the experiments on man and those on isolated muscle confirmed and threw light on one another. (66)

Skeletal Muscle Mass

Whether the muscle mass engaged in exercise could play a role was not a major concern in the initial studies of oxygen transport. As mentioned earlier, Hill and others noted that the highest $\dot{V}O_2max$ was commonly reported during cross-country skiing. However, the question of whether adding upper-body muscles was the cause of the observed higher $\dot{V}O_2max$ was not discussed or answered. Henry Longstreet Taylor and colleagues were the first to determine the role of adding arm to leg exercise in reaching $\dot{V}O_2max$ (179). They established that arm cranking while running on a treadmill increased the value of maximal oxygen uptake (figure 11.4).

In later research by Åstrand and Saltin (8), Bergh and colleagues (16), and Secher and colleagues (169), who more systematically evaluated the effect of adding arm exercise to leg exercise, a difference in $\dot{V}O_2max$ in the two exercises was also found. However, it was not as large as that reported by Taylor and others (179). Arm-plus-leg exercise compared with leg exercise alone most commonly produced a difference of only 0% to 10%. It is noteworthy that when a difference existed it was primarily explained by a wider arteriovenous (a-v)O_2 difference and not by elevated cardiac output (174). The critical question that emerged was what minimum amount of muscle mass must be engaged in exercise to obtain a true measurement of $\dot{V}O_2max$. As shown by Andersen and Saltin (1), peak O_2 uptake by limb skeletal muscle is on the order of 0.2 (sedentary) to 0.4 (well trained) $L \cdot min^{-1} \cdot kg^{-1}$. Thus, a muscle mass of only 10 to 15 kg needs to be involved in order for the exercise to elicit $\dot{V}O_2max$. In healthy, normal humans, and especially in trained cyclists, this is accomplished when running or cycling. However, it has been repeatedly demonstrated that in an individual without cycling training, $\dot{V}O_2max$ on a cycle ergometer is lower than $\dot{V}O_2max$ on a treadmill by as much as 25%. Presumably, local muscle fatigue prevents subjects from activating sufficient muscle mass (64).

Submaximal Methods

The measurement of $\dot{V}O_2max$ is technically difficult and requires the subject to exercise to exhaustion, which is unpleasant and potentially dangerous for elderly subjects. This has led to the development of submaximal tests, which are based on the linear relationship between $\dot{V}O_2$ and heart rate from light exercise to maximal values. The submaximal $\dot{V}O_2$ for any given power output or speed on the treadmill is highly predictable, allowing the test to be conducted without expensive equipment. Steady-state heart rates are measured at two levels of exercise—moderate and moderately hard—and are plotted against the estimated $\dot{V}O_2$ for these two levels. The line connecting these points is then extrapolated to the age-related maximal heart rate. A series of lines of varying slopes can be generated for any population. Fitter individuals have the lower heart rate for any given power output.

Herman Hellerstein (1916-1993) was the first person to recognize that a single equation for all individuals (including heart patients, athletes, and sedentary people) accurately describes this relationship when submaximal $\dot{V}O_2$ and submaximal heart rate (HR) are expressed as a percentage of their maximal values. His

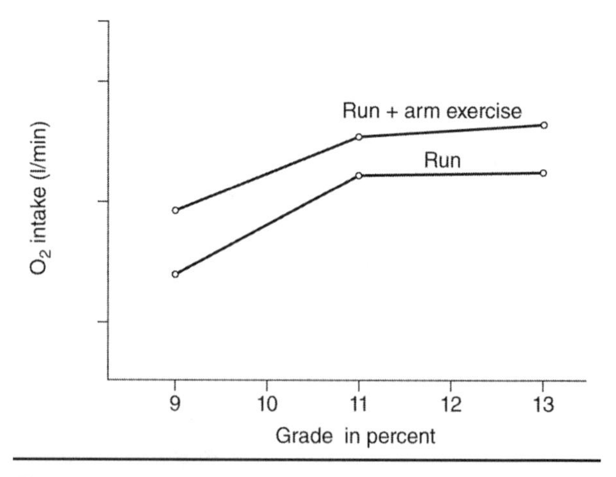

Figure 11.4 Oxygen intake (L/min) during running [7 miles (11.3 km)/h] at increasing grades (percentage) on a treadmill (Run). When arm cranking was added (Run + Arm Exercise), a greater oxygen intake occurred at each level.

Reprinted from H.L. Taylor, E. Buskirk, and A. Henschel, 1955, "Maximal oxygen intake as an objective measure of cardio-respiratory performance," *Journal of Applied Physiology* 8: 73-80, 1955. With permission of American Physiological Society.

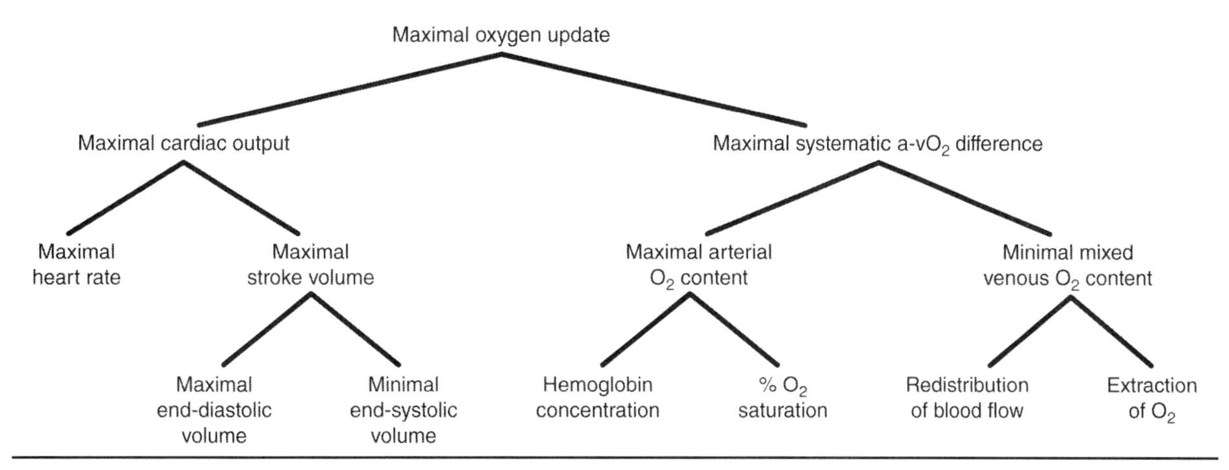

Figure 11.5 Determinants of maximal oxygen uptake. The key variables for oxygen transport and utilization are outlined.

Reprinted from J.H. Mitchell and B. Saltin, 2003, The oxygen transport system and maximal oxygen uptake. In *Exercise physiology,* edited by C.H. Tipton (Oxford: Oxford University Press). ©American Physiological Society.

original formula, published in 1973 (61), was as follows: % $\dot{V}O_2$max = 1.41 · % HRmax − 42. This equation allows one to predict $\dot{V}O_2$max from a single steady-state bout of submaximal exercise during which the heart rate is measured and expressed as a percentage of the age-predicted maximal heart rate.

Oxygen Transport System

Determinants of Maximal Oxygen Uptake

In Hill's 1924 article with Long and Lupton (67), the dominant discussion relates to the factors that determine maximal oxygen uptake (figure 11.5). They considered primarily the role of breathing in saturating the arterial blood with oxygen and the output of the heart, but they also touched on the oxygen capacity of the blood and the degree of desaturation of the venous blood. In their experiments, they saw consistently higher $\dot{V}O_2$max values when subjects breathed O_2-enriched air than when they breathed air only. Much later, Welch and colleagues (194) proposed that the variation and magnitude of the effect of hyperoxia suggested a methodological limitation. However, an elevation in performance was observed in the early work. Hill and colleagues concluded from their work, "It is necessary, however, to assume that a rich oxygen mixture works primarily by increasing the saturation of the blood with oxygen; there is no other way it may work" (67). Not having measured cardiac output, Hill and colleagues estimated what may be the flow rate. Because they assumed a rather low oxygen-carrying capacity of the blood (it appears that they were unaware of hemoconcentration during severe exercise), a low arterial O_2 saturation (−85%; possibly correct because the respiratory valve and tubes added resistance to breathing), and a low O_2 extraction (~60%), the calculated demand on the heart was high. Stroke volume was estimated to be 170 to 220 ml and cardiac output to be 30 to 40 L/min. Although these values are in the current expected range for well-trained people with high maximal oxygen uptakes (>5.0 L/min), they are too high for the subjects studied by Hill and others (67).

Herbst also discussed the reason for the variation in $\dot{V}O_2$max (63). He cites Hill and colleagues (67) and points to the heart and its peak output; he also cites a study by Himwich and Barr (70) suggesting that O_2 saturation may become reduced during exercise. Rowell and colleagues in 1964 (154) were the first to report desaturation in intense exercise. Carter and colleagues observed a decrease in saturation to 80% during an exhaustive run on the current world record holder for the mile in 1965 (22). Later, Dempsey and colleagues (31) confirmed that this occurred in some well-trained subjects, and it appears from a wealth of more recent studies that it was a common phenomenon in many well-trained individuals (94). Moreover, those with the very highest $\dot{V}O_2$max values did not always desaturate; this appears to be a function of pulmonary oxygen diffusing capacity. The phenomenon was most common in athletes (141) when the exercise engaged a large fraction of the muscles of the body and breathing was coupled to the cadence of the movements of the extremities, as in running and rowing.

Herbst highlighted another important point (63). Based on Hill's finding of an effect of O_2-enriched breathing on improved performance and $\dot{V}O_2$, Herbst concluded that skeletal muscle metabolism cannot be at its peak in normoxic exhaustive exercise and thus cannot limit $\dot{V}O_2max$. This last conclusion was reiterated by Bainbridge, Boch, and David Bruce Dill (1891-1986) in a chapter (10) on the limits of muscular exertion in which they say that "this limit is imposed by the supply of oxygen to the muscles and brain rather the functional capacity of skeletal muscles." They also base their conclusion on experiments, such as that performed by Hill and Flack in 1910 (69), in which performance is enhanced when pure oxygen is given in connection with all-out exertion. A review article by Christensen, Krogh, and Lindhard in 1934 also addressed the question of what limits maximal oxygen uptake (25). The article essentially summarized all relevant studies in the field and included a section called "The Oxygen Supply in Work of Maximum Intensity." The article was a report to the predecessor of the World Health Organization in Geneva. They listed the following factors as important in an individual's exercise capacity:

1. Pulmonary ventilation
2. The rate of diffusion of oxygen from alveolar air to the blood
3. The minute volume of the heart
4. The rate of blood flow through the muscles
5. The conditions for O_2 diffusion from the muscle capillaries to the tissue

Points 3 to 5 reiterate what others had concluded before this list was published, but some of the remarks under points 1 and 2 were novel and deserve to be mentioned. Liljestrand used estimates of cost of breathing based on experiments in which CO_2 rebreathing was used to elevate ventilation (108). From rest to a ventilation approaching 100 L/min, O_2 cost increased in proportion to ventilation with a factor of 4 to 5. Another argument for a possible limitation of the lungs in severe exercise came with the demonstration by August and Marie Krogh that the estimated maximum rate of diffusion of O_2 from alveoli to blood was 4.0 to 4.5 L/min, or possibly 5 L/min in exceptional cases (112). In a theoretical and experimental analysis, the value for peak O_2 diffusion in the lungs was later increased to 6.0 L/min (28). Thus, the points that were addressed in those early days in regard to the physiological meaning of and limits to $\dot{V}O_2max$ were the same as those addressed today, as were the arguments. The capacity of the lungs and the heart was thought to be the most likely explanation for the limitation.

Cardiac Output and Systemic a-vO_2 Difference

As outlined in figure 11.6, maximal oxygen uptake is a function of maximal cardiac output and maximal systemic a-vO_2 difference. Foreign gas methods were used early in the 20th century to estimate cardiac output at rest and during exercise up to exhaustive work (18, 113). Although the method became heavily criticized, especially during exercise, the early data have proven to be reasonably similar to what was later found using invasive techniques when the proper solubility constant for acetylene was applied (23, 156). The first to use dye dilution with arterial samples of blood during exercise were Erling Asmussen (1907-1991) and Marius Nielsen (1903-2000) in the early 1950s (3). They measured cardiac output during very intense bicycle exercise, although maximal exertion may not have been achieved. Their subjects reached cardiac outputs of approximately 22 L/min at an oxygen uptake of 3.2 L/min, giving an a-vO_2 difference of 150 ml/L. Similar values were provided by Freedman and colleagues some years later (45).

Mitchell and colleagues (125) were the first to provide data on the missing determinants at maximal oxygen uptake depicted in figure 11.6. In their 1958 article titled "The Physiological Meaning of the Maximal Oxygen Intake Test," they determined the values for cardiac output (heart rate and stroke volume) and total-body a-vO_2 difference at maximal oxygen uptake in untrained subjects. Peak cardiac output was 23.4 L/min at a maximal oxygen uptake of 3.22 L/min. From the Fick principle, they estimated that a-vO_2 difference was 143 ml/L, arterial oxygen content was 193 ml/L, and mixed venous blood O_2 content was 50 ml/L. The latter value was slightly greater than the oxygen content found in blood from the femoral vein during exercise, which was 35 ml/L. Thus, definite answers were given to questions that had been discussed for several decades. For a person with a moderate $\dot{V}O_2max$, oxygenation of blood was not impaired in going from rest to exhaustive exercise because Pa O_2 remained unchanged (87-88 mmHg). The small reduction in O_2 saturation from 97.1% to 94.7% was a function of a shift in acid–base balance and pH decreasing to 7.19. A 9.5-fold elevation in oxygen uptake was brought about by a 2.3-fold widening of the a-vO_2 difference and a 4.3-fold increase in cardiac output. Heart rate and stroke volume contributed about equally to the elevation in cardiac output when subjects were in an upright position both at rest and during exercise. Of particular note was that although O_2 extraction was markedly increased in these studies, blood draining from an exercising limb still contained appreciable O_2, an indication of incomplete

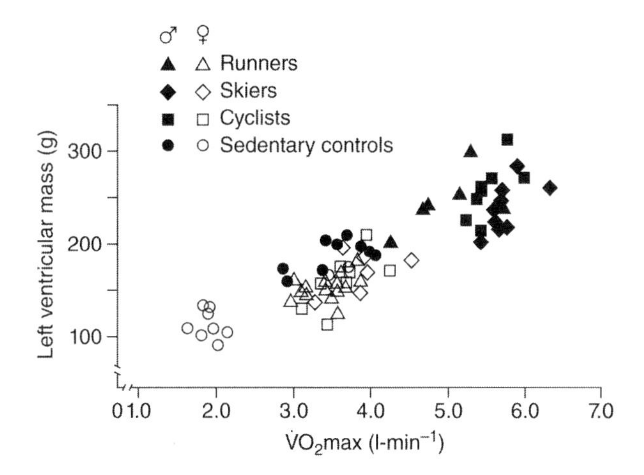

Figure 11.6 The relation of left ventricular mass (g) to maximal oxygen uptake (L/min) of male and female sedentary controls and endurance athletes.

Reprinted from *The American Journal of Cardiology*, Vol. 62, M.C. Milliken, J. Stray-Gundersen, R.M. Peshock, J. Katz, and J.H. Mitchell, "Left ventricular mass as determined by magnetic resonance imaging in male endurance athletes," pgs. 301-305, copyright 1988, with permission of Elsevier; data from Richardson et al. 1993.

O_2 extraction by contracting muscle at $\dot{V}O_2$max. The large widening of the a-vO_2 difference in exercise is primarily a function of a larger fraction of the cardiac output being directed to the exercising limbs, which are able to more completely extract the O_2. These findings have been confirmed in later studies with similar exercise paradigms and measurements; that is, the absolute values for maximal cardiac output and arterially transported O_2 vary among people and are the primary variables most closely related to $\dot{V}O_2$max and exercise capacity (122).

Mitchell and colleagues (125) questioned whether there was a fixed relationship between $\dot{V}O_2$max and maximal cardiac output. In the study, cardiac output was also measured at an exercise intensity higher than that which elicited $\dot{V}O_2$max. At this intensity, the same maximal $\dot{V}O_2$ was achieved but with a 2.8 L/min lower cardiac output and a widening of the a-vO_2 from 14.3 to 15.7 ml/L (125). The explanation is that the more intense exercise elicits a pronounced increase in sympathetic activity, which reduces blood flow to skin and other nonactive tissues, thereby directing a larger fraction of the cardiac output to the contracting muscles. Additional support for this concept is obtained from measurements early (~3-4 min) and late (~7 min) during running (160). At the end of the run, $\dot{V}O_2$max is the same whereas cardiac output is 2 L/min higher. Moreover, if the exercise is performed in a hot environment, cardiac output is also approximately 3 L/min higher at peak exercise, eliciting $\dot{V}O_2$max (156). Thus, there does not appear to be a fixed relationship between $\dot{V}O_2$max and maximal cardiac output in an individual. This should not detract from the early findings of a tight relationship between cardiac output and $\dot{V}O_2$; the ratio is 5:1. That is, for each liter of O_2 uptake above the resting value, there is an increase in cardiac output of about 5 L/min.

Cardiac Dimensions and Function

There is a very close relationship between the size of the heart and the cardiac output. This was suggested by very early studies on heart volume (43). In 1899, Henschen in Sweden used the palpation technique to estimate heart size of cross-country skiers and found a positive relationship between this variable and success in skiing (62). This observation has since been confirmed with more direct measurements. Keys and Friedell (88) used X-ray in the posterior–anterior projection and observed larger diastolic and stroke areas in athletes compared with sedentary subjects, whereas systolic area was similar in the two groups. Further refinements were made by Jonsell (79) and Nylin (136) in Sweden in the 1930s and 1940s, respectively. They developed a method for measuring the size of the heart in diastole in two planes simultaneously at a 90° angle. This method, which was later used in Sweden but used most extensively by Herbert Reindell (1908-1990) (143) in Germany, more precisely linked heart size with the performance of the heart and exercise capacity than was previously possible. Interestingly, Reindell in collaboration with running coach Woldemar Gerschler, developed the concept of interval running (the Gerschler-

Reindell law), which was associated with many famous athletes of the 1940s and 1950s, including Rudolf Harbig, Emil Zátopec, and Gordon Pirie. They claimed that the maximal expansion stimulus on the left ventricle occurred during the immediate postexercise recovery phase. Stroke volume of the heart is larger in the best-trained subjects (17, 190). Although none of these groups measured $\dot{V}O_2max$, they all demonstrated very close relationships between heart size and measures of exercise capacity. This relationship was first confirmed in the 1960s, when $\dot{V}O_2max$ was determined together with other circulatory variables (72). It has now been well established by echocardiography that eccentric hypertrophy is present in endurance-trained athletes (87). Also, as shown in figure 11.6, there is a close correlation between the magnitude of this adaptation of the left ventricle detected using magnetic resonance imaging and the level of maximal oxygen uptake in both male and female endurance athletes (120, 147).

Given that the maximum heart rate of athletes is, if anything, lower than that of nonathletes (152), it follows that the primary distinguishing feature of athletes that allows a large cardiac output is their large stroke volume. Because end systolic volume has never been reported to be smaller in athletes than in nonathletes, the single most important factor allowing this large stroke volume is a large end diastolic volume. Research in Levine's laboratory in Dallas in the United States in 1991 demonstrated the mechanism for this unique characteristic (104) using direct, invasive techniques. Endurance athletes have a markedly greater ability to use the Starling mechanism to increase stroke volume. Contractility was not different between athletes and nonathletes, so virtually all the difference in stroke volume was due to a large end diastolic volume. Athletes were able to achieve such a large end diastolic volume by virtue of markedly enhanced cardiac chamber compliance (figure 11.7). Both static compliance determined from pressure (*P*)–volume (*V*) curves and operational compliance determined from d*V*/d*P* of the pressure–volume curve were substantially larger in the endurance athletes (104). In these studies, the largest hearts for male athletes showed an end diastolic volume in the supine position during volume infusion of around 250 ml, which generated a stroke volume between 130 and 150 ml for peak heart rate of 200 beats/min; this gives a peak cardiac output of about 30 L. Studies in dogs (175) and pigs (58) provided evidence that the pericardium was a critical restraint to maximal left ventricular (LV) filling. When the pericardium was removed in these studies, maximal LV end diastolic volume was significantly increased, leading to increased cardiac output and $\dot{V}O_2max$. Thus, the key distinguishing characteristic of elite endurance athletes is a large end diastolic volume due to a compliant heart and a distensible pericardium.

In order to fill to these large end diastolic volumes during maximal exercise with very high heart rates, the heart must have very rapid diastolic relaxation with vigorous suction. Work by Ferguson and colleagues (42) has shown that athletes do indeed have hearts that fill more rapidly at high levels of exercise intensity, which allows endurance athletes to continue to increase stroke volume at all levels of exercise (49). Diastolic suction develops because the remodeling of the athlete's heart (137) increases the equilibrium volume of the left ventricle, which is the volume in the heart when transmural filling pressure is 0 mmHg (129, 198). When the heart contracts below the equilibrium volume in systole, it engages mechanical restorative forces that markedly augment the transmitral intraventricular pressure gradients that literally suck blood from the left atrium across the mitral valve into the apex of the left ventricle (129, 198). This active process is particularly important during upright exercise when gravitational gradients must be overcome to maximize venous return (106).

Blood Volume and Hemoglobin

The role of blood volume and the total amount of hemoglobin in determining $\dot{V}O_2max$ has been long debated. In 1965, Grande and Taylor (52) summarized matters in stating that "at submaximal exercise, plasma

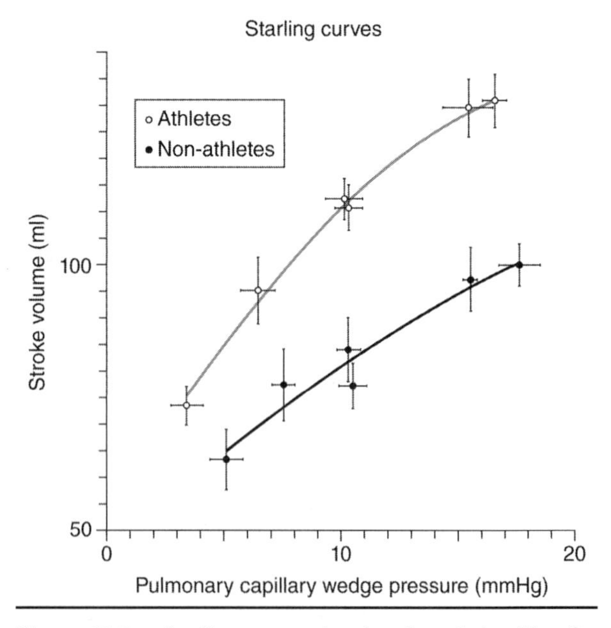

Figure 11.7 Starling curves showing the relationship of stroke volume (ml) to pulmonary capillary wedge pressure (mmHg) in athletes and nonathletes.

Reprinted, by permission, from B.D. Levine, 1991, "Left ventricular pressure-volume and Frank-Starling relations in endurance athletes. Implications for orthostatic tolerance and exercise performance," *Circulation* 84(3): 1016-1023.

volume affected the stroke volume and heart rate response, but it was more questionable whether this was the case also at peak exercise rates." They based this on the finding that plasma volume could be reduced acutely up to 20% without a decrease in cardiac output and $\dot{V}O_2$max (156). In contrast, previous studies in the 1940s, 1950s, and early 1960s found close relationships between both blood volume and total hemoglobin and maximal values for stroke volume, cardiac output, and maximal oxygen uptake (38, 72). Of note was that only a few training bouts may be needed to expand the plasma volume enough to enlarge peak stroke volume, as demonstrated by Green and colleagues (53). Training also stimulates red cell production and thus increases the total amount of hemoglobin, a process that takes considerable time (months), as first convincingly shown in the late 1940s (89). However, there is little or no change in the hemoglobin concentration of the blood.

An elevation in hemoglobin concentration with only minor changes in blood volume also increases $\dot{V}O_2$max, as most clearly demonstrated in the experiments of Ekblom and colleagues in which they gave their subjects a red cell infusion (39). Conversely, a lowering of the hemoglobin concentration, as with anemia, brings about a low exercise capacity and $\dot{V}O_2$max (11, 173). This is in line with systemic oxygen delivery at peak exercise being most closely related to $\dot{V}O_2$max (82).

Factors Influencing the Level of Maximal Oxygen Uptake

How high was the maximal oxygen uptake of our ancestors? A precise answer to this question cannot be given because such measurement has been performed for less than 100 yr. However, there are different approaches to providing a reasonable estimate. One estimate was made during the International Biological Programme in the 1960s when various ethnic groups were physically characterized and the key measure was their maximal aerobic power (100). These studies demonstrated that $\dot{V}O_2$max varied markedly among ethnic groups of the world and appeared to be a function of the demands of their daily lives. People with a sedentary lifestyle, such as the young adults on Easter Island, achieved a $\dot{V}O_2$max of approximately 35 $ml \cdot kg^{-1} \cdot min^{-1}$ (38). In contrast, as long as nomadic Lapps took care of their reindeer, they had a value of 50 $ml \cdot kg^{-1} \cdot min^{-1}$ in $\dot{V}O_2$max; the range around this mean value was very narrow (100). This latter value is of interest because it may represent what the normal $\dot{V}O_2$max was in ancient time for active, healthy adults. Whipp's note on the estimated aerobic capacity of Roman legionnaires suggested that it may have been approximately 50 $ml \cdot kg^{-1} \cdot min^{-1}$ (195). Whipp's estimate was based on reported values of body size of the soldiers, the distances they marched, and the speeds they maintained. Similar estimates of soldiers in the army of Alexander the Great who marched as far as India indicate that $\dot{V}O_2$max was also likely in the range of 50 to 55 $ml \cdot kg^{-1} \cdot min^{-1}$.

In 1938, Sid Robinson (1902-1981) was the first to study the effect of age on $\dot{V}O_2$max (148). Boys from the age of 7 yr increased their $\dot{V}O_2$max as a function of growth and reached the highest level in young adulthood. Normalizing for body weight, the increase was small and the peak $\dot{V}O_2$max reached just below 50 $ml \cdot kg^{-1} \cdot min^{-1}$. The oldest subjects, who were 75 yr of age, reached half the $\dot{V}O_2$max of the young adults (i.e., a reduction of 0.5 $ml \cdot kg^{-1} \cdot min^{-1}$ per year).

Robinson's investigations were cross-sectional and did not take into account modifying influences such as regular physical activity. More recent longitudinal studies (44, 84, 85, 117, 139) have demonstrated that a high $\dot{V}O_2$max may be sustained well into the fourth decade, at which time an accelerated decline occurs.

Dill and colleagues (34) were the first to report on 16 well-trained runners who were followed for 24 yr. They found an average decline of 1.1 $ml \cdot kg^{-1} \cdot min^{-1}$ per year (or 0.06 $L \cdot min^{-1} \cdot yr^{-1}$). This was a larger reduction with age than was found later, where the decline was more in the range of 0.4 to 0.7 $ml \cdot kg^{-1} \cdot min^{-1}$ for both sedentary subjects (118) and physically active people who maintained their activity level (7, 54, 60, 86). Thus, the large decrease in $\dot{V}O_2$max observed by Dill and others (34) was most likely explained by a reduced level of physical activity adding to the effect of age.

Per-Olof Åstrand (1922–) was the first to include girls and young women in his investigations (6). He found that up to puberty, increases in $\dot{V}O_2$max in girls were similar to those in boys (figure 11.8). However, he saw a graduated increase in the difference (~15%) between the sexes after puberty, primarily due to a relatively low hemoglobin concentration and high percentage of body fat in girls. The first to perform studies on the effect of aging on $\dot{V}O_2$max in women was Irma Åstrand in Sweden, who confirmed that an up to 20% difference in $\dot{V}O_2$max existed between women and men (4). She also showed that $\dot{V}O_2$max declined with age at the same rate in women and men.

None of the previously cited studies used measurements from a random sample of subjects in a healthy population. During the last decades of the 20th century, such measurements of $\dot{V}O_2$max were performed in subsamples of large epidemiological studies. In these studies, mean values for $\dot{V}O_2max$ were generally 5% to 10% lower than those observed in previous studies of exercise capacity (20). This could be anticipated. Robinson (148) as well as Åstrand (6) recruited healthy volunteers with a keen interest in being studied. In addition to this explanation for the difference, there could in more recent years be a general trend for lower aerobic power in the population coupled with an increase in fatness. Numerous follow-up studies performed throughout the world during the past three decades that have evaluated the $\dot{V}O_2max$ status of various populations demonstrate mean values for young adult men in the range of 40 to 48 $ml \cdot kg^{-1} \cdot min^{-1}$; values for females are some 10% lower. More pronounced and more important from a general health perspective, however, is the trend for a larger range in $\dot{V}O_2max$ observed in these population studies, especially in the age groups from adolescence to the middle aged (4). The focus on fitness in a small fraction of the population performing regular physical training yields results reflecting the fact that they maintain quite high values into old age (55, 60). However, the possibility of living an almost completely physically inactive life at work and in leisure time is an option today in many countries, and it is the activity level preferred by most people. This adds a large group of people at the lower end of the range, which increases the risk for premature morbidity and mortality, as first shown by Morris and colleagues (128).

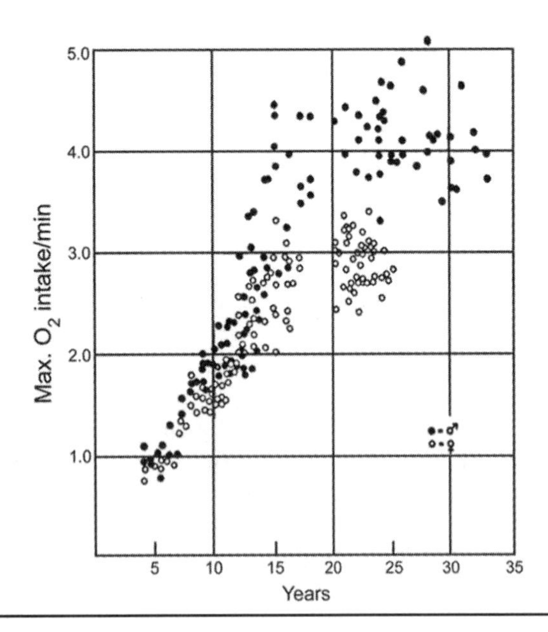

Figure 11.8 The relationship of maximal O_2 intake (L/min) to age (yr) in males and females. Subjects were school children in Stockholm and students at the Gymnastiska Central Institutet.

Adapted from P-O. Åstrand, 1952, *Experimental studies of physical working capacity in relation to sex and age* (Copenhagen: Munksgaard). By permission of P-O. Åstrand.

Level of Habitual Activity

Inactivity

The first study to include accurate measurements of $\dot{V}O_2max$ after bed rest in healthy volunteers was published in 1945 by Taylor and colleagues (180). $\dot{V}O_2max$ was reduced by 17%, or from 3.85 L/min to 3.18 L/min, after 20 d of bed rest. No detailed circulatory measurements were performed to determine the mechanism of this decline; however, such data are available in the Dallas Bed Rest and Training Study of 1968 (160). These findings are shown in figure 11.9. The $\dot{V}O_2max$ of the five subjects in the Dallas study was reduced by 26%, from 3.3 L/min to 2.4 L/min. This reduction, which was slightly larger than that observed by Taylor and colleagues (180), was probably due to a more strict enforcement of inactivity. The reduction in $\dot{V}O_2max$ was fully explained by a lowering of maximal cardiac output, which in turn was due entirely to a reduction in stroke volume (160). It is well known that bed rest caused people to become orthostatic, and it could be that this change contributed to the lowering of stroke volume. In this regard, it is noteworthy that the reduction in stroke volume in the upright position was identical to that found during supine exercise and suggested that an altered size and function of the heart may be the cause of the smaller stroke volume. This has recently been demonstrated in a series of elegant experiments that estimated changes in ventricular compliance with bed rest and aging (2).

Physical Activity (Training)

Cross-sectional studies of $\dot{V}O_2max$ in untrained and trained subjects and the effect of training were performed by Erik Hohwü-Christensen (1904-1996), as was an evaluation of the submaximal $\dot{V}O_2$, cardiac output, heart rate, and stroke volume responses (24). However, a true longitudinal study with $\dot{V}O_2max$ measurements was published in 1941 when Robinson and Harmon (150) reported the changes in $\dot{V}O_2max$ for 9 men as a result of regular training for 26 wk. The subjects were initially fit and had an average $\dot{V}O_2max$ of 53 $ml \cdot kg^{-1} \cdot min^{-1}$, which increased 17% to just above 60 $ml \cdot kg^{-1} \cdot min^{-1}$. A parallel study performed at the Harvard Fatigue Laboratory by Knehr and colleagues and published in 1942 demonstrated an increase in

$\dot{V}O_2max$, but only by 4% (92). No circulatory variables were measured in these two studies.

In the 1960s, three investigations of exercise training were performed that included the measurement of circulatory variables. Those studies had similar designs, duration of training, and methods applied to evaluate, in detail, the cardiovascular response to exhaustive exercise before and after training. One was performed in Minneapolis by Loring B. Rowell (1930–) (153). Another study (figure 11.9), performed in Dallas, included a period of inactivity (bed rest) before the training started (160). Ekblom and colleagues (37) performed the third study in Stockholm. Although the percentage increase in $\dot{V}O_2max$ varied from 11% to 33% in the three studies, the mechanism by which the increase in $\dot{V}O_2$ was brought about was quite similar. In all three studies maximal cardiac output increased and maximal a-vO_2 difference widened. In the Dallas study, these two variables contributed equally; this was also the case in the other two studies when considering the mean values of changes. Stroke volume alone contributed to the elevated maximal cardiac output because maximal heart rate was essentially unchanged in all three studies. A slightly more complete O_2 extraction by the exercising limb explained the enlarged systemic a-vO_2 difference. The larger heart volume contributed to the change in stroke volume.

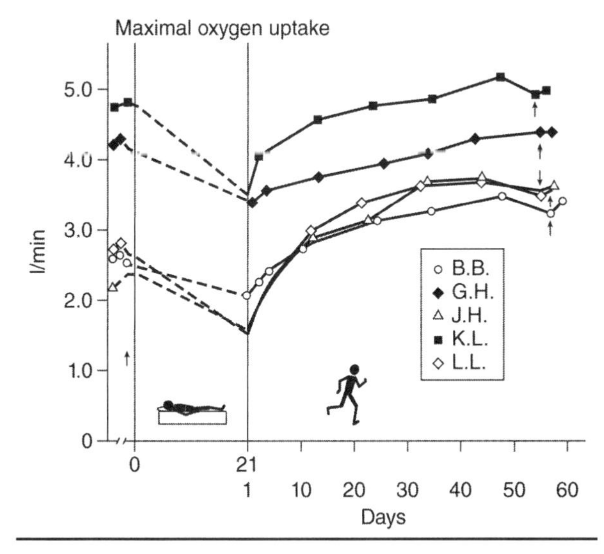

Figure 11.9 Dallas Bed Rest and Training Study, in which researchers examined the effect of 20 d of strict bed rest and 53 to 55 d of intense training on three sedentary and two athletic subjects.

Reprinted by permission, from B. Saltin et al., 1968, "A longitudinal study of adaptive changes in oxygen transport and body composition," *Circulation* 38(suppl 5): VII-1-VII-78.

Endurance Athletes

Hill had an interest in the superb athlete. The highest $\dot{V}O_2max$ value he observed was 65.7 $ml \cdot kg^{-1} \cdot min^{-1}$ in a runner (67), a value similar to that of a Danish bicyclist (63.3 $ml \cdot kg^{-1} \cdot min^{-1}$) studied by Christensen (24). It was Robinson and colleagues, however, who in 1937 made human power a special subject and reported that the best 2 mile runner at the time, Don Lash, had a $\dot{V}O_2max$ of 5.35 L/min, or 81.5 $ml \cdot kg^{-1} \cdot min^{-1}$ (149). This value is comparable with the best runners today, yet his world record of 8 min 58.4 s for the 2 mile now stands at less than 8 min. Higher values have been found in some bicyclists and cross-country skiers. Ingjer measured values higher than 90 $ml \cdot kg^{-1} \cdot min^{-1}$ in elite Norwegian skiers (74). The first to report $\dot{V}O_2max$ values for female athletes engaged in various sports were Saltin and Åstrand (159), who observed in the early 1960s values up to 70 $ml \cdot kg^{-1} \cdot min^{-1}$. Female elite athletes today approach 80 $ml \cdot kg^{-1} \cdot min^{-1}$; cross-country skiers have the highest values (75). Ultimately, $\dot{V}O_2max$ becomes a physiological characteristic bounded by the parametric limits of the Fick equation: (LV end diastolic volume − LV end systolic volume) × heart rate × a-vO_2 difference. Some simple calculations help highlight these boundaries. First, assume a maximum hemoglobin concentration for competitive male athletes of 17 g/dl, the upper limit established by the Fédération Internationale de Ski for allowing athletes to begin a race. If we assume 100% arterial oxygen saturation (i.e., no diffusion limitation or ventilation–perfusion mismatch, certainly an overestimation in competitive athletes), and the lowest mixed venous oxygen saturations measured −14% near the summit of Mount Everest (176), then the largest possible a-vO_2 difference would be 200 ml/L (20 volumes percentage). To the best of our knowledge, the highest published cardiac output during exercise and the largest stroke volume may be those reported in a world-class orienteer of 42.3 L/min and a world champion cyclist of 212 ml (40), respectively, though anecdotal unpublished comments have suggested the possibility of higher values (http://indurain.chez-alice.fr/). Assuming maximal possible a-vO_2 difference and the highest reported exercise cardiac output, this absolute upper limit for $\dot{V}O_2max$ is approximately 8 L/min. The highest reported $\dot{V}O_2max$ of which we are aware is 7.48 L/min in a large, elite skier (158), suggesting that modern elite athletes are approaching these limits of oxygen transport.

One feature of the cardiovascular system is the close matching of oxygen transport to extraction. Early research on training and performance emphasized adaptation of the heart and cardiac output. Measurements of a-vO_2 difference suggested that as long as enough skeletal muscle was recruited a-vO_2 difference could be maintained at approximately 17 ml/dl in the face of a wide range of O_2 delivery. It is now appreciated that adaptations in muscle such as increased capillary and mitochondrial density are key to improved extraction through greater perfusion and reduced capillary transit time coupled with greater flux of substrates through the energy-releasing pathways. In concert with these changes are the reduced resistance to muscle blood flow through increased vasodilator capacity and an accompanying reduced total peripheral resistance, which augment the pumping capacity of the heart.

The training-induced changes in muscle are the result of signaling pathways in muscle fibers being activated to induce transcription of hundreds of genes that code for proteins in muscle filaments, glycolytic and Krebs cycle enzymes, and a variety of receptors responsible for facilitating entry of hormones and fuel into the muscle fiber. It is therefore clear that the more muscle fibers that can participate in the contractile activity, the greater the total amount of new proteins synthesized. This concept, which has important implications for training of elite athletes, is also important for health. Greater muscle fiber recruitment is a countermeasure for sarcopenia and insulin resistance in the elderly and has led to the adoption of interval training. In general, muscular activity can be sustained beyond 10 min at a level demanding about 75% to 80% $\dot{V}O_2max$, called the lactate or ventilatory threshold. Interval training allows short periods of muscular contraction for 1 to 2 min at 100% to 110% $\dot{V}O_2max$.

Ongoing research is attempting to define with greater precision the dose for favorable adaptations in muscle. This whole area is complicated by issues of overtraining and maladaptation. Muscle activity that is too strenuous has many negative effects. These include the activation of stress hormones such as cortisol and excess stress on body structures.

$\dot{V}O_2max$ is not performance, although it clearly is one of the major characteristics that determines performance in endurance sport (32, 138) and frequently this relationship may be obscure when only elite athletes with similar $\dot{V}O_2max$ values are considered (15, 130, 172). However, an elite athlete with a $\dot{V}O_2max$ of 80 $ml \cdot kg^{-1} \cdot min^{-1}$ can run 5,000 m faster than a recreational athlete with a $\dot{V}O_2max$ of 40 $ml \cdot kg^{-1} \cdot min^{-1}$ can, so this characteristic clearly is closely tied to endurance performance (15, 32, 172). Of course, how much $\dot{V}O_2max$ contributes to performance depends on the distance being covered, which determines the rate of optimal or possible energy utilization and the substrate used to produce ATP (32, 80, 138). Moreover, it has been widely recognized for decades that $\dot{V}O_2max$ is not the only characteristic that determines how fast an athlete can travel, especially because the differences in outcome at a world-class level are measured in fractions of a second. Other traits such as sport-specific economy, anaerobic capacity, and, for longer distances, fuel utilization and the speed or oxygen uptake at the maximal steady state all contribute to the final count on the stopwatch (32, 80, 138).

Because maximal oxygen delivery has little to do with most of these factors, athletes can of course perform at work rates higher than $\dot{V}O_2max$ for brief periods of time, regardless of the specific limitations to $\dot{V}O_2max$. This duration and intensity is not unlimited (i.e., a sprinter cannot maintain a speed of 10 m/s for a marathon distance) because human skeletal muscle cannot produce ATP at a high enough rate for a sustained period of time (or regenerate it on a sustained basis via oxidative metabolism) to support this kind of external work. It seems clear that the muscle, not the brain, fatigues over these brief bursts of extremely high levels of motor recruitment, though such local signals are communicated to the brain and influence the athlete's sense of how fast they can continue to run. Motivation and voluntary motor recruitment influenced by sensations coming from skeletal muscle, not the vague action of a central governor, clearly play a role in exercise performance independent of $\dot{V}O_2max$ and cardiovascular performance.

Genetics

In Grande and Taylor's chapter in *Handbook of Physiology* (52), they comment on a possible genetic explanation for the huge individual variation that was observed in $\dot{V}O_2max$ at the time (30-80 $ml \cdot kg^{-1} \cdot min^{-1}$). Based on the available investigation of physical conditioning with 4% to 17% (~9 $ml \cdot kg^{-1} \cdot min^{-1}$) increases in $\dot{V}O_2max$, they concluded that "the individual with a large $\dot{V}O_2max$ per unit of body weight owes this characteristic primarily to the traits of the heart and circulatory system at the time when he reached his full growth."

Studies of twins further highlight the role of genetic factors that influence $\dot{V}O_2max$ and its response to training. To what extent genetics play a role, however, is still debated. In the early research of Klissouras, the inheritance factor was high (90), but it was found to be less strong in later studies (20, 41). Some further support for a role for hereditary factors is found in the ongoing studies (HEITAGE Family Study) by Claude Bouchard (1939–) and colleagues. Among many physiological variables, they are investigating the response of $\dot{V}O_2max$ to training (19). Although the magnitude of the elevation in $\dot{V}O_2max$ with a physical conditioning program varied in earlier studies, very seldom do the participants not improve at all (35, 150, 153, 160). A major drawback with many longitudinal training studies is that they are not based on random samples of subjects but rather use individuals who have an interest in improving fitness. This may, at least in part, explain the difference between the HERITAGE Family Study and other training studies. Another possibility is that the method of applying the training dose may have resulted in an exercise intensity that was too low in the subjects who did not improve. Bouchard and colleagues used cycle ergometers in which the resistance (power output) was automatically adjusted to maintain a preset heart rate. The training heart rate corresponded to 75% $\dot{V}O_2max$ on the initial maximal test. In the experience of one of the authors (PGS) with this equipment, some individuals have a slow heart rate response to the ramp increase in power output, causing the work rate to increase too high initially. This results in an overshoot of the target heart rate, which remains high as the resistance decreases to levels of work well below the desired percentage of $\dot{V}O_2max$.

Another aspect of the response to training has been highlighted. It has been found that the largest improvement in $\dot{V}O_2max$ is observed in people with a low initial pretraining $\dot{V}O_2max$; this finding applies to a wide range of age groups (155, 162). Those with a very high initial $\dot{V}O_2max$ achieve a smaller improvement, percentage wise, in $\dot{V}O_2max$ than those with low initial levels of $\dot{V}O_2max$. The proposed explanation is that the influence of training would be more dominant when the genetic potential is sufficient for a response to be expressed. It has not been defined where that limit in $\dot{V}O_2max$ is and how it differs, most likely, among individuals. Based on a multitude of population studies, a $\dot{V}O_2max$ of 45 to 50 $ml \cdot kg^{-1} \cdot min^{-1}$ appears to be obtainable for healthy young individuals (155). The huge influence of physical activity below that level is underscored by the findings from the three sedentary subjects in the Dallas Bed Rest and Training Study (160). After a 20 d period of inactivity (bed rest), these three subjects lost 33% of their $\dot{V}O_2max$ and reached a level of 24 $ml \cdot kg^{-1} \cdot min^{-1}$. This value then doubled to 48 $ml \cdot kg^{-1} \cdot min^{-1}$ with 2 mo of subsequent training.

Maximal Oxygen Uptake and Endurance Performance

In 1928, Hill discussed the steady-state concept emphasizing "the equilibrium between processes building up versus breaking down" (65). If the latter was in excess of the former, there was no steady state and the exercise had to stop due to exhaustion. Hill also pointed out that the higher the $\dot{V}O_2max$, the higher the exercise intensity that could be performed under a steady-state condition. The first researcher to test a possible coupling between $\dot{V}O_2max$ and performance was Herbst (63), whose subjects performed a 3 mile run. Herbst noted an inverse relationship: The higher the subject's $\dot{V}O_2max$ (expressed in $ml \cdot kg^{-1} \cdot min^{-1}$), the shorter the running time. Herbst also plotted the running performance in relation to $\dot{V}O_2max$ as a percentage of the estimated energy demand for the running speed. Again, the higher the percentage, the shorter the running time.

The concept of percentage of maximal oxygen uptake also appeared in Robinson's 1938 study of males of various ages (148), but he did not apply this concept to the capacity for endurance performance. However, in the parallel study performed at the Harvard Fatigue Laboratory, Knehr and colleagues (92) discussed the huge discrepancy that may exist in how much $\dot{V}O_2max$ may change in comparison with changes in performance. They demonstrated an increase in $\dot{V}O_2max$ of 4%, but Karpovich and Perstrecov (83) concomitantly demonstrated that time to exhaustion at a given submaximal intensity increased markedly after several weeks of training. The explanation is that time to exhaustion is exponentially related to the relative exercise intensity expressed as a percentage of $\dot{V}O_2max$ (165). Thus, at a given workload, which may represent 90% $\dot{V}O_2max$ before training, exhaustion may come after 20 min. With a 20% elevation in $\dot{V}O_2max$, the same exercise intensity now represents approximately 75% $\dot{V}O_2max$ and time to exhaustion is elongated at least four- to fivefold to 2 to 3 h.

A relevant question is whether changes in activity level also affect the time one can perform at a given relative exercise intensity. P.-O. Åstrand and K. Rodahl have promoted this possibility (5), in part based on Irma Åstrand's studies published in 1960 (4). This is in contrast to Hill's explanation of the steady-state concept and accepts that expressing a given aerobic energy yield as a percentage of $\dot{V}O_2max$ is a way to normalize people with different exercise capacities (65). Thus, performance should be quite similar at a given percentage of $\dot{V}O_2max$ regardless of training status. David Costill and colleagues have worked extensively with these problems and found that elite marathon runners exercised at a similar percentage of $\dot{V}O_2max$ (80%-90%) during a race as those finishing in the middle or at the end of the field (29).

The numerous studies highlighting the interplay between $\dot{V}O_2max$ and relative exercise intensity, with endurance performance primarily being a function of the latter variable, should not overshadow the fact that endurance performance can be markedly different among people with the same $\dot{V}O_2max$. Coyle and colleagues performed a study on this particular problem, with intriguing results (30). They argued that the explanation for a difference could be the metabolic capacity of the skeletal muscle engaged in the exercise. In bicyclists with a $\dot{V}O_2max$ of 67 $ml \cdot kg^{-1} \cdot min^{-1}$, they used the blood lactate response to submaximal exercise to categorize well-trained bicyclists into two groups: low-lactate responders and high-lactate responders. Endurance performance was markedly different in the two groups. The high-lactate responders could bicycle only half the time at 88% $\dot{V}O_2max$ compared with the low-lactate responders. Leg muscles were studied for mitochondrial marker enzymes, fiber, and size as well as capillaries. Most surprisingly, there was no difference between groups in regard to mitochondrial enzyme activity and capillaries per muscle fiber. Thus, neither the lactate response nor the performance could be related to the indexes of muscle adaptation. Both variables are easily affected by changes in physical activity level and are proposed to be the link to a more efficient metabolism of the muscle (48, 71, 127), although it may vary despite whether the exercise is performed at the same relative workload (12). Two other muscle indexes differed. Those with a poor performance had fewer Type I muscle fibers than those with a good performance (47% vs. 67%), and their mean muscle fiber size was 15% larger. Type II muscle fibers are not easily fully changed to Type I muscle fibers in humans with training, although more IIa fibers coexpress the myosin heavy chain I isoform (91). Nevertheless, it is difficult to relate the observed differences in lactate and performance to responses to an effect of training. Rather, heredity may be critical because fiber type composition is influenced by genetic endowment (93). It is difficult to envision, however, how differences in myosin heavy chain composition can explain the very different metabolic responses in the two groups because their mitochondrial enzyme profiles are so similar, taking into account the finding by Jansson and Kaijser that in well-trained endurance athletes, Type II fibers are equally as oxidative as Type I fibers (76). Larger fiber size in the group performing poorly probably plays a role because diffusion distances are large, but it can hardly be the whole explanation. We are then left with the conclusion that, generally speaking, expressing exercise intensity as a percentage of $\dot{V}O_2max$ normalizes the endurance performance; however, there is substantial individual variation. Very well-trained people are in the upper range of performance, whereas less trained or untrained people are in the lower range of variation. The prevailing explanation for the variation that exists in time to exhaustion at a given percentage of $\dot{V}O_2max$ is then not only differences in capillaries and mitochondrial capacity of skeletal muscle but also fiber type composition, which in turn has an effect on the metabolic response to exercise.

Limitations to Maximal Oxygen Uptake

The view that Hill expressed in 1923 and 1924 on the meaning of the $\dot{V}O_2max$ measurements became the unchallenged dogma (68). On the topic of what limits maximal oxygen uptake, no consensus has been reached, although almost a century of vigorous debate has been stimulated by the results obtained in many (occasionally brilliant) experiments. A voice in these ongoing discussions has argued that every single link in the oxygen transport system is limiting. Others have argued that no single link can be pinpointed.

The strongest argument that skeletal muscle capacity is the limiting factor in $\dot{V}O_2max$ comes from Ewald Weibel (1929–) of Bern and Richard C. Taylor (1939-1995) at Harvard (178, 192). Their impressive data comprise studies relating $\dot{V}O_2max$ and total volume of muscle mitochondria in various species of animals with weights ranging from 1 g to approximately 400 to 500 kg (horses). Based on a close relationship in a log-log plot of these two variables, they argue that the muscle mitochondrial content sets the ultimate limit for $\dot{V}O_2max$ (73, 193). This may be the case in certain quadrupeds but is hardly the case in humans (151). Abrupt elevation in the maximal amount of oxygen delivered to the muscle results in elevated oxygen consumption. Several interventions have been used to achieve this goal. In addition to O_2-enriched gas and red cell infusion, reduction in blood volume or Hb concentration as well as changing heart rate in a subject with a pacemaker alters $\dot{V}O_2max$ as a function of arterial oxygen delivery (164, 170). Another approach to demonstrating that mitochondrial respiration has a marked overcapacity to utilize O_2 comes from experiments in humans who perform exercise with a small muscle mass. Muscle blood flow is several times higher when a small muscle mass is engaged in the exercise than when a large muscle mass is engaged (e.g., bicycling, running). Muscle oxygen uptake follows the increase in blood flow and arterial oxygen delivery, as does the power output. Peak muscle oxygen uptake per kilogram of muscle can be as high as 0.3 to 0.5 L/min (1, 146). Measurements of Vmax of aerobic enzymes such as pyruvate dehydrogenase and a-keto oxygluterate dehydrogenase, which have the lowest activity of all mitochondria enzymes, also provide evidence for this conclusion (22).

Further evidence that skeletal muscle capacity is the limiting factor comes from the work of Jan Pretorius Clausen (1939-1976) in Copenhagen, which focused on the interplay between cardiac output and peripheral vasodilation during maximal exercise (26). These studies included both a cross-sectional examination of untrained and well-trained subjects and a longitudinal investigation of subjects before and after training. The main finding was that the lower the total peripheral resistance, the higher the $\dot{V}O_2max$. Moreover, a low mixed venous oxygen saturation was related to a low peripheral resistance. At the time of these studies, skeletal muscle adaptation to endurance training (increased capillarization and mitochondrial capacity) (71, 163) was paramount and influenced the interpretation of the finding by Clausen and colleagues (26, 27). Thus, they concluded that the training-induced changes at the muscle level were a prerequisite for a larger muscle blood flow, an enhanced O_2 extraction by the muscle, and increased $\dot{V}O_2max$ (27). A one-leg training study by Saltin and colleagues in 1976 strengthened this view (163). They found that peak $\dot{V}O_2$ and endurance increased more in the trained leg than in the untrained leg. This was compatible with the observed muscle adaptation in regard to increased capillarization and mitochondrial capacity, which was highest in the endurance-trained leg. Thus, in the late 1970s, the prevailing view of what limited maximal oxygen uptake, especially in trained subjects, shifted from the output of the heart to the capacity of skeletal muscle to accommodate high blood flows and to extract O_2. This view was supported by the research of Snell and colleagues (171), who showed in a cross-sectional study that the conductance of the calf muscle was strongly related to $\dot{V}O_2max$ in trained runners and only modestly related in sedentary individuals. The role of peripheral limitations has also been extensively elaborated on, primarily by Wagner and colleagues (145, 187).

Further studies in Copenhagen by Niels H. Secher (1946–), however, swung the pendulum back toward oxygen delivery being the limiting factor. He demonstrated that adding arm exercise to ongoing heavy leg exercise induced vasoconstriction and reduced leg blood flow (168). The conclusion from this research was that the heart could not provide sufficient cardiac output to maintain blood pressure and provide adequate perfusion of all capillary beds and, thus, adequate oxygen delivery. Blood pressure at peak exercise was unaffected by adding active muscle mass, which implied that total peripheral resistance was reduced (168). Further support for a key role of increased cardiac output in explaining the effect of endurance training on $\dot{V}O_2max$ came from subjects who trained with either their arms or their legs (27). Arm training elevated peak arm $\dot{V}O_2$ but not leg $\dot{V}O_2max$, whereas leg training increased both peak and leg $\dot{V}O_2max$. The explanation proposed by Clausen and colleagues was that arm training was not adequate to increase cardiac output whereas leg training was adequate (27). Thus, when exercising with the untrained arms after leg training, a higher blood pressure was generated due to an elevation in peak cardiac output and a strong vasoconstriction of vessels in nonexercising limbs, forcing a larger perfusion through the arm muscle with an elevated peak arm $\dot{V}O_2$ as a result.

When the concept of a peripheral limitation to $\dot{V}O_2max$, in part overcome by training, was re-examined, researchers proposed considering the role of improved cardiac capacity from another perspective. A greater cardiac output would be a lesser threat to a decrease in blood pressure and thus cause less of an effect of vasoconstrictor activity in the active muscle during the intense exercise, resulting in a lowering of peripheral resistance. This would explain the findings of Clausen that individuals with the highest $\dot{V}O_2max$ have the lowest peripheral resistance (26). This interpretation should be viewed in the light of the concept of functional sympatholysis, presented in 1962 by Remensnyder and colleagues (144). During heavy or maximal muscular work with a small muscle mass, locally produced metabolites in the muscle induce vasodilation by relaxing the smooth muscles and inhibit the effect of an elevated sympathetic activity by partially or fully blocking the action of norepinephrine on the α_1 receptor of vascular smooth muscle (81, 167). Such a regulatory control mechanism could play an important role in optimizing perfusion of skeletal muscle during exercise and, at peak exercise, could ensure that skeletal muscle conductance can be matched to the pumping capacity of the heart in order to maintain blood pressure. This effect may become attenuated in intense exercise involving a large muscle mass, especially in the untrained individual with a low cardiac capacity. Increasing the pump capacity of the heart with training would lessen the need for sympathetically mediated vasoconstriction, which could be brought about by lowering the sympathetic discharge to contracting skeletal muscle but also by a more potent functional sympatholysis. Moreover, in this perspective, the primary purpose of enlarging the bed in capillary skeletal muscle with training is to maintain or possibly increase mean transit time, which allows for a more complete O_2 extraction when O_2 delivery is elevated (157). The role for the enlarged mitochondrial volume and enhanced mitochondrial enzyme capacity in this process is still not fully clarified, but, as several have argued, it is of less importance for $\dot{V}O_2max$ than it is for the choice of more lipid as substrate rather than carbohydrate in the metabolism of the skeletal muscle (192).

Continuing discussions of what limits maximal oxygen uptake commonly try to identify one specific link as the limitation. As advocated by di Prampero, such an approach is nonphysiological and overshadows the close interaction that exists between the links in securing a continuous oxygen supply to tissues, most especially to contracting muscle during exercise (33), without compromising blood pressure (169). He has analyzed the oxygen transport system by considering the various links in the chain as resistances in Ohm's law. By using cross-sectional as well as longitudinal (inactivity–activity) data, he has been able to assign a resistance value to each link. Instead of one link being the limiting factor, he has estimated the relative role of each link. The largest roles (i.e., lowest resistance) were assigned to cardiac output and muscle blood flow. Other links such as the lungs, Hb concentration, capillarization, and mitochondrial capacity also had an influence, but their resistances are considerably higher. di Prampero's estimations provide information on the relative contribution and thereby the functional significance of a link rather than whether it limits the maximal oxygen uptake. From a physiological perspective, di Prampero's approach is more constructive than trying to identify a single limiting factor.

From this perspective, one can examine the various steps in the oxygen transport process. In a subject who becomes desaturated during exhaustive exercise, even if the reduction is small and could be explained by several reasons, the lungs to some extent limit maximal oxygen uptake. Thus, without this desaturation, maximal oxygen uptake would have been higher, as proven by Powers and colleagues (140). However, the magnitude of reduction in $\dot{V}O_2max$ due to incomplete oxygen saturation of the hemoglobin molecules may be 0.1 to 0.3 L/min, which could represent approximately 5% of the $\dot{V}O_2max$ in most humans. A similar point can be made when examining the transport of O_2 that occurs from the red blood cells in the capillaries to the myoglobin and the mitochondria in the muscle cell. All the O_2 in the blood is not extracted in exercising muscle; 5 to 30 ml/L remains in the blood draining from the muscle at $\dot{V}O_2max$. As in the lungs, there could be many explanations for the nonextracted oxygen. However, if extraction were complete, $\dot{V}O_2$ would have been higher, as elaborated on by Wagner and colleagues (145, 188). They consider extraction to be a diffusion limitation in contracting skeletal muscles at $\dot{V}O_2max$, which may amount to approximately 0.1 to 0.4 L/min of O_2 (or less than 10% of $\dot{V}O_2max$) that could have been extracted. Thus, using di Prampero's approach, it is possible to evaluate the extent to which differences in an individual's $\dot{V}O_2max$ can be explained by variation in incomplete oxygenation of the blood in the lungs, hemoglobin content, cardiac output, distribution of the systemic blood flow, and incomplete O_2 extraction by active muscle (33). Such an analysis demonstrated that the conclusions reached by the pioneers in the field were correct: Peak cardiac output was the variable varying the most (two- to threefold) in healthy subjects and thus can be said to have the largest effect on the achieved maximal oxygen uptake.

Of course, the factors determining oxygen uptake are not independent, and where the limits to $\dot{V}O_2max$ lie depends on the nuance of how the question is asked (185). Thus, increasing cardiac output by itself may increase diffusion limitation at the lung or at the skeletal muscle and therefore not increase oxygen transport. This problem is especially true at high altitude, where the gradients driving diffusion of oxygen from the alveoli into the pulmonary capillary are greatly reduced (78, 189, 191). Indeed, maximal rates of diffusion in the periphery are so limited at high altitude that even increasing oxygen content of the blood by transfusion (199) or erythropoietin (115, 199) does not increase $\dot{V}O_2max$, even though increasing O_2 content by increasing the fraction of inspired oxygen—emphasizing the importance of the pressure gradient from blood to muscle cell—obviously does (21, 166). Many detailed and scholarly discussions of the limitations to $\dot{V}O_2max$ have been reported by others (e.g., 32, 161, 164, 185, 186, 188).

It may be instructive to examine how $\dot{V}O_2max$ changes with hypoxia at altitude in order to demonstrate the tight relationship between $\dot{V}O_2max$ and performance, even in athletes with relatively uniform $\dot{V}O_2max$. It has long been known that $\dot{V}O_2max$ decreases with high-altitude exposure or hypoxia (for review, see 47) and that this decrease is evident in athletes at altitudes as low as a few hundred meters (51, 101, 181, 191). The mechanism for this reduction is related to diffusion limitation in the lung, which is exaggerated in athletes with high pulmonary blood flows (78, 105, 183) and who may develop exercise-induced hypoxemia even at sea level (31, 154). Exercise performance at altitude also clearly deteriorates at all running distances greater than 800 m (events lasting longer than 2 min) (138). Noakes (135) suggested that this decrease in $\dot{V}O_2max$ and performance is a function of reduced motor recruitment and argued that it provides evidence in support of the central governor model. However, this speculation has been convincingly proven to be incorrect. Some insight on this topic can be obtained from an early study by Medbo and colleagues (119), who elaborated on Krogh and Lindhard's concept of the accumulated oxygen deficit as a measure of anaerobic capacity. In this study, the investigators performed supramaximal exercise on a treadmill after carefully assessing individual running economy. During the uphill run, the total energy expended during the test was divided into aerobic and anaerobic components. The anaerobic capacity was defined as the difference between the predicted cost of the total work if all energy had been derived from oxidative sources and the directly measured accumulated oxygen uptake. That this measure is truly representative of anaerobic capacity was proven by the demonstration that the measure is independent of oxygen uptake and unaffected by hypoxia. Consistent with this hypothesis, the data showed that there was no difference in the anaerobic capacity measured under normoxic and hypoxic conditions, equivalent to an altitude of 3,500 m; 100% of the reduction in performance (slower speed, lower grade) was attributable to a reduction in accumulated oxygen uptake. However, in order to keep the duration of the test constant in normoxia and hypoxia, the speed and grade of the treadmill had to be reduced. More recently, Wehrlin and Hallen (191) extended this research by performing repeated supramaximal running tests in a group of trained athletes at multiple low to moderate altitudes ranging from 300 m to 2,800 m. In order to ensure that motor recruitment and power output were the same in all tests, each supramaximal test was performed at exactly the same speed at 107% normoxic velocity at $\dot{V}O_2$max. Despite keeping the speed absolutely constant at all altitudes, $\dot{V}O_2$max was reduced progressively and linearly by 0.6%/100 m of altitude in direct proportion to the reduction in oxygen saturation (191). Performance was reduced by 1.4%/100 m of altitude in direct proportion to the decrease in $\dot{V}O_2$max. This study provides strong evidence that $\dot{V}O_2$max is closely tied to oxygen transport, even when the differences are quite small and especially in well-trained endurance athletes, and that the reduction in $\dot{V}O_2$max at altitude is not likely attributable to decreased motor unit recruitment because running speed was constant at all altitudes studied.

Conclusions

Many important discoveries were made by early investigators, including Lavoisier, Zuntz, Benedict, Krogh, and Liljestrand, who described the mechanisms involved in oxygen transport from the ambient air to the active muscles during exercise. However, starting with the work of Hill and Lupton, the concept of maximal oxygen uptake in humans was enunciated. Studies during the past 100 yr have confirmed this concept and provided quantification of the important factors and links in the process. Some of the significant conceptual advances are shown in table 11.1. However, to reach a more fundamental understanding of the molecular and integrative aspects of the movement of oxygen from inspired air to energy-yielding mitochondria, major contributions still have to be made. They range from identifying genes of importance for $\dot{V}O_2$max and how they are activated to the very subtle and precise interplay between central nervous factors and reflexes to optimally match and distribute the available cardiac output to active muscle and other organs during maximal exercise.

Table 11.1 Maximal Oxygen Uptake Milestones

Year	Investigator	Significant advance	Reference
1923	Hill and Lupton	Were the first to define the meaning of maximal oxygen uptake.	68
1928	Herbst	Used the concept of leveling off (plateau) to determine $\dot{V}O_2$max; was the first to express this value in milliliters per kilogram per minute.	63
1938	Robinson	Was the first to publish a cross-sectional study on the effects of age on maximal oxygen uptake and found a linear decline.	148
1938	Keys and Friedell	Showed using X-rays that end diastolic and stroke volume were larger in athletes than in sedentary subjects.	88

(continued)

Table 11.1 *(continued)*

Year	Investigator	Significant advance	Reference
1952	Åstrand	Published a cross-sectional study on the effects of gender on maximal oxygen uptake and found higher values in males than in females.	6
1955	Taylor et al.	Were the first to show the important effect of exercising muscle mass on maximal oxygen uptake.	179
1958	Mitchell et al.	Measured and derived the major physiological factors in determining maximal oxygen uptake.	125
1961	Åstrand and Saltin	Showed the increased level of maximal oxygen uptake in male and female endurance athletes.	8
1962	Remensnyder et al.	Showed that sympathetically induced vasoconstriction was inhibited in contracting skeletal muscle (functional sympatholysis).	144
1963	Bevegård et al.	Showed that the high maximal oxygen uptake in endurance athletes is attributable to a large stroke volume during exercise.	17
1964	Rowell et al.	Were the first to show that in some subjects arterial desaturation occurred at maximal exercise.	154
1968	Saltin et al.	Showed that bed rest caused a marked decrease in maximal oxygen uptake and that intense exercise training caused a marked increase in maximal oxygen uptake.	160
1971	Klissouras	Was the first to study maximal oxygen uptake in monozygotic and dizygotic twins and showed that it was largely genetically determined.	90
1973	Clausen et al.	Showed that the lower total-body peripheral resistance was during exercise, the higher maximal oxygen uptake was.	27
1974	Secher et al.	Examined the role of active muscle mass on maximal oxygen uptake in a more systematic study.	169
1981	Heath et al.	Showed that continued intense exercise training markedly blunted the age-related decrease in maximal oxygen uptake.	60
1982	Andersen and Saltin	Showed in humans that blood flow to maximally exercised skeletal muscle can be as high as $2.5\ L \cdot kg^{-1} \cdot m^{-1}$.	1
1984	Dempsey et al.	Confirmed that arterial desaturation occurred in some subjects during heavy exercise and studied the mechanism by which it occurred.	31
1985	di Prampero	Analyzed the links in oxygen transport and assigned a relative role to each in determining maximal oxygen uptake.	33
1986	Bouchard et al.	Studied monozygotic and dizygotic twins and found that genetic factors accounted for about 40% of the value of maximal oxygen uptake.	20
2005	Fleg et al.	Showed an accelerated decline in maximal oxygen uptake with age in a longitudinal study involving numerous male and female subjects.	44
2007	Hawkins et al.	Confirmed that $\dot{V}O_2$ plateaus even in subjects who can run at very high oxygen demands for sufficient periods to stabilize oxygen uptake.	59

References

1. Andersen P, Saltin B. Maximal perfusion of skeletal muscle in man. *J Physiol* 366: 233-249, 1985.
2. Arbab-Zadeh A, Dijk E, Prasad A, Fu Q, Torres P, Zhang R, Thomas JD, Palmer D, et al. Effect of aging and physical activity on left ventricular compliance. *Circulation* 110: 1799-1805, 2004.
3. Asmussen E, Nielsen M. The cardiac output in rest and work determined simultaneously by the acetylene and the dye injection methods. *Acta Physiol Scand* 27: 217-230, 1952.
4. Åstrand I. Aerobic work capacity in men and women with special reference to age. *Acta Physiol Scand* 49(Suppl.): 1-92, 1960.
5. Åstrand P-O, Rodahl K. *Textbook of Work Physiology. Physiological Bases of Exercise*. New York: McGraw-Hill, 1986, 420-422.
6. Åstrand PO. *Experimental Studies of Physical Working Capacity in Relation to Sex and Age*. Coenhagen: Munksgaard, 1952.
7. Åstrand PO, Bergh U, Kilbom A. A 33-yr follow-up of peak oxygen uptake and related variables of former physical education students. *J Appl Physiol* 82: 1844-1852, 1997.
8. Åstrand PO, Saltin B. Maximal oxygen uptake and heart rate in various types of muscular activity. *J Appl Physiol* 16: 977-981, 1961.
9. Atwater WO, Benedict FG. *Experiments on the Metabolism of Matter and Energy in the Human Body*. Washington, DC: Government Printing Office, republished 1988.
10. Bainbridge FA, Bock AV, Dill DB. *The Physiology of Muscular Exercise*. London: Longmans Green, 1923.
11. Balke B, Grillo GP, Konecci EB, Luft UC. Work capacity after blood donation. *J Appl Physiol* 7: 231-238, 1954.
12. Barnard J, Holloszy JO. The metabolic systems: Aerobic metabolism and substrate utilization in exercising skeletal muscle. In: Tipton CM, ed. *Exercise Physiology: People and Ideas*. Oxford: Oxford University Press, 2003, 292-321.
13. Benedict FG. An apparatus for studying respiratory exchange. *Am J Physiol* 24: 345-374, 1909.
14. Benedict FG, Cathcart EP. *A Metabolic Study With Special Reference to the Efficiency of the Human Body as a Machine*. New York: Blanchard Co., 1913, 345-374.
15. Bergh U, Ekblom B, Åstrand PO. Maximal oxygen uptake "classical" versus "contemporary" viewpoints. *Med Sci Sports Exerc* 32: 85-88, 2000.
16. Bergh U, Kanstrup IL, Ekblom B. Maximal oxygen uptake during exercise with various combinations of arm and leg work. *J Appl Physiol* 41: 191-196, 1976.
17. Bevegård S, Holmgren A, Jonsson B. Circulatory studies in well-trained athletes at rest and during heavy exercise, with special reference to stroke volume and the influence of body position. *Acta Physiol Scand* 57: 26-50, 1963.
18. Bock AV, Dill DB, Talbott JH. Studies in muscular activity: I. Determination of the rate of circulation of blood in man at work. *J Physiol* 66: 121-132, 1928.
19. Bouchard C, An P, Rice T, Skinner JS, Wilmore JH, Gagnon J, Perusse L, Leon AS, et al. Familial aggregation of VO2max response to exercise training: Results from the HERITAGE Family Study. *J Appl Physiol* 87: 1003-1008, 1999.
20. Bouchard C, Lesage R, Lortie G, Simoneau JA, Hamel P, Boulay MR, Perusse L, Theriault G, et al. Aerobic performance in brothers, dizygotic and monozygotic twins. *Med Sci Sports Exerc* 18: 639-646, 1986.
21. Boushel R, Calbet JA, Radegran G, Sondergaard H, Wagner PD, Saltin B. Parasympathetic neural activity accounts for the lowering of exercise heart rate at high altitude. *Circulation* 104: 1785-1791, 2001.
22. Carter JEL, Kasch FW, Boyer JL, Phillips WH, Ross WD, Susec A. Structural and functional assessments on a champion runner—Peter Snell. *Res Q* 38: 355-365, 1967.
23. Chapman CB, Taylor HL, Borden C, Ebert RV, Keys A. Simultaneous determinations of the resting arteriovenous oxygen difference by the acetylene and direct Fick methods. *J Clin Invest* 29: 651-659, 1950.
24. Christensen EH. Betrage zur Phsiologie Schwerer Korperlicher Arbeit. V. Minutenvolumen und Schlagvolumen des Herzens Wahrend Schwerer Korperlicher Arbeit. *Arbeitsphysiologie* 4: 470-502, 1932.
25. Christensen EH, Krogh A, Lindhard J. Recherches sur l'effort musculaire intense. Bulleting trimestriel de l'organisation d'hygiene. *Societe des Nations* 3: 407-439, 1934.
26. Clausen JP. Circulatory adjustments to dynamic exercise and effect of physical training in normal subjects and in patients with coronary artery disease. *Prog Cardiovasc Dis* 18: 459-495, 1976.
27. Clausen JP, Klausen K, Rasmussen B, Trap-Jensen J. Central and peripheral circulatory changes after training of the arms or legs. *Am J Physiol* 225: 675-682, 1973.
28. Cohn JE, Carroll DG, Armstrong BW, Shepard RH, Riley RL. Maximal diffusing capacity of the lung in normal male subjects of different ages. *J Appl Physiol* 6: 588-597, 1954.
29. Costill DL, Thomason H, Roberts E. Fractional utilization of the aerobic capacity during distance running. *Med Sci Sports* 5: 248-252, 1973.

30. Coyle EF, Coggan AR, Hopper MK, Walters TJ. Determinants of endurance in well-trained cyclists. *J Appl Physiol* 64: 2622-2630, 1988.
31. Dempsey JA, Hanson PG, Henderson KS. Exercise-induced arterial hypoxaemia in healthy human subjects at sea level. *J Physiol* 355: 161-175, 1984.
32. di Prampero PE. Factors limiting maximal performance in humans. *Eur J Appl Physiol* 90: 420-429, 2003.
33. di Prampero PE. Metabolic and circulatory limitations to VO2 max at the whole animal level. *J Exp Biol* 115: 319-331, 1985.
34. Dill DB, Robinson S, Ross JC. A longitudinal study of 16 champion runners. *J Sports Med Phys Fitness* 7: 4-27, 1967.
35. Douglas EG, Haldane JS. The causes of absorption of oxygen by the lungs. *J Physiol (Lond)* 44: 305-354, 1912.
36. du Bois-Reymond E. Uber angeblich saure Reaction des Muskelfleisches. *Gesammelte Abhandl Zur allg Muskel U Nervenphysik* 2-36, 1877.
37. Ekblom B, Astrand P-O, Saltin B, Stenberg J, Wallstrom R. Effect of training on circulatory response to exercise. *J Appl Physiol* 24: 518-528, 1968.
38. Ekblom B, Gjessing E. Maximal oxygen uptake of the Easter Island population. *J Appl Physiol* 25: 124-129, 1968.
39. Ekblom B, Goldbarg AN, Gullbring B. Response to exercise after blood loss and reinfusion. *J Appl Physiol* 33: 175-180, 1972.
40. Ekblom B, Hermansen L. Cardiac output in athletes. *J Appl Physiol* 25: 619-625, 1968.
41. Fagaard R, Beilen E, Avery A. Heritability of aerobic power and anaerobic energy generation during exercise. *J Appl Physiol* 70: 357-362, 1991.
42. Ferguson S, Gledhill N, Jamnik VK, Wiebe C, Payne N. Cardiac performance in endurance-trained and moderately active young women. *Med Sci Sports Exerc* 33: 1114-1119, 2001.
43. Fick A. Uber die messung des blutquantums in den herzentrikeln. *Physmed Ges Worzburg* 1870.
44. Fleg JL, Morrell CH, Bos AG, Brant LJ, Talbot LA, Wright JG, Lakatta EG. Accelerated longitudinal decline of aerobic capacity in healthy older adults. *Circulation* 112: 674-682, 2005.
45. Freedman ME, Snider GL, Brostoff P, Kimelblot S, Katz LN. Effect of training on response of cardiac output to muscular exercise in athletes. *J Appl Physiol* 8: 37-47, 1955.
46. Fu Q, Levine BD. Cardiovascular response to exercise in women. *Med Sci Sports Exerc* 37: 1433-1435, 2005.
47. Fulco CS, Rock PB, Cymerman A. Maximal and submaximal exercise performance at altitude. *Aviat Space Environ Med* 69: 793-801, 1998.
48. Gollnick PD, Armstrong RB, Saubert CWI, Piehl K, Saltin B. Enzyme activity and fiber composition in skeletal muscle of untrained and trained men. *J Appl Physiol* 33: 312-319, 1972.
49. Gonzalez-Alonso J. Point: Stroke volume does/does not decline during exercise at maximal effort in healthy individuals. *J Appl Physiol* 104: 275-276, discussion 279-280, 2008.
50. Goodwin GM, McCloskey DI, Mitchell JH. Cardiovascular and respiratory responses to change in central command during isometric exercise of constant muscle tension. *J Physiol* 226: 173-190, 1972.
51. Gore CJ, Hahn AG, Scroop GC, Watson DB, Norton KI, Wood RJ, Campbell DP, Emonson DL. Increased arterial desaturation in trained cyclists during maximal exercise at 580 m altitude. *J Appl Physiol* 80: 2204-2210, 1996.
52. Grande F, Taylor HL. Adaptive changes in the heart, vessels and patterns of control under chronically high loads. In: Hamilton WF, Dow P, eds. *Handbook of Physiology*. Washington, DC: American Physiological Society, 1965, 2615-2678.
53. Green HJ, Thomson JA, Ball ME, Hughson RL, Houston ME, Sharratt MT. Alterations in blood volume following short-term supramaximal exercise. *J Appl Physiol* 56: 145-149, 1984.
54. Grimby G, Nilsson NJ, Saltin B. Cardiac output during submaximal and maximal exercise in active middle-aged athletes. *J Appl Physiol* 21: 1150-1156, 1966.
55. Grimby G, Saltin B. Physiological analysis of physically well-trained middle-aged and old athletes. *Acta Med Scand* 179: 513-526, 1966.
56. Haldane JS. Some improved methods of gas analysis. *J Physiol (Lond)* 22: 465, 1897-1898.
57. Haller RG, Henriksson KG, Jorfeldt L, Hultman E, Wibom R, Sahlin K, Areskog NH, Gunder M, et al. Deficiency of skeletal muscle succinate dehydrogenase and aconitase. Pathophysiology of exercise in a novel human muscle oxidative defect. *J Clin Invest* 88: 1197-1206, 1991.
58. Hammond HK, White FC, Bhargava V, Shabetai R. Heart size and maximal cardiac output are limited by the pericardium. *Am J Physiol* 263: H1675-H1681, 1992.
59. Hawkins MN, Raven PB, Snell PG, Stray-Gundersen J, Levine BD. Maximal oxygen uptake as a parametric measure of cardiorespiratory capacity. *Med Sci Sports Exerc* 39: 103-107, 2007.
60. Heath GW, Hagberg JM, Ehsani AA, Holloszy JO. A physiological comparison of young and older endurance athletes. *J Appl Physiol* 51: 634-640, 1981.
61. Hellerstein H, Franklin B. Exercise testing and prescription. In: Wenger N, Hellerstein H, eds. *Rehabilitation of the Coronary Patient*. New York: Wiley, 1978, 149.

62. Henschen ES. *Skiddlauf und Skidwettlauff: Eini medizinische Sportstudie. Mit. Med. Klin.* Upsala: Jena Fischer Verlag, 1899.
63. Herbst R. Der Gasstoffwechsel als Mass der Korperlichen Leistungsfahigkeit. I. Mitteilung: Die Bestimmung des Sauerstoffaufnahmevermogens beim Gesunden. *Deut Arch Klin Med* 162: 33-50, 1928.
64. Hermansen L, Saltin B. Oxygen uptake during maximal treadmill and bicycle exercise. *J Appl Physiol* 26: 31-37, 1969.
65. Hill AV. The role of oxidation in maintaining the dynamic equilibrium of the muscle cell. *Proc Royal Soc Lond* 103: 138-162, 1928.
66. Hill AV. *Trails and Trials in Physiology* Baltimore, MD: Williams & Wilkins, 1966.
67. Hill AV, Long CNH, Lupton H. Muscular exercise, lactic acid and the supply and utilization of oxygen, IV-VI and VII-VIII. *Proc Royal Soc Lond* 97: 84-138, 155-176, 1924.
68. Hill AV, Lupton H. Muscular exercise, lactic acid, and the supply and utilization of oxygen. *Q J M* 16: 135-171, 1923.
69. Hill L, Flack M. The influence of oxygen inhalations on muscular work. *J Physiol* 40: 347-372, 1910.
70. Himwich HE, Barr DP. Studies in the physiology of muscular exercise. Relationships in the arterial blood. *J Biol Chem* 57: 363-378, 1923.
71. Holloszy JO. Biochemical adaptations in muscle. Effects of exercise on mitochondrial oxygen uptake and respiratory enzyme activity in skeletal muscle. *J Biol Chem* 242: 2278-2282, 1967.
72. Holmgren A, Astrand PO. DL and the dimensions and functional capacities of the O2 transport system in humans. *J Appl Physiol* 21: 1463-1470, 1966.
73. Hoppeler H, Kayar SR, Claassen H, Uhlmann E, Karas RH. Adaptive variation in the mammalian respiratory system in relation to energetic demand. III. Skeletal muscles: Setting the demand for oxygen. *Respir Physiol* 69: 27-46, 1987.
74. Ingjer F. Development of maximal oxygen uptake in young elite male cross-country skiers: A longitudinal study. *J Sports Sci* 10: 49-63, 1992.
75. Ingjer F. Maximal oxygen uptake as a predictor of performance ability in women and men elite cross-country skiers. *Scand J Med Sci Sports* 1: 25-30, 1991.
76. Jansson E, Kaijser L. Muscle adaptation to extreme endurance training in man. *Acta Physiol Scand* 100: 315-324, 1977.
77. Johansson JE. Uber die einwirkung der muscel auf die atmung und die hertz. *Skandinaviesches Archiv fur Physiologie* 5: 20-66, 1895.
78. Johnson RL. Pulmonary diffusion as a limiting factor in exercise stress. *Circ Res* 20: S154-S160, 1967.
79. Jonsell S. A method for the determination of the heart size by teleroentgenographie. *Acta Radiol* 20: 325-340, 1939.
80. Joyner MJ, Coyle EF. Endurance exercise performance: The physiology of champions. *J Physiol* 586: 35-44, 2008.
81. Joyner MJ, Nauss LA, Warner MA, Warner DO. Sympathetic modulation of blood flow and O2 uptake in rhythmically contracting human forearm muscles. *Am J Physiol* 263: H1078-H1083, 1992.
82. Kanstrup IL, Ekblom B. Blood volume and hemoglobin concentration as determinants of maximal aerobic power. *Med Sci Sports Exerc* 16: 256-262, 1984.
83. Karpovich PV, Perstrecov K. Effect of gelatin upon muscular work in man. *Am J Physiol* 134: 1941.
84. Kasch FW, Boyer JL, Schmidt PK, Wells RH, Wallace JP, Verity LS, Guy H, Schneider D. Ageing of the cardiovascular system during 33 years of aerobic exercise. *Age Ageing* 28: 531-536, 1999.
85. Kasch FW, Boyer JL, Van Camp S, Nettl F, Verity LS, Wallace JP. Cardiovascular changes with age and exercise. A 28-year longitudinal study. *Scand J Med Sci Sports* 5: 147-151, 1995.
86. Kasch FW, Wallace JP. Physiological variables during 10 years of endurance exercise. *Med Sci Sports* 8: 5-8, 1976.
87. Keul J, Dickhuth HH, Simon G, Lehmann M. Effect of static and dynamic exercise on heart volume, contractility, and left ventricular dimensions. *Circ Res* 48: I162-1170, 1981.
88. Keys A, Friedell HL. Size and stroke of the heart in young men in relation to athletic activity. *Science* 88: 456-458, 1938.
89. Kjellberg SR, Rudhe U, Sjostrand T. Increase of the amount of hemoglobin and blood volume in connection with physical training. *Acta Physiol Scand* 19: 146-151, 1949.
90. Klissouras V. Heritability of adaptive variation. *J Appl Physiol* 31: 338-344, 1971.
91. Klitgaard H, Bergman O, Betto R, Salviati G, Schiaffino S, Clausen T, Saltin B. Co-existence of myosin heavy chain I and IIa isoforms in human skeletal muscle fibres with endurance training. *Pflugers Arch* 416: 470-472, 1990.
92. Knehr CA, Dill DB, Neufeld W. Training and its effects on man at rest and work. *Am J Physiol* 136: 148-156, 1942.
93. Komi PV, Karlsson J. Physical performance, skeletal muscle enzyme activities, and fibre types in monozygous and dizygous twins of both sexes. *Acta Physiol Scand* 462(Suppl.): 1-28, 1979.
94. Koskolou MD, McKenzie DC. Arterial hypoxemia and performance during intense exercise. *Eur J Appl Physiol Occup Physiol* 68: 80-86, 1994.

95. Krogh A. The number and distribution of capillaries in muscles with calculations of the oxygen pressure head necessary for supplying the tissue. *J Physiol* 52: 409-415, 1919.

96. Krogh A, Krogh M. On the tensions of gases in the arterial blood. *Skand Arch Physiol* 23: 179-192, 1910.

97. Krogh A, Lindhard J. The changes in respiration at the transition from work to rest. *J Physiol* 53: 431-439, 1920.

98. Krogh A, Lindhard J. The regulation of respiration and circulation during the initial stages of muscular work. *J Physiol* 47: 112-136, 1913.

99. Krogh A, Lindhard J. The relative value of fat and carbohydrate as sources of muscular energy: With appendices on the correlation between standard metabolism and the respiratory quotient during rest and work. *Biochem J* 14: 290-363, 1920.

100. Lange-Andersen L, Elsner R, Saltin B, Hermansen L. *Physical Fitness in Terms of Maximal Oxygen Uptake of Nomadic Lapps*. Alaska: Arctic Aeromedical Laboratory, 1961.

101. Lawler J, Powers SK, Thompson D. Linear relationship between VO2max and VO2max decrement during exposure to acute hypoxia. *J Appl Physiol* 64: 1486-1492, 1988.

102. Leonard B, Mitchell JH, Mizuno M, Rube N, Saltin B, Secher NH. Partial neuromuscular blockade and cardiovascular responses to static exercise in man. *J Physiol* 359: 365-379, 1985.

103. Levine BD. VO2max: What do we know, and what do we still need to know? *J Physiol* 586: 25-34, 2008.

104. Levine BD, Lane LD, Buckey JC, Friedman DB, Blomqvist CG. Left ventricular pressure-volume and Frank-Starling relations in endurance athletes. Implications for orthostatic tolerance and exercise performance. *Circulation* 84: 1016-1023, 1991.

105. Levine BD, Stray-Gundersen J. Exercise at high altitude. In: Melton M, ed. *Sports Medicine Secrets*. Philadelphia: Hanley & Belfus, 1999, 91-96.

106. Levine BD, Stray-Gundersen J. "Living high-training low": Effect of moderate-altitude acclimatization with low-altitude training on performance. *J Appl Physiol* 83: 102-112, 1997.

107. Lewis SF, Haller RG, Blomqvist CG. Neuromuscular diseases as models of cardiovascular regulation during exercise. *Med Sci Sports Exerc* 16: 466-471, 1984.

108. Liljestrand G. Untersuchungen uber die Atmungsarbeit. *Skand Arch Physiol* 35: 199-293, 1918.

109. Liljestrand G, Lindhard J. Uber das Minutvolumen des Herzens beim Schwimmen. Studien uber die Physiologie des Schwimmens. *Skand Arch Physiol* 39: 64-77, 1920.

110. Liljestrand G, Lindhard J. Zur Physiologie des Ruderns. *Skand Arch Physiol* 39: 215-235, 1920.

111. Liljestrand G, Stenstrom N. Respirationsversuche beim Gehen, Laufen, Ski- und Schlittschuhlaufen. *Skand Arch Physiol* 39: 167-206, 1920.

112. Liljestrand G, Stenstrom N. Studien uber die Physiologie des Schwimmens. *Skand Arch Physiol* 39: 1-63, 1920.

113. Lindhard J. Uber das Minutenvolumen des Herzens bei Ruhe und bei Muskelabeit. *Pflugers Arch* 161: 233-383, 1915.

114. Lohmann K. Darstellung der Adenylphosphorsaure aus Muskulatur. *Biochem Z* 233: 460-472, 1931.

115. Lundby C, Damsgaard R. Exercise performance in hypoxia after novel erythropoiesis stimulating protein treatment. *Scand J Med Sci Sports* 16: 35-40, 2006.

116. Lundsgaard E. Untersuchungen uber Muskelkontraktionen ohne Milchshaurebildung. *Biochem Z* 217: 162-177, 1930.

117. McGavock JM, Hastings JL, Snell PG, McGuire DK, Pacini EL, Levine BD, Mitchell JH. A forty-year follow-up of the Dallas Bed Rest and Training study: The effect of age on the cardiovascular response to exercise in men. *J Gerontol A Biol Sci Med Sci* 64: 293-299, 2009.

118. McGuire DK, Levine BD, Williamson JW, Snell PG, Blomqvist CG, Saltin B, Mitchell JH. A 30-year follow-up of the Dallas Bedrest and Training Study: I. Effect of age on the cardiovascular response to exercise. *Circulation* 104: 1350-1357, 2001.

119. Medbo JI, Mohn AC, Tabata I, Bahr R, Vaage O, Sejersted OM. Anaerobic capacity determined by maximal accumulated O2 deficit. *J Appl Physiol* 64: 50-60, 1988.

120. Milliken MC, Stray-Gundersen J, Peshock RM, Katz J, Mitchell JH. Left ventricular mass as determined by magnetic resonance imaging in male endurance athletes. *Am J Cardiol* 62: 301-305, 1988.

121. Mitchell JH. Neual control of the circulation during exercise. *Med Sci Sports* 22: 141-154, 1990.

122. Mitchell JH, Blomqvist CG. Maximal oxygen uptake. *N Engl J Med* 284: 1018-1022, 1971.

123. Mitchell JH, Reeves DR Jr., Rogers HB, Secher NH, Victor RG. Autonomic blockade and cardiovascular responses to static exercise in partially curarized man. *J Physiol* 413: 433-445, 1989.

124. Mitchell JH, Saltin, B. The oxygen transport system and maximal oxygen uptake. In: Tipton CM, ed. *Exercise Physiology*. Oxford: Oxford University Press, 2003.

125. Mitchell JH, Sproule BJ, Chapman CB. The physiological meaning of the maximal oxygen intake test. *J Clin Invest* 37: 538-547, 1958.

126. Mitchell JH, Victor RG. Neural control of the cardiovascular system: Insights from muscle sympathetic

nerve recordings in humans. *Med Sci Sports Exerc* 28: S60-S69, 1996.

127. Morgan T, Cobb L, Short F, Ross R, Gunn D. Effects of long-term exercise on human muscle mitochondria. In: Pernow B, Saltin B, eds. *Muscle Metabolism During Exercise*. New York: Plenum Press, 1971, 87-96.
128. Morris JN, Heady JA, Raffle PA, Roberts CG, Parks JW. Coronary heart-disease and physical activity of work. *Lancet* 265: 1053-1057, 1953.
129. Nikolic S, Yellin EL, Tamura K, Vetter H, Tamura T, Meisner JS, Frater RW. Passive properties of canine left ventricle: Diastolic stiffness and restoring forces. *Circ Res* 62: 1210-1222, 1988.
130. Noakes TD. 1996 J.B. Wolffe Memorial Lecture. Challenging beliefs: Ex Africa semper aliquid novi. *Med Sci Sports Exerc* 29: 571-590, 1997.
131. Noakes TD. The central governor model of exercise regulation applied to the marathon. *Sports Med* 37: 374-377, 2007.
132. Noakes TD. How did AV Hill understand the VO2max and the "plateau phenomenon"? Still no clarity? *Br J Sports Med* 42: 574-580, 2008.
133. Noakes TD. Time to move beyond a brainless exercise physiology: The evidence for complex regulation of human exercise performance. *Appl Physiol Nutr Metab* 36: 23-35, 2011.
134. Noakes TD, Marino FE. Does a central governor regulate maximal exercise during combined arm and leg exercise? A rebuttal. *Eur J Appl Physiol* 104: 757-759, 2008.
135. Noakes TD, Peltonen JE, Rusko HK. Evidence that a central governor regulates exercise performance during acute hypoxia and hyperoxia. *J Exp Biol* 204: 3225-3234, 2001.
136. Nylin G. On the amount of, and changes in, the residual blood of the heart. *Am Heart J* 25: 598-608, 1943.
137. Pelliccia A, Culasso F, Di Paolo FM, Maron BJ. Physiologic left ventricular cavity dilatation in elite athletes. *Ann Intern Med* 130: 23-31, 1999.
138. Peronnet F, Thibault G, Cousineau DL. A theoretical analysis of the effect of altitude on running performance. *J Appl Physiol* 70: 399-404, 1991.
139. Pollock ML, Mengelkoch LJ, Graves JE, Lowenthal DT, Limacher MC, Foster C, Wilmore JH. Twenty-year follow-up of aerobic power and body composition of older track athletes. *J Appl Physiol* 82: 1508-1516, 1997.
140. Powers SK, Dodd S, Woodyard J, Beadle RE, Church G. Haemoglobin saturation during incremental arm and leg exercise. *Br J Sports Med* 18: 212-216, 1984.
141. Rasmussen J, Hanel B, Diamant B, Secher NH. Muscle mass effect on arterial desaturation after maximal exercise. *Med Sci Sports Exerc* 23: 1349-1352, 1991.
142. Raven PB. Neural control of the circulation during exercise. *Exp Physiol* 91: 10-13, 2012.
143. Reindell H. Ober den Kreislauf der Trainierten. Uber den Restblutmenge des Herzens und uber die besondere Bedeutung rontgenologischer (kymographischer) hamodynamischen Beobachtungen in Ruhe und nach Belastung. *Arch Kreislaufforsch* 12: 265, 1943.
144. Remensnyder JP, Mitchell JH, Sarnoff SJ. Functional sympatholysis during muscular activity. Observations on influence of carotid sinus on oxygen uptake. *Circ Res* 11: 370-380, 1962.
145. Richardson RS, Noyszewski EA, Kendrick KF, Leigh JS, Wagner PD. Myoglobin O2 desaturation during exercise. Evidence of limited O2 transport. *J Clin Invest* 96: 1916-1926, 1995.
146. Richardson RS, Poole DC, Knight DR, Kurdak SS, Hogan MC, Grassi B, Johnson EC, Kendrick KF, et al. High muscle blood flow in man: Is maximal O2 extraction compromised? *J Appl Physiol* 75: 1911-1916, 1993.
147. Riley-Hagan M, Peshock RM, Stray-Gundersen J, Katz J, Ryschon TW, Mitchell JH. Left ventricular dimensions and mass using magnetic resonance imaging in female endurance athletes. *Am J Cardiol* 69: 1067-1074, 1992.
148. Robinson S. Experimental studies of physical fitness in relation to age. *Arbeitsphysiologie* 10: 251-323, 1938.
149. Robinson S, Edwards HT, Dill DB. New records in human power. *Science* 85: 409-410, 1937.
150. Robinson S, Harmon PM. The effect of training and of gelatin upon certain factors which limit muscular work. *Am J Physiol* 133: 161-169, 1941.
151. Rose RJ, Cluer D, Saltin B. Some comparative aspects of the camel as a racing animal. *Acta Physiol Scand* 150(Suppl. 617): 87-95, 1984.
152. Rowell L. Circulatory adjustments to dynamic exercise. In: *Human Circulation: Regulation During Physical Stress*. New York: Oxford University Press, 1986.
153. Rowell L. Factors affecting the prediction of maximal oxygen intake from measurements made during submaximal work with observations related to factors which may limit maximal oxygen uptake. University of Minnesota, 1962.
154. Rowell LB, Taylor HL, Wang Y, Carlson WS. Saturation of arterial blood with oxygen during maximal exercise. *J Appl Physiol* 19: 284-286, 1964.
155. Saltin B, ed. *Cardiovascular and Pulmonary Adaptation to Physical Activity*. Champaign, IL: Human Kinetics, 1990, 187-204.

156. Saltin B. Circulatory response to submaximal and maximal exercise after thermal dehydration. *J Appl Physiol* 19: 1125-1132, 1964.

157. Saltin B. Malleability of the system in overcoming limitations: Functional elements. *J Exp Biol* 115: 345-354, 1985.

158. Saltin B. The physiology of competitive cross country skiing across a four decade perspective. In: Muller E, Schwameder H, Kornxl E, Raschner C, eds. *Science and Skiing*. London: E & FN Spon, 1996, 435-469.

159. Saltin B, Åstrand PO. Maximal oxygen uptake in athletes. *J Appl Physiol* 23: 353-358, 1967.

160. Saltin B, Blomqvist G, Mitchell JH, Johnson RL Jr., Wildenthal K, Chapman CB. Response to exercise after bed rest and after training. *Circulation* 38(5 Suppl.): VII1-78, 1968.

161. Saltin B, Calbet JA. Point: In health and in a normoxic environment, VO2 max is limited primarily by cardiac output and locomotor muscle blood flow. *J Appl Physiol* 100: 744-745, 2006.

162. Saltin B, Hartley LH, Kilborn A, Astrand I. Physical training in sedentary middle-aged and older men. II. Oxygen uptake, heart rate, and blood lactate concentration at submaximal and maximal exercise. *Scand J Clin Lab Invest* 24: 323-334, 1969.

163. Saltin B, Nazar K, Costill DL, Stein E, Jansson E, Essen B, Gollnick PD. The nature of the training response: Peripheral and central adaptations to one-legged exercise. *Acta Physiol Scand* 96: 289-305, 1976.

164. Saltin B, Strange S. Maximal oxygen uptake: "Old" and "new" arguments for a cardiovascular limitation. *Med Sci Sports Exerc* 24: 30-37, 1992.

165. Savard G, Kiens B, Saltin B. Central cardiovascular factors as limits to endurance. In: Macleod D, Maughan R, Nimmo M, Reilly T, Williams C, eds. *Exercise: Benefits, Limits and Adaptations*. London: E and FN Spon, 1987, 162-180.

166. Savard GK, Areskog NH, Saltin B. Cardiovascular response to exercise in humans following acclimatization to extreme altitude. *Acta Physiol Scand* 154: 499-509, 1995.

167. Savard GK, Richter EA, Strange S, Kiens B, Christensen NJ, Saltin B. Norepinephrine spillover from skeletal muscle during exercise in humans: Role of muscle mass. *Am J Physiol* 257: H1812-H1818, 1989.

168. Secher NH, Clausen JP, Klausen K, Noer I, Trap-Jensen J. Central and regional circulatory effects of adding arm exercise to leg exercise. *Acta Physiol Scand* 100: 288-297, 1977.

169. Secher NH, Ruberg-Larsen N, Binkhorst RA, Bonde-Petersen F. Maximal oxygen uptake during arm cranking and combined arm plus leg exercise. *J Appl Physiol* 36: 515-518, 1974.

170. Snell PG. *Metabolic and Cardiovascular Responses to Exercise in Man with Fixed Heart Rate*. Pullman, WA: 1982.

171. Snell PG, Martin WH, Buckey JC, Blomqvist CG. Maximal vascular leg conductance in trained and untrained men. *J Appl Physiol* 62: 606-610, 1987.

172. Snell PG, Mitchell JH. The role of maximal oxygen uptake in exercise performance. *Clin Chest Med* 5: 51-62, 1984.

173. Sproule BJ, Mitchell JH, Miller WF. Cardiopulmonary physiological responses to heavy exercise in patients with anemia. *J Clin Invest* 39: 378-388, 1960.

174. Stenberg J, Åstrand PO, Ekblom B, Royce J, Saltin B. Hemodynamic response to work with different muscle groups, sitting and supine. *J Appl Physiol* 22: 61-70, 1967.

175. Stray-Gundersen J, Musch TI, Haidet GC, Swain DP, Ordway GA, Mitchell JH. The effect of pericardiectomy on maximal oxygen consumption and maximal cardiac output in untrained dogs. *Circ Res* 58: 523-530, 1986.

176. Sutton JR, Reeves JT, Wagner PD, Groves BM, Cymerman A, Malconian MK, Rock PB, Young PM, et al. Operation Everest II: Oxygen transport during exercise at extreme simulated altitude. *J Appl Physiol* 64: 1309-1321, 1988.

177. Taivassalo T, Jensen TD, Kennaway N, DiMauro S, Vissing J, Haller RG. The spectrum of exercise tolerance in mitochondrial myopathies: A study of 40 patients. *Brain* 126: 413-423, 2003.

178. Taylor CR, Karas RH, Weibel ER, Hoppeler H. Adaptive variation in the mammalian respiratory system in relation to energetic demand. *Respir Physiol* 69: 1-127, 1987.

179. Taylor HL, Buskirk E, Henschel A. Maximal oxygen intake as an objective measure of cardio-respiratory performance. *J Appl Physiol* 8: 73-80, 1955.

180. Taylor HL, Erickson L, Henschel A, Keys A. The effect of bed rest on the blood volume of normal young men. *Am J Physiol* 144: 227-232, 1945.

181. Terrados N, Mizuno M, Andersen H. Reduction in maximal oxygen uptake at low altitudes: Role of training status and lung function. *Clin Physiol* 5(Suppl. 3): 75-79, 1985.

182. Tissot J. A new method for measuring and registering the respiration of humans and animals [in French]. *J Physiol Path Gen* 6: 688, 1904.

183. Torre-Bueno JR, Wagner PD, Saltzman HA, Gale GE, Moon RE. Diffusion limitation in normal humans during exercise at sea level and simulated altitude. *J Appl Physiol* 58: 989-995, 1985.

184. Von Dobeln W. A simple bicycle ergometer. *J Appl Physiol* 7: 222-224, 1954.

185. Wagner PD. Counterpoint: In health and in normoxic environment VO2max is limited primarily by

cardiac output and locomotor muscle blood flow. *J Appl Physiol* 100: 745-747, discussion 747-748, 2006.

186. Wagner PD. Determinants of maximal oxygen transport and utilization. *Ann Rev Physiol* 58: 21-50, 1996.
187. Wagner PD. Gas exchange and peripheral diffusion limitation. *Med Sci Sports Exerc* 24: 54-58, 1992.
188. Wagner PD. New ideas on limitations to VO2max. *Exerc Sport Sci Rev* 28: 10-14, 2000.
189. Wagner PD. A theoretical analysis of factors determining VO2 max at sea level and altitude. *Respir Physiol* 106: 329-343, 1996.
190. Wang Y, Shepherd JT, Marshall RJ, Rowell LB, Taylor HL. Cardiac response to exercise in unconditioned young men and in athletes. *Circulation* 24: 1064, 1961. (Abstract).
191. Wehrlin JP, Hallen J. Linear decrease in VO2max and performance with increasing altitude in endurance athletes. *Eur J Appl Physiol* 96: 404-412, 2006.
192. Weibel E. *Symmorphosis. On Form and Function in Shaping Life.* Cambridge, MA: Harvard University Press, 2000.
193. Weibel ER, Taylor CR, Hoppeler H. The concept of symmorphosis: A testable hypothesis of structure-function relationship *Proc Nat Acad Sci* 88: 10357-10361, 1991.
194. Welch HG. Hyperoxia and human performance: A brief review. *Med Sci Sports Exerc* 14: 253-262, 1982.
195. Whipp BJ, Ward SA, Hassall M. Estimating the metabolic rate of marching Roman Legionaries. *J Physiol (Lond)* 491: 60P, 1996.
196. Williamson JW, Fadel PJ, Mitchell JH. New insights into central cardiovascular control during exercise in humans: A central command update. *Exp Physiol* 91: 51-58, 2006.
197. Williamson JW, McColl R, Mathews D, Mitchell JH. Activation of the insular cortex is affected by the intensity of exercise. *J Appl Physiol* 87: 1213-1219, 1999.
198. Yellin EL, Nikolic S, Frater RW. Left ventricular filling dynamics and diastolic function. *Prog Cardiovasc Dis* 32: 247-271, 1990.
199. Young AJ, Sawka MN, Muza SR, Boushel R, Lyons T, Rock PB, Freund BJ, Waters R, et al. Effects of erythrocyte infusion on VO2max at high altitude. *J Appl Physiol* 81: 252-259, 1996.
200. Zuntz N, Schumburg W. Studien zu einer Physiologie des Marsches. *Zbl Physiol* 15: 327-330, 1901.

CHAPTER 12

The Cardiovascular System: Central Influences

Charles M. Tipton, PhD

Introduction

Historically, the response of the cardiovascular system to exercise has resulted in extensive literature, which is best presented in separate sections that are devoted to central and peripheral considerations. Thus, chapter 13 is devoted to peripheral influences on the cardiovascular system. Due to page constraints and the voluminous literature on the subject, the major emphasis of this chapter is the acute and chronic responses of young, healthy subjects to light (~25% $\dot{V}O_2$ max), moderate (~50% $\dot{V}O_2$ max), heavy (~75% $\dot{V}O_2$ max), very heavy (~85%-95% $\dot{V}O_2$ max), and maximal exercise (194); select animal findings are cited as well. With the exception of results pertaining to recovery blood pressure, resting results will not be emphasized. In addition, the narrative includes qualitative details of representative findings but avoids using the term *significant* unless stated by the author(s). Table 12.1 lists the abbreviations used in this chapter.

Cortical Irradiation and Central Command Considerations

Julius Geppert (1856-1937) and Nathan Zuntz (1847-1920) in 1888 proposed that exertion elicited neural stimuli that would activate the respiratory center (54). However, it was the 1893 studies on passive, active, and electrical stimulation using rabbits by E. Johan Johansson (1862-1938) that suggested the rapid increase in heart rate (HR) with exercise was related to the activation of neural centers in the brain (91). According to Secher and Ludbrook (170), in 1908 Aulo proposed the term *cortical irradiation* to explain the rapid HR response with exercise. Five years later, August Krogh (1862-1949) and Johannes Lindhard (1870-1947) (100) indicated that cortical irradiation was involved in the HR results. In fact, they (100, p. 117) reported that seven subjects exercising on a bicycle ergometer experienced within 0.5 s an increase in HR of 19 beats (34%). Nearly 60 yr later, the term *central command* (CC) was introduced by Goodwin and colleagues, including Jere Mitchell (1928–; figure 12.1), to describe a feed-forward command from higher neural centers to exercising muscles (61). In a human experiment that included isometric muscle contractions coupled with vibration of muscle spindle afferents, they effectively demonstrated that muscle tension, HR, and arterial blood pressure (ABP) were increased when CC was initiated and that the opposite occurred when a reduction in CC was noted.

However, animal studies were necessary for identifying the neural pathways that "must act ultimately through medullary neurons" that would affect parasympathetic and sympathetic neural activity while simultaneously eliciting locomotor and cardiovascular responses (201, p. 339). This process began in 1960 with Smith and colleagues, who stimulated the diencephalon of dogs and observed responses similar to those characteristic of volitional exercise (175). They were followed in the 1980s by Eldridge and colleagues (42, 43), who demonstrated that electrical stimulation of hypothalamic locomotor and mesencephalic regions would elicit increased HR and ABP responses. They also effectively demonstrated in paralyzed animals that electrical stimulation of these same regions would produce fictive locomotion, which was accompanied by an elevation in ABP (42). Several years later, Waldrop and colleagues electrically stimulated the hypothalamic locomotor regions in cats and observed elevations in ventricular pressure and contractility along with increased blood flow to skeletal muscles and to the myocardium while observing a reduced blood flow to the kidneys (202). A 2006 study with cats indicated that cells of the nucleus tractus solitarius of the medulla oblongata were involved with CC signals that also facilitated the responses of baroreceptors (33).

Beginning in 1997 and continuing into the 2000s, the research of Jon Williamson (1963–; figure 12.2) and

Table 12.1 Abbreviations Used in This Chapter

ABP	arterial blood pressure
CBF	coronary blood flow
CC	central command
CO	cardiac output
DBP	diastolic blood pressure
EDV	end diastolic volume
EF	ejection fraction
EPR	exercise pressor reflex
ESV	end systolic volume
F	female
HR	heart rate
LBNP	lower-body negative pressure
LVM	left ventricular mass
M	male
MBP	mean blood pressure
MSNA	muscle sympathetic nerve activity
MVC	maximum voluntary contraction
NT	nontrained
PCWP	pulmonary capillary wedge pressure
SBP	systolic blood pressure
SNS	sympathetic nervous system
SV	stroke volume
TPR	total peripheral resistance

colleagues demonstrated that the insular cortex was activated with dynamic exercise (209, 210, 212). When they conducted a CC and hypnotic study with an emphasis on perception of effort using single-photon-emission computed tomography and magnetic resonance imaging techniques, they found that HR, mean blood pressure (MBP), and cerebral blood flow were increased when the insular cortex, anterior cingulate cortex, and thalamic regions were activated (208, 211).

The HR and ABP results of Green and colleagues in 2007 demonstrated that the periaqueductal grey region of the midbrain was an essential integrating region for the feed-forward signals of the CC and for the feedback signals from exercising muscles (64). Results from neuromuscular blockage studies conducted between 1977 and 1990 that included curarization (63), epidural anesthesia (45), and peripheral sensory anesthesia (102) combined with findings from a 1992 investigation by Innes and colleagues (88) with subjects who exhibited unilateral leg weakness effectively demonstrated during exercise that a positive relationship existed between an increase in CC activation and elevations in HR and ABP.

Training and Central Command

Since 1972, a plethora of studies on the relationship between acute exercise and CC have been conducted. However, investigations on chronic exercise, especially as it pertains to the central influences of the cardiovascular system, have been lacking. In 1986, Talan and Engel trained three monkeys to lift weights (light stimulus and food reward) to either increase or decrease cardiovascular responses (operant conditioning by electrical shocks to the tail) (190). After conditioning, the investigators recorded a reduction in mean HR of approximately 21 beats/min, consistent decreases in diastolic blood pressure (DBP), and a lower product of HR × ABP. They concluded that it was possible to demonstrate the influences of CC and that CC could dissociate select cardiovascular effects of exercise (190). Eleven years later, they essentially repeated the experiment when they implanted electrodes in 24 sites in the brains of four monkeys. These sites were electrically stimulated when animals exercised in response to CC or when the animal exercised to avoid receiving an electrical shock (25). The brains were removed and evaluated after the termination of the experiment. From their analysis the investigators concluded that the mediodorsal nucleus, nucleus ventralis anterior, and cingulate cortex were the anatomical regions that demonstrated that CC had initiated HR increases or had exhibited reduced HRs with lifting (25). Interestingly, the blunting of a CC response with exercising animals has a parallel in biathlon participants in that these competitors have learned to reduce their HRs before running or firing (85).

Figure 12.1 Jere H. Mitchell, MD (1928–), esteemed investigator and author, is currently located at the University of Texas Southwestern Medical Center in Dallas in the United States. His contributions to this chapter are found in references 61, 93, 102, 126, 130 to 135, 176, 202, 208, 210 to 212.

This 1990 photograph is courtesy of Charles M. Tipton.

Figure 12.2 Jon Williamson, PhD (1963–), is a distinguished investigator who began his research career in the laboratory of Jere H. Mitchell and has emerged as an authority on the central command with exercise. He currently is affiliated with the University of Texas Southwestern School of Health Professions in Dallas in the United States. His contributions to this chapter are found in references 208 to 212.

Photo courtesy of Jon Williamson.

Since the 1970s the exercise pressor response has been associated with CC activation and muscle mechanoreflex involvement (30). Fisher and White in 1999 investigated the pressor response in subjects who participated in a 6 wk training program of the triceps surae muscles in which the dominant leg was trained and the nondominant, contralateral leg was not (nontrained; NT) (48). The study included a 30% maximum voluntary contraction (MVC) by both legs as well as an electrically evoked contraction of muscles in both legs to elicit mechanoreflex activation and exclude CC involvement. Finally, the muscles were subjected to a postexercise circulatory occlusion test during which changes in HR and ABP were monitored in order to better understand the contributions from chemoreflexes and mechanoreflexes in skeletal muscles. In the trained leg, MVC reduced DBP by 28% and exhibited significant attenuation in HR. Occlusion per se had no statistical influence. However, DBP declined by 27% when these muscles were electrically stimulated. Again, occlusion had no meaningful influence. In the NT leg, a MVC contraction was associated with a 24% attenuation in DBP and a significant reduction in the elevation of HR with exercise. Again, electrical stimulation or occlusion resulted in no meaningful changes in either HR or DBP. To the investigators, these collective responses of the trained and control legs indicated that training had induced a change in CC; however, the changes in DBP lacked statistical significance. Therefore, their conclusion concerning the benefits of training must be interpreted as suggestive and intriguing (48).

Summary

The ideas of Johansson plus Krogh and Lindhard and the experiments of Goodwin and colleagues established the existence of a CC with acute and chronic exercise. Insightful animal experiments and innovative human investigations have effectively established that the neural pathways include medullary neurons, hypothalamic locomotor areas, thalamic regions, and sites in the insular and cingulate cortexes. HR and ABP exhibit physiological changes with exercise that demonstrate CC influence and involvement. Moreover, there is sufficient evidence to indicate that trained populations initiate CC involvement and responses. However, it has yet to be demonstrated that the characteristics or intent of a CC have changed because of training.

Heart Rate and Exercise

As discussed in chapter 1, Galen indicated more than 2,000 yr ago that the rate of the heart increases with exercise. Between 1895 and 1914, Henrich Ewald Herring (1896-1948) (81), Wilbur Bowen (20), and Herbert Gasser and Walter Meek (53) attributed the increase in exercise HR to accelerator nerve activity and increased vagal activity, an increase in the restraining activity of the inhibitory center, and an inhibition of vagal withdrawal, respectively. In 1966, Robinson and colleagues conducted a progressive double-blockade study during incremental dynamic exercise using atropine sulfate and propranolol. They reported HR changes that exhibited a decrease in parasympathetic nervous system activity and a progressive increase in sympathetic nervous system (SNS) activity (160). With the advent of microneurographic recordings of muscle sympathetic nerve activity (MSNA), it has been documented that involvement of the SNS does not occur instantaneously with the initiation of CC (197); rather, moderate exercise must be performed in order to activate the SNS (24). It is the opinion of Loring B. Rowell (1930–; figure 12.3) that CC initiates vagal withdrawal (163). Moreover, Rowell and O'Leary state that SNS activation occurs when vagal withdrawal approaches 100 beats/min and that the subsequent availability of norepinephrine, epinephrine, plasma renin activity, and angiotensin II elevates HR to higher limits (165). In addition, Rowell and O'Leary believe that a mismatch between cardiac output (CO) and vascular conductance will cause a pressor error to initiate an increase in sympathetic nerve activity (165).

Results reported by Sid Robinson (1902-1981) in 1938 (161), Per-Olaf Åstrand (1922–) and Erik Hohwü-Christensen (1904-1996) in 1964 (4), and Pollock and colleagues in 1978 (151) demonstrated that a linear relationship existed between HR and $\dot{V}O_2max$ until very heavy or maximal exercise was performed (194). Maximal HR results from 350 male (M) and female (F) subjects reported by Åstrand and Christensen in 1964 showed a mean value for the 10-yr-old groups (>200 beats/min), after which the HR means progressively declined to lower values (4). A similar response pattern was observed in the data collected by Robinson in 1938 (161). These results are of interest because in 1919 Brainbridge felt that maximal HR with exercise would be approximately 180 beats/min (21).

Training and Heart Rate

Maximal Heart Rate

In 1954, Åstrand and Rhyming (9) stated that the maximal HR for 62 trained Fs between 20 and 30 yr of age was 195 beats/min. Several years later, Hartley and colleagues summarized the results of three training studies involving 17 young M and indicated that the differences between maximal values for NT and T subjects were 3 beats/min lower in T subjects (74). Meanwhile, Ekblom and colleagues conducted a 16 wk study of moderate to heavy training with M subjects and reported that maximal HR declined from 200 beats/min to 198 beats/min (39). During 2000, Zavorsky summarized the results from 20 studies conducted between 1957 and 1998 involving 187 Ms and 29 Fs of a wide range of ages who were trained on assorted modalities for 6 to 365 d. Before training, the mean maximal HR was 186 beats/min. After training, the mean maximal HR was significantly lowered by 6 beats/min (3%) (215). Swine and rodent training studies have demonstrated maximal

Figure 12.3 Loring (Larry) B. Rowell, PhD (1930–), an esteemed investigator and author, is currently emeritus professor of physiology at the University of Washington in Seattle. His contributions to this chapter are found in references 145 and 163 to 166.

Photo courtesy of Charles M. Tipton.

HR changes of 50 beats/min, which represented significant reductions of 10% and 18%, respectively (120, 195). Although it can be concluded that training will not increase maximal HR values and that reductions are expected, the mechanism(s) responsible remains uncertain. Although alterations in the autonomic nervous system are most frequently cited (215), a change in the intrinsic HR continues to be emphasized (19).

Submaximal Heart Rate (Exercise Bradycardia)

Christensen indicated in 1931 that submaximal HRs at specific workloads were lower after training but that the training stimulus had to be increased for HR to remain reduced (27). In 1968 Ekblom and colleagues reported that training reduced mean HR by 26 beats/min (15%) (39). In 1982, Klausen and colleagues conducted a one-leg and two-leg training study with 6 young subjects for 8 wk. Each leg was trained separately but tested together. After training, submaximal HR was reduced by 19 beats/min (11%) when one leg exercised and by 4 beats/min (3%) when two legs were exercised. The investigators suggested that the differences noted were related to the vasoconstricting effect with a large amount of inactive mass (97).

According to Åstrand and Rodahl (8), during 9 mo of moderate-intensity training and submaximal testing mean HR decreased 32 beats/min (23%) after 1 mo, when it plateaued and remained for the next 3 mo (p. 369). However, it was the study of Hickson and colleagues in 1981 that effectively demonstrated the bradycardia concept of Christensen. Specifically, Hickson and colleagues conducted a 9 wk training study with 9 subjects. In the first phase (4 wk) of the study the training load was held constant, whereas in the second phase it was elevated to a higher level of intensity. Weekly testing at the same workload revealed that maximal HRs were significantly reduced by 21 beats/min (11%) after 2 wk, when they plateaued and exhibited a half-time value of 5 d. When the training stimulus was subsequently increased, submaximal values were further reduced by 8 beats/min (5%) (83). Representative training studies conducted since 1969 have repeatedly demonstrated the presence of exercise bradycardia in exercising subjects of different ages and sexes (65, 139, 140, 169). A similar trend has been observed with the training of mongrel dogs (193). The usual explanation given for the decreased HR with submaximal exercise is that activation of the SNS is decreased at specific intensity levels because less work is performed by the heart (19). (Chapter 9 provides more information on this subject.) Interestingly, HR is increased more by arm exercise than by leg exercise at the same submaximal O_2 consumption because of the increased involvement of the SNS (19).

Intrinsic Heart Rate

Intrinsic HR is the HR that is recorded after blockade of the adrenergic and cholinergic receptors by select pharmacological agents (e.g., atropine sulfate and propranolol) and relates to the activity of the sinoatrial (SA) node. Although the intrinsic HR has been extensively investigated with regard to the resting bradycardia of training, investigations pertaining to submaximal exercise are limited. Lewis and colleagues conducted a cross-sectional study with highly trained cyclists and controls. They intravenously infused combinations of atropine sulfate, propranolol, or metoprolol at 50%, 75%, and 100% $\dot{V}O_2max$ and found that significant reductions of 8 beats/min (6%), 16 beats/min (10%), and 11 beats/min (6%), respectively, occurred in the trained subjects. They suggested that cardiac hypertrophy was the mechanism responsible for any reduction in the intrinsic rate (107). Boushel and colleagues also indicated that cardiac enlargement could affect intrinsic HR due to the influence of stretch on the SA node (19). Investigations conducted by Bolter and colleagues (18) and by Hughson and colleagues (86), with isolated atrial preparation from NT and T rats, have reported that atrial preparation from the T rats had lower rates.

Summary

HR increases with dynamic exercise because the CC initiates a rapid vagal withdrawal that is augmented by SNS activation when HRs approach 100 beats/min. As the intensity of exercise increases, central and reflex influences supplement SNS dominance, as manifested by increased concentrations of norepinephrine, plasma renin activity, and epinephrine from the adrenal gland as well as by an increasing core temperature. Maximal HR is not enhanced by training; in fact, it exhibits a modest decrease, possibly due to a change in the intrinsic HR. Submaximal exercise bradycardia is a consistent and reproducible training effect that has been explained by a combination of mechanisms, including reduced cardiac work, decreased SNS activation, less-responsive beta adrenergic receptors, and enhanced stroke volume (SV) that occurs because of an expanded plasma volume.

Arterial Blood Pressure and Exercise

Dynamic Exercise

Stephen Hales (1667-1761) was the first to measure arterial blood pressure (ABP) in a living animal (72) and is regarded as "the father of hemodynamics" (163). According to the German sport physician Hebert Herxheimer, Grebner and Graunbum in 1900 were the first to measure ABP pressure in humans during dynamic exercise and reported an increase of 60 mmHg (82). In 1904, Wilbur Bowen had M subjects perform moderate to heavy exercise on a cycle ergometer. Bowen measured systolic blood pressure (SBP) with an Erlanger sphygmomanometer connected to a revolving smoke drum and reported that SBP approached 190 mmHg in 5 to 10 min (20).

During 1965 Åstrand and colleagues had 14 young subjects perform light to moderate and moderate to heavy dynamic exercise with the arms and legs on different occasions (6). In both exercise situations, MBP for the arms was significantly higher than that for the legs by 16 mmHg (17%) and 35 mmHg (36%), respectively (6). These results were later confirmed by Bevegard and colleagues using direct measurements from the aorta and were attributed to the increased vasoconstriction from SNS activation, which also elevated vascular total peripheral resistance (TPR) (13). However, when Volianitis and colleagues in 2004 had 12 young M subjects perform moderate to heavy exercise with the arms and legs, the MBP was reset to a lower pressure (−5 mmHg) than previously recorded for arm exercise. They suggested that the reduction in pressure was related to a shift in central blood volume by the muscle pump because of the competition between the arms and the legs for an increase in blood flow (199).

Gleim and colleagues in 1990 assessed SBP responses of 10 young M and 10 F subjects who performed incremental dynamic exercise on four modalities (treadmill running, rowing, arm and leg ergometers) at fixed percentages of their peak HR values. Ms had significantly higher pressures than did Fs (30-45 mmHg) at each stage (50%, 75%, and 100% peak HR). However, this sex difference in SBP did not exist when evaluated according to body surface area and was markedly reduced after body weight was considered. When evaluated per kilogram of lean body mass, Fs had significantly higher SBP values than Ms. Urinary catecholamine results were similar for both M and F subjects and were unable to explain the absolute M results or the lean body mass findings with the F participants (59).

In 2001, Gallagher and colleagues measured changes in intramuscular pressure with dynamic exercise that combined thigh cuff occlusion and lower-body negative pressure (LBNP). They observed that parallel increases occurred between ABP and intramuscular pressures at the higher work rates (50).

Although postexercise hypotension is regarded as a contemporary finding with dynamic exercise, it was first reported by Leonard Hill in 1898 (84). Hill's subject ran 400 yards and exhibited a reduction in ABP of 10 to 15 mmHg after 10 min and a reduction of 30 to 35 mmHg after 1 h. He stated, "The arterial pressure becomes depressed below the normal resting level after severe muscular work" (p. xxvii). Lowsley (112) was cognizant of the existence of postexercise hypotension but regarded it as a subnormal response associated with heart strain. In 1911 he examined the effects of five types of exercise (brief, rapid, vigorous, fatiguing, and exhausting) on ABP. Runs of 5 to 9 miles (8-14.5 km) were included in the "fatiguing" category and runs of 10, 13, and 20 miles (16.1, 20.9, and 32.2 km) were included in the "exhausting" category. SBP and DBP were recorded before and at the end of the run and during the time required to return to baseline. Decreased pressures were observed in all five categories that averaged −16 mmHg for SBP and ranged from 7 to 10 mmHg for DBP with the recovery time to baseline resting values being variable (112).

Nearly a century later, Polito and Farinatti reported significant reductions of 13 mmHg for SBP and 7 mmHg for DBP after heavy leg-resistance exercise during a 60 min recovery period (150). Results pertaining to responsible mechanisms (especially from animals) have implicated the actions of endogenous opioids (19), involvement of serotonergic pathways, reduced sympathetic activity, the presence of endothelium relaxing factors (191), and the interaction between substance P and the gamma-aminobutyric acid (GABAergic) system in the nucleus tractus solitarii (26).

Dynamic- and Static-Resistance Exercise

James H. McCurdy (1866-1940) was among the first to record ABP responses from subjects lifting weights. In 1901, he used the modified Hill and Riva-Rocci sphygmomanometer with 11 young M students performing back and leg lifts and reported that group ABP values increased from 111 mmHg to 180 mmHg with several subjects exceeded 200 mmHg. McCurdy attributed these pressure results to changes in intrapulmonic (~60 mmHg) and intra-abdominal (~87 mmHg) pressures (124). It is disappointing that this explanation is seldom mentioned in many textbooks and that many elementary textbooks ascribe all pressure changes to the activity of the SNS.

According to Herxheimer, Lindhard in 1920 was the first to investigate the effects of static (isometric) exercise on the cardiovascular and respiratory systems by having subjects hang from a beam (82). More than four decades later, Lind and colleagues used sustained handgrip contractions at various percentages of MVC to assess cardiovascular responses. The results reaffirmed that isometric contractions can elicit greater increments in ABP than dynamic exercise can. The rate of increase in ABP was linear with the isometric contraction force, which was expressed as a percentage of MVC, and mean aortic blood pressure was markedly elevated at 50% MVC whereas TPR continued to decrease (108). Static exercise results from subjects at the University of Massachusetts in the United States showed that peak SBP progressively increased from 172 mmHg at 25% MVC to 225 mmHg with maximal effort with peak DBP being increased from 106 mmHg to 156 mmHg under the same conditions (121). These results support the concept that pressor responses are proportional to the percentage of the MVC performed.

During 1980, Mitchell and colleagues demonstrated that the increase in MBP and HR with static exercise (40% MVC) was related to the magnitude of the muscle mass involved in the contraction process and was not proportional to the MVC percentage (131). One year later, Mitchell and his group reaffirmed the importance of muscle mass in eliciting ABP responses while investigating the role of neural control in static exercise. Besides recording ABP responses, they secured measures of EMG activity with and without holding force constant in order to show the involvement of the CC. From their collective results the authors concluded that both CC and reflex activity, initiated by muscle afferents, were contributing mechanisms for the elevation in pressure (133).

Using cats, Petrofsky and colleagues in 1981 examined the relationship between muscle fiber types and pressure responses at various MVC percentages. For Type I or slow-twitch muscles (soleus), there was no relationship between ABP responses and fiber type stimulation and recruitment profiles. In contrast, when Type II or fast-twitch muscles (medial gastrocnemius) were activated, both SBP and DBP exhibited marked elevations (147).

In 1992, Sullivan and colleagues (188) recruited 10 young M subjects to perform submaximal and maximal isometric deadlift exercise. After the brachial artery was cannulated for direct pressure measurements, they found that MAP was significantly elevated by 23 mmHg at 50% maximal effort; this was primarily due to a significant elevation in HR because no significant changes in CO or TPR occurred. However, group MAP results significantly increased by 56 mmHg (52%) with 100% maximal effort due to significantly higher (90%) CO values and a 17% reduction in TPR. Intrathoracic pressures were assessed by esophageal pressure measurements, and no significant results were reported (188).

Training and Arterial Blood Pressure

Dynamic Exercise and Maximum Arterial Blood Pressure Results

The results of Ekblom and colleagues' 12 wk study of young M subjects demonstrated that training increased maximal SBP values by 32 mmHg (18%) to 213 mmHg and increased MBP by 9 mmHg (8%) and DBP by 6 mmHg (8%) (39). The increase in maximal MBP was attributed to a significant increase in CO (8%); change in TPR was minimal (1%). In the cross-sectional investigation of Karjalainen and colleagues, maximal dynamic exercise was associated with an SBP of 203 mmHg in the controls whereas a significant value of 228 mmHg was obtained in elite athletes. They attributed the elevation in pressure to an increased left ventricular mass (LVM) (94). Interestingly, when they assigned athletes to either a high or a low LVM group, the SBP values with maximal dynamic exercise were significantly higher in the subjects in the high group than in those in the low group. Furthermore, their statistical analysis indicated a high association between maximum pressures and LVM.

Submaximal Arterial Blood Pressure Results

In the 1969 training study of Hartley and colleagues with older M subjects, moderate exercise was identified with a 6 mmHg (5%) reduction in MBP, which was associated with a 4% reduction in CO and a 1% decrease in TPR (74). In the 1982 investigation of Klausen and colleagues in which subjects performed either one-leg or two-leg exercise, one-leg exercise reduced MBP by 14 mmHg (14%) and two-leg exercise reduced MBP by 7 mmHg (7%). TPR units were reduced by 12% and 7% in the one-leg and two-leg exercise groups, respectively. The authors were unable to explain the higher resistance results in the two-leg group (97). During 1986, Jennings and colleagues had 10 young subjects train at a heavy intensity level either 3 or 7 times/wk for 4 wk.

When subjects were tested at 50% of their maximal workload, the calculated decrease in SBP was 7 mmHg for both groups, which was explained by the significant reduction in the TPR units. The investigators also reported that the exercise norepinephrine spillover rate was significantly reduced in 80% of the subjects and was a major factor for the reduction in ABP, which was best demonstrated with resting results. Moreover, training 3 times/wk and 7 times/wk yielded results that were very similar (90).

When the elite athletes and control subjects mentioned previously in the Karjalainen and colleagues investigation (94) were compared while exercising at 100 W, mean SBP results for the athletes were 7 mmHg lower than those for the controls. Contrary to expectation, this pressure difference had a minimal relationship with LVM, in part, because the work of the heart was lower at that specific workload (19).

Maximal Arterial Blood Pressure With Dynamic- and Static-Resistance Exercise

In 1926, Wilhelm Ewig, a German physician, stated that the maximal ABP obtainable by athletes was 250 mmHg (82, p. 18). In a 1985 study by McDougall and colleagues, direct pressure measurements were obtained from the brachial artery of 5 highly trained and experienced bodybuilders who performed one-arm curls and two-leg presses. Their training and genetic endowment enabled them to achieve mean peak SBP and DBP results of 255/190 mmHg with one-arm curls and from 255/190 to 480/350 mmHg with two-leg presses (125). They used the Valsalva maneuver to increase intrathoracic pressures and were capable of increasing MBP by 60 mmHg even when no weights were lifted. The investigators attributed these extraordinary increases in ABP to the mechanical compression of tissues, the results of the Valsalva maneuver in increasing intrathoracic pressures, and a potent pressure response (125). For such pressures to occur, training likely increased muscle mass, enhanced vascular adaptability to withstand the elevated compressional forces and the vasoconstricting effects, enhanced the resetting of baroreceptors associated with the exercise pressor reflex (EPR), and improved the skill and efficiency of lifting (62, 176, 213).

In contrast to ABP results with subjects performing dynamic exercise (74), McCartney and collaborators conducted a 12 wk dynamic resistance-training program with older M subjects that included one-arm curls and two-leg presses. Direct brachial arterial pressures were measured during both exercises. After training, maximum SBP values (228 mmHg) were significantly decreased by 9% whereas DBP values (144 mmHg) were significantly decreased by 19%. With the two-leg press, a significant reduction of 8% occurred at 60% and 80% of 1 repetition maximum, respectively. The investigators felt that a feed-forward CC mechanism was primarily responsible for the training effect, although they did not minimize the importance of possible reduced intramuscular compression forces being involved (122).

In a 1997 cross-sectional study involving athletes and controls, comparisons were made after subjects performed a 3 min static handgrip test. The athletes had higher SBP results (21%) and significantly higher DBP values (28%), which were explained, in part, by a greater LVM (94). Hagerman and colleagues in 2000 conducted a dynamic resistance-training study with older M subjects resulting in significant strength and anatomical changes in the Type II muscle fibers. However, these training effects had no statistical relationship with ABP measurements obtained during lifting at 60%, 80%, or 100% of the 1 repetition maximum (70).

Summary

ABP increases with acute dynamic exercise because of the influence of CC, activation of the SNS, and reflexes from active muscles that result in an augmentation of CO and reduction in TPR. Both arm and leg exercise elevates ABP, but arm exercise elicits higher pressures due to a greater vasoconstricting SNS effect with the arm vasculature. However, combined arm and leg exercise of moderate to heavy intensity can be associated with a lower ABP, presumably due to competition between the limbs for adequate blood flow. Ms generally have higher pressures than Fs during exercise, but the explanation is complicated by body composition. Differences in SNS activation and responsiveness are apparently not major considerations. Postexercise hypotension has been a consistent finding over the years. Its explanation in humans remains elusive, although the influence of opiates must be considered.

Mass, compressional forces, reflex responses, and changes in TPR units are important considerations for an increase in ABP with resistance exercise. Tissues of elite athletes are able to safely withstand the extraordinary muscular forces generated with resistance exercises only because adaptive vascular changes have occurred with long-term training. Chapter 13 offers insights on responsible mechanism. In animals and humans the percentage of Type II fibers has relevance to ABP, but careful comparative studies concerning the effects of chronic exercise need to be undertaken.

In the majority of studies, dynamic exercise training was associated with significantly elevated exercise ABP values that were attributed to an increased CO and augmented SNS activation. Athletes had elevated pressures with exercise when compared with nonathletic controls. However, whether LVM was a primary determinant remains uncertain. Trained subjects consistently demonstrate reduced ABP values with submaximal dynamic exercise, which has been explained by diminished SNS activity and decreased vascular resistance due to a reduced afterload.

The explanation for the extraordinary ABP elevations that occur in elite athletes with dynamic resistance exercise is unknown, but genetic endowment is a major consideration. Athletes also have higher ABP values with static exercise than nonathletes do. In general, older trained subjects exhibit significantly attenuated maximal and submaximal responses with dynamic resistance exercise. It is uncertain whether this observation is a result of the aging process.

Select Reflex Activity and Cardiovascular Responses

Background Information

Marey's law (119) was operational during the early decades of the 20th century even though Zuntz (68) and Brainbridge (21) had conducted experiments indicating it was an untenable concept concerning HR and ABP relationships during exercise. Also known but seldom investigated was the Geppert and Zuntz concept that "unknown blood absorbing substance(s) from the contracting muscles during exertion could elicit cardiorespiratory response" (54).

During 1923, Heinrich E. Herring (1886-1948) of Prague demonstrated that mechanical stimulation of the carotid sinus in animals resulted in a reflex decrease in both HR and ABP (80). With time, the reflex became identified with afferent input to the nucleus tractus solitarii and activation of the autonomic nervous system (ANS) leading to changes in HR and ABP with alterations in atrioventricular conduction and in TPR (38).

Alam and Smirk of Cairo, Egypt, in 1937 incorporated limb occlusion with their exercising subjects and observed marked increases of MBP (~30 mmHg) that remained elevated, in contrast to the controls, for approximately 10 min (3). Later, in a series of innovative experiments, Rowell and colleagues confirmed these findings and attributed the result to Geppert and Zuntz's concept that active muscles were reflexly responsible for the increase in MBP because a mismatch had occurred between muscle blood flow and oxygen demand. In addition, Rowell considered the blood pressure-increasing effect to be the result of a chemoreflex (163). Subsequently, limb occlusion effects became identified as a chemoreflex, ischemic muscle pressor reflex, muscle metaboreflex, or a pressor reflex (192). In 1983, Mitchell and colleagues introduced the term *exercise pressor reflex* (EPR) for the increased ABP resulting from muscle contractions (130) and reminded readers that muscle contractions can reflexly influence ANS outputs to the cardiovascular system to increase ABP, HR, ventricular contractility, CO, and changes in CO distribution (130).

During 1966 Bevegard and Shepherd used neck suction and pressure in a neck collar in humans to alter the transmural pressure forces in the carotid sinus in order to influence the neural input from the carotid sinus during moderate exercise (15). They wrote, "Thus, the carotid sinus mechanism continues to oppose, through negative feedback, the rise in blood pressure during exercise, but is overcome by the exercise stimulus so that the net result is an increase in both HR and blood pressure" (p. 142). The results of this human study and one conducted by Bristow et al in 1971 (22) followed by animal experiments (71, 127, 196) led to concepts that during exercise the baroreceptors were overcome, turned off, not involved, or suppressed at the onset of exercise. This concept prevailed until the later part of the 20th century.

Two ideas emerged from the animals studies. The first was that the elevation in ABP at the onset of exercise, and to some extent during exercise, required the presence of baroreceptors. The second was that baroreceptors initiated vasoconstriction of the vascular beds to cause an elevation in ABP (192). However, it was the animal studies conducted by David Donald (1921-2004; figure 12.4) and colleagues of the Mayo Clinic (128, 181, 203) that provided the essential information for understanding the functioning of baroreceptors during exercise. In 1983, Walgenbach and Donald, using dogs, perfected and employed the technique known as the reversible vascular isolation of the carotid sinus, which involved selective denervation of the aortic arch and the contralateral carotid sinus that isolated one carotid sinus. This approach allowed evaluation of baseline and changes from baseline in MBP, HR, blood flow, and sinus pressures (203). The authors concluded that baroreceptors "serve to maintain arterial blood pressure in the face of metabolic vasodilation in the active muscle; as exercise increases in severity, they prevent an abnormal increase in arterial pressure caused by excessive sympathetic outflow to systemic resistance vessels" (203, p. 261). This change indicated that carotid baroreceptors regulate ABP during exercise and have the capability of resetting during exercise, a possibility suggested by Robinson and colleagues in 1966 (160) and reinforced by the animal investigation of Ludbrook and

Graham in 1985 (113). According to Raven (154) and Potts (152), resetting is a result of neural signals from the cortex, skeletal muscles, and baroreceptors that are integrated in the nucleus tractus solitaries (NTS) and the rostral ventrolateral medulla.

Carotid Baroreflex

Studinger and coworkers, with the aid of ultrasonographic techniques and radial tonometry, recently measured the effect of ABP deformation on arterial barosensory regions. Diameter measurements were made near the bifurcation before and during peak exercise and significant changes (6%) in diameter (mm) were recorded (187).

Emerging from the 1983 study of Walgenbach and Donald (203) was the realization that a sigmoid relationship existed between changes in pressure of the isolated carotid sinus and systemic ABP, which meant that an increase in HR and systemic ABP occurred when sinus pressure was reduced. On the other hand, decreases in HR and systemic ABP were observed when pressures in the carotid sinus were elevated. They found that baroreceptor functions were best understood by analyzing sigmoid stimulus–response curves that allowed identification of thresholds, saturation, operating points, gains, or sensitivity and their shifts on the stimulus–response curves (166). According to Peter Raven (1940–; figure 12.5) and colleagues, significant increases in baroreflex threshold and saturation points for both HR and ABP along with no statistical evidence for changes in maximum baroreflex gain (or sensitivity) were convincing evidence for resetting (figure 12.6). It should be noted in figure 12.6 that the resetting shown in panel D will shift the operating point (OP) and facilitate the optimal buffering of the EPR (156).

During 1993, Potts and colleagues had subjects perform light to moderate dynamic exercise on a cycle ergometer. Subjects wore neck collars to activate the carotid baroreflex by using incremental negative and positive pressures with HR, MBP, and central venous pressures being measured (153). With moderate exercise they found significant increases in pressure for threshold, saturation, and the operating point with no changes in baroreceptor gain. Thus, resetting had occurred. Later, Papelier and colleagues conducted a related study with subjects performing heavy exercise and found higher HR and MBP values at any given carotid sinus pressure with no changes in gain, a result that also indicated that resetting had occurred. However, they did not report data pertaining to threshold or saturation changes (145).

As noted previously, reports demonstrated that arm exercise alone elicited higher ABP results than leg exercise did. However, when both arm and leg movements were combined while exercising at the workload for arms alone, a reduction in MBP and a reduction in CO occurred (6, 171). In 2002, Volianitis and Secher observed a 20% reduction in arm blood flow when heavy exercise was performed by both arms and legs (198). Two years later, these investigators along with others used the neck collar method (neck suction coupled with neck pressure) with subjects who per-

Figure 12.4 David E. Donald, PhD (1921-2004), was a creative and innovative investigator at the Mayo Clinic in Rochester, Minnesota. His contributions to this chapter are found in references 35, 128, 181, and 203.
Photo courtesy of Charles M. Tipton.

Figure 12.5 Peter B. Raven, PhD (1940–), an esteemed investigator, author, and editor, is currently located at the University of North Texas Health Science Center in Fort Worth. His contributions to this chapter are found in references 50 to 52, 143, 144, 146, 153 to 157, 172, and 182.
Photo courtesy of Peter B. Raven.

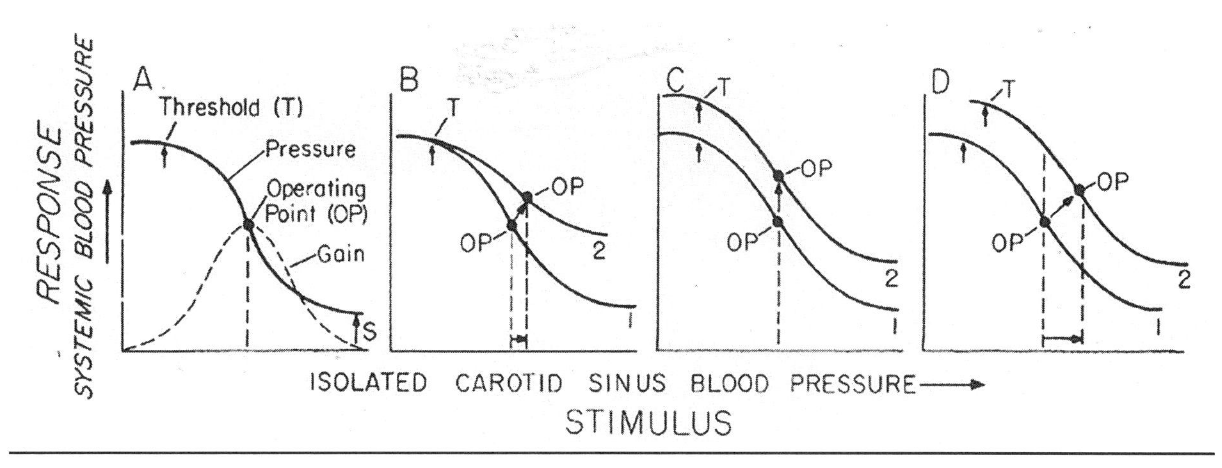

Figure 12.6 Stimulus–response or baroreflex function curves for the carotid sinus baroreflex. In panel A, dashed lines represent gain or sensitivity. The point of maximum gain is considered to be the baroreflex operating point. T = threshold, OP = operating point, S = saturation. Panel B represents a decrease in baroreflex gain in which the OP and systemic pressure curve are shifted upward and no change occurs in the threshold. Panel C demonstrates an upward shift in the systemic pressure curve and no changes in T, gain (OP), or S point. Panel D exhibits an upward shift in T and gain (OP) to a higher carotid sinus and systemic pressure and serves as an example of baroreceptor resetting.

Reprinted from L.B. Rowell et al., Integration of cardiovascular control systems in dynamic exercise. In *Handbook of physiology. Exercise: Regulation and integration of multiple systems,* edited by L.B. Rowell and J.T. Shephard (New York: Oxford University Press), 770-838, 1996. With permission of American Physiological Society.

formed upright and supine exercise with only the arms. The stimulus–response curves showed a marked shift up and to the right in MBP, which indicated that resetting had occurred with the various exercise conditions. They found that the carotid baroreflex operation point during exercise, compared with resting conditions, exhibited elevations that ranged from 9% to 23% with upright arm exercises being the highest and supine leg exercises being the lowest. Carotid baroreflex threshold and saturation pressures followed the same pattern with gain being similar for all conditions. The downward resetting of baroreceptors with the combined exercise was associated with a 6 mmHg decrease in MBP, which was attributed to the increase in central blood volume facilitated by the activity of the muscle pump during leg exercise (199).

During 2003, Ogoh and colleagues investigated the relationships between the carotid baroreflex, CO, SV, total vascular conductance, and MBP in subjects who performed mild to heavy exercise on a bicycle ergometer (143). They reported that the activity of the carotid baroreflex had no significant influence on CO or SV but that it was associated with upward resetting of the stimulus–response curves for HR and MBP without inducing changes in gain. With heavy exercise, total vascular conductance was markedly increased. They concluded that the SV did not contribute to the reflex-induced increase in MBP, that HR increase was reflex driven, and that the increased total vascular conductance was more responsible for the elevations in MBP than CO. They also believed that changes in vasomotion were the primary means by which the carotid baroreflex regulated ABP during mild to heavy exercise (143).

Aortic–Cardiac Baroreflex

Human research, predominately by Raven's group, has demonstrated that aortic baroreceptors dominate carotid baroreceptors in the reflex control of HR under resting conditions (172). To determine whether exercise alters this relationship, they combined light dynamic exercise, LBNP, and phenylephrine infusions and used the ratio of change in HR to MAP to assess baroreflex functioning. They also measured changes in gain during the two conditions. Because the ratio changes and the gain values were similar in both conditions, they concluded that aortic baroreceptors continued to dominate carotid baroreceptors during exercise. How heavy or maximal exercise would alter this relationship was not pursued.

In 2003, Komine and colleagues conditioned three cats to perform static exercise for a food reward and found a rapid and significant increase in HR that began before exercise was initiated. After surgically implanting electrodes to excite the aortic depressor nerve, they found that stimulation elicited a profound baroreflex. When the nerve was stimulated just before bar pressing, they observed a significant attenuated bradycardia before the onset of static exercise that prevailed during the initial period. The attenuation suggested that the cardiac component of the aortic baroreflex before and during the early period was inhibited by a central mechanism (e.g., CC) rather than by the EPR. Because there were no significant changes with ABP, they felt that the central property of the reflex had been preserved (98).

In an attempt to identify the roles of the aortic and carotid baroreflexes in influencing changes in MSNA and DBP during isometric exercise (DBP is significantly correlated with MSNA whereas SBP is not), Ichinose and colleagues in 2006 evaluated select characteristics of the MSNA response and related them to changes in DBP during 30% MVCs combined with postexercise ischemia. They found with isometric exercise that these reflexes modulated, in a time-dependent manner, a progressive increase in MSNA plus a change in baroreflex gain (87).

Cardiopulmonary Baroreflex

Walgenbach and Shepherd in a 1984 Mayo Clinic publication summarized exercise studies conducted with intact, sinoaortic, denervated, vagotomized dogs as well as dogs with isolated carotid sinus preparations and concluded that cardiopulmonary baroreceptors were not important for the control of blood pressure during exercise (192). Later, they combined with Daskalopoulos (32) to study chronic sinoaortic, denervated dogs during exercise before and after vagotomy and again reported minimal changes with HR and MBP. This finding reinforced their conclusion that cardiopulmonary mechanoreflexes had minimal influence on the regulation of blood pressure during exercise (192).

In humans, Mack and colleagues investigated the role of the cardiopulmonary baroreflex during moderate supine dynamic exercise using systematic reductions in LBNP (−10 to −40 mmHg) (116). Because of results from previous LBNP investigations (2, 217), they assumed that results at −10 mmHg would selectively unload cardiopulmonary baroreceptors whereas the carotid baroreceptors would be responsible at negative pressures of −20 mmHg and more. Compared with exercise control conditions, they found no significant changes in MBP or HR at −10 mmHg but observed significant changes in CO (−10%), SV (−9%), forearm blood flow (−30%), TPR (+10%), and forearm vascular resistance (+66%). From these findings they concluded that the cardiopulmonary baroreflex was selectively unloaded during exercise and that it reflexly defended the elevated MBP (116).

After the turn of the 21st century, Ogoh and collaborators investigated the influence of changes in central blood volume (measured indirectly) with exercise on cardiopulmonary baroreceptor functions (144). Exercise was performed in three situations (two upright and one supine) that were similar in $\dot{V}O_2$consumption but different in ergometer revolutions in the supine position to activate the muscle pump. Stimulus–response curves were analyzed for functional changes. The researchers found that MBP was increased with upright dynamic exercise and associated with a shift in the stimulus–response curve that indicated that the baroreceptors had been reset to a higher pressure. However, with supine exercise, MBP was significantly lower than in control conditions (10%) and was associated with the operating point exhibiting a significant shift downward and to the left when central blood volume was elevated. These changes were coupled with significant reductions in threshold and saturation value and indicated that cardiopulmonary baroreceptors could be reset depending on nature of the exercise and the magnitude of change in the central blood volume (144).

Exercise Pressure Reflex

After Mitchell and colleagues defined the EPR in 1983, they suggested that a subset of group III and group IV muscle afferents may be involved as ergoreceptors and capable of responding to mechanical or metabolic stimuli, including bradykinin, serotonin, potassium, and capsaicin, as well as hypermolar concentrations of lactate and phosphate (130). (In later years, mechanical stimuli activated mechanoreceptors and the metabolic stimuli affected the metaboreceptors.) Mitchell and colleagues indicated that little was known about the site of integration or brain stem involvement but implicated the lateral reticular nucleus, fastigial nucleus, and raphe nuclei (130). Using anaesthetized cats to study the effect of static and rhythmic (dynamic) twitch contractions on the discharge profile of group III and IV muscle afferents, Kaufman and colleagues in 1984 found that group III afferents were more responsive to static than to rhythmic contractions and were likely a major contributor for HR changes with vagal withdrawal (95).

From their muscle contraction experiments in animals in 1988, Rotto and Kaufman reported that lactic acid and select cyclooxygenase products such as prostaglandins and thromboxanes were the most promising for stimulating group III and IV afferents (162). During 1994, Sinoway and colleagues used a design that included animals as well as young subjects performing ischemic static exercise (50% MVC). ^{31}P-nuclear magnetic resonance imaging procedures were used to measure muscle H^+ and $H_2PO_4{}^-$ (dipronated phosphate) concentrations and microneurography techniques were used to record MSNA. Sinoway and colleagues concluded that $H_2PO_4{}^-$ was an important stimulant of muscle afferents and of MSNA during exercise (174). Although the specific mechanism of action of lactic acid in activating EPR continues to be unclear, a 2008 study by Hayes and colleagues that included acid-sensitive ion channels, arterial lactic acid injections, static contractions by decerebrate cats, and injections of an inhibitor of channel activity indicated that infusions of lactic acid that were combined with static contractions reduced the EPR response by approximately 50% (76).

To determine the contribution of EPR in increasing MBP during dynamic exercise, Collins and colleagues in 2001 used bilateral carotid occlusion techniques with instrumented dogs performing light to heavy exercise and measured blood flow in active and inactive beds. Their data indicated that the vasoconstriction initiated by the EPR in the active vascular beds was responsible for the elevated MBP that occurred with graded exercise (29). It is known that static exercise will stimulate both mechanoreceptors and metaboreceptors to induce changes in blood volume and flow. During 2007, Stewart and colleagues examined the effect of exhaustive static exercise (35% MVC) on baroreflex responses in 16 young M and F subjects. They found that central blood volume, CO, and TPR were elevated whereas the HR component of the baroreflex in the exercise pressure reflex had been uncoupled from regulation. They concluded that the EPR response was associated with a reduction in baroreflex cardiovagal control and that the reflex was driven by increased CO, preload, cardiac contractility, and by an elevated blood volume that had originated from the splanchnic bed (183).

In 2006, Gallagher and colleagues had a small number of young M subjects perform static leg exercise (20% MVC) to determine the interaction between the EPR and CC in resetting the baroreflex (51). Their experimental design was (a) to examine changes in the carotid baroreflex–HR stimulus–response curve and (b) to determine the carotid baroreflex–vasomotor (MBP) stimulus–response curve after conditions of control, neuromuscular blockage, and after activation of the EPR by use of antishock trousers inflated to 100 mmHg. Their results supported the concept that CC controlled the HR resetting response whereas both CC and the EPR contributed to the carotid baroreflex–vasomotor reflex resetting response during exercise (51).

Training and Responses of Select Reflexes

Background Information

As discussed previously, during maximal exercise well-trained populations have the capacity to increase ABP beyond the limits achieved by NT subjects. However, the explanation for the increase is far from simple because, as demonstrated by the elegant resting experiments of Ogoh and colleagues concerning ABP changes with baroreceptor stimulation, approximately 33% of the changes noted were attributed to HR and CO influences whereas the remaining 67% of changes were associated with vascular resistance alterations (143).

Dynamic Exercise and Arterial Baroreflexes

In the 1970s, reports from two cross-sectional studies in which trained subjects were subjected to progressive increases in LBNP demonstrated that trained individuals exhibited attenuated baroreceptor responses and less tolerance to orthostatic challenges (114, 180). Raven and colleagues stated in 1984 that individuals with elevated fitness levels had significantly reduced HR and SBP responses to more negative LBNPs whereas the less-fit subjects had higher TPR results (157). In a 1992 study by Stevens and colleagues, subjects endurance trained for 8 mo before being measured at four LBNP stages (between −5 to −45 mmHg) (182). The trained subjects were 23% faster in achieving presyncope conditions, exhibited impaired arterial baroreflex control of ABP, and, at −45 mmHg, showed primarily a reduced vasoconstrictor response and secondarily a reduced HR (182). When animal results (rabbits and rats) were considered, Raven and colleagues concluded that there was convincing evidence that daily exercise training attenuates arterial baroreflex control of the cardiovascular system (156, p. 17). In explaining training effects, especially aspects that related to the incidence of orthostatic intolerance, Raven and Pawelczyk listed decreased responsiveness of the aortic and carotid baroreceptors, increased total blood volume, enhanced limb compliance, and the presence of eccentric ventricular hypertrophy (155).

Citing predominately animal results, Komine and colleagues stated that trained populations have increased arterial baroreflex sensitivity (or gain), although they were uncertain whether it was the result of neural influences or changes in vascular compliance. Hence, they conducted a cross-sectional investigation in 2009 with healthy trained runners and sedentary subjects and, based on their results, proposed that changes in sensitivity had occurred in the neural component of the baroreflex arc with little or no alterations occurring in vascular compliance (99).

Exercise Pressor Reflex

A major contribution of the EPR is its role in the activation of MSNA during both static and dynamic exercise. In addition, EPR and CC are actively involved in the resetting of ABP to a higher level (51, 52, 176). To determine whether arm isometric training would affect the chemoreflex component (metabolites) that activate SNS responses, Somers and colleagues in 1992 had 8 subjects train at 33% MVCs for 6 wk. They found significant gains (1,146%) in endurance performance but a marked decrease (73%) in microneurographic recordings of SNS activity, which indicated that training had attenuated the chemoreflex component of the EPR. Using data from other isometric-training studies, Somers and colleagues speculated that a reduction in muscle pH had affected muscle afferent activity, which in turn was responsible for the attenuated findings (177).

In 1998, Mostoufi-Moab and colleagues conducted a 4 wk dynamic exercise-training study with the forearm at 35% MVCs under ischemic (positive pressure) and nonischemic conditions in order to elicit an EPR response (138). They reported that training increased the positive pressure threshold necessary to initiate reflex responses while inducing significant reductions (11 mmHg) in mean ABP (10%) and in venous lactate concentrations (38%). In addition, the elicited pH value after training (7.25) was significantly elevated by 0.04 units. Their investigation demonstrated that training could reset baroreceptors and supported the concept of Somers and colleagues that alterations in the chemoreflex component were major contributors to the attenuated response (138).

Fadel and colleagues in 2001 conducted a cross-sectional investigation involving fit and nonfit subjects that employed neck suction with thigh ischemia. They found that the fit subjects had significantly reduced MSNA responses with significant decreases in mean ABP when compared with the less-fit subjects. In addition, they felt that fitness levels affected measures of vascular reactivity (44).

Emerging from the late years of the 20th century and the first decade of the 21st century was the clinical awareness that individuals with chronic heart failure also had exercise intolerance, elevated sympathoexcitation, and exaggerated EPR responses (12, 141, 176). Wang and colleagues in 2010 reported that exercise training in rats prevented an exaggerated EPR response (204). Whether a similar effect occurs in humans is unknown and at the present time is the subject of intensive speculation and future research (12, 141).

Cardiopulmonary Baroreflex

It is disappointing that few published exercise studies with young and healthy subjects exist on this topic. Between 1986 and 1993, 4 cross-sectional studies involving 53 fit or highly trained athletes were conducted. These studies assessed forearm blood flow and vascular resistance when the athletes were subjected to low negative pressures generated by LBNP procedures (55, 115, 117, 189). The collective results indicated that fit subjects, when compared with their less-fit counterparts, exhibited lower forearm vascular resistance, lower plasma norepinephrine concentrations, significantly reduced baroreflex gain values, and an attenuated baroreflex response, which Mack and colleagues believed were the primary result of training-induced hypervolemia (115). Raven and colleagues were also of the opinion that an increased plasma volume would inhibit the cardiopulmonary baroreflex due to the increased tonic influences from cardiac afferent nerves (156).

The single longitudinal study conducted during this time period pertained to 14 subjects who trained for 10 wk (115). Endurance training significantly elevated $\dot{V}O_2max$ (20%) and total blood volume (9%). When compared to the NT group, who exhibited no meaningful changes, the T subjects exhibited significant differences when measurements of forearm vascular resistance (FVR) and central venous pressure (CVP were made. Specifically, the slope of the linear relationship between FVR and CVP, an index of the responsiveness of the cardiopulmonary reflex, was significantly decreased by 32% in the T subjects. Furthermore, the reduction in slope was inversely related to the increase in blood volume (115). When the results were combined with a previous cross-sectional study(116), the authors concluded that exercise training had decreased the sensitivity of the cardiopulmonary reflex because of an increase in central blood volume.

Summary

With alterations in ABP, baroreceptors located in the cardiovascular system respond by inducing changes in HR, CO, and TPR. For many decades, the role of baroreceptors during exercise was unknown or thought to be of minimal importance. From the results of many creative investigators, notably David Donald and colleagues at Mayo Clinic, it became accepted that baroreceptors maintain ABP despite the vasodilation influences of metabolites and the vasoconstricting influences of an activated SNS. The capability of baroreceptors to reset during exercise has been well described but are not completely understood. Resetting is a complex process that involves interactions between the CC and baroreflexes in which afferent input is initiated from select sites throughout the body, especially the skeletal muscles. Baroreflex stimulus–response curves have been characterized by a logistic function with properties of threshold, saturation, operation point, and gain (sensitivity), which have been incorporated into explanations that characterize baroreflex function curves. Recall that for the carotid baroreflex, Raven and colleagues interpreted significant increases in saturation and threshold values and minimal changes in gain as convincing evidence that resetting had occurred.

The various baroreflexes mentioned are operational during exercise and are excellent examples of functional redundancy. The carotid baroreflex is likely the most investigated because of its role in the resetting of HR and ABP. Although it has minimal influence on SV or CO, it can affect vascular conductance. Keep in mind that the aortic–cardiac baroreflex, especially in resting conditions, has more influence on the control of HR than the carotid baroreflex does. Although exercise research has minimized the importance of the cardiopulmonary baroreflex in animals, this does not appear to be the situation in humans. The EPR is a key reflex during exercise because its responses are determined by mechanoreceptors or metaboreceptors and transmitted to the central nervous system by group III or IV afferents. More importantly, the EPR activates MSNA involvement when vagal withdrawal occurs during exercise. Although not extensively documented, some authorities are of the opinion that the carotid and pulmonary baroreflexes have an inhibitory influence on the activation of the SNS during exercise.

A major limitation of the data pertaining to the effects of training is that the majority of the published record has been obtained under resting conditions, with the exception being the aortic–cardiac baroreflex. In addition, the record has more cross-sectional data than longitudinal data. With this background, the collective results indicate that the responsiveness of the various baroreflexes is attenuated with training, of which a reduction in SNS activity via the chemoreflex component of the EPR is a major effect. Because heart failure produces an exaggerated EPR, the animal data showing that dynamic-exercise training will blunt the reflex are an important stimulus for more human studies on the topic. The finding that training appears to change the neural component of the baroreflex arc associated with an enhanced sensitivity (gain) of the arterial baroreflex has created similar interest for the same reason.

Cardiac Output During Exercise

Insights on the Measurement of Cardiac Output

Measurement of cardiac output (CO) during exercise occurred after Adolph Eugen Fick (1829-1901) in 1870 established the principle that the flow of blood to an organ was proportional to the difference in concentration of a substance in the blood as it entered and as it left an organ in the body (46). It became recognized as the Fick principle but was never used until 1898, when Nathan Zuntz (1847-1920) and Hagemann meticulously used the principle to measure CO in horses during light to moderate exercise on a horizontal surface. They reported mean values in excess of 50 L/min (218). Subsequently, Zuntz began to perfect a foreign gas rebreathing method (nitrous oxide) that was used by Johannes Lindhard (1870-1947) in 1915 with four humans (109). Not surprisingly, methodologies that measured CO were subsequently investigated, notably by Grollman (67) and Hamilton and colleagues (73). Grollman compared the Fick principle with the foreign gas principle (acetylene) (66), whereas Hamilton and colleagues evaluated the Fick principle with the dye-injection method (73). Both accepted the procedure for scientific investigations. (Readers are encouraged to become familiar with Rowell's informative discussion on this subject in reference 164). Since 1948, multiple methodologies for measuring CO have emerged. It is beyond the scope of this chapter to discuss these methodologies; however, it is important to note that the Fick principle continues to reign as the gold standard for measuring CO (110). Moreover, since 1997, Liu and colleagues (110) plus Jere Mitchell (figure 12.1) and Bengt Saltin (1935–) have emphasized that careful evaluations of the early research results have given credence to the merits of the foreign gas methodology (132).

Background Information

As reviewed by Rowell, CO is the product of HR (beats/min) and SV (ml/beat) or, in accordance with the Fick principle, the volume of O_2 consumed (L/min) divided by the arteriovenous (a-v) difference in O_2 content per liter of blood that was simultaneously obtained from either the pulmonary vein or a peripheral artery that contained oxygenated blood and from the pulmonary artery, which contained deoxygenated blood (163). Recall that Janicki and colleagues emphasized with upright dynamic exercise that HR is responsible for approximately 63% of the increase in CO and that SV changes are associated with the remainder (89).

Cardiac Outputs and Arteriovenous O_2 Differences

Although the initial CO results during exercise were obtained from animals (horses and hounds), this section relates essentially to studies in humans. In 1915, Johannes Lindhard (1870-1947) performed the first exercise study pertaining to CO that clearly demonstrated a close and linear relationship between CO and O_2 consumption (109, p. 359). In addition, Lindhard reported that subject JJ performed 1,452 kgm of work, utilized 3.948 L/min of O_2, and had a CO of 43.0 L/min (109, p.

352), which represented a maximum ratio of 10.92 L of CO delivered for every liter of O_2 consumed; this vastly exceeded the range (5.9-7.5) cited by Saltin and Calbet as being associated with trained subjects or athletes (168). In 1928, Bock and colleagues measured the CO of 4 NT and trained individuals (DeMar was included) who exercised on a bicycle ergometer (17). The highest CO achieved in a NT subject was 26.8 L/min.

As detailed in chapter 11, Mitchell and colleagues in 1958 published the first meaningful physiological explanation of the maximum O_2 intake that verified the relationship between CO and O_2 consumption. They also reported a maximum ratio of 7.26 L of CO/L of O_2 consumed for 15 normal M subjects (135). Interestingly, when 6 subjects performed a test beyond their maximal oxygen intake level, the CO decreased by 2.8 L (13%) (135). Although an explanation was not provided, it is possible that the increased workload impaired the ability of the circulatory system to sustain O_2 delivery, which affected performance. In fact, in 1997, Harms and colleagues in the laboratory of Jerome Dempsey (1938–; figure 12.7) reported that when the respiratory muscles associated with the work of breathing were experimentally loaded during inspiration, a redistribution of CO occurred that significantly decreased blood flow to the locomotor muscles (legs) as well as their significantly reducing the oxygen consumption of the legs (75). In 2001, McCole and colleagues reported that NT subjects who performed prolonged maximal-exercise tests experienced significant reductions in CO and SV and suggested that a dissociation had occurred between CO and $\dot{V}O_2$ (123). In 2005, Mortensen and colleagues had trained subjects perform incremental maximal exercise and observed that the circulatory system was unable to maintain a linear oxygen delivery, a linear increase in CO, and an increased SV that had reached a plateau. The authors concluded their results provided evidence for a central limitation to aerobic power and capacity in humans (137).

In a frequently cited 1964 study, Åstrand and colleagues had 23 active young M and F subjects perform submaximal and maximal exercise on a bicycle ergometer. They reported increased CO and a-v O_2 differences, with maximal CO values of 24.1 L/min for the Ms and 18.5 L/min for the Fs with maximal a-v O_2 differences being 170 ml/L for the Ms and 143 ml/L for the Fs. At 50% $\dot{V}O_2$max, CO was essentially 50% of the maximum values for both sexes whereas for the a-v O_2 differences between 70% and 80% of the available O_2 had been extracted by both M and F subjects (5). HR and SV changes explained the increased CO results, and lower hemoglobin concentrations were mentioned to account for the observed a-v O_2 differences in the F subjects (5). However, not mentioned were changes in hemoconcentration, reductions in venous PO_2 values, and shifts in the hemoglobin dissociation curve, which would help to explain changes in oxygen extraction (146). In a 1997 study with young M subjects, dynamic exercise elicited a mean CO of 24 L/min and a maximum a-v O_2 difference of 150 ml/L. The oxygen extracted at 65% peak $\dot{V}O_2$ was 87% whereas the oxygen extracted at 85% peak $\dot{V}O_2$ was 97% (129).

Stroke Volume and Cardiac Considerations

According to Janicki and colleagues, SV contributions to CO are directly influenced by the filling volume of the left ventricle (preload)—the physical force that the ventricle must develop and maintain to inject blood into the aorta (afterload)—and the contractility properties of the myocardium and indirectly by the factors that either modify venous return and filling of the left ventricle (or preload) or impede the systemic circulation (or afterload). They also note, especially with upright exercise, the importance of the length–tension or Frank-Starling relationship with cardiac fibers (89). These selective aspects are discussed in subsequent sections.

The SV that enters the aorta is the difference between left ventricle end diastolic volume (EDV) and left ventricle end systolic volume (ESV). However, body position is important. In the 1960 study of Bevegard and colleagues, the SV of nonathletic subjects in

Figure 12.7 Jerome (Jerry) A. Dempsey (1938–), an esteemed investigator, author, and editor, is currently located at the University of Wisconsin in Madison in the United States. His important contribution to this chapter is found in reference 75.

Photo courtesy of Jerome A. Dempsey.

the transition from supine to an upright sitting position was reduced by 10% with moderate to heavy exercise (14). Poliner and colleagues reported that a reduction in SV of 7% occurred in the transition from a supine position to peak upright exercise (n = 7) (149). Compared with supine subjects, upright subjects had lower EDV values (14%) and demonstrated a progressive and significant decline in ESV (34%) values with peak exercise. The ejection fraction (EF; the difference between EDV and SV, expressed as the percentage of blood ejected per beat) with peak exercise exhibited a significant increase of 8%. In NT subjects, EF and its changes generally paralleled those of SV (93). Consequently, a change in position definitely affects the contractile state and the Frank-Starling relationship (149).

With the measurements of CO, it was assumed SV would increase with exercise. The 1928 SV results of Bock and colleagues (16, 17) showed hyperbolic changes with plateaus occurring between 88% and 95% of maximum (17). For SV changes, 1964 was an interesting year because Åstrand and colleagues demonstrated that upright cycling was associated with an increased SV that plateaued at approximately 40% of maximal aerobic capacity (5). Later, Hartley and colleagues (74) published SV results with exercise that were similar to those of Åstrand and colleagues. Although Rushmer was of the opinion that SV was fairly constant during exercise (167), the results from the 1960s plus the influence of the textbooks by Åstrand and Rodal (7, 8) helped create a perception that SV would increase with exercise and then plateau at submaximal levels of aerobic capacity. However, as demonstrated Gledhill and colleagues in 1994 (58) and later emphasized by Janicki and colleagues, not all published research pertaining to SV supports this perception (89).

Cardiac Contractility, the Frank-Starling Relationship, Constraints, and Ventricular Performance

Cardiac contractility, an intrinsic property of myocardial cells, is influenced by HR, increased activity of the cardiac sympathetic nerves, and by plasma norepinephrine concentrations (163). Assessment of contractility includes the SBP and ESV relationship (31, 78, 163) and the EF percentage (93, 163). The Frank-Starling law or relationship began with Starling, who in 1918 stated, "The relation between the length of the heart fiber and its power of contraction I have called 'the law of the heart'" (179, p. 266). Measurements pertaining to EDV and ESV are emphasized in the Frank-Starling relationship and in ventricular performance (89).

Cognizant that body position had a profound influence on the SV response, Weiss and colleagues in 1979 investigated with 12 M and F NT subjects (mean age = 36 yr) the influence of incremental moderate to heavy semisupine exercise to exhaustion (bicycle ergometer) on ventricular performance. With peak exercise, group HR was increased by 98 beats/min or by 213%. Peak exercise was also associated with a significant increase in end diastolic diameter (16%), no significant changes in end systolic diameter, significant increases (41%) in stroke dimensions (end diastolic diameter – end systolic diameter), and with a significant increase (85%) in the mean velocity of circumferential shortening (normalized for end diastolic diameter). The authors concluded that increases in indices of left ventricular fiber shortening and rates of lengthening appear early in exercise before changes in end diastolic fiber length, which suggested an elevation in CO occurred as a result of an increase in HR, especially during light to moderate exercise (206). Another interpretation of their semisupine exercise results is that HR is responsible for SV changes with light to moderate exercise, whereas with heavy or peak exercise, the Frank-Starling relationships become more important for SV and ventricular performance changes. In 1985, Mahler and colleagues used upright bicycle ergometry in 9 M subjects to determine the contributions of ventricular contractility and the Frank-Starling relationship in moderate exercise (50% $\dot{V}O_2max$) (118). The SBP and ESV relationship and EF were used to assess contractility with EDV changes in both ventricles being used to appraise the Frank-Starling involvement. The investigators found that SBP was significantly increased (28%) and that the left ventricle ESV exhibited a nonsignificant (7%) reduction with both SV and EF being increased by 13%. Both ventricles experienced significant increases in EDV measurements (15% and 24% respectively), which lead to the conclusion that moderate exercise was associated with enhanced ventricular contractility and an improved Frank-Starling relationship (118).

In 1986, Plotnick and colleagues examined these relationships with upright progressive bicycle ergometry in an older M population. They demonstrated with light to moderate exercise an increased EF, decreased ESV, elevated SV, and augmented EDV, which indicated that ventricular contractility and the Frank-Starling relationship were active and contributing factors. However, with heavy and peak exercise, EF and SBP continued to increase and ESV and EDV became reduced while SV had plateaued. These findings demonstrated that contractility was augmented while the influence of the Frank-Starling relationship had diminished with heavy and peak exercise. Although SNS activity was implicated for the decrease in the Frank-Starling influence, there was no supporting data for the interpretation (148). Their concept for upright exercise was substantiated by the 1991 radionuclide study of Goodman and colleagues, who made ventricular performance measurements in a mixed-gender population at moderate and maximal exercise intensities and concluded that augmented contractility prevailed throughout exercise because of the Frank-Starling relationship (60).

Sullivan and colleagues measured continuous left ventricular performance during maximal isometric deadlift exercise in 10 young M subjects. In the transition from rest to 100% maximal exercise, the SBP and ESV relationship increased (9%) while the EF percentage remained constant; none of the changes had statistical significance. Because the ESV value exhibited a significant decrease (37%), it is likely that a modest degree of contractility was present at 100% maximal effort. If correct, it would be similar to other reports with dynamic exercise (60, 148). Because the SV and EDV results from rest to exercise demonstrated significant increases of 28% and 32%, respectively, it would appear that the Frank-Starling relationship was operative with isometric exercise throughout the transition to maximal exercise (188).

Animal experiments have been invaluable in learning the constraining functions of the pericardium. Investigations with dogs that experienced sutured and opened pericardiums have demonstrated that ventricular dimensions, pressures, volumes, and their respective SV results were altered by the process (57, 185). In 1986, Stray-Gundersen and colleagues performed pericardiectomies on NT and sham-operated dogs. All dogs were instrumented with aorta and pulmonary artery catheters and familiarized with treadmill running before performing submaximal and maximal exercise before and after pericardiectomy (186). Although the surgical procedure had minimal influence on the cardiac responses with submaximal exercise, removal of the pericardium with maximal exercise was associated with significant increases in CO (20%) and SV (17%). In addition, pericardiectomy resulted in a significantly higher $\dot{V}O_2max$ (7%). Because the sham-operated animals exhibited no meaningful changes in these parameters with exercise, the collective effect was that the pericardium had a constraining influence on ventricular performance during exercise (186).

Experiments with dogs have also provided information on the capability of the myocardium to respond to maximal exercise when HR has been compromised. Specifically, Donald and Shepherd reported that after recovery from cardiac denervation, SV achieved 93% of its maximal value and contributed 67% to the maximal CO within 15 to 20 s of maximal treadmill exercise (35).

Mode, Arm, and Leg Exercise

In 1961, Åstrand and Saltin evaluated the effect of mode on $\dot{V}O_2max$ and reported that treadmill exercise elicited higher values (6%) than leg cycling on an ergometer. No CO results were included (10). Later, Hermansen and colleagues (79) and Miyamura and Honda (136) measured maximal CO changes in a small number of NT subjects on both modalities and found either no differences or a modest 4% reduction in CO when exercising on a bicycle ergometer. However, the results of Miyamura and Honda were identified with significantly lower HR and a-v O_2 differences (136).

Although the 1961 investigation of Åstrand and Saltin (10) had a limited number of subjects, it was apparent that the O_2 consumption results with arm-cycling exercise were markedly lower (~26% and ~23%, respectively) than either leg-cycling exercise or arm-and-leg-cycling exercise. Nearly a decade later, Simmons and Shepherd measured CO during arm or leg exercise in 5 M subjects and reported that maximum arm work elicited lower (26%) CO results than maximum leg work; however, the a-v O_2 differences (144 and 150 ml/L for arms and legs, respectively) were similar. The maximum CO to O_2 consumption ratio was 6.56 for arms and 11.20 for legs (173). Using an indirect impedance method to assess CO, Haennel and colleagues in 1992 had 5 subjects perform maximal arm and leg isometric and isokinetic exercises at 3 velocities that ranged from 0.52 to 2.62 rads/s. During leg isometric exercise CO was increased (38%) nearly twofold over arm exercise, and leg isokinetic exercise was again associated with essentially a twofold increase in CO when compared with arm exercise. The authors recommended using arm mass rather than arm velocity when evaluating CO results from isokinetic exercise (69). In 2007, Calbet and collaborators had 9 M subjects perform incremental arm and leg exercise to exhaustion (Wmax). The percentage of CO being delivered to the legs increased curvilinearly up to 84% Wmax before plateauing and experiencing a significant reduction in the rate of increase. Arm exercise at 20% Wmax demonstrated both a reduction and a plateau that remained until Wmax had been obtained. These specific CO results during exercise suggested that sympathetic nervous system vasoconstrictor signals originating during leg exercise had effectively opposed the vasodilatory signals being elicited with arm exercise (23).

Training and Cardiac Output

Response With Exercise

Maximal Exercise

Ekblom and Hermansen reported in 1968 that a member of the Swedish national team had a CO of 42.3 L/min during maximal exercise (40). Although this result continues to be listed in prominent textbooks, readers should be aware that in 1980 an American Olympic athlete recorded a maximal volume of 56.6 L/min (159).

Cross-sectional studies consistently demonstrate that athletes or highly trained individuals have significantly higher values than their nonathletic or sedentary counterparts. This relationship was demonstrated in the 1992 study of Ogawa and colleagues with young M and F trained subjects whose CO values were 29% and 21% higher, respectively, than their controls (142). In a study in which 8 young M subjects underwent moderate to heavy dynamic training for 16 wk, Ekblom and colleagues found significant increases in CO (1.4 L/min or 8%) and SV (15 ml/beat or 13%) and found that the a-v O_2 difference widened by 50 ml/L or 4% (39). Rerych and colleagues had 17 young M and F subjects swim train at a heavy intensity for 6 mo. They noted a significant elevation in CO of 5.5 L/min (25%) and an increase in SV of 28%. In addition, total blood volume in the subjects was increased by 35% (159). Recently, Murias and colleagues conducted a 12 wk dynamic heavy-exercise study in 8 young M subjects and observed significant incremental increases in group CO results (10%) until 9 wk, after which CO plateaued. Significant increases were also observed for SV (11%) and a-v O_2 differences (7%) (140). Interestingly, none of these previously mentioned studies exhibited increased HR means, suggesting that the increased CO was the result of an elevated SV and the widening of the a-v O_2 differences. Murias and colleagues also conducted a related study with the same design with 8 young F subjects and reported similar training results, although the M subjects had higher absolute results (139).

Few long-term longitudinal training studies concerning CO have been published. One of note was the 2001 Dallas Bed Rest and Training Study, which included 5 middle-aged Ms who were subjects in the 1966 investigation that had a similar purpose. They participated in a 6 mo dynamic exercise-training study at 75% of their maximal HR. The group maximal HR decreased by 9 beats/min and was markedly lower than the 1966 value whereas maximal CO increased by 300 ml/min and was close to the 1966 pretraining levels. Compared with the 1966 posttraining results, maximal SV was 9 ml/beat higher (8%) and a-v O_2 differences were 19 ml/L lower (11%). However, the maximal TPR units were 24% higher when measured in 2001. Although the results of this study were from a small number of subjects, they reinforced the concept that dynamic exercise training can attenuate the influence of inactivity (126).

In a noteworthy study, minipigs were trained for 10 wk at high exercise intensities. The animals exhibited significant increases in CO (31%) and SV (40%) with no meaningful changes in either HR or a-v O_2 differences (207).

Submaximal Exercise

Murias and colleagues measured their subjects after 3, 6, and 9 wk of the training program at a fixed power output that represented approximately 33% of the pretraining maximum. In Ms, CO was virtually unchanged at these designated times whereas HR was significantly decreased after 3 wk. At this time period, SV was significantly elevated (12%) and remained at that level throughout the program. There were no significant changes in a-v O_2 differences (140). The results for the Fs were nearly identical to those for the Ms except SV was significantly increased after 9 and 12 wk by 8% and 9%, respectively (139). Hartley and colleagues conducted an 8 to 10 wk training study with 15 middle-aged sedentary M who were assessed during moderate exercise. CO was virtually unchanged, HR was reduced by 10%, SV was elevated by 7%, and a-v O_2 differences were nonexistent (74). Thus, their training profile was similar to results obtained from younger subjects.

Stroke Volume Considerations

As with CO, select cross-sectional studies show predominately M athletes and fit subjects exhibiting markedly higher SV results (~21%-73%) than their designated controls (31, 142, 184). In 1992, Spina and colleagues conducted a study in which 12 young M and F subjects performed moderate to heavy dynamic exercise for 12 wk and demonstrated a mean SV that plateaued at 50% $\dot{V}O_2max$ and did not change with maximal conditions. This result supported the earlier findings of Åstrand and colleagues (5). In 1994, Gledhill and colleagues conducted a cross-sectional study with young trained and sedentary subjects (n = 7/group). They reported that the sedentary individuals exhibited an SV plateau at 48% $\dot{V}O_2max$ (HR at 120 beats/min) whereas the SV for the trained individuals continued to increase throughout exercise with each target HR and at $\dot{V}O_2max$. Although these findings for trained individuals were explained by significantly enhanced ventricular filling and emptying times, the researchers were uncertain whether the results were due to genetic endowment, to training per se, or to both (58).

In the recent training study conducted with young M subjects, Murias and colleagues reported significant SV increases with maximal exercise of 7 ml/beat (5%), 11 ml/beat (8%), and 17 ml/beat (13%) after 3, 6, and 9 wk, respectively, after which a plateau was reached (11%) (140). These findings effectively demonstrated that training can elevate SV beyond 40% to 60% $\dot{V}O_2$max before a plateau occurs, but not to the magnitude achieved by elite athletes. The 1985 study by Crawford and colleagues with competitive and noncompetitive runners (31) and the 2001 investigation by Zhou and colleagues with M elite distance runners, university distance runners, and NT university students (216) provide credence for this concept.

Although no results from inexperienced lifters existed for comparative purposes, the SV profile of highly trained and experienced weightlifters performing concentric, eccentric, and isometric muscle contractions in the lifting process is interesting because of the contrasting changes. As demonstrated by the five highly trained weightlifters in the study by Lentini and colleagues, SV decreased (~19%) during the concentric lifting phase, increased (~25%) during the lockout or isometric phase, and decreased (~20%) during the eccentric or lowering phase. The reductions in SV were associated with an increase in TPR and a decrease in EDV, whereas the elevations in SV were attributed to a reduction in TPR, an increase in EDV, improved activity of the muscle pump, and an enhanced venous return (103). They were also associated with contractility changes and the Frank-Starling relationships (emphasized later).

End Diastolic Volume and End Systolic Volume

Select cross-sectional studies conducted between 1985 and 2006 consistently demonstrated that athletic, fit, and highly trained subjects exhibit higher (19%-49%)—and frequently significant—EDV values (31, 34, 49, 56, 185). Explanations favored enhanced ventricular compliance and more rapid filling but also included enlarged chambers, increased blood volume, and improved utilization of the Frank-Starling relationship. Results pertaining to ESV were not as consistent or available and were uncertain in demonstrating whether training was associated with either increased or decreased volumes. Seldom cited is the 1980 longitudinal investigation by Rerych and colleagues in which M and F athletes participated in a 6 mo heavy swim-training program and tested on a cycle ergometer. The researchers found that EDV was significantly higher (23%), as was total blood volume (35%), and felt that the increase in EDV occurred because of better utilization of the Frank-Starling relationship and an enhanced blood volume (159). However, they did not report ESV values.

Myocardial Contractility, Frank-Starling Relationship, and Ventricular Filling

In 1985, Crawford and colleagues compared the dynamic exercise responses of marathon runners with those of noncompetitive runners (NT) using changes in the SBP and ESV ratio and EF to evaluate myocardial contractility and shifts in EDV values to assess the Frank-Starling relationship (31). SV progressively increased in the marathon runners throughout the exercise bout whereas SV plateaued at 70% of maximal HR in the NT subjects. The SBP and ESV ratio for the NT subjects was consistently higher than that of the trained subjects throughout the exercise bout, especially during maximal conditions. EDV results for the marathon runners were elevated until 70% maximal HR was reached, after which it remained constant. The NT subjects had a similar profile but their absolute volumes were markedly lower than those of the marathon runners. However, there were no statistical differences in EF between the groups. The authors suggested that myocardial contractility was utilized more by the NT runners than by the trained runners and that the trained runners utilized the Frank-Starling relationship more effectively because it was a less energy utilizing mechanism (31).

In search of an explanation for the orthostatic intolerance of well-trained athletes, Levine and colleagues in 1991 investigated the interrelationships between left ventricle EDV, SV, and pulmonary capillary wedge pressures in well-trained endurance athletes and their sedentary controls. The experiment included saline infusions and exposure to conditions of LBNP (105). From their data they prepared Frank-Starling curves (pulmonary capillary wedge pressure vs. SV) and observed that the curve for the athletes had shifted up and to the left, which enhanced SV and facilitated ventricular filling pressures. In the relationship between pulmonary capillary wedge pressure and EDV, they observed that the curve for the athletes had shifted far to the right, which indicated that the ventricles of the athletes had become more compliant and distensible than those of the nonathletes, thus enabling a larger left ventricle EDV to occur over a wider range of filling pressures. Two years later, Levine in a review article (104) indicated that highly trained athletes with an elevated EDV effectively used the Frank-Starling relationship to produce an elevated SV during exercise. However, this "mechanical property of the heart" placed the athlete at a disadvantage during orthostasis because large decreases in SV occurred when filling pressures were reduced (104).

In 2002, Warburton and colleagues conducted a study in which highly trained M cyclists performed supine and upright incremental exercise to determine the contributions of myocardial contractility and the Frank-Starling relationship to SV and CO (205). SV exhibited a linear 120% increase with upright exercise and a 60% increase with supine exercise. The marked increase in EDV accounted for 93% and 83% of the SV change with upright and supine exercise, respectively. There were no significant changes in ESV, although the EF percentage increased throughout exercise with minimal differences being found between upright and supine exercise. The SBP to ESV ratio percentage demonstrated higher values for the upright position, although the most marked differences occurred with submaximal exercise. These findings from trained subjects reaffirmed that postural positions have an effect on SV changes and indicate that the Frank-Starling relationship prevailed throughout the exercise period. Warburton and colleagues stated, "Thus it appears that there is a progressive utilization of the Frank-Starling mechanism throughout incremental to maximal exercise" (205, p. 620). In addition, they felt that contractility contributions occurred when HRs were between 110 and 130 beats/min and during maximal dynamic exercise conditions.

Resistance exercise, such as the leg press, has components of both dynamic and static exercise. As demonstrated by SBP:ESV ratio changes in the study by Lentini and colleagues with experienced lifters, estimated myocardial contractility decreases (~16%) during the lifting or the concentric contraction phase (103, p. 2705) whereas it increased by 6% during the lockout or the isometric contraction phase. Myocardial contractility decreases by approximately 25% during the lowering or the eccentric contraction phase (103).

Cross-sectional studies have also demonstrated (58) or suggested (104) that ventricular diastolic filling times are faster in highly fit subjects or well-conditioned athletes than in their sedentary or NT controls. In a study of 17 young M subjects, Levy and colleagues reported that peak early filling rate was significantly faster after 6 mo of training (8%) compared with pretraining levels. Although they were concerned that ventricular hypertrophy might impede filling times, this issue did not materialize. They were also uncertain about the mechanism responsible and suggested that a change in the time constant for relaxation could be a factor for the faster rate (106).

In a study with athletes, Levine indicated that training had increased ventricular compliance and distension and was responsible for the increased filling pressures. The Frank-Starling relationship was utilized during exercise by both NT and trained populations; however, it appeared that trained populations had more effectively maximized the Frank-Starling relationship so that SV had been augmented (104).

Cardiac Mass

According to Duncker and Bache, training can be associated with a 30% increase in cardiac mass (36). Mass, myocardial contractility, and the Frank-Starling relationship are important determinants for increases in both CO and SV. Readers should refer to chapter 11 for further information on cardiac mass.

Arteriouvenous Considerations

Like in other CO parameters, young and well-trained or highly fit populations exhibit higher (and frequently significant) a-v O_2 difference values when compared to sedentary and less-active populations (1, 142, 215). In the 1973 study of Clausen and colleagues, a marked increase in a-v O_2 differences occurred with submaximal and supramaximal exercise during arm training but not during leg training. The differences were attributed to a reduction in blood flow to the arm muscles, increased O_2 extraction, and less vasoconstriction by nonexercising tissues (28). The 12 wk training study of Ogawa and colleagues with 12 young M and F subjects demonstrated that the program significantly increased a-v O_2 differences by 7% (142). Although the recent training programs conducted by Murias and colleagues exhibited significant a-v O_2 differences (139, 140), they did not address relevant aspects pertaining to the oxygen content of mixed venous blood, changes in hemoconcentration or saturation, shifts in the hemoglobin oxygen dissociation curve, or changes in the redistribution of blood flow (146).

Select Redistributions of Cardiac Output

Coronary Blood Flow

Coronary blood flow (CBF) is under central cardiovascular control and represents between 4% and 5% of the exercise CO (213). Because of its high energy requirements, the heart at rest extracts approximately 70% to 80% of the arterially delivered oxygen. Therefore, to meet the myocardial oxygen requirements during incremental exercise, CBF must increase to match the demand (36). For multiple reasons, recent exercise studies with direct measurements from coronary arteries that pertained to young and healthy subjects are limited.

Moreover, when mechanisms were investigated, studies with animals have been preferred.

One of the early investigators of exercising humans was the cardiologist Richard Bing (1909-2010), who in the late 1940s used accepted medical procedures and placed catheters into the coronary sinus of supine patients and used nitrous oxide to assess CBF and the Fick principle to determine myocardial oxygen consumption ($M\dot{V}O_2$) of patients (111). When healthy, normal subjects were measured during the next two decades, the subjects performed light exercise. An example is the 1961 study of Regan and colleagues using 8 normal subjects. The investigators observed a 38% increase in CBF, 32% elevation in $M\dot{V}O_2$, and 10% increase in O_2 extraction (158). To determine the effects of upright exercise, Kitamura and colleagues in 1972 used procedures similar to those mentioned previously and had 10 healthy, normal subjects perform light, medium, and heavy exercise. However, they failed to include baseline data (96). The transition from light to heavy exercise was associated with increases in CBF and M $\dot{V}O_2$ of 102% and 131%, respectively, that were coupled with a 10% increase in O_2 extraction. They correlated CBF and $M\dot{V}O_2$ with different variables and found that HR × aortic SBP × 10^{-2} had the highest r values with CBF (.89) and with $M\dot{V}O_2$ (.90). Unexpectedly, the tension time index results were lower; namely, the r values were .83 and .77for CBF and M $\dot{V}O_2$, respectively (96). During 1972, Ekstrom-Jodal and colleagues used Xe^{133} to measure CBF in 6 young M subjects who performed light exercise in a beta receptor blockade study. Exercise was associated with a 177% elevation in CBF, a 200% increase in $M\dot{V}O_2$, a 66% reduction in coronary vascular resistance, a 17% decline in coronary venous O_2 content, and a 9% increase in O_2 extraction. In addition, CO was augmented by 54%. Reductions occurred in CBF, $M\dot{V}O_2$, and CO when a beta receptor blocker was infused, denoting the influence of the SNS (41). Two decades later, Kaijser and Berglund conducted a coronary sinus blood flow and lactate investigation with 13 young M subjects who performed supine moderate and maximal exercise (92). CBF was increased by 260% and 325% at both intensities, respectively, whereas $M\dot{V}O_2$ demonstrated increases of 349% and 443% at the two conditions, respectively (92). At maximal exercise, coronary oxygen saturation decreased from 34% to 25% while oxygen extraction increased by 33%. CO was not reported, but HR was elevated by 101 beats/min. Of the substrates measured, lactate extraction was the highest of what was recorded (92).

Other than the influence of the SNS, few authors discussed mechanisms responsible for their results. According to Kaijser and Kanstrup (93), the heart has the capacity to increase myocardial O_2 consumption faster than activation factors responsible for increasing local blood flow.

The regulation of coronary artery blood flow is a complex process because flow must match oxygen demand. Moreover, during exercise, there is a limited extraction reserve to assist local metabolic influences or sympathetic feed-forward mechanisms in the matching process that enables vasodilator and vasoconstrictor influences to function. While the contributions of sympathetic α and β components are well known; this is not the case for mediators of local metabolic control (37). This topic is discussed in more detail in chapter 13.

Training and Coronary Blood Flow

The 2008 *Physiological Reviews* article by Duncker and Bache includes 627 citations concerning the regulation of CBF during exercise. Yet the lack of longitudinal studies involving healthy young subjects is striking, especially with regard to whether increases in CBF occur with maximal exercise. This situation has not been aided by animal studies because when the article was published nine were in the affirmative and eight had negative results (36). In fact, the percentage of studies reporting physiological advantages from training was remarkably low. A human study of note was a cross-sectional investigation conducted in 1976 by Heiss and colleagues with 5 sedentary NT subjects and 6 athletes. Under medical supervision, catheters were placed in the coronary sinus, pulmonary artery, and the abdominal aorta. When measurements were obtained at 65% $\dot{V}O_2max$, CO increased 3.5-fold in the NT subjects and 2.7-fold in the athletes (78). Furthermore, myocardial blood flow values were significantly lower in the athletes (103%) than in the NT subjects (215%). The trained group also had significantly lower $M\dot{V}O_2$ results (102%) and used 46% less myocardial O_2 per heart stroke. Athletes were also able to extract more O_2 (19%) than their NT (13%) counterparts. Heiss and colleagues concluded that trained subjects had significantly lower MBF and $M\dot{V}O_2$ values at rest than their sedentary controls and that these values become exaggerated with submaximal exercise. This situation occurs, in part, because the trained heart muscle requires less energy at any specific workload than nontrained hearts (78). Two well-conducted training studies with dogs provide supporting evidence for the concept that training will significantly reduce CBF and $M\dot{V}O_2$ with submaximal exercise (11, 200). In the investigation by von Restorff and colleagues, trained dogs increased the myocardial O_2 extraction rate increased from 75% to 93% by increasing myocardial a-v O_2 differences from 15 volumes to 22 volumes percent (200).

Using positron emission tomography technology with $[^{15}O]H_2O$ to measure myocardial perfusion, Laaksonen and colleagues conducted a cross-sectional investigation with young athletic and nonathletic populations. The subjects exercised at the same absolute workloads (150 W) and at 70% Wmax (101). At 150 W, the NT individuals exhibited a 292% increase in perfusion units ($ml \cdot g^{-1} \cdot min^{-1}$) whereas the trained subjects exhibited a significantly lower increase of 98%. The same trend occurred at 70% Wmax, but the differences (~35%) had no statistical significance. The authors came to the same conclusion as Heiss and colleagues (78), namely that the myocardiums of athletes consume less O_2 than those of NT subjects during submaximal exercise.

During 2008, Heinonen and colleagues examined the relationship between left ventricle adenosine A_{2A} receptors and MBF in T and NT subjects. The total volume of A_{2A} was higher (~75%) in the T subjects than in the NT subjects; however, the density results were similar. The MBF results were significantly higher in the T subjects (77).

Summary

CO can be determined as the product of HR and SV or, in accordance with the Fick principle, by dividing the oxygen consumption by the a-v O_2 differences. Although the principle was formulated in 1870, it was not until Zuntz perfected and tested a foreign gas rebreathing method (nitrous oxide) that it was used in human experimentation. Despite challenges over the years, the methodology continues to reign as the gold standard.

In 1915, Lindhard demonstrated that CO increased in humans with exercise and that a close relationship existed between CO and oxygen consumption. As an approximation, 63% of the CO is due to the HR and SV is responsible for the remainder. Although HR has been discussed previously and the focus is on CO changes associated with SV and a-v O_2 differences, we must not forget that SNS activation is essential for a maximal CO response. With dynamic exercise by most NT or sedentary subjects, SV increases in a linear relationship with $\dot{V}O_2$ until approximately 40% to 50% $\dot{V}O_2max$ is obtained, after which a plateau occurs. The magnitude of the SV is dependent on the difference between EDV and ESV; the former systematically increases during exercise whereas the latter, when measured and reported, does not always exhibit an expected decrease. In most situations, the EF demonstrates a response profile similar to that exhibited by the SV. The ventricular contributions during exercise are influenced by its contractile properties, which appear to be operational during light and heavy exercise, whereas the Frank-Starling relationship is functioning throughout the exercise bout. Based on animal studies, it is evident that the pericardium is a constraint on CO being achievable during exercise. CO values are higher when the subject is in the supine position and modality favors the treadmill. Training affects CO responses, although the plethora of cross-sectional studies and the paucity of longitudinal investigators complicate the interpretation. Because maximal HR does not increase with chronic exercise, dynamic exercise training markedly increases maximal CO results due to its influences on SV, a-v O_2 differences, and promoting a decrease in TPR. The CO changes with submaximal exercise are different because HR is reduced, in part, to diminished SNS activity while SV is elevated. It is evident in trained and athletic populations that SV does not plateau with moderate exercise and may not plateau at all. An elevated SV is attributed to an enlarged EDV that has been augmented by an expanded blood volume and, at times, by a reduced ESV. Although it is unclear whether training improves contractility, it is evident that training changes chamber dimensions, enhances filling pressures, improves the compliance and distension of the ventricles, and shifts the Frank-Starling relationship so that SV is increased. Training is also associated with significantly reduced CBF and $M\dot{V}O_2$ measures with submaximal dynamic exercise and, in trained dogs, an extraordinary ability to extract oxygen from arterial blood.

Closing Remarks

The summaries in this chapter give historical insight of the central cardiovascular influences on acute and chronic effects of exercise. Keep in mind that the central influences are inextricably linked with functions of the central and autonomic nervous systems and are integrative with the peripheral components (discussed in chapter 13). Table 12.2 identifies discoveries that have enhanced our understanding of the central influences. The information gleaned from the in-depth investigations of the CC, blood pressure regulation, reflex contributions from baroreceptors and active muscles, and the role of the heart in the distribution of the CO has enhanced our understanding of the acute effects of exercise. However, our knowledge of the mechanisms associated with the continuum of training lacks completeness due, in part, to the reliance on cross-sectional investigations rather than careful and extended longitudinal investigations. We are hopeful that this situation will change in the decades ahead.

Table 12.2 Select Exercise Physiology Milestones for Central Influences on the Cardiovascular System (1910-2010)

Year	Investigators	Milestone	Reference
1913	Krogh and Lindhard	Proposed that cortical irradiation was also responsible for the sudden increase in heart rate with exercise.	100
1915	Lindhard	Was the first to measure cardiac output in exercising human subjects using the nitrous oxide method and the first to demonstrate the close, linear relationship between cardiac output and oxygen consumption during exercise.	109
1937	Alam and Smirk	Conducted a limb-occlusion procedure in an exercise experiment that provided the first evidence that muscle activity could initiate cardiovascular reflex responses.	3
1972	Goodwin, McClosky, and Mitchell	Conducted a vibration experiment that stimulated or inhibited muscle spindles while subjects performed isometric contractions so that heart rate and arterial blood pressure responded accordingly to the intent of a central command.	61
1976	Heiss et al.	Secured coronary blood flow and myocardial oxygen consumption values from athletic and nonathletic subjects during submaximal exercise; reported that coronary artery blood flows were reduced and myocardial oxygen consumption values were lower in athletic subjects when compared to nonathlete subjects because the energy requirements of the heart were less at a specific workload.	78
1983	Walgenbach and Donald	Surgically perfected an isolated carotid sinus preparation in exercising dogs in which arterial pressures remained constant; indirectly demonstrated the occurrence of resetting in baroreceptors.	203
1985	McDougall et al.	With medical supervision, secured direct arterial blood pressure measurements from highly trained and experienced weightlifters performing two-leg presses, which elevated pressures to the extraordinary value of 480/450 mmHg.	125
1985	Eldridge at al.	Demonstrated in unanesthetized, decorticated cats that electrical stimulation of hypothalamic and mesencephalic regions produced normal and fictive locomotion while elevating arterial blood pressure.	42
1990	Rowell and O'Leary	Proposed the concept that with the onset of dynamic exercise in humans, vagal withdrawal is near completeness at 100 beats/min whereas the activation of the sympathetic nervous system begins around 90 beats/min as manifested by changes in plasma NE, plasma renin activity, and muscle sympathetic nerve activity.	165
1993	Levine	Compared Frank-Starling pressure–volume curves from athletic and nonathletic populations and found markedly different profiles, which indicated that training produced more compliant and distensible ventricles and augmented stroke volume.	104

1994	Papelier et al.	Measured subjects who performed 4 stages of cycling exercise (highest at 75% $\dot{V}O_2max$) while using a neck suction chamber capable of delivering pulsatile positive and negative pressures to the carotid sinus region. Their human results indicated that resetting of the baroreceptors occurred with heavy exercise.	145
1994	Gledhill et al.	Demonstrated in well-conditioned athletic subjects performing maximal exercise that stroke volume did not plateau (as expected) during submaximal stages but rather continued to increase throughout the progressive exercise bout.	58
2001	Williamson et al.	Used single-photon-emission computed tomography in exercising subjects who were hypnotically motivated to manipulate their sense of effort. As a result, the subjects exhibited increased heart rates, elevated arterial blood pressures, and activation of thalamic and insular cortical regions when confronted with a perceived uphill climbing task.	211
2005	Mortensen et al.	Reported that endurance-trained or recreationally active subjects who performed incremental maximal exercise demonstrated impaired systemic oxygen delivery to the tissues, which was attributed to a markedly reduced stroke volume that occurred due to a plateau in cardiac output. The authors suggested that the results were a prime example of a central limitation.	137

References

1. Andrew GM, Guzman CA, Becklake MR. Effect of athletic training on exercise cardiac output. *J Appl Physiol* 21: 603-608, 1966.
2. Abboud FM, Eckberg DL, Johannsen UJ, Mark AL. Carotid and cardiopulmonary baroreceptor control of splanchnic and forearm vascular resistance during venous pooling in man. *J Physiol* 286: 173-184, 1979.
3. Alam M, Smirk FH. Observations in man upon a blood raising reflex arising from the voluntary muscles. *J Physiol* 89: 372-383, 1937.
4. Åstrand P-O, Christensen EH. Aerobic work capacity. In: Dickens F, Neil E, Widdas WF, eds. *Oxygen in the Animal Organism.* New York: Pergamon Press, 1964, 295-303.
5. Åstrand P-O, Cuddy TE, Saltin B, Stenberg J. Cardiac output during submaximal and maximal work. *J Appl Phsiol* 19: 268-274, 1964.
6. Åstrand P-O, Ekblom B, Messin R, Saltin B, Stenberg J. Intra-arterial blood pressure during exercise with different muscle groups. *J Appl Physiol* 20: 253-256, 1965.
7. Åstrand P-O, Rodal K. *Textbook of Work Physiology.* New York: McGraw-Hill, 1970.
8. Åstrand P-O, Rodahl K. *Textbook of Work Physiology.* 3rd ed. New York: McGraw-Hill, 1986.
9. Åstrand P-O, Ryhming I. A nomogram for calculation of aerobic capacity (physical fitness) from pulse rate during submaximal work. *J Appl Physiol* 7: 218-221, 1954.
10. Åstrand P-O, Saltin B. Maximal oxygen uptake and heart rate in various types of muscular activity. *J Appl Physiol.* 16: 977-981, 1961.
11. Barnard RJ, Duncan HW, Baldwin KM, Grimditch G, Buckberg GD. Effects of extensive exercise training on myocardial performance and coronary blood flow. *J Appl Physiol* 49: 444-449, 1980.
12. Belardinelli R. Exercise training in chronic heart failure. How to harmonize oxidative stress, sympathetic outflow and angiotension II. *Circulation* 115: 3042-3044, 2007.
13. Bevegard S, Freyschuss U, Strandell T. Circulatory adaptation to arm and leg exercise in supine and sitting position. *J Appl Physiol* 21: 37-46, 1966.
14. Bevegard S, Holmgren A, Jonsson B. The effect of body position on the circulation at rest and during exercise, with special reference to the influences on the stroke volume. *Acta Physiol Scand* 49: 279-298, 1960.
15. Bevegard BS, Shepherd JT. Circulatory effects of stimulating the carotid arterial stretch receptors in man at rest and during exercise. *J Clin Invest* 45: 132-142, 1966.
16. Bock AV, Dill DB. *The Physiology of Muscular Exercise.* 3rd Ed. London: Longmans, Green & Co., 1931.
17. Bock AV, Vancaulaert C, Dill DB, Folling A, Hurxthal LM. Studies in muscular activity. III Dynamical changes occurring in man at work. *J Physiol* 66: 136-161, 1928.
18. Bolter CP, Banister EW, Sigh AK. Intrinsic rates and adrenergic responses of atria from rats on sprinting. endurance and walking programmes. *Aust J Exp Biol Med Sci* 64(Pt 3): 251-256, 1986.
19. Boushel R, Snell P, Saltin B. Cardiovascular regulation with endurance training. In: Saltin B, Bouchel R, Secher N, Mitchell J, eds. *Exercise and Circulation in Health and Disease.* Champaign, IL: Human Kinetics, 2000, 225-243.

20. Bowen WP. Changes in heart-rate, blood-pressure, and duration of systole resulting from bicycling. *Am J Physiol* 11: 59-77, 1904.
21. Brainbridgc FA. *The Physiology of Muscular Exercise*. London: Longsman, Green & Co., 1919.
22. Bristow JD, Brown EB Jr., Cunningham DJC, Howson MG, Strange-Peterson E, Pickering TG, Sleight P. Effect of bicycling on the baroreflex regulation of the pulse interval. *Circ Res* 28: 582-592, 1971.
23. Calbet JAL, Gonzalez-Alonso J, Helge JW, Sondergaard H, Munch-Andersen T, Boushel R, Saltin B. Cardiac output and leg and arm blood flow during incremental exercise to exhaustion on the cycle ergometer. *J Appl Physiol* 103: 969-978, 2007.
24. Callister R, Ng AV, Seals DR. Arm muscle sympathetic nerve activity during preparation for and initiation of leg-cycling exercise in humans. *J Appl Physiol* 77: 1403-1410, 1994.
25. Chefer SI, Talan MI, Engel BT. Central neural correlates of learned heart rate control during exercise: Central command demystified. *J Appl Physiol* 83: 1448-1453, 1997.
26. Chen C-Y, Bonham AC. Postexercise hypotension: Central mechanisms. *Exerc Sports Sci Rev* 38: 122-127, 2010.
27. Christensen EH. Beitrage zur Physiologie schwerer korperlicher Arbeit: Minutenvolumen und Schlagvolumen des Herzens wahrend schwerer korperlicher Arbeit. *Arbeitsphysiol* 4: 453-470, 1931.
28. Clausen JP, Klausen K, Rasmussen B, Trap-Jensen J. Central and peripheral circulatory changes after training of the arms or legs. *Am J Physiol* 225: 675-682, 1973.
29. Collins HL, Augustyniak RA, Ansorge EJ, O'Leary DS. Carotid baroreflex pressor responses at rest and during exercise: Cardiac output vs. regional vasoconstriction. *Am J Physiol* 280: H642-H648, 2001.
30. Coote JH, Hilton SM, Perez-Gonzalez IF. The reflex nature of the pressor response to muscular exercise. *J Physiol* 215: 789-804, 1971.
31. Crawford MH, Petru MA, Rabinowitz CR. Effect of isotonic exercise training on left ventricular volume during upright exercise. *Circulation* 72: 1237-1243, 1985.
32. Daskalopoulos DA, Shepherd JT, Walgenbach SC. Cardiopulmonary reflexes and blood pressure in exercising sinoaortic-denervated dogs. *J Appl Physiol* 57: 1417-1421, 1984.
33. Degtyarenko AM, Kaufman MP. Barosensory cells in the nucleus tractus solitarius receive convergent input from group III muscle afferents and central command. *Neurosci* 140: 1041-1050, 2006.
34. Di Bello V, Santoro G, Talarico L, Di Muro C, Caputo MT, Giorgi D, Bertini D, et al. Left ventricular function during exercise in athletics and in sedentary man. *Med Sci Sports Exerc* 28: 190-196, 1996.
35. Donald DE, Shepherd JT. Response to exercise in dogs with cardiac denervation. *Am J Physiol* 205: 393-400, 1963.
36. Duncker DJ, Bache RJ. Regulation of coronary blood flow during exercise. *Physiol Revs* 88: 1009-1086, 2008.
37. Duncker DJ, Merkus D. Acute adaptations of the coronary circulation to exercise. *Cell Biochem Biophys* 43: 17-35, 2005.
38. Eckberg DL. High- and low-pressure baroreceptors. In: Robertson D, Low PA, Polinsky RJ, eds. *Primer on the Autonomic Nervous System*. San Diego: Academic Press, 1966, 59-65.
39. Ekblom B, Åstrand P-O, Saltin B, Stenberg J, Wallstrom B. Effect of training on circulatory response to exercise. *J Appl Physiol* 24: 518-528, 1968.
40. Ekblom B, Hermansen L. Cardiac output in athletes. *J Appl Physiol* 25: 619-625, 1968.
41. Ekstrom-Jodal B, Haggendal E, Malmberg R, Svedmyr N. The effect of adrenergic β receptor blockade on coronary circulation in man during work. *Acta Med Scand* 191: 245-248, 1972.
42. Eldridge FL, Milhorn DE, Kiley JP, Waldrop TG. Stimulation by central command of locomotion, respiration and circulation during exercise. *Respir Physiol* 59: 313-337, 1985.
43. Eldridge FL, Milhorn DE, Waldrop TG. Exercise hypernea and locomotion: Parallel activation from the hypothalamus. *Science* 211: 844-846, 1981.
44. Fadel PJ, Stromstad M, Hansen J, Sander M, Horn K, Ogoh S, Smith ML, et al. Arterial baroreflex control of sympathetic nerve activity during acute hypotension: Effect of fitness. *Am J Physiol* 280: H2524-H2532, 2001.
45. Fernandes A, Galbo H, Kjaer M, Mitchell JH, Secher NH, Thomas SN. Cardiovascular and ventilatory responses to dynamic exercise during epidural anaesthesia in man. *J Physiol* 420: 281- 293, 1990.
46. Fick A. Ueber die Messung des Blutquantums in des Herzventrikeln. *S B Phys Ges*. Wurzburg, 1870, p. 16.
47. Field H, Bock AV, Gildea EF, Lathrop FL. The rate of the circulation of the blood in normal resting individuals. *J Clin Invest* 1: 65-85, 1924.
48. Fisher WJ, White MJ. Training-induced adaptations in the central command and peripheral reflex component of the pressor response to isometric exercise of the human triceps surae. *J Physiol* 520: 621-628, 1999.
49. Fisman EZ, Frank G, Ben-Ari E, Kessler G, Pines A, Drory Y, Kellermann JJ. Altered left ventricular volume and ejection fraction responses to supine dynamic exercise in athletes. *J Am Coll Cardiol* 15: 582-588, 1990.
50. Gallagher KM, Fadel PJ, Smith SA, Norton KH, Querry RQ, Olivencia-Yurvati A, Raven PB. Increases in intramuscular pressure raise arterial blood

pressure during dynamic exercise. *J Appl Physiol* 91: 2351-2358, 2001.

51. Gallagher KM, Fadel PJ, Smith SA, Stromstad M, Ide K, Secher NH, Raven PB. The interaction of central command and the exercise reflex in mediating baroreflex resetting during exercise in humans. *Exp Physiol* 91: 79-87, 2006.
52. Gallagher KM, Fadel PJ, Stromstad M, Ide K, Smith SA, Querry RQ, Raven PB, Secher NH. Effects of exercise pressor reflex activation on carotid baroreflex function during exercise in humans. *J Physiol* 533: 871-880, 2001.
53. Gasser HS, Meek WJ. A study of the mechanisms by which muscular exercise produces acceleration of the heart. *Am J Physiol* 34: 48-71, 1914.
54. Geppert J, Zuntz N. Uber die Regulation der Athmung. *Pflugers Archiv* 42: 189-245, 1888.
55. Giannattasio C, Seravalle G, Bolla GB, Cattaneo BM, Cleroux J, Cuspidi C, Sampieri L, et al. Cardiopulmonary receptor reflexes in normotensive athletes with cardiac hypertrophy. *Circulation* 82: 1222-1229, 1990.
56. Ginzton LE, Conant R, Brizendine M, Laks MM. Effect of long-term high intensity aerobic training on left ventricular volume during maximal upright exercise. *J Am Coll Cardiol* 14: 364-371, 1989.
57. Glantz SA, Misbach GA, Moores WY, Mathey DG, Lekven J, Stowe DF, Parmley WW, Tyberg JV. The pericardium substantially affects the left ventricular diastolic pressure-volume relationship in the dog. *Cir Res* 42: 433-441, 1978.
58. Gledhill N, Cox D, Jamnik R. Endurance athletes' stroke volume does not plateau: Major advantage is diastolic function. *Med Sci Sports Exerc* 26: 1116-1121, 1994.
59. Gleim GW, Stachenfeld NS, Coplan NL, Nicholas JA. Gender differences in the systolic blood pressure response to exercise. *Am Heart J* 121: 524-530, 1990.
60. Goodman JM, Lefkowitz CA, Liu PP, McLaughlin PR, Plyley MJ. Left ventricular functional response to moderate and intense exercise. *Can J Sport Sci* 16: 204-209, 1991.
61. Goodwin GM, McCloskey DI, Mitchell JH. Cardiovascular and respiratory responses to changes in central command during isometric exercise at constant muscle tension. *J Physiol* 226: 173-190, 1972.
62. Gould AG, Dye JA. *Exercise and Its Physiology*: New York: A.S. Barnes and Co., 1932.
63. Grandevia SC, McCloskey DI. Changes in motor commands, as shown by changes in perceived heaviness, during partial curarization and anaesthesia in man. *J Physiol* 272: 673-689, 1977.
64. Green AL, Wang S, Purvis S, Owen SLF, Bain PG, Stein JR, Guz A, et al. Identifying cardiorespiratory neurocircuitry involved in central command during exercise in humans. *J Physiol* 578: 605-612, 2007.
65. Grimby G, Saltin B. Physiological analysis of physical well-trained middle-aged and old athletes. *Acta Med Scand* 179: 513-526, 1966.
66. Grollman A. A comparison of the triple extrapolation (Fick Principle) and the acetylene (Foreign Gas Principle) methods for the determination of the cardiac output of man. *Am J Physiol* 93: 116-123, 1930.
67. Grollman A. The determination of the cardiac output by the use of acetylene. *Am J Physiol* 88: 432-445, 1929.
68. Gunga HC. Biography. In: *Nathan Zuntz: His Life and Work in the Fields of High Altitude Physiology and Aviation Medicine.* New York: American Physiological Society, 2009, 20-21.
69. Haennel RG, Snydmiller GD, Teo KK, Greenwood PV, Quinney HA, Kappagoda CT. Changes in blood pressure and cardiac output during maximal isokinetic exercise. *Archiv Phys Med Rehab* 73: 150-155, 1992.
70. Hagerman FC, Walsh SJ, Staron RS, Hikida RS, Gilders RM, Murray K, Toma K, Ragg KE. Effects of high-intensity resistance training on untrained older man. I. Strength, cardiovascular and metabolic responses. *J Gerontol A Biol Sci Med Sci* 55: B 336-B346, 2000.
71. Hales JRS, Ludbrook J. Baroreflex participation in redistribution of cardiac output at onset of exercise. *J Appl Physiol* 64: 627-634, 1988.
72. Hales S. *Statistical Essays. Vol. II. Haemastatiks.* London: Innys & Manby, 1733.
73. Hamilton WF, Riley RL, Attyah AM, Cournand A, Fowell DM, Himmelstein A, Noble RP, et al. Comparison of the Fick and dye injection methods of measuring the cardiac output in man. *Am J Physiol* 153: 309-321, 1948.
74. Hartley LH, Grimby G, Kilbom A, Nilsson NJ, Åstrand I, Bjure J, Ekblom G, Saltin B. Physical training in sedentary middle-aged and older men. III. Cardiac output and gas exchange at submaximal and maximal exercise. *Scand J Clin Lab Invest* 24: 335-344, 1969.
75. Harms CA, Bablock MA, McClaren SR, Pegelow DF, Nickele GA, Nelson WB, Dempsey JA. Respiratory muscle work comprises leg blood flow during maximal exercise. *J Appl Physiol* 82: 1573-1583, 1997.
76. Hayes SG, McCord JL, Rainier J, Liu Z, Kaufman MP. Role played by acid-sensitive ion channels in evoking the exercise pressor reflex. *Am J Physiol* 295: H1720-H1725, 2008.
77. Heinonen I, Nesterov SV, Liukko KK, Kemppainen J, Nagren K, Luotolahti M, Virsu P, et al. Myocardial blood flow and adenosine A_{2A} R receptor density in endurance athletes and untrained men. *J Physiol* 586: 5193-5202, 2008.
78. Heiss HW, Barmeyer J, Wink K, Hell G, Cerny FJ, Keul J, Reindell H. Studies of the regulation of myo-

cardial blood flow in man. I: Training effects on blood flow and metabolism of the healthy heart at rest and during standardized heavy exercise. *Basic Res Cardiol* 71: 658-675, 1976.

79. Hermansen L, Ekblom B, Saltin B. Cardiac output during submaximal and maximal treadmill and bicycle exercise. *J Appl Physiol* 29: 82-86, 1970.
80. Herring HE. Der Karotisdruckversuch. *Munch Med Wochschr* 70: 1287-1290, 1923.
81. Herring HE. Ueber die Beziehung der extracardialen Herznerven zur Steigeruiig der Herzschlagenzal dei Muskelthdtigkeit. *Arch gesamte Physiol* 40: 429-492, 1895.
82. Herxheimer H. *The Principles of Medicine in Sport for Physicians and Students* [translated by Tuttle WW, Knowlton GC]. Leipzig: George Thieme, 1933.
83. Hickson RC, Hagberg JM, Ehsani AA, Holloszy JO. Time course of the adaptive responses of aerobic power and heart rate to training. *Med Sci Sports Exerc* 13: 17-20, 1981.
84. Hill L. Arterial pressure in man while sleeping, resting, working, bathing. *J Physiol* 22: xxvi-xxx, 1898.
85. Hoffman MD, Street GM. Characterization of the heart rate response during biathlon. *Int J Sports Med* 13: 390-394, 1992.
86. Hughson RL, Sutton JR, Fitzgerald JD, Jones NL. Reduction of intrinsic sinoatrial frequency and norepinephrine response of the exercised rat. *Can J Physiol Pharmacol* 55: 813-820, 1977.
87. Ichinose M, Saito M, Kondo N, Nishiyasu T. Time-dependent modulation of arterial baroreflex control of sympathetic nerve activity during isometric exercise in humans. *Am J Physiol* 290: 1419-1426, 2006.
88. Innes JA, DeCort SC, Evans PJ, Guz A. Central command influences cardiorespiratory response to dynamic exercise in humans with unilateral weakness. *J Physiol* 448: 551-563, 1992.
89. Janicki JS, Sheriff DD, Robotham JL, Wise RA. Cardiac output during exercise: Contributions of the cardiac, circulatory, and respiratory systems. In: Rowell LB, Shepherd JT, eds. *Handbook of Physiology, Section 12. Exercise: Regulation and Integration of Multiple Systems*. Bethesda, MD: American Physiological Society, 1996, 649-704.
90. Jennings G, Nelson L, Nestel P, Esler M, Korner P, Borton D, Bazelmans J. The effects of changes in physical activity on major cardiovascular risk factors, hemodynamics, sympathetic function, and glucose utilization in man: A controlled study of four levels of activity. *Circulation* 73: 30-40, 1986.
91. Johansson JE. Ueber die Einwirkung der Muskelthatigkeit auf die Athmung und Herzthatigkeit. *Skand Arch Physiol* 5: 20-66, 1893.
92. Kaijser L, Berglund B. Myocardial lactate extraction and release at rest and during heavy exercise in healthy men. *Acta Physiol Scand* 144: 39-45, 1992.
93. Kaijser L, Kanstrup I-L. Coronary blood flow and cardiac hemodynamics. In: Saltin B, Boushel R, Secher N, Mitchell J, eds. *Exercise and Circulation in Health and Disease*. Champaign, IL: Human Kinetics, 2000, 67-78.
94. Karjalainen JK, Mantysaari M, Viitasalo M, Kujala U. Left ventricular mass, geometry, and filling in endurance athletes: Association with exercise blood pressure. *J Appl Physiol* 82: 531-537, 1997.
95. Kaufman MP, Waldrop TG, Rybicki KJ, Ordway GA, Mitchell JH. Effects of static and rhythmic twitch contractions on the discharge of group III and IV afferents. *Cardiovasc Res* 18: 663-668, 1984.
96. Kitamura K, Jorgensen CR, Gobel FL, Taylor HL, Wang Y. Hemodynamic correlates of myocardial oxygen consumption during upright exercise. *J Appl Physiol* 32: 516-522, 1972.
97. Klausen K, Secher NH, Clausen JP, Hartling O, Trap-Jensen J. Central and regional circulatory adaptations to one-leg training. *J Appl Physiol* 52: 976-983, 1982.
98. Komine H, Matsukawa K, Tsuchimachi H, Murata J. Central command blunts the baroreflex bradycardia to aortic nerve stimulation at the onset of voluntary static exercise in cats. *Am J Physiol* 285: H516-H526, 2003.
99. Komine H, Sugawara J, Hayashi K, Yoshizawa M, Yokoi T. Regular endurance exercise in young men increases arterial baroreflex sensitivity through neural alteration of baroreflex arc. *J Appl Physiol* 106: 1499-1505, 2009.
100. Krogh A, Lindhard J. The regulation of respiration and circulation during the initial stages of muscular work. *J Physiol* 47: 112-136, 1913.
101. Laaksonen MS, Kalliokoski KK, Luotolahti M, Kemppalnen J, Teras M, Kyrolainen H, Nuutila P, Knuuti J. Myocardial perfusion during exercise in endurance-trained and untrained humans. *Am J Physiol* 293: R837-R843, 2007.
102. Lassen A, Mitchell JH, Reeves DR Jr., Rogers HB, Secher NH. Cardiovascular responses to brief static contractions in man with topical nervous blockade. *J Physiol* 409: 333-341, 1989.
103. Lentini AC, McKelvie RS, McCartney N, Tomlinson CW, MacDougall JD. Left ventricular response in healthy young men during heavy-intensity weight-lifting exercise. *J Appl Physiol* 75: 2703-2710, 1993.
104. Levine BD. Regulation of central blood volume and cardiac filling in endurance athletes: The Frank-Starling mechanism as a determinant of orthostatic tolerance. *Med Sci Sports Exerc* 25: 727-732, 1993.
105. Levine BD, Lane LD, Buckey JC, Friedman DB, Blomqvist CG. Left ventricular pressure volume and Frank-Starling relations in endurance athletes: Im-

plications for orthostatic tolerance and exercise performance. *Circulation* 84: 1016-1023, 1991.

106. Levy WC, Cerqueira MD, Abrass IB, Schwartz RS, Stratton JR. Endurance exercise training augments diastolic filling at rest and during exercise in healthy young and older men. *Circulation* 88: 116-126, 1993.
107. Lewis SL, Nylander E, Gad P, Areskog NH. Non-autonomic component in bradycardia of endurance trained men at rest and during exercise. *Acta Physiol Scand* 109: 207-305, 1980.
108. Lind AR, Taylor SH, Humphreys PW, Kennelly BM, Donald KM. The circulatory effects of sustained voluntary muscular contraction. *Clin Sci* 27: 229-244, 1964.
109. Lindhard J. Uber das Minutenvolum des Herzens bei Ruhe und bei Muskelarbeit. *Pflugers Archiv* 161: 233-383, 1915.
110. Liu Y, Menold E, Dullenkkopf A, Reibnecker S, Lormes W, Lehmann M, Steinacker JM. Validation of the acetylene rebreathing method for measurement of cardiac output at rest and during high intensity exercise. *Clin Physiol* 17: 171-182, 1997.
111. Lombardo TA, Rose L, Taeschler M, Tuluy S, Bing RJ. The effect of exercise on coronary blood flow, myocardial oxygen consumption and cardiac efficiency in man. *Circulation* 7: 71-78, 1953.
112. Lowsley O. The effects of various forms of exercise on systolic, diastolic, and pulse pressures and pulse rate. *Am J Physiol* 27: 446-466, 1911.
113. Ludbrook J, Graham WF. Circulatory responses to onset of exercise: Role of arterial and cardiac baroreceptors. *Am J Physiol* 248: H457-H467, 1985.
114. Luft UC, Myrhe LG, Leoppky JA, Venters AM. A study of factors affecting tolerance of gravitational stress stimulated by lower body negative pressure. Research report on specialized physiology studies in support of manned flight. NASA contract NAS 59-14472. Albuquerque, NM: Lovelace Foundation, 1976, 1-60.
115. Mack GW, Convertino VA, Nadel ER. Effect of exercise training on cardiopulmonary baroreflex control of forearm vascular resistance in humans. *Med Sci Sports Exerc* 25: 722-726, 1993.
116. Mack GW, Nose H, Nadel ER. Role of cardiopulmonary baroreflexes during dynamic exercise. *J Appl Physiol* 65: 1827-1832, 1988.
117. Mack GW, Shi X, Nose H, Tripathi A, Nadel ER. Diminished baroreflex control of forearm vascular resistance in physically fit humans. *J Appl Physiol* 63: 105-110, 1987.
118. Mahler DA, Matthay RA, Synder PE, Pytlik L, Zaret BL, Loke J. Volumetric responses of right and left ventricles during upright exercise in normal subjects. *J Appl Physiol* 58: 1818-1822, 1985.
119. Marey EJ. *Physiologie Medicale de la Circulation du Sang.* Paris: Delahaye, 1863.
120. McAllister RM, Kimani JK, Webster JL, Parker JL, Laughlin MH. Effects of exercise training on responses of peripheral and visceral arteries in swine. *J Appl Physiol* 80: 216-225, 1996.
121. McArdle WA, Katch FL, Katch VL. *Exercise Physiology.* 5th ed. Philadelphia: Lippincott Williams & Wilkins, 2001.
122. McCartney N, McKelvie RS, Martin J, Sale DG, MacDougall JD. Weight-training-induced attenuation of the circulatory response of older males to weight lifting. *J Appl Physiol* 74: 1056-1060, 1993.
123. McCole SD, Davis AM, Fueger PT. Is there a dissociation of maximal oxygen consumption and maximal cardiac output? *Med Sci Sports Exerc* 33: 1265-1269, 2001.
124. McCurdy JH. The effect of maximum muscular effort on blood-pressure. *Am J Physiol* 5: 95-103, 1901.
125. McDougall JD, Tuxen D, Sale DG, Moroz JR, Sutton JR. Arterial blood pressure response to heavy resistance exercise. *J Appl Physiol* 58: 785-790, 1985.
126. McGuire DK, Levine BD, Williamson JW, Snell PG, Blomqvist CG, Saltin B, Mitchell JH. A 30-year follow-up of the Dallas bed rest and training study. *Circulation* 104: 1358-1364, 2001.
127. McRitchie RJ, Vatner SF, Boettcher D, Heyndrickx GR, Patrick TA, Braunwald E. Role of the arterial baroreceptors in mediating cardiovascular response to exercise. *Am J Physiol* 230: 85-89, 1976.
128. Melcher A, Donald DE. Maintained ability of carotid reflex to regulate arterial blood pressure during exercise. *Am J Physiol* 241: H838-H849, 1981.
129. Minson CT, Kenney WL. Age and cardiac output during cycle exercise in thermoneutral and warm environments. *Med Sci Sports Exerc* 29: 75-81, 1997.
130. Mitchell JH, Kaufman MP, Iwamoto GA. The exercise pressor reflex: Its cardiovascular effects, afferent mechanisms, and central pathways. *Ann Rev Physiol* 45: 229-242, 1983.
131. Mitchell JH, Payne FC, Saltin B, Schibye G. The role of muscle mass in the cardiovascular response to static contractions. *J Physiol* 309: 45-54, 1980.
132. Mitchell JH, Saltin B. The oxygen transport system and maximal oxygen uptake. In: Tipton CM, ed. *Exercise Physiology: People and Ideas.* New York: Oxford University Press, 2003, 255-291.
133. Mitchell JH, Schibye B, Payne FC III, Saltin B. Response of arterial blood pressure to static exercise, in relation to muscle mass, force development, and electromyographic activity. *Cir Res* 48(6 Pt. 2): 170-175, 1981.
134. Mitchell JH, Smith SA. Unraveling the mysteries of the exercise pressor reflex at the cellular level. *J Physiol* 586: 3025-3026, 2008.

135. Mitchell JH, Sproule BJ, Chapman CB. The physiological meaning of the maximum oxygen intake test. *J Clin Invest* 37: 538-547, 1958.
136. Miyamura M, Honda Y. Oxygen intake and cardiac output during maximal treadmill and bicycle exercise. *J Appl Physiol* 32: 185-188, 1972.
137. Mortensen SP, Dawson EA, Yoshiga CC, Dalsgaard MK, Damsgaard R, Secher NH, Gonzalez-Alonso J. Limitations to systemic and locomotor limb muscle oxygen delivery and uptake during maximal exercise in humans. *J Physiol* 566: 273-285, 2005.
138. Mostoufi-Moab S, Widmaier EJ, Cornett JA, Gray K, Sinoway LI. Forearm training reduces the exercise pressor reflex during ischemic rhythmic handgrip. *J Appl Physiol* 84: 277-283, 1998.
139. Murias JM, Kowalchuk JM, Paterson DH. Mechanisms for increases in VO_2max with endurance training in older and young women. *Med Sci Sports Exerc* 42: 1891-1898, 2010.
140. Murias JM, Kowalchuk JM, Paterson DH. Time course and mechanisms of adaptations in cardiorespiratory fitness with endurance training in older and young men. *J Appl Physiol* 108: 621-627, 2010.
141. Negrao CE, Middlekauf HR. Exercise training in heart failure: Reduction in angiotensin II, sympathetic nerve activity, and baroreflex control. *J Appl Physiol* 104: 577-578, 2008.
142. Ogawa T, Spina RJ, Martin WH III, Kohrt WM, Schechtman KB, Holloszy JO, Ehsani AA. Effects of aging, sex, and physical training on cardiovascular responses to exercise. *Circulation* 86: 494-503, 1992.
143. Ogoh S, Fadel PJ, Nissen P, Jans O, Selmer C, Secher NH, Raven PB. Baroreflex-mediated changes in cardiac output and vascular resistance in response to alterations in carotid sinus pressure during exercise in humans. *J Physiol* 550: 317-324, 2003.
144. Ogoh S, Fisher JP, Fadel P, Raven PB. Increase in central blood volume modulate carotid baroreflex resetting during dynamic exercise in humans. *J Physiol* 581: 405-418, 2007.
145. Papelier Y, Escourrou P, Gauthier JP, Rowell LB. Carotid baroreflex control of blood pressure and heart rate in men during dynamic exercise. *J Appl Physiol* 77: 502-506, 1994.
146. Perego GB, Marenzi GC, Guazzi M, Sganzerla P, Assanelli E, Palermo P, Conconi B, et al. Contribution of PO_2, P_{50}, and Hb to changes in arteriovenous O_2 content during exercise in heart failure. *J Appl Physiol* 80: 623-631, 1996.
147. Petrofsky JS, Phillips CA, Sawka MN, Hanpeter D, Lind AR, Stafford D. Muscle fiber recruitment and blood pressure response to isometric exercise. *J Appl Physiol* 50: 32-37, 1981.
148. Plotnick GD, Becker LC, Fisher ML, Gerstenblith G, Renlund DG, Fleg JL, Weisfeldt ML, Lakatta EG. Use of the Frank-Starling mechanism during submaximal versus maximal upright exercise. *Am J Physiol* 251: H1101-H1105, 1986.
149. Poliner LR, Dehmer GJ, Lewis SE, Parkey, RW, Blomqvist CG, Willerson JT. Left ventricular performance in normal subjects: A comparison of the responses to exercise in the upright and supine positions. *Circulation* 62: 528-534, 1980.
150. Polito MD, Farinatti PT. The effects of muscle mass and number of sets during resistance exercise on postexercise hypotension. *J Strength Cond Res* 23: 2351-2357, 2009.
151. Pollock ML, Wilmore JH, Fox SM III. Health and fitness through physical activity. In: *American College of Sports Medicine Series.* New York: Wiley, 1978.
152. Potts JT. Inhibitory neurotransmission in the nucleus tractus solitarii: Implications for baroreceptor resetting during exercise. *Exp Physiol* 91: 59-72, 2006.
153. Potts JT, Shi X, Raven PB. Carotid baroreflex responsiveness during dynamic exercise in man. *Am J Physiol* 265: H1928-H1938, 1993.
154. Raven P. Neural control of the circulation during exercise. *Exp Physiol* 91: 25-26, 2006.
155. Raven PB, Pawelczyk JA. Chronic endurance exercise training: A condition of inadequate blood pressure regulation and reduced tolerance to LBNP. *Med Sci Sports Exerc* 25: 713-721, 1993.
156. Raven PB, Potts JT, Shi Z, Pawelczyk J. Baroreceptor-mediated reflex regulation of blood pressure during exercise. In: Saltin B, Boushel R, Secher N, Mittchell J., eds. *Exercise and Circulation in Health and Disease.* Champaign, IL: Human Kinetics, 2000, 3-24.
157. Raven PB, Rohm-Young D, Blomqvist CG. Physical fitness and cardiovascular response to lower body negative pressure. *J Appl Physiol* 56: 138-144, 1984.
158. Regan TJ, Timmis G, Gray M, Binak K, Hellems HK. Myocardial oxygen consumption during exercise in fasting and lipemic subjects. *J Clin Invest* 40: 624-630, 1961.
159. Rerych SK, Scholz PM, Sabiston DC Jr., Jones RH. Effects of exercise training on left ventricular function in normal subjects: A longitudinal study by radionuclide angiography. *Am J Cardiol* 45: 244-252, 1980.
160. Robinson BF, Epstein SE, Beiser GD, Braunwald E. Control of heart rate by the autonomic nervous system: Studies in man on the interrelations between baroreceptor mechanisms and exercise. *Cir Res* 19: 400-411, 1966.
161. Robinson S. Experimental studies of physical fitness in relation to age. *Arbeitsphysiol* 10: 251-323, 1938.
162. Rotto DM, Kaufman MP. Effects of metabolic products of muscular contraction on the discharge of group III and IV afferents. *J Appl Physiol* 64: 2306-2313, 1988.
163. Rowell LB. *Human Cardiovascular Control.* New York: Oxford University Press, 1993.

164. Rowell LB. The cardiovascular system. In: Tipton CM, ed. *Exercise Physiology: People and Ideas*. New York: Oxford University Press, 2003, 98-137.
165. Rowell LB, O'Leary DS. Reflex control of the circulation during exercise: Chemoreflexes and mechanoreflexes. *J Appl Physiol* 69: 407-418, 1990.
166. Rowell LB, O'Leary DS, Kellog DL Jr. Integration of cardiovascular control systems in dynamic exercise. In: Rowell LB, Shepherd JT, eds. *Handbook of Physiology. Exercise: Regulation and Integration of Multiple Systems. Control of Respiratory and Cardiovascular Systems.* Bethesda, MD: American Physiological Society, 1996, 770-838.
167. Rushmer RF. Constancy of stroke volume in ventricular responses to exertion. *Am J Physiol* 196: 745-750, 1959.
168. Saltin B, Calbet JAL. Point: In health and in a normoxic environment, VO_2max is limited primarily by cardiac output and locomotor muscle flow. *J Appl Physiol* 100: 744-746, 2006.
169. Saltin B, Hartley LH, Kilbom A, Åstrand I. Physical training in sedentary middle-aged and older men. II. Oxygen uptake, heart rate, and blood lactate concentration at submaximal and maximal exercise. *Scand J Clin Lab Invest* 24: 323-334, 1969.
170. Secher NH, Ludbrook J. Introduction. In: Saltin B, Boushel R, Secher N, Mitchell J, eds. *Exercise and Circulation in Health and Disease*. Champaign, IL: Human Kinetics, 2000, vii-xiv.
171. Secher NH, Volianitis S. Are the arms and legs in competition for cardiac output? *Med Sci Sports Exerc* 38: 1797-1803, 2006.
172. Shi X, Potts JT, Raven PB, Foresman BH. Aortic-cardiac reflex during dynamic exercise. *J Appl Physiol* 78: 1569-1574, 1995.
173. Simmons R, Shepherd RJ. Measurements of cardiac output in maximum exercise. Application of an acetylene rebreathing method to arm and leg exercise. *Int Z angew Physiol* 29: 159-172, 1971.
174. Sinoway LL, Smith MB, Enders B, Leuenberger U, Dzwonczyk T, Gray K, Whisler K, Moore RL. Role of diprotonated phosphate in evoking muscle reflex responses in cats and humans. *Am J Physiol* 267: H770-H778, 1994.
175. Smith OA Jr., Rushmer RE, Lasher EP. Similarity of cardiovascular responses to exercise and to diencephalic stimulation. *Am J Physiol* 198: 1139-1142, 1960.
176. Smith SA, Mitchell JH, Garry MG. The mammalian exercise pressor reflex in health and disease. *Exp Physiol* 91: 89-102, 2006.
177. Somers VK, Leo KC, Schields R, Clary M, Mark AL. Forearm endurance training attenuates sympathetic nerve response to isometric handgrip in normal individuals. *J Appl Physiol* 72: 1039-1043, 1992.
178. Spina RJ, Ogawa T, Martin WH III, Coggan AR, Holloszy JO, Ehsani AA. Exercise training prevents decline in stroke volume in young healthy subjects. *J Appl Physiol* 72: 2458-2462, 1992.
179. Starling EH. On the circulatory changes associated with exercise. *J Roy Army Med Corps* 34: 258-272, 1919.
180. Stegemann J, Busert A, Brock D. Influence of fitness on blood pressure control system in man. *Aerosp Med* 45: 45-48, 1974.
181. Stephenson RB, Donald DE. Reversible vascular isolation of carotid sinuses in conscious dogs. *Am J Physiol* 238: H809-H814, 1980.
182. Stevens GHJ, Foresman BH, Shi X, Stern SA, Raven PB. Reduction in LBNP tolerance following prolonged endurance exercise training. *Med Sci Sports Exerc* 24: 1235-1244, 1992.
183. Stewart JM, Montgomery LD, Glover JI, Medow MS. Changes in regional blood volume and blood flow during static handgrip. *Am J Physiol* 292: H215-H223, 2007.
184. Stickland MK, Welsh RC, Petersen SR, Tyberg JV, Anderson WD, Jones RL, Taylor DA, et al. Does fitness level modulate the cardiovascular hemodynamic response to exercise. *J Appl Physiol* 100: 1895-1901, 2006.
185. Stokland O, Miller MM, Lekven J, Ilebekk A. The significance of the intact pericardium for cardiac performance in the dog. *Cir Res* 47: 27-32, 1980.
186. Stray Gundersen, J, Musch TI, Haidet GC, Swain DP, Ordway GA, Mitchell JH. The effect of pericardiectomy on maximal oxygen consumption and maximal cardiac output in untrained dogs. *Cir Res* 58: 523-530, 1986.
187. Studinger P, Lenard Z, Kovats Z, Kocsis L, Kollai M. Static and dynamic changes in carotid artery diameter in humans during and after strenuous exercise. *J Physiol* 550: 575-583, 2003.
188. Sullivan J, Hanson P, Rahko PS, Folts JD. Continuous measurement of left ventricular performance during and after maximal isometric deadlift exercise. *Circulation* 85: 1406-1413, 1992.
189. Takeshita A, Jingu S, Imaizumi T, Kunihiko Y, Koyanagi S, Nakamura M. Augmented cardiopulmonary baroreflex control of forearm vascular resistance in young athletes. *Cir Res* 59: 43-48, 1986.
190. Talan MI, Engel BT. Learned control of heart rate during dynamic exercise in nonhuman primates. *J Appl Physiol* 61: 545-553, 1986.
191. Tipton CM. Exercise, training and hypertension: An update. *Exerc Sports Sci Rev* 19: 447- 505, 1991.
192. Tipton CM. The autonomic nervous system. In: Tipton CM, ed. *Exercise Physiology: People and Ideas.* New York: Oxford University Press, 2003, 188-254.
193. Tipton CM, Carey RA, Eastin WC, Erickson HH. A submaximal test for dogs: Evaluation of effects of training, detraining, and cage environment. *J Appl Physiol* 37: 271-275, 1974.

194. Tipton CM, Franklin BA. The language of exercise. In: Tipton CM, ed. *ACSM's Advanced Exercise Physiology*. Philadelphia: Lippincott Williams & Wilkins, 2006, 3-10.
195. Tipton CM, Matthes RD, Callahan A, Tcheng T-K, Lais L. The role of chronic exercise on resting blood pressures of normotenisve and hypertensive rats. *Med Sci Sports* 9: 168-177, 1977.
196. Vanhoutte P, Lacroix E, Leussen I. The cardiovascular adaptation of the dog to muscular exercise. Role of the arterial pressoreceptors. *Arch Int Physiol Biochem* 74: 201-222, 1966.
197. Victor RG, Seals DR, Mark AL. Differential control of heart rate and sympathetic nerve activity during dynamic exercise. Insight from intraneural recordings in humans. *Clin Invest* 79: 508-516, 1987.
198. Volianitis S, Secher NH. Arm blood flow and metabolism during arm and combined arm and leg exercise in humans. *J Physiol* 544: 977-984, 2002.
199. Volianitis S, Yoshiga CC, Vogelsang T, Secher NH. Arterial blood pressure and carotid baroreflex function during arm and combined arm and leg exercise in humans. *Acta Physiol Scand* 181: 289-295, 2004.
200. von Restorff W, Holtz J, Bassenge E. Exercise induced augmentation of myocardial oxygen extraction in spite of normal coronary dilatory capacity in dogs. *Pflugers Arch* 372: 181-185, 1977.
201. Waldrop TG, Eldridge FL, Iwamoto GA, Mitchell JH. Central neural control of respiration and circulation during exercise. In: Rowell LB, Shepherd JT, eds. *Handbook of Physiology. Exercise: Regulation and Integration of Multiple Systems. Control of Respiratory and Cardiovascular Systems.* Bethesda, MD: American Physiological Society, 1996, 333-380.
202. Waldorp TG, Henderson MC, Iwamoto GA, Mitchell JH. Regional blood flow responses to stimulation of the subthalamic region. *Respir Physiol* 64: 93-102, 1986.
203. Walgenbach SC, Donald DE. Inhibition by carotid baroreflex of exercise-induced increase in arterial pressure *Circ Res* 52: 253-262, 1983.
204. Wang H-J, Pan Y-X, Wang W-Z, Gao L, Zimmerman MC, Zucker IH, Wang W. Exercise training prevents the exaggerated exercise pressor reflex in rats with chronic heart failure. *J Appl Physiol* 108: 1365-1375, 2010.
205. Warburton DER, Haykowsky MJ, Quinney HA, Blackmore D, Teo KK, Humen DP. Myocardial response to incremental endurance-trained athletes: Influence of heart rate, contractility and the Frank-Starling effect. *Exp Physiol* 87: 613-622, 2002.
206. Weiss JL, Weisfeldt ML, Mason SJ, Garrison JB, Livengood SV, Fortuin NJ. Evidence of Frank-Starling effect in man during severe semisupine exercise. *Circulation* 59: 655-661, 1979.
207. White FC, McKirnan D, Breisch EA, Guth BD, Liu Y-M, Bloor CM. Adaptation of the left ventricle to exercise-induced hypertrophy. *J Appl Physiol* 62: 1097-1110, 1987.
208. Williamson JW, Fadel PJ, Mitchell JH. New insights into central cardiovascular control during exercise in humans: A central command update. *Exp Physiol* 91: 51-58, 2006.
209. Williamson JW, McColl R, Matthews D. Evidence for central command activation of the human insular cortex during exercise. *J Appl Physiol* 94: 1726-1734, 2003.
210. Williamson JW, McColl R, Matthews D, Ginsburg M, Mitchell JH. Activation of the insular cortex is affected by the intensity of exercise. *J Appl Physiol* 87: 1213-1219, 1999.
211. Williamson JW, McColl R, Matthews D, Mitchell JH, Raven PR, Morgan WP. Hypnotic manipulation of effort sense during dynamic exercise: Cardiovascular responses and brain activation. *J Appl Physiol* 90: 1392-1399, 2001.
212. Williamson JW, Nobrega ACL, McColl R, Matthews D, Winchester P, Friberg L, Mitchell JH. Activation of the insular cortex during dynamic exercise in humans. *J Physiol* 503: 277-283, 1997.
213. Wilmore JH, Costill DL, Kenney WL. *Physiology of Sport and Exercise.* 4th ed. Champaign, IL: Human Kinetics.
214. Wolfe LA, Cunningham DA, Davis GM, Rosenfeld H. Relationship between maximal oxygen uptake and left ventricular function in exercise. *J Appl Physiol* 44: 44-49, 1978.
215. Zavorsky GS. Evidence and possible mechanisms of altered maximum heart rate with endurance training and tapering. *Sports Med* 29: 13-26, 2000.
216. Zhou B, Conlee RK, Jensen R, Fellingham GW, George JD, Fisher AG. Stroke volume does not plateau in elite male distance runners. *Med Sci Sports Exerc* 33: 1849-1854, 2001.
217. Zoller RD, Mark AL, Abboud FM, Schmid PG, Heistad DD. The role of low pressure baroreceptors in reflex vasoconstriction responses in man. *J Clin Invest* 51: 2967-2972, 1972.
218. Zuntz O, Hagemann O. Untersuchungen uber den Stoffwechsel des Pferedes bei Ruhe und Arbeit. *Landwirtschaftliche Jarbucher* 27 Erganzunband 3: 371-412, 1898.

CHAPTER 13

The Cardiovascular System: Peripheral Circulation

Grant H. Simmons, PhD

Bruno Roseguini, MS

Jaume Padilla, PhD

M. Harold Laughlin, PhD

Introduction

During dynamic exercise the energy cost of sustained rhythmic skeletal muscle contractions is paid by oxidative metabolism. As a result, large increases in oxygen consumption of the muscle tissue lead to the requirement for a sustained supply of oxygen from the cardiovascular system. Oxygen requirements in this setting are met by increases in cardiac output and changes in the regional distribution of blood flow; up to 90% of cardiac output is directed to the dynamically active skeletal muscles during maximal aerobic effort. As exercise progresses for extended periods, additional peripheral circulatory adjustments are necessary in order to achieve thermal equilibrium despite increased metabolic heat production. A salient feature of the response to increased heat load during prolonged dynamic exercise is an increase in cutaneous blood flow and venous volume. The combination of these cutaneous vascular responses and increases in skeletal muscle vascular conductance presents a challenge to the maintenance of blood pressure, particularly if exercise is performed in a hot environment (which exacerbates the increase in cutaneous blood volume). Thus, the metabolic and thermoregulatory demands of dynamic exercise are now recognized to produce the greatest stress on the cardiovascular system that is encountered in normal life. The purpose of this chapter is to summarize the historical development (table 13.1) of our current understanding of the effects of acute and chronic exercise on the cardiovascular system. In view of recent historical summaries that address reflex control of the cardiovascular system during dynamic exercise, this chapter focuses on the development of ideas regarding the peripheral vascular responses to acute and chronic dynamic exercise and specifically discusses the skeletal muscle and cutaneous circulations.

Acute Responses of the Cardiovascular System to Dynamic Exercise

Skeletal Muscle Exercise Hyperemia

This section presents a historical overview of the field of exercise hyperemia. In the first part we review the literature that initially described the basic features of muscle blood flow responses to exercise, including magnitude and temporal profile, influence of muscle fiber type, and the integrated responses of the muscle microcirculation. In the second part we focus on the development of ideas regarding the mechanistic basis of exercise hyperemia.

Initial Observations

The seminal observations on the characteristics and nature of the circulatory adjustments to muscle contraction date back to the end of the 19th century. Not mentioned in chapter 1 were the observations of Sadler in 1869 (204) and Gaskell in 1877 (92), who reported the first descriptions of changes in venous outflow before, during, and after electrically induced tetanic contractions in dog muscle. These authors, working in

Table 13.1 Milestones in the Study of Peripheral Circulation and Exercise

Year	Investigator	Key finding
1930	Rein et al.	Described a reduction in sympathetic vasoconstriction during exercise.
1935	Anrep and von Saalfeld	Described changes in arterial blood flow to muscle during contractions in dog gracilis muscle.
1938	Grant	Refined water-filled plethysmograph to study local exercise hyperemia in the human forearm.
1960	D'Silva and Fouche	Demonstrated capacity of conduit arteries to increase diameter (i.e., vasodilation) in response to increased flow.
1968	Folkow and Halicka	Described the influence of oxidative capacity on the circulation to skeletal muscle at rest and during exercise.
1975	Johnson and Rowell	Demonstrated that prolonged dynamic exercise causes cutaneous vasodilation driven by the thermal load of increased metabolic activity.
1978	Gorczynski et al.	Described the localized effect of muscle fiber activity on arterioles in a muscle.
1980	Furchgott and Zawadzki	Described the obligatory role of endothelial cells in acetylcholine-induced smooth muscle relaxation.
1983	Armstrong and Laughlin	Recorded muscle blood flows >350 $ml \cdot min^{-1} \cdot 100\ g^{-1}$ in rats performing treadmill exercise, challenging the dogma that maximal in vivo muscle blood flow values were <100 $ml \cdot min^{-1} \cdot 100\ g^{-1}$.
1985	Andersen and Saltin	Recorded muscle blood flows of 250 $ml \cdot min^{-1} \cdot 100\ g^{-1}$ in humans performing one-leg knee-extensor exercise.
1986	Segal and Duling	Demonstrated the existence of cell-to-cell conducted vasodilation in the microvasculature.
1992	Celermajer et al.	Established the technique of blood flow-induced dilation for the noninvasive study of conduit artery endothelial vasomotor function in human limbs.
1993	Delp et al.	Demonstrated that endothelium-dependent vasodilation is increased after exercise training.
1997	Asahara et al.	Isolated endothelial progenitor cells from human peripheral blood.
1998	Sheriff and Van Bibber	Demonstrated that isolated skeletal muscle can perfuse itself via contractile action (muscle pump) when its circulation is isolated from the heart.
2000	Hambrecht et al.	Established that exercise training improves coronary artery endothelial vasomotor function in patients with overt coronary artery disease.

Carl Ludwig's laboratory at the Institute of Physiology at Leipzig, used a special device (washing-out apparatus) (92) to measure the quantity of blood draining from a cannulated vein. Despite the technical constraints, Gaskell's description of the behavior of venous outflow during and after contractions highlighted some of the essential features of the hyperemic response to exercise and profoundly influenced subsequent work (14). He summarized the findings of these initial experiments as follows:

> With the commencement of the tetanus there appears a spurt-like quickening of the stream, followed on further contraction by an almost complete cessation of flow, which then again steadily and gradually increases in volume; at the end of the

tetanus, the rate of flow sinks a second time for a moment and then rises again in the course of some seconds to a new maximum, from which it falls gradually and much more slowly than it arose to near the value it possessed before the stimulation. (92, p. 391)

Similar findings were reported by Burton-Optiz from Columbia University in 1903 (43) in New York. He used a stromuhr to measure venous outflow during tetanic contractions in dog muscle and observed that the magnitude of the blood flow response depended on the intensity of the current stimulating the muscle. In a series of experiments in the early 1930s, Gleb Anrep (figure 13.1) and colleagues (8, 9) from the University of Cairo described for the first time the changes in arterial inflow to the muscle during rhythmic and tetanic contractions. Flow was measured by a hot-wire anemometer. These authors confirmed the notion advanced by Gaskell that contractions mechanically impede arterial blood flow, causing a back thrust and flow arrest followed by the hyperemic event during muscle relaxation (8, 9). In these experiments it also became clear that the hyperemic response was proportional to the contraction strength (8, 9). The close relationship between muscle metabolism and blood flow was elegantly demonstrated in a series of experiments by Kurt Kramer and colleagues (131-133) at the University of Heidelberg between 1937 and 1939. Using the dog gastrocnemius preparation and continuous measurements of venous outflow, these authors showed that blood flow responses were related to the magnitude of work performed as well as to oxygen consumption and lactate output from the muscles (131-133).

Figure 13.1 Gleb Anrep during his tenure at the University of Cambridge, circa 1928.

Reprinted, by permission, from S.M. Yentis, 1998, "Minds and hearts: Themes in the life of Gleb von Anrep," *Journal of Royal Society of Medicine* 91: 211.

The first demonstrations of exercise hyperemia in humans appeared in the late 1930s after the development of venous occlusion plethysmography. This technique was first described in 1905 by Thomas Brodie and A.E. Russell and was later modified by Sir Thomas Lewis and Roland Grant at University College London in 1925 (156). By using a water-filled forearm plethysmograph (156), Grant in 1938 performed the first measurements of blood flow in the forearm muscles during and after sustained gripping of an iron bar for 1 min (99). He observed that blood flow increased slightly during contractions followed by a robust hyperemia during relaxation (99). A year later, Henry Barcroft (figure 13.2) and J. Millen at Queen's University of Belfast reported measurements of blood flow through the calf during contractions (19). These authors estimated blood flow by measuring heat clearance from limb muscle using copper-constantan thermocouples inserted into the calf. They observed an intensity-dependent response in which low-intensity contractions evoked a significant hyperemia whereas blood flow was arrested during strong contractions when the subjects were standing on tiptoe (19). After these initial studies, Barcroft started using the water-filled plethysmograph to study muscle blood flow and with Dornhorst in 1949 reported measurements of calf blood flow during rhythmic contractions (17). In agreement with the aforementioned studies, they observed that forceful rhythmic contractions caused blood flow to intermittently cease; the hyperemic response was evident only during the relaxation phase (17). The relationship between exercise intensity and the magnitude of the blood flow response, previously observed in animal studies, was demonstrated by John Black in 1959 (29) and later by Corcondilas and colleagues at Mayo Clinic in 1964 (53). In the latter study, Corcondilas and colleagues used strain-gauge plethysmography to measure the blood flow to the forearm muscles after a brief grasp (~0.3 s) of a metallic ring. They found that the hyperemic response to a single contraction was tightly related to force and was remarkably fast irrespective of the contraction strength (53).

Fiber Type Differences in Muscle Blood Flow Responses to Contractions

Histochemical, biochemical, and functional studies of skeletal muscle in the late 1950s and early 1960s demonstrated that the metabolic and contractile properties of red and white muscles differ substantially. The concept that the vascular supply differed between red and white muscle emerged at this time as well (96, 180, 199), but it was not until the late 1960s that the effect of fiber type on muscle blood flow at rest and during contractions began to be unraveled. Donald Reis and colleagues (195) at Cornell University and Sidney Hilton (114) from Cambridge University were among the first to demonstrate that red muscles have higher blood flows at rest compared with predominantly white muscle. After initial reports by Hilton (114), Bjorn Folkow and Halina Halicka in 1968 performed experiments in cats in which they compared the characteristics of the predominantly red soleus muscle with those of the predominantly white gastrocnemius in response to muscle contractions (83). They demonstrated that the soleus muscle had larger maximal blood flow and capillary surface area and a 35% lower oxygen cost of contraction per unit of weight when compared with the gastrocnemius. Of note, they also observed that the soleus muscle had a much smaller vasoconstrictor response during stimulation of sympathetic nerves compared with the white gastrocnemius (83).

In the 1970s, more sophisticated classification systems based on metabolic and physiologic properties were proposed for muscle fiber types (42, 187). These concepts are covered in more detail in chapter 9. An important development during this era was the division of muscle fibers by Peter and colleagues (187) into three major types: fast-twitch oxidative glycolytic (FOG), fast-twitch glycolytic (FG), and slow-twitch oxidative (SO). This nomenclature was rapidly incorporated in the blood flow studies, and it became clear that the magnitude and profile of flow responses to exercise varied greatly among these three types of muscle (162).

Muscle Blood Flow Capacity

Initial estimates of maximal blood flow capacity (BFC) in feline skeletal muscle during electrical stimulation were in the range of 36 to 40 $ml \cdot min^{-1} \cdot 100\ g^{-1}$ (83, 125) in the predominantly white gastrocnemius and approximately 120 $ml \cdot min^{-1} \cdot 100\ g^{-1}$ in the red soleus (83). In humans, Dornhorst and Whelan (69) recorded a peak blood flow as measured by a plethysmograph of around 20 $ml \cdot min^{-1} \cdot 100\ g^{-1}$ after rhythmic calf exercise. Later, Black (29) reported peak flows of 40 $ml \cdot min^{-1} \cdot 100\ g^{-1}$ in the calf after running. In subsequent studies that used xenon washout, values between 50 and 60 $ml \cdot min^{-1} \cdot 100\ g^{-1}$ were reported (101, 232). These reports led most authors at the time to conclude that even during vigorous exercise in healthy subjects the pumping capacity of the heart could attend to the demands of contracting muscles without compromising the control of blood pressure. This concept was challenged in the early 1980s with the reports of Armstrong and Laughlin (figure 13.3) in exercising rats (10) and Andersen and Saltin (figure 13.4) in humans (5) that the maximal perfusion of skeletal muscle far exceeded the existing estimates. Armstrong and Laughlin determined that in muscle sections such as the red portion of the gastrocnemius of the rat, maximal blood flow could be as high as 395 $ml \cdot min^{-1} \cdot 100\ g^{-1}$ with the animal running at 60 m/min (10). Using the knee-extensor model of exercise and the thermodilution technique to measure blood flow, Andersen and Saltin detected blood flows of 250 $ml \cdot min^{-1} \cdot 100\ g^{-1}$ in the human quadriceps (5). Also in the 1980s, Musch and colleagues measured muscle blood flows as high as 385 $ml \cdot min^{-1} \cdot 100\ g^{-1}$ during maximal treadmill exercise in trained foxhounds (177). Collectively, these remarkable values gave rise to the notion that skeletal muscle is a sleeping giant and that blood pressure control may become compromised during high-intensity dynamic exercise involving large muscle mass if the vasoconstrictor action of the sympathetic nervous system does not limit skeletal muscle vascular conductance (208, 212).

Figure 13.2 Henry Barcroft in 1990 at a meeting of the Physiological Society in Belfast.

Segmental Differences in the Skeletal Muscle Vascular Bed

Until the early 1960s most animal studies investigating the hemodynamic responses to muscle contractions used pump-perfused preparations in which either flow or pressure to the limb is kept constant during the intervention. Although this model allows for calculations of total resistance changes in the vasculature, it was not possible to determine how the different segments of the vascular network behave during muscle contractions. To gain insight into this problem, Bjorn Folkow, Ingemar Kjellmer, Stefan Mellander, and colleagues at the University of Göteborg in Sweden started combining the constant pressure and variable flow method for recording resistance with plethysmographic techniques for recording capacitance responses, capillary fluid transfer, and estimations of the capillary surface area (50, 125). Using this approach in cats, these authors observed heterogeneous responses in the skeletal muscle vascular network to a number of different stimuli, including exercise (50, 125). In 1965, Kjellmer reported that precapillary arterioles (precapillary sphincters) could partially escape from the sympathetic vasoconstrictor action during muscle contractions and concomitant sympathetic stimulation, an effect that was minimal in larger, proximal vessels (126). Based on these initial observations, Folkow and colleagues (84) adapted a method for determining pressure in large and small vessels during muscle contractions in the cat. They found that electrical stimulation of the muscle evoked an initial decrease in vascular resistance in distal vessels only. With the progression of the exercise bout, this vasodilator response slowly ascended to the larger vessels (84). This insightful study provided the first experimental evidence that the site for control of skeletal muscle perfusion seems to shift from precapillary segments to larger, proximal segments during exercise. It also highlighted the idea that distal arterioles were more prone to metabolic-induced vasodilation evoked by muscle contractions when compared with larger arterioles (84, 126).

Figure 13.3 Robert (Bob) Armstrong and M. Harold Laughlin (left) at the annual meeting of the American College of Sports Medicine in 1996.

Photo courtesy of M. Harold Laughlin.

Figure 13.4 Bengt Saltin in Copenhagen, Denmark, where he served as director of the Copenhagen Muscle Research Center.

Photo courtesy of Bengt Saltin.

Intravital Studies

August Krogh (figure 13.5) is inarguably the pioneer of the field of oxygen transport and exchange in the skeletal muscle microcirculation. He was among the first to perform direct observations of the living microcirculation before and after muscle contractions in superficial muscles of frogs and guinea pigs (135, 137). The idea of capillary recruitment during exercise advanced by his initial experiments has since received considerable attention. Interestingly, it was not until the late 1970s that models of intravital microscopy were adapted to study the microcirculatory adjustments to muscle contractions. In their seminal report, Gorczynski and colleagues at the University of Virginia in the United States described a model in the hamster cremaster muscle for direct observation of the effect of electrically induced contractions on the diameter of arterioles (98). These authors elegantly illustrated the localized nature of arteriolar vasodilation during exercise. By stimulating single fibers to contract, they observed that arteriolar dilation was noticeable only in the region of the arteriole that intersected with the active fiber (98). Subsequent work from Duling's laboratory in the hamster cremaster (60, 128) and experiments by Ingrid Sarelius at the University of Rochester in the United States (23, 209, 210) strengthened the concept that terminal arterioles control capillary blood flow and that recruitment during exercise occurs through perfusion of groups of capillaries, called modules (23) or units (60). In 1997, using a preparation similar to that used by Gorczynski and colleagues, Berg and colleagues demonstrated that stimulation of a few muscle fibers signals a rapid increase in red cell flux in the capillary module overlying the active muscle and evokes dilation of the module inflow arteriole. This study highlighted the concept that capillaries can sense changes in tissue metabolism during contractions and that this information was transmitted to the upstream arterioles (22).

Recently there has been renewed interest in the issue of capillary recruitment during contractions and its implication for oxygen transfer in skeletal muscle (47, 193). Poole and colleagues compiled evidence derived mostly from intravital microscopy studies and advanced the notion that no recruitment occurs during muscle contractions because the vast majority of capillaries support red blood cell flux at rest. Instead, these investigators suggest that increased muscle oxygen diffusion capacity during contractions arises from increased capillary hematrocit as well as increased length over which oxygen is exchanged (192). Unfortunately, it is still not known whether vascular control in thin muscles that are suitable for visualization under a microscope resembles that of load-bearing muscles that are normally recruited during exercise. Resolution of this dispute awaits the availability of techniques that allow for continuous monitoring of the terminal vascular network in locomotor muscles at rest and during voluntary exercise.

Mechanisms of Control in Exercise Hyperemia

Metabolic Factors

In his initial report in 1877 (92), Walter Gaskell hypothesized that changes in blood flow during contractions were the result of vasodilator nerve fibers that were stimulated simultaneously with the motor nerves. In subsequent perfusion studies in frogs, during which he found that lactic acid caused a dilatory effect in the muscle, Gaskell advanced the hypothesis that byproducts of tissue metabolism could diffuse to the vasculature and directly cause vasodilation (93). This hypothesis rapidly gained widespread acceptance and prompted several attempts to determine what metabolic factors could mediate or determine the hyperemia of exercise. Initial attention was given to hallmark byproducts of tissue metabolism, including CO_2, hydrogen ion, and lactic acid (20, 77). This list of metabolites grew longer as new substances were discovered to have dilator effects in perfused skeletal muscle. In the 1960s and 1970s, several proposals emerged from independent laboratories as described by Barcroft (15): Dawes (57) and later Ingemar Kjellmer (127) proposed the role of potassium ion, Stefan Mellander and colleagues argued in favor of osmolarity (170), Sidney Hilton and colleagues proposed the role of inorganic phosphate (115), and Forrester and Lind focused on the action of adenosine triphosphate (85, 86). In the 1980s, with the discovery of the importance of the endothelial layer in vasomotor control (88), new candidates such as nitric oxide and prostaglandins emerged. Since then, endothelium-derived substances have been a major target of studies on exercise hyperemia, as reviewed recently by Joyner and Wilkins (122) and Saltin and colleagues (205). Although the list of metabolic mediators continues to grow, a definite conclusion of what factors mediate this response remains to be reached. As recently emphasized by Davis and colleagues (56), the disparate findings and lack of agreement on which factors contribute to exercise hyperemia stems from several confounding factors and differences in experimental preparations, including the following: the stimulus protocol used to evoke contraction, the muscle mass stimulated, the type of contraction, the muscle fiber type, the method of tissue perfusion used, species differences, and the quality of the experimental preparation. Currently, the consensus among investigators is that blood flow is linked to metabolism in striated muscle through the orchestration of a number of metabolites and that no single factor is essential. It is also generally accepted that substantial redundancy exists in the action of these metabolites and that combinations of several agents are necessary to evoke the hyperemic response to exercise.

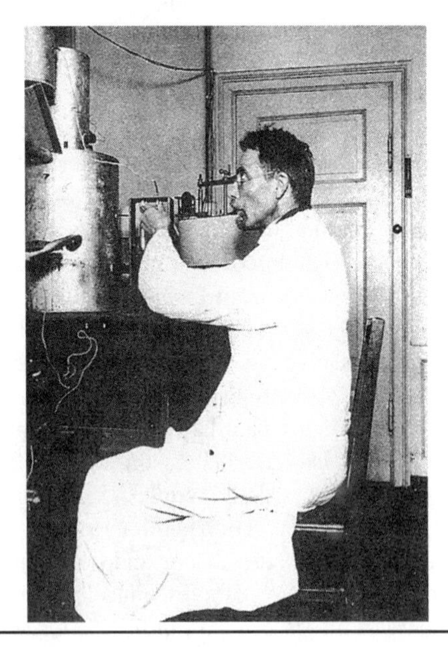

Figure 13.5 August Krogh in his laboratory in Ny Vestergade, circa 1925.

Reprinted from B. Schmidt-Nielsen, 1995, *August & Marie Krogh: Lives in science* (New York: American Physiological Society). With permission of American Physiological Society.

Conducted Vasodilation

By 1920, August Krogh had observed in experiments performed in the tongue of the frog that local mechanical and thermal stimulation and application of agents such as weak acid, cocaine, and adrenaline caused a dilatory response that "spread to an area greater than that stimulated" (136, p. 418). In the 1930s, Schretzenmayr (211) and later Fleisch (76) reported that electrically induced contractions of the calf muscles in the cat and dog triggered an increase in the diameter of the femoral artery. Fleisch also noticed that intra-arterial injections of metabolites, acetylcholine, and histamine provoked a similar response (76). These observations were later confirmed in experiments performed by Sydney Hilton in cats (113). He found that the vasodilator response of the femoral artery to muscle contractions was remarkably similar to the one evoked by administration of a number of agents (i.e., acetylcholine, histamine, bradykinin, and nicotine) and concluded that "both vascular reactions involve the same peripheral conducting mechanism" (113, p. 110). He reasoned that a conducting system in the wall of the artery was responsible for the observed responses because these responses were not abolished after the nerves to the limb had been cut (113). Although insightful, a major confounding factor in these experiments was the associated changes in flow, which later were shown to independently induce vasodilation. The notion that conducted responses can occur even in the absence of changes in flow grew from experiments performed in Brian Duling's laboratory (figure 13.6). In 1970, Duling and Robert Berne observed that application of acetylcholine to an arteriole in the hamster cheek pouch evoked vasodilation that spread rapidly upstream and downstream from the point of application (70). In their seminal report in 1986, Steve Segal (figure 13.7) and Duling (215) showed that acetylcholine-induced propagated vasodilation in hamster cheek pouch arterioles was not affected by arteriolar occlusions, thus suggesting that this response occurs independent of flow changes. These authors also proposed that conducted responses were necessary for the proper coordination of microvascular adjustments to an increase in tissue metabolic demand (214, 215). This concept, later strengthened by the aforementioned observations of Berg and colleagues (22), suggested that a conducted response initiated in arterioles and capillaries embedded in the contracting muscles helps recruit feed arteries that are located outside the tissue and were not exposed to the vasodilator action of byproducts of muscle contraction (214, 215).

Myogenic Mechanisms

In 1901, Sir William Bayliss from University College in London described what later became known as the myogenic mechanism (20). He performed experiments in dogs, cats, and rabbits as well as in excised carotid arteries and observed that increases in intravascular pressure were associated with a vasoconstrictor response and that, likewise, sudden decreases in pressure caused relaxation of the vessel wall (20). This concept later gained widespread acceptance as an important mediator of blood flow autoregulation in vascular beds (78). Based on the notion that skeletal muscle contractions cause intramuscular pressure to increase considerably, David Morhman and Harvey Sparks from the University of Michigan in the United States conducted a study to verify whether the sudden alteration in transmural pressure during contractions could independently contribute to exercise hyperemia (172). These authors measured arterial inflow and venous outflow in the dog

Figure 13.6 Brian Duling in his laboratory at the University of Virginia in the United States, circa 2009.
Photo courtesy of Brian Duling.

Figure 13.7 Steve Segal in his laboratory at the University of Missouri in the United States, circa 2010.
Photo courtesy of Steve Segal.

calf during and after a brief tetanus and during externally applied mechanical compressions using a pressure-cuff system. They observed that intramuscular pressure elevation by external cuff inflation caused an immediate increase in vascular conductance and thus advanced the notion that compressive forces contribute, at least partially, to exercise hyperemia (172). This idea gained momentum recently with the demonstration by Clifford and colleagues that external mechanical compression of isolated skeletal muscle arterioles evokes a fast dilatory response that is partially endothelial independent (49). According to these authors, the vasodilator response induced by external deformation is a major contributor to the rapid-onset vasodilation observed at the initiation of muscle contractions (49).

Muscle Pump

The powerful effect of oscillations in intramuscular pressure on muscle perfusion was recognized from the initial observations of the profile of hyperemic response to exercise (19, 92). In 1949, Barcroft and Dornhorst noticed that rhythmic calf contractions caused the calf volume to reduce by almost 30 ml in individuals lying on a couch, a finding that was attributed to venous emptying evoked by the pumping action of calf muscles (18). In the following years, a number of studies documented the behavior of superficial and deep venous pressure during locomotion. Starting with the work of Albert Pollack and Earl Wood in 1949 (191), several investigations demonstrated that pressure in the veins of the dependent limb was reduced during locomotion (116, 226). As a result of this reduction, the pressure difference between the arterial and venous side is enhanced, therefore favoring arterial inflow. This concept was elegantly demonstrated in a series of experiments by Bjorn Folkow and colleagues (80-82). In their first series of experiments in cats they observed that positioning the calf muscles in a dependent position 35 cm below the heart level increased flow through rhythmically working muscles when compared with positioning the limb at heart level (80). The same group subsequently demonstrated in humans the effect of increased transmural pressure on the blood flow response to contractions (82). These authors measured calf blood flow with the xenon clearance technique in healthy subjects performing heavy plantar flexion exercise (82). In accordance with the results in cats, blood flow in the calf muscle during rhythmic exercise increased markedly when the subjects were tilted from supine to the leg-down position (82). Since these initial experiments the muscle pump hypothesis has received substantial attention. One particularly ingenious experiment that proved the effectiveness of the muscle pump was performed by Don Sheriff (figure 13.8) and Richard Van Bibber in pigs (218). Using an experimental design analogous to the heart–lung preparation, these authors demonstrated that the muscle could perfuse itself during rhythmic contractions when its circulation was short-circuited and isolated from the remainder of the circulation (218). Despite being effective as a peripheral heart, the contribution of the muscle pump to exercise hyperemia remains debatable. As proposed by Laughlin and Schrage, one source of confusion is that the instrumentation necessary for properly investigating the action of this mechanism may prevent its normal function and therefore preclude decisive conclusions about its role in exercise hyperemia in intact subjects (153).

Flow-Induced Vasodilation

D'Silva and Fouche were among the first to propose that flow itself could induce dilation of large arteries during exercise (55). Based on the initial observations of dilation of the femoral artery during muscle contractions by Schretzenmayr (211), Fleisch (76), and Hilton (113), these authors evaluated the effect of creating an arteriovenous fistula on the caliber of large arteries perfusing the calf of cats. They created a connection between the anterior tibial artery of one leg and a vein of the other leg and observed widening of the femoral artery of about 30% when the fistula was opened (55). The magnitude of dilation was remarkably similar to the magnitude evoked by repeated contractions of the calf in the same model, which led these authors to propose that increased flow and not a conducted dilation (113) was responsible for the exercise-induced vasodilation of the conduit arteries (55). In the ensuing years, the dilatory response to flow in large arteries was well characterized and was shown to depend on intact endothelium (220, 222). The importance of this mechanism to microcirculatory control in skeletal muscle

Figure 13.8 Don Sheriff in his laboratory at the University of Iowa in the United States, circa 2010.
Photo courtesy of Don Sheriff.

was later demonstrated in intravital studies of the rat cremaster muscle by Akos Koller and Gabor Kaley from New York Medical College (130). They occluded an arteriole with a micro-occluder and studied changes in red blood cell velocity and diameter in one of the parallel coupled arterioles before and after light-dye treatment, which impairs the endothelium. They found that the increased flow in the coupled nonoccluded arteriole induced a fast, significant dilatory response that was completely abolished after light-dye treatment. These authors hypothesized that rapid shear stress-induced dilation would be especially meaningful for the coordination of microvascular adjustments between proximal and distal elements of the arteriolar network to an increase in metabolic demand (e.g., during exercise) (130). Thus, during increases in tissue metabolism the precapillary arterioles would be the first to dilate and the consequent increase in flow in the upstream vessels could lead to their secondary dilation, therefore allowing for an increase in bulk flow to the tissue (130).

Interaction Between Metabolic Vasodilation and Sympathetically Induced Vasoconstriction

By the end of the 1940s, the notion that the sympathetic nervous system exerts tonic vasoconstrictor activity in resting muscles was well established (16, 79). In the ensuing years several reports indicated that sympathetic activation also occurs during muscle contractions, as evidenced by reductions in blood flow to inactive tissues during exercise (30, 38, 39). These observations gave rise to the idea that the contracting muscles could also be a target for sympathetically mediated vasoconstriction and, by inference, that a competition between vasodilator systems and nervous control would take place in the skeletal muscle vasculature. The first insights into this apparent paradox emerged in 1930 from the work of Hermann Rein and colleagues from Gottingen, Germany. These investigators studied the effect of sympathetic activation and intravenous infusion of adrenaline on vascular resistance in resting and exercising skeletal muscle (194). They observed for the first time that sympathetic activation caused less efficient vasoconstriction in exercising muscle than in resting muscle (194). The ability of the skeletal muscle vasculature to escape vasoconstriction during contractions was later confirmed by Remensnyder and colleagues, who referred to this effect as functional sympatholysis (196). The characteristics and functional significance of this effect became better understood after a series of elegant experiments conducted by Ingemar Kjellmer (126) and later by Bjorn Folkow and colleagues (84). Using a combination of measurements that allowed for the study of different segments of the vasculature, these authors observed that the distal portion of the arteriolar network seemed to be particularly susceptible to the sympatholytic effect of contractions whereas the proximal branches had a sustained vasoconstriction throughout the exercise bout (84, 126). More recent studies that used intravital microscopy have confirmed that the preferential locus of metabolic inhibition of sympathetically mediated vasoconstriction is the terminal segment of the network (66, 237). As originally proposed, the preferential escape of distal segments ensures appropriate capillary perfusion and matching between oxygen delivery and utilization in the microcirculation (84). At the same time, the proximal feed vessels, which seem to show preserved vasoconstriction, serve as key sites for regulation of cardiac output, blood flow distribution, and, ultimately, blood pressure control (237).

Observations on the Cutaneous Vascular Response to Dynamic Exercise

Heat is the most abundant byproduct of cellular metabolism. As such, dynamic exercise in which a significant percentage of muscle mass is engaged generates thermoregulatory demands that, in humans, are met in part by cutaneous vascular adjustments. Anecdotes indicate a general appreciation for this concept by early 20th century physiologists. For example, in his 1924 text titled *The Physiology of Exercise*, McCurdy (describing the work of Bowen, 1903) states:

> The stopping of the rise in blood-pressure during work while the heart-rate is still increasing is due to decreased peripheral resistance in the working muscles and in the skin. The fall in pressure is due to the dilatation of the skin vessels, which is affected by the skin temperature due to heat output . . . ." (167, p. 81)

Clearly, by 1924 there was an appreciation for the circulatory changes occurring in the skin during dynamic exercise and for the thermoregulatory mechanisms driving these changes. However, the details of these responses would not be elucidated experimentally for a number of decades.

Brief Periods of Dynamic Exercise

At the beginning of the 1900s, scattered reports suggested that dynamic exercise of the lower limbs caused a reduction in skin temperature of the hands (Adae, 1876, referenced in 28). In 1911, Stewart (227), working in the H.K. Cushing Laboratory of Experimental Medicine at Western Reserve University (now Case Western Reserve University), measured the heat eliminated from one hand (using calorimetry) while the subject performed exercise with the contralateral hand. Stewart noted a reduction in heat elimination to the calorimeter bath from the resting hand and concluded that heat delivery—and thus blood flow—to the hand must have been reduced. In 1942, Christensen and Nielson at the University of Copenhagen used the plethysmograph to extend Stewart's observations to dynamic exercise of the legs, showing a highly reproducible vasoconstriction in the fingers at the onset of cycle exercise (46). The vasoconstriction was abolished if exercise was prolonged. Because the circulation to the hand perfuses mostly cutaneous tissue, these findings led to the belief that the cutaneous vasculature was constricted at the onset of exercise. However, in the 1950s it became clear that hand (acral) and forearm (nonacral) skin segments are controlled differently by the autonomic nervous system, at least during passive thermal stress. Thus, questions remained about whether the findings in the acral skin of the hand could be extrapolated to the nonacral circulation of the forearm.

In 1957, Bishop and colleagues (28) at the University of Birmingham in the United Kingdom conducted the first systematic study investigating perfusion of the upper limb during lower-limb exercise (lasting 10 min). Vasoconstriction of the hand circulation was confirmed via calorimetry, and forearm blood flow was estimated by measuring oxygen saturation of the venous effluent draining the forearm. Estimates of forearm blood flow indicated that flow was reduced at the onset of exercise, in agreement with the blood flow data in the hand. However, blood flow appeared to increase toward the end of the exercise period in both the hand and the circulation to the whole arm. The authors also measured the clearance of radioactive sodium injected into the muscle to estimate blood flow to the skeletal muscle vasculature. Because the radioactive sodium measurements indicated no change in muscle blood flow during exercise, the authors concluded that forearm vasoconstriction, and the subsequent return of forearm blood flow to baseline, was occurring in the nonacral skin.

In 1961, Blair and colleagues (30) at Queen's University in Belfast showed constriction of the hand and forearm at the onset of leg exercise using plethysmography, confirming the observations of Bishop and colleagues (28). Flow increased back toward baseline during the final minutes of exercise (5 min bout) in the forearm and increased above resting in the hand at a higher workload. Blocking the deep nerves above the elbow (which innervate muscle) with local anesthetic eliminated the forearm vasoconstriction during exercise whereas blocking the cutaneous nerves had no effect on forearm vasoconstriction at exercise onset. Thus, the authors concluded that forearm vasoconstriction at the onset of exercise takes place in the muscle circulation and not the skin of the forearm. It is possible that the small proportion of limb blood flow circulating through the skin was responsible for the failure to detect an effect of cutaneous nerve block on the limb vasoconstrictor response. Nevertheless, Blair and colleagues (30) considered the significance of the tendencies for forearm blood flow to return to baseline as exercise progressed beyond the initial minutes and for blood flow in the hand to surpass baseline levels. The authors indicate the persistence of a concept regarding the role of the cutaneous circulation in the dissipation of heat generated by muscular work:

> The finding that the blood flow rose to a higher level in the normal side than in the nerve-blocked side indicated that the vasodilatation was not due to release of vasoconstrictor tone in skin, but rather to an active vasodilator mechanism mediated through fibers running with the cutaneous nerves. Body heating causes such a reflex vasodilatation in forearm skin, and it is likely that the increase in heat production during exercise was responsible for the vasodilatation in the present experiments. The fact that the subjects often felt hot and were sweating after the period of exercise would support this. (30, p. 272)

Although the role of the cutaneous circulation in body temperature regulation was well established by 1961, it would be 14 yr before the magnitude of cutaneous hyperemia with prolonged exercise (sufficient to elevate core temperature) was demonstrated (121).

In 1966, Bevegard and Shepherd (26) at Mayo Clinic estimated blood flow to the forearm muscle and skin separately during 5 min of leg cycling. In addition to measuring total forearm blood flow with plethysmography, they estimated blood flow in the forearm by measuring oxygen saturation of the venous blood draining a deep vein and a superficial vein, the latter presumably draining primarily cutaneous circulation. The oxygen saturation measurements from the deep vein followed the pattern of the forearm blood flow response (vasoconstriction). In contrast, oxygen saturation measurements from the superficial vein indicated an initial transient vasoconstriction in skin; that is, oxygen saturation decreased. This was followed by vasodilation, particularly if the exercise load was heavy (exercise duration 5 min). In the hand, vasoconstriction at exercise onset was more marked, but flow similarly began increasing toward baseline as exercise progressed. These results clearly indicated that skin blood flow progressively increases during dynamic exercise after the initial cutaneous vasoconstriction.

In 1969, Zelis and colleagues (250) at the National Institutes of Health took advantage of a technique developed by Barcroft and colleagues (16) whereby the cutaneous circulation of the forearm could be arrested and thus the circulation to the noncutaneous tissue alone (mostly muscle) could be investigated. The technique involved iontophoresis of epinephrine into the skin, which stimulates alpha adrenergic receptors and, when applied at high enough doses, causes complete cessation of skin blood flow. Indeed, after topical iontophoresis of epinephrine, the skin does not bleed when cut with a blade (52). In the study of Zelis and colleagues (250), subjects exercised on a cycle ergometer at three different workloads for 6 min per load. Comparison of blood flow measurements (plethysmography) between control and experimental (skin circulation arrested) arms indicated that both muscle and skin constricted during the first 3 min of exercise. Three workloads were investigated. In agreement with the findings of Bevegard and Shepherd, results indicated that the muscle circulation remained constricted during the last 3 min of exercise whereas skin blood flow returned to baseline in mild exercise and further increased with moderate exercise.

Prolonged Dynamic Exercise

Until 1969, all investigations of the cutaneous vascular response to dynamic exercise were done in the context of short-duration exercise protocols lasting 10 min or less. This undoubtedly delayed development of an appreciation for the magnitude and cardiovascular consequences of sustained cutaneous hyperemia during exercise. In 1969, Kamon and Belding (123) at the University of Pittsburgh in the United States provided the first assessment of skin blood flow (albeit indirect) during prolonged exercise. Blood flow was estimated by the technique of heat uptake through the skin. The subject's arm and hand were submerged in a water bath maintained at approximately 41 °C, and a change in the amount of heat necessary to maintain a constant bath temperature was measured to estimate changes in skin blood flow (or heat uptake by the limb). Authors showed initial adjustments in skin blood flow early in exercise that were similar to those in previous studies. During prolonged (1 h) exercise blood flow in the forearm increased considerably, eventually demonstrating a plateau after 40 min of exercise. Although the methods used by these investigators were indirect, these results represent the first demonstration of a potent cutaneous vasodilation occurring during sustained (>10 min) dynamic exercise involving large muscle groups.

In 1975, Wenger and colleagues (243) in the John B. Pierce Foundation Laboratory at Yale University tested subjects during cycle ergometer exercise over a range of exercise intensities and ambient temperatures. The duration of each exercise bout was 30 min. In agreement with the findings of Kamon and Belding, this study demonstrated a massive forearm hyperemia during sustained dynamic exercise. Importantly, blood flow was measured via plethysmography, but no attempt was made to distinguish skeletal muscle from cutaneous components of the hyperemia. Nevertheless, previous studies that suggested a sustained vasoconstriction in muscle and a developing vasodilation in the skin led the authors to speculate that the forearm vasodilation might be confined to the skin. In addition, by this time it had been demonstrated that the forearm vasodilation during passive body heating is confined to the skin. Wenger and colleagues speculated that a similar mechanism was responsible for the forearm vascular response to exercise heat stress. Later that year, the speculation of Wenger and colleagues was confirmed in a landmark study performed by John Johnson (figure 13.9) and Loring Rowell (figure 13.10) at the University of Washington in Seattle (121). Johnson and Rowell exercised subjects on a cycle ergometer for 1 h while measuring total forearm blood flow with plethysmography and estimating muscle blood flow by radioisotope clearance (injected into the brachioradialis muscle). During prolonged exercise, total forearm blood flow increased by approximately 8 $ml \cdot 100\ ml^{-1} \cdot min^{-1}$ and estimated muscle blood flow decreased. Thus, the findings of Johnson and Rowell (121) demonstrated that cutaneous vasodilation of considerable magnitude occurs during prolonged dynamic exercise whereas the skeletal muscle circulation is the target of vasoconstriction. An important implication of this finding was that the cutaneous circulation could be studied by measuring forearm blood flow during prolonged lower-limb exercise.

Recently, the divergent responses in the cutaneous and muscle circulations during sustained dynamic exercise were confirmed using modern techniques to measure blood flow. In 2008, Ooue and colleagues (188) at Kobe University in Japan used Doppler ultrasound to measure blood flow in the brachial artery (perfusing the whole limb), a deep (muscle) vein, and a superficial (cutaneous) vein. During leg exercise, forearm perfusion through the brachial artery increased and flow in the cutaneous vein concomitantly increased. In contrast, blood flow through the deep muscle vein was unchanged during exercise. Thus, modern methods of blood flow assessment have confirmed the results of Johnson and Rowell (and those before them), which indicate that forearm hyperemia during leg exercise is confined to the skin circulation and that a considerable cutaneous vasodilation occurs during prolonged dynamic exercise in order to subserve thermoregulation.

Figure 13.9 John Johnson near his laboratory at the University of Texas Health Science Center in San Antonio in the United States.

Photo courtesy of John Johnson.

Figure 13.10 Loring (Larry) Rowell, professor emeritus of physiology and biophysics at the University of Washington in Seattle.

From the collection of Charles M. Tipton.

Cardiovascular Responses to Isometric Exercise

According to Loring Rowell, "The cardiovascular response to isometric contractions is characterized by its failure to secure adequate blood flow for the contracting muscles" (201, p. 303). The concept that mechanical impedance to muscle blood flow at high levels of contractile force cannot be overcome by increased perfusion pressure has been appreciated since early in the 20th century. For example, Lindhard measured cardiovascular and respiratory responses to isometric contractions in humans in 1920 (159). He noted that the increase in oxygen uptake was small during strong contractions of the arms and that oxygen uptake increased dramatically upon relaxation. These observations were supported by the subsequent work of Dolgin and Lehmann (67), who demonstrated that the duration of a strong isometric contraction was unaffected by occlusion of arterial inflow. In contrast, the duration of weaker contractions was shortened during arterial occlusion, confirming that isometric contractions cause local circulatory arrest when performed at high intensities (67). Later, Barcroft and Millen (19) noted that local blood flow in calf muscles was occluded at isometric contractions of 20% of maximal voluntary contraction (MVC) or greater. These findings appear to be consistent with subsequent work conducted by Lind and McNichol in 1967 (158) in which the effect of isometric handgrip contraction on forearm blood flow was evaluated at different levels of force development. These authors clearly demonstrated that inadequate matching of blood flow and metabolism occurred at 20% MVC and greater, as evidenced by signs of fatigue as well as postcontraction hyperemia.

To summarize, static contractions in which force development surpasses approximately 20% of maximal are characterized by a reduction and, at high levels of force production, a cessation of blood flow to the contracting muscle. In this setting, reflex increases in arterial blood pressure caused in part by peripherally stimulated metaboreflexes can increase blood flow and oxygen delivery to the working muscle only if the intraluminal blood pressure surpasses extraluminal (intramuscular) pressure. Some evidence for this has been presented, primarily during contractions of relatively modest effort (10% MVC) (244).

Adaptations to Chronic Dynamic Exercise and Exercise Training

In the past 100 yr it has become well established that aerobic-exercise training improves cardiovascular function and increases the vascular transport capacity of skeletal muscle (26, 48, 201). Research in the 1980s and 1990s by a number of laboratories demonstrated that exercise hyperemia in skeletal muscle and vasomotor reactivity of skeletal muscle arteries are altered by exercise training (11, 177, 245). It is established that adaptations induced by chronic exercise training in the skeletal muscle vasculature can be grouped into two general categories: structural vascular adaptation (remodeling of existing vessels and angiogenesis of capillaries) and adaptive changes in the control of blood flow (neurohumoral control, local control, and the effects of structural adaptations on vascular resistance). In this section we trace the historical development of advances in the understanding of these vascular adaptations induced by chronic dynamic exercise. We begin by discussing structural vascular adaptation induced by exercise training and then summarize discoveries pertaining to alterations in the phenotype of vascular cells that contribute to reported alterations in vasomotor reactivity of skeletal muscle arteries after exercise training (11, 245). These discoveries resulted from the application of vascular cell biology techniques and concepts to the study of exercise-induced adaptations in the vasculature. Improvements in BFC to striated muscle and alterations in mechanisms controlling blood flow to skeletal muscle appear to be key components in training-induced increases in maximal oxygen consumption and endurance in normal subjects (145, 198, 207, 238). A growing body of evidence indicates that exercise training alters the endothelium and vascular smooth muscle of the arteries supplying blood flow to skeletal muscle as well as endothelial progenitor cells. Before discussing alterations in vascular control, we summarize the historical development of our understanding of structural adaptations.

Structural Vascular Adaptation

Capillary Density

The initial studies on the effects of increased muscle activity or exercise on skeletal muscle capillarity emerged in the mid-1930s. Using the rather crude techniques of staining of erythrocytes and perfusion with different materials, Vanotti and Magiday (236) in 1934 and Petren and colleagues in 1936 (188) demonstrated that capillary density is increased in trained muscles. In the 1960s, it became established that capillary density varies in muscles comprising different fiber types. The seminal work of Flaviu Romanul and colleagues from Harvard University advanced the notion that a tight relationship exists between capillarity and oxidative capacity in different fiber types (199). This idea, coupled with knowledge of muscle recruitment patterns during exercise, prompted investigators in the field to evaluate the effects of exercise training on capillarity in muscles comprising different fiber types. Carrow and colleagues (44) were among the first to determine the effects of training (wheel running and swimming) on capillarity in predominantly red or white areas of the rat gastrocnemius. After 35 d of exercise they perfused the rat hind limb with India ink and formalin to determine capillarity. Interestingly, they found that the capillary-to-fiber ratio increased predominantly in the white portion of the muscle in rats exposed to both swimming and voluntary wheel running (44). In 1970, Mai and colleagues (163) exposed guinea pigs to a program of 18 wk of endurance running (40 m/min) on a treadmill and determined muscle capillarity and oxidative capacity changes in the soleus (red) and medial gastrocnemius (white). The authors anticipated that running on the treadmill at these intensities would recruit predominantly red fibers, and, accordingly, they expected that the changes in capillarity would occur mainly around red fibers. Confirming their prediction, the capillary-to-fiber ratio increased by approximately 28% in red fibers only (163). This study provided the first evidence that vascular adaptations to exercise in skeletal muscle can be spatially different depending on muscle recruitment characteristics. An important series of experiments in the late 1980s and 1990s using different types of exercise training (e.g., intensity, duration) in rats revealed that exercise training-induced angiogenesis of capillaries in and among skeletal muscles is greatest in the muscle tissue with the greatest relative increase in fiber activity during training bouts (102, 103, 146, 217). That is, training-induced adaptations of capillarity and mitochondrial content in the muscle are spatially coupled so that the regions that exhibited increased oxidative capacity also exhibited increased capillarity (102, 103, 200). In the case of endurance training, increases in capillarity were seen in high-oxidative skeletal muscle (SO and FOG) but not in low-oxidative muscle (FG) (102, 200). In contrast, capillarity was increased in FG muscle by sprint training only (103). Of note, the available results from these studies indicate that the magnitude of the adaptations in and among skeletal muscles, induced by different forms of exercise training, is not coupled. This series of studies suggested that BFC must be increased by structural changes in arteries and that stimuli for arterial remodeling (functional or structural) are different than those for capillary angiogenesis.

After the development of a method for percutaneous muscle biopsy in 1962 by Jonas Bergstrom (24), it became possible to study the effects of training on muscle capillarization in humans. Bengt Saltin and colleagues in the well-known Dallas Bed Rest and Training Study (206) were the first to determine the effects of inactivity and exercise training on capillary density in healthy subjects. In that study, 8 wk of intensive dynamic-exercise training (after 3 wk of bed rest) resulted in no changes in muscle capillarity. In contrast, Per Andersen and Jan Henriksson from Copenhagen in 1977 provided unequivocal evidence that prolonged intense-exercise training does increase capillarity in skeletal muscle. After 8 wk of training on a cycle ergometer at 80% $\dot{V}O_2max$, they detected a 20% increase in capillary density that was evident in all three fiber types studied (4). Thus, aerobic-exercise training is currently considered to increase capillarity of muscle in humans; these increases are greatest in SO (Type I) and FOG (Type IIA) skeletal muscle (40, 51, 94).

Arteriolar Density

In 1989, Martin and colleagues quantified resistance arterioles (10-100 μm) in the soleus, gastrocnemius, and vastus lateralis muscles of sedentary Sprague-Dawley rats (165). They observed that arteriolar density increased progressively with muscle depth and was much greater in areas comprising slow-twitch and FG fibers (165). Julia Lash and Glenn Bohlen were the first to study the effect of prolonged endurance-exercise training on arteriolar density. In this study, young rats were exposed to 8 to 10 wk of treadmill running, after which histological techniques were used to assess arteriolar density in the gracilis and spinotrapezius muscles (140). They found that training did evoke a small but significant increase in the diameter of larger arterioles and the density of smaller arterioles in the spinotrapezius (140). These results are surprising given the notion that both muscles are not actually recruited at the running intensities chosen in this study (143, 178), which raised the possibility that structural changes in the microcirculation can occur in muscles that are not active during the exercise bouts. Olga Hudlicka's group from Birmingham also provided valuable evidence for the understanding of arteriolar growth with increased muscle activity. These authors used chronic electrical stimulation of the lower-limb flexor muscles of rats as a model of increased muscle activity. In this model, the density of preterminal and terminal arterioles begins to increase after only 2 d of stimulation, reaching densities fivefold higher than preintervention levels within 7 d (107). A major limitation of this approach, however, is that it fails to mimic the normal recruitment patterns associated with locomotion, which precludes a direct extrapolation of the results to the more physiological modes of exercise.

Work completed over the past 10 yr suggests that whether increases in skeletal muscle arteriolar density or arteriolar size are observed in response to exercise training is determined by which muscles are examined and by the type of exercise training performed (152). For example, Laughlin and colleagues in 2006 showed that aerobic-endurance training increased arteriolar density throughout fast-twitch skeletal muscle (FOG and FG) but that the arteriolar density of the soleus muscle was not significantly altered by aerobic training (144). When rats were trained with interval sprint training there were no changes in arteriolar density, percentage artery area, total artery area, or arteriolar wall thickness in either fast-twitch or slow-twitch muscle tissue (144). Vascular casting techniques and computational network model analysis that examined mechanisms responsible for increased skeletal muscle BFC in mixed fiber type muscle of sprint-trained rats revealed that arteriolar densities and dimensions increased after sprint training (27). Importantly, computational model analysis indicated that these changes in the tissue of muscle with mixed fiber composition would be sufficient to create the doubling of flow capacity reported for this muscle of sprint-trained rats (27, 146). These combined results indicated that structural arteriolar remodeling may be the primary adaptation responsible for the training-induced increase in BFC in muscle with mixed fiber composition (27).

To summarize, over the past 50 yr the understanding of the role of structural vascular adaptation in exercise-induced adaptations has improved dramatically. At this time it appears that angiogenesis of capillaries improves capillary exchange capacity in skeletal muscle and that structural vascular adaptations contribute importantly to training-induced increases in BFC. However, these effects are dependent on fiber-type composition and recruitment patterns of muscle fibers during training bouts. The fact that increases in flow capacity are not fully explained by structural adaptations (i.e., increased capillarity, increased arteriolar number or sizes) indicates that control of vascular resistance and vasomotor responses mediated by endothelial cells or smooth muscle cells must also contribute to increased BFC in exercise-trained skeletal muscle (103, 146, 217).

Functional Vascular Adaptation: Changes in Control of Conductance

Vascular Smooth Muscle Cell Biology

Early evidence that muscular activity (exercise training) could alter sensitivity of vascular smooth muscle was presented in 1992 by Hudlická and Fronek (118). They demonstrated that chronic electrical stimulation (10 Hz) of rabbit skeletal muscles produced alterations in the reactivity of arteries supplying these muscles to various stimuli. Also in the 1990s, several laboratories reported that exercise training altered reactivity of vascular smooth muscle of abdominal aorta (62-64, 225) and coronary arteries (184) and altered the cellular and molecular control of intracellular free Ca^{2+} in coronary vascular smooth muscle cells (35, 36). More recently it has been demonstrated that exercise training alters vascular smooth muscle responsiveness in various resistance arteries of the soleus muscle and gastrocnemius muscle of rats (154, 166). It is now well established that vasomotor reactivity and a number of electrophysiological characteristics of vascular smooth muscle cells are altered by exercise training (37, 147, 241).

The recognition that factors such as cyclic stretch of the vessel wall, which is acutely altered during an exercise bout, can lead to pronounced modifications in the phenotype of vascular smooth muscle cells (104) favors the concept that repeated exposure to physical activity could promote beneficial adaptations in vascular smooth muscle (37). There is now substantial evidence that exercise training induces adaptations that cause changes in the responsiveness of vascular smooth muscle to a number of local control signals. Space allows us to summarize only a few of these studies to illustrate the progress in this area.

Adenosine

The adenosine (ADO) hypothesis of blood flow control emerged from the work of Berne, who proposed that ADO is released extracellularly during muscle contraction and accumulates in the muscle interstitium in a parallel fashion with the muscle blood flow response (25, 71, 164). In the 1990s, evidence that exercise training can alter vascular responsiveness to ADO was reported from studies on coronary arteries of exercise-trained pigs that exhibited increased sensitivity to ADO (184). Subsequently, divergent findings were reported in animal studies after treadmill training: Studies showed reduced (229) or unaltered (139) responsiveness to ADO in the gracilis and spinotrapezius muscles, respectively. Given that both the gracilis and spinotrapezius muscles are not believed to be recruited during low to moderate running intensities in rats, these differing results could be the result of making measurements in arteries of muscles only modestly recruited during training bouts. More recently, work by McCurdy and colleagues demonstrated that decreased physical activity (perhaps detraining) resulted in decreased sensitivity of isolated gastrocnemius arterioles to the dilatory actions of ADO (168) and did not alter responses of arterioles from soleus muscle. These results may suggest that, similar to the reported effect in coronary artery, exercise training increases ADO sensitivity in muscles known to be active during exercise (184).

Adrenergic Vasomotor Reactivity

In humans, the limited evidence available suggests that exercise training does not affect sympathetically mediated constriction at rest (181, 221). Consistent with this conclusion, a recent study examined sympathetic constriction during moderate exercise in sedentary subjects and competitive cyclists and reported no differences between groups (248).

Animal experiments conducted in the 1990s indicated that exercise training decreases the reactivity of vascular smooth muscle to vasoconstrictor agonists in the abdominal aorta (63, 64, 225). Over the past 10 yr, several interesting studies have indicated that exercise training also alters the responsiveness of arterioles and resistance arteries of rat skeletal muscle tissue. For example, Laughlin and colleagues (154) reported that sprint training decreased the constrictor responses of gastrocnemius feed arteries to phenylephrine but increased constrictor responses of white gastrocnemius arterioles. Donato and colleagues (68) also reported that training decreased the constrictor responses of gastrocnemius arterioles in old and young rats. In contrast to these observations in gastrocnemius arterioles, Jasperse and Laughlin (119) reported that endurance-exercise training did not alter the constrictor response of soleus feed arteries to phenylephrine. Similarly, Donato and colleagues (68) reported that exercise training did not alter constrictor responses of soleus arterioles of young rats but did decrease constrictor responses in soleus arterioles of old rats. They attributed this finding to changes in the endothelium, not smooth muscle (68). Thus, at this time it appears that the effects of training on smooth muscle adrenergic reactivity in skeletal muscle arterioles may be determined by the fiber-type composition of the muscle, the types of exercise training, and age of the animal.

Myogenic Response

As discussed in "Acute Responses of the Cardiovascular System to Dynamic Exercise," the myogenic response appears to be a fundamental property of vascular smooth muscle of arterioles throughout the body. The first report that exercise training may alter the myogenic response in resistance arteries was that of Muller and colleagues (176), who reported in the early 1990s that exercise training increased spontaneous tone and the myogenic responsiveness of coronary arterioles. Of interest, other vasoconstrictor responses of similar coronary arterioles were found to be uninfluenced by exercise training (149). Subsequently, Laughlin and colleagues (154) reported that sprint training increased spontaneous tone in arterioles isolated from white gastrocnemius but not arterioles isolated from red gastrocnemius in rats. Equally important, studies that have addressed the effect of inactivity induced by hind-limb unweighting on the myogenic response in rat skeletal muscle arterioles report heterogeneous adaptations depending on the composition of muscle fiber type surrounding the arteriole as well as a branch order effect. Decreased myogenic responsiveness was reported in second-order arterioles isolated from the white gastrocnemius muscle of rats after hind-limb unloading (61), whereas Heaps and Bowles (109) found no effect of hind-limb unloading in first-order arterioles of the same muscle. In the mixed gastrocnemius, however, Heaps and Bowles found both increased spontaneous tone and increased myogenic responsiveness after endurance training (109). At this time it appears that increased myogenic responsiveness is associated with exercise training in arterioles of muscle tissue active during training bouts.

Vascular Endothelial Cell Biology

The transformation in understanding the biology of vascular smooth muscle cells that occurred between 1909 and 2009 is astonishing; however, the transformation of understanding endothelial cell biology is even more striking. By 1908, *Gray's Anatomy* used the term *endothelium* to refer to the cells that line the blood vessels and that form the capillaries connecting the arteries and veins (141). However, the physiology of these cells and their role in vascular homeostasis remained largely unappreciated for much of the 20th century. In the early 1980s a major paradigm shift occurred in understanding the biology of endothelial cells with revelation of three processes in which endothelium play a central role. First came the observation that many of the metabolic functions attributed to the lung actually occurred in pulmonary endothelial cells (75). Also, largely from studies of endothelial cells in culture, investigators found that endothelial cells underwent phenotypic changes in response to inflammatory mediators involving expression of adhesion factors and other markers; this response was called endothelial activation (141). Third was the discovery that endothelium could be stimulated to release a number of dilators, constrictors, and molecules that can modulate vascular structure. The first report was that of Moncada and colleagues (173) in 1976, which demonstrated that microsomes of endothelium produce prostacyclin. This was followed by the Nobel Prize-wining discovery of Furchgott (figure 13.11) and Zawadzki in 1980 in which they demonstrated the existence of what came to be called endothelium-dependent relaxation to acetylcholine. They reported that acetylcholine induced relaxation of vascular smooth muscle of carefully isolated arteries and that this response was abolished by removal of the endothelium (91). This observation started a major shift in the understanding of the role of the vascular endothelium in health and disease (89, 90). Over the next 20 yr, intense investigation of processes involved in endothelium-dependent phenomena dramatically increased the appreciation of the importance of the endothelium to all phases of vascular function (2, 3, 203). It is now known that the vascular endothelium plays an important role in homeostasis, inflammation, lipid metabolism, vascular growth, cell migration, formation of (and interactions with) extracellular matrix molecules, and control of vascular permeability and vascular resistance (both vasodilator and vasoconstrictor responses). The endothelium is able to detect chemical substances in the blood as well as physical forces (i.e., shear stress and distention) imparted to blood vessel walls and initiate responses to these chemical or physical signals by releasing substances that modulate vascular tone or blood vessel structure (2, 3). The endothelium is the primary sensory tissue of the arterial wall and serves as a gatekeeper against circulating molecules.

The discovery that exercise training produces enhanced endothelium-dependent dilation represents a major advance in understanding the mechanisms for vascular adaptation stimulated by chronic exercise training (64, 216, 242). Any alteration in endothelial

Figure 13.11 Nobel laureate Robert Furchgott in his laboratory (date unknown).

Photo courtesy of Robert F. Furchgott Society.

function may be important because of the pivotal role the endothelium plays in normal cardiovascular function in health and because improved endothelial function holds the potential to provide significant improvement of cardiovascular function in disease (95, 239, 240).

In the past decade it has been demonstrated that exercise-induced adaptations in endothelium of skeletal muscle arteries are related to muscle fiber type and recruitment patterns so that the distribution of training-induced adaptations in the arteries that perfuse the gastrocnemius muscle were dramatically different after sprint training (154) and endurance training (166). These results are consistent with the proposition that exercise-induced vascular adaptations occur, spatially, in the area of muscle tissue with the greatest relative increase in fiber activity during exercise-training bouts. Also, blood flow in the previously discussed studies increased in the same areas that demonstrated vascular adaptations. Therefore, these data are consistent with the hypothesis that adaptations in the endothelium produced by exercise training result from changes in shear stress during exercise bouts (97, 100, 150).

Flow-induced dilation, a vascular-control mechanism that was intensively investigated in the late 1990s (discussed above), is now considered to be important as a mechanism for exercise hyperemia and, in conduit arteries, appears to be an important marker of vascular health (233). Flow-mediated dilation of resistance arterioles in skeletal muscle is increased by roughly 20% after exercise training (223-225, 228). Noteworthy is that there is substantial heterogeneity for training-induced changes in responsiveness to flow in and between muscles. For example, in the rat gastrocnemius, McAllister and colleagues observed that although second-order arterioles showed an approximate 25% increase in dilatory responses to flow after endurance training, first-order arterioles dilated less compared with the control group (166). Moreover, Jasperse and Laughlin (119) failed to find any significant difference in the responsiveness to flow in soleus muscle arterioles after endurance training.

As discussed in "Acute Responses of the Cardiovascular System to Dynamic Exercise," another endothelium-mediated process, conducted vasodilation, has been discovered during the past 50 yr. The dilatory response to muscle contraction is not uniform along the arteriolar tree; it occurs faster and to a greater extent in the small arterioles (66). From distal remote sites, it is believed that dilation ascends to proximal larger segments, predominantly thorough endothelial cell-to-cell communication (213). It is possible that this mechanism is fundamental for the coordination of the dilatory response and, given the velocity the signal travels, important for the early-onset dilatory response seen during contractions. This endothelium-mediated vascular-control mechanism does not appear to be altered by exercise training (21). Bearden and colleagues in 2007 showed that voluntary wheel running for 8 wk did not increase the magnitude of the conducted response in second-order arterioles of the gluteus maximus muscle in mice (21).

Endothelial Progenitor Cells

In the early portion of this century it was believed that vasculogenesis could occur only during development of the embryo and that postnatal formation of new arteries was the result of arteriolarization of capillaries. It was considered that only capillaries could sprout and form entirely new vessels, a process called angiogenesis, and that they could only form capillaries, not other vessel types. However, recent evidence indicates that circulating endothelial progenitor cells (EPCs) make it possible for vasculogenesis and angiogenesis to occur during postnatal life as well (197).

The idea that circulating cells may give rise to endothelial cells was proposed in the 1930s (197), but it is generally agreed that Asahara and colleagues in 1997 were the first to describe EPCs, which were isolated from human blood and have been shown to be derived of bone marrow stem and progenitor cells (12). Since this time an explosion of research has revealed that EPCs play an important role in maintaining a healthy endothelium and in healing the endothelium after damage and that insufficient EPCs may play an important role in a number of vascular diseases, including peripheral arterial disease and myocardial ischemia or infarction (171, 234). In 2004 it was demonstrated that a single bout of exercise can mobilize EPCs from bone marrow (1) and that exercise training for 3 to 4 wk resulted in an approximate doubling of EPCs circulating in the blood of both humans and mice (142). These results have been confirmed in healthy humans, patients on hemodialysis, and individuals with heart failure, peripheral vascular disease, and coronary artery disease (171). This new area of vascular cell biology promises to provide important new insight into mechanisms for exercise-induced vascular adaptation to physical activity.

Arteries Are Not All the Same and Do Not Respond to Exercise Training the Same

Before the late 1990s the endothelium was generally considered to consist of very similar endothelial cells throughout the vascular tree. However, in the late 1990s it started to be reported that fundamental differences exist among vasomotor properties of arteries of different tissues (160) and in and among arteries in the same tissue (7, 157). Over the past 15 yr results have demonstrated that the relative importance of vascular-control mechanisms is not constant along the length of the arterial network (i.e., vasomotor properties of arterioles are different among the branch orders, such as 1A versus 3A) (138, 145, 148). Also, it has become apparent that substances released from the endothelial cells differ among vessels in the vascular tree. For example, nitric oxide (NO) is the primary vasodilator substance released by the endothelium in larger arteries (72, 111, 161) whereas prostaglandin production also appears to be important in determination of vascular tone in smaller arteries and arterioles (129). Ando and colleagues (7) reported that (eNOS) expression and eNOS enzyme activity were higher in cultured aortic endothelial cells than in cultured microvascular endothelial cells, consistent with the concept of variable expression of eNOS in endothelium of different-size arteries. In addition, recent evidence indicated that exercise training increased eNOS protein content nonuniformly throughout the coronary arterial tree (151) and skeletal muscle vascular beds (154, 166).

The concept that vasomotor properties of vascular smooth muscle vary along the arterial tree has developed with a similar time course over the past 20 yr. For example, L-type calcium channel density is inversely related to arterial diameter in the coronary arterial tree (34). Also, adrenoreceptor-mediated vasoconstrictor mechanisms vary throughout the vascular tree in skeletal muscle (189). Adrenergic-induced constriction is mediated by both α_1 and α_2 receptors in larger resistance arteries but primarily through α_2 receptors in small resistance arteries (73, 74). Differences in the spatial distribution of adrenergic-receptor subtypes are important because the susceptibility of constriction to interference by metabolic (6, 145, 148), hormonal (145, 148), and endothelium-derived (145, 148, 182, 183) factors varies between the two alpha receptor subtypes. Also, it is believed that arterioles in small distal skeletal muscle may be more responsive to several vasoactive mediators than are proximal large segments, possibly due to nonuniform expression of receptors or ion channels throughout the arteriolar tree (112, 213). Similarly, it has been suggested that both the release of metabolites during muscle contraction and the arteriolar responsiveness to metabolites might be different depending on the fiber-type composition of skeletal muscle (164, 249).

In summary, present available information indicates that training-induced increases in skeletal muscle BFC are the result of combinations of structural and functional vascular adaptation. Exercise training increases BFC through increases in capillarity, growth and remodeling of arterioles, alterations in endothelial and smooth muscle phenotypes, and altered control of vascular resistance. Importantly, the relative contributions of each of these adaptations, and the spatial distribution of these adaptations in the musculature are not uniform throughout the arteriolar network of skeletal muscle. Depending on the mode of training, the muscle fiber recruitment patterns associated with it, and the duration and intensity of the training, these mechanisms can all be activated to differing degrees to accomplish vascular adaptation. Thus, the consensus supports the hypothesis that increases in BFC in skeletal muscle are the result of a mixture of increased arteriolar density, increased capillarity, and changes in control of resistance.

In the past century, advances in the understanding of skeletal muscle fiber recruitment patterns during exercise, muscle fiber type composition, and vascular adaptations have revealed an interaction among these factors and exercise intensity or duration, indicating that differing combinations of exercise intensity and duration stimulate graded adaptations throughout the arterial vasculature of skeletal muscle. In light of these results it seems possible that each type of skeletal muscle (fast red, fast white, slow) can adapt via graded changes in capillarity, arteriolar density, and vascular control processes (145). Clearly, more research is needed in this area because the biological strategies of the adaptations produced in response to altered activity in skeletal muscle of different phenotypes are only partially characterized at this time. It will be exciting to watch the further development of this field over the next 10 yr.

Exercise Training and Endothelial Function in Humans

The observation in animals that blood flow-induced dilation (FMD) is abolished after removal of the endothelial lining (190) led Celermajer and colleagues (45) in 1992 to develop a technique for assessing conduit artery flow-mediated vasodilation in humans. This landmark paper (45) introduced an approach for noninvasively assessing endothelium-dependent dilation in peripheral conduit arteries, a technique that rapidly became popular in the research community. In brief, this technique uses a transient (5 min) limb occlusion to induce reactive hyperemia. The relative increase in diameter from baseline, caused by augmented shear stress, is measured via high-resolution ultrasound and used as a marker of endothelial function (percentage FMD) (54). This noninvasive measure of conduit artery endothelial function is an extremely useful method that physiologists can use to evaluate the influence of exercise on vascular function in humans. The first study to apply the FMD technique to exercise training was conducted by Hornig and colleagues (117) in 1996 at the University of Freiburg. The authors evaluated radial artery FMD in patients with chronic heart failure at baseline, after 4 wk of daily handgrip training, and 6 wk after cessation of the training program. Handgrip exercise training—an adaptation that was abolished 6 wk after the intervention—locally restored FMD in patients with heart failure. Importantly, the portion of FMD that was attenuated by nitric oxide synthase blockade was significantly greater after exercise training, thus demonstrating that the improvement in FMD was mediated by an augmented ability to produce NO. In 1997, Katz and colleagues (124) performed a similar study investigating the effects of handgrip training on the function of skeletal muscle resistance vessel using intra-arterial infusions of acetylcholine before and after exercise training (also in congestive heart failure patients). Results indicated an approximate 90% increase in forearm vascular reactivity to acetylcholine, suggesting that exercise training-induced adaptations also occur in the skeletal muscle resistance vessels of humans. In 2000, for the first time, Hambrecht and colleagues (figure 13.12) eloquently evaluated the effect of exercise training on coronary endothelial function in patients with coronary artery disease (105). Nineteen patients were randomly assigned to an exercise-training group (n = 10) or a control group (n = 9). The exercise training consisted of six 10 min bouts/d of supervised exercise at 80% of peak heart rate for 4 wk. The control group continued their sedentary lifestyle. Before and after the 4 wk intervention, changes in arterial diameter and flow velocity in response to intracoronary infusion of increasing doses of acetylcholine were assessed. The exercise-training group exhibited a marked improvement in endothelium-dependent dilation of coronary conduit and resistance vessels, whereas coronary function in the control group was unaltered. Today, Hambrecht's study (105) continues to represent the best evidence that exercise training has a favorable effect on coronary endothelial function in the presence of atherosclerotic vascular disease.

Adaptations to Physical Inactivity

During the past century, extensive research has been devoted to enhancing our understanding of the effect of restricted or reduced natural physical activity. Fascinatingly, the motivation driving research of physical inactivity has evolved considerably over the years and, as a result, research designs and models have reflected these developments. For example, World War II appeared to stimulate an interest in bed rest research in order to reveal the potential adverse effects of inactivity and deconditioning in the treatment of hospitalized patients (87). At that time, bed rest was considered a standard therapeutic approach for many disorders. A symposium titled "Abuse of Rest in the Treatment of Disease" during the annual meeting of the American Medical Association (108) in 1944 set the foundation for the general practice of early ambulation for hospitalized patients (87); this was a revolutionary school of thought at that point in time. In 1965, Norman Browse, an English physician, wrote a book titled *The Physiology and Pathology of Bed Rest* that contained all the facts about physiology of bed rest known at that time. The principal purpose of the book was "to lay bare our ignorance of the whole subject (bed rest), to stimulate research and to emphasize the absurdity of using a nonspecific treatment for specific disease without reason or proven value" (41, p. v).

In recent years, another health hazard has emerged in Western societies: sedentary death syndrome (155). It is now accepted that physical inactivity in sedentary societies directly contributes to multiple chronic health disorders. In 2004, Frank Booth at the University of Missouri in the United States stated:

Figure 13.12 Rainer Hambrecht during his tenure as head of the departments of cardiology and angiology at Bremen Heart Center, Bremen, Germany.

Photo courtesy of Rainer Hambrecht.

> Genetic evolution has not changed as rapidly as social evolution and trends of physical inactivity over the last 40 to 100 yr. The human genome requires physical activity to lower the prevalence of many chronic diseases. (155, p. 451)

Novel inactivity research models in both animals and humans have recently been applied in order to further understand the underlying mechanism by which reduction in physical activity promotes chronic disease. This section briefly and chronologically highlights the most prominent advancements in our understanding of the deleterious consequences of physical inactivity that are pertinent to the cardiovascular system.

In 1948, Deitrick and colleagues (59) from Cornell University studied 4 normal, healthy men before, during, and after a 6 wk period of bed rest. Activity in bed was standardized by immobilization of the pelvic girdle and legs in plaster casts. The authors reported an average increase in heart rate of 3.8 beats/min after a 6 wk period of rest. A year later, Taylor and colleagues (230) from the University of Minnesota in the United States studied the effects of 3 to 4 wk of bed rest in healthy young men. Bed rest produced a 17% decrease in heart volume, an increase in resting heart rate that averaged 0.5 beat/min for each day of bed rest, an increase in heart rate of 40 beats/min during submaximal exercise, and a decrease in $\dot{V}O_2max$ of 17%. In 1968, the first data set from the aforementioned Dallas Bed Rest and Training Study in 5 young men became available (206) from the University of Texas Southwestern Medical Center in the United States. Saltin and colleagues (206) found that $\dot{V}O_2max$ decreased by 26%, which was similar to the 26% reduction in maximal cardiac output. The slight increase in maximal heart rate from 193 to 197 beats/min did not compensate for the decline in maximal stroke volume from 104 to 74 ml. The arteriovenous oxygen difference was unaltered, thus suggesting that compromised stroke volume was the main mechanism for bed rest-induced reduction in $\dot{V}O_2max$. Interestingly, this finding was in contrast to the finding that stroke volume and arteriovenous oxygen difference contributed equally to an increase in $\dot{V}O_2max$ associated with exercise training in healthy subjects (202). The detrimental effects of inactivity in the Dallas Bed Rest and Training Study recently became more apparent after data from the 30 yr follow-up visit were published (169). The most remarkable finding of the follow-up study was that 3 wk of bed rest in 1966 caused a greater deterioration in cardiovascular and physical work capacity than did 30 yr of aging in these 5 men (169).

A head-down hind-limb suspension model in rodents was developed by Morey at the National Aeronautics and Space Administration Ames Research Center in the late 1970s (174, 175). Many laboratories around the world used this model to simulate weightlessness and to study various aspects of the skeletal muscle. This model was later used by researchers to investigate the effect of inactivity on the cardiovascular system. Consistent with the observations made in humans with bed rest, exposure of rats to head-down hind-limb suspension elicited a decrease in aerobic capacity characterized by a reduction in maximal oxygen consumption and cardiac output (186, 247). The use of this model also became valuable when investigating the effects of inactivity on the vasculature. Data from the Health Activity Center at the University of Missouri indicated that chronic reduction in soleus blood flow associated with reduced physical activity during head-down hind-limb suspension decreased expression of eNOS and SOD-1 messenger ribonucleic acid and protein in soleus feed arteries and impaired endothelium-dependent vasodilation (120, 246).

With the development of the FMD technique in humans (45) (described previously) and the continued interest in the effect of inactivity on conduit artery function, a number of researchers began to study the influence of inactivity on endothelial function. Counterintuitively, studies of bed rest and unilateral lower-limb immobilization and comparisons between patients with spinal cord injury and able-bodied controls appeared to suggest that endothelium-dependent dilation is preserved in peripheral conduit arteries (58, 231). A plausible explanation for this conservation of vascular function may be related to a compensatory decrease in conduit artery dimensions (235). The inward remodeling associated with decreased blood flow may create a new equilibrium characterized by chronic increased shear stress in which vascular function is normalized. This decrease in conduit artery diameter and subsequent increase in shear stress was noted in a number of human studies evaluating the effect of inactivity (58, 231). In the microvasculature, however, the effect of inactivity on endothelial function appears to be more prominent. A number of bed rest studies, ranging from 7 to 56 d, have demonstrated an impairment of microvascular function (31, 65, 106, 110, 179, 219).

As described by Frank Booth and colleagues (33), the human genome was selected in an environment of high physical activity. Hunting and foraging for food and other necessities in the wild was the condition of human life for millions of years (13). Our urbanized, highly technological society has substituted much of the physical activity that selected optimal gene expression, and the consequences of such transformations are now manifesting through an explosion of modern chronic diseases (32). In the attempt to replicate the rapid shift of activity levels experienced by our society, a relatively recent rat model referred to as "wheel lock" was developed by Booth (33). Rats are given access to voluntary wheel running, which mimics the intermittent physical activity of our ancestors. The sudden lock of running wheels permits the study of the biological consequences of transitioning from a physically active life to a sedentary life. Rats retain the ambulatory cage activity after the wheel is locked and do not become fully immobile, as occurs in other models of extreme inactivity such as hind-limb suspension in animals and bed rest in humans (33). In the human equivalent of the wheel-lock rat model, participants are instructed to significantly reduce the number of daily steps (e.g., from 10,000 to 1,500) for a period of time (e.g., 2 wk) and biological measurements are performed before and after this decline in physical activity. This reduction in activity was recently shown to induce a 7% decline in maximal oxygen consumption (134). Although recent models of inactivity in both the rat and human have been used primarily in the context of metabolic dysfunction, the application of these approaches in the study of the cardiovascular system is extremely promising. We look forward to the upcoming developments in this area of research and hope that new discoveries will inspire society to improve its physical activity habits.

Overall Conclusions

The past century of cardiovascular research has produced more advancement in the basic knowledge of cardiovascular regulation during exercise than was achieved in all of human history before this period. In general, major advancements in the area of blood flow control and distribution were made in the first half of the 20th century, whereas biochemical and molecular advances (i.e., molecular mechanisms) were concentrated toward the second half of the century. There is no doubt that the electronic revolution that began in the last quarter of the 20th century has accelerated the pace of discovery. Much in the same way that the past 100 yr saw more accomplishment (with respect to cardiovascular physiology) than did the previous 10,000 yr, it is possible that the magnitude of discoveries that occurred between 1909 and 2009 will be surpassed in the next decade.

It is the view of the authors of this chapter that the next 100 yr of research in cardiovascular exercise physiology will witness a shift toward the integration of physical activity and inactivity research with the goal of preventing or reversing the societal burden of chronic disease. Human behavior and lifestyles have turned in a direction that is contrary to that which our biology evolved to fulfill. It is the responsibility of current and future exercise and cardiovascular physiologists to discover and educate society on how to strike a balance between technological progress and maintenance—or increased levels—of physical activity and dynamic exercise.

References

1. Adams V, Lenk K, Linke A, Lenz D, Erbs S, Sandri M, Tarnok A, Gielen S, Emmrich F, Schuler G, Hambrecht R. Increase of circulating endothelial progenitor cells in patients with coronary artery disease After exercise-induced ischemia. *Arterioscler Thromb Vasc Biol* 24: 684-690, 2004.
2. Aird WC. Phenotypic heterogeneity of the endothelium: I. Structure, function, and mechanisms. *Circ Res* 100: 158-173, 2007.
3. Aird WC. Phenotypic heterogeneity of the endothelium: II. Representative vascular beds. *Circ Res* 100: 174-190, 2007.
4. Andersen P, Henriksson J. Capillary supply of the quadriceps femoris muscle of man: Adaptive response to exercise. *J Physiol* 270: 677-690, 1977.
5. Andersen P, Saltin B. Maximal perfusion of skeletal muscle in man. *J Physiol* 366: 233-249, 1985.
6. Anderson KM, Faber JE. Differential sensitivity of arteriolar a_1- and a_2-adrenoceptor constriction to metabolic inhibition during rat skeletal muscle contraction. *Circ Res* 69: 174-184, 1991.
7. Ando H, Kubin T, Schaper W, Schaper J. Cardiac microvascular endothelial cells express alpha-smooth muscle actin and show low NOS III activity. *Am J Physiol* 276: H1755-H1768, 1999.
8. Anrep GV, Cerquia S, Samaan A. The effect of muscular contraction on the blood flow in skeletal muscle. *Proc Roy Soc London* 114: 223-244, 1934.
9. Anrep GV, von Saalfeld E. The blood flow through the skeletal muscle in relation to its contraction. *J Physiol* 85: 375-399, 1935.
10. Armstrong RB, Laughlin MH. Blood flows within and among rat muscles as a function of time during high speed treadmill exercise. *J Physiol* 344: 189-208, 1983.
11. Armstrong RB, Laughlin MH. Exercise blood flow patterns within and among rat muscles after training. *Am J Physiol* 246: H59-H68, 1984.
12. Asahara T, Murohara T, Sullivan A, Silver M, van der Zee R, Li T, Witzenbichler B, Schatteman G, Isner JM. Isolation of putative progenitor endothelial cells for angiogenesis. *Science* 275: 964-966, 1997.

13. Astrand PO, Rodahl K. *Textbook of Work Physiology*. New York: McGraw-Hill, 1970.
14. Barcroft H. Circulation in skeletal muscle. In: *Handbook of Physiology, Sec 2*. Bethesda, MD: American Physiological Society, 1963, 1353-1385.
15. Barcroft H. An enquiry into the nature of the mediator of the vasodilatation in skeletal muscle in exercise and during circulatory arrest. *J Physiol* 222: 99P-118P, 1972.
16. Barcroft H, Bonnar WM, Edholm OG, Effron AS. On sympathetic vasoconstrictor tone in human skeletal muscle. *J Physiol* 102: 21-31, 1943.
17. Barcroft H, Dornhorst AC. The blood flow through the human calf during rhythmic exercise. *J Physiol* 109: 402-411, 1949.
18. Barcroft H, Dornhorst AC. Demonstration of the muscle pump in the human leg. *J Physiol* 108: Proc 39, 1949.
19. Barcroft H, Millen JL. The blood flow through muscle during sustained contraction. *J Physiol* 97: 17-31, 1939.
20. Bayliss W. The action of carbon dioxide on blood vessels. *J Physiol* 26: 32P-33P, 1901.
21. Bearden SE, Linn E, Ashley BS, Looft-Wilson RC. Age-related changes in conducted vasodilation: Effects of exercise training and role in functional hyperemia. *Am J Physiol Regul Integr Comp Physiol* 293: R1717-R1721, 2007.
22. Berg BR, Cohen KD, Sarelius IH. Direct coupling between blood flow and metabolism at the capillary level in striated muscle. *Am J Physiol* 272: H2693-H2700, 1997.
23. Berg BR, Sarelius IH. Functional capillary organization in striated muscle. *Am J Physiol* 268: H1215-H1222, 1995.
24. Bergstrom J. Percutaneous needle biopsy of skeletal muscle in physiological and clinical research. *Scand J Clin Lab Invest* 35: 609-616, 1975.
25. Berne RM. Regulation of coronary blood flow. *Physiol Rev* 44: 1, 1964.
26. Bevegard BS, Shepherd JT. Reaction in man of resistance and capacity vessels in forearm and hand to leg exercise. *J Appl Physiol* 21: 123-132, 1966.
27. Binder KW, Murfee WL, Song J, Laughlin MH, Price RJ. Computational network model prediction of hemodynamic alterations due to arteriolar remodeling in interval sprint trained skeletal muscle. *Microcirculation* 14: 181-192, 2007.
28. Bishop JM, Donald KW, Taylor SH, Wormald PN. The blood flow in the human arm during supine leg exercise. *J Physiol (Lond)* 137: 294-308, 1957.
29. Black JE. Blood flow requirements of the human calf after walking and running. *Clin Sci (Lond)* 18: 89-93, 1959.
30. Blair DA, Glover WE, Roddie IC. Vasomotor responses in the human arm during leg exercise. *Circ Res* 9: 264-274, 1961.
31. Bleeker MW, De Groot PC, Rongen GA, Rittweger J, Felsenberg D, Smits P, Hopman MT. Vascular adaptation to deconditioning and the effect of an exercise countermeasure: Results of the Berlin Bed Rest study. *J Appl Physiol* 99: 1293-1300, 2005.
32. Booth FW, Gordon SE, Carlson CJ, Hamilton MT. Waging war on modern chronic diseases: Primary prevention through exercise biology. *J Appl Physiol* 88: 774-787, 2000.
33. Booth FW, Laye MJ, Lees SJ, Rector RS, Thyfault JP. Reduced physical activity and risk of chronic disease: The biology behind the consequences. *Eur J Appl Physiol* 102: 381-390, 2008.
34. Bowles DK, Hu Q, Laughlin MH, Sturek M. Heterogeneity of L-type calcium current density in coronary smooth muscle. *Am J Physiol* 273: H2083-H2089, 1997.
35. Bowles DK, Laughlin MH, Sturek M. Exercise training alters the Ca2+ and contractile responses of coronary arteries to endothelin. *J Appl Physiol* 78: 1079-1087, 1995.
36. Bowles DK, Laughlin MH, Sturek M. Exercise training increases K+-channel contribution to regulation of coronary arterial tone. *J Appl Physiol* 84: 1225-1233, 1998.
37. Bowles DK, Wamhoff BR. Coronary smooth muscle adaptation to exercise: Does it play a role in cardioprotection? *Acta Physiol Scand* 178: 117-121, 2003.
38. Bradley SE. Variations in hepatic blood flow in man during health and disease. *N Engl J Med* 240: 456-461, 1949.
39. Bradley SE, Childs AW, Combes B, Cournand A, Wade OL, Wheeler HO. The effect of exercise on the splanchnic blood flow and splanchnic blood volume in normal man. *Clin Sci (Lond)* 15: 457-463, 1956.
40. Brodal P, Ingjer F, Hermansen L. Capillary supply of skeletal muscle fibers in untrained and endurance-trained men. *Am J Physiol* 232: 705-712, 1977.
41. Browse NL. *The Physiology and Pathology of Bed Rest*. Springfield, IL: Charles C. Thomas, 1965.
42. Burke RE, Levine DN, Zajac FE III. Mammalian motor units: Physiological-histochemical correlation in three types in cat gastrocnemius. *Science* 174: 709-712, 1971.
43. Burton-Opitz R. Muscular contraction and the venous blood-flow. *Am J Physiol* 9: 161-185, 1903.
44. Carrow RE, Brown RE, Van Huss WD. Fiber sizes and capillary to fiber ratios in skeletal muscle of exercised rats. *Anat Rec* 159: 33-39, 1967.
45. Celermajer DS, Sorensen KE, Gooch VM, Spiegelhalter DJ, Miller OI, Sullivan ID, Lloyd JK, Deanfield JE. Non-invasive detection of endothelial dys-

function in children and adults at risk of atherosclerosis. *Lancet* 340: 1111-1115, 1992.

46. Christensen EH, Nielsen M. Investigations of the circulation in the skin at the beginning of muscular work. *Acta Physiol Scand* 4: 162-170, 1942.
47. Clark MG, Rattigan S, Barrett EJ, Vincent MA. Point: There is capillary recruitment in active skeletal muscle during exercise. *J Appl Physiol* 104: 889-891, 2008.
48. Clausen JP, Trap-Jensen J. Effects of training on the distribution of cardiac output in patients with coronary artery disease. *Circulation* 42: 611-624, 1970.
49. Clifford PS, Kluess HA, Hamann JJ, Buckwalter JB, Jasperse JL. Mechanical compression elicits vasodilatation in rat skeletal muscle feed arteries. *J Physiol* 572: 561-567, 2006.
50. Cobbold A, Folkow B, Kjellmer I, Mellander S. Nervous and local chemical control of pre-capillary sphincters in skeletal muscle as measured by changes in filtration coefficient. *Acta Physiol Scand* 57: 180-192, 1963.
51. Coggan AR, Spina RJ, King DS, Rogers MA, Brown M, Nemeth PM, Holloszy JO. Skeletal muscle adaptations to endurance training in 60- to 70-yr-old men and women. *J Appl Physiol* 72: 1780-1786, 1992.
52. Cooper KE, Edholm OG, Mottram RF. The blood flow in skin and muscle of the human forearm. *J Physiol* 128: 258-267, 1955.
53. Corcondilas A, Koroxenidis GT, Shepherd JT. Effect of a brief contraction of forearm muscles on forearm blood flow. *J Appl Physiol* 19: 142-146, 1964.
54. Corretti MC, Anderson TJ, Benjamin EJ, Celermajer D, Charbonneau F, Creager MA, Deanfield J, Drexler H, Gehard-Herman M, Herrington D, Vallance P, Vita J, Vogel R. Guidelines for the ultrasound assessment of endothelial-dependent flow-mediated vasodilation of the brachial artery. *J Am College Cardiol* 39: 257-265, 2002.
55. D'Silva JL, Fouche RF. The effects of changes in blood flow on the calibre of large arteries. *J Physiol Lond* 150: 23-24, 1960.
56. Davis MJ, Hill MA, Kuo L. Local regulation of microvascular perfusion. In: Tuma RF, Duran WN, Klaus L, eds. *Handbook of Physiology: Microcirculation*. San Diego: Academic Press, 2008, 161-284.
57. Dawes GS. The vaso-dilator action of potassium. *J Physiol* 99: 224-238, 1941.
58. de Groot PCE, Bleeker MWP, Hopman MTE. Magnitude and time course of arterial vascular adaptations to inactivity in humans. *Exerc Sport Sci Rev* 34: 65-71, 2006.
59. Deitrick JE, Whedon GD, Shorr E. Effects of immobilization upon various metabolic and physiologic functions of normal men. *Am J Med* 3: 1948.
60. Delashaw JB, Duling BR. A study of the functional elements regulating capillary perfusion in striated muscle. *Microvasc Res* 36: 162-171, 1988.
61. Delp MD. Myogenic and vasoconstrictor responsiveness of skeletal muscle arterioles is diminished by hindlimb unloading. *J Appl Physiol* 86: 1178-1184, 1999.
62. Delp MD, Laughlin MH. Regulation of skeletal muscle perfusion during exercise. *Acta Physiol Scand* 162: 411-419, 1998.
63. Delp MD, Laughlin MH. Time course of enhanced endothelium-mediated dilation in aorta of trained rats. *Med Sci Sports Exerc* 29: 1454-1461, 1997.
64. Delp MD, McAllister RM, Laughlin MH. Exercise training alters endothelium-dependent vasoreactivity of rat abdominal aorta. *J Appl Physiol* 75: 1354-1363, 1993.
65. Demiot C, Gnat-George F, Fortrat JO, Sabatier F, Gharib C, Larina I, Gauquelin-Koch G, Hughson R, Custaud MA. WISE 2005: Chronic bed rest impairs microcirculatory endothelium in women. *Am J Physiol Heart Circ Physiol* 293: H3159-H3164, 2007.
66. Dodd LR, Johnson PC. Diameter changes in arteriolar networks of contracting skeletal muscle. *Am J Physiol* 260: H662-H670, 1991.
67. Dolgin P, Lehmann G. Ein Beitrag zur Physiologie der statischen Arbeit. *Arbeitsphysiologie* 2: 248-252, 1930.
68. Donato AJ, Lesniewski LA, Delp MD. Ageing and exercise training alter adrenergic vasomotor responses of rat skeletal muscle arterioles. *J Physiol* 579: 115-125, 2007.
69. Dornhorst AC, Whelan RF. The blood flow in muscle following exercise and circulatory arrest; The influence of reduction in effective local blood pressure, of arterial hypoxia and of adrenaline. *Clin Sci (Lond)* 12: 33-40, 1953.
70. Duling BR, Berne RM. Propagated vasodilation in the microcirculation of the hamster cheek pouch. *Circ Res* 26: 163-170, 1970.
71. Duncker DJ, Bache RJ. Regulation of coronary blood flow during exercise. *Physiol Rev* 88: 1009-1086, 2008.
72. Ekelund U, Mellander S. Role of endothelium-derived nitric oxide in the regulation of tonus in large-bore arterial resistance vessels, arterioles and veins in cat skeletal muscle. *Acta Physiol Scand* 140: 301-309, 1990.
73. Faber JE. In situ analysis of alpha-adrenoceptors on arteriolar and venular smooth muscle in rat skeletal muscle microcirculation. *Circ Res* 62: 37-50, 1988.
74. Faber JE, Meininger GA. Selective interaction of alpha-adrenoceptors with myogenic regulation of microvascular smooth muscle. *Am J Physiol* 259: 1126-1133, 1990.

75. Fishman AP. Endothelium: A distributed organ of diverse capabilities. *Ann NY Acad Sci* 401: 1-8, 1982.
76. Fleisch A. Les reflexes nutritifs ascendants producteurs de dilatation arterielle. *Arch Int Physiol* 41: 1935.
77. Fleisch A, Sibul I. Über nutritive Kreislauf regulierung. II. Die Wirkung von pH, intermediären Stoffwechselprodukten und anderen biochemischen Verbindungen. *Pflügers Arch* 231: 787-804, 1933.
78. Folkow B. Description of the myogenic hypothesis. *Circ Res* 15(Suppl.): 279-287, 1964.
79. Folkow B. Nervous control of the blood vessels. *Physiol Rev* 35: 629-663, 1955.
80. Folkow B, Gaskell P, Waaler BA. Blood flow through limb muscles during heavy rhythmic exercise. *Acta Physiol Scand* 80: 61-72, 1970.
81. Folkow B, Gaskell P, Waaler BA. Blood flow through muscles during heavy rhythmic exercise. *Acta Physiol Scand* 76: 22A-23A, 1969.
82. Folkow B, Haglund U, Jodal M, Lundgren O. Blood flow in the calf muscle of man during heavy rhythmic exercise. *Acta Physiol Scand* 81: 157-163, 1971.
83. Folkow B, Halicka H. A comparison between "red" and "white" muscle with respect to blood supply, capillary surface area and oxygen uptake during rest and exercise. *Microvasc Res* 1: 1-14, 1968.
84. Folkow B, Sonnenschein RR, Wright DL. Loci of neurogenic and metabolic effects on precapillary vessels of skeletal muscle. *Acta Physiol Scand* 81: 459-471, 1971.
85. Forrester T, Lind AR. Adenosine triphosphate in the venous effluent and its relationship to exercise. *Fed Proc* 28: 1280-1283, 1969.
86. Forrester T, Lind AR. Identification of adenosine triphosphate in human plasma and the concentration in the venous effluent of forearm muscles before, during and after sustained contractions. *J Physiol* 204: 347-364, 1969.
87. Fortney SM, Schneider VS, Greenleaf JE. *Handbook of Physiology. Environmental Physiology. The Physiology of Bed Rest.* New York: Oxford University Press, 1996, 889-939.
88. Furchgott RF. Role of endothelium in responses of vascular smooth muscle. *Circ Res* 53: 557-573, 1983.
89. Furchgott RF, Carvalho MH, Khan MT, Matsunaga K. Evidence for endothelium-dependent vasodilation of resistance vessels by acetylcholine. *Blood Vessels* 24: 145-149, 1987.
90. Furchgott RF, Vanhoutte PM. Endothelium-derived relaxing and contracting factors. *FASEB J* 3: 2007-2018, 1989.
91. Furchgott RF, Zawadzki JV. The obligatory role of endothelial cells in the relaxation of arterial smooth muscle by acetylcholine. *Nature* 288: 373-376, 1980.
92. Gaskell WH. The changes of the blood-stream in muscles through stimulation of their nerves. *J Anat Physiol* 11: 360-363, 1877.
93. Gaskell WH. On the tonicity of the heart and blood vessels. *J Physiol* 3: 48-92, 1880.
94. Gavin TP, Ruster RS, Carrithers JA, Zwetsloot KA, Kraus RM, Evans CA, Knapp DJ, Drew JL, McCartney JS, Garry JP, Hickner RC. No difference in the skeletal muscle angiogenic response to aerobic exercise training between young and aged men. *J Physiol* 585: 231-239, 2007.
95. Gielen S, Hambrecht R. Effects of exercise training on vascular function and myocardial perfusion. *Cardiol Clin* 19: 357-368, 2001.
96. Giovacchini RP, Smith RD. The vascularity of some red and white muscles of the rabbit. *Acta Anat (Basel)* 28: 342-358, 1956.
97. Gokce N, Vita JA, Bader DS, Sherman DL, Hunter LM, Holbrook M, O'Malley C, Keaney JF, Balady GJ. Effect of exercise on upper and lower extremity endothelial function in patients with coronary artery disease. *Am J Cardiol* 90: 124-127, 2002.
98. Gorczynski RJ, Klitzman B, Duling BR. Interrelations between contracting striated muscle and precapillary microvessels. *Am J Physiol* 235: H494-H504, 1978.
99. Grant RT. Observations on the blood circulation in voluntary muscle in man *Clin Sci* 3: 157-173, 1938.
100. Green DJ, Maiorana A, O'Driscoll G, Taylor R. Effect of exercise training on endothelium-derived nitric oxide function in humans. *J Physiol (Lond)* 561: 1-25, 2004.
101. Grimby G, Haggendal E, Saltin B. Local xenon 133 clearance from the quadriceps muscle during exercise in man. *J Appl Physiol* 22: 305-310, 1967.
102. Gute D, Fraga C, Laughlin MH, Amann JF. Regional changes in capillary supply in skeletal muscle of high-intensity endurance-trained rats. *J Appl Physiol* 81: 619-626, 1996.
103. Gute D, Laughlin MH, Amann JF. Regional changes in capillary supply in skeletal muscle of interval-sprint and low intensity, endurance trained rats. *Microcirculation* 1: 183-193, 1994.
104. Haga JH, Li Y-SJ, Chien S. Molecular basis of the effects of mechanical stretch on vascular smooth muscle cells. *J Biomech* 40: 947-960, 2007.
105. Hambrecht R, Wolf A, Gielen S, Linke A, Hofer J, Erbs S, Schoene N, Schuler G. Effect of exercise on coronary endothelial function in patients with coronary artery disease. *N Engl J Med* 342: 454-460, 2000.
106. Hamburg NM, McMackin CJ, Huang AL, Shenouda SM, Widlansky ME, Schulz E, Gokce N, Ruderman NB, Keaney JF, Vita JA. Physical inactivity rapidly

induces insulin resistance and microvascular dysfunction in healthy volunteers. *Atheroscler Thromb Vasc Biol* 27: 2650-2656, 2007.

107. Hansen-Smith F, Egginton S, Hudlicka O. Growth of arterioles in chronically stimulated adult rat skeletal muscle. *Microcirculation* 5: 49-59, 1998.
108. Harrison TR. Abuse of rest as a therapeutic measure for patients with cardiovascular disease. *J Am Med Assoc* 125: 1075-1077, 1944.
109. Heaps CL, Bowles DK. Nonuniform changes in arteriolar myogenic tone within skeletal muscle following hindlimb unweighting. *J Appl Physiol* 92: 1145-1151, 2002.
110. Hesse C, Siedler H, Luntz SP, Arendt BM, Goerlich R, Fricker R, Heer M, Haefeli WE. Modulation of endothelial and smooth muscle function by bed rest and hypoenergetic, low-fat nutrition. *J Appl Physiol* 99: 2196-2203, 2005.
111. Hester RL, Eraslan A, Saito Y. Differences in EDNO contribution to arteriolar diameters at rest and during functional dilation in striated muscle. *Am J Physiol* 265: 146-151, 1993.
112. Hill CE, Phillips JK, Sandow SL. Heterogeneous control of blood flow amongst different vascular beds. *Med Res Rev* 21: 1-60, 2001.
113. Hilton SM. A peripheral arterial conducting mechanism underlying dilatation of the femoral artery and concerned in functional vasodilatation in skeletal muscle. *J Physiol* 149: 93-111, 1959.
114. Hilton SM. The search for the cause of functional hyperemia in skeletal muscle. In: Hudlicka O, ed. *Symposium on Circulation in Skeletal Muscle*. Oxford: Pergamon Press, 1967.
115. Hilton SM, Vrbova G. Inorganic phosphate—A new candidate for mediator of functional vasodilatation in skeletal muscle. *J Physiol* 206: 29P-30P, 1970.
116. Hojensgard IC, Sturup H. Static and dynamic pressures in superficial and deep veins of the lower extremity in man. *Acta Physiol Scand* 27: 49-67, 1952.
117. Hornig B, Maier V, Drexler H. Physical training improves endothelial function in patients with chronic heart failure. *Circulation* 93: 210-214, 1996.
118. Hudlicka O, Fronek K. Effect of long-term electrical stimulation of rabbit fast muscles on the reactivity of their supplying arteries. *J Vasc Res* 29: 13-19, 1992.
119. Jasperse JL, Laughlin MH. Vasomotor responses of soleus feed arteries from sedentary and exercise trained rats. *J Appl Physiol* 86: 441-449, 1999.
120. Jasperse JL, Woodman CR, Price EM, Hasser EM, Laughlin MH. Hindlimb unweighting decreases ecNOS gene expression and endothelium-dependent dilation in rat soleus feed arteries. *J Appl Physiol* 87: 1476-1482, 1999.
121. Johnson JM, Rowell LB. Forearm skin and muscle vascular responses to prolonged leg exercise in man. *J Appl Physiol* 39: 920-924, 1975.
122. Joyner MJ, Wilkins BW. Exercise hyperaemia: Is anything obligatory but the hyperaemia? *J Physiol* 583: 855-860, 2007.
123. Kamon E, Belding HS. Dermal blood flow in the resting arm during prolonged leg exercise. *J Appl Physiol* 26: 317-320, 1969.
124. Katz SD, Yuen J, Bijou R, Lejemtel TH. Training improves endothelium-dependent vasodilation in resistance vessels of patients with heart failure. *J Appl Physiol* 82: 1488-1492, 1997.
125. Kjellmer I. The effect of exercise on the vascular bed of skeletal muscle. *Acta Physiol Scand* 62: 18-30, 1964.
126. Kjellmer I. On the competition between metabolic vasodilatation and neurogenic vasoconstriction in skeletal muscle. *Acta Physiol Scand* 63: 450-459, 1965.
127. Kjellmer I. The role of potassium ions in exercise hyperaemia. *Med Exp Int J Exp Med* 5: 56-60, 1961.
128. Klitzman B, Damon DN, Gorczynski RJ, Duling BR. Augmented tissue oxygen supply during striated muscle contraction in the hamster. Relative contributions of capillary recruitment, functional dilation, and reduced tissue PO2. *Circ Res* 51: 711-721, 1982.
129. Koller A, Huang A, Sun D, Kaley G. Exercise training augments flow-dependent dilation in rat skeletal muscle arterioles. Role of endothelial nitric oxide and prostaglandins. *Circ Res* 76: 544-550, 1995.
130. Koller A, Kaley G. Endothelium regulates skeletal muscle microcirculation by a blood flow velocity-sensing mechanism. *Am J Physiol* 258: H916-H920, 1990.
131. Kramer K, Obal F, Quensel W. Unter-suchungen über den Muskelstoffwechsel des Warmblüters. III. Mitteilung. Die Sauerstoffaufnahme des Muskels während rhythmischer Tätigkeit. *Pflüg Arch Ges Physiol* 241: 717, 1939.
132. Kramer K, Quensel W. Untersuchungen uber den Muskelstoffwechsel des Warmbluters. I. Mitteilung. Der Verlauf der Muskeldurchblutung wahrend der tetanischen Kontraktion. *Pflug Arch Ges Physiol* 239: 621, 1937.
133. Kramer K, Quensel W, Schäfer KE. Untersuchungen über den Muskelstoffwechsel des Warmblüters. IV. Mitteilung. Beziehungen zwischen Sauerstoffaufnahme und Milchsäureabgabe des Muskels während der Tätigkeit. *Pflüg Arch Ges Physiol* 241: 730, 1939.
134. Krogh-Madsen R, Thyfault JP, Broholm C, Mortensen OH, Olsen RH, Mounier R, Plomgaard P, Hall GV, Booth FW, Pedersen BK. A 2-wk reduction of ambulatory activity attenuates peripheral insulin sensitivity. *J Appl Physiol* 108: 1034-1040, 2010.

135. Krogh A. Reminiscences of work on capillary circulation; A lecture to the students in the Harvard Medical School, 1946. *Isis* 41: 14-20, 1950.
136. Krogh A. Studies on the capillariometer mechanism: I. The reaction to stimuli and the innervation of the blood vessels in the tongue of the frog. *J Physiol* 53: 399-419, 1920.
137. Krogh A. The supply of oxygen to the tissues and the regulation of the capillary circulation. *J Physiol* 52: 457-474, 1919.
138. Kuo L, Davis MJ, Chilian WM. Longitudinal gradients for endothelium-dependent and -independent vascular responses in the coronary microcirculation. *Circulation* 92: 518-525, 1995.
139. Lash JM. Exercise training enhances adrenergic constriction and dilation in the rat spinotrapezius muscle. *J Appl Physiol* 85: 168-174, 1998.
140. Lash JM, Bohlen HG. Functional adaptations of rat skeletal muscle arterioles to aerobic exercise training. *J Appl Physiol* 72: 2052-2062, 1992.
141. Laubichler MD, Aird WC, Maienschein J. The endothelium in history. In: Aird WCMD, ed. *Endothelial Biomedicine*. New York: Cambridge University Press, 2007, 5-19.
142. Laufs U, Werner N, Link A, Endres M, Wassmann S, Jurgens K, Miche E, Bohm M, Nickenig G. Physical training increases endothelial progenitor cells, inhibits neointima formation, and enhances angiogenesis. *Circulation* 109: 220-226, 2004.
143. Laughlin MH, Armstrong RB. Muscle blood flow during locomotory exercise. *Exerc Sport Sci Rev* 13: 95-136, 1985.
144. Laughlin MH, Cook J, Tremble R, Ingram D, Colleran PN, Turk JR. Exercise training produces nonuniform increases in arteriolar density of rat soleus and gastrocnemius muscle. *Microcirculation* 13: 175-186, 2006.
145. Laughlin MH, Korthuis RJ, Duncker DJ, Bache RJ. *Control of Blood Flow to Cardiac and Skeletal Muscle During Exercise*. New York: Oxford University Press, 1996, 705-769.
146. Laughlin MH, Korthuis RJ, Sexton WL, Armstrong RB. Regional muscle blood flow capacity and exercise hyperemia in high intensity trained rats. *J Appl Physiol* 64: 2420-2427, 1988.
147. Laughlin MH, Korzick DH. Vascular smooth muscle: Integrator of vasoactive signals during exercise hyperemia. *Med Sci Sports Exerc* 33: 81-91, 2001.
148. Laughlin MH, McAllister RM, Delp MD. Heterogeneity of blood flow in striated muscle. In: *The Lung*. Philadelphia: Raven, 1997, 1945-1955.
149. Laughlin MH, Muller JM. Vasoconstrictor responses of coronary resistance arteries in exercise-trained pigs. *J Appl Physiol* 84: 884-889, 1998.
150. Laughlin MH, Newcomer SC, Bender SB. Importance of hemodynamic forces as signals for exercise-induced changes in endothelial cell phenotype. *J Appl Physiol* 104: 588-600, 2008.
151. Laughlin MH, Pollock JS, Amann JF, Hollis ML, Woodman CR, Price EM. Training induces nonuniform increases in eNOS content along the coronary arterial tree. *J Appl Physiol* 90: 501-510, 2001.
152. Laughlin MH, Roseguini B. Mechanisms for exercise training-induced increases in skeletal muscle blood flow capacity: Differences with interval sprint training versus aerobic endurance training. *J Physiol Pharmacol* 59: 71-88, 2008.
153. Laughlin MH, Schrage WG. Effects of muscle contraction on skeletal muscle blood flow: When is there a muscle pump? *Med Sci Sports Exerc* 31: 1027-1035, 1999.
154. Laughlin MH, Woodman CR, Schrage WG, Gute D, Price EM. Interval sprint training enhances endothelial function and eNOS content in some arteries that perfuse white gastrocnemius muscle. *J Appl Physiol* 96: 233-244, 2004.
155. Lees SJ, Booth FW. Sedentary death syndrome. *Can J Appl Physiol* 29: 447-460, 2004.
156. Lewis T, Grant RT. Observations upon reactive hyperaemia in man. *Heart* 12: 73-120, 1925.
157. Li S, Fan YS, Chow LH, Van Den Diepstraten C, van Der Veer E, Sims SM, Pickering JG. Innate diversity of adult human arterial smooth muscle cells: Cloning of distinct subtypes from the internal thoracic artery. *Circ Res* 89: 517-525, 2001.
158. Lind AR, McNichol GW. Local and central circulatory responses to sustained contractions and the effect of free and restricted arterial inflow on post-exercise hyperaemia. *J Physiol* 192: 575-593, 1967.
159. Lindhard J. Untersuchungen uber statischer Muskelarbeit I. *Skand Arch Physiol* 40: 145-195, 1920.
160. Lindner V, Maciag T. The putative convergent and divergent natures of angiogenesis and arteriogenesis. *Circ Res* 89: 747-749, 2001.
161. Luscher TF, Vanhoutte PM. *The Endothelium: Modulator of Cardiovascular Function*. Boca Raton, FL: CRC Press, 1990, 1-162.
162. Mackie BG, Terjung RL. Blood flow to different skeletal muscle fiber types during contraction. *Am J Physiol* 245: H265-H275, 1983.
163. Mai JV, Edgerton VR, Barnard RJ. Capillarity of red, white and intermediate muscle fibers in trained and untrained guinea pigs. *Experientia* 26: 1222-1223, 1970.
164. Marshall JM. The roles of adenosine and related substances in exercise hyperaemia. *J Physiol* 583: 835-845, 2007.
165. Martin WH III, Coggan AR, Spina RJ, Saffitz JE. Effects of fiber type and training on beta-adreno-

ceptor density in human skeletal muscle. *Am J Physiol* 257: E736-E742, 1989.

166. McAllister RM, Jasperse JL, Laughlin MH. Nonuniform effects of endurance exercise training on vasodilation in rat skeletal muscle. *J Appl Physiol* 98: 753-761, 2005.
167. McCurdy JH, ed. *The Physiology of Exercise*. Philadelphia: Lea & Febiger, 1924.
168. McCurdy MR, Colleran PN, Muller-Delp J, Delp MD. Effects of fiber composition and hindlimb unloading on the vasodilator properties of skeletal muscle arterioles. *J Appl Physiol* 89: 398-405, 2000.
169. McGuire DK, Levine BD, Williamson JW, Snell PG, Blomqvist CG, Saltin B, Mitchell JH. A 30-year follow-up of the Dallas Bed Rest and Training Study. *Circulation* 104: 1350-1357, 2001.
170. Mellander S, Johansson B, Gray S, Jonsson O, Lundvall J, Ljung B. The effect of hyperosmolarity on intacet and isolated vascular smooth muscle. Possible role in exercise hyperemia. *Angiologica* 4: 310-322, 1967.
171. Möbius-Winkler S, Höllriegel R, Schuler G, Adams V. Endothelial progenitor cells: Implications for cardiovascular disease. *Cytometry Part A* 75A: 25-37, 2009.
172. Mohrman DE, Sparks HV. Myogenic hyperemia following brief tetanus of canine skeletal muscle. *Am J Physiol* 227: 531-535, 1974.
173. Moncada S, Gryglewsik R, Bunting S, Vane JR. An enzyme isolated from arteries transforms prostaglandin endoperoxides to an unstable substance that inhibits platelet aggregation. *Nature* 263: 663-665, 1976.
174. Morey-Holton ER, Globus RK. Hindlimb unloading rodent model: Technical aspects. *J Appl Physiol* 92: 1367-1377, 2002.
175. Morey ER. Space flight and bone turnover: Correlation with a new rat model of weightlessness. *Bioscience* 29: 168-172, 1979.
176. Muller JM, Myers PR, Laughlin MH. Exercise training alters myogenic responses in porcine coronary resistance arteries. *J Appl Physiol* 75: 2677-2682, 1993.
177. Musch TI, Haidet GC, Ordway GA, Longhurst JC, Mitchell JH. Training effects on regional blood flow response to maximal exercise in foxhounds. *J Appl Physiol* 62: 1724-1732, 1987.
178. Musch TI, Poole DC. Blood flow response to treadmill running in the rat spinotrapezius muscle. *Am J Physiol* 271: H2730-H2734, 1996.
179. Navasiolava NM, Gnat-George F, Sabatier F, Larina IM, Demiot C, Fortrat JO, Gauquelin-Koch G, Kozlovskaya IB, Custaud MA. Enforced physical inactivity increases endothelial microparticle levels in healthy volunteers. *Am J Physiol Heart Circ Physiol* 299: H248-H256, 2010.
180. Nishiyama A. Histochemical studies on the red, white and intermediate muscle fibers of some skeletal muscles. II. The capillary distribution on three types of fibers of some skeletal muscles. *Acta Med Okayama* 19: 191-198, 1965.
181. O'Sullivan SE, Bell C. Training reduces autonomic cardiovascular responses to both exercise-dependent and -independent stimuli in humans. *Auton Neurosci* 91: 76-84, 2001.
182. Ohyanagi M, Faber JE, Nishigaki K. Differential activation of alpha 1- and alpha 2-adrenoceptors on microvascular smooth muscle during sympathetic nerve stimulation. *Circ Res* 68: 232-244, 1991.
183. Ohyanagi M, Nishigaki K, Faber JE. Interaction between microvascular a_1- and a_2-adrenoceptors and endothelium-derived relaxing factor. *Circ Res* 71: 188-200, 1992.
184. Oltman CL, Parker JL, Adams HR, Laughlin MH. Effects of exercise training on vasomotor reactivity of porcine coronary arteries. *Am J Physiol* 263: 372-382, 1992.
185. Ooue A, Ichinose T, Inoue Y, Nishiyasu T, Koga S, Kondo N. Changes in blood flow in conduit artery and veins of the upper arm during leg exercise in humans. *Eur J Appl Physiol* 103: 367-373, 2008.
186. Overton JM, Woodman CR, Tipton CM. Effect of hindlimb suspension on VO_2max and regional blood flow responses to exercise. *J Appl Physiol* 66: 653-659, 1989.
187. Peter JB, Barnard RJ, Edgerton VR, Gillespie CA, Stempel KE. Metabolic profiles of three fiber types of skeletal muscle in guinea pigs and rabbits. *Biochemistry* 11: 2627-2633, 1972.
188. Petren T, Sjöstrand T, Sylven B. Der Einfluss des Trainings auf die Häufigkeit der Capillaren in Herz- und Skeletmuskulatur. *Arbeitsphysiol* 9: 376-386, 1936.
189. Ping P, Faber JE. Characterization of alpha-adrenoceptor gene expression in arterial and venous smooth muscle. *Am J Physiol* 265: H1501-H1509, 1993.
190. Pohl U, Holtz J, Busse R, Bassenge E. Crucial role of endothelium in the vasodilator response to increased flow in vivo. *Hypertension* 8: 37-44, 1986.
191. Pollack AA, Wood EH. Venous pressure in the saphenous vein at the ankle in man during exercise and changes in posture. *J Appl Physiol* 1: 649-662, 1949.
192. Poole DC, Behnke BJ, Padilla DJ. Dynamics of muscle microcirculatory oxygen exchange. *Med Sci Sports Exerc* 37: 1559-1566, 2005.
193. Poole DC, Brown MD, Hudlicka O. Counterpoint: There is not capillary recruitment in active skeletal muscle during exercise. *J Appl Physiol* 104: 891-893, 2008.

194. Rein H, Kreller J, Loeser A, Rein E, Schneiden M. Die Interferenz der Vasomotorischen Regulationen. *Klinische Wochenschrift* 9: 1485-1489, 1930.

195. Reis DJ, Wooten GF, Hollenberg M. Differences in nutrient blood flow of red and white skeletal muscle in the cat. *Am J Physiol* 213: 592-596, 1967.

196. Remensnyder JP, Mitchell JH, Sarnoff SJ. Functional sympatholysis during muscular activity. Observations on influence of carotid sinus on oxygen uptake. *Circ Res* 11: 370-380, 1962.

197. Ribatti D. The discovery of endothelial progenitor cells: An historical review. *Leuk Res* 31: 439-444, 2007.

198. Roca J, Agust AGN, Alonso A, Poole DC, Viegas C, Barbera JA, Rodriguez-Roisin R, Ferrer A, Wagner PD. Effects of training on muscle O2 transport at VO2max. *J Appl Physiol* 73: 1067-1076, 1992.

199. Romanul FC. Capillary supply and metabolism of muscle fibers. *Arch Neurol* 12: 497-509, 1965.

200. Rossiter HB, Howlett RA, Holcombe HH, Entin PL, Wagner HE, Wagner PD. Age is no barrier to muscle structural, biochemical and angiogenic adaptations to training up to 24 months in female rats. *J Physiol* 565: 993-1005, 2005.

201. Rowell LB. *Human Cardiovascular Control.* Oxford: Oxford University Press, 1993, 500.

202. Rowell LB. *Human Circulation: Regulation During Physical Stress.* New York: Oxford University Press, 1986, 227-286.

203. Rubanyi GM. *Cardiovascular Significance of Endothelium-Derived Vasoactive Factors.* Mount Kisco, NY: Futura, 1991.

204. Sadler W. *Arbeiten aus dem Physiologischen.* Leipsig: Institute zu Leipsig, 77-100.

205. Saltin B. Exercise hyperaemia: Magnitude and aspects on regulation in humans. *J Physiol* 583: 819-823, 2007.

206. Saltin B, Blomqvist G, Mitchell JH, Johnson RL, Wildenthal K, Chapman CB. Response to exercise after bed rest and after training: A longitudinal study of adaptive changes in oxygen transport and body composition. *Circulation* 37/38(Suppl. VII): VII1-VII78, 1968.

207. Saltin B, Gollnick PD. Skeletal muscle adaptability: Significance for metabolism and performance. *Handbook of Physiology: Skeletal Muscle* 10: 555-631, 1977.

208. Saltin B, Radegran G, Koskolou MD, Roach RC. Skeletal muscle blood flow in humans and its regulation during exercise. *Acta Physiol Scand* 162: 421-436, 1998.

209. Sarelius IH. Cell flow path influences transit time through striated muscle capillaries. *Am J Physiol* 250: H899-H907, 1986.

210. Sarelius IH, Damon DN, Duling BR. Microvascular adaptations during maturation of striated muscle. *Am J Physiol* 241: H317-H324, 1981.

211. Schhetzenmayr A. Ober kreislaufregulatorische Vorgange an den grossen Arterien bei der Muskelarbeit. *Pfliiger Arch Ges Physiol* 232: 743, 1933.

212. Secher NH, Clausen JP, Klausen K, Noer I, Trap-Jensen J. Central and regional circulatory effects of adding arm exercise to leg exercise. *Acta Physiol Scand* 100: 288-297, 1977.

213. Segal SS. Regulation of blood flow in the microcirculation. *Microcirculation* 12: 33-45, 2005.

214. Segal SS, Duling BR. Communication between feed arteries and microvessels in hamster striated muscle: Segmental vascular responses are functionally coordinated. *Circ Res* 59: 283-290, 1986.

215. Segal SS, Duling BR. Flow control among microvessels coordinated by intercellular conduction. *Science* 234: 868-870, 1986.

216. Sessa WC, Pritchard K, Seyedi N, Wang J, Hintze TH. Chronic exercise in dogs increases coronary vascular nitric oxide production and endothelial cell nitric oxide synthase gene expression. *Circ Res* 74: 349-353, 1994.

217. Sexton WL, Korthuis RJ, Laughlin MH. High intensity exercise training increases vascular transport capacity of rat hindquarters. *Am J Physiol Heart Circul Physiol* 254: H274-H278, 1988.

218. Sheriff DD, Van Bibber R. Flow-generating capability of the isolated skeletal muscle pump. *Am J Physiol* 274: H1502-H1508, 1998.

219. Shoemaker JK, Hogeman CS, Silber DH, Gray K, Herr M, Sinoway LI. Head-down-tilt bed rest alters forearm vasodilator and vasoconstrictor responses. *J Appl Physiol* 84: 1998.

220. Smiesko V, Kozik J, Dolezel S. The control of arterial diameter by blood flow velocity is dependent upon intact endothelium. *Physiol Bohemoslov* 32: 558, 1983.

221. Smith ML, Graitzer HM, Hudson DL, Raven PB. Baroreflex function in endurance- and static exercise-trained men. *J Appl Physiol* 64: 585-591, 1988.

222. Sparks HV, Hull SS. Role of the endothelium in blood flow induced arterial vasodilation. *Int J Microcirc* 3: 313, 1984.

223. Spier SA, Delp MD, Meininger CJ, Donato AJ, Ramsey MW, Muller-Delp JM. Effects of ageing and exercise training on endothelium-dependent vasodilatation and structure of rat skeletal muscle arterioles. *J Physiol* 556: 947-958, 2004.

224. Spier SA, Delp MD, Stallone JN, Dominguez JM II, Muller-Delp JM. Exercise training enhances flow-induced vasodilation in skeletal muscle resistance arteries of aged rats: Role of PGI2 and nitric oxide. *Am J Physiol Heart Circ Physiol* 292: H3119-H3127, 2007.

225. Spier SA, Laughlin MH, Delp MD. Effects of acute and chronic exercise on vasoconstrictor responsiveness of rat abdominal aorta. *J Appl Physiol* 87: 1752-1757, 1999.
226. Stegall HF. Muscle pumping in the dependent leg. *Circ Res* 19: 180-190, 1966.
227. Stewart GN. Studies on the circulation in man. I. The measurement of the blood flow in the hands. *Heart* 3: 33-75, 1911.
228. Sun D, Huang A, Koller A, Kaley G. Adaptation of flow-induced dilation of arterioles to daily exercise. *Microvasc Res* 56: 54-61, 1998.
229. Sun D, Huang A, Koller A, Kaley G. Short-term daily exercise activity enhances endothelial NO synthesis in skeletal muscle arterioles of rats. *J Appl Physiol* 76: 2241-2247, 1994.
230. Taylor HL, Henschel A, Brozek J, Keys A. Effects of bed rest on cardiovascular function and work performance. *J Appl Physiol* II: 223-239, 1949.
231. Thijssen DHJ, Maiorana AJ, O'Driscoll G, Cable NT, Hopman MTE, Green DJ. Impact of inactivity and exercise on the vasculature in humans. *Eur J Appl Physiol* 108: 845-875, 2010.
232. Tonnesen KH. Blood-flow through muscle during rhythmic contraction measured by 133-xenon. *Scand J Clin Lab Invest* 16: 646-654, 1964.
233. Tschakovsky ME, Joyner MJ. Nitric oxide and muscle blood flow in exercise. *Appl Physiol Nutr Metab* 33: 151-161, 2008.
234. Umemura T, Higashi Y. Endothelial progenitor cells: Therapeutic target for cardiovascular diseases. *J Pharmacol Sci* 108: 1-6, 2008.
235. van Duijnhoven NTL, Green DJ, Felsenberg D, Belavy DL, Hopman MTE, Thijssen DHJ. Impact of bed rest on conduit artery remodeling. Effect of exercise countermeasures. *Hypertension* 56: 2010.
236. Vanotti A, Magiday M. Untersuchungen zum Studium Muskulatur.Uber die Capillarisierung der trainierten. *Arbeitsphysiologie* 7: 615-622, 1934.
237. VanTeeffelen JW, Segal SS. Interaction between sympathetic nerve activation and muscle fibre contraction in resistance vessels of hamster retractor muscle. *J Physiol* 550: 563-574, 2003.
238. Varnauskas E, Bjorntorp P, Fahlen M, Prerovsky I, Stenberg J. Effects of physical training on exercise blood flow and enzymatic activity in skeletal muscle. *Cardiovasc Res* 4: 418-422, 1970.
239. Vita JA, Keaney JF. Exercise—Toning up the endothelium? *N Engl J Med* 342: 503-505, 2000.
240. Vita JA, Loscalzo J. Shouldering the risk factor burden: Infection, atherosclerosis, and the vascular endothelium. *Circulation* 106: 164-166, 2002.
241. Wamhoff BR, Bowles DK, Dietz NJ, Hu Q, Sturek M. Exercise training attenuates coronary smooth muscle phenotypic modulation and nuclear Ca2+ signaling. *Am J Physiol Heart Circ Physiol* 283: H2397-H2410, 2002.
242. Wang J, Wolin MS, Hintze TH. Chronic exercise enhances endothelium-mediated dilation of epicardial coronary artery in conscious dogs. *Circ Res* 73: 829-838, 1993.
243. Wenger CB, Roberts MF, Stolwijk JA, Nadel ER. Forearm blood flow during body temperature transients produced by leg exercise. *J Appl Physiol* 38: 58-63, 1975.
244. Wesche J. The time course and magnitude of blood flow changes in the human quadriceps muscles following iscometric contraction. *J Physiol* 377: 445-462, 1986.
245. Wiegman DI, Harris PD, Joshua IG, Miller FN. Decreased vascular sensitivity to norepinephrine following exercise. *J Appl Physiol* 52: 282-287, 1981.
246. Woodman CR, Schrage WG, Rush JWE, Ray CA, Price EM, Hasser EM, Laughlin MH. Hindlimb unweighting decreases endothelium-dependent dilation and eNOS expression in soleus not gastrocnemius. *J Appl Physiol* 91: 1091-1098, 2001.
247. Woodman CR, Sebastian LA, Tipton CM. Influence of simulated microgravity on cardiac output and blood flow distribution during exercise. *J Appl Physiol* 79: 1762-1768, 1995.
248. Wray DW, Donato AJ, Nishiyama SK, Richardson RS. Acute sympathetic vasoconstriction at rest and during dynamic exercise in cyclists and sedentary humans. *J Appl Physiol* 102: 704-712, 2007.
249. Wunsch SA, Muller-Delp J, Delp MD. Time course of vasodilatory responses in skeletal muscle arterioles: Role in hyperemia at onset of exercise. *Am J Physiol Heart Circ Physiol* 279: H1715-H1723, 2000.
250. Zelis R, Mason DT, Braunwald E. Partition of blood flow to the cutaneous and muscular beds of the forearm at rest and during leg exercise in normal subjects and in patients with heart failure. *Circ Res* 24: 799-806, 1969.

CHAPTER 14

The Muscular System: Muscle Plasticity

Kenneth M. Baldwin, PhD

Fadia Haddad, PhD

Introduction

Skeletal muscles collectively compose the organ system of movement. A pivotal property of this system is its plasticity, which can be defined as the muscle's ability to change its mass, metabolic properties, and contractile phenotype in response to alterations in physical activity, inactivity, and certain hormonal stimuli. This field has matured as a critical research topic in the past century, especially in the past 40 to 50 yr. Before this time, the science of muscle structure and function evolved through a long history of discovery, initially dating back to the late 1600s. It laid dormant for approximately 300 yr and then suddenly blossomed as new biochemical and molecular biology techniques and technologies came to the forefront.

This chapter traces the evolution of scientific inquiry on skeletal muscle plasticity during the past century (table 14.1) along three central, integrated topics: the fundamental properties of skeletal muscle centered on the structural and functional features of the motor unit, which provides the gradation in movement and force-generation capacity of living organisms; differential properties of fast versus slow skeletal muscle motor units and their adaptive capacity in response to acute and chronic endurance exercise; and the evolution of studies concerning alterations in loading state in the regulation of muscle mass as well as the features of both the contractile and metabolic phenotypes in contributing to the plasticity of skeletal muscle.

Key Discoveries From 1910 to 1950: The Origin of Motor Units and Intrinsic Contractile Properties of Skeletal Muscle

Motor Unit

Sir Charles Sherrington (1857-1952) in 1925 established the concept of the motor unit as a colony of fibers linked to a single motor nerve (120). He considered the motor unit to be the functional entity of movement and believed that all activities along with the forces generated by the muscles must be graded to the number of motor units being activated by the nervous system (36, 134). During this same period, John Eccles (1903-1997), a medical student (and future Nobel laureate) from Melbourne, Australia, joined Sherrington's laboratory and discovered that two distinct types of motor neurons—one large and one small—innervated skeletal muscle (59). Later studies reported that the large neurons innervated the myofibers whereas the small neurons innervated the muscle spindles (107, 118). Importantly, this work triggered intense research on motor unit structure, contractile, and biochemical properties, which continue to dominate the field of exercise physiology.

Fast- and Slow-Type Muscle: Connecting the Muscle Fiber to Its Motor Neuron

In 1929 Derek Denny-Brown (1901-1981), a member of Sherrington's laboratory at Oxford University (1925-1927), described the relationship of the histological features of skeletal muscle with function. He de-

Table 14.1 Chronological Timeline of Discoveries in Skeletal Muscle Plasticity

Year	Discovery	Reference
1925	Sherrington provides the concept of a motor unit as a colony of muscle fibers linked to a common nerve.	120
1929	Denny-Brown differentiates the contractile properties of fast and slow muscles.	51
1938	A.V. Hill derives the force–velocity equation to mathematically define fundamental contractile properties of skeletal muscle.	100
1960-1970	Several laboratories develop histochemical tools for defining metabolic and functional properties of 3 types of motor unit.	34, 55, 56, 61, 191
1960	The cross-innervation study of Buller and colleagues demonstrates the power of the motor nerve in modulating contractile speed in fast versus slow muscle.	35
1967	Barany demonstrates that myosin ATPase activity is correlated with contractile speed of skeletal muscles spanning many animal species.	16
1967	Holloszy discovers that endurance running induces mitochondrial biogenesis as a means of increasing muscle oxidative capacity and endurance.	102
1968-1971	Lesch and colleagues, Gutmann and colleagues, and Guth and Yellin independently show that muscle overload via ablation of synergists results in compensatory hypertrophy and transformation of fast muscle into slow muscle.	85-87, 119
1969	Salmons and Vrbová show that chronic low-frequency electrical stimulation can transform a fast muscle into a slow muscle.	175
1969-1979	Several groups develop technology for separating myosin subunits; in particular, myosin light chain isoforms can be separated by polyacrylamide gel electrophoresis under dissociation conditions.	49, 124, 156, 209
1972	Peter and colleagues provide new physiological nomenclature to describe the properties of the 3 classical motor units.	152
1972	Gollnick and colleagues show that trained athletes and nontrained subjects have divergent fiber typing profiles along with specific functions involving strength and endurance in different types of athletes.	78
1972	Baldwin and colleagues show that the oxidative profiles of all 3 types of muscle fiber can increase in response to endurance training to expand the spectrum of exercise capability.	11
1973-1975	D'Albis and Gratzer and Hoh independently develop technology for separating isomyosin on native pyrophosphate gels by electrophoresis.	48, 101
1975	Fitts and colleagues demonstrate that increments in exercise endurance capacity are closely correlated to the adaptive increase in mitochondrial biogenesis.	67
1982-1993	Several laboratories contribute to the discovery of the myosin heavy gene family in which specific genes encode protein products that regulate the contractile properties of skeletal muscle in the various motor unit pools.	117, 126, 188, 211, 221
1985	Collaborative studies from the Edgerton and Baldwin laboratories demonstrate that fast-to-slow contractile and metabolic phenotype switching in response to functional overload originates primarily from the red fast-oxidative fibers.	172

1985	Reiser correlates MHC composition and contraction speed in single fibers.	167
1989	Schiaffino and colleagues discover and characterize monoclonal antibodies for identifying developmental and adult MHC isoforms and discover the Type IIx MHC.	181
1990, 1993	Methodology improves for separating all MHC isoforms by SDS denaturing polyacrylamide gel electrophoresis.	115, 195
1991	Bottinelli and colleagues differentiate the contractile speed of single fibers and correlate them to type of MHC isoform expression.	33
1992	Rosenblatt and Parry report that gamma irradiation prevents muscle hypertrophy in mice, implicating proliferation of satellite cells as a requirement for muscle hypertrophy.	170
1994	Smerdu and colleagues discover that human skeletal muscle Type II fibers express IIx, not IIb, MHC.	186
1995, 1998, 1999	Coleman and colleagues, Barton-Davis and colleagues, and Adams and colleagues use different approaches to demonstrate that insulin-like growth factor 1 induces muscle hypertrophy.	5, 20, 21, 44
1998	Caiozzo and colleagues demonstrate the power of thyroid hormone and unloading stimuli in repressing slow MHC genes and de novo activating the fastest MHC genes in rodents. Soleus fibers are shown to express all four MHC isoforms, and many belong to I/IIx hybrid, nulling out the dogma stating that MHC transitions have to occur in this order: I↔IIa↔IIx↔IIb.	38
1998	Chin and colleagues are the first to implicate calcineurin nuclear factor of activated T-cell in fiber type regulation.	43
1999	Speigelman's group identifies an inducible coactivator (PGC-1) that activates nuclear respiratory factors necessary for mitochondrial deoxyribonucleic acid synthesis as well as several transcription factors to orchestrate the regulation of many nuclear and mitochondrial genes encoding mitochondrial proteins.	220
2001	Bodine and colleagues identify pathways involved in driving mechanisms of muscle hypertrophy and atrophy.	26, 27
2003	Haddad and colleagues discover antisense ribonucleic acid complementary to the cardiac beta (slow Type I) MHC gene in cardiac muscle, implicated in antithetical regulation of beta and alpha cardiac MHC.	91
2004	Sandri and colleagues and Stitt and colleagues independently implicate FOXO transcription factors with muscle atrophy.	177, 192
2006, 2008	Pandorf and colleagues and Rinaldi and colleagues discover that long noncoding antisense ribonucleic acid is expressed in skeletal muscle and is implicated in Type II MHC gene switching.	147, 168
2007	Holloszy's group identifies the signaling pathway and coactivators for inducing mitochondrial biogenesis in skeletal muscle.	215, 216
2009	Pandorf and colleagues implicate histone modifications with MHC gene switching.	148

MHC = myosin heavy chain.

veloped a recording system to expand the findings of Ranvier (164, 165), who in 1873 was the first to show that red muscles contracted slowly and were more resistant to fatigue compared with pale muscles. Denny-Brown (51) examined the mechanical properties of cat skeletal muscle and obtained results for fast- and slow-twitch muscles that were remarkably similar to later recordings made using more sophisticated instruments (128). He observed that the pale gastrocnemius muscle produced a faster twitch time than the red soleus muscle did and that this difference was unrelated to fiber diameter or the amount of fatty granules (implicating mitochondria) in the fibers.

Contributions of Archibald Vivian Hill to Fundamental Muscle-Contraction Processes

Archibald Vivian Hill (1896-1977), the Nobel Prize recipient in physiology and medicine in 1922 for his findings on lactic acid metabolism, made several milestone discoveries in exercise physiology (23). One of his key studies involved the characterization of muscle mechanics during contraction. In 1938, Hill published a paper titled "The Heat of Shortening and the Dynamic Constants of Muscle" (100) in which he proposed that skeletal muscles possess two distinctive structural components in the contractile machinery that are aligned in series with one another: a contractile component that shortens during contraction and an elastic component that lengthens in response to the force generated during the shortening process of contraction. From these studies, Hill mathematically derived the force–velocity relationship (figure 14.1). This equation is still commonly used as an analytical tool because it enables scientists to determine the intrinsic mechanical properties of a given muscle (or a single muscle fiber) when activated.

Key Discoveries From 1950 to 1970: Building a Foundation in Muscle Plasticity Via Histochemical and Biochemical Techniques

Muscle Histochemistry and the Biochemistry of Myosin

During the early 1960s several investigative teams independently developed a variety of histochemical techniques for characterizing different types of muscle fibers. Using a variety of markers such as oxidative and glycolytic enzyme staining intensity and actomyosin ATPase staining intensity, two fiber types were classified in human skeletal muscles (55, 61) and at least three fiber types were identified in most animal species. Several classification schemes were identified: Type I (slow, red, high oxidative, low phosphorylase) and Type II (fast, white, low oxidative, high phosphorylase) (55, 56); A, B, and C, as defined on the basis of the cytochemical distribution of succinate dehydrogenase activity (SDH) activity (191); and Types I, IIA, and IIB based on the actomyosin ATPase staining intensity using a pH lability protocol ranging from pH 4.3 to pH 10 (34). A major breakthrough came with the use of glycogen depletion as a marker for fibers that were subjected to repetitive electrical stimulation. As a result, Edstrom and Kugelberg demonstrated that the fibers in a motor unit are homogeneous based on SDH staining. Also, they demonstrated that there are two types of fast motor units: a fatigue-resistant fast type where the

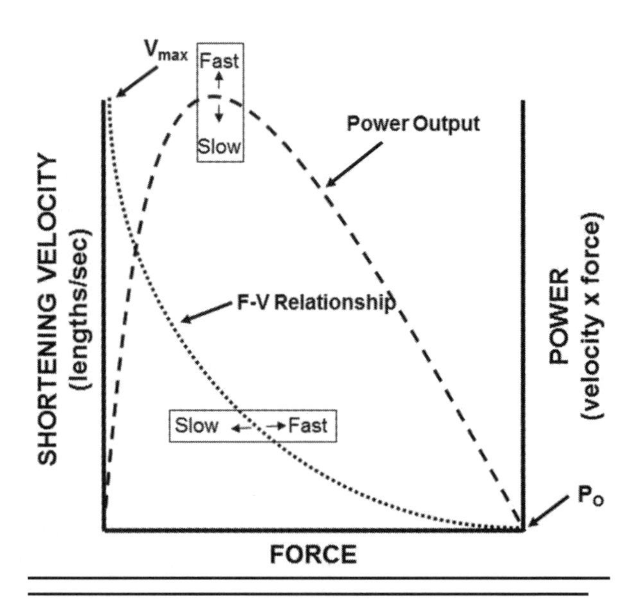

Figure 14.1 Force–velocity relationship and power output. Hill's force–velocity relationship has been used to describe the mechanical behavior of contracting skeletal muscle. The force–velocity relationship identifies the velocity of shortening against a fixed load and therefore dictates the power output (force × velocity) that can be achieved with any velocity–load combination. The represented curve is based on the Hill equation: $(\text{force} + a)(\text{velocity} + b) = b(\text{force}_{max} + a) = C$, where a is a constant with a dimension of force, b is a constant with a dimension of velocity, and C is a constant with a dimension of power. P_o = maximal isometric tension; V_{max} = maximal shortening velocity. The x-intercept in the force–velocity relationship represents the point where the force is so great that the muscle fiber cannot shorten and therefore represents the maximal isometric force (P_o). The y-intercept represents an extrapolated value for the maximal velocity (V_{max}) that would be achieved if no force was produced. The value is extrapolated because it cannot be measured experimentally. This relationship demonstrates that the velocity of lifting the load decreases as the load imposed on a muscle is increased; this inverse relationship follows a hyperbolic curve (100). Also shown are directional shifts as muscles transform for fast to slow or slow to fast properties.

fibers have high oxidative capacity and a fatigable fast type with low oxidative capacity (60). Based on this foundation, additional classification schemes (discussed later) eventually evolved.

In 1967, Michael Barany published a seminal paper demonstrating that actin and calcium activation of purified myosin ATPase were highly correlated to the speed of shortening of their respective muscles (16). This was the first investigation to illustrate that the intrinsic speed of the muscle contraction is a characteristic of the muscle's myosin ATPase activity and that this phenomenon is independent of the actin moiety. Barany's data indicated that the actin-binding properties are similar in myosins that show great differences in ATPase activity (16). Importantly, this breakthrough justified the use of ATPase as a primary marker for fiber typing and provided a biochemical mechanism of A.V. Hill's equation. Thus, Barany's observations provided a foundation for later studies on the role of myosin heavy chain (MHC) isoforms as the regulator of the contractile properties across different motor unit types.

Early Science of Muscle Plasticity: Adaptive Responses of Muscle Fibers to Simulated Physical Activity

In 1960, the classical cross-innervation study of Buller and colleagues (35) showed that the nerve itself dictated the intrinsic contractile properties (biochemical, metabolic, and functional) of mammalian skeletal muscles. These manipulations of motor nerve switching between intrinsic fast and slow muscles revealed that terminally differentiated muscle is not static but rather is a highly versatile organ system capable of changing phenotype from fast to slow and vice versa. The critical question concerning these unique findings is whether such alterations involve trophic or chemical influences (originating via the neurons), alter patterns of activation of the muscles via slow- and fast-type neurons, and require a critical volume of loading affecting the target muscle. Overriding this critical question is the issue of identifying molecular mechanisms by which the nerve regulates the physiological properties of the muscle fibers it innervates.

In the late 1960s Stanly Salmons (1939–) and Gerta Vrbová (175) first addressed some of these issues by developing an electrical nerve-stimulation technique whereby the motoneurons are stimulated chronically at preprogrammed frequencies and durations in order to alter neuromuscular function. They showed that chronic low-frequency stimulation at approximately 10 Hz involving a rabbit fast muscle such as the external digital longus (EDL) can transform the muscle's fast-type properties into slow-type properties. These findings set in motion a series of studies largely directed by Dirk Pette and colleagues at the University of Konstanz in Germany (155), who examined the effect of chronic low-frequency stimulation in different animal models. Their findings clearly illustrated that the biochemical and metabolic properties of fast muscles could be transformed into slower phenotypes. The degree of these transformations depended on the targeted animal species (complete in rabbits but apparently incomplete in rodents and humans).

Despite these remarkable accomplishments, a major concern of this particular muscle plasticity model lies in its lack of physiological context and significance because chronic low-frequency stimulation is associated with a simultaneous activation of all motor units in the target muscle rather than the graded recruitment of motor units that normally occurs during activities of different intensities. In many cases the electrically stimulated muscle atrophied, which is not a typical outcome with normal chronic physical activity (54, 157). Thus, many investigators began to seek more physiological paradigms such as running, swimming, and resistance training. Also, the muscle overload model was introduced as a way to evaluate the contribution of excessive use on the muscle (86, 119).

Early Studies on Exercise-Induced Adaptations in Skeletal Muscle

In the 1930s, Russian biochemist Alexander V. Palladin (1895-1972) studied the effects of physical activity on skeletal muscle (19). He and his colleagues at the Institute of Biochemistry at Harkov used electrical stimulation 2 times/d for a period of 15 d; they called this stimulation paradigm "training." They found that training increased the levels of phosphocreatine and glycogen in skeletal muscle and that fatiguing work decreased the levels of lactic acid and increased work duration. These alterations have become markers of aerobic endurance exercise that remain a centerpiece of training paradigms. Unfortunately, most texts on exercise physiology have failed to recognize the early contributions of Palladin and other Russian scientists.

In the early 1950s, investigators used comparative biology approaches to compare muscles of animals that are normally active with those of animals that are less active. For example, active animals generally have a higher proportion of oxidative enzymes in their leg muscles (116, 150). These observations provided the foreshadowing that activity-related factors are important in accounting for the marked difference in the inherent metabolic properties across different muscle fibers.

The first studies in the United States to examine the effects of exercise training on the oxidative capacity of skeletal muscle used a swimming paradigm involving rats. Two independent studies found that 30 min/d of swimming spanning 5 to 8 wk did not induce increases in oxidative enzymes in the leg muscle of rodents (82, 97). In retrospect, this is not surprising given that later studies showed that rodents could swim acutely for as long as 6 h/d. Thus, these early swimming studies did not generate a sufficient exercise stress; the activity intensity was well within the endurance limits of the limb muscles.

However, in 1967 John Holloszy at Washington University School of Medicine in the United States (102) demonstrated that high-intensity running (in contrast to high-intensity swimming) for progressively longer durations up to 2 h/d 5 d/wk for 12 wk increased the oxidative capacity of rodent limb muscles. The exercise induced an increase in the concentration of specific mitochondrial enzymes and in total mitochondria number, suggesting induced mitochondrial biogenesis. Additionally, he showed that mitochondria of exercised animals exhibited a high level of respiratory control and a tightly coupled oxidative phosphorylation profile. Most important, Holloszy's findings spearheaded a marked interest in subsequent studies on muscle plasticity.

Key Discoveries From 1970 to 1980: Contributions of Exercise Biochemistry to the Study of Muscle Adaptations to Physical Activity

Fiber Type Characterization of Mammalian Skeletal Muscle: Linking Biochemistry to Muscle Function

In 1971, Robert E. Burke (1934–) provided a thorough physiological and histochemical correlation involving three types of mammalian motor units in the cat gastrocnemius (37). It was confirmed that the fast Type IIA motor unit was more resistant to fatigue compared with the fast Type IIB unit. In 1972, a study by James B. Peter and colleagues at the University of California, Los Angeles (UCLA) provided several lines of evidence that the three types of fibers in the lower leg muscles of guinea pigs could be classified based on the combined variables of myosin and actomyosin ATPase activity, contractile properties, and the oxidative and glycolytic metabolic profile of the muscle fiber clusters (152). These analyses provided a physiological and functional identification scheme based on comparisons of fast red and white fiber clusters and slow red muscle types. These biological descriptive properties still hold true when classifying the three primary motor unit pools defining skeletal muscle in animals, although additional properties of MHC isoforms have embellished these important descriptions.

Adaptive Responses of Motor Units to Endurance Exercise

Animal Studies

In 1970, James Barnard, V. Reggie Edgerton, and James Peter reported the effects of long-term training (20 wk of treadmill running) on the metabolic, biochemical, histochemical, and functional properties of gastrocnemius and plantaris muscles in guinea pigs (17, 18). Their findings showed that the muscles adapted metabolically by expressing higher densities of mitochondria and oxygen-utilization capacity, which substantiated the findings reported by Holloszy in 1967 (102). Further, they showed via histochemical analyses that there were more red oxidative fibers in the muscles of the trained groups than in the muscles of the sedentary groups (17) whereas there were no changes in the myosin ATPase properties and contractile intensity of the target muscles of the trained group compared with the muscles of the control group (18). The authors concluded that the exercise paradigm stimulated the muscles to convert low-oxidative fast fibers into high-oxidative fast fibers without necessarily inducing a change in fiber type. Other laboratories using histochemical techniques arrived at conclusions similar to the conclusions of the UCLA group (summarized previously) (65).

In 1971, Paul Molé (1939-2001) and colleagues (139) demonstrated that homogenates obtained from trained rodent skeletal muscle have a twofold greater capacity than untrained skeletal muscle to metabolize any given concentration of long-chain fatty acid substrate. Expanding this analytical approach, Kenneth Baldwin and colleagues (11) assessed the effects of endurance training on the biochemical properties of different fiber types of skeletal muscle. In addition to performing histochemical analyses across the different muscle types, they assessed the capacity of the muscles to metabolize different substrates (e.g., pyruvate and the long-chain fatty acid palmitate) and quantified other marker oxidative enzymes. Although the histochemical analyses were similar to those in previous studies, their biochemical assessments of muscle homogenates clearly showed that the oxidative capacities of all three muscle types to metabolize substrates doubled and that cytochrome c and citrate synthase activity increased to the same extent. The upshot was that the oxidative capacity of all types of skeletal muscle has the ability to increase. In another study published in 1973 (8), the Holloszy group demonstrated that the divergent muscle types differentially expressed glycolytic enzymes and that these particular enzymes were affected differently by chronic endurance running such that the glycolytic capacity of slow muscle increased whereas the opposite occurred in the fast white and fast red muscle types. These collective findings demonstrated a pattern of adaptation such that endurance-trained skeletal muscle took on properties similar to those of cardiac muscle, which is the epitome of endurance-conditioned striated muscle.

Finally, in the context of this topic on fiber type adaptations, the authors would be remiss to not mention another seminal paper that has had a major effect on modern endurance-training protocols. The late Gary Dudley, William Abraham, and Ronald Terjung (57) trained different groups of rats with 19 combinations of intensity and duration of running speed and examined mitochondrial adaptation across the three fibers types. Their collective findings demonstrated the following: Time-independent outcomes were dependent on running intensity; the duration of training to achieve the optimal outcome could be shortened as running intensity increased; and, with regard to the low-oxidative fiber type, the adaptive response increased exponentially as intensity was increased. This study on rodents laid the groundwork for current training paradigms that invoke high-intensity, short-duration training stimuli in order to optimize the endurance capacity and the sustainability of running at very high speeds.

Human Studies

In 1972, a key seminal paper was published by Philip Gollnick (1934-1991), Bengt Saltin, and colleagues (78). They obtained biopsies from quadriceps and deltoid muscles of a wide spectrum of subjects that either were untrained or were trained for their lower-limb adaptations (i.e., cycling, sprinting, long-distance running) or upper arm and shoulder muscle adaptations (i.e., canoeing, swimming, weightlifting). The biopsy material was analyzed for all the established histochemical and metabolic enzyme markers for aerobic and anaerobic metabolic pathways. Confirming previous findings (55, 61), these analyses showed a two fiber type system (e.g., Type I and Type II) based chiefly on histochemical-derived myosin ATPase activity. This study (78) provided several key results. First, fast-oxidative and slow-oxidative fibers in leg muscle were more prevalent in endurance-running athletes than in either sedentary subjects or athletes specializing in arm or upper-body activities. Second, fast fibers were typically larger than slow fibers when comparing all the groups of subjects. Third, although slow fibers typically demonstrated high levels of oxidative enzyme activity, fast fibers in the endurance runners showed equivalent or higher activity compared with that typically observed for slow fibers. Fourth, unlike oxidative enzymes, the glycolytic enzyme marker phosphofructokinase showed little differentiation between the two fiber types and across the different training profiles. Finally, muscle glycogen storage was typically higher in the muscle fibers of trained subjects, and there was no obvious difference between fast and slow fibers. This study provided a foundation for a variety of training studies that unfolded in the 1970s using both animal and human subjects.

In 1973, Gollnick and colleagues (77) performed what is considered one of the most intensive endurance-training studies (5 mo) of its time using high-intensity cycling exercise targeting the vastus lateralis muscle of untrained adult subjects. Novel observations included significant increases in whole-organism capacity for maximal oxygen consumption, marked increases in both SDH and phosphofructokinase (PFK) enzyme levels that were attributed to both fast and slow fiber types, and increases in the relative cross-sectional area of slow versus fast fibers without any evidence of fiber type switching. In particular, the relative increase in slow fiber cross-sectional area suggests that submaximal exercise of these subjects could be sustained better in the trained versus the untrained state without the need to recruit the fast fiber motor unit pool during submaximal exercise intensities. This likely accounted for the remarkable increment in the ability of these subjects to sustain exercise in the trained state at 85% to 90% of maximal oxygen consumption capacity.

In another key study, David Costill (1936–) and colleagues at Ball State University (47) compared groups of highly trained male and female runners who typically ran approximately 60 miles (96.6 km)/wk spanning a period of 7 to 8 yr with nontrained, sedentary individuals as well as moderately trained individuals who ran 3.5 miles (5.6 km)/d 4 d/wk over a period of 10 wk. The results indicated that there were no differences in the percentage of slow Type I fibers among the four groups. However, the highly trained male and female subjects had more fast Type IIa fibers relative to fast Type IIb (now referred to as Type IIx) fibers compared with the sedentary and moderately trained subjects. Interestingly, the trained male and female runners had equivalent maximal oxygen consumption capacities.

Jan Henriksson performed a training study in which only one leg was trained via bicycle exercise for 1 mo (98). The results demonstrated that the metabolic enzyme adaptations of the quadriceps muscle in the trained leg were higher compared with those of the contralateral muscle in the untrained leg. When both legs were exercised simultaneously, the trained leg had a lower respiratory quotient indicative of a greater utilization of fatty acids; this was consistent with a greater uptake of fatty acid substrate by the trained muscle. This study clearly demonstrated that the adaptations in the muscle have a direct effect on its metabolic capacity independent of the delivery processes of the circulatory system.

Effect of Training on Skeletal Muscle Fiber Types During Acute Bouts of Exercise

Animal Studies

The studies previously reviewed conclusively demonstrated that one of the hallmark adaptations to endurance training affecting skeletal muscle involves an increase in the density of mitochondria and, hence, a greater capacity to perform substrate oxidation. This adaptation is reflected across all muscle types. Therefore, studies were carried out to determine the role that this adaptation plays in enhancing exercise capacity. Two studies were conducted by the Holloszy laboratory. First, Baldwin and colleagues (7) studied a large group of rodents that were trained to sustain moderate-intensity running lasting at least 2 h. Then, subgroups subsequently ran for 15, 60, or 120 min using 3 intensity paradigms that were typically used in training rodents. One of the regimens involved interval sprint exercise, which was of higher intensity than the other two regimens.

The findings were quite surprising. Only the fast, high-oxidative and slow-oxidative motor units appeared to contribute to the sustained activity across the three intensity paradigms, and little evidence indicated that the low-oxidative fast Type IIb fibers were recruited. The glycogen storage pools of both oxidative fast and slow fibers were markedly depleted, yet only the fast-oxidative fibers utilized significant quantities of muscle-stored triglycerides. Interestingly, the pattern of liver glycogen utilization was extensive across all three exercise paradigms, and at the end of 120 min approximately 85% of the liver glycogen pool was depleted. In absolute terms it was calculated that glycogen in the liver contributed more calories for exercise than did glycogen in the skeletal muscles. Given the effect of liver glycogen depletion on the hypoglycemic state, it was speculated that the various exercise tests would have come to a halt once the liver glycogen was fully depleted.

One paradox of this study was the striking pattern of energy utilization in the slow-oxidative fibers given their high reliance on glycogen and the little evidence for using endogenous muscle-derived triglycerides as a source of energy. Interestingly, ketone bodies are known to accumulate in blood during long-duration exercise, as reported by William Winder and colleagues (213). Of note, later investigations in the Baldwin laboratory showed that slow-oxidative muscle fibers, compared with the fast fiber types, possess the greatest capacity for metabolizing this substrate (105). Consequently, the slow type of muscle appears to preferentially use ketone bodies as a substrate in place of other substrates such as fatty acids.

In a similar study, Robert Fitts and colleagues (67) studied groups of rats that were treadmill trained for different durations (10, 30, 60, and 120 min) during each training session. Sessions were conducted 5 d/wk for 13 wk, and subgroups were subsequently analyzed for skeletal muscle oxidative capacity. The results showed that the longer the training sessions, the greater the increase in oxidative capacity of the muscles. Two types of exercise tests—a moderate-intensity endurance run to exhaustion and a 30 min test of moderately high intensity—were performed to compare trained groups against nontrained groups. Results of the run to exhaustion indicated that a progression improvement in of running time correlated with the degree of muscle oxidative capacity such that the animals that trained the longest during each training session had the greatest running time to exhaustion and vice versa. In the 30 min endurance-run test, the utilization of both muscle and liver glycogen was inversely proportional to the oxidative capacity of skeletal muscle among the groups. Thus, these animal studies clearly established the importance of oxidative capacity of skeletal muscle in regulating the ability of individuals to exercise until exhaustion by utilizing fuels other than carbohydrate sources.

Human Studies

In 1971, Jan Karlsson and Bengt Saltin (114) published a fascinating study in which a group of subjects ran a 30 km race twice 3 wk apart—once after consuming a normal mixed diet and once after consuming a special high-carbohydrate diet that increased muscle glycogen. The results showed that vastus muscle glycogen concentration was about twofold higher after consuming the high-carbohydrate diet than after consuming the mixed diet (34 vs. 17 g/kg). Importantly, when on the high-carbohydrate diet, each participant ran faster times, maintained optimal pace from start to finish, and maintained higher muscle glycogen stores at the end of the race (19 vs. 5.2 g/kg). The authors hypothesized that the high pace of the race was sufficiently demanding to necessitate the recruitment of a large number of glycogen-dependent fibers (fast Type II) and that the high content of glycogen in these fibers enabled the fibers to contract longer before fatiguing. These findings catalyzed a variety of studies dealing with the role of fiber types and substrate utilization. Because this chapter lacks the space to address this important topic in further detail, the reader is referred to references 79, 80, and 81 for additional information.

Can Physical Activity Paradigms Convert Fast-Type Fibers Into Slow-Type Fibers?

In the research noted previously on skeletal fiber adaptation to endurance exercise, little evidence suggests that animal and human muscle fibers are capable of shifting the properties of the fast types to a true slow contractile phenotype. Instead, the findings suggested that shifts occurred within the pool of Type II fibers (11). In the early 1970s, another approach was adopted in which skeletal muscle was mechanically stressed by surgically removing the synergistic muscles to the target muscle of interest. This manipulation forced the overloaded muscle to take over the functional needs of antigravity support and routine locomotion. It was shown that the overload stimulus transforms the muscles from a faster phenotype to a slower phenotype based on contractile and histochemical properties (85-87). Thus, this new approach demonstrated that muscle fiber types are indeed dynamic in nature and respond to both neural input and the degree of loading that is imposed. Additional studies on this model were published by several laboratories using chiefly histochemical observations, which clearly suggested that more slow-appearing fibers could be induced when the target muscle was a fast muscle such as the plantaris (9, 180, 199). However, it was not until strategic biochemical and functional analyses were performed that a clearer picture evolved. Collaborations between the laboratories of Baldwin and V. Reggie Edgerton and Roland Roy (9, 13, 14, 172, 173) showed that when the plantaris muscle was functionally overloaded for several weeks, the muscle hypertrophied by expanding the myofibril protein pool in the overloaded fibers. Glycogenolytic enzymatic activity, myofibrillar ATPase-specific activity, and the contractile properties were shifted to a slower phenotype. Interestingly, the muscle's capacity to oxidize pyruvate and fatty acids decreased but the capacity to metabolize ketones increased. Finally, the endurance properties of the overloaded muscle were enhanced, which is typical of slow fibers. In additional experiments, both the medial gastrocnemius and plantaris muscles were overloaded and their fast red and fast white fiber clusters were isolated and analyzed. These data (13, 172, 173) provided strong evidence that the fast red regions in both muscles contained fibers that were likely transformed to express the slow phenotype properties. As a result, there appears to be both a specificity about and a limitation to the extent that the fundamental properties of a given fiber type can transform in response to mechanical stimuli.

Polymorphism of Myofibril Proteins and Role of Myosin

In the 1970s, the understanding of muscle structure and function became significantly enhanced at the molecular level through the efforts of several investigators. By this time, it was well established that force production requires the interaction between actin and myosin as well as the presence of adenosine triphosphate (ATP). Also, research showed that myosin is a hexameric molecule containing two pairs of light chains (essential and regulatory) and two heavy chains. In addition, other myofibrillar proteins, including actin, troponin, and tropomyosin, were identified (see 194 for review). Furthermore, gel electrophoresis enabled the separation of multiple forms of myosin molecules (48, 49, 101) as well as multiple forms of light chains (49, 124, 156, 209) in a single muscle or muscle fiber, and the pattern of their expression was different in fast versus slow muscle types. These multiple forms were referred to as isoforms, isomyosins, or isozymes, which refer to proteins of the same species that play a very similar biological role but exhibit differences in their primary structures that can subtly alter their biological properties. Importantly, evidence demonstrated that myosin subunit expression may be transformed as a result of altered nerve activity based on the findings of cross-innervation experiments (101, 176). These findings placed the myosin isoform expression profile as a central theme in studying muscle plasticity and adaptation to various physiological, hormonal, and pathological stimuli. In 1977, a report using immunocytochemical analyses determined that more than one form of myosin is expressed in skeletal muscle (69). In addition to whole-muscle analyses, there were some observations of heterogeneity of myosin found in single fibers using immunocytochemical approaches and specific polyclonal antibodies against fast and slow myosins (69, 70).

By this time, understanding of the role of myosin in striated muscle structure and function had increased. Through the research of both Hugh Huxley (1934–) and Andrew Huxley (1917–; unrelated), it became clear that the thick filaments of the sarcomere comprised myosin whereas the thin filaments primarily consisted of actin (as well as its associated regulatory proteins). Also, the muscle contraction process at the molecular level evolved to the point that the cross-bridge cycle, calcium, and ATP were considered to be the essential players (108, 194). The MHC subunits possess the ATPase activity. The dogma that evolved stated that MHC binding to actin in the presence of ATP transforms chemical energy via terminal phosphate cleavage into free energy to drive both force development and movement of the thin filament (e.g., the sliding filament theory).

Discoveries From 1980 to 2000: Myosin Isoform Gene Discovery, Advancements in Analytical Technology, and Expansion of Activity Models to Overcome the Atrophy of Inactivity

Advancing Biotechnologies and Identification of the MHC Gene Family

Studies in the 1980s on muscle plasticity were mostly descriptive in terms of better characterizing muscle fiber types (based on myosin heavy chain [MHC] and myosin light chain [MLC] profiles). Also, a working hypothesis was established that any given muscle fiber existing in the broad pool of motor units has the capability to express more than one type of MHC isoform. Thus, studies were geared to characterizing the contractile phenotype alterations during development and in response to altered conditions such as loading state, nerve activity, and hormonal status. However, the central question concerning the mechanism of phenotype change is how isoform expression is regulated and at what level of gene expression the alteration occurs. Thus, there became a need to integrate molecular biology and physiology in order to better understand mechanisms of muscle plasticity. An explosion in biotechnology information occurred in the early 1980s as molecular biology methods for identifying nucleic acids became available. Complementary deoxyribonucleic acid (DNA) and genomic DNA materials were cloned into libraries, and clones for specific genes were identified and sequenced.

Given this backdrop, several groups of investigators, including Bernardo Nadal-Ginard (Harvard University), Leslie Leinwand, (Albert Einstein College of Medicine), Murray Rabinowitz (University of Illinois in the United States), Margaret Buckingham (Pasteur Institute, France), and their respective colleagues, spearheaded the discovery and characterization of the MHC gene family in rodents and humans (117, 126, 188, 211, 221). These discoveries provided important insight into sarcomeric MHC genes, their tandem organization into clusters (figure 14.2), and the high degree of similarity in their respective messenger ribonucleic acid (RNA) coding sequences. This understanding added more depth to the understanding of muscle plasticity and transformed the direction of MHC gene regulation at the messenger RNA expression level as well as promoter activity assessments and their regulation by hormones, nerve activity, and altered loading state. Also, there was increasing evidence that the MHC gene regulation in skeletal muscle occurs at the transcriptional level. Frank Booth, in the authors' opinion, led the way for the exercise science field with his focus on the infrastructure of gene expression (29). These early studies served as a catalyst for other investigators working in the field of muscle plasticity.

New Approaches for Identifying MHC Proteins and Fiber Typing At the Protein and Molecular Levels

Between 1986 and 1989, Stefano Schiaffino and colleagues (50, 181) discovered and characterized a series of monoclonal antibodies that allowed researchers to identify with better precision different fiber types based on the expression of specific MHC protein isoforms. This technical achievement confirmed the existence of hybrid MHC isoforms in single fibers and led to the discovery of a new MHC isoform, IIx, which has fast properties and is a major isoform expressed in human Type II fibers (50, 181, 186). Also, denaturing electrophoretic techniques were improved in order to enable separation of specific MHC isoforms (195), which was applied to both the whole muscle and single fibers (see figure 14.2).

As depicted in figure 14.2, individual native isoforms are made up of different combinations of both the

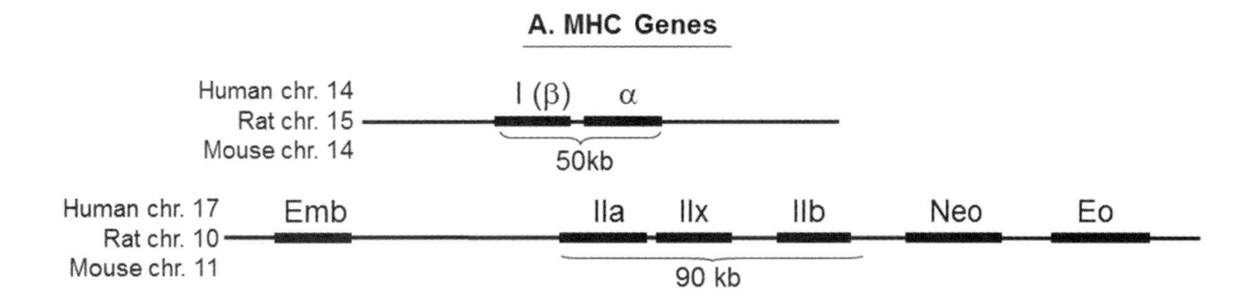

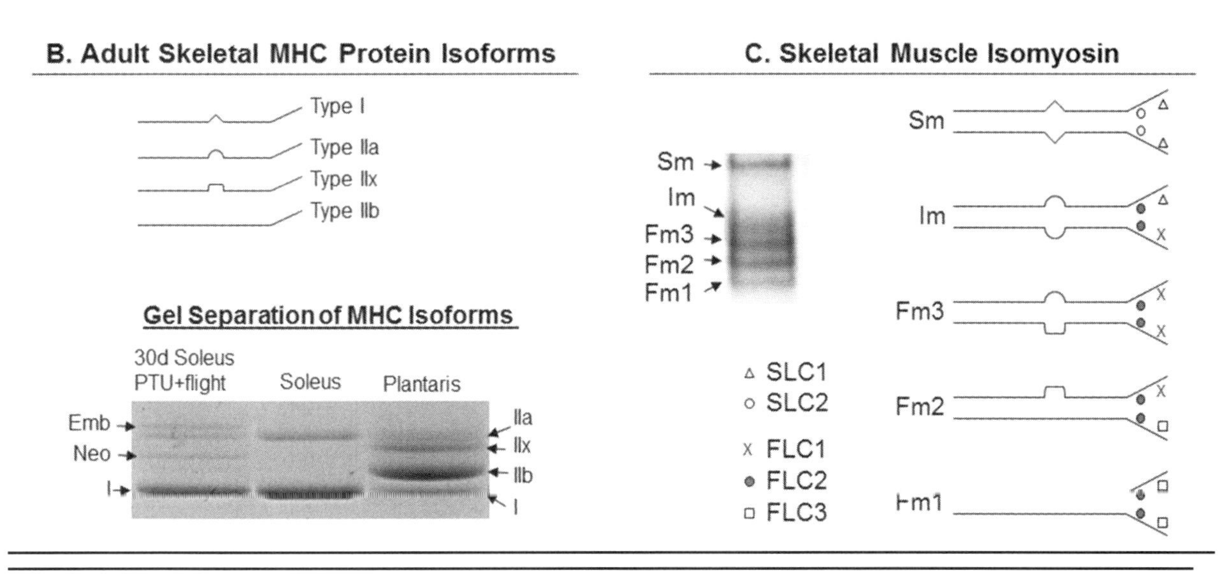

Figure 14.2 The sarcomeric myosin heavy chain (MHC) gene family and protein isoforms. *(a)* MHC gene organization. At least eight MHC genes are expressed in striated muscle and are found in two clusters. The first is the cardiac MHC gene cluster on rat chromosome 15, which consists of the Type I (also called beta) and the alpha cardiac MHC genes. Type I is the slow MHC expressed in slow skeletal muscle fibers. The second is the skeletal MHC gene cluster on rat chromosome 10. The embryonic (Emb), fast IIa, IIx, IIb, neonatal (Neo), and extraocular (Eo) genes are located in tandem in the order depicted. This MHC gene organization, order, head-to-tail orientation, and spacing has been conserved through millions of years of evolution and could be of great significance to the way these genes are regulated in response to various stimuli. Human and mouse cardiac MHCs are found on chromosome 14, whereas human skeletal MHCs are found on chromosome 17 and mouse skeletal MHCs are found on chromosome 11. *(b)* MHC isoform separation by SDS polyacrylamide gel electrophoresis under denaturing conditions according to the method described by Talmadge and colleagues (195). Shown are an adult plantaris (expresses all four MHC isoforms), an adult soleus (rich in Type I; also expresses IIa), and a soleus from 30d old animal that was made hyporthyroid and exposed to space flight (shows some expression of developmental Emb and neonatal MHCs in addition to Type I and IIa). *(c)* Separation of native isomyosins by the method described by Hoh using native pyrophosphate gels (101). We present an illustration depicting the five primary native myosin isoforms that have been identified in adult rodent skeletal muscle. These molecules consist of different combinations of MHC and light chain isoform. Thus, the native myosin protein exists as a hexameric molecule comprising two MHCs and two pairs of light chains, the latter of which are thought to modulate the heavy chain ATPase component (232).

MHC and the light chain gene products. For example, the slow native MHC is made up of both the slow MHC isoform and slow light chain isoform pairs. The intermediate native MHC isoform is thought to consist of the Type IIa MHC and specific fast light chain pairs, whereas the FM3, FM2, and FM1 isoforms comprise combinations of the Type IIx and IIb isoforms along with the fast MHC light chain combinations. It turns out that the expression of these native myosins is different across a variety of skeletal muscles as well as in specific regions of a given muscle (197, 203). Moreover, such expression is very plastic and subject to altered patterns of expression depending on the activity or inactivity pattern of the motor units in which they exist.

Functional Properties of the MHC Isoforms

Beginning with the classical myosin study of Michael Barany in 1967 (16), it was generally assumed that differences in specific activity of the myosin ATPase enzyme was the key factor responsible for differentiating the intrinsic functional properties of fast versus slow skeletal muscle fibers. With the new understanding that myosin is expressed in many isoforms, it became important to see how these isoforms relate to ATPase activity and muscle fiber contractile function. Peter Reiser and colleagues (167) were among the first to examine the relationship of MHC composition and contraction speed in single fibers. In contrast, Staron and Pette established a relationship between MHC composition and ATPase activity in single fibers (190). However, in 1991, Roberto Bottinelli and colleagues published a seminal paper comparing the force–velocity properties and MHC isoform composition of isolated skinned rat fibers from slow and fast skeletal muscles (33). They identified four groups of fibers based on the specific MHC antibody staining pattern and designated these fibers Types I, IIa, IIx, and IIb. The analyses included the force–velocity relationship based on the Hill equation. Single-fiber analyses created a continuum of contraction velocity kinetics in the ascending order of Type I < IIa < IIx < IIb. Also they observed that the force per cross-sectional area was lower in the slow fiber type (I) relative to the three fast types (IIa, IIx, and IIb). These velocity differences also were correlated with power curves that were generated for each fiber type and showed that the two fastest fiber types (IIb and IIx) generated greater power output than the Type I and IIa fibers.

Based on these observations it became generally accepted that MHC isoform expression patterns are a major determinant of the mechanical properties of a muscle fiber. Additional studies by Bottinelli and colleagues (31, 32), Robert Fitts and colleagues (212), and Steve Harridge and colleagues (96) focused on human skeletal muscle fiber types, which have been categorized into three primary types: I, IIa, and IIx/b. Similar to observations made in rat muscle, the contractile velocity in the slow Type I fibers is lower than that in the Type IIa and IIx/b fibers in human muscle. At this time it is important to point out that the human genome contains an intact IIb MHC gene. However, its product is not detected at the protein level (186, 210) and is rarely detected at the messenger RNA level. Thus, three types of myofibers are generally recognized for human limb muscle: slow Type I and fast Types IIa and IIx. Interestingly, Robert Fitts and colleagues (212) discovered that slow and fast fibers in small animals have faster kinetics than those in larger animals, including humans. Despite this scaling phenomenon, the adaptive plasticity of animal and human skeletal muscle displays remarkable consistency in fiber type adaptations in response to mechanical activity and hormonal interventions.

New Activity and Inactivity Paradigms Involving Animal Models

During this time period many investigators explored muscle plasticity using different paradigms of either increased or decreased muscle activation in order to induce either fast-to-slow or slow-to-fast phenotype transitions, respectively. Thus, many animal models were developed for this purpose. Plasticity analyses focused on describing changes in gene expression (mainly myosin, MHC, and light chain isoforms) at both the protein and messenger RNA levels. At this time functional promoter analyses were also unfolding, which represented transcriptional analyses of the gene's regulation. Further, it became evident that skeletal MHC genes are regulated by thyroid hormone in a fashion specific to muscle fiber type (109). Therefore, alterations in thyroid hormone levels were studied separately or in combination with altered activity models (see 10 and 40 for review).

Functional Overload in Combination With Running

William Stirewalt and colleagues (84) were the first to determine the effects of functional overload in combination with treadmill running on the expression of native myosin isoforms. This study showed that the contractile phenotype of a typical fast muscle such as the plantaris could be transformed toward a slower phenotype by increasing the loading state on the muscle. In 1989, the Stirewalt group (151) analyzed messenger RNA expression for the target MHC and light chain genes. Their findings showed that messenger RNA expression increased for the Type I and IIa genes and decreased in the Type IIb gene. It is important to point out that the fast Type IIx gene had not yet been discovered when this study was conducted. Nevertheless, this study demonstrated that the regulation of the MHC gene family was largely regulated by pretranslational processes. It would take another 10 yr or more to establish that the MHC gene family is regulated by signaling pathways and transcription factors that directly act on the MHC promoters (e.g., regulatory sequences that program transcription of the target genes).

Role of Loading in Spinal Transection and Spinal Isolation

At the same time, another model for unloading and thus altering muscle homeostasis gained notoriety. This model involved the technique of transecting (severing) the spinal cord (ST) in the thoracic area to paralyze the lower extremities while keeping the upper limbs intact. This technique was developed by V. Reggie Edgerton and Roland Roy at UCLA (12, 123, 187). The same group also developed a companion model to ST that involved severing the spinal cord in the midthoracic area and the high sacral area along with bilaterally deafferenting all sensory input into the spinal cord between the two transections. This technique created a near-complete silencing of the skeletal muscles in the lower limbs while maintaining an intact nerve–muscle connection. This technique is referred to as spinal isolation (SI) (83, 111). These studies were initially performed on cats and were then refined for use on rats, which enabled more mechanistic studies on the properties of the affected skeletal muscles.

The ST studies demonstrated that extensor muscle fibers of the hind limbs underwent both atrophy and shifts to a faster phenotype in both slow-type (soleus) and mixed-type (gastrocnemius muscle fibers (12). Larger diameter fibers of slow and fast fiber types atrophied to a common size, suggesting that all types of fibers were sensitive to atrophy. Although static weight-bearing activity did not prevent slow-to-fast MHC shifts, it was successful in maintaining the oxidative enzyme activity of the fibers.

Similar types of metabolic and contractile phenotype shifts were also observed with SI, especially in the soleus muscle (83, 111). With the evolution to the rat SI model, it became possible to ascertain the cellular processes linked to the marked atrophy that occurs during the initial 15 d of SI, in which the muscles typically atrophy by approximately 50% (92). It was determined that degradation of the myofibril fraction was the primary contributor to the loss of muscle size. There was also a reduction in muscle total RNA and DNA content as well as messenger RNA for the slow myosin and actin genes. The loss in DNA is consistent with the decrease in the number of nuclei in the atrophying SI soleus muscle (223).These observations suggested that both the messenger RNA substrate and the translational machinery (ribosomal RNA) were reduced, making it difficult to maintain protein translation. Additional observations suggested that protein degradation processes were upregulated (93), further suggesting that the marked atrophy seen in this unique model is likely the result of several processes affecting transcription, translation, and protein degradation processes.

Resistance Exercise as a Countermeasure to Limb Unloading

The unloading model involves applying traction to the rodent (or mouse) tail with a bandage system linked to a swivel device that is connected to a hook at the top of a cage housing a single animal. With its hind limbs raised off the floor, the animal cannot bear weight; this causes muscle atrophy. This model is called hind limb suspension (HS) and has evolved to be an important ground-based analogue for studying the effects of space flight on the homeostasis of the skeletal muscle system. In the early 1980s, the U.S. National Aeronautics and Space Administration was gearing up for both animal and human experiments performed on the space shuttle. However, due to a number of delays in launching the shuttle program, especially the Challenger disaster in the mid-1980s, meaningful research on the space shuttle was delayed until the early 1990s. Gary Diffee and colleagues (53) were probably the first to ascertain whether a resistance-exercise countermeasure paradigm could ameliorate the muscle atrophy and the shift to fast MHC gene expression in unloaded slow skeletal muscle. The outcomes were partially successful but pointed to the difficulty of maintaining normal homeostasis of slow-type skeletal muscle in the face of HS (and likely space flight). This issue remains a problem for maintaining muscle homeostasis in astronauts while they are living at the international space station and performing a variety of countermeasure paradigms.

What Happens When Functional Overload is Interacted With Either Hind Limb Suspension or Thyroid State?

As discussed previously, functional overload induces a net anabolic, hypertrophic state and a fast-to-slow contractile phenotype whereas HS induces a net catabolic, atrophic state and fast contractile phenotype. Richard Tsika and colleagues ascertained how these contrasting stimuli affected one another by simultaneously imposing them on adult rats (202). Their findings demonstrated that the hind limb suspension state nulls out the functional overload effect on hypertrophic processes and partially blunts the fast-to-slow phenotype shifts. Similarly, when functional overload was carried out and then followed by the HS stimulus, the anabolic and phenotype shifts associated with functional overload could be reversed. Interestingly, the kinetics for altering muscle mass and phenotype are much faster in the unloading condition than in the overloading condition (204). These responses further emphasized the importance of the daily loading state in either maintaining or enhancing skeletal muscle homeostasis in different types of muscle fibers.

Given the effect of earlier studies suggesting that thyroid state can modulate the muscle contractile phenotype, Dan Fitzsimons and colleagues (68) and Steve Swoap and colleagues (193) independently investigated the effects of thyroid state on the contractile phenotype and isomyosin expression of skeletal muscle. Findings by Fitzsimons and colleagues (68) suggested that elevated levels of thyroid hormone had the same effect as the unloading state on the expression of the isomyosin profile. That is, a hyperthyroid state induced repression of slow MHC expression in the soleus muscle and induced upregulation of the Type II MHCs whereas a hypothyroid state caused the opposite effect. Furthermore, Swoap and colleagues (193) examined the effects of hyperthyroidism (which normally induces a fast-to-slow isomyosin shift, as noted previously) on functional overload responses. Although thyroid hormone treatment did not affect the hypertrophic response induced by the overload stimulus, it completely blocked the upregulation of the slow Type I MHC expression at both the messenger RNA and protein levels of regulation. In contrast, Type IIb MHC expression was affected mostly by the loading state and was repressed in the overloaded plantaris regardless of thyroid state (193). The mechanism causing these phenomena has not been investigated.

Single-Fiber MHC Polymorphism: Patterns and the Role of Loading Conditions

As discussed previously, it has been reported that activity with high loading paradigms can shift the contractile phenotype type to a slower pattern. However, less is known about the opposite when unloading conditions such as bed rest and space flight prevail. In the early 1990s, independent efforts were successful in separating all four adult MHC isoforms by denaturing gel electrophoresis (115, 195). Vincent Caiozzo and colleagues (39) demonstrated the power of using such a technique on single-fiber MHC analyses. They found that multiple MHC isoforms exist simultaneously in most of the fast muscle fibers of adult rats and that this polymorphism can be extended further during postbirth development (52) and with hind limb suspension and hyperthyroidism with interventions to simultaneously impose hind limb suspension and hyperthyroidism to alter altered loading and thyroid states combined. For example, the slow soleus muscle can be transformed into a true fast muscle that expresses predominantly a mixture of fast Type II MHCs when animals are simultaneously treated with HS and thyroid hormone. Type I MHCs provide only a minor fraction of the total MHC pool (38). Type I MHCs contribute a minor fraction (38). This study demonstrated for the first time that true slow fibers possess the genetic milieu to express the entire spectrum of adult MHCs. This observation spurred researchers to investigate the mechanisms of MHC gene switching. Importantly, these studies showed that some of the hybrid fibers identified showing Type I/IIx coexpression do not comply with the scheme of MHC transitions proposed by Pette and Staron (158, 159). These findings suggested that some other mechanism (perhaps involving antisense noncoding RNA, which is discussed later) must exist to allow for such hybrid or polymorphic MHC gene regulation.

2000 to Present: Mechanisms Regulating Protein Balance and Muscle Mass, Mitochondrial Biosynthesis, and Contractile Phenotype Switching

Mechanisms of Altered Protein Balance Affecting Muscle Mass

In 2001, Susan Bodine and colleagues (27) were some of the first to identify and characterize signaling and effector molecules that regulate protein synthesis affecting myofibril protein accumulation in functionally overloaded rat plantaris skeletal muscle. They specifically characterized phosphorylation and activation of the ribosomal P70 S6 kinase (S6K1) protein as a key regulator of ribosomal function along with the pathway leading to its activation. The cascade to activate S6K1 is referred to as the insulin-like growth factor 1 (IGF-1)–IRS1–Akt–mammalian target of rapamycin (mTOR) pathway (simplified as the Akt pathway; figure 14.3*a*). Thus, studies implicated mTORC1 and S6K1 as the principal mediators of general muscle growth and loading-induced hypertrophy. Additional studies involving both normal and growth hormone-deficient rats (3, 222) further suggested that mechanical stimuli enhance expression of endogenous IGF-1 messenger RNA and concomitant elevation of IGF-1 peptide levels in the target muscle as a crucial driver of the Akt cascade. This event was correlated with accumulation of total RNA, DNA, and myofibril protein in the overloaded muscles, further implicating growth factor(s) as a trigger in response to the mechanical stress signal (3). Also, studies demonstrated that over-

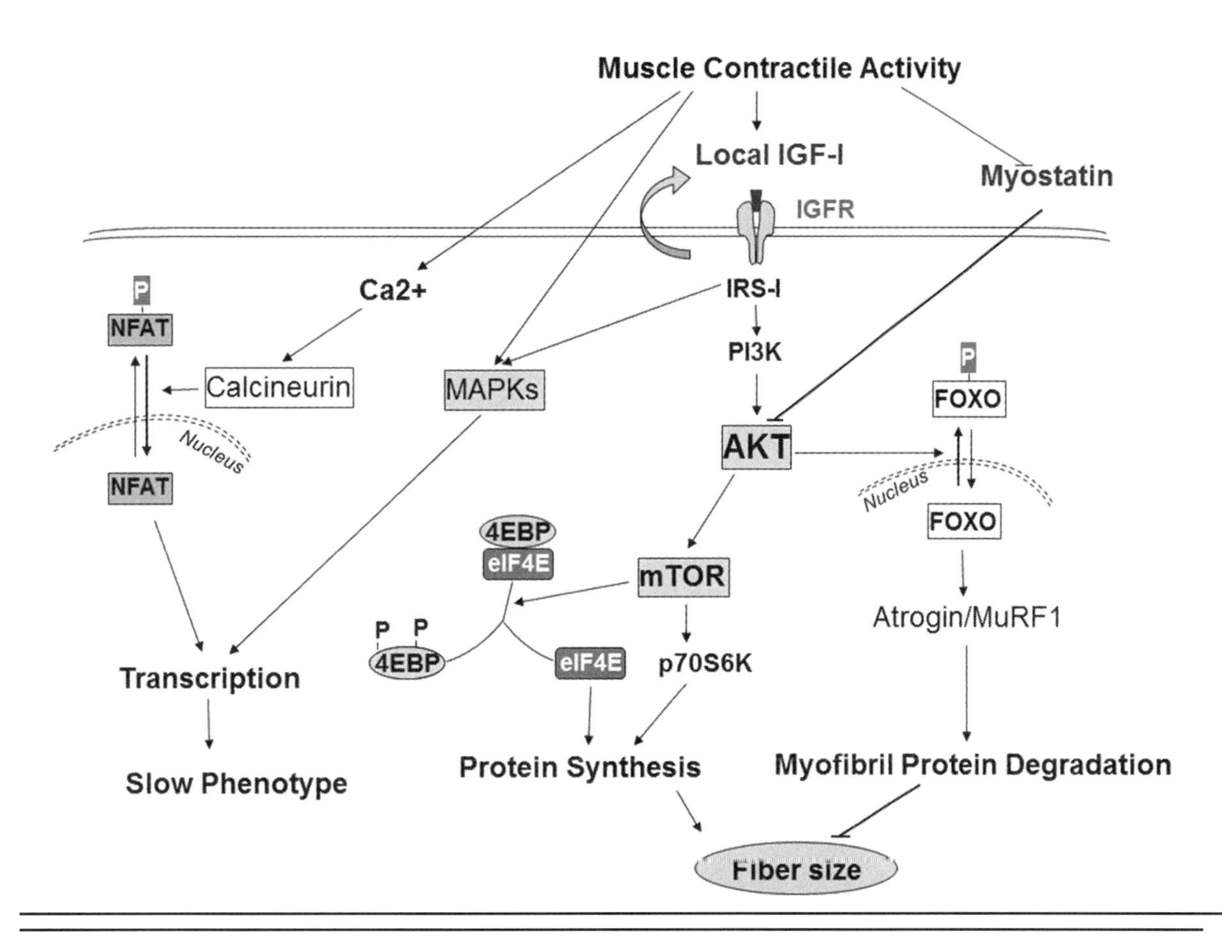

Figure 14.3 Muscle contractile activity leading to muscle plasticity that affects fiber size and mitochondria content. *(a)* Signaling pathways leading to altered protein balance affecting muscle fiber size. A schematic of anabolic cascade signaling for insulin-like growth factor 1 affecting protein synthesis via mammalian target of rapamycin–Akt signals, protein transcription via mitogen activator protein kinase ERKs, and protein degradation through FOXO (a transcription factor of the forkhead family)–atrogin–muscle RING finger protein 1 (MuRF-1) action (25, 27,169,177,192, 200). *(continued)*

expressing IGF-1 in transgenic mice (44), directly infusing IGF-1 (5) into the myofiber compartment, or injecting a plasmid vector into the target muscle to overexpress IGF-1 could essentially mimic the compensatory growth paradigm (20, 21). As a result, there appears to be two components to the anabolic–hypertrophic process: a protein synthetic response mediated by the Akt–mTORC1–S6K1 pathway (which also includes activation or additional regulators such as 4E binding protein) (figure 14.3*a*), and an additional pathway that causes mitogenic-induced satellite cell proliferation and differentiation processes in order to promote myonuclei incorporation into the myofiber domain to maintain regulatory control per unit volume of myofibril protein (45, 89). This latter pathway has been identified as the mitogen activator protein kinase (MAPK)–ERK pathway (figure 14.3*a*). Thus, it has been proposed in both animal and human models of hypertrophy that there is an obligatory requirement to incorporate new myonuclei into the enlarging fiber domain in order to significantly increase myofiber mass (21).

Are Satellite Cells Required for Skeletal Muscle Hypertrophy?

Since muscle satellite cells were discovered in 1961 (127) they have been implicated in muscle repair, growth, and hypertrophy (see 1 and 66 for reviews). However, some findings and interpretations do not support the idea that satellite cells play an obligatory role in muscle hypertrophy, and the issue of the role of satellite cells in muscle hypertrophy has recently become the subject of strong debate (131, 143). Based on various perspectives, it is clear that some of the confusion has to do with the experimental design of various

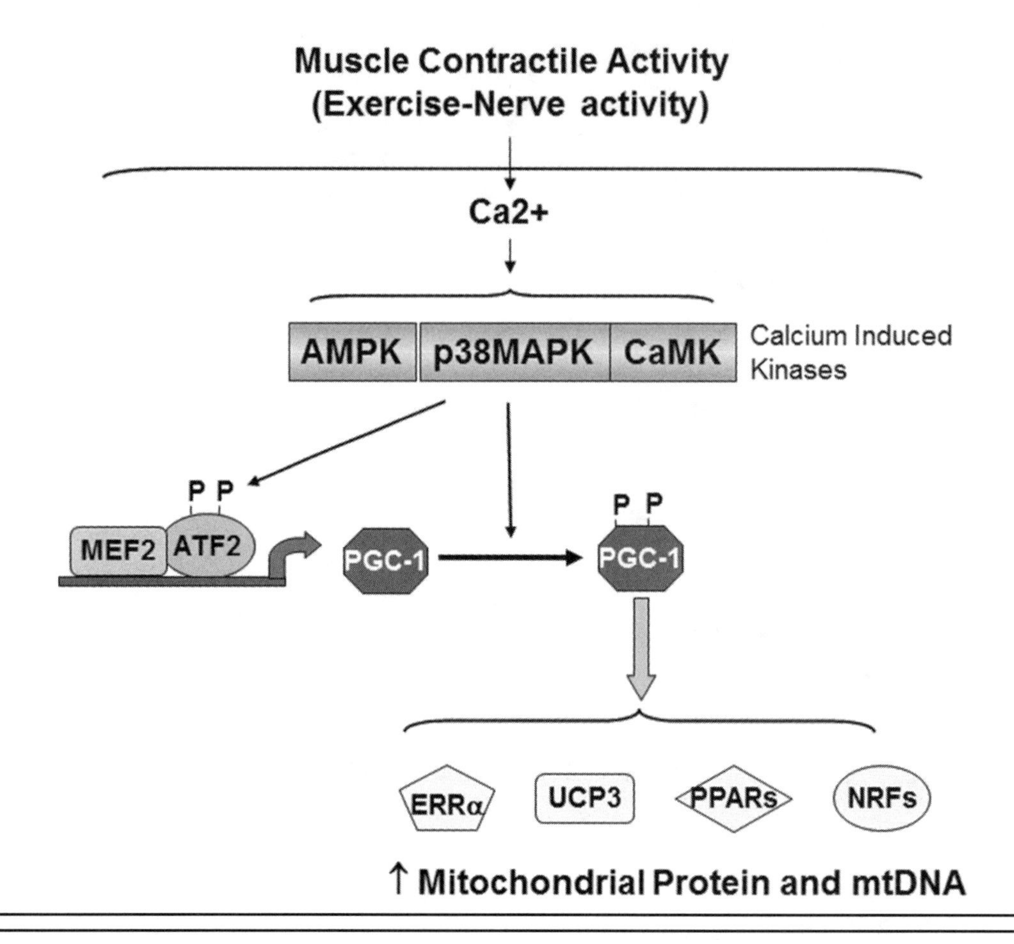

Figure 14.3 *(continued) (b)* Mitochondria biogenesis. (PGC-1α) is central in mitochondria biogenesis in response to increased muscle activity. Physical activity causes increased intracellular calcium. This event serves as the trigger to induce many kinases to phosphorylate PGC-1α. This response induces PGC-1α to translocate to the nucleus whereby it serves as a key coregulating transcription factor to regulate mitochondrial biogenesis (215, 216).

studies (144). For example, muscle growth may be induced via either hormone treatment (e.g., clenbuterol, testosterone) or mechanical overload, which appear to involve different mechanisms. Also, the age of the animal, the duration of the stimulus, and the choice of species may confound the interpretation of results. Rehfeldt and colleagues (166) suggested that in order to resolve the debate one needs to use mechanistic approaches that selectively disable the regenerative capacity of satellite cells without altering other systems.

Supporting Evidence

Studies by Rosenblatt and colleagues (170, 171) and Adams and colleagues (2) tested the hypotheses that interfering with satellite cell proliferation or differentiation processes via gamma irradiation treatment would prevent significant increments in muscle hypertrophy in response to functional overload. In both studies, irradiation was associated with the lack of muscle hypertrophy, supporting the need for satellite cells. In the study by Adams and colleagues (2), temporal analyses of protein, DNA, and myonuclei accumulation and concentration as well as markers of the Akt signaling pathway were performed. Results showed that myofibril, DNA, and myonuclei accumulation were equivalent between the irradiated and control overloaded groups during the first few days of overload. However, during the long-term course of the study (i.e., d 15-90), only the control muscle truly hypertrophied and accumulated DNA, total RNA (ribosomal), and new myonuclei compared with the irradiated group. Importantly, markers of growth factor stimulation and of satellite cell proliferation (cyclin D1) and differentiation (myogenin) were equivalent among the two groups, suggesting that the irradiated muscle was attempting to proliferate and incorporate potential differentiated satellite cells into the nuclear domain of the irradiated muscle. These collective findings suggest that there is indeed a requirement to maintain the nuclear domain in the face of muscle fiber myofibril protein expansion during extensive muscle enlargement in the compensatory hypertrophy model.

Recent studies by Marcas Bamman and colleagues (15, 153, 154) have addressed the issue of whether myonuclear addition to myofibers is required during skeletal muscle hypertrophy in humans. These investigators used K-means cluster analysis to classify 66 human subjects after 16 wk of knee-extension resistance training into three response categories based on the degree of myofiber hypertrophy: extreme, modest, and nonresponders. Mean fiber hypertrophy averaged 58%, 28%, and 0% across the three groups, respectively. The working hypothesis was that the robust hypertrophy seen in the extreme group was caused by a high level of satellite cell activation and incorporation into the myofibers. Their findings strongly support the idea that myonuclear addition via satellite cell recruitment is required in order to achieve substantial myofiber hypertrophy in humans. Individuals with a greater basal presence of satellite cells demonstrated with training a remarkable ability to expand the satellite cell pool, incorporate new nuclei, and achieve robust growth (154).

Opposing Evidence

In recent years the transgenic mouse has become a key model for studies on muscle plasticity given the technical achievements in generating transgenic lines to either knock out or overexpress a specific gene likely involved in muscle structure, function, and adaptation to physical activity. Two recent studies of relevance to the issue of hypertrophy mechanisms are noteworthy. Espan Spangenburg and colleagues (189) used a transgenic mouse model (called MKR) that expresses a dominant negative IGF-1 receptor specifically targeting skeletal muscle. The objective of the study was to ascertain whether muscle hypertrophy could be induced in the plantaris muscle by functional overload in the absence of IGF-1 activation of the Akt signaling pathway via its upstream target receptor. Wild-type control and MKR mouse subgroups were subject to 0, 7, and 35 d of functional overload. In the control state, the plantaris muscle mass of the wild type group was initially 11% greater than that of the MKR group. However, both groups responded to the overload stimulus with significant increases in muscle mass that averaged approximately 62%. Examination revealed marked activation of the Akt–mTORC1 signaling pathway in both experimental groups, suggesting that its activation was pivotal to the hypertrophy process. Based on these results, it was concluded that mechanical loading can induce muscle hypertrophy and activate the appropriate signaling pathways independent of a functional IGF-1 receptor. The implication is that IGF-1 may not be the only signaling trigger to induce the muscle hypertrophy cascade.

In another study of muscle hypertrophy involving mice, Blaauw and colleagues (24) performed a different type of transgenic manipulation by examining whether inducible activation of the Akt pathway in the absence of mechanical stimuli can generate muscle hypertrophy and result in satellite cell activation and incorporation into the growing muscle fibers. After 3 wk of Akt activation, the gastrocnemius muscle hypertrophied by 62% and muscle strength was proportionally increased, suggesting that the hypertrophy response was physiological. Other muscles such as the soleus and extensor digitorum longus (EDL) were hypertrophied as well. Interestingly, there was no evidence of myonuclear incorporation, suggesting that a major expansion of the myonuclear domain occurred without loss of function. The authors concluded that satellite cell incorporation into myofibers is not obligatory to the hypertrophy process.

These findings using transgenic mouse models are at odds with those findings observed in conventional studies. Also, studies using clenbuterol demonstrated muscle hypertrophy without the addition of myonuclei (131). Thus, the authors agree with O'Connor and colleagues (144) that better-controlled studies are needed to solve these discrepancies.

Role of Activity in Reversing Atrophy Responses to Unloading Stimuli: Importance of Resistance Exercise

Atrophy of skeletal muscle fibers occurs in response to states of unloading such as space flight and ground-based analogues such as hind-limb suspension in rodents and bed rest in humans. Unloading-induced atrophy appears to involve both sides of the protein-balance equation, which is defined as the ratio of protein synthesis rate divided by the protein degradation rate (i.e., synthesis rate/degradation rate) (198). When the degradation rate exceeds the synthesis rate, net protein loss (atrophy) occurs. Although the equation is simple, the mechanism is rather complex. Research in the past decade focused heavily on understanding the upstream events regulating these protein processes.

Alterations in Protein Synthesis

Accumulated evidence shows that the IGF-1 signaling pathway involving Akt–mTOR–p70S6K (figure 14.3*a*) is repressed in unloaded muscle (90, 106, 112). To make matters worse, evidence shows that transcription of myofibrillar protein (actin and myosin) is severely repressed in unloaded muscle. For instance, Julia Giger and colleagues (71) recently showed that one of the early responses contributing to the rapid atrophy of slow skeletal muscle during unloading involves the transcriptional repression of both the slow MHC and actin genes that manifests as a marked reduction in expression of both their respective premessenger RNA and messenger RNA pools. It is apparent that this response contributes to the very rapid loss of muscle mass (35%) that occurs in slow muscle during the first 7 d of unloading.

Alterations in Protein Degradation

The net atrophy response of a muscle fiber primarily involves targeting the myofibril fraction for degradation. In order to degrade the contractile proteins, which account for approximately 50% to 60% of the total protein pool, two events must occur. First, the myofibril machinery must undergo an initial process of proteolysis to disassemble the contractile machinery. This process is thought to be regulated by calcium-activated proteolytic enzymes such as the calpains and caspases (63, 133, 184). Next, these naked proteins become targeted for ubiquitination by a three-enzyme step reaction that involves polyligating the target protein with ubiquitin molecules. It is known that specific E3 ligase isozymes are responsible for the specificity of targeting a particular protein for destruction (26, 76, 141). Once ubiquitinated, the target protein is transported to the complex proteasomal machinery in the cytosol, in which the protein is progressively broken down into small peptides and eventually to free amino acids, which can be recycled. This degradation process is highly dependent on ATP as the energy source.

In recent years considerable insight has been gathered about the regulation of this degradation cascade. As presented in figure 14.3*a*, when the IGF-1 Akt–mTOR–P70S6K pathway is sufficiently activated, Akt phosphorylates a transcription factor of the forkhead family referred to as FOXO1. FOXO1 regulates gene expression of key ubiquitin E3 ligases such as muscle RING-finger protein 1 (MuRF-1) and atrogin 1/muscle atrophy F box (177, 192). When FOXO1 is phosphorylated it is transferred out of the nucleus and is no longer capable of activating transcription of atrogin 1 and MuRF-1, which target degradation of myosin (and actin) proteins (177, 192). Thus, when the loading state is reduced, as during hind-limb suspension involving the rat, Akt signaling pathways are reduced, FOXO1 becomes activated, and the E3 ligases become upregulated to elevate protein degradation processes. Also facilitating this degradation cascade is the activation of expression of myostatin, an antigrowth factor that is a member of the transforming growth factor beta family of proteins. In certain cattle species that possess a nonfunctional myostatin gene, the muscles become huge due to the loss of function of this important gene. Interestingly, myostatin is upregulated during unloading, and part of its action involves the ability to inhibit the Akt pathway, as noted in figure 14.3*a* (140, 200).

Muscle protein loss leading to atrophy is the result of closely interacting signals simultaneously affecting protein synthesis and degradation. A key question remains about whether any practical countermeasure can ameliorate the large disruption in the protein-balance equation. To address this question, Adams and colleagues (4) simultaneously initiated both a robust resistance-exercise paradigm and the unloading stimulus and trained one limb daily during this steep decline in muscle mass. The results demonstrated that the atrophy process involved total-muscle and myofibril fraction reduction, a reduction in total RNA, and an upregulation of the atrogin, MuRF-1, and myostatin genes. However, the program of heavy resistance exercise blunted these alterations and increased activity-mediated IGF-1 Akt–mTOR signaling responses that inhibit protein degradation mechanisms. Thus, it is apparent that loading state and ancillary intermittent high loading stimuli are the foundation for maintaining a positive protein balance in skeletal muscle.

Mechanisms of Mitochondrial Biosynthesis Regulation Impacting Muscle Performance

It is well documented that repetitive endurance exercise is a primary inducer of mitochondrial biogenesis. Investigators have been interested in the mechanism of this important adaptation given its broad physiological effect on exercise performance in general and its effect on general health (104). In early studies, long-term training was used to induce muscle mitochondria biogenesis, which provided the impression that the adaptive process was very slow (103). However, later studies (57) have shown that adaptation occurs rapidly and involves a complex coordinated expression and assembly of approximately 1,500 proteins encoded by both the nuclear and mitochondrial genomes (42, 179). Due to the complexity of processes in mitochondrial biogenesis, progress toward understanding the mechanisms has been slow. Two important discoveries were critical in understanding mitochondria biogenesis in the past decade. First, Scarpulla and colleagues discovered nuclear respiratory factors 1 and 2 (64, 206, 207), which are transcription factors for several nuclear genes encoding mitochondrial proteins such as the enzymes of oxidative phosphorylation (cytochrome c, succinic acid dehydrogenase, cytochrome oxidase [COX], ATP synthase) as well as factors involved with mitochondrial DNA replication factors (Tfam) and heme biogenesis (figure 14.3*b*). The second discovery, led by Bruce Speigelman's group (110), identified an inducible coactivator (PGC-1) that activates nuclear respiratory factors necessary for mitochondrial DNA synthesis as well as several transcription factors that orchestrate the regulation of many nuclear and mitochondrial genes encoding mitochondrial proteins. Several studies, as reviewed by Holloszy (103), showed that either a single bout of exercise or repetitive muscle contraction can induce a rapid increase in PGC-1α transcription in skeletal muscle in both humans and animals.

In subsequent studies by members of John Holloszy's research team (215, 216), several key advances were achieved concerning the mechanisms of exercise-induced mitochondrial biogenesis. First, they demonstrated that exercise induction of mitochondrial biogenesis occurs before any upregulation of the PGC-1α gene occurs, suggesting that other scenarios are operating in the induction process (216). Using an in vitro model of mitochondria biogenesis in C2C12 muscle cells, the Holloszy group showed that p38-MAPK is first activated and, in turn, phosphorylates PGC-1α, which is located primarily in the cytosol. This phosphorylation process enables PGC-1α to enter the nucleus to become a transcription coactivator. In the same year, Speigelman's group reported a similar mechanism of PGC-1α activation via AMPK (110).

Another paper by the Holloszy group (215) demonstrated that p38-MAPK phosphorylates and activates another transcription factor, referred to as activating transcription factor 2, that interacts with myocyte enhancing factor 2 (MEF2) on the PGC-1α promoter to activate its transcription. It was further demonstrated that activating transcription factor 2, when activated, interacts with another transcription factor. Thus, the role of PGC-1α in mitochondrial biogenesis is biphasic; it involves an initial phase of phosphorylation and translocation and a later phase that increases PGC-1α levels via increased transcriptional regulation.

What happens if muscle-specific PGC-1α expression is knocked out? Handschip and colleagues (94, 95) generated a muscle-specific PGC-1α knockout mouse model and showed that these animals presented with reduced locomotive activity, low maximal exercise tolerance, and reduced muscle function, which are consistent with reduced oxidative capacity of skeletal muscle. However, these mice exhibited normal voluntary running activity along with shift from Type IIb to IIa fibers in skeletal muscle in response to the exercise training. However, expression of mitochondrial enzymes (cytochrome c and cytochrome oxidase IV [COXIV]) induced by endurance exercise was attenuated. These findings on PGC-1α muscle knock out animals illustrate the importance of PGC-1α in mitochondrial biogenesis but not fiber type transformations involving Type II MHC transformations. The reader is encouraged to examine excellent review articles on the broader issues of PGC-1α in muscle metabolism and organism function (28, 103, 104).

Transcriptional Regulation of Contractile Phenotype Switching in Response to Altered Activity and Loading States

Muscle plasticity involves changes in gene expression. Therefore, it is important to study how gene expression is regulated. Gene regulation can be achieved at many levels, including the transcriptional, posttranscriptional, pretranslational, translational, and posttranslational levels. Transcription—the first step in gene expression in which the gene (DNA) is transcribed into RNA (pre-messenger RNA) in the nucleus—is a key regulator of mRNA availability as substrate for protein translation. In this section we primarily discuss MHC gene transcription because MHC phenotype is a major determinant of contractile function. Based on the inherent expression of messenger RNA and protein expression, there is strong evidence that transcriptional regulation underlies MHC plasticity. We discuss various layers of transcription regulation that are thought to be involved in response to altered activity.

Approaches in Studying Gene Transcription in Response to Altered Activity Paradigms

In eukaryotic cells, transcription is very complex and involves interactions between RNA polymerase II and the gene promoter as well many protein–protein and protein–DNA interactions. The key questions pertaining to muscle plasticity in response to altered loading state are how transcription regulation occurs and what factors are involved. Traditionally, the nuclear run-on assay provides a direct measure of transcription acting as a snapshot in time. Darryl Neufer and colleagues used this approach in determining the increase in transcription of genes of metabolic function during exercise and recovery (99, 160, 161). In contrast to nuclear run on-assays, promoter analyses via a reporter assay system can provide an indirect measure of gene transcription. In this approach, a DNA fragment of the promoter (containing regulatory elements) of a specific gene is linked to a reporter gene complementary DNA. Upon transfection into cells, the promoter becomes transcriptionally active and transcribes the reporter gene, which is translated into protein. A variety of reporter proteins—initially chloramphenicol acetyl transferase, then the firefly and Renilla luciferases and various fluorescent proteins—have been used. These readouts are not normally expressed in muscle tissue, their activities represent the activity of the linked promoter, and they can be assayed using highly sensitive assay systems. Two approaches have been used in promoter studies: transgenic animals and a transient assay system in cultured cells or in intact muscle after transfection with plasmid DNA. The use of transgenic animals to study regulation of muscle gene promoters in response to overloading and loading was pioneered by Richard Tsika's group. They studied the muscle creatine kinase, beta MHC, and slow myosin light chain 1 gene promoters in transgenic mice in overloaded plantaris and soleus muscles of rats suspended from the hind limb as compared to normal control mice (113, 132, 201, 205, 208). These studies implicated many transcription factors in the regulation of fast-to-slow and slow-to-fast phenotype transitions. For example, they show that transcription enhancing factor-1(TEF-1), palustrisedoxin beta (Pur β), and transcription factor specificity protein 3(SP3) promote slow Type I MHC gene expression.

The use of direct gene transfection in muscle was made possible after the breakthrough discovery by Wolff and colleagues that striated muscles have the capability to take up plasmid DNA (214). The transfected plasmid DNA locates into the nucleus in an episomal fashion; that is, it does not integrate into the genome, as in the case of transgenic animals. Because muscle fibers are postmitotic, the plasmid DNA remains in the nucleus of muscle fibers for long durations (>1 mo). This discovery was attractive for developing vaccines or gene therapy strategies, and muscle molecular biologists took advantage of it in order to study promoter activity via reporter systems. This method is faster and much less expensive to develop compared with the transgenic approach. For example, using this approach it was found that the activity of the alpha skeletal actin promoter is not specific to fiber type (46). Members of the Baldwin group have used this approach to study the slow Type I MHC gene promoter activity in response to decreased loading (72, 74, 75) and overloading (73). More recently, Gary McCall and colleagues studied the IIb promoter activity in slow muscle in response to decreased loading (130). Although these gene transfection studies can be useful in giving specific information on the function of cis acting (binding site) or transacting factors (transcription factors) in regulating promoter activity, the promoter in this approach does not represent its native state in the nuclear milieu. Consequently, the promoter often fails to be active at all or to represent muscle type specificity or regulation (129, 146). It was speculated that certain promoters may not succeed because they lack the chromatin assembly; these genes are found in the chromosomes in their native nuclear milieu. Furthermore, another limitation of studying promoter fragments is that the DNA does not encompass all potential regulatory elements for the corresponding gene. In some cases, a regulatory element might be several kilobases away in upstream or downstream sequences relative to the transcription start site.

In a more recent approach, premessenger RNA expression has been used as a marker for gene transcription. It has been shown that premessenger RNA levels correlate well with transcription of genes, as measured by nuclear run-on assays (62). The premessenger RNA is determined based on reverse transcription polymerase chain reaction (RT-PCR) using primers targeting intronic sequences for amplification (91, 147). The expansion of genomic DNA databases in the 21st century has facilitated the primer design for specific target genes. Analyzing premessenger RNA has an advantage in determining rapid change in gene expression that could be missed when studying protein or even messenger RNA, both of which possess significantly longer half-life.

Signaling Pathways Controlling MHC Gene Transcription During Exercise

The use of transgenic mice, promoter analyses, muscle gene transfer, and modern approaches such as gain and loss of function of a gene have collectively enriched the knowledge on the regulatory mechanisms involved with the regulation of MHC gene transcription. As a result, it appears that the nuclear factor of activated T-cell (NFAT) calcineurin (Cn) signaling is widely accepted to be a major influence in activity-induced altered fiber type (see 22 and 182 for reviews).

Cn Signaling and Slow MHC Gene During Altered Activity Patterns

Cn is a calcium- and calmodulin-regulated serine–threonine phosphatase that acts on the NFAT family of transcription factors. Muscle intracellular calcium titer increases in response to mechanical activation, which activates Cn activity. Activated Cn dephosphorylates NFATs, which promotes their translocation into the nucleus to regulate target genes (MHCs and others). NFAT is also thought to interact with other transcription factors such as MEF2 involved with muscle gene transcription (6, 217, 218). The role of the Cn signaling pathway in regulating muscle gene expression was intensely investigated in the past decade (43, 58, 135, 145, 183). Most data support the role of Cn–NFAT in activity-induced gene regulation; however, there is some disagreement about its role in Type I MHC gene regulation (72). Recently, a study by Stefano Schiaffino's group (183) suggested that Cn–NFAT signaling acts as a nerve activity sensor in skeletal muscle in vivo and controls activity-dependent MHC switching.

Clay Pandorf and colleagues (149) took a different approach to delineating the role of Cn in phenotype remodeling, particularly its role in driving expression of Type I MHC, by using a novel strategy in which a profound transition from fast to slow was examined in the absence and presence of cyclosporine A, a Cn inhibitor. To induce the fast-to-slow transition, rats were first subjected to 7 d of hind-limb unloading plus thyroid hormone treatment. This intervention represses the Type I MHC gene transcription while de novo activating the Type IIx and IIb MHCs. HS and thyroid treatment then were withdrawn and the rats resumed normal weight bearing and ambulation, during which either vehicle or optimal doses of cyclosporine A were administered for 7 and 14 d. The findings demonstrated that, despite significant inhibition of Cn, premessenger RNA, messenger RNA, and protein abundance of Type I MHC increased markedly during reloading relative to the prereloading conditions. Type I MHC expression was, however, attenuated compared with vehicle treatment. In addition, Type IIa and IIx MHC premina, messenger RNA, and protein levels were increased in cyclosporine A-treated rats relative to vehicle-treated rats. These findings suggest that in adult slow prototype muscle, Cn has a modulatory role in MHC transcription rather than a major role as the primary regulator of the slow MHC in adult skeletal muscle. These findings are in line with a large number of studies that have been addressed and summarized in this chapter concerning Cn–NFAT regulation in the past decade (149).

Epigenetics and Muscle Gene Regulation in Response Unloading and to Exercise

Epigenetics is a new and rapidly growing research field that investigates heritable alterations in chromosome function and gene expression caused by mechanisms other than changes in DNA sequence. Epigenetic mechanisms are diverse but can be classified into three interacting areas involving modulation of the chromatin–histone structure (methylation, acetylation, phosphorylation), DNA methylation, and noncoding RNA such as microRNA and long noncoding intergenic and antisense RNA.

Posttranslational modification of histones and DNA methylation can alter the chromatin structure to become either active or repressive for transcription (see figure 14.4). Recent studies have shown that histone modifications at specific genes can also occur dynamically and rapidly in response to environmental changes to alter gene expression. For example, in carrying out recent studies on the plasticity of the MHC gene family in response to altered loading state, the Baldwin group discovered two types of epigenetic phenomena. The first involves the expression of antisense RNAs in the fast MHC gene locus in which the MHC genes are organized in tandem on the same chromosome. These antisense RNAs allow adjacent genes to cross-talk and to coordinate regulation of neighboring MHC genes (147, 168). Second, repression of slow MHCs and activation of fast MHCs (and vice versa) in a given muscle involve altered patterns of acetylation and methylation of the histones that regulate expression of MHCs (e.g., slow to fast and fast to slow depending on the loading conditions) (148). These studies suggest that histone acetyltransferases (HATs) and histonedeacetylases (HDACs) associate with the chromatin of the MHC promoters as presented in the model in figure 14.4; however, the identity of these HATs and HDACs remains unknown. Also, the state of repressive histone methylation and associated histone-modifying enzymes are unknown.

In mammals, more than a dozen HDAC family members have been identified and can be categorized into three classes based on their structure, complex formation, and expression pattern. Members of the HDAC class II group (HDAC 4, 5, 7, and 9) are highly expressed in the heart and skeletal muscle (88). Recent studies implicated class II HDACs with coordinated muscle gene expression in response to exercise (136, 137). Furthermore, there is evidence that HDACs interact with myogenin (196), NFAT (122, 185), and MEF2 (163). MEF2 has been implicated in fiber type-specific gene expression (162, 219).

Gene manipulations in transgenic mice have identified the involvement of class II HDACs with the regulation of fiber phenotype (162, 163). Also, it was found that these class II HDAC effects are mediated via myogenin and MEF2 and their downstream targets (138). Other studies have shown that NFATs and MyoD are involved with recruitment of HATs on gene promoters (not necessarily MHC) (121, 122). Even though these studies show a role for class II HDACs in skeletal muscle remodeling via specific transcription factors (TFs), it has not been demonstrated how these HDACs and HATs affect the chromatin and coordinate the regulation of the four adult MHCs (I, IIa, IIx, IIb). Furthermore, little is known about the role of repressive

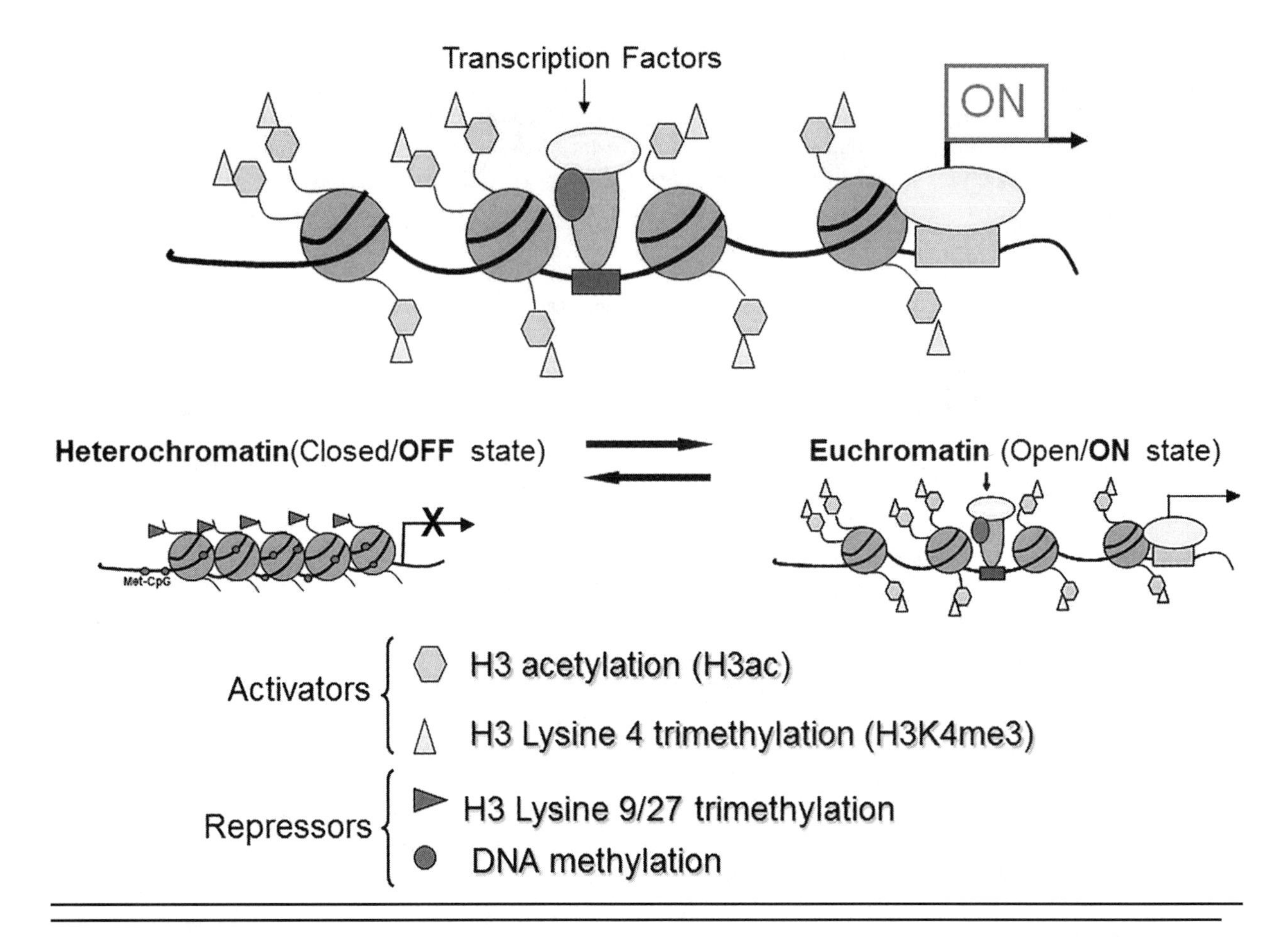

Figure 14.4 Chromatin state and gene transcription. Model for chromatin factors interacting with transcription factors to regulate transcription of a gene. Histone modifications and DNA methylation are important factors in regulating the chromatin from active to repressed and vice versa. Histone H3 acetylation and histone H3 methylation and lysine 4 are associated with an active chromatin state. In contrast, histone H3 methylation at lysine 9 or lysine 27 as well as DNA methylation are associated with repressive chromatin state. Chromatin is in a dynamic equilibrium between the two states.

histone methylation in the regulation of muscle MHC genes. Is there any interplay among the regulatory networks for each MHC isoform? It is likely that a complex network of interactions exists among various histone-modifying enzymes and transcription factors in the regulation of MHC isoforms to confer specificity of fiber type and adaptability. Discovering interplay among factors will be critical for better understanding the regulation of MHC genes in different conditions.

The notion that environmentally induced epigenetic traits have an effect on future generations has important ramifications for future research involving diet and exercise. For example, can diet and exercise induce specific epigenetic modulations that serve as a countermeasure for some disorders and thus help in overcoming human genetic weaknesses and predisposition to certain diseases (30)?

Role of Noncoding Antisense RNA During Altered Loading States

The observation that genomic organization of the MHC genes and their tandem arrangement has been conserved for millions of years raised questions about whether the gene arrangement is of functional significance in their regulation. Recent evidence implicated a noncoding antisense RNA in the coordinated regulation of two tandemly positioned genes, which emphasizes the importance of the genomic organization of these MHC genes in their coordinated regulation.

In 2003, Fadia Haddad and colleagues (91) made the novel discovery that a naturally occurring antisense RNA to the cardiac β-MHC gene is involved in cardiac MHC gene regulation in cardiac muscle. Cardiac α- and β-MHC isoforms are the products of two distinct genes that are organized in tandem in a head-to-tail position on the chromosome in the order of β → α (125) and are separated by a 4.5 kb intergenic space (figure 14.2). A long noncoding antisense RNA is transcribed from the DNA strand that is opposite to the MHC genes, creating a beta antisense RNA (91). This beta antisense transcript was implicated with the MHC isoform gene switching in the heart in response to diabetes and hypothyroidism (91). Given this observation, a series of studies was subsequently carried out on skeletal muscle to ascertain whether antisense expression in slow and fast skeletal muscle contributes to the patterns of MHC expression in response to unloading and overloading stimuli.

In 2006, Clay Pandorf and colleagues (147) published a paper that investigated Type II MHC gene regulation in slow Type I soleus muscle fibers undergoing a slow-to-fast MHC transformation in response to unloading of the target muscle 7 d after SI in rats. Transcriptional products of both the sense and antisense strands across the IIa-IIx-IIb MHC gene locus were examined. Results showed that the messenger RNA and premessenger RNA of each MHC had a similar response to the SI, suggesting regulation of these genes at the transcriptional level. In addition, detection of previously unknown antisense strand transcription occurred that produced natural antisense transcripts (NATs). RT-PCR mapping of the RNA products revealed that the antisense activity resulted in the formation of three major products: aII, xII, and bII NATs (i.e., antisense products of the IIa, IIx, and IIb genes, respectively). The upshot of this experiment is that the SI-induced inactivity caused marked inhibition of both the slow Type I and Type IIa genes along with upregulation of both the IIx and IIb genes. Thus, the inactivity model negatively affects transcription of the Type I MHC gene directly by inhibiting its promoter and uses antisense expression to primarily repress the IIa MHC gene promoter, thereby creating a slow-to-fast fiber switch of the normally slow soleus muscle. Importantly, this observation explains the existence of Type I/IIx fibers in the study by Caiozzo and colleagues (38) and in nulling out the transition schemes proposed by Pette and Staron that state that MHC transition in muscle fibers occurs in the precise order of I↔IIa↔IIx↔IIb (158, 159).

What about exercise? In 2008, Chiara Rinaldi and colleagues (168) investigated the effect of resistance-exercise training on regulation of fast MHCs in the white gastrocnemius muscle of rats. Resistance exercise causes a rapid transition from IIb to IIx expression in this muscle, which made it a good model to test the hypothesis whether an antisense RNA to the IIx MHC is involved. Results show that when IIb expression is greater than 90% in the normal white gastrocnemius, there is a strong antisense RNA to IIx (xII NAT) that starts in the intergenic region between the IIx and IIb genes and extends to overlap the IIx gene. The xII NAT expression becomes downregulated in the trained muscle, along with IIb, whereas the expression of IIx becomes upregulated. Data analyses implicated the xII antisense with tight coordinated and antithetical regulation between IIx and IIb genes in this model. These results (147, 168) suggest that noncoding RNA expression is a critical regulator of Type II MHC expression in both fast and slow muscles undergoing alterations in mechanical loading states.

Role of MicroRNA

Another epigenetic regulator that is worth discussing is microRNA. MicroRNAs are small noncoding RNAs that regulate gene expression at the posttranscriptional level by controlling the level of available messenger RNA in influencing their stability (178). These highly conserved, approximately 21-mer RNAs regulate the expression of genes by binding to the 3' untranslated regions of specific messenger RNAs. Each microRNA could act posttranscriptionally to target hundreds of messenger RNAs for translational repression, degradation, or destabilization. They are involved in many aspects of cell function and play a significant role in disease development. Research suggests that microRNAs are major regulators of gene expression and thus are part of the adaptive response (41). MicroRNAs together with transcription factors generate a complex combinatorial code regulating gene expression. Thus, identifying and targeting microRNA transcription factor gene networks may be a potent approach in future research in exercise science as applied to therapy and disease prevention. Recently, microRNA-mediated regulation was suggested to be involved in the complex regulatory networks that govern skeletal adaptation to endurance exercise in both mice (174) and humans (142).

Concluding Comments

During the past century, especially the past 50 yr, great strides have been achieved in understanding plasticity of skeletal muscle fiber types in response to alterations in physical activity, inactivity and unloading, and hormonal state. A large body of information has been gathered and we now have clear insight into the mechanisms that drive important phenomena such as muscle hypertrophy and atrophy, mitochondrial biogenesis and substrate energy metabolism, and contractile gene switching. As we point to the future, the importance of physical activity as medicine and the need to enhance muscle health as an ingredient for general health and longevity in order to ameliorate a variety of degenerative disorders is rapidly emerging. The challenge will be to establish and understand the prescriptions that are needed as we turn to a new chapter in the exercise sciences.

References

1. Adams GR. Satellite cell proliferation and skeletal muscle hypertrophy. *Appl Physiol Nutr Metab* 31: 782-790, 2006.
2. Adams GR, Caiozzo VJ, Haddad F, Baldwin KM. Cellular and molecular responses to increased skeletal muscle loading after irradiation. *Am J Physiol* 283: C1182-C1195, 2002.
3. Adams GR, Haddad F. The relationships among IGF-1, DNA content, and protein accumulation during skeletal muscle hypertrophy. *J Appl Physiol* 81: 2509-2516, 1996.
4. Adams GR, Haddad F, Bodell PW, Tran PD, Baldwin KM. Combined isometric, concentric, and eccentric resistance exercise prevents unloading-induced muscle atrophy in rats. *J Appl Physiol* 103: 1644-1654, 2007.
5. Adams GR, McCue SA. Localized infusion of IGF-I results in skeletal muscle hypertrophy in rats. *J Appl Physiol* 84: 1716-1722, 1998.
6. Allen DL, Sartorius CA, Sycuro LK, Leinwand LA. Different pathways regulate expression of the skeletal myosin heavy chain genes. *J Biol Chem* 276: 43524-43533, 2001.
7. Baldwin K, Reitman J, Terjung R, Winder W, Holloszy J. Substrate depletion in different types of muscle and in liver during prolonged running. *Am J Physiol* 225: 1045-1050, 1973.
8. Baldwin K, Winder W, Terjung R, Holloszy J. Glycolytic enzymes in different types of skeletal muscle: Adaptation to exercise. *Am J Physiol* 225: 962-966, 1973.
9. Baldwin KM, Cheadle WG, Martinez OM, Cooke DA. Effect of functional overload on enzyme levels in different types of skeletal muscle. *J Appl Physiol* 42: 312-317, 1977.
10. Baldwin KM, Haddad F. Effects of different activity and inactivity paradigms on myosin heavy chain gene expression in striated muscle. *J Appl Physiol* 90: 345-357, 2001.
11. Baldwin KM, Klinkerfuss GH, Terjung RL, Mole PA, Holloszy JO. Respiratory capacity of white, red, and intermediate muscle: Adaptative response to exercise. *Am J Physiol* 222: 373-378, 1972.
12. Baldwin KM, Roy RR, Sacks RD, Blanco C, Edgerton VR. Relative independence of metabolic enzymes and neuromuscular activity. *J Appl Physiol* 56: 1602-1607, 1984.
13. Baldwin KM, Valdez V, Herrick RE, MacIntosh AM, Roy RR. Biochemical properties of overloaded fast-twitch skeletal muscle. *J Appl Physiol* 52: 467-472, 1982.
14. Baldwin KM, Valdez V, Schrader LF, Herrick RE. Effect of functional overload on substrate oxidation capacity of skeletal muscle. *J Appl Physiol* 50: 1272-1276, 1981.
15. Bamman MM, Petrella JK, Kim J-S, Mayhew DL, Cross JM. Cluster analysis tests the importance of myogenic gene expression during myofiber hypertrophy in humans. *J Appl Physiol* 102: 2232-2239, 2007.
16. Barany M. ATPase activity of myosin correlated with speed of muscle shortening. *J Gen Physiol* 50: 197-216, 1967.
17. Barnard RJ, Edgerton VR, Peter JB. Effect of exercise on skeletal muscle. I. Biochemical and histochemical properties. *J Appl Physiol* 28: 762-766, 1970.
18. Barnard RJ, Edgerton VR, Peter JB. Effect of exercise on skeletal muscle. II. Contractile properties. *J Appl Physiol* 28: 767-770, 1970.
19. Barnard RJ, Holloszy JO. The metabolic systems: Aerobic metabolism and substrate utilization in exercising skeletal muscle. In: Tipton CM, ed. *Exercise Physiology: People and Ideas*. New York: Oxfrord University Press, 2003, 292-321.
20. Barton-Davis ER, Shoturma DI, Musaro A, Rosenthal N, Sweeney HL. Viral mediated expression of insulin-like growth factor I blocks the aging-related loss of skeletal muscle function. *Proc Natl Acad Sci* 95: 15603-15607, 1998.
21. Barton-Davis ER, Shoturma DI, Sweeney HL. Contribution of satellite cells to IGF-I induced hypertrophy of skeletal muscle. *Acta Physiol Scand* 167: 301-305, 1999.
22. Bassel-Duby R, Olson EN. Signaling pathways in skeletal muscle remodeling. *Ann Rev Biochem* 75: 19-37, 2006.
23. Bassett DR. Scientific contributions of A.V. Hill: Exercise physiology pioneer. *J Appl Physiol* 93: 1567-1582, 2002.
24. Blaauw B, Canato M, Agatea L, Toniolo L, Mammucari C, Masiero E, Abraham R, et al. Inducible activation of Akt increases skeletal muscle mass and force without satellite cell activation. *FASEB J* 23: 3896-3905, 2009.
25. Bodine SC. mTOR signaling and the molecular adaptation to resistance exercise. *Med Sci Sports Exerc* 38: 1950-1957, 2006.
26. Bodine SC, Latres E, Baumhueter S, Lai VK, Nunez L, Clarke BA, Poueymirou WT, et al. Identification of ubiquitin ligases required for skeletal muscle atrophy. *Science* 294: 1704-1708, 2001.
27. Bodine SC, Stitt TN, Gonzalez M, Kline WO, Stover GL, Bauerlein R, Zlotchenko E, et al. Akt/mTOR pathway is a crucial regulator of skeletal

muscle hypertrophy and can prevent muscle atrophy in vivo. *Nat Cell Biol* 3: 1014-1019, 2001.

28. Bonen A. PGC-1α-induced improvements in skeletal muscle metabolism and insulin sensitivity. *Appl Physiol Nutr Metab* 34: 307-314, 2009.
29. Booth FW. Perspectives on molecular and cellular exercise physiology. *J Appl Physiol* 65: 1461-1471, 1988.
30. Booth FW, Laye MJ. Lack of adequate appreciation of physical exercise's complexities can pre-empt appropriate design and interpretation in scientific discovery. *J Physiol (Lond)* 587: 5527-5539, 2009.
31. Bottinelli R, Canepari M, Pellegrino MA, Reggiani C. Force-velocity properties of human skeletal muscle fibres: Myosin heavy chain isoform and temperature dependence. *J Physiol (Lond)* 495: 573-586, 1996.
32. Bottinelli R, Pellegrino MA, Canepari M, Rossi R, Reggiani C. Specific contributions of various muscle fibre types to human muscle performance: An in vitro study. *J Electromyogr Kinesiol* 9: 87-95, 1999.
33. Bottinelli R, Schiaffino S, Reggiani C. Force-velocity relations and myosin heavy chain isoform compositions of skinned fibres from rat skeletal muscle. *J Appl Physiol* 437: 655-672, 1991.
34. Brooke MH, Kaiser KK. Three "myosin adenosine triphosphatase" systems: The nature of their pH lability and sulfhydryl dependence. *J Histochem Cytochem* 18: 670-672, 1970.
35. Buller AJ, Eccles JC, Eccles RM. Interactions between motoneurones and muscles in respect of the characteristic speeds of their responses. *J Physiol (Lond)* 150: 417-439, 1960.
36. Burke RE. Sir Charles Sherrington's *The integrative action of the nervous system*: A centenary appreciation. *Brain* 130: 887-894, 2007.
37. Burke RE, Levine DN, Zajac FE. Mammalian motor units: Physiological-histochemical correlation in three types in cat gastrocnemius. *Science* 174: 709-712, 1971.
38. Caiozzo VJ, Baker MJ, Baldwin KM. Novel transitions in MHC isoforms: Separate and combined effects of thyroid hormone and mechanical unloading. *J Appl Physiol* 85: 2237-2248, 1998.
39. Caiozzo VJ, Baker MJ, McCue SA, Baldwin KM. Single-fiber and whole muscle analyses of MHC isoform plasticity: Interaction between T3 and unloading. *Am J Physiol* 273: C944-C952, 1997.
40. Caiozzo VJ, Haddad F. Thyroid hormone: Modulation of muscle structure, function, and adaptive responses to mechanical loading. *Exerc Sport Sci Rev* 24: 321-361, 1996.
41. Callis TE, Deng Z, Chen J-F, Wang D-Z. Muscling through the microRNA world. *Exp Biol Med* 233: 131-138, 2008.
42. Calvo S, Jain M, Xie X, Sheth SA, Chang B, Goldberger OA, Spinazzola A, et al. Systematic identification of human mitochondrial disease genes through integrative genomics. *Nat Genet* 38: 576-582, 2006.
43. Chin ER, Olson EN, Richardson JA, Yang Q, Humphries C, Shelton JM, Wu H, et al. A calcineurin-dependent transcriptional pathway controls skeletal muscle fiber type. *Genes Dev* 12: 2499-2509, 1998.
44. Coleman ME, DeMayo F, Yin KC, Lee HM, Geske R, Montgomery C, Schwartz RJ. Myogenic vector expression of insulin-like growth factor I stimulates muscle cell differentiation and myofiber hypertrophy in transgenic mice. *J Biol Chem* 270: 12109-12116, 1995.
45. Coolican SA, Samuel DS, Ewton DZ, McWade FJ, Florini JR. The mitogenic and myogenic actions of insulin-like growth factors utilize distinct signaling pathways. *J Biol Chem* 272: 6653-6662, 1997.
46. Corin S, Levitt L, O'Mahoney J, Joya J, Hardeman E, Wade R. Delineation of a slow-twitch-myofiber-specific transcriptional element by using in vivo somatic gene transfer. *Proc Natl Acad Sci* 92: 6185-6189, 1995.
47. Costill DL, Fink WJ, Getchell LH, Ivy JL, Witzmann FA. Lipid metabolism in skeletal muscle of endurance-trained males and females. *J Appl Physiol* 47: 787-791, 1979.
48. d'Albis A, Gratzer WB. Electrophoretic examination of native myosin. *FEBS Lett* 29: 292-296, 1973.
49. d'Albis A, Pantaloni C, Bechet J-J. An electrophoretic study of native myosin isozymes and of their subunit content. *Eur J Biochem* 99: 261-272, 1979.
50. DeNardi C, Ausoni S, Moretti P, Gorza L, Velleca M, Buckingham M, Schiaffino S. Type 2X-myosin heavy chain is coded by a muscle fiber type-specific and developmentally regulated gene. *J Cell Biol* 123: 823-835, 1993.
51. Denny-Brown DE. The histological features of striped muscle in relation to its functional activity. *Proc R Soc Lond B* 104: 371-411, 1929.
52. Di Maso NA, Caiozzo VJ, Baldwin KM. Single-fiber myosin heavy chain polymorphism during postnatal development: Modulation by hypothyroidism. *Am J Physiol* 278: R1099-R1106, 2000.
53. Diffee GM, Caiozzo VJ, McCue SA, Herrick RE, Baldwin KM. Activity-induced regulation of myosin isoform distribution: Comparison of two contractile activity programs. *J Appl Physiol* 74: 2509-2516, 1993.
54. Donselaar Y, Eerbeek O, Kernell D, Verhey BA. Fibre sizes and histochemical staining characteristics in normal and chronically stimulated fast muscle of cat. *J Physiol (Lond)* 382: 237-254, 1987.
55. Dubowitz V, Pearse AGE. A comparative histochemical study of oxidative enzyme and phosphorylase activity in skeletal muscle. *Histochem Cell Biol* 2: 105-117, 1960.
56. Dubowitz V, Pearse AGE. Reciprocal relationship of phosphorylase and oxidative enzymes in skeletal muscle. *Nature* 185: 701-702, 1960.

57. Dudley GA, Abraham WM, Terjung RL. Influence of exercise intensity and duration on biochemical adaptations in skeletal muscle. *J Appl Physiol* 53: 844-850, 1982.
58. Dunn SE, Simard AR, Bassel-Duby R, Williams RS, Michel RN. Nerve activity-dependent modulation of calcineurin signaling in adult fast and slow skeletal muscle fibers. *J Biol Chem* 276: 45243-45254, 2001.
59. Eccles JC, Sherrington CS. Numbers and contraction-values of individual motor-units examined in some muscles of the limb. *Proc Royal Soc Lond Series B* 106: 326-357, 1930.
60. Edström L, Kugelberg E. Histochemical composition, distribution of fibres and fatiguability of single motor units. Anterior tibial muscle of the rat. *J Neurol Neurosurg Psychiat* 31: 424-433, 1968.
61. Edstrom L, Nystrom BO. Histochemical types and sizes of fibers in normal human muscles. *Acta Neurol Scand* 45: 257-269, 1969.
62. Elferink CJ, Reiners JJ Jr. Quantitative RT-PCR on CYP1A1 heterogeneous nuclear RNA: A surrogate for the in vitro transcription run-on assay. *Biotechniques* 20: 470-477, 1996.
63. Enns D, Raastad T, Ugelstad I, Belcastro A. Calpain/calpastatin activities and substrate depletion patterns during hindlimb unweighting and reweighting in skeletal muscle. *Eur J Appl Physiol* 100: 445-455, 2007.
64. Evans MJ, Scarpulla RC. NRF-1: A trans-activator of nuclear-encoded respiratory genes in animal cells. *Genes Dev* 4: 1023-1034, 1990.
65. Faulkner J, Maxwell L, Brook D, Lieberman D. Adaptation of guinea pig plantaris muscle fibers to endurance training. *Am J Physiol* 221: 291-297, 1971.
66. Favier F, Benoit H, Freyssenet D. Cellular and molecular events controlling skeletal muscle mass in response to altered use. *Pflungers Arch* 456: 587-600, 2008.
67. Fitts R, Booth F, Winder W, Holloszy J. Skeletal muscle respiratory capacity, endurance, and glycogen utilization. *Am J Physiol* 228: 1029-1033, 1975.
68. Fitzsimons DP, Herrick RE, Baldwin KM. Isomyosin distributions in rodent muscles: Effects of altered thyroid state. *J Appl Physiol* 69: 321-327, 1990.
69. Gauthier G, Lowey S. Polymorphism of myosin among skeletal muscle fiber types. *J Cell Biol* 74: 760-779, 1977.
70. Gauthier GF, Lowey S. Distribution of myosin isoenzymes among skeletal muscle fiber types. *J Cell Biol* 81: 10-25, 1979.
71. Giger JM, Bodell PW, Zeng M, Baldwin KM, Haddad F. Rapid muscle atrophy response to unloading: Pretranslational processes involving MHC and actin. *J Appl Physiol* 107: 1204-1212, 2009.
72. Giger JM, Haddad F, Qin AX, Baldwin KM. Effect of cyclosporin A treatment on the in vivo regulation of type I MHC gene expression. *J Appl Physiol* 97: 475-483, 2004.
73. Giger JM, Haddad F, Qin AX, Baldwin KM. Functional overload increases b-MHC promoter activity in rodent fast muscle via the proximal MCAT (be3) site. *Am J Physiol* 282: C518-C527, 2002.
74. Giger JM, Haddad F, Qin AX, Baldwin KM. In vivo regulation of the beta-myosin heavy chain gene in soleus muscle of suspended and weight-bearing rats. *Am J Physiol* 278: C1153-C1161, 2000.
75. Giger JM, Haddad F, Qin AX, Zeng M, Baldwin KM. Effect of unloading on type I myosin heavy chain gene regulation in rat soleus muscle. *J Appl Physiol* 98: 1185-1194, 2005.
76. Glass DJ. Skeletal muscle hypertrophy and atrophy signaling pathways. *Int J Biochem Cell Biol* 37: 1974-1984, 2005.
77. Gollnick PD, Armstrong RB, Saltin B, Saubert CW, Sembrowich WL, Shepherd RE. Effect of training on enzyme activity and fiber composition of human skeletal muscle. *J Appl Physiol* 34: 107-111, 1973.
78. Gollnick PD, Armstrong RB, Saubert CW, Piehl K, Saltin B. Enzyme activity and fiber composition in skeletal muscle of untrained and trained men. *J Appl Physiol* 33: 312-319, 1972.
79. Gollnick PD, Armstrong RB, Sembrowich WL, Shepherd RE, Saltin B. Glycogen depletion pattern in human skeletal muscle fibers after heavy exercise. *J Appl Physiol* 34: 615-618, 1973.
80. Gollnick PD, Karlsson J, Piehl K, Saltin B. Selective glycogen depletion in skeletal muscle fibres of man following sustained contractions. *J Physiol (Lond)* 241: 59-67, 1974.
81. Gollnick PD, Piehl K, Saubert CW, Armstrong RB, Saltin B. Diet, exercise, and glycogen changes in human muscle fibers. *J Appl Physiol* 33: 421-425, 1972.
82. Gould MK, Rawlinson WA. Biochemical adaptation as a response to exercise. Effect of swimming on the levels of lactic dehydrogenase, malic dehydrogenase and phosphorylase in muscles of 8-, 11- and 15-week-old rats. *Biochem J* 73: 41-44, 1959.
83. Graham SC, Roy RR, Navarro C, Jiang B, Pierotti D, Bodine-Fowler S, Edgerton VR. Enzyme and size profiles in chronically inactive cat soleus muscle fibers. *Muscle Nerve* 15: 27-36, 1992.
84. Gregory P, Low RB, Stirewalt WS. Changes in skeletal-muscle myosin isoenzymes with hypertrophy and exercise. *Biochem J* 238: 55-63, 1986.
85. Guth L, Yellin H. The dynamic nature of the so-called "fiber types" of mammalian skeletal muscle. *Exp Neurol* 31: 277-300, 1971.
86. Gutmann E, Hájek I, Horský P. Effect of excessive use on contraction and metabolic properties of cross-striated muscle. *J Physiol (Lond)* 203: 46P-47P, 1969.
87. Gutmann E, Schiaffino S, Hanzliková V. Mechanism of compensatory hypertrophy in skeletal muscle of the rat. *Exp Neurol* 31: 451-464, 1971.

88. Haberland M, Montgomery RL, Olson EN. The many roles of histone deacetylases in development and physiology: Implications for disease and therapy. *Nat Rev Genet* 10: 32-42, 2009.
89. Haddad F, Adams GR. Inhibition of MAP/ERK kinase prevents IGF-I-induced hypertrophy in rat muscles. *J Appl Physiol* 96: 203-210, 2004.
90. Haddad F, Adams GR, Bodell PW, Baldwin KM. Isometric resistance exercise fails to counteract skeletal muscle atrophy processes during the initial stages of unloading. *J Appl Physiol* 100: 433-441, 2006.
91. Haddad F, Bodell PW, Qin AX, Giger JM, Baldwin KM. Role of antisense RNA in coordinating cardiac myosin heavy chain gene switching. *J Biol Chem* 278: 37132-37138, 2003.
92. Haddad F, Roy RR, Zhong H, Edgerton VR, Baldwin KM. Atrophy responses to muscle inactivity: I. Cellular markers of protein deficit. *J Appl Physiol* 95: 781-790, 2003.
93. Haddad F, Roy RR, Zhong H, Edgerton VR, Baldwin KM. Atrophy responses to muscle inactivity: II. Molecular markers of protein deficit. *J Appl Physiol* 95: 791-802, 2003.
94. Handschin C, Chin S, Li P, Liu F, Maratos-Flier E, LeBrasseur NK, Yan Z, et al. Skeletal muscle fiber-type switching, exercise intolerance, and myopathy in PGC-1α muscle-specific knock-out animals. *J Biol Chem* 282: 30014-30021, 2007.
95. Handschin C, Choi C, Chin S, Kim S, Kawamori D, Kurpad A, Neubauer N, et al. Abnormal glucose homeostasis in skeletal muscle-specific PGC-1alpha knockout mice reveals skeletal muscle-pancreatic beta cell crosstalk. *J Clin Invest* 117: 3463-3474, 2007.
96. Harridge S, Bottinelli R, Canepari M, Pellegrino M, Reggiani C, Esbjörnsson M, Saltin B. Whole-muscle and single-fibre contractile properties and myosin heavy chain isoforms in humans. *Pflügers Arch* 432: 913-920, 1996.
97. Hearn GR, Wainio WW. Succinic dehydrogenase activity of the heart and skeletal muscle of exercised rats. *Am J Physiol* 185: 348-350, 1956.
98. Henriksson J. Training induced adaptation of skeletal muscle and metabolism during submaximal exercise. *J Physiol (Lond)* 270: 661-675, 1977.
99. Hildebrandt AL, Neufer PD. Exercise attenuates the fasting-induced transcriptional activation of metabolic genes in skeletal muscle. *Am J Physiol* 278: E1078-E1086, 2000.
100. Hill AV. The heat of shortening and the dynamic constants of muscle. *Proc R Soc Lond (Biol)* 126: 136-195, 1938.
101. Hoh JF. Neural regulation of mammalian fast and slow muscle myosins: An electrophoretic analysis. *Biochem* 14: 742-747, 1975.
102. Holloszy JO. Biochemical adaptations in muscle. Effects of exercise on mitochondrial oxygen uptake and respiratory enzyme activity in skeletal muscle. *J Biol Chem* 242: 2278-2282, 1967.
103. Holloszy J. Regulation by exercise of skeletal muscle content of mitochondria and GLUT4. *J Physiol Pharmacol* 59(Suppl. 7): 5-18, 2008.
104. Hood DA. Mechanisms of exercise-induced mitochondrial biogenesis in skeletal muscle. *Appl Physiol Nutr Metab* 34: 465-472, 2009.
105. Hooker AM, Baldwin KM. Substrate oxidation specificity in different types of mammalian muscle. *Am J Physiol* 236: C66-C69, 1979.
106. Hornberger TA, Hunter RB, Kandarian SC, Esser KA. Regulation of translation factors during hindlimb unloading and denervation of skeletal muscle in rats. *Am J Physiol* 281: C179-C187, 2001.
107. Hunt CC. The reflex activity of mammalian small-nerve fibres. *J Physiol (Lond)* 115: 456-469, 1951.
108. Huxley HE. Muscular contraction and cell motility. *Nature* 243: 445-449, 1973.
109. Izumo S, Nadal-Ginard B, Mahdavi V. All members of the MHC multigene family respond to thyroid hormone in a highly tissue-specific manner. *Science* 231: 597-600, 1986.
110. Jäger S, Handschin C, St.-Pierre J, Spiegelman BM. AMP-activated protein kinase (AMPK) action in skeletal muscle via direct phosphorylation of PGC-1α. *Proc Natl Acad Sci* 104: 12017-12022, 2007.
111. Jiang BA, Roy RR, Navarro C, Nguyen Q, Pierotti D, Edgerton VR. Enzymatic responses of cat medial gastrocnemius fibers to chronic inactivity. *J Appl Physiol* 70: 231-239, 1991.
112. Kandarian SC, Jackman RW. Intracellular signaling during skeletal muscle atrophy. *Muscle Nerve* 33: 155-165, 2006.
113. Karasseva N, Tsika G, Ji J, Zhang A, Mao X, Tsika R. Transcription enhancer factor 1 binds multiple muscle MEF2 and A/T-rich elements during fast-to-slow skeletal muscle fiber type transitions. *Mol Cell Biol* 23: 5143-5164, 2003.
114. Karlsson J, Saltin B. Diet, muscle glycogen, and endurance performance. *J Appl Physiol* 31: 203-206, 1971.
115. LaFramboise WA, Daood MJ, Guthrie RD, Moretti P, Schiaffino S, Ontell M. Electrophoretic separation and immunological identification of type 2X myosin heavy chain in rat skeletal muscle. *Biochim Biophys Acta* 1035: 109-112, 1990.
116. Lawrie RA. The activity of the cytochrome system in muscle and its relation to myoglobin. *Biochem J* 55: 298-305, 1953.
117. Leinwand LA, Saez L, McNally E, Nadal-Ginard B. Isolation and characterization of human myosin heavy chain genes. *Proc Natl Acad Sci* 80: 3716-3720, 1983.
118. Leksell L. The action potential and excitatory effects of the small ventral root fibers to skeletal muscle. *Acta Physiol Scand* 10(Suppl. 31): 1-84, 1945.

119. Lesch M, Parmley W, Hamosh M, Kaufman S, Sonnenblick E. Effects of acute hypertrophy on the contractile properties of skeletal muscle. *Am J Physiol* 214: 685-690, 1968.
120. Liddell EGT, Sherrington CS. Recruitment and some other features of reflex inhibition. *Proc Roy Soc London B* 97: 488-518, 1925.
121. Liu Y, Randall W, Schneider M. Activity-dependent and -independent nuclear fluxes of HDAC4 mediated by different kinases in adult skeletal muscle. *J Cell Biol* 168: 887-897, 2005.
122. Liu Y, Shen T, Randall W, Schneider M. Signaling pathways in activity-dependent fiber type plasticity in adult skeletal muscle. *J Muscle Res Cell Motil* 26: 13-21, 2005.
123. Lovely RG, Gregor RJ, Roy RR, Edgerton VR. Effects of training on the recovery of full-weight-bearing stepping in the adult spinal cat. *Exp Neurol* 92: 421-435, 1986.
124. Lowey S, Risby D. Light chains from fast and slow muscle myosins. *Nature* 234: 81-85, 1971.
125. Mahdavi V, Chambers AP, Nadal-Ginard B. Cardiac alpha- and beta-myosin heavy chain genes are organized in tandem. *Proc Natl Acad Sci USA* 81: 2626-2630, 1984.
126. Mahdavi V, Izumo S, Nadal-Ginard B. Developmental and hormonal regulation of sarcomeric myosin heavy chain gene family. *Circ Res* 60: 804-814, 1987.
127. Mauro A. Satellite cell of skeletal muscle fibers. *J Biophys Biochem Cytol* 9: 493-495, 1961.
128. Mayer RF. The motor unit and electromyography—The legacy of Derek Denny-Brown. *J Neurol Sci* 189: 7-11, 2001.
129. McCall GE, Allen DL, Haddad F, Baldwin KM. Transcriptional regulation of IGF-I expression in skeletal muscle. *Am J Physiol Cell Physiol* 285: C831-C839, 2003.
130. McCall GE, Haddad F, Roy RR, Zhong H, Edgerton VR, Baldwin KM. Transcriptional regulation of the myosin heavy chain IIB gene in inactive rat soleus. *Muscle Nerve* 40: 411-419, 2009.
131. McCarthy JJ, Esser KA. Counterpoint: Satellite cell addition is not obligatory for skeletal muscle hypertrophy. *J Appl Physiol* 103: 1100-1102, 2007.
132. McCarthy JJ, Vyas DR, Tsika GL, Tsika RW. Segregated regulatory elements direct beta-myosin heavy chain expression in response to altered muscle activity. *J Biol Chem* 274: 14270-14279, 1999.
133. McClung JM, Kavazis AN, Whidden MA, DeRuisseau KC, Falk DJ, Criswell DS, Powers SK. Antioxidant administration attenuates mechanical ventilation-induced rat diaphragm muscle atrophy independent of protein kinase B (PKB Akt) signalling. *J Physiol (Lond)* 585: 203-215, 2007.
134. McComas AJ, ed. *The Neuromuscular System*. New York: Oxford University Press, 2003, 39-97.
135. McCullagh KJA, Calabria E, Pallafacchina G, Ciciliot S, Serrano AL, Argentini C, Kalhovde JM, et al. NFAT is a nerve activity sensor in skeletal muscle and controls activity-dependent myosin switching. *Proc Natl Acad Sci* 101: 10590-10595, 2004.
136. McGee SL, Fairlie E, Garnham AP, Hargreaves M. Exercise-induced histone modifications in human skeletal muscle. *J Physiol (Lond)* 587: 5951-5958, 2009.
137. McGee SL, Hargreaves M. Histone modifications and exercise adaptations. *J Appl Physiol* 110: 258-263, 2011.
138. McKinsey TA, Zhang CL, Olson EN. Control of muscle development by dueling HATs and HDACs. *Curr Opin Gen Dev* 11: 497-504, 2001.
139. Molé PA, Oscai LB, Holloszy JO. Adaptation of muscle to exercise. Increase in levels of palmityl CoA synthetase, carnitine palmityltransferase, and palmityl CoA dehydrogenase, and in the capacity to oxidize fatty acids. *J Clin Invest* 50: 2323-2330, 1971.
140. Morissette MR, Cook SA, Buranasombati C, Rosenberg MA, Rosenzweig A. Myostatin inhibits IGF-I-induced myotube hypertrophy through Akt. *Am J Physiol* 297: 1124-1132, 2009.
141. Myung J, Kim KB, Crews CM. The ubiquitin-proteasome pathway and proteasome inhibitors. *Med Res Rev* 21: 245-273, 2001.
142. Nielsen S, Scheele C, Yfanti C, Åkerström T, Nielsen AR, Pedersen BK, Laye M. Muscle specific microRNAs are regulated by endurance exercise in human skeletal muscle. *J Physiol (Lond)* 588: 4029-4037, 2010.
143. O'Connor RS, Pavlath GK. Point/counterpoint: Satellite cell addition is/is not obligatory for skeletal muscle hypertrophy. *J Appl Physiol* 103: 1099-1100, 2007.
144. O'Connor RS, Pavlath GK, McCarthy JJ, Esser KA. Last word on point/counterpoint: Satellite cell addition is/is not obligatory for skeletal muscle hypertrophy. *J Appl Physiol* 103: 1107, 2007.
145. Olson EN, Williams RS. Remodeling muscles with calcineurin. *Bioessays* 22: 510-519, 2000.
146. Pandorf CE, Haddad F, Qin AX, Baldwin KM. IIx myosin heavy chain promoter regulation cannot be characterized in vivo by direct gene transfer. *Am J Physiol* 293: C1338-C1346, 2007.
147. Pandorf CE, Haddad F, Roy RR, Qin AX, Edgerton VR, Baldwin KM. Dynamics of myosin heavy chain gene regulation in slow skeletal muscle: Role of natural antisense RNA. *J Biol Chem* 281: 38330-38342, 2006.
148. Pandorf CE, Haddad F, Wright C, Bodell PW, Baldwin KM. Differential epigenetic modifications of histones at the myosin heavy chain genes in fast and slow skeletal muscle fibers and in response to muscle unloading. *Am J Physiol* 297: C6-C16, 2009.
149. Pandorf CE, Jiang WH, Qin AX, Bodell PW, Baldwin KM, Haddad F. Calcineurin plays a modulatory role in loading-induced regulation of type I

myosin heavy chain gene expression in slow skeletal muscle. *Am J Physiol* 297: R1037-R1048, 2009.

150. Paul MH, Sperling E. Cyclophorase system. XXIII. Correlation of cyclophorase activity and mitochondrial density in striated muscle. *Proc Soc Exp Biol Med* 79: 352-354, 1952.
151. Periasamy M, Gregory P, Martin BJ, Stirewalt WS. Regulation of myosin heavy-chain gene expression during skeletal-muscle hypertrophy. *Biochem J* 257: 691-698, 1989.
152. Peter JB, Barnard RJ, Edgerton VR, Gillespie CA, Stempel KE. Metabolic profiles of three fiber types of skeletal muscle in guinea pigs and rabbits. *Biochem* 11: 2627-2633, 1972.
153. Petrella JK, Kim J-S, Cross JM, Kosek DJ, Bamman MM. Efficacy of myonuclear addition may explain differential myofiber growth among resistance-trained young and older men and women. *Am J Physiol* 291: E937-E946, 2006.
154. Petrella JK, Kim J-S, Mayhew DL, Cross JM, Bamman MM. Potent myofiber hypertrophy during resistance training in humans is associated with satellite cell-mediated myonuclear addition: A cluster analysis. *J Appl Physiol* 104: 1736-1742, 2008.
155. Pette D. Plasticity in skeletal, cardiac, and smooth muscle: Historical perspectives: Plasticity of mammalian skeletal muscle. *J Appl Physiol* 90: 1119-1124, 2001.
156. Pette D, Henriksson J, Emmerich M. Myofibrillar protein patterns of single fibres from human muscle. *FEBS Lett* 103: 152-155, 1979.
157. Pette D, Müller W, Leisner E, Vrbová G. Time dependent effects on contractile properties, fibre population, myosin light chains and enzymes of energy metabolism in intermittently and continuously stimulated fast twitch muscles of the rabbit. *Pflügers Arch* 364: 103-112, 1976.
158. Pette D, Staron RS. Myosin isoforms, muscle fiber types, and transitions. *Microsc Res Tech* 50: 500-509, 2000.
159. Pette D, Staron RS. Transitions of muscle fiber phenotypic profiles: Myosin isoforms, muscle fiber types, and transitions. *Histochem Cell Biol* 115: 359-372, 2001.
160. Pilegaard H, Ordway GA, Saltin B, Neufer PD. Transcriptional regulation of gene expression in human skeletal muscle during recovery from exercise. *Am J Physiol* 279: E806-E814, 2000.
161. Pilegaard H, Osada T, Andersen LT, Helge JW, Saltin B, Neufer PD. Substrate availability and transcriptional regulation of metabolic genes in human skeletal muscle during recovery from exercise. *Metabolism* 54: 1048-1055, 2005.
162. Potthoff MJ, Olson EN, Bassel-Duby R. Skeletal muscle remodeling. *Curr Opinion Rheumatol* 19: 542-549, 2007.
163. Potthoff MJ, Wu H, Arnold MA, Shelton JM, Backs J, McAnally J, Richardson JA, et al. Histone deacetylase degradation and MEF2 activation promote the formation of slow-twitch myofibers. *J Clin Invest* 117: 2459-2467, 2007.
164. Ranvier L. De quelques traits relatifs á l'histologie et á la physiologie des muscles striés. *Arch Physiol Norm Pathol* 6: 1-15, 1874.
165. Ranvier L. Propriétés et structures différents des muscles rouges et des muscles blancs, chez les lapins et chez les raies. *CR Acad Sci (Paris)* 77: 1030-1034, 1873.
166. Rehfeldt C, Mantilla CB, Sieck GC, Hikida RS, Booth FW, Kadi F, Bodine SC, et al. Satellite cell addition is/is not obligatory for skeletal muscle hypertrophy. *J Appl Physiol* 103: 1104-1106, 2007.
167. Reiser PJ, Moss RL, Giulian GG, Greaser ML. Shortening velocity in single fibers from adult rabbit soleus muscles is correlated with myosin heavy chain composition. *J Biol Chem* 260: 9077-9080, 1985.
168. Rinaldi C, Haddad F, Bodell PW, Qin AX, Jiang W, Baldwin KM. Intergenic bidirectional promoter and cooperative regulation of the IIx and IIb MHC genes in fast skeletal muscle. *Am J Physiol* 295: 208-218, 2008.
169. Rommel C, Bodine SC, Clarke BA, Rossman R, Nunez L, Stitt TN, Yancopoulos GD, et al. Mediation of IGF-1-induced skeletal myotube hypertrophy by PI(3)K/Akt/mTOR and PI(3)K/Akt/GSK3 pathways. *Nat Cell Biol* 3: 1009-1013, 2001.
170. Rosenblatt JD, Parry DJ. Gamma irradiation prevents compensatory hypertrophy of overloaded mouse extensor digitorum longus muscle. *J Appl Physiol* 73: 2538-2543, 1992.
171. Rosenblatt JD, Yong D, Parry DJ. Satellite cell activity is required for hypertrophy of overloaded adult rat muscle. *Muscle Nerve* 17: 608-613, 1994.
172. Roy RR, Baldwin KM, Martin TP, Chimarusti SP, Edgerton VR. Biochemical and physiological changes in overloaded rat fast- and slow-twitch ankle extensors. *J Appl Physiol* 59: 639-646, 1985.
173. Roy RR, Meadows ID, Baldwin KM, Edgerton VR. Functional significance of compensatory overloaded rat fast muscle. *J Appl Physiol* 52: 473-478, 1982.
174. Safdar A, Abadi A, Akhtar M, Hettinga BP, Tarnopolsky MA. miRNA in the regulation of skeletal muscle adaptation to acute endurance exercise in C57Bl/6J male mice. *PLoS ONE* 4: e5610, 2009.
175. Salmons S, Vrbová G. The influence of activity on some contractile characteristics of mammalian fast and slow muscles. *J Physiol (Lond)* 201: 535-549, 1969.
176. Samaha FJ, Guth L, Albers RW. The neural regulation of gene expression in the muscle cell. *Exp Neurol* 27: 276-282, 1970.
177. Sandri M, Sandri C, Gilbert A, Skurk C, Calabria E, Picard A, Walsh K, et al. Foxo transcription factors induce the atrophy-related ubiquitin ligase atrogin-1 and cause skeletal muscle atrophy. *Cell* 117: 399-412, 2004.

178. Saunders M, Lim L. (micro)Genomic medicine: MicroRNAs as therapeutics and biomarkers. *RNA Biol* 6: 324-328, 2009.
179. Scarpulla RC. Transcriptional paradigms in mammalian mitochondrial biogenesis and function. *Physiol Rev* 88: 611-638, 2008.
180. Schiaffino S, Bormioli SP. Adaptive changes in developing rat skeletal muscle in response to functional overload. *Exp Neurol* 40: 126-137, 1973.
181. Schiaffino S, Gorza L, Sartore S, Saggin L, Ausoni S, Vianello M, Gundersen K, et al. Three myosin heavy chain isoforms in type 2 skeletal muscle fibres. *J Muscle Res Cell Motil* 10: 197-205, 1989.
182. Schiaffino S, Sandri M, Murgia M. Activity-dependent signaling pathways controlling muscle diversity and plasticity. *Physiology* 22: 269-278, 2007.
183. Serrano AL, Murgia M, Pallafacchina G, Calabria E, Coniglio P, Lomo T, Schiaffino S. Calcineurin controls nerve activity-dependent specification of slow skeletal muscle fibers but not muscle growth. *Proc Natl Acad Sci* 98: 13108-13113, 2001.
184. Servais S, Letexier D, Favier R, Duchamp C, Desplanches D. Prevention of unloading-induced atrophy by vitamin E supplementation: Links between oxidative stress and soleus muscle proteolysis? *Free Radic Biol Med* 42: 627-635, 2007.
185. Shen T, Liu Y, Randall W, Schneider M. Parallel mechanisms for resting nucleo-cytoplasmic shuttling and activity dependent translocation provide dual control of transcriptional regulators HDAC and NFAT in skeletal muscle fiber type plasticity. *J Muscle Res Cell Motil* 27: 405-411, 2006.
186. Smerdu V, Karsch-Mizrachi I, Campione M, Leinwand L, Schiaffino S. Type IIx myosin heavy chain transcripts are expressed in type IIb fibers of human skeletal muscle. *Am J Physiol* 267: C1723-C1728, 1994.
187. Smith JL, Edgerton VR, Eldred E, Zernicke RF. The chronic spinalized cat: A model for neuromuscular plasticity. *Birth Defects Orig Artic Ser* 19: 353-373, 1983.
188. Soussi-Yanicostas N, Whalen RG, Petit C. Five skeletal myosin heavy chain genes are organized as a multigene complex in the human genome. *Hum Mol Genet* 2: 563-569, 1993.
189. Spangenburg EE, Le Roith D, Ward CW, Bodine SC. A functional insulin-like growth factor receptor is not necessary for load-induced skeletal muscle hypertrophy. *J Physiol (Lond)* 586: 283-291, 2008.
190. Staron RS, Pette D. Correlation between myofibrillar ATPase activity and myosin heavy chain composition in rabbit muscle fibers. *Histochem Cell Biol* 86: 19-23, 1986.
191. Stein JM, Padykula HA. Histochemical classification of individual skeletal muscle fibers of the rat. *Am J Anat* 110: 103-123, 1962.
192. Stitt TN, Drujan D, Clarke BA, Panaro F, Timofeyva Y, Kline WO, Gonzalez M, et al. The IGF-1/PI3K/Akt pathway prevents expression of muscle atrophy-induced ubiquitin ligases by inhibiting FOXO transcription factors. *Mol Cell* 14: 395-403, 2004.
193. Swoap SJ, Haddad F, Caiozzo VJ, Herrick RE, McCue SA, Baldwin KM. Interaction of thyroid hormone and functional overload on skeletal muscle isomyosin expression. *J Appl Physiol* 77: 621-629, 1994.
194. Szent-Györgyi AG. The early history of the biochemistry of muscle contraction. *J Gen Physiol* 123: 631-641, 2004.
195. Talmadge RJ, Roy RR. Electrophoretic separation of rat skeletal muscle myosin heavy-chain isoforms. *J Appl Physiol* 75: 2337-2340, 1993.
196. Tang H, Macpherson P, Marvin M, Meadows E, Klein WH, Yang X-J, Goldman D. A histone deacetylase 4/myogenin positive feedback loop coordinates denervation-dependent gene induction and suppression. *Mol Biol Cell* 20: 1120-1131, 2009.
197. Thomason DB, Baldwin KM, Herrick RE. Myosin isozyme distribution in rodent hindlimb skeletal muscle. *J Appl Physiol* 60: 1923-1931, 1986.
198. Thomason DB, Booth FW. Atrophy of the soleus muscle by hindlimb unweighting. *J Appl Physiol* 68: 1-12, 1990.
199. Tomanek RJ. A histochemical study of postnatal differentiation of skeletal muscle with reference to functional overload. *Dev Biol* 42: 305-314, 1975.
200. Trendelenburg AU, Meyer A, Rohner D, Boyle J, Hatakeyama S, Glass DJ. Myostatin reduces Akt/TORC1/p70S6K signaling, inhibiting myoblast differentiation and myotube size. *Am J Physiol* 296: C1258-C1270, 2009.
201. Tsika G, Ji J, Tsika R. Sp3 proteins negatively regulate {beta} myosin heavy chain gene expression during skeletal muscle inactivity. *Mol Cell Biol* 24: 10777-10791, 2004.
202. Tsika RW, Herrick RE, Baldwin KM. Interaction of compensatory overload and hindlimb suspension on myosin isoform expression. *J Appl Physiol* 62: 2180-2186, 1987.
203. Tsika RW, Herrick RE, Baldwin KM. Subunit composition of rodent isomyosins and their distribution in hindlimb skeletal muscles. *J Appl Physiol* 63: 2101-2110, 1987.
204. Tsika RW, Herrick RE, Baldwin KM. Time course adaptations in rat skeletal muscle isomyosins during compensatory growth and regression. *J Appl Physiol* 63: 2111-2121, 1987.
205. Tsika RW, McCarthy J, Karasseva N, Ou Y, Tsika GL. Divergence in species and regulatory role of beta-myosin heavy chain proximal promoter muscle-CAT elements. *Am J Physiol* 283: C1761-C1775, 2002.
206. Virbasius JV, Scarpulla RC. Activation of the human mitochondrial transcription factor A gene by nuclear respiratory factors: A potential regulatory link between nuclear and mitochondrial gene ex-

pression in organelle biogenesis. *Proc Natl Acad Sci* 91: 1309-1313, 1994.

207. Virbasius JV, Virbasius CA, Scarpulla RC. Identity of GABP with NRF-2, a multisubunit activator of cytochrome oxidase expression, reveals a cellular role for an ETS domain activator of viral promoters. *Genes Dev* 7: 380-392, 1993.
208. Vyas DR, McCarthy JJ, Tsika GL, Tsika RW. Multiprotein complex formation at the beta myosin heavy chain distal muscle CAT element correlates with slow muscle expression but not mechanical overload responsiveness. *J Biol Chem* 276: 1173-1184, 2001.
209. Weeds AG. Light chains of myosin. *Nature* 223: 1362-1364, 1969.
210. Weiss A, McDonough D, Wertman B, Acakpo-Satchivi L, Montgomery K, Kucherlapati R, Leinwand L, et al. Organization of human and mouse skeletal myosin heavy chain gene clusters is highly conserved. *Proc Natl Acad Sci USA* 96: 2958-2963, 1999.
211. Weydert A, Daubas P, Lazaridis I, Barton P, Garner I, Leader DP, Bonhomme F, et al. Genes for skeletal muscle myosin heavy chains are clustered and are not located on the same mouse chromosome as a cardiac myosin heavy chain gene. *Proc Natl Acad Sci* 82: 7183-7187, 1985.
212. Widrick JJ, Romatowski JG, Karhanek M, Fitts RH. Contractile properties of rat, rhesus monkey, and human type I muscle fibers. *Am J Physiol* 272: R34-R42, 1997.
213. Winder WW, Baldwin KM, Holloszy JO. Exercise-induced adaptive increase in rate of oxidation of beta-hydroxybutyrate by skeletal muscle. *Proc Soc Exp Biol Med* 143: 753-755, 1973.
214. Wolff JA, Malone RW, Williams P, Chong W, Acsadi G, Jani A, Felgner PL. Direct gene transfer into mouse muscle in vivo. *Science* 247: 1465-1468, 1990.
215. Wright DC, Geiger PC, Han D-H, Jones TE, Holloszy JO. Calcium induces increases in peroxisome proliferator-activated receptor γ coactivator-1α and mitochondrial biogenesis by a pathway leading to p38 mitogen-activated protein kinase activation. *J Biol Chem* 282: 18793-18799, 2007.
216. Wright DC, Han D-H, Garcia-Roves PM, Geiger PC, Jones TE, Holloszy JO. Exercise-induced mitochondrial biogenesis begins before the increase in muscle PGC-1α expression. *J Biol Chem* 282: 194-199, 2007.
217. Wu H, Naya FJ, McKinsey TA, Mercer B, Shelton JM, Chin ER, Simard AR, et al. MEF2 responds to multiple calcium-regulated signals in the control of skeletal muscle fiber type. *Embo J* 19: 1963-1973, 2000.
218. Wu H, Rothermel B, Kanatous S, Rosenberg P, Naya FJ, Shelton JM, Hutcheson KA, et al. Activation of MEF2 by muscle activity is mediated through a calcineurin-dependent pathway. *EMBO J* 20: 6414-6423, 2001.
219. Wu Y, Koenig RJ. Gene regulation by thyroid hormone. *Trends Endocrinol Metab* 11: 207-211, 2000.
220. Wu Z, Puigserver P, Andersson U, Zhang C, Adelmant G, Mootha V, Troy A, et al. Mechanisms controlling mitochondrial biogenesis and respiration through the thermogenic coactivator PGC-1. *Cell* 98: 115-124, 1999.
221. Wydro RM, Nguyen HT, Gubits RM, Nadal-Ginard B. Characterisation of sacromeric myosin heavy chain genes. *J Biol Chem* 258: 670-678, 1983.
222. Yamaguchi A, Ikeda Y, Hirai T, Fujikawa T, Morita I. Local changes of IGF-I mRNA, GH receptor mRNA, and fiber size in rat plantaris muscle following compensatory overload. *Jap J Physiol* 53: 53-60, 2003.
223. Zhong H, Roy RR, Siengthai B, Edgerton VR. Effects of inactivity on fiber size and myonuclear number in rat soleus muscle. *J Appl Physiol* 99: 1494-1499, 2005.

CHAPTER 15

The Endocrine System: Actions of Select Hormones

Peter A. Farrell, PhD

Henrik Galbo, MD, DMS

Introduction

Meeting the demands of prolonged heavy exercise requires the coordinated effort of essentially all the systems of the body. It is clear that the endocrine system has important functions. During exercise, hormones function to regulate the following:

- Fuel mobilization and utilization
- Growth in response to overload
- Blood flow distribution
- Bone stability (balance between synthesis and degradation of bone)
- Select psychological adjustments to exercise
- Select aspects of temperature regulation
- Select aspects of reproductive function
- Chronotropic and ionotropic regulation of heart rate
- Muscle contractility
- Kidney filtration and reabsorption

Those functions require the action of most hormones in the body. However, due to space limitations, we have chosen to cover insulin, epinephrine, growth hormone, and cortisol.

This chapter highlights many of the most notable accomplishments (table 15.1) by exercise scientists who have sought to determine the effects of exercise on endocrinology and, perhaps more important, how endocrinology allows successful human movement and adaptations to that movement when exercise is repeated on a regular basis. The chapter, which covers the period from 1910 to 2010, has several limitations: A length limitation was necessarily invoked, we did not have full access to research published in languages other than English, and, perhaps most important, we made a judgment call about the relative importance of particular lines of investigation. Consequently, we fully recognize that some outstanding scientists and their work have not been adequately reviewed in this chapter.

Most of the chapter is organized according to specific hormones. Within that organization, we concentrate on the historical record which provided an understanding of the importance of a hormone for a specific physiological function. We present, in mostly chronological order, important studies on the role of a hormone in a function during exercise. Before starting the body of the chapter, we feel it is informative to highlight one scientist who used organ ablation and replacement strategies.

In the absence of detailed knowledge of the various hormones and absence of purified hormones and methods for selectively suppressing their secretion and blocking their receptors, much of the work in the early 20th century used organ limitation and replacement strategies. Endocrinology lends itself to this strategy because an entire endocrine gland can be removed (in most cases without immediate fatality to the animal) and the effects of exercise with or without that gland could be determined. The drawback to this strategy is that all endocrine glands secrete more than one hormone and therefore it is impossible to assign a specific change in function during exercise to a specific hormone. In an attempt to overcome this limitation, most organ elimination studies in the 1930s and 1940s used replacement strategies in which a specific (or not) organ extract was replaced.

Dwight J. Ingle from the Division of Experimental Medicine at the Mayo Foundation, and later from the UpJohn Company in Michigan and the Ben May Laboratory at the University of Chicago, developed an anesthetized rat model in the early 1930s with which he could document the strength of gastrocnemius muscle

Table 15.1 Important Endocrine Findings (1910-2010)

Year	Investigators	Finding	Reference
1922	Hartman et al.	Used a bioresponse (pupil dilation) to show that exercise stimulates release of epinephrine.	98
1926	Lawrence	Demonstrated the importance of exercise for people with type 1 diabetes; stated that regular exercise helps reduce the amount of insulin needed to control glucose and suggested that exercise enhances the function of insulin.	167
1936	Ingle	Found that leg contractions stimulated by electrical current are diminished when part or all of the adrenal gland is removed from rats.	124
1940	Asmussen et al.	Determined the importance of adrenal hormones in fuel metabolism during exercise.	5
1950	Ingle and Nezamis	Used glucose tolerance as an indicator of glucose uptake to show that contractions stimulate glucose uptake when no insulin is present.	126
1955	Ingle	Summarized 20 yr of work on the influence of hormones on the ability to sustain muscle contractions.	122
1965	Hunter et al.	Showed that even low levels of exertion stimulated increased plasma concentrations of growth hormone.	118
1967	Goldberg	Demonstrated that hypophysectomized rats can hypertrophy muscle due to exercise.	79
1970	Pruett	In this and other studies, she demonstrated that plasma insulin concentrations decline during exercise.	193
1970	Pruett	Demonstrated the importance of diet before exercise in the insulin response.	192
1972	Tipton et al.	Demonstrated that adrenalectomized rats could exercise train and adapt fairly normally to prolonged exercise.	238
1972	Hartley et al.	Showed that exercise training altered the response of many hormones to subsequent acute maximal exercise.	96
1972	Hartley et al.	Showed that exercise training altered the response of many hormones to prolonged exercise.	97
1973	Hansen	Demonstrated that obese subjects have blunted or no growth hormone response to exercise.	94
1974	Rennie and Johnson	One of the first training studies to document reduced insulin responses to exercise and glucose stimulation.	198
1975	Brandenberger and Follenius	Showed that background pattern of cortisol must be known before the effects of exercise can be accurately assessed because cortisol is released in a circadian rhythm.	16
1977	Galbo et al.	Showed reduced hormonal responses to exercise after training for many hormones not previously studied; introduced the concept of a general adapatation to training.	76
1977	Sutton	Demonstrated that the endocrine response to exercise is exaggerated under hypoxic conditions.	230

1977	Cashmore et al.	Demonstrated that the increase in plasma cortisol is due to increased secretion with minimal changes in degradation.	23
1978	Winder et al.	Provided the time course of changes in horomonal responses to acute exercise in response to endurance training.	256
1979	Järhult and Holst	Showed that adrenergic innervation to the pancreatic islets is important for the reduced insulin secretion found during exercise.	135
1980	Mondon et al.	Demonstrated that the increase in insulin sensitivity with exercise training is primarily due to increases in the activated muscle.	180
1981	Galbo et al.	This paper presented many seminal concepts on exercise endocrinology. Galbo proposed that neural feed-forward mechanisms control endocrine secretion, circulatory, respiratory, and thermoregulatory responses to exercise.	65
1982	Wallberg-Henriksson et al.	Showed that insulin sensitivity in people with type 1 diabetes mellitus increases in response to endurance training.	251
1982	Richter et al.	Used precise measures of glucose uptake to establish that acute exercise increased insulin sensitivity for at least 24 h after exercise.	201
1983	Vinten and Galbo	Demonstrated that regular exercise training in rats increases insulin sensitivity in adipocytes.	244
1984	Ploug et al.	Showed that muscle contractions increase glucose uptake with no need for insulin.	188
1986	Hoelzer et al.	Demonstrated that a multitude of redundant endocrine systems interact during exercise to ensure glucose availability.	105
1986	Hoelzer et al.	Demonstrated in humans that epinephrine is not critical in preventing hypoglycemia during exercise.	106
1987	Kjaer et al.	Used curarae in humans to demonstrate that central command is important to hormonal reponses to exercise.	154
1988	Mikines et al.	Used sequential euglycemic hyperinsulinemic clamps to demonstrate the effect of physical exercise on sensitivity and responsiveness to insulin in humans.	176
1993	Yarasheski et al.	Showed that short-term growth hormone treatment does not increase muscle protein synthesis in experienced weightlifters.	262
1995	Engdahl et al.	Showed that insulin pulse profile adapts to endurance-exercise training by decreasing pulse amplitude but not pulse frequency.	43
1998	Gosselink et al.	Showed that specific skeletal muscle afferents regulate release of bioassayable growth hormone during exercise in rats.	85
1998	Goodyear and Kahn	Summarized many of the intracellular insulin signaling mechanisms that were studied between the late 1980s and 1998.	83

contractions over prolonged periods of time (hours to days). He used work performance as an index of normality of function for many endocrine organs.

Some of the experimental procedures used by Ingle and colleagues were quite extraordinary. For example, they could keep rats anesthetized for 17 continuous d using phenobarbital and cyclopal while studying the effects of electrical stimulation with or without ablation of selected endocrine glands and hormone replacement. Much of his early research centered on the experimental paradigm in which work performance was assessed in adrenalectomized rats with or without supplementation

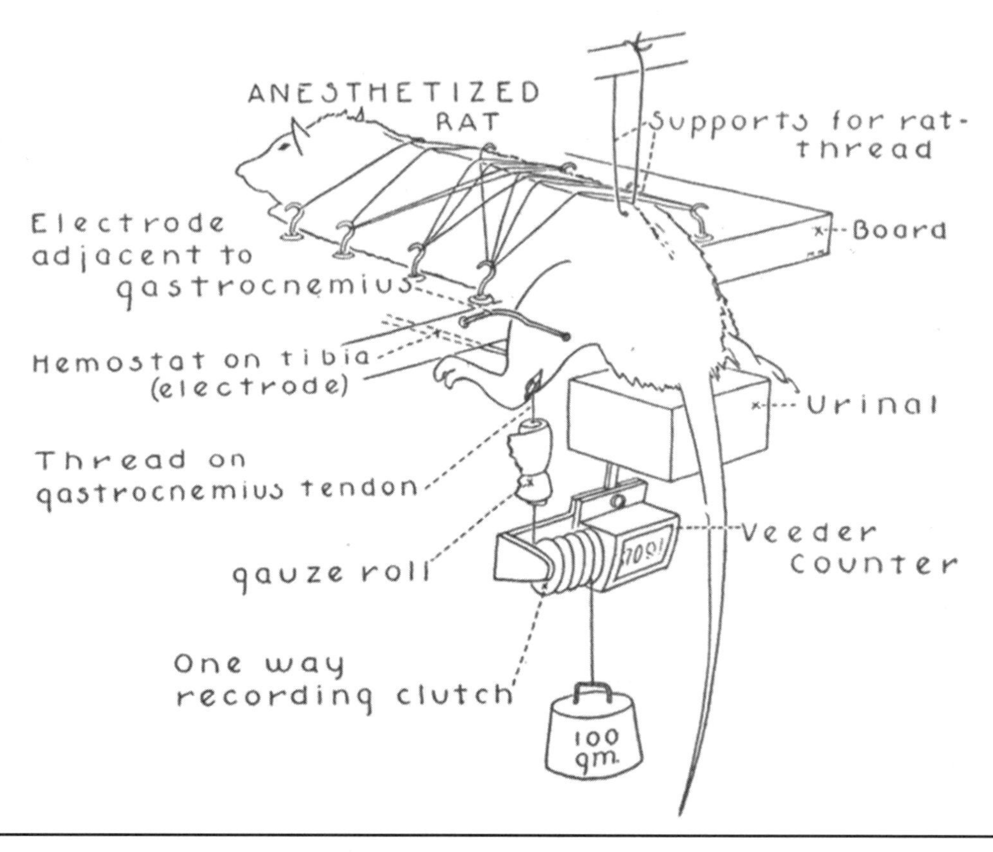

Figure 15.1 The rat model of functional performance used by D.J. Ingle in the 1930s through 1950s. Anesthetized rats could contract the hind limb for many days while various hormones were infused.

of adrenal cortex extracts or pure substances such as cortisone, hydrocortisone, or corticosterone. Ingle and colleagues repeatedly showed that most but not all of the decrement in work performance due to adrenalectomy could be reversed by injections or infusions of specific glucocorticoids. Ingle also used adrenal demedullation and found that work capacity was reduced very quickly after the operation. Similar results were observed after hypophysectomy. Initially, Ingle believed that loss of function after adrenalectomy was due to hypocorticalism, but later he showed using the adrenalectomized–hypophysectomized rat that substances other than corticotropin were involved in what he referred to as a work factor. His group continued to work on the unidentified work factor and turned their attention to the posterior pituitary of the hypophysis and provided some evidence for the importance of vasopressin in maintaining work capacity due to its effect on maintaining blood pressure. Ingle was doing this work as individual hormones were just being discovered and his group wasted no time in securing pure hormones and testing them in his model. Some of this work was exquisite in that they infused biochemicals that differed only by the position of one element, such as oxygen, in the molecule. They found that glucocorticoids must contain oxygen in position 11 to have any influence on work performance. Finally, Ingle's research was a testament to perseverance. Rarely did the number of rats per group decrease below 15 to 20 rats, and in some cases these experiments continued nonstop for a week (24 h/7 d). Figure 15.1 shows Ingle's animal model and figure 15.2 shows some of his results (101, 125).

In figure 15.2 (123, 125), the terms *growth, thyrotropic*, and so on refer to extracts taken from areas of the anterior pituitary that had been shown to have some importance to that function. The value of periodically reviewing older literature is exemplified when we note work done by Ingle in 1948 (128). He and James E. Nezamis used the eviscerated rat model (most of the internal organs were removed and vessels were ligated, followed by whole-body glucose perfusion) to determine whether insulin was needed for glucose tol-

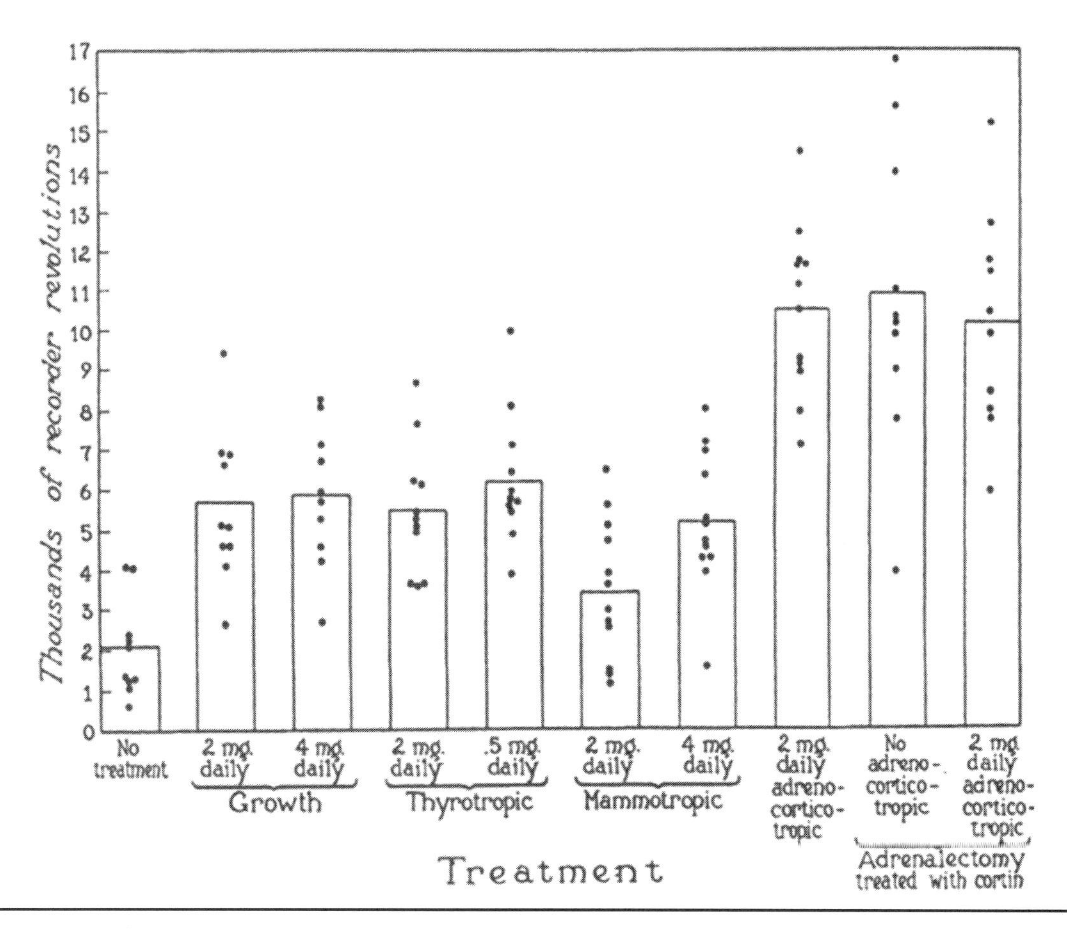

Figure 15.2 Data from the anesthetized rat model showing leg contractions while various extracts from the anterior pituitary were infused.

Reprinted from D.J. Ingle, H.D. Moon, and H.M. Evans, 1938, "Work performance of hypophysectomized rats treated with anterior pituitary extracts," *American Journal of Physiology* 123: 620-624. With permission of American Physiological Society.

erance during muscle contractions. The design used one-leg contraction protocols that included the following: no work, no insulin; work, no insulin; no work, insulin; and work with insulin. Over a 2 h period of stimulation glucose tolerance was not dependent on insulin, and when insulin was added to the perfusate, insulin and contractions were fully additive for glucose tolerance. Ingle wrote, "It seems reasonable to draw the tentative conclusion that the stimulation of muscle causes a marked increase in the utilization of glucose without any change in the insulin content of the body fluids" (p. 17). These same conclusions were drawn by two groups 37 yr later (182, 188) using actual measures of glucose uptake into muscle, which were not performed by Ingle. In 1955 (122), Ingle summarized 25 yr of research using this model.

Insulin

The major function of insulin is to promote storage of ingested fuels that can be used when needed. As such, it makes eminent sense now that the plasma concentration of this hormone should decrease during exercise when fuel mobilization is the priority. One overarching concept that started in 1948 with Ingle's work is that muscle contractions stimulate glucose uptake into skeletal muscle during exercise, providing energy for contractions without a need for insulin action. However, when this process is repeated on a regular basis (e.g., training or habitual physical activity), the ability of insulin to stimulate glucose uptake into resting muscle is improved. Therefore, an adaptation occurs in response to contractions that is measurable during parts of the day when the organism is inactive.

One of the earliest physician scientists to take an interest in the effects of exercise on insulin was R.D. Lawrence at Kings College Hospital in London. Although others had noted the ability of exercise to reduce plasma glucose concentration in patients with diabetes, Lawrence and colleague G.A. Harrison in the early 1920s made fundamental observations on the effects of exercise and insulin injections on diabetic patients. (This was, of course, more than 30 yr before we had the ability to actually measure insulin in body fluids.) In case studies (167) showed that the combination of exercise and insulin injections caused a marked reduction in glucose that was severe enough to require treatment with exogenous glucose. They also observed what is probably the first documented incidence of late-onset postexercise hypoglycemia. Their diabetic patients required significant exogenous glucose to avoid hypoglycemia in the evening and early-morning hours after exercise. Their conclusion was that muscular activity greatly enhances the action of insulin P649. This case study report also concluded that exercise, when repeated for many days and weeks, causes a reduction in the insulin dose required to control glucose. Data on weight stability, calorie intake, and other factors known to affect insulin sensitivity were not presented in these early studies. However, the earliest investigators noticed improved glucose control due to regular exercise in people with diabetes. Early work on the value of exercise for people with diabetes continued and was summarized by Soskin and colleagues in 1934 (220).

A main function of insulin is to stimulate glucose uptake, and it was known that exercise requires glucose uptake for continued muscle contraction. However, Ingle in 1950 (121) showed, using pancreatectomized rats, that insulin was not necessary for glucose uptake during muscle contractions. Using the isolated frog sartorius muscle, John Holloszy (1933–) and Narahara in 1965 (108) provided a quantitative measure of this insulin-independent glucose uptake during contractions. This idea of insulin-independent glucose uptake during exercise was solidified in the mid-1980s by two groups headed by Nesher (182) and Henrik Galbo (1946–) (188).

In 1960, Rosalyn S. Yalow and Solomon A. Berson published a seminal paper in endocrinology (260) in which they described a radioimmunoassay for endogenous plasma insulin in humans. Yalow received the Nobel Prize in physiology and medicine in 1977 for her research contributions. Several modifications to this assay were made in subsequent years. However, the ability to quantify peptide hormones in tissues was fundamentally important to the advancement of the understanding of peptide hormones and exercise.

Do Plasma Concentrations of Insulin Change During Exercise?

Devlin (40) in 1963 was among the first to measure plasma insulin during exercise using a bioassay procedure. He reported declines in insulin-like activity during exercise. In 1966 Cochran and colleagues (18) used a radioimmunoassay to measure circulating insulin and showed a decrease during step-climbing exercise in the majority of subjects and no change in others. The lower limit of detection for their assay was 5 μU/ml, and several of the subjects who showed no change were already at this lower limit before exercise. Later, Don S. Schalch (207) reported no significant change in insulin during a competitive game of basketball or women's lacrosse. These first reports using a radioimmunoassay and showing conflicting results established the tone for a controversy that existed for many years. At least six papers in the 1960s and 1970s showed no significant change in plasma insulin during mild exercise of short duration but an equal number of reports showed declines in insulin. Interestingly, no papers reported elevations in insulin during exercise.

Several of these early observations hypothesized that insulin should decline with exercise to allow mobilization of free fatty acids for use in muscle metabolism. This same rationale was mentioned for GH (discussed later). This was the rationale put forward by Hunter and Sukkar in 1968 (119) when they reported a decline in circulating insulin during 2 h of treadmill exercise in five of six subjects (figure 15.3).

In 1970, Esther D.R. Pruett from the Institute of Work Physiology in Oslo, Norway, published a series

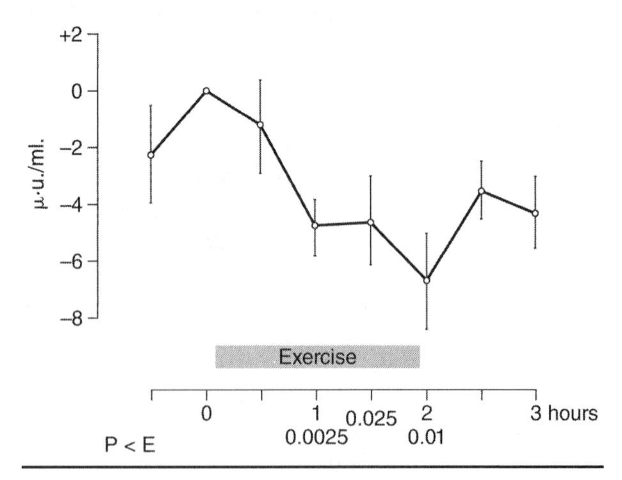

Figure 15.3 Data from Hunter and Sukkar (1968) showing declines in circulating plasma insulin during prolonged exercise in humans.

Reprinted, by permission, from W.M. Hunter and M.Y. Sukkar, 1968, "Changes in plasma insulin levels during muscular exercise," *Journal of Physiology* 196 (Suppl): 110P-111P.

of four studies (191-194; three of these were single authored) that consistently reported a decline in insulin during exercise. One of the major points in these studies was that the rate of glucose transport into cells (interpreted from oxygen consumption data) was dependent on the exercise intensity relative to $\dot{V}O_2max$ rather than the total absolute exercise intensity. Her studies clearly demonstrated that exercise reduces plasma insulin concentrations and showed that a very rapid increase in insulin above basal levels occurred after the cessation of exercise. This fact (along with high lower limits of detection for radioimmunoassays at that time) might explain some of the controversy about changes in insulin concentrations due to exercise because most early studies took postexercise samples at varying times after exercise. Her studies also showed that endogenous mechanisms other than plasma glucose itself regulated insulin concentrations during exercise. She also conducted research on the effect of high-fat or high-carbohydrate diets given before prolonged exercise and found that dietary manipulations did not alter the decline in plasma insulin during exercise. This suggested that powerful mechanisms mandate reduced insulin during exercise. After each of the exercise sessions Pruett also administered glucose tolerance tests and found improved tolerance of ingested glucose after heavy exercise. She concluded, "Taken together, these findings might indicate an increase in insulin activity when the glucose concentration gradient between the blood and the muscle and liver cells is very great as during a glucose tolerance test following prolonged severe exercise" (192, p. 207). Further support for the robustness and importance of the decline in insulin during exercise was provided by Henrik Galbo and colleagues (68), who fasted individuals for 59 h before exercise and then measured differences in hormone responses. The fast reduced the insulin concentrations before exercise by 50%, but exercise still reduced insulin concentrations by an additional 50%.

Pruett's finding that a rapid increase in circulating insulin occurred immediately after exercise was verified in 1979 by Mladen Vranic (1930–) and Kawamori in Vranic's laboratory (249), Galbo and colleagues in 1981 (68, 70), and Engdahl and colleagues in 1995 (43).

Changes in Insulin Secretion

In 1968, Esko A. Nikkila (1926-1986) and colleagues (183) estimated insulin secretion rates by infusing I^{131} insulin into humans during exercise and found no significant decline in plasma insulin with exercise and no significant change of insulin secretion rate. However, in several cases glucose was infused to maintain hyperglycemia during exercise. Secretion declined in subjects who exercised but did not receive a glucose infusion of insulin. Nikkila and colleagues speculated that hypoxia in the islets may have reduced insulin secretion. Because there was no change in circulating insulin at a time when glucose uptake increased during exercise, they emphasized that enhanced glucose uptake during exercise was not due to circulating insulin. This paper was later criticized because the catabolic rate of naturally occurring insulin is markedly different from that of the iodinated insulin they used. Nevertheless, from a historical perspective these early experiments using radiolabeled insulin in humans provided valuable insights. Using a rat model in 1968, Wright and Malaisse (259) showed that prolonged strenuous exercise resulted in suppressed insulin secretion. J.R.M. Franckson (63) and colleagues followed up on the observation of Nikkila and colleagues using a better tracer (their claim) and, in contrast to the results of Nikkila and colleagues, found that mild to moderate exercise caused a reduction in insulin secretion, which they attributed to an elevation in epinephrine concentrations.

Further evidence that the reduction in insulin during exercise was due to reduced secretion was provided by Mladen Vranic (1933–) and colleagues in 1976 (250) using dogs. In nondiabetic dogs, there is a rapid and pronounced decrease in circulating insulin at the start of exercise. They also showed that insulin concentrations during exercise did not change in insulin-infused depancreatized dogs. Furthermore, very low levels of insulin were not changed during exercise when insulin was withheld from depancreatized dogs. These data suggested that a small but essential amount of insulin must be present for proper glucose regulation and glucose uptake in exercising muscle. This suggestion was debated because others felt and later demonstrated (182, 188) that glucose uptake could be stimulated by contractions in the complete absence of insulin. The studies by Vranic (248, 249) demonstrated that a reduction in insulin secretion and perhaps a small increase in insulin clearance were responsible for reduced insulin during exercise. This suggestion was later partially validated in humans by Hilsted and colleagues in 1980 (104) when they demonstrated that the decrease in insulin was due to decreased secretion while clearance also decreased. This interpretation was based on changes in the ratio of plasma c-peptide to insulin.

Mechanisms That Explain Reduced Insulin During Exercise

In 1971, Brisson and colleagues (18) demonstrated in swimming rats that the reduction in circulating insulin could be blocked using the alpha adrenergic antagonist phentolamine. Similar results were provided by Leclercq-Meyer and Malaisse in 1975 (168) using rats. In 1977, Galbo and colleagues (66) showed in humans that the decline in insulin during exercise occurred to a lesser extent during infusions of phentolamine while glucagon was not influenced. In contrast, cholinergic blockade by atropine did not affect pancreatic hormones or lipolysis during exercise. These results were extended by Järhult and Holst (135), who showed that insulin declined identically in adrenal-intact and biadrenalectomized subjects during exercise. Two biadrenalectomized subjects exercised with or without prior phentolamine injections that markedly blunted the decline in insulin during exercise. These data suggested that alpha adrenergic nerves innervating the pancreas (beta cells) were activated during exercise and were responsible for much of the decline in circulating insulin. These data also demonstrated that circulating catecholamines were not necessary for this response.

As noted previously, Nikkila and colleagues proposed that hypoxia in the pancreatic islets may be a reason for reduced insulin secretion during exercise. The concept was investigated in 1977 by John Sutton (230), who found markedly lower insulin concentrations during exercise in hypoxic conditions. This finding was verified by Michael Kjaer (1957–) and colleagues in 1988 (145); however, they emphasized the close relationship between hormonal responses and relative exercise intensity, the latter being increased in hypoxia.

More recently, investigators examined neural mechanisms to explain the declines in insulin concentrations. In 1995 Redmon and colleagues (197) studied individuals who had received a pancreas transplant. Glucose decreased slightly in both control subjects and transplant patients during light exercise. As expected, plasma insulin decreased significantly in the control subjects whereas there was only a small and insignificant decline in both insulin and c-peptide in the transplant patients. This again supports the idea that sympathetic nerves innervating the pancreas were important for the exercise-induced decreases in plasma insulin. At the same time, Houwing and colleagues (114) studied rats that had been made diabetic before being given islet transplants. They found that circulating epinephrine and norepinephrine were not responsible for the decrease in insulin concentrations during exercise but that sympathetic neural influences were important. In 1999, Coker and colleagues (27) in David Wasserman's (1957–) laboratory used dogs to determine whether pancreatic innervation was necessary for insulin and glucagon responses to exercise. All nerves and plexuses to the pancreas were sectioned in one group of dogs. The results indicated that changes in both glucagon and insulin in the portal vein were similar between nerve-ablated dogs and sham-operated dogs. Thus, the issue was not settled.

Glucose-Stimulated Insulin Secretion After Acute Exercise

In 1970 Pruett showed that insulin secretion due to an intravenous glucose tolerance test was reduced immediately after exercise (192) but then increased later. However, this test may reflect changes in both beta cell function and insulin sensitivity. Avoiding such ambiguity, Kari Mikines and colleagues (175) showed in 1988 that a single bout of bicycle exercise does not alter glucose-stimulated insulin secretion at low, moderate, or high levels of sustained (90 min) hyperglycemia at either 2 h or 48 h after exercise. David James and colleagues (132) showed in 1983 that glucose-stimulated insulin secretion was reduced in untrained rats 30 min after a single bout of exercise but not 24 h after.

Insulin Sensitivity After Acute Exercise

During the years when plasma insulin responses and insulin secretion were being studied, many groups investigated whether exercise altered the effectiveness of insulin for glucose uptake. As mentioned, several of the earliest investigators (e.g., Ingle in 1948) noted that glucose uptake during exercise occurred at a higher rate than at rest even when insulin was reduced (e.g., diabetic rat model; 126) or absent (e.g., eviscerated rat model; 128).

In 1975, Michael Berger (1944-2002) and colleagues in Neil Ruderman's (1937–) laboratory (8) demonstrated that acute muscle stimulation (isometric contractions) increased insulin-stimulated glucose uptake in isolated perfused rat hind limbs to a higher level than is found with comparable amounts of insulin in a nonexercised leg. In 1982, Erik Richter (201) in the same laboratory used a hind-limb perfusion in rats after treadmill running to demonstrate that glucose utilization due to insulin was higher after acute exercise. They showed that the Km (insulin concentration that stimulated half-maximal rates of glucose uptake) was reduced (i.e., insulin sensitivity increased) after exercise whereas the maximal effect of insulin (insulin responsiveness) was increased. This increase in the effect of insulin lasted for 4 h but not 24 h. The increase in glucose uptake occurred predominantly in the fast-twitch red fibers of the red gastrocnemius whereas the elevation in fast-twitch white fibers was minimal. In 1983, John Ivy (1945–)

and colleagues (131) in Holloszy's laboratory expanded Richter's finding by using hind-limb perfusion to show that the increased insulin sensitivity lasts for approximately 2 d. In 1984, Garetto and colleagues (77) from Neil Ruderman's laboratory described two phases of glycogen repletion after exercise. In the first phase immediately after exercise and for approximately 2.5 h, glycogen is repleted even if no additional insulin is available. After this time, insulin-mediated glycogen repletion is dominant. The idea that short-term and long-term increases in insulin sensitivity occur after acute exercise was the subject of intense investigation for many years.

Kari Mikines (1956–) and colleagues in 1988 (176) performed an important series of studies using sequential multistep euglycemic hyperinsulinemic clamps after rest, immediately after exercise, and 48 h after aerobic exercise. Immediately (2 h postexercise) after exercise the Km was significantly reduced and the maximal conversion rate of glucose to glycogen was higher than in a resting state; this was maintained for 48 h (figure 15.4). They also studied the same subjects 5 d after exercise and found that the effects of acute exercise were absent. The value of this particular study was that it investigated a dose–response relationship between insulin and glucose disposal in humans using multilevel clamps, which was necessary due to the sigmoidal shape (see figure 15.4) of the in vivo insulin–glucose disposal relationship. The findings of Mikines and colleagues relative to the effects of the last bout of exercise were extended in 1988 by King and colleagues (142), who found that trained people who refrained from exercise for 10 d had a significant reduction in insulin sensitivity but no change in insulin responsiveness. Both studies (142,176) measured whole-body glucose disposal. However, in 1989 Erik Richter and colleagues (203) used one-leg exercise in healthy untrained humans and reported that enhanced insulin-stimulated glucose uptake after exercise was limited to the active muscle.

Effects of Training on Insulin Secretion

Michael J. Rennie in 1974 conducted a training study of 4 wk duration and measured plasma insulin at the same absolute running speed before and after training (198). Resting concentrations of insulin in the fasted state were significantly lower after training. When subjects exercised at the same absolute work level after training, the plasma levels of insulin were lower throughout the exercise session. In 1977, Galbo and colleagues (76) found that 12 wk of swim training in rats resulted in a blunted hormone (including insulin) response to acute exercise. This indicated that endocrine responses to exercise are less marked in the trained state.

David James and colleagues in 1983 (132) showed that exercise training reduced glucose-stimulated insulin secretion, but this effect was present for at least 48 h after exercise (132). In 1987, Farrell (1947–) (48) used hyperglycemic glucose clamps to show that treadmill training in rats resulted in a markedly reduced insulin secretion. In 1989, Mikines (175) used multilevel hyperglycemic clamps to demonstrate that at both low and very high glucose concentrations the c-peptide and insulin responses were lower in the trained state. Dela and colleagues in 1990 demonstrated that the insulin secretion response to arginine is also reduced by regular endurance training (37). All of these observations on training adaptations at the whole body level of investigation required follow-up because the beta cells do not function in isolation.

To determine the anatomical location(s) of training-induced decreases in glucose-stimulated insulin se-

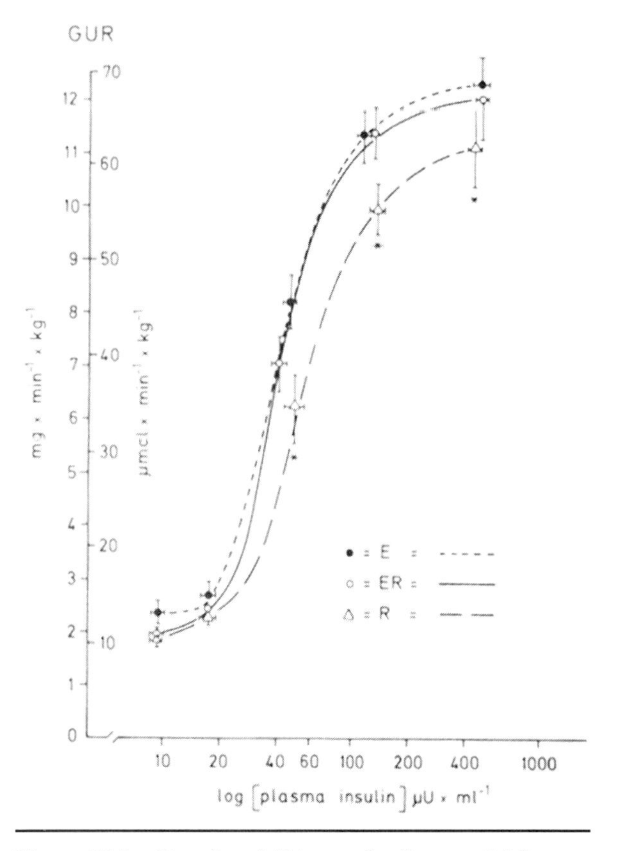

Figure 15.4 Data from Mikines and colleagues (176) exhibiting rates of glucose uptake during sequential euglycemic hyperinsulinemic glucose clamps in humans 1 and 48 h after endurance exercise.

cretion, Shima and colleagues in 1987 (212) used the isolated perfused whole-pancreas preparation and showed that insulin secretion stimulated by either glucose or arginine was reduced after 8 wk of voluntary wheel running in rats. Using isolated islets of Langerhans, Reaven and Reaven in 1981 (196), Galbo and colleagues in 1981 (69), and Zawalich and colleagues in 1982 (263) demonstrated that regular exercise reduced glucose-stimulated insulin secretion at the islet level of inquiry. Farrell and colleagues (50) in 1992 demonstrated that endurance training reduced glucose-stimulated insulin secretion from single isolated (via laser-enhanced flow cytometry) pancreatic beta cells. Beta cell size and degree of granularity (an estimate of beta cell content) were not altered by exercise training. Surprisingly, the effects of acute endurance exercise have not been studied at the islet level. However, Fluckey and colleagues reported in 1995 that arginine-stimulated insulin secretion (but not glucose-stimulated secretion) was elevated from islets after acute resistance exercise in rats (61).

Changes in Insulin Secretion Due to Training in Insulin-Deficient Models

The early observations by Lawrence (167) in 1926 suggested that chronic exercise could allow people with diabetes to use less insulin. In 1991 Farrell and colleagues (49) used partially pancreatectomized rats (mild, moderate, and severely diabetic groups) and 9 wk of treadmill-exercise training to show that mildly and moderately diabetic animals could actually increase insulin secretion during hyperglycemic clamps. Severely diabetic rats had no such adaptation. In 2004, Dela and colleagues (38) studied people with type 2 diabetes who were characterized (using Oral Glucose Tolerance Tests [OGTT] results) as low or moderate insulin secretors to determine whether exercise training would alter beta cell responses to glucose. They found that the insulin response to hyperglycemia or arginine stimulation did not change in the low secretors after training. However, beta cell responses significantly increased in the moderate secretors after endurance training. The difference in pancreatic adaptations between insulin-deficient models (increased secretion due to endurance training) and insulin-sufficient models (decreased secretion due to endurance training) is quite fascinating. In addition to studies on insulin secretion, several groups have determined whether insulin-mediated glucose uptake is altered by training in insulin-deficient models. Regular endurance-exercise training increases insulin sensitivity in people with type 1 diabetes, as demonstrated by Wallberg-Henriksson and colleagues in 1982 (251). In 1995, Dela and colleagues (35) demonstrated in people with type 2 diabetes that regular one-leg endurance-exercise training augments insulin-mediated glucose clearance in the trained leg only.

Insulin Pulsatility

Insulin is secreted from the pancreas in a pulsatile manner. Each burst occurs at approximately 11 min intervals. The effectiveness of a given amount of most hormones (luteinizing hormone [LH], GH, and insulin have been studied from this perspective) is enhanced when the hormone is presented to the target tissue in a pulsatile (on–off) manner rather than a constant presentation. Engdahl and colleagues in 1995 (43) demonstrated that regular exercise training results in a pulse profile where the pulse frequency is not different from that in untrained subjects but the pulse amplitude (mass of insulin per burst) is about half of that measured in the untrained subjects. The authors suggested that less insulin per burst of insulin is needed in the trained state due to increased peripheral insulin sensitivity.

Effects of Training on Insulin Sensitivity

In 1979, Soman and colleagues in Ralph DeFronzo's laboratory (216) showed that 6 wk of endurance training in humans improved insulin-stimulated glucose uptake (determined by the insulin clamp technique). The increase in insulin sensitivity was proportional to the improvement in physical fitness.

Mondon and colleagues (180) in 1980 used in situ organ-perfusion techniques and showed that insulin sensitivity was increased in hind-limb muscles in rats after endurance training. Interestingly, they found that the rate of glucose uptake in the hind limbs was faster in trained rats both with and without insulin in the perfusate. They also measured glucose uptake in liver and found that after an overnight fast the liver of trained rats took up significantly less glucose than that of control rats. They made the fundamental observation that skeletal muscle is the organ primarily responsible for increased insulin sensitivity after endurance training (spontaneous running). In 1985, David James and colleagues (133) used the euglycemic hyperinsulinemic glucose clamps along with glucose tracers in nonanesthetized rats to show that whole-body insulin sensitivity is increased by endurance training. They found a training-induced increase in insulin-mediated glucose uptake in many leg muscles and the diaphragm and showed that adipose tissue increased insulin-stimulated glucose uptake. This finding complemented data from

Vinten and Galbo in 1983 (244), who demonstrated that insulin-stimulated glucose uptake into isolated fat cells was enhanced by training. This finding was later verified by Ferrara and colleagues in 1998 (56) when they studied rat adipocyctes after short-term training and found increased insulin stimulated glucose uptake.

In 1983, Heath and colleagues in Holloszy's laboratory demonstrated that the effects of training on improved glucose tolerance test (100) were short lived in that 10 d of detraining led to a complete reversal of this adaptation. Noteworthy was that one bout of exercise after 10 d of detraining returned the glucose and insulin response during an oral glucose tolerance test (OGGT) to levels almost reaching those of the trained state. In 1990, Nagasawa and colleagues (181) also reported that although insulin sensitivity is reduced very quickly after the cessation of training, insulin responsiveness remains elevated for a 3 wk period in rats. In 2009, Bajpeyi and colleagues (6) in Joe Houmard's (1960–) laboratory reported some interesting findings relative to how long insulin responsiveness remains elevated after the cessation of training in humans. Three groups exercised for 8 mo using protocols that were described as follows: low volume, moderate intensity; low volume, vigorous intensity; and high volume, vigorous intensity. All programs increased insulin sensitivity when studied 16 to 24 h after exercise. However, the low-volume, moderate-intensity and high-volume, vigorous-intensity groups maintained elevated insulin responsiveness for 15 d after the cessation of training. Thus, there is consistency in the finding that exercise-induced increases in insulin sensitivity are lost very quickly after cessation of exercise. However, the effects of exercise on insulin responsiveness are retained for prolonged periods of time after cessation of training.

Many other reports from the 1980s to the present firmly established that insulin sensitivity increases (at least transiently) with exercise training in all species investigated. It would be impractical to list those studies or even the reviews dealing with this topic. However, comparisons between trained, detrained, and untrained subjects have revealed that the training-induced decrease in beta cell secretory capacity generally matches the accompanying increase in insulin action because glucose disposal during hyperglycemic clamping was identical in the groups despite widely differing plasma insulin concentrations (177). However, oral glucose tolerance is enhanced in extremely well-trained subjects compared with untrained subjects, indicating that enhancement of insulin action is not fully compensated by reduced beta cell function (36). At the other end of the activity spectrum, insulin secretion and insulin action are not well matched. During bed rest, glucose tolerance deteriorates and glucose disposal is diminished in the face of augmented insulinemia during hyperglycemic clamps. Thus, the reduction in insulin sensitivity is not fully overcome by enhanced glucose-stimulated insulin secretion. Consequently, a minimal level of physical activity is necessary for normal glucose homeostasis. Taken together, in terms of glucose homeostasis, the adaptive effect of physical activity on beta cell function is somewhat governed by the changes in insulin action; this may be considered a compensatory phenomenon on the part of the pancreas.

An important concept that had not been fully delineated before 1986 was whether changes in insulin sensitivity due to training were attributable to alterations in insulin binding to its receptor or to changes in intracellular mechanisms. In 1986 Grimditch and colleagues (89) found that binding affinity (number of either high or low affinity insulin binding sites) was not changed in skeletal muscle of trained rats compared to control rats.

Resistance Exercise

Most of the work done to describe the effects of exercise on insulin action, secretion, or adaptations to training has used endurance-type exercise. Very similar results are found when resistance exercise is used. Resistance-exercise training (178) as well as an acute bout of resistance exercise (158) improves insulin sensitivity for glucose uptake into muscle.

Aging

In 1984, Doug Seals (1954–) and colleagues in Holloszy's laboratory demonstrated that older athletes who performed regular endurance-exercise training had better responses of both glucose and insulin during oral glucose tolerance tests (210). In 1990, Kahn and colleagues (138) studied older men during 6 mo of intensive exercise training and found significantly lower concentrations of fasting insulin after training. In 2003, Short and colleagues (214) found that aerobic-exercise training programs are not effective for increasing insulin sensitivity in older individuals. This finding was verified in 2005 by Goulet and colleagues (87); however, many studies have shown that insulin sensitivity can increase in the elderly. For example, Bloem and Chang (11) in 2008 found that 7 consecutive d of aerobic training significantly increased insulin sensitivity by 59% in older people. Thus, there is some controversy about the effects of exercise training on insulin sensitivity in the elderly.

Functions of Insulin Other Than Glucose Uptake

Farrell and others in his laboratory conducted a series of studies (53, 54, 159) demonstrating that insulin has a permissive role in allowing rates of protein synthesis to increase after resistance exercise in rats, a finding later verified in humans (10). In 1988, Mikines and colleagues (176) demonstrated that previous acute exercise significantly enhances the inhibitory effect of insulin on lipolysis. Dela and colleagues in 1990 demonstrated that insulin-mediated suppression of lipolysis is enhanced by both acute and chronic exercise (37). Stallknecht in 2000 (224) confirmed the influence of training on insulin inhibition of lipolysis.

Intracellular Signalling

After these studies in the late 1980s and early 1990s, attention turned away from insulin secretion and action per se and toward the regulation of insulin-regulatable glucose transporters with exercise, as exemplified by the research of Joe Houmard (1960–) and colleagues in the early 1990s (111-113). Due to space limitations, we do not review those studies. Later reviews (83, 99, 206) capture the important studies that have defined intracellular pathways for insulin that are common or not common to pathways activated by contractions (206).

Epinephrine

The first studies on epinephrine and exercise used approaches that were similar to those used by Ingle in terms of endocrine gland removal and the determination of exercise responses or capacity with or without hormone replacement. For example, Stewart and Rogoff in 1922 (226) removed various amounts of one or two adrenal glands from cats and then had them exercise over many months. In most cases the cats could sustain significant levels of running once they recovered from the surgery and seemed to be normal. They found that cats could engage in regular exercise with very little deficiency even when a large portion of the adrenal gland was removed. It should be noted that the data are reported as descriptions of individual responses for each cat and not as group responses. Also, the researchers made no attempt to require all cats to exercise at the same intensity or for similar durations. This complicates the interpretation of the heart rate, respiratory, and body temperature data that are reported. Probably the most important information gained in the study was that a very large portion of the adrenal glands (~80%) had to be removed before significant decrements were observed. Similar results were reported by Schweitzer (209) in 1945.

Bioresponse of Epinephrine

In 1922 Hartman and colleagues (98) published a paper using a bioresponse to document changes in epinephrine with exercise. Epinephrine causes a dilation of the pupils (figure 15.5). Dilation in a sympathetically denervated eye is most likely attributable to epinephrine secretion. Essentially, removal of the adrenal gland can completely ablate the dilation of eye pupils during the rapid removal of intense light. When epinephrine increases in plasma due to a stress (exercise), there is a dose–response relationship between the increase in epinephrine and the degree of pupil dilation. Cats ran on a hand-driven running wheel. The researchers found that epinephrine secretion was directly proportional to the exercise intensity. Figure 15.5 shows individual responses of cats under various workloads, as indicated in the legend. In each photo, the cat's left eye is sympathetically denervated.

Many of the early noted physiologists contributed to our understanding of epinephrine responses to exercise. For example, after Hartman's initial report in 1922, Walter B. Cannon and colleagues the finding that exercise stimulates epinephrine release in 1924 (22). Cannon later formulated the fight or flight concept for the role of epinephrine during extreme stress. In 1936, L. Brouha, W.B. Cannon, and D.B. Dill published a paper (19) using sympathectomized dogs and the effects of adrenalin (epinephrine) injection on cardiac performance. They suggested that the acceleration in heart rate during exercise is not attributable to secreted adrenalin but rather to withdrawal of cardiac inhibition by the vagus nerve.

Urinary Catecholamines

In the early 1950s Euler published data (45, 46) that reported significant increases in urinary catecholamines after muscular work. Many have used a similar approach, but urinary excretion of catecholamines is a poor measure of sympathoadrenal activity. This was illustrated 20 yr later by findings of constant epinephrine excretion over a wide range of submaximal workloads (116).

Changes in Plasma Concentrations

Gray and Beetham in 1957 (88) used a fluorescent assay and reported threefold increases in plasma epinephrine and norepinephrine after both indoor treadmill exercise and an 18 mile run outside. Vendsalu (242) in

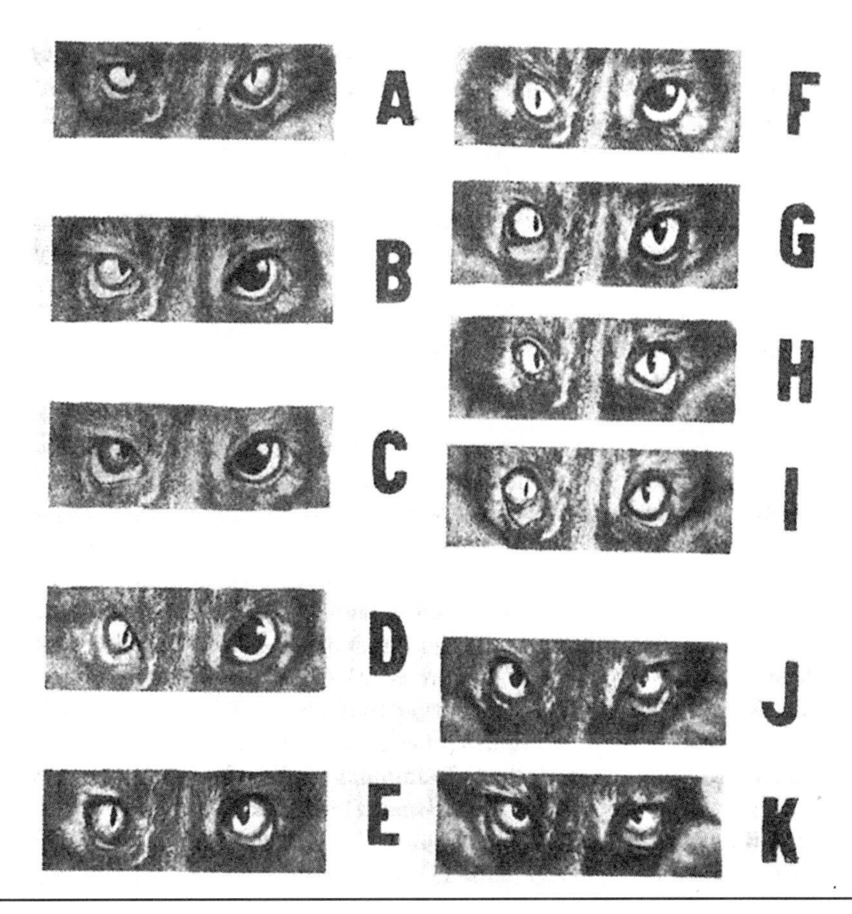

Figure 15.5 Data demonstrating pupillary responses of cats at various times after stopping treadmill exercise. Dilation of a denervated eye (the cat's left eye in the figure) in response to intense light was interpreted as indicating epinephrine release into the circulation due to exercise. These are responses of the denervated pupil of cats to in vivo exercise with the left adrenal gland denervated. (*a*) Before exercise. (*b*) 292 m in 3 min; observed 12 s after stopping. (*c*) 595 m in 23 min; observed 20 s after stopping. (*d*) Same run but observed 50 s after stopping. (*e*) Same run but observed 4 min 17 s after stopping. (*f*) 810 m in 42 min; observed 17 s after stopping. (*g*) Same run but observed 4 min 22 s after stopping. (*h*) Same run but observed 9 min 33 s after stopping. (*i*) Same run but observed 15 min 5 s after stopping. (*j*) With the right adrenal gland removed; before treadmill exercise. (k) With the right adrenal gland removed; 280 m in 6 min.

1960 also showed that both plasma and urine catecholamines increase with exercise.

In the early 1970s there was some controversy about whether exercise increased plasma levels of epinephrine. Some studies failed to show a change in epinephrine with exercise whereas most studies showed an increase as long as the exercise intensity was relatively high. The increase in plasma concentrations with exercise intensity is exponential in nature (figure 15.6). This was first shown in 1970 when Haggendal and colleagues (91) reported an exponential increase in arterial plasma norepinephrine, which was further characterized as being most related to exercise intensity when expressed relative to $\dot{V}O_2max$. This fundamentally important observation was verified for epinephrine by Kotchen and colleagues in 1971 (160). Later, Galbo and colleagues (75) demonstrated the importance of duration of exercise; a slow increase of several hormones occurs when exercise is prolonged.

Using a fluorescent assay, Banister and Griffiths in 1972 (7) also showed that epinephrine concentrations increased as workload increased and that there was a direct linear relationship between plasma epinephrine and norepinephrine over various workloads from very light to very heavy. However, probably reflecting inadequate sensitivity of the assay and marked variation in epinephrine recovery between samples, in the early 1970s there was still controversy about whether exercise increases plasma levels of epinephrine. Reliable measurements of plasma catecholamines—epinephrine in particular—first became possible after 1970 with the introduction of the radioenzymatic assay (71). Using

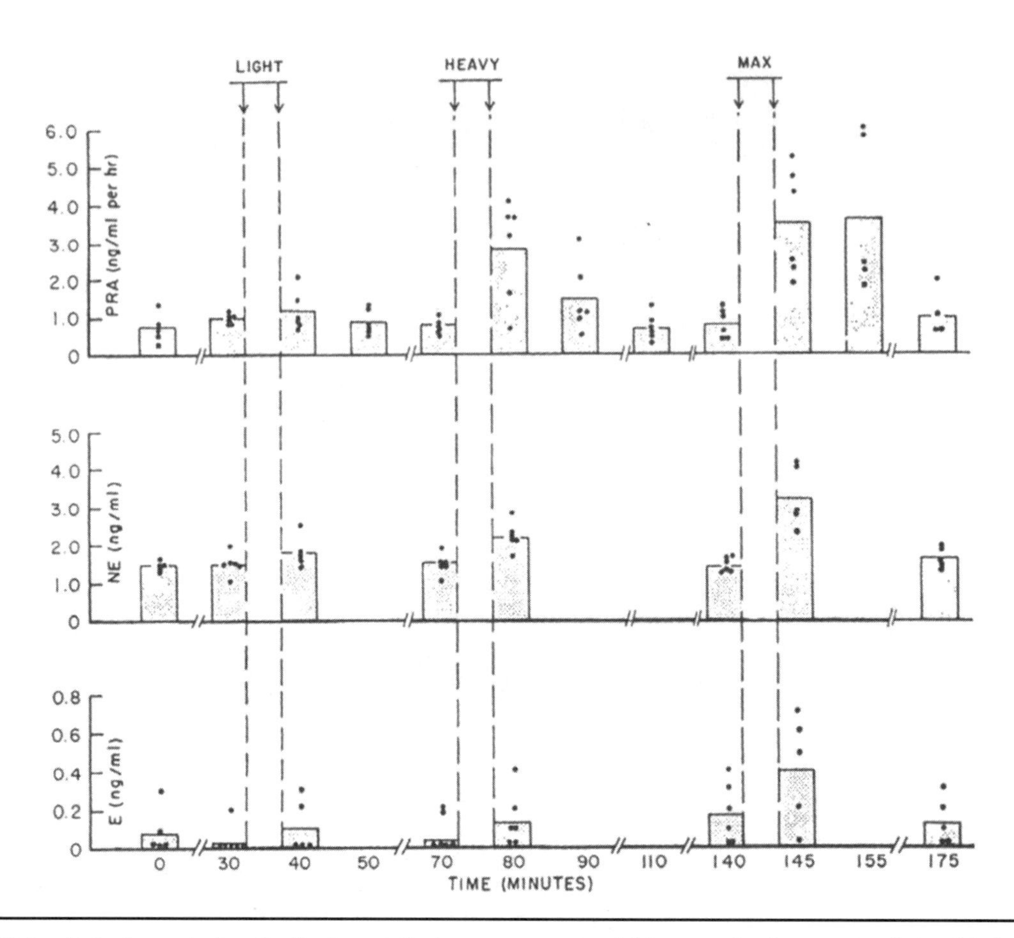

Figure 15.6 Data demonstrating the fundamentally important exponential increase in plasma epinephrine (E) during short duration exercise of increasing intensity relative to maximal oxygen consumption.

Reprinted from T.A. Kotchen et al. 1971, "Renin, norepinephrine, and epinephrine responses to graded exercise," *Journal of Applied Physiology* 31: 178- 84. With permission of American Physiological Society.

this method, Galbo and colleagues (71) were the first to show that the plasma epinephrine concentration increases exponentially with exercise intensity greater than 30% of $\dot{V}O_2max$ and that the response is closely related to relative workload. Furthermore, they found that during prolonged moderate exercise the plasma epinephrine concentration increases gradually from onset and throughout. These findings indicated a meticulously regulated secretion and, accordingly, a physiological role in everyday physical activities rather than just in extreme stress, as previously thought.

Epinephrine Secretion Versus Clearance

Gordon and colleagues in 1966 (84) used a combination of α-methyl tyrosine, which is a potent inhibitor of the rate-limiting step in catecholamine synthesis, and radiolabeled tyrosine to measure catecholamine synthesis in rats. They showed under conditions of severe exercise and α-methyl tyrosine infusion that the epinephrine concentration in the adrenal glands was markedly reduced whereas during exercise alone it was only slightly reduced. The researchers interpreted these data as demonstrating that a simultaneous increase occurs during exercise in both the synthesis of epinephrine and the release of epinephrine from the adrenal medulla. The study also showed that there was a marked increase in the synthesis of brain epinephrine as a result of exercise. Warren and colleagues in 1984 (252) infused unlabeled epinephrine in order to calculate epinephrine clearance. Probably reflecting the weaknesses of this method, they found unchanged epinephrine concentration in plasma at a heart rate of 100 beats/min despite a 46% decrease in clearance, indicating that secretion was diminished under these conditions compared with resting conditions. Using tracer technology, Kjaer and colleagues in 1985 (146) found that the clearance of epinephrine was closely related to relative workload; it increased 15% above basal levels at 30% $\dot{V}O_2max$ and was 22% below basal levels at 76% $\dot{V}O_2max$. Because changes in clearance were relatively small, the changes in plasma epinephrine concentrations seen with exercise essentially reflect changes in epinephrine secretion.

What Causes the Increase in Epinephrine During Exercise?

In 1974, Irving and colleagues (130) used several adrenergic blockers to study whether they had an effect on the catecholamine response to exercise. They used the beta adrenergic antagonists propranolol, oxprenolol, and pindolol, and in all cases the epinephrine responses to exercise were enhanced. Davies and colleagues in 1976 (33) used practolol (β adrenergic blocker) or atropine (a cholinergic blocker described in this study as a parasympathetic blocker) to study the change in plasma catecholamine concentrations after exercise. Atropine always caused an increase in catecholamine response; however, the beta blocker had no effect. These data conflict with those of Galbo and colleagues (72), who showed that propranolol increased the epinephrine response to prolonged exercise. In 1977, Galbo and colleagues (67) extended that observation by infusing glucose as plasma glucose decreased during exercise, which resulted in a reduction of the epinephrine response.

Role of Central Command

In 1981 Galbo (65) introduced the unifying concept that neural feedforward mechanisms are as important in the control of endocrine secretion as they are for circulatory, respiratory, and thermoregularoty responses to exercise. In a highly cited landmark paper, Kjaer and colleagues (154) in 1987 demonstrated a role for central command in the regulation of endocrine responses to exercise. Briefly, the investigators used tubocurarine in humans to weaken peripheral muscles so that greater activity of the motor centers in the brain must be elicited to perform a certain amount of work. During submaximal exercise at the same absolute oxygen consumption, epinephrine, norepinephrine, GH, adrenocorticotropic hormone (ACTH), and beta endorphin responses were higher during curarization than under control conditions. However, at maximal effort and, accordingly, lower absolute workload, the epinephrine response was lower after curare compared with control conditions. This is in line with the view that factors originating in the working muscles that depend on force or metabolism (by humoral or afferent nervous mechanisms) contribute to the epinephrine response. In agreement with this, the same authors found that epinephrine concentrations increased in parallel with increasing absolute workload when continuous maximal exercise—and, accordingly, maintained maximal central command—was carried out during fading away of neuromuscular blockade with tubocurarine (73).

Other studies using direct manipulation of certain regions of the brain have also shown a role for central command. In 1989, Vissing and colleagues (247) demonstrated that anesthetizing the ventral medial hypothalamus of rats with marcaine resulted in lower plasma catecholamines as well as of overall rates of hepatic glycogenolysis and initial hepatic glucose production during exercise. Later, van Dijk and colleagues (241) used bupivacaine to anesthetize the paraventricular nucleus of the hypothalamus and found that the epinephrine response to exercise was markedly reduced during treadmill running. However, the most convincing evidence for a direct role of motor centers in hormonal and metabolic responses to exercise was provided by Vissing and colleagues in 1994 (246). They electrically stimulated the posterior hypothalamic locomotor region in unanesthetized, decorticated cats in which feedback from contracting muscle was prevented by tubocurarine. They found that electrical stimulation elicited a hormonal and metabolic response, including an increase in plasma epinephrine, mimicking that seen in exercise. This group established the importance of central and afferent somatic neural mechanisms for the epinephrine response to exercise by stimulating muscle afferents in humans and rats (246), performing epidural blockade in healthy men (155), and studying spinal cord-injured tetraplegic humans during electrically induced cycling (153). Several of these control mechanisms for hormones are similar to those found for the cardiovascular and ventilatory systems (64), a fact implying coherence of the major physiological systems during exercise.

Effects of Endogenous Opioid Peptides

In 1986 Farrell and colleagues (52) demonstrated that endogenous opioid peptides have an inhibitory effect on the plasma epinephrine response to exercise. They suggested that this regulation is important for appropriate endocrine responses to the stress of moderate exercise. Gullestad and colleagues (90) in 1987 did not verify these results, but Staessen and colleagues (222) in 1988, Armstrong and Hatfield in 2006 (3), and Angelopoulos and colleagues in 1995 (2) found that naloxone treatment augmented epinephrine responses to exercise.

Plasma Glucose

The fast nervous mechanisms related to motor activity, which govern epinephrine secretion, operate throughout exercise. At high work rates these mechanisms dominate. However, sensing of error signals also plays a role and may dominate at low exercise intensities as they do at rest. Based on exercise studies with various drugs and diets that allow variation in rates of decrease in plasma glucose concentrations as well as studies with glucose infusion, Galbo and colleagues first pointed out that the most important metabolic error signal in exercise is a decrease in plasma glucose concentration and that epinephrine secretion is very sensitive to even small declines in glucose (67, 70, 72).

Effect of Training

In 1977, Galbo and colleagues (76) demonstrated that swim training in rats resulted in reduced plasma epinephrine responses to exercise and that adipose tissue sensitivity to catecholamines is increased. In 1982, Will Winder (1942–) and colleagues (255) found similar results using rats and treadmill exercise. In 1978, Winder and colleagues (256) reported important results that continue to be highly cited today. They showed the time course of changes in plasma catecholamines with endurance training over a 7 wk period (figure 15.7). The epinephrine responses to standard exercise (same absolute exercise intensity) were reduced after only 3 wk of training. A year later, Winder and colleagues (257) performed a similar study that included glucagon, which also decreased after only 3 wk of training. In that study, they also included a workload after training that was the same as the relative workload before training (a higher absolute workload) and found that the epinephrine response was still lower in the trained state than in the untrained state but was higher (but still lower than the untrained state) when the same absolute workload was used.

In the mid-1980s, Kjaer performed a series of studies using trained and untrained subjects (143, 144, 152) in which he compared the catecholamine (and other hormones) response to several stresses. These studies led to the concept of a sport adrenal medulla, suggesting that endurance training increases the secretory capacity for epinephrine during very strenuous exercise. Substantiating this, Stallknecht in the same laboratory (223) found in 1990 that endurance training (swimming) in rats resulted in heavier adrenal glands, higher catecholamine content in the glands, and larger adrenal medulla volumes. Nevertheless, during insulin-induced hypoglycemia the epinephrine response was much smaller in trained rats. This finding probably reflects that the repeated hypoglycemia, which occurred in the rats during regular training, resulted in less-acute, hypoglycemia-induced nervous output to the adrenal medulla (224). Similar effects on the volume of the adrenal medulla were presented by Schmidt and colleagues in 1992 (208) using treadmill running as the training stimulus.

Role of Epinephrine in Glycogenolysis

During exercise, muscle glycogen is an important energy source that is mobilized when motor nerve stimulation elicits a Ca++-mediated conversion of glycogen phosphorylase b to phosphorylase a. However, as early as 1935, Dill and colleagues (41) and later Asmussen and colleagues (5) injected epinephrine in human subjects and found an exaggerated increase in blood lactate during exercise, suggesting that epinephrine may en-

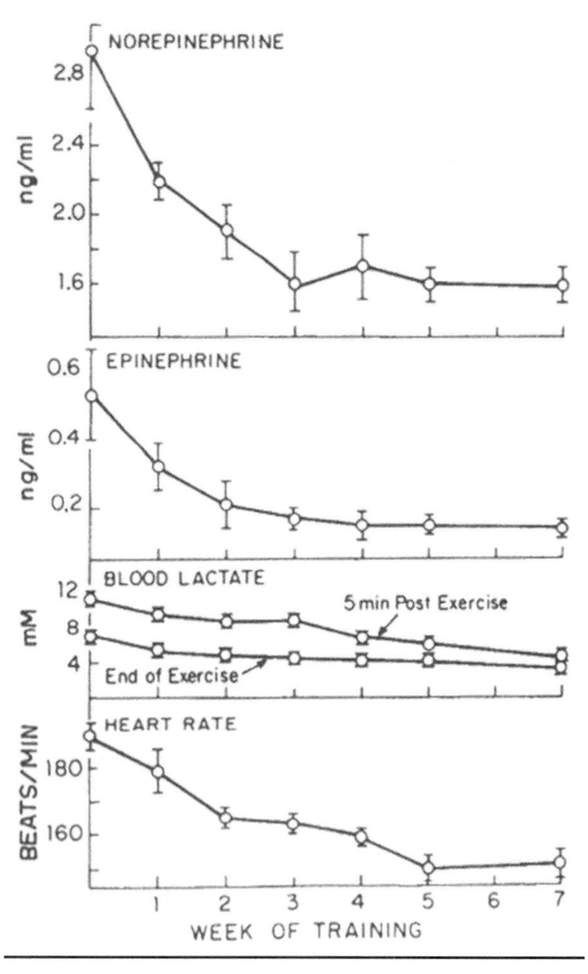

Figure 15.7 Data from Winder and colleagues (256) demonstrating the lower response of plasma epinephrine to exercise during 7 wk of endurance training.

Reprinted from W.W. Winder et al., 1978, "Time course of sympathoadrenal adaptation to endurance exercise training in man," *Journal of Applied Physiology* 45: 370 -374. With permission of American Physiological Society.

hance muscle glycogenolysis in exercise. However, epinephrine was administered in supraphysiological amounts and glycogen was not directly measured. Furthermore, no effect of adrenodemedullation on glycogenolysis was found in exercising rats (82). Richter and Galbo in the following years, however, described more extensive effects of adrenodemedullation (74, 202, 219). In adrenodemedullated rats only slightly reduced glycogen concentrations were found at exhaustion after prolonged exercise (118, 200); this substantiates an essential role of adrenomedullary hormones for muscle glycogenolysis in vivo. Furthermore, epinephrine infusion in similar rats completely restored glycogenolysis in muscle (199). The same authors perfused isolated muscle and found support for the view that the effect of epinephrine seen in vivo is exerted at least partially on muscle (204). Epinephrine, in concentrations that had no effect on glycogenolysis in resting muscle, enhanced glycogenolysis during contractions by prolonging the transient effect of contractions on phosphorylase a activity and by inhibiting glycogen synthase activity.

In humans a physiological role of epinephrine for glycogenolysis in exercising muscle is less apparent. Galbo and colleagues in 1976 (72) found that during exercise glycogenolysis was inhibited in experiments with beta-receptor blockade compared with experiments in which lipolysis was similarly blocked by nicotinic acid. However, Kjaer and colleagues in 1991 (151) showed that the rate of glycogen breakdown as well as lactate release from contracting leg muscle were unaffected by addition of arm cranking, which increased plasma epinephrine threefold. Kjaer found that muscle glycogen breakdown and lactate accumulation in contracting muscle were identical in healthy subjects and adrenalectomized patients regardless of whether the latter had epinephrine infused 149. The finding that plasma lactate levels were higher in experiments with increases in plasma epinephrine than in experiments without increases (149) may reflect that epinephrine may enhance glycogenolysis in nonexercising muscle during exercise (1).

In the 1980s, Richter, Sonne, and Galbo used adrenodemedullated rats, either swimming (200, 202) or running (217), and found that epinephrine may enhance hepatic glycogen breakdown during exercise. Furthermore, these authors (218) as well as Winder's laboratory (4, 258) emphasized that epinephrine may also promote hepatic glucose production during exercise by enhancing muscle glycogenolysis, thereby providing lactate for gluconeogenesis. Similarly, Wasserman's laboratory used adrenalectomized dogs to show that epinephrine may play a role in hepatic glucose production late during prolonged exercise (179). In humans, probably reflecting the existence of redundant mechanisms, a clear physiological role of epinephrine for hepatic glucose production in exercise has not been demonstrated. Philip Cryer's laboratory (106) together with John Holloszy in 1986 reported that the exercise-induced increase in hepatic glucose production is inhibited during alpha and beta adrenergic blockade (105), but they could not reproduce the findings in 1991 (173). In this later report, they also showed that the increase in glucose production during exercise is the same in adrenalectomized and healthy subjects (60, 106). Kjaer also reported that glucose homeostasis is maintained during exercise in adrenalectomized humans (115). When epinephrine was infused in the patients a transient enhancement of glucose production was seen early in exercise, but epinephrine concentrations were higher than in exercising healthy subjects. The conclusion that epinephrine is not normally a major regulator of glucose production in exercising humans was also drawn in another study by these authors (147). Thus, glucose production increased identically in exercise experiments with and without blockade of the celiac ganglion and, in turn, with and without impairment of epinephrine secretion. Furthermore, glucose production increased identically regardless of whether epinephrine was infused in substituting amounts in the former experiments. Hepatic glucose production was enhanced when epinephrine was infused to concentrations higher than those corresponding to the actual exercise intensities.

Lipolysis-Related Studies

Lithell and colleagues in 1981 (169) were among the first to show that there is a positive relationship between muscle lipoprotein lipase activity and the excretion of urinary epinephrine after heavy exercise. In 1984, Martin and colleagues (174) showed that there was a significant reduction in epinephrine-induced lipolysis 4 d after the cessation of training in humans compared with the enhanced levels apparent in the trained state. In 1995, Klein (157) performed epinephrine infusions in endurance-trained athletes to determine whether a single bout of exercise before epinephrine infusion altered the lipolytic rate. They found that previous acute exercise did not alter the rate of lipolysis in these trained athletes. Horowitz and colleagues (110) conducted a longitudinal training study in 1999 for 16 wk and found no change in the lipolytic rate or adipose tissue blood flow sensitivity to epinephrine infusions during resting conditions. Crampes and colleagues (31) in 1989 reported that adipose tissue responses to epinephrine in adipocytes taken from trained humans are markedly higher than those found in untrained people and that a beta adrenergic pathway seems to be involved.

Effects of Catecholamines on Heart Function

Although much of the attention concerning epinephrine and exercise has centered on metabolism, it is clear that catecholamines have important effects on the circulation and heart. In 1986, Chilian and colleagues (24) used dogs to show that alpha adrenergic coronary vasoconstrictor tone during submaximal exercise was primarily due to circulating catecholamines rather than direct neural effects. In 1986, Svendenhag and colleagues (233) determined that both the vasodilator and systolic pressure responses to epinephrine were greater in endurance-trained subjects. However, the cardiac chronotropic and inotropic effects of epinephrine were independent of the training state.

Growth Hormone

In the mid-1960s, W.M. Hunter and colleagues (117-119) at the University of Edinburgh conducted much of the initial research on GH responses to muscular activity. These studies used very long walks or games of squash as the exercise stimuli and measurements of GH were made with recently developed radioimmunoassays. The data (117) published in *Science* showed that even with relatively slow walking (6.4 km/h) there was a large increase in plasma GH in six subjects. If the walking continued for 2 to 3 h there was a marked decrease in circulating GH toward basal levels. Fonseka and colleagues in 1966 (62) emphasized the importance of intensity of exercise on the secretion of GH. They reported several studies such as monitoring GH in physicians during normal ward duty and never found a significant elevation of GH, yet when these subjects exercised there was a marked elevation in GH. Some of their research was also unique in that they measured GH in the same subjects on both exercise and nonexercise days. The results showed an early recognition that exercise responses must be made relative to the concentrations found on a control day. This is especially true for hormones such as GH that are secreted in a pulsatile manner. As is typical in many of these early studies, an observation was sometimes made on a single subject. This early research on GH centered on whether GH was responsible for stimulating fat mobilization during exercise because previous work had shown that injections of GH resulted in elevations in nonesterified free fatty acids in the plasma.

Effect of Hypophysectomy

In 1966, C.M. Tipton and colleagues (239) at the University of Iowa in the United States studied hypophysectomized rats with GH replacement in an attempt to determine the effects of exercise training on spleen weights and body weight with or without replacement manipulations. Rats were trained or remained sedentary and some were supplemented with GH. Manipulations included vagotomy, diencephalon lesioning, immunological sympathectomy, hypophysectomy, and hypophysectomy plus GH supplement. Training had no effect on spleen weights under most manipulations, with the exception that hypophysectomized rats that were trained had markedly lower spleen weights.

R.H. Johnson and colleagues (136) at the University of Glasgow in 1971 studied patients who had panhypopituitarism (i.e., part of the pituitary was removed because of tumors) (figure 15.8). They found that marked elevations in GH occurred during exercise in control subjects but not patients. In contrast, free fatty acid, glycerol, and ketone body concentrations were higher or similar in patients compared with control subjects. This led to the conclusion that GH is not needed for mobilization of free fatty acids during exercise. This finding was supported by Federspil and colleagues (55) in 1975 when they showed that elevations in free fatty acids after exercise in rats were not affected by hypophysectomy. However, a purported role for GH-stimulated lipolysis remained controversial for several decades.

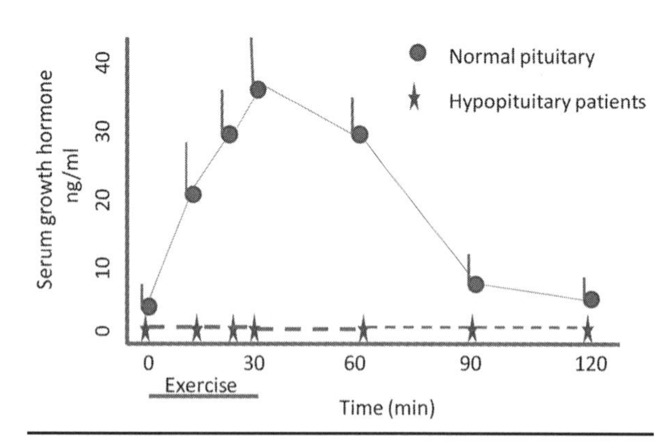

Figure 15.8 The response of plasma growth hormone (GH) in humans with a normal pituitary and those in humans with reduced pituitary function due to surgery for removal of tumors

Reproduced with permission from R.H. Johnson et al., 1971, *Clinical Science* 40: 127-136. "The effect of moderate exercise on blood metabolites in patients with hypopituitarism," © the Biochemical Society and the Medical Research Society.

Factors Potentially Responsible for Elevations in GH During Exercise

In 1977, John R. Sutton (1941-1996) (230) at the Garvan Institute in Sydney, Australia, demonstrated that the GH response during hypoxic exercise was greater than that during normoxic exercise. The remarkable magnitude of the difference in the GH responses demonstrated the importance of multiple stresses to endocrine responses. Sutton reported other data on the importance of exercise intensity, acid–base imbalance, and other factors that affect GH responses to exercise (229, 231, 232). In 1994, Engfred and colleagues (44) showed that training at hypoxia versus normoxia does not alter the GH response to acute exercise. Almost 30 yr after Sutton's first report, J.R. Hoffman and colleagues (107) measured muscle oxygenation and GH responses to two types of resistance exercise: low intensity, high volume and high intensity, low volume. They found no difference in muscle oxygenation between these two regimens. The GH response in the postexercise period was higher for the low-intensity, high-volume routine compared with the high-intensity, low-volume routine, which could be due to differences in postexercise reoxygenation. There was better reoxygenation in the low-intensity, high-volume protocol, which is when they observed a greater GH response.

In 2006, Collier and colleagues (28) provided arginine before resistance exercise and found that arginine significantly inhibited the GH response to exercise. Jill Kanaley (1958–) published a review paper in 2008 (140) in which she stated that arginine alone increases resting GH, as does exercise. However, arginine blunts the GH response when it is given during or before exercise.

Effects of Exercise Intensity on Postexercise GH Levels

GH is primarily an anabolic hormone despite some effects on stimulating lipolysis. Thus, its actions after exercise are important. Blood sampling at frequent intervals and for prolonged periods is necessary for an accurate assessment of responses due to GH pulsatility.

In 2000, Pritzlaff and colleagues (189) in Art Weltman's (1950–) laboratory performed a series of studies (190) in which individuals exercised at many exercise intensities (normalized to the lactate threshold) on different days. The intention of these studies was to determine whether 30 min of exercise at these various intensities resulted in the same integrated GH concentrations during a 4 h recovery period. They found that the integrated area of GH increased linearly with exercise intensity. The mass of GH secreted per pulse increased with no change in GH pulse frequency.

Specific Antagonists and the GH Response to Exercise

In 1980, Few and Davies (59) at the London School of Hygiene and Tropical Medicine studied the effect of atropine on the GH response to exercise at low work levels. Atropine had no effect on cortisol but it greatly reduced the GH response to exercise. However, at high workloads atropine enhanced the increase in cortisol and again the GH response was diminished. They concluded that a cholinergic mechanism inhibited the GH response to exercise. These results were verified in 1986 by Brillon and colleagues (17), who showed that phentolamine blocks the GH response to moderate exercise whereas propranolol increases the GH response. Serotonergic mechanisms are probably not involved.

In 1986, Struthers and colleagues (227) reported that alpha adrenergic antagonists augmented the GH response to exercise. This finding was counter to the thinking at that time and to the data of Brillon and colleagues. In 1993, Thompson and colleagues (236) determined the effect of cholinergic and opioid mechanisms on GH released during and after exercise. Naltrexone (a general opioid antagonist) did not affect the GH response. This finding did not agree with that of Farrell and colleagues in 1986 (52) or Armstrong and Hatfield in 2006 (3), who found that opioid blockade enhanced GH responses during exercise. Speculation at that time centered on differences in results that were attributable to when the antagonists was given relative to the start of exercise. Thompson and colleagues (236) further showed that cholinergic mechanisms potentiate the GH response during recovery from exercise and that this may be due to inhibition of somatostatin release inhibiting factor. Kjaer and colleagues in 1999 (148) combined hypoxia with epidural anesthesia and found that hypoxia caused a greater increase in GH during exercise. Epidural blockade blocks the neural feedback from working muscle. When epidural blockade was added to the hypoxia, there was no additional effect on GH responses to exercise.

GH in Special Populations

It was not long after initial studies on plasma concentrations of GH before investigators took an interest in whether GH responses to exercise were normal in people with various diseases. In 1971, Hansen (93) from Aarhus, Denmark, showed that young people with type 1 diabetes had greater elevations in GH during both moderate and intense exercise. In 1988, Sundkvist and colleagues (228) furthered this research by showing that the GH response to exercise is augmented in type 1 diabetic patients with retinopathy compared with diabetic patients who do not have retinopathy.

Hansen (94) also showed that a glucose infusion in normal subjects during exercise abolishes the GH response during moderate work but not during heavy work. A glucose infusion during exercise did not alter the GH response in people with type 1 diabetes. Hansen noted that the effect of glucose was unique to exercise because the GH response to sleep is not suppressed by glucose. (*Note:* GH concentrations increase markedly a couple hours after the onset of sleep.) It is interesting that the discussion in this paper includes the speculation that "humoral factors may be liberated locally from the working muscles and carried by the blood stream to hypothalamic growth releasing centers or neural reflex mechanisms may be involved as with many circulatory and respiratory adaptations to exercise" (93, p. 204).

In 1973, Hansen (94) showed that obese young adults have a blunted or nonexistent GH response to moderate-intensity exercise even though their exercise heart rate was higher than that of lean control subjects. Garlaschi and colleagues (78) verified this finding in 1975. It should be noted that the obese individuals in Hansen's work were hyperinsulinemic but not diabetic.

More than 30 yr later, Irving and colleagues (129) investigated the influence of the metabolic syndrome on overnight GH secretion in obese people. Subjects engaged in 16 wk of endurance exercise of low or high intensity. Compared with controls, both exercise-training regimens resulted in increased nocturnal GH. The increase in nocturnal GH was inversely related to the change in fat mass. It seems that people with the metabolic syndrome can still increase GH responses with training.

GH in Organ-Transplanted or Paralyzed People

The importance of nervous system innervation to the GH response was also tested using both organ transplant and nerve blockage strategies in the 1990s. Much of this research was accomplished by Michael Kjaer.

In 1989, Kjaer and colleagues (156) used epidural blockade as a means to eliminate or reduce afferent feedback during exercise. Epidural blockade did not alter hormonal or metabolic profiles at rest. During exercise the GH response was the same with or without blockade. In contrast, epidural blockade completely blocked the ACTH and beta endorphin response to exercise. The data indicate that impulses from the working muscle are essential for the ACTH and beta endorphin responses to submaximal exercise. They also show that afferent activity is less important than efferent activity from motor centers for the GH response (and catecholamines and insulin). Testing a similar hypothesis concerning the importance of neural feedback for immunoassayable GH release, Gosselink and colleagues (85) in 1998 conducted an important study on hypophysectomized rats. This study demonstrated that low-threshold nerve afferents from the proximal end of severed nerves are important for the release of bioassayable GH from the pituitary. Such stimulation did not affect immunoassayable GH, a finding compatible with Kjaer's data. The study by Gosselink and colleagues makes the fundamentally important point that bioassayable GH can be neurally regulated.

Kjaer and colleagues (150) in 1995 used liver- or kidney-transplant patients and found that the GH response to exercise was similar in liver- and kidney-transplant patients compared with controls. The norepinephrine and insulin responses also were similar. In 1996, Kjaer and colleagues studied tetraplegics (153) during functional electrical stimulation and the result was an increase in GH. Thus, activity in the motor centers and afferent muscle nerves are important for GH, catecholamines, insulin, glucose production, and lipolysis. Kjaer also used epidural blockade in healthy humans. In this case complete temporary paralysis of the legs was induced and then individuals exercised in response to functional electrical stimulation. This design allowed these subjects to be compared when they were normal (voluntary contractions) and when they were paralyzed. While exercising at the same oxygen consumption, epinephrine, GH, ACTH, and cortisol levels were higher during voluntary exercise than during functional electrical stimulation. Kjaer concluded that humoral feedback alone is not sufficient to trigger normal endocrine responses, some of which are important to mobilization of extra muscular fuels.

Role for GH in Muscle Growth

A role for GH and exercise-stimulated muscle growth was not explored to a great extent until the mid-1960s. One exception was a study completed in 1952 by Brenda Bigland and Barbara Jehring (9) at University College, London, who studied female rats with an intact pituitary by injecting them with GH for 21 d and then measuring their quadriceps strength. Interestingly, previous research had shown that the quadriceps muscles were more sensitive to GH treatment than was the gastrocnemius. Importantly, throughout the period of GH injection, food intake of the GH-treated rats was restricted to that of their age-matched controls. The results are most intriguing in that the researchers found that even though there was greater muscle mass (15%-40%) and muscle fiber area (6%-20%) in the GH-treated female rats, the ability to generate force was not increased. When the amount of tension that could be developed was made relative to the muscle mass, the ability to develop tension was actually decreased in GH-treated animals.

Several important studies on the role of GH in muscle growth were conducted by A.L. Goldberg at Harvard Medical School. In 1967, Goldberg (79) demonstrated that work-induced muscle growth can occur in hypophysectomized rats. Two years later, Goldberg and Goodman (80) used hypophysectomized rats with or without GH replacement combined with muscle growth induced by a synergist ablation model. They reported that GH treatment can reduce or abolish inactivity-related muscle loss. An important finding in this early study was that GH stimulated muscle mass both in muscles that hypertrophied during the synergist ablation model and in control GH-treated muscles. After making this observation, the investigators demonstrated wonderful insight by testing the effect of GH in the same animal in which one leg muscle exhibited hypertrophy and the other leg muscle demonstrated atrophy. They used tentonomy on one leg to produce hypertrophy and sciatic nerve ablation on the other leg to cause atrophy. Treatment with GH caused significant growth in the atrophying muscle (GH caused less atrophy) and enhanced muscle hypertrophy in the overloaded muscles.

In the early 1970s, Katarina Tomljenovic Borer (1940–) at the University of Michigan in the United States showed that ponderal and skeletal growth increased in female golden hamsters engaging in voluntary running activity. In 1978, Borer, Kelch, and Brown (12, 20) reported a complicated but important study that was among the first to include all of the following: food restriction, voluntary exercise, short-term fasting, early- and late-phase compensatory growth, and precise measurements of food intake and serum levels of anabolic hormones. They sought to determine whether GH or insulin regulated increased growth due to exercise. GH responses after exercise depended on the food intake immediately (first 4 h) after exercise. Food restriction of animals during this time period abolishes the exercise-induced elevations in GH. This study also showed a relationship between serum insulin and GH, and the results suggested that exercise-induced GH release was a stimulus for insulin release after exercise. This study along with Goldberg's research may have contributed to a shift away from studying the GH responses to exercise relative to fuel metabolism and toward studying GH's role as a promoter of growth. In 1986, Borer and colleagues (13) found, again in the hamster model of running, that rapid growth was related to the elevated GH pulse amplitude and frequency and inversely related to low troughs of GH. In 1987, Shapiro and colleagues (211) in Borer's laboratory found in spontaneously running golden hamsters that the concentration of somatostatin-like immunoreactivity was increased in the group that exercised and had adequate food in comparison with other groups that were food restricted or did not exercise. Borer's research solidified the concept that multiple factors coalesce to regulate GH responses to exercise as well as growth consequent to that exercise.

GH and Resistance Exercise

The earliest work on resistance exercise and GH in humans appears to have been conducted in the mid-1970s by Lukaszewska and colleagues (170). These studies were published in Polish journals and therefore have not been frequently cited in English-language journals. These studies showed that acute resistance exercise elevates serum GH concentrations. Lukaszewska and colleagues also made initial observations about the importance of exercise intensity to the magnitude of this increase.

Starting in the late 1980s and continuing until the current time, William Kraemer (1953–) at several institutions, including the Natick Laboratory, the University of Connecticut in the United States, and Penn State University in the United States, conducted numerous studies on the endocrine response to resistance exercise. A significant portion of his early studies defined the GH response to various types of resistance exercise. These parameters included intensity, volume, rest intervals, single set versus multiset, skill level, periodicity, gender, and aging. In 1988, Kraemer and colleagues (161) wrote a review article that was among the first on the topic of endocrine responses to resistance exercise. In 1990, Kraemer and colleagues (163) reported that the highest GH response occurred during a session that required the greatest workload and the shortest rest interval. He also made the important observation that changes in insulin-like growth factor 1 (called somatomedin-C at that time) did not follow the same pattern as GH. This study initiated many subsequent investigations in which his group tried to determine the optimal design of resistance-exercise sessions for stimulating the release of GH.

In 1990, Kraemer and colleagues (162) documented the GH responses to several types of heavy resistance exercise and concluded that not all resistance-exercise protocols result in elevated levels of GH. In the late 1990s he conducted several investigations focused on resistance exercise and muscle fiber hypertrophy (165). He found a significant correlation between absolute mean elevations in GH and the increase in muscle fiber size with training (165). By 2005, Kraemer and Ratamess concluded that "protocols high in volume, moderate to high intensity, using short rest intervals and stressing a large muscle mass, tend to produce the greatest acute hormonal elevations" (166).

Kraemer's group later focused on the fact that there are many (>100) isoforms of GH in the plasma and that resistance exercise alters these isoforms in multiple ways. Immunoassayable GH and bioassayable GH are not the same and show different profiles after heavy resistance exercise. In 2000, Nindl and colleagues (185) demonstrated that GH increased during exercise whether it was measured by functional (bioactive GH assay using the rat tibial line) or immunoradiological assays. However, in the postexercise period immunoreactive GH increased approximately twofold when compared with functionally assayed GH. These studies stimulated numerous other investigations during the next decade that have provided insight into changes in GH variants in response to resistance exercise. Nindl and colleagues (184) showed that an acute bout of heavy resistance exercise reduced GH pulse amplitude early in the evening after exercise but that a significant elevation in GH occurred for the rest of the night. In 2001, Wes Hymer (120) in collaboration with Kraemer and colleagues found that plasma obtained after resistance exercise shows an increase in immunoassayable GH, but not when measured by the rat tibial line bioassayable GH. Also, most of the GH released after exercise was able to dimerize the GH receptor in vitro. Thus, there should be stimulation at the target tissues. Kraemer's group used chromatography to fraction plasma samples by molecular mass and then measured GH using three assays, including the rat tibial line bioassay. In this study (184), women were studied before and after a 24 wk resistance-exercise program and another group remained sedentary. They found no consistent change in bioactive GH after acute exercise, but chronic exercise increased bioactive GH. This occurred across all fractions, suggesting that exercise training increases many bioactive forms of GH. Some of this early research was summarized by Brad Nindl (one of Kraemer's former students) and colleagues in 2003 (186). In 2006, Kraemer and colleagues (164) made the important observation that chronic resistance-exercise training results in an increase in biologically active GH molecular variants. This is true for several types of training regimens.

Training and GH Responses to Exercise

In 1972, Hartley and colleagues (96, 97) published two heavily cited papers that documented the endocrine responses of many hormones to endurance exercise training. GH responses before and during submaximal exercise were lower after training than during pretraining; however, this difference was statistically significant only for very high workloads. Chronic resistance-exercise training does not seem to alter resting concentrations of GH. This suggests that the response of GH after an acute bout of resistance exercise may be most important for tissue remodeling, a point made by Kraemer and Ratamess (166) in 2005.

Role of GH in Free Fatty Acid Mobilization

As discussed previously, one of the initial reasons for investigating GH and exercise in the early 1960s was to examine a role for GH in the metabolic response to acute exercise. In 1975, Federspil and colleagues (55) conducted a study using hypophysectomized rats in which they showed that the pituitary gland plays an essential role in promoting the increase in exercise plasma free fatty acids immediately after exercise (swimming). However, this paper reported that in nonhypophysectomized rats swimming did not induce an increase in GH but rather induced a decrease.

Jorgensen and colleagues (137) in 2003 studied GH-deficient individuals and healthy individuals to determine the effect of ghrelin (known to stimulate GH) on GH responses to moderate exercise. They found no change in ghrelin levels in both groups but the normal increase in GH occurred only in control subjects. They also studied the effects of GH treatment during exercise in GH-deficient patients and found that GH stimulated the turnover of free fatty acids in the recovery period from exercise. Jill Kanaley and colleagues (141) reported in 2004 that GH released during aerobic exercise plays a significant role in regulating fatty acid metabolism. The possibility of GH stimulating free fatty acid availability for muscle oxidation was the original stimulus for the first studies by Hunter and Fonseka in the 1960s and it is still of interest a half-century later.

Exercise and GH Pulsatility

The effectiveness of most hormones is enhanced when that hormone is secreted and delivered to target tissues in a pulsatile manner as opposed to a constant delivery. Therefore, it was logical to investigate whether exercise training alters the GH profile. These studies involve having subjects exercise (aerobic or resistance) and then taking blood samples at short intervals (10-20 min) for longer periods of time (12-24 h). Manning and Bronson in 1991 (171) showed that female rats lose their GH pulsatility during prolonged forced exercise. In contrast, GH pulsatility is enhanced when animals are given voluntary running wheels. Finally, any negative effects of forced running are reversed by gonadotropin releasing hormone (GnRH) infusions. In 1992, Weltman and colleagues (253) showed that GH pulsate amplitude was significantly greater in women who exercise trained by running for 1 yr at intensities above, not below, the lactate threshold. GH pulse frequency was not altered by the training schedule.

Kraemer's group determined the effect of resistance exercise on GH pulsatility. In 2006, Tuckow and colleagues (240) had people perform resistance exercise and they measured GH by both immunoreactive assays and bioassays on plasma obtained for prolonged periods after the exercise. The results relative to GH pulsatility seem to contrast with other studies in that they found higher burst frequency and lower burst mass of GH after exercise. Approximate entropy (an estimate of orderliness of burst occurrence) was higher after exercise, which suggests a less-orderly GH release pattern. The average GH concentration was similar between control and exercise conditions. Pierce and colleagues in 2009 (187) conducted a study using acute resistance exercise before and after 8 wk of training. They found that acute exercise leads to a greater appearance of disulfide-linked immunoreactive GH aggregates. This increase is not affected by training. They speculated that structural alterations in the GH molecule after acute exercise may help in the regulation of biological activity.

Exercise Plus GH Administration

Because GH is an anabolic hormone, it was logical to investigate the effect of combining exogenous GH with resistance exercise. In 1993, Yarasheski and colleagues (262) showed that giving GH in addition to resistance exercise in young men did not further augment protein accumulation, muscle strength, or rates of protein synthesis in skeletal muscle. Yarasheski along with other colleagues (261) later demonstrated the same lack of effect in older people.

Hansen and colleagues (95) in 2005 had people exercise with or without supplemental GH. They found that the extra GH increased lipolysis but that there was no further augmentation of whole-body fat oxidation during exercise. In 2009, Goto and colleagues (86) in Michael Kjaer's laboratory determined the effects of administering a GH receptor antagonist for 16 d along with exercise. Subjects preformed either short-term or long-term exercise in the same session. Basal GH was elevated in the group receiving the receptor antagonist treatment whereas insulin-like growth factor 1 was reduced. The GH receptor antagonist did not alter metabolism during acute exercise.

Effects of Aging

The effect of aging on GH responses to exercise also gathered interest in the 1990s. In 1992, Pyka and colleagues (195) showed that the GH response to acute resistance exercise is greatly diminished in the elderly; there was no sex difference in this regard. Marcell and colleagues in 1999 (172) confirmed this result and extended it by suppressing somatostatin release inhibiting factor with a drug. They found that the response of the elderly was not re-established to the level observed in the young. Thus, the age deficit in releasing GH is not solely attributable to somatostatin release inhibitor mechanisms. Weltman's group in 2006 (254) showed that the GH pulsatile response to exercise is approximately 50% lower in older people than in younger people and that exercise selectively augmented the mass of GH secreted per burst in younger subjects but not older subjects.

Adrenal Cortex: Cortisol

Some of the first studies on cortisol (F) centered on whether the adrenal gland was necessary for changes in heart rate during exercise. In 1922, Stewart and Rogoff (226) performed studies using exercise in adrenalectomized cats. Most importantly, they found that no major decrement occurred in the ability to exercise or in heart rate, temperature, and other general physiological responses when even a large portion of the adrenal gland was removed. Others subsequently found marked decreases in work performance in adrenalectomized dogs. In 1924, W.B. Cannon and colleagues (22) used denervated heart preparations and reported responses to hind-limb stimulation in cats. In these studies, the investigators tested a previous hypothesis made by Johansson in 1895 that muscle activity sets into motion the release of a substance into the plasma that causes acceleration of the heart and that this is true even when nerves to the heart to have been eliminated. Cannon and colleagues confirmed earlier work that also showed that heart rate does not increase normally during muscle contractions when the adrenal glands are rendered inactive.

After these early studies, some attention switched to the effect of adrenal gland extracts on work performance. In 1932, Eagle and Britton (42) used a limited number of normal dogs (named Brownie, Snookie, and Lady) but studied them repeatedly over many months. Running ability increased by about 90% during trials in which the extract was injected. In 1936, Dwight J. Ingle (124) used anesthetized adrenalectomized rats in which the leg muscles were stimulated to contract. Adrenalectomy caused muscular work performance to decrease markedly. A substance called cortin caused a dose–response increase in muscle work performance when injected into the rats. This stimulation was continuously applied until the animal died or 120 h had elapsed. Over the next 20 yr Ingle refined the observations (75) on the adrenal cortex by replacing various corticosterone-like substances (which increased in specificity as the years went by) and noted improvements in performance. As noted previously, Ingle performed an extensive amount of excellent research on the adrenal cortex in the 1930s and 1940s. In 1972, Tipton and colleagues (238) showed that adrenalectomized rats can adapt to 80 d of treadmill training with some differences noted but no differences in muscle work performance or in various hematological measures.

Urinary and Plasma Concentrations

After the initial observations with adrenal cortex extracts, studies began to center on changes in circulating or urinary F in humans in response to exercise. In 1946, Venning and Kazmin (243) reported that there was an increase in urinary corticosteroids during exercise in humans. The accuracy and validity of the steroid measurements, although the best at the time, were questioned later. This may explain the discrepancy between this initial finding and several subsequent studies that showed a decline in urinary 17-ketosteroids during and after exercise (92). In 1958, Robinson and Macfarlane (205) determined that urinary 17-ketosteroids decrease during exercise in the heat. Also in 1958, Connell and colleagues (29) reported reduced urinary 17-ketosteroid levels for 24 h after intense exercise. This study was the first to report changes in urinary 17-ketosteroids due to exercise training. They found a significant and uniform decrease in urinary steroids during a 10 d training period.

Staehelin and colleagues (221) in 1955 may have been the first to measure plasma 17'21-hydroxysteroids during and after exercise in humans. They found an immediate increase in 17'21-hydroxysteroids by 15 min of exercise; however, if the exercise was continued for 2 h, the concentration of 17'21-hydroxysteroids significantly declined. This was true for both intense treadmill exercise and prolonged marching. In 1965, Cornil and colleagues reported a significant decrease in F and corticosterone in human plasma during and after 20 min of high-intensity bicycle ergometer exercise and speculated that the decline in plasma F was due to an "increased utilization of cortisol" (30, p. 167). The last sentence of their discussion section states, "The problem is not resolved; the determination of the production rate of F during exercise by means of the isotope dilution method would probably allow an approach to the solution" (30, p. 167). In 1974, J.D. Few published relevant data using isotope dilution (reviewed in "Secretion of Cortisol During Exercise"). During the 1970s, G.C. Cashmore, J.D. Few, and C.T.M. Davies from the Environmental Physiology Unit, London School of Hygiene and Tropical Medicine, provided numerous important insights into the control of cortisol during exercise.

In 1970, Davies and Few (34) showed for the first time that the F response was dependent on the intensity of the treadmill exercise. Low levels of exercise intensity (less than ~60% of the individual's maximal aerobic capacity) result in decreases in plasma cortisol; however, at higher work intensity there was a marked increase in cortisol. As a precursor to an important study published in 1974 by Few, they injected tritiated cortisol (a single injection) into two subjects, and plasma cortisol concentration increased during high-intensity work. They believed that this increase was attributable to an increased production of cortisol (57).

Secretion of Cortisol During Exercise

In 1974, Few (57) infused tritiated cortisol into humans to determine whether the increase in plasma concentration of F during exercise was due to increased secretion. He found significant increases in cortisol production and marked decline in specific radioactivity of the injected cortisol. These data also showed that exercise causes an increase in rate of uptake of cortisol by peripheral tissues. Thus, the speculation by Cornil and colleagues in 1965 (30) was confirmed. In 1977, Cashmore, Davies, and Few (23) reviewed data from several protocols that were similar to that used in Few's 1974 paper and showed that measuring only plasma levels of F gave a good estimate of changes in secretion rate. Few, Davies, and Cashmore continued to provide significant insight into the adrenocortical responses to exercise for many years. In 1980, Few (60) had subjects ingest 11 α^3H 4-^{14}C cortisol before running a 100 km race. Twenty-four hour urine collection again showed that the running significantly increased the rate of F secretion.

Factors That Might Influence Cortisol Secretion During Exercise

Some of the controversy in the early 1970s concerning whether F increased with exercise may have been due to marked differences in experimental design in terms of exercise intensity and duration. The first measure of both plasma ACTH and F under controlled laboratory conditions was made by Farrell and colleagues in 1983 (51). They showed that treadmill running at 65% or 80% of the maximal oxygen consumption resulted in elevations of both F and ACTH and that these changes were correlated at both intensities. He also showed a correlation between lactic acid accumulation and both plasma ACTH and F. In 1984, Tabata and colleagues (234) showed that both ACTH and F increased when plasma glucose decreased during exercise. Later, the same group (235) showed that both ACTH and F responses to prolonged exercise are abolished when plasma glucose is maintained by a glucose infusion. Hypoxia has also been considered as a contributor to F responses to exercise. For example, in 1988 Bouissou and colleagues (14) studied men under both normoxic and hypoxic conditions and verified Farrell's finding that 60 min of exercise at 65% $\dot{V}O_2max$ resulted in elevations in both ACTH and F in normoxia. However, under hypoxic conditions the F response was reduced but the ACTH response was not affected.

More recently some other factors that may contribute to F secretion have been investigated. In 2003, Steensberg and colleagues (225) demonstrated that the myokine interleukin-6 enhanced plasma F in humans. An interaction by interleukin-6 with the adrenal cortex may prove to be quite important because it could indicate a solid connection between the endocrine and immune system that was not previously appreciated. In 2007, Viru and colleagues (245) showed that beta adrenergic blockade resulted in enhanced F response to short-duration exhaustive exercise. In 2007, Coiro and colleagues (26) reported that lipid infusion resulted in a significant reduction in both ACTH and F during exercise. In 2009, Jankord and colleagues (134) found that chronic inhibition of nitric oxide synthase caused an exaggerated ACTH response to exercise in pigs. In 2009, Kamimori and colleagues (139) found that atropine when given before intermittent exercise caused a significantly higher increase in F than a placebo did. The effect of exercise intensity on F levels during exercise was still being investigated in 2008 by Hill and colleagues (103) when they confirmed that moderate- and high-intensity exercise resulted in the elevations in F whereas lower-intensity exercise (40% maximal oxygen consumption) resulted in a reduction of F levels. All of these recent studies point to the fact that, despite efforts from the 1930s, we still do not understand the complete mechanisms regulating F secretion during exercise.

Pulsatile Nature of F

Gabrielle Brandenberger and colleagues (15, 16) in the mid-1970s and early 1980s made the important point that we need to know the complete hormone pattern of plasma cortisol concentrations in order to conclude whether exercise or a meal stimulates the release of cortisol. They demonstrated that plasma F concentrations do not change when the start of exercise coincides with a circadian peak of F in the early afternoon. Exercise in the morning when F is low results in a significant elevation. She also made the observation that exercise-induced elevations in F are not additive to meal-induced increases in F. In figure 15.9, plasma F values during exercise (solid line) are compared with a control day (dotted line) when the same subjects were at rest. The arrows indicate when a meal was taken.

In 1982, Brandenberger and colleagues (15) showed that exercise-related increases in F inhibit subsequent stimulation of the adrenal cortex. Ignoring the background circadian rhythm can lead to significant errors in estimating exercise responses. In 1995, Thuma and

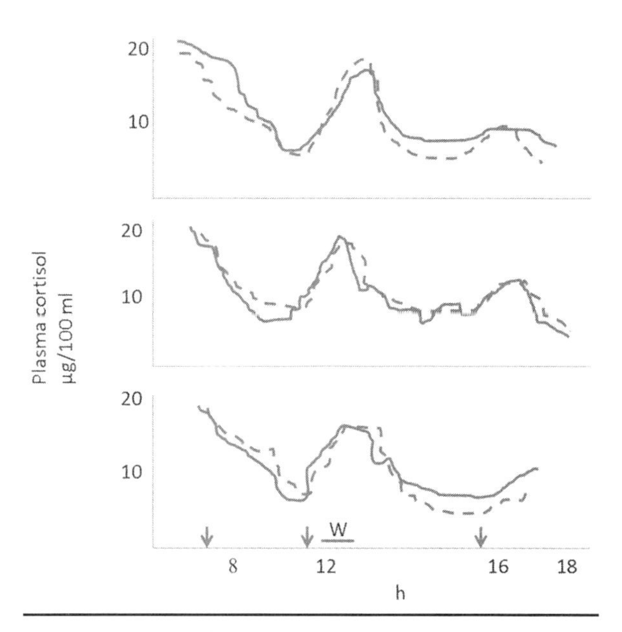

Figure 15.9 The importance of measuring hormones that have a natural daily rhythm when assessing the effects of exercise on that hormone—in this case, cortisol. The solid line shows the cortisol concentrations when exercise (W) occurs, and the dotted lines show the cortisol concentrations when no exercise occurs.

Reprinted, by permission, from G. Brandenberger and M. Follenius, 1975, "Influence of timing and intensity of muscular exercise on temporal patterns of plasma cortisol levels," *Journal of Clinical Endocrinology and Metabolism* 40(5): 845-849.

colleagues from Anne Loucks' laboratory (237) found that areas under the curve for exercise responses underestimated F secretion by as much as 93% when they were calculated based on pre-exercise concentrations (as the baseline) compared with areas under the curve based on measured circadian baselines (multiple samples taken on a rest day at the same time points as those taken on the exercise day). This research, along with previous results by Brandenberger and colleagues, clearly shows the imprecision in interpreting hormone data when only pre–post samples are obtained and the natural circadian rhythms are not evaluated.

Effect of Activated Muscle Mass on Plasma F Responses

In 1980, Few and colleagues (58) compared F responses in one- and two-leg exercise. Plasma F levels increased with one-leg exercise. However, the plasma F response was much lower when the same absolute workload was accomplished using two legs. This group also reported that some subjects had no increase in F levels during exercise, but each of these individuals had elevated resting F before exercise. Deschenes and colleagues (39) in 2000 demonstrated that when subjects exercised at the same relative oxygen consumption (50% $\dot{V}O_2max$) by maintaining either 40 or 80 revolutions/min on a bicycle ergometer, the F response was elevated only at the 40 revolutions/min cadence. These data suggest that the muscle activation pattern is important to F responses.

Effect of F on Decreasing Muscle Mass

Excess F is known to cause skeletal muscle atrophy primarily by stimulating proteolysis of myosin heavy chain proteins. Using rats, Hickson and colleagues (102) in 1984 found that the protective aspects of exercise were not attributable to any change in androgen or glucocorticoid cytosol receptor levels during regular endurance exercise. In 1989, Czerwinski and colleagues (32) in Bob Hickson's laboratory determined the effect of regular exercise on glucocorticoid-induced muscle atrophy. Specifically, they measured myosin heavy chain breakdown rates as well as synthesis rates and found that regular exercise attenuated the effect of glucocorticoids on myosin heavy chain breakdown but that exercise had no effect on the rate of myosin heavy chain synthesis. In 1992, Falduto (47) from Bob Hickson's laboratory demonstrated that regular endurance exercise inhibited several of the important steps in the pathway that leads to glucocorticoid-induced muscle atrophy.

Effect of Training on F Responses and Adrenal Cortex Volume

The measurement of changes in the volume and density of the adrenal cortex has been used since the time of Hans Selye as an indicator of chronic stress. Several scientists, such as Buuck and Tharp in 1971 (21), used rats and found that the zona fasciculata of the adrenal cortex was significantly enlarged due to endurance training. Like most other hormones, the F response to submaximal exercise is reduced by endurance training when humans exercise at the same absolute workload before and after training. However, if exercise is intense and prolonged, the F response in trained individuals is greater than that in untrained individuals. This is similar to the adaptation observed for epinephrine (reviewed earlier). If athletes overtrain, F is reduced during exercise. Dexamethasone is a competitive feedback inhibitor of F secretion that causes less of a suppression in F in trained individuals than in untrained individuals, as shown by Silva and colleagues in 2008 (215), suggesting a desensitization of the adrenal cortex due to prolonged training.

Effects on Immunology

As noted previously, F has long been known to regulate immune function. In 1996, Shinkai and colleagues (213) had male subjects perform bicycle exercise and then separated the subjects into two groups: responders who had a high F response to exercise and nonresponders who had little response. They then monitored changes in several classes of immune cells. Their data suggested that in the group of responders F may contribute to postexercise suppression of helper cytotoxic T-cells but that F did not seem to be involved in natural killer cell or beta cell function. Horne and colleagues in 1997 (109) used exercise training in humans (separated into a concurrent resistance and endurance-exercise group and resistance or endurance-training only groups) for 12 wk. The main outcome was that there was no correlation between changes in resting F or tumor necrosis factor (TNFα) plasma concentrations in any of the interventions.

Summary and Future Directions

During the period from 1910 to 2010, investigators went from not knowing that a certain hormone existed to understanding how exercise altered the circulating concentrations of most hormones. The next logical steps have been taken to understand intracellular signaling for these hormones and, unfortunately, space limitations for this chapter do not allow us to review those important achievements. There is, of course, still much to be learned about adaptations of hormones to regular exercise, especially from the important view that many of those changes have positive health implications. The ability of regular exercise to reduce insulin resistance stands out as a powerful beacon that should be followed.

References

1. Ahlborg G. Mechanism for glycogenolysis in nonexercising human muscle during and after exercise. *Am J Physiol* 248: E540-E545, 1985.
2. Angelopoulos TJ, Denys BG, Weikart C, Dasilva SG, Michael TJ, Robertson RJ. Endogenous opioids may modulate catecholamine secretion during high intensity exercise. *Eur J Appl Physiol Occup Physiol* 70: 195-199, 1995.
3. Armstrong DW, Hatfield, B. Hormonal responses to opioid receptor blockade during rest and exercise in cold and hot environements. *Europ J Appl Physiol* 97: 43-51, 2006.
4. Arnall DA, Marker JC, Conlee RK, Winder WW. Effect of infusing epinephrine on liver and muscle glycogenolysis during exercise in rats. *Am J Physiol* 250: E641-E649, 1986.
5. Asmussen E, Wilson JW, Dill DB. Hormonal influences on carbohydrate metabolism during work. *Am J Physiol* 130: 600-607, 1940.
6. Bajpeyi S, Tanner CJ, Slentz CA, Duscha BD, McCartney JS, Hickner RC, et al. Effect of exercise intensity and volume on persistence of insulin sensitivity during training cessation. *J Appl Physiol* 106: 1079-1085, 2009.
7. Banister EW, Griffiths J. Blood levels of adrenergic amines during exercise. *J Appl Physiol* 33: 674-676, 1972.
8. Berger M, Hagg S, Ruderman NB. Glucose metabolism in perfused skeletal muscle: Interaction of insulin and exercise on glucose uptake. *Biocheml J* 146: 231-238, 1975.
9. Bigland B, Jehring, B. Muscle performance in rats, normal and treated with growth hormone. *J Physiol* 116: 129-136, 1952.
10. Biolo G, Williams BD, Fleming RYD, Wolfe RR. Insulin action on muscle protein kinetics and amino acid transport during recovery after resistance exercise. *Diabetes* 48: 949-957, 1999.
11. Bloem CJ, Chang AM. Short-term exercise improves beta-cell function and insulin resistance in older people with impaired glucose tolerance. *J Clin Endocrinol Metab* 93: 387-392, 2008.
12. Borer KT, Kelch RP. Increased serum growth hormone and somatic growth in exercising adult hamsters. *Am J Physiol* 234: E611-E616, 1978.
13. Borer KT, Nicoski DR, Owens V. Alteration of pulsatile growth hormone secretion by growth-inducing exercise: Involvement of endogenous opiates and somatostatin. *Endocrinology* 118: 844-850, 1986.
14. Bouissou P, Fiet J, Guezennec CY, Pesquies PC. Plasma adrenocorticotrophin and cortisol responses to acute hypoxia at rest and during exercise. *Eur J Appl Physiol Occup Physiol* 57: 110-113, 1988.
15. Brandenberger G, Follenius M, Hietter B, Reinhardt B, Siméoni M. Feedback from meal-related peaks determines diurnal changes in cortisol response to exercise. *J Clin Endocrinol Metab* 54: 592-596, 1982.
16. Brandenberger G, Follenius, M. Influence of timing and intensity of muscular exercise on temporal patterns of plasma cortisol levels. *J Clin Endocrinol Metabol* 40: 845-849, 1975.
17. Brillon D, Nabil N, Jacobs LS. Cholinergic but not serotonergic mediation of exercise-induced growth hormone secretion. *Endocr Res* 12: 137-146, 1986.
18. Brisson GR, Malaisse-Lagae F, Malaisse WJ. Effect of phentolamine upon insulin secretion during exercise. *Diabetologia* 7: 223-226, 1971.
19. Brouha L, Cannon WB, Dill DB. The heart rate of symphthectomized dog in rest and exercise. *J Physiol* 612: 345-359, 1936.
20. Browne SA, Borer KT. Basis for the exercise-induced hyperphagia in adult hamsters. *Physiol Behav* 20: 553-557, 1978.
21. Buuck RJ, Tharp GD. Effect of chronic exercise on adrenocortical function and structure in the rat. *J Appl Physiol* 31: 880-883, 1971.
22. Cannon WB, Linton JR, Linton RR. Conditions of activity in endocrine glands XIV. The effects of muscle metabolites on adrenal secretion. *Am J Physiol*: 153-162, 1924.
23. Cashmore GC, Davies CT, Few JD. Relationship between increases in plasma cortisol concentration and rate of cortisol secretion during exercise in man. *J Endocrinol* 72: 109-110, 1977.
24. Chilian WM, Harrison DG, Haws CW, Snyder WD, Marcus ML. Adrenergic coronary tone during submaximal exercise in the dog is produced by circulating catecholamines. Evidence for adrenergic denervation supersensitivity in the myocardium but not in coronary vessels. *Circ Res* 58: 68-82, 1986.
25. Cochran B Jr., Marbach EP, Poucher R, Steinberg T, Gwinup G. Effect of acute muscular exercise on serum immunoreactive insulin concentration. *Diabetes* 15: 838-841, 1966.
26. Coiro V, Casti A, Rubino P, Manfredi G, Maffei ML, Melani A, Saccani J, et al. Free fatty acids inhibit adrenocorticotropin and cortisol secretion

stimulated by physical exercise in normal men. *Clin Endocrinol* 66: 740-743, 2007.

27. Coker RH, Koyama Y, Lacy DB, Williams PE, Rheaume N, Wasserman DH. Pancreatic innervation is not essential for exercise-induced changes in glucagon and insulin or glucose kinetics. *Am J Physiol* 277: E1122-E1129, 1999.
28. Collier SR, Collins E , Kanaley JA. Oral arginine attenuates the growth hormone response to resistance exercise. *J Appl Physiol* 101: 848-852, 2006.
29. Connell AM, Cooper J, Redfearn JW. The contrasting effects of emotional tension and physical exercise on the excretion of 17-ketogenic steroids and 17-ketosteroids. *Acta Endocrinol (Copenh)* 27: 179-194, 1958.
30. Cornil A, Decoster A, Copinschi G, Franckson JR. Effect of muscular exercise on the plasma level cortisol in man. *Acta Endocrinol (Copenh)* 48: 163-168, 1965.
31. Crampes F, Riviere D, Beauville M, Marceron M, Garrigues M. Lipolytic response of adipocytes to epinephrine in sedentary and exercise-trained subjects: Sex-related differences. *Eur J Appl Physiol Occup Physiol* 59: 249-255, 1989.
32. Czerwinski SM, Zak R, Kurowski TT, Falduto MT, Hickson RC. Myosin heavy chain turnover and glucocorticoid deterrence by exercise in muscle. *J Appl Physiol* 67: 2311-2315, 1989.
33. Davies CT, Brotherhood JR, Few JD, Zeidifard E. Effects of beta blockade and atropinisation on plasma catecholamine concentration during exercise. *Eur J Appl Physiol Occup Physiol* 36: 49-56, 1976.
34. Davies CTM, Few JD. Adrenalcortical activity and exercise. *PS EBM* 95: 35P-36P, 1970.
35. Dela F, Larsen JJ, Mikines KJ, Ploug T, Petersen LN, Galbo H. Insulin-stimulated muscle glucose clearance in patients with NIDDM. Effects of one-legged physical training. *Diabetes* 44: 1010-1020, 1995.
36. Dela F, Mikines KJ, Linstow MV, Galbo H. Effect of training on response to a glucose load adjusted for daily carbohydrate intake. *Am J Physiol* 260: E14-E20, 1991.
37. Dela F, Mikines KJ, Tronier B, Galbo H. Diminished arginine-stimulated insulin secretion in trained men. *J Appl Physiol* 69: 261-267, 1990.
38. Dela F, von Linstow ME, Mikines KJ, Galbo H. Physical training may enhance beta-cell function in type 2 diabetes. *Am J Physiol* 287: E1024-E1031, 2004.
39. Deschenes MR, Kraemer WJ, McCoy RW, Volek JS, Turner BM, Weinlein JC. Muscle recruitment patterns regulate physiological responses during exercise of the same intensity. *Am J Physiol* 279: R2229-R2236, 2000.
40. Devlin JG. The effect of training and acute physical exercise on plasma insulin-like activity. *Ir J Med Sci* 14: 423-425, 1963.
41. Dill DB, Edwards HT, de Meio RH. Effects of adrenalin injection in moderate work. *Am J Physiol* 31: 9-20, 1935
42. Eagle E, Britton SW. The influence of cortio-adrenal extract on energy output. *Science* 75: 221-222, 1932.
43. Engdahl JH, Veldhuis JD, Farrell PA. Altered pulsatile insulin secretion associated with endurance training. *J Appl Physiol* 79: 1977-1985, 1995.
44. Engfred K, Kjaer M, Secher NH, Friedman DB, Hanel B, Nielsen OJ, Bach FW, et al. Hypoxia and training-induced adaptation of hormonal responses to exercise in humans. *Eur J Appl Physiol Occup Physiol* 68: 303-309, 1994.
45. Euler US. *Hormones in Blood.* New York: Academic Press, 1961.
46. Euler US, Hellner, S. Noradrenaline excretion in musclar work. *Acta Physiol Scand* 26: 183, 1952.
47. Falduto MT, Hickson RC Exercise interrupts ongoing glucocorticoid-induced muscle atrophy and glutamine synthetase induction. *Am J Physiol* 263: E1157-E1163, 1992.
48. Farrell PA. Decreased insulin response to sustained hyperglycemia in exercise trained rats. *Med Sci Sports Exerc* 20: 469-473, 1987.
49. Farrell PA, Caston AL, Rodd AD. Changes in insulin response to glucose after exercise training in partially pancreactomized rats. *J Appl Physiol* 70: 1563-1569, 1991.
50. Farrell PA, Caston AL, Rodd D, Engdahl J. Effect of training on insulin secretion from single pancreatic beta cells. *Med Sci Sports Exerc* 24: 426-433, 1992.
51. Farrell PA, Garthwaite TL, Gustafson AB. Plasma adrenocorticotropin and cortisol responses to submaximal and exhaustive exercise. *J Appl Physiol* 55: 1441-1444, 1983.
52. Farrell PA, Gustafson AB, Garthwaite TL, Kalkhoff RK, Cowley AW Jr., Morgan WP. Influence of endogenous opioids on the response of selected hormones to exercise in humans. *J Appl Physiol* 61: 1051-1057, 1986.
53. Farrell PA, Hernandez JM, Fedele MJ, Vary TC, Kimball SR, Jefferson LS. Eukaryotic initiation factors and protein synthesis after resistance exercise in rats. *J Appl Physiol* 88: 1036-1042, 2000.
54. Fedele MJ, Hernandez JM, Lang CH, Vary TC, Kimball SR, Jefferson LS, Farrell PA. Severe diabetes prohibits elevations in skeletal muscle protein synthesis following acute resistance exercise in rats. *J Appl Physiol* 88: 102-108, 2000.
55. Federspil G, Udeschini, G, De Palo C, Sicolo N. Role of growth hormone in lipid mobilization stimulated by prolonged muscular exercise in the rat. *Hormone Metab Res* 7: 484-488, 1975.
56. Ferrara CM, Reynolds TH, Zarnowski MJ, Brozinick JT, Cushman SW. Short-term exercise enhances insulin-stimulated GLUT-4 translocation and glucose transport in adipose cells. *J Appl Physiol* 85: 2106-2111, 1998.

57. Few JD. Effect of exercise on the secretion and metabolism of cortisol in man. *J Endocrinol* 62: 341-353, 1974.
58. Few JD, Cashmore GC, Turton G. Adrenocortical response to one-leg and two-leg exercise on a bicycle ergometer. *Eur J Appl Physiol Occup Physiol* 44: 167-174, 1980.
59. Few JD, Davies CT. The inhibiting effect of atropine on growth hormone release during exercise. *Eur J Appl Physiol Occup Physiol* 43: 221-228, 1980.
60. Few JD, Thompson MW. Investigation of the secretion and metabolism of cortisol during a 100 km race using [4-14C] and [11 alpha-3H] cortisol. *Acta Endocrinol (Copenh)* 94: 204-212, 1980.
61. Fluckey JD, Kraemer WJ, Farrell PA. Pancreatic islet insulin secretion is increased after resistance exercise in rats. *J Appl Physiol* 79: 1100-1105, 1995.
62. Fonseka CC, Hunter WM, Passmore R. Effects of long continued exercise on plasma growth hormone levels in human adults. *J Physiol* 182(Suppl.): 26P-27P, 1966.
63. Franckson VR Jr., Leclercq R, Brunengraber H, Ooms HA. Labelled insulin catabolism and pancreatic responsiveness during long-term exercise in man. *Horm Metabol Res* 3: 366-373, 1971.
64. Galbo H. *Hormonal and metabolic adaptations to exercise*. Stuggart, FRG, Thieme, 1983.
65. Galbo H. Endocrinology and metabolism in exercise. *Int. J. Sports. Med.* 2:203-211, 1981.
66. Galbo H, Christensen NJ, Holst JJ. Catecholamines and pancreatic hormones during autonomic blockade in exercising man. *Acta Physiol Scand* 101: 428-437, 1977.
67. Galbo H, Christensen NJ, Holst JJ. Glucose-induced decrease in glucagon and epinephrine responses to exercise in man. *J Appl Physiol* 42: 525-530, 1977.
68. Galbo H, Christensen NJ, Mikines KJ, Sonne B, Hilsted J, Hagen C, Fahrenkrug J. The effect of fasting on the hormonal response to graded exercise. *J Clinl Endocrinol Metab* 52: 1106-1111, 1981.
69. Galbo H, Hedeskov CJ, Capito K, Vinten J. The effect of physical training on insulin secretion of rat pancreatic islets. *Acta Physiol Scand* 111: 75-79, 1981.
70. Galbo H, Holst JJ, Christensen NJ. The effect of different diets and of insulin on the hormonal response to prolonged exercise. *Acta Physiol Scand* 107: 19-32, 1979.
71. Galbo H, Holst JJ, Christensen NJ. Glucagon and plasma catecholamine responses to graded and prolonged exercise in man. *J Appl Physiol* 38: 70-76, 1975.
72. Galbo H, Holst JJ, Christensen NJ, Hilsted J. Glucagon and plasma catecholamines during beta-receptor blockade in exercising man. *J Appl Physiol* 40: 855-863, 1976.
73. Galbo H, Kjaer M, Secher NH. Cardiovascular, ventilatory and catecholamine responses to maximal dynamic exercise in partially curarized man. *J Physiol* 389: 557-568, 1987.
74. Galbo H, Richter EA, Christensen NJ, Holst JJ. Sympathetic control of metabolic and hormonal responses to exercise in rats. *Acta Physiol Scand* 102: 441-449, 1978.
75. Galbo H, Richter EA, Hilsted J, Holst JJ, Christensen NJ, Henriksson J. Hormonal regulation during prolonged exercise. *Ann NY Acad Sci* 301: 72-80, 1977.
76. Galbo H, Richter EA, Holst JJ, Christensen NJ. Diminished hormonal responses to exercise in trained rats. *J Appl Physiol* 43: 953-958, 1977.
77. Garetto LP, Richter EA, Goodman MN, Ruderman NB. Enhanced muscle glucose metabolism after exercise in the rat: The two phases. *Am J Physiol* 246: E471-E475, 1984.
78. Garlaschi C, di Natale B, del Guercio MJ, Caccamo A, Gargantini L, Chiumello G. Effect of physical exercise on secretion of growth hormone, glucagon, and cortisol in obese and diabetic children. *Diabetes* 24: 758-761, 1975.
79. Goldberg AL. Work-induced growth of skeletal muscle in normal and hypophysectomized rats. *Am J Physiol* 213: 1193-1198, 1967.
80. Goldberg AL, Goodman HM. Relationship between growth hormone and muscular work in determining muscle size. *J Physiol* 200: 655-666, 1969.
81. Goldsmith SR, Iber C, McArthur CD, Davies SF. Influence of acid-base status on plasma catecholamines during exercise in normal humans. *Am J Physiol* 258: R1411-R1416, 1990.
82. Gollnick PD, Soule RG, Taylor AW, Williams C, Ianuzzo CD. Exercise-induced glycogenolysis and lipolysis in the rat: Hormonal influence. *Am J Physiol* 219: 729-733, 1970.
83. Goodyear L, Kahn BB. Exercise, glucose transport, and insulin sensitivity. *Annu Rev Med* 49: 235-261, 1998.
84. Gordon R, Spector S, Sjoerdsma A, Udenfriend S. Increased synthesis of norepinephrine and epinephrine in the intact rat during exercise and exposure to cold. *J Pharmacol Exp Ther* 153: 440-447, 1966.
85. Gosselink KL, Grindeland RE, Rou RR, Zhong H, Bigbee AJ, Grossman EJ, Edgerton VR. Skeletal muscle afferent regulation of bioassayable growth hormone in the rat pituitary. *J Appl Physiol* 84: 1425-1430, 1998.
86. Goto K, Doessing S, Nielsen RH, Flyvbjerg A, Kjaer M. Growth hormone receptor antagonist treatment reduces exercise performance in young males. *J Clin Endocrinol Metab* 94: 3265-3272, 2009.
87. Goulet ED, Melancon MO, Aubertin-Leheudre M, Dionne IJ. Aerobic training improves insulin sensitivity 72-120 h after the last exercise session in

younger but not in older women. *Eur J Appl Physiol* 95: 146-152, 2005.
88. Gray I, Beetham WP Jr. Changes in plasma concentration of epinephrine and norepinephrine with muscular work. *PSEBM* 96: 636-638, 1957.
89. Grimditch GK, Barnard RJ, Kaplan SA, Sternlicht E. Effect of training on insulin binding to rat skeletal muscle sarcolemmal vesicles. *Am J Physiol* 250: E570-E575, 1986.
90. Gullestad L, Dolva LO, Kjeldsen SE, Eide I, Kjekshus J. The effects of naloxone and timolol on plasma catecholamine levels during short-term dynamic exercise. *Scand J Clin Lab Invest* 47: 847-851, 1987.
91. Haggendal J, Hartley LH, Saltin B. Arterial noradrenaline concentration during exercise in relation to the relative work levels. *Scand J Clin Lab Invest* 26: 337-342, 1970.
92. Hamerman D, Hatch FT, Ziporin ZZ. Urinary excretion of mucoproteins and adrenal corticosteroids during rest and exercise; Relation to urine volume and physical fitness. *J Appl Physiol* 9: 465-468, 1956.
93. Hansen AP. The effect of intravenous glucose infusion on the exercise induced serum growth hormone rise in normals and juvenile diabetics. *Scand J Clin Lab Invest* 28: 195-205, 1971.
94. Hansen AP. Serum growth hormone response to exercise in non-obese and obese normal subjects. *Scand J Clin Lab Invest* 31: 175-178, 1973.
95. Hansen M, Morthorst R, Larsson B, Dall R, Flyvbjerg A, Rasmussen MH, Orskov H, et al. No effect of growth hormone administration on substrate oxidation during exercise in young, lean men. *J Physiol* 567: 1035-1045, 2005.
96. Hartley LH, Mason JW, Hogan RP, Jones LG, Kotchen TA, Mougey EH, Wherry FE, et al. Multiple hormonal responses to graded exercise in relation to physical training. *J Appl Physiol* 33: 602-606, 1972.
97. Hartley LH, Mason JW, Hogan RP, Jones LG, Kotchen TA, Mougey EH, Wherry FE, et al. Multiple hormonal responses to prolonged exercise in relation to physical training. *J Appl Physiol* 35: 607-610, 1972.
98. Hartman FA, Waite RH, McCordock, HA. The liberation of epinephrin during muscular exercise. *Am J Physiol* 62: 225-241, 1922.
99. Hayashi T, Wojtaszewski JFP, Goodyear LJ. Exercise regulation of glucose transport in skeletal muscle. *Am J Physiol* 273: E1039-E1051, 1997.
100. Heath GW, Gavin JR III, Hinderliter JM, Hagberg JM, Bloomfield SA, Holloszy JO. Effects of exercise and lack of exercise on glucose tolerance and insulin sensitivity. *J Appl Physiol* 55: 512-517, 1983.
101. Heron WT, Hales WM, Ingle DJ. Capacity of skeletal muscle in rats to maintain work output. *Am J Physiol* 110: 357-361, 1934.
102. Hickson TTK, Capaccio JA, Chatterton RT Jr. Androgen cytosol binding in exercise-induced sparing of muscle atrophy. *Am J Physiol* 247: E597-E603, 1984.
103. Hill EE, Zack E, Battaglini C, Viru M, Viru A, Hackney AC. Exercise and circulating cortisol levels: The intensity threshold effect. *J Endocrinol Invest* 31: 587-591, 2008.
104. Hilsted J, Galbo H, Sonne B, Schwartz T, Fahrenkrug J, de Muckadell OB, Lauritsen KB, Tronier B. Gastroenteropancreatic hormonal changes during exercise. *Am J Physiol* 239: G136-G140, 1980.
105. Hoelzer DR, Dalsky GP, Clutter WE, Shah SD, Holloszy JO, Cryer PE. Glucoregulation during exercise: Hypoglycemia is prevented by redundant glucoregulatory systems, sympathochromaffin activation, and changes in islet hormone secretion. *J Clin Invest* 77: 212-221, 1986.
106. Hoelzer DR, Dalsky GP, Schwartz NS, Clutter WE, Shah SD, Holloszy JO, Cryer PE. Epinephrine is not critical to prevention of hypoglycemia during exercise in humans. *Am J Physiol* 251: E104-E110, 1986.
107. Hoffman JRI, Joohee X, Rundell KW, Kang JIE, Nioka S, Speiring BA, Kime R, Chance B. Effect of muscle oxygenation during resistance exercise on anabolic hormone response. *Med Sci Sports Exerc* 35: 1929-1934, 2003.
108. Holloszy JO, Narahara HT. Studies of tissue permeability. X. Changes in permeability to 3-methylglucose associated with contraction of isolated frog muscle. *J Biol Chem* 240: 3493-3500, 1965.
109. Horne LBG, Fisher B, Warren S, Janowska-Wieczorek RA. Interaction between cortisol and tumour necrosis factor with concurrent resistance and endurance training. *Clinl J Sport Med* 7: 247-253, 1997.
110. Horowitz JF, Braudy RJ, Martin WH III, Klein S. Endurance exercise training does not alter lipolytic or adipose tissue blood flow sensitivity to epinephrine. *Am J Physiol* 277: E325-E331, 1999.
111. Houmard JA, Egan PC, Neufer PD, Friedman JE, Wheeler WS, Israel RG, Dohm GL. Elevated skeletal muscle glucose transporter levels in exercise-trained middle-aged men. *Am J Physiol* 261: E437-E443, 1991.
112. Houmard JA, Hortobagyi T, Neufer PD, Johns RA, Fraser DD, Israel RG, Dohm GL. Training cessation does not alter GLUT-4 protein levels in human skeletal muscle. *J Appl Physiol* 74: 776-781, 1993.
113. Houmard JA, Shinebarger MH, Dolan PL, Leggett-Frazier N, Bruner RK, McCammon MR, Israel RG, Dohm GL. Exercise training increases GLUT-4 protein concentration in previously sedentary

middle-aged men. *Am J Physiol* 264: E896-E901, 1993.

114. Houwing H, Fränkel KMA, Strubbe JH, Van Suylichem PTR, Steffens AB. Role of the sympathoadrenal system in exercise-induced inhibition of insulin secretion: Effects of islet transplantation. *Diabetes* 44: 565-571, 1995.
115. Howlett K, Galbo H, Lorentsen J, Bergeron R, Zimmerman-Belsing T, Bulow J, Feldt-Rasmussen U, Kjaer M. Effect of adrenaline on glucose kinetics during exercise in adrenalectomized humans. *J Physiol* 519(Pt. 3): 911-921, 1999.
116. Howley E, Skinner JS, Mendez J, Buskirk ER. Effect of different intensities of exercise on catecholamine excretion. *Med Sci Sports Exer* 2: 193-196, 1970.
117. Hunter WM, Fonseka CC, Passmore R. Growth hormone: Important role in muscular exercise in adults. *Science* 150: 1051-1053, 1965.
118. Hunter WM, Fonseka CC, Passmore R. The role of growth hormone in the mobilization of fuel for muscular exercise. *Q J Exp Physiol Cogn Med Sci* 50: 406-416, 1965.
119. Hunter WM, Sukkar MY. Changes in plasma insulin leveles during muscular exercise. *J Physiol* 196(Suppl.): 110P-111P, 1968.
120. Hymer WC, Kraemer WJ, Nindl BC, Marx JO, Benson DE, Welsch JR, Mazzetti SA, et al. Characteristics of circulating growth hormone in women after acute heavy resistance exercise. *Am J Physiol* 281: E878-E887, 2001.
121. Ingle DJ. The effect of a stress upon the glycosuria of partially depancreatized force-fed rats. *Endocrinology* 46: 67-71, 1950.
122. Ingle DJ. Effect of endocrine glands on normal muscle work. *Am J Med* 19: 724-728, 1955.
123. Ingle DJ. The time for the occurrence of cortico-adrenal hypertrophy in rats during continued work. *Am J Physiol* 124: 627-630, 1938.
124. Ingle DJ. Work capacity of the adrenalectomized rat treated with cortin. *Am J Physiol* 116: 622-625, 1936.
125. Ingle DJ, Moon HD, Evans HM. Work performance of hypophysectomized rats treated with anterior pituitary extracts. *Am J Physiol* 123: 620-624, 1938.
126. Ingle DJ, Nezamis JE. Effect of stress upon glycosuria of force-fed depancreatized and of adrenalectomized-depancreatized rats. *Am J Physiol* 162: 1-4, 1950.
127. Ingle DJ, Prestrud MC, Nezamis JE. Effect of cortisone acetate upon plasma amino acids in the eviscerated rat. *PSEBM* 75: 801-803, 1950.
128. Ingle DW, Nezamis JE. Effect of muscle work upon tolerance of eviscerated rat for glucose. *Am J Physiol* 155: 15-17, 1948.
129. Irving BA, Weltman JY, Patrie JT, Davis CK, Brock DW, Swift D, Barrett EJ, et al. Effects of exercise training intensity on nocturnal growth hormone secretion in obese adults with the metabolic syndrome. *J Clin Endocrinol Metab* 94: 1979-1986, 2009.
130. Irving MH, Britton BJ, Wood WG, Padgham C, Carruthers M. Effects of beta adrenergic blockade on plasma catecholamines in exercise. *Nature* 248: 531-533, 1974.
131. Ivy JL, Young JC, McLane JA, Fell RD, Holloszy JO. Exercise training and glucose uptake by skeletal muscle in rats. *J Appl Physiol* 55: 1393-1396, 1983.
132. James DE, Burleigh KM, Kraegen EW, Chisholm DJ. Effect of acute exercise and prolonged training on insulin response to intravenous glucose in vivo in rat. *J Appl Physiol* 55: 1660-1664, 1983.
133. James DE, Kraegen EW, Chisholm DJ. Effects of exercise training on in vivo insulin action in individual tissues of the rat. *J Clin Invest* 76: 657-666, 1985.
134. Jankord R, McAllister RM, Ganjam VK, Laughlin MH. Chronic inhibition of nitric oxide synthase augments the ACTH response to exercise. *Am J Physiol* 296: R728-R734, 2009.
135. Järhult J, Holst J. The role of the adrenergic innervation to the pancreatic islets in the control of insulin release during exercise in man. *Pflügers Arch* 383: 41-45, 1979.
136. Johnson RH, Rennie MJ, Walton JL, Webster MH. The effect of moderate exercise on blood metabolites in patients with hypopituitarism. *Clin Sci* 40: 127-136, 1971.
137. Jorgensen JO, Krag M, Kanaley J, Moller J, Hansen TK, Moller N, Christiansen JS, Orskov H. Exercise, hormones, and body temperature regulation and action of GH during exercise. *J Endocrinol Invest* 26: 838-842, 2003.
138. Kahn SE, Larson VG, Beard JC, Cain KC, Fellingham GW, Schwartz RS, Veith RC, et al. Effect of exercise on insulin action, glucose tolerance, and insulin secretion in aging. *Am J Physiol* 258: E937-E943, 1990.
139. Kamimori GH, Bellar D, Fein HG, Smallridge RC. Hormonal and cardiovascular response to low-intensity exercise with atropine administration. *Mil Med* 174: 253-258, 2009.
140. Kanaley JA. Growth hormone, arginine and exercise. *Curr Opin Clin Nutr Metab Care* 11: 50-54, 2008.
141. Kanaley JA, Dall R, Moller N, Nielsen SC, Christiansen JS, Jensen MD, Jorgensen JO. Acute exposure to GH during exercise stimulates the turnover of free fatty acids in GH-deficient men. *J Appl Physiol* 96: 747-753, 2004.
142. King DS, Dalsky GP, Clutter WE, Young DA, Staten MA, Cryer PE, Holloszy JO. Effects of exercise and lack of exercise on insulin sensitivity and responsiveness. *J Appl Physiol* 64: 1942-1946, 1988.
143. Kjaer M. Adrenal medulla and exercise training. *Eur J Appl Physiol Occup Physiol* 77: 195-199, 1998.

144. Kjaer M. Sports adrenal medulla [Sports medicine]. *Ugeskr Laeger* 160: 5623, 1998.
145. Kjaer M, Bangsbo J, Lortie G, Galbo H. Hormonal response to exercise in humans: Influence of hypoxia and physical training. *Am J Physiol* 254: R197-R203, 1988.
146. Kjaer M, Christensen NJ, Sonne B, Richter EA, Galbo H. Effect of exercise on epinephrine turnover in trained and untrained male subjects. *J Appl Physiol* 59: 1061-1067, 1985.
147. Kjaer M, Engfred K, Fernandes A, Secher NH, Galbo H. Regulation of hepatic glucose production during exercise in humans: Role of sympathoadrenergic activity. *Am J Physiol* 265: E275-E283, 1993.
148. Kjaer M, Hanel B, Worm L, Perko G, Lewis SF, Sahlin K, Galbo H, Secher NH. Cardiovascular and neuroendocrine responses to exercise in hypoxia during impaired neural feedback from muscle. *Am J Physiol* 277: R76-R85, 1999.
149. Kjaer M, Howlett K, Langfort J, Zimmerman-Belsing T, Lorentsen J, Bulow J, Ihlemann J, et al. Adrenaline and glycogenolysis in skeletal muscle during exercise: A study in adrenalectomized humans. *J Physiol* 528(Pt. 2): 371-378, 2000.
150. Kjaer M, Keiding S, Engfred K, Rasmussen K, Sonne B, Kirkgård P, Galbo H. Glucose homeostasis during exercise in humans with a liver of kidney transplant. *Am J Physiol* 268: E636-E644, 1995.
151. Kjaer M, Kiens B, Hargreaves M, Richter EA. Influence of active muscle mass on glucose homeostasis during exercise in humans. *J Appl Physiol* 71: 552-557, 1991.
152. Kjaer M, Farrell PA, Christensen NJ, Galbo H. Increased epinephrine responses and inaccurate glucoregulation in exercising athletes. *J Appl Physiol* 61: 1693-1700, 1986.
153. Kjaer M, Pollack SF, Mohr T, Weiss H, Gleim GW, Bach FW, Nicolaisen T, et al. Regulation of glucose turnover and hormonal responses during electrical cycling in tetraplegic humans. *Am J Physiol* 271: R191-R199, 1996.
154. Kjaer M, Secher NH, Bach FW, Galbo H. Role of motor center activity for hormonal changes and substrate mobilization in humans. *Am J Physiol* 253: R687-R695, 1987.
155. Kjaer M, Secher NH, Bach FW, Galbo H, Reeves DR Jr., Mitchell JH. Hormonal, metabolic, and cardiovascular responses to static exercise in humans: Influence of epidural anesthesia. *Am J Physiol* 261: E214-E220, 1991.
156. Kjaer M, Secher NH, Bach FW, Sheikh S, Galbo H. Hormonal and metabolic responses to exercise in humans: Effect of sensory nervous blockade. *Am J Physiol* 257: E95-E101, 1989.
157. Klein S, Coyle EF, Wolfe RR. Effect of exercise on lipolytic sensitivity in endurance-trained athletes. *J Appl Physiol* 78: 2201-2206, 1995.
158. Koopman R, Manders RJ, Zorenc AH, Hul GB, Kuipers H, Keizer HA, van Loon LJ. A single session of resistance exercise enhances insulin sensitivity for at least 24 h in healthy men. *Eur J Appl Physiol* 94: 180-187, 2005.
159. Kostyak J, Kimball SR, Jefferson LS, Farrell PA. Severe diabetes inhibits the resistance exercise-induced increase in eukaryotic initiation factor 2B activity. *J Appl Physiol* 91: 79-84, 2001.
160. Kotchen TA, Hartley LH, Rice TW, Mougey EH, Jones LG, Mason JW. Renin, norepinephrine, and epinephrine responses to graded exercise. *J Appl Physiol* 31: 178-184, 1971.
161. Kraemer WJ, Deschenes MR, Fleck SJ. Physiological adaptations to resistance exercise. Implications for athletic conditioning. *Sports Med* 6: 246-256, 1988.
162. Kraemer WJ, Dziados JE, Gordon SE, Marchitelli LJ, Fry AC, Reynolds KL. The effects of graded exercise on plasma proenkephalin peptide F and catecholamine responses at sea level. *Eur J Appl Physiol Occup Physiol* 61: 214-217, 1990.
163. Kraemer WJ, Marchitelli L, Gordon SE, Harman E, Dziados JE, Mello R, Frykman P, et al. Hormonal and growth factor responses to heavy resistance exercise protocols. *J Appl Physiol* 69: 1442-1450, 1990.
164. Kraemer WJ, Nindl BC, Marx JO, Gotshalk LA, Bush JA, Welsch JR, Volek JS, et al. Chronic resistance training in women potentiates growth hormone in vivo bioactivity: Characterization of molecular mass variants. *Am J Physiol* 291: E1177-E1187, 2006.
165. Kraemer WJ, Patton JF, Gordon SE, Harman EA, Deschenes MR, Reynolds K, Newton RU, et al. Compatibility of high-intensity strength and endurance training on hormonal and skeletal muscle adaptations. *J Appl Physiol* 78: 976-989, 1995.
166. Kraemer WJ, Ratamess NA. Hormonal responses and adaptations to resistance exercise and training. *Sports Med* 35: 339-361, 2005.
167. Lawrence RD. The effect of exercise on insulin action in diabetes. *Br Med J* 1: 648-650, 1926.
168. Leclercq-Meyer V, Malaise WJ. Participation of catecholamines in the inhibition of insulin secretin and stimulation of glucagon release during exercise and stress. *Diabetes Metab* 1: 119-122, 1975.
169. Lithell H, Cedermark M, Froberg J, Tesch P, Karlsson J. Increase of lipoprotein-lipase activity in skeletal muscle during heavy exercise. Relation to epinephrine excretion. *Metabolism* 30: 1130-1134, 1981.
170. Lukaszewska J, Biczowa B, Bobilewicz D, Wilk M, Obuchowicz-Fidelus B. Effect of physical exercise on plasma cortisol and growth hormone levels in young weight lifters. *Endokrynol Pol* 27: 149-157, 1976.

171. Manning JM, Bronson FH. Suppression of puberty in rats by exercise: Effects on hormone levels and reversal with GnRH infusion. *Am J Physiol* 260: R717-R723, 1991.
172. Marcell TJ, Wiswell RA, Hawkins SA, Tarpenning KM. Age-related blunting of growth hormone secretion during exercise may not be soley due to increased somatostatin tone. *Metabolism* 48: 665-670, 1999.
173. Marker JC, Hirsch IB, Smith LJ, Parvin CA, Holloszy JO, Cryer PE. Catecholamines in prevention of hypoglycemia during exercise in humans. *Am J Physiol* 260: E705-E712, 1991.
174. Martin WH III, Coyle EF, Joyner M, Santeusanio D, Ehsani AA, Holloszy JO. Effects of stopping exercise training on epinephrine-induced lipolysis in humans. *J Appl Physiol* 56: 845-848, 1984.
175. Mikines KJ, Farrell PA, Sonne B, Tronier B, Galbo H. Postexercise dose-response relationship between plasma glucose and insulin secretion. *J Appl Physiol* 64: 988-999, 1988.
176. Mikines KJ, Sonne B, Farrell PA, Tronier B, Galbo H. Effect of physical exercise on sensitivity and responsiveness to insulin in humans. *Am J Physiol* 254: E248-E259, 1988.
177. Mikines KJ, Sonne B, Tronier B, Galbo H. Effects of training and detraining on dose-response relationship between glucose and insulin secretion. *Am J Physiol* 256: E588-E596, 1989.
178. Miller JP, Pratley RE, Goldberg AP, Gordon P, Rubin M, Treuth MS, Ryan AS, Hurley BF. Strength training increases insulin action in healthy 50-65 yr-old men. *J Appl Physiol* 77: 1122-1127, 1994.
179. Moates JM, Lacy DB, Goldstein RE, Cherrington AD, Wasserman DH. Metabolic role of the exercise-induced increment in epinephrine in the dog. *Am J Physiol* 255: E428-E436, 1988.
180. Mondon CE, Dolkas CB, Reaven GM. Site of enhanced insulin sensitivity in exercise-trained rats at rest. *Am J Physiol* 239: E169-E177, 1980.
181. Nagasawa J, Sato Y, Ishiko T. Effect of training and detraining on in vivo insulin sensitivity. *Int J Sports Med* 11: 107-110, 1990.
182. Nesher R, Karl IE, Kipnis DM. Dissociation of effects of insulin and contraction on glucose transport in rat epitrochlearis muscle. *Am J Physiol* 249: C226-C232, 1985.
183. Nikkila EA, Taskinen M, Miettinen TA, Pelkonen R, Poppius H. Effect of muscular exercise on insulin secretion. *Diabetes* 17: 209-218, 1968.
184. Nindl BC, Kraemer WJ, Hymer WC. Immunofunctional vs. immunoreactive growth hormone responses after resistance exercise in men and women. *Growth Horm IGF Res* 10: 99-103, 2000.
185. Nindl BC, Kraemer WJ, Marx JO, Arciero PJ, Dohi K, Kellogg MD, Loomis GA. Overnight responses of the circulating IGF-I system after acute, heavy-resistance exercise. *J Appl Physiol* 90: 1319-1326, 2001.
186. Nindl BC, Kraemer WJ, Marx JO, Tuckow AP, Hymer WC. Growth hormone molecular heterogeneity and exercise. *Exerc Sport Sci Rev* 31: 161-166, 2003.
187. Pierce JR, Tuckow AP, Alemany JA, Rarick KR, Staab JS, Harman EA, Nindl BC. Effects of acute and chronic exercise on disulfide-linked growth hormone variants. *Med Sci Sports Exerc* 41: 581-587, 2009.
188. Ploug T, Galbo H, Richter EA. Increased muscle glucose uptake during contractions: No need for insulin. *Am J Physiol* 247: E726-E731, 1984.
189. Pritzlaff CJ, Wideman L, Blumer J, Jensen M, Abbott RD, Gaesser GA, Veldhuis JD, Weltman A. Catecholamine release, growth hormone secretion, and energy expenditure during exercise vs. recovery in men. *J Appl Physiol* 89: 937-946, 2000.
190. Pritzlaff CJ, Wideman L, Weltman JY, Abbott RD, Gutgesell ME, Hartman ML, Veldhuis JD, Weltman A. Impact of acute exercise intensity on pulsatile growth hormone release in men. *J Appl Physiol* 87: 498-504, 1999.
191. Pruett EDR. FFA mobilization during and after prolonged severe muscular work in men. *J Appl Physiol* 29: 809-815, 1970.
192. Pruett EDR. Glucose and insulin during prolonged work stress in men living on different diets. *J Appl Physiol* 28: 199-208, 1970.
193. Pruett EDR. Plasma insulin concentrations during prolonged work at near maximal oxygen uptake. *J Appl Physiol* 29: 155-158, 1970.
194. Pruett EDR, Oseid S. Effect of exercise on glucose and insulin response to glucose infusion. *Scand J Clin Lab Invest* 26: 277-285, 1970.
195. Pyka G, Wiswell RA, Marcus R. Age-dependent effect of resistance exercise on growth hormone secretion in people. *J Clin Endocrinol Metab* 75: 404-407, 1992.
196. Reaven EP, Reaven GM. Structure and function changes in the endocrine pancreas of aging rats with reference to the modulating effects of exercise and caloric restriction. *J Clin Invest* 68: 75-84, 1981.
197. Redmon JB, Kubo SH, Robertson RP. Glucose, insulin, and glucagon levels during exercise in pancreas transplant recipients. *Diabetes Care* 18: 457-462, 1995.
198. Rennie MJ, Johnson RH. Alteration of metabolic and hormonal responses to exercise by physical training. *Eur J Appl Physiol Occup Physiol* 33: 215-226, 1974.
199. Richter EA, Sonne B, Christensen NJ, Galbo H. Role of epinephrine for muscular glycogenolysis and pancreatic hormonal secretion in running rats. *Am J Physiol* 240: E526-E532, 1981.
200. Richter EA, Galbo H, Christensen NJ. Control of exercise-induced muscular glycogenolysis by adrenal

medullary hormones in rats. *J Appl Physiol* 50: 21-26, 1981.
201. Richter EA, Garetto LP, Goodman MN. Muscle glucose metabolism following exercise in the rat: Increased sensitivity to insulin. *J Clin Invest* 69: 785-793, 1982.
202. Richter EA, Galbo H, Sonne B, Holst JJ, Christensen NJ. Adrenal medullary control of musclular and hepatic glycogenolysis and of pancreatic hormonal secretion in exercising rats. *Acta Physiol Scand* 108: 235-242, 1980.
203. Richter EA, Mikines KJ, Galbo H, Kiens B. Effect of exercise on insulin action in human skeletal muscle. *J Appl Physiol* 66: 876-885, 1989.
204. Richter EA, Ruderman NB, Gavras H, Belur ER, Galbo H. Muscle glycogenolysis during exercise: Dual control by epinephrine and contractions. *Am J Physiol* 242: E25-E32, 1982.
205. Robinson KW, Macfarlane WV. Urinary excretion of adrenal steroids during exercise in hot atmospheres. *J Appl Physiol* 12: 13-16, 1958.
206. Rockl KS, Witczak CA, Goodyear LJ. Signaling mechanisms in skeletal muscle: Acute responses and chronic adaptations to exercise. *IUBMB Life* 60: 145-153, 2008.
207. Schalch D. The influence of physical stress and exercise on growth hormone and insulin secretion in man. *J Lab Clin Med* 69: 256-268, 1967.
208. Schmidt KN, Gosselin LE, Stanley WC. Endurance exercise training causes adrenal medullary hypertrophy in young and old Fischer 344 rats. *Horm Metab Res* 24: 511-515, 1992.
209. Schweitzer A. The effect of adrenalectomy on the contractile power of skeletal muscle. *J Physiol* 104: 21-31, 1945.
210. Seals DR, Hagberg JM, Allen WK, Hurley BF, Dalsky GP, Ehsani AA, Holloszy JO. Glucose tolerance in young and older athletes and sedentary men. *J Appl Physiol* 56: 1521-1529, 1984.
211. Shapiro B, Borer KT, Fig LM, Vinik AI. Exercise-induced hyperphagia in the hamster is associated with elevated plasma somatostatin-like immunoreactivity. *Regul Pept* 18: 85-92, 1987.
212. Shima KHM, Sato M, Iwami T, Oshima I. Effect of exercise training on insulin and glucagon release from perfused rat pancreas. *Horm Metab Res* 19: 395-399, 1987.
213. Shinkai S, Watanabe S, Asai H, Shek PN. Cortisol response to exercise and post-exercise suppression of blood lymphocyte subset counts. *Int J Sports Med* 17: 597-603, 1996.
214. Short KR, Vittone JL, Bigelow ML, Proctor DN, Rizza RA, Coenen-Schimke JM, Nair SK. Impact of aerobic exercise training on age-related changes in insulin sensitivity and muscle oxidative capacity. *Diabetes* 52: 1888-1896, 2003.
215. Silva TS, Faria LC, Rocha CD, Melo MN, Faria MR, de Souza TG, Almeida JA, et al. Impact of prolonged physical training on the pituitary glucocorticoid sensitivity determined by very low dose intravenous dexamethasone suppression test. *Horm Metab Res* 40: 718-721, 2008.
216. Soman VR, Koivisto VA, Deibert D, Felig P, DeFronzo RA. Increased insulin sensitivity and insulin binding to monocytes after physical training. *New Engl J Med* 301: 1200-1204, 1979.
217. Sonne B, Galbo H. Carbohydrate metabolism during and after exercise in rats: Studies with radioglucose. *J Appl Physiol* 59: 1627-1639, 1985.
218. Sonne B, Mikines KJ, Galbo H. Glucose turnover in 48-hour-fasted running rats. *Am J Physiol* 252: R587-R593, 1987.
219. Sonne B, Mikines KJ, Richter EA, Christensen NJ, Galbo H. Role of liver nerves and adrenal medulla in glucose turnover of running rats. *J Appl Physiol* 59: 1640-1646, 1985.
220. Soskin S, Strouse S, Molander CO, Vidgoff B, Henner RI. Value of muscular exercise in the treatment of diabetes mellitus. *J Am Med Assoc* 103: 1767-1768, 1934.
221. Staehelin D, Labhart A, Froesch R, Kagi HR. The effect of muscular exercise and hypoglycemia on the plasma level of 17-hydroxysteroids in normal adults and in patients with the adrenogenital syndrome. *Acta Endocrinol (Copenh)* 18: 521-529, 1955.
222. Staessen J, Fiocchi R, Bouillon R, Fagard R, Hespel P, Lijnen P, Moerman E, Amery A. Effects of opioid antagonism on the haemodynamic and hormonal responses to exercise. *Clin Sci (Lond)* 75: 293-300, 1988.
223. Stallknecht B, Kjaer M, Mikines M, Maroun L, Plough T, Ohkuwa T, Vinten J, Galbo H. Diminished epinephrine response to hypoglycemia despite enlarged adrenal medulla in trained rats. *Am J Physiol* 259: R998-R1003, 1990.
224. Stallknecht B, Larsen JJ, Mikines KJ, Simonsen L, Bulow J, Galbo H. Effect of training on insulin sensitivity of glucose uptake and lipolysis in human adipose tissue. *Am J Physiol* 279: E376-E385, 2000.
225. Steensberg A, Fischer CP, Keller C, Møller K, Pedersen BK. IL-6 enhances plasma IL-1ra, IL-10, and cortisol in humans. *Am J Physiol* 285: E433-E437, 2003.
226. Stewart GN, Rogoff JM. The influence of muscular exercise on normal cats compared with cats deprived of the greater part of the adrenals, with special reference to body temperature, pulse and respiratory frequency. *J Pharm Exp Therap* 19: 87-95, 1922.
227. Struthers AD, Burrin JM, Brown MJ. Exercise-induced increases in plasma catecholamines and growth hormone are augmented by selective alpha 2-adrenoceptor blockade in man. *Neuroendocrinology* 44: 22-28, 1986.
228. Sundkvist G, Lilja B, Almer LO. Absent elevations in growth hormone, factor VIII related antigen, and plasminogen activator activity during exercise in di-

abetic patients resistant to retinopathy. *Diabetes Res* 7: 25-30, 1988.

229. Sutton JR, Lazarus L. Growth hormone in exercise: Comparison of physiological and pharmacological stimuli. *J Appl Physiol* 41: 523-527, 1976.
230. Sutton JR. Effect of acute hypoxia on the hormonal response to exercise. *J Appl Physiol* 42: 587-592, 1977.
231. Sutton JR, Jones NL, Toews CJ. Growth hormone secretion in acid-base alterations at rest and during exercise. *Clin Sci Mol Med* 50: 241-247, 1976.
232. Sutton JR, Young JD, Lazarus L, Hickie JB, Maksvytis J. The hormonal response to physical exercise. *Australas Ann Med* 18: 84-90, 1969.
233. Svendenhag J, Martinsson A, Ekblom B, Hjemdahl P. Altered cardiovascular responsiveness to adrenalin in endurance trained subjects. *Acta Physiol Scand* 126: 539-550, 1986.
234. Tabata I, Atomi Y, Miyashita M. Blood glucose concentration dependent ACTH and cortisol responses to prolonged exercise. *Clin Physiol* 4: 299-307, 1984.
235. Tabata I, Ogita F, Miyachi M, Shibayama H. Effect of low blood glucose on plasma CRF, ACTH, and cortisol during prolonged physical exercise. *J Appl Physiol* 71: 1807-1812, 1991.
236. Thompson DL, Weltman JY, Rogol AD, Metzger DL, Veldhuis JD, Weltman A. Cholinergic and opioid involvement in release of growth hormone during exercise and recovery. *J Appl Physiol* 75: 870-878, 1993.
237. Thuma JR, Gilders R, Verdun M, Loucks AB. Circadian rhythm of cortisol confounds cortisol responses to exercise: Implications for future research. *J Appl Physiol* 78: 1657-1664, 1995.
238. Tipton CM, Struck PJ, Baldwin KM, Matthes RD, Dowell RT. Response of adrenalectomized rats to chronic exercise. *Endocrinology* 91: 573-579, 1972.
239. Tipton CM, Tharp GD, Schild RJ. Exercise, hypophysectomy, and spleen weight. *J Appl Physiol* 21: 1163-1167, 1966.
240. Tuckow AP, Rarick KR, Kraemer WJ, Marx JO, Hymer WC, Nindl BC. Nocturnal growth hormone secretory dynamics are altered after resistance exercise: Deconvolution analysis of 12-hour immunofunctional and immunoreactive isoforms. *Am J Physiol* 291: R1749-R1755, 2006.
241. van Dijk G, Vissing J, Steffens AB, Galbo H. Effect of anaesthetizing the region of the paraventricular hypothalamic nuclei on energy metabolism during exercise in the rat. *Acta Physiol Scand* 151: 165-172, 1994.
242. Vendsalu A. Studies on adrenaline and noradrenaline in human plasma. *Acta Physiol Scand Suppl* 173: 57-69, 1960.
243. Venning EH, Kazmin V. Excretion of urinary corticoids and 17-ketosteroids in the normal individual. *Endocrinology* 39: 131-139, 1946.
244. Vinten J, Galbo H. Effect of physical training on transport and metabolism of glucose in adipocytes. *Am J Physiol* 244: E129-E134, 1983.
245. Viru A, Viru M, Karelson K, Janson T, Siim K, Fischer K, Hackney AC. Adrenergic effects on adrenocortical cortisol response to incremental exercise to exhaustion. *Eur J Appl Physiol* 100: 241-245, 2007.
246. Vissing J, Iwamoto GA, Fuchs IE, Galbo H, Mitchell JH. Reflex control of glucoregulatory exercise responses by group III and IV muscle afferents. *Am J Physiol* 266: R824-R830, 1994.
247. Vissing J, Wallace JL, Scheurink AJ, Galbo H, Steffens AB. Ventromedial hypothalamic regulation of hormonal and metabolic responses to exercise. *Am J Physiol* 256: R1019-R1026, 1989.
248. Vranic M, Engerman R, Doi K, Morita S, Yip CC. Extrapancreatic glucagon in the dog. *Metabolism* 25: 1469-1473, 1976.
249. Vranic M, Kawamori R. Essential roles of insulin and glucagon in regulating glucose fluxes during exercise in dogs. *Diabetes* 28(Suppl. 1): 45-52, 1979.
250. Vranic M, Kawamori R, Pek S, Kovacevic N, Wrenshall GA. The essentiality of insulin and the role of glucagon in regulating glucose utilization and production during strenuous exercise in dogs. *J Clin Invest* 57: 245-255, 1976.
251. Wallberg Henriksson H, Gunnarsson R, Henriksson J, DeFronzo RA, Felig P, Östman J, Wahren J. Increased peripheral insulin sensitivity and muscle mitochondrial enzymes but unchanged blood glucose control in type I diabetics after physical training. *Diabetes* 31: 1044-1050, 1982.
252. Warren J, Dalton N, Turner C, Clark TJ, Toseland PA. Adrenaline secretion during exercise. *Clin Sci* 66: 87-90, 1984.
253. Weltman A, Weltman JY, Schurrer R, Evans WS, Veldhuis JD, Rogol AD. Endurance training amplfies the pulsatile release of growth hormone: Effects of training intensity. *J Appl Physiol* 72: 2188-2196, 1992.
254. Weltman A, Weltman JY, Roy CP, Wideman L, Patrie J, Evans WS, Veldhuis JD. Growth hormone response to graded exercise intensities is attenuated and the gender difference abolished in older adults. *J Appl Physiol* 100: 1623-1629, 2006.
255. Winder WW, Beattie MA, Holman RT. Endurance training attenuates stress hormone responses to exercise in fasted rats. *Am J Physiol* 243: R179-R184, 1982.
256. Winder WW, Hagberg JM, Hickson RC, Ehsani AA, McLane JA. Time course of sympathoadrenal adaptation to endurance exercise training in man. *J Appl Physiol* 45: 370-374, 1978.
257. Winder WW, Hickson RC, Hagberg JM, Eshani AA, McLane J. Training-induced changes in hormonal and metabolic responses to submaximal exercise. *J Appl Physiol* 46: 766-771, 1979.

258. Winder WW, Terry ML, Mitchell VM. Role of plasma epinephrine in fasted exercising rats. *Am J Physiol* 248: R302-R307, 1985.
259. Wright PH, Malaisse WJ. Effects of epinephrine, stress, and exercise on insulin secretion by the rat. *Am J Physiol* 214: 1031-1034, 1968.
260. Yalow RS, Berson SA. Immunoassay of endogenous plasma insulin in man. *J Clin Invest* 39: 1157-1175, 1960.
261. Yarasheski K, Zachwieja J, Campbell J, Bier D. Effects of growth hormone and resistance exercise on muscle growth and strength in elderly men. *Am J Physiol* 268: E268-E276, 1995.
262. Yarasheski KE, Zachwieja JJ, Angelopoulos TJ, Bier DM. Short-term growth hormone treatment does not increase muscle protein synthesis in experienced weight lifters. *J Appl Physiol* 74: 3073-3076, 1993.
263. Zawalich WS, Maturo S, Felig P. Influence of physical training on insulin release and glucose utilization by islet cells and liver glucokinase activity in the rat. *Am J Physiol* 243: E464-E469, 1982.

CHAPTER 16

The Gastrointestinal System

G. Patrick Lambert, PhD

Introduction

Of all the organ systems in the human body, the gastrointestinal (GI) system has likely received the least attention from exercise physiologists. After the experiments of Beaumont in the early 1800s (88), there were no publications on GI function during exercise until the American physiologist Anton J. Carlson (1875-1956) published his text in 1916 (14). After Carlson's research, there were few publications in this area until the 1960s. In the mid- to late 1960s, however, two major phenomena emerged that brought more attention to the importance of GI function during exercise: the invention of the first sport drink (i.e., Gatorade) and the running boom in the 1970s. As millions of people ran longer and longer distances, they became aware of the profound effects that dehydration, hypoglycemia, glycogen depletion, and loss of electrolytes had on their ability to perform optimally. Because the GI system was the organ system responsible for the intake and absorption of required fluid and nutrients during running, it began to receive more attention from the general public and the scientific community.

Research inevitably led to studies on gastric emptying and intestinal absorption. Possibly the most influential study was published in 1967 by John Fordtran (1931–) and Bengt Saltin (1935–) (28). Their experiments, discussed in detail later, set the stage for further research in this area, mainly by four laboratories—two in the United States and two in Europe.

In the early 1970s, David Costill (1936–) from Ball State University began conducting experiments on gastric emptying during exercise with colleagues in his laboratory. These investigators performed numerous studies into the early 1990s. In the 1980s, laboratories led by Carl V. Gisolfi (1942-2000) at the University of Iowa in the United States, Ronald J. Maughan (1951–) at Aberdeen University, and Wim H.M. Saris (1949–) at the University of Limburg (i.e., Maastricht University) became interested in the study of GI function during exercise. Research led by Costill, Gisolfi, Maughan, and Saris has had the most influence on the subject.

The majority of research conducted on GI function during exercise has been focused on fluid and energy provision for endurance activity. The two processes most important in this regard are gastric emptying and intestinal absorption. In addition, exercise has been found to affect GI blood flow, motility, transit time, barrier function, and symptoms. These subjects are discussed in subsequent sections of this chapter. Although most of the discussion pertains to human experiments, animal results are mentioned when appropriate.

GI Blood Flow

As with all organs in the body, adequate blood flow to the organs of the GI tract is essential for proper function. The earliest studies in this area were conducted in dogs. In 1929, Barcroft and Florey (2) at the Physiological and Pathological Laboratories in Cambridge examined the color change of an exteriorized patch of colon in a dog while the dog performed 17 s runs in the laboratory passage. It was reported that the mucous membrane became intensely pale during the run. However, within 60 s after the run, the intestine became "as red as, if not more red than, it was when the animal was at rest." The authors commented that "a little, but only a little, vasoconstriction" occurred. Furthermore, they indicated that some of the pallor observed in the mucosa after running was likely due to excitement because the pallor subsided as the dog became disinterested in the experiments. These findings stimulated a later study by Herrick and colleagues (42), who examined superior mesenteric artery (SMA) blood flow in dogs after surgical application of a thermostromuhr to the vessel. After 17 experiments in 3 dogs, the authors reported that 6 experiments showed no change in flow, 10 showed increased flow, and only 1 showed decreased flow. They concluded that blood flow in the SMA was not decreased during exercise and discussed that vasoconstriction may be offset by increased blood pressure during exercise.

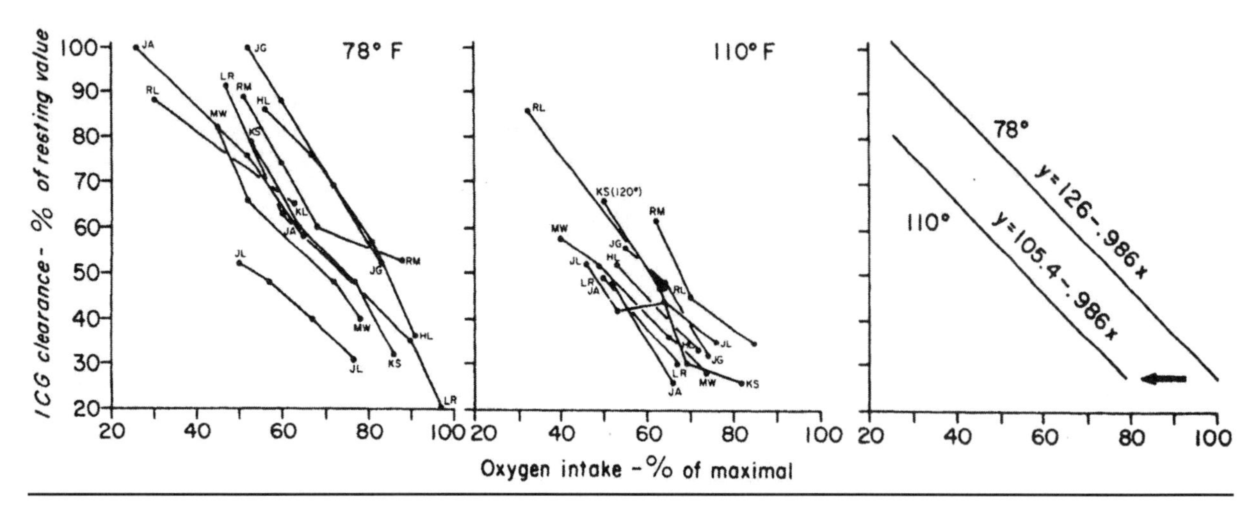

Figure 16.1 Splanchnic blood flow (ICG clearance − percentage of resting value) as a function of exercise intensity (percentage $\dot{V}O_2$max) and heat stress.

Reprinted from L.B. Rowell et al., 1965, "Hepatic clearance of indocyanine green in man under thermal and exercise stresses," *Journal of Applied Physiology* 20: 384-394. With permission of American Physiological Society.

Studies on GI blood flow in humans began in the 1940s. The initial studies in this area did not directly measure blood flow to specific organs of the GI tract. Instead, they estimated total splanchnic blood flow by measuring clearance of a tracer such as bromophthalein sodium (BSP) or indocyanine green (ICG) by the liver. In 1949, Bradley (8) at Columbia University reported a reduction in splanchnic blood flow during recumbent cycling exercise using the BSP method. However, in 1952, Lowenthal and colleagues (67) did not observe any change in splanchnic blood flow (also using BSP) during 15 min of moderate-intensity (average heart rate 155 beats/min) supine leg exercise (weightlifting with a pulley system). Four years later, Wade and colleagues (including Bradley) at Columbia University (135) supported Bradley's previous findings when they observed a 20% reduction in splanchnic blood flow during recumbent leg exercise (7-8 min of weightlifting); this reduction was sustained for up to 18 min postexercise. The reduction in blood flow to the region was attributed to "increased splanchnic vascular resistance, presumably as a result of vasoconstriction." The controversy persisted through 1957 when Harpuder and colleagues (including Lowenthal) (36) again reported no reduction in splanchnic (i.e., liver) blood flow during dynamic exercise in upright subjects.

No further studies were conducted in this area until 1964, when Loring B. Rowell (1930–) and colleagues at the University of Washington in Seattle began a series of studies on peripheral blood flow responses during exercise. Using ICG (the authors believed it was more sensitive than BSP because it was exclusively cleared by the liver), they observed that total splanchnic blood flow was reduced by as much as 80% during moderate to severe upright exercise in humans (106). This result, which supported the findings of Bradley (8) and Wade and colleagues (135), confirmed that blood flow to the splanchnic organs (including the GI tract) is most certainly reduced during exercise. A year later, Rowell and colleagues (107) demonstrated that the imposition of heat stress (43.3 °C) with heavy dynamic exercise further reduces splanchnic blood flow by approximately 20% (figure 16.1).

Further studies documenting the effect of exercise on GI blood flow in various animal models were conducted from the late 1960s through the early 1990s. In 1969, Burns and Schenk (11), using an electromagnetic flow probe placed around the SMA in dogs during severe exercise, did not observe a significant reduction in blood flow. In 1976, using the radioactive microsphere technique in dogs during exhaustive exercise, Sanders and colleagues (113) observed an initial decrease (~33%) in intestinal blood flow and a subsequent increase to resting levels. The authors commented that blood flow to the splanchnic organs was not homogenous during exercise and that assessing only splanchnic blood flow without accounting for the heterogeneity of the different organs it serves could be misleading. This concept was supported in 1987 by the findings of Musch and colleagues (81), who, also using radioactive microspheres in dogs, observed a significant reduction in blood flow to the stomach and large intestine at even low workloads (30% $\dot{V}O_2$max), but not to the small intestine, until the animal reached 100% $\dot{V}O_2$max. Using the microsphere technique during aquatic and treadmill exercise in rats, Flaim and colleagues (26) found in 1979 that blood flow was significantly reduced to the stomach (47% aquatic, 48% treadmill), jejunum (31% aquatic), and colon (47% aquatic, 64% treadmill). In 1983, Bell and colleagues (5), also using the microsphere method, observed significant reductions (21%) in blood flow to the small intestine and large intestine in sheep during treadmill running (30 min). Manohar (68) in 1986 studied GI blood flow in ponies exercising to exhaustion and observed, using microsphere technique, large reductions in blood flow to the small intestine (82%) and large intestine (74%).

Further studies in humans on GI blood flow during exercise essentially supported the predominant finding of the earlier studies that exercise reduced GI blood flow. In 1987, Qamar and Read (95) used transcutaneous Doppler ultrasound to assess SMA blood flow during 15 min of treadmill exercise (5 km/h and 20% gradient). In accordance with Rowell's findings (106), these authors reported a 43% decrease in SMA blood flow in the fasted state. Puvi-Rajasingham and colleagues (94) reported similar results in 1997 during supine cycling at 25, 50, and 75 W (3 min at each workload). Using both ultrasound Doppler and ICG clearance techniques, Perko and colleagues (92) in 1998 reported a 32% decrease in mesenteric flow, a 30% decrease in celiac flow, and a 43% reduction in splanchnic flow during submaximal cycling in fasted subjects. In all these studies, feeding before exercise attenuated but did not abolish the reduction in GI blood flow. One study that did not observe a reduction in splanchnic blood flow was that of Eriksen and Waaler (24) from the University of Oslo. These investigators found no changes in SMA blood flow using Doppler ultrasound with short (4 min) bouts of cycling in the reclined position at intensities reaching 75% $\dot{V}O_2$max. The authors indicated that the discrepancy in results with other studies was likely attributable to species differences or differences in exercise duration or intensity during the studies.

Portal venous blood flow has also been used as an index of GI blood flow. Iwao and colleagues (44) showed in 1995 that portal flow was reduced after maximal exercise. In 2001, Nancy Rehrer and colleagues (101) in New Zealand reported reductions in portal vein blood flow of up to 80% in subjects cycling at 70% $\dot{V}O_2$max. In a follow-up study published in 2005 (100), this group found that the ingestion of neither a carbohydrate (CHO) beverage nor water was able to maintain portal blood flow at pre-exercise levels (~50% reduction for both beverages) after 60 min of cycle ergometry at 70% $\dot{V}O_2$max. However, both beverages did appear to keep blood flow higher than the values reported in their first study.

In summary, research over the past century has indicated that GI blood flow in most nonhuman species was reduced with increasing exercise intensity. The exception was dogs, in which blood flow changes to the GI tract were not always reduced during exercise. In humans, almost all studies found that GI blood flow decreased during exercise and that this reduction was exacerbated by heat stress.

Gastric Emptying

With regard to exercise, the primary function of the stomach is to provide the intestine with ingested fluids and nutrients. Thus, the rate of gastric emptying can significantly influence fluid homeostasis and performance.

After the observations of Beaumont (88) on the effect of exercise on gastric function (see chapter 1), it was not until 1916 that another physiologist took interest in the topic. At this time, Anton J. Carlson (figure 16.2) from the University of Chicago published a book titled *The Control of Hunger in Health and Disease* (14). Carlson described experiments conducted in both dogs and humans. In dogs with simple gastric fistulas, treadmill running resulted in reduced gastric tonus and hunger contractions. The degree of inhibition became greater with increased intensity and persisted for 20 to 40 min after running. In humans running in situ,

Figure 16.2 Anton J. Carlson (1875-1956), professor of physiology at the University of Chicago.
Courtesy of American Physiological Society.

Carlson similarly reported that exercise inhibited gastric tonus and hunger contractions. However, walking in situ did not have any effect. Of note, the human subjects were a gastric fistula patient, three of Carlson's assistants, and Carlson himself.

It was more than a decade before another report on the effect of exercise on gastric function was published. In 1928, J.M.H. (Maurice) Campbell (1891-1973) and colleagues (13) from Guy's Hospital in London studied the effect of exercise on gastric emptying. The test meal used was porridge in which "the constant composition and consistency which did not readily block the stomach tube were great practical advantages." Describing the exercise protocol, the authors wrote, "[The subjects] ran round the laboratory for part of the hour after the meal, the distance run (from one to four miles) depending on the degree of training. The object being to take moderately easy exercise, rests were taken as required and no attempt was made to run fast, the conditions being made less tedious by some humorous rivalry." In these studies, it was found that running led to greater gastric residue of the test meal 1 h after ingestion. In other words, exercise slowed the gastric emptying rate. It was also observed that the more-fit subjects were able to empty more at a given level of exercise and that mild exercise, such as walking and talking, actually enhanced gastric emptying. As stated by the authors, "Exercise which did not produce discomfort of any sort increased the rate of digestion, but the opposite effect followed exercise which produced any discomfort" (13, pp. 290-291).

In 1934, a pioneer in exercise physiology, Frances A. Hellebrandt (1901-1992), and her colleagues in the department of physiology at the University of Wisconsin in the United States published a series of three papers (39-41) under the main title "Studies on the Influence of Exercise on the Digestive Work of the Stomach." The effect of exercise on gastric secretions, gastric emptying, and motility after a meal was studied using gastric aspiration, fluoroscopic visualization, and manometry. From these studies, the general conclusions were that intense exercise reduces initial secretion of gastric acid, initial inhibition of gastric acid secretion may be followed by hyperacidity, mild exercise did not affect peak acidity, mild exercise enhanced gastric emptying rate, "violent and exhaustive" exercise initially inhibited gastric emptying (which may be enhanced later to compensate for the initial reduction), and motility and secretory functions were coordinated and the mechanisms appeared to be the same.

After the work of Hellebrandt and colleagues (39-41), the effect of exercise on GI function was infrequently investigated for the next 30 yr. However, in the 1960s and 1970s the running boom exploded and an interest in fluid, CHO, and electrolyte replacement was on the rise. In 1967, John Fordtran (figure 16.3), a gastroenterologist from Dallas, and Bengt Saltin (see figure 16.4), from Stockholm, collaborated on a classic study in this field (28). They reported that treadmill exercise at approximately 70% $\dot{V}O_2max$ for 60 min did not adversely affect gastric emptying of water or a 13.3% glucose solution, nor did it alter gastric acid secretion. Moreover, water emptied much faster from the stomach during exercise than the concentrated glucose solution did. Thus, they were the first to demonstrate that prolonged moderate-intensity exercise did not affect gastric emptying and that a high concentration of CHO in a beverage did slow gastric emptying. As discussed later, these findings were confirmed by others.

In 1970, David Costill (figure 16.4) and colleagues at Ball State University published their first study on gastric emptying during exercise (19). Highly trained marathon runners ran for 2 h at approximately 70% $\dot{V}O_2max$ while drinking 100 ml of water or a glucose–electrolyte beverage every 5 min. Gastric aspiration of the stomach contents revealed that gastric residue volumes were nearly identical between conditions, although all subjects experienced "extreme sensations of fullness." It was calculated that the subjects sweat at an average rate of 1,668 ml/h but replaced (i.e., emptied) an average of only 822 ml/h from the stomach. From these data the authors concluded that "it is highly un-

Figure 16.3 John S. Fordtran (1931–), MD, gastroenterologist and professor of medicine at the University of Texas Southwestern Medical School and Baylor University Medical Center in the United States.

Photo courtesy of John S. Fordtran.

likely that fluid replacement could keep pace with fluid loss" (19, p. 524) during a competitive marathon event.

Costill and Saltin (20) in 1974 examined the influences of exercise intensity, beverage volume, temperature, osmolarity, and glucose concentration on gastric emptying. Using the gastric-aspiration technique, they found that as the volume of fluid increased, up to approximately 600 ml, the rate of emptying also increased; that as beverage glucose concentration (and osmolarity) increased above 139 mM (209 mOsm/L), gastric emptying decreased compared with water; and that warmer solutions tended to empty more slowly than cooler ones. Also, in accordance with the previous findings by Fordtran and Saltin (28), there was no effect of exercise on gastric emptying at intensities below approximately 70% $\dot{V}O_2$max during short-term (15 min) or prolonged (120 min) exercise. However, it was found that the rate of gastric emptying was slowed when cycling at 80% to 90% $\dot{V}O_2$max (figure 16.5).

Costill's laboratory, still using the gastric-aspiration method, then began to study the effects of the compositions of many different beverages on gastric emptying rate. Specifically, Coyle and colleagues in 1978 studied the gastric emptying of commercially available athletic drinks (21). In 1980, Foster and colleagues compared gastric emptying rates among solutions containing different forms of CHO (i.e., glucose and glucose polymers) (29). The main finding from these studies was that both higher osmolality and higher CHO (i.e., glucose) concentrations delay gastric emptying. This finding confirmed previous results (20, 28) on the negative effects of increasing osmolality and CHO content on gastric emptying rate.

In 1986, Neufer and colleagues in Costill's laboratory reported that 15 min of running at 50% to 70% $\dot{V}O_2$max enhanced the emptying rates of fluids compared with resting conditions (82). They also observed that water emptied faster during rest or running than did solutions containing maltodextrin or combinations of maltodextrin, glucose, and fructose, even though all were hypotonic. Another interesting finding was that the addition of glucose to the various solutions inhibited emptying during exercise whereas the addition of fructose did not. Later (1989), Neufer and colleagues at the U.S Army Research Institute of Environmental Medicine (84) found that increasing treadmill exercise intensity (up to ~74% $\dot{V}O_2$max) did not inhibit the gastric emptying rate of water. In fact, it had an enhancing effect. The same authors also found that 15 min of moderate treadmill exercise (~50% $\dot{V}O_2$max; three 15 min bouts) in a hot environment (49 °C) reduced gastric emptying of fluids, as did exercise in a warm environment (35 °C) when hypohydrated (83).

During this same time period, Mitchell and colleagues in Costill's laboratory studied the effect of repeated ingestion of fluids on gastric emptying during cycling exercise. Initially (73), they reported that CHO solutions no higher than 7.5% can empty as rapidly as water from the stomach and that more than 95% of the ingested fluid was emptied during the exercise period (seven 12 min rides at 70% $\dot{V}O_2$max and a final 12 min

Figure 16.4 David L. Costill (1936–; far right), professor of exercise science and director of the Human Performance Laboratory at Ball State University. Bengt Saltin (1935–; left of Costill), professor at the Karolinska Institute in Stockholm, professor at the August Krogh Institute at the University of Copenhagen, and director of the Copenhagen Muscle Research Center. Also pictured are Ball State graduate students Rich Cote (to the left of Saltin) and Ed Burke (with headset attached).

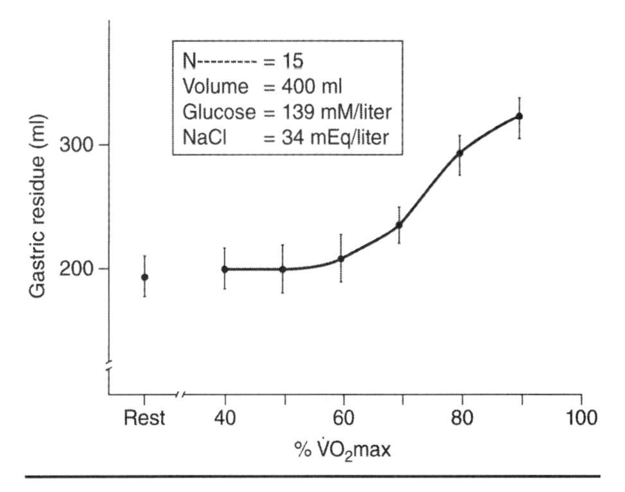

Figure 16.5 Volume of gastric residue remaining in the stomach 15 min after ingestion of 400 ml of fluid at different exercise intensities (percentage $\dot{V}O_2$max).

self-paced performance ride). Later (72), they demonstrated that concentrated solutions (12% and 18% CHO) emptied much more slowly than did less-concentrated solutions (i.e., 6% CHO) even when ingested repeatedly. In addition, they found that gastric emptying was not hindered by prolonged moderate-intensity (70% $\dot{V}O_2max$) exercise in a neutral (22 °C; 50% relative humidity) environment.

From the mid-1980s through the early 1990s, Carl Gisolfi (figure 16.6) at the University of Iowa began devoting a great deal of his laboratory attention to the gastric emptying of fluids during exercise. Two PhD students, Mallard Owen and Alan Ryan, and their colleagues conducted the first studies on gastric emptying of fluids that combined prolonged, continuous exercise with heat stress. Using a gastric-aspiration technique similar to that used by Costill and Saltin (20), Owen and colleagues (89) observed that gastric emptying of a hypertonic (586 mOsm/kg of H_2O) 10% glucose solution was slower (41% of drink emptied) during 2 h of running at 65% $\dot{V}O_2max$ in the heat (35 °C) compared with gastric emptying of a hypotonic (193 mOsm/kg of H_2O) 10% glucose polymer solution (51% emptied) or sweetened water (59% emptied). Ryan and colleagues (108) further observed that a 5% glucose solution (300 mOsm/kg of H_2O) emptied more slowly (93% emptied) compared with sweetened water (99% emptied) during a 3 h bout of cycling (60% $\dot{V}O_2max$) in the heat (33 °C). Hypotonic glucose polymer (82 mOsm/kg of H_2O; 94% emptied) and glucose polymer–fructose (156 mOsm/kg of H_2O; 98% emptied) solutions, however, were found to empty as rapidly as sweetened water.

Figure 16.6 Carl V. Gisolfi (1942-2000; left), professor of exercise science and physiology and biophysics at the University of Iowa, with friend and colleague Charles M. Tipton (1927–), professor and chair of exercise and sport sciences, University of Arizona.

Photo courtesy of Mrs. Joan Seye, University of Iowa.

In the late 1980s, a group of researchers in Wim Saris' (figure 16.7) laboratory at the University of Limburg in Maastricht published a series of studies dealing with gastric emptying during exercise. An early paper by Beckers and colleagues (4) provided the methodology for determining gastric residue volumes and secretions at any given time during an experiment. Using this technique, Rehrer and colleagues conducted a number of experiments on gastric emptying during exercise. These studies addressed factors such as training status, exercise intensity, exercise mode, repeated drinking, dehydration, and drink composition. Novel findings from these studies were that gastric emptying of fluids was not affected by level of training (97), gastric emptying was initially similar between running and cycling but then appeared to decline with running after approximately 40 min of exercise at 70% $\dot{V}O_2max$ (99), initial bolus size affects the initial rate of emptying of all solutions (97), and a high emptying rate was maintained with repeated consumption of water or an isotonic drink (99). Of these results, the most influential was that a relatively high ingested volume and repeated drinking enhanced the rate of emptying. These findings were confirmed by Mitchell and Voss (74) and subsequently emphasized by Noakes and colleagues (85). The finding that training status had no effect on emptying of fluids during exercise was also novel. However, during the same time, Carrio and colleagues (15) found that a solid meal (i.e., egg omelet) was emptied more rapidly at rest by trained runners compared with sedentary individuals. To date, these studies have provided the only evidence concerning the effect of chronic exercise training on gastric emptying of fluids or solid meals.

In the early 1990s, Zachwieja and colleagues from Costill's laboratory conducted two studies on the ef-

Figure 16.7 Wim H.M. Saris (1949–), professor of human nutrition at Maastricht University (formerly the University of Limburg).

Photo courtesy of Wim Saris.

fects of carbonation in sport drinks on gastric emptying at rest (138) and during exercise (137) and reported that carbonation did not influence gastric emptying of solutions with the same CHO (or no CHO) content. During the same time period, similar results were obtained by Ryan and colleagues (111) and by Lambert and colleagues (51) at the University of Iowa. Another study conducted at this time by Houmard and colleagues documented that there was no difference in gastric emptying rate of fluids between running and cycling (each at 75% $\dot{V}O_2max$ for 1 h) (43).

The effect of CHO concentration and osmolality on gastric emptying rate received further attention in the mid-1990s. Two studies from the laboratory of Ronald J. Maughan (figure 16.8) at Aberdeen University, using the double-sampling gastric-aspiration method and authored by Vist and Maughan (133, 134), found that glucose concentrations higher than 2% inhibited the emptying rate of a single 600 ml bolus of solution (133). Furthermore, a 4% glucose solution (230 mOsm/kg of H_2O) emptied more slowly than did a glucose polymer solution with the same energy content but lower osmolality (42 mOsm/kg of H_2O). However, a 4% glucose solution was found to empty more rapidly than a glucose polymer solution with the same osmolality but much higher glucose content (134). These findings indicated that both osmolality and CHO content of ingested solutions affect gastric emptying rate. CHO content exerted the greatest effect, and osmolality was more influential at higher CHO concentrations.

In 1995, Fred Brouns and colleagues (10) in Maastricht also systematically studied the effect of osmolarity and CHO content on gastric emptying using the modified double-sampling method previously described by Beckers and colleagues (4). They concluded that it was the rate of CHO delivery to the intestine and not the osmolarity of the fluid-replacement solutions (moderately hypotonic to moderately hypertonic) that determined the gastric emptying rate. These findings supported the concept that the CHO content of dilute solutions has more influence on the rate of gastric emptying than does the osmolarity of the solution, as indicated by Vist and Maughan (134) during the same time period.

Figure 16.8 Ronald J. Maughan (1951–), professor from Aberdeen University and Loughborough University.
Photo courtesy of Ronald Maughan.

With heightened interest in the formulation of sport drinks, Robert Murray (1949–) and colleagues at the Gatorade Sports Science Institute published studies in the mid-1990s on the gastric emptying rates of various solutions. They used the repeated double-sampling method and observed that gastric emptying rates could be interpreted differently depending on how they were assessed (80). In their study, five solutions (including pure water) differing in CHO concentration, CHO type, and osmolality were tested. When the gastric emptying rates were expressed as milliliters per minute during the 30 min test period, no differences among solutions were found. However, when the results were expressed as the percentage of original beverage remaining in the stomach during the first 10 or 20 min time periods, or as the gross gastric volumes at these time points, statistically significant differences were observed (i.e., 6% glucose and 6% maltodextrin solutions emptied more slowly than did water, 6% sucrose, or 4% sucrose/2% glucose solutions) (80). A further observation from this study was that osmolality did not appear to affect gastric emptying rate because the maltodextrin solution, with an osmolality of only 53 mOsm/kg of H_2O, emptied more slowly than did solutions with higher osmolality values (80).

Murray and colleagues reported in 1999 that repeated ingestion of an 8% CHO beverage resulted in slower gastric emptying rates when compared with ingestion of water, 4% CHO, or 6% CHO beverages, which highlighted the negative effect of higher CHO concentrations on gastric emptying rates (79). In 2000, Shi and colleagues (117) found that the gastric emptying rate of 6% CHO solutions was not affected by osmolality (250-434 mOsm/kg of H_2O) or by the number, types, or combination of CHO (i.e., glucose, fructose, glucose and fructose, sucrose).

In 2001, John Leiper and colleagues from Maughan's laboratory published two studies examining gastric emptying rates during intermittent high-intensity cycling (62) and a soccer match (66). In each activity, emptying rates were delayed when compared with either resting or walking trials. Similar results were subsequently obtained by Leiper and colleagues in 2005 (65) when they compared high-intensity shuttle running with walking. These studies have added valuable information to the body of knowledge in this area.

From the mid-1990s through the first decade of the 21st century, researchers at the University of Iowa published a series of studies (31, 34, 54, 55, 105, 110) simultaneously examining gastric emptying and intestinal absorption of fluids during exercise. These studies followed the concept that repeated fluid consumption maintains a relatively high and constant gastric emptying rate. Lambert and colleagues (54) developed a procedure for these simultaneous determinations that used the double-sampling aspiration technique for determining gastric emptying and the triple-lumen tube segmental perfusion method for measuring intestinal absorption.

A key finding from these studies was that solutions that differed markedly in CHO content (55, 105, 110) emptied from the stomach at similar rates when provided an equilibration period (35 min) to reach a relatively constant gastric volume from repeated drinking. Although this finding initially appeared to be in contrast to other results, the authors observed that they had to rely on a greater steady-state stomach volume with the solutions of higher concentrations in order to achieve the same emptying rate found with the solutions of lower concentration. Thus, the more highly concentrated solutions did not empty as rapidly in the early stages of the experiments (i.e., during the equilibration period), but did so later by relying on a higher gastric volume. This result supports the concept that gastric volume has a major influence on gastric emptying rate (85). In practical terms, repeated consumption of solutions with higher CHO (e.g., 9% CHO) can eventually produce relatively high gastric emptying rates (110). However, they do so at the cost of maintaining higher stomach volumes and potentially a greater chance of GI distress.

As the 20th century came to a close, Michiel van Nieuwenhoven and colleagues from the Saris laboratory examined a number of GI variables during exercise, including gastric emptying. These investigators, using the ^{13}C-acetate breath test, reported that prolonged intense cycling (70% maximal W for 90 min) did not affect gastric emptying rate (129) or rates recorded during a rest–exercise–rest protocol when ingesting water or an approximately 7% CHO beverage (with or without caffeine) (130). However, a 3% loss in body mass due to sweating before cycling at 70% maximal W for 90 min was shown to delay gastric emptying rate (131). These results corresponded with findings from most previous studies (20, 28, 83, 84). However, Ryan and colleagues (110), using the double-sampling gastric-aspiration technique, did not observe an effect of hypohydration (~2.7% body mass) on gastric emptying of a water placebo solution during prolonged cycling (90 min at 65% $\dot{V}O_2$max) with repeated drinking. The discrepancy with the findings of van Nieuwenhoven and colleagues (131) could be explained by differences in the method used to determine gastric emptying rate or in the ingested fluid.

To determine whether the gastric emptying rate can be enhanced by repeated sessions of drinking large volumes during exercise, Lambert and colleagues (56) had runners perform five 90 min treadmill runs, each separated by approximately 1 wk. During each run, the subjects repeatedly drank a glucose–electrolyte solution to match their predetermined sweat rates. After the first and last runs, the gastric emptying rate was determined by gastric aspiration. No statistical differences were observed between these trials, indicating that meaningful adaptations did not occur to enhance gastric emptying with time, although stomach comfort did significantly improve.

Table 16.1 summarizes the earliest or most novel findings in the area of gastric emptying during exercise. These findings include the following: Mild to moderate exercise did not likely affect gastric emptying rate or enhanced it (13, 20, 41, 72, 82, 84); strenuous, continuous, or intermittent exercise inhibited gastric emptying (13, 14, 41, 62); gastric emptying of fluids slows with increasing energy (e.g., CHO) content and, thus, osmolality (10, 20, 21, 28, 29, 72, 89, 108, 133, 134); severe heat stress or hypohydration inhibited gastric emptying (83); and large fluid volumes and repeated drinking maintained a rapid gastric emptying rate (74, 99).

Intestinal Absorption

Experiments in the area of intestinal absorption during exercise were not conducted in humans until Fordtran and Saltin's groundbreaking study in 1967 (28). In addition to gastric emptying, these authors used the triple-lumen tube steady-state perfusion technique (i.e., segmental perfusion method) to measure net water and solute fluxes from a glucose–electrolyte solution in the proximal jejunum and ileum during treadmill exercise. The key finding was that exercise at approximately 70% $\dot{V}O_2$max did not significantly affect absorption of water, glucose, or sodium.

No further studies on intestinal absorption during exercise were published until 1988. At that time, Barclay and Turnberg (1) at the University of Manchester also used the segmental perfusion technique during moderate-intensity cycling. In contrast to the findings by Fordtran and Saltin (28), they found significant reductions in jejunal water and electrolyte absorption. The conflicting results were likely attributable to methodological differences in exercise mode, exercise intensity, perfused solution, and perfusion rate.

Table 16.1 Novel Findings Related to Gastric Emptying During Exercise (1910-2010)

Author	Year (reference)	Finding
Carlson Campbell et al. Hellebrandt et al.	1916 (14) 1928 (13) 1934 (41)	Mild to moderate exercise enhanced gastric emptying of a solid meal; strenuous exercise inhibited gastric emptying of solid meal.
Fordtran and Saltin	1967 (28)	Prolonged exercise at ~70% $\dot{V}O_2$max did not inhibit gastric emptying of fluids; water emptied faster than 13.3% glucose solution did.
Costill and Saltin	1974 (20)	Increasing beverage volume (up to 600 ml) increased gastric emptying rate; increasing glucose concentrations in beverage reduced gastric emptying rate; exercise at ~80%-90% $\dot{V}O_2$max reduced gastric emptying rate of ingested fluid.
Neufer et al.	1989 (83)	Treadmill exercise in a hot (49 °C) environment, or a warm environment (35 °C) when hypohydrated, slowed gastric emptying rate.
Owen et al. Ryan et al.	1986 (89) 1989 (108)	Prolonged running or cycling (60%-65% $\dot{V}O_2$max) in heat (35 °C) did not affect gastric emptying rate with repeated drinking.
Rehrer et al. Mitchell and Voss	1990 (99) 1991 (74)	Large fluid volume and repeated drinking enhanced gastric emptying rate during exercise.
Leiper et al.	2001 (66)	Intermittent high-intensity exercise inhibited gastric emptying rate.

During the mid- to late 1980s, a number of intestinal perfusion studies examined the effect of solution composition on water and solute absorption in resting humans. Although the studies were not conducted during exercise, the results influenced the formulation of fluid-replacement beverages used by athletes. For instance, Leiper and Maughan at the University of Aberdeen studied the effects of solution composition on jejunal fluid absorption and reported that an isotonic glucose electrolyte solution was absorbed at a higher rate than plain water (63) and that moderately hypotonic glucose–electrolyte solutions were absorbed more rapidly than relatively isotonic solutions (64). These findings supported the concept that the presence of glucose stimulated water absorption and that hypotonicity created a greater osmotic gradient with the blood, thus promoting fluid absorption. At approximately the same time, Wheeler and Banwell in Cleveland in the United States reported that CHO-containing solutions perfused into the jejunum could be absorbed as well as, but not necessarily better than, plain distilled water (136). A subsequent study by Rehrer and colleagues (102) indicated that a dilute (4.5%) glucose solution was absorbed more rapidly than water in the perfused jejunum.

Numerous studies related to intestinal absorption during exercise using the segmental perfusion technique were conducted in the 1990s in the Gisolfi laboratory at the University of Iowa. Early findings indicated that isotonic CHO–electrolyte solutions were absorbed more rapidly than water or electrolyte-only solutions when perfused into the duodenojejunum and that cycling for 90 min at 70% $\dot{V}O_2$max did not affect absorption of either water or a CHO–electrolyte solution (32) (figure 16.9). Shi and colleagues reported that higher glucose concentrations (i.e., 8%, in the form of either glucose or maltodextrin) inhibited water absorption in the duodenojejunum (120). However, multiple forms of CHO enhanced water absorption of even hypertonic solutions at rates not different from those of isotonic or hypotonic solutions (119, 120). This important finding was attributed to multiple forms of CHO enhancing total solute absorption, which promoted greater water absorption in the intestinal segment studied—provided that the osmolality of the solution was not markedly increased. It was also during this time that Gisolfi and colleagues (33) observed that 6% CHO solutions with no sodium were absorbed at rates similar to those containing sodium (i.e., 25 or 50 mEq/L). They concluded that CHO (e.g., glucose) was much more important for stimulating water absorption than was sodium and that sodium secretion into the intestine provided adequate sodium to low-sodium solutions without negatively affecting overall water flux (33).

As noted, Lambert and colleagues described a method for the simultaneous determination of both gastric emptying and intestinal absorption during exercise (54). This study allowed the researchers to measure the intestinal absorption of the actual solution leaving the stomach at the actual gastric emptying rate rather than perfusing a solution at a predetermined flow rate. Several studies were subsequently conducted using this technique.

In 1997, Lambert and colleagues (55) reported that absorption in the proximal small intestine depended on the segment studied and the solution composition in a given segment. They found that an ingested water placebo solution was absorbed more rapidly than an ingested isotonic 6% CHO–electrolyte solution in the "leaky" duodenum (first 25 cm of small intestine), where the osmotic gradient for the water placebo solution was the highest. However, in the proximal jejunum (first 50 cm), where glucose transporters were in abundance, the isotonic 6% CHO–electrolyte solution was absorbed more rapidly because increased solute absorption enhanced water absorption. The authors concluded that over the first 75 cm of proximal small intestine there is no difference in fluid absorption rate between water and the CHO–electrolyte solution. However, the different solutions rely on different mechanisms to achieve their respective overall rates. Another important finding using this technique was made in

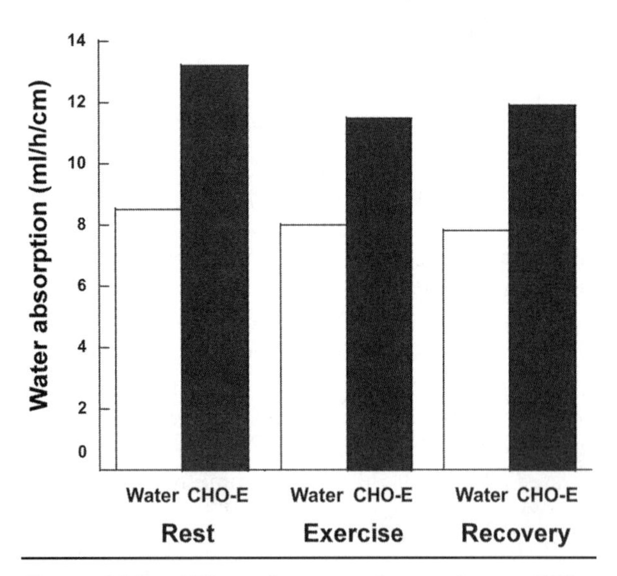

Figure 16.9 Effect of rest, cycle exercise at 70% $\dot{V}O_2$max , and recovery on duodenojejunal absorption of distilled water or a CHO–electrolyte solution (CHO-E). There were no differences between rest, exercise, and recovery conditions in a solution or between solutions in a condition. However, pooled data for all three conditions revealed a significant ($P < 0.05$) increase in water absorption from CHO-E compared with distilled water.

Data from Gisolfi et al. 1991.

1998, when Ryan and colleagues (110) demonstrated that hypohydration did not affect water absorption. In their study they also observed that high CHO concentration (≥8%), resulting in high osmolality, reduced water absorption. The latter finding confirmed the results of Shi and colleagues (120) and was later supported by Lambert and colleagues (58). Also at this time, Gisolfi and colleagues (34) reported that ingested hypotonic, isotonic, and moderately hypertonic 6% CHO–electrolyte solutions could be absorbed as rapidly as a water placebo solution in the proximal small intestine. However, similar to the results of Lambert and colleagues (55), the solutions differed in the segment at which the majority of fluid was absorbed.

Select investigators have estimated CHO absorption during exercise using urinary excretion of ingested, nonmetabolized CHO such as 3-O-methyl-D-glucose (actively absorbed) and D-xylose (passively absorbed). Findings by van Nieuwenhoven and colleagues in Maastricht in 1999 (129) and Lang and colleagues in Iowa in 2006 (61) indicated that exercise at 70% $\dot{V}O_2$max reduced both active (61, 129) and passive (61) glucose absorption. Although these findings did not agree with results using the segmental perfusion methodology (28, 32), the differences between studies were likely attributable to different methods for determining absorption or the exercise mode employed (running versus cycling).

In terms of intestinal adaptations to exercise, Harris and colleagues reported in 1991 that chronically active persons with elevated energy intakes have rapid orocecal transit rates (37). However, absorption of CHO was not compromised (as measured by D-xylose absorption), indicating that small intestinal adaptations likely occurred to allow efficient assimilation of nutrients in such individuals.

Table 16.2 summarizes the earliest or most novel findings in the area of intestinal absorption during exercise. These findings included the following: Prolonged exercise below approximately 70% $\dot{V}O_2$max did not affect fluid or solute absorption (28); hypotonic solutions containing CHO were absorbed more rapidly at rest than water, isotonic solutions, or hypertonic solutions in the jejunum (63, 64); an isotonic 6% CHO–electrolyte solution was absorbed more rapidly during exercise than water in the duodenojejunum (32); intestinal fluid absorption of CHO solutions at rest decreased with increasing CHO concentration and osmolality (58, 120); multiple forms of CHO in a solution enhanced total solute and water absorption at rest (120); absorption of an ingested water placebo occurred at a high rate during exercise in the duodenum with a significant reduction in the jejunum, whereas absorption of

Table 16.2 Novel Findings Related to Intestinal Absorption During Exercise (1910-2010)

Author	Year (reference)	Finding
Fordtran and Saltin	1967 (28)	Prolonged exercise at ~70% $\dot{V}O_2max$ did not affect fluid or solute absorption in the jejunum.
Leiper and Maughan	1986 (63, 64)	Isotonic glucose solutions absorbed more rapidly than water in the jejunum at rest; hypotonic glucose solutions absorbed faster than isotonic solutions in the jejunum at rest.
Gisolfi et al.	1991 (32)	CHO–electrolyte solution absorbed more rapidly than water during rest, exercise, and recovery in the duodenojejunum.
Shi et al.	1994 (120)	Multiple forms of CHO in a solution enhanced total solute and water absorption in the duodenojejunum.
Lambert et al.	1996 (54)	Developed method for determining intestinal absorption rates from ingested fluids during exercise.
Lambert et al.	1997 (55)	Developed method for determining absorption rates of ingested fluids in different segments of the small intestine.
Ryan et al.	1998 (110)	Hypohydration did not affect absorption rate of ingested water during exercise; high CHO concentration (≥8%) reduced rate of water absorption from ingested solutions.
Gisolfi et al.	1998 (34)	Moderately hypertonic CHO–electrolyte solutions with multiple transportable CHO absorbed as rapidly as hypotonic solutions when ingested.

CHO = carbohydrate.

an ingested CHO–electrolyte solution occurred at a more constant rate in the duodenojejunum (i.e., intestinal fluid absorption during exercise differed depending on the intestinal segment studied) (55); hypohydration did not affect intestinal water absorption during exercise (110); and moderately hypertonic solutions with multiple transportable CHO ingested during exercise were absorbed as rapidly as hypotonic solutions in the duodenojejunum (34).

GI Transit and Motility

GI transit and motility involves all segments of the GI tract. The preponderance of research in this area, in terms of exercise, has been related to gastric emptying of fluids for hydration purposes. However, research in this area has also been important for those interested in possibly altering overall GI transit and motility via exercise.

The first study in the past century to examine the acute effect of exercise on GI motility was conducted by DeYoung and colleagues (23) in 1931. In their study on dogs with chronic cecostomies, brief (~6 min) bouts of treadmill running resulted in a temporary increase in tone and motility of the colon, which was followed by a period of hypomotility. The next study dealing with acute exercise was published in 1956 by Stickney and colleagues (126). These investigators studied both rats and dogs and found no effect of treadmill running on propulsion through the small intestine in either species.

The only study to assess both gastric emptying and small intestinal transit during exercise in humans was performed in 1982 by Cammack and colleagues (12) at the University of Sheffield. These investigators labeled a solid meal with radioactive technetium sulfur colloid and measured gastric radioactivity with a single crystal scintillation detector. Small bowel transit time was determined by measuring hydrogen concentrations in the breath. These investigators reported that although gastric emptying was accelerated, small bowel transit time during acute, intermittent, light cycling exercise was unchanged compared with resting conditions. In contrast, Keeling and Martin (49) in 1987 and Keeling and colleagues (48) in 1990 observed more rapid orocecal transit times of a 360 kcal liquid meal during acute, mild treadmill exercise than during resting conditions. However, Meshkinpour and colleagues (71) observed slower orocecal transit time during treadmill walking compared with sitting in subjects after ingestion of a relatively dilute liquid meal (i.e., 10 g of lactulose in 150 ml of water). Soffer and colleagues (123) at Iowa also did not observe faster orocecal transit time during high-intensity (20 min at 80% $\dot{V}O_2$max) cycling. It should be noted that the latter studies did not differentiate between gastric emptying and small bowel transit time because they measured only hydrogen concentrations in the breath as the ingested meal reached the cecum.

In terms of the effect of chronic exercise training on GI transit time and motility, the earliest study in the past century was conducted in animals and was published in 1954 by Van Liere and colleagues (128). In this study, groups of adult albino rats either ran on a treadmill (2 h/d, 6 d/wk, for approximately 1 mo) or remained untrained. After the training period, both the trained and untrained rats were administered a powdered charcoal solution and the distance it travelled through the GI tract was measured after 40 min. The powdered charcoal was found to travel a significantly farther distance in the trained rats than in the untrained rats, indicating more rapid transit and greater motility. The authors reported that the trained animals ate more than their untrained counterparts and therefore may have had greater strength of the intestinal musculature due to increased work. The authors further commented that they had heard from a surgeon that the intestinal musculature of soldiers in World War I was "thicker and in far better tone than the musculature of the intestines of men of sedentary habits." In addition, the authors noted that parasympathetic tone increases with training and "would, of course, cause increased propulsive motility of the small intestine."

In 1986, Cordain and colleagues (18) were the first to study the effect of chronic exercise training on GI transit time in humans. These authors reported more rapid (23%) whole-bowel transit time after a 6 wk running program. In contrast, Bingham and Cummings (6) in 1989 found no consistent effect of a running program on large intestinal transit time. Thus, faster whole-bowel transit time with exercise training appeared to reflect changes occurring proximal to the colon (i.e., stomach or small intestine). This finding was confirmed by Harris and colleagues (37), who found in 1991 a more rapid orocecal transit time in physically active individuals with increased dietary intakes. In 1991, Oettle (86) showed faster whole-bowel transit in healthy subjects after 1 wk of daily running or cycling. Conflicting results subsequently appeared throughout the 1990s as Lampe and colleagues (60), Coenen and colleagues (17), Robertson and colleagues (103), Sesboue and colleagues (116), and Kayaleh and colleagues (46) reported no effect of chronic exercise on bowel transit time. Only one study during this period, by Koffler and colleagues (50), was supportive of faster whole-bowel transit time with exercise training; however, this study involved strength training rather than aerobic exercise. Most recently, De Schryver and colleagues (22) reported in 2005 that regular physical activity improved total colonic transit time in patients with constipation.

Few studies have directly measured motility of the GI tract. The first study in humans was conducted in trained cyclists by Soffer and colleagues (123) at Iowa in 1991. The authors reported an increase in small intestinal motility in the postprandial state during high-intensity cycling (80%-90% $\dot{V}O_2$max). Cheskin and colleagues (16) subsequently documented in 1992 that distal colonic motility did increase during running at 70% to 80% of maximal heart rate in trained runners, especially in those with a history of runner's diarrhea. In contrast, Rao and colleagues (96) found in 1999 that colonic motility was reduced during cycle ergometry at 25, 50, and 75% $\dot{V}O_2$max and increased after exercise in untrained subjects.

Soffer and colleagues (122, 124) were the first to examine esophageal motility. They reported that esophageal motility (duration, amplitude, and frequency of contractions) decreased with increasing cycling intensity in both trained and untrained subjects. In 1999, van Nieuwenhoven and colleagues (129) in Maastricht also examined esophageal motility during cycling at 70% of maximal workload. They found that the number of peristaltic contractions decreased whereas peristaltic velocity increased. Furthermore, lower esophageal pressure was reduced whereas there was no evidence of a change in gastroesophageal reflux.

In summary, acute mild-intensity exercise was shown to have no effect on small bowel transit of a solid meal, but overall orocecal transit time was shown to increase (48, 49). Acute high-intensity exercise was shown to have no effect on orocecal transit time (123). Chronic exercise training enhanced whole-bowel transit time (18, 86), likely due to enhanced gastric and small intestinal transit time (37). Small intestinal (123) and colonic (123) motility was also found to increase in trained subjects; however, this was not the case, in terms of colonic motility, in untrained subjects (96). Esophageal motility was found, in general, to be reduced with exercise (122, 124, 129).

GI Barrier Function

The GI barrier prevents unwanted substances in the lumen of the GI tract from entering the internal environment of the body. Research has indicated that prolonged strenuous exercise (77, 87, 90), especially when combined with heat stress (59) or nonsteroidal anti-inflammatory drug (NSAID) use (52, 53, 109, 121), causes dysfunction of this barrier. This possibly leads to increased permeation of substances such as endotoxin (7, 9), a highly pathogenic molecule derived from the walls of gram-negative bacteria. Endotoxemia can possibly exacerbate the problems of heat stress (112). The first indication that exercise affects GI barrier function was reported in 1988 by Bosenberg and colleagues (7) and Brock-Utne and colleagues (9) from the laboratory of Steve Gaffin at the University of Natal. These investigators found that participation in a long-endurance triathlon (7) or an ultraendurance run (9) resulted in significant elevations in plasma endotoxin concentrations. Further human research by Moses and colleagues (77) in 1991 and Oktedalen and colleagues (87) in 1992 indicated that prolonged strenuous running was associated with GI barrier dysfunction as determined by urinary excretion of ingested, nonmetabolizable probes. It also became apparent that running created more GI barrier problems than cycling did. For instance, Moore and colleagues (75), under the direction of James Knochel in Dallas, did not observe significant endotoxemia in cyclists admitted to the medical tent during a long-endurance ride [100 miles (160.9 km)] in Texas in 1995. However, Ryan and colleagues (109) showed in 1996 that aspirin use before running greatly increased small intestinal permeability. A subsequent study by Pals and colleagues (90) demonstrated that prolonged high-intensity running (i.e., 80% $\dot{V}O_2max$ for 60 min) also increased GI permeability to large molecules. Later, two more studies indicated that prolonged exercise with or without NSAID use can cause gut barrier dysfunction. First, Smetanka and colleagues (121) reported that runners in the 1996 Chicago Marathon (42 km) had increased gut permeability if they used ibuprofen before or during the race. Second, Lambert and colleagues (59) found that participation in a long-endurance triathlon (1998 Hawaii Ironman: 3.9 km swim, 180 km bike, 42 km run) increased small intestinal permeability (figure 16.10). In 2000, Jeukendrup and colleagues (45) confirmed the results of Lambert and colleagues (59). Subsequent studies in the first decade of the 21st century by Lambert and colleagues examined GI barrier function during exercise (52, 53, 57). These studies supported the concept that NSAIDs do exacerbate GI barrier dysfunction during prolonged distance running (52, 53) and the idea that fluid ingestion attenuated select effects of both aspirin and dehydration on gut permeability (53, 57).

In summary, GI barrier dysfunction was shown to occur during prolonged strenuous exercise based on increased plasma endotoxin levels (7, 9, 45) and increased urinary excretion of ingested permeability probes (59, 77, 87). Furthermore, use of NSAIDs was shown to increase gut permeability during distance running (52, 109, 121), and fluid restriction promoted greater GI barrier dysfunction (57).

GI Symptoms During Exercise

Probably the most obvious indication that gut barrier function was compromised during strenuous exercise was the recurrent GI complaints of runners during the running boom of the 1970s. Richard Fogoros first described this phenomenon in 1980 (27). He presented two case studies of male distance runners suffering from abdominal pain, diarrhea, and bloody stools. These symptoms were found to be associated only with running and subsided with time or with reductions in training, but did recur in one patient after a marathon race. Fogoros speculated that the likely explanation for the symptoms was gut ischemia and that the imposition of heat stress and hyperthermia could predispose a runner to ischemic necrosis. These explanations were later found to be correct as cases of running-associated ischemic colitis (38) and hemorrhagic colitis (78) were

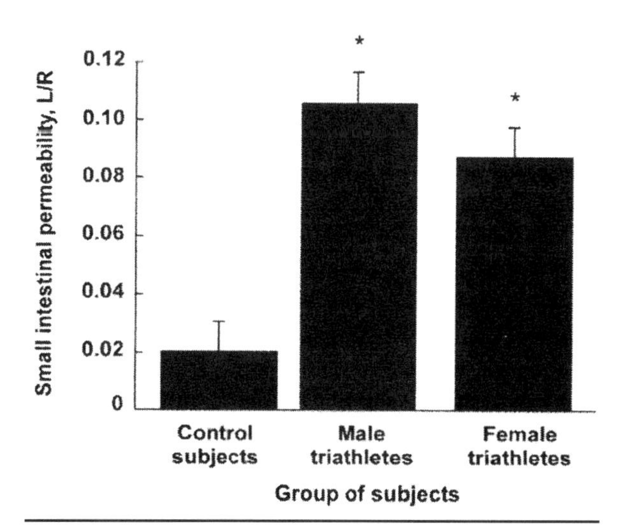

Figure 16.10 Small intestinal permeability after participation in the Hawaii Ironman triathlon (3.9 km swim, 180 km bike, 42 km run). Asterisk (*) denotes significant difference ($P < 0.05$) from control subjects.

Reprinted, by permission, from G.P. Lambert, 2009, "Stress-induced gastrointestinal barrier dysfunction and its inflammatory effects," *Journal of Animal Science* 87(14 Suppl): E101-E108.

documented. A number of studies in the 1980s and 1990s further described cases of GI erosions and lesions (30, 87, 115), blood loss (3, 25, 35, 69, 70, 104, 114, 115, 125), and hemorrhage (91) in distance runners.

Keefe and colleagues (47) in 1984 were the first to comprehensively document GI symptoms in distance runners. In that study the reported symptoms were heartburn, abdominal cramps, nausea, vomiting, urge to defecate, diarrhea, and bloody stools. Subsequent survey-type investigations on runners and other endurance athletes have supported the findings of Keefe and colleagues (47) for abdominal cramps, nausea, vomiting, diarrhea, and bloody stools (127) and for heartburn, nausea, vomiting, abdominal cramps, urge to defecate, and diarrhea (93).

Studies aimed at determining the cause of GI complaints have also been conducted. As previously noted, numerous studies indicated that GI ischemia was a main culprit leading to GI bleeding. However, in 1988, Brock-Utne and colleagues (9) reported that 80.6% of runners with high plasma endotoxin values after the Comrades Marathon (89.4 km) suffered from nausea, vomiting, or diarrhea, implicating endotoxemia as a possible cause of some GI symptoms. With regard to fluid consumption, Rehrer and colleagues (98) reported in 1990 that dehydration, rather than the consumption of large volumes of fluid, increased the frequency of GI disorders during endurance events. Furthermore, van Nieuwenhoven and colleagues (132), Shi and colleagues (118), and Morton and colleagues (76) observed that GI distress increased as the CHO concentration of the ingested solution was elevated, possibly due to the reduced gastric emptying rates or intestinal absorption rates of the more highly concentrated solutions.

In summary, case studies and surveys revealed that prolonged strenuous exercise, especially distance running, promoted GI symptoms. Such symptoms included nausea, vomiting, abdominal cramps, urge to defecate, diarrhea, and bloody stools. Cases of ischemic colitis, hemorrhagic colitis, and GI lesions were also documented.

Summary

Even though the GI system has received less attention over the past 100 yr than other organ systems with regard to exercise, we still have learned a great deal about its function. Research has reported that most GI processes were not greatly affected by mild to moderate exercise. However, prolonged strenuous exercise, especially when combined with heat stress or dehydration, negatively affected gastric emptying, intestinal absorption, GI barrier function, and GI symptoms.

Research has provided much information regarding the optimal formulation of fluid-replacement beverages for exercise. Thus, the importance of the GI system in maintaining proper fluid balance and thermoregulation during exercise is more fully appreciated. A great deal is yet to be learned regarding how GI function during exercise may ultimately affect the functions of other organ systems, such as the cardiovascular, neuromuscular, renal, and immune systems. Future decades of research will hopefully provide much new information in this regard.

References

1. Barclay GR, Turnberg LA. Effect of moderate exercise on salt and water transport in the human jejunum. *Gut* 29: 816-820, 1988.
2. Barcroft J, Florey H. The effects of exercise on the vascular conditions in the spleen and the colon. *J Physiol* 68: 181-189, 1929.
3. Baska RS, Moses FM, Graeber G, Kearney G. Gastrointestinal bleeding during an ultramarathon. *Dig Dis Sci* 35: 276-279, 1990.
4. Beckers EJ, Rehrer NJ, Brouns F, Hoor FT, Saris WHM. Determination of total gastric volume, gastric secretion and residual meal using the double sampling technique of George. *Gut* 29: 1725-1729, 1988.
5. Bell AW, Hales JRS, King RB, Fawcett AA. Influence of heat stress on exercise-induced changes in regional blood flow in sheep. *J Appl Physiol: Respirat Environ Exercise Physiol* 55: 1916-1923, 1983.
6. Bingham SA, Cummings JH. Effect of exercise and physical fitness on large intestinal function. *Gastroenterology* 97: 1389-1399, 1989.
7. Bosenberg A, Brock-Utne J, Gaffin S, Wells M, Blake G. Strenuous exercise causes systemic endotoxemia. *J Appl Physiol* 65: 106-108, 1988.
8. Bradley SE. Variations in hepatic blood flow in man during health and disease. *N Engl J Med* 240: 456-461, 1949.
9. Brock-Utne J, Gaffin S, Wells M, Gathiram P, Sohar E, James M, Morrell D, Norman R. Endotoxemia in exhausted runners after a long-distance race. *S African Med J* 73: 533-536, 1988.
10. Brouns F, Senden J, Beckers EJ, Saris WHM. Osmolarity does not affect the gastric emptying rate of oral rehydration solutions. *J Parent Ent Nutr* 19: 403-406, 1995.
11. Burns GP, Schenk WG. Effect of digestion and exercise on intestinal blood flow and cardiac output. *Arch Surg* 98: 790-794, 1969.
12. Cammack J, Read NW, Cann PA, Greenwood B, Holgate AM. Effect of prolonged exercise on the passage of a solid meal through the stomach and small intestine. *Gut* 23: 957-961, 1982.

13. Campbell JMH, Mitchell GO, Powell ATW. The influence of exercise on digestion. *Guy's Hosp Rep* 78: 279-293, 1928.
14. Carlson AJ. *The Control of Hunger in Health and Disease*. Chicago: The University of Chicago Press, 1916.
15. Carrio I, Estorch M, Serra-Grima R, Ginjaume M, Notivol R, Calabuig R, Vilardell F. Gastric emptying in marathon runners. *Gut* 30: 152-155, 1989.
16. Cheskin LJ, Crowell MD, Kamal B, Rosen B, Schuster MM, Whitehead WE. The effects of acute exercise on colonic motility. *J Gastrointest Motil* 4: 173-177, 1992.
17. Coenen C, Wegener M, Wedmann B, Schmidt G, Hoffmann S. Does physical exercise influence bowel transit time in healthy young men? *Am J Gastroenterol* 87: 292-295, 1992.
18. Cordain L, Latin RW, Behnke JJ. The effects of an aerobic running program on bowel transit time. *J Sports Med* 26: 101-104, 1986.
19. Costill D, Kammer W, Fisher A. Fluid ingestion during distance running. *Arch Environ Health* 21: 520-524, 1970.
20. Costill DL, Saltin B. Factors limiting gastric emptying during rest and exercise. *J Appl Physiol* 37: 679-683, 1974.
21. Coyle EF, Costill DL, Fink WJ, Hoopes DG. Gastric emptying rates for selected athletic drinks. *Res Quart* 49: 119-124, 1978.
22. De Schryver AM, Keulmans YC, Peters HP, Akkermans LM, Smout AJ, DeVries WR, vanBerge-Henegouwen GP. Effects of regular physical activity on defecation pattern in middle aged-patients complaining of chronic constipation. *Scand J Gastroenterol* 40: 422-429, 2005.
23. DeYoung VR, Rice HA, Steinhaus AH. Studies in the physiology of exercise. VII. The modification of colonic motility induced by exercise and some indications for a nervous mechanism. *Am J Physiol* 99: 52-63, 1931.
24. Eriksen M, Waaler BA. Priority of blood flow to splanchnic organs in humans during pre-and postmeal exercise. *Acta Physiol Scand* 150: 363-372, 1994.
25. Fisher RL, McMahon LF, Ryan MJ, Larson D, Brand M. Gastrointestinal bleeding in competitive runners. *Dig Dis Sci* 31: 1226-1228, 1986.
26. Flaim SF, Minteer WJ, Clark DP, Zelis R. Cardiovascular response to acute aquatic and treadmill exercise in the untrained rat. *J Appl Physiol: Respirat Environ Exercise Physiol* 46: 302-308, 1979.
27. Fogoros R. "Runners trots": Gastrointestinal disturbances in runners. *JAMA* 243: 17431744, 1980.
28. Fordtran JS, Saltin B. Gastric emptying and intestinal absorption during prolonged severe exercise. *J Appl Physiol* 23: 331-335, 1967.
29. Foster C, Costill DL, Fink WJ. Gastric emptying characteristics of glucose and glucose polymer solutions. *Res Quart Exer Sport* 51: 299-305, 1980.
30. Gaudin C, Zerath E, Guezennec CY. Gastric lesions secondary to long-distance running. *Dig Dis Sci* 35: 1239-1243, 1990.
31. Gisolfi CV, Lambert GP, Summers RW. Intestinal fluid absorption during exercise: Role of sport drink osmolality and [Na+]. *Med Sci Sports Exerc* 33: 907-915, 2001.
32. Gisolfi CV, Spranger KJ, Summers RW, Schedl HP, Bleiler TL. Effects of cycle exercise on intestinal absorption in humans. *J Appl Physiol* 71: 2518-2527, 1991.
33. Gisolfi CV, Summers RD, Schedl HP, Bleiler TL. Effect of sodium concentration in a carbohydrate-electrolyte solution on intestinal absorption. *Med Sci Sports Exerc* 27: 1414-1420, 1995.
34. Gisolfi CV, Summers RW, Lambert GP, Xia T. Effect of beverage osmolality on intestinal fluid absorption during exercise. *J Appl Physiol* 85: 1941-1948, 1998.
35. Halvorsen F-A, Lyng J, Ritland S. Gastrointestinal bleeding in marathon runners. *Scand J Gastroenterol* 21: 493-497, 1986.
36. Harpuder K, Lowenthal M, Blatt SD. Peripheral and visceral effects of exercise in the erect position. *J Appl Physiol* 11: 185-188, 1957.
37. Harris A, Lindeman AK, Martin BJ. Rapid orocecal transit in chronically active persons with high energy intake. *J Appl Physiol* 70: 1550-1553, 1991.
38. Heer M. Acute ischaemic colitis in a female long distance runner. *Gut* 28: 896-899, 1987.
39. Hellebrandt FA, Dimmitt LL. Studies on the influence of exercise on the digestive work of the stomach. III. Its effect on the relation between secretory and motor function. *Am J Physiol* 107: 364-369, 1934.
40. Hellebrandt FA, Hoopes SL. Studies on the influence of exercise on the digestive work of the stomach. I. Its effect on the secretory cycle. *Am J Physiol* 107: 348-354, 1934.
41. Hellebrandt FA, Tepper RH. Studies on the influence of exercise on the digestive work of the stomach. II. Its effect on emptying time. *Am J Physiol* 107: 355-363, 1934.
42. Herrick JF, Grindlay JH, Baldes EJ, Mann FC. Effect of exercise on the blood flow in the superior mesenteric, renal and common iliac arteries. *Am J Physiol* 128: 338-344, 1939.
43. Houmard JA, Egan PC, Johns RA, Neufer PD, Chenier TC, Israel RG. Gastric emptying during 1 h of cycling and running at 75% VO2max. *Med Sci Sports Exerc* 23: 320-325, 1991.
44. Iwao T, Toyonaga A, Ikegami M, Sumino M, Oho K, Sakaki M, Shigemori H, Tanikawa K, Iwao J. Effects of exercise-induced sympathoadrenergic activation on portal blood flow. *Dig Dis Sci* 40: 48-51, 1995.
45. Jeukendrup AE, Vet-Joop K, Sturk A, Stegen JH, Senden J, Saris WH, Wagenmakers AJ. Relationship between gastro-intestinal complaints and endotoxemia, cytokine release and the acute-phase reaction during and after a long-distance triathlon in highly trained men. *Glin Sci (Lond)* 98: 47-55, 2000.
46. Kayaleh RA, Meshkinpour H, Avinashi A, Tamadon A. Effect of exercise on mouth-to-cecum transit in trained athletes: A case against the role of runners'

abdominal bouncing. *J Sports Med Phys Fitness* 36: 271-274, 1996.

47. Keefe EB, Lowe DK, Goss JR, Wayne R. Gastrointestinal symptoms of marathon runners. *West J Med* 141: 481-484, 1984.
48. Keeling WF, Harris A, Martin BJ. Orocecal transit during mild exercise in women. *J Appl Physiol* 68: 1350-1353, 1990.
49. Keeling WF, Martin BJ. Gastrointestinal transit during mild exercise. *J Appl Physiol* 63: 978-981, 1987.
50. Koffler KH, Menkes A, Redmond RA, Whitehead WE, Pratley RE, Hurley BF. Strength training accelerates gastrointestinal transit in middle-aged and older men. *Med Sci Sports Exerc* 24: 415-419, 1992.
51. Lambert GP, Bleiler TL, Chang R-T, Johnson AK, Gisolfi CV. Effects of carbonated and noncarbonated beverages at specific intervals during treadmill running in the heat. *Int J Sport Nutr* 3: 177-193, 1993.
52. Lambert GP, Boylan MW, Laventure J-P, Bull AJ, Lanspa SJ. Effect of aspirin and ibuprofen on GI permeability during exercise. *Int J Sports Med* 28: 722-726, 2007.
53. Lambert GP, Broussard LJ, Mason BL, Mauermann WJ, Gisolfi CV. Gastrointestinal permeability during exercise: Effects of aspirin and energy-containing beverages. *J Appl Physiol* 90: 2075-2080, 2001.
54. Lambert GP, Chang RT, Joensen D, Shi X, Summers RW, Schedl HP, Gisolfi CV. Simultaneous determination of gastric emptying and intestinal absorption during cycle exercise in humans. *Int J Sports Med* 17: 48-55, 1996.
55. Lambert GP, Chang RT, Xia T, Summers RW, Gisolfi CV. Absorption from different intestinal segments during exercise. *J Appl Physiol* 83: 204-212, 1997.
56. Lambert GP, Lang JA, Bull AJ, Eckerson JM, Lanspa S, O'Brien J. Fluid tolerance while running: Effect of repeated trials. *Int J Sports Med* 29: 878-882, 2008.
57. Lambert GP, Lang JA, Bull AJ, Pfeifer PC, Eckerson JM, Moore GA, Lanspa SJ, et al. Fluid restriction increases GI permeability during running. *Int J Sports Med* 29: 194-198, 2008.
58. Lambert GP, Lanspa S, Welch R, Shi X. Combined effects of glucose and fructose on fluid absorption from hypertonic carbohydrate-electrolyte beverages. *J Exerc Physiol online* 11: 46-55, 2008.
59. Lambert GP, Murray R, Eddy D, Scott W, Laird R, Gisolfi CV. Intestinal permeability following the 1998 Ironman triathlon. *Med Sci Sports Exerc* 31: S318, 1999.
60. Lampe JW, Slavin JL, Apple FS. Iron status of active women and the effect of running a marathon on bowel function and gastrointestinal blood loss. *Int J Sports Med* 12: 173-179, 1991.
61. Lang JA, Gisolfi CV, Lambert GP. Effect of exercise intensity on active and passive glucose absorption. *Int J Sport Nutr Exerc Metab* 16: 485-493, 2006.
62. Leiper JB, Broad N, Maughan R. Effects of intermittent high-intensity exercise on gastric emptying in man. *Med Sci Sports Exerc* 33: 1270-1278, 2001.
63. Leiper JB, Maughan RJ. Absorption of water and electrolytes from hypotonic, isotonic, and hypertonic solutions. *J Physiol* 373: 90P, 1986.
64. Leiper JB, Maughan RJ. The effect of luminal tonicity on water absorption from a segment of the intact human jejunum. *J Physiol* 378: 95P, 1986.
65. Leiper JB, Nicholas CW, Ali A, Williams C, Maughan RJ. The effect of intermittent high-intensity running on gastric emptying of fluids in man. *Med Sci Sports Exerc* 37: 240-247, 2005.
66. Leiper JB, Prentice AS, Wrightson C, Maughan RJ. Gastric emptying of a carbohydrate-electrolyte drink during a soccer match. *Med Sci Sports Exerc* 33: 1932-1938, 2001.
67. Lowenthal M, Harpuder K, Blatt SD. Peripheral and visceral vascular effects of exercise and postprandial state in supine position. *J Appl Physiol* 4: 689-694, 1952.
68. Manohar M. Blood flow to the respiratory and limb muscles and to abdominal organs during maximal exertion in ponies. *J Physiol* 377: 25-35, 1986.
69. McCabe ME, Peura DA, Kadakia SC, Bocek Z, Johnson LF. Gastrointestinal blood loss associated with running a marathon. *Dig Dis Sci* 31: 1229-1232, 1986.
70. McMahon LF, Ryan MJ, Larson D, Fisher RL. Occult gastrointestinal blood loss in marathon runners. *Ann Int Med* 100: 846-847, 1984.
71. Meshkinpour H, Kemp C, Fairshter R. Effect of aerobic exercise on mouth-to-cecum transit time. *Gastroenterology* 96: 938-941, 1989.
72. Mitchell JB, Costill DL, Houmard JA, Fink WJ, Robergs RA, Davis JA. Gastric emptying: Influence of prolonged exercise and carbohydrate concentration. *Med Sci Sports Exerc* 21: 269-274, 1989.
73. Mitchell JB, Costill DL, Houmard JA, Flynn MG, Fink WJ, Beltz JD. Effects of carbohydrate ingestion on gastric emptying and exercise performance. *Med Sci Sports Exerc* 20: 110-115, 1988.
74. Mitchell JB, Voss KW. The influence of volume on gastric emptying and fluid balance during prolonged exercise. *Med Sci Sports Exerc* 23: 314-319, 1991.
75. Moore GE, Holbein MEB, Knochel JP. Exercise-associated collapse in cyclists is unrelated to endotoxemia. *Med Sci Sports Exerc* 27: 1238-1242, 1995.
76. Morton DP, Aragon-Vargas LF, Callister R. Effect of ingested fluid composition on exercise-related transient abdominal pain. *Int J Sport Nutr Exerc Metab* 14: 197-208, 2004.
77. Moses F, Singh A, Smoak B, Hollander D, Deuster P. Alterations in intestinal permeability during prolonged high intensity running. *Gastroenterology* 100: A472, 1991.
78. Moses FM, Brewer TG, Peura DA. Running-associated proximal hemorrhagic colitis. *Ann Int Med* 108: 385-386, 1988.
79. Murray R, Bartoli W, Stofan J, Horn M, Eddy D. A comparison of the gastric emptying characteristics of

selected sports drinks. *Int J Sport Nutr* 9: 263-274, 1999.

80. Murray R, Eddy DE, Bartoli WP, Paul GL. Gastric emptying of water and isocaloric carbohydrate solutions consumed at rest. *Med Sci Sports Exerc* 26: 725-732, 1994.
81. Musch TI, Friedman DB, Pitetti KH, Haidet GC, Stray-Gunderson J, Mitchell JH, Ordway GA. Regional distribution of blood flow of dogs during graded dynamic exercise. *J Appl Physiol* 63: 2269-2277, 1987.
82. Neufer PD, Costill DL, Fink WJ, Kirwan JP, Fielding RA, Flynn MG. Effects of exercise and carbohydrate composition on gastric emptying. *Med Sci Sports Exerc* 18: 658-662, 1986.
83. Neufer PD, Young AJ, Sawka MN. Gastric emptying during exercise: Effects of heat stress and hypohydration. *Eur J Appl Physiol* 58: 433-439, 1989.
84. Neufer PD, Young AJ, Sawka MN. Gastric emptying during walking and running: Effects of varied exercise intensity. *Eur J Appl Physiol* 58: 440-445, 1989.
85. Noakes TD, Rehrer NJ, Maughan RJ. The importance of volume in regulating gastric emptying. *Med Sci Sports Exerc* 23: 307-313, 1991.
86. Oettle GJ. Effect of moderate exercise on bowel habit. *Gut* 32: 941-944, 1991.
87. Oktedalen O, Lunde OC, Opstad PK, Aabakken L, Kvernebo K. Changes in the gastrointestinal mucosa after long distance running. *Scand J Gastroenterol* 27: 270-274, 1992.
88. Osler W. William Beaumont. A pioneer American physiologist. *XIIIth International Physiological Congress*, Boston, 1929.
89. Owen MD, Kregel KC, Wall PT, Gisolfi CV. Effects of ingesting carbohydrate beverages during exercise in the heat. *Med Sci Sports Exerc* 18: 568-575, 1986.
90. Pals KL, Chang RT, Ryan AJ, Gisolfi CV. Effect of running intensity on intestinal permeability. *J Appl Physiol* 82: 571-576, 1997.
91. Papaioannides D, Giotis C, Karagiannis N, Voudouris C. Acute upper gastrointestinal hemorrhage in long-distance runners. *Ann Int Med* 101: 719, 1984.
92. Perko MJ, Nielsen HB, Skak C, Clemmesen JO, Schroeder TV, Secher NH. Mesenteric, coeliac, splanchnic blood flow in humans during exercise. *J Physiol* 513: 907-913, 1998.
93. Peters HPF, Bos M, Seebregts L, Akkermans LMA, van Berge Henegouwen GP, Bol E, Mosterd WL, et al. Gastrointestinal symptoms in long-distance runners, cyclists, and triathletes: Prevalence, medication, and etiology. *Am J Gastroenterol* 94: 1570-1581, 1999.
94. Puvi-Rajasingham S, Wijeyekoon B, Natarajan P, Mathias CJ. Systemic and regional (including superior mesenteric) haemodynamic responses during supine exercise while fasted and fed in normal man. *Clin Autonomic Res* 7: 149-154, 1997.
95. Qamar MI, Read AE. Effects of exercise on mesenteric blood flow in man. *Gut* 28: 583587, 1987.
96. Rao SSC, Beaty J, Chamberlain M, Lambert PG, Gisolfi C. Effects of acute graded exercise on human colonic motility. *Am J Physiol* 276: G1221-G1226, 1999.
97. Rehrer NJ, Beckers E, Brouns F, TenHoor F, Saris WHM. Exercise and training effects on gastric emptying of carbohydrate beverages. *Med Sci Sports Exerc* 21: 540-549, 1989.
98. Rehrer NJ, Beckers EJ, Brouns F, TenHoor F, Saris WHM. Effects of dehydration on gastric emptying and gastrointestinal distress while running. *Med Sci Sports Exerc* 22: 790-795, 1990.
99. Rehrer NJ, Brouns F, Beckers EJ, TenHoor F, Saris WHM. Gastric emptying with repeated drinking during running and bicycling. *Int J Sports Med* 11: 238-243, 1990.
100. Rehrer NJ, Goes E, DuGardeyn C, Reynaert H, DeMeirleir K. Effect of carbohydrate on portal vein blood flow during exercise. *Int J Sports Med* 26: 171-176, 2005.
101. Rehrer NJ, Smets A, Reynaert H, Goes E, DeMeirleir K. Effect of exercise on portal vein blood flow in man. *Med Sci Sports Exerc* 33: 1533-1537, 2001.
102. Rehrer NJ, Wagenmakers AJM, Beckers EJ, Halliday D, Leiper JB, Brouns F, Maughan RJ, et al. Gastric emptying, absorption, and carbohydrate oxidation during prolonged exercise. *J Appl Physiol* 72: 468-475, 1992.
103. Robertson G, Meshkinpour H, Vandenberg K, James N, Cohen A, Wilson A. Effects of exercise on total and segmental colon transit. *J Clin Gastroenterol* 16: 300-303, 1993.
104. Robertson JD, Maughan RJ, Davidson RJL. Faecal blood loss in response to exercise. *Br Med J* 295: 303-305, 1987.
105. Rogers J, Summers RW, Lambert GP. Gastric emptying and intestinal absorption of a low-carbohydrate sport drink during exercise. *Int J Sport Nutr Exerc Metab* 15: 220-235, 2005.
106. Rowell LB, Blackmon JR, Bruce RA. Indocyanine green clearance and estimated hepatic blood flow during mild to maximal exercise in upright man. *J Clin Invest* 43: 1677-1690, 1964.
107. Rowell LB, Blackmon JR, Martin RH, Mazzarella JA, Bruce RA. Hepatic clearance of indocyanine green in man under thermal and exercise stresses. *J Appl Physiol* 20: 384-394, 1965.
108. Ryan AJ, Bleiler TL, Carter JE, Gisolfi CV. Gastric emptying during prolonged cycling in the heat. *Med Sci Sports Exerc* 21: 51-58, 1989.
109. Ryan AJ, Chang R-T, Gisolfi CV. Gastrointestinal permeability following aspirin intake and prolonged running. *Med Sci Sports Exerc* 28: 698-705, 1996.
110. Ryan AJ, Lambert GP, Shi X, Chang RT, Summers RW, Gisolfi CV. Effect of hypohydration on gastric emptying and intestinal absorption during exercise. *J Appl Physiol* 84: 1581-1588, 1998.
111. Ryan AJ, Navarre AE, Gisolfi CV. Consumption of carbonated and noncarbonated sports drinks during prolonged treadmill exercise in the heat. *Int J Sport Nutr* 1: 225-239, 1991.

112. Sakurada S, Hales JRS. A role for gastrointestinal endotoxins in enhancement of heat tolerance by physical fitness. *J Appl Physiol* 84: 207-214, 1998.
113. Sanders TM, Werner RA, Bloor CM. Visceral blood flow distribution during exercise to exhaustion in conscious dogs. *J Appl Physiol* 40: 927-931, 1976.
114. Schoch DR, Sullivan AL, Grand RJ, Eagan WF. Gastrointestinal bleeding in an adolescent runner. *J Pediatr* 111: 302-304, 1987.
115. Scobie BA. Recurrent gut bleeding in five long-distance runners. *NZ Med J* 98: 966, 1985.
116. Sesboue B, Arhan P, Devroede G, Lecointe-Besancon I, Congard P, Bouchoucha M, Fabre J. Colonic transit in soccer players. *J Clin Gastroenterol* 20: 211-214, 1995.
117. Shi X, Bartoli W, Horn M, Murray R. Gastric emptying of cold beverages in humans: Effect of transportable carbohydrates. *Int J Sport Nutr* 10: 394-403, 2000.
118. Shi X, Horn MK, Osterberg KL, Stofan JR, Zachwieja JJ, Horswill CA, Passe DH, Murray R. Gastrointestinal discomfort during intermittent high-intensity exercise: Effect of carbohydrate-electrolyte beverage. *Int J Sport Nutr Exerc Metab* 14: 673-683, 2004.
119. Shi X, Summers RW, Schedl HP, Chang RT, Lambert GP, Gisolfi CV. Effects of solution osmolality on absorption of select fluid replacement solutions in human duodenojejunum. *J Appl Physiol* 77: 1178-1184, 1994.
120. Shi X, Summers RW, Schedl HP, Flanagan SW, Chang RT, Gisolfi CV. Effects of carbohydrate type and concentration and solution osmolality on water absorption. *Med Sci Sports Exerc* 27: 1607-1615, 1995.
121. Smetanka RD, Lambert GP, Murray R, Eddy D, Horn M, Gisolfi CV. Intestinal permeability in runners in the 1996 Chicago marathon. *Int J Sports Nutr* 9: 426-433, 1999.
122. Soffer EE, Merchant RK, Duethman G, Launspach J, Gisolfi C, Adrian TE. Effect of graded exercise on esophageal motility and gastroesophageal reflux in trained athletes. *Dig Dis Sci* 38: 220-224, 1993.
123. Soffer EE, Summers RW, Gisolfi C. Effect of exercise on intestinal motility and transit in trained athletes. *Am J Physiol* 260(*Gastrointest Liver Physiol* 23): G698-G702, 1991.
124. Soffer EE, Wilson J, Duethman G, Launspach J, Adrian TE. Effect of graded exercise on esophageal motility and gastroesophageal reflux in nontrained subjects. *Dig Dis Sci* 39: 193-198, 1993.
125. Stewart JG, Ahlquist DA, McGill DB, Ilstrup DM, Schwartz S, Owen RA. Gastrointestinal blood loss and anemia in runners. *Ann Int Med* 100: 843-845, 1984.
126. Stickney JC, Northup DW, Van Liere EJ. Resistance of the small intestine (motility) against stress. *J Appl Physiol* 9: 484-486, 1956.
127. Sullivan SN, Wong C. Runners' diarrhea. Different patterns and associated factors. *J Clin Gastroenterol* 14: 101-104, 1992.
128. Van Liere E, Hess HH, Edwards JE. Effect of physical training on the propulsive motility of the small intestine. *J Appl Physiol* 7: 186-187, 1954.
129. van Nieuwenhoven MA, Brouns F, Brummer RJ. The effect of physical exercise on parameters of gastrointestinal function. *Neurogastroenterol Motil* 11: 431-439, 1999.
130. van Nieuwenhoven MA, Brummer R-JM, Brouns F. Gastrointestinal function during exercise: Comparison of water, sports drink, and sports drink with caffeine. *J Appl Physiol* 89: 1079-1085, 2000.
131. van Nieuwenhoven MA, Vriens BEPJ, Brummer R-JM, Brouns F. Effect of dehydration on gastrointestinal function at rest and during exercise in humans. *Eur J Appl Physiol* 83: 578-584, 2000.
132. van Nieuwenhoven MA, Brouns F, Kovacs EM. The effect of two sports drinks and water on GI complaints and performance during an 18-km run. *Int J Sports Med* 26: 281-285, 2005.
133. Vist GE, Maughan RJ. Gastric emptying of ingested solutions in man: Effect of beverage glucose concentration. *Med Sci Sports Exerc* 26: 1269-1273, 1994.
134. Vist GE, Maughan RJ. The effect of osmolality and carbohydrate content on the rate of gastric emptying of liquids in man. *J Physiol* 486: 523-531, 1995.
135. Wade OL, Combes B, Childs AW, Wheeler HO, Cournand A, Bradley SE. The effect of exercise on the splanchnic blood flow and splanchnic blood volume in normal man. *Glin Sci* 15: 457-463, 1956.
136. Wheeler KB, Banwell JG. Intestinal water and electrolyte flux of glucose-polymer electrolyte solutions. *Med Sci Sports Exerc* 18: 436-439, 1986.
137. Zachwieja JJ, Costill DL, Beard GC, Robergs RA, Pascoe DD, Anderson DE. The effects of a carbonated carbohydrate drink on gastric emptying, gastrointestinal distress, and exercise performance. *Int J Sport Nutr* 2: 239-250, 1992.
138. Zachwieja JJ, Costill DL, Widrick JJ, Anderson DE, McConnell GK. Effects of drink carbonation on the gastric emptying characteristics of water and flavored water. *Int J Sport Nutr* 1: 45-51, 1991.

CHAPTER 17

Metabolic Systems: Substrate Utilization

Andrew R. Coggan, PhD

This chapter traces advances made during the past century in our understanding of the effects of acute and chronic exercise on substrate metabolism. The chapter emphasizes the integrative physiology of substrate metabolism in exercising humans, although selected studies using animal models are also considered. Because chapter 1 did not cover the topic of substrate utilization in great detail, the following section is presented first.

Substrate Utilization During Exercise: Understanding Before 1910

As discussed by previous authors (4, 9, 19), many of the salient features of the metabolic response to exercise were initially described well before 1910. For example, in 1807 Berzelius reported that lactate concentrations were higher in the fully functioning muscles of hunted deer than in the partially paralyzed muscles (15). In 1845, von Helmholtz observed a decrease in alcohol-soluble substances and an increase in water-soluble substances in muscle with activity, a finding consistent with the breakdown of glycogen to form lactate and other glycolytic intermediates (179). In 1859, du Bois-Reymond determined that muscle contractions were associated with an increase in H^+ concentration, which he concluded was due to the production of lactic acid from glycogen (62). In 1871 Weiss was the first to directly demonstrate that muscle glycogen was utilized during contractile activity (183), and in 1887 Chauveau and Kauffman found in arteriovenous balance experiments that glucose uptake by muscle increased several-fold during exercise (27). Before this, in 1862 Smith reported that treadmill exercise did not increase the daily loss of nitrogen in the urine of prisoners consuming a constant diet (168), indicating that, contrary to what was previously thought, protein was not an important energy source during exercise. That exercise did not significantly increase protein oxidation was confirmed a few years later in experiments by Fick and Wislicenus (70) and von Pettenkofer and Voit (180).

According to Nathan Zuntz (1847-1920) writing in 1911 (191), findings such as these led to the belief that carbohydrate was the sole source of energy used by exercising muscle. However, based on measurements in his own laboratory of the respiratory exchange ratio (RER) in exercising horses and humans starting in the late 1880s (reviewed in 191), Zuntz believed that both fat and carbohydrate could be oxidized by muscle during exercise and that the relative contribution from each depended on the intensity and duration of the exercise as well as the availability of bodily carbohydrate stores. Zuntz also recognized that, in addition to muscle glycogen, liver glycogen was an important source of fuel for contracting muscle and that liver glycogenolysis served to maintain blood glucose concentration during moderate exercise and elevate it during strenuous exercise. Finally, on the basis of his own experiments as well as those of others, Zuntz understood that the oxidation of carbohydrate yielded only slightly more energy per unit of O_2 consumed than the oxidation of fat did. Indeed, as emphasized by Barnard and Holloszy (9), perhaps the only point on which Zuntz's understanding of substrate metabolism during exercise differed substantially from that of the present day was his mistaken belief that the production of lactate by exercising muscle was due to hypoxia.

Substrate Utilization During Exercise: Advances From 1910 to World War II

Despite Zuntz's quite modern view as expressed in his 1911 review, it took approximately another 30 yr of research before a widely agreed-upon (albeit flawed, as shall be discussed) picture of substrate utilization during exercise fully emerged. During these years, investigation in this area focused primarily on clarifying two important issues:

- What substrate(s) are in fact oxidized during exercise?
- What is the immediate source of energy fueling muscle contraction?

These two interrelated questions remained unsettled largely because of the influence of two individuals—J.B. Auguste Chauveau (1837-1917) and especially Archibald V. Hill (1886-1977) of England—and because basic research in biochemistry was not at the level of physiological measurements.

On the basis of experiments in which RER was measured during intense exercise leading to exhaustion in a little more than 1 h, Chauveau held that carbohydrate was the only fuel directly oxidized during exercise and that fat could be used only if it was first converted into carbohydrate, a process that he estimated would reduce the net energy yield by approximately 30% (27). Although this 1896 hypothesis was inconsistent with the results of Nathan Zuntz's experiments performed at around the same time, Chauveau's intellectual influence was such that the first question listed previously remained a point of controversy for quite some time.

The concept that carbohydrate was the sole fuel used by exercising muscle was further reinforced by the publication in 1907 of Walter M. Fletcher (1873-1910) and Frederick G. Hopkins' (1861-1947) now-classic studies definitively demonstrating the production of lactate by contracting muscle (72). This was followed closely thereafter by the publication of a series of papers between 1910 and 1914 by Fletcher's student Hill, who demonstrated that muscle contraction was accompanied by the liberation of the same amount of heat regardless of whether O_2 was present (94-96). If O_2 was present, however, additional heat was released during the recovery process. These observations led Hill, as well as Fletcher and Hopkins, to believe that formation of lactate was directly responsible for muscle contraction and that O_2 was used to rebuild lactate into its parent molecule during recovery. This parent molecule was originally believed to be Hermann's inogen (92) but was later accepted to be glycogen after experiments by Parnas and Wagner in 1914 (148) and Otto F. Meyerhof (1884-1951) in 1920 (137).

The controversy over whether carbohydrate alone or both fats and carbohydrate were oxidized during exercise led Francis G. Benedict (1870-1957) and Edward P. Cathcart (1877-1954) to readdress this issue in 1913 (11). Using a professional cyclist as their primary research subject, Benedict and Cathcart conducted an extensive series of experiments in which they quantified the rates of O_2 uptake ($\dot{V}O_2$) and CO_2 production ($\dot{V}CO_2$) as well as heart rate, rate of respiration, and, in some experiments, rectal temperature before, during, and after exercise performed at various intensities on a carefully calibrated cycle ergometer. These experiments clearly demonstrated that although RER generally increased with the transition from rest to exercise, it typically remained well below unity even at a $\dot{V}O_2$ as high as 45 $ml \cdot min^{-1} \cdot kg^{-1}$. Furthermore, they revealed that both gross and net efficiency were essentially constant during exercise performed at a somewhat lower intensity despite a large difference in RER (i.e., 0.77 vs. 0.88-0.90) induced by the subject's preceding diet. This led Benedict and Cathcart to at least tentatively reject Chauveau's hypothesis, concluding that "although the evidence in all of (our) experiments . . . is strikingly in favor of the view that during periods of muscular work there is an increased draft upon carbohydrate material in the body, the data by no means indicate that muscular work is performed exclusively by the combustion of carbohydrate" (11, p. 94) and that "the fact that the experiments with the diet poor in carbohydrates showed not the slightest indication of an increase in the energy output per unit of work is strongly suggestive of the absence of the transformation during work of fat to glycogen with a consequent liberation of unavailable heat" (11, p. 146). That Benedict and Cathcart did not dismiss Chauveau's idea completely despite their completely contradictory findings speaks to the sway Chauveau held over the field at the time.

Benedict and Cathcart's experiments were essentially replicated in 1920 by August Krogh (1874-1949) and Johannes Lindhard (1870-1947), who, distrustful of the intermittent collection of expired air during exercise using a mouthpiece and closed-circuit system as employed by Benedict and Cathcart, used an elaborately constructed and extensively validated Jacquet-type respiration chamber that completely enclosed the subject and the cycle ergometer to continuously measure respiratory gas exchange using the open-circuit approach (figure 17.1) (118). Regardless of this difference in methodology, however, Krogh and Lindhard's results were quite similar to those obtained by Benedict and Cathcart. RER tended to be higher during exercise than at rest (especially after several days of a high-carbohydrate diet) but still remained well below unity. Also, although the subjects' energy expenditure appeared to be slightly greater when the rate of fat oxidation was higher, the difference was only 11%—much smaller than the 30% suggested by Chauveau or the minimum of 24% recalculated by Zuntz.

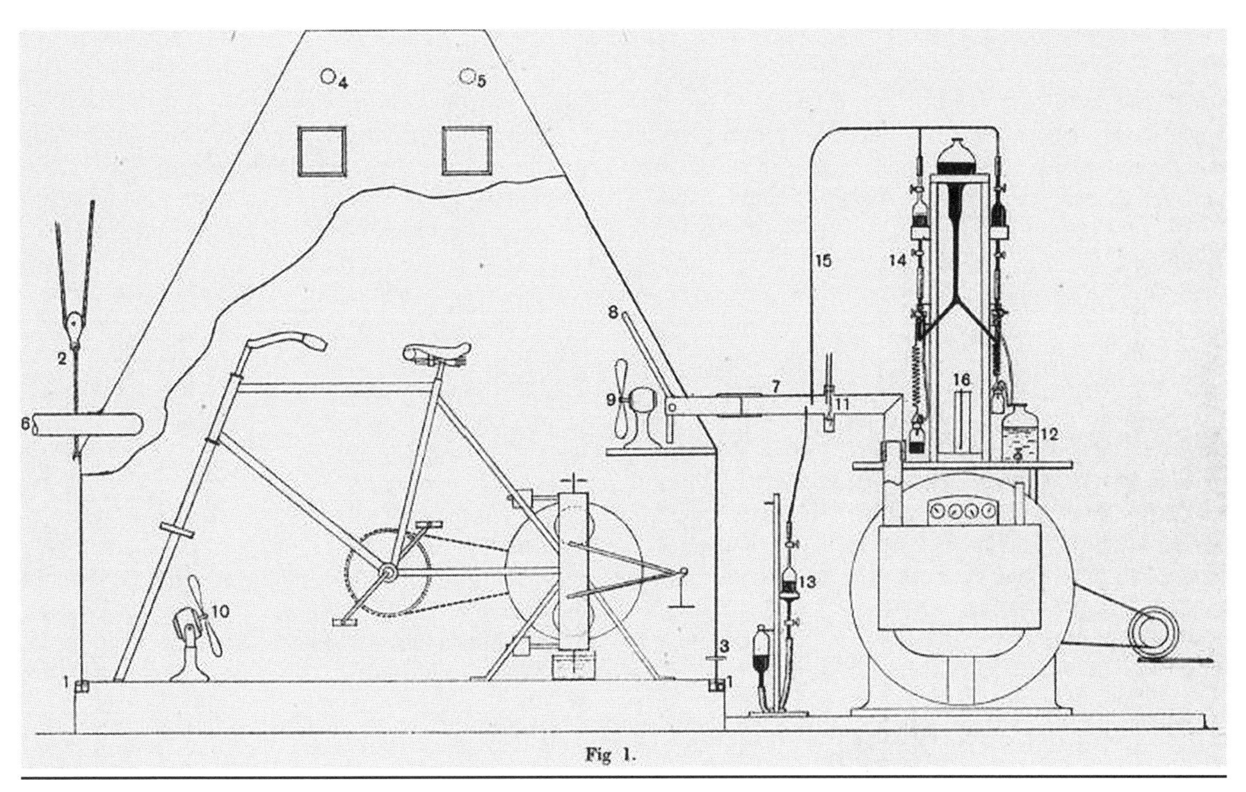

Figure 17.1 The cycle ergometer and Jacquet-type respiration chamber used by Krogh and Lindhard.
Reproduced, with permission, from A. Krogh and J. Lindhard, 1920, *Biochemical Journal* 14: 290-363. © the Biochemical Society.

Despite such data, Hill continued to believe that muscle was wholly dependent on carbohydrate as an energy source during exercise. This idea was further established by a series of studies of exercising humans performed in the early 1920s by Hill along with Lupton and C.N. Hugh Long (1901-1970) (100-102). In these experiments, Hill and colleagues related changes in blood lactate concentration to changes in $\dot{V}O_2$ during and especially after exercise at various intensities and while breathing various percentages of O_2. In particular, Hill was interested in reconciling the results of these human studies with the results of his and Meyerhof's previous experiments with isolated frog muscle, for which they shared a Nobel Prize in physiology or medicine in 1922. This was achieved via Hill's now well-known O_2 debt hypothesis, which proposed that the initial rapid decline in $\dot{V}O_2$ after exercise represented the oxidation of lactate in the muscles where it was formed whereas the later slower decline in $\dot{V}O_2$ represented the oxidation of lactate after it had escaped from such muscles via diffusion. However, on the basis of the recovery coefficient (i.e., the ratio of CO_2 retained—assumed to equal the amount of lactate oxidized—to O_2 consumed after exercise), it was held that much of the lactate formed during exercise was in fact not oxidized. This dovetailed the results of these human studies with Hill's and Meyerhof's prior calculations that only somewhere between one sixth to one fifth (Meyerhof) or one fourth to one third (Hill) of the lactate produced was oxidized after exercise and that the remainder was resynthesized into its precursor. Hill and Lupton (103) therefore concluded the following:

> The muscle is to be regarded as an accumulator of energy, energy available for rapid nonoxidative discharge, stored during previous oxidations. The transformation of glycogen into lactic acid, the action of the lactic acid on the muscle proteins, and the neutralization of the lactic acid by the alkaline buffers of the muscle, are the vehicle by which this stored energy is made manifest; during recovery the process is reversed at the expense of a portion of lactic acid oxidized. The accumulator has been recharged at the expense of oxidations required to run the dynamo. We must regard the muscle, therefore, as possessing two mechanisms: (a) the anaerobic one of discharge and (b) the oxidative one of recovery. (103, p. 139)

It is not entirely clear why Hill (and many others) continued to believe so strongly that exercising muscle could use only carbohydrate as substrate when the whole-body data of Zuntz, Benedict and Cathcart, and Krogh and Lindhard clearly suggested otherwise. In part, it may have been because studies of rabbit muscle in 1904 by Leathes (122) and frog muscle in 1915 by Winfield (189) failed to demonstrate any reduction in total fatty acid content as a result of electrical stimulation—although Leathes in fact concluded, "It is doubtful whether the utilization of fat in muscular activity can be proved by stimulating the muscles . . . but in other ways it has been established beyond doubt that muscles can and do make use of fat as a source of energy" (122). As well, Hill's misunderstanding may have simply stemmed from his differing conceptualization of the problem at hand. For example, in a lecture delivered to the Mayo Foundation and five universities in the autumn of 1924, he stated:

> It has long been discussed whether the breakdown of carbohydrate, rather than of other substances, is primarily responsible for the provision of energy in muscular contraction. It is known and accepted that work may be done, in the general melting-pot of the body, by the use of any kind of foodstuff. We are now concerned, however, specifically with the primary process of muscular contraction. In the complete chain of processes involved in long-continued exercise, this primary process may be disguised, or even apparently obliterated, by simultaneous transformations that take place between the different food constituents. (97, p. 505)

Thus, Hill's focus was clearly on the molecular events underlying the contractile process itself, which had yet to be elucidated but which he felt were intimately connected to the production of lactate from glycogen. As a result, he (mis)interpreted Krogh and Lindhard's observation of a slightly greater energy cost associated with the oxidation of fat as evidence that "the primary breakdown is of carbohydrate, and that fat is used only in a secondary manner, e.g., to restore the carbohydrate which has disappeared" (97, p. 506), even though the magnitude of the difference that was observed was much less than that predicted by Chauveau or even Zuntz.

The Hill-Meyerhof lactic acid theory of muscle contraction was finally displaced in the late 1920s by a series of breakthroughs that Hill himself described in 1932 as "the revolution in muscle physiology" (98). First among these was the almost simultaneous discovery in 1927 of phosphagen by Philip Eggleton (1903-1954) and Grace Palmer Eggleton (1901-1970) (64) and its identification as creatine phosphate (CrP) by Cyrus H. Fiske and Yellapragada SubbaRow (1895-1948) (68). Both sets of authors found that the compound decreased during exercise and increased during recovery. This, along with its high heat of hydrolysis as soon measured by Meyerhof and Suranyi (138), suggested that it might be an important energy source in muscle. In 1929, Karl Lohman (1898-1978) (124) and Fiske and SubbaRow (69) discovered adenosine triphosphate (ATP), which Lohman correctly hypothesized was the direct source of energy for muscle contraction, with CrP resynthesizing ATP via the famous Lohman reaction: CrP + adenosine diphosphate (ADP) $\leftrightarrow$ Cr + ATP) (125). Finally, in the early 1930s, Einar Lundsgaard (1899-1968) demonstrated that the muscles of frogs poisoned with iodoacetic acid were able to contract despite being unable to produce lactate (126-129), thus driving the final stake into the heart of the Hill-Meyerhof theory of muscle contraction. In 1887, Pohl had reported that the pH of frog muscles treated with bromoacetic acid did not decline during the resultant contracture (150), which in 1924 Schwartz and Oschmann had demonstrated was attributable to the absence of lactate production (162). However, due to the comprehensive nature and timing of Lundsgaard's experiments as well as the fact that they involved electrically induced contractions rather than chemically induced contractures, he is generally credited with bringing the end to the lactic acid era.

These findings, especially Lundsgaard's observations, finally laid to rest the belief that the lactate molecule was directly involved in the contractile process. Other aspects of the O_2 debt hypothesis, however, remained unrefuted. Thus, in 1933, Rodolfo Margaria (1901-1983), Harold T. Edwards (1897-1937), and David Bruce Dill (1891-1986) proposed that the more rapid, or alactacid, part of the biphasic decline in $\dot{V}O_2$ after exercise represented the oxidative resynthesis of ATP and CrP using energy derived from the combustion of "ordinary fuels" whereas the slower, or lactacid, part represented the O_2 cost of oxidizing approximately one tenth of the lactate formed to provide the energy required to resynthesize the remainder into glycogen (131). Margaria and colleagues assumed a value of one tenth rather than one fifth, as suggested by Hill and Meyerhof, because the smaller value was more consistent with their own data and fit with their inherent belief that resynthesis of glycogen from lactate was a highly energetically efficient process. This hypothesis was based on the newly available data described in the previous paragraph as well as the observation of a close linear relationship between the postexercise lactate concentration and the magnitude of the O_2 debt (but only once the latter exceeded approximately 3 L). Although Margaria and colleagues somewhat pointedly declined to speculate on the nature of the "ordinary fuels" oxidized to repay the alactacid portion of the O_2 debt, they concluded that "the lactacid mechanism has to be considered more like a mechanism of emergency" (131, p. 698) that is called on "only when there may be reasons to believe that the work is carried on in anaerobic conditions" (131, p. 714). The paper by Margaria and colleagues did much to perpetuate the mistaken belief that muscle relied extremely heavily, if not exclusively, on carbohydrate (glycogen) as an energy source during exercise and reinforced the misconception, based in part on studies of the Pasteur effect in yeast, that skeletal muscle produced lactate only when it was hypoxic.

Studies of the role of blood glucose availability in metabolism and fatigue also contributed during this time to the emphasis on carbohydrate as a source of fuel for exercising muscle. For example, in 1924 Levine and colleagues observed that hypoglycemia often developed during long-distance (i.e., marathon) running (123). The following year the same group reported that this hypoglycemia could be prevented, and performance seemingly improved, by ingesting additional carbohydrate before and during exercise (83). Similarly, in 1932 Dill and colleagues reported that when a dog was provided with only water to drink during prolonged treadmill exercise, circulating glucose levels decreased markedly and the animal was unable to continue after 3 to 6.5 h of running (57). However, when fed 20 g of glucose every hour throughout exercise or 40 g at the point of fatigue, the dog could continue running for at least 13 h. Dill and colleagues therefore concluded that fatigue during prolonged exercise was due to lack of bloodborne glucose as a muscular fuel.

Finally, in 1939, E. Hohwü-Christensen (1904-1996) and O. Hansen published a series of papers [which in fact was a supplement to a prior publication by Christensen, Krogh, and Lindhard (35)] in which they readdressed, in part, the issue of how variations in carbohydrate availability induced by different diets influence the mixture of fuels oxidized during exercise (30-34). As previously found by Zuntz (191), Benedict and Cathcart (11), Krogh and Lindhard (118), and others (17, 63), Christensen and Hansen observed that consuming a high-carbohydrate diet for 3 to 7 d beforehand enhanced the rate of carbohydrate oxidation and postponed the onset of fatigue during moderate-intensity exercise. On the other hand, consuming a low-carbohydrate diet for the same period of time markedly reduced the rate of carbohydrate oxidation during exercise and impaired exercise performance; fatigue was accompanied by ketosis and hypoglycemia that were severe enough to result in symptoms of neuroglucopenia. To differentiate between these possible causes of premature fatigue, additional experiments were performed in which two subjects were fed 200 g of glucose at the point of almost complete exhaustion. In response, blood glucose levels quickly increased, the neuroglucopenic symptoms disappeared, and the men were able to exercise for an additional hour. The rate of carbohydrate oxidation, on the other hand, did not change much either before the initial onset of fatigue or after glucose ingestion. In another experiment, one of the men ingested 200 g of glucose 3 h before exercise and then fatigued prematurely due to hypoglycemia and symptoms of neuroglucopenia even though the rate of carbohydrate oxidation was quite high. Although these findings led Christensen and Hansen to conclude, as did Bøje before them (18), that hypoglycemia must cause fatigue by way of its effects on the central nervous system and not by affecting muscle metabolism, the results of these experiments nonetheless dramatically demonstrated the effect of prolonged muscular exercise on bodily carbohydrate stores and especially glucose homeostasis.

By the onset of World War II the following picture had emerged. On the basis of experiments in which RER was measured, it was understood that although the contribution of carbohydrate oxidation to overall energy production increased curvilinearly with increasing exercise intensity, fat oxidation still played a significant role, at least at low to moderate exercise intensities. However, it was generally believed that skeletal muscle itself was heavily, if not exclusively, dependent on carbohydrate as an energy source during exercise and that this need was met by circulating glucose during prolonged aerobic exercise that required much less than 100% of maximal O_2 uptake ($\dot{V}O_2max$) and by glycogen during "violent" exercise that required a greater fraction of $\dot{V}O_2max$. Fatty acids could not be oxidized directly by muscle but could be utilized only after being somehow converted into carbohydrate in other tissues (e.g., the liver). For example, in 1941 Arthur H. Steinhaus (1897-1970) wrote in *Annual Review of Physiology* that "if fats are used as fuel in muscles, it must be by some indirect route" (170, p. 702). Similarly, based on his own research (77) as well as a survey of the literature, in 1942 Gemmill wrote in *Physiological Reviews* that "the results of experiments on fat utilization during muscular work have demonstrated that this substance is used indirectly. There is no experimental evidence at the present time for the direct utilization of fat by mammalian muscle" (78, p. 49). Even Lundsgaard himself adhered to this view, stating in his lecture to the Harvey Society of New York in 1937 that "it is probable that the high-molecular fatty acids are not attacked, or not readily attacked oxidatively in the muscles" (130, p. 180). Instead, Lundsgaard believed that fatty acids were first converted to ketone bodies in the liver before being metabolized by exercising muscle (discussed later).

Lundsgaard's belief was largely based on experiments performed by his colleague in Copenhagen, Blixenkrone-Møller, in which the clearance of β-hydroxybutyrate by perfused cat hindquarters was measured (16). These studies demonstrated that even resting muscle had a large capacity to metabolize ketone bodies, which increased several-fold during electrical stimulation. Experiments by Barnes and Drury at around the same time also revealed a significant extraction of ketone bodies by dog, rat, and human muscle (10), whereas follow-up studies a few years later by Drury and colleagues in ketotic rats and humans (61) and by Neufeld and Ross in ketotic guinea pigs and humans (143) showed that exercise resulted in a temporary decline in ketone levels in the blood, which was followed by a subsequent increase. This pattern was assumed to reflect an exercise-induced increase in ketone utilization by muscle and ketone production by the liver, the latter becoming evident only during recovery. As a result of these and other studies, Lundsgaard's view of substrate metabolism during exercise became the prevailing wisdom for the next 15 to 20 yr—that is, until the recognition of the importance of nonesterified (free) fatty acids (FFAs) as a source of fuel for muscle (discussed later).

Substrate Utilization During Exercise: Advances From World War II to the 1960s

World War II and its aftermath temporarily slowed the pace of research in exercise physiology. Many budding or already notable scientists took leave of absence from their positions to contribute to the war effort, were forced to relocate, or lost their lives in the conflict. As well, the physical demands of military activity led to a strong shift by exercise physiologists away from studies of substrate metabolism and toward studies of fitness testing, cardiovascular function, heat and altitude acclimatization, ergogenic aids, and so on, as emphasized by C. Taylor and Peter V. Karpovich in their review articles on exercise in *Annual Review of Physiology* in 1945 (174) and 1947 (114), respectively. Notably, Karpovich also lamented that so few physiological studies were conducted in schools of physical education, which he considered the logical place for such research. Although he expressed hope that this situation would soon be remedied such that physical education programs would have a sound scientific basis, it appears to this author that Karpovich's hope has to this day been only partially fulfilled.

On the other hand, research in the more basic fields of biochemistry and muscle physiology forged ahead during the immediate postwar years. It was during this time period, for example, that the details of major metabolic pathways such as the tricarboxylic acid cycle were fully elucidated, aided in part by the availability, in the new nuclear era, of ^{14}C-labeled compounds as described by Sir Hans Krebs (1900-1981) in his 1953 Nobel lecture (117). This increased understanding of basic biochemistry paved the way for an explosion in knowledge regarding metabolism during exercise in the late 1960s and beyond (discussed later). As well, the enhanced availability of radioactively labeled compounds and advances in methods for their analysis and data interpretation saw the use of such materials steadily spread from studies of cell extracts and isolated cells to studies of whole organisms, including exercising humans. This pattern essentially repeated itself in the 1990s with the increased availability of stable isotopically labeled compounds and improved methods for their analysis.

It was also during the decade or so after World War II that biochemists definitively answered the second of two questions posed at the outset of this chapter: What is the immediate source of energy fueling muscle contraction? Since the work of Lohman in the 1920s and 1930s, it had become generally accepted that ATP served this role. Due to the rapid rephosphorylation of ADP by CrP, however, it had not been possible to irrefutably demonstrate a decrease in ATP in response to a single twitch or even in response to repeated muscle contractions except at the point of extreme fatigue. This led Hill to publish his famous paper titled "A Challenge to Biochemists" in 1950 (99) in which he pointedly emphasized this fact and suggested that yet another revolution might still occur and that some other substrate might displace ATP in our understanding in precisely the same way that ATP had displaced phosphagens (CrP) and phosphagens had displaced lactate previously.

To address this issue, Hill proposed that biochemists attempt to measure changes in ATP in response to a tetanus of tortoise muscle in which the very slow speed of contraction (i.e., one fifteenth that of frog muscle) and hence ATP turnover could be further reduced by lowering its temperature to 0 °C to 5 °C. Utilizing precisely this approach, however, neither Münoh-Petersen in 1953 (142) nor Mommaerts and colleagues in 1955 (139) were able to provide absolute proof that ATP was directly utilized. On the other hand, that same year

Table 17.1 Changes in ATP, ADP, and AMP Concentrations in Response to a Single Contraction of Frog Muscle Treated with 1-Fluoro-2,4-dinitrobenzene to Inhibit Creatine Kinase

Condition	ATP (μmol/g)	ADP (μmol/g)	AMP (μmol/g)
Mean value at rest	1.25	0.64	0.10
Mean value after contraction	0.81	0.90	0.24
Change ± SEE*	−0.44 ± 0.05	+0.26 ± 0.02	+0.14 ± 0.03

*n = 9 experiments.
ATP = adenosine triphosphate; ADP = adenosine diphosphate; AMP = adenosine monophosphate.
Adapted from Cain and Davies 1962.

Lange reported decreases in ATP and CrP and increases in ADP during potassium chloride- and acetylcholine-induced contractures of frog rectus abdominis muscle that had been treated with iodoacetic acid and mustard gas (119). These results, however, were questioned by Davies and colleagues because the contractions took several seconds and significant amounts of adenosine monophosphate (AMP) were found, suggesting that damage to the muscle occurred while it was being removed from its support and processed in a high-speed blender (56). Thus, in 1962 Cain and Davies used 1-fluoro-2,4-dinitrobenzene to potently and irreversibly inhibit creatine kinase in frog muscle, thus enabling them to demonstrate small but significant reductions in ATP and increases in ADP as a result of a single contraction (table 17.1) (22). This study therefore conclusively established, as had been believed for quite some time, that "ATP is the primary energy source and CrP is the secondary energy source for contraction of working muscle" (22, p. 366).

The late 1950s and early 1960s also saw renewed interest in the overall pattern of substrate metabolism during exercise, driven in part by the realization that the FFAs long known to be present in plasma were potentially important energy sources for many tissues rather than artifacts arising from in vitro lipolysis of plasma triglycerides (TG) during sample processing. For example, using measurements of the arteriocoronary sinus difference in O_2 content and FFA concentration in cardiac patients, Gordon and Cherkes in 1956 estimated that up to 70% of the O_2 uptake of the heart was used to oxidize FFAs (84). Using samples provided by Andres and colleagues (3), Gordon and colleagues also measured the arteriovenous difference in FFAs across the forearm of normal volunteers (85). Their analyses revealed a significant extraction of FFAs in 15 out of 22 sample pairs. These observations, coupled with the findings of Andres and colleagues (3) that the uptake of glucose by the forearm could account for only 7% of simultaneously measured O_2 uptake and studies of isolated rat skeletal muscle (e.g., that by Fritz and colleagues in 1958) in which high rates of $^{14}CO_2$ production from ^{14}C-labeled FFAs were observed under a wide variety of conditions, including electrical stimulation (76), finally cemented the importance of FFAs as an energy source for muscle, at least in the basal state.

It was not until 1960, however, that data first emerged that provided clear-cut evidence of increased utilization of FFAs during exercise in humans. In that year, Friedberg and colleagues reported the results of experiments in which a bolus of [1-^{14}C]palmitate was administered to healthy men performing 15 min of vigorous cycling exercise (74). The exercise led to a rapid reduction in plasma FFA concentration, which promptly rebounded during recovery. However, this was not due to a reduction in release of FFAs into the circulation but rather to an increased rate of utilization, as indicated by a near-doubling in the rate of [1-^{14}C]palmitate clearance. Taking into consideration the reduction in plasma FFA pool size, it was estimated that the exercise increased the flux of FFAs through plasma by 25% to 45%. Friedberg and colleagues obtained comparable results during lower-intensity exercise that could be maintained for up to 45 min (75) and used the forearm arteriovenous balance model to demonstrate that contractions increased the rate of conversion of [1-^{14}C]palmitate to $^{14}CO_2$ in human muscle (73). The findings of Friedberg and colleagues were soon confirmed using constant infusion of [1-^{14}C]palmitate in studies reported by Carlson and Pernow in Stockholm (25) and Havel and colleagues in San Francisco (88). Havel then spent 1962 and 1963 on sabbatical in Stockholm, working with Carlson and comparing the use of various tracers (i.e., [1-^{14}C]palmitate, [U-^{14}C]palmitate, [9,10-^{3}H]palmitate, [1-^{14}C]oleate, and [1-^{14}C]linoleate]) to measure FFA turnover during exercise (87). With the exception of [1-^{14}C]linoleate, similar results were obtained regardless of the tracer employed, further establishing the relevance of the prior (and subsequent) results obtained using palmitate labeled in the first position with ^{14}C (or, later, ^{13}C).

Just as importantly, the results of this and previous studies by Carlson and colleagues and later by others (e.g., 149) using ^{14}C-labeled FFA tracers indicated that although exercise markedly increased the turnover and oxidation of this energy source, FFAs still accounted for only approximately 50% of the fat that was being oxidized, based on data obtained from indirect calorimetry. In an attempt to account for the "missing" fat, Carlson and colleagues first determined the effects of exercise on the rate of clearance of a commercial chylomicron-like TG emulsion from the circulation (24). Observing no change in this rate, Carlson and Fröberg measured the TG and phospholipid contents of the muscles of rats before and after both electrical stimulation and treadmill running (23). Although the phospholipid content of skeletal muscle remained constant in both cases, the TG content was significantly reduced, thus demonstrating the importance of this intramuscular source of lipids during contractile activity. However, due to the difficulty of accurately quantifying small changes in muscle TG levels [as previously recognized by Leathes (122)] or an inadequate increase in metabolic demand [as concluded by Barclay and Stainsby (8)], in 1966 Masoro and colleagues found no change in the TG content of the muscles of monkeys subjected to 5 h of electrical stimulation at 3 Hz (133). The contribution of such lipid stores to overall energy metabolism therefore remained controversial for many years, and only in the past decade or so has any sort of consensus been reached (182).

Thus, with the end of the 1960s approaching the pendulum had seemingly swung so far that FFAs had supplanted not only ketone bodies but even carbohydrate as the primary fuel thought to be used by exercising muscle, at least during aerobic exercise that did not result in a significant increase in lactate production. For example, in 1963 Havel and colleagues (88) concluded that FFAs "are the major circulating metabolites burned by working muscle in the postabsorptive state" (p. 1054), whereas in 1965 Rowell and colleagues (158) wrote, "Although for many years investigators were reluctant to accept the phenomenon of fatty acid oxidation by skeletal muscle, recent evidence has allotted this substrate a major role in skeletal muscle metabolism. Indeed fatty acid oxidation is considered the primary if not the sole energy source in exercising men" (p. 1032). Glycogen was still held to be important during high-intensity exercise but was thought to be mobilized primarily in response to tissue hypoxia, and the resultant production of lactate was thought to limit the contribution of FFAs to energy production by inhibiting lipolysis. On the other hand, other substrates (plasma-borne glucose in particular) were relegated to a relatively minor role. For example, based on experiments in which [U-^{14}C]glucose (which underestimates glucose flux due to tracer recycling) was infused into dogs during prolonged treadmill exercise, Paul and Issekutz (149) concluded that "the rates of turnover and oxidation of plasma glucose play only a minor role in exercise metabolism, and they cannot make up the difference between the energy required to perform the work and the energy supplied by the adipose tissue in form of FFA" (p. 621). Along the same lines, based on studies using the arteriovenous balance approach, Havel and colleagues (89) estimated that "uptake of blood glucose in leg tissues could account for no more than 16% [of total energy expenditure] and a quantity of carbon equivalent to one-third of glucose uptake [is] released as lactate" (p. 95).

Substrate Utilization During Exercise: Advances From the late 1960s to Approximately 1990

The mid- to late 1960s proved to be yet another important watershed in the understanding of exercise metabolism. This was due in part to the reintroduction earlier in the decade of the percutaneous needle biopsy procedure by the Swedish nephrologist Jonas Bergström (1929-2001). Charrière and Duchenne first described the percutaneous needle muscle biopsy technique in 1865 (26). This technique enabled Bergström and his colleagues, primarily Eric Hultman but also others such as Lars Hermansen (1933-1984) and Bengt Saltin (1935–), to directly assess, for the very first time, the effects of exercise on muscle glycogen and high-energy phosphate metabolism in humans (12-14, 108). In one early experiment, for example, they obtained biopsy samples from the vastus lateralis muscles of healthy young men before and after 30 min of moderate-intensity cycle ergometer exercise while also measuring glucose exchange across the leg using the arteriovenous balance method (13). Muscle glycogen content decreased, on average, 29% during exercise, but the arteriovenous glucose difference decreased to 0 or even became negative. These results led Bergström and Hultman (13) to conclude that "muscle glycogen is the main carbohydrate source for muscle activity" whereas "glucose uptake from the blood [is] negligible" (p. 20). In another highly influential study, Bergström and colleagues examined the effects of dietary carbohydrate intake on muscle glycogen levels and exercise per-

Table 17.2 Pre-Exercise Muscle Glycogen Concentration and Time to Fatigue During Prolonged Exercise in Men Consuming Different Diets

Diet	Muscle glycogen (g/100 g)	Time to fatigue (min)
High protein and fat	0.63 ± 0.10	56.9 ± 1.7
Normal mixed	1.75 ± 0.15	113.6 ± 5.3
High carbohydrate	3.31 ± 0.30	166.5 ± 17.8

Adapted from Bergström et al. 1967.

formance (14). Muscle glycogen stores were found to be markedly increased when strenuous exercise was followed by several days of rest and consumption of a high-carbohydrate diet but were significantly reduced when a low-carbohydrate, high-fat diet was consumed instead. Changes in exercise performance (time to fatigue) closely paralleled these changes in initial muscle glycogen levels on both an individual (i.e., $R = .92$; $P < .001$) and group mean (table 17.2) basis. This frequently cited study provided the basis for the common practice of carbohydrate loading among endurance athletes and did much to establish the primacy of muscle glycogen as an energy source and glycogen depletion as an important factor in fatigue during prolonged exercise.

The window into human muscle metabolism during exercise provided by the biopsy approach also helped catalyze the further growth of exercise physiology research in the United States and helped fuel the careers of investigators such as Philip Gollnick (1934-1991) at Washington State University and David Costill (1932–) at Ball State University. Beginning in 1969, Gollnick and various colleagues used the biopsy method to address questions such as the effects of prolonged exercise on skeletal muscle ultrastructure (80), the pattern of motor unit recruitment during exercise (81), and the effects of endurance training on muscle enzyme activities and fiber type composition (79, 82). Gollnick was also an early leader in the burgeoning field of equine exercise physiology (157). Costill used the biopsy approach to document the metabolic demands of, and adaptations to, endurance exercise, especially the responses to distance running (52, 53). In a key study, Costill and colleagues demonstrated that, in subjects consuming a normal mixed diet, muscle glycogen concentration declined progressively in response to running 16.1 km/d for 3 d and in some subjects did not fully recover after 5 d of rest (53). Combined with the prior observations of Bergström and colleagues (14), the results of Costill and colleagues' study reinforced the importance of muscle glycogen as an energy source and demonstrated the need for endurance athletes to consume adequate dietary carbohydrate during routine training as well as when tapering for an important competition.

Paralleling the expansion of human research that started in the 1960s was a greatly increased interest in the biochemical responses and adaptations of muscle to exercise as studied in animal models, especially rats. This line of research was largely spearheaded by John O. Holloszy (1933–) (figure 17.2), who initially trained as a physician specializing in internal medicine, endocrinology, and metabolism. However, after 2 yr in the U.S. Public Health Service, during which he worked in the Physical Fitness Research Laboratory at the University of Illinois, Holloszy became fascinated by the improvements in performance ability and by the metabolic changes (e.g., reduction in plasma TG) resulting from repeated bouts of exercise. Holloszy therefore returned to Washington University in St. Louis in 1963 to pursue additional training in biochemistry. He joined the faculty in 1965 and for next 10 yr focused on elucidating the biochemical adaptations of skeletal (e.g., 6, 104, 106, 140) and cardiac (147, 177) muscles to chronic exercise. These studies revealed the phenomenal plasticity of skeletal muscle in response to increases in metabolic demand and established the critical importance of muscle mitochondrial content in determining the pattern of substrate utilization during exercise (described in more detail later). Holloszy's research endeavors later expanded to include studies of glucose transport in mammalian (rat) muscle and studies of the metabolic, hormonal, and cardiovascular responses and adaptations to exercise in humans, especially in patient populations such as the elderly or those with type 2 diabetes. For his research, in 2000 Holloszy was awarded the International Olympic Committee's Olympic Prize on Sports Sciences.

The remarkable adaptations observed in rodent skeletal muscle and shortly thereafter in human skeletal muscle (79, 82, 141, 178) in response to increases in physical activity led to renewed attention to the effects of such training on the pattern of substrate utilization during exercise. Based largely on cross-sectional studies in which RER or blood lactate were measured, it had been generally accepted since early in the century that trained individuals did not rely as heavily on bodily carbohydrate stores during exercise (see 38 for review). However, few if any formal studies of this issue appear to have been performed; most investigations of training focused instead on, for example, the cardiovascular responses to exercise. Indeed, before the 1970s the only longitudinal study of the effects of training on substrate metabolism during exercise appears to be the 1936 report of McNelly (135), who measured the RER of three athletes cycling at a constant intensity before, during, and after a period of training. In 1967, however, Hermansen working with Hultman and Saltin observed that the initial rate of glycogen utilization was 24% lower in trained men than in untrained men during exercise at 77% of $\dot{V}O_2$max (93). This glycogen-sparing effect of training was confirmed in subsequent cross-sectional and longitudinal studies of both rats (5, 71) and humans (66, 112, 113, 161), thus explaining, at least in part, the already well-known training-induced reductions in RER and blood lactate levels during exercise. Importantly, studies of rats demonstrated that the rate of muscle glycogen utilization during exercise was significantly and inversely correlated with muscle respiratory capacity when the latter varied over a twofold range in groups of rats that were sedentary or had been trained 10, 30, 60, or 120 min/d (73). Conversely, time to fatigue during moderate-intensity treadmill exercise was found to be positively associated with muscle respiratory capacity. These data added further evidence that muscle glycogen is by far the most important energy substrate for exercising muscle and that fatigue is often related to glycogen depletion.

Figure 17.2 John Otto Holloszy (1933–) of Washington University in St. Louis.
Photo courtesy of John O. Holloszy.

Notably, however, the study by Fitts and colleagues (71) as well as other studies published around the same time (5) also demonstrated that training reduces the rate of utilization of liver glycogen in rats during exercise. In fact, the glycogen-sparing effect of training was even more prominent in the liver than in muscle and was more closely correlated with muscle respiratory capacity (71). Such observations clearly implied that training reduced the dependence of rat muscle on plasma-borne glucose as well as intramuscular glycogen during contractile activity. This hypothesis was confirmed in 1983 by Brooks and Donovan, who used [U-^{14}C]glucose to quantify the turnover and oxidation of glucose in untrained and trained rats during treadmill running (58). On the other hand, previous one-leg training studies conducted in humans failed to show any difference in glucose uptake between the trained and untrained legs during two-leg exercise at 65% to 70% of $\dot{V}O_2$max (161). Presumably, however, these results reflected the difficulty in accurately quantifying small changes in glucose uptake using the arteriovenous balance method. Thus, in 1990 the present author along with Kohrt, Spina, Bier, and Holloszy used a primed constant infusion of the nonrecycling tracer [U-^{13}C]glucose to measure glucose turnover and oxidation in human subjects before and after 3 mo of endurance-exercise training (43). Consistent with prior studies in rats, this study demonstrated that training reduced the rates of glucose turnover and oxidation by approximately one third (figure 17.3); this decrease in glucose use accounted for roughly one half of the reduction in overall carbohydrate oxidation resulting from training. This preferential sparing of glucose was confirmed in subsequent studies (50, 136), which also demonstrated that this adaptation (a) develops quite rapidly (i.e., within the first 10 d of training) (136) and (b) is evident not only during exercise performed at the same absolute intensity (i.e., the same power output or $\dot{V}O_2$) before and after training but also during exercise performed at the same relative intensity (i.e., the same percentage of $\dot{V}O_2$max) in the untrained and trained states (45). In addition, it has been shown that the training-induced reduction in glucose production during exercise in humans is the result of an attenuated rate of gluconeogenesis as well as a decline in the rate of glycogenolysis (50). This diminished reliance on plasma glucose for energy after training contributes to the training-induced improvement in endurance by protecting against the development of hypoglycemia.

Due to the emphasis at the time on the importance of plasma-borne FFAs as an energy source for muscle, it was initially assumed that the previously described

Table 17.3 Effect of Endurance-Exercise Training on Utilization of Intramuscular Triglycerides

Time of biopsy sample	Before training	After training
Before exercise	59.2 ± 7.7	63.3 ± 17.7
After exercise	46.4 ± 8.9	37.2 ± 12.3*
Decrease	12.7 ± 5.5	26.1 ± 9.3†

Values are mean ± standard deviation for n = 9.
Significant difference (after training versus before training): *$P < .05$; † $P < .001$.
Adapted from Hurley et al. 1986.

training-induced reduction in carbohydrate utilization during exercise was compensated for by, and was at least partially due to, an increased uptake and oxidation of FFAs (105, 112). In fact, Issekutz and colleagues reported in 1965 that the rate of FFA turnover in exercising mongrel dogs was inversely related to their blood lactate level and, hence, positively related to their aerobic fitness (111). Similarly, when comparing the results they obtained in untrained men (87) with what they had previously measured in athletes exercising at a similar absolute intensity (88), Havel and colleagues also noted that the trained subjects derived more of their energy from FFAs. Somewhat paradoxically, however, in longitudinal experiments plasma FFA concentrations were observed to increase less during exercise performed at the same absolute intensity after training compared with before training (187, 188). Although in theory this could have been due to a training-induced increase in the rate of FFA clearance from plasma, circulating glycerol concentrations were also

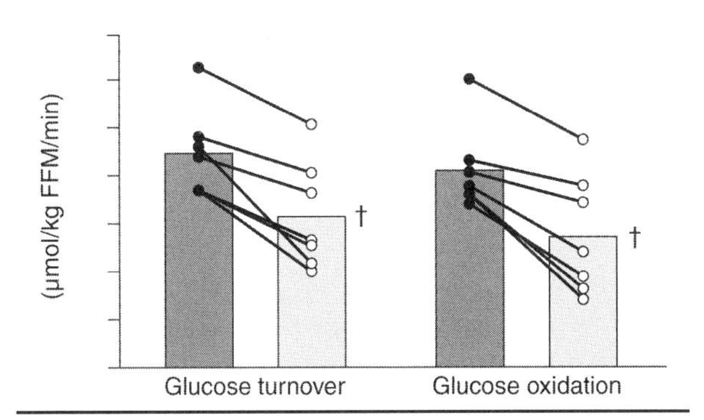

Figure 17.3 Effect of endurance training on plasma glucose turnover and oxidation during prolonged exercise. FFM = fat-free mass. Bars depict group means (dark bars = pretraining; light bars = posttraining); circles (open = pretraining; closed = posttraining) and lines depict individual subjects. †$P < .001$.

Reprinted from A.R. Coggan, 1990, "Endurance training decreases plasma glucose turnover and oxidation during moderate intensity exercise in men," *Journal of Applied Physiology* 68: 990-996. With permission of American Physiological Society.

observed to be lower after training (187, 188), implying that the reduction in FFA concentration was due to a reduced rate of lipolysis. The authors therefore hypothesized "on the basis of the apparently lower rates of lipid mobilization after training, that the trained individual obtains a greater proportion of his energy requirement during long-term submaximal exercise from intramuscular lipid stores" (188, p. 770). Thus, in 1986, Hurley working in Holloszy's laboratory measured the TG content of carefully dissected (to remove possible intramuscular adipocytes) muscle biopsy samples obtained from subjects before and after a bout of prolonged exercise performed in the untrained and trained states (109). The results of this study revealed an approximate doubling in the net utilization of this important energy source with training (table 17.3). Other longitudinal and cross-sectional studies using stable isotopic tracers also support the conclusion that training increases the utilization of intramuscular TG during exercise at the same absolute (115, 132, 166) and even the same relative (44) intensity. Under the latter condition, however, the overall rate of lipolysis (as judged from the rate of appearance of glycerol) is greater in the trained state (44, 116), resulting in higher FFA concentrations and hence a greater utilization of this lipid source (44).

Along with the effects of training, another major theme of metabolic research during the period between the late 1960s and approximately 1990, especially during the 1980s, was the effect of the ingestion of carbohydrate (and other substrates) during exercise on metabolism and performance. As previously discussed, it had long been accepted that providing glucose to exercising animals or humans could delay fatigue. Based on the studies of Levine and colleagues (123), Dill and colleagues (57), and Christensen and Hansen (33), this delay was generally believed to be due to prevention of hypoglycemia and symptoms of neuroglucopenia. In 1982, however, Felig and colleagues of Yale University reported in *New England Journal of Medicine* that glucose ingestion did not delay fatigue in untrained men exercising at 60% of $\dot{V}O_2max$ (67). This stimulated Coyle (figure 17.4) and colleagues working in Holloszy's laboratory to reinvestigate the effects of carbohydrate feedings during exercise on metabolism and performance (54). Using trained cyclists exercising at a higher intensity (i.e., 75% of $\dot{V}O_2max$) more representative of athletic competition, Coyle and colleagues found that consumption of glucose polymers delayed the onset of fatigue in 7 out of the 10 subjects tested. Because only 2 of the subjects exhibited symptoms of neuroglucopenia, it was hypothesized that postponement of fatigue was due to a muscle glycogen-sparing effect of the feedings. Contrary to this hypothesis, however, in a subsequent study carbohydrate ingestion was found to not alter the rate of muscle glycogen utilization during exercise to fatigue at 70% to 75% of $\dot{V}O_2max$ (55). Instead, it appeared that such feedings delayed fatigue by maintaining plasma glucose availability late in exercise, when muscle glycogen concentrations were very low. Thus, for his dissertation under Coyle's supervision, the present author essentially replicated Christensen and Hansen's classic experimental design by restoring glucose concentrations to normal via either carbohydrate ingestion (40, 41) or glucose infusion (40) late in prolonged exercise or at the point of fatigue.

Carbohydrate ingestion at the point of fatigue often failed to re-establish and maintain euglycemia. On the other hand, both carbohydrate feeding late in exercise and glucose infusion at the point of fatigue did so. This increase in glucose availability was accompanied by an increase in the rate of carbohydrate oxidation and postponement or even reversal of fatigue. Notably, the rate of glucose infusion required to maintain plasma glucose concentrations during continued exercise averaged more than 1.1 g/min and apparently provided more than three quarters of total carbohydrate oxidation during this time period. These data, along with the careful and detailed arteriovenous balance experiments by Felig and Wahren in collaboration with Ahlborg and colleagues at the Karolinska Institute in Stockholm in which the exchange of glucose and other substrates was determined across the legs and splanchnic bed of exercising men (1, 2, 181), helped illustrate the importance of circulating glucose as a source of fuel for muscle, as originally reported by Chauveau and Kaufmann approximately 100 yr before (28). Indeed, it is now well established that plasma-borne glucose normally provides 20% to 50% of total oxidative energy production and 25% to 100% of the total carbohydrate oxidized during prolonged exercise (see 37 for review). Moreover, the studies of Coyle and colleagues also clearly demonstrated that fatigue during prolonged exercise is normally the result of both depletion of muscle glycogen and a decline in plasma glucose availability and is not due to muscle glycogen depletion alone.

Figure 17.4 Edward F. Coyle (seated) of the University of Texas in the United States along with former and current doctoral students Andrew R. Coggan, Scott J. Montain, Marc T. Hamilton, Ricardo Mora-Rodriguez, Ted W. Zederic, and Matthew D. Pahnke (standing, left to right).
Photograph courtesy of Edward F. Coyle.

Substrate Utilization During Exercise: Advances from the Mid-1990s to Present

During the 19th and much of the 20th century, metabolic research in exercise physiology was focused primarily on the acute responses and chronic adaptations of healthy young men. With the approach of the new millennium, however, greater consideration was given to such issues in other subject populations (e.g., women, the elderly, those who are obese or diabetic). As well, ever-increasing attention was paid to unraveling the mechanisms responsible for regulating substrate use in these contexts. These trends were driven in large part by societal issues (i.e., the growing participation of women in sport, the aging of the overall population, and especially the dramatically increasing rates of diseases of inactivity such as obesity and type 2 diabetes) in Western countries and advances in the more basic biological sciences—in the fields of molecular biology and genetics in particular—that led to a more reductionist approach.

As an example of the former trend, in 1990 Tarnopolsky working in MacDougall's laboratory initiated an extensive series of investigations examining the effects of sex on substrate metabolism during and after acute exercise and in response to chronic training (173). Tarnopolsky's studies were not the first in this area but were better controlled than previous research. These studies established that, compared with men, women tended to rely slightly less on intramuscular glycogen and plasma glucose and slightly more on intramuscular TG and plasma FFAs during endurance exercise. Leucine turnover and oxidation were also found to be lower in women than in men during exercise. Similar results have been obtained by others (159). This difference between men and women in the pattern of substrate utilization during exercise appears to be due in part to a sex hormone-mediated enhancement of the biochemical pathways of lipid mobilization and utilization because short-term administration of 17β-estradiol to men tends to induce similar changes (173). In keeping with this interpretation, in women the rate of plasma glucose utilization and the overall rate of carbohydrate oxidation have been found to be lower during the luteal phase of the menstrual cycle than during the follicular phase (190). Conversely, aging has been found to result in a greater dependence on muscle glycogen and plasma glucose for energy during exercise along with a corresponding reduction in the rate of fat oxidation (163). This appeared to be the result of an age-related reduction in muscle mitochondrial respiratory capacity (39, 48), which impaired the ability of aged muscle to maintain energetic homeostasis during exercise (39). However, both the age-related decrease in muscle enzyme levels (49) and the age-related alterations in substrate metabolism during exercise (164) were at least partially reversible with endurance training, suggesting that they were due to a reduction in habitual physical activity with age and did not reflect a primary effect of aging.

Another area of research that has received considerable attention from exercise physiologists in recent decades has been the effect of obesity or type 2 diabetes on substrate metabolism. Although much of the focus in this area has been on the effects of physical activity on metabolic responses at rest, a number of studies (reviewed in 107) have examined the response of obese or diabetic individuals to exercise. In general, these studies have revealed a marked reduction in the overall capacity of muscle to oxidize fatty acids both at rest and during exercise; this reduction was not reversible upon weight loss (176). At the same time, the uptake of FFAs and their incorporation into intramuscular TG or conversion into downstream metabolites such as ceramide or diacylglycerol appeared to be enhanced in obese persons and type 2 diabetics, likely contributing to the insulin resistance exhibited by such individuals (107).

With regard to mechanisms, in recent years much effort has been expended attempting to elucidate how muscle contractions (or insulin) increase the permeability of muscle to glucose and thereby increase glucose uptake. This has facilitated advances in our understanding of the signaling pathways involved, as recently reviewed by Röckl and colleagues of Harvard University (156). Although much of the interest in this area has originated from its potential importance in insulin-resistant states, the enhanced knowledge regarding the regulation of glucose transport into muscle has greatly increased our understanding of substrate metabolism during exercise. For example, demonstration in 1998 by Hayashi and colleagues (90) of an important role of AMP-activated protein kinase (AMPK) in mediating contraction-induced translocation of glucose transporter 4 (GLUT4) to the sarcolemma provided a potential explanation for the curvilinear relationship between exercise intensity and glucose uptake; both AMPK activity and glucose uptake generally parallel changes in the AMP:ATP ratio in muscle (29). Similarly, attenuation of the exercise-induced increase in AMPK activity with training (134) may explain the somewhat paradoxical finding that even though training increases total muscle GLUT4 content, fewer GLUT4 are translocated to the sarcolemma during exercise in the trained state (155), thus accounting for a slower rate of glucose utilization after training. Training-induced changes in plasma glucose metabolism during exercise are therefore now mechanistically linked to the accompanying alterations in cellular energetics and hence to the increase in mitochondrial respiratory capacity that occurs with training, as originally hypothesized (47). The increase in total GLUT4 that accompanies training therefore appears to be an adaptation aimed more at enhancing the rate of glycogen resynthesis in the postexercise state than at increasing the rate of glucose utilization during exercise itself.

Significant advances have also occurred in the past couple decades in our understanding of the regulation of fatty acid metabolism at the cellular level. For many years it was assumed that FFAs were taken up by muscle and other tissues via simple diffusion across the lipid bilayer of the plasma membrane. This assumption led to the widespread belief that the rate of FFA oxidation during exercise was determined largely, if not entirely, by the plasma concentration of this substrate. When the exercise intensity was low to moderate, adipose tissue lipolysis released FFAs at ever-increasing rates during prolonged exercise, leading to a progressive increase in plasma FFA levels and hence in their rate of oxidation. When the exercise intensity was higher and the duration shorter, inhibition of lipolysis due to an increase in lactate or the accompanying change in pH or entrapment of FFAs in adipose tissue due to redirection of blood flow led to a decline in FFA levels and hence limited their contribution to energy metabolism during exercise. The perspective that exercising muscle is largely a passive consumer of whatever FFAs are presented to it was reinforced in 1963 by the famous glucose–fatty acid cycle hypothesis of Randle and colleagues (151), which proposed that an increase in FFA availability and hence oxidation inhibited glucose utilization via accumulation of citrate, which inhibited phosphofructokinase and hence caused an elevation of glucose-6-phosphate, which in turn inhibited glucose phosphorylation by hexokinase. This was especially true after Rennie and Holloszy reported that the glucose–fatty acid cycle was operative not only in cardiac and diaphragm muscle as originally reported by Randle and colleagues but also in adequately oxygenated perfused skeletal muscle (154). As a consequence, numerous putative FFA-elevating and hence glycogen-sparing treatments (e.g., heparin injection to activate lipoprotein lipase activity, caffeine ingestion) were studied by various investigators, albeit with decidedly mixed results. The glucose–fatty acid cycle was also often invoked to explain the training-induced reduction in carbohydrate utilization during exercise (105, 112).

Perspectives on the factors regulating lipid utilization by muscle during exercise began to change in the 1990s. This was due in part to the demonstration by Turcotte and colleagues that, as had previously been found in cultured adipocytes, hepatocytes, and cardiac myocytes, uptake of palmitate by perfused skeletal muscle appeared to be saturable when plotted as a function of the unbound instead of the total palmitate concentration (figure 17.5) (177). This clearly implied some sort of transport-mediated process, leading Bonen and colleagues (20) to investigate this issue using a giant sarcolemmal vesicle preparation. As in the perfused hindquarter, vesicular uptake of palmitate was found to follow Michaelis-Menten kinetics (K_m), with a K_m of approximately 6 nM in vesicles prepared from either red or white muscle. On the other hand, the V_{max} of palmitate uptake was approximately twice as high in vesicles from red muscle as in vesicles from white muscle, in keeping with the vesicles from red muscle's approximately twofold greater content of the putative fatty acid transporters fatty acid translocase (FAT/CD36) and plasma membrane fatty acid binding protein ($FABP_{PM}$). Subsequent studies firmly established that the uptake of FFA by skeletal muscle is acutely regulated in a manner quite similar to the uptake of glucose (i.e., via the translocation of transport proteins from intracellular storage pools to the sarcolemma and back again, as reviewed by Bonen and colleagues; 19). Moreover, endurance training has been recently shown to enhance the capacity of skeletal muscle to metabolize FFAs, in part by increasing the expression of FAT/CD36 and $FABP_{PM}$ (172). Thus, rather than being an entirely passive process, FFA uptake by muscle is now recognized as being under both short-term and long-term control. At the same time, other studies have shown that although muscle citrate levels are higher during exercise in the trained state (46, 112), glucose-6-phosphate concentrations are actually lower (46, 51), demonstrating that "the training-induced reduction in carbohydrate utilization during prolonged submaximal exercise occurs by modification of the glycogenolytic/glycolytic flux before the phosphofructokinase step and is not due to inhibition of phosphofructokinase by citrate or by other metabolites" (46, p. E220). Thus, although the glucose–fatty acid cycle appears to operate in exercising skeletal muscle under at least some conditions (154), this mechanism does not explain the lesser dependence on muscle glycogen and plasma glucose during exercise in the trained state.

Finally, data obtained in the past two decades have revealed new pathways by which carbohydrate and lipid metabolism interact in various tissues, including exercising skeletal muscle. In particular, significant attention has been focused on the role of malonyl-CoA, the product of acetyl-CoA carboxylase and a potent inhibitor of carnitine palmityltransferase I (CPT-I). Based on studies demonstrating that CPT-I in skeletal muscle is especially sensitive to inhibition by malonyl-CoA, in 1989 Winder and colleagues assessed changes in muscle malonyl-CoA levels in response to acute treadmill exercise in rats (185). Malonyl-CoA levels were found to decrease by nearly two thirds, thus presumably disinhibiting CPT-I and contributing to the exercise-induced increase in fat oxidation. Subsequent studies from Winder's group revealed that the decline in malonyl-CoA occurs in a time- (186) and intensity- (152) dependent manner during exercise but was attenuated by the infusion of glucose (65). Observations such as these led to the proposal that a reverse glucose–fatty acid cycle is operative during exercise (i.e., that increases or decreases in the rate of carbohydrate oxidation decrease or increase the rate of fat oxidation via changes in malonyl-CoA) (152, 165, 184). This hypothesis is seemingly supported by the results of studies in which [1-^{14}C]octanoate and [1-^{13}C]palmitate were infused simultaneously into exercising humans to quantify the relative oxidation of medium- and long-chain fatty acids, respectively (166, 167). In these studies, the percentage of the long-chain fatty acid tracer that was oxidized decreased with increasing exercise intensity in untrained individuals (167) but was higher in trained subjects than in untrained subjects (166). On the other hand, the percentage oxidation of the medium-chain fatty acid tracer did not vary significantly across conditions, which was interpreted to mean that the differences in long-chain fatty acid oxidation were due to inhibition of CPT-I, presumably by malonyl-CoA. In contrast to the [1-^{13}C]palmitate, however, it now seems likely that the vast majority of

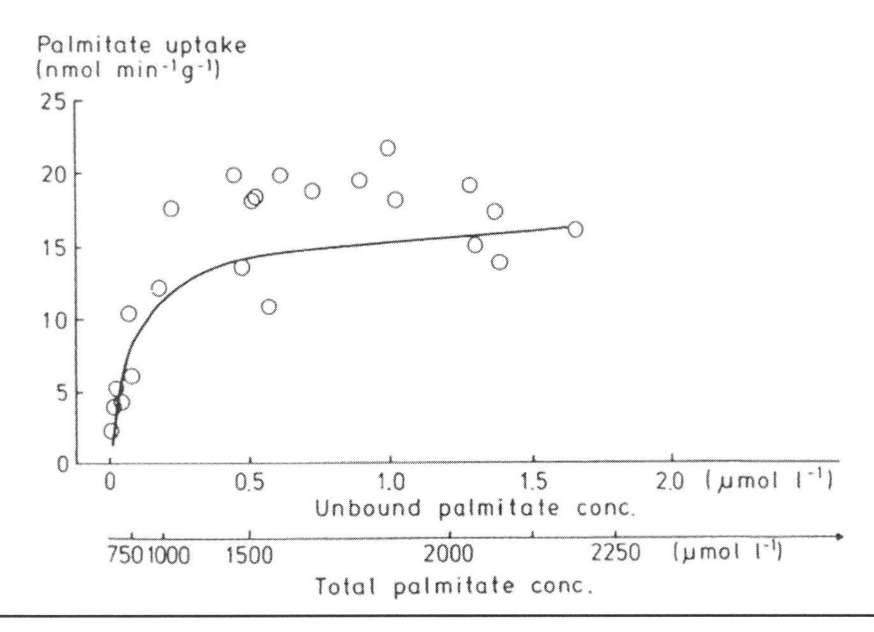

Figure 17.5 Relationship of palmitate uptake to unbound palmitate concentration in perfused rat hindquarters.

Reprinted from *FEBS Letters*, Vol. 279, L.P. Turcotte, B. Kiens, and E.A. Richter, "Saturation kinetics of palmitate uptake in perfused skeletal muscle," pgs. 279-327, copyright 1991, with permission of Elsevier.

the [1-^{14}C]octanoate was oxidized by tissues other than skeletal muscle (presumably the liver), making the results of these whole-body studies difficult to interpret (42). Furthermore, other studies have demonstrated that malonyl-CoA levels in human muscle change only transiently, if at all, during exercise (144, 145). Finally, the sensitivity of muscle CPT-I to inhibition by malonyl-CoA is even higher in trained persons than in untrained persons (169). Hence, in humans the reciprocal relationship that exists between carbohydrate and fat utilization during exercise both before and after training must be due to some mechanism other than malonyl-CoA-mediated changes in CPT-I activity. As recently reviewed by Sahlin (160), a number of alternative hypotheses have been proposed: inhibition of CPT-I by lactate or H^+; reduced flux through CPT-I due to a decline in free carnitine levels, stemming from overproduction of acetyl-CoA via pyruvate dehydrogenase and hence an increase in acetyl carnitine; a reduction in beta oxidation due to the resultant increase in the acetyl-CoA:free CoA ratio; and inhibition of beta oxidation by NADH produced via metabolism of pyruvate-derived acetyl-CoA in the tricarboxylic acid cycle. Further research is required to fully elucidate the precise mechanisms involved. Regardless, it now seems clear that the rate of carbohydrate utilization during exercise largely dictates the rate of fat utilization, not the reverse as originally believed. In turn, this implies that the increase in fat oxidation with training is primarily the result of the reduction in carbohydrate use, not vice versa as initially hypothesized (105, 112).

Summary

Table 17.4 lists some of the most significant studies of substrate metabolism during acute and chronic exercise that have been published during the past century. These studies, along with the other investigations mentioned in this chapter as well as experiments far too numerous to discuss, have yielded a highly detailed understanding of the pattern of substrate utilization during exercise. Consequently, the first question posed at the outset of this chapter (i.e., what substrates are in fact oxidized during exercise?) seems to have been largely answered. In contrast to what had been widely believed early in the century, it is now well established that no single substrate provides the energy to support muscle contraction. Rather, exercising muscle has been revealed to be a metabolic omnivore, utilizing both carbohydrate and fat derived from intramuscular stores and delivered via the plasma. The precise mix depends on a wide variety of factors, including the intensity, duration, and mode of exercise; the environmental conditions; and the habitual diet, nutritional state (i.e., fasted vs. fed), training status, sex, age, and health status of the individual in question. Carbohydrate in general (and muscle, especially in untrained individuals or during high-intensity exercise, with fatigue muscle glycogen in particular), however, still seems to be the preferred substrate of exercising during prolonged intense exercise that often corresponds to depletion of bodily carbohydrate stores.

Table 17.4 Historically Significant Studies of the Effects of Acute and Chronic Exercise on Substrate Metabolism (1910-Present)

Year	Authors and reference	Findings and significance
1913	Benedict and Cathcart (11)	Demonstrated that respiratory exchange ratio varied markedly with pre-exercise diet but that efficiency did not, thus providing evidence against Chauveau's hypothesis that fat must first be converted into carbohydrate before being oxidized by muscle.
1920	Krogh and Lindhard (118)	Confirmed the results of Benedict and Cathcart.
1923	Hill and Lupton (103)	Proposed the O_2 debt hypothesis linking changes in postexercise O_2 consumption to lactate oxidation.
1924	Levine at al. (123)	Observed that hypoglycemia often developed during prolonged running.
1927	Eggleton and Eggleton (64); Fiske and SubbaRow (68)	Eggleton and Eggleton discovered phosphagen, which Fiske and SubbaRow almost simultaneously identified as creatine phosphate.
1929	Lohman (124); Fiske and SubbaRow (69)	Lohman and Fiske and SubbaRow identified adenosine triphosphate, which Lohman correctly hypothesized was the direct source of energy for muscle contraction.
1930-1932	Lundsgaard (126-129)	Demonstrated that frog muscles poisoned with iodoacetic acid were able to contract despite being unable to produce lactate, thus finally refuting the Hill-Meyerhof theory of muscle contraction.
1932	Dill et al. (57)	Demonstrated in dogs that prolonged exercise led to hypoglycemia and fatigue, which could be prevented or reversed by feeding glucose throughout exercise or at the point of exhaustion, thus cementing the link between carbohydrate availability and exercise performance.
1933	Margaria et al. (131)	Modified Hill and Lupton's original O_2 debt hypothesis to include an alactacid component reflecting creatine phosphate resynthesis as well as a lactacid component due to lactate oxidation.
1936	McNelly (135)	Was the first to demonstrate, using a longitudinal study design, that endurance training reduces the rate of carbohydrate oxidation and increases the rate of fat oxidation during exercise at the same absolute intensity.
1938	Blixenkrone-Møller (16)	Demonstrated that resting muscle had a large capacity to metabolize ketone bodies, which increased several-fold during contractions. Combined with prior observations, this contributes to the widespread belief that muscle utilizes fatty acids only after they are partially oxidized in the liver.
1939	Christensen and Hansen (30-34)	Published a series of papers confirming and extending the earlier research of Benedict and Cathcart, Krogh and Lindhard, and Dill and colleagues.
1950	Hill (99)	Issued his famous challenge to biochemists that questioned whether adenosine triphosphate is in fact the direct source of energy during muscle contractions.
1957	Gordon et al. (85)	Demonstrated significant extraction of FFAs across the forearm of resting humans.

1958	Fritz et al. (76)	Using [1-^{14}C]palmitate, demonstrated increased oxidation of FFAs by isolated rat muscle in response to electrical stimulation.
1960	Friedberg et al. (74)	Using [1-^{14}C]palmitate, demonstrated that exercise increases the flux of FFAs through plasma in humans.
1961-1964	Carlson and Pernow (25); Havel et al. (88); Havel et al. (87)	Performed extensive studies using constant infusion of [1-^{14}C]palmitate and other ^{14}C-labeled fatty acid tracers during exercise, thus firmly establishing the importance of FFAs as an energy source.
1962	Cain and Davies (22)	Used 1-fluoro-2,4-dinitrobenzene to inhibit creatine kinase in frog muscle, making it possible to demonstrate changes in adenosine triphosphate during a single contraction and thus answering Hill's challenge.
1963	Randle et al. (151)	Proposed the glucose–fatty acid cycle by which oxidation of FFAs inhibits glycolysis and, hence, glucose uptake and oxidation via citrate-mediated inhibition of phosphofructokinase activity. This concept strongly influenced beliefs regarding the regulation of substrate metabolism during exercise for several decades.
1966-1971	Bergström and Hultman (13); Bergström et al. (14); Holloszy (106); Molé and Holloszy (140)	Bergström and colleagues used the muscle biopsy technique to quantify changes in muscle glycogen concentration during exercise in humans and demonstrated a very close relationship between initial glycogen levels and time to fatigue during prolonged exercise performed after low, moderate, or high carbohydrate intake. These findings emphasized the critical importance of muscle glycogen even during submaximal exercise and provided the scientific basis for the practice of carbohydrate loading among athletes. Holloszy and colleagues performed detailed studies of the effects of endurance training of rats on muscle mitochondrial respiratory capacity and enzyme activities, thus revealing the remarkable plasticity of muscle in response to alterations in chronic use.
1974-1975	Karlsson et al. (113); Baldwin et al. (5); Fitts et al. (71)	Using a longitudinal study design, Karlsson and colleagues demonstrated the muscle glycogen-sparing effect of endurance training in humans, thus providing a mechanistic explanation for the reduction in respiratory exchange ratio first observed by McNelly. Working in Holloszy's laboratory, Baldwin and colleagues demonstrated that training of rats reduces utilization of both muscle glycogen and liver glycogen during exercise. This diminished reliance on such stores is closely correlated with the training-induced increase in muscle respiratory capacity.
1977	Rennie and Holloszy (154)	Confirmed operation of Randle's glucose–fatty acid cycle in skeletal muscle, thus reinforcing the perspective that FFA availability was a primary determinant of the pattern of substrate utilization during exercise.
1986	Hurley et al. (109); Coyle et al. (55)	Hurley and colleagues were the first to demonstrate that the training-induced increase in fat oxidation during exercise is due, at least in part, to an increased reliance on intramuscular triglycerides. Coyle and colleagues reaffirmed the results of prior studies demonstrating that carbohydrate feedings during prolonged exercise can delay fatigue; however, they demonstrated that this was not due to a sparing of muscle glycogen.

(continued)

Table 17.4 *(continued)*

Year	Authors and reference	Findings and significance
1987	Coggan and Coyle (40); Constable et al. (51)	Coggan and Coyle confirmed Christensen and Hansen's classic observations that restoring plasma glucose availability late in exercise can reverse fatigue and showed that this was associated with an increase in carbohydrate oxidation. They also demonstrated using the euglycemic clamp technique that the rate of glucose uptake can exceed 1 g/min during the latter stages of prolonged intense exercise, thus helping reestablish the importance of this energy source. Constable and colleagues demonstrated that the training-induced reduction in carbohydrate utilization during electrical stimulation of rat muscle was accompanied by a reduction in glucose-6-phosphate levels, thus undermining the hypothesis that endurance training affects substrate use via Randle's glucose–fatty acid cycle.
1989-1990	Winder et al. (185, 186)	Reported that muscle malonyl-CoA levels decrease during exercise in rat muscle, thus relieving inhibition of carnitine palmitoyltransferase I and thereby increasing fatty acid oxidation.
1991	Turcotte et al. (177)	Demonstrated that FFA uptake by perfused muscle exhibits saturation kinetics when the rate of uptake is plotted as a function of the free instead of the total (predominantly albumin bound) FFA concentration, implying that FFAs are taken up via a specific transport mechanism instead of via simple diffusion as previously believed.
1993	Coggan et al. (46)	Demonstrated that muscle glucose-6-phosphate concentrations are lower in humans during prolonged exercise after training compared with before training even though muscle citrate concentrations are higher. This confirmed Constable's prior observations in electrically stimulated rat muscle and definitively ruled out the glucose–fatty acid cycle as the mechanism by which training reduces carbohydrate use during exercise.
1996	Odland et al. (144)	Reported that, in contrast to the responses observed in rat muscle, malonyl-CoA does not decrease during exercise in humans, thus indicating that other mechanisms must be responsible for regulating fatty acid oxidation in exercising humans.
1998	Hayashi et al. (90)	Demonstrated the importance of adenosine monophosphate-activated protein kinase in regulating glucose transport during contractions, thus providing a mechanistic link between alterations in cellular energetic and glucose uptake.
1998	Bonen et al. (20)	Used giant sarcolemmal vesicle preparation to demonstrate that palmitate transport into skeletal muscle follows Michaelis-Menten kinetics: The twofold difference in V_{max} between red and white muscle paralleled the twofold difference in fatty acid translocase (FAT/CD36) and plasma membrane fatty acid binding protein ($FABP_{PM}$). This study, along with the previous work of Turcotte and colleagues (177) and subsequent experiments, helped establish that the transport of FFAs into muscle is actively regulated in a manner similar to glucose rather than being due to passive diffusion.

FFA = free fatty acid.

References

1. Ahlborg G, Felig P. Lactate and glucose exchange across the forearm, legs, and splanchnic bed during and after prolonged exercise. *J Clin Invest* 69: 45-54, 1982.
2. Ahlborg G, Felig P, Hagenfeldt L, Hendler R, Wahren J. Splanchnic turnover during prolonged exercise in man. Splanchnic and leg metabolism of glucose, free fatty acids, and amino acids. *J Clin Invest* 53: 1080-1090, 1974.
3. Andres R, Cader G, Zierler KL. The quantitatively minor role of carbohydrate in oxidative metabolism by skeletal muscle in intact man in the basal state. Measurements of oxygen and glucose uptake and carbon dioxide and lactate production in the forearm. *J Clin Invest* 35: 671-682, 1956.
4. Asmussen E. Muscle metabolism during exercise in man. A historical survey. In: Pernow B, Saltin B, eds. *Muscle Metabolism During Exercise*. New York: Plenum Press, 1971, 1-12.
5. Baldwin KM, Fitts RH, Booth FW, Winder WW, Holloszy JO. Depletion of muscle and liver glycogen during exercise. Protective effect of training. *Pflügers Arch* 354: 203-212, 1975.
6. Baldwin KM, Klinkerfuss GH, Terjung RL, Molé PA, Holloszy JO. Respiratory capacity of white, red, and intermediate muscle: Adaptive response to exercise. *Am J Physiol* 222: 363-378, 1972.
7. Bang O. The lactate content of the blood during and after muscular exercise in man. *Skand Arch Physiol* 74(Suppl. 10): 49-82, 1936.
8. Barclay JK, Stainsby WN. Intramuscular lipid store utilization by contracting dog skeletal muscle in situ. *Am J Physiol* 223: 115-119, 1972.
9. Barnard RJ, Holloszy JO. The metabolic systems: Aerobic metabolism and substrate utilization in exercising muscle. In: Tipton CM, ed. *Exercise Physiology: People and Places*. New York: Oxford University Press, 2003, 292-321.
10. Barnes RH, Drury DR. Utilization of ketone bodies by the tissues in ketosis. *Proc Soc Exp Biol Med* 36: 350-352, 1937.
11. Benedict FG, Cathcart EP. Muscular work. A metabolic study with special references to the efficiency of the human body as a machine. Washington, DC: Carnegie Institution, 1913.
12. Bergström J, Hultman E. Muscle glycogen synthesis after exercise: An enhancing factor localized to the muscle cells in man. *Nature* 210: 309-310, 1966.
13. Bergström J, Hultman E. The effect of exercise on muscle glycogen and electrolytes in normals. *Scand J Clin Lab Invest* 18: 16-20, 1966.
14. Bergström J, Hermansen L, Hultman E, Saltin B. Diet, muscle glycogen and physical performance. *Acta Physiol Scand* 71: 140-150, 1967.
15. Berzelius M. XX. On the existence of lactic acid in living bodies. *Philosophical Magazine Series 3* 33: 128-132, 1848.
16. Blixenkrone-Møller N. Über den Abbau von Ketonkörpern. *Z Physiol Chem* 253: 261-275, 1938.
17. Bock AV, Vancaulaert C, Dill DB, Folling A, Hurxthal LM. Studies in muscular activity. III. Dynamical changes occurring in man at work. *J Physiol (Lond)* 66: 136-161, 1928.
18. Bøje O. Der Blutsucker während und nach körperlicher Arbeit. *Skand Arch Physiol* 74(Suppl. 10): 1-46, 1936.
19. Bonen A, Campbell SE, Benton CR, Chabowski A, Coort SL, Han XX, Koonen DP, Glatz JF, Luiken JJ. Regulation of fatty acid transport by fatty acid translocase/CD36. *Proc Nutr Soc* 63: 245-249, 2004.
20. Bonen A, Luiken JJ, Liu S, Dyck DJ, Kiens B, Kristiansen S, Turcotte LP, Van Der Vusse GJ, Glatz JF. Palmitate transport and fatty acid transporters in red and white muscles. *Am J Physiol* 275: E471-E478, 1998.
21. Brooks GA, Gladden B. The metabolic systems: Anaerobic metabolism (glycolytic and phosphagen). In: Tipton CM, ed. *Exercise Physiology: People and Places*. New York: Oxford University Press, 2003, 322-359.
22. Cain DF, Davies RE. Breakdown of adenosine triphosphate during a single contraction of working muscle. *Biochem Biophys Res Commun* 8: 361-366, 1962.
23. Carlson LA. Lipid metabolism and muscular work. *Fed Proc* 26: 1755-1759, 1967.
24. Carlson LA, Ekelund L-G, Fröberg SO, Hallberg D. Effect of exercise on the elimination of exogenous triglycerides from blood in man. *Acta Med Scand Suppl* 472: 245-252, 1967.
25. Carlson LA, Pernow B. Studies on blood lipids during exercise. II. The arterial plasma-free fatty acid concentration during and after exercise and its regulation. *J Lab Clin Invest* 58: 673-681, 1961.
26. Charrière M, Duchenne GB. Emporte pièce histologique. *Bull Acad Natl Med* 30: 1050-1051, 1865.
27. Chauveau A. Source et nature du potential directment utilisé dans la travail musculairs d'asprès les exchanges respiratories, chez l'hormme en état d'abstinence. *CR Acad Sci (Paris)* 122: 1163-1221, 1896.
28. Chauveau A, Kaufmann M. Expériences pour la détermination du coefficient de l'activité nutritive et respiratoire des muscles en respos et en travail. *CR Acad Sci (Paris)* 104: 1126-1132, 1887.
29. Chen ZP, Stephens TJ, Murthy S, Canny BJ, Hargreaves M, Witters LA, Kemp BE, McConell GK. Effect of exercise intensity on skeletal muscle AMPK signaling in humans. *Diabetes* 52: 2205-2212, 2003.

30. Christensen EH, Hansen O. I. Zur Methodik der respiratorischen Quotient-Bestimmungen in Ruhe and bei Arbeit. *Skand Arch Physiol* 81: 137-151, 1939.
31. Christensen EH, Hansen O. II. Untersuchungen fiber die Verbrennungsvorgange bei langdauernder, schwerer Muskelarbeit. *Skand Arch Physiol* 81: 152-159, 1939.
32. Christensen EH, Hansen O. III. Arbeitsfähigkeit und Ernährung. *Skand Arch Physiol* 81: 160-171, 1939.
33. Christensen EH, Hansen O. IV. Hypoglykamie, Arbeitfähigkeit und Ermudung. *Skand Arch Physiol* 81: 172-179, 1939.
34. Christensen EH, Hansen O. V. Respiratorisher Quotient und O_2-Aufnahme. *Skand Arch Physiol* 81: 180-189, 1939.
35. Christensen EH, Krogh A, Lindhard J. Investigations on heavy muscular work. *Quart Bull Health Org League Nations* 3: 338-417, 1934.
36. Costill DL, Sparks K, Gregor R, Turner C. Muscle glycogen utilization during exhaustive running. *J Appl Physiol* 31: 353-356, 1971.
37. Coggan AR. Plasma glucose metabolism during exercise in humans. *Sports Med* 11: 102-124, 1991.
38. Coggan AR. The glucose crossover concept is not an important new concept in exercise metabolism. *Clin Exper Pharmacol Physiol* 24: 896-900, 1997.
39. Coggan AR, Abduljalil AM, Swanson SC, Earle MS, Farris JW, Mendenhall LA, Robitaille P-M. Muscle metabolism during exercise in young and older untrained and endurance-trained men. *J Appl Physiol* 75: 2125-2133, 1993.
40. Coggan AR, Coyle EF. Reversal of fatigue during prolonged exercise by carbohydrate infusion or ingestion. *J Appl Physiol* 63: 2388-2395, 1987.
41. Coggan AR, Coyle EF. Metabolism and performance following carbohydrate ingestion late in exercise. *Med Sci Sports Exerc* 21: 59-65, 1989.
42. Coggan AR, del Corral P, Lynch NA, Wolfe JE. Oxidation of ^{14}C-octanoate during moderate intensity exercise: Muscle vs. other tissues. *FASEB J* 14: A429, 2000.
43. Coggan AR, Kohrt WM, Spina RJ, Bier DM, Holloszy JO. Endurance training decreases plasma glucose turnover and oxidation during moderate intensity exercise in men. *J Appl Physiol* 68: 990-996, 1990.
44. Coggan AR, Raguso CA, Gastaldelli A, Sidossis LS, Yeckel CW. Fat metabolism during high intensity exercise in endurance-trained and untrained men. *Metabolism* 49: 122-128, 2000.
45. Coggan AR, Raguso CA, Williams BD, Sidossis LS, Gastaldelli A. Glucose kinetics during high-intensity exercise in endurance-trained and untrained humans. *J Appl Physiol* 78: 1203-1207, 1995.
46. Coggan AR, Spina RJ, Kohrt WM, Holloszy JO. Effect of prolonged exercise on muscle citrate concentration before and after endurance training in men. *Am J Physiol* 264: E215-E220, 1993.
47. Coggan AR, Spina RJ, Kohrt WM, Kirwan JP, Bier DM, Holloszy JO. Plasma glucose kinetics during exercise in subjects with high and low lactate thresholds. *J Appl Physiol* 73: 1873-1880, 1992.
48. Coggan AR, Spina RJ, Rogers MA, King DS, Brown M, Nemeth PM, Holloszy JO. Histochemical and enzymatic comparison of the gastrocnemius muscle of young and elderly men and women. *J Geront* 47: B71-B76, 1992.
49. Coggan AR, Spina RJ, Rogers MA, King DS, Brown M, Nemeth PM, Holloszy JO. Skeletal muscle adaptations to endurance training in 60-70 year old men and women. *J Appl Physiol* 72: 1780-1786, 1992.
50. Coggan AR, Swanson SC, Mendenhall LA, Habash DL, Kien CL. Effect of endurance training on hepatic glycogenolysis and gluconeogenesis during prolonged exercise in men. *Am J Physiol* 268: E375-E383, 1995.
51. Constable SH, Favier RJ, McLane JA, Fell RD, Chen M, Holloszy JO. Energy metabolism in contracting rat skeletal muscle: Adaptation to exercise training. *Am J Physiol* 253: C316-C322, 1987.
52. Costill DL, Gollnick PD, Jansson ED, Saltin B, Stein EM. Glycogen depletion pattern in human muscle fibres during distance running. *Acta Physiol Scand* 89: 374-383, 1973.
53. Costill DL, Bowers R, Branam G, Sparks K. Muscle glycogen utilization during prolonged exercise on successive days. *J Appl Physiol* 31: 834-838, 1971.
54. Coyle EF, Hagberg JM, Hurley BF, Martin WH, Eshani AA, Holloszy JO. Carbohydrate feeding during prolonged strenuous exercise can delay fatigue. *J Appl Physiol* 55: 230-235, 1983.
55. Coyle EF, Coggan AR, Hemmert MK, Ivy JL. Muscle glycogen utilization during prolonged strenuous exercise when fed carbohydrate. *J Appl Physiol* 61: 165-172, 1986.
56. Davies RE, Cain D, Delluva AM. The energy supply for muscle contractions. *Ann NY Acad Sci* 81: 468-476, 1959.
57. Dill DB, Edwards HT, Talbott JH. Studies in muscular activity. VII. Factors limiting the capacity for work. *J Physiol* 77: 49-62, 1932.
58. Brooks GA, Donovan CM. Effect of endurance training on glucose kinetics during exercise. *Am J Physiol* 244: E505-E512, 1983.
59. Douen AG, Ramlal T, Klip A, Young DA, Cartee GD, Holloszy JO. Exercise-induced increase in glucose transporters in plasma membranes of rat skeletal muscle. *Endocrinol* 124: 449-454, 1989.
60. Douglas CG. Coordination of respiration and circulation with variations in bodily activity. *Lancet* ii: 213-218, 1927.
61. Drury DR, Wick AN, MacKay EM. The action of exercise on ketosis. *Am J Physiol* 134: 761-768, 1941.

62. du Bois-Reymond EH. On the supposed acid reaction of muscular fibre. *Philosophical Magazine Series 4* 18: 544-546, 1859.
63. Edwards HT, Margaria R, Dill DB. Metabolic rate, blood sugar, and the utilization of carbohydrate. *Am J Physiol* 108: 203-209, 1934.
64. Eggleton P, Eggleton GP. XXV. The inorganic phosphate and a labile form of organic phosphate in the gastrocnemius of the frog. *Biochem J* 21: 190-195, 1927.
65. Elayan IM, Winder WW. Effect of glucose infusion on muscle malonyl-CoA during exercise. *J Appl Physiol* 70: 1495-1499, 1991.
66. Evans WJ, Bennett AS, Costill DL, Fink WJ. Leg muscle metabolism in trained and untrained men. *Res Q* 50: 350-359, 1979.
67. Felig P, Cherif A, Minigawa A, Wahren J. Hypoglycemia during prolonged exercise in normal men. *N Eng J Med* 306: 895-900, 1982.
68. Fiske CH, SubbaRow Y. The nature of the "inorganic phosphate" in voluntary muscle. *Science* 65: 401-403, 1927.
69. Fiske CH, SubbaRow Y. Phosphorous compounds of muscle and liver. *Science* 70: 381-382, 1929.
70. Fick A, Wislicenus J. LXX. On the origin of muscular power. *Philosophical Magazine Series 4* 31: 485-503, 1866.
71. Fitts RH, Booth FW, Winder WW, Holloszy JO. Skeletal muscle respiratory capacity, endurance, and glycogen utilization. *Am J Physiol* 228: 1029-1033, 1975.
72. Fletcher WM, Hopkins FG. Lactic acid in amphibian muscle. *J Physiol (Lond)* 35: 247-309, 1907.
73. Friedberg SJ, Estes EH. Direct evidence for the oxidation of free fatty acids by peripheral tissues. *J Clin Invest* 41: 677-681, 1961.
74. Friedberg SJ, Harlan WR Jr., Trout DL, Estes EH. The effect of exercise on the concentration and turnover of plasma nonesterified fatty acids. *J Clin Invest* 39: 215-220, 1960.
75. Friedberg SJ, Sjer PB, Bogdonoff MD, Estes EH. The dynamics of plasma free fatty acid metabolism during exercise. *J Lipid Res* 4: 34-38, 1963.
76. Fritz IB, Davis DG, Holtrop RH, Dundee H. Fatty acid oxidation by skeletal muscle during rest and activity. *Am J Physiol* 194: 379-386, 1958.
77. Gemmill CL. The effect of stimulation on the fat and carbohydrate content of the gastrocnemius muscle of the phlorizinized rat. *Bull Johns Hopkins Hosp* 66: 71-89, 1940.
78. Gemmill CL. The fuel for muscular exercise. *Physiol Rev* 22: 32-53, 1942.
79. Gollnick PD, Armstrong RB, Saubert CW IV, Piehl K, Saltin B. Enzyme activity and fiber composition in skeletal muscle of untrained and trained men. *J Appl Physiol* 33: 312-319, 1972.
80. Gollnick PD, Ianuzzo CD, Williams C, Hill TR. Effect of prolonged, severe exercise on the ultrastructure of human skeletal muscle. *Int Z Angew Physiol* 27: 257-265, 1969.
81. Gollnick PD, Armstrong RB, Sembrowich WL, Shepherd RE, Saltin B. Glycogen depletion pattern skeletal muscle fibers after heavy exercise. *J Appl Physiol* 34: 615-618, 1973.
82. Gollnick PD, Armstrong RB, Saltin B, Saubert CW IV, Sembrowich CL, Shepherd RE. Effect of training on enzyme activity and fiber composition of human skeletal muscle. *J Appl Physiol* 34: 107-111, 1973.
83. Gordon B, Kohn LA, Levine SA, Matton M, Scriver WM, Whiting WB. Sugar content of blood in runners following a marathon race, with especial reference to the prevention of hypoglycemia: Further observations. *JAMA* 85: 508-509, 1925.
84. Gordon RS Jr., Cherkes A. Unesterified fatty acid in human blood plasma. *J Clin Invest* 35: 206-212, 1956.
85. Gordon RS Jr., Cherkes A, Gates H. Unesterified fatty acid in human blood plasma. II. The transport function of unesterified fatty acid. *J Clin Invest* 36: 810-815, 1957.
86. Grimmel CL. The fuel for muscular exercise. *Physiol Rev* 22: 32-53, 1942.
87. Havel RJ, Carlson LA, Ekelund L-G, Holmgren A. Turnover rate and oxidation of different free fatty acids in man during exercise. *J Appl Physiol* 19: 613-618, 1964.
88. Havel RJ, Naimark A, Borchgrevink CF. Turnover rate and oxidation of free fatty acids of blood plasma in man during exercise: Studies during continuous infusion of palmitate-1-C14. *J Clin Invest* 42: 1054-1063, 1963.
89. Havel RJ, Pernow B, Jones NL. Uptake and release of free fatty acids and other metabolites in the legs of exercising men. *J Appl Physiol* 23: 90-99, 1967.
90. Hayashi T, Hirshman MF, Kurth EJ, Winder WW, Goodyear LJ. Evidence for 5' AMP-activated protein kinase mediation of the effect of muscle contraction on glucose transport. *Diabetes* 47: 1369-1373, 1998.
91. Henricksson J. Training-induced of skeletal muscle and metabolism during submaximal exercise. *J Physiol (Lond)* 270: 661-675, 1977.
92. Hermann L. *Untersuchungen über den Stoffwechsel der Muskeln ausgehend von des Gaswechsel derselben*. Berlin: August Hirschwald, 1867, 79.
93. Hermansen LE, Hultman E, Saltin B. Muscle glycogen during prolonged severe exercise. *Acta Physiol Scand* 71: 129-139, 1967.
94. Hill AV. The heat produced in contracture and muscle tone. *J Physiol (Lond)* 40: 389-403, 1910.
95. Hill AV. The energy degraded in recovery processes of stimulated muscles. *J Physiol (Lond)* 46: 28-80, 1913.
96. Hill AV. The oxidative removal of lactic acid. *J Physiol (Lond)* 48: x-xi, 1914.

97. Hill AV. Muscular activity and carbohydrate metabolism. *Science* 60: 505-514, 1924.
98. Hill AV. The revolution in muscle physiology. *Physiol Rev* 12: 56-67, 1932.
99. Hill AV. A challenge to biochemists. *Biochim Biophys Acta* 4: 4-11, 1950.
100. Hill AV, Long CNH, Lupton H. Muscular exercise, lactic acid and the supply and utilization of oxygen. Pt. 1-III. *Proc Royal Soc B* 96: 438-475, 1924.
101. Hill AV, Long CNH, Lupton H. Muscular exercise, lactic acid and the supply and utilization of oxygen. Pt. IV-VI. *Proc Royal Soc B* 97: 84-138, 1924.
102. Hill AV, Long CNH, Lupton H. Muscular exercise, lactic acid and the supply and utilization of oxygen. Pt. VII-IX. *Proc Royal Soc B* 97: 155-176, 1924.
103. Hill AV, Lupton H. Muscular exercise, lactic acid, and the supply and utilization of oxygen. *Q J Med* 16: 135-171, 1923.
104. Holloszy JO. Biochemical adaptations in muscle. Effects of exercise on mitochondrial oxygen uptake and respiratory enzyme activity in skeletal muscle. *J Biol Chem* 242: 2278-2282, 1967.
105. Holloszy JO. Biochemical adaptations to exercise: Aerobic metabolism. In: Wilmore JH, ed. *Exercise and Sport Sciences Reviews*. Vol 1. New York: Academic, 1973, 45-71.
106. Holloszy JO, Oscai LB, Don IJ, Molé PA. Mitochondrial citric acid cycle and related enzymes: Adaptive response to exercise. *Biochem Biophys Res Commun* 40: 1368-1373, 1970.
107. Houmard JA. Intramuscular lipid oxidation and obesity. *Am J Physiol* 294: R111-R116, 2008.
108. Hultman E, Bergström J, Anderson NM. Breakdown and resynthesis of phosphorylcreatine and adenosine triphosphate in connection with muscular work in man. *Scand J Clin Lab Invest* 19: 56-66, 1967.
109. Hurley BF, Nemeth PM, Martin WH III, Hagberg JM, Dalsky GP, Holloszy JO. Muscle triglyceride utilization during exercise: Effect of training. *J Appl Physiol* 60: 562-567, 1986.
110. Issekutz B Jr., Paul P. Intramuscular energy sources in exercising normal and depancreatectomized dogs. *Am J Physiol* 215: 197-204, 1968.
111. Issekutz B Jr., Miller HI, Paul P, Rodahl K. Aerobic work capacity and plasma FFA turnover. *J Appl Physiol* 20: 293-296, 1965.
112. Jansson E, Kaijser L. Substrate utilization and enzymes in skeletal muscle of extremely endurance-trained men. *J Appl Physiol* 62: 999-1005, 1987.
113. Karlsson J, Nordesjo L-O, Saltin B. Muscle glycogen utilization during exercise after physical training. *Acta Physiol Scand* 90: 210-217, 1974.
114. Karpovich PV. Exercise. *Ann Rev Physiol* 9: 149-162, 1947.
115. Klein S, Coyle EF, Wolfe RR. Fat metabolism during low-intensity exercise in endurance-trained and untrained men. *Am J Physiol* 267: E934-E940, 1994.
116. Klein S, Weber J-M, Coyle EF, Wolfe RR. Effect of endurance training on glycerol kinetics during strenuous exercise in humans. *Metabolism* 45: 357-361, 1996.
117. Krebs HA. The citric acid cycle. In: *Nobel Lectures, Physiology or Medicine 1942-1962*. Amsterdam: Elsevier, 1964, 399-410.
118. Krogh A, Lindhard J. The relative value of fat and carbohydrate as sources of muscular energy. *Biochem J* 14: 290-363, 1920.
119. Lange G. Über die Dephosphorylierung von Adenosintriphosphat zu Adenosindihosphat während der Kontraktionsphase von Froschrectusmuskeln. *Biochem Z* 326: 172-186, 1955.
120. Lange G. Methoden zur Differenzierung der Adeninucleotide von Froschrectusmuskeln. *Biochem Z* 326: 225-230, 1955.
121. Lange G. Untersuchungen an glatten Muskeln über Anderungen im Gehalt an Adenosintriphosphat, Adenosindiphosphat, Adenosinmonophosphat und Kreatinphosphat bei der Kontraktion und bei der Erschlaffung. *Biochem Z* 326: 369-379, 1955.
122. Leathes JB. *Problems in Animal Metabolism. A Course of Lectures Given in the Physiological Laboratory of the London University at South Kensington in the Summer Term, 1904.* Philadelphia: Blakiston's Son & Co., 1906, 99.
123. Levine SA, Gordon B, Derick CL. Changes in chemical constituents of blood following a marathon race. *JAMA* 82: 1778-1779, 1924.
124. Lohman K. Über die Pyrophosphatfraktion im Muskel. *Die Naturwissenschaftenn* 17: 624-625, 1929.
125. Lohman K. Über die enzymatische Aufspaltung der Kreatinphosphorsäure zugleich ein Beitrag zum Chemismis der Muskelkontraktion. *Biochem Z* 271: 264-277, 1934.
126. Lundsgaard E. Untersuchungen über Muskel-Kontraktionen ohne Milchsäurebildung. *Biochem Z* 217: 162-177, 1930.
127. Lundsgaard E. Weitere Untersuchungen über Muskelkontraktionen ohne Milchsäurebildung. *Biochem Z* 227: 51-83, 1930.
128. Lundsgaard E. Über die Energetik der anaeroben Muskelkontraktion. *Biochem Z* 233: 322-343, 1931.
129. Lundsgaard, E. Betydningen af fænomenet "mælkesyrefri muskelkontraktioner for opfattelsen af muskelkontraktionens kemi. *Hospitalstidende* 75: 84-95, 1932.
130. Lundsgaard E. The Pasteur-Meyerhof reaction in muscle metabolism. *Bull NY Acad Sci* 14: 163-182, 1938.
131. Margaria R, Edwards HT, Dill DB. The possible mechanisms of contracting and paying the oxygen debt and the role of lactic acid in muscular contraction. *Am J Physiol* 106: 689-715, 1933.
132. Martin WH III, Dalsky GP, Hurley BF, Matthews DE, Bier DM, Hagberg JM, Rogers MA, King DS,

Holloszy JO. Effect of endurance training on plasma free fatty acid turnover and oxidation during exercise. *Am J Physiol* 265: E708-E714, 1993.
133. Masoro EJ, Rowell LB, McDonald RM, Steiert B. Skeletal muscle lipids. II. Nonutilization of intracellular lipid esters as an energy source for contractile activity. *J Biol Chem* 241: 2626-2634, 1966.
134. McConell GK, Lee-Young RS, Chen ZP, Stepto NK, Huynh NN, Stephens TJ, Canny BJ, Kemp BE. Short-term exercise training in humans reduces AMPK signaling during prolonged exercise independent of muscle glycogen. *J Physiol* 568: 665-676, 2005.
135. McNelly WC. Some effects of training on the respiratory response to exercise. *Am J Physiol* 116: 100-101, 1936.
136. Mendenhall LA, Swanson SC, Habash DL, Coggan AR. Ten days of exercise training reduces glucose production and utilization during moderate-intensity exercise. *Am J Physiol* 266: E136-E143, 1994.
137. Meyerhof O. Die Engergieumwandlungen im Muskel. III. Kohlenhydrat und milchsaureumsatz im Froschmuskel. *Pflugers Arch Physiol Mensch Tiere* 185: 11-32, 1920.
138. Meyerhof O, Suranyi J. Über die Wärmetönungen der chemische Reaktionsphasen im Muskel. *Biochem Z* 191: 106-124, 1927.
139. Mommaerts WFHM. Investigation of the presumed breakdown of adenosine-triphosphate and phosphocreatine during a single muscle twitch. *Am J Physiol* 182: 585-593, 1955.
140. Molé PA, Holloszy JO. Adaptation of muscle to exercise. Increase in the levels of palmityl CoA synthetase, carnitine palmityltransferase, and palmityl CoA dehydrogenase, and in the capacity to oxidize fatty acids. *J Clin Invest* 50: 2323-2330, 1971.
141. Morgan TE, Cobb LA, Short FA, Ross R, Gunn DR. Effect of long-term exercise on human muscle mitochondria. In: Pernow B, Saltin B, eds. *Muscle Metabolism During Exercise.* New York: Plenum, 1971, 87-95.
142. Münoh-Petersen A. Dephosphorylation of adenosine triphosphate during the rising phase of a twitch. *Acta Physiol Scand* 29: 202-219, 1953.
143. Neufeld AH, Ross WD. Blood ketone bodies in relation to carbohydrate metabolism in muscular exercise. *Am J Physiol* 138: 747-752, 1943.
144. Odland LM, Heigenhauser GJ, Lopaschuk GD, Spriet LL. Human skeletal muscle malonyl-CoA at rest and during prolonged submaximal exercise. *Am J Physiol* 270: E541-E544, 1996.
145. Odland LM, Howlett RA, Heigenhauser GJ, Hultman E, Spriet LL. Skeletal muscle malonyl-CoA content at the onset of exercise at varying power outputs in humans. *Am J Physiol* 274: E1080-E1085, 1998.
146. Olwes WH. Alterations in the lactic acid content of the blood as a result of light exercise, and associated changes in the CO_2-combining power of the blood and in the alveolar CO_2 pressure. *J Physiol* 69: 214-237, 1930.
147. Oscai LB, Molé PA, Holloszy JO. Effect of exercise on cardiac weight and mitochondria in male and female rats. *Am J Physiol* 220: 1944-1948, 1971.
148. Parnas JK, Wagner R. Uber den Kohlenhydratumsatz isolierter amphibien Muskeln und uber die Beziehungenzwichen Kohlenhydratschwund und Milchsaurebildung im Muskel. *Biochem Z* 61: 387-427, 1914.
149. Paul P, Issekutz B Jr. Role of extramuscular energy sources in the metabolism of exercising dogs. *J Appl Physiol* 22: 615-622, 1967.
150. Pohl J. Zur Lehre von der Wirkung substituirter Fettsäuren. *Arch Exp Pathol Pharmakol* 24: 142-150, 1887.
151. Randle PJ, Garland PB, Hales CN, Newsholme EA. The glucose fatty-acid cycle. Its role in insulin sensitivity and the metabolic disturbances of diabetes mellitus. *Lancet* 1(7285): 785-789, 1963.
152. Rasmussen BB, Winder WW. Effect of exercise intensity on skeletal muscle malonyl-CoA and acetyl-CoA carboxylase. *J Appl Physiol* 83: 1104-1109, 1997.
153. Rasmussen BB, Wolfe RR. Regulation of fatty acid oxidation in skeletal muscle. *Annu Rev Nutr* 19: 463-484, 1999.
154. Rennie MJ, Holloszy JO. Inhibition of glucose uptake and glycogenolysis by availability of oleate in well-oxygenated perfused skeletal muscle. *Biochem J* 168: 161-170, 1977.
155. Richter EA, Kristiansen KS, Wojtaszewski J, Daugaard JR, Asp S, Hespel P, Kiens B. Training effects on muscle glucose transport during exercise. *Adv Exp Biol Med* 441: 107-116, 1998.
156. Röckl KSC, Witczak CA, Goodyear LJ. Signaling mechanisms in skeletal muscle: Acute responses and chronic adaptations to exercise. *IUBMB Life* 60: 145-153, 2008.
157. Rose RJ, Hodgson DR, Kelso TB, McCutcheon LJ, Bayly WM, Gollnick PD. Maximum O_2 uptake, O_2 debt and deficit, and muscle metabolites in Thoroughbred horses. *J Appl Physiol* 64: 781-788, 1988.
158. Rowell LR, Masoro EJ, Spencer MJ. Splanchnic metabolism in exercising man. *J Appl Physiol* 20: 1032-1037, 1965.
159. Ruby BC, Coggan AR, Zderic TW. Gender differences in glucose kinetics and substrate oxidation during exercise near the lactate threshold. *J Appl Physiol* 92: 1125-1132, 2002.
160. Sahlin K. Control of lipid oxidation at the mitochondrial level. *Appl Physiol Nutr Metab* 34: 382-388, 2009.
161. Saltin B, Nazar K, Costill DL, Stein E, Jansson E, Essén B, Gollnick PD. The nature of the training response; peripheral and central adaptations to one-

legged exercise. *Acta Physiol Scand* 96: 289-305, 1976.

162. Schwartz A, Oschmann A. Contribution au problèm du mécanisme des contractures musculaires. Le taux de l'acide phosphorique musculaire libre dans les contractures des animaux empoisonnés par l'acide monobromique. *C R Soc Seances Soc Biol Fil* 91: 275-277, 1924.
163. Sial S, Coggan AR, Carroll R, Goodwin JS, Klein S. Fat and carbohydrate metabolism during exercise in elderly and young subjects. *Am J Physiol* 271: E983-E989, 1996.
164. Sial S, Coggan AR, Hickner RC, Klein S. Training-induced alterations in fat and carbohydrate metabolism during exercise in elderly subjects. *Am J Physiol* 274: E785-E790, 1998.
165. Sidossis LS. The role of glucose in the regulation of substrate interaction during exercise. *Can J Appl Physiol* 23: 558-569, 1998.
166. Sidossis LS, Wolfe RR, Coggan AR. Regulation of fatty acid oxidation in untrained versus trained men during exercise. *Am J Physiol* 274: E510-E515, 1998.
167. Sidossis LS, Gastaldelli A, Klein S, Wolfe RR. Regulation of plasma fatty acid oxidation during low- and high-intensity exercise. *Am J Physiol* 272: E1065-E1070, 1997.
168. Smith E. On the elimination of urea and urinary water. *Philos Trans R Soc London* 151: 747-834, 1862.
169. Starritt EC, Howlett RA, Heigenhauser GJ, Spriet LL. Sensitivity of CPT I to malonyl-CoA in trained and untrained human skeletal muscle. *Am J Physiol* 278: E462-E468, 2000.
170. Steinhaus AH. Exercise. *Ann Rev Physiol* 3: 695-716, 1941.
171. Sternlicht E, Barnard RJ, Grimditch GK. Mechanism of insulin action on glucose transport in rat skeletal muscle. *Am J Physiol* 254: E633-E638, 1988.
172. Talanian JL, Holloway GP, Snook LA, Heigenhauser GJ, Bonen A, Spriet LL. Exercise training increases sarcolemmal and mitochondrial fatty acid transport proteins in human skeletal muscle. *Am J Physiol* 299: E180-E188, 2010.
173. Tarnopolsky MA. Sex differences in exercise metabolism and the role of 17-beta estradiol. *Med Sci Sports Exerc* 40: 648-654, 2008.
174. Taylor CL, Karpovich PV. Exercise. *Ann Rev Physiol* 7: 599-622, 1945.
175. Terjung RL, Klinkerfuss GH, Baldwin KM, Winder WW, Holloszy JO. Effect of exhausting exercise on rat heart mitochondria. *Am J Physiol* 225: 300-305, 1973.
176. Thyfault JP, Kraus RM, Hickner RC, Howell AW, Wolfe RR, Dohm GL. Impaired plasma fatty acid oxidation in extremely obese women. *Am J Physiol* 287: E1076-E1081, 2004.
177. Turcotte LP, Kiens B, Richter EA. Saturation kinetics of palmitate uptake in perfused skeletal muscle. *FEBS Lett* 279: 327-329, 1991.
178. Varnauskas E, Björntorp P, Fahlén M, Přerovský I, Stenberg J. Effects of physical training on exercise blood flow and enzymatic activity in skeletal muscle. *Cardio Res* 4: 418-422, 1970.
179. von Helmoltz, H. Ueber den Stoffverbrauch bei der Muskelaction. *Arch Anat Physiol* 72-83, 1845.
180. von Pettenkofer M, Voit C. Untersuchungen über den Stoffwechselverbrauch des normalen Menschen. *Z Biol* 2: 459-573, 1866.
181. Wahren J, Felig P, Ahlborg G, Jorfeldt L. Glucose metabolism during leg exercise in man. *J Clin Invest* 50: 2715-2725, 1971.
182. Watt MJ, Heigenhauser GJF, Spriet LL. Intramuscular triacylglycerol utilization in human skeletal muscle during exercise: Is there a controversy? *J Appl Physiol* 93: 1185-1195, 2002.
183. Weiss S. Verbunden von Glycogen zu Muskelaktivität. *Wien Acad Bericht* 64: 284-291, 1871.
184. Winder WW. Malonyl-CoA—Regulator of fatty acid oxidation in muscle during exercise. *Exerc Sport Sci Rev* 26: 117-132, 1998.
185. Winder WW, Arogyasami J, Barton RJ, Elayan IM, Vehrs PR. Muscle malonyl-CoA decreases during exercise. *J Appl Physiol* 67: 2230-2233, 1989.
186. Winder WW, Arogyasami J, Elayan IM, Cartmill D. Time course of exercise-induced decline in malonyl-CoA in different muscle types. *Am J Physiol* 259: E266-E271, 1990.
187. Winder WW, Baldwin KM, Holloszy JO. Exercise-induced increase in the capacity of rat skeletal muscle to oxidize ketones. *Can J Physiol Pharmacol* 53: 86-91, 1975.
188. Winder WW, Hickson RC, Hagberg JM, Ehsani AA, McLane JA. Training-induced changes in hormonal and metabolic responses to submaximal exercise *J Appl Physiol* 46: 766-771, 1979.
189. Winfield G. The fate of fatty acids in the survival processes of muscle. *J Physiol* 49: 171-179, 1915.
190. Zderic TW, Coggan AR, Ruby BC. Glucose kinetics and substrate oxidation during exercise in the follicular and luteal phases. *J Appl Physiol* 90: 447-453, 2001.
191. Zuntz VN. Umsatz der nahrstoffe. XI. Betrachtungen uber die beziehungen zwischen nahrstoffen und leistungen des korpers. In: Oppenheimer K, ed. *Handbuch der Biochemie des Menschen und der Tierre.* Jena, FRG: Fischer, 1911, 826-855.

CHAPTER 18

Metabolic Systems: The Formation and Utilization of Lactate

George A. Brooks, PhD

In a previous review, L. Bruce Gladden (1951–, currently on the faculty of Auburn University) and I addressed the history of studies of anaerobic metabolism during exercise (21). Our emphasis was on the early origins and progression of studies on phosphagen and glycolytic metabolism, and readers are referred to that chapter as well as others in the same work (143). These referrals are necessary because it is now realized that most lactate disposal via oxidation or gluconeogenesis involves vascular conductance of lactate and oxygen to mitochondria in anatomically distributed tissue beds. Hence, knowledge of pulmonary, cardiovascular, respiratory, and endocrine physiology is necessary to develop a contemporary integrative understanding of lactate metabolism. For information on studies of lactate metabolism and scientists before approximately 1900, Gladden and I relied heavily on Fletcher and Hopkins (46), Keilin (79), Kleiber (83), Leicester (90), Rothschuh (130), von Muralt (147), Williams (156), and Zuntz (163) as well as the primary sources that could be found and interpreted. Studies of lactate metabolism make up a large part of the history of studies of exercise and muscle physiology and metabolic biochemistry. Now, less than a decade after our review (21), and despite significant progress, the field is as controversial as ever. In fact, so contentious is the field that aspects of this chapter will likely be in dispute with major tenets of other chapters in this work. Hence, for students of the field, this is a great time to see the process of science in action and learn how ideas come to be and how they are disseminated and acculturated, only to be overturned. Science is strong. However, it has its politics and forms that typically lead to major advances but sometimes also impede progress. Indeed, the lack of communication that exists and persists between and within fields can be a serious trap that students need to be aware of and avoid.

Today, knowledge of lactate metabolism is important because lactate is a major energy source, the major gluconeogenic precursor, and a signaling molecule (a "lactormone"). These three, and possible more, functions of lactate are encapsulated in the lactate shuttle hypothesis (11-14), which stands in stark contrast to earlier ideas traceable to the end of the 19th century that pointed to lactate as a metabolic waste and poison. Still, despite balkanization in science, ideas of the lactate shuttle hypothesis are so powerful that they have crossed over into several fields. Today, the effect of the lactate shuttle and its components is such that lactate polymers are used to provide nutritive support to athletes (1) and are engineered into scaffolds for stem cell proliferation and growth used in tissue repair (87). Because lactate is the preferred fuel for neurons, lactate salts and esters are being evaluated for the management of traumatic brain injury when glycolysis is inhibited (57). In cancer treatment, blocking major lactate shuttle elements, such as monocarboxylate transporter (MCT) and lactate dehydrogenase (LDH) isoforms, offers the possibility of killing tumor cells (32, 75, 80, 94, 111, 136). Thus, although emanating from exercise physiology, lactate shuttle ideas have found the way to fields as far ranging as tissue repair, neurosurgery, and oncology.

For readers unfamiliar with the history of lactic acid metabolism, an abbreviated appraisal of early research (21) is helpful in understanding the mistaken history that metabolism is either anaerobic or aerobic. A contemporary view is that lactate is the link between glycolytic and mitochondrial enzymatic networks because lactate is the product of the former and a substrate for the latter. (In this review the terms *lactate* and *lactic acid* are used interchangeably.) Some (e.g., 126) may assert that glycolysis produces lactate, an anion, and that other processes associated with muscle contraction and other forms of cell work give rise to hydrogen ions. This is an important discussion, but for another time (15).

Prelactic Acid and Lactic Acid Eras

Fermentation and the Pasteur Effect

The notion that lactic acid is formed as the result of lack of oxygen can be traced to alcohol-fermentation technology of the 18th century. The key observations of Louis Pasteur (1822-1895), the great French scientist and father of modern biochemistry and immunology, were that some microorganisms could live and proliferate in the absence of air and cannot use oxygen (113, 114). He also found that some facultative cells are capable of living in both the presence and absence of oxygen. Moreover, Pasteur found that these facultative cells respire normally in the presence of oxygen and cause very little fermentation, but in anoxia they show very active fermentation (79, p. 68). Not only did Pasteur recognize the existence of aerobic and anaerobic organisms (90), he also recognized that different types of the organisms he called yeast formed very different products (113, 114). For instance, one kind of organism fermented sugar to lactic acid whereas another organism produced alcohol.

Lactic Acid

According to Leicester (90), in 1808 Jöns Jacob Berzelius (1779-1848) found an elevated concentration of lactate in "the muscles of hunted stags" (107). In 1845, the great German scientist and philosopher Hermann von Helmholtz (1821-1894) reported findings that were consistent with lactic acid formation at the expense of glycogen (147). Shortly thereafter (1859), Emil H. Du Bois-Reymond (1818-1896) (45, 147) noted that activity caused muscles to become acidic and actually related this finding to the increased lactic acid reported by Berzelius. Soon after (1864), Rudolph P.H. Heidenhain (1834-1897) (147) reported that the amount of lactic acid increased with the amount of muscle work done.

Despite all of the evidence of acid accumulation in isolated muscles over time, others found little or no increase in muscle acid over time, particularly if the circulation remained intact (45, 90). Nevertheless, considerable confusion remained until the classic studies of Walter Morley Fletcher (1873-1910) and Frederick Gowland Hopkins (1861-1947) in 1907 (46). Fletcher was mentor to A.V. Hill, and in 1929 Hopkins was awarded the Nobel Prize for his discovery of "growth-stimulating vitamins" (108).

Glycolysis

In 1912 Gustav Georg Embden (1874-1933) and colleagues (43) showed that yeast and working muscle produced the same intermediate, which was believed to be a glyceraldehyde. Subsequently, in 1914 Embden and Laquer (44) found a phosphorus compound in muscle that caused production of lactic acid. They called this substance lacticidogenen and afterward identified it as a hexose monophosphate (45). As reviewed by Leicester (90, pp. 203-204), this "embden ester" turned out to be a mixture of hexose monophosphates that were subsequently identified: fructose-6-P, glucose-6-P, and, eventually, glucose-1-P. In 1920, using frog muscle preparations, Otto Fritz Meyerhof (1884-1951) (98-101) identified glycogen as the precursor of lactic acid. Meyerhof's apparatus was a nonperfused and nonoxygenated setup that allowed for repeated electrically stimulated contractions (figure 18.1*a*) (98). In addition to relating glycogen and lactic acid masses before and after contraction, Meyerhof (98-101) observed that when lactate disappeared after contractile activity, glycogen reappeared in a corresponding amount less a quantity that, calculated from heat production and O_2 consumption, approximated the enthalpy of a fraction (one fourth to one third) of the lactate that disappeared. Independently of Meyerhof, Archibald Vivian Hill (1886-1977) found that the recovery heats of isolated frog muscles could account for combustion of about one sixth to one fifth of the lactate that disappeared in recovery (67-69) (figure 18.1*b*). Even though the results of Hill and Meyerhof disagreed on the combustion coefficient (i.e., fraction of frog muscle lactate oxidized in recovery to restore the remainder to glycogen in situ), in 1922 Meyerhof and Hill shared the Nobel Prize in physiology and medicine. Meyerhof was recognized "for his discovery of the fixed relationship between the consumption of oxygen and the metabolism of lactic acid in the muscle," whereas Hill was recognized "for his discovery relating to the production of heat in the muscle" (108). The awards were richly deserved, but the studies were on ex vivo preparations of unperfused and poorly oxygenated muscles (figure 18.1) not designed for sustained activity through evolution. Linking the experimental results to human physiology was a challenge that Hill took up in the early 1920s, although, as noted later, he and others either were unaware of or ignored the results of investigators such as Ole Bang. In retrospect, it is important to note that the basis of the one fifth to four fifths theory of O_2 debt (i.e., oxidation of one fifth of the lactate formed during exercise to provide energy for recon-

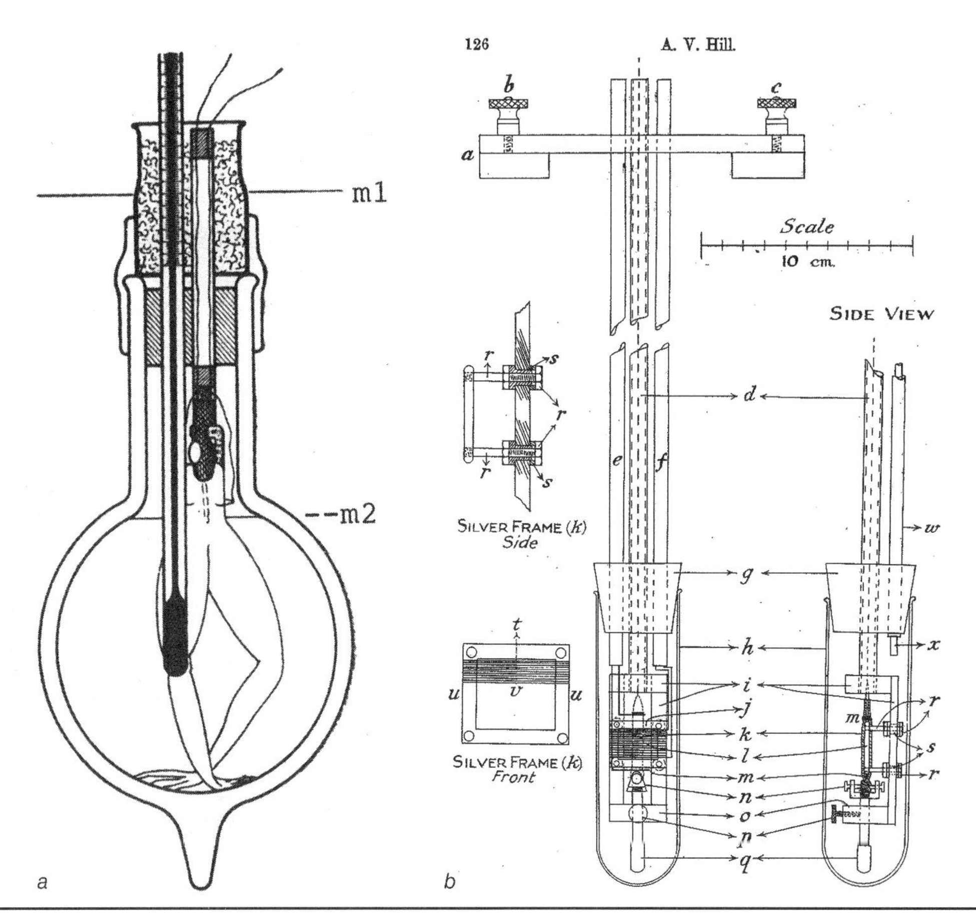

Figure 18.1 The apparatus used by Otto Meyerhof and A.V. Hill in their Nobel Prize-winning experiments that framed the idea of oxidative and glycolytic metabolism for much of the 20th century. (*a*) Meyerhof's calorimeter. The original legend is translated as "Calorimeter with electrode for indirect electrical stimulation of a hemicorpus. The first horizontal mark (ml, inserted) shows the level of the thermostatically controlled water bath. The mark in the inside of the vessel (m2, added) shows the level of the Ringers solution" (96). (*b*) Hill's thermopile device for measuring heat release from isolated frog muscles. From the original text: "All-metal thermopile, for a pair of frog sartorius muscles. Centre, front view; right, side view; left, enlarged front and side views of silver frame. Scale refers to centre and right only. a, vulcanite carrier; b, brass terminals to electrodes; c, copper terminals to thermopile; d, brass tube admitting muscle connection; e, glass tube carrying electrode leads; f, glass tube carrying thermopile leads; g, rubber stopper; h, glass cover; i, brass carriage; j, silver frame; k, muscle; 1, thermopile; m, electrodes; n, muscle clamp; o, brass carriage; p, screw; q, vulcanite rod; r, brass rods and nuts; s, "Elo" bushes; t, hot junctions; u, cold junctions; v, wire windings; w, gas exit-pipe; x, gas inlet-pipe" (69).

(a) Reprinted from O. Myerhoff, 1920, "Energieumwandlungen im muskel III. Kohlenhydrat- und milchsaureumsatz im froschmuskel," [Energy conversions in the muscles IIITH carbohydrate and milk sour revenue in the frog muscle] *Pflugers Acchiv Ges Physiol Mensch Tiere* 185: 11-32, 1920. *(b)* Reprinted from A.V. Hill, 1928, "Myothermic apparatus," *Proceedings of the Royal Society of London. Series B* 103:117-137.

version of the remaining four fifths) to glycogen is the heat released from frog muscles ex vivo (figure 18.1*b*) (68). Surprisingly, Hill never measured lactate levels in the isolated muscles he studied, and the combustion coefficient of one fifth was never verified in humans or other mammalian species or preparations (vide infra). Indeed, the idea stands in contrast to data obtained on rodents in which most lactate is removed by oxidation and only a minor fraction is converted to glycogen in postexercise recovery (16, 20).

Elaboration of the full Embden-Meyerhof (glycolytic) pathway took another two decades (101). However, in the early 20th century, the essence of the phenomena described by Pasteur was confirmed on lactic acid-producing cells and tissues from vertebrates. When yeast cells were incubated or muscles made to contract without oxygen, lactic acid accumulated. Even though 1931 Nobel laureate Otto Heinrich Warburg (1883-1970) (108) reported extensively on glycolysis leading to lactate accumulation in some types of well-oxygenated cells (e.g., cancer cells) (149), the phenomenon describing glucose–oxygen–lactic acid interactions came to be known as the Pasteur effect in textbooks of biochemistry (e.g., 89, p. 408). Retrospectively, because lactate removal is accomplished primarily by oxidation, we now understand that in the absence of oxygen, cultured facultative cells and isolated frog muscles could only produce but not remove protons, lactate, and other glycolytic or fermentation products. It was in this context of nonperfused glycolytic muscle made to contract ex vivo that the notion of a causative relationship between hypoxia and lactate accumulation evolved.

Current Knowledge

Today we know that some types of facultative cells can be cultured with lactate as a preferred fuel because their mitochondria possess means to consume and oxidize lactate directly without conversion to pyruvate in the cytosol. For instance, mitochondria of yeast (*Saccharomyces cerevisiae*) contain flavocytochrome b_2, a lactate–cytochrome c oxidoreductase (33) that couples lactate dehydrogenation with reduction of cytochrome c (34). In fact, the association between cytochrome b_2 and LDH in yeast can be traced to the 1940s (79, p. 274). A similar phenomenon occurs in mammalian muscle (19, 35, 42, 91, 112, 141). Contemporary studies have reported a balance of lactic acid production and oxidation in exercising men at sea level and 4,300 m altitude; such data were simply not obtainable with the technology of the 1920s (123). Similarly, the advent of nuclear magnetic resonance (NMR) technology that allows us to know that contractions stimulate glycolysis in muscle independently of O_2 availability (8, 29, 30, 105, 124) and that lactate is oxidized in working skeletal (8) and cardiac (26, 54) muscle are recent developments.

O_2 Debt and Meaning of Blood Lactate Accumulation During Exercise

As already noted, using frog muscle preparations (figure 18.1*a*), in 1920 Meyerhof (98-101) identified glycogen as the precursor of lactic acid in frog muscle. Further, using the thermopile, Hill (figure 18.1*b*) (67, 69) found a recovery or latent heat release from frog muscles after stimulation. The problem then was to make surrogate measures on intact persons during exercise and recovery that would be representative of lactate formation in humans during exercise. Attempts to resolve the problem led to studies of O_2 debts and deficits.

In 1920, Danes Shack August Steenberg Krogh (1874-1949) and Jens Peter Johannes Lindhard (1870-1947) were the first to report the exponential decline in O_2 consumption in men after exercise (86). Krogh (1874-1949) was a 1920 Nobel Prize winner "for his discovery of the capillary motor regulating mechanism" (108). After the report of Krogh and Lindhard, Hill and colleagues turned their attention to studies of humans in an attempt to integrate the new knowledge of muscle biochemistry and human metabolism. In 1923 Hill and Lupton (73) articulated the O_2 debt hypothesis, and the following year Hill and colleagues (70-72) published a series of noteworthy reports. The O_2 debt was defined (72, p. 142) as the "total amount of oxygen used after cessation of exercise in recovery therefrom." Recognizing that there was a deficit in oxygen consumption during exercise onset and maximal exercise conditions, Hill and colleagues sought to measure the excess postexercise O_2 consumption (O_2 debt) (51) to obtain an energy equivalent of the anaerobic lactate-producing work done during exercise. Hill and colleagues attributed the first (fast) phase of decline in postexercise $\dot{V}O_2$ to oxidative removal of lactate in the previously active muscles and attributed the second (slow) phase of decline in postexercise $\dot{V}O_2$ to oxidation of lactate that had escaped from muscles by diffusion.

In formulating their hypothesis, Hill and colleagues assumed that the energetics and metabolism in muscles of healthy humans were the same as in isolated muscles of frogs. That metabolic pathways, activities, and strategies differ significantly between mammals and lower vertebrates, particularly with regard to pathways of lactate disposal, is more recent knowledge provided by comparative physiologists Steven J. Wickler (1952-2007) and Todd T. Gleeson (1952–; currently dean of the College of Arts and Sciences, University of Colorado, Boulder in the United States) (155). Lactate is actively oxidized in red mammalian muscle; however, lactate-to-glycogen conversion is the favored pathway of disposal in poorly perfused muscles of carp, lizards, frogs, and other lower vertebrates. In mammals, glyconeogenesis from lactate involves the liver and kidneys, but there persists a vestigial pathway in fast-twitch white muscles, as seen in rabbits (110).

Ole Bang (1901-1988) was the first to question the fundamental assumptions and conclusions of O_2 debt theory (3, 4). Through the study of exercise of varied intensities and durations, Bang showed that the results of Hill and colleagues (70-73) and Margaria and colleagues (93) were fortuitous consequences of the duration of their experiments (4).

In two letters (dated September 10, 1972 and December 18, 1972) to George A. Brooks (1944–), Bang offered opinions on why his work appeared to have been overlooked. First, A.V. Hill did not acknowledge the work. Second, Bang's clinical commitments and the events of World War II deflected his efforts from exercise physiology. Also, the fact that the German-language journal (*Skandinavisches Archiv für Physiologie*) selected for publication of results ceased publication was likely important. Nonetheless, O_2 debt theory dominated scientific thinking for decades and still persists in some clinical fields, notable with regard to the concept of anaerobic threshold (21, 150, 151).

To juxtapose O_2 debt theory with contemporary data, the results of Molé and colleagues (105) and Richardson and colleagues (124, 125) suggest that lactate is produced always, including in muscles working under fully aerobic conditions. The observation that lactate production occurs under fully aerobic conditions can be traced to work of Otto Warburg on cancer cells. Accordingly, the contemporary idea is that, because it is the product of one energy system and the substrate or another, lactate links glycolytic and oxidative energy systems.

In subsequent sections ideas that exercise and exercise training increase lactate clearance during exercise is elaborated. Roughly defined, the lactate metabolic clearance rate is the rate of disposal from the blood (Rd) divided by the blood lactate concentration: lactate MCR = Rd/[La$^-$]. In this view, then, an increase in blood lactate concentration is not necessarily the fault of overproduction but rather of limited lactate clearance. Numerous groups are exploring the utility of determining the maximal lactate steady state on subjects exercising in health and disease (e.g., 159).

Mitochondrial Lactate Oxidation

With the necessary contextual background of separate and independent aerobic and anaerobic metabolic systems aside, in the early literature there were several threads to contemporary understanding of lactate oxidation during exercise and recovery. Fletcher and Hopkins (46) showed that lactic acid disappeared when frog muscles fatigued by electrical stimulation were placed in oxygen-rich environments. Hence, oxidative removal of lactate without reconversion to glucose in liver could be suspected.

Similarly, in humans blood lactate would decline to basal levels during mild- to moderate-intensity exercise (4). As well, Carl V. Gisolfi (1942-2000) and colleagues (55) among others showed that mild exercise during recovery hastened the decline in blood lactate level after intense exercise. Hence, a role for oxygen in lactate removal was always part of the story.

Most importantly, two sets of findings in the 20th century—observations that mitochondria could oxidize lactate directly, and evidence from isotope tracer studies of oxidative lactate disposal in mammals and humans in vivo—led to the realization of oxidative lactate disposal. These two lines of evidence are addressed sequentially.

Contemporary textbooks of biochemistry physiology abound with the 19th century concept that metabolism is either anaerobic (without O_2) or aerobic (with O_2). Typically, glycolytic flux from glucose and glycogen is depicted as progressing to pyruvate and then to the TCA cycle. However, the textbooks assert that glycolysis progresses to lactate if oxygen is absent. This is a convenient motif that typically is copied by one author from another and then through serial editions of texts. Amazingly, some textbook authors are biochemists who work with cells in culture under fully aerobic conditions in which lactate production from glucose containing-media requires daily changing to maintain glucose concentration and pH. Other textbook authors are physiologists who measure lactate:pyruvate concentration ratios of 10 in muscles and blood in resting mammals, including humans, and further observe the lactate:pyruvate concentration ratio to increase to more than 100 during submaximal exercise (66). Did anyone ever look to determine whether isolated mitochondria oxidize lactate? Yes, some have looked for mitochondrial LDH (mLDH) and mitochondrial lactate oxidation, but when successful their results were overlooked, castigated as being controversial, and dismissed for failing to fit with established dogma. Now that the MitoCarta is published (109), mLDH is a recognized constituent of the mitochondrial proteome. Acknowledging that lactate, not pyruvate, is the main mitochondrial fuel upsets many—indeed, most—concepts of the organization of intermediary metabolism. Still, despite the evidence, many ignored their own observations and what was in the literature.

The first investigator to evaluate mitochondrial lactate metabolism was probably Mario Umberto Dianzani (1925–), who in 1951 published a paper, the translation of which is "Distribution of Lactic Acid Oxidase in Liver and Kidney Cells of Normal Rats and Rats With Fatty Degeneration of the Liver" (38). One section translates as follows: "In conclusion the [cellular] lactic acid oxidizing systems are localized 80% to 100% in the mitochondria" (p. 182). Although Dianzani had a prolific career in hepatic toxicology and only recently retired from the department of experimental medicine and oncology at the University of Turin, he apparently never followed up on his finding of mitochondrial lactate oxidation. However, others independently observed the presence of mLDH and lactate oxidation capability.

Given the filamentous nature of the mitochondrial reticulum (81, 82) and the presence of dynamic mitochondrial morphological transitions (160), it is not surprising that some mitochondrial constituents (e.g., cytochrome c and LDH) are labile during mitochondrial vesicle isolation (25). Consequently, depending on the tissue and conditions, some investigators can obtain mitochondrial preparations that oxidize lactate and others cannot. Among those who found mLDH were Baba and Sharma (2), Brandt and colleagues (10, 84), Brooks and colleagues (19), Nakae and colleagues (106), Taylor and colleagues (142), Lemire and colleagues (91), and Pagliarini and colleagues (109). In fact, based on their observations of mLDH using electron microscopy, Baba and Sharma were probably the first to use the term *lactate shuttle*. Beyond Dianzani (38), among those who isolated mitochondria that were capable of oxidizing lactate were Kline and colleagues (84), Brandt and colleagues (10), Szczesna-Kaczmarek (141), Brooks and colleagues (19), Lemire and colleagues (91), and de Bari and colleagues (35). Among those who failed in attempts to isolate mitochondrial preparations that contain LDH and oxidize lactate were Rasmussen and colleagues (122), Sahlin and colleagues (132), and Yoshida and colleagues (161). As well, others routinely found LDH in mitochondrial preparations but regarded their findings as artifact and took steps to block LDH, thus permitting the measured rate of exogenous pyruvate oxidation to increase (28). The lesson to take from this is that mLDH is necessary for lactate oxidation. When mLDH is lost in the isolation of mitochondrial vesicles (25), or if oxamate is used to block mLDH (19), the preparation will be able to oxidize pyruvate but not lactate. In comparison with mLDH, the presence of mitochondrial MCT isoform 1 (MCT1) is less controversial. MCT1 is the only lactate–pyruvate transporter represented in the MitoCarta, the mammalian mitochondrial proteome (109, 142). That the mitochondrial reticulum can respire lactate is fundamental to the operation of lactate shuttles because oxygen consumption establishes the lactate concentration gradients down which lactate fluxes. Hence, highly oxidative (red) fiber types, cardiac tissue, the liver, and the kidneys are sites of net lactate disposal. In this model, lactate fluxes (shuttles) from fast-twitch glycolytic fibers to oxidative fibers in a working muscle bed, from working muscle to heart and liver, and from skin and subcutaneous adipose tissue to working red muscle, liver, and heart (11-14).

Tracers and Lactate Oxidation During Recovery from Exercise: Testing and Disproving a Fixed Ratio Theory

In science, as in other fields, the authority of accomplishment leads to wide acceptance of ideas, sometimes even in the face of contrary findings and conclusions. So it was with the Hill-Meyerhof theories of glycogen–lactic acid interactions and O_2 debt, aspects of which we now know to be incorrect. To reiterate, by varying the duration and intensity of exercise, Bang was able to disassociate the presence of lactate from the O_2 debt in human subjects. Also, from the late 1920s (31) to well into the 1960s, the work of Nobel Prize winners Carl Ferdinand (1896-1984) and Gerty Theresa Cori (1896-1957) emphasized the disposal of lactate via hepatic gluconeogenesis. Although the Hill-Meyerhof theories of in situ muscle glyconeogenesis held sway, and despite the presence of contrary ideas, technical limitations prevented early investigators from evaluating the pathways of lactate disposal in resting humans or mammalian models during exercise or recovery from exercise. It took the advent of isotope tracer technology to advance the field of lactate metabolism.

In hindsight, the mid-1950s was a landmark time in the history of understanding lactate metabolism. ^{14}C-glucose became commercially available to investigators, but the investigators had to ferment tracer glucose to lactate and isolate and purify the compound of interest in order to study lactate metabolism. With those capabilities, in 1955 Douglas S. Drury, Arne N. Wick, and Toshiko Morita (41) used both continuous-injection and pulse-injection approaches to administer [U-^{14}C]lactate tracer to eviscerated rabbits. The investigators were impressed by their surprising findings of very high rates of extrahepatic lactate disposal by oxidation. In 1956 Drury and Wick (40) repeated the experiments on intact and awake rabbits. In addition to monitoring blood lactate turnover and oxidation, the investigators measured the incorporation of ^{14}C from lactate into blood glucose and liver glycogen. They wrote, "Our results indicate that lactate has a very rapid turnover. It is disposed of largely by oxidation; relatively little is converted to glycogen and glucose." In discussing their results further, the investigators continued, "The very high lactate oxidation rates found by us would suggest a widespread participation by the tissues of the body." Noting the relatively sluggish utilization of glucose compared with lactate, the investigators went on to discuss oxidative disposal of lactate in tissues such as heart, brain, and skeletal muscle. From a historical perspective these interpretations were prophetic, but in their contemporary context the results were largely ignored, and despite the prominence of Drury and Wick, their writings did not become citation

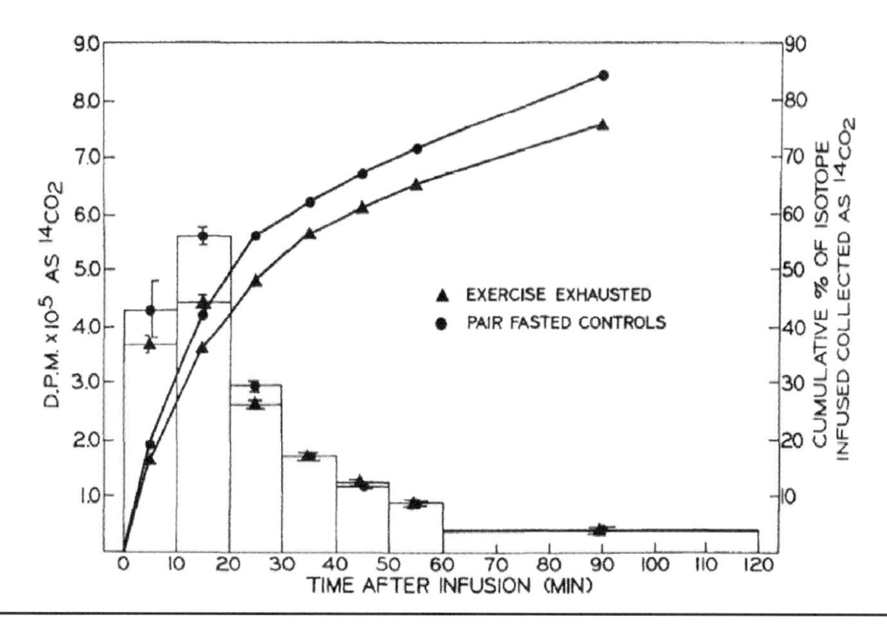

Figure 18.2 Expiration of labeled CO_2 (means ± standard error) as a function of time after infusion of 1 μCi of l-^{14}C-labeled lactate in exercise-exhausted and pair-fasted control rats (bar graph on left ordinate). Expiration of labeled CO_2 is also expressed as a cumulative percentage of activity infused (right ordinate). n = 10 animals/group.

Reprinted from G.A. Brooks, K.E. Brauner, and R.G. Cassens, 1973, "Glycogen synthesis and metabolism of lactic acid after exercise," *American Journal of Physiology* 224: 1162-1166. With permission of American Physiological Society.

classics. The writings went unrecognized by textbook authors and did not have the deserved effect of changing widespread views of the organization of intermediary metabolism. Instead, on individual and independent bases, others were destined to corroborate the results of Drury, Wick, and Morita and come to similar conclusions. Until recently the Hill-Meyerhof concepts of separate and distinct anaerobic and aerobic metabolic pathways held sway, but general understanding of the role of lactate in intermediary metabolism and gene regulation is finally changing with findings of lactate shuttling within and between muscle cells as well as in the brain and in tumors.

The first challenge to the one fifths to four fifths theory of lactate oxidation and glyconeogenesis came from Brooks and colleagues (16), who exercised rats to exhaustion, injected [1-^{14}C]lactate, and took samples of expired air and tissue intermittently in recovery. Figure 18.2 displays results showing interval and cumulative $^{14}CO_2$ excretion after tracer lactate injection. Over 2 h of recovery, 75% of the injected dose appeared as $^{14}CO_2$, compared with 80% recovery in pair-fed but unexercised control animals. Blood glucose was restored to pre-exercise levels within 20 min, but neither muscle glucose nor liver glucose levels were restored over the course of 2 h of recovery. The authors concluded that lactate contributed to gluconeogenesis for the restoration of blood sugar in recovery but that the primary metabolic fate of lactate after exercise was to oxidation and not restitution of muscle glycogen.

More definitive results on the pathways of lactate disposal and glycogen restitution from lactate were to follow when Brooks was appointed to a faculty position at the University of California, Berkeley. There, with Timothy P. White (1949–; currently chancellor at the University of California, Riverside) and major assistance from Donner Laboratory machinists, Brooks built an indirect calorimetry system that was used by White and others in their dissertation studies (22). Among those was Glenn A. Gaesser (1950–; currently at Arizona State University in the United States). Gaesser and Brooks used [U-^{14}C]lactate and glucose tracers, indirect calorimetry, and two-dimensional chromatography to trace the paths of lactate and glucose disposal during recovery from exhausting exercise (20, 50). In fact, the chromatography techniques and apparatus (figure 18.3) they used were the same as those used by Melvin Calvin (1911-1997) and Alvin Bassham for the work that gave rise to Calvin's Nobel Prize in chemistry for tracking the path of carbon in photosynthesis (108). Again, Gaesser and Brooks found little incorporation of lactate-derived carbon into muscle, liver, kidney, or heart glycogen but major disposal as $^{14}CO_2$. In mammals, oxidation—not reconversion to glycogen—is the major metabolic fate of lactate after exercise (figure 18.4). Simply, a one fifths to four fifths model provided an approximate fit to the data, but the frac-

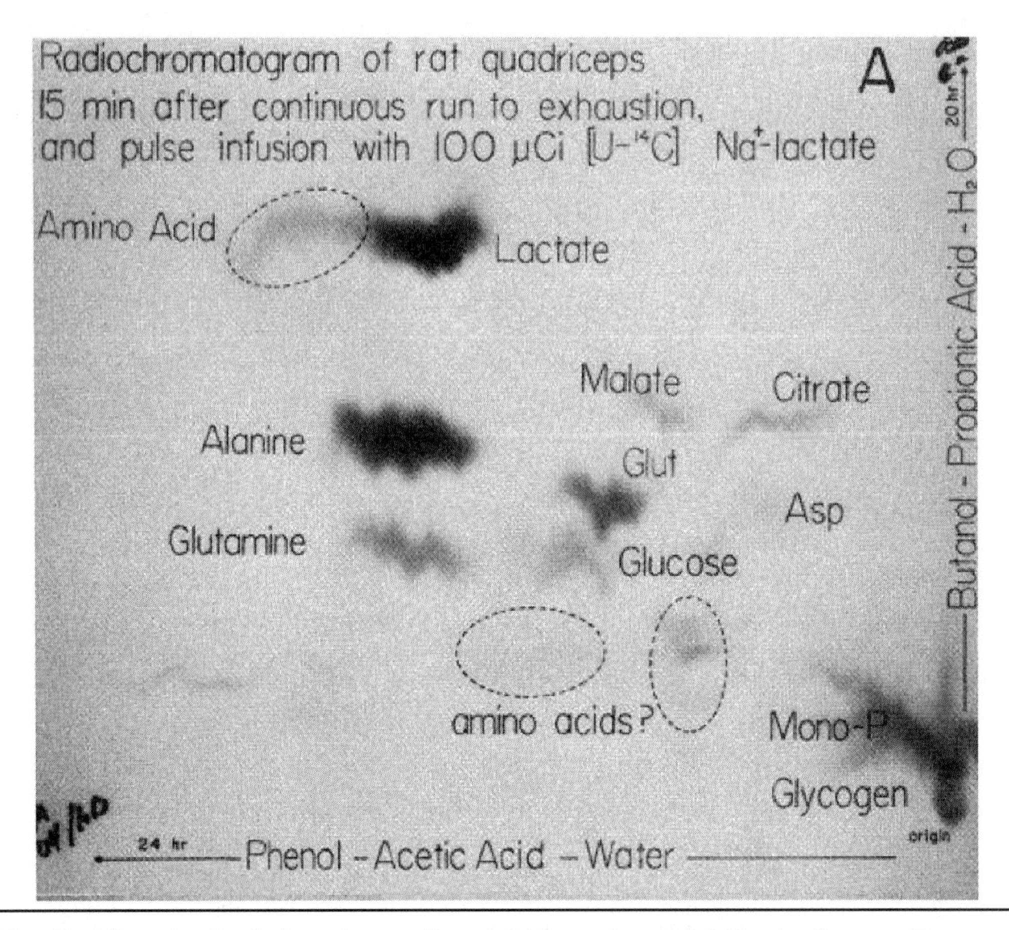

Figure 18.3 Two-dimensional radiochromatograms from skeletal muscle sampled 15 min after a continuous run to exhaustion and injection of [U-^{14}C]lactate. Incorporation of label into amino acids (including alanine and glutamine) is greater after lactate injection than after glucose injection. Glucose injection, however, resulted in greater incorporation into glycogen.

Reprinted from G.A. Brooks and G.A. Gaesser, 1980, "End points of lactate and glucose metabolism after exhausting exercise," *Journal of Applied Physiology* 49: 1057-1069. With permission of American Physiological Society.

tions obtained varied and, if anything, the oxidation–glyconeogenesis relationships were reversed from classical ideas of lactate oxidation and glyconeogenesis. The data do not fit the idea of fixed combustion and glyconeogenesis coefficients. Rather, substrate partitioning during exercise and recovery depend on individual needs and the availability of alternative substrates.

Tracers and Lactate Turnover During Exercise

Pioneer work in the field of lactate kinetics in exercising mammals was conducted by Florent Depocas (1923-2004) and colleagues. Using continuous infusion of [U-^{14}C]lactate into dogs during rest and continuous steady-state exercise, these investigators (36) (figure 18.5) made several key, fundamental findings regarding lactate metabolism. They discovered that there is active lactate turnover during the resting postabsorptive condition; that a large fraction (approximately one half) of lactate formed during rest is removed through oxidation; that the turnover rate of lactate increases during exercise compared with rest even if only a minor change occurs in blood lactate concentration; that the fraction of lactate disposal through oxidation increases to approximately three fourths during exercise; and that a minor fraction (one tenth to one fourth) of lactate removed is converted to glucose via the Cori cycle during exercise. Though the fractions are subject to species and experimental variations, the essential results have been reproduced in rats (39), rabbits (40), dogs (76, 77), horses (153), and humans (7, 23, 24, 95, 138, 139). Depocas' work on lactate metabolism was but one small part of his work at the Canadian NRC, where he had wide-ranging interest in the metabolic and endocrine responses of mammals to hypothermia.

Over the span of two decades commencing in the mid-1980s, Brooks, staff research associates Michael

Horning, Gretchen Casazza, and Jill Fattor, and their students and collaborators made concerted efforts to describe the effect of exercise intensity and training state on whole-body and working muscle lactate oxidation and gluconeogenesis from lactate in humans. The effects of exercise intensity on whole-body lactate turnover and oxidation were reported by William C. Stanley and colleagues (1957–; currently on the faculty at the University of Maryland Medical School, Baltimore) (138, 139) and Robert S. Mazzeo and colleagues (1952–; currently on the faculty at the University of Colorado, Boulder) (95), who showed that tracer-measured lactate turnover and oxidation scaled to metabolic rate during exercise and that oxidation accounted for approximately 50% of lactate disposal in resting men and 75% to 80% of lactate disposal during continuous hard exercise (figure 18.6).

Studies of lactate metabolism in working skeletal and cardiac muscle in men during exercise were reported by Stanley and colleagues in 1986 (139) and Gertz and colleagues in 1988 (54), respectively. The advantage of using tracer infusions along with simultaneous measurements of arteriovenous difference across working muscle beds is that lactate extraction, net release, production (extraction + net release), and oxidation can be determined simultaneously. Figure 18.7 depicts results showing simultaneous lactate extraction (uptake) and release (production) by working human leg muscles. Due to intramuscular lactate extraction, turnover, and oxidation, measurements of net chemical balance underestimate true rates of lactate production and oxidation.

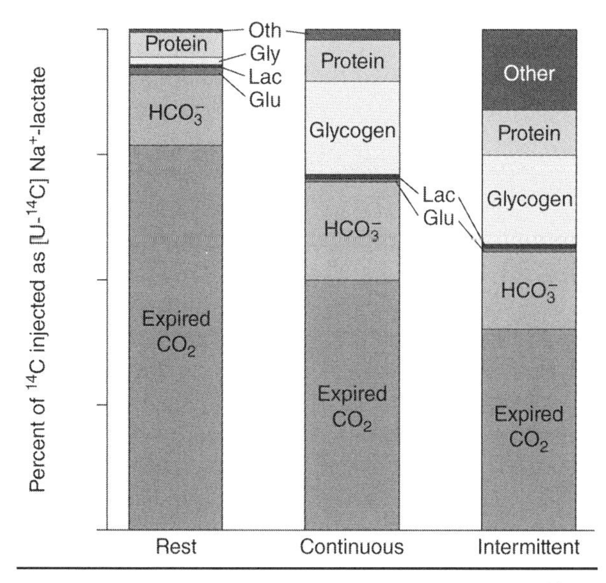

Figure 18.4 Histograms representing the fraction of ^{14}C injected as [U-^{14}C]lactate in rats that could be accounted for as CO_2, HCO_3^-, glycogen, protein, glucose, and lactate after 4 h at rest and after continuous and intermittent exercise to exhaustion. Data represent the combined results of whole-body oxidation and tissue studies.

Reprinted from G.A. Brooks and G.A. Gaesser, 1980, "End points of lactate and glucose metabolism after exhausting exercise," *Journal of Applied Physiology* 49: 1057-1069. With permission of American Physiological Society.

Another relevant but underappreciated and often unmentioned aspect of the issue of blood lactate accumulation during exercise is the assumption that the circulating lactate level increases because of net release from active muscle. Surprisingly, although net lactate release from resting muscle is common, net release from working muscle is usually transient if power output is submaximal and held constant. As shown first by Stainsby and colleagues in 1966 (137) and Welch and Stainsby in 1967 (137, 154) using dog muscle preparations contracting in situ, the Stainsby effect of transient muscle net lactate release at exercise onset followed by a switch to net uptake from the blood by working muscle has been confirmed in exercising humans (24). In this regard, the studies of Gladden and colleagues on dog muscles contracting in situ (56) are noteworthy. Gladden clearly showed that lactate uptake is concentration (substrate) dependent and not O_2 dependent, a finding that also appears to be true in human muscle (7) vide infra. Thus, it is certain that working skeletal muscle is not the sole source of blood lactate in humans during whole-body exercise. Epinephrine is more likely to signal glycolysis and lactate production in noncontracting tissues than working muscle is. In working muscle epinephrine augments glycolysis, leading to increased lactate accumulation (60, 152).

Endurance Training and Lactate Metabolism

The first measurements of the effect of endurance training on lactate kinetics using radioactive tracers in rats were performed by PhD student Casey M. Donovan (1952–; currently on the faculty at the University of Southern California) at the University of California, Berkeley. Donovan and Brooks (18, 39) showed that endurance training produced the classic result: Training lowered circulating lactate levels during exercise at a given intensity. However, those studies on rats showed that circulating lactate concentration was not reduced by changing production but rather by increasing lactate clearance. More recently, using a longitudinal training design and a combination of stable isotope tracers for lactate and glucose, arteriovenous difference measurements across working limbs, and muscle biopsies on humans, Berkeley PhD student Bryan Bergman (1971–; currently on the faculty at the University of Colorado, Denver in the United States) and colleagues showed that training has a small but significant effect on blood

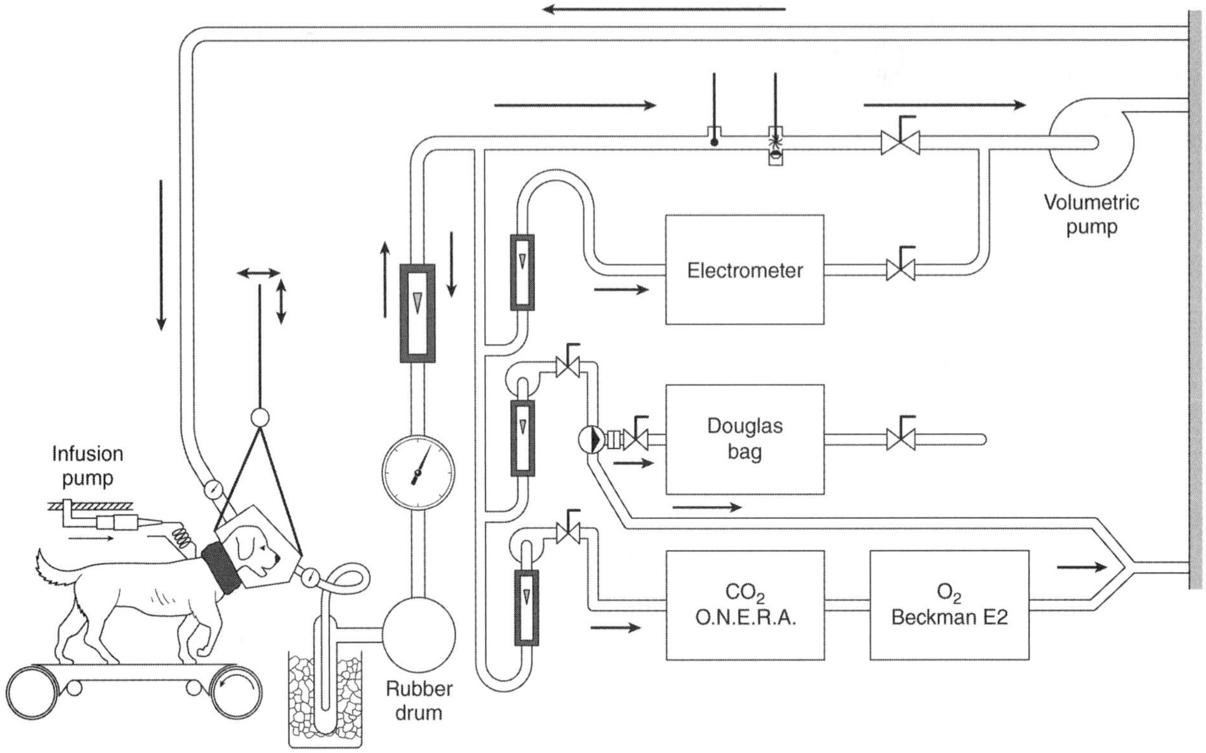

Figure 18.5 Experimental setup used to study lactic acid metabolism in resting and running dogs. Fresh air is drawn with a volumetric pump through the plastic mask enclosing the animal's head. The air is then partly dehumidified, metered, and monitored for water content by dry and wet thermometry. A rubber drum before the dry-test meter dampens pressure changes due to respiration of the animal. Three metered continuous air samples from the main duct are taken for radioactivity monitoring with an ion-chamber electrometer, for collection into Douglas bags for determination of respiratory CO_2-specific activity, and for continuous monitoring of CO_2 and O_2 content.

Reprinted, by permission, from F. Deposes, Y. Minaire, and J. Chatonnet, 1969, "Rates of formation and oxidation of lactic acid in dogs at rest and during moderate exercise," *Canadian Journal of Physiology and Pharmacology* 47: 603-610. ©2008 Canadian Science.

lactate appearance and a major effect on improving lactate clearance, especially during hard exercise (figure 18.8) (7). Importantly, by measuring limb blood flow, the arteriovenous differences for lactate and CO_2 contents, and lactate isotopic enrichments, Bergman also showed that training improves the capacity for intramuscular lactate oxidation (figure 18.9). From determining the secondary labeling of blood glucose with ^{13}C from infused [3-^{13}C]lactate, Bergman and colleagues also showed that endurance training improves gluconeogenesis from lactate (6); this was shown previously in rats (39). The role of lactate in maintaining blood glucose homeostasis during prolonged exercise is often overlooked but vitally important. The major effect of training on improving lactate oxidation in working muscles (7), an effect that is attributable to increasing expression of lactate transporters in muscle sarcolemmal (9, 17, 41, 42, 59, 119-121, 128, 129) and mitochondrial (17, 42) membranes, was shown in the Brooks laboratory by postdoctoral fellow Hervé Dubouchaud (1969–; currently on the faculty at the Université Joseph Fourier, Grenoble). These adaptations facilitate operation of cell–cell and intracellular lactate shuttles (i.e., exchange of lactate between glycolytic and oxidative fibers and the oxidative disposal of lactate within oxidative fibers, respectively).

In summary, results of studies using isotopic tracers have shown the following: Lactate turnover is prominent at rest and scales exponentially to metabolic rate during exercise; lactate is actively exchanged both within and between cells, organs, and tissues; lactate disposal through oxidation increases with exercise and is the dominant pathway; only a minor fraction of lactate removed is converted to glucose, but gluconeogenesis is an important means for maintaining glycemia during sustained exercise; net release from exercising muscle is usually transient at a constant power output; contracting skeletal muscle is not the sole source of blood lactate; and training lowers tissue lactate levels during exercise mainly by increasing clearance.

Lactate Shuttle and Membrane Lactate Transport and Lactate–Pyruvate Exchange

With knowledge of glucose and lactate fluxes and oxidation rates in resting and exercising rats (18, 39), in 1984 George Brooks took a unique approach to explaining phenomena related to lactate responses to exercise when he articulated the lactate shuttle hypothesis (11). In describing key elements of the hypothesis, Brooks stated, "The shuttling of lactate through the interstitium and vasculature provides a significant carbon source for oxidation and gluconeogenesis during rest and exercise." As such, the lactate shuttle hypothesis represented a model of how the formation and distribution of lactate is a central means by which the coordination of intermediary metabolism in diverse tissues and different cells in tissues can be accomplished. The initial hypothesis was developed from results of original isotope tracer studies conducted on laboratory rats in Brooks' own laboratory along with numerous other studies (many of which are cited in the previous section). Thus, the working hypothesis was developed that much of the glycolytic flux during exercise passed through lactate.

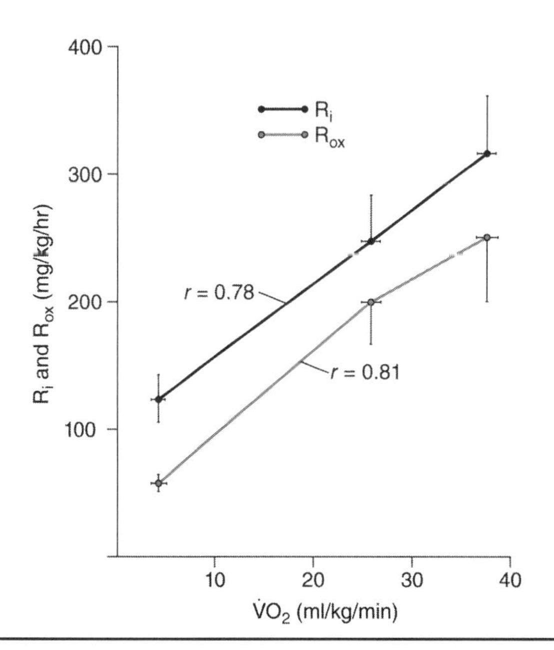

Figure 18.6 Lactate disposal (R_i) and oxidation (R_{ox}) rates plotted as functions of oxygen consumption rate ($\dot{V}O_2$) in six men at rest and at exercise power outputs eliciting 50% and 75% of $\dot{V}O_2$max. Values are means ± SEM.

Reprinted from R.S. Mazzeo et al., 1986, "Disposal of blood [1-13C]lactate in humans during rest and exercise," *Journal of Applied Physiology* 60: 232-241. With permission of American Physiological Society.

According to the cell–cell lactate shuttle hypothesis, lactate is a metabolic intermediate rather than an end product (11-14). Lactate is continuously formed in and released from diverse tissues such as skeletal muscle, skin, and erythrocytes; it also serves as an energy source in highly oxidative tissues such as the heart and is a gluconeogenic precursor for the liver. Lactate exchanges among these tissues appear to occur under various conditions ranging from postprandial to sustained exercise (11-14).

If lactate does serve as a key metabolic intermediate that shuttles into and out of tissues at high rates, particularly during exercise, then transmembrane movement is critical. For many years lactate was assumed to move across membranes by simple diffusion. However, by 1980 facilitated protein carrier-mediated transport of lactate across erythrocyte membranes had been well documented (37). It was not until 1990 that the characteristics of sarcolemmal membrane lactate transport were described by David A. Roth (1953–) and Brooks (128, 129). The study of cell membrane lactate transport proteins took another major leap in 1994 when, looking for the mevalonate transporter gene in Chinese hamster ovary (CHO) cells, Christine Kim Garcia (1968–), Michael. S. Brown (1941–), Joseph L. Goldstein (1940–), and colleagues (53) cloned and sequenced an MCT that they termed MCT1. In 1985, Goldstein and Brown shared the Nobel Prize for their discoveries concerning the regulation of cholesterol metabolism (108). They found that MCT1 was abundant in erythrocytes, heart, and basolateral intestinal epithelium. MCT1 was detectable only in oxidative muscle fiber types and not in liver. With an interest in describing a role for MCT isoforms in the Cori cycle, Garcia and colleagues (52) subsequently described the isolation of a second isoform (MCT2) by screening a Syrian hamster liver library. MCT2 was initially found in the liver and testes and was subsequently found in the brain (62, 118) and some tumor cell lines (75, 91, 136).

Independently of Garcia and colleagues (52, 53), in 1998 Andrew P. Halestrap (1949–) (59) and colleagues (157) identified another MCT isoform that is now known as MCT4. In terms of physiology, Halestrap and colleagues continued to hold traditional Hill-Meyerhof views of an association between lactate transporter expression and oxidative metabolism even though their data showed a high correlation between MCT1 expression and mitochondrial markers (9). However, with a different concept based on knowledge that lactate was formed and oxidized continuously in muscle and the heart in vivo, Brooks and colleagues sought to identify the molecular basis of coupling glycolytic pathways with oxidative pathways. In that pursuit, Brooks and

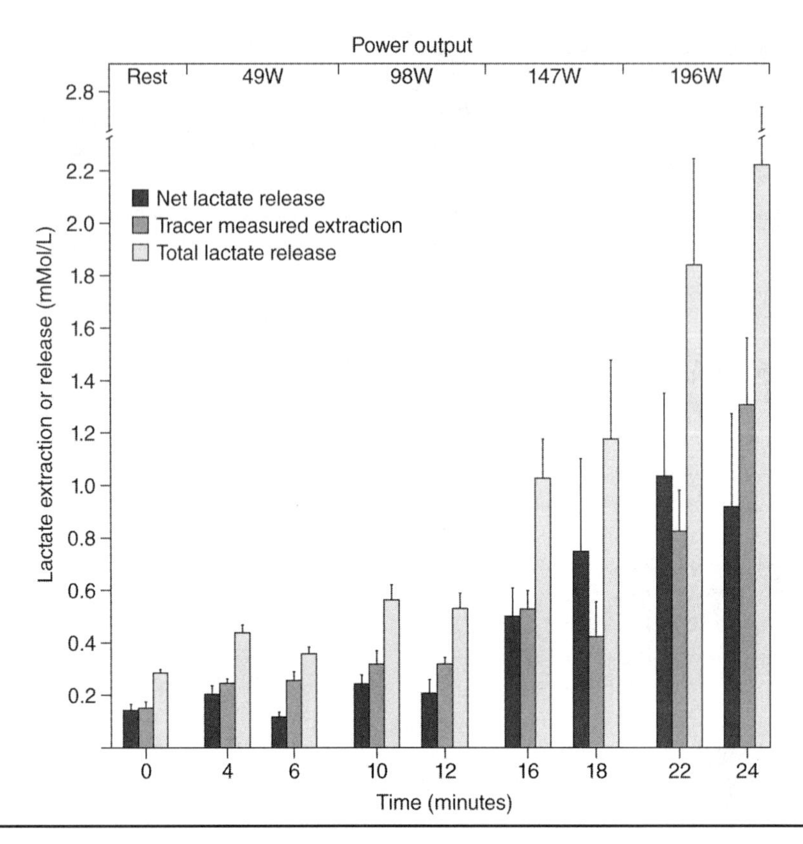

Figure 18.7 Net lactate release, tracer-measured lactate extraction, and total lactate release (extraction + net release) in working leg muscles as a function of time. Net lactate release underestimates total intramuscular turnover at all times. Values are means ± SEM; n = 6 for all but the last sample, where n = 3.

Reprinted from W.C. Stanley et al., 1986, "Lactate extraction during net lactate release in legs of humans during exercise," *Journal of Applied Physiology* 160: 1116-1120. With permission of American Physiological Society.

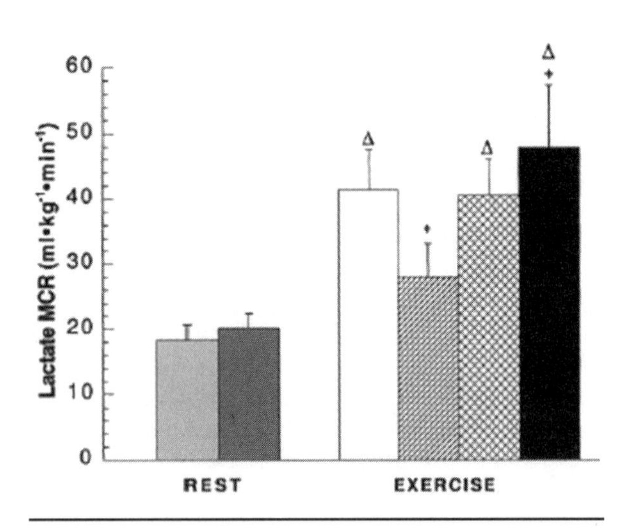

Figure 18.8 Effects of exercise intensity and training on lactate metabolic clearance rate (MCR). Values are means 6 SE for eight to nine subjects. Values are means ± SEM for nine subjects.

Reprinted from B.C. Bergman et al., 1999, "Active muscle and whole body lactate kinetics after endurance training in men," *Journal of Applied Physiology* 87: 1684-96. With permission of American Physiological Society.

colleagues first observed the presence of LDH in rat liver, cardiac, and skeletal muscle mitochondria (19). Subsequently, they also showed MCT1 to be in mitochondria of the same tissues (17), thus allowing mitochondria to directly oxidize lactate (19). With such knowledge, in 1998 Brooks extended the cell–cell lactate shuttle concept to include an intracellular component: the intracellular lactate shuttle. As already noted, several of the observations such as the presence of mLDH (2) and mitochondrial lactate oxidation (38) were swamped by the mediocrity of authority and were buried and forgotten in the literature.

Compartmentation Issues and Controversies: The Intracellular Lactate Shuttle

Although there is growing agreement on aspects of the cell–cell lactate shuttle in exercise physiology and other fields such as neurobiology (117, 118, 134, 135, 191) and cancer research (75, 136), aspects of the intracellular lactate shuttle remain controversial. In essence, disagreement exists over where the first step in lactate oxidation (i.e., conversion to pyruvate) occurs. This is why the discussion over whether mitochondria contain LDH and, if they do, where the LDH resides to be able to oxidize lactate (vide supra) is important. Results of light and electron micrographic studies conducted to date are unclear on where LDH resides in the mitochondrial reticulum. Both morphometry and susceptibility to loss during mitochondrial isolation suggest that LDH resides in the intermembrane space or, like cytochrome c, is loosely fixed to the inner membrane. On the other hand, if it is acknowledged that cardiac and red skeletal muscle take up and oxidize lactate but the presence of mLDH and a mitochondrial lactate oxidation complex (mLOC) is denied, opponents of the intracellular lactate oxidation complex must rely on cytosolic oxidation of lactate to pyruvate. Cytosolic oxidation of lactate is unlikely (14) and contrary to the evidence available from Peter Hochachka (74) or Bradley Zinker and colleagues (162) working in David H. Wasserman's laboratory at Vanderbilt University.

The first depiction of an intracellular lactate shuttle is attributable to the comparative physiologist Peter W. Hochachka (1937-2002) (74), who linked the presence of a unique LDH isoform (LDH-C) to the ability of sperm mitochondria to oxidize lactate. Hochachka fully recognized the physiological and evolutionary significance of lactate oxidation by sperm mitochondria, but he did not have opportunity to extend these findings to other cell systems.

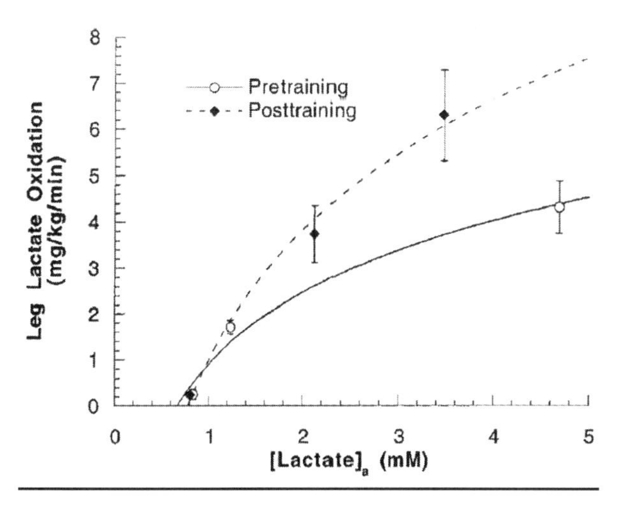

Figure 18.9 Relationship between tracer-measured leg lactate oxidation and arterial lactate concentration before and after endurance training. Values are means ± SEM for seven to nine subjects.

In their studies, Zinker and colleagues (162) infused ^{3}H- and ^{14}C-glucose tracers into chronically instrumented dogs during continuous submaximal treadmill running. Consistent with other work on exercising mammals, including humans, Zinker and colleagues found that exercise increased whole-body turnover and uptake rates of working muscle glucose and lactate during normoxic (21% O_2) and hypoxic (11% O_2) breathing. Notably, during normoxia, half the tracer glucose taken up by working limb muscles was released as labeled lactate while the muscle took up and oxidized lactate on a net basis. This extensive report was one of a series on a complex set of studies that provided many intriguing results, including the possibilities of compartmental phenomena. One idea to consider is that tracer glucose is taken up and metabolized to lactate in one compartment while carbon flux from glycogen is equilibrated with tracer lactate in a different pool. For the present, the findings of net glucose and lactate uptake and the release of 50% of glucose as lactate argues against cytosolic oxidation of lactate to pyruvate. Hence, in working dog (162) and human (7, 23, 24, 139) muscle, the results point to direct mitochondrial lactate oxidation in vivo.

Beyond a cytosol-to-mitochondria lactate shuttle, other intercellular lactate shuttles likely exist (e.g., between cytosol and peroxisomes, where it is known that a system for the reoxidation of NADH is essential for the functioning of β-oxidation). In this context it is noteworthy that lactate for pyruvate exchange across peroxisomal membranes must be accomplished in order to control peroxisomal redox balance (96).

Results of studies using proton and ^{13}C-NMR support the contention of lactate shuttles in vivo, but the data suggest that knowledge of cell–cell and intracellular lactate exchange and metabolism are in their infancy. For instance, although results from NMR spectroscopy show preferential lactate oxidation in skeletal muscle (8) and the heart (26, 88), the pathways are not necessarily as expected. Pyruvate tracer given into the circulation is rapidly converted to lactate, likely through uptake via the action of LDH in erythrocytes and cell lactate–pyruvate exchange mediated by MCT1 (140). With tracers injected directly into the myocardial circulation, ^{13}C-pyruvate exchanges with lactate and alanine in cytosol; all three peaks are detected in spectra (88). However, when ^{13}C-lactate is injected, cytosolic pyruvate and alanine were not labeled (88). More recently, Chatham and colleagues (26) elaborated on this apparent compartmentation of lactate metabolism. The results of their studies indicate preferential oxidation of exogenous lactate in the heart with glycolytically derived lactate exported from the heart.

Mitochondrial Lactate Oxidation Complex

Knowledge that mitochondria contained MCT and LDH isoforms, that the scaffold for MCT1 was basigin (CD147), and that LDH and MCT isoforms were necessary for mitochondrial lactate oxidation led to the hypothesis that those proteins were associated with some constituent of the mitochondrial electron transport chain that proved to be cytochrome oxidase (COx). Hence, in the Brooks laboratory at Berkeley, Takeshi Hashimoto (1975–; currently a faculty member at Ritsumeikan University) used confocal laser scanning microscopy, immunocoprecipitation, and cell fraction techniques followed by western blotting to verify the presence of the purported mLOC in cultured myocytes as well as in thin sections of adult rat tissues. Colocalization of MCT1, COx, and LDH in mitochondria of adult rat skeletal muscle was visualized by means of combinations of primary and fluorescent-labeled secondary antibodies plus MitoTracker Red and dual-wavelength scanning confocal microscopy (figure 18.10) (61, 64). Subsequently, similar techniques allowed the investigators to show colocalization of mLOC components in cultured L6 (rat muscle-derived) cells (figure 18.11). Again, those results were confirmed by immunocoprecipitation and western blotting of isolated cell fractions (61-64).

A pictorial representation of how the mLOC is organized on the inner mitochondrial membrane with projection into the intermembrane space is constructed from results of immunocytochemistry and immunoprecipitation studies (figure 18.12).

Lactate in the Brain

As reviewed by Schurr (134, 135), lactate as a neuronal fuel has long been of interest to neuroscientists. Over decades, numerous investigators showed lactate use by brain slices and brain cell cultures. In 1998, 14 yr after articulation of the lactate shuttle concept for working muscle, MCT messenger ribonucleic acids (117) were found in rodent brains and an astrocyte-neuron lactate shuttle was proposed. Then, after a 10 yr lag, Niels Secher (1946–) and colleagues showed cerebral lactate uptake and oxidation in exercising men (146).

The issue of universality of the presence of mLOC components in mammalian tissues other than muscle has been addressed, in part, by findings of brain cell MCT1 and MCT2 in 2005 (117) and LDH in 2008 (91). Colocalization of LDH and MCT1 (or MCT2, depending on the brain region) with COx was demonstrated in 2008 by Takeshi Hashimoto and colleagues using histological sections of rat brains and primary rat neuron cultures (62). The investigators used immunoco-precipitation and cell fractionation followed by western blotting techniques to support the results of visualization studies. Hence, a conclusion from the morphometric and other studies of Hashimoto and colleagues is that although an astrocyte-neuron lactate shuttle may operate in vivo, neurons are capable of taking up and oxidizing vascularly supplied lactate because the cerebral capillary endothelium is rich with MCTs for lactate to cross the blood–brain barrier (62).

How Lactate Production Dominates Metabolism

During rest or exercise, glycolysis produces lactate (66, 85, 125). Glycolysis is stimulated by carbohydrate nutrition, and significant postprandial glycogen synthesis occurs via the indirect pathway (47). The pathway is referred to as indirect because, paradoxically, some hepatic portal glucose from CHO nutrition bypasses the liver and enters the systemic circulation. In the direct pathway of hepatic glycogen synthesis, portal vein glucose from dietary carbohydrate is taken up and converted to hepatic glycogen. In the indirect pathway in muscle and other tissues, glucose undergoes glycolysis that results in the release of lactate into the systemic circulation. Arterial perfusion provides the liver with an ample supply of lactate that serves as a precursor for glycogen synthesis by this indirect or glucose paradox pathway. Estimates vary depending on species, but the relative contribution of the direct pathway to postprandial hepatic glycogen synthesis in humans (~25%) (97, 158) is less than that in animal models (47).Ordinarily, after consumption of dietary carbohydrate, lactate clearance mechanisms—whether by oxidation, hepatic gluconeogenesis (GNG), or glycogen synthesis —are adequate to minimize excursions in arterial lactate concentration, maintaining levels around 1 mM whether lactate appears from exogenous or endogenous sources. However, energy substrate partitioning is affected when clearance mechanisms cannot prevent an increase in circulating lactate. As already noted, arterial lactate concentration increases after CHO nutrition; this result supports liver glycogen synthesis (144, 145). Other metabolic consequences stem from increasing circulating lactate. Among these are increased gluconeogenesis, lactate substitution for glucose oxidation, decreased free fatty acid (FFA) release from adipose and lower plasma FFA levels, and decreased mitochondrial FFA uptake and oxidation.

Hepatic plus renal glucose production in overnight (10-12 h) fasted humans is typically estimated from the

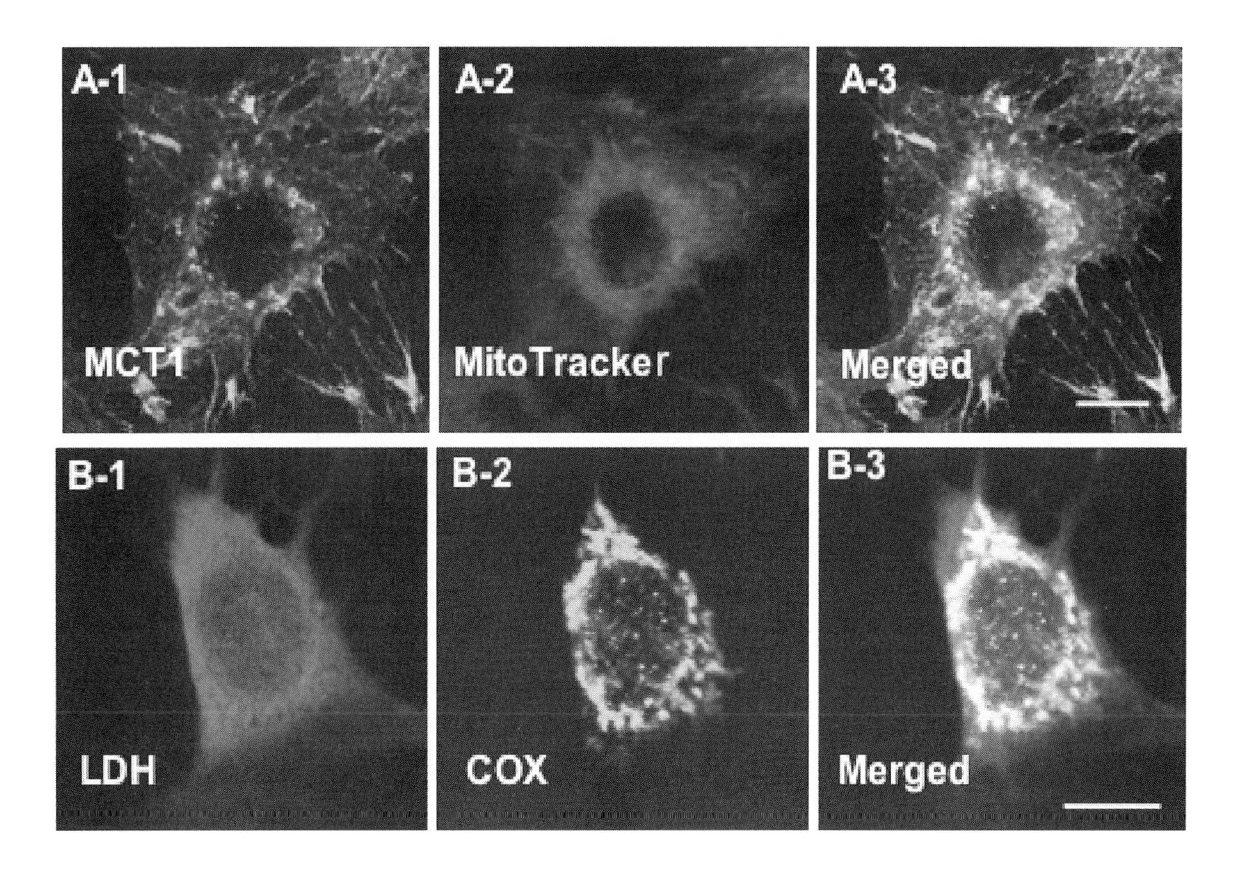

Figure 18.10 Immunohistochemical images demonstrating some components of the lactate oxidation complex in cultured L6 (rat muscle-derived) muscle cells. This complex involves the mitochondrial constituent cytochrome oxidase (COx), the lactate–pyruvate transport protein monocarboxylate transporter 1 (MCT1), lactate dehydrogenase (LDH), and other constituents. (*a*) Colocalization of MCT1 and the mitochondrial reticulum. MCT1 was detected at both sarcolemmal and intracellular domains (A-1). Using MitoTracker, the mitochondrial reticulum was extensively elaborated and detected at intracellular domains throughout L6 cells (A-2). When signals from probes for the lactate transporter MCT1 (green, A-1) and mitochondria (red, A-2) were merged, superposition of the signals (yellow) showed colocalization of MCT1 and components of the mitochondrial reticulum, particularly at perinuclear cell domains (A-3). (*b*) LDH (B-1) and mitochondrial COx (B-2). Superposition of signals for LDH (red, B-1) and COx (green, B-2) shows colocalization of LDH in the mitochondrial reticulum (yellow) of cultured L6 rat muscle cells (D-3). Depth of field is approximately 1 μm; scale bar = 10 μm.

Reprinted from T. Hashimoto, R. Hussien, G.A. Brooks, "Colocalization of MCT1, CD147 and LDH in mitochondrial inner membrane of L6 cells: Evidence of a mitochondrial lactate oxidation complex," *American Journal of Physiology Endocrinology and Metabolism* 290: 1237-1244. With permission of American Physiological Society.

isotopic dilution of a continuously infused tracer such as [6,6-^{2}H]glucose, the doubly deuterium-labeled glucose commonly referred to as D2-glucose. In humans, arterial and hepatic vein catheterizations for directly determining hepatic glucose production are rare and computations depend on estimates of hepatic arterial and portal vein blood flow to the liver. Accordingly, from tracer studies we know that endogenous hepatic and renal glucose production autoregulates to maintain glycemia during fasting in humans adapted to a balanced diet (65, 102, 103, 144, 145). In support of glycemia, lactate is well established as the main hepatic and renal gluconeogenic precursor in resting postabsorptive humans; the contribution of lactate is one or two orders of magnitude greater than that of pyruvate and alanine (97). The importance of lactate as a gluconeogenic precursor in exercising humans is even more prominent during exercise (6, 102, 103, 145) when the arterial lactate:pyruvate concentration ratio increases from 10 at rest to more than 100 during exercise (66). In resting postabsorptive individuals, the contribution of gluconeogenesis to hepatic and renal glucose production depends on the time interval since carbohydrate nutrition and ranges from essentially no endogenous glucose production from GNG immediately after eating

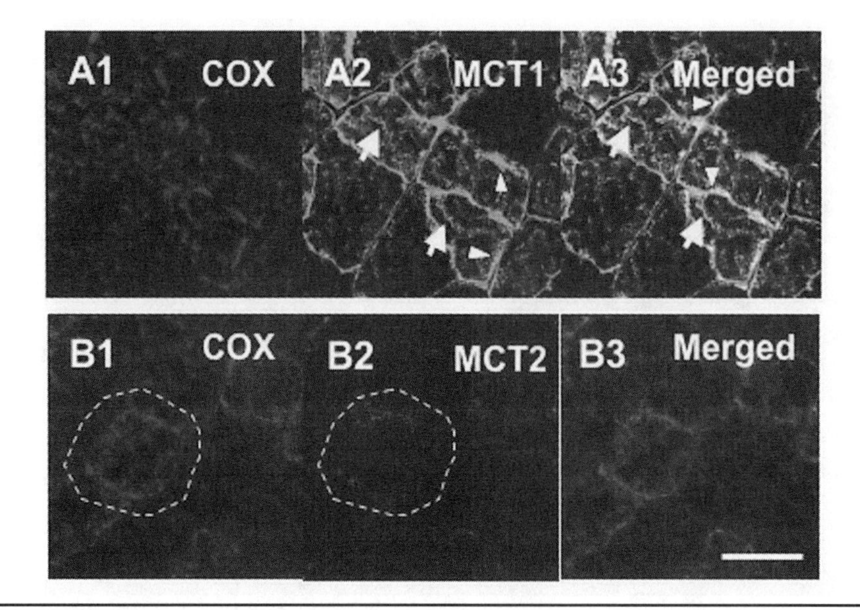

Figure 18.11 Cellular locations of monocarboxylate transporter (MCT) 1 and 2 lactate transporter isoforms and the mitochondrial reticulum cytochrome oxidase (COx) in adult rat plantaris muscle determined using confocal laser scanning microscopy and fluorescent probes for the respective proteins. Comparisons for MCT1 in the first row (plates A1-A3) and MCT2 in the second row (plates B1-B3). The localization of COx was detected in rat plantaris muscle (plates A1 and B1). MCT1 was detected throughout the cells including subsarcolemmal (arrow heads) and interfibrillar (arrows) domains (plate A2). MCT1 abundance was greatest in oxidative fibers where COx is abundant and the signal strong. When MCT1 (green) and COx (red) were merged, superposition of the two probes was clear (yellow), a finding prominent at interfibrillar (arrows) as well as sarcolemmal (arrow heads) cell domains (plate A3). In contrast, the signal for MCT2 (plate B2) was weak and relatively more noticeable in fibers denoted by strong staining for COx (plates B1 and B3; broken line is delineated around oxidative fiber to distinguish the faint signal for MCT2). Overlap of MCT2 and COx is insignificant and is denoted by absence of yellow in plate B3. Scale bar = 50 µm. Sections are from the same animal.

to approximately 50% of endogenous glucose production from GNG after 24 h of fasting (65, 145).

Endogenous glucose production increases in postabsorptive men (6, 49) and women (48) during exercise. As well, total gluconeogenesis from all precursors and gluconeogenesis from lactate increase in humans during exercise (6) (figure 18.13). The effect of training is to decrease glucose Rd (6, 48, 49), but as part of the post-training enhancement in lactate clearance capacity, GNG from lactate increases during exercise at a given exercise power output after training (figure 18.13). Also, endurance training increases GNG from lactate during exercise at a given (≥65% $\dot{V}O_2$max) relative exercise intensity after training (figure 18.13) (6).

Perhaps no body of work better illustrates the dominance of lactate on energy substrate partitioning than the lactate clamp experiments of Berkeley PhD student Benjamin F. Miller (1972–; currently on the faculty at Colorado State University, Fort Collins in the United States). In those studies, Miller, Brooks, and colleagues infused a sodium lactate–lactic acid cocktail to increase arterial lactate concentration to 4 mM during rest and exercise (102, 103). In the first experiment (102), deuterated and ^{13}C-labeled glucose tracers were given during rest and exercise. Results showed autoregulation of endogenous glucose production to be independent of precursor lactate supply during rest or exercise. However, exogenous lactate infusion decreased glucose oxidation rate. Thus, results showed preferential lactate over glucose oxidation during exercise. In the second experiment (103), deuterated glucose and ^{13}C-labeled lactate tracers were given during rest and exercise. Results confirmed autoregulation of hepatic and renal glucose production and the effect of exogenous lactate competing for glucose as an energy source. With the lactate clamp procedure more ^{13}C from infused lactate was incorporated into blood glucose, indicating GNG from lactate and suppression of liver glycogen degradation.

Relative carbohydrate (CHO) oxidation is decreased after training and lipid oxidation is enhanced during exercises eliciting power outputs that require less than 65% $\dot{V}O_2$max (5). However, regardless of training state, exercises eliciting power outputs that require relatively hard intensities (i.e., ≥65% $\dot{V}O_2$max) are CHO dependent (5). Increased lactate oxidation and GNG

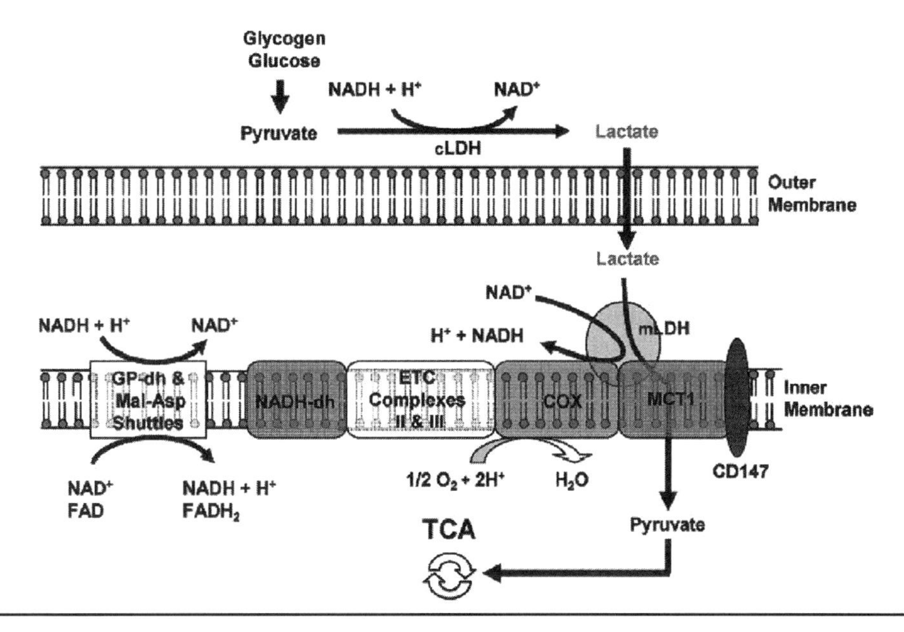

Figure 18.12 The putative lactate oxidation complex. Lactate is oxidized to pyruvate via mitochondrial lactate dehydrogenase in association with cytochrome oxidase (COx). This endergonic lactate oxidation reaction is coupled to the exergonic redox change in COx during mitochondrial electron transport. Transport of pyruvate across the inner mitochondrial membrane is facilitated by monocarboxylate transporter 1. GP = glycerol phosphate; Mal–Asp = malate–aspartate; ETC = electron transport chain; TCA = tricarboxylic acid.

Reprinted from T. Hashimoto, R. Hussien, G.A. Brooks, "Colocalization of MCT1, CD147 and LDH in mitochondrial inner membrane of L6 cells: Evidence of a mitochondrial lactate oxidation complex," *American Journal of Physiology Endocrinology and Metabolism* 290: 1237-1244. With permission of American Physiological Society.

from lactate occur after training exercises that require 65% or more $\dot{V}O_2$max. However, because the metabolic fate of glucose produced by GNG is mainly oxidative whether lactate clearance is by direct oxidation or oxidation after conversion to glucose, most lactate is oxidized during or immediately after exercise.

In addition to supporting glycemia and overall CHO oxidation during hard exercise, lactate suppresses lipid mobilization and oxidation during hard exercise. The inverse relationship between elevated lactate and suppressed plasma FFA concentration has long been recognized but is underappreciated. In the 1960s a group of talented pioneers in the field of metabolic regulation—including Bella Issekutz (1912-1999), Kårre Rodahl (1917–), Harvey I. Miller, and Pavle Paul working in the division of research at Lankenau Hospital in Philadelphia—subjected human and canine models to stress and thereby studied metabolic regulation. These investigators noted the effect of lactacidemia on diminishing circulating FFA concentration in individuals during hard exercise (76, 104, 127), and lactate infusion into running dogs caused circulating FFA concentration to decline (58, 76, 77, 127). These investigators clearly observed an effect of lactate on adipose FFA release, but whether the mechanism was an inhibition of lipolysis or a stimulation of re-esterification was unresolved.

Beyond decreasing availability of FFA concentration, lactate has perhaps even greater effects on mitochondrial lipid oxidation during exercise. When glycolysis is accelerated during muscle contraction, concentrations of the glycolytic products lactate and pyruvate increase the lactate:pyruvate concentration ratio. At rest, the lactate:pyruvate concentration ratio in muscle and venous effluent from a muscle bed approximates 10, but the ratio increases an order of magnitude or more during moderate-intensity exercise (66). The monocarboxylate pair floods entry into the mitochondrial reticulum (131), giving rise to formation of acetyl-CoA and, thereby, malonyl-CoA. The increase in malonyl-CoA inhibits the entry of activated FFA into the mitochondrial matrix by inhibiting carnitine–palmitoyl transferase-1 (131). Also, by allosteric and redox regulation of β-ketothiolase (the terminal and rate-limiting enzyme of the mitochondrial beta oxidation pathway), the accumulation of acetyl-CoA limits mitochondrial lipid oxidation irrespective of lipid availability.

Lactate as a Signaling Molecule and the Autocrine, Paracrine, and Hormonal Effects of Lactate

Terms typically used in describing cell-signaling messengers such as cytokines do not fully and adequately describe the roles of lactate as encompassed in lactate shuttle concepts, but by analogy with those terms, concepts may be illustrated so far as differences may be appreciated. Lactate differs from other signaling moieties in that lactate is highly abundant and is a substrate for the process that it regulates.

Magnitude of the Lactate and Lactate:Pyruvate Concentration Ratio Change Signals

In resting individuals the circulating lactate level ranges from 0.1 to 1.0 mM; these levels are comparable with interleukin-6 levels of 1 to 2 pg/mL. During strenuous exercise, blood lactate levels can increase one to two orders of magnitude; this range is comparable with that of cytokines such as interleukin-6 (115), meaning that lactate concentration and lactate:pyruvate concentration ratio signals are huge (66) (two to three orders of magnitude greater than those of other purported myokines) (115).

Lactate as a Substrate

Lactate released from working muscle and other tissue beds of production enter blood, the heart, red skeletal muscle, and the liver, moving down concentration and pH gradients (7, 37, 128, 129). Lactate entry into cytosolic and mitochondrial compartments changes cell redox status in those compartments (14) and results in significant metabolic effects. Increasing arterial lactate concentration results in increased tissue lactate oxidation (7) and substitution of lactate for glucose in muscle (102, 103) and the brain (146). Increasing arterial lactate concentration also suppresses FFA release from adipose (104) and mitochondrial FFA uptake and oxidation (vide supra). In terms of gluconeogenesis, increasing lactate supports hepatic glucose production (6), thus making substrate available to cells and tissues with obligate glucose requirements. Hence, in numerous ways, lactate influences energy substrate use both directly and indirectly.

Lactate as a Hormone

The recent observations of Liu and colleagues (92) showing that lactate inhibits lipolysis in fat cells through activation of an orphan G-protein coupled receptor (GPR81) is the first example of lactate signaling by allosteric regulation. Whether this is the first or the

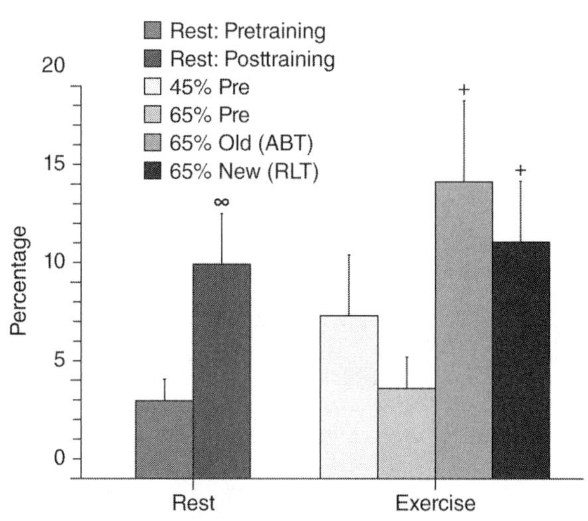

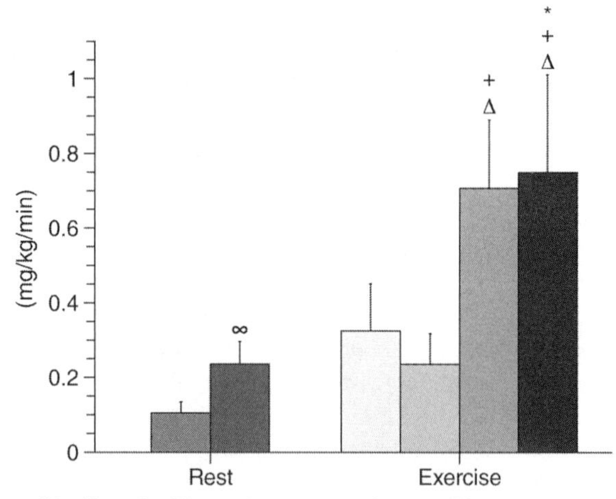

Figure 18.13 Gluconeogenesis (GNG) from lactate during rest and exercise, before and after training. (*a*) Percentage endogenous glucose production from GNG and (*b*) gluconeogenesis from lactate. Values are means ± SEM; n = 8 or 9.

Reprinted from B.C. Bergman et al., 2000, "Endurance training increases gluconeogenesis during rest and exercise in men," *American Journal of Physiology Endocrinology and Metabolism* 278: E244-E251. With permission of American Physiological Society.

first and last such observation of lactate signaling is yet to be determined.

Lactate as an Antifatigue Agent

Another major twist in the 100 yr tale of the role of lactate in physical exercise is its role as an antifatigue agent (116). Recent results of studies by Ole B. Nielsen and colleagues at the University of Aarhus, Denmark, on isolated rat muscle preparations show that high extracellular lactate, as would appear from release into the systemic circulation from metabolically active and epinephrine-responsive tissue beds, could block chloride channels, thus relieving the effects of extracellular potassium accumulation. From a historical perspective, could it have been that lactate accumulation in the experiments of Meyerhof (figure 18.1*a*) reflected a muscular stress response to the strain of supporting ATP homeostasis in the absence of perfusion oxygenation? Consequently, from the standpoint of cellular regulation the influence of lactate on muscle performance and fatigue is autocrine (from lactate produced in contracting muscle fibers), paracrine (from lactate accumulated in the tissue bed), and hormonal (from lactate in the arterial circulation).

Lactate as a Signaling Molecule

Recently, Liu and colleagues (92) showed that lactate inhibits lipolysis in fat cells through activation of an orphan G-protein coupled receptor (GPR81). In mouse, rat, and human adipocytes GPR81 appears to act as a lactate sensor that inhibits lipolysis. In comparing the recent paper of Liu and colleagues (92) with the classic papers of Issekutz, Miller, Paul, and Rodahl of the 1960s, it is interesting to observe the difference in perspective. The data showing that lactate inhibit lipolysis are largely consistent, but the emerging view of lactate sensing with lactate acting as a signaling molecule stands in stark contrast to the view that lactate is a metabolic waste that serves little purpose except to interfere with essential processes such as muscle contraction and lipolysis.

Autocrine and Paracrine Roles of Lactate

Cell and tissue self-signaling roles of lactate have been unsuspected and only recently studied, initially in the context of MCT1 expression. By studying rat muscle-derived L6 cells in typical high-glucose media, it was found that both MCT1 protein levels and lactate concentraton increased continuously due to aerobic glycolysis (63). That observation led to a study of genome-wide responses of L6 cells to elevated (10 and 20 mM) sodium–lactate added to buffered media. Those high but physiological lactate incubation levels changed cell redox and increased reactive oxygen species (ROS) production, which in turn upregulated 673 genes, many known to be responsive to ROS and Ca^{2+} (63). The genes responding could be assigned to six distinct groups according to function; these were genes encoding for mLOC components, mitochondrial biogenesis, ROS generation, ROS neutralization, mitochondrial dynamics, and calcium signaling. Indeed, by itself, lactate represents a physiological signal that is associated with many of the responses associated with exercise training.

Most recently, working in the laboratory of Salvatore Passarella in Bari and Università del Molise, Campobasso, Italy, produced evidence of the presence of an unknown lactate oxidase in the mitochondrial intermembrane space (35, 112). This work on the presence of a rotenone-insensitive mitochondrial ROS generator provides important evidence on the mechanism of mitochondrial ROS generation from lactate. Investigation of mechanisms by which lactate affects cellular redox state, ROS generation, and gene transcriptions offers a new frontier in understanding the physiological functioning of lactate, an ancient and misunderstood metabolite.

The Newest Frontier: Lactate and the mLOC in Cancer

Data from the ongoing Cancer Genome Atlas project and other cancer genome projects (32) show that all cancers display mutations in multiple genes and that the number of mutations and the affected genes vary during tumor growth and development (111). Despite the molecular heterogeneity in and between tumors, the genetic program encoding intermediary metabolism to support growth, proliferation, metastasis, and lactate production is robust in cancers. High concentrations of lactate (up to 40 nmol/g of tumor) have been detected in malignant cancer cells (148), and it has been suggested that the Warburg effect, which produces copious amounts of lactate, supplies approximately 50% of the ATP requirement of most tumors. During the past 30 yr we have studied and published extensively on lactate metabolism in vivo in healthy young humans and rodents. Consequently, it is not surprising that investigators have commenced studying lactate shuttling not only in the brain (57, 117) but also in tumors (75, 80, 94, 136). Of the three important functions of lactate shuttles (fuel energy source, gluconeogenic precursor, and signaling molecule), the first and third are suspected to be involved in cancer cell metabolism.

To date, most noteworthy have been the efforts of Sonveaux and colleagues (136), who noted that whereas tumor cells are highly glycolytic under fully aerobic conditions, lactate is the preferred energy fuel source in rapidly proliferating cells. Incubation of several human tumor cell lines in which MCT1 has been diminished by use of siRNAs or by treatment with the lactate transport protein blocker cinnamate (19, 28, 128, 129) caused death of those cells. As well, they found that when tumor cells were xenotransplanted into healthy nude (immunosuppressed) mice, growth of the resulting tumors was repressed when animals were injected with cinnamate. Significance of the experimental findings for cancer treatment resulted in a commentary in *Journal of Clinical Investigation* on how to disrupt lactate shuttling to kill cancer cells.

Seeing the results of Sonveaux and colleagues and knowing our experiments in which incubation of L6 myocytes with lactate augmented intracellular H_2O_2 and was associated with the induction of approximately 600 genes (63), we explored the literature and found that lactate also produced a robust transcriptomic response (~1,700 genes modulated) in the MCF-7 breast cancer cell line (27). Hence, it appears that in addition to metabolic role of lactate, simultaneous modulation of transcriptional response by lactate contributes to cell growth and proliferation.

Having previously worked on muscle, a postmitotic tissue, we commenced work on breast cancer cells (75) that display active mitosis. Initial efforts were on known mLOC proteins. These included lactate–pyruvate; MCT1, MCT2, and MCT4; the scaffold glycoprotein CD147; LDH (isoforms A and B); and mitochondrial COx. Our primary hypothesis was that the mLOC complex expression and assembly are dysregulated in breast cancer, and detailed results have recently appeared (75).

We tested our primary hypothesis in breast cancer cell lines MCF-7, MDA-MB-231, and HMEC-184. The former two are transformed breast cell lines that are cancerous whereas HMEC-184 is a nontransformed primary breast cell line used as a control. Protein expression of MCTs, CD147, and LDHs was determined by immunoblot analysis of whole-cell extracts from the three cell lines grown under standard conditions. The expression of the seven proteins in the two transformed cell lines was compared with that of those in the primary, untransformed cell line HMEC-184. Relative to HMEC-184, in the transformed cells some proteins (MCT2, LDH, and CD147) are upregulated. Remarkably, the expression of MCT1 was low in MCF-7 cells and undetectable in MDA-MB-231 cells. A notable discovery from our study is the reciprocal relationship between the protein expression levels of MCT1 and CD147. Because MCT1 and CD147 expressions are coregulated, our data in breast cancer cell lines suggest a deregulation of the coordinated expression of these two genes and its contribution to the carcinogenic process in breast cancer.

We also examined the subcellular localization of MCTs and LDH isoforms in the three cell lines studied. Confocal laser scanning microscopy and related immunohistochemical techniques showed the presence of MCT, LDH isoforms, and COx. LDH isoforms, MCT2, and MCT4 were colocalized with mitochondrial protein marker COx, but, distinct from the situation in muscle, MCT1 was not mitochondrial and was localized to the plasma membrane exclusively. The localization of MCT and LDH isoforms in the two cancer cell lines (MCF-7 and MDA-MB-231) and the normal cell line (HMEC-184) was the same. MCT2, MCT4, and LDH were localized in mitochondria, plasma membrane, and cytosol whereas MCT1 was localized mainly in plasma membrane. Our data show that breast cancer appears to change the expression of lactate shuttle proteins but not their subcellular localizations. The reasons for these changes need elucidation.

Considering results on cancer cells to date, it seems that the history of studies of lactate metabolism have come full circle. Once perceived as a lowly waste product formed as the result of lack of oxygen, as embodied in the shuttle hypothesis, lactate is now realized to fulfill three essential metabolic functions: It is an energy source, gluconeogenic precursor, and signaling molecule. Lactate was originally conceived of in the context of exercise physiology and supporting short- and long-term adaptive processes. Now, on one end of the spectrum, ranging from metabolic waste and fatigue agent to the lactate shuttle concept, some investigators think of using lactate esters and salts to sustain athletes in prolonged bouts of hard exercise (1) and to provide nutritive support to humans after traumatic brain injury (57), and on the other end of the spectrum investigators are giving thought to killing cancer cells by blocking lactate shuttling (75, 80, 94, 136).

Summary

Once thought to be the consequence of lack of oxygen in contracting skeletal muscle, we now know that lactate is formed and utilized continuously in diverse cells under fully aerobic conditions. In fact, as the product of one metabolic pathway (glycolysis) and the substrate for a downstream pathway (mitochondrial respiration), lactate can be regarded as the link between glycolytic and aerobic pathways. Importantly, this linkage can transcend compartment barriers and occur within and between cells, tissues, and organs. Today, knowledge of lactate metabolism is important because lactate is a major energy source, the major gluconeogenic precursor, and a signaling molecule with autocrine, paracrine, and endocrine-like effects; it is a "lactormone." Cell–cell and intracellular lactate shuttle concepts describe the roles of lactate in the delivery of oxidative and gluconeogenic substrates as well as in cell signaling. Examples of the cell–cell shuttles include lactate exchanges between white glycolytic and red oxidative fibers in a working muscle bed, working skeletal muscle, and the heart, brain, liver, and kidneys. Examples of intracellular lactate shuttles include lactate exchange among plasma, cytosol, and mitochondria as well as pyruvate for lactate exchange in peroxisomes. Pyruvate for lactate exchange balances redox status, thus allowing peroxisomal β-oxidation. Lactate for pyruvate exchanges affects cell redox, and by itself lactate generates ROS that have signaling functions. As well, lactate may affect metabolic regulation, for instance, by binding to G-protein receptors in adipocytes inhibiting lipolysis, thus decreasing plasma FFA availability. In cultured myocytes, lactate accumulation upregulates expression of *MCT1* and genes coding for other components of the mitochondrial reticulum in skeletal muscle. The muscle mitochondrial reticulum and mitochondrial networks in other aerobic tissues establish concentration and proton gradients necessary for cells with high oxidative capacities to utilize lactate. In contrast to its early portrayal as a metabolic waste product and poison, lactate is part of a classic feedback loop in which short-term challenges to ATP supply stimulate production of glycolytic ATP as well as a rapidly used oxidative substrate that also signals tissue adaptations to stimulate lactate clearance in support of ATP homeostasis over the long term. Over the course of 100 yr the understanding of lactate production and utilization processes has changed dramatically in diverse fields ranging from exercise physiology to tissue-repair technology, neurosurgery, and oncology. Table 18.1 provides a chronology of important discoveries in the field of lactate metabolism including confirmation of key features of the intracellular lactate shuttle hypothesis (164, 165).

Table 18.1 Timeline of Milestone Events in Lactate Metabolism

Dates	Events
1907-1950	The lactic acid era. Archibald Vivian (A.V.) Hill and the Cambridge School of Physiologists believe that the "processes of muscle contraction are due to the liberation of lactic acid from some precursor" (70-73). Otto Meyerhof quantifies the relationship between glycogen and lactic acid formation in isolated, nonperfused frog hemicorpus preparations (98-101). A.V. Hill develops the O_2 debt concept in human studies in the 1920s (70-73). August Krogh and Johannes Lindhard (86) provide evidence supporting Zuntz's assertion that both fat and carbohydrate are substrates for energy during exercise in humans (163), but those studies are outside the prevailing theory of the Cambridge School of A.V. Hill and colleagues. In 1933, Rodolfo Margaria, Harold T. Edwards, and David Bruce Dill of the Harvard Fatigue Laboratory apply the new knowledge of the phosphagens to O_2 debt theory in humans and segment the phenomenon into lactacid and alactacid components (93). Commencing in 1934 and again in 1936, Ole Bang (3, 4) challenges established concepts of lactacid O_2 debt. However, his results are ignored, World War II intervenes, and O_2 debt ideas become entrenched.
1951	Mario Umberto Dianzani shows that rat liver mitochondria oxidize lactate (38).
1955-1956	Douglas S. Drury, Arne N. Wick, and Toshiko Morita use ^{14}C-lactate tracers to show extrahepatic oxidative disposal of lactate in rabbits (40, 41).
1964	Karlman Wasserman and Malcolm B. McIlroy assume the Hill-Meyerhof O_2 debt theory of oxygen-limited lactate production and coin the term *anaerobic threshold* (151).
1966-1968	Wendell Stainsby and Hugh Welch demonstrate the transient nature of muscle lactic acid output in canine muscle in situ and present evidence that argues for the O_2 independence of lactic acid formation during muscle contractions (137, 154). In 1998 Brooks and colleagues observe the same phenomenon in humans and name it the Stainsby effect (23). Franz F. Jöbsis and Wendel Stainsby demonstrate oxidation of complex 1 in the mitochondrial electron transport chain of dog gastrocnemius muscles contracting in situ of sufficient intensity to produce and release lactate (78).

(continued)

Table 18.1 *(continued)*

Dates	Events
1969	Florent Depocas and colleagues use radioactive tracers to study lactate turnover and oxidation in resting and exercising dogs. Fundamental discoveries of Depocas and colleagues showing active lactate turnover in resting and exercising individuals have been replicated on numerous species, including humans (36).
1971	Nobuhisa Baba and Hari M. Sharma visualize mitochondrial lactate dehydrogenase using electron microscopy. They were the first to postulate a lactate shuttle, but did not follow up on their observation (2).
1973	Brooks and colleagues provide ^{14}C-lactate to rats after exhausting exercise and, contrary to classic O_2 debt theory, find little incorporation into glycogen but major disposal as $^{14}CO_2$. This is the first challenge to the classic Hill-Meyerhof concept of a one fifth to four fifths ratio of lactate to lactate conversion to glycogen (16).
1980	Glenn Gaesser and George Brooks use bolus injections of [U-^{14}C]glucose and lactate tracers, indirect calorimetry, and two-dimensional chromatography to trace the paths of lactate and glucose disposal during recovery from exhausting exercise. Again, they find little incorporation of lactate-derived carbon into glycogen but major disposal as $^{14}CO_2$. In mammals, oxidation—not reconversion to glycogen—is the major metabolic fate of lactate after exercise (20, 50).
1983	Casey M. Donovan and George A. Brooks use primed continuous infusions [U-^{14}C]glucose and lactate tracers, indirect calorimetry, and two-dimensional chromatography to determine the flux, oxidation, and conversion rates of lactate to glucose in endurance-trained and untrained rats during exercise. Training results in the classic finding of lowered arterial lactate concentration that is due to increased clearance via oxidation and gluconeogenesis (18, 39).
1984	Richard Connett, Tom Gayeski, and Carl Honig observe lactate production in canine muscle in situ when intramuscular PO_2 is apparently above the critical value for mitochondrial oxidative phosphorylation (30). Based on tracer-measured glucose and lactate fluxes and bochemical evidence, George Brooks proposes the lactate shuttle in a meeting of comparative physiology in Liege, Belgium. Meeting proceedings are published the following year (11). Daniel Foster presents the annual Banting lecture revealing the importance of lactate to hepatic glycogen synthesis (the indirect pathway) after carbohydrate nutrition (47).
1990	David Roth and George Brooks describe the characteristics of sarcolemmal lactic acid transport (128, 129).
1994	Christine Kim Garcia, Michael S. Brown, Joseph L. Goldstein, and colleagues sequence and clone the gene encoding for a muscle cell membrane MCT (53). Subsequently, they identify a second isoform, MCT2, found mainly in liver (52).
1998	George Brooks proposes the intracellular lactate shuttle (13). Russ Richardson, Peter Wagner, and colleagues use magnetic resonance spectroscopy to show lactate production and net release from fully aerobic, working human skeletal muscle (124, 125). Andrew Halestrap and colleagues clone and sequence four new MCT isoforms and describe tissue variability in MCT isoform expression (121). Pierre Magistretti, Luc Pellerin, and colleagues propose an astrocyte-neuron lactate shuttle (117, 118).
1999	Paul Molé and colleagues use magnetic resonance spectroscopy to confirm results of Richardson and colleagues showing lactate production and net release from fully aerobic, working human skeletal muscle (105). Henriette Pilegaard, Andrew Halestrap, and Carsten Juel and colleagues show MCT1 and MCT4 distribution in human skeletal muscle (119, 120).
1999-2000	George Brooks, Marcy Brown, Hervé Dubouchaud, and colleagues show lactate dehydrogenase and MCT1 in muscle mitochondria of rats and humans (17, 19). Bryan Bergman, Eugene Wolfel, Gail Butterfield, and colleagues show that endurance training improves lactate clearance by intramuscular oxidation and gluconeogenesis in humans (6, 7, 42).

2002	Benjamin Miller, George Brooks, and colleagues use exogenous lactate infusion (lactate clamp) and stable isotope tracer technology to test lactate clearance mechanisms and show preferential lactate over glucose oxidation in exercising men (102, 103).
2006	Takeshi Hashimoto, George A. Brooks, and colleagues provide evidence of a mitochondrial lactate oxidation complex comprising mitochondrial lactate dehydrogenase, mMCT1, basigin (CD147), and cytochrome oxidase in rodent skeletal muscle (61).
2007	Takeshi Hashimoto, George Brooks, and colleagues provide evidence of gene regulation by lactate. In L6 (rat muscle-derived) cells, the upregulated genes include those encoding for MCT1, mitochondrial proteins, proteins involved in reactive oxygen species generation and quenching, and calcium response elements (63).
2008	Garret van Hall, Niels Secher, and colleagues show cerebral lactate uptake and oxidation in exercising humans (146). Takeshi Hashimoto, George Brooks, and colleagues provide evidence of a mitochondrial lactate oxidation complex comprising mitochondrial lactate dehydrogenase, mMCT1, basigin (CD147), and cytochrome oxidase in rat neurons (62). MCTs become targets for cancer treatment when Pierre Sonveaux and colleagues find MCTs and cell–cell lactate shuttling to be prevalent among tumor cells. They use siRNAs and MCT pharmacological to kill cancer cells in vitro and in mice in vivo (136). Lidia de Bari, Daniela Valenti, Anna Atlante, and Salvatore Passarella provide evidence of the presence of an intermembrane lactate oxidase that generates H_2O_2 sufficient to activate the known reactive oxygen species response elements signaling mitochondrial adaptation and other adaptations to exercise. This work provides a mechanism by which lactate generation in muscle exercise participates in the feedback loop in which lactate generation in exercise leads to adaptations facilitating high rates of lactate disposal in exercise (35).
2013	Among others, Daniel Kane and colleagues and Robert Jacobs, Carsten Lundby, and colleagues confirm the ability of rodent (164) and human skeletal muscle mitochondria (165), thus confirming a key feature of the intracellular lactate shuttle hypothesis.

MCT = monocarboxylate transport protein.

Acknowledgment

Supported by National Institutes of Health grants DK019577, AR043906, and AR050459 and a gift from CytoSport Incorporated of Benecia, California.

References

1. Azevedo JL, Tietz E, Two-Feathers T, Paull J, Chapman K. Lactate, fructose and glucose oxidation profiles in sports drinks and the effect on exercise performance. *PLoS One* 2(9): e927, 2007.
2. Baba N, Sharma HN. Histochemistry of lactic dehydrogenase in heart and pectoralis muscles of rat. *J Cell Biol* 51: 621-635, 1971.
3. Bang O. Blutmilchsaure und Sauerstoffaufnahme wahrend und nach Muskelarbeit beim Menschen. *Arbeitsphysiologie* 7: 544-554, 1934.
4. Bang O. The lactate content of the blood during and after muscular exercise in man. *Skand Arch Physiol* 74: 46-82, 1936.
5. Bergman BC, Brooks GA. Respiratory gas-exchange ratios during graded exercise in fed and fasted trained and untrained men. *J Appl Physiol* 86: 479-487, 1999.
6. Bergman BC, Horning MA, Casazza GA, Wolfel EE, Butterfield GE, Brooks GA. Endurance training increases gluconeogenesis during rest and exercise in men. *Am J Physiol Endocrinol Metab* 278: E244-E251, 2000.
7. Bergman BC, Wolfel EE, Butterfield GE, Lopaschuk GD, Casazza GA, Horning MA, Brooks GA. Active muscle and whole body lactate kinetics after endurance training in men. *J Appl Physiol* 87: 1684-1696, 1999.
8. Bertocci LA, Lujan B. Incorporation and utilization of [3-13C]lactate and [1,2-^{13}C]acetate by rat skeletal muscle. *J Appl Physiol* 86: 2077-2089, 1999.
9. Bonen A, Tonouchi M, Miskovic D, Heddle C, Heikkila JJ, Halestrap AP. Isoform-specific regulation of the lactate transporters MCT1 and MCT4 by contractile activity. *Am J Physiol Endocrinol Metab* 279: E1131-E1138, 2000.
10. Brandt RB, Laux JE, Spainhour SE, Kline ES. Lactate dehydrogenase in rat mitochondria. *Arch Biochem Biophys* 259: 412-422, 1987.
11. Brooks GA. Lactate: Glycolytic end product and oxidative substrate during sustained exercise in

mammals—The "lactate shuttle." In: Gilles R, ed. *Proceedings In Life Sciences. Circulation, Respiration, and Metabolism: Current Comparative Approaches; First International Congress Of Comparative Physiology and Biochemistry, Liege, Belgium, Aug. 1984.* Berlin: Springer-Verlag, 1985, 208-218.

12. Brooks GA. Current concepts in lactate exchange. *Med Sci Sports Exerc* 23: 895-906, 1991.
13. Brooks GA. Mammalian fuel utilization during sustained exercise. *Comp Biochem Physiol B Biochem Molec Biol* 120: 89-107, 1998.
14. Brooks GA. Cell–cell and intracellular lactate shuttles. *J Physiol* 587: 5591-5600, 2009.
15. Brooks GA. What does glycolysis make and why is it important? *J Appl Physiol* 108: 1450-1451, 2010.
16. Brooks GA, Brauner KE, Cassens RG. Glycogen synthesis and metabolism of lactic acid after exercise. *Am J Physiol* 224: 1162-1166, 1973.
17. Brooks GA, Brown MA, Butz CE, Sicurello JP, Dubouchaud H. Cardiac and skeletal muscle mitochondria have a monocarboxylate transporter MCT1. *J Appl Physiol* 87: 1713-1718, 1999.
18. Brooks GA, Donovan CM. Effect of endurance training on glucose kinetics during exercise. *Am J Physiol Endocrinol Metab* 244: E505-E512, 1983.
19. Brooks GA, Dubouchaud H, Brown M, Sicurello JP, Butz CE. Role of mitochondrial lactate dehydrogenase and lactate oxidation in the intracellular lactate shuttle. *Proc Natl Acad Sci USA* 96: 1129-1134, 1999.
20. Brooks GA, Gaesser GA. End points of lactate and glucose metabolism after exhausting exercise. *J Appl Physiol* 49: 1057-1069, 1980.
21. Brooks GA, Gladden LB. The metabolic systems: Anaerobic metabolism (Glycolytic and phosphagen). In: Tipton CM, ed. *Exercise Physiology: People and Ideas*. Oxford: Oxford University Press, 2003, 322-360.
22. Brooks GA, White TP. Determination of metabolic and heart rate responses of rats to treadmill exercise. *J Appl Physiol* 45: 1009-1015, 1978.
23. Brooks GA, Wolfel EE, Butterfield GE, Cymerman A, Roberts AC, Mazzeo RS, Reeves JT. Poor relationship between arterial [lactate] and leg net release during exercise at 4,300 m altitude. *Am J Physiol Endocrinol Metab* 275: R1192-R1201, 1998.
24. Brooks GA, Wolfel EE, Groves BM, Bender PR, Butterfield GE, Cymerman A, Mazzeo RS, Sutton JR, Wolfe RR, Reeves JT. Muscle accounts for glucose disposal but not blood lactate appearance during exercise after acclimatization to 4,300 m. *J Appl Physiol* 72: 2435-2445, 1992.
25. Butz CE, McClelland GB, Brooks GA. Confirmation of MCT1 in mitochondria of rat striated muscle. *J Appl Physiol* 97: 1059-1066, 2004.
26. Chatham JC, Des Rosiers C, Forder JR. Evidence of separate pathways for lactate uptake and release by the perfused rat heart. *Am J Physiol Endocrinol Metab* 281: E794-E802, 2001.
27. Chen JL, Merl D, Peterson CW, Wu J, Liu PY, Yin H, Muoio DM, Ayer DE, West M, Chi JT. Lactic acidosis triggers starvation response with paradoxical induction of TXNIP through MondoA. *PLoS Genet* 6.
28. Chretien D, Pourrier M, Bourgeron T, Sene M, Rotig A, Munnich A, Rustin P. An improved spectrophotometric assay of pyruvate dehydrogenase in lactate dehydrogenase contaminated mitochondrial preparations from human skeletal muscle. *Clin Chim Acta* 240: 129-136, 1995.
29. Conley KE, Kushmerick MJ, Jubrias SA. Glycolysis is independent of oxygenation state in stimulated human skeletal muscle in vivo. *J Physiol (Lond)* 511: 935-945, 1998.
30. Connett RJ, Gayeski TE, Honig CR. Lactate accumulation in fully aerobic, working, dog gracilis muscle. *Am J Physiol Heart Physiol* 246: H120-H128, 1984.
31. Cori CF, Cori GT. Glycogen formation in the liver from d- and l-lactic acid. *J Biol Chem* 81: 389, 1929.
32. Cowin PA, Anglesio M, Etemadmoghadam D, Bowtell DD. Profiling the cancer genome. *Annu Rev Genomics Hum Genet* 11: 133-159, 2010.
33. Daff S, Ingledew WJ, Reid GA, Chapman SK. New insights into the catalytic cycle of flavocytochrome b2. *Biochem* 35: 6351-6357, 1996.
34. Daum G, Bohni PC, Schatz G. Import proteins into mitochondria. Cytochrome b2 and cytochrome c peroxidase are located in the intermembrane space of yeast mitochondria. *J Biol Chem* 252: 13028-13033, 1982.
35. de Bari L, Valenti D, Atlante A, Passarella SL. L-lactate generates hydrogen peroxide in purified rat liver mitochondria due to the putative L-lactate oxidase localized in the intermembrane space. *FEBS Lett* 584: 2285-2290, 2010.
36. Depocas F, Minaire Y, Chatonnet J. Rates of formation and oxidation of lactic acid in dogs at rest and during moderate exercise. *Can J Physiol Pharmacol* 47: 603-610, 1969.
37. Deuticke B. Monocarboxylate transport in erythrocytes. *J Membr Biol* 70: 89-103, 1982.
38. Dianzani MU. La ripartizione del sistema ossidante l'acido lactico nelle cellule del pagato e del rene di ratti normali e di ratti con degenerazione grassa del fegato. *Archivo di Physiologica* 50: 181-186, 1951.
39. Donovan CM, Brooks GA. Endurance training affects lactate clearance, not lactate production. *Am J Physiol Endocrinol Metab* 244: E83-E92, 1983.
40. Drury DR, Wick AN. Metabolism of lactic acid in the intact rabbit. *Am J Physiol* 184: 304-308, 1956.

41. Drury DR, Wick AN, Morita TN. Metabolism of lactic acid in extrahepatic tissues. *Am J Physiol* 180: 345-349, 1955.
42. Dubouchaud H, Butterfield GE, Wolfel EE, Bergman BD, Brooks GA. Endurance training, expression, and physiology of LDH, MCT1, and MCT4 in human skeletal muscle. *Am J Physiol Endocrinol Metab* 278: E571-E579, 2000.
43. Embden G, Baldes K, Schmitz E. Über den Chemismus der Milchsaurebildung aus Traubenzucker im Teirkorper. *Biochemishe Zeitung* 45: 108-133, 1912.
44. Embden G, Laquer F. Über die Chemi des Lactocidogens. I. Isolierungsversuche. *Hoppe-Seylers Zeitschrift fur Physiologische Chemie* 93: 94-123, 1914.
45. Embden G, Laquer F. Über die Chemie des Lactacidogens. II. *Hoppe-Seylers Zeitschrift fur Physiologische Chemie* 98: 181, 1917.
46. Fletcher WM, Hopkins F. Lactic acid in amphibian muscle. *J Physiol (Lond)* 35: 247-309, 1907.
47. Foster D. Banting lecture 1984. From glycogen to ketones—and back. *Diabetes* 33: 1188-1199, 1984.
48. Friedlander AL, Casazza GA, Horning MA, Huie MJ, Piacentini MF, Trimmer KK, Brooks GA. Training-induced alterations of carbohydrate metabolism in young women: Women respond differently from men. *J Appl Physiol* 85: 1175-1186, 1998.
49. Friedlander AL, Casazza GA, Huie MJ, Horning MA, Brooks GA. Endurance training alters glucose kinetics in response to the same absolute, but not the same relative workload. *J Appl Physiol* 82: 1360-1369, 1997.
50. Gaesser GA, Brooks GA. Glycogen depletion following continuous and intermittent exercise to exhaustion. *J Appl Physiol* 49: 722-728, 1980.
51. Gaesser GA, Brooks GA. Metabolic bases of excess post-exercise oxygen consumption: A review. *Med Sci Sports Exerc* 16: 29-43, 1984.
52. Garcia CK, Brown MS, Pathak RK, Goldstein JL. cDNA cloning of MCT2, a second monocarboxylate transporter expressed in different cells than MCT1. *J Biol Chem* 270: 1843-1849, 1995.
53. Garcia CK, Goldstein JL, Patha RK, Anderson RG, Brown MS. Molecular characterization of a membrane transporter for lactate, pyruvate, and other monocarboxylates: Implications for the Cori cycle. *Cell* 76: 865-873, 1994.
54. Gertz EW, Wisneski JA, Stanley WS, Neese RA. Myocardial substrate utilization during exercise in humans. Dual carbon-labeled carbohydrate isotope experiments. *J Clin Invest* 82: 2017-2025, 1988.
55. Gisolfi C, Robinson S, Turrell ES. Effects of aerobic work performed during recovery from exhausting work. *J Appl Physiol* 21: 1767-1777, 1966.
56. Gladden LB, Crawford RE, Webster MJ. Effect of lactate concentration and metabolic rate on net lactate uptake by canine skeletal muscle. *Am J Physiol Reg Integ Comp Physiol* 266: R1095-R1101, 1994.
57. Glenn TC, Kelly DF, Boscardin WJ, McArthur DL, Vespa P, Oertel M, Hovda DA, Bergsneider M, Hillered L, Martin NA. Energy dysfunction as a predictor of outcome after moderate or severe head injury: Indices of oxygen, glucose, and lactate metabolism. *J Cereb Blood Flow Metab* 23: 1239-1250, 2003.
58. Gold M, Miller HI, Issekutz B, Spitzer JJ. Effect of exercise and lactic acid infusion on individual free fatty acids of plasma. *Am J Physiol* 205: 902-904, 1963.
59. Halestrap AP. Transport of pyruvate and lactate into human erythrocytes. Evidence for the involvement of the chloride carrier and a chloride-independent carrier. *Biochem J* 156: 193-207, 1976.
60. Hamann JJ, Kelley KM, Gladden LB. Effect of epinephrine on net lactate uptake by contracting skeletal muscle. *J Appl Physiol* 91: 2635-2641, 2001.
61. Hashimoto T, Hussien R, Brooks GA. Colocalization of MCT1, CD147 and LDH in mitochondrial inner membrane of L6 cells: Evidence of a mitochondrial lactate oxidation complex. *Am J Physiol Endocrinol Metab* 290: 1237-1244, 2006.
62. Hashimoto T, Hussien R, Cho H-S, Kaufer D, Brooks GA. Evidence for a mitochondrial lactate oxidation complex in rat neurons: A crucial component for a brain lactate shuttle. *PloS-One* 3(8): e2915, 2008.
63. Hashimoto T, Hussien R, Oommen S, Gohil K, Brooks GA. Lactate sensitive transcription factor network in L6 myocytes: Activation of MCT1 expression and mitochondrial biogenesis. *FASEB J* 21: 2602-2612, 2007.
64. Hashimoto T, Masuda S, Taguchi S, Brooks GA. Immunohistochemical analysis of MCT1, MCT2 and MCT4 expression in rat plantaris muscle. *J Physiol (Lond)* 567: 121-129, 2005.
65. Hellerstein MK, Neese RA, Linfoot P, Christiansen M, Turner S, Letscher A. Hepatic gluconeogenic fluxes and glycogen turnover during fasting in humans. A stable isotope study. *J Clin Invest* 100: 1305-1319, 1997.
66. Henderson GC, Horning MA, Lehma SL, Wolfel EE, Bergman BC, Brooks GA. Pyruvate shuttling during rest and exercise before and after endurance training in men. *J Appl Physiol* 97: 317-325, 2004.
67. Hill AV. The energy degraded in recovery processes of stimulated muscles. *J Physiol (Lond)* 46: 28-80, 1913.
68. Hill AV. The oxidative removal of lactic acid. *J Physiol (Lond)* 48: x-xi, 1914.
69. Hill AV. Myothermic apparatus. *Proc Royal Soc London B* 103: 117-137, 1928.

70. Hill AV, Long CNH, Lupton H. Muscular exercise, lactic acid and the supply and utilisation of oxygen. Pt I-III. *Proc Royal Soc B* 96: 438-475, 1924.
71. Hill AV, Long CNH, Lupton H. Muscular exercise, lactic acid and the supply and utilisation of oxygen. Pt IV-VI. *Proc Royal Soc B* 97: 84-138, 1924.
72. Hill AV, Long CNH, Lupton H. Muscular exercise, lactic acid and the supply and utilisation of oxygen. Pt VII-VIII. *Proc Royal Soc B* 97: 155-176, 1924.
73. Hill AV, Lupton H. Muscular exercise, lactic acid and the supply and utilization of oxygen. *Quar J Med* 16: 135-171, 1923.
74. Hochachka PW. *Living Without Oxygen.* Cambridge, MA: Harvard University Press, 1980.
75. Hussien R, Brooks GA. Mitochondrial and plasma membrane lactate transporter and lactate dehydrogenase isoform expression in breast cancer cell lines. *Phys Genom* 43: 255-264, 2011.
76. Issekutz B. Effect of beta-adrenergic blockade on lactate turnover in exercising dogs. *J Appl Physiol* 57: 1754-1759, 1984.
77. Issekutz B, Miller H. Plasma free fatty acids during exercise and the effect of lactic acid. *Proc Soc Exp Biol Med* 110: 237, 1962.
78. Jöbsis FF, Stainsby WN. Oxidation of NADH during contractions of circulated mammalian skeletal muscle. *Respir Physiol* 4: 292-300, 1968.
79. Keilin D. *The History of Cell Respiration and Cytochrome.* Cambridge: Cambridge University Press, 1966, xx, 416.
80. Kennedy KM, Dewhirst MW. Tumor metabolism of lactate: The influence and therapeutic potential for MCT and CD147 regulation. *Future Oncol* 6: 127-148, 2010.
81. Kirkwood SP, Munn EA, Packer L, Brooks GA. Mitochondrial reticulum in limb skeletal muscle. *Am J Physiol* 251: C395-C402, 1986.
82. Kirkwood SP, Munn EA, Packer L, Brooks GA. Effects of endurance training on mitochondrial reticulum in limb skeletal muscle. *Arch Biochem Biophys* 255: 80-88, 1987.
83. Kleiber M. *The Fire of Life: An Introduction to Animal Energetics.* New York: Wiley, 1961, 116-128, 291-311.
84. Kline ES, Brandt RB, Laux JE, Spainhour SE, Higgins ES, Rogers KS, Tinsley SB, Waters MG. Localization of L-lactate dehydrogenase in mitochondria. *Arch Biochem Biophys* 246: 673-680, 1986.
85. Krebs HA, Kornberg HL, Burton K. A survey of the energy transformations in living matter. *Ergeb Physiol* 49: 212-298, 1957.
86. Krogh A, Lindhard J. The relative value of fat and carbohydrate as sources of muscular energy. *Biochem J* 14: 290-363, 1920.
87. Lam CX, Hutmacher DW, Schantz T, Woodruff MA, Teoh SH. Evaluation of polycaprolactone scaffold degradation for 6 months in vitro and in vivo. *J Biomed Mater Res A* 90: 906-919, 2009.
88. Laughlin MR, Taylor J, Chesnick AS, DeGroot M, Balaban RS. Pyruvate and lactate metabolism in the in vivo dog heart. *Am J Physiol* 264: H2068-H2079, 1993.
89. Lehninger AL. *Biochemistry: The Molecular Basis of Cell Structure and Function.* New York: Worth, 1970, xiii, 833.
90. Leicester HM. *Development of Biochemical Concepts from Ancient to Modern Times.* Cambridge, MA: Harvard University Press, 1974, 286.
91. Lemire J, Mailloux RJ, Appanna VD. Mitochondrial lactate dehydrogenase is involved in oxidative-energy metabolism in human astrocytoma cells (CCF-STTG1). *PLoS One* 3(2): e1550, 2008.
92. Liu C, Wu J, Zhu J, Kuei C, Yu J, Shelton J, Sutton SW, Li X, Yun SJ, Mirzadegan T, Mazur T, Kamme F, Lovenberg TW. Lactate inhibits lipolysis in fat cells through activation of an orphan G-protein-coupled receptor, GPR81. *J Biol Chem* 284: 2811-2822, 2009.
93. Margaria R, Edwards RHT, Dill DB. The possible mechanisms of contracting and paying the oxygen debt and the role of lactic acid in muscular contraction. *Am J Physiol* 106: 689-715, 1933.
94. Mazan-Mamczarz K, Hagner P, Dai B, Corl S, Liu Z, Gartenhaus RB. Targeted suppression of MCT-1 attenuates the malignant phenotype through a translational mechanism. *Leuk Res* 33: 474-482, 2009.
95. Mazzeo RS, Brooks GA, Schoeller DA, Budinger TF. Disposal of blood [1-^{13}C]lactate in humans during rest and exercise. *J Appl Physiol* 60: 232-241, 1986.
96. McClelland GB, Khanna S, González GF, Butz CE, Brooks GA. Peroxisomal membrane monocarboxylate transporters: Evidence for a redox shuttle system? *Biochem Biophys Res Comm* 203: 130-135, 2003.
97. Meyer C, Stumvoll M, Dostou J, Welle S, Haymond M, Gerich JE. Renal substrate exchange and gluconeogenesis in normal postabsorptive humans. *Am J Physiol Endocrinol Metab* 282: E428-E434, 2002.
98. Meyerhof O. Die Energieumwandlungen im Muskel I. Über die Beziehungen der Milchsaure zur Warmebildung and Arbeitsleistung des Muskels in der Anaerobiose. *Pflügers Archiv ges Physiol Mensch Tiere* 182: 232-283, 1920.
99. Meyerhof O. Die Energieumwandlungen im Muskel II. Das Schicksal der Milchsaure in der Erholungsperiode des Muskels. *Pflügers Archiv ges Physiol Mensch Tiere* 182: 284-317, 1920.
100. Meyerhof O. Die Energieumwandlungen im Muskel III. Kohlenhydrat- und Milchsaureumsatz im

Froschmuskel. *Pflügers Archiv ges Physiol Mensch Tiere* 185: 11-32, 1920.

101. Meyerhof O. University of Chicago, and University of Wisconsin. *A Symposium on Respiratory Enzymes*. Madison, WI: University of Wisconsin Press, 1942, xii, 281.
102. Miller BF, Fattor JA, Jacobs KA, Horning MA, Suh S-H, Navazio F, Brooks GA. Metabolic and cardiorespiratory responses to an exogenous lactate infusion during rest and exercise: "The lactate clamp." *Am. J. Physiol Endocrinol Metab* 283: E889-E898, 2002.
103. Miller BF, Fattor JA, Jacobs KA, Horning MA, Suh S-H, Navazio F, Brooks GA. Lactate-glucose interaction in men during rest and exercising using lactate clamp procedure. *J Physiol (Lond)* 544: 963-975, 2002.
104. Miller HI, Issekutz B, Rodahl K, Paul P. Effect of lactic acid on plasma free fatty acids in pancreatectomized dogs. *Am J Physiol* 207: 1226-1230, 1964.
105. Molé PA, Chung Y, Tran TK, Sailasuta N, Hurd R, Jue T. Myoglobin desaturation with exercise intensity in human gastrocnemius muscle. *Am J Physiol Reg Integ Comp Physiol* 277: R173-R180, 1999.
106. Nakae Y, Stoward PJ, Shono M, Matsuzaki T. Localisation and quantification of dehydrogenase activities in single muscle fibers of mdx gastrocnemius. *Histochem Cell Biol* 112: 427-436, 1999.
107. Needham J, Hill R. *The Chemistry of Life: Eight Lectures on the History of Biochemistry*. Cambridge, England: University Press, 1970, xxix, 213.
108. Nobel Prize. *The First 100 Years*. London: World Scientific, 2001.
109. Pagliarini DJ, Calvo SE, Chang B, Sheth SA, Vafai SB, Ong SE, Walford GA, Sugiana C, Boneh A, Chen WK, Hill DE, Vidal M, Evans JG, Thorburn DR, Carr SA, Mootha VK. A mitochondrial protein compendium elucidates complex I disease biology. *Cell* 134: 112-123, 2008.
110. Pagliassotti MJ, Donovan CM. Glycogenesis from lactate in rabbit skeletal muscle fiber types. *Am J Physiol* 258: R903-R911, 1990.
111. Parsons DW, Li M, Zhang X, Jones S, Leary RJ, Lin JC, et al. The genetic landscape of the childhood cancer medulloblastoma. *Science* 331: 435-439, 2011.
112. Passarella S, de Bari L, Valenti D, Pizzuto R, Paventi G, Atlante A. Mitochondria and L-lactate metabolism. *FEBS Lett* 582: 3569-3576, 2008.
113. Pasteur L. Memorie sur la fermentation appelee lactique. 52: 404-418, 1858.
114. Pasteur L. Recherches sur putrefaction. 56: 1189-1194, 1863.
115. Pedersen BK, Edward F. Adolph distinguished lecture: Muscle as an endocrine organ: IL-6 and other myokines. *J Appl Physiol* 107: 1006-2014, 2009.
116. Pedersen TH, de Paoli F, Nielsen OB. Increased excitability of acidified skeletal muscle: Role of chloride conductance. *J Gen Physiol* 125: 237-246, 2005.
117. Pellerin L, Pellegri G, Bittar PG, Charnay Y, Bouras C, Martin JL, Stella N, Magistretti PJ. Evidence supporting the existence of an activity-dependent astrocyte-neuron lactate shuttle. *Dev Neurosci* 20: 291-299, 1998.
118. Pierre K, Pellerin L. Monocarboxylate transporters in the central nervous system: Distribution, regulation and function. *J Neurochem* 94: 1-14, 2005.
119. Pilegaard H, Bangsbo J, Richter EA, Juel C. Lactate transport studied in sarcolemmal giant vesicles from human muscle biopsies: Relation to training status. *J Appl Physiol* 77: 1858-1862, 1994.
120. Pilegaard H, Terzis H, Halestrap A, Juel C. Distribution of the lactate/H+ transporter isoforms MCT1 and MCT4 in human skeletal muscle. *Am J Physiol Endocrinol Metab* 276: E843-E848, 1999.
121. Price NT, Jackson VN, Halestrap AP. Cloning and sequencing of four new mammalian monocarboxylate transporter (MCT) homologues confirms the existence of a transporter family with an ancient past. *Biochem J* 329: 321-328, 1998.
122. Rasmussen HN, van Hal G, Rasmussen UF. Lactate dehydrogenase is not a mitochondrial enzyme in human and mouse vastus lateralis muscle. *J Physiol* 541, 575-580, 2002.
123. Reeves JT, Wolfel EE, Green JJ, Mazze RS, Young AJ, Sutton JR, Brooks GA. Oxygen transport during exercise at altitude and the lactate paradox: Lessons from Operation Everest II and Pikes Peak. *Exer Sport Sci Rev* 20: 275-296, 1992.
124. Richardson RS, Noyszewski EA, Kendrick KF, Leigh JS, Wagner PD. Myoglobin O_2 desaturation during exercise: Evidence of limited O_2 transport. *J Clin Invest* 96: 1916-1926, 1995.
125. Richardson RS, Noyszewski EA, Leigh JS, Wagner PD. Lactate efflux from exercising human skeletal muscle: Role of intracellular PO_2. *J Appl Physiol* 85: 627-634, 1998.
126. Robergs RA, Ghiasvand F, Parker D. Biochemistry of exercise-induced metabolic acidosis. *Am J Physiol Regul Integr Comp Physiol* 287: R502-R516, 2004.
127. Rodahl K, Miller H, Issekutz B. Plasma free fatty acids in exercise. *J Appl Physiol* 19: 489-492, 1964.
128. Roth DA, Brooks GA. Lactate transport is mediated by a membrane-bound carrier in rat skeletal muscle sarcolemmal vesicles. *Arch Biochem Biophys* 279: 377-385, 1990.
129. Roth DA, Brooks GA. Lactate and pyruvate transport is dominated using a pH gradient-sensitive

carrier in rat skeletal muscle sarcolemmal vesicles. *Arch Biochem Biophys* 279: 386-394, 1990.
130. Rothschuh KE. *History of Physiology*. Huntington, NY: Krieger, 1973, xxi, 379.
131. Saddik M, Gamble J, Witters J, Lopaschuk GD. Acetyl-CoA carboxylase regulation of fatty acid oxidation in the heart. *J Biol Chem* 268: 25836-25845, 1993.
132. Sahlin K, Fernstrom M, Svensson M, Tonkonogi M. No evidence of an intracellular lactate shuttle in rat skeletal muscle. *J Physiol* 541, 569-574, 2002.
133. Schoenheimer R. *The Dynamic State of Body Constituents*. New York: Hafner, 1964, x, 78.
134. Schurr A. Lactate: The ultimate cerebral oxidative energy substrate? *J Cereb Blood Flow Metab* 26: 142-152, 2006.
135. Schurr A. Lactate: A major and crucial player in normal function of both muscle and brain. *J Physiol* 586: 2665-2666, 2008.
136. Sonveaux P, Végran F, Schroeder T, Wergin MC, Verrax J, Rabbani ZN, De Saedeleer CJ, Kennedy KM, Diepart C, Jordan BF, Kelley MJ, Gallez B, Wahl MJ, Feron O, Dewhirst MW. Targeting lactate-fueled respiration selectively kills hypoxic tumor cells in mice. *J Clin Invest* 118: 3930-3942, 2008.
137. Stainsby WN, Welch HG. Lactate metabolism of contracting dog skeletal muscle in situ. *Am J Physiol* 211: 177-183, 1966.
138. Stanley WC, Gertz EW, Wisneski JA, Morris DL, Neese RA, Brooks GA. Systemic lactate kinetics during graded exercise in man. *Am J Physiol Endocrinol Metab* 249: E595-E602, 1985.
139. Stanley WC, Gertz EW, Wisneski JA, Morris DL, Neese RA, Brooks GA . Lactate extraction during net lactate release in legs of humans during exercise. *J Appl Physiol* 160: 1116-1120, 1986.
140. Sumeji B, Podanyi B, Forgo P, Kovel KE. Metabolism of [3-^{13}C]pyruvate and [3-^{13}C]-propionate in normal and ischaemic rat heart in vivo: ^{1}H- and ^{13}C-NMR studies. *Biochem J* 312: 75-81, 1995.
141. Szczesna-Kaczmarek A. L-lactate oxidation by skeletal muscle mitochondria. *Int J Biochem* 22: 617-620, 1990.
142. Taylor SW, Fahy E, Zhang B, Glenn GM, Warnock DE, Wiley S, Murphy AN, Gaucher SP, Capaldi RA, Gibson BW, Ghosh SS. Characterization of the human heart mitochondrial proteome. *Nat Biotechnol* 21: 281-286, 2003.
143. Tipton CM. *Exercise Physiology: People and Ideas*. Oxford: Oxford University Press, 2003.
144. Trimmer JK, Casazza GA, Horning MA, Brooks GA. Autoregulation of glucose production in men with a glycerol load during rest and exercise. *J Physiol Endocrinol Metab* 280: E657-E668, 2001.
145. Trimmer JK, Schwarz J-M, Casazza GA, Horning MA, Rodriguez N, Brooks GA. Measurement of gluconeogenesis in exercising men by mass isotopomer distribution analysis. *J Appl Physiol* 93: 233-241, 2002.
146. van Hall G, Strømstad M, Rasmussen P, Jans O, Zaar M, Gam C, Quistorff B, Secher NH, Nielsen HB. Blood lactate is an important energy source for the human brain. *J Cereb Blood Flow Metab* 29: 1121-1129, 2008.
147. von Muralt A. The development of muscle chemistry: A lesson in neurophbysiology. *Biochim Biophys Acta* 4: 126-129, 1950.
148. Walenta S, Schroeder T, Mueller-Klieser W. Lactate in solid malignant tumors: Potential basis of a metabolic classification in clinical oncology. *Curr Med Chem* 11: 2195-2204, 2004.
149. Warburg O. The origin of cancer cells. *Science* 124: 269-270, 1956.
150. Wasserman, K. Coupling of external to cellular respiration during exercise: The wisdom of the body revisited. *Am J Physiol Endocrinol Metab* 266: E519-E539, 1994.
151. Wasserman K, McIlroy MB. Detecting the threshold of anaerobic metabolism in cardiac patients during exercise. *Am J Card* 14: 844-852, 1964.
152. Watt MJ, Howlett KF, Febbraio MA, Spriet LL, Hargreaves M. Adrenaline increases skeletal muscle glycogenolysis, pyruvate dehydrogenase activation and carbohydrate oxidation during moderate exercise in humans. *J Physiol (Lond)* 534: 269-278, 2001.
153. Weber JM, Parkhouse WS, Dobson GP, Harman JC, Snow DH, Hochachka PW. Lactate kinetics in exercising Thoroughbred horses: Regulation of turnover rate in plasma. *Am J Physiol Reg Integ Comp Physiol* 253: R896-R903, 1987.
154. Welch HG, Stainsby WN. Oxygen debt in contracting dog skeletal muscle in situ. *Resp Physiol* 3: 229-242, 1967.
155. Wickler SJ, Gleeson TT. Lactate and glucose metabolism in mouse (Mus musculus) and reptile (Anolis carolinensis) skeletal muscle. *Am J Physiol Reg Integ Comp Physiol* 264: R487-R491, 1993.
156. Williams TI. *A Biographical Dictionary of Scientists*. New York: Wiley, 1974, xv, 641.
157. Wilson MC, Jackson VN, Heddle C, Price NT, Pilegaard H, Juel C, Bonen A, Montgomery I, Hutter OF, Halestrap AP. Lactic acid efflux from white skeletal muscle is catalyzed by the monocarboxylate transporter isoform MCT3. *J Biol Chem* 273: 15920-15926, 1998.
158. Woerle HJ, Meyer C, Dostou JM, Gosmanov NR, Islam N, Popa E, Wittlin SD, Welle SL, Gerich JE. Pathways for glucose disposal after meal ingestion in humans. *Am J Physiol Endocrinol Metab* 284: E716-E725, 2003.

159. Wonisch M, Hofmann P, Fruhwald FM, Hoedl R, Schwaberger G, Pokan R, von Duvillard SP, Klein W. Effect of beta(1)-selective adrenergic blockade on maximal blood lactate steady state in healthy men. *Eur J Appl Physiol* 87: 66-71, 2002.

160. Yoon Y. Sharpening the scissors: Mitochondrial fission with aid. *Cell Biochem Biophys* 41: 193-206, 2004.

161. Yoshida Y, Holloway GP, Ljubicic V, Hatta H, Spriet LL, Hood DA, Bonen A. Negligible direct lactate oxidation in subsarcolemmal and intermyofibrillar mitochondria obtained from red and white rat skeletal muscle. *J Physiol* 582: 1317-1335, 2007.

162. Zinker BA, Namdaran K, Wilson R, Lacy DB, Wasserman DH. Acute adaptation of carbohydrate metabolism to decreased arterial PO_2. *Am J Physiol* 266: E921-E929, 1994.

163. Zuntz N. Betrachtungen über die beziehungen zwischen nahrstoffen und leistungen des korpers. In: Oppenheimer K, ed. *Handbuch der Biochemie des Menschen und der Tiere*. Jena, FRG: Fischer.1911, 826-863.

164. Elustondo PA, White AE, Hughes ME, Brebner K, Pavlov E, Kane DA. Physical and functional association of lactate dehydrogenase (LDH) with skeletal muscle mitochondria. *J Biol Chem.* 288: 25309-25317, 2013.

165. Jacobs RA, Meinild AK, Nordsborg NB, Lundby C. Lactate oxidation in human skeletal muscle mitochondria. *Am J Physiol Endocrinol Metab.* 304: E686-694, 2013.

CHAPTER 19

The Temperature Regulatory System

Suzanne Schneider, PhD

Pope Moseley, MD

This chapter presents a select overview of advances in our understanding of exercise thermoregulation. In the early 1900s, John Scott Haldane was one of the first to document in field and controlled laboratory investigations many of the factors that influence the increase in body temperature during exertion. However, it was not until 1938 that Marius Nielson demonstrated that the exercise-induced increase in core temperature is tightly regulated. In the early 1960s, a debate ensued about whether core or skin temperature is the primary controller for thermoregulation. This issue was finally settled when Theodore Benzinger developed a new sensitive method for measuring core temperature. He showed that tympanic temperature correlates closer with changes in sweating and skin blood flow during exercise than skin temperature does. Richard Hellon and Alexander Lind next described a prescriptive zone in which the ambient temperature does not affect core temperature during moderate exercise. Exercise outside of this prescriptive zone results in a continued rise in core temperature until a critical temperature is reached, where the subject either fatigues or risks severe heat illness. Fitness and heat acclimation widen this prescriptive zone and increase the critical temperature for heat exhaustion. Artificial heat acclimation was first used by Aldo Dreosti in 1935 to reduce heat-related deaths in South African miners. The traditional methods of heat acclimation we use today were developed by Sid Robinson to prepare soldiers for desert or jungle warfare during World War II. Grover C. Pitts from the Harvard Fatigue Laboratory recommended that soldiers replace all fluids lost when working in the heat to avoid dehydration and heat exhaustion. The cardiovascular and thermal consequences of dehydration were explicitly described by Edward F. Adolph in his book *Physiology of Man in the Desert*, published in 1947. It wasn't until the late 1980s that Carl Gisolfi challenged the early dogma of Adolph and others that water is the most effective hydration fluid. Today, the debate continues about the best methods of hydrating and rehydrating during exertion. Humans have greater difficulty adapting to cold than to heat. Whether humans can acclimate to cold depends on the characteristics of the environmental stress and the individual. Cold acclimatization has been documented only in special populations, and it may involve insulative, hypothermic, or metabolic adaptations. In the 21st century we are discovering the processes of thermoadaptation at the cellular and molecular levels. Exciting new evidence suggests that cellular thermotolerance induces a shared resistance against not only thermal stress but other environmental and disease stressors.

1910 to 2010: A Century of Discovery

The goal of this chapter is to provide a colorful and accurate view of select advances in the understanding of exercise thermoregulation from 1910 to 2010 (table 19.1). The authors apologize for omissions; only a fraction of the deserving investigations conducted during this century of discovery can be mentioned. Readers are encouraged to obtain original papers (many available through the American Physiological Society legacy series) and to read excellent reviews on thermoregulation (16, 18, 30, 70, 100).

Table 19.1 Milestones of Discovery in Thermoregulation (1910-2010)

Year	Observation	Reference
Thermoregulation, central controller, and pathways		
1912	A second thermal control center in the hypothalamus is discovered	105
1913	Temperature regulation is a balance between two thermal control centers	128
1926	The concept of homeostasis is illustrated using temperature regulation as an example	32
1935	Thermosensitivity of the spinal cord is shown	199
1936	The proportional control theory of thermoregulation with core and skin temperature inputs is described	28
1949	Anterior hypothalamus heating causes a vasomotor response	60
1950	Temperature-dependent potentials from cells in the hypothalamus are measured	56
1951	Electrical properties of thermoreceptors in the skin are characterized	97
1953	The effect of electrolytes ions on thermoregulation is noted	4
1958	The shell–core model of thermoregulation is presented	8
1963	The set point theory of thermoregulation is advocated by Hammel	86
1964	The amine theory for central thermal regulation is proposed	57
1966	Vasodilation and sweating are independently regulated in nonpalmar regions	218
1967	3 types of thermosensitive cells are identified in the anterior hypothalamus identified	95
1970	The ionic theory of central thermoregulation is proposed	58
1977	Thermosensitive neurons are shown to be influenced by hormones, glucose, and osmolality	25
1983	Muscle thermosensitivity is noted	109
1995	The heat flux theory is extended from the Burton and Snellen model	211
2002	Molecular mechanisms for receptor thermosensitivity are described	125, 156
2006	The reciprocal innervation theory of thermoregulation is advocated by Bligh	23
2008	Dual pathways for thermosensation and thermoregulation are proposed	132, 138
Physiological responses to exercise and heat		
1912	The importance of fluid intake for heat tolerance is shown	103
1923	Caucasians can acclimate to heat	11
1923	High ambient temperatures reduce productivity in workers	208
1923	Exposure to high temperature increases blood volume and improves heat responses	14
1924	Exposure to heat increases skin blood flow and radiant heat loss	13
1933	The benefits of salt intake on reducing heat injuries is shown in Hoover Dam workers	197
1933	Sweating responses during heat acclimation are characterized	45
1935	Heat acclimation is used by Dreosti to reduce heat illness in miners	113
1938	Core temperature is regulated during exercise	143

1941	Sex differences in sweating are noted	89
1943	A rapid heat acclimation protocol is developed and classic responses are developed	167
1944	The importance of water and salt in the diet to allow for prolonged work in heat is noted	158
1947	Voluntary dehydration and the importance of drinking to replace all fluid losses is described	1
1958	Exercise core temperature is independent of ambient temperature in a prescriptive zone	96
1960	Core temperature during exercise is related to the relative intensity of exercise	9
1961	The Hardy engineering model of thermoregulation with zones of regulation is presented	88
1963	The controlled hyperthermia method for heat acclimation is shown	65
1964	The effects of dehydration on maximal and submaximal work is characterized	174
1965	Artificial heat acclimation reduces mortality in miners by 50%	113
1965	A lower sweating rate in women does not make them heat intolerant	221
1966	Physical training alone is not sufficient to induce all the benefits of heat acclimation	196
1967	Ice cream study: The predominance of core temperature for thermoregulation is shown	18
1970	Human sweating involves sympathetic cholinergic innervation	64
1980	Input from osmoreceptors and baroreceptors are added to thermoregulation models	136
1986	The effects of exercise and heat on cardiac output distribution during exercise are quantified	170
1987	The critical temperature theory of fatigue is proposed	26
1988	The importance of sodium in drinks for rehydration and rapid recovery after exercise is demonstrated	146, 147
1989	The benefits of electrolyte and glucose in drinking solutions during prolonged exercise in heat is shown	33
2009	The role of nitric oxide in active skin vasodilation is recognized	112
Thermoregulation in cold		
1924	Evidence of central cold receptors is found	186
1930	Insulative cold acclimation is shown in Australian Aborigines	98
1930	The hunting reaction is described in human skin	120
1943	The afterdrop phenomena is noted and attributed to a circulatory response	115
1950	The critical temperature concept is described	182
1958	Ama divers confirm metabolic cold acclimation in humans	99
1959	Hypothermic acclimation is noted in bushmen	85
1961	Shivering efficiency at rest and during exercise in cold water is measured	88
1961	Nonshivering thermogenesis is shown in adult humans	40
1963	The importance of subcutaneous fat in heat conservation is quantified	31
1967	The role of central versus peripheral cold reception in thermoregulation is debated	52
1986	The conductive theory of the afterdrop is proposed	210

(continued)

Table 19.1 *(continued)*

Year	Observation	Reference
Molecular changes with thermoregulation		
1962	Heat shock proteins are discovered in *Drosophila*	163
1989	Heat exposure reduces inflammation by inhibiting the nuclear factor kappa B pathway	80
1991	Heat shock protein increases in human leukocytes after exercise and heat exposure	173
1995	Acute heat stress increases heat shock protein 72 in liver and gut tissues	59
1997	Heat shock protein 72 reduces gut permeability and inflammation	133
1997	A crossover between the heat shock response and bacterial resistance is shown	114
2007	Heat shock protein is upregulated in humans during heat acclimation	222

Physical Processes and Measurement of Heat Exchange

Thermoregulation involves two processes of heat exchange: passive and active thermoregulation. Passive thermoregulation, discussed only briefly in this section, involves the physical transfer of heat. The remainder of this chapter focuses on the physiological processes of active thermoregulation.

Before the 18th century, heat loss from living animals could be quantified using only the method of direct calorimetry, which involved placing an animal in a temperature-controlled container and measuring the increase in the temperature of the container or its walls. Near the end of the 18th century, Lavoisier discovered a relationship between the amount of heat produced by a living organism and the amount of CO_2 in its expired air (126). This formed the basis for the method of indirect calorimetry, which relates heat production to the chemical energy released when food is metabolized and excreted. In 1842 Robert Meyer formulated the first law of thermodynamics, which states that heat production must equal heat loss (126). By the early 1900s Rubner and Atwater had shown agreement between measurements obtained by direct and indirect calorimetry (126). Max Rubner, who was known as the energy physiologist, validated energy production of various foods and quantified the heat loss of dogs at rest and during exercise (171). From the heat balance equation it was known that heat exchange occurs through the avenues of radiation, convection, conduction, and evaporation. In 1937 Winslow and colleagues (216) constructed a partitional calorimeter, which measured heat exchange by each of the avenues of the heat balance equation. A disadvantage of their device, however, was that it could be used only in resting subjects. Because of the complexity of the method of direct calorimetry, indirect calorimetry became the primary method for measuring energy exchange during exercise in humans. In the early 1900s, indirect calorimetry was performed on humans only in specialized chambers. However, this method was soon modified so that respiratory measurements could be obtained in a normal room by collecting the expired gases in a metal tissot tank (203) and later in rubberized (Douglas) bags (figure 19.1) (51). In 1906, Nathan Zuntz constructed the first portable respirometer, in which ventilation was measured with a dry gas meter and expired gas samples were collected using a mercury tonometer and stored for later chemical analysis. This portable system allowed researchers to measure heat production in ambulatory subjects, even during exercise and at altitude. Zuntz also is credited with developing the first treadmill with controlled grade and speed, which allowed heat production to be measured during exact levels of exercise (81). In 1933, hot-wire and magnetic measurements of O_2 and CO_2 from expired gases became possible; these methods allowed continuous measurement of heat production during exercise (16). Such technological advances allowed for advances in the understanding of thermoregulation during exercise.

Temperature Regulating System

From ancient times physicians have understood that animal heat is a sign of life and that fluctuations are associated with disease. The physiological system that regulates body temperature can be divided into three parts: a central controller and integrator, thermosensors and afferent neural pathways, and efferent pathways and effectors.

Central Controller

In 1867, Tshcetschichin concluded from his spinal transaction studies that the brain is the site of temperature control (18). In 1887, Isaac Ott used a sharp probe to pique a small region of the brain just above the optic chiasma and observed an immediate increase in body temperature (150). However, Aronsohn and Sachs (7) are credited with first discovering this thermoregulatory center. They applied electrical stimulation to this region of the brain and observed an immediate increase in the rectal temperatures of rabbits, dogs, and guinea pigs. Aronsohn and Sachs concluded that this supraoptic region of the hypothalamus causes heat production, which is limited by input from other regions of the brain. Further support for their theory was provided by Henry Gray Barbour in 1912 when he inserted cannulas into the supraoptic region, which he perfused with hot or cold fluids. Each perfusion provoked an appropriate heat-loss or heat-production response (12). During the same year, Isenschmid and Krehl (105) discovered a second heat-control center. These two medical students transected the brains of rabbits at various levels and identified a region posterior to the first heat center that caused a decrease in body temperature upon stimulation and a loss of thermoregulation after transection. Isenschmid and Krehl discounted the works of Aronson and Sachs and Ott as effects of local infection and concluded that they had discovered the true heat-control center of the body. They failed to consider the possibility that there could be two thermoregulatory centers.

In 1913, H.H. Meyer suggested for the first time that body temperature is regulated by a balance between two centers: one functioning to prevent heat gain (the anterior hypothalamus) and the other functioning to prevent heat loss (the posterior hypothalamus) (128). When Walter Cannon (figure 19.2) presented his classic concept of homeostasis in 1926, he used the reciprocal regulation of body temperature to illustrate this concept (32).

In 1964 Feldberg and Meyers (57) proposed the amine theory to describe the chemical regulation of body temperature. They found that a microinjection of serotonin in the brain of monkeys caused body temperature to increase after activation of heat production pathways. After an injection of norepinephrine, body temperature decreased as heat production was inhibited and heat loss responses were activated. These investigators later found that the hypothalamic control of body temperature can be modulated by a variety of circulating and neural factors. For example, an injection of calcium ions into the lateral cerebral ventricle of a cat or monkey produced hypothermia, but the mechanism and site of action for calcium was different from the hypothermic effect of norepinephrine. The ionic effect acted in the posterior hypothalamus whereas the amine effect acted in the anterior hypothalamus. Myers and Feldberg also showed that a central injection of a fluid containing excess sodium ions caused an increase in body temperature, calcium caused a decrease (129), and other cations in equivalent concentrations had little

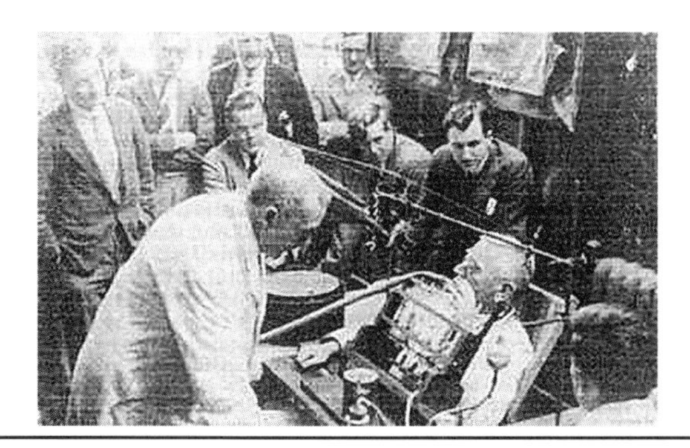

Figure 19.1 C.G. Douglas (of the Douglas bag method used in indirect calorimetry; seated) acting as the subject for J.G. Priestly as they demonstrate the acetylene rebreathing method for determining cardiac output in 1937.

Reprinted from H.W. Davenport, 1993, "The life and death of laboratory teaching of medical physiology: A personal narrative. Part I," *American Journal of Physiology* 264: S16-S23. Photo courtesy of American Physiological Society.

Figure 19.2 Walter Bradford Cannon, who defined the concept of homeostasis using temperature regulation as an example.

Reprinted with permission of American Physiological Society.

effect. These findings led Feldberg and colleagues to propose the ionic theory of thermoregulation, which states that body temperature is determined by the balance between sodium and calcium ions in the posterior hypothalamus (58). More recently, Jack Boulant and colleagues found that in addition to cations, the firing frequency of thermosensitive neurons in the hypothalamus is influenced by glucose or reproductive hormones as shown in figure 19.3, local temperature, and the osmolality of the cerebrospinal fluid (187, 188). These warm-sensitive neurons in the preoptic hypothalamus receive neural input from other regions of the brain, such as the hippocampus, which links the limbic system (emotion and stress) with thermoregulation (25), and from the suprachiasmic nucleus, which provides input concerning arousal state (75) and circadian and sleep rhythms (42).

Thermoreceptors and Afferent Pathways of Thermoregulation

Thermosensitive cells that provide sensory input to the hypothalamic thermoregulatory center are located throughout the body. Thermosensitive neurons have now been identified in many regions of the brain, including the ventromedial hypothalamus, midbrain, and medulla oblongata (217). In 1967, Thauer identified thermosensitive cells in the spinal cord (116). In the 1970s, cold (82) and warm (162) receptors were identified in splanchnic and abdominal regions. By taking advantage of a unique feature of the goat skeleton, Jessen and colleagues (109) first reported thermal sensitivity near skeletal muscles. These authors perfused the wide cavities of the upper-limb bones with cold water while maintaining a constant body temperature using intravascular heat exchangers. With local leg cooling, they found a small but significant reduction in respiratory water loss, the primary avenue for heat loss in the goat. Jessen and colleagues proposed that thermal input from either active muscles themselves or from the temperature of venous blood perfusing the muscles is integrated with other temperature input to determine thermoregulatory responses during exercise.

Thermoreceptors were characterized in the skin by Hensel and Zotterman in 1951 (98) based on the electrophysiological responses of myelinated (cold) and primarily unmyelinated (warm) small nerve endings during changes in local temperature. Recently, the molecular mechanism of thermodetection has been determined and ascribed to a transient receptor potential (TRP) family of cation channel proteins. In 2002, McKemy and colleagues (125) reported that the TRPM8 channel is the dominant mediator for cold and is sensitive to even a modest reduction (<27 °C air temperature) in ambient temperature. TRPM8 channels are found in the cell bodies, axons, and peripheral free nerve endings of somatosensory neurons that have cell

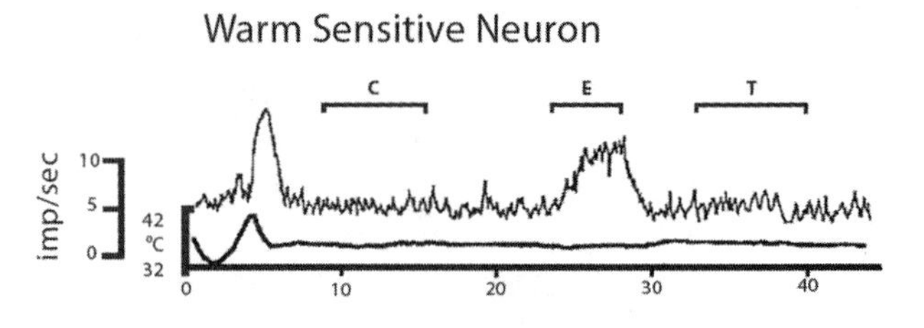

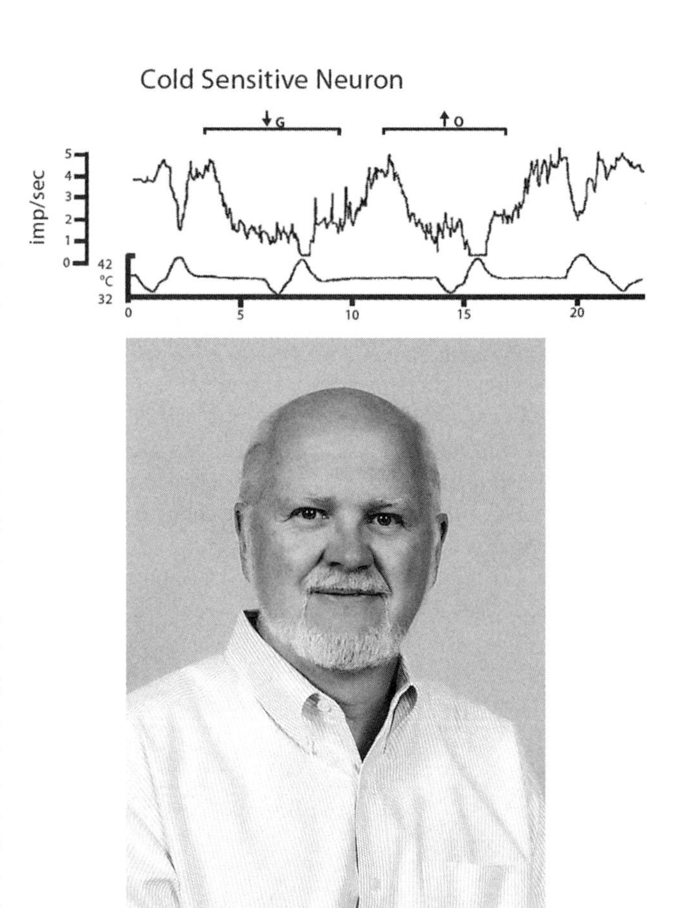

Figure 19.3 (*a*) Effects of ethanol control (C), estrogen (E), testosterone (T), low glucose (1 mM; G), and hyperosmolality (309 mOsm; O) on neuron recordings from rat preoptic tissue slices. (*b*) Jack Boulant, who first demonstrated the sensitivity of rat preoptic temperature-sensitive neurons to a variety of hormonal and ionic substances.

(*a*) Reprinted, by permission, from N.L. Silva and J.A. Boulant, 1984, "Effects of osmotic pressure, glucose, and temperature on neurons in preoptic tissue slices," *American Journal of Physiology - Regulatory, Integrative and Comparative Physiology* 247: R335-R345; Reprinted, by permission, from N.L. Silva and J.A. Boulant, 1986, "Effects of testosterone, estradiol, and temperature on neurons in preoptic tissue slices," *American Journal of Physiology - Regulatory, Integrative and Comparative Physiology* 250: R625-R632.

(*b*) Photo courtesy of Jack Boulant.

bodies in the dorsal root and the trigeminal ganglia. Also in 2002, Peier and colleagues (156) identified TRPV3 and TRPV4 channels as the warm thermosensors. However, the expression of TRPV3 and TRPV4 is prominent in skin keratinocytes, and it is not known how the thermal information detected by these TRP channels is transmitted to the local sensory nerves.

It was first recognized that the spinothalamocortical tract provides afferent transmission of thermal sensations up the dorsal horn of the spinal cord to the brain (20). The dorsal horn neurons synapse with neurons in the thalamus, which then project to the primary somatosensory cortex, allowing the perception and discrimination of peripheral body temperature. This pathway, which initiates behavior thermoregulatory responses, was originally thought to be the primary afferent pathway for thermoregulation. However, a connection between the spinothalamocortical pathway and hypothalamic temperature-control centers has not been documented.

Therefore, as proposed by Morrison and colleagues (131, 132, 138), it is unlikely that the spinothalamocortical pathway triggers the afferent pathways (which activate the physiological responses that defend body temperature) to increase heat production or facilitate heat loss. Instead, these authors proposed a dual thermoregulatory pathway model (figure 19.4). First, the spinothalamocortical pathway (on the left side of the figure) functions as described previously for thermal perception and localization. Second, a spinoparabrachiopreoptic pathway drives the physiological thermoregulatory reflexes. This second pathway shares the same thermosensors and neurons that project up the spinal dorsal horn. However, secondary somatosensory neurons branch off from the dorsal horn pathway and synapse with neurons in the lateral parabrachial nucleus in the pons. Upon cold stimulation, interneurons from the external lateral parabrachial nucleus (LPBel) activate interneurons in the median preoptic hypothalamus (MnPO) that inhibit the tonic inhibitory output of warm-sensitive neurons in the medial preoptic area (MPO). This disinhibition of thermogenesis-promoting neurons in the hypothalamus, and possibly the disinhibition of sympathetic and somatic premotor neurons in the medulla and spinal cord, activates efferent excitatory neurons that increase heat production. A similar reciprocal pathway involving warm thermosensors and a different area of the lateral parabrachial nucleus (LPBd) has been proposed to inhibit heat production and stimulate heat-loss responses (132). At this time, the central pathways that control panting and saliva production in animals or sweating in humans remain undefined.

Efferent Pathways and Effectors

As described previously, the hypothalamic integration and efferent pathways for thermoregulation are incompletely understood, especially in humans. Figure 19.4 illustrates some of the proposed efferent pathways for heat production that have been developed using an animal model. It is believed that the activity of warm-sensitive neurons in the medial preoptic hypothalamus (MPO) controls both the heat-loss and heat-gain responses. In a thermoneutral environment, the warm-sensitive neurons are tonically active and inhibit efferent heat-production pathways. With increases in body temperature, the activity of the warm-sensitive neurons increases and stimulates sympathetically mediated heat-loss responses. With body cooling, the input from cold thermosensors through the spinoparabrachiopreoptic tract inhibits the activity of the medial preoptic warm-sensitive neurons, disinhibiting their effect on cold-defense thermogenic responses (20). This disinhibition releases thermogenic activity of neurons in the dorsomedial hypothalamic nucleus (DMH) and of sympathetic and somatic premotor neurons, including neurons in the rostral raphe pallidus nucleus (rRPa). Separate efferent pathways (see figure 19.4) descend from the preoptic area (POA) and excite spinal sympathetic and somatic motor circuits activating cutaenous vasoconstriction (cvc) and stimulating brown adipose tissue (BAT) thermogenesis and shivering (132).

Effector responses to body heating were first quantified in the early 1900s by Max Rubner (17), who measured the volume of sweat secreted with heating and determined the amount of heat removed during its evaporation from the skin. Thermal sweating in humans is unique among animal species because it is activated by sympathetic cholinergic pathways (164). This was first shown by Foster and Weiner when they placed small doses of atropine on the forearm and blocked the neural activation of the sweat glands (64). Sympathetic adrenergic sweating also occurs in humans, although heat loss through this pathway is minimal. Cholinergic sweating is the primary avenue for heat loss during exercise (164).

Rubner (17) also identified a second primary mechanism for heat loss in humans: heat loss from the skin caused by increasing cutaneous blood flow as body temperature increases. Burton and Bazett (28) measured heat loss from various skin regions using a bath calorimeter. They found that skin blood flow varies regionally with body heating and that separate responses occur in the extremities (hand, foot, ears, lips, nose), the trunk and proximal limbs, and the head and brow. In 1966, Wurster and colleagues (218) quantified the regional blood flow responses with heating and found that

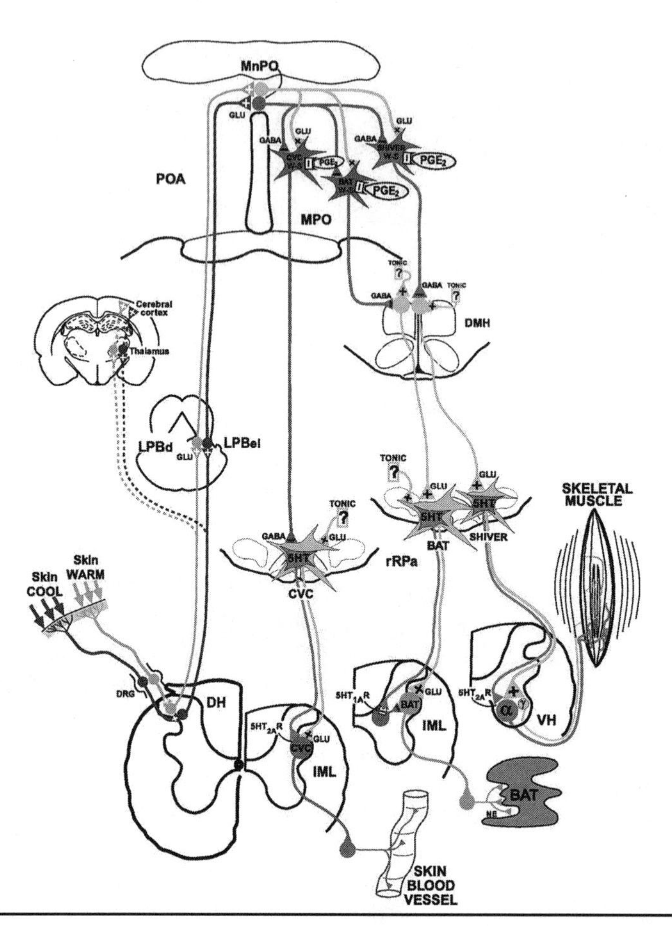

Figure 19.4 Current concept of the dual thermoregulatory pathways as developed from animal studies. The spinothalamocortical pathway begins with the activation of thermal receptors in the skin, which transmit signals to respective primary sensory neurons in the dorsal root ganglia (DRG), which then relay this information to second-order sensory neurons in the dorsal horn (DH). Thermosensory signals from DH neurons are relayed to the thalamus and then to the cortex for conscious thermal perception and localization. A second spinoparabrachial pathway is postulated to provide the excitatory drive that controls heat-loss and heat-production effector responses. This pathway shares the same skin receptors, DRG, and DH neurons. However, DH neurons synapse with third-order sensory neurons in the external lateral subnucleus of the lateral parabrachial nucleus (LPBel) with cold stimulation or the dorsal subnucleus of the lateral parabrachial nucleus (LPBd) with warm stimulation. Cool signals are glutamatergically transmitted from the LPBel to the preoptic area (POA), where GABAergic interneurons in the median preoptic (MnPO) subnucleus inhibit warm-sensitive (W-S) neurons in the medial preoptic (MPO) subnucleus that tonically inhibit cutaneous vasoconstriction (cvc), brown adipose thermogenesis (BAT), and shivering (activating heat-loss responses). Cold stimulation thus disinhibits the inhibition of cold responses by the W-S POA neurons, resulting in heat production. In contrast, with a warm skin stimulus, interneurons in the MnPO excite W-S neurons. Prostaglandin (PG) binds to PG3 receptors on W-S neurons, further inhibiting heat-production pathways. Preoptic W-S neurons inhibit cvc sympathetic premotor neurons in the rostral ventromedial medulla, including the rostral raphe pallidus (rRPa), that project to sympathetic preganglionic neurons in the spinal intermediolateral nucleus (IML). Similarly, preoptic W-S neurons controlling BAT thermogenesis inhibit BAT sympathetic premotor neurons in the rRPa that project to the IML. Finally, preoptic W-S neurons inhibit shivering-promoting neurons in the dorsomedial hypothalamus (DMH), which excite shivering premotor neurons in the rRPa, which then project to the spinal ventral horn to excite alpha and gamma motor neurons during shivering.

Reprinted from S. Morrison, 2011, "2010 Carl Ludwig distinguished lectureship of the APS neural control and autonomic regulation section: Central neural pathways for thermoregulatory cold defense," *Journal of Applied Physiology* 110: 1137-1149. With permission of American Physiological Society.

blood flow to the palm increases twofold and is controlled by sympathetic noradrenergic fibers whereas blood flow to the trunk and proximal limbs increases sevenfold and is controlled by a separate active vasodilatory pathway. An important issue in human thermoregulation has been the identification of the neurotransmitter(s) involved in active vasodilation. This question has been pursued for more than 50 yr, and it now appears that there may be two independent mechanisms, both of which involve nitric oxide. Recent work by Kellogg and colleagues suggests that vascular endothelial release of nitric oxide may mediate the local heating response whereas neural release of nitric oxide may mediate the whole-body active vasodilatory response (112).

Models of Thermoregulation

Physiologists have attempted to understand the process of thermoregulation by using models, which can be described as verbal, pictorial, analog, mathematical, and, most recently as in figure 19.4, neuronal. Each model presents a hypothetical framework by which the three components of the thermoregulatory system interact and control body temperature in response to environmental, disease, or exercise stressors. Early modeling approaches are reviewed by Hardy (87) and Buskirk (30).

Aschoff and Wever (8) in 1958 proposed a pictorial core–shell model. According to this model, only core temperature is regulated whereas shell temperature varies with the ambient conditions and provides insulation to protect the core. Consistent with this view, Marius Nielsen (143) reported that core temperature during exercise is regulated at a level proportional to exercise intensity, independent of the ambient conditions. Hellon and Lind (96) narrowed this effect and determined that core temperature is independent of ambient conditions only in a prescriptive zone of ambient temperature, which becomes more limited as the work intensity increases.

Engineering models of temperature regulation were prevalent in the 1960s. For example, J.D. Hardy presented thermoregulation as a servosystem in which the output from a central controller is modified by feedback loops from central and peripheral inputs (87). This approach also was adopted by Theodore Benzinger, but a lively debate ensued regarding the relative importance of the peripheral versus core inputs. Stolwijk and Hardy proposed that core and skin temperature inputs were important in feedback control at a ratio of approximately 4 to 1, respectively (195). Benzinger disagreed with this weighting and argued that the core to skin ratio should be closer to 9 to 1 (19). This debate led to the famous ice cream experiments in which Benzinger placed subjects in a gradient layer calorimeter that permitted instantaneous recording of sweating and vasodilation during rapid changes in core and skin temperatures. Benzinger argued that it was critical to develop a method for assessing core temperature that was more accurate than the slow-responding rectal temperature method. Initially, he placed probes intracranially in sinuses in his head. Eventually, he and his wife developed a tympanic membrane probe that provided a rapid assessment of changes in core temperature. He found that when hyperthermic subjects ate ice, their tympanic temperatures rapidly decreased while skin temperature paradoxically increased. Because sweating also abruptly decreased, Benzinger concluded that heat-loss responses are driven primarily by core temperature (18). Hardy, on the other hand, argued that temperature regulation varies according to ambient conditions (88). He described three zones of regulation: a vasomotor zone (in neutral ambient conditions), a metabolic zone (cold conditions), and an evaporative and vasomotor zone (hot conditions). Hardy proposed that temperature regulation in cold and neutral zones is influenced strongly by skin thermoreceptors whereas regulation in a hot environment becomes more strongly influenced by core thermoreceptors. Both perspectives are held today: Core temperature is considered the primary input but core–shell weightings of 0.9:0.1, 0.8:0.2, or 0.67:0.33 are used in hot, warm, and cool conditions, respectively (177).

The idea of a set point controlling body temperature was suggested as early as 1875 when Liebermeister described the upward shift in temperature during fever as a resetting of a biological set point (122). In 1941, Burton described his proportional control theory whereby temperature regulation was controlled as a linear function of the load error from temperatures received from the periphery and core as compared with the body's internal standard value (27). However, the set point theory of temperature regulation has been ascribed to Ted Hammel. According to Hammel's theory, thermal inputs from peripheral and core regions are integrated to form a single signal that is compared with an internal set point (86). An error signal above or below this set point then elicits an appropriate heat-loss or heat-production response. Hammel proposed that there are two sets of thermosensitive neurons in the anterior hypothalamus; this is in agreement with data concurrently published by Nakayama and colleagues in anesthetized cats (139). One set of neurons is sensitive to heat and facilitates heat-loss responses while inhibiting heat production. The other set of neurons is less responsive to heat and facilitates heat production while inhibiting heat loss. Hammel proposed that the balance between the firing of these two sets of neurons determines the set point, which is adjusted by input from neurons of the reticular activating system to alter the set point throughout the day (86). In 1967, Hellen described an interaction of warm- and cold-sensitive neurons in the anterior hypothalamus of unanesthetized rabbits. About 10% of the cells in this region are thermosensitive, but the most interesting are the C cells, which have a threshold for activation that is within 0.5 °C of normal hypothalamic temperature. Hellen proposed that the C cells are responsible for the internal set point. The validity of this set point theory is still hotly debated today. Supporters of this theory allow multiple avenues for adjustment of the set point, including interactions with other temperature-sensitive neurons in the brain and body (25), effects of pyrogens during fever (22), and nonthermal factors such as hormones, ions, baroreceptors, or osmoreceptors (178).

An alternative theory of temperature regulation is the reciprocal inhibition theory, described by John Bligh (23). This theory includes the same neuronal architecture as the set point theory but it does not require a comparative central reference. Instead, it relies on a separate response coefficient for cold and warm sensors and a reciprocal inhibitory influence for each sensor-to-effector pathway. Thus, as body temperature increases, warm sensors are activated, which activates heat-loss responses and inhibits heat gain. When body temperature decreases, heat sensor activation is reduced and heat-production responses are eventually activated. Body temperature is therefore maintained by a reciprocal balance between the two effector systems and does not require a central set point. Because there is a temperature gap between the onset of heat-loss and heat-gain responses, a null zone exists where body temperature can be maintained efficiently by minimal vasomotor regulation. This theory has been expanded to include excitatory or inhibitory influences from nonthermal factors (127).

A third theory of thermoregulation is quite different. According to the heat regulation theory described by Paul Webb (211), heat flow—not temperature—is the controlled variable. Heat flow in this model is sensed by thermal gradients in the body. This idea of temperature sensors located at various depths to detect thermal gradients was first proposed by Bazett in 1951 (15). According to this theory, as internal temperature increases at onset of exercise, the increasing core-to-skin thermal gradient is detected, resulting in activation of heat-loss responses in order to minimize heat storage. Heat loss increases until heat balance is attained, although at an elevated core temperature. In this theory, the heat content of the body varies in relation to the heat load. Rather than sensing and integrating peripheral and central temperatures, the body senses heat flow from temperature gradients across the skin surface and between the skin and core. The central integration of heat balance then operates from feedback of heat loss and possible feed-forward control from heat production. One advantage of this theory is that it can easily account for an altered regulation of core temperature under conditions of varying metabolic rate (e.g., exercise, circadian rhythm, or the menstrual cycle) without requiring the resetting of a central controller.

Exercise and Thermoregulation

Before 1900, an increase in body temperature with exertion had been described in the occupational, physiological, and biochemical literatures. The 1900s saw the beginning of quantitative measurements and the development of more precise equipment for controlling work and assessing energy costs.

Acute Effects

In the 1900s, John Scott Haldane was famous for his self-experimentations, including assessment of the effects of physical work and environmental stressors on core temperature, sweating, and hydration. One of his primary goals was to reduce accidental deaths in miners, who in an emergency had to climb ladders under extreme conditions of heat and noxious gases. Therefore, Haldane designed his laboratory experiments to quantify heart rate and core temperature while subjects climbed up and down a ladder (83). He quickly discarded oral temperature measurement and reported that rectal temperature more accurately agreed with heat-related symptoms. His field studies in mines and sewers, during diving, and at altitude represent some of the first physiological field studies in extreme environments (77).

At the University of Copenhagen in 1920, August Krogh and Johannes Lindhard performed critical experiments to characterize gas exchange and cardiac responses during exercise (119). They were the first to explain how energy production varies with the macronutrients metabolized and how reliance on carbohydrate increases as exercise intensity increases. August Krogh and his wife Marie published their "seven critical devils"—seven papers (180) that proved unequivocally that oxygen diffuses from the lungs to arterial blood and is not secreted. Once it was accepted that the lungs cannot secrete oxygen, oxygen consumption during exercise could be viewed as an accurate method for calculating energy cost.

In the United States, the Harvard Fatigue Laboratory was founded by L.J. Henderson in 1926. The history of this laboratory as well as descriptions of key studies and investigators are presented in chapter 3. Under the direction of David Bruce Dill more than 300 peer-reviewed articles on exercise and environmental stress, including key studies in the field of exercise thermoregulation, were published during the 19 yr this laboratory was in existence (101).

In 1938 Marius Nielsen attributed the controlled increase in core temperature during exercise to an upward shift thermoregulatory in set point (143). Irma Åstrand refined this observation to show that interindividual differences in core temperature were reduced when male subjects of varying fitness worked at the same relative exercise level (~50% of maximal oxygen consumption) (9). Her finding was further expanded by Saltin and Hermansen in 1966 to include male and female subjects exercising at intensities from 26% to 69% $\dot{V}O_2max$ (175). However, the debate to explain exactly how core temperature is regulated during exercise continues. Nielsen had proposed that exercise elevates the set point (143). However, if heat-loss responses are determined by a load error between the current core temperature and an increasing set point, heat-loss responses should be inhibited during exercise. This led investigators such as Ted Hammel (107) to suggest that the set point is adjustable and that it decreases during exercise. Others have argued that the relationship between exercise intensity and core temperature can be explained without invoking a change in set point (192). For example, Jan Snellen in 1972 (192) found using direct calorimetry that sweat rate correlated best with the increase in average body temperature, which he calculated by dividing heat storage by the water equivalent of the body. He proposed that heat-loss responses are proportional to mean body temperature and therefore that "the regulating mechanism operates as if there is heat regulation instead of temperature regulation." This argument is similar to that promoted by Paul Webb in the heat regulation theory (211). A complicating factor in this model is determining a practical method for assessing mean body temperature, as recently reviewed by Jay and colleagues (108).

During strenuous exercise, heat production often exceeds the capacity for heat loss, resulting in a continuing increase in body temperature. Another current debate is how much central fatigue limits exercise capacity. A critical temperature theory was first proposed by Bruck in 1987 (26) and has been advocated by Nielsen (142), Nybo (148), and Gonzalez-Alonso and colleagues (76), who observed that cyclists fatigue at the same core temperature regardless of their initial resting temperature. According to this theory, once core temperature exceeds approximately 40 °C, the central motor center is inhibited and the ability to recruit motor units is impaired, resulting in muscular fatigue. This effect is viewed as a protective response to limit heat storage and prevent tissue thermal injury. Nielsen and colleagues (142) have shown that at the critical point of fatigue muscle blood flow is not reduced and glycogen stores are not depleted, suggesting the fatigue is not attributable to peripheral muscle effects.

Chronic Effects

Physical training and either heat acclimatization (natural adaptation to a hot environment) or heat acclimation (artificial adaptation to a hot environment, such as using heat chambers) each produce similar circulatory and thermoregulatory adaptations, including increased blood volume, reduced heart rate and core temperature, and increased sweating and skin vasodilation. Roberts and colleagues (165) observed that physical conditioning improves heat-loss responses by decreasing the core temperature thresholds for sweating and skin blood flow, an effect that also occurs with heat acclimation. In the early 1960s, Sid Robinson (figure 19.5) and colleagues (2, 159, 166) suggested that heat acclimation may be achieved by physical training without exposure to a hot environment. In a study by Piwonka and colleagues (159), five distance runners performed moderate exercise for 85 min in a hot environment and their responses were typical of heat-acclimated men. In another study by Allan and colleagues (2), soldiers who trained in a cold environment appeared to tolerate a prolonged heat test as well as soldiers who were heat acclimated. However, the conclusions of such studies were suspect because the work intensities were not always normalized between trained and untrained subjects, the state of heat acclimation was uncertain, and, in the study by Allan and colleagues, oral temperatures were used to assess core temperature.

What is certain is that vigorous physical training, especially when it includes repeated intervals of high-intensity exercise, provides a superior heat tolerance over

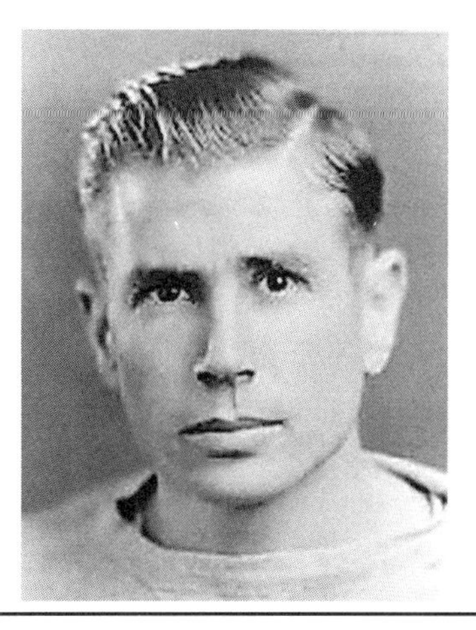

Figure 19.5 Sid Robinson, a respected athlete and exercise physiologist, proposed that physical training even in a cool environment improves heat tolerance. Robinson is known for his development of the traditional method of inducing heat acclimation.

Photo used with permission from the Indiana University Athletics. Available: http://iuhoosiers.cstv.com/trads/ind-trads-hof-1985.html.

nontraining (73, 166). In 1945, Eichna and colleagues (54) studied 64 young Army recruits during hot–humid and hot–dry acclimations. They concluded that although fit men acclimated more rapidly, fitness did not replace heat acclimation. In a definitive study by Strydom and colleagues in 1966 (196), two groups of subjects performed 5 h training sessions for 12 d under either hot or cool conditions. On the first and last days of training, both groups performed a 5 h heat test in which they exercised until their core temperatures reached 40 °C and then rested in the heat for the remaining time. During the first hour of the posttraining tests, the cool-trained group had responses similar to those of the heat-acclimated group even though their sweat rates were less than 50% of the rates of the acclimated group. However, three of the five cool-trained subjects could not complete the 5 h exposure and the two that did complete had much higher heart rates and core temperatures compared with the heat-acclimated subjects. Strydom and colleagues concluded that physical conditioning can improve heat tolerance during short heat exposures only.

Exercise in a Hot Environment

By 1910, the role of high environmental temperature in reducing work productivity and increasing accidents had been documented for a variety of industries (154). It is interesting that most of the early observations on the effect of heat stress on human performance were driven by industrial hygiene concerns rather than the exercise physiology viewpoint of today.

Acute Effects

As mentioned previously, Haldane obtained some of the earliest measurements of core temperature in heat-stressed subjects, including workers in Turkish baths and miners (77, 83). He noted that visitors to the mines were greatly affected by the heat and experienced mouth temperatures exceeding 39.2 °C, shortness of breath, and had great difficulty climbing out of the mine whereas the acclimated miners bore the heat well. Haldane also noted the effects of wind velocity, clothing, and hydration on heat tolerance.

The construction of the Hoover Dam that began in 1931 afforded the opportunity for members of the Harvard Fatigue Laboratory to directly study the effects of living and working in a desert environment (43). In the first year of construction, 17 deaths were attributed to the heat. The next year, there were no deaths and only 7 men were hospitalized for heat prostration or muscle cramps. The major changes made during this time, as documented by Talbott and Michelsen (197), included moving the workers to a cooler sleeping area, making cold drinking water more readily available, and encouraging the workers to add salt to their meals.

World War II provided another impetus for research on the responses to exertion in a hot environment. As described by Newburgh in the preface of his book *Physiology of Heat Regulation*, this was a new type of war in which generals planned their campaigns regardless of the weather, aviators were exposed to extreme cold and hypoxia for the first time while tank drivers and seamen experienced extreme heat, and the foot soldier often carried all of his equipment on his person for the sake of mobility. Civilians were recruited to find solutions for such environmental hazards, and much of the early work was performed by personnel from laboratories such as the Harvard Fatigue Laboratory and the department of physiology at the University of Rochester (140). In his seminal book *Physiology of Man in the Desert*, Adolph described the changes that occur in the body fluids during dehydration, documented the phenomena of voluntary dehydration, and emphasized the importance of water ingestion in the survival of men marching in the desert or stranded in life rafts (1). Adolph's work dispelled the myth that humans are able to adapt to water restriction.

Early observations of sex differences in thermoregulation included the observation that women have less heat production and sweat less than men. In 1941, Hardy measured rectal temperatures and heat balance in two male and six female subjects while resting in the Russell Sage calorimeter during exposure to a wide range of ambient temperatures (89). Although the women's mean rectal temperature was similar to that of the men, they had greater variation in body temperature. Their skin temperatures were warmer, they had less heat loss per unit of surface area, and they began to sweat at a higher internal temperature and maintained a lower sweating rate. Hardy did not conclude that women were more heat intolerant but rather that they had lower heat production and therefore required less heat dissipation.

However, by the 1960s most investigators concluded that the lower sweating rates in women caused lower heat tolerance (67, 213, 221, 224). In these early studies men and women exercised at a similar absolute level without consideration for differences in aerobic capacity or heat acclimation (221, 224). In later studies in which these factors were equated, women still sweated less than men; however, a lower sweat rate was not interpreted as indicating lower heat tolerance (1, 10, 46, 152). For example, Dill and colleagues concluded that young men and women equally tolerated sustained exercise in desert heat but that women were able to regulate body temperature with greater economy by sweating less than men (46). Wyndham and colleagues arrived at a similar conclusion that women are better able to adjust their sweat rate in hot, humid conditions than are men, who are prolific, wasteful sweaters (221).

In 1904, van de Velde noted that the body temperature of young women varied as a function of their menstrual cycles (207). We now know that core temperature is regulated higher during the luteal phase, accompanied by an upward shift in core temperature thresholds for thermal sensation (38), sweating (193), and vasodilation (90, 193). Although several authors have reported no effect of menstrual function on exercise thermotolerance (69, 185, 215), when the menstrual cycle is well documented and circadian effects are controlled, subtle differences in sweating and skin blood flow (193) are shown to reduce exercise tolerance in nonacclimated women during the luteal phase compared with the follicular phase (10).

Chronic Effects

Until the past century, climatic heat was considered a detriment to the health of white men and women. This idea originated during the earliest European expeditions to tropical regions and was undoubtedly influenced by the hardships and tropical diseases experienced by white settlers. It became a creed that Europeans should avoid muscular activity and rely on the manual labor of "colored races." Women in particular were thought to suffer from deleterious effects of tropical sunrays, which were thought to damage their reproductive organs. Andrew Balfour, considered an expert on this subject in the early 1920s, concluded that "so far as race is concerned, I am persuaded that the hot and humid tropics are not suited to white colonization and never will be with our present knowledge, even if they were rendered as free from disease as in England" (161).

However, sporadic observations from white individuals living in the tropics offered a more optimistic perspective. In 1789, a physician attending British soldiers in India noted that although exertions of a single day may be harmful, even in the hottest weather bad effects were rare after a campaign had continued for a few days (100). Also, in a less flattering way, E.R. Stitt appreciated the benefits of manual work for adapting to the tropics. He described that female settlers had greater difficulties than men because "as a rule, they [white women] have no serious employment and considerably less domestic work than at home." Price, in his monograph *White Settlers in the Tropics*, concurred with Stitt's viewpoint and went further to write, "Almost every tropical settlement has its quota of lazy, bored, card-playing, spirit-drinking women, who would be far healthier and happier if financial circumstances forced them to do their own housework, if not some labor out of doors" (160). Although many believed that continuing residence in the tropics would result in a gradual deterioration in health, some tropical residents made repeated observations on themselves and noted no health concerns. For example, John Davy (41) in 1850 reported that although his body temperature was higher during his 3.5 yr in the West Indies, his skin was more active in transpiration and his kidneys less active. He noted that although his body temperature increased during exercise, it rapidly recovered to its normal condition after exercise. J.M. Crombie obtained more than 1,288 body temperature measurements during his stay in Bengal and reported that body temperature gradually decreased after the first couple weeks. This concern about whether Caucasians can adapt to heat persisted until the 1920s when Balfour, Ciliento, and others described successful adaptations that they attributed to physiological changes. Eijkman and others at first dismissed any improvements as attributable to improved hygiene, learning to drink more fluids, and performing exertions during a cooler time of day. This early controversy and its resolution is humorously summarized by Horvath (100).

Vernon in 1923 was one of the first to obtain physiological measurements in an environmental chamber to document the effects of heat acclimation. He observed with repeated heat exposures that his heart rate decreased from 131 to 120 beats/min and his rate of sweating increased (208). Also in 1923, Joseph Barcroft reported a progressive increase in blood volume as ambient temperatures increased while he and his colleagues traveled from England to Peru. He concluded that this expanded blood volume contributed to their ability to adjust to working in a tropical environment (14). Dill and colleagues in 1933 documented a progressive increase in sweat rate and decline in sweat sodium and chloride during the first 6 d of desert heat exposure in men maintained on a constant diet (1, 43, 45). Thus, by the mid-1930s it was finally accepted that significant cardiovascular and thermal adaptations occur so that even Caucasians can survive and thrive in hot–wet or hot–dry environments.

A strong impetus for developing methods for rapid heat acclimation began in the 1930s when the gold mines in South Africa exceeded depths of 1,000 m. With rock temperatures approaching 60 °C and water spray being used to contain dust in the mines, heat stress became extreme and 26 heat stroke fatalities occurred in a single year (113). In 1935 Aldo Dreosti, a young physician in the medical service of the Rand Mines Limited, developed a method for artificially inducing heat acclimation, and he used a standardized heat-tolerance protocol to screen prospective miners (205). By the 1950s, heat acclimation was a standard procedure that involved two stages of 6 d each during which the miners worked in areas of progressively increasing temperature in the mine. However, the uncontrolled environmental and work conditions sometimes resulted in suboptimal levels of heat acclimation. Therefore, in the 1960s with 300,000 men/yr undergoing artificial acclimation, surface climatic chambers began to be used in which work rate was controlled by simple block-stepping protocols (205). During the 1960s and 1970s, studies were performed by Wyndham, Strydom, Mitchell, and others to understand the physiological and psychological responses to heat and the effects of hydration, fitness, race, and sex on thermal tolerance and heat acclimation (196, 219, 221). The artificial acclimation and standardized heat-tolerance tests they developed resulted in an approximately 50% reduction in heat stroke mortality in the miners. However, the heat chamber acclimation procedure had several disadvantages, including the onerous nature of the process and the number of unproductive shifts. Therefore, in the 1980s microclimate cooling was incorporated whereby miners are acclimated by working in hot areas of the mines while wearing a cooling jacket that covers about one third of the body surface area. In this way, an excess increase in body temperature can be prevented while the miners perform productive work. This method is reported to induce heat acclimation while increasing productivity and improving morale (13).

In the United States a flurry of research to develop methods for inducing heat acclimation occurred in the 1940s as U.S. soldiers were deployed to desert and jungle battle zones. In 1943, Sid Robinson and colleagues performed a classic study that documented a protocol and physiological responses that are now considered the traditional method for acclimating to working in heat (167). Five subjects walked at a constant speed on a treadmill for 1 to 1.5 h in a simulated desert environment for 24 d. The rapid decline in heart rate and rectal and skin temperatures at a set level of work became the standard indicators of heat acclimation. Eichna and colleagues (54) reported a similar pattern of improvement during acclimation to humid heat. They also reported that not exercising during heat exposure, sleeplessness, intake of alcohol, and inadequate intake of fluids impeded the process (54). They noted that aerobically fit subjects acclimated faster and that forced fluid intake resulted in better results than did allowing the soldiers to voluntarily hydrate. Taylor and colleagues (198) described the thermal and cardiovascular responses of six men who lived continuously in a chamber maintained at 110 °F to 120 °F during the day and 85 °F at night for 8 d. They noted that most of the improvement in thermal responses and work tolerance occurred during the first 3 d; while the sweating rate increased only 4%. The sweat rate continued to increase and became about 30% greater on d 8 compared with d 1 of heat exposure. Thus, these authors attributed the early improvements of heat acclimation almost entirely to cardiovascular adaptations, such as the increased blood volume and enhanced cutaneous blood flow as documented by Fox (66), and later improvements to thermal adaptations, such as an increased sweat output. Nielsen recently attributed the increased sweating response to an increased sensitivity of the sweat glands for thermal and humoral stimuli and also to an increase in size and number of active sweat glands (141).

Conn and colleagues (35) proposed that the initial expansion of blood volume was caused by endocrine changes that cause retention of salt and water. However, with continued heat exposure using the traditional heat acclimation protocol, blood volume returns toward preacclimation levels yet the cardiovascular improvements remain. Senay (184) proposed that an improved efficiency of the cardiovascular system occurs during the later stages of heat acclimation, which results from an increased mobility of proteins between the interstitial and vascular compartments. Thus, after acclimation subjects are better able to defend their plasma volume during exposure to heat and exercise.

As opposed to the traditional heat acclimation protocol, Fox and colleagues in the 1960s described a method of controlled hyperthermic heat acclimation (65). Using this method, during each day of heat acclimation the subject's core temperature is elevated rapidly to approximately 39 °C while the subject sits in a reclining chair wearing vapor-barrier clothing to maintain this elevated core temperature for at least 1 h. Using this approach, the core temperature, cardiovascular responses, and thermal strain remain elevated during each session of heat acclimation rather than decrease with each acclimation session, as occurs during the traditional acclimation protocol. Fox and colleagues (65) proposed that heat acclimation can be induced more rapidly—10 d instead of approximately 21—with their controlled hyperthermia protocol. Improvements in sweating occurred sooner using this method and predicted the degree of improvement in a heat-tolerance test. Although this method was moderately effective, other investigators noted that this passive heat acclimation protocol provides less protection against conditions involving exercise in heat than a traditional heat and exercise protocol does (54, 219). Recent modifications of the controlled hyperthermia protocol require that subjects exercise during acclimation sessions in which the exercise level or ambient conditions are altered to maintain a controlled hyperthermic response during all days of heat acclimation (155).

Hydration and Thermoregulation

Humans are unusual among animal species in our strong reliance on evaporative heat loss for thermoregulation and, therefore, our total dependence on adequate hydration (164). The hydration status of the body is tightly regulated by feedback from cardiovascular baroreceptors and from osmotic and ionic receptors. Neural and hormonal responses are triggered by these hydration sensors, which modify central temperature integration and efferent temperature responses. The type and amount of fluids required to maintain hydration during exercise in the heat have been researched extensively.

Effects of Blood Volume and Osmolality

Hunt in 1912 (103) recognized an absolute requirement for free access to drinking water during prolonged work in the heat. In 1924, Barbour and Tolstoi (13) observed that drinking alters the water content of the blood, which is essential for maintaining body temperature. They noted that hemodilution and increased peripheral blood flow enhanced radiative heat loss in response to warm ambient temperatures. Rietschel and Beck in 1928 documented a greater increase in rectal temperature during stair stepping in the heat when men were dehydrated compared with when they were euhydrated. These authors suggested this thirst fever was due to heat liberated by osmotic reactions in the body (78).

The term *voluntary dehydration* was first coined by Rothstein (1) in the classic book *Physiology of Man in the Desert*. Adolph and colleagues noted that most soldiers marching in the desert with water freely available do not drink an adequate amount of fluid. Yet, to preserve their ability to march, every bit of sweat loss must be replaced sooner or later. This obligatory volume of drinking was not substantially altered after training or with repeated heat exposure. Failure to replace sweat losses resulted in dehydration, which was characterized by a reduced blood volume and increased plasma osmolality.

A serious consequence of dehydration is an increased incidence of heat illness under conditions of extreme heat, such as that observed in workers on the Hoover Dam (43), miners exposed to hot and humid conditions (219), or soldiers marching in the desert (1). The effects of dehydration on cardiovascular and exercise responses were characterized by Saltin in 1964 (174) when 10 men performed submaximal and maximal exercise before and 90 min after dehydration caused by thermal (sauna), metabolic (cycle exercise), or combined thermal and metabolic stressors. In each condition, a similar level of dehydration (weight loss of 1.7-4.6 kg) produced a similar impairment in work capacity. At a given submaximal level of stress, the subjects cycled with the same oxygen consumption but with a markedly reduced stroke volume, increased heart rate, and reduced blood lactate. At maximal exercise there were no differences in maximal oxygen consumption, heart rate, or blood lactate compared with the hydrated condition. However, the subjects could not sustain exercise as long when dehydrated. Saltin concluded that because muscle strength and maximal aerobic capacity were not impaired, the reduced endurance capacity after dehydration must be related to changes in the muscle at the cellular level.

Ekblom and colleagues in 1970 (55) found that only a 1% loss of body weight from dehydration resulted in a notable elevation in rectal temperature during moderate exercise (60% $\dot{V}O_2max$) in the heat. These authors attributed the higher core temperature to impaired sweating. However, because the sweat rate was not near maximal they concluded that the reduced sweating was most likely due to less central stimulation of the sweat glands rather than a failure of the sweat glands themselves. Greenleaf and Castle (79) quantified the effect of dehydration on core temperature when 8 men exercised at 49% $\dot{V}O_2max$ under three hydration conditions: euhydrated, hyperhydrated (+1.2% body weight), and dehydrated (−5.2% body weight). They determined that rectal temperature is elevated by 0.1 °C for each 1% of body weight lost. They agreed that the greater increase in rectal temperature during dehydration is due to inadequate sweating.

Plasma osmolality also is a critical determinant of thermal homeostasis. The Scottish physiologist Archibald Byron Macallum in 1926 noted the conserved ionic concentration of body fluids across a wide variety of species (123). According to the ionic theory proposed by Amatruda and Welt in 1953 (4), the tonicity of the body fluids alters the threshold at which the body thermostat responds to heat: Hypertonicity increases and hypotonicity decreases the threshold for heat loss. Feldberg and colleagues later demonstrated that specific ions, such as sodium and calcium, can directly alter the activity of hypothalamic neurons (58).

Greenleaf in 1979 (78) quantified the ionic effects where intravenous or oral ingestion of electrolyte solutions resulting in an excess plasma sodium of 5 mEq/L or an excess plasma calcium of 1 mEq/L each increases (for sodium) or decreases (for calcium) core temperature during exercise by about 0.1 °C.

The independent actions of hypovolemia and hyperosmolality on thermoregulation were shown in a series of experiments by Nadel and colleagues. When subjects ingested a diuretic producing iso-osmotic hypovolemia, the threshold for cutaneous vasodilation was elevated, maximal skin blood flow was reduced by 50% (137), and the sensitivity of the sweating response was reduced (62). In another study, isovolemic hyperosmolality was induced by infusion of either an isotonic or a 3% saline solution with a plasma expander. With an increase in plasma osmolality to approximately 303 mOsm/kg, the thresholds for both vasodilation and sweating were elevated without change in blood flow sensitivity or maximal skin blood flow (63). Nadel and colleagues integrated these nonthermal inputs (osmolality, sodium concentration, blood volume) into the set point theory of thermoregulation (136).

Larry Rowell summarized the cardiovascular responses during intense or prolonged exercise in the heat and the effects of hydration in the book *Human Circulation: Regulation During Physical Stress* (170). When skin and muscle vascular beds compete for limited cardiac output during exercise in the heat, the pumping capacity of the heart can easily be exceeded. Initially during prolonged exercise, skin perfusion increases and cardiac stroke volume is reduced, resulting in a progressive cardiovascular drift as heart rate increases to maintain cardiac output. With continued exercise, vasoconstrictor reflexes are invoked to maintain cardiac filling, resulting in progressive reductions in blood flow to the splanchnic organs, kidneys, and nonactive muscle vascular beds. Under extreme exercise and heat conditions—especially if dehydration occurs—cardiac filling is compromised and cardiac output decreases, triggering reductions in skin blood flow and sweating responses. At this point, body temperature increases more rapidly and heat exhaustion is likely without hydration or cessation of exercise. Thus, with severe dehydration, the cardiovascular system defends blood pressure first, at the expense of maintaining body temperature.

Hydration Solutions

As Europeans migrated to the tropics it was soon realized that noncontaminated drinking water was required for successful adaptation to heat. E.H. Hunt (103), for example, recommended that at least 6 L/d of water is required for an able-bodied European leading an active outdoor life in Deccan, India. He described the ill effects of prolonged work without copious fluid intake and the futility of attempting to restrict drinking to conserve water. Adolph (1) observed that men walking briskly at 3 or 4 miles (4.8-6.4 km)/h over hot desert trails lose more than 2 lb (0.9 kg)/h of body mass. Pitts (158) and colleagues from the Harvard Fatigue Laboratory also studied the water requirements of men marching up grades at 3.5 miles (5.6 km)/h for 6 h in the heat. They concluded that "the more nearly water intake approximates sweat loss, the better off the subject remains." They also concluded that electrolyte supplementation was not needed as long as adequate salt was available ad libitum during meals.

The possible benefit of adding glucose or electrolytes to a hydration solution during repeated days of exercise in the heat was first examined by Costill and colleagues in 1975 (36). Two groups rehydrated with either water or a 13% glucose–electrolyte solution during repeated days of cycling in the heat while attaining a 3% loss of body weight. Both groups expanded their plasma volumes similarly during the repeated heat exposures, and there was no additional benefit of adding salt or sugar to the rehydration solution because the excess salt was excreted. This crucial study influenced athletes and scientists to believe that water is the optimum hydration solution for maintaining performance during exercise in the heat.

In the 1980s scientists began to question whether water is always the best hydration or rehydration solution. Electrolytes are lost during prolonged exercise in the heat and must be replaced even though sweat is hypotonic. Also, muscle glycogen stores become depleted with prolonged intense exercise. Carbohydrate intake could spare muscle glycogen and delay fatigue. Theoretically, salt and glucose added to a hydration solution should enhance endurance for prolonged intense exercise. However, it was known that hypertonic solutions delay stomach emptying, reduce intestinal absorption of electrolytes and water (176), and, from studies such as that of Costill and colleagues (36), appeared to offer no performance benefit.

In the 1960s, significant progress was made in the treatment of cholera, a disease associated with severe diarrhea and dehydration. Although intravenous infusions of isotonic electrolyte solutions were highly effective for rehydrating cholera patients, pyrogen-free intravenous fluids often were not available in the regions where cholera was prevalent. Therefore, an oral rehydration solution was desperately sought. In 1968, Fordtran and colleagues (61) found that adding a small quantity of glucose to a rehydration fluid enhances reabsorption of sodium and water in the human small intestine. In 1969, Pierce and colleagues (157) confirmed the effectiveness an oral rehydration solution containing 100 mEq/L of sodium and 120 mmol/L of glucose; the solution was only slightly hypertonic (327 mOsm/L). The small quantity of sugar in the fluid enhanced sodium uptake in the gut, resulting in 87% retention of the administered fluid.

This clinical background was known to Carl Gisolfi (figure 19.6), an exercise physiologist who was working to develop an exercise rehydration solution (179). By inserting triple-lumen tubes into the gastrointestinal tracts of his students and collaborators, he could deliver and sample various drinking solutions as they moved through the gut at rest, during exercise, and after exercise. Gisolfi postulated that an effective sport drink should provide carbohydrate to delay depletion of muscle glycogen, replace electrolytes to prevent hyponatremia, and provide fluid to the vascular compartment to prevent dehydration. Such a drink should improve endurance for prolonged exercise and prevent hyperthermia. Gisolfi and colleagues, however, had to address several challenges. First, during exercise blood is shunted away from the gastrointestinal tract, thus slowing intestinal absorption of water and electrolytes. Gisolfi and colleagues in 1991 (74) studied the effect of exercise on fluid absorption and found that as long as the exercise intensity was less than approximately 70% $\dot{V}O_2max$, exertion had only a minor effect on absorption from the small intestine. Another challenge in developing a hydration solution is that stomach emptying is delayed if drinking solutions are hypertonic. Gisolfi and colleagues found that dilute solutions of glucose (<6%) and electrolytes empty from the stomach almost as fast as water (172); that they are effective in reducing heart rate, maintaining plasma volume, and preventing hyperthermia; and that they often delay fatigue during prolonged running in the heat (151).

Figure 19.6 Carl Gisolfi performed extensive studies on the effect of hydration solutions on exercise responses. His findings led to the current explosion of sport drinks.
Photo courtesy of the personal collection of Dr. Charles M. Tipton.

In recent studies, Noakes and colleagues challenged the early findings of Hunt, Pitts, and others (1, 103, 158) that subjects should completely replace their fluids lost due to sweating during prolonged exercise (>4 h) in the heat. Noakes argued that the classic recommendation to drink to maintain body mass increases the risk of hyponatremia in some individuals (144) and that body mass losses of up to 3% are tolerated without alteration in exercise core temperature or cardiovascular responses. Noakes and colleagues suggested that there are body water reserves in the gastrointestinal tract and bound with glycogen in muscle and that these reserves can be mobilized to the vascular compartment when a well-hydrated individual performs prolonged exercise in the heat (145). This theory of a water reserve in humans requires further investigation.

Rehydration after exercise is critical for preventing hyperthermia during repeated exercise in the heat. As Hunt observed in 1912 (103), humans are slow to rehydrate and take hours to restore body water after exposure to severe heat. Carter and Gisolfi (33) noted that subjects drink more fluid, plasma volume increases to a higher level, and body weight increases more rapidly when subjects rehydrate using a dilute carbohydrate–electrolyte solution compared with water during recovery from exercise in the heat. The mechanism for this quicker recovery was proposed by Hiroshi Nose and colleagues. These authors (147) studied six subjects during recovery from exercise in the heat, during which their body weights were reduced by 2.3%. The subjects rested 1 h without fluids and then over the next 3 h were offered either tap water or a dilute salt solution (0.45 g of NaCl/100 ml of water). Subjects who rehydrated with water recovered only 68% of their fluid deficit whereas subjects who rehydrated with the salt solution recovered 82% of their deficit. Nose and colleagues concluded that the less-complete rehydration with water was attributable to removal of the osmotic drive to drink and to greater diuresis. In a subsequent paper, Nose and colleagues (146) found that plasma volume was selectively expanded when rehydrating with water, which diminished the drive to increase plasma renin and aldosterone. Therefore, less sodium is retained, resulting in a slower recovery of total body water.

Exercise in a Cold Environment

Thermoregulation in a cold environment follows pathways for heat exchange that are similar to those followed in a hot environment, as described in several excellent reviews (29, 88, 194, 200, 202, 204). Cold receptors in central and peripheral body regions trigger afferent impulses that ascend in the spinoparabrachial pathway to inhibit heat-loss responses from the anterior hypothalamus and to activate heat-production and heat-conservation responses as shown in figure 19.4 (132, 138). In addition, heat production is influenced by alterations in thyroid hormone, although this effect is less in humans than in most animals (34).

Acute Effects of Cold

The concept of a critical temperature during exposure to cold air was first proposed by Scholander and colleagues in 1950 and was defined as the lowest ambient temperature before an increase in metabolism above the resting level occurs (182). For individuals this value varies from approximately 20 °C to 27 °C in air but is considerably higher—30 °C to 34 °C—in a person immersed in water (204). Before this critical temperature is reached, core temperature is defended primarily by vasoconstriction to increase body insulation.

Body insulation is increased by veno- and vasoconstriction. Webb and Shepherd (212) described how veins constrict more rapidly than arteries because of a greater cold sensitivity. This venoconstriction reduces blood volume in the periphery, increasing the body shell (8). Similar to heating, changes in skin blood flow during cold exposure involve both local and whole-body reflex responses. John Johnson's group (3) used liquid-cooling garments to separately manipulate whole-body and local skin temperatures to identify their respective roles in cold-induced vasoconstriction. Whole-body cooling triggers a generalized sympathetic constrictor response. However, local cooling induces a more complex response that produces local vasoconstriction and inhibits the generalized reflex vasoconstriction. Under normal circumstances, when a decrease in environmental temperature causes whole-body cooling, the generalized reflex response dominates the overall skin blood flow response.

Shell thickness and composition are important variables in maintaining body temperature during cold exposure. Buskirk and colleagues (31) characterized the linear positive relationship between rectal temperature and percentage body fat for men and women exposed to 10 °C air where rectal temperature (Tr) = 36.33 + 0.011 × (% body fat). Keatinge (111) reported a direct and inverse relationship between skinfold thickness and the decrease in rectal temperature of men immersed in cold water. However, others have shown that the insulation of a tissue is not solely explained by the amount of subcutaneous fat. For example, Hayward and Keatinge (93) found that subcutaneous fat in the limbs accounts for only a small portion of limb insulation. Toner and McArdle (204) ascribed the additional heat conservation of limbs to a more complete vasoconstriction, the countercurrent arrangement of blood vessels, and insulation provided by a constricted muscle mass. However, this protective effect of muscle applies only to a resting limb. Once the limb moves (e.g., if swimming is required), blood flow to the muscle increases, significantly reducing the insulation of the limb. In situations of rescue swimming, a decision must be made about whether the metabolic heat produced by swimming is greater than the increased heat loss from the exercising limbs. Ducharme and Lounsbury (53) reviewed this dilemma and injected a further complication: The prominent reason for drowning in cold water often is not hypothermia per se but rather the effect of local cooling in causing muscle incapacitation.

When the ambient air or water temperature decreases below the critical temperature, oxygen consumption increases, initially in relation to the decrease in air or water temperature. This initial increase in metabolism is caused primarily by an involuntary shivering reflex; the increased cost of breathing, muscle tensing, and voluntary movements make smaller contributions. J.D. Hardy in 1961 (88) characterized heat production during shivering as the "efficiency of shivering," which he defined as the percentage of increased heat production stored in the body. Harvey found that the efficiency of shivering for men resting in a cold environment was approximately 48%. However, when they exercised, their efficiency of shivering declined to only 20%. Hayward and Eckerson (91) measured the metabolic cost of shivering in subjects immersed in ice water. They found that the metabolic rate increased to about 419 W within 15 to 20 min of immersion in average-fit subjects; values were even higher in their most-fit subjects (93). Iampietro and colleagues (104) studied men during exposure to cold air and high wind velocity and found a maximum metabolic response (370 W) only slightly lower than what was reported during water immersion. When shivering was insufficient, Dill and Forbes (44) described how patients become hypothermic and how shivering becomes sporadic and eventually ceases once core temperature decreases below 30 °C. Cardiac arrhythmias and death may occur when core temperature plunges below approximately 25 °C. However, exceptions have occurred under conditions of very rapid cooling. In 2000, Gilbert and Skagseth described a case report where a 29-yr-old woman fell in a waterfall and was resuscitated despite reaching a core temperature of 13.7 °C (72). Such cases may be possible with rapid cooling and when using slow internal-rewarming procedures.

Another interesting phenomenon and possible cause of death in accidental hypothermia is the afterdrop. An afterdrop occurs when core temperature decreases suddenly during active rewarming; it can be caused by limb movement or by warming of the periphery. Although the afterdrop has been noted for more than 200 yr, a vigorous debate over its mechanism began in the 1940s when Konig (115) attributed this effect to a circulatory response. A circulatory mechanism presumably occurs during external rewarming or with sudden movement when chilled peripheral blood returns to the heart, causing a sudden further decrease in core temperature. Based on this circulatory theory, the clinical treatment for accidental hypothermia would be to slowly warm the skin and avoid movement of the limbs (92). In the 1980s, however, this theory was challenged by the conduction theory. As explained by Webb (210), a number of observations did not support the circulatory theory. First, an afterdrop was not observed during the rewarming of small animals and it did occur when rewarming dead pigs without an intact circulation. Second, when humans underwent rewarming in a direct calorimeter (210), the afterdrop could be explained using simple rules of heat transfer without a circulatory explanation. That is, when rewarming an object by applying heat to its surface, the central temperature will continue to decrease as heat moves by conduction from the deeper tissues to warmer surface tissues. This afterdrop debate continues today, and some authors suggest that both circulatory and conductive mechanisms are involved (169).

Chronic Effects

Cold acclimation is a more subtle and varied response than heat acclimation in humans, leading some physiologists to even doubt its existence (99). Some confusion exists because cold adaptations depend on whether the cold exposure induces only local versus whole-body cooling and on the degree and duration of cold exposure (223). Buskirk, in his review in 1978 (29), summarized possible adaptive changes that could improve man's survival in the cold. The adaptations included long-term morphological changes in body shape and composition as well as regulatory changes in the threshold and gain of heat production and conservation responses.

Early studies of cold acclimatization were performed by assessing the cold responses of indigenous peoples. In 1930, a field study by Sir (Woodley) Cedric Hicks (figure 19.7) (98) examined the metabolic and skin responses of central Australian Aborigines. His report portrayed the deprivations of field work in the Australian outback, which relied on labor-intensive instrumentation including Douglas bags, a Haldane apparatus, a Benedict metabolimeter, and a bolograph to measure radial pulse volume as an index of hand blood flow. An early and intense peripheral vasoconstriction in the Aborigines reduced their body heat loss at night, allowing them to sleep comfortably with little clothing or bedding, unlike the unacclimatized European investigators. Such field studies were halted during World War II and then revived in the 1950s by Scholander, Hammel, and others. This interesting work was summarized by Ted Hammel in the 1964 *Handbook of Physiology* (84) and more recently by Andrew Young in the 1996 *Handbook of Physiology* (223).

Bittle and colleagues in 1989 (21) synthesized results from field and laboratory studies to characterize four types of cold adaptation. They described the first pattern as a hypothermic adaptation. Examples of this pattern included Ted Hammel's work with the Kalahari bushmen and Elsner and Bolstad's work with Peruvian Indians (181). In this pattern of cold adaptation, core temperature is allowed to decrease at night because the subjects have blunted shivering and a normal skin vasoconstrictor response. The subjects simply tolerate the lower body temperatures.

A second pattern of insulative adaptation was described by Hammel and colleagues (84), who studied coastal tribes in tropical northern Australia, and by Davis and Johnston, who studied Europeans who naturally adapted to a cold climate (84, 181). These subjects exhibit an exaggerated skin vasoconstrictor response that maintains core temperature during cold exposures.

The third pattern, called insulative hypothermic adaptation, was described by Hicks in the Western Australian Aborigines discussed previously (98). Subsequent studies by Scholander and Hammel in this same population (84) confirmed the insulative response of the Aborigines and documented a decrease in rectal temperature and blunted shivering response. Thus, this

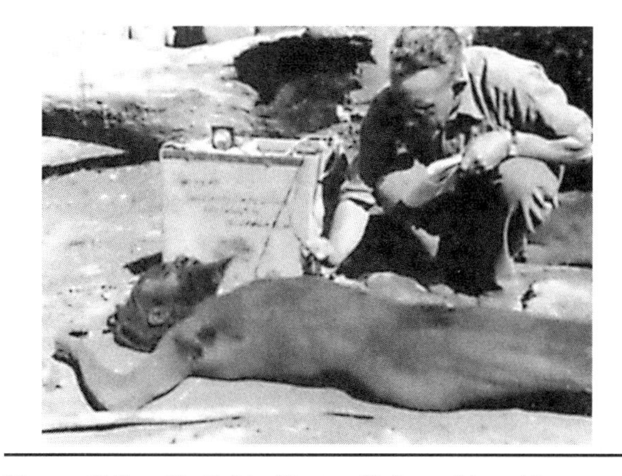

Figure 19.7 Sir Cedric Stanton Hicks making skin temperature measurements on an Australian Aborigine in 1934.

Used with courtesy of the American Physiological Society. Hicks, CS. Terrestial animals in cold: Exploratory studies of primitive man. In *Handbook of Physiology, Adaptation to the Environment* Bethesda, MD Am Physiol. Soc., 1964. Section 4, chapter 25, p407.

group exhibited both insulative and hypothermic adaptations.

The fourth cold adaptation pattern, called metabolic adaptation, was observed in Europeans living with bushmen (181), in the Alacaluf Indians of Terra del Fuego, in Arctic Indians, and in Eskimos (84). In the Alacaluf subjects, for example, Hammel observed that metabolic heat production was more than 50% higher than that of nonadapted subjects. The subjects also had higher mean skin temperatures and maintained normal core temperatures during exposure to cold (84).

The most compelling documentation of human cold acclimatization was made between 1958 and 1977 by Suk Ki Hong (figure 19.8), Donald Rennie, and Yan Park (99) in female Korean Ama divers, who dove in sea temperatures of 10 °C in winter to 27 °C in summer while wearing only cotton swimsuits. During a dive, these women experienced a decrease in rectal temperature to approximately 35 °C, a delayed onset of shivering, and an increased subcutaneous insulation due to an intense vasoconstriction of the limbs. These responses are characteristic of an insulative-hypothermic adaptation. In addition, however, these women experienced seasonal variations in basal metabolic rate. In summer their basal metabolic rate was similar to that of nondiving Korean women but in winter it was 30% greater (110). When the authors returned to study this same population from 1980 to 1982, the divers had begun wearing wetsuits and these adaptations to cold had mostly disappeared (153).

What determines the pathway an individual will use to acclimate to cold? Bittle and colleagues (21) suggested that this is determined by specific environmental

Figure 19.8 Suk Ki Hong, who studied the thermal responses of the Ama pearl divers. Before the 1980s when the Ama began to wear wetsuits, these women had evidence of insulative and metabolic cold acclimatization.

Photo courtesy of Dr. Harold Strauss of the Department of Physiology and Biophysics at SUNY and with permission of the Suk Ki Hong family.

conditions, the individual's diet, and individual factors such as body fat or fitness. More recently, Andrew Young (223) developed a flowsheet (figure 19.9) to describe the adaptation process. In the first step of the flowsheet, intermittent cold exposure, even without whole-body cooling, induces habituation responses such as delayed shivering and less-intense vasoconstriction. In 1930, Lewis (120) described a skin response that reduced local pain by increasing peripheral blood flow with repeated cold exposure. After about 5 to 10 min of hand immersion in cold water, the blood vessels in the fingertips vasodilate and then alternate between constriction and dilation. This alternating local blood flow despite continuing cold exposure is called a hunting reaction. It is more pronounced in persons born in cold regions or who expose their hands to cold for prolonged periods. This reaction was thought to be a local reflex that reduced tissue injury due to cold (39).

The next step in Young's flowsheet (223) occurs if significant amounts of body heat are lost during the cold exposure. If the cold exposure is not severe and an increase in metabolism is sufficient to maintain core temperature, the metabolic acclimation pathway will be adopted. However, if the cold exposure is severe enough to cause a decrease in core temperature despite increased heat production, an insulative approach develops that includes intense vasoconstriction and possibly a reduced core temperature to minimize conductive and convective heat losses.

In animals, an important response to increased heat production during cold exposure is nonshivering thermogenesis. In 1961, Thomas Davis (40) reported that 8 h cold exposures for 31 d in sedentary men resulted in an increase in metabolic rate despite a gradual decrease in shivering; this supported the existence of nonshivering thermogenesis in humans. Lichtenbelt and Daanen (121) described two possible mechanisms for nonshivering thermogenesis in man. The first involves brown adipose tissue thermogenesis, in which catecholamines released in response to cold activate an uncoupling protein, UCP-1, located in the mitochondria of brown adipose tissue to increase heat production. Early work by Cramer in 1920 (37) noted the presence of small amounts of brown adipose tissue in humans. Morphological studies from cadavers by Heaton in 1972 (94) quantified the reduction in brown adipose tissue that occurs with age but noted that some is present around the deeper organs of the body even in the eighth decade. In 1997, Oberkolfer and colleagues (149) developed a sensitive assay for detecting UCP-1 messenger ribonucleic acid. They found that UCP-1 is interspersed throughout the adipose tissues of adult humans. The second form of nonshivering thermogenesis involves several subtypes of UCPs located in

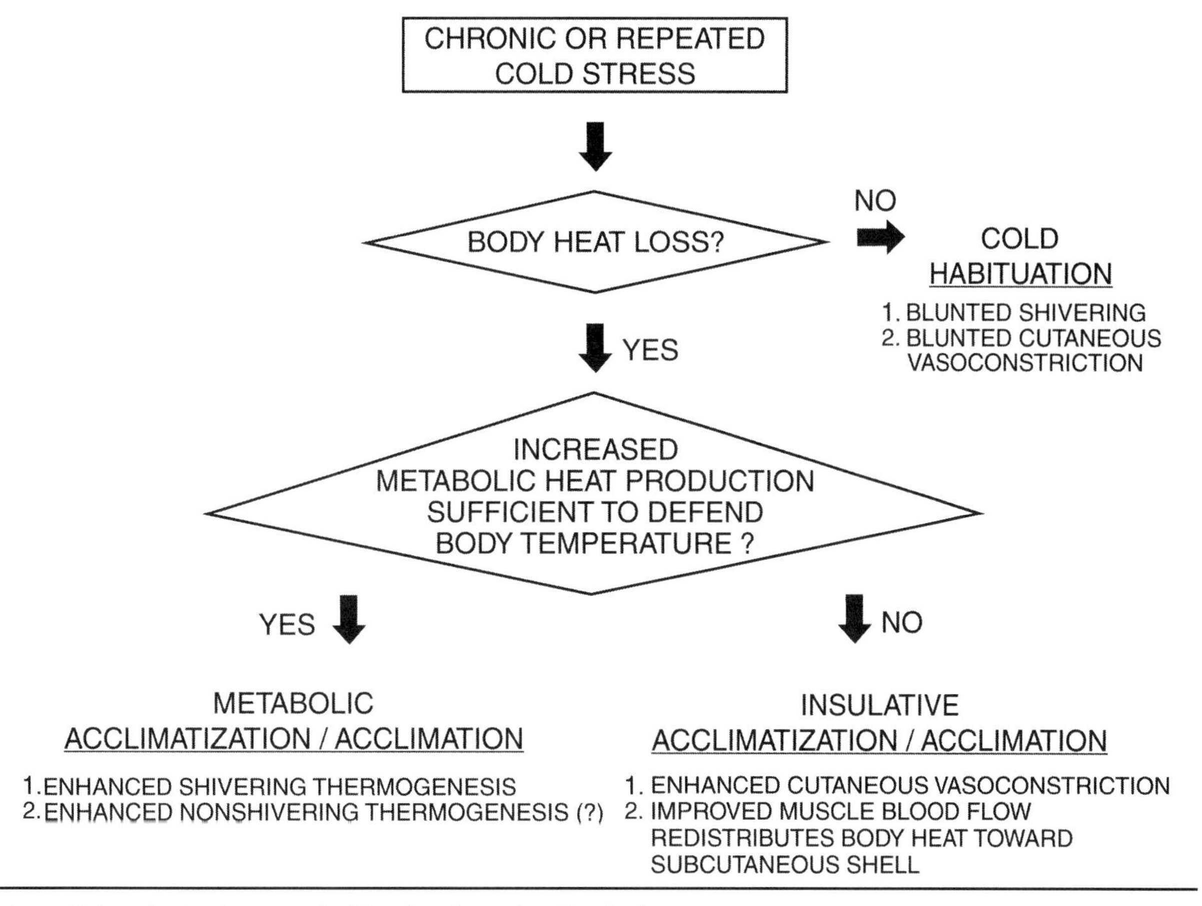

Figure 19.9 The development of cold acclimation and acclimatization.

Adapted from A.J. Young, 1996, *Handbook of physiology* (Bethesda, MD: The American Physiological Society), 435. With permission from American Physiological Society and Andrew Young.

various tissues of the body. Simonsen and colleagues (189) measured thermogenesis after infusion of epinephrine and estimated that 40% of the increased metabolism occurred in skeletal muscle. They proposed that during an acute exposure to severe cold, catecholamines may be released and induce UCP thermogenesis by uncoupling adenosine triphosphate synthesis from respiratory chain oxidation in brown adipose tissue and by uncoupling and altering calcium handling in the mitochondria of muscle cells. Nonshivering thermogenesis is considered a minor source of heat production in humans compared with the thermogenesis caused by shivering. It may play another indirect role in thermoregulation by an effect on overall energy balance and therefore through its effect in altering body morphology and fat content (121).

Molecular Thermoregulation

During the past 50 yr, understanding of thermoregulation at the molecular level has exploded. Temperature, a critical stressor at the cellular level, activates molecular pathways that affect the thermotolerance of individual cells and stimulate generalized stress responses that influence the entire organism.

Molecular Responses of the Cell to Heat

The investigation of thermoregulation-related changes at the cellular and tissue levels centers around the heat shock response and the cellular accumulation of heat shock proteins (HSPs). This ubiquitous response was first noted by Ritossa in 1962 (163) when *Drosophila* cells exposed to heat demonstrated a characteristic pattern of chromosomal puffing. These changes were associated with the state of thermotolerance, or the ability of a cell conditioned by a single stress to resist a subsequent (otherwise lethal) heat stress. Subsequent research defined the biology of the HSPs and included the following findings: The common factor in induction was the impairment of protein translation (214) or the presence of unfolded proteins (6); the cellular HSP accumulation is titrated to the degree of stress (130); and HSP transcription is dependent on a set of highly regulated transcription factors (particularly heat shock factor 1), which are themselves regulated by binding to HSPs (102). Removal of HSPs during stress results in heat shock factor 1 trimerization, translocation to the nucleus, and transcriptional activation.

In cell and animal systems the HSP response serves as a highly reliable biomarker of heat stress. Flanagan and colleagues used this property as a means to define the tissues most susceptible to heat stress (59). This study demonstrated that the Hsp70 response to heat stress was greatest in the liver and gut; these results supported earlier physiologic data implicating that the gut and liver are critical target tissues of heat stress (117). These findings, and others from the literature, supported a model of heat injury as a form of the systemic inflammatory response syndrome driven by heat and ischemic injury to the gut, portal vein endotoxemia, and resultant cytokine production (135).

HSPs were found to be more than biomarkers in that they conferred protection to tissues and the intact organism through a variety of mechanisms that were not identical in their ability to confer thermotolerance. Tissues and animals rely on mechanisms that allow multicellular systems to continue to function as units (as opposed to single-cell survival) in order to tolerate severe heat challenges. This phenomenon, which has been termed physiological thermotolerance (133), has been shown to be a critical factor in heat tolerance. Given the model of the gut as the critical target tissue in heat-related illness, research that focused on cellular adaptations demonstrated the ability of the heat shock response to enhance epithelial barrier integrity and block cytokine production. Early studies demonstrated a heat-associated inhibition of tumor necrosis factor-driven cytotoxicity (80); target cells expressing HSPs were resistant to cytokines (106). This finding was extended to animal models (47, 114). More recent studies in animals using single-gene Hsp70 expression demonstrated that Hsp70 alone could confer this effect through the inhibition of the transcription factor nuclear factor kappa light chain enhancer of activated B cells (NF-κB) (48). Interestingly, the HSP transcription machinery, specifically heat shock factor 1, also was shown to exert independent inhibitory control on inflammatory cytokine production through inhibition of NF-κB as well as direct interference with tumor necrosis factor alpha transcription (190).

Epithelial barrier integrity also is modulated by the heat shock response. This discovery extends from observations in 1990 demonstrating that heat conditioning sufficient to result in cellular Hsp70 accumulation resulted in improved barrier resistance to a reversible heat injury (134). This protective effect lasted as long as elevated cellular Hsp70 could be demonstrated. Inhibition of the heat shock response blocked this effect (49). Hsp70 expression alone was sufficient to confer barrier resistance. However, as was shown for cytokine resistance, heat shock factor 1 itself was found to have important effects independent of HSP accumulation because it was shown to directly modulate the transcription of the epithelial tight junction protein occludin 1 (50).

Molecular Responses of the Cell to Exercise

Animal studies conducted in the 1990s demonstrated the ability of the whole organism to generate a heat shock response in response to severe passive heating and to exercise (118, 168, 191). Although severe stresses in cell or animal models could clearly induce the heat shock response, studies with passive heating or in febrile humans failed to detect a change in cellular HSPs. Ryan and colleagues (173) used a novel system to detect HSP alterations and were the first to demonstrate that the cellular HSP response could be activated in humans. Moreover, these studies, using an exercise stimulus, demonstrated that cellular HSP accumulation could occur under the physiologic condition of exercise in the heat. This foundational study was followed by others demonstrating an HSP response in muscle after eccentric stress (201), exhaustive exercise (209), or exercise-induced heat acclimation (222).

Molecular Effects of Heat Acclimation

One of the central and elusive quests of thermoregulation and heat adaptation is the link between thermotolerance and acclimation. Although both are clear adaptive responses, they have clearly different characteristics. Thermotolerance is induced by a single severe stress and is generally defined in terms of survival of a subsequent (otherwise lethal) heat challenge. Acclimation is induced by multiple less-severe heat challenges or by exercise resulting in temperature elevations. Thus, thermotolerance is a state of tolerance of higher temperatures. The adaptations of acclimation are largely those that result in a lower temperature for a given level of work. These adaptations include earlier onset of sweating, increased sweat rate, and increased plasma volume. However, several studies have noted that individuals are able to continue to work at higher core temperatures in the acclimated state than in the unacclimated state; this mirrors the state of thermotolerance. Interestingly, studies of evolutionary biology in ants (71), lizards (206), and hydra (24) have noted differences in HSPs between hot- and cool-environment species. These tantalizing and provocative studies offer the possibility that thermotolerance and both acclimation and acclimatization are linked through cellular mechanisms such as the heat shock response. And, while we await demonstration of this link, clear associations exist: activation of the cellular heat stress machinery by exercise acclimation protocols (5, 222), the ability of acclimated animals (68) and humans (124, 183) to work at higher temperatures, and the understanding that gut barrier function and inflammation are crucial in heat-related illness and are clearly modulated by the heat shock response (48-50).

Summary

This chapter presents brief snapshots of some of the discoveries made in human exercise thermoregulation during the past century. As such, it may be useful as an introduction for searching historical articles and books on this topic. An appreciation of the past offers a perspective for current and new investigators to evaluate the importance of their own work and to more clearly identify what information is required for the next major scientific breakthroughs. Also, by recognizing the mistakes and successes of the past, we may gain insight into what techniques are needed to advance most efficiently and effectively. For the student, we hope this chapter has brought to life some of the important concepts and interesting people who have contributed to the field of thermoregulation.

Acknowledgments

The authors thank John Greenleaf and Charlie Senay for their helpful comments and suggestions and Joan Iverson and Rebecca Gustaf from the college of education at the University of New Mexico in the United States for their help in the preparation of this manuscript.

References

1. Adolph EF. *Physiology of Man in the Desert*. New York: Interscience, 1947.
2. Allan JR. The effects of physical training in a temperate and hot climate on the physiological responses to heat stress. *Ergonomics* 8: 445-453, 1965.
3. Alvarez GE, Zhao K, Kosiba WA, Johnson JM. Relative roles of local and reflex components in cutaneous vasoconstriction during skin cooling in humans. *J Appl Physiol* 100: 2083-2088, 2006.
4. Amatruda TT Jr., Welt LG. Secretion of electrolytes in thermal sweat. *J Appl Physiol* 5: 759-772, 1953.
5. Amorim FT, Yamada PM, Robergs R, Schneider SM, Moseley PL. The effect of the rate of heat storage on serum heat shock protein 72 in humans. *Eur J Appl Physiol* 104: 965-972, 2008.
6. Ananthan J, Goldberg AL, Voellmy R. Abnormal proteins serve as eukaryotic stress signals and trigger the activation of heat shock genes. *Science* 232: 522-524, 1986.
7. Aronsohn E, Sachs J. Die beziehungen des gehirns zur korperwarme und zum fieber. *Pflugers Arch Physiol* 37: 232-249, 1885.
8. Aschoff J, Wever R. Kern and schale im warmehauschalt des menshen. *Naturwissenchaften* 45: 47-485, 1958.
9. Astrand I. Aerobic work capacity in men and women with special reference to age. *Acta Physiol Scand* 49: 1-92, 1960.
10. Avellini BA, Kamon E, Krajewski JT. Physiological responses for physically fit men and women to acclimation to humid heat. *J Appl Physiol* 49: 254-261, 1980.
11. Balfour A. Sojourners in the tropics and problems with acclimation. *Lancet* 204. 1329-1334, 1923.
12. Barbour H. Die wirkung unmittelbarer erwarmung und abkuhlung der warmezentra auf die korpertemperatur. *Arch Exp Path Pharm* 70: 1-16, 1912.
13. Barbour H, Tolstoi E. Heat regulation and water exchange. II The role of water content of the blood, and its control by the central nervous system. *Am J Physiol* 67: 378-387, 1924.
14. Barcroft J. On the relation of external temperature to blood volume. *Philos Trans R Soc London Ser B* ccxi: 417, 455, 1923.
15. Bazett HC. Theory of reflex controls to explain regulation of body temperature at rest and during exercise. *J Appl Physiol* 4: 245-262, 1951.
16. Benzinger TH. Heat regulation: Homeostasis of central temperature in man. *Phys Rev* 49: 671-759, 1969.
17. Benzinger TH. On physical heat regulation and the sense of temperature in man. *Proc Nat Acad Sci* 45: 645-659, 1959.
18. Benzinger TH. *Temperature Part II Thermal Homeostasis*. Stroudsburg, PA: Dowden, Hutchinson & Ross, 1977, 1-414.
19. Benzinger TH. The thermal homeostasis of man. In: Benzinger TH, ed. *Temperature*. Stroudsburg, PA: Dowden, Hutchinson & Ross, 1977, 289-377.
20. Bernarroch EE. Thermoregulation. Recent concepts and remaining questions. *Neurology* 69: 1293-1297, 2007.

21. Bittle JHM, Livecchi-Gonnot GH, Hanniquet AM, Poulain C, Etienne J-L. Thermal changes observed before and after J.-L. Etienne's journey to the North Pole. *Eur J Appl Physiol* 58: 646-651, 1989.
22. Blatteis CM. The neurobiology of endogenous pyrogens. In: Braun HA, Bruck K, Heldmaier G, eds. *Thermoregulation and Temperature Regulation*. Berlin: Springer-Verlag, 1990, 257-272.
23. Bligh JA. A theoretical consideration of the means whereby the mammalian core temperature is defended at a null zone. *J Appl Physiol* 100: 1332-1337, 2006.
24. Bosch TC, Krylow SM, Bode HR, Steele RE. Thermotolerance and synthesis of heat shock proteins: These responses are present in *Hydra attenuate* but absent in *Hydra oligactis*. *Proc Nat Acad Sci* 85: 7927-7931, 1988.
25. Boulant JA, Demieville HN. Responses of thermosensitive preoptic and septal neurons to hippocampal and brain stem stimulation. *J Neurophysiol* 40: 1356-1368, 1977.
26. Bruck K, Olschewski H. Body temperature related factors diminishing the drive to exercise. *Can J Physiol Pharmacol* 65: 1274-1280, 1987.
27. Burton AC. The operating characteristics of the human thermoregulatory mechanism. In: Hardy JD, ed. *Temperature: Its Measurement and Control in Science and Industry*. New York: Reinhold, 1941, 522-528.
28. Burton AC, Bazett HC. The study of the average temperature of the exchanges of heat and vasomotor responses in man by means of a bath calorimeter. *Am J Physiol* 117: 36-54, 1936.
29. Buskirk ER. Cold stress: A selective review. In: Folinsbee LJ, Wagner JA, Borgia JF, Drinkwater BL, Gliner JA, Bedi IF, eds. *Environmental Stress: Individual Human Adaptations*. New York: Academic Press, 1978, 249-266.
30. Buskirk ER. The temperature regulatory system. In: Tipton CM, ed. *Exercise Physiology: People and Ideas*. Oxford: Oxford University Press, 2003, 423-451.
31. Buskirk ER, Thompson RH, Whedon GD. Metabolic response to cooling in the human: Role of body composition and particularly of body fat. In: Hardy JD, ed. *Temperature: Its Measurement and Control in Science and Industry*. New York: Reinhold, 1963, 429-442.
32. Cannon W. Physiological regulation of normal states: Some tentative postulates concerning biological homeostatics. In: Pettit A, ed. *A Charles Richet: Ses Amis, Aes Collegues, Ses Eleves*. Paris: Les Editions Medicales, 1926, 91-93.
33. Carter JE, Gisolfi CV. Fluid replacement during and after exercise in the heat. *Med Sci Sports Exerc* 21: 532-539, 1989.
34. Collins KJ, Weiner JS. Endocrinological aspects of exposure to high environmental temperatures. *Physiol Rev* 48: 785-839, 1968.
35. Conn JW, Johnston MW, Louis HL. Acclimatization to humid heat: A function of adrenal cortical activity. *J Clin Invest* 25: 912, 1946.
36. Costill DL, Cote R, Miller E, Miller T, Wynder S. Water and electrolyte replacement during repeated days of work in the heat. *Aviat Space Environ Med* 46: 795-800, 1975.
37. Cramer W. On glandular adipose tissue and its relation to the endocrine organs and the vitamin problem. *Br J Exp Pathol* 1: 184-196, 1920.
38. Cunningham DJ, Cabanac M. Evidence from behavioral thermoregulatory responses of a shift in setpoint temperature related to the menstrual cycle. *J Physiol (Paris)* 63: 236-238, 1971.
39. Daanen HAM. Finger cold-induced vasodilation: A review. *Eur J Appl Physiol* 89: 411-426, 2003.
40. Davis TRA. Chamber cold acclimatization in man. *J Appl Physiol* 16: 1011-1015, 1961.
41. Davy J. On the temperature of man within the tropics. *Phil Trans (London)* 140: 437-466, 1850.
42. Derambure PS, Boulant JA. Circadian thermosensitive characteristics of suprachiasmic neurons in vitro. *Am J Physiol* 266: R1876-R1884, 1994.
43. Dill DB. *The Hot Life of Man and Beast*. Springfield, IL: Charles C. Thomas, 1985, 177.
44. Dill DB, Forbes WH. Respiratory and metabolic effects of hypothermia. *Am J Physiol* 132: 685-697, 1941.
45. Dill DB, Jones BF, Edwards HT, Oberg SA. Salt economy in extreme dry heat. *J Biol Chem* 100: 755, 1933.
46. Dill DB, Soholt LF, McLean DC, Drost TF Jr., Loughran MT. Capacity of young males and females for running in desert heat. *Med Sci Sports Exerc* 9: 137-142, 1977.
47. Dokladny K, Kozak A, Wachulec M, Wallen ES, Menache MG, Kozak W, Kluger MJ, Moseley PL. Effect of heat stress on LPS-induced febrile response in D-galactosamine-sensitized rats. *Am J Physiol* 280: R338-R344, 2001.
48. Dokladny K, Lobb R, Wharton W, Ma TY, Moseley PL. LPS-induced cytokine levels are repressed by elevated expression of HSP70 in rats: Possible role of NF-kappaB. *Cell Stress Chaperones* 15: 153-163, 2010.
49. Dokladny K, Wharton W, Lobb R, Ma TY, Moseley PL. Induction of physiological thermotolerance in MDCK monolayers: Contribution of heat shock protein 70. *Cell Stress Chaperones* 11: 268-275, 2006.
50. Dokladny K, Ye D, Kennedy JC, Moseley PL, Ma TY. Cellular and molecular mechanisms of heat stress-induced up-regulation of occludin protein ex-

pression: Regulatory role of heat shock factor-1. *Am J Pathol* 172: 659-670, 2008.

51. Douglas C. A method for determining the total respiratory exchange in man. *J Physiol (London)* 42: xvii-xviii, 1911.
52. Downey JA, Chiodi HP, Darling RC. Central temperature regulation in the spinal man. *J Appl Physiol* 22: 91-94, 1967.
53. Ducharme MB, Lounsbury DS. Self-rescue swimming in cold water: The latest advice. *Appl Physiol Nutr Metab* 32: 799-807, 2007.
54. Eichna LW, Bean WB, Ashe WF, Nelson N. Performance in relation to environmental temperature. *Bull Johns Hopkins Hosp* 76: 25-58, 1945.
55. Ekblom B, Greenleaf CJ, Greenleaf JE, Hermansen L. Temperature regulation during exercise dehydration in man. *Acta Physiol Scand* 79: 475-483, 1970.
56. von Euler C. Slow temperature potentials in the hypothalamus. *J Cellular Comp Physiol* 36: 333-350, 1950.
57. Feldberg W, Meyers RD. Effects on temperature of amines injected into the cerebral ventricles. A new concept of temperature regulation. *J Physiol* 173: 226-237, 1964.
58. Feldberg W, Myers RD, Veale WL. Perfusion from cerebral ventricle to ciserna magna in the unanaesthetized cat. Effect of calcium on body temperature. *J Physiol* 207: 403-416, 1970.
59. Flanagan SW, Ryan AJ, Gisolfi CV, Moseley PL. Tissue-specific HSP70 response in animals undergoing heat stress. *Am J Physiol* 268: R28-R32, 1995.
60. Folkow B, Strom G, Uvnas B. Efferent nervous pathway involved in cutaneous vaodilation induced by activation of hypothalamic heat loss mechanisms. *Acta Physiol Scand* 17: 327-338, 1949.
61. Fordtran JS, Rector FC, Carter NW. The mechanism of sodium reabsorption in the human small intestine. *J Clin Invest* 47: 884-900, 1968.
62. Fortney SM, Nadel ER, Wenger CB, Bove JR. Effect of blood volume on sweating rate and body fluids in exercising humans. *J Appl Physiol* 51: 1594-1600, 1981.
63. Fortney SM, Wenger CB, Bove JR, Nadel ER. Effect of hyperosmolality on control of blood flow and sweating. *J Appl Physiol* 57: 1688-1695, 1984.
64. Foster KG, Weiner JS. Effects of cholinergic and adrenergic blocking agents on the activity of the eccrine sweat glands. *J Physiol* 210: 883-895, 1970.
65. Fox RH, Goldsmith R, Kidd DJ, Lewis HE. Acclimatization to heat in man by controlled elevation of body temperature. *J Physiol* 166: 530-547, 1963.
66. Fox RH, Goldsmith R, Kidd DJ, Lewis HE. Blood flow and other thermoregulatory changes with acclimatization to heat. *J Physiol* 166: 548-562, 1963.
67. Fox RH, Lofstedt BE, Woodward PM, Erickson E, Werkstrom B. Comparison of thermoregulatory function in men and women. *J Appl Physiol* 26: 444-453, 1969.
68. Fruth JM, Gisolfi CV. Work-heat tolerance in endurance-trained rats. *J Appl Physiol* 54: 249-253, 1983.
69. Frye AJ, Kamon E, Webb M. Responses of menstrual women, amenorrheal women, and men to exercise in a hot, dry environment. *Eur J Appl Physiol* 48: 279-288, 1982.
70. Fulton J, Wilson L. Homeostasis and the regulation of body temperatures. In: *Selected Readings in the History of Physiology*. Springfield, IL: Charles C. Thomas, 1966.
71. Gehring WJ, Wehner R. Heat shock protein synthesis and thermotolerance in *Cataglyphis*, an ant from the Sahara desert. *Proc Nat Acad Sci* 92: 2994-2998, 1995.
72. Gilbert M, Skagseth A. Resuscitation from accidental hypothermia of 13.7°C with circulatory arrest. *Lancet* 355: 375-376, 2000.
73. Gisolfi CV, Cohen JS. Relationships among training, heat acclimation, and heat tolerance in men and women: The controversy revisited. *Med Sci Sports Exerc* 11: 56-59, 1979.
74. Gisolfi CV, Spranger KJ, Summers RW, Schedl HP, Bleiler TL. Effects of cycle exercise on intestinal absorption in humans. *J Appl Physiol* 71: 2518-2527, 1991.
75. Glotzbach SF, Heller HC. Changes in the thermal characteristics of hypothalamic neurons during sleep and wakefulness. *Brain Res* 309: 17-26, 1984.
76. Gonzalez-Alonzo J, Teller C, Andersen SL, Jensen FB, Hyldig T, Nielsen B. Influence of body temperature on the development of fatigue during prolonged exercise in the heat. *J Appl Physiol* 86: 1032-1039, 1999.
77. Goodman M. *Suffer and Survive*. London: Simon & Schuster UK, 2007, 422.
78. Greenleaf JE. Hyperthermia and exercise. In: Robershaw D, ed. *Environmental Physiology III*. Baltimore, MD: University Park Press, 1979, 157-208.
79. Greenleaf JE, Castle BL. Exercise temperature regulation in man during hypohydration and hyperhydration. *J Appl Physiol* 30: 847-853, 1971.
80. Gromkowski SH, Yagi J, Janeway CAJ. Elevated temperature regulates tumor necrosis factor-mediated immune killing. *Eur J Immunol* 19: 1709-1714, 1989.
81. Gunga, HC. *Nathan Zuntz*. Amsterdam: Elsevier, 2009, 1-252.
82. Gupta BN, Nier K, Hensel H. Cold-sensitive afferents from the abdomen. *Pflugers Arch* 380: 203-204, 1979.

83. Haldane J. The influence of high air temperatures. *J Hygiene* 12: 494-513, 1905.
84. Hammel HT. Terrestrial animals in cold: Recent studies of primitive man. In: Dill DB, Adolph EF, Wilber CG, eds. *Handbook of Physiology*. Washington, DC: American Physiological Society, 1964, 413-433.
85. Hammel HT, Elsner RW, Le Messurier DH, Milan FA. Thermal and metabolic responses of the Australian Aborigines to moderate cold in summer. *J Appl Physiol* 14: 605-615, 1959.
86. Hammel HT, Jackson DC, Stolwijk JA, Hardy JD, Stromme SB. Temperature regulation by hypothalamic proportional control with an adjustable set point. *J Appl Physiol* 18: 1146-1154, 1963.
87. Hardy JD. Models of temperature regulation—A review. In: Bligh J, Moore RE, eds. *Essays on Temperature Regulation*. New York: Elsevier, 1972, 163-186.
88. Hardy JD. Physiology of temperature regulation. *Physiol Rev* 41: 521-606, 1961.
89. Hardy JD, Milhorat AT, DuBois EF. Heat loss and heat production in women under basal conditions at temperatures from 23 to 35°C. In: Hardy JD, ed. *Temperature: Its Measurement and Control in Science and Industry*. New York: Reinhold, 1941, 529-536.
90. Haslag SWM, Hertzman AB. Temperature regulation in young women. *J Appl Physiol* 20: 1283-1288, 1965.
91. Hayward JS, Eckerson JD. Physiological responses and survival time prediction for humans in ice-water. *Aviat Space Environ Med* 53: 206-212, 1984.
92. Hayward JS, Eckerson JD, Kemna D. Thermal and cardiovascular changes during three methods of resuscitation from mild hyperthermia. *Resuscitation* 11: 21-33, 1984.
93. Hayward MG, Keatinge WR. Roles of subcutaneous fat and thermoregulatory reflexes in determining ability to stabilize body temperature in water. *J Physiol (Lond)* 320: 229-251, 1981.
94. Heaton JM. The distribution of brown adipose tissue in the human. *J Anat* 112: 35-39, 1972.
95. Hellon RF. Thermal stimulation of hypothalamic neurones in unanesthetized rabbits. *J Physiol (Lond)* 193: 381-395, 1967.
96. Hellon RF, Lind AR. The influence of age on peripheral vasodilation in a hot environment. *J Physiol (Lond)* 141: 262-272, 1958.
97. Hensel H, Zotterman Y. Quantitative beziehungen zwischen der entladung einzelner kaltefasern und der temperatur. *Acta Physiol Scand* 23: 291-319, 1951.
98. Hicks CS. Terrestrial animals in cold: Exploratory studies of primitive man. In: Dill DB, Adolph EF, Wilber CG, eds. *Handbook of Physiology*. Washington, DC: American Physiological Society, 1964, 405-412.
99. Hong SK, Rennie DW, Park YS. Humans can acclimatize to cold: A lesson from Korean women divers. *NIPS* 2: 79-82, 1987.
100. Horvath SM. Historical perspectives of adaptation to heat. In: Horvath SM, Yousef MK, eds. *Environmental Physiology: Aging, Heat and Altitude*. New York: Elsevier North Holland, 1981, 11-25.
101. Horvath SM, Horvath EC. *The Harvard Fatigue Laboratory: Its History and Contributions*. Englewood Cliffs, NJ: Prentice-Hall, 1973, 182.
102. Hu Y, Mivechi NF. HSF-1 interacts with Ral-binding protein 1 in a stress-responsive multiprotein complex with HSP90 in vivo. *J Biol Chem* 278: 17299-17306, 2003.
103. Hunt EH. The regulation of body temperature in extremes of dry heat. *J Hygiene* 12: 479-488, 1912.
104. Iampietro PF, Vaughan JA, Goldman RF, Kreider MB, Masucci F, Bass DE. Heat production during shivering. *J Appl Physiol* 15: 632-634, 1960.
105. Isenschmid R, Krehl L. Uber den einflub des gehirns auf die warmeregulation. *Archiv Exp Pathol Pharmakol* 70: 109-134, 1912.
106. Jaattela M, Wissing D. Heat-shock proteins protect cells from monocyte cytotoxicity: Possible mechanism of self-protection *J Exp Med* 177: 231-236, 1993.
107. Jackson DC, Hammel HT. Hypothalamic "set" temperature decreased in exercising dog. *Life Sci* 8: 554-563, 1963.
108. Jay O, Reardon FD, Webb P, DuCharme MB, Ramsay T, Nettlefold L, Kenny GP. Estimating changes in mean body temperature for humans during exercise using core and skin temperatures is inaccurate even with a correction factor. *J Appl Physiol* 103: 443-451, 2007.
109. Jessen C, Feistkorn G, Nagel A. Temperature sensitivity of skeletal muscle in the conscious goat. *J Appl Physiol* 54: 880-886, 1983.
110. Kang BS, Song SH, Suh CS, Hong SK. Changes in body temperature and basal metabolic rate of the Ama. *J Appl Physiol* 18: 483-488, 1963.
111. Keatinge WR. The effects of subcutaneous fat and of previous exposure to cold on the body temperature, peripheral blood flow and metabolic rate of men in cold water. *J Physiol (Lond)* 153: 166-178, 1960.
112. Kellogg DL, Zhao JL, Wu Y. Roles of nitric oxide synthase isoforms in cutaneous vasodilation induced by local warming of the skin and whole body heat stress in humans. *J Appl Physiol* 107: 1438-1444, 2009.
113. Kielblock AJ, Schutte PC. The evolution of heat acclimatization practices and procedures in the South African gold mining industry with special reference

to fundamental considerations. In: Samueloff S, Yousef MK, eds. *Adaptive Physiology to Stressful Environments*. Boca Raton, FL: CRC Press, 1987, 67-74.

114. Kluger MJ, Rudolph K, Soszynski D, Conn CA, Leon LR, Kozak W, Wallen ES, Moseley PL. Effect of heat stress on LPS-induced fever and tumor necrosis factor. *Am J Physiol* 273: R858-R863, 1997.
115. Konig FH. Bemerkungen zur physikalischen behandlung von unterkuhlten. *Klin Wochenschr* 22: 45-50, 1943.
116. Kosaka M, Simon E, Thauer R. Shivering in intact and spinal rabbits during spinal cord cooling. *Experientia* 23: 385-387, 1967.
117. Kregel KC, Gisolfi CV. Circulatory responses to vasoconstrictor agents during passive heating in the rat. *J Appl Physiol* 68: 1220-1227, 1990.
118. Kregel KC, Moseley PL, Skidmore R, Gutierrez JA, Guerriero V Jr. HSP70 accumulation in tissues of heat-stressed rats is blunted with advancing age. *J Appl Physiol* 79: 1673-1678, 1995.
119. Krogh A, Lindhard J. The relative value of fat and carbohydrate as sources of muscular energy. *Biochem J* 14: 290-362, 1920.
120. Lewis T. Observations upon the reactions of the vessels of the human skin to cold. *Heart* 15: 177 208, 1930.
121. Lichtenbelt WDM, Daanen HAM. Cold-induced metabolism. *Curr Opinion Clin Nutr Metabol Care* 6: 469-475, 2003.
122. Liebermeister C. *Hanbuch der Pathologie und Therapie des Fiebers*. Leipzig, Germany: Vogelwelt, 1875.
123. Macallum AB. The paleochemistry of the body fluids and tissues. *Physiol Rev* 6: 316-357, 1926.
124. Maron MB, Wagner JA, Horvath SM. Thermoregulatory responses during competitive marathon running. *J Appl Physiol* 42: 909-914, 1977.
125. McKemy DD, Neuhausser WM, Julius D. Identification of a cold receptor reveals a general role for TRP channels in thermosensation. *Nature* 416: 52-58, 2002.
126. McLean J, Tobin G. Historical. In: *Animal and Human Calorimetry*. Cambridge, UK: Cambridge University Press, 1987, 6-23.
127. Mekjavic IB, Eiken O. Contribution of thermal and nonthermal factors to the regulation of body temperature in humans. *J Appl Physiol* 100: 2065-2072, 2005.
128. Meyer H. Theorie des fiebers und seiner behandlung. *Zentr Inn Med* 6: 385-386, 1913.
129. Meyers RD, Veale WL. Possible ionic mechanism in the hypothalamus controlling the set-point. *Science* 170: 95-97, 1970.
130. Mizzen LA, Welch WJ. Characterization of the thermotolerant cell. I. Effects on protein synthesis activity and the regulation of heat-shock protein 70 expression. *J Cell Biol* 106: 1105-1116, 1988.
131. Morrison SF. 2010 Carl Ludwig Distinguished Lectureship of the APS Neural Control and Autonomic Regulation Section: Central neural pathways for the thermoregulatory cold defense. *J Appl Physiol* 110: 1137-1149, 2011.
132. Morrison SF, Nakamura K, Madden CJ. Central control of thermogenesis in mammals. *Exp Physiol* 93: 773-797, 2008.
133. Moseley PL. Heat shock proteins and heat adaptation of the whole organism. *J Appl Physiol* 83: 1413-1417, 1997.
134. Moseley PL, Gapen C, Wallen ES, Walter ME, Peterson MW. Thermal stress induces epithelial permeability. *Am J Physiol* 267: C425-C434, 1994.
135. Moseley PL, Gisolfi CV. New frontiers in thermoregulation and exercise. *Sports Med* 16: 163-167, 1993.
136. Nadel ER. Circulatory and thermal regulations during exercise. *Fed Proc* 39: 1491-1497, 1980.
137. Nadel ER, Fortney SM, Wenger CB. Effect of hydration state on circulatory and thermal regulations. *J Appl Physiol* 49: 715-721, 1980.
138. Nakamura K, Morrison SF. A thermosensory pathway that controls body temperature. *Nat Neurosci* 11: 62-71, 2008.
139. Nakayama T, Eisenman JS, Hardy JD. Single unit activity of anterior hypothalamus during local heating. *Science* 134: 560-561, 1961.
140. Newburgh L. *Physiology of heat regulation and the science of clothing*. Philadelphia: W.B. Saunders Comp., 1949, p. v-vi.
141. Nielsen B. Heat acclimation-mechanisms of adaptation to exercise in the heat. *Int J Sports Med* 19(Suppl. 2): S154-S156, 1998.
142. Nielsen B, Savard G, Richter EA, Hargreaves M, Saltin B. Muscle blood flow and muscle metabolism during exercise and heat stress. *J Appl Physiol* 69: 1040-1046, 1990.
143. Nielsen M. Die regulation der korpertemperatur bec muskelarbeit. *Skan Arch Physiol* 79: 193-230, 1938.
144. Noakes TD, Goodwin N, Rayner BL, Branken T, Taylor RKN. Water intoxication: A possible complication during endurance exercise. *Med Sci Sports Exerc* 17: 370-375, 1985.
145. Nolte H, Noakes TD, van Vuuren B. Ad libitum fluid replacement in military personnel during a 4-h route march. *Med Sci Sports Exerc* 42: 1675-1680, 2010.
146. Nose H, Mack GW, Shi X, Nadel ER. Involvement of sodium retention hormones during rehydration in humans. *J Appl Physiol* 65: 332-336, 1988.
147. Nose H, Mack GW, Shi X, Nadel ER. Role of osmolality and plasma volulme during rehydration in humans. *J Appl Physiol* 65: 325-331, 1988.

148. Nybo L. Hyperthermia and fatigue. *J Appl Physiol* 104: 871-878, 2008.

149. Oberkolfer H, Dallinger G, Liu Y-M, Hell E, Krempler F, Patsch W. Uncoupling protein gene: Quantification of expression levels in adipose tissue of obese and non-obese humans. *J Lipid Res* 38: 2125-2133, 1997.

150. Ott I. The heat-centre in the brain. *J Nerv Ment Dis* 14: 152-162, 1887.

151. Owen MD, Kregel KC, Wall PT, Gisolfi CV. Effects of ingesting carbohydrate beverages during exercise in the heat. *Med Sci Sports Exerc* 18: 568-575, 1986.

152. Paolone AM, Wells CL, Kelly GT. Sexual variations in thermoregulation during heat stress. *Aviat Space Environ Med* 49: 715-719, 1978.

153. Park YS, Rennie DW, Lee IS, Park YD, Paik KS, Kang DH, Suh DJ, Lee SH, Hong SY, Hong SK. Time course of deacclimatization to cold water immersion in Korean women divers. *J Appl Physiol* 54: 1708-1716, 1883.

154. Parsons K. Interference with activities, performance and productivity. In: Parsons K, ed. *Human Thermal Environments: The Effects of Hot, Moderate and Cold Environments on Human Health, Comfort and Performance*. London: Taylor & Francis, 1993.

155. Patterson MJ, Stocks JM, Taylor NA. Sustained and generalized extracellular fluid expansion following heat acclimation. *J Physiol* 559(Pt. 1): 327-334, 2004.

156. Peier AM, Reeve AJ, Andersson DA, Moqrich A, Earley TJ, Heregarden AC, Story GM, Colley S, Hogenesch JB, McIntryre P, Bevan S, Patapoutian A. A heat-sensitive TRP channel expressed in keratinocytes. *Science* 296: 2046-2049, 2002.

157. Pierce NF, Bradley R, Mitra RC, Banwell JG, Brigham KL, Fedson DS, Mondal A. Replacement of water and electrolyte losses in cholera by an oral glucose-electrolyte solution. *Ann Int Med* 70: 1173-1181, 1969.

158. Pitts GC, Johnson RE, Consolazio FC. Work in the heat as affected by intake of water, salt and glucose. *Am J Physiol* 142: 253-259, 1944.

159. Piwonka RW, Robinson S, Gay VL, Manalis RS. Preacclimatization of men to heat by training. *J Appl Physiol* 20: 379-384, 1965.

160. Price A. Diet, clothing, exercise. In: Price A, ed. *White Settlers in the Tropics*. New York: American Geographical Society, 1939, 217-227.

161. Price A. The problem of white settlement in the tropics. In: Price A, ed. *White Settlers in the Tropics*. New York: American Geographical Society, 1939, 3-12.

162. Riedel W. Warm receptors in the dorsal abdominal wall of the rabbit. *Pflugers Arch* 361: 205-206, 1976.

163. Ritossa F. A new puffing pattern induced and temperature shock and DBP in *Drosophila. Experientia* 18: 571-573, 1962.

164. Robershaw D. The evolution of thermoregulatory sweating in men and animals. *Intern J Biometeorol* 15: 263-267, 1971.

165. Roberts MF, Wenger CB, Stolwijk JA, Nadel ER. Skin blood flow and sweating changes following exercise training and heat acclimation. *J Appl Physiol* 43: 133-137, 1977.

166. Robinson S. Training, acclimatization and heat tolerance. *Can Med Assoc J* 96: 795-799, 1967.

167. Robinson S, Turell ES, Belding HS, Horvath SM. Rapid acclimatization to work in hot environments. *J Appl Physiol* 140: 168-176, 1943.

168. Rollet E, Lavoie JN, Landry J, Tanguay RM. Expression of *Drosophila's* 25 kDa heat shock protein into rodent cells confers thermal resistance. *Biochem Biophys Res Commun* 185: 116-120, 1992.

169. Romet TT. Mechanism of afterdrop after cold water immersion. *J Appl Physiol* 65: 1535-1538, 1988.

170. Rowell LB. Circulatory adjustments to dynamic exercise and heat stress: Competing controls. In: Rowell LB, ed. *Human Circulation: Regulation During Physical Stress*. New York: Oxford University Press, 1986, 363-406.

171. Rubner M. *Die Gesetze des Energieverbrauchs bei der Ernahrung*. Leipzig: 1902.

172. Ryan AJ, Bleiler TL, Carter JE, Gisolfi CV. Gastric emptying during prolonged cycling exercise in the heat. *Med Sci Sports Exerc* 21: 51-58, 1989.

173. Ryan AJ, Gisolfi CV, Moseley PL. Synthesis of 70K stress protein by human leukocytes: Effect of exercise in the heat. *J Appl Physiol* 70: 466-471, 1991.

174. Saltin B. Aerobic and anaerobic work capacity after dehydration. *J Appl Physiol* 19: 1114-1118, 1964.

175. Saltin B, Hermansen L. Esophageal, rectal, and muscle temperature during exercise. *J Appl Physiol* 21: 1757-1762, 1966.

176. Saunders PU, Telford RD, Pyne DD, Gore CJ, Hahn AG. Improved race performance in elite middle-distance runners after cumulative altitude exposure. *Int J Sports Physiol Perform* 4: 134-138, 2009.

177. Sawka MN, Castellani JW. How hot is the human body? *J Appl Physiol* 103: 419-420, 2007.

178. Sawka MN, Wenger CB, Pandolf KB. Thermoregulatory responses to acute exercise-heat stress and heat acclimation. In: Fregley MJ, Blatteis CM, eds. *Handbook of Physiology*. New York: Oxford University Press, 1996, 157-185.

179. Schedl HP, Maughan RJ, Gisolfi CV. Intestinal absorption during rest and exercise: Implications for formulating an oral rehydration solution (ORS). *Med Sci Sports Exerc* 26: 267-280, 1994.

180. Schmidt-Nielsen B. August and Marie Krogh and respiration physiology. *J Appl Physiol* 57: 293-303, 1984.
181. Scholander PF, Hammel HT, Andersen KL, Lovning Y. Mebabolic acclimation to cold in man. *J Appl Physiol* 12: 1-8, 1958.
182. Scholander PF, Hock R, Walters V, Johnson F, Irving L. Heat regulation in some arctic and tropical mammals and birds. *Biol Bull* 99: 237-258, 1950.
183. Selkirk GA, McLellan TM, Wright HE, Rhind SG. Mild endotoxemia, NF-kappaB translocation, and cytokine increase during exertional heat stress in trained and untrained individuals. *Am J Physiol* 295: R611-R623, 2008.
184. Senay LC. Changes in plasma volume and protein content during exposures of working men to various temperatures before and after acclimatization to heat: Separation of the roles of cutaneous and skeletal muscle circulation. *J Physiol* 224: 61-81, 1972.
185. Shapiro Y, Pandolf KB, Avellini BA, Pimental NA, Goldman RF. Physiological responses of men and women to humid and dry heat. *J Appl Physiol* 49: 1-8, 1980.
186. Sherrington CS. Notes on temperature after spinal transection, with some observations on shivering. *J Physiol (Lond)* 58: 405-421, 1924.
187. Silva NL, Boulant JA. Effects of osmotic pressure, glucose, and temperature on neurons in preoptic tissue slices. *Am J Physiol* 247: R335-R345, 1984.
188. Silva NL, Boulant JA. Effects of testosterone, estradiol, and temperature on neurons in preoptic tissue slices. *Am J Physiol* 250: R625-R632, 1986.
189. Simonsen L, Bulow J, Madsen J, Christensen NJ. Thermogenic response to epinephrine in the forearm and abdominal subcutaneous adipose tissue. *Am J Physiol* 263: E850-E855, 1992.
190. Singh IS, He JR, Hester L, Fenton MJ, Hasday JD. Bacterial endotoxin modifies heat shock factor-1 activity in RAW 264.7 cells: Implications for TNF-alpha regulation during exposure to febrile range temperatures. *J Endotoxin Res* 10: 175-184, 2004.
191. Skidmore R, Gutierrez JA, Guerriero V Jr., Kregel KC. HSP70 induction during exercise and heat stress in rats: Role of internal temperature. *Am J Physiol* 268: R92-R97, 1995.
192. Snellen J. Set-point and exercise. In: Bligh J, Moore RE, eds. *Essays on Temperature Regulation*. Amsterdam: North-Holland, 1972, 139-148.
193. Stephenson LA, Kolka MA. Menstrual cycle phase and time of day alter reference signal controlling arm blood flow and sweating. *Am J Physiol* 249: R186-R191, 1985.
194. Stocks JM, Taylor NAS, Tipton MJ, Greenleaf JE. Human physiological responses to cold exposure. *Aviat Space Environ Med* 75: 444-457, 2004.
195. Stolwijk JA, Hardy JD. Partional calorimetric studies of responses of man to thermal transients. *J Appl Physiol* 21: 967-977, 1966.
196. Strydom NB, Wyndham CH, Williams CG, Morrison JF, Bredell GAG, Benade AJS, von Rahden M. Acclimatization to humid heat and the role of physical conditioning. *J Appl Physiol* 21: 636-642, 1966.
197. Talbott JH, Michelsen J. Heat cramps. A clinical and chemical study. *J Clin Invest* 12: 533-549, 1933.
198. Taylor H, Henschel A, Keys A. Cardiovascular adjustments of man in rest and work during exposure to dry heat. *Am J Physiol* 139: 583-591, 1943.
199. Thauer R. Warmeregulation und fieberfahigkeit nach operativen eingriffen am nervensystem homoiothermer saugetiere. *Arch Ges Physiol* 236: 102-147, 1935.
200. Thompson GE. Physiological effects of cold exposure. In: Robertshaw D, ed. *Environmental Physiology II*. Baltimore, MD: University Park Press, 1977, 29-69.
201. Thompson HS, Clarkson PM, Scordilis SP. The repeated bout effect of heat shock proteins: Intramuscular HSP27 and HSP70 expression following two bouts of eccentric exercise in humans. *Acta Physiol Scand* 174: 47-56, 2002.
202. Tipton MJ. The intial responses to cold water immersion in man. *Clin Sci* 77: 581-588, 1989.
203. Tissot J. Nouvelle methode de mesure et d'inscription du debit et des mouvements respiratoires de l'homme et des animaux. *J de Physiol et de Pathol Generale* 6: 688-700, 1904.
204. Toner MM, McArdle WD. Human thermoregulatory responses to acute cold stress with special reference to water immersion. In: Fregley MJ, Blatteis CM, eds. *Handbook of Physiology*. New York: Oxford University Press, 1996, 379-397.
205. Turk J, Thomas IR. Artificial acclimatization to heat. *Ann Occup Hyg* 17: 271-278, 1975.
206. Ulmasov KA, Shammakov S, Karaev K, Evgen'Ev MB. Heat shock proteins and thermoresistance in lizards. *Proc Nat Acad Sci* 89: 1666-1670, 1992.
207. van de Velde TH. Ueber den zusammenhang zwischen ovarial function, wellenbewegung, und menstrualblutung, und uber die entstehung des sogenannten mittelschmerzes. *Haarlem,* 1904.
208. Vernon HM. The index of comfort at high atmospheric temperatures. *Med Res Coun Rep Series (London)* 73: 116-149, 1923.
209. Walsh RC, Koukoulas I, Garnham A, Moseley PL, Hargreaves M, Febbraio MA. Exercise increases serum HSP72 in humans. *Cell Stress Chaperones* 6: 386-393, 2001.
210. Webb P. Afterdrop of body temperature during rewarming: An alternate explanation. *J Appl Physiol* 60: 385-390, 1986.

211. Webb P. The physiology of heat regulation. *Am J Physiol* 268: R838-R850, 1995.
212. Webb PM, Shepard JT. Responses of superfiscial limb veins of the dog to changes in temperature. *Circ Res* 22: 737-746, 1968.
213. Weinman KP, Slabochova Z, Bernauer EM, Morimoto T, Sargent F II. Reactions of men and women to repeated exposure to humid heat. *J Appl Physiol* 22: 533-538, 1967.
214. Welch WJ, Mizzen LA. Characterization of the thermotolerant cell. II. Effects on the intracellular distribution of heat-shock protein 70, intermediate filaments, and smaller nuclear ribonucleoprotein complexes. *J Cell Biol* 106: 1117-1130, 1988.
215. Wells CL, Horvath SM. Heat stress responses related to the menstrual cycle. *J Appl Physiol* 35: 1-5, 1973.
216. Winslow C-EA, Herrington LP, Gagge AP. Physiological reactions of the human body to varying environmental temperatures. *Am J Physiol* 120: 288-299, 1937.
217. Witte JD, Sessler DI. Perioperative shivering. *Anesthesiology* 96: 467-484, 2002.
218. Wurster RD, McCook RD, Randall WC. Cutaneous vascular and sweating responses to tympanic and skin temperatures. *J Appl Physiol* 31: 617-622, 1966.
219. Wyndham CH. The physiology of exercise under heat stress. *Ann Rev Physiol* 35: 193-220, 1973.
220. Wyndham CH, Benade AJS, Williams CG, Strydom NB, Goldin A, Heyns AJA. Changes in central circulation and body fluid spaces during acclimatization to heat. *J Appl Physiol* 25: 586-593, 1968.
221. Wyndham CH, Morrison JF, Williams CG. Heat reactions of male and female Caucasians. *J Appl Physiol* 20: 357-364, 1965.
222. Yamada PM, Amorim FT, Moseley PL, Robergs R, Schneider SM. Effect of heat acclimation on heat shock protein 72 and interleukin-10 in humans. *J Appl Physiol* 103: 1196-1204, 2007.
223. Young AJ. Homeostatic responses to prolonged cold exposure: Human cold acclimatization. In: Fregley MJ, Blatteis CM, eds. *Handbook of Physiology*. New York: Oxford University Press, 1996, 419-438.
224. Yousef MK, Dill DB. Sweat rate and concentration of chloride in hand and body sweat in desert walks: Male and female. *J Appl Physiol* 36: 82-85, 1974.

CHAPTER 20

The Renal System

Jacques R. Poortmans, PhD

Edward J. Zambraski, PhD

Introduction

Kidneys are important organs that contribute to the homeostasis of mammalian organisms under resting conditions. They filter plasma (180 L/24 h), reabsorb most (99%) of the water and other elements from the tubule fluid, and eliminate waste products such as urea and creatinine. Therefore, it is quite understandable that about 25% of total cardiac output passes through these two organs (150 g each) each day under resting conditions. Strenuous and prolonged exercise induces changes in renal hemodynamics, excretory function, and hormone release. All have profound effects and implications for total-body homeostasis, mediated by alterations in renal function. Many of these changes could alter or limit exercise performance

Given the essential role of the kidney in the homeostasis of mammalian organisms at rest and the diversified and potentially important changes in renal function that occur with exercise, it is surprising that the field of exercise physiology has placed relatively little or no emphasis on the issue of the kidneys and exercise. *The Physiology of Muscular Exercise*, the first meaningful textbook on the subject, was published in 1919 by F.A. Brainbridge (9). None of the 214 pages were devoted to the kidney and its functions.

In the 1930s four textbooks were devoted to exercise physiology. Of these, three did contain some information about the kidney. The first, written by Adrian G. Gould and Joseph A. Dye, was titled *Exercise and its Physiology* (40). Of its 312 pages, approximately 1 page was related to kidney function. The second book was published by Percy M. Dawson in 1935 (21). About 7 pages of this textbook, which was likely used by most physical education departments, dealt with the kidney and changes in renal function during exercise. The third book of that period, published by Edward C. Schneider in 1933, contained 366 pages and included no information whatsoever on the kidney (115). However, the third edition of this book, published in 1959 with Peter V. Karpovich as a coauthor, included 1 page (out of 313) that indicated that lactic acid was found in urine after exercise (55). The fourth book, a human general physiology textbook published in 1936 by Evans, included a few sentences on the composition of the urine after severe exercise. Specifically, it stated, "[a] small trace of albumin will often be found in the urine which is passed shortly after taking muscular exercise, but this has no pathognomonic significance" (31).

In the 1950s, Sarah R. Reidman wrote a textbook titled *The Physiology of Work and Play*; about 5 pages out of 565 were devoted to urinary changes with exercise (105). By 1959, when Karpovich became the sole author of the fifth edition of Schneider's text, it still contained only 1 page (out of 368) on the kidney (55). He wrote that "changes in the urine after excessive exertion cannot represent pathological conditions of the kidneys."

Information on renal function during exercise is almost absent from most exercise physiology textbooks published in the 1960s. However, a 13 page chapter titled "Kidney Function in Exercise" emphasized the participation of the kidney in the general adjustments that decrease renal plasma flow and glomerular filtration rate (132). Nevertheless, Ernst Simonson did not mention the word *kidney* in his textbook *Physiology of Work Capacity and Fatigue* (118). In contrast, David Lamb devoted a whole chapter (11 pages) to renal responses and adaptations to exercise in his 1978 book titled *Physiology of Exercise: Responses and Adaptations*. He emphasized "the complex nature of the responses and a lack of research and contributions to our poor understanding of the function of this system under exercise conditions" (59). Around this time, some other textbooks referred to the kidneys and exercise. In the book *Exercise Physiology*, Phillip Rasch and I. Dodd Wilson wrote a chapter titled "The Kidney" (9 pages), which mentioned postexercise proteinuria, hemoglobinuria, and myoglobinuria (103). In 1982, Roy Shephard

Table 20.1 Pivotal Human Studies Involving Renal Exercise Physiology Over the Past Century

Year	Investigator	Topic
1910	Barach	Renal function: Effects of a marathon
1947	Barclay	Exercise and renal plasma flow
1949	Radigan	Exercise and heat stress: Effects on renal function
1962	Poortmans	Characterization of exercise-induced protein excretion
1965	Grimby	Exercise and glomerular filtration
1967	Castenfors	Assessment of renal hemodynamics and excretory function with exercise
1988	Hasking	Renal norepinephrine spillover with exercise
1991	Meredith	Effects of exercise training on renal sympathetic nerve activity
1994	Painter	Effects of exercise training in patients with end-stage renal disease

included 4 pages on renal responses during exercise in his book (672 pages) titled *Physiology and Biochemistry of Exercise* (117). Historically, sports medicine textbooks have also given little attention to the kidney and exercise. In a 1933 publication, Herbert Herxheimer related that "kidney function during exercise and in recovery has been explained in a very incomplete manner" (47). More recent sports medicine textbooks describe only some of the renal abnormalities that occur during exercise, namely hematuria and proteinuria observed in athletes (36, 39).

It is quite surprising that recent textbooks on exercise physiology include meager information on kidney function, with the possible exception of a statement concerning the redistribution of blood flow from the kidney to skeletal muscle. Scott K. Powers and Edward T. Howley included 0.5 page (out of 552) on the regulation of acid–base balance via the kidneys (100), and only 0.5 page of information (out of 1,068) was included in the 2007 edition of a textbook on exercise physiology by William D. McArdle, Frank I. Katch, and Victor L. Katch (65). Nevertheless, the second issue of *Physiology of Sport and Exercise*, published by Jack H. Wilmore and David L. Costill in 1999, did contain 2.5 pages (out of 710) on erythropoietin doping and 0.5 page on electrolyte loss in urine during exercise (134). The most recent textbook of Georges A. Brooks and colleagues titled *Exercise Physiology* provided 2 pages (out of 830) on renal hemodynamics and fluid volume regulation by the kidney during acute exercise (11).

When one examines the past it is evident that the history of nephrology and renal physiology is limited compared with other fields of medicine (table 20.1). These early days of information, from Hippocrates (460-370 BC), are covered by Charles Tipton in chapter 1 of this text. Despite these many early observations and insights, it is surprising that there have been so few descriptions of abnormal constituents in the urine of individuals engaged in strenuous physical exercise. We know that bloody urine can occur as a consequence of exercise. Perhaps one of the first papers to describe the collection of bloody urine from athletes was that of Barach, who studied the renal effects of the marathon run, in 1910 (5).

Renal Plasma and Renal Plasma Flow During Exercise

Up to 1940, the literature on renal plasma flow (RPF) changes during exercise was limited. This lack of information may be attributed to the absence of an adequate method for measuring RPF or renal blood flow (RBF) in humans. Nevertheless, in 1935 Edwards and colleagues suggested that a diversion of blood flow from the kidney during exercise could provide more blood flow for working muscles (26).

Human Observations

In the early 1940s the application of Fick's first law was extended to renal physiology for the measurement of RPF. It was found that the clearance of diodrast or p-aminohippurate (PAH) could be used to determine RPF or RBF. For practical reasons (e.g., to minimize interference or trauma to the subjects), this clearance technique was performed almost exclusively under resting conditions when it was first used to measure RBF and RPF in humans.

The first report on the effects of exercise on RPF came from the department of physiology at the University of Birmingham in England (6). Barclay and colleagues measured diodrast clearance in 10 healthy students immediately after cessation of a 0.25 mile run at full speed. They found that this short-term exercise reduced RPF 30% from the resting value and that 10 to 40 min were required for RPF to return to resting levels. Barclay and colleagues also investigated the effect of 15 min of sustained bicycle exercise (intensity not specified) in two subjects. They found a steady decline in RPF during the exercise, and there was no correlation between changes in RPF and in urine flow. In 1948, additional reports on the effect of exercise on RPF appeared. White and Rolf investigated the renal responses of five subjects subjected to relatively light exercise in the form of jogging (6.8-11.5 km/h for 13-22 min). They noted little change in RPF during this workload but found an 80% decrease during heavy exercise in two subjects (133). The authors postulated that the renal vascular bed had constricted in response to the normal stimuli of exercise and that the degree of renal vasoconstriction is proportional to the intensity of exercise. Renal vascular resistance increased at least fivefold during heavy exercise.

A more detailed and controlled study was conducted that same year by Chapman and colleagues, who measured PAH clearance in nine healthy young males. These tests involved constant intravenous infusion of PAH while the subjects were at rest, while the subjects walked on a motor-driven treadmill, and during recovery (16). Each subject was investigated during two 20 min walking exercise periods (5 km/h) at three slopes on a motor-driven treadmill, resulting in exercise at different intensities. The results of this study were in accord with those obtained by other groups but demonstrated more clearly the relationship between the decline in RPF and the intensity of the exercise. Moreover, they emphasized that the recovery of RPF after exercise was considerably slower than the recovery of pulse rate and blood pressure. RPF remained depressed even after 40 min of rest after all three levels of exercise.

Once studies demonstrated the effects of exercise on RPF, the effects of various environmental conditions were then examined in healthy subjects. Knowing that exposure to heat may cause profound circulatory stress, Radigan and Robinson in 1949 investigated the effects of moderate exercise (walking 5 km/h up a 5% grade for 80 min) on RBF in cool (21 °C) and hot (50 °C) environments (102). The decreases in RPF were significantly greater when exercising in the heat (50 °C) than in a cool environment (21 °C). Together with Smith and Pearcy, Robinson extended his studies of environmental conditions and the effect of exercise on kidney function to include dehydration (119). Moderate exercise (walking 5 km/h up a 5% grade) in a 50 °C environment while in a dehydrated state induced a 56% decrease in RPF; the subjects became fatigued or syncopal and were unable to complete the exercise period. These results were later confirmed by German researchers in 1960 (120).

All previous investigators restricted their studies to the effects of absolute power on renal hemodynamics. The next step was to relate changes in RPF to exercise quantified as relative power outputs. Thus, Grimby from Göteborg exercised 15 healthy young males on a bicycle ergometer at different power requirements expressed as a percentage of their maximal oxygen uptake ($\dot{V}O_2max$) (41). Using PAH clearance, it was shown that RPF decreased linearly as exercise intensity increased from 18% to 89% of $\dot{V}O_2max$; the lowest value of RPF was 33% of the resting value. Castenfors extended this observation in his PhD thesis and measured the glomerular filtration rate (GFR) and the filtration fraction as well (figure 20.1).

These reports indicated that the degree of renal vasoconstriction is proportional to the intensity of exercise. The first experiment to attempt to alter this response was reported in 1967 by Jan Castenfors from the Karolinska Institute in Sweden (15) (figure 20.2).

Castenfors used dihydralazine, a vasodilator agent that causes smooth-muscle relaxation. When dihydralazine was injected intravenously at the start and in the middle of a 45 min session of bicycle exercise when the subject's heart rate reached approximately 170 beats/min, the decrease in RPF was attenuated.

Both the afferent and efferent glomerular arterioles are densely innervated by sympathetic adrenergic nerves. Neurally induced renal vasoconstriction during exercise could also be supplemented by the effects of elevated levels of the circulating catecholamines norepinephrine and epinephrine.

In the 1970s and 1980s, sensitive assays for the measurement of plasma and urine catecholamines became available. Such measurements, however, represent only a generalized or global measure of sympathetic nervous system activation both at rest and during perturbations such as exercise. As discussed in more detail in chapter 9 of this text, the measurement of plasma norepinephrine (NE) spillover rates in the 1980s emerged as a more rigorous method for evaluating sympathetic responses from select organs. In 1991, Tidgren and colleagues (123) demonstrated that the differences in RPF at several exercise intensities were closely re-

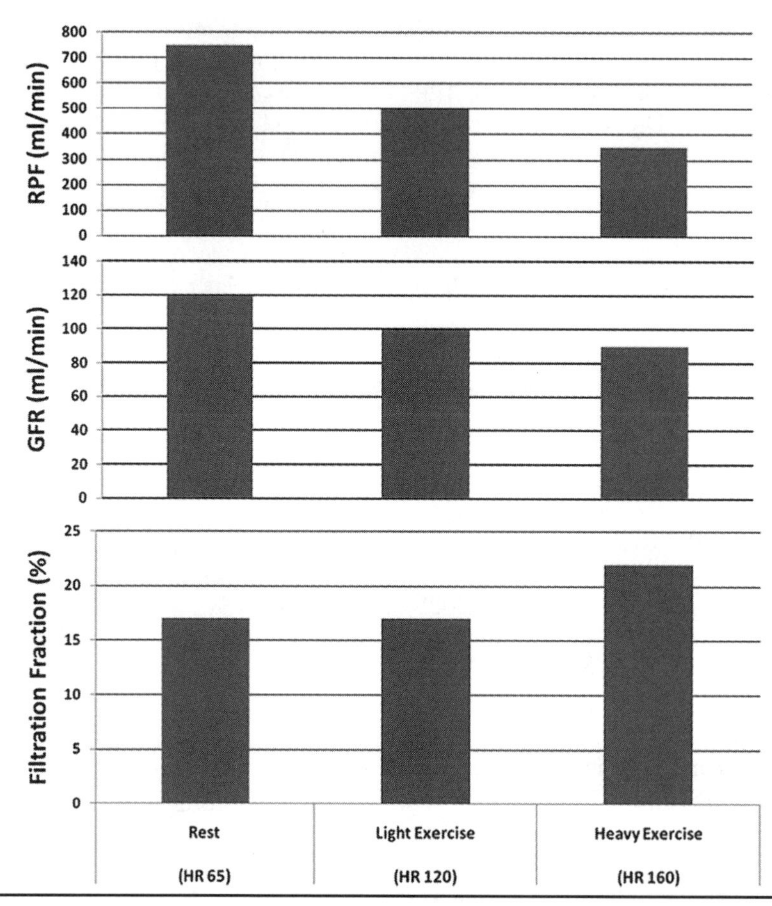

Figure 20.1 Effect of light and heavy exercise on renal plasma flow (measured by clearance of para-aminohippuric acid), glomerular filtration rate (measured by insulin clearance), and filtration fraction (FF) in bicycling human subjects. Heart rates (HR) are shown at rest and during light and heavy exercise.
Adapted from Castenfors 1967.

lated to the renal NE spillover rate, which indicated that renal vasoconstriction during exercise was related to increased renal sympathetic activity. In 1988, Hasking and colleagues measured renal NE spillover both at rest and during exercise in healthy men (44). Their work demonstrated that at rest renal NE spillover, as a contribution to total-body NE spillover, was 22%, an amount that was similar to the contribution from muscle. With exercise, renal NE spillover increased 3.6-fold.

Figure 20.2 Jan Castenfors.
Photo courtesy of Jan Castenfors.

Animal Studies

In contrast to studies in humans using indirect clearance techniques, direct measurements of RPF and RBF have been made during exercise in dogs, miniature swine, and baboons. Measurements made in conscious, healthy dogs indicate that RBF does not decrease during exercise as it does in humans (7, 51, 126). In contrast, the renal hemodynamic response to exercise in miniature swine appears to be similar to that observed in man. Bloor and colleagues in the late 1980s conducted experiments in miniature swine that were trained to run on a motor-driven treadmill at various power intensities (8). Radioactive microspheres were used to measure both total RBF and the distribution of RBF in the kidney (i.e., cortex versus medulla) (8, 112). Severe exercise reduced RBF by 70% of control, whereas the intrarenal blood flow distribution was unchanged throughout the rest and exercise periods. These results clearly showed that exercising swine, like humans, increases blood flow to skeletal muscle in association with a decrease in RBF. In contrast, these changes have not been observed in the baboon. Vatner implanted Doppler ultrasonic flow probes in the left renal arteries of conscious baboons that were permitted to exercise (i.e., climbing, running, jumping) in a large outdoor enclosure (127). Moderate exercise (heart rate of ~179 beats/min) did not decrease RBF.

Exercise and Renal Sympathetic Nerve Activity

The animal studies helped quantify changes in exercise and renal sympathetic nerve activity (RSNA) during exercise in conscious animals. To obtain renal nerve recordings from awake animals, chronic implantation of recording electrodes in the renal nerve bundle is required. Successful recording of RSNA in animals during exercise was achieved by Schad and Seller in 1975 using cats (113). These results have been confirmed by others using rabbits (24, 77). All three studies demonstrated that RSNA was increased by acute exercise. Also, the increase in RSNA was related to the intensity of exercise, and RSNA remained elevated well into the postexercise recovery period. This direct evidence showing increased RSNA during acute exercise was extremely important and has profound implications regarding the mechanisms responsible for changes in various elements of renal function during exercise (i.e., changes in renal hemodynamics, water and electrolyte excretion, hormone release). A clear example is renin release. With acute exercise there is an increase in renal renin release. Although there are multiple mechanisms that could potentially be responsible for the increased renin secretion (e.g., change in perfusion pressure, prostaglandin release, macula densa feedback), Zambraski and colleagues showed that the increased renin release during exercise is mediated by the renal sympathetic nerves (140) (figure 20.3).

GFR During Exercise

Before the 1930s, methodology for measuring GFR had not been developed. Consequently, the initial studies that examined the effects of exercise on renal excretion function were limited to the examination of excretion rates of water and solutes.

In 1921, James Campbell and Thomas Webster were the first to report rates of creatinine excretion in subjects at rest, performing daily activities, and after exercising on a bicycle ergometer. Their results varied and no consistent conclusions could be drawn concerning the effects of exercise on glomerular filtration (12, 13). Twenty years later, Eggleton reported that in a group of 60 of his biochemistry class students, full-speed running over a 0.25 mile distance resulted in a significant decrease in creatinine excretion, a measure used to estimate GFR (28).

The first attempt to directly measure changes in GFR with exercise was made in 1936 by F.G. Covian and P.B. Rehberg on two healthy young men who performed bicycle exercise on a Krogh ergometer at power intensities varying from 120 to 240 W for a duration of 12 to 40 min (19). They used endogenous creatinine clearance (creatinine rate excretion divided by plasma creatinine concentration) before and after exercise. The authors observed that exercise at low and moderate intensity did not affect GFR whereas heavy or intense exercise reduced GFR by nearly 40%. This result was later confirmed by several authors using inulin clearance, a more accurate method for measuring GFR (6, 68, 133). A further step in the evaluation of the effect of exercise on glomerular filtration was introduced by Radigan and Robinson (102) and by Smith and colleagues (119), who exercised their subjects in a hot environment (50 °C) or in a dehydrated state. Under heat and dehydration conditions, GFR showed earlier and more exaggerated decreases.

Figure 20.3 Edward J. Zambraski.

Photo courtesy of Edward J Zambraski.

Experiments conducted between 1965 and 1993 confirmed the important effect of exercise intensity and duration on GFR. Grimby (41), Kachadorian and Johnson (52), and Freund and colleagues (35) reported that GFR changes are dependent on exercise intensity. GFR is constant, or even slightly elevated, when heart rates are under 150 beats/min, an exercise intensity representing 50% to 60% of $\dot{V}O_2$max. Above these values, GFR decreases gradually as the power output is increased. Thus, the kidney maintains GFR relatively unchanged over low to moderate power outputs even though RPF is decreased. This indicates that filtration fraction (the ratio of GFR to RPF) is increased during exercise (6, 15). Prolongation of exercise duration does not produce further changes. Severe exercise may cause a doubling of the filtration fraction because the decrease in RPF is greater than the decrease in GFR (15).

Both endocrine and neural factors influence GFR. In 1981, atrial natriuretic peptide (ANP) was discovered by de Bold and colleagues (22). This endogenous peptide is a potent renal vasodilator that increases both GFR and sodium excretion. Plasma levels of ANP increase during acute exercise, and it has been speculated that ANP may be important in maintaining or even increasing GFR during low-intensity exercise (35). In 1990, Zambraski (138) speculated that in moderate to severe exercise the effects of ANP are likely overridden by elevated RSNA, angiotensin II, and circulating catecholamines; these factors are probably responsible for the observed decreased GFR.

The role of ANP as a determinant of renal function, both at rest and during exercise, however, is likely to be approved by the discovery of a similar peptide, urodilatin, in 1998 (116). Urodilatin is of renal origin. Its properties are similar to those of ANP, and it has been shown that the excretion of urodilatin is increased with exercise (114). The exact roles that ANP or urodilatin play in mediating changes in kidney function with exercise will not be known until such time that pharmacological antagonists are available for each of these compounds and applied during an exercise test.

Water, Electrolyte, and Metabolite Excretion During Exercise

A previous review (91) described an early story of Mr. Weston, who walked for 100 miles (160.9 km) in England under the supervision of F.W. Pavy, a physician at Guy's Hospital in London. The urine analyses were extensively reported in 10 issues of *The Lancet* and *British Medical Journal* (84). Weston's daily urine volume was reported as 1,700 ml before the race and an average of 1,810 ml after 12 d of walking. In summary, Pavy contradicted the conclusion reached by Flint. He stated that "although the elimination of urinary nitrogen is increased by muscular exercise, yet the increase is nothing nearly sufficient to give support to the proposition that the source of the power manifested in muscular action is due to the oxidation of muscular tissue" (84).

Water and Electrolytes

Historically, all of the studies that have evaluated the effects of exercise on creatinine excretion or the renal clearance of compounds to assess changes in GFR or RPF confirm that, in the absence of excessive fluid loading, strenuous exercise results in decreased urine output. The urine excreted is reduced in volume and its composition is changed such that it is more concentrated (i.e., higher osmolarity) and more acidic. Resting urine flow rates in humans of approximately 1.0 to 1.5 ml/min are reduced by approximately 50% with strenuous exercise. One of the first papers to actually comment on the mechanisms of this exercise-induced reduction in urine volume was that of Castenfors (15), who in 1967 described that the reduction in urine flow during exercise was largely a function of renal vasoconstriction and decreased GFR. Also, in the 1960s and 1970s, new assays for measuring hormones such as aldosterone, renin–angiotensin, and vasopressin (antidiuretic hormone) became available. Numerous groups observed that plasma concentrations of these hormones increased during moderate to heavy exercise, and these would be predicted to contribute to the conservation of fluid by the kidney during exercise. A specific early example of plasma antidiuretic hormone analysis is that of the group of Kozslowski in Warsaw, which showed the activation of the antidiuresis system by the hypothalamo-hypophyseal axis during exercise.

One interesting finding concerning urine output and exercise was first made by Refsum and Stromme in 1977 (104) and Wade and Claybaugh in 1980 (130). These two groups reported that the kidney's capacity to concentrate urine was impaired during heavy exercise. Despite elevated levels of vasopressin, there may actually be a decrease in urine osmolarity and an increase in free water clearance. This response is not seen in all cases and the consequences of this impairment in terms of fluid balance during exercise are minimal. The mechanism responsible for this renal resistance to vasopressin has not been determined.

Sodium is the major extracellular osmotically active electrolyte, and its homeostasis is tightly regulated to maintain plasma and total-body tonicity at a fixed level.

Therefore, the renal excretion of sodium is a major factor in the regulation and control of extracellular volume (including plasma volume). During the stress of exercise, especially when coupled with dehydration due to excess water and sodium loss through sweating, the renal handling of sodium is important in terms of maintaining or re-establishing body fluid homeostasis. In the exercise setting, the renal conservation of sodium and water (i.e., decreased excretion) is essential.

It is quite interesting to observe that sodium and chloride excretion during exercise are concordant. Koriakina and colleagues (57) observed that running 3 to 28 km or playing soccer reduced chloride excretion. From the 1930s through 1947, these findings were confirmed by several researchers (6, 29, 45, 63, 135). More recent studies have evaluated the effects of exercise on sodium excretion, both in the laboratory and during athletic competitions such as the marathon (138). In general, sodium excretion decreases by 30% to 50% or more with prolonged intense exercise. In 1967 Castenfors (15) suggested that the decrease in sodium excretion could not be fully accounted for by changes in GFR or decreased filtered sodium load. Hence, with exercise there is an increase in renal tubular sodium reabsorption. In 1991 Freund and colleagues demonstrated that the antinatriuretic effect of exercise is dependent on the intensity of the exercise (35).

Because the control of sodium handling by the kidney is of paramount importance, there are several mechanisms that can influence sodium excretion during exercise stress. These include changes in GFR (filtered sodium load); the renal effects of renin–angiotensin and aldosterone, both of which may independently increase renal tubular sodium reabsorption; and ANP, which decreases renal tubular sodium reabsorption. In addition, it has been conclusively shown that renal sympathetic nerves directly increase renal tubular sodium reabsorption, an effect that is independent of changes in renal hemodynamics or other hormonal systems (23).

The identification of the mechanism responsible for increased renal tubular sodium reabsorption during exercise is an important issue. Because increased renal tubular sodium reabsorption parallels changes in circulating renin–angiotensin and aldosterone with moderate to heavy exercise, it has been widely presumed (and most textbooks espouse) that the antinatriuretic effect of exercise is attributable to increases in these hormones. These associations, however, are probably incorrect. The association between the changes in urinary sodium excretion and changes in renin–angiotensin and aldosterone during exercise was challenged by Wade and colleagues in 1987 (129). They demonstrated that the changes in urinary sodium excretion with exercise were not altered in subjects who were treated with drugs to block the angiotensin II production. This response was confirmed by Mittleman and colleagues in 1996 (71). In the exercised miniature swine, the antinatriuretic response to exercise is extremely fast, suggesting a neural effect. Also, the antinatriuretic response was unaltered by treatment with an aldosterone antagonist (138). It has been suggested that there is significant neurogenic control of tubular sodium reabsorption during exercise (138). However, experiments involving exercise in the presence of renal denervation have not been performed.

An important question concerning the renal excretory response to exercise involves the problem of water intoxication and the resultant hyponatremia (76). Subjects who drink copious amounts of pure water during a long-term exercise event, such as a marathon, may develop dilutional hyponatremia. Questions have arisen about why the kidney fails to excrete the excess free water. In 1984, Poortmans (92) from the Université Libre de Bruxelles demonstrated that the antidiuretic effects of strenuous exercise were profound in fluid-loaded subjects with urine flow rates of 10 to 12 ml/min. In these subjects, exercise reduced urine flow rates to 2 to 3 ml/min despite continued fluid loading. It is likely that the observed renal hemodynamic changes, especially the decrease in GFR and filtered loads of sodium and water, substantially reduce the delivery of filtrate to the diluting segment of the nephron. The result is an antidiuresis rather than an anticipated diureris.

Urine Metabolites as Evidence of Muscle Energetics

Historically, the examination of urine metabolites was used to indirectly assess changes in muscle energetics during exercise. Considering urinary urea during exercise, Justus von Liebig (1803-1873) from the University of Giessen (Germany) was convinced that "the sum of the mechanical effects produced is proportional to the amount of nitrogen in the urine." According to von Liebig, muscle protein breakdown was necessarily involved in muscular contraction (60). Subsequent studies have shown that this does not occur to an appreciable extent. This issue was clarified when the concept of renal clearance was introduced by Moller, McIntosh, and Van Slyke in 1928 (74). In 1932, D.D. Van Slyke and colleagues from the Rockefeller Institute for Medical Research (125) reported a decrease of urea clearance during several types of physical activities: tennis, weightlifting, vigorous exercise, football, soccer, basketball, and running on a treadmill. The decrease was 25% to 75% of the original clearance, depending on the intensity and duration of the exercise. They concluded that during severe exercise it is probable that "some dehydration of the blood by sweating occurred, and perhaps also some diversion of blood flow from the kidneys to the heavily working muscles" (125).

Regarding the renal metabolism of lactic acid, small amounts of lactate have been found in the urine of humans at rest. The origin of lactic acid excretion came from the work of Archibald V. Hill (1886-1977), which showed that mild or moderate exercise did not lead to a continuous accumulation of lactate in the plasma. This may account for the fact that only small quantities of lactate have been found in the urine after long periods of exercise, such as mountain climbing (49) or hard work for 5 h (13). In 1925, Liljestrand and Wilson made a thorough study of urinary lactate excretion after muscular exercise (61). They were the first to actually isolate d(-)lactate (sarcolactate) from exercise urine and construct a complete curve for rates of lactate excretion during recovery from exercise.

In 1937, Johnson and Edwards addressed the problem of evaluating blood and urine lactate (50). They calculated that the total excretion of excess lactate and pyruvate accounts for about 2% of the total excess that the body has dissipated after short-term intense exercise. This figure is remarkably close to present-day estimates. Meanwhile, Miller and Miller examined urinary lactate excretion in two subjects who ran on a treadmill for 2 to 15 min at speeds of 10 to 16 miles (16.1-25.7 km)/h (69). A threshold plasma lactate of 5 to 6 mmol/L was associated with increased rates of lactate clearance (20 ml/min versus 1 ml/min at rest). Dies and colleagues in 1969 concluded that lactate was actively reabsorbed in the renal proximal tubule (25).

What was the fate of the lactate reabsorbed by the kidney? Biochemical investigations by Hans A. Krebs and T. Yoshida demonstrated that rat kidney cortex slices synthesize glucose from lactate under resting conditions (five times less than in liver). They also collected cortex slices 35 min after swimming (58). Gluconeogenesis from lactate increased by 55% immediately after exercise. This increase was interpreted as an adaptative reaction effecting a more rapid removal of lactate from blood. In order to understand this enhanced gluconeogenesis, Sanchez-Medina investigated the activities of the enzymes responsible for these biosynthetic pathways and showed that the activity of kidney phosphoenolpyruvate carboxykinase doubled when rats were forced to swim for 2 h (109-111). In 1971, Wahren and colleagues determined arterial concentrations and net substrate exchange across the kidney in two human subjects who bicycled for 40 min at increasing intensities up to 200 W (131). Renal uptake and production of glucose were inconsistent during the exercise period. A limitation of this study was that the mean arterial lactate was less than 4 mmol/L. Thus, the question of whether excess lactate produced during heavy exercise is oxidized versus converted to glucose has not yet been answered.

High Molecular Weight Substances in Urine During Exercise

Postexercise Proteinuria

The presence of proteins in the urine of healthy subjects was first recognized in 1878 by W. von Leube in soldiers who underwent 1 to 3 d of strenuous battalion march (see review in 91). von Leube was convinced that this phenomenon was a "physiologischen albuminurie" and suspected that its origin could be attributable to filtration through the pores of the glomerular membrane. A few other colleagues observed that young men performing exercise activities had protein, blood cells, and casts in their urine. The authors claimed that these findings were almost identical to an acute parenchymatous nephritis and that functional albuminuria disappeared after a night of rest. The two major statements made from these observations were that "no one should any longer advise young men who pass large quantities of albumin in the urine after severe muscular exercise to give up all hard athletic competition" and that "insurance companies should not continue to refuse to consider the acceptance of the lives of young men between the ages of 18 and 30 whose urines are found to contain albumin after exercise."

In 1910, a study by J.H. Barach investigated the effects on the kidney of running a marathon race (5). He confirmed the presence of urinary protein, red cells, and casts after the race and pointed out that the subjects with the highest maximum blood pressures at the end of the race were excreting the largest amounts of albumin.

While several studies confirmed the quantitative observation of postexercise proteinuria, which may amount to several grams of protein per liter, other investigators tried to find an explanation for the occurrence of exercise proteinuria. Isaac Starr from the University of Pennsylvania in the United States introduced the concept of induced renal vasoconstriction (by emotional excitement in cats, stimulation of renal nerves in dogs, or subcutaneous injection of adrenaline in men) to explain a concomitant excretion of albumin in urine (121). He ended his summary by saying that "this concept may apply to those albuminurias in man which result from excessive muscular exercise."

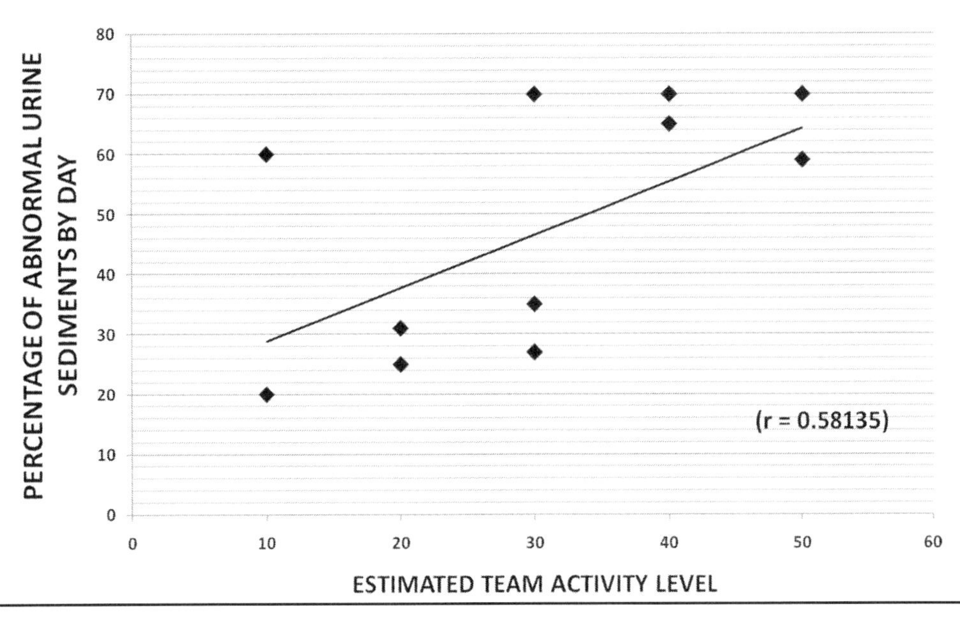

Figure 20.4 Influence of football practice on production of urinary sediment abnormalities (albumin, casts, red blood cells). Estimated team physical activity: 0 = light, 30 = moderate, 60 = maximal.

Adapted from Gardner 1956.

In 1931, Edwards, Richards, and Dill from the Harvard Fatigue Laboratory examined protein in urine as estimated by the degree of turbidimetry produced in the nitric acid test (27). Of the 42 football players studied after competition, 60% had proteinuria. The researchers related these findings to the duration of exercise. In 1932, Frances A. Hellebrandt from the University of Wisconsin in the United States was convinced that albuminuria was related to the subnormal phase in blood pressure that follows intense exercise (46). She tested this relationship in 47 women after bicycling, running, or rowing exercise and concluded that the greatest amount of albumin excretion occurred during the maximal depression in postexercise blood pressure. Moreover, Hellebrandt claimed that strenuous exercise redirected blood flow to the working muscles and the skin, affecting the circulation of the kidney in such a way that caused "asphyxiation of the renal cells beyond that compatible with normal function." She suspected that abnormal accumulation of acids in the renal tissue alters its permeability to the blood proteins and induces albuminuria. This conclusion was also reached in 1944 by Guy Viarnaud of Lyon, France, when he concluded that albuminuria after soccer and rugby games may indicate underlying renal disease (128).

Rise of the Athletic Pseudonephritis Concept

As indicated previously, the concept of pathological urine induced by exercise originated with Baldes and colleagues from Frankfort (Germany) in 1906 (see review in 91). In 1950, Erik Hohwü-Christensen and P. Högberg reported that postexercise proteinuria occurred in practically all of the best Swedish skiers after horizontal track and uphill skiing (17). In 1952, Javitt and Miller suggested three possible causes of exercise proteinuria: increased glomerular filtration, increased acidity, or decreased tubular reabsorption of protein (48). This suggestion was revisited and extended by Kenneth D. Gardner 4 yr later (37). Gardner collected urine specimens from 47 members of a football team. Of the 424 specimens, 44.8% met the criteria established for abnormal urine (red blood cell casts, broad granular casts). Gardner also noted that the occurrence of protein was related to the estimated team activity. This study was intended to confirm and re-emphasize the fact that formed elements heretofore considered almost pathognomonic of parenchymal renal disease may appear in the urine of players during the athletic season. However, Gardner stated that these elements usually do not indicate serious renal disease because they promptly disappear from the urinary sediment when the athlete withdraws from daily exertion. The name *athletic pseudonephritis* was proposed for this phenomenon "with the intent that the sediment abnormalities that appear in the urine of an athlete will not lead to the incorrect diagnosis of nephritis in candidates for life insurance or induction into the armed forces" (37) (figure 20.4).

Since 1956, the concept of athletic pseudonephritis has been widely used to characterize the urinary responses of healthy subjects to heavy exercise. It has been observed in many types of exercise (4, 20, 53), including nontraumatic exercise (2). An association between postexercise proteinuria and the intensity of the exercise was mentioned in a several reports (54, 85, 97, 124) that hypothesized a link between degree of proteinuria, heart rate, blood lactate, and intensity of the exercise load imposed on the subjects.

Identification of Macromolecules

In order to characterize urinary protein excretion, one has to identify the origin of these macromolecules. Before 1958, incomplete information was available. Upon the introduction of paper electrophoresis, further insight was obtained by Nedbal and Seliger from Charles University in Prague. In their studies, albumin remained the major constituent (67%) of postexercise proteinuria; four globulin fractions were present as well (75). They concluded that the ratio of the different protein fractions in postexercise proteinuria was practically identical with the protein spectrum in blood serum. In 1961, British scientists Rowe and Soothill studied exercise proteinuria. Using immune antisera, they identified plasma albumin, transferrin, ceruloplasmin, and immunoglobulin G in urine (106). They suggested that exercise urine proteins might be the product of a selective filtration through the glomerular basement membrane. This heterogeneity of postexercise urinary proteins was confirmed by Jacques R. Poortmans in 1962 (90) and subsequently by other investigators (34, 66) (figure 20.5). The introduction of sensitive and accurate immunochemical techniques allowed Poortmans and colleagues to suggest that postexercise proteinuria was in fact of both glomerular and tubular origin (87, 93, 98) (figure 20.6).

In 1977, Mogensen and Solling (73) reported that certain amino acids (i.e., lysine) inhibit, either partially or completely, renal tubular protein reabsorption. They showed that amino acids that were positively charged interfere with tubular protein reabsorption. Under those conditions, the results obtained under lysine perfusion may be considered as an index of glomerular passage of proteins. Using this information, Poortmans administered lysine to healthy subjects submitted to severe exercise to confirm that exercise proteinuria involved both an increase in glomerular permeability and impaired renal tubular protein reabsorption (94).

There are probably several mechanisms for exercise proteinuria (99). As indicated in the studies cited, the magnitude of exercise proteinuria is related to exercise intensity and duration (97, 122) (figure 20.7).

Changes in renal hemodynamics (i.e., renal vasoconstriction) are also likely to be involved. One of the most definitive studies dealing with the mechanism of exercise proteinuria in humans was published by Mittleman and Zambraski in 1992 (72). In 8 subjects undergoing 30 min of steady-state treadmill exercise at 75% of $\dot{V}O_2max$, the prostaglandin (PG) inhibitor indomethacin significantly attenuated the exercise proteinuria observed during the placebo tests. This effect of PG inhibition was not associated with any change in renal hemodynamics, excretory function, or systemic arterial pressure. In these same subjects, angiotensin-converting enzyme inhibition did not alter exercise proteinuria. These data suggest that PG contributes to an increase in glomerular membrance permeability to protein during exercise.

Acute Renal Failure Induced by Exercise

Acute renal failure (ARF) is defined as a sudden decline in GFR over hours or days. With regards to exercise, ARF could be the result of renal tubular obstruction, as observed with exertional rhabdomyolysis, in which the muscle breakdown and release of intracellular enzymes and myoglobin result in myoglobinuria and the possibility of renal tubular obstruction. ARF may also occur during exercise due to excessive renal vasoconstriction, which would decrease renal perfusion and GFR. When combined with exercise, environmental factors such as dehydration, sodium depletion, and heat stress would be predicted to increase renal vasoconstriction and, thereby, the likelihood of ARF.

Historically, research or research interest concerning the issue of exercise-induced ARF has been limited,

Figure 20.5 Jacques R. Poortmans.
Photo courtesy of Jacques R Poortmans.

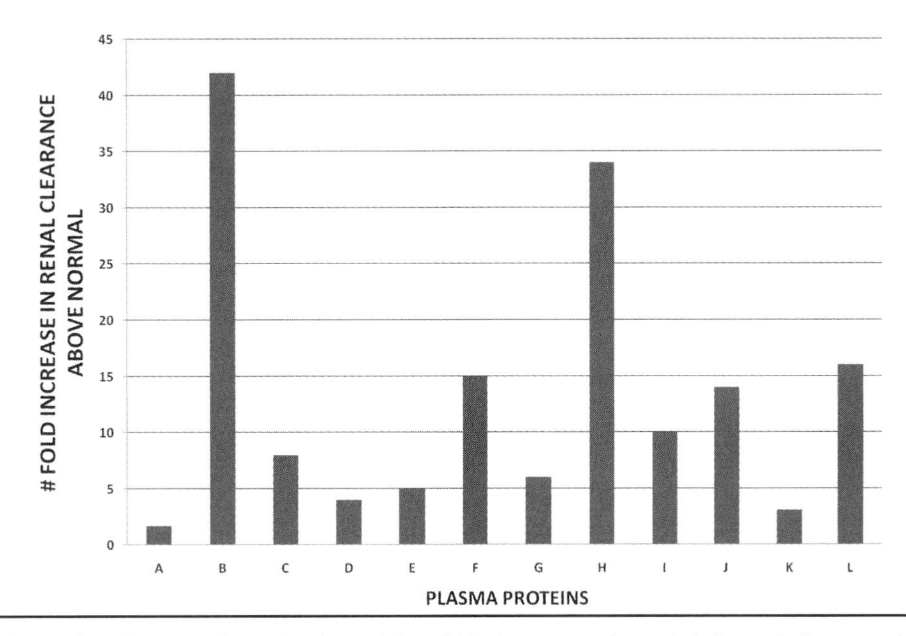

Figure 20.6 Comparison between the molecular weight of 12 plasma proteins and their renal clearances in resting urine (normal) and urine collected 1 h after the 1966 Boston marathon race (exercise urine). The letter under each column identifies the protein tested and their molecular weight in kdaltons. A: α_2-glycoprotein I; B: α_1-acid glycoprotein; C: α_1-antitrypsin; D: α_2-Gc globulin; E: tryptophan-rich prealbumin; F: albumin; G: hemopexin; H: transferrin; I: haptgloblin; J: ceruloplasmin, K: IgA-immunoglobin A, L: IgG-immunoglobin G. Results have been modified from reference 87. Those responses emphasize the specific enhanced glomerular membrane permeability and the saturation of the tubular reabsorption of different plasma proteins.

Data from Poortmans and Jeanloz 1968.

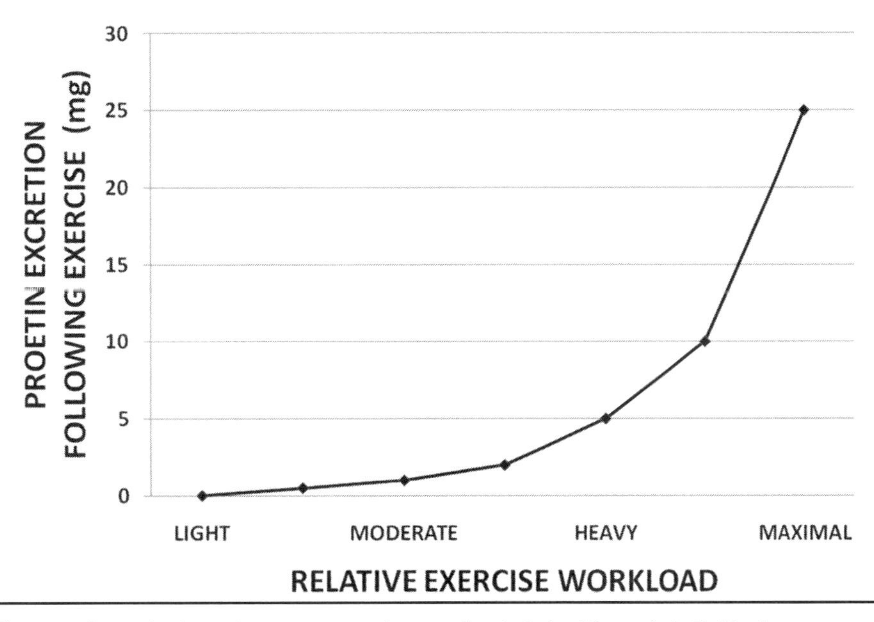

Figure 20.7 Influence of exercise intensity on postexercise proteinuria in healthy male individuals.

Adapted from Taylor 1960.

mainly because this complication of exercise is relatively rare. The first attempt to quantitate the frequency of ARF in an athletic event, such as a marathon, was made by MacSearraigh and colleagues (64), who evaluated the number of cases of ARF that occurred in the 90 km South African Comrades Marathon. In this race with more than 20,000 competitors, only 10 cases of ARF were reported over a 9 yr period. Besides the low frequency of ARF induced by exercise, the individual's level of training or other environmental factors (e.g., high external temperature, reduced fluid intake) might contribute to this pathological condition.

Rather than address the issue of what caused ARF during exercise, in a 1996 review Zambraski (139) approached this question from a different perspective and asked what prevented ARF during exercise. It was emphasized that during exercise high levels of RSNA, angiotensin II, endothelin, neuropeptide Y, and circulating catecholamines would actually be predicted to cause excessive renal vasoconstriction and potentially cause ARF. It was hypothesized that these renal vasoconstriction factors must be counterbalanced by some renal vasodilatory influence that is yet to be identified. Possibilities include ANF, PG, and nitric oxide. Studies that have evaluated the effects of PG inhibition during exercise in humans, tested under conditions that mimic those that occur in an event such as a marathon, suggest that PG is not responsible for this renal vasodilatory protective effect. Moreover, PG inhibition does not prevent postexercise proteinuria after short-term high-intensity exercise (33). Definitive studies evaluating the role of nitric oxide in kidney function during exercise have yet to be conducted. Finally, the suggestion has been made that the effects of exercise training on RSNA may be an important factor in preventing exercise-induced ARF (139).

Acute Renal Failure Induced by Nutritional Supplements During Exercise

Concerns about the deleterious consequences of oral creatine supplementation were initiated in the spring of 1998 when two British nephrologists suggested that there was "strong circumstantial evidence that creatine was responsible for the deterioration in renal function" (101). The Association of Professional Team Physicians and the American College of Sports Medicine (3) also concluded that much more long-term research needs to be done before one can issue a verdict on the health status of creatine. In 2004, a scientific panel of the European Food Safety Agency reported that "the safety and bioavailability of creatine, creatine monohydrate, in food for particular nutritional uses, is not a matter of concern provided there is adequate control of purity of this source of creatine" (30). Nevertheless, that same year, a report of the Agence Française de Sécurité Sanitaire et Alimentaire claimed that "one should not encourage publicity of creatine in order to protect sport participants to any potential pathological consequences" (1).

The first investigations on renal function in healthy individuals who consumed exogenous creatine was published in 2002. Poortmans and colleagues compared renal clearances of creatinine and urea in three groups of active subjects who consumed creatine for a duration of 5 d, 9 wk, and up to 5 yr with those of control groups (95). They did not observe statistical differences in renal function between the control groups and the creatine consumers. From these experimental protocols it can be stated that GFR and tubular reabsorption processes were not affected by oral supplementation with the usual daily amount of creatine (20 g/d for 5 d and <10 g/d thereafter).

As reported by Wyss and Kaddurah-Daouk, the excess conversion of creatine to sarcosine may result in cytotoxic agents such as methylamine (137). The latter has been found to be deaminated by semicarbazide-sensitive amine oxidase (EC 1.4.3.6) to produce formaldehyde and hydrogen peroxide. Under special conditions, the presence of methylamine and formaldehyde—two well-known cytotoxic agents—can be revealed by urine analyses. Recently, Poortmans and colleagues investigated 20 healthy young males who were supplemented daily with 21 g of creatine monohydrate for a duration of a couple weeks (96, 107). In this study, 24 h urine collections were made before and after creatine supplementation and urine determination of creatine, creatinine, methylamine, formate, and formaldehyde was made. These investigations showed that short-term and heavy-load oral creatine supplementation (up to 21 g/d) stimulates the production of an excess of methylamine and formaldehyde in the urine of healthy humans. However, the levels of those compounds did not reach the normal upper-limit values for healthy humans (96, 107).

Adaptations to Chronic Exercise Training

The various functions of the kidney are largely dictated by extrarenal regulating factors, such as perfusion pressure, neural influences, and endocrines. With chronic exercise, renal function would be predicted to change to the extent that these regulatory factors may be altered by exercise training. A clear example is RSNA. As discussed previously, during exercise the renal sympathetic nerves appear to influence renal hemodynamics, renin release, and possibly tubular sodium reabsorption and proteinuria. It had been suggested that in humans who were exercise trained the peripheral sympathetic nerve activity response to a given absolute exercise power output was decreased (136). In fact, an important human study by Meredith and colleagues in 1991 (67) showed that of the significant decrease in resting total-body NE spillover induced by chronic exercise training, two thirds of the reduction in total-body peripheral sympathetic activity is attributable to a decrease in renal NE spillover RSNA or RSNA. Thus, there is a selective and disproportionate reduction in RSNA with chronic exercise training. This may explain the reduction of postexercise proteinuria observed in trained subjects submitted to submaximal exercise (14, 99). A decrease in RSNA with exercise training may also be a major factor preventing ARF during exercise. This training effect has profound implications (yet to be tested) for the acute renal response to exercise as well as for changes in renal function that may be important in certain disease states, such as hypertension.

Exercise Training and Select Kidney Disease Conditions

In the last quarter of the 20th century, several researchers (38, 42, 70, 79, 81, 82) recognized the positive rehabilitation effect of exercise training on end-stage renal disease patients and kidney transplant recipients. They were convinced that exercise training would have positive effects on the social and health care of those patients. However, one must make a distinction between an individual with end-stage renal disease—an irreversible situation—and a kidney transplant recipient, who may become a healthy individual after transplantation.

End-Stage Renal Disease

The importance of exercise training in the rehabilitation of patients with end-stage renal disease has been confirmed by different authorities (18, 80, 88, 89). Stationary cycling during hemolysis treatment has been used as an alternative to exercise training (79, 82, 108). Hemodialysis patients can safely participate in a variety of exercise programs with no evidence of adverse effects (10, 83). However, more education of health care providers on the importance of exercise is needed (78). Among the research teams, Patricia Painter from the United States made a major contribution (figure 20.8).

Kidney Transplant Recipients

Whether exercise training can positively affect associated outcomes such as physical functioning, kidney function, and metabolism in transplant recipients has been addressed in a limited number of investigations (62). In a study published in 1985, Evans and colleagues reported that 79% of transplant patients functioned at nearly normal levels compared with 48% to 59% of those treated with various forms of dialysis (32). Almost 75% of transplant patients were able to work compared with 25% to 59% of dialysis patients. Kempeneers and colleagues investigated 16 renal transplant recipients (mean age: 33 yr) during a 24 wk exercise-training program (56). The training routine consisted of 60 min sessions of supervised exercise (including walking, jogging, aerobic exercise, and ball games) 3 times/wk. The subjects were able to increase their peak $\dot{V}O_2$ from 29 $ml \cdot kg^{-1} \cdot min^{-1}$ to 38 $ml \cdot kg^{-1} \cdot min^{-1}$; the latter value represents 78% of a control healthy group. Muscle biopsies (vastus lateralis) showed a lower oxidative capacity (64%) in the transplant recipients compared with healthy, moderately trained subjects. Recently, Hamiwka and colleagues reported a study on children aged 6 to 18 yr 4.8 ± 3.2 yr after kidney transplantation (43). Questionnaires regarding physical activity frequency and health-related quality of life were collected for recipients and control subjects. The results of this study suggested that kidney-recipient children and teenagers were less active than the control peers. Moreover, they had a mean GFR of 79 ± 21 ml/min expressed per 1.73 m^2, a value slightly under the normal range for this age.

An important fundamental question was how a transplanted kidney functions during the stress of exercise. Only a few papers have addressed this issue. Poortmans and colleagues investigated the GFR and albumin excretion rate in adult kidney recipients, who were regularly involved in exercise practice, when submitted to maximal work load on a bicycle ergometer (86). They observed that the GFR was slightly reduced by maximal exercise to an extent similar to reduction found in control subjects. The albumin excretion rate was

Figure 20.8 Patricia Painter.
Photo courtesy of Patricia Painter.

slightly increased (about 50%) in the recipient subjects whereas it was majorly increased (20 times the resting level) in the control subjects. The difference between the two groups could be related to the lack of increase in plasma renin activity normally seen during strenuous exercise in the transplant recipients. The absence of an increase in renin with exercise in the kidney transplant patient may be attributable to the denervation of the transplanted kidney.

Concluding Remarks

The understanding and investigation of kidney function during exercise have lingered behind that of the other organ systems. This may be explained by two fundamental reasons. First, renal function during exercise is usually not considered to be essential or critical in terms of limiting or determining exercise performance. Second, and more important, renal function is more difficult to assess during exercise. Although a urine analysis provides some information, it is extremely difficult to apply invasive techniques to asses kidney function in an exercising individual. Noninvasive methodologies for evaluating various elements of renal function (e.g., hemodynamics, tubular transport, hormone release) are not available.

In the past decade or so there has been an increased interest in and acknowledgment of the importance of kidney function during exercise. Individuals studying exercise and environmental stressors and the associated changes in extracellular or plasma volumes have begun to look at changes in kidney function as the mediator of some of these changes. Changes in cardiovascular and thermoregulatory function with aging are pointing toward important alterations in kidney function with age. The issue of exercise or exercise and dehydration-induced ARF has had its continued presence. The fact that many athletes are using anti-inflammatory drugs or nutritional supplements, which might be nephrotoxic under certain conditions, has generated interest in the potential deleterious combined effects of exercise and these agents on renal function. Finally, the role of the kidneys as an endocrine organ, for the release of both renin–angiotensin and NE, has profound implications to possibly explain the effects of chronic exercise on certain disease states. This may be particularly true for diseases associated with elevated peripheral sympathetic nerve activity, such as hypertension, congestive heart failure, or obesity.

References

1. Agence Française de Sécurité Sanitaire et Alimentaire. Avis relatif à la publicité portant sur des substances de développement musculaire et de mise en forme contenue dans un magazine spécialisé (Saisines 2003-SA-0385 & 2003-SA-0386).
2. Alyea EP, Boone AW. Urinary findings resulting from nontraumatic exercise. *Southern Med J* 50: 905-910, 1957.
3. American College of Sports Medicine. The physiological and health effects of oral creatine supplementation. *Med Sci Sports Exerc* 32: 706-717, 2000.
4. Bailey RR, Dann E, Gillies AHB, Lynn KL, Abernethy MH, Neale TJ. What the urine contains following athletic competition. *New Z Med J* 83: 309-313, 1976.
5. Barach JH. Physiological and pathological effects of severe exertion (the marathon race) on the circulatory and renal systems. *Arch Intern Med* 5: 382-405, 1910.
6. Barclay J, Cooke W, Kenney R, Nutt M. The effects of water diuresis and exercise on the volume and composition of the urine. *Am J Physiol* 148: 327-337, 1947.
7. Blake W. Effect of exercise and emotional stress on renal hemodynamics, water and sodium excretion in the dog. *Am J Physiol* 165: 149-157, 1951.
8. Bloor C, McKirnan M, Guth B, White F. Renal blood flow during exercise in trained and untrained swine. In: Tumbleson M, ed. *Swine in Biomedical Research*. New York: Plenum Press, 1986, 1789-1794.
9. Brainbridge FA. *The Physiology of Muscular Exercise*. London: Longmans, Green & Co., 1919, 214.
10. Brenner I. Exercise performance by hemodialysis patients: A review of the literature. *Phys Sportsmed* 37: 84-96, 2009.
11. Brooks GA, Fahey TD. *Exercise Physiology*. London: Macmillan, 2000, 830.
12. Campbell J, Webster T. LXXX. Day and night urine during complete rest, laboratory routine, light muscular work and oxygen administration. *Biochem J* 15: 660-664, 1921.
13. Campbell J, Webster T. XV. Effect of severe muscular work on the composition of the urine. *Biochem J* 16: 106-110, 1922.
14. Cantone A, Cerretelli P. Effect of training on proteinuria following muscular exercise. *Int Z Angew Physiol Arbeitsphysiol* 18: 324-329, 1960.
15. Castenfors J. Renal function during exercise. *Acta Physiol Scand* 70(Suppl. 293): 43 pages, 1967.
16. Chapman C, Henschel A, Minckler J, Forsgren A, Keys A. The effect of exercise on renal plasma flow in normal male subjects. *J Clin Invest* 27: 639-644, 1948.

17. Christensen E, Högberg P. Physiology of skiing. *Arbeitsphysiol* 14: 292-303, 1950.
18. Clyne N. Physical working capacity in uremic patients. *Scand J Urol Nephrol* 30: 247-252, 1996.
19. Covian F, Rehberg P. Uber die nierenfunktion während schwerer muskelarbeit. *Skand Arch Physiol* 75: 21-37, 1936.
20. Coye RD, Rosandich RR. Proteinuria during the 24-hour period following exercise. *J Appl Physiol* 15: 592-594, 1960.
21. Dawson PM. *The Physiology of Physical Education*. Baltimore, MD: Williams & Wilkins Co., 1935, 913.
22. de Bold AJ, Borenstein HB, Veress AT, Sonnenberg H. A rapid and potent natriuretic response to intravenous injection of atrial myocardial extract in rats. *Life Sci* 28: 89-94, 1981.
23. Dibona GF, Kopp UC. Neural control of renal function. *Physiol Rev* 77: 75-197, 1997.
24. Dicarlo S, Bishop V. Onset of exercise shifts operating point of arterial baroreflex to high pressures. *Am J Physiol* 262: H303-H307, 1992.
25. Dies E, Ramos G, Avelar E, Lennhoff M. Renal excretion of lactic acid in the dog. *Am J Physiol* 216: 106-111, 1969.
26. Edwards H, Cohen M, Dill D. Renal function in exercise. *Arbeitsphysiol* 9: 610-618, 1935-1937.
27. Edwards HT, Richards TK, Dill DB. Blood sugar, urine sugar and urine protein in exercise. *Am J Physiol* 98: 352-356, 1931.
28. Eggleton M. A class experiment on urinary changes after exercise. *J Physiol* 101: 1-2P, 1942.
29. Eggleton MG. The effect of exercise on chloride excretion in man during water diuresis and during tea diuresis. *J Physiol (London)* 102: 140-154, 1943.
30. European Food Safety Agency. Creatine monohydrate for use in foods for particular nutritional uses (Question number EFSA-Q-2003-125). *EFSA J* 36: 1-6, 2004.
31. Evans L. *Starling's Principles of Human Physiology*. London: Churchill, 1936, 926.
32. Evans R, Manninen D, Garrison Jr. L, Hart G, Blagg C, Gutman R, Hull R, Lowrie E. The quality of life of patients with end-stage renal disease. *N Engl J Med* 312: 553-559, 1985.
33. Farquhar W, Morgan A, Zambraski E, Kenney W. Effects of acetaminophen and ibuprofen on renal function in the stressed kidney. *J Appl Physiol* 86: 598-604, 1999.
34. Freedman MH, Connell GE. The heterogeneity of gamma-globulin in post-exercise urine. *Can J Biochem* 42: 1065-1097, 1964.
35. Freund B, Shizuru E, Hashiro G, Claybaugh J. Hormonal, electrolyte, and renal responses to exercise are intensity dependent. *J Appl Physiol* 70: 900-906, 1991.
36. Gardner K. Exercise and the kidney. In: Appenzeller O, ed. *Sports Medicine*. Baltimore, MD: Urban & Schwarzenberg, 1988, 189-195.
37. Gardner KD. "Athletic pseudonephritis"—Alteration of urine sediment by athletic competition. *JAMA* 161: 1613-1617, 1956.
38. Goldberg A, Geltman E, Hagberg JM, Gavin I, Delmez J, Carney R, Naumowicz A, Oldfield M, Harter H. Therapeutic benefits of exercise training for hemodialysed patients. *Kidney Int* 24 (Suppl. 16): S303-S309, 1983.
39. Goldszer R, Siegel A. Renal abnormalities during exercise. In: Strauss R, ed. *Sports Medicine*. Philadelphia: Saunders, 1984, 130-139.
40. Gould AG, Dye JA. *Exercise and its Physiology*. New York: Barnes and Co., 1932, 312.
41. Grimby G. Renal clearances during prolonged supine exercise at different loads. *J Appl Physiol* 20: 1294-1298, 1965.
42. Hagberg JM, Goldberg A, Ehsani A, Heath G, Delmez J, Harter H. Exercise training improved hypertension in hemodialysed patients. *Am J Nephrol* 3: 209-212, 1983.
43. Hamiwka L, Cantell M, Crawford S, Clark C. Physical activity and health related quality of life in children following kidney transplantation. *Pediatr Transplantation* 13: 861-867, 2009.
44. Hasking G, Esler M, Jennings G, Dewar E, Lambert G. Norepinephrine spillover to plasma during steady-state supine bicycle exercise. *Circulation* 78: 516-521, 1988.
45. Havard RE, Reay GA. The influence of exercise on the inorganic phosphates of the blood and urine. *J Physiol (London)* 61: 35-48, 1926.
46. Hellebrandt FA. Studies on albuminuria following exercise. I. Its incidence in women and its relationship to the negative phase in pulse pressure. *Am J Physiol* 101: 357-364, 1932.
47. Herxheimer H. *The Principles of Medicine in Sport for Physicians and Students*. Leipzig: George Thieme, 1933.
48. Javitt NB, Miller AT. Mechanisms of exercise proteinuria. *J Appl Physiol* 4: 834-839, 1952.
49. Jerusalem E. Ueber ein neuer verfahren zur quantitativen bestimmung der milchsäure in organen und tierischen flüssigkeiten. *Biochem Ztschr Berlin* xii: 361-389, 1908.
50. Johnson RE, Edwards HT. Lactate and pyruvate in blood and urine after exercise. *J Biol Chem* 118: 427-432, 1937.
51. Joles JA. Renal function and exercise in dogs. Utrecht: University of Utrecht, 1984, 131.
52. Kachadorian W, Johnson R. Renal responses to various rates of exercise. *J Appl Physiol* 28: 748-752, 1970.

53. Kachadorian WA, Johnson RE. Athletic pseudonephritis in relation to rate of exercise. *Lancet* 1: 472, 1970.
54. Kachadorian WA, Johnson RE, Buffington RE, Lawler L, Serbin JJ, Woodall T. The regularity of "athletic pseudonephritis" after heavy exercise. *Med Sci Sports* 2: 142-145, 1970.
55. Karpovich P. *Physiology of Muscular Activity*. Philadelphia: Saunders, 1959, 232-233.
56. Kempeneers G, Noakes T, van Zyl-Smit R, Myburgh K, Lambert M, Adams B, Wiggins T. Skeletal muscle limits the exercise tolerance of renal transplant recipients. *Am J Kidney Dis* 16: 57-65, 1990.
57. Koriakina AF, Kossowskaja EB, Krestownikoff AN. Uber die schwankungen des chloridgehaltes im blut, harn und schweib bei muskeltätigkeit. *Arbeitsphysiol* 2: 461-473, 1930.
58. Krebs HA, Yoshida T. Muscular exercise and gluconeogenesis. *Biochem Z* 338: 241-244, 1963.
59. Lamb D. *Physiology of Exercise: Responses and Adaptations*. New York: Macmillan, 1978, 287-298.
60. Liebig VJ. *Animal Chemistry, or Chemistry in its Application to Physiology and Pathology*. London: Taylor & Walton, 1843, 384.
61. Liljestrand SH, Wilson DW. The excretion of lactic acid in the urine after muscular exercise. *J Biol Chem* 65: 773-782, 1925.
62. Macdonald J, Kirkman D, Jibani M. Kidney transplantation: A systematic review on interventional and observational studies of physical activity on intermediate outcomes. *Adv Chronic Kidney Dis* 16: 482-500, 2009.
63. MacKeith NW, Pembrey MS, Spurell WR, Warner EC, Westlake HJWJ. Observations on the adjustment of the human body to muscular work. *Proc Roy Soc B* 95: 413-439, 1923.
64. MacSearraigh ETM, Kallmeyer JC, Schiff HB. Acute renal failure in marathon runners. *Nephron* 24: 236-240, 1979.
65. McArdle WD, Katch FI, Katch VL. *Exercise Physiology*. Baltimore, MD: Williams & Wilkins, 2007, 1068.
66. McKay E, Slater RJ. Studies on human proteinuria. II. Some characteristics of the gamma globulins excreted in normal, exercise, postural, and nephrotic proteinuria. *J Clin Invest* 41: 1638-1652, 1962.
67. Meredith I, Friberg P, Jennings G, Dewar E, Fazio V, Lambert G, Esler M. Exercise training lowers resting renal but not cardiac sympathetic activity in humans. *Hypertension* 18: 575-582, 1991.
68. Merrill A, Cargill W. The effect of exercise on the renal plasma flow and filtration rate of normal and cardiac subjects. *J Clin Invest* 27: 272-277, 1948.
69. Miller AT, Miller JO. Renal excretion of lactic acid in exercise. *J Appl Physiol* 1: 614-618, 1949.
70. Miller T, Squires R, Gerald T, Ilstrup D, Frohnert P, Sterioff S. Graded exercise testing and training after renal transplantation: A preliminary study. *Mayo Clin Proc* 62: 773-777, 1987.
71. Mittelman K. Influence of angiotensin II blockade during exercise in the heat. *Eur J Appl Physiol* 72: 542-547, 1996.
72. Mittleman KD, Zambraski EJ. Exercise-induced proteinuria is attenuated by indomethacin. *Med Sci Sports Exerc* 24: 1069-1074, 1992.
73. Mogensen CE, Solling K. Studies on renal tubular absorption:partial and near complete inhibition by certain amino acids. *Scand J Clin Lab Invest* 37: 477-486, 1977.
74. Moller E, McIntosh JF, Van Slyke DD. Studies of urea excretion. *J Clin Invest* 6: 427-465, 1928.
75. Nedbal J, Seliger V. Electrophoretic analysis of exercise proteinuria. *J Appl Physiol* 13: 244-246, 1958.
76. Noakes TD, Goodwin N, Rayer BL, Branken T, Taylor RKN. Water intoxication: A possible complication during endurance exercise. *Med Sci Sports Exerc* 17: 370-375, 1985.
77. O'Hagen K, Bell L, Mittelstadt S, Clifford P. Effect of dynamic activity on renal sympathetic nerve activity in conscious rabbits. *J Appl Physiol* 74: 2099-2104, 1993.
78. Painter P. Exercise in chronic disease: Physiological research needed. *Exerc Sport Sci Rev* 36: 83-90, 2008.
79. Painter P. Exercise training during hemodialysis: Rates of participation. *Dial Transplant* 17: 165-168, 1988.
80. Painter P. The importance of exercise training in rehabilitation of patients with end-stage renal disease. *Am J Kidney Dis* 24 (Suppl. 1): S2-S9, 1994.
81. Painter P, Hanson P, Messer-Rehak D, Zimmerman S, Glass N. Exercise tolerance changes following renal transplantation. *Am J Kidney Dis* 10: 452-456, 1987.
82. Painter P, Nelson-Worel J, Hill M, Thornberry D, Shelp W, Harrington A, Weinstein A. Effects of exercise training during hemodialysis. *Nephron* 43: 87-92, 1986.
83. Parsons T, King-Vanvlack C. Exercise and end-stage kidney disease: Functional exercise capacity and cardiovascular outcomes. *Adv Chronic Kidney Dis* 16: 459-481, 2009.
84. Pavy FW. Effect of prolonged muscular exercise upon the urine in relation to the source of muscular power. *Lancet* II: 815-818, 1876.
85. Perlman LV, Cunningham D, Montoye H, Chiang B. Exercise proteinuria. *Med Sci Sports* 2: 20-23, 1970.
86. Poortmans J, Hermans L, Vandervliet A, Niset G, Godefroid C. Renal responses to exercise in heart and kidney transplant patients. *Transplant Int* 10: 323-327, 1997.

87. Poortmans J, Jeanloz RJ. Quantitative immunological determination of 12 plasma proteins excreted in human urine collected before and after exercise. *J Clin Invest* 47: 386-393, 1968.

88. Poortmans J, Niset G, Godefroid C, Lamotte M. Responses to exercise and limiting factors in hemodialysis and renal transplant patients. In: Rieu M, ed. *Physical Work Capacity in Organ Transplantations*. Basel: Karger, 1998, 113-133.

89. Poortmans J, Painter P. Renal disease. In: LeMura L, von Duvillard S, eds. *Clinical Exercise Physiology*. Hagerstown, MD: Lippincott Williams & Wilkins, 2004, 331-346.

90. Poortmans J, Van Kerchove E, Jaumain P. Aspects physiologiques et biochimiques de la protéinurie d'effort. *Int Z angew Physiol Einschl Arbeitsphysiol* 19: 337-354, 1962.

91. Poortmans J, Zambraski E. The renal system. In: Tipton C, ed. *Exercise Physiology: People and Ideas*. Oxford: Oxford University Press, 2003, 452-474.

92. Poortmans JR. Exercise and renal function. *Sports Med* 1: 125-153, 1984.

93. Poortmans JR, Blommaert E, Baptista M, De Broe ME, Nouwen EJ. Evidence of differential renal dysfunctions during exercise in men. *Eur J Appl Physiol* 76: 88-91, 1997.

94. Poortmans JR, Brauman H, Staroukine M, Verniory A, Decaestecker C, Leclercq R. Indirect evidence ofglomerular mixed-type postexercise proteinuria in healthy humans. *Am J Physiol* 254: F277-F283, 1988.

95. Poortmans JR, Francaux M. Renal implications of exogenous creatine monohydrate supplementation. *Am J Med Sports* 4: 212-216, 2002.

96. Poortmans JR, Kumps A, Duez P, Fofonka A, Carpentier A, Francaux M. Effect of oral creatine supplementation on urinary methylamine, formaldehyde, and formate. *Med Sci Sport Exerc* 37: 1717-1720, 2005.

97. Poortmans JR, Labilloy D. The influence of work intensity on postexercise proteinuria. *Eur J Appl Physiol* 57: 260-263, 1988.

98. Poortmans JR, Vancalck B. Renal glomerular and tubular impairment during strenuous exercise in young women. *Eur J Clin Invest* 8: 175-178, 1978.

99. Poortmans JR, Vanderstraeten J. Kidney function during exercise in healthy and diseased humans. *Sports Med* 18: 419-437, 1994.

100. Powers SK, Howley ET. *Exercise Physiology*. Boston: McGraw-Hill, 1996, 552.

101. Pritchard N, Kalra P. Renal dysfunction accompanying oral creatine supplementations. *Lancet* 351: 1252-1253, 1998.

102. Radigan L, Robinson S. Effects of environmental heat stress and exercise in renal blood flow and filtration rate. *J Appl Physiol* 2: 185-191, 1949.

103. Rasch P, Dodd Wilson I. Other body systems and exercise. In: Falls H, ed. *Exercise Physiology*. New York: Academic Press, 1978, 129-151.

104. Refsum HE, Stromme SB. Renal osmol clearance during prolonged heavy exercise. *Scand J Clin Lab Invest* 38: 19-22, 1977.

105. Reidman SR. *The Physiology of Work and Play*. Dryden Press, 1950, 565.

106. Rowe DS, Soothill JF. The proteins of postural and exercise proteinuria. *Clin Sci* 21: 87-91, 1961.

107. Sale C, Harris RC, Florance J, Kumps A, Sanvura R, Poortmans JR. Urinary creatine and methylamine excretion following 4 x 5 $g \cdot day^{-1}$ or 20 g x 1 $g \cdot day^{-1}$ of creatine monohydrate for 5 days. *J Sports Sci* 27: 759-766, 2009.

108. Sams B, Thompson J, Stray-Gudersen J. Three months of cycling while on hemodialysis improves work capacity to improving peripheral extraction in chronic renal failure patients. *Med Sci Sport Exerc* 28: S55, 1996.

109. Sanchez-Medina F, Sanchez-Urrutia L, Medina JM, Mayor F. Effect of muscular exercise and glycogen depletion on rat liver and kidney phosphoenolpyruvate carboxykinase. *FEBS Letters* 19: 128-130, 1971.

110. Sanchez-Medina F, Sanchez-Urrutia L, Medina JM, Mayor F. Effect of short-term exercise on gluconeogenesis by rat kidney cortex. *FEBS Letters* 26: 25-26, 1972.

111. Sanchez-Urrutia L, Garcia-Ruiz JP. Lactic acidosis and renal phosphoenolpyruvate carboxykinase activity during exercise. *Biochem Med* 14: 355-367, 1975.

112. Sanders M, Rasmussen S, Cooper D, Bloor C. Renal and intrarenal blood flow distribution in swine during severe exercise. *J Appl Physiol* 40: 932-935, 1976.

113. Schad H, Seller H. A method for recording autonimic activity in unanesthetized, freely moving cats. *Brain Res* 100: 425-430, 1975.

114. Schmidt W, Bub A, Meyer M, Weiss T, Schneider G, Maasen N, Forssmann W. Is urodilatin the missing link in exercise-dependent renal sodium retention? *J Appl Physiol* 84: 123-128, 1998.

115. Schneider EC. *Physiology of Muscular Exercise*. London: Saunders, 1933, 366.

116. Schulz-Knappe P, Forssmann K, Herbst F, Hock D, Pipkorn R, Forsmann WG. Isolation and structural analysis of "urodilatin," a new peptide of the cardiodilatin-[ANP]-family, extracted from human urine. *Klin Wochenschr* 66: 752-759, 1988.

117. Shephard R. *Physiology and Biochemistry of Exercise*. New York: Praeger, 1982, 312-316.

118. Simonson E, Keys A. *Physiology of Work Capacity and Fatigue*. Springfield, IL: Charles C. Thomas, 1971, 571.
119. Smith J, Robinson S, Pearcy M. Renal responses to exercise, heat and dehydration. *J Appl Physiol* 4: 659-665, 1952.
120. Starlinger H, Bandino R. Der einfluss von arbeit bei hoher temperatur auf die inulin- und PAH-clearance sowie die elektrolyt- und 17-21-dihydroxy-20-ketosteroidausscheidung im harn. *Arbeitsphysiol* 18: 285-305, 1960.
121. Starr I. The production of albuminuria by renal vasoconstriction in animals and in man. *J Exp Med* 43: 31-51, 1926.
122. Taylor A. Some characteristics of exercise proteinuria. *Clinical Science* 19: 209-217, 1960.
123. Tidgren D, Hjemdahl P, Theodorsson E, Nussberger J. Renal neurohormonal and vascular responses to dynamic exercise in humans. *J Appl Physiol* 70: 2279-2286, 1991.
124. Todorovic B, Nikolic B, Brdaric R, Nikolic V, Stojadinovic B, Pavlovic-Kentera V. Proteinuria following submaximal work of constant and variable intensity. *Int Z angew Physiol* 30: 151-160, 1972.
125. Van Slyke DD, Alving A, Rose WC. Studies of urea excretion. VII. The effects of posture and exercise on urea excretion. *J Clin Invest* 11: 1053-1064, 1932.
126. Vatner S, Pagani M. Cardiovascular adjustments to exercise: Hemodynamics and mechanisms. *Progr Cardiovasc Dis* 19: 91-108, 1976.
127. Vatner SF. Effects of exercise and excitement on mesenteric and renal dynamics in conscious, unrestrained baboons. *Am J Physiol* 234: H210-H214, 1978.
128. Viarnaud G. *L'albuminurie d'effort*. Lyon: Université de Lyon, 1944.
129. Wade C, Ramee S, Hunt M, White C. Hormonal and renal responses to converting enzyme inhibition during maximal exercise. *J Appl Physiol* 63: 1796-1800, 1987.
130. Wade CE, Claybaugh J. Plasma renin activity, vasopressin concentration, urinary excretory responses to exercise in men. *J Appl Physiol* 49: 930-936, 1980.
131. Wahren J, Felig P, Ahlborg G, Jorfeldt L. Glucose metabolism during leg exercise in man. *J Clin Invest* 50: 2715-2725, 1971.
132. Wesson LG. Kidney function in exercise. In: Johnson WR, ed. *Science and Medicine of Exercise and Sports*. New York: Harper & Brothers, 1960, 270-284.
133. White H, Rolf D. Effects of exercise and of other influences on the renal circulation in man. *Am J Physiol* 152: 505-516, 1948.
134. Wilmore JH, Costill DL. *Physiology of Sport and Exercise*. Champaign, IL: Human Kinetics, 1999, 710.
135. Wilson DW, Long HC, Thompson HC, Thurlow S. Changes in the composition of the urine after muscular exercise. *J Biol Chem* 65: 755-771, 1925.
136. Winder WW, Hagberg JM, Hickson RC, Ebashi AA, McLane JA. Time course of sympatho-adrenal adaptation to endurance exercise. *J Appl Physiol* 45: 370-374, 1978.
137. Wyss M, Kaddurah-Daouk R. Creatine and creatinine metabolism. *Physiol Rev* 80: 1107-1213, 2000.
138. Zambraski E. Renal regulation of fluid homeostasis during exercise. In: Gisolfi C, Lamb D, eds. *Perspectives in Exercise Science and Sports Medicine*. Carmel, IN: Benchmark Press, 1990, 247-280.
139. Zambraski E. The kidney and body fluid balance during exercise. In: Buskirk E, Puhl S, eds. *Body Fluid Balance Exercise and Sport*. Boca Raton, FL: CRC Press, 1996, 75-95.
140. Zambraski E, Tucker M, Lakas C, Grassl S, Scanes C. Mechanism of renin release in exercising dog. *Am J Physiol* 246: E71-E78, 1984.

CHAPTER 21

The Immune System

Roy J. Shephard, MD, PhD, DPE, LLD (Hon. Caus.)

Introduction

The history of exercise immunology is relatively brief. This account traces its origins to early clinical studies of the interactions between strenuous bouts of physical activity and an individual's resistance to bacterial and viral infections and provides detailed comments on strenuous dynamic exercise and the risk of upper respiratory infections. The text then discusses major technical advances that have spurred rapid progress in the scientific understanding of cellular and humoral response to various types and intensities of exercise. The chapter briefly notes areas where animal physiologists have made contributions to exercise immunology as well as potential modifications of the exercise response induced by nutritional factors and pharmacological agents. Practical applications in the exercise sciences and sports medicine are then discussed. Given the specific scientific orientation of this text, details of the personalities behind the development of exercise immunology as a specific subdiscipline with its own international society (International Society for Exercise Immunology) and journal (*Exercise Immunology Review*) are not considered here, although this information can be found in a recent publication (179). The present chapter concludes by looking briefly at future research challenges.

Strenuous Exercise in Relation to Acute Bacterial and Viral Illnesses

The earliest studies of interactions between bouts of strenuous physical activity and immune function occurred in the context of recurrent epidemics of infectious disease during the first half of the 20th century. Given the lack of specific remedies for many of these diseases, physicians commonly advised bed rest. This usually left patients in a weakened state after their recovery from the disease; this was a particular concern in military hospitals. Physicians thus began to enquire whether a bout of strenuous physical activity was a factor predisposing to a number of bacterial and viral conditions (particularly anterior poliomyelitis and viral hepatitis) and to examine whether the continuation of normal or more vigorous physical activity would adversely affect the course of the disease process.

Anterior Poliomyelitis

Early studies in monkeys suggested that cold exposure and fatiguing exercise both increased the severity of paralysis after deliberate inoculation of an animal with the poliomyelitis virus (106). Russell presented case histories on 100 human cases of acute poliomyelitis, choosing individuals who were old enough to give a good personal account of their symptoms. He noted that an unusual amount of physical activity had preceded the attack by 1 to 3 d in 14 of these cases; in these individuals, paralysis was either severe or fatal and was generally localized to the body parts that had been involved in the strenuous physical activity (173). Although localization to the active limb is suggestive of causation, there was no control group, and one obvious weakness of such retrospective questioning is that it can cause a patient to remember a bout of physical activity that a healthy person might forget. Hargreaves (75) and Russell (172) reported that serious paralysis was most likely to develop in individuals who continued to engage in vigorous exercise in the face of early symptoms. Horstmann (86) studied 411 cases of poliomyelitis from 3 epidemics and concluded that an adverse effect was most likely to occur if strenuous exercise occurred in the second acute phase of the disease.

It was suspected that exercise increased blood flow to the anterior horn cells involved in a given type of physical activity, thus exposing these cells to a higher concentration of the virus, although an adverse effect on the immune response may also have been involved. Further exploration of this issue was discouraged by the emergence of effective antibiotics and vaccines for the treatment and prevention of anterior poliomyelitis: the bacteriostatic antimicrobial agent chloramphenicol (isolated by David Gottlieb and introduced into clinical practice in 1949), Jonas Salk's inactivated vaccine (licensed in 1955), and Albert Sabin's attenuated oral vaccine (licensed in 1962).

Viral Hepatitis

Type A viral hepatitis was a common problem consequent to the disruption of urban water supplies during World War II. Physicians generally agreed that rest was needed during the acute stages of the disease; indeed, this view prevails to the present day. However, there was less agreement on an optimal regimen during convalescence. One case report noted a sharp increase of icteric index in a patient who elected to engage in moderate exercise during convalescence (6). However, a detailed review of World War II data for U.S. troops in the Mediterranean suggested that normal moderate physical activity had an adverse effect on convalescing soldiers only if their serum bilirubin still exceeded 3 mg/100 ml (194). Chalmers and colleagues (33, 131) studied 188 U.S. naval personnel serving in the Korean War. Delays in hospitalization and bed rest during the early stages of the disease were associated with an increased duration of illness. However, even if the patient still showed some residual jaundice, light *ad libitum* activity could be resumed during convalescence without increasing the risk of a relapse or delaying recovery.

Subsequent case reports (19) and more detailed research have generally confirmed that strenuous exercise exacerbates the early stages of type A viral hepatitis. Krinkler and Zilberg (96) commented that five patients in whom the hepatitis had taken a virulent form had all engaged in some form of vigorous physical activity during the early stages of their infection (although the question remains whether the onset of disease heightened recollections of unusual physical activity). Edlund (52) studied 23 cases of viral hepatitis, mostly of the long incubation type. All of the patients studied undertook moderate, controlled dynamic exercise on a cycle ergometer during their convalescence, apparently without any adverse effect on outcomes.

Ringertz and colleagues (13, 166) noted an epidemic of 600 cases of hepatitis among Swedish orienteers between 1957 and 1966. It was originally thought to be a distinct infection (hepatitis sylvatica), but serological tests disclosed hepatitis B infection. Subsequent research showed that exposure of skin injuries to contaminated water (34), rather than the heavy physical demands of orienteering, was responsible for the high incidence of disease. The seroprevalence of infection decreased to that of the general population once orienteers were required to wear protective clothing (82). Likewise, a 1980 epidemic of type B hepatitis among Sumo wrestlers (121) was probably transmitted through the exposure of injured skin to contaminated mats.

Miscellaneous Viral Infections

Athletic teams have occasionally experienced epidemics of other acute infections. In one instance, 41 of 63 high school football players were affected by an enteroviral illness, and 33 (80% of those infected) developed aseptic meningitis. In contrast, students who were not on the football team had a lower attack rate, and only 10% of those who were infected developed aseptic meningitis (8). Outbreaks of herpes gladiatorum have also been described among those involved in contact sports. In most of these studies it seems that sport participation increased exposure to infection rather than reduced the individual's resistance to disease.

Elevated titers of the Epstein-Barr virus and related viruses have been described in chronic fatigue syndrome (24, 130), but their clinical significance is doubtful because high titers can also occur in healthy individuals. Strenuous physical activity may exacerbate fatigue in affected individuals, although it is unclear whether this reflects disturbed immune function, concomitant psychological issues, or a combination of the two. Moderate, graded dynamic exercise seems to pose no danger to the patient once the acute stage of the disease has passed (53).

Bacterial Infections

An early report from Cowles (40) noted an apparent association between recent participation in intensive exercise and the incidence of pneumonia among pupils at a boys' school. Others reported that bouts of exhausting exercise increased susceptibility to experimental pneumococcal infections in both rabbits (4) and guinea pigs (137), particularly if physical activity was undertaken during the infectious stage of the disease. However, 16 to 18 d of preliminary exercise-wheel training enhanced the survival of mice that were given a subsequent experimental *Salmonella Typhimurium* infection (29).

As with viral infections, the introduction of potent antibiotics greatly reduced interest in this type of experimentation. More recent reviewers (influenced in part by the popularity of the open-window or natural killer cell hypothesis) have argued that although heavy dynamic exercise adversely affects susceptibility to viral infections, it does not compromise resistance to bacteria (21, 144). Seemingly in support of this view, researchers found no change in the response to subcutaneous vaccines when a group of recent infantry recruits undertook 3 mo of rigorous initial training. Likewise, over the 2 to 4 wk period after participation in a triathlon event, the antibody response to pneumococcal, tetanus, and diphtheria vaccines did not differ from that seen in control subjects (23).

One continuing concern has been the spread of methicillin-resistant *Staphylococcus aureus* infections among athletes (164). As with other diseases, the problem seems to be mainly a contamination of cutaneous injuries during either competition or subsequent hospital treatment, although the possible effect of a bout of exhausting exercise on the resistance to this specific group of organisms remains to be studied.

Summary

Early clinical observations suggested that unusually vigorous physical activity exacerbated the initial stages of both anterior poliomyelitis and viral hepatitis infections but that recovery was unaffected or even speeded by moderate physical activity. A number of epidemics among sport participants were traced to an increased exposure to viruses and bacteria. However, it remains unclear whether strenuous physical activity directly affects an individual's susceptibility to bacterial infection.

Strenuous Dynamic Exercise and the Risk of Upper Respiratory Infections

Given apparent interactions between fatiguing activity and vulnerability to various diseases, sport physicians became concerned that periods of strenuous athletic training might increase an athlete's susceptibility to viral infections of the respiratory tract, thereby compromising performance and possibly exposing the competitor to the risk of a fatal myocarditis. Doctors providing medical supervision at major competitions had long noted a high prevalence of upper respiratory infections among participants (1, 181), and the possible causal roles of intensive training and fatiguing exercise continue to be a major interest of exercise immunologists. Issues have included a strengthening of epidemiological evidence, the establishment of dose–response relationships, examination of implications for the athlete's performance, and the possible prevention of immunosuppression by the use of dietary supplements or medications.

Infections around the time of international competition are a major discouragement to athletes who have spent many years preparing for their chosen event. Moreover, in commercial sport, millions of dollars may hang on the health of a single player. Early papers on strenuous exercise and upper respiratory infection were based mainly on accumulated clinical experience. The 1974 International Union of Physiological Sciences meeting in India included a satellite symposium on exercise in the city of Patiala. Ernst Jokl (figure 21.1) delivered the keynote address at this symposium, arguing that participation in an athletic event often depressed the immune response, a view that he had recently published (89). At the end of his lecture he was vigorously questioned by Bengt Eriksson, an exercise physiologist from the University of Göteborg. The latter made three successive demands for scientific evidence supporting Jokl's views and remained unconvinced at the end of a heated exchange. An article in the 1984 Olympics issue of *Journal of the American Medical Association* expressed equal skepticism, commenting, "There is no clear experimental or clinical evidence that exercise will alter the frequency or severity of human infections" (192, p. 2738). After reviewing 10 reports examining the effects of exercise on various host-defense factors, Simon concluded, "Further studies will be needed before it can be concluded that exercise affects the host response to infection in any clinically meaningful way" (192, p. 2735).

Figure 21.1 Ernst Jokl was one of the founders of the American College of Sports Medicine and an early pioneer of sports medicine. During the 1970s, he suggested (mainly on the basis of clinical observations) that athletic training and sport participation could modify an individual's immune response (89).

The photograph was obtained through the courtesy of his son, Dr. Peter Jokl.

Risk of Myocarditis

In experimental animals, an acute infectious myocarditis leads to an acute phase reaction and degradation of myocardial protein. If animals are exercised early during infection, viral replication and damage to cardiac muscle are both increased. Between 1979 and 1992, 16 Swedish orienteers died suddenly while racing; there were fears that a combination of upper respiratory infection and the prolonged bouts of exercise demanded by their sport had exacerbated a myocardial infection (25, 208, 209). Wesslen and colleagues (209) noted that, perhaps because of the substantial skill component involved in orienteering, performance was not greatly affected and many athletes competed even when suffering from such infections.

Many anecdotal reports have linked viral infections to deaths that have occurred while engaged in exercise. An early analysis of data from the Toronto Rehabilitation Centre noted that one of 688 postcoronary patients had died suddenly while undertaking moderate dynamic exercise (jogging). He was recovering from influenza and, despite advice to the contrary, had continued his normal exercise regimen (180). Among young athletes, myocarditis normally accounts for 5% to 20% of sudden exercise-induced deaths; parvovirus B19 and herpes virus 6 are the most commonly identified microorganisms (58, 120). However, in the sample of fatalities among Swedish orienteers, about three fourths of postmortems showed histological evidence of myocarditis. High titers of *Chlamydia pneumoniae* were found in four of five individuals where serological information was available. This organism has previously been linked to the development of myocarditis (70). Nevertheless, the problem seems to have been a chronic rather than an acute infection in most of the orienteers; as judged from symptoms such as arrhythmia, the cardiac abnormality had been present from 4 mo to 2 yr before death. A further puzzling feature is that no incidents have occurred subsequent to 1992; this raises suspicions that some factor other than infection—possibly blood doping—was responsible.

A comprehensive comparison of lymphocyte subsets, immunoglobulin, cytokine, and complement levels disclosed no important differences between 119 top-ranked orienteers and a control group of 36 highly trained distance runners and cross-country skiers (101). This further reduces the likelihood that an exercise-induced disturbance of immune function was responsible for the fatal episodes.

Upper Respiratory Infections and Competitive Performance

No strong evidence links a sustained deterioration of an athlete's physical performance to serological evidence of a recent infection. A prospective study of 68 elite track and field athletes found that 54% developed elevated titers of antibody to Coxsackie B 1-5 virus over their winter training, but such findings were not associated with poor competitive outcomes (167). In laboratory volunteers, experimental injections of either pyrogens (72) or live microorganisms (42, 59) reduced the capacity for dynamic aerobic exercise (as estimated from submaximal testing) and isometric and dynamic strength and endurance exercise during the period of fever. However, the immediate functional deterioration occurred before any significant degradation of muscle protein had occurred and seemed correlated with symptoms such as myalgia rather than with physiological or biochemical changes. Longer-term deteriorations in muscle strength and endurance have been hard to dissociate from the bed rest that typically accompanies acute infections.

Performance might be adversely affected if a respiratory infection increased the competitor's liability to exercise-induced bronchospasm. One study compared 19 nonasthmatic cross-country skiers with 22 sedentary controls (80). Of these subjects, 12 of the skiers and 10 of the controls contracted an upper respiratory tract infection over the course of the study. No restrictions were placed on training. One week after the onset of symptoms, the response to a methacholine challenge was increased in the skiers but not in the sedentary controls, suggesting that the infected athletes would have been more vulnerable to exercise or cold-induced bronchospasm during competition.

Strengthening of Epidemiological Data

The first epidemiological evidence suggesting that bouts of vigorous dynamic exercise increased susceptibility to upper respiratory infections came in a briefly reported cross-sectional comparison of 61 university rowers and 126 cadets (48). The rowers reported more frequent and more severe respiratory symptoms than the cadets did, although it could be argued that their desire to continue undertaking vigorous exercise made them more conscious of the effects of infections. The effect of a competitive event was examined in a case-control study of 150 participants in the Two Oceans ultramarathon in Cape Town (153). In the 2 wk after this event, symptoms suggestive of upper respiratory infection developed in 33.3% of the runners compared with 15.3% of controls; moreover, a fast race time or the development of muscle pain after the event appeared to be factors predisposing the runners to infection. Nevertheless, the implied frequency of infection in controls (once every 13 wk) was high, underlining the problem in relying on subjective reports of respiratory disease.

Subsequent studies have suggested that prolonged bouts of dynamic endurance training also increase the likelihood of developing respiratory symptoms (77, 112, 145). Our questioning of masters athletes indicated that most of the group was conscious of a critical weekly training distance (commonly around 50 km/wk). If this threshold was exceeded, the athletes became more vulnerable to upper respiratory infections (182, 186). Likewise, David Nieman (figure 21.2) and colleagues found that runners training 97 km/wk had twice the risk of a self-reported infection seen in those who were training only 32 km/wk (142). Adverse effects from a single race were not seen if the distance covered was less than 42 km, but the risk of infection after marathon participation was six times that found in controls (142, 143, 145).

To this point, studies had relied on the reporting of symptoms of the type expected with an upper respiratory tract infection. Further progress depended on drawing a careful distinction between infections and similar symptoms arising from inflammatory or allergic responses to exercise (160). One important epidemiological landmark was the application of modern quality criteria to 28 published studies (183). Only 7 of the 28 reports met current standards of epidemiological excellence in terms of a careful categorization of physical activity, adequate experimental design and methodology, and a confirmed diagnosis of upper respiratory infection (table 21.1). Even these 7 papers did not satisfy all of Bradford Hill's criteria of a causal relationship. Causality demands a strong, consistent, temporally correct and specific relationship; a clear dose–response curve; a plausible hypothesized mechanism; a coherent explanation of the data; experimental verification; and demonstration of analogous processes in animals or in vitro (183). However, the likelihood of reporting an increased risk of infection after exercising seemed just as great in weaker studies as in well-designed investigations.

Figure 21.2 David Nieman, professor of health, leisure, and exercise science at Appalachian State University, has been one of the main driving forces behind the growth of interest in exercise immunology in North America. Particular contributions have included the demonstration of a sixfold increase in self-reported upper respiratory infections after participation in a marathon run, exploration of the changes in blood leukocyte levels associated with dynamic endurance exercise, and development of the hypothesis of a J-shaped relationship between exercise dose and susceptibility to infections of the upper respiratory tract.

Photo courtesy of David Nieman.

Dose–Response Relationships

As noted previously, a period of very heavy training or a very prolonged bout of strenuous dynamic exercise is needed for an adverse effect. Available information suggests that moderate dynamic exercise may even lead to some improvement in a person's resistance to infection. Although detailed data are still lacking, the hypothesis of a J-shaped relationship between the amount of exercise performed and the risk of infection has been repeatedly proposed by David Nieman and colleagues (138, 139). The concept gained credibility in part because it seemed consonant with the general public health finding that moderate physical activity improves many aspects of health whereas excessively prolonged or strenuous activity—particularly in those who are ill prepared—can have adverse health consequences. Nevertheless, caution is needed in inferring details of the shape of the relationship from an amalgam of studies that differ in subject characteristics, methods of investigation, and rigor of analysis (183).

Nutritional and Pharmacological Factors Modifying Risk

The potential for athletes to modify their risk of developing upper respiratory infections through the use of dietary supplements or specific medication is discussed later in this chapter.

Summary

In some way, strenuous dynamic physical activity appears to modify the susceptibility of the body to the

Table 21.1 Critical Epidemiological Analysis of Studies Examining the Influence of Sport, Physical Activity, and Training on Susceptibility to Upper Respiratory Infections

Author	Population and study type	Type and intensity of exercise	Susceptibility to infection	Study ratings (numbers in parenthesis indicate maximum possible score for a given rating)			
				PA (14)	URI (8)	M (16)	O (36)
Weidner et al. (1997, 1998)	34 M and F, 18-29 yr	40 min at 70% heart rate reserve 3 d/wk (M)	No change in duration or severity of experimental URI	13	7	10	30
Douglas and Hanssen (1978)	61 M rowers and 126 unfit M cadets, 18-26 yr Prospective study	University rowing	Validated checklist Increased frequency and severity of URI in rowers	12	5	6	23
Verde et al. (1992)	10 M Prospective study	Running, 3 wk 38% increase of training	Respiratory symptoms in 3 out of 10	14	2	7	23
Nieman et al. (1990a)	36 mildly obese F, 25-45 yr Prospective study	45 min at 60% heart rate reserve 5 d/wk for 15 wk	Reduced duration of URI symptoms compared with controls (3.6 vs. 7.0 d/infection); duration vs. Δ fitness ($r = .37$)	13	5	4	22
Nieman et al. (1990b)	2,311 M and F, 35-37 yr Retrospective and prospective studies	Marathon runners	Twofold increase in URI with training; sixfold increase in URI in faster runners 1 wk after race (12.9% vs. 2.2%)	8	4	8	20
Schouten et al. (1988)	92 M and 107 F, 20-23 yr Retrospective study	Questionnaire: activity level and fitness	No effect of activity in males; URI less in active females	12	2	3	17
Nieman et al. (1993)	32 sedentary F, 12 active F, 67-85 yr Cross-sectional study	Walking 37 min 5 d/wk; fit vs. callisthenic controls	Lower incidence of URI in fit subjects (9%) and walkers (21%) relative to controls (50%)	11	2	2	15
Shephard et al. (1995)	551 M and 199 F, 40-81 yr Retrospective study	Masters athletes	URI increased in 16% if distance >70-80 km/wk	9	1	3	13

Heath et al. (1991)	447 M and 83 F, 13-75 yr Prospective study	Distance running Logistic regression	URI (monthly log) increased if distance >15 km/wk	4	5	4	13
Kingsbury et al. (1998)	14 M and 38 F Retrospective study	39 track and field, 12 judo, 1 rower; all international	10 out of 19 with chronic fatigue and reduced training had URIs; only 1 out of 33 able to maintain training	8	3	2	13
Karper and Boschen (1993)	6 M and 10 F, 60-72 yr Prospective study	Moderate exercise 3 d/wk for 9-12 mo	Reduced number of infections relative to initial state	6	3	2	11
Nieman et al. (1989a)	183 M and 90 F Retrospective study	Distance runners	No increase in URI 1 wk after run (5, 10, 21.1 km) compared with URI 2 mo previously; flu in 6.8% of 21.1 km runners and 17.9% of 5 and 10 km runners; training 42 km/wk led to lower URI than training 12 km/wk	4	2	5	11
Peters and Bateman (1983)	145 M and 5 F, 18-65 yr Prospective study	Ultramarathoners vs. nonrunners	URI symptoms increased compared with controls (33% vs. 15%) 2 wk postrace, especially in faster runners (questionnaire and interview)	4	0	7	11
Peters (1997)	Pooling of 3 studies; 567 subjects	Ultramarathoners	Postrace URI symptoms related to training status and race intensity; greater if no vitamin A or C supplementation				
Nieman et al. 1989b)	9 M and 1 F, 27-44 yr Retrospective study	Marathon runners	7 perceived increased risk of URI after competition	8	1	0	9
Green et al. (1981)	20 M, 23-46 yr Retrospective study	Marathon runners	9 out of 20 perceived decreased risk of URI; 1 out of 20 perceived increased risk	7	1	0	8
Österback and Qvamberg (1987)	76 M and 61 F, 11-14 yr Prospective study	Interview: school sport involvement vs. controls	No difference from controls (nurse interviews)	2	3	2	7
Budgett and Fuller (1989)	69 M, 18-33 yr Retrospective study	30 international rowers, 39 other rowers	Incidence of URI 1.4/yr	5	1	1	7
Linde (1987)	55 M and 28 F, 19-34 yr (44 orienteers) Prospective study	Orienteers vs. nonorienteers	Incidence (2.5 vs. 1.7) and duration (7.9 d vs. 6.4 d, ns) of URI greater in orienteers	3	0	4	7

(continued)

Table 21.1 *(continued)*

Author	Population and study type	Type and intensity of exercise	Susceptibility to infection	Study ratings (numbers in parenthesis indicate maximum possible score for a given rating)			
				PA (14)	URI (8)	M (16)	O (36)
Mackinnon et al. (1996)	8 M and 16 F, 15-26 yr Prospective study	Elite Australian swimmers	Risk of URI increased at peak training (10 out of 24 infected in 4 wk)	5	1	1	7
Strauss et al. (1988)	87 M Prospective study	University athletes (wrestling, swimming, gymnastics)	86% report URI over 8 wk (weekly interviews)	2	2	2	6
Lee et al. (1992)	96 Air Force cadets Prospective study	After initial training vs. recruitment	Immune function depressed but no change in URI (questionnaire and medical records)	2	2	1	5
Berglund and Hemmingson (1990)	121 M and 53 F, 16-21 yr Prospective study	38 national cross-country skiers vs. 136 at sport colleges	Diary; no difference in frequency or distribution of infections	2	0	2	4
Linenger et al. (1993)	482 M trainee months Prospective study	Special warfare training	High incidence of physician-diagnosed URI	2	1	1	4
Seyfried et al. (1985)	8,000 subjects, all ages, 0-70 yr Prospective study	Recreational swimmers vs. nonswimmers	Increased URI relative to nonswimmers	2	1	0	3
Moolla (1996)	88 subjects	Ultramarathoners	No increase of URI postrace, but incidence reduced by beta carotene and vitamin C	Details not available			

For details on references, consult 183.

M = Males; F = Females.

Studies have been arbitrarily ranked in terms of four criteria: precision of categorization of physical activity (PA), diagnosis of upper respiratory tract infections (URI), experimental design and methodology (M), and overall quality (O).

Reprinted from R.J. Shephard, 2000, "Special feature for the Olympics: Effects of exercise on the immune system: Overview of the epidemiology of exercise immunology," *Immunology & Cell Biology* 78: 485-495. By permission of R.J. Shephard.

acute stage of infection (27). The underlying mechanisms remain under debate. As transient exercise-related changes in circulating immune cell counts became defined, it was argued that a decrease in natural killer (NK) cell count might have created an open window during which it was easier for microorganisms to penetrate the body's defenses. However, the changes in circulating NK cell count are short lived (often no more than a few hours), and it seems doubtful that they could account for a two- to eightfold increase in the risk of respiratory infection over a period of several weeks. Immunologists have more recently considered changes in mucosal immunoglobulin levels and changes in cluster of differentiation (CD-4) (helper) T-cell populations with severe exercise. Plainly, nonimmunological factors also potentially make contributions. Mouth breathing might lead to a bypassing of nasal filtration mechanisms during vigorous exercise. Exercise also causes local changes in blood flow, depletes body stores of some nutrients, and increases concentrations of reactive species, prostaglandins, and sometimes endotoxins. Finally, there may have been exposure to a common source of infection: Swimmers may have used an infected pool or athletes may have shared water bottles on the sport field. In one specific episode in which 90 of 97 football players developed viral hepatitis, all were found to have drunk from a contaminated water supply (129).

Technology and Leukocyte Responses to Exercise

In many areas of biological science, the progress of knowledge has been propelled not simply by human curiosity or serendipity but also by the advent of new technology. Thus, cardiorespiratory physiology made enormous strides during World War II. High-altitude research laboratories replaced the traditional straw levers and smoked drums by recording galvanometers that were linked to sophisticated electronic transducers of respiratory and intravascular pressures, and the tedious chemical analysis of gas samples was replaced by continuous infrared, paramagnetic, and mass-spectroscopic determination of gas concentrations. Similar forces have played an important role in shaping the history of exercise immunology. Important technical developments for the cellular immunologist have included automated cell counters and sorters as well as radioactive techniques for the accurate measurement of cell toxicity and applications of molecular biology for examining the metabolism of immune cells.

In her introduction to the second meeting of the International Society of Exercise Immunology, Bente Klarlund Pedersen (figure 21.3) drew attention to an early publication by Schulte (1893), who had commented that a bout of exercise gave rise to a leukocytosis. Larrabee also noted a three- to fivefold increase of blood leukocyte count in four runners after completion of a marathon run (104). These early studies relied heavily on differential cell counts that had painstakingly used a light microscope and a graticule to count 100 to 200 cells as seen in eosin-stained blood smears. The idea of automated cell counting dates back to 1934 (125). By 1947, Wallace Coulter had developed a practical device for detecting the number and size of particles suspended in a fluid. The apparatus, now known as the Coulter counter, was patented in 1953 (39) and became widely available in clinical laboratories beginning in the late 1960s.

In brief, the apparatus detects changes in electrical conductance as a cell-containing fluid is drawn through a small aperture. Because the cells are essentially nonconducting, they alter the effective cross-section of the aperture and thus its conductance. The critical technical issue is to allow the particles to pass through the orifice in a reasonable time. Electrophoresis and electrolysis proved to be unsuccessful approaches, but the problem was overcome by using two chambers separated by multiple apertures. Results obtained by the automated

Figure 21.3 Bente Klarlund Pedersen is professor in the department of infectious diseases and the Muscle Research Centre at the University of Copenhagen. She served as president of the International Society for Exercise Physiology from 1995 to 1997 and is one of the leading investigators in exercise immunology in Scandinavia. Her particular contributions include the concept that the immune response to exercise provides a useful general model of immune responses to stress and her demonstration that muscle production of a cytokine (interleukin-6) plays a major role in the regulation of metabolism during exercise.
Photo courtesy of Bente Pedersen.

counter correlate well with manual counts of total leukocytes, lymphocytes, and neutrophils (7).

Automatic sorting of cell subtypes became possible with the development of the fluorescence-activated cell sorter and appropriate monoclonal antibodies. The idea of measuring the fluorescence of particles in a fluid stream dates back to 1969 (46). Use of a mercury lamp allowed cell markers to be detected in the ultraviolet spectrum. Electronic devices that sorted cells by volume soon made their appearance (60), and absorbing dyes were quickly replaced by fluorescent markers linked to monoclonal antibodies. By the early 1970s, the first versions of the Becton-Dickinson fluorescence-activated cell sorter (FACScan) were introduced (61, 87, 124). In current equipment, light from one or more laser beams impinges on the stream of suspended cells. The forward scatter of the light beam reflects cell size, and lateral scatter provides information about the surface form of the cells, their membrane folding and granularity, and the intensities of green, orange, and red fluorescence. After photomultiplication, data for scatter and fluorescence can be gated, plotted against each other, and compared with responses to standard calibration beads (figure 21.4). Over the past 20 yr, an ever-growing range of monoclonal antibodies has been formulated. This has allowed very precise study of the surface antigens on various leukocytes and the manner in which surface characteristics change during and after exercise (61, 124, 171). However, procedures remain quite time consuming and costly in terms of reagents, so studies are still based on relatively small samples of subjects, often without controls. Typically, pre-exercise data are compared with two to six blood samples collected during and after an exercise bout. Figure 21.5 illustrates one such study in which young adults performed 1 h of moderate dynamic exercise on a cycle ergometer at 60% of maximal aerobic power. A number of early reports also neglected to make allowance for the 5% to 10% hemoconcentration associated with sustained vigorous exercise. Initial studies focused on cell numbers, but subsequent investigations have examined changes in the functional activity of the cells and their adhesion molecules as well as changes in resting characteristics and responses of the immune system to an acute bout of exercise after several months of endurance training.

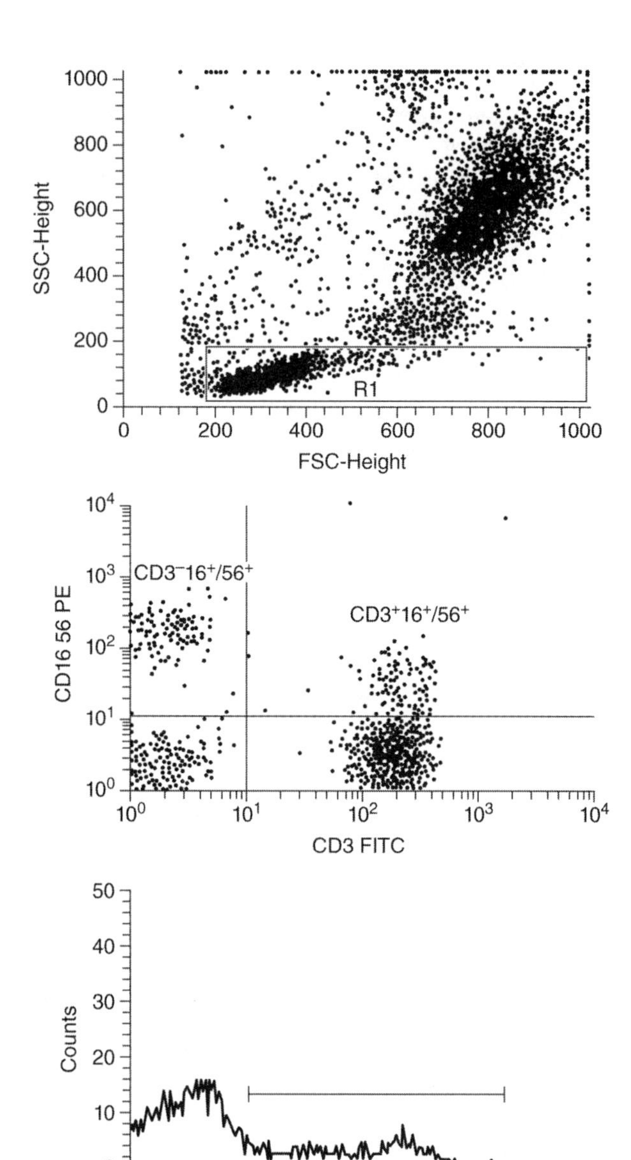

Figure 21.4 Introduction of the FACScan apparatus greatly facilitated identification and counting of various types of leukocyte. In this example, it is determining the natural killer cell count. Panel 1: Cells with a substantial forward scatter (horizontal axis) but little side scatter (vertical axis) are identified as box R1. These are the lymphocytes. Panel 2: CD3 fluorescence is plotted for the lymphocytes, allowing the investigator to distinguish the $CD3^+$ population (right side of diagram). Panel 3: A histogram of CD16 fluorescence is plotted in order to count the number of $CD3^+CD16^+CD56^+$ (natural killer) cells (to left of marker bar).

Reprinted, by permission, from R.J. Shephard, 1997, *Physical activity, training and the immune response* (Carmel, IN: Cooper Publishing Group), 40.

Neutrophils

Because the neutrophils form the largest fraction of the circulating leukocyte population (40%-75% of total cells), exercise-induced increases in neutrophil count—a 60% increase with 5 min of moderate dynamic exercise on a cycle ergometer (2) and a somewhat smaller increase with 25 min of treadmill walking (2, 44)—were noted even before the introduction of automated counters. It was quickly recognized that such changes reflected a demargination of cells from reservoir sites, particularly in the lungs and bone marrow, in response to changes in cortisol and complement concentrations (43, 170).

The past 20 yr have seen many reports on changes in the activity of neutrophils during and after various types of exercise (activation, adherence and demargination, chemotaxis, phagocytosis, and microbicidal activity) (182), but little consensus has been reached on the extent or even the direction of change. One issue leading to intraobserver disagreement has been the timing of blood samples. Any increase in circulating neutrophil count seems a late phenomenon, possibly a reaction to tissue injury. Studies have also varied in the type, intensity, and duration of exercise bouts, resulting in variations in the cortisol response. Strenuous bouts of exercise may also spur some increase in neutrophil activity. The intracellular generation of superoxide anions can now be monitored by FACscan equipment; this may provide clearer information on this issue (159).

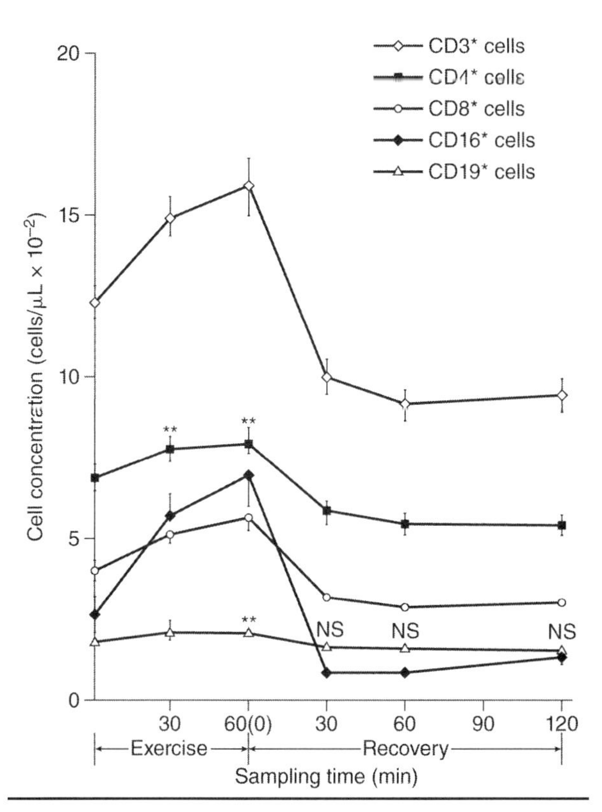

Figure 21.5 Changes in blood concentrations of specific leukocyte populations during and after 1 h of dynamic cycle ergometer exercise at 60% of maximal aerobic power.

Reprinted, by permission, from S. Shinkai, S. Shore, P.N. Shek and R.J. Shephard, 1992, "Acute exercise and immune function: Relationship between lymphocyte activity and changes in subset counts," *International Journal of Sports Medicine* 13: 452-461.

Eosinophils

Early manual counting found a decrease of eosinophils after a marathon run (104), increases in response to moderate sustained dynamic exercise on a cycle ergometer (2, 44) and a 100 m run (30), and no change of cell numbers in response to a basketball game (92). More recent data have continued to show both increases and decreases in eosinophil counts with no consistent pattern. One study of 15 healthy adults showed a 250% increase after participation in a triathlon run (190). Because eosinophils are associated with the dampening of allergic reactions, an increase in eosinophil count may reflect in part an allergic response to exercise.

Basophil Counts

One early study found a 207% increase in basophil count with 5 min of moderate dynamic exercise (2), but most subsequent authors have found little change in basophil counts in response to exercise.

Monocytes and Macrophages

Early reports that dynamic exercise increased the monocyte count (2, 3, 44, 50) have been supported by more recent investigations. FACScan studies of surface antigens have confirmed the view that the increase in cell numbers reflects a trafficking of cells from tissue depots. Circulating counts subsequently decline as the cells migrate to sites of tissue injury (62).

Macrophage activity is important to both the ingestion of bacteria and the secretion of various cytokines. Early studies measured activity in terms of insulin binding (95) or glucose metabolism (15). More recent approaches have exploited new techniques for detecting adherence, the proportion of mature cells, and cytokine production and have made direct observation of phagocytosis (particularly in animal studies). Almost all techniques point to the main conclusion: Monocyte activity is increased with short bouts of moderate dynamic exercise but tends to be depressed by very strenuous or very prolonged exercise (182).

Lymphocytes

Study of the various lymphocyte subsets—T-helper and T-suppressor cells, cytotoxic T-cells, B-cells, and NK cells—began as FACScan equipment became available to exercise scientists. Lymphocyte proliferation rates have generally been assessed by the incorporation of tritiated thymidine (122). Exercise-induced changes in total activity tend to match changes in cell count (191).

Overall T-Cell Count

Large increases in overall T-cell count have generally been seen and are greater with intense or prolonged exercise than with moderate dynamic exercise (191). FACScan data generally confirm older results based on blood smears (67). Disparate reports probably reflect limited or inappropriate timing of blood sampling. More recent research has focused on identification of various subpopulations, including CD4 (helper) and CD8 (suppressor) cells and variants in T-cell antigen receptor expression (normally alpha–beta chains, but gamma–delta in 5% of T-cells).

CD4:CD8 Cell Ratio

The first investigators to examine the separate responses of CD4 (T-helper and memory) cells and CD8 (T-cytotoxic and suppressor) cells found a greater increase in CD8 cells than CD4 cells immediately after dynamic exercise (14, 100). More detailed analyses confirmed this pattern, but in the 2 h after a bout of strenuous exercise the CD4:CD8 ratio increased to greater than its resting value. In one study this increase persisted for as long as 7 d (176).

The classical T helper cells (now termed T_{h1} cells) have a proinflammatory action, but advances in the identification of surface markers have revealed a second population of T helper cells (termedT_{h2} cells) that have an anti-inflammatory action. Examination of differential responses to exercise began with Rabin and colleagues (161). Studies of T-cell proliferation and cytokine production have offered one insight into how moderate dynamic exercise may enhance resistance to infection. Access to an exercise wheel has been shown to increase both T-cell proliferation and cytokine production in mice in response to subcutaneous or intranasal administration of a model antigen (ovalbumin) (168). Current work also suggests that exercise-induced changes in the relative proportions of T_{h1} and T_{h2} cells may increase susceptibility to infection and contribute to the picture of overtraining.

B-Cells

Early studies of B-cells used the rosetting technique (15, 78). Unfortunately, this did not distinguish clearly between B and NK cells. FACScan methodology has thus demonstrated much smaller exercise-induced increases in B-cell count than were reported by early authors (45).

Beginning with Hedfors and colleagues (79), exercise immunologists have studied exercise-induced changes in the functional activity of B-cells in terms of the in vitro pokeweed mitogen-stimulated production of immunoglobulins. As understanding of interactions has increased, investigators have appreciated that a decrease of immunoglobulin production could reflect either a decrease in B-cell activity or a decrease in CD4:CD8 cell ratio and thus a lesser availability of regulatory cytokines.

NK and Other Cytotoxic Cells

Developments in FACScan technology have allowed the study of both NK cells and a subset of cytotoxic T-cells capable of lysing microorganisms without antigen priming (45). Although circulating concentrations of NK cells were increased during most types and intensities of exercise (45), multiple sampling showed a substantial decrease immediately after a bout of moderate or vigorous dynamic exercise (191). At first, this was thought to offer a simple immunological explanation for why such exercise increased vulnerability to upper respiratory infections. However, in most studies, blood concentrations of NK cells were depressed for no more than 6 to 24 h—a rather short time to cause any substantial increase in the risk of viral infection. One study did report a depression of cell count that persisted for at least 7 d (176), but these observations have not yet been replicated.

Assessments of cytotoxicity depend on success in isolating the NK cells. The original rosetting technique was essentially a form of negative selection: B-cells were removed from the suspension by adherence to nylon wool and T-cells were removed by panning with anti-CD3 (38). More recently, this approach has been replaced by a form of positive selection. The NK cells are coated with biotinylated specific antibody and are then passed through a column containing the biotin-binding substance avidin linked to polyacrylamide beads (12). Ferric oxide coating and a magnet are sometimes used to complete the cell separation (163).

The cytotoxicity of isolated NK cells was traditionally assessed by the release of ^{51}Cr from myeloid tumor cells (148, 158). However, this method was time consuming, and flow cytometric methods are now preferred (32). Although depressed cytotoxic activity has been seen immediately after most forms of exercise, this seems to largely reflect the decrease in circulating NK cell count noted previously. Other suggested bases for the modulation of cell activity include cell trafficking (110), prostaglandin secretion (152), and the release of endogenous opioids (56).

Receptor Structures

As understanding of the immune system developed, it became apparent that the changes in cell count observed during and after exercise largely reflected a modulation of cell adhesion molecules and a resulting sequestration or demargination of cells (204). Heinrich Liesen and German colleagues were quick to observe that cell redistribution was correlated with norepinephrine levels, which were increased by strenuous exercise and other forms of stress (193). They speculated that the spleen might be involved in cell sequestration. However, a study of patients who had undergone splenectomy because of trauma found no alteration in immune responses to exercise (10). The early German investigations, led by Holger Gabriel, gave rise to an extensive study of hormonal modulation of adhesion molecules (185). An ever-growing number of cytokines were identified, leading to an expanding study of the manner in which their respective receptor sites were modified during and after a bout of exercise (108, 165).

Summary

Many forms of physical activity lead to substantial immediate changes in various components of the circulating leukocyte count. However, multipoint sampling has established that most of these changes are reversed quite quickly after cessation of exercise. They largely reflect a temporary demargination of cells from reservoir sites in response to the release of stress hormones during vigorous exercise.

Humoral Responses to Exercise

The first of the humoral elements of the immune system to attract the attention of exercise immunologists was the production of immunoglobulins, but an increasing range of cytokines drew intensive scrutiny as they were identified. The complement cascade is also modified by exercise, but it has as yet received much less attention. As with cellular elements of the immune system, technical developments spurred these investigations and allowed a progressively more specific examination of exercise and training responses. Methods for identifying and quantifying concentrations of specific proteins have included gas–liquid chromatography, enzyme-linked immunoabsorbent assay (ELISA), and western blot technologies. Unfortunately, the elements under investigation have often remained close to the detection thresholds of the available assay methods.

Immunoglobulins

Early studies noted that physical training or sport participation enhanced the production of antigens in response to specific antibodies such as diphtheria (49) and tetanus toxoid (54). Observations on overall antigen levels focused on responses to intradermal injections (110) and the injection of foreign erythrocytes (69) or neutrophils (107). More recently, globulin concentrations in plasma, saliva, and mucosal secretions have generally been assessed by ELISA.

One early study found small increases in serum immunoglobulins IgA and IgG but not IgM in response to a progressive cycle ergometer test (157). More commonly, no change in serum immunoglobulin levels has been reported after acute bouts of vigorous submaximal dynamic exercise (71), although early Russian reports of decreases during periods of particularly strenuous endurance training (155) have been confirmed by more recent observations on swimmers attending the Australian Institute of Sport in Canberra (114).

Attention was subsequently directed to the mucosal immune system, which offers the primary physical, biochemical, and immunological barrier to invasion by most microorganisms. A study of nasal IgA in cross-country skiers found low levels before an event and a further decrease after competition (201). Subsequent research has generally confirmed a depression of nasal IgA concentrations in response to periods of heavy training and competition (196); this has offered the most convincing explanation to date of any exercise-induced increase in vulnerability to upper respiratory infections (table 21.2).

Cytokines

Cytokines act mainly on closely adjacent cells, and plasma concentrations thus tend to be minute. Their accurate assay is challenging due to complicating factors such as their strong binding to receptors, short half-life, and possible neutralization by circulating inhibitors. Assay methods have included radioimmunoassay, ELISA, competitive binding to a receptor molecule, and, most recently, the transcription of cytokine messenger ribonucleic acids using ribonucleic acid isolation kits and the molecular biology technique of reverse transcription polymerase chain reaction assay (126, 165). This last approach has raised the issue that circulating cytokines may arise from sites other than circulating leukocytes. The cellular response often depends on the simultaneous presence of other cytokines or cell stimulants. Therefore, the in vitro production of cytokines by mitogen- or phytohemagglutinin-stimulated cells has not always mirrored the changes observed in vivo. Detailed investigations by Bente Klarlund Pedersen (151) demonstrated that a bout of strenuous exercise typically induces a cascade response involving a substantial number of interacting cytokines (figure 21.6).

Factors identified as influencing the in vivo pattern of cytokine production include the relative numbers of T_{h1} and T_{h2} cells, body temperature, and any prosta-

Table 21.2 Changes of Salivary Immunoglobulin A Concentration After an Acute Bout of Exercise

Author	Population	Duration and type of exercise	Change in immunoglobulin
Sustained moderate-intensity dynamic exercise			
McDowell et al. (1991)	9 fit M 9 unfit M 9 unfit M	20 min treadmill, 50% $\dot{V}O_2max$ 15-45 min treadmill, 60% $\dot{V}O_2max$ 20 min treadmill, 65% $\dot{V}O_2max$	No change per milliliter No change per milliliter No change per milliliter
Mackinnon and Hooper (1994)	18 M and F	40-90 min treadmill, 55% $\dot{V}O_2max$	No change per milligram of protein
Sustained vigorous dynamic exercise			
Cameron and Priddle (1990)	13 fit M	10.5 km run	Decreased
Housh et al. (1991)	9 fit M	30 min treadmill, 80% $\dot{V}O_2max$	No change per milliliter
McDowell et al. (1991)	9 unfit M	20 min treadmill, 75% $\dot{V}O_2max$	No change per milliliter
McDowell et al. (1992)	24 M	Submaximal treadmill running	Decreased
Mackinnon et al. (1992)	12 fit F	Hockey competition or training	Decreased per milligram of protein
Mackinnon and Hooper (1994)	18 M and F	40-90 min treadmill, 75% $\dot{V}O_2max$	Decreased concentration, no change per milligram of protein
Gleeson et al. (1995)	26 swimmers 18 controls	Routine training session Routine exercise session	Decreased No change
Very prolonged dynamic exercise			
Tomasi et al. (1982)	8 M and F	Cross-country skiing (50 km M, 20 km F)	IgA −40%
Mackinnon et al. (1989)	6 M	120 min cycle ergometer, 70%-80% $\dot{V}O_2max$	IgA −65%; IgM −55%/mg of protein
Muns et al. (1989)	33 M 15 M	31 km run Marathon run	Decreased Decreased
Tharp and Barnes (1990)	21 M	120 min routine swim training	Decreased
Steinenberg et al. (1997)	42 triathletes	Triathlon competition	Decreased
Nehlsen-Cannarella et al. (2000)	20 F rowers	120 min training session	Decreased

Brief maximal or submaximal dynamic exercise			
Schouten et al. (1988)	84 fit M, 91 fit F	15 min treadmill, 100% $\dot{V}O_2max$	Increase in M; decrease in F
McDowell et al. (1992)	29 fit M	16-40 min treadmill, 100% $\dot{V}O_2max$	−17% to 24%/ml
Other types of exercise			
Tharp (1991)	27 prepubescent M 23 postpubescent M	Basketball game Basketball game	+2% +3%
McDowell et al. (1992)	9 M	3 sets, 6-10 reps of resistance exercise	No change in concentration; IgA flow +25%
Mackinnon and Jenkins (1993)	12 M	5 × 1 treadmill intervals, 110% $\dot{V}O_2max$	−21% IgA; −23% IgG
Mackinnon et al. (1993)	8 M	30 min interval kayak training	−27% to −38%

For details on references, consult 68 and 182.
M = Males; F = Females.

glandins secreted in response to tissue injury. Joe Cannon and colleagues noted that 1 h of dynamic exercise on a cycle ergometer at 60% of maximal aerobic power increased monocyte secretion of interleukin (IL)-1, an important mediator of defense reactions to stresses imposed by the environment and disease (28). Based on these observations, the idea emerged that excessive physical activity, particularly eccentric exercise, could mirror to a lesser degree the serious changes associated with trauma and sepsis. The work of Camus and colleagues in Belgium (26) and Northoff and colleagues in Germany (147) demonstrated that such a reaction, initiated by an increased macrophage production of IL-1, could temporarily depress the immune response to infections (figure 21.7). A parallel to sepsis was seen in the penetration of the gut endothelium by gram-negative microorganisms during intensive exercise (18); this led to increased plasma concentrations of lipopolysaccharides. The hypothesis proved sufficiently intriguing that in 1997 Toronto investigators hosted an entire colloquium devoted to this theme (84).

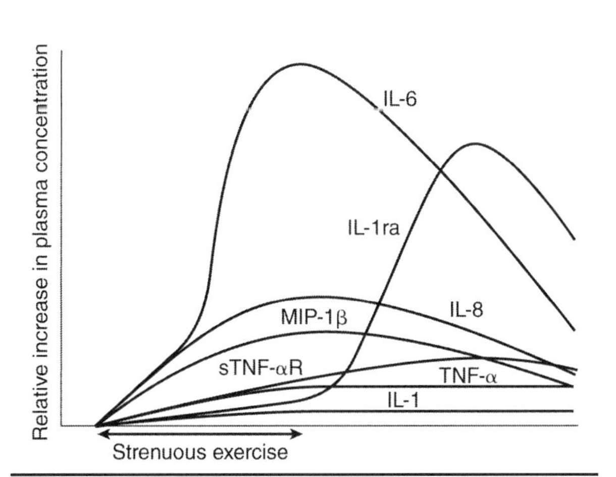

Figure 21.6 Sequential changes in plasma cytokine concentrations in response to strenuous dynamic exercise. IL-1ra = interleukin receptor antagonist; MIP-1β = macrophage inflammatory protein 1β; sTNF-αR = soluble tumor necrosis factor-α receptor.

Reprinted, by permission, from B.K. Pedersen and A.D. Toft, 2000, "Effects of exercise on lymphocytes and cytokines," *British Journal of Sports Medicine* 34: 246-251.

Complement

An early report found an increase of total plasma complement titer in response to 20 min of dynamic exercise on a cycle ergometer at a relatively strenuous power output averaging 216 W (51). This study neglected to correct for possible exercise-induced changes in plasma volume, but the majority of more recent studies have confirmed that strenuous exercise activates the classical complement cascade.

Acute Phase Proteins

An increase of acute phase reactants with both strenuous short-distance exercise and endurance training was an early finding of German investigators (74, 110). This pointed to at least a temporary exercise-induced enhancement of lytic activity and spurred interest in associated production of the cytokine IL-6 and an aseptic inflammatory response to exercise (146). More recently, moderate endurance training has been found to reduce inflammatory reactions in young

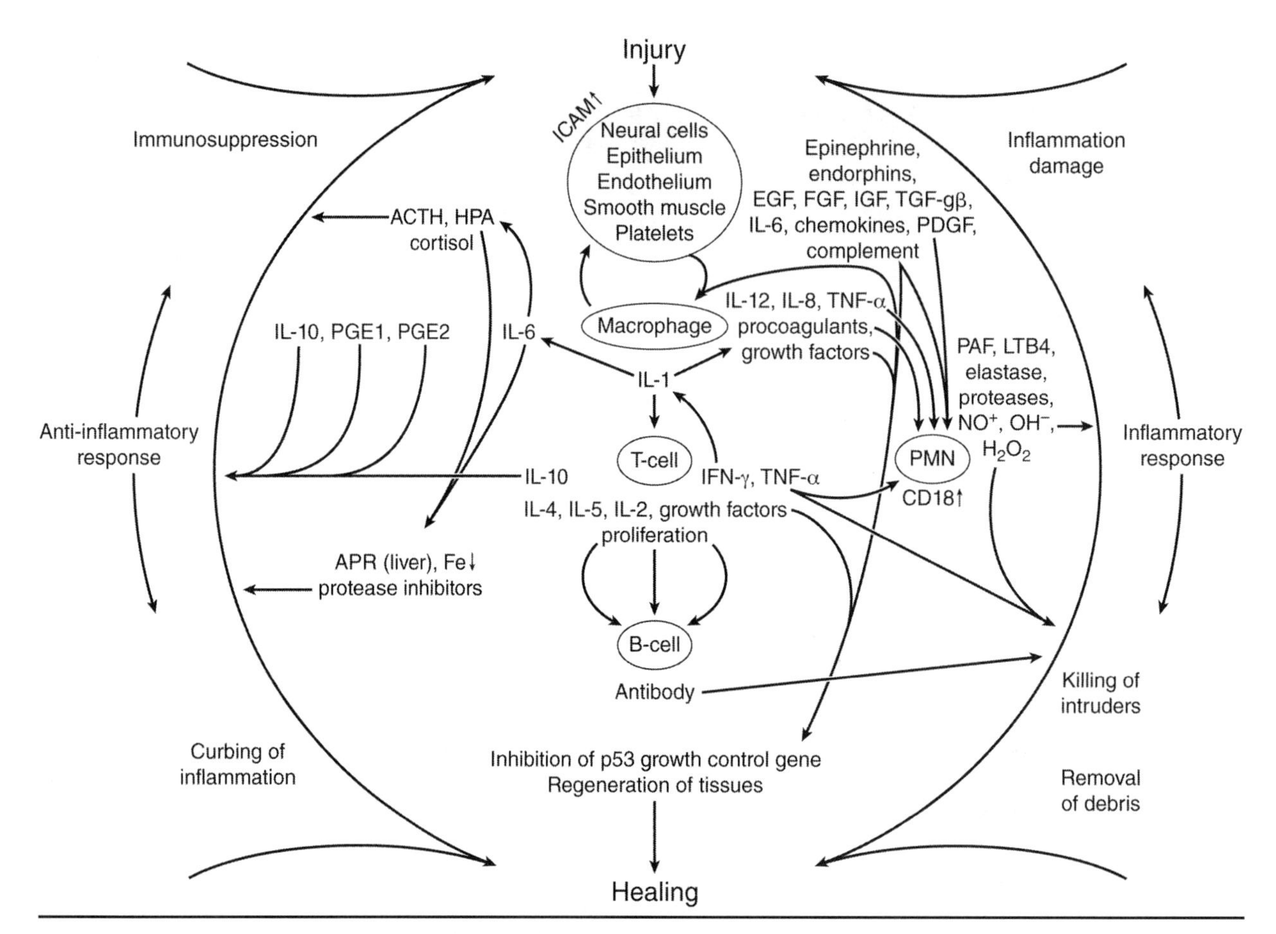

Figure 21.7 Some of the multiple mechanisms involved in the healing process.

Reprinted from H. Northoff, S. Enkel, and C. Weinstock 1995, "Exercise, injury and immune function," *Exercise Immunology Review* 1: 26-48. By permission of H. Northoff.

people (197) and the acute phase response to exercise in patients with intermittent claudication (200).

Summary

Strenuous dynamic exercise induces increases in circulating levels of complement, acute phase reactants, and a cascade of cytokines. The overall response to such exercise shows some parallels to a septic reaction. Recent research suggests that cytokines may also play a role in metabolic regulation. Very strenuous exercise and training depress serum and mucosal immunoglobulin levels; any impairment of mucosal immunity probably has the greatest significance for health.

Contribution of Animal Experimentation

A surprisingly large fraction of exercise immunology research has been based on studies of athletes and healthy volunteers. One reason for this choice is that it is difficult to involve laboratory animals in controlled amounts of voluntary exercise. Mice can exercise on activity wheels, but the cumulative effort is often relatively light and the energy expenditure is usually not precisely known. Rats can be persuaded to run on treadmills in response to electric shock or can engage in enforced swimming with weighted tails, but in either of these situations there is an element of stress that may affect the immune response as much as the exercise. Indeed, laboratory caging in itself imposes a substantial stress on most animals. With patient conditioning, animals can engage in resistive exercise that is rewarded by the delivery of food pellets. If small mammals are used, the volume of blood and other tissue samples that can be collected is quite limited. Finally, there are difficulties related to differences in life span and body size between humans and most animal models. Nevertheless, one advantage of experimental animals is that they can be readily assigned in random fashion to pre-selected exercise regimens; this is often difficult to arrange with human volunteers.

Certainly, animal experimentation has proven attractive and even essential to investigators in some important areas of exercise immunology. It is possible to examine the responses of an animal to the injection of various experimental toxoids and bacteria (4, 29, 49, 137), although it may be objected that such techniques as intratracheal injection of bacteria bypass primary mucosal defense mechanisms. Immune responses can be studied in parts of the body other than the blood stream, such as the spleen (83). Those with access to laboratory sheep have also been able to examine the distribution of lymphocytes among the blood, thymus, bone marrow, and lymph ducts (76). This is an important piece of information because samples taken from the blood stream represent only a very small (and often a biased) fraction of the total cell population.

Perhaps the most important topic of animal investigation for exercise immunologists is the study of interactions among physical activity, cancer, and the immune system. Laurie Hoffmann-Goetz of the University of Waterloo has for many years played a leading role in this area. Human studies have been almost entirely epidemiological, examining associations between an individual's physical activity history and the risk of neoplasm (184). The progress of human work has been hampered by the slow process of carcinogenesis, uncertainty about the relative importance of recent versus lifetime physical activity, and the continuing absence of any unifying theory about why such activity might reduce the risk of developing this disease (187, 210). Animal studies have allowed tumor cells to be transplanted (5) and tumors to be induced by injection of a potent carcinogen (198, 202) or a massive number of tumor cells (118) at various points before and after the onset of regular physical activity. Early studies noted a parallel between antibody levels (judged by a plaque-forming response to foreign erythrocytes) and a significant delay of tumor growth in exercised animals (69). Advances in the use of animal treadmills have also allowed some exploration of dose–response relationships. Current evidence suggests protection (a 37% reduction of incidence and a 60% decrease of multiplicity) against chemically induced tumors at a treadmill speed corresponding to 70% of the animal's maximal oxygen intake but no response at 35% of maximal oxygen intake. Surprisingly, the benefit is similar for daily exercise sessions that are 20 and 40 min in duration (199).

Although such findings are interesting, there are continuing problems in comparing such massive chemical and cellular challenges with the gradual process of spontaneous carcinogenesis that seems likely in most human tumors. Attempts have been made to counter some of these problems through the breeding of transgenic mice that have a high incidence of spontaneous tumors (47, 109). Research with such animals has recently led to the controversial conclusion that moderate or strenuous treadmill running for 22 or 45 min/d reduced survival times; strenuous exercise also shortened tumor latency and increased the multiplicity of tumors. The authors of this report hypothesized that either exercise had a negative effect on this particular strain of animal or *ad libitum* feeding countered any benefit from the exercise regimen (36). Another potential application of transgenic mice that will undoubtedly contribute to future knowledge of exercise immunology is the exploration of responses to physical activity in animals with deficiencies in parts of their immune system (17, 97).

Summary

Animal models have to date been helpful in providing tissue samples from areas other than the circulating blood and in allowing examination of potential interactions between exercise and exposure to dangerous bacteria, toxoids, and tumor cells. The study of transgenic animals holds promise as a method for exploring individual components of complex and closely linked immune reactions.

Nutritional and Pharmaceutical Implications

Attempts to avert an exercise-induced suppression of immune function have been hampered by lack of clear evidence on the underlying mechanisms and mediators (188). Proposed treatments have included immunoglobulins, nutritional supplements, vitamins, and prostaglandin antagonists.

Immunoglobulins

The first purified preparations of immunoglobulin for intravenous administration were developed by the Behring Company in 1979. Noting decreased circulating levels of the immunoglobulins IgM and IgG in athletes after stressful exercise (207), treatment by the administration of polyvalent immunoglobulins was advocated (205, 206). Beriglobin (Aventis-Behring, Marburg, Germany) was thus given to the German national boxing team before the 1992 Olympic Games (205-207). Although the study was uncontrolled, it was claimed that competitors who had been treated in this manner did not show the minor respiratory infections that had plagued many of their peers in previous major competitions. A study of Austrian athletes found that the oral administration of thymomodulin prevented the decline of IgG2 in response to strenuous training (64), and Scandinavian skiers who received a nasal instillation of IgA remained free of infection in the Alberville Olympics (in contrast with 9 of 19 coaches who developed infections) (81). However, a study of world-class canoeists found that although the nasal administration of IgA increased salivary IgA concentrations during 17 d of strenuous training, it did not lead to any significant decrease in respiratory symptoms relative to untrained and untreated control officials and students (111). Further progress on this issue may need more objective evidence of protection than subjective reports of respiratory symptomatology.

Diet

Newsholme aroused an early interest in dietary problems by advancing the hypothesis that the suppression of immune function was due to an exercise-induced decrease in plasma glutamine concentration (135). Glutamine is certainly needed for leukocyte metabolism, and a lack of glutamine has been accepted as a factor in the impairment of immune function in malnourished African populations. However, a glutamine deficiency seems unlikely in athletes who are consuming a well-balanced diet with protein content that is sufficient to maintain nitrogen balance (1.6 g/kg of body mass in many categories of athletes).

Studies from the Oxford laboratory of Newsholme found an average 20% decrease of plasma glutamine concentrations after a marathon run, and a controlled trial found a substantial reduction in the reporting of respiratory infections in marathoners who consumed two drinks each containing 5 g of glutamine relative to those given a maltodextrin placebo (19% vs. 51% of runners reported infections) (31). However, more recent research from other laboratories has generally failed to support either a substantial decrease in plasma glutamine levels in athletes during periods of strenuous training or changes in NK cell count and activity after the administration of glutamine supplements (116, 169).

Another plausible dietary mediator might be an exercise-induced depletion of muscle and liver glycogen reserves; such depletion might indeed lead to a depletion of plasma glutamine. Carbohydrate depletion in the endurance athlete was found to increase blood levels of stress hormones and appeared to exacerbate immunosuppression, possibly through the corticosteroid mechanism described in an early review by Cupps and Fauci (41). Nieman and colleagues thus tested the effects of providing a 6% carbohydrate supplement during 2.5 h of high-intensity treadmill running. Immediately and 1.5 h postexercise, larger increases of IL-6 and IL-1 receptor antagonist were observed in the placebo controls, suggesting that the carbohydrate drink had attenuated the release of cytokines in the inflammatory cascade (132). These observations on the effects of exogenous carbohydrate seem linked to the more recent discovery that the cytokine cascade is important to the release of glucose from intramuscular glycogen stores.

Studies in other areas of immunology have suggested an important role for essential fatty acids. However, there has to date been little study of the possible benefits of administering omega fatty acid supplements to athletes. A German study of 63 triathletes noted that plasma levels of cortisol and IL-6 were lower in competitors with high levels of polyunsaturated fatty acids during a period of intensive training (99). However, a heavy intake of fat could have had negative effects simply because it reduced the athlete's carbohydrate reserves.

Vitamins

One factor contributing to a depression of immune function during strenuous exercise could be an accumulation of reactive species. The source might be the oxidative process itself or muscle damage along with invasion of the injured tissues by neutrophils, or conceivably it could arise in the immune system as exercise increased both the numbers and the phagocytic activity of neutrophils to a self-destructive level. However, the depression of immune function seems to be unrelated to changes in neutrophil count or activity, and because it is usually most marked after eccentric exercise, minor muscle injury seems to be the likely explanation.

A double-blind placebo controlled trial conducted by Edith Peters in South Africa suggested that if a daily 600 mg supplement of vitamin C was administered during the supposed open-window period (3 wk after a marathon race), the incidence of upper respiratory tract symptoms was halved (154). However, the study has been criticized on the basis that the incidence of respiratory symptoms in the control group was unrealistically high. A subsequent study showed that the administration of a larger (1,000 mg/d) vitamin C supplement for 8 d did not modify the response to 2.5 h of treadmill running (140), although it could be objected that the environmental stress was less in the laboratory situation than in a marathon run. Other vitamin supplements have been proposed from time to time, but none have as yet shown any consistent benefit to the immune systems of individuals who are exercising intensely. Megadoses of vitamins A and E may even have toxic effects.

Table 21.3 Effects of Heavy Training on Select Immune Parameters

Parameter	Effect
Natural killer cell counts	Decreased
Natural killer cell activity	Decreased
CD4+:CD8+ cell ratio	Decreased
Lymphocyte count	Unchanged
Lymphocyte proliferation	Increased
Granulocyte count	Increased
Serum immunoglobulins	Decreased
Salivary immunoglobulins	Decreased

For supporting research, see references 115 and 182.

Prostaglandin Antagonists

The minor tissue injury associated with strenuous exercise releases prostaglandin E2. Early researchers appreciated that this substance reduced the response of mononuclear cells to mitogens (66). Prostaglandin release also appears to be a factor in the exercise-induced reduction in NK cell activity because this change can be averted both in vivo and in vitro by the administration of therapeutic doses of indomethacin (91). However, it seems that researchers have not conducted further therapeutic trials of prostaglandins, perhaps because the decrease in NK cell count is now thought to be of limited significance to health.

Summary

At various times, glutamine, carbohydrate, vitamins, and various mineral supplements together with prostaglandin antagonists have been proposed as possible methods for countering exercise-induced immunosuppression. However, to date the evidence supporting such tactics remains unconvincing.

Practical Applications in Exercise Science

What practical applications have the exercise scientist found for the growing understanding of relationships between physical activity and the immune response? Topics of particular interest include overtraining, new concepts of metabolic regulation, modeling of the inflammatory response, and interactions between exercise and environmental stressors.

Overtraining

Attempts have long been made to relate the exercise-induced suppression of immune function to the syndrome of athletic overtraining. Given the combination of very heavy training and the psychological distress associated with waning performance, a major disturbance of immune function might certainly be anticipated with overtraining. However, in practice any disturbances of immune function (table 21.3) have often been small and inconsistent. Evaluations of mood state and performance have provided a clearer basis of diagnosis and a better indication of the severity of the disorder (63, 123, 203).

One practical problem for investigators has been that the deliberate induction of overtraining is not ethically acceptable in human subjects. In experimental studies, the level of training adopted (although heavy) may thus have lacked the intensity needed to modify immune function. More has been learned by following teams of international athletes during periods of very intensive training leading up to major competition. Laurel Mackinnon and colleagues at the Australian National Institute for Sport have played a leading role in these investigations. Such studies have shown decreased leukocyte (105), lymphocyte (73, 117), and NK cell (65) counts as well as reduced levels of serum immunoglobulins (114).

New Concepts of Metabolic Regulation

Bente Klarlund Pedersen and colleagues have been active in innovative explorations of the associations between cytokine secretion and metabolic regulation.

They demonstrated that the IL-6 gene was activated in muscle during exercise and that this cytokine was released into the blood, particularly as glycogen stores became depleted. There was an associated stimulation of lipolysis and cortisol production as well as an inhibition of tumor necrosis factor (TNF) production (150). Infusion of small doses of recombinant human TNF-α also led to systemic lipolysis and an increased plasma level of free fatty acids (156).

Modeling of the Inflammatory Response

Interest in exercise as a model of immune responses to inflammation and trauma began in the early 1990s (26, 146) and peaked in 1996 with an international colloquium titled "Immune Responses to Inflammation and Trauma: A Physical Training Model" (84). This conference brought many of the world's leading exercise immunologists to Toronto. Researchers believed that the physical training model was important to investigate because it might throw new light on the clinical immunologist's problem of sepsis (26). Although there are some striking similarities between strenuous exercise and sepsis, important differences have also emerged (177). It is usually not possible to push the exercise model to the point of inducing a septic reaction, at least in humans; the inflammatory response seen to date has remained subclinical and well ordered. Therefore, the early promise of an experimental approach to sepsis has yet to be realized. There are continuing practical difficulties (i.e., low concentrations, short half-lives, and neutralizing factors in the circulation) in determining the cytokines concerned. Possibly, there may be a resurgence of interest in this question as molecular biologists facilitate the intracellular assay of cytokine secretions. Many clinical investigators have been attracted to more moderate intensities of dynamic exercise because of its apparent potential to enhance immune function and control unwanted inflammation in such conditions as autoimmune disease, aging, and cancer.

Interactions with Other Environmental Stressors

One growing strand of exercise immunology research has argued that exercise affects the immune system much like a variety of other stressors. There are many points of similarity between strenuous exercise (particularly athletic competition) and responses to other forms of environmental stress, including modulations of hypothalamic activity and plasma levels of the stress hormones epinephrine and cortisol (35, 149). It has thus been argued that exercise provides a useful experimental model of the immune response to stress (149), leading to important interactions between exercise immunology and the growing area of psychoneuroimmunology (the complex two-way interaction between the pysche and immune function that modulates many chronic diseases and conditions) (93, 175). Psychoneuroimmunologists have progressively underlined the close interrelationships between the immune system, the endocrine system, and the brain (90). An expanding volume of research has shown that regular physical activity has positive effects on mental health and on neuroendocrine and immune functioning (103, 113). A mechanism has thus been suggested whereby the psyche modulates exercise responses both in high-performance sport and in the treatment of a variety of chronic diseases, including cancer and autoimmune disorders (figure 21.8). The concept seems applicable to humans as well as animals; factors such as housing density and environmental enrichment modify the immune responses of laboratory mice (94).

In addition to the potent influence of the sympathetic nervous system, a growing range of neuropeptides (including vasoactive intestinal peptide and neuropeptide Y), pituitary neuroendocrine hormones (including adrenocorticotropin, growth hormone, prolactin, and endogenous opioids), and monoamines (e.g., serotonin) have all been found to modify immune function. Moreover, receptors for these compounds have been demonstrated on many types of leukocytes, and these cells themselves have been shown to produce neuropeptides, thus allowing a bidirectional communication between the brain and the immune system (16, 119). Interregulation appears to be highly redundant, and much further research is needed to define the relative importance of these various potential mediators.

Exercise is particularly stressful under conditions of international competition or military combat, and the resulting changes in the concentrations of various hormones inevitably have a major effect on immune function. Military laboratories have had a particular interest in this topic because troops may need to undertake very prolonged exercise in the face of combat as well as other stressors such as extreme heat or cold, hyperbaric environments, food shortages, and sleep deprivation (20). This has led to specific studies of interactions between exercise and the immune response at high altitude, during deep diving, and in very hot and very cold conditions (182).

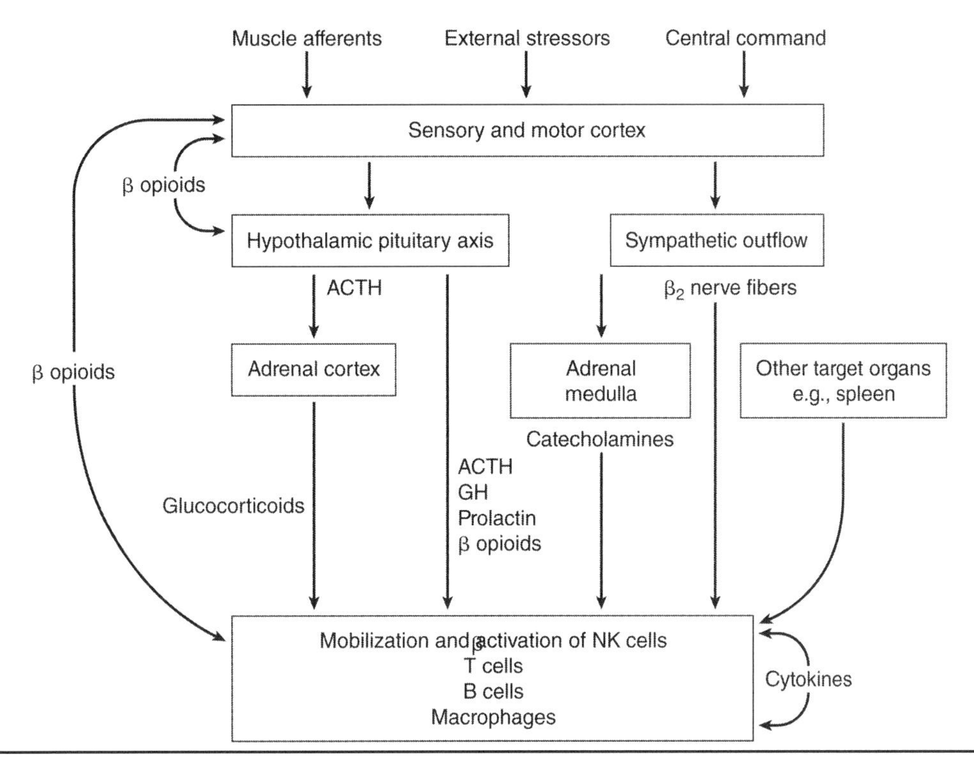

Figure 21.8 Potential pathways involved in the exercise-induced modulation of natural immunity include both the sympathetic nervous system and the neuroendocrine system. ACTH = adrenocorticotrophic hormone; GH = growth hormone.

Adapted from Jonsdottir 2000.

Practical Applications in Medicine

During the past 20 yr the findings of exercise immunologists have had many important applications in medical practice, and it is likely that more applications will be discovered in the future. We briefly discuss issues of aging, the prevention of upper respiratory infections, protection against cancer, and human immunodeficiency virus infections.

Aging

Many countries currently face the problem of a rapidly aging population, and it is surprising that there have been so few studies of interactions among aging, exercise, and immune function in either animals or elderly people (57, 189). Investigators have been hampered by the many concomitants of aging that can affect immune function, including poor nutrition, chronic depression, and a buildup of chronic infection and degenerative changes in various body organs. The elderly are certainly more susceptible to many diseases than their younger peers and show a reduced ability to adapt to a variety of stressors. It is now recognized that the aging immune system is challenged by a combination of impaired cellular function and increased inflammatory activity in many parts of the body. It remains less clear how far a deterioration of immune function is responsible for the increased prevalence of disease, although there is current speculation that an excessive inflammatory response may contribute to such problems as sarcopenia and Alzheimer's disease.

Characteristics of the exercise response noted in elderly individuals include an unchanged ability to recruit T-cells and NK cells, a reduced recruitment of neutrophils after eccentric exercise, less stimulation of lymphocyte proliferation by moderate dynamic exercise, and less suppression of proliferation by strenuous dynamic exercise. Highly trained individuals show greater cytokine production and (in some studies) show an enhancement of NK cell function (189). However, any possible benefits of rigorous training have yet to be clearly dissociated from the healthy lifestyle typical of the very active senior. Although one report suggested that moderate dynamic training reduced the risk of upper respiratory infection in seniors, other studies reported that immune function was not enhanced by either a 6 mo program of moderate dynamic exercise (211) or 12 wk of resistance training (162). Indeed, there have even been suggestions that exercise may actually decrease NK cell function in frail and very elderly individuals.

The biological factors involved in immunosenescence have yet to be fully elucidated. Likely candidates include an involution of the thymus gland, dysregulation of cytokine secretion, and cumulative cell damage due to free radicals, a mitotic clock, and apoptosis. A further understanding of these processes may help in devising appropriate tactics for countering immunosenescence.

Autoimmune Diseases

Although rheumatoid arthritis is an autoimmune disease, evidence is accumulating that moderate dynamic and resistive exercise is clinically beneficial except at times of acute inflammation. However, it is less certain that any benefit from such exercise is due to immunomodulation. Studies to date have not discerned any associated changes in leukocyte counts or cytokine production (9). One growing area of inquiry is the role of regular dynamic exercise in reducing both the muscular generation of reactive oxygen species and the activation of transcription factors for inflammatory cytokines such as nuclear factor kappa-β and other inflammation-related genes (22, 88).

Susceptibility to Upper Respiratory Infections

Although strenuous exercise may have negative effects on the immune response, growing empirical evidence shows that an appropriately chosen dose of moderate dynamic exercise can enhance both cellular and humoral resistance to upper respiratory infections. Moderate dynamic exercise enhanced NK cell activity and reduced the risk of upper respiratory symptoms in middle-aged and older women relative to controls (141, 145, 186), and a moderate level of basketball training increased salivary levels of the immunoglobulin IgA (195). Thus, it may prove possible to commend regular moderate physical activity in terms of its long-term health benefits as well as its ability to reduce susceptibility to acute infections.

Cancer

Good evidence is now available that exercise reduces the risk of several types of neoplasm (184); consistent decreases have been shown in all-cause, colon, and breast cancer mortality. Many possible mechanisms have been discussed, but the collection of evidence has been hampered by the long disease latency. Investigators need to attempt the very difficult task of assessing an individual's pattern of physical activity over 10 or even 20 yr. If an exercise-induced enhancement of some aspect of immune function (e.g., increased activity of macrophages, NK cells, and related cytokines) is involved, it is puzzling why exercise protects against only some forms of cancer. Nevertheless, evidence from animal experiments shows that an exercise-induced stimulation of NK cell function can decrease risk from the more immediate danger of metastasis in those with established cancers (118).

Human Immunodeficiency Virus Infections

Human immunodeficiency virus (HIV) infections have been widely studied by immunologists since the early 1980s. Moderate dynamic and resistance exercise was proposed as an element in therapy for its potential to stimulate NK cell function, its ability to counter depression (which itself depresses immune function), and its potential to reverse the poor general physique seen in many patients with established HIV infections. It has been necessary to examine carefully potential interactions between such exercise and the compromised immune systems of patients infected with this virus. One continuing area of concern is that once acquired immune deficiency syndrome (AIDS) has developed, a high proportion of patients have evidence of myocarditis (with a potential for sudden, exercise-induced death, as discussed previously). Reports have generally noted a favorable response to a moderate training program: Increases in lean body mass, decreases in anxiety and depression, and possibly a slowing of the progression toward AIDS were achieved without further depression of T-cell numbers. However, studies to date have generally been on a small scale and often have lacked a control group. A detailed evaluation of resulting changes in immune response has also generally been lacking.

Arthur LaPerriere, the first president of the International Society of Exercise Immunology, has been particularly interested in this area of investigation. One of his studies found that a moderate dynamic-exercise program increased counts of both CD4+ and CD4+CD45+ cells; the latter have the potential to activate CD8+ cytotoxic cells. The response that he observed was at least as good as that anticipated from zidovudine (AZT) treatment (102). Another report claimed that moderate dynamic exercise enhanced skin responses to the Multitest Mérieux antigen (174), but immune benefits were less clearly demonstrated in other studies (182).

Future Challenges and Conclusions

In terms of future challenges, brief mention is made of three specific areas that are currently attracting attention: the likelihood that vigorous exercise can induce deoxyribonucleic acid (DNA) damage and apoptosis, the possibility that exercise-related modification of immune responses can slow aging of the brain, and the development of methods for detecting athletes who have engaged in various forms of doping.

DNA Damage and Apoptosis

Study of the DNA damage caused by strenuous dynamic exercise, and particularly the relationship of such damage with the release of reactive species and immune potential, is likely to increase (134, 171). DNA damage could accelerate the aging process and increase the susceptibility of the individual to neoplasia. The problem of apoptosis, first described in rat thymocytes in 1993 (37) and possibly a part of immunosenescence, is currently attracting considerable attention. Mooren noted that apoptosis of lymphocytes occurred when humans engaged in treadmill running to exhaustion at 80% of maximal oxygen intake. However, he pointed out the need to determine whether this represented an elimination of harmful autoreactive cells or a suppression of immune function (127). The extent of any such apoptosis during a single bout of exercise is decreased by training (128), but differing mechanisms in different parts of the lymphatic system remain to be explained (98). Another puzzling feature of the phenomenon is that wheel training reduces splenic lymphocyte apoptosis when samples are collected at rest but not when they are taken after an acute bout of exercise (85). Assuming that the changes are indeed harmful, it will be important to follow up on indications of protection from heat shock proteins (55), training, and antioxidant medication.

Aging as an Expression of Disturbed Immune Function

Aging has multiple causes (178), but one likely factor is cell damage caused by exposure to reactive species, including peroxidation of lipids, damage to mitochondrial enzymes, and inactivation of one or more of the 3,000 genes that are essential to cell function. Although strenuous exercise increases the production of reactive species if practiced regularly, it also increases blood levels of antidotes such as the enzymes superoxide dismutase and catalase. This could be helpful both in reducing the adverse effects of acute exercise and slowing the overall aging process.

Possibly related to this, aging is associated with an increasing level of inflammation. Increased concentrations of the proinflammatory cytokines IL-1β and TNF-α are known to contribute to neurodegeneration in the hippocampal region of transgenic mice models of Alzheimer's disease and may lead to such problems as sarcopenia. A promising avenue for future inquiry is suggested by the effect of regular wheel exercise in reducing the level of these cytokines and delaying one component of the aging process of the brain (136). If proinflammatory agents play a significant role in aging, there will also be much scope for investigation of the benefit of anti-inflammatory agents.

Detection of Doping

Over the past 20 yr, sport physicians have waged an increasingly sophisticated campaign to prevent doping among high-level athletes, and exercise immunologists have increasingly been recruited to help in this task. One troublesome form of such cheating is homologous red blood cell transfusion. A flow cytometric technique of detecting such transfusions based on the presence of foreign antigens has been developed (133). There are also growing concerns that some athletes are abusing gene transfer technology (gene doping). The detection of this practice has to date been difficult, but a polymerase chain reaction technique that allows detection of transgenic DNA in whole blood has recently been described (11).

Conclusions

Since the beginning of the modern era of exercise immunology some 20 yr ago, a small group of investigators have together made remarkable progress in developing the understanding of this discipline. Table 21.4 summarizes some important milestones in the process. There have been many technical challenges to overcome, particularly the extremely low active concentrations of messenger substances in the circulating blood stream. Much remains that is still unclear. However, application of the methods of molecular biology seem destined to allow striking advances in exercise immunology over the next two decades as well as a corresponding growth in the comprehension of this important topic.

Table 21.4 Important Milestones in the History of Exercise Immunology

Year	Finding	Reference
Infections and strenuous dynamic exercise		
1918-1924	Strenuous exercise increased susceptibility to bacterial infections	4, 40, 137
1945-1950	Strenuous exercise exacerbated acute anterior poliomyelitis	75, 86, 106, 173
1950-1963	Strenuous exercise exacerbated acute viral hepatitis	33, 131, 194
1983-1988	Prolonged distance running predisposes runners to upper respiratory infections	142, 153
1992-1995	Moderate dynamic exercise may increase resistance to infection; hypotheses of J-shaped curve and open window for infection	139, 141, 144, 186
2000	Application of current epidemiological evaluation to studies of strenuous exercise and infection	183
Technological innovations important to exercise immunology		
1947-1953	Development of automated cell counter	183
1965-1973	Automated cell sorting and sizing	46, 60, 87
Cellular responses to various forms of strenuous exercise		
1893	Description of leukocytosis by Schulte	No reference available
1902-1960	Increase of neutrophil counts	3, 44, 50, 104
1935-1960	Increase of monocyte and macrophage counts	3, 44, 50, 92
1978-1994	Recognition of importance of cell migration and adhesion molecules	62, 78, 110, 204
1982	Corticosteroid-mediated decrease in monocytes	41
1983-1988	Increase of T-cell counts	14, 79, 100
1988	Clear differentiation of B-cell responses from natural killer cell responses	45
1992	Description of biphasic natural killer cell response	191
1995	Differentiation of T helper 1 and T helper 2 responses	161
Humoral responses to various forms of strenuous exercise		
1969	Increase of acute phase reactants	74
1971	Increased production of complement	51
1982	Decrease of mucosal immunoglobulins	201
1986	Increased secretion of interleukin-1	28
1994-1995	Parallels noted between strenuous exercise and sepsis	26, 147
1995	Application of reverse transcription technology to cytokine detection	126
2000	Description of cascade of humoral secretions	149
2004	Role of cytokines in metabolic regulation noted	150

References

1. Aguirre G, Frontera WR, Colon LR, Amy E, Micheo W, Correa JJ, Camunas JF. Profile of health services utilization during the XVII Central American and Caribbean Sport Games: Delegation of Puerto Rico, Ponce 1993. *Puerto Rico Health Sci J* 13: 273-278, 1994.
2. Ahlborg B, Ahlborg G. Exercise leukocytosis with and without beta-adrenergic blockade. *Acta Med Scand* 187: 241-246, 1970.
3. Anderssen KL. Leukocyte response to brief, severe exercise. *J Appl Physiol* 7: 671-674, 1955.
4. Bailey GW. The effect of fatigue upon the susceptibility of rabbits to intratracheal injection of type I pneumococcus. *Am J Hygiene* 5: 176-295, 1924.
5. Baracos VE. Exercise inhibits progressive growth of the Morris hepatoma 7777 in male and female rats. *Can J Physiol Pharmacol* 67: 864-870, 1989.
6. Barker MH, Capps RB, Allen FW. Acute infectious hepatitis in the Mediterranean Theater. *J Am Med Assoc* 128: 997, 1945.
7. Barnard DF, Barnard SA, Carter AB, Machin SJ, Patterson KG, Yardumian A. Detection of important abnormalities of the differential count using the Coulter STKR blood counter. *J Clin Pathol* 42: 772-776, 1989.
8. Baron RC, Hatch MH, Kleeman K, MacCormack JN. Aseptic meningitis among members of a high school football team. An outbreak associated with echovirus 16 infection. *J Am Med Assoc* 248: 1724-1727, 1982.
9. Baslund B, Lyngberg K, Andersen V, Halkjaer-Kristensen J, Hansen M, Klokker M, Pedersen BK. Effect of 8 wk of bicycle training on the immune system of patients with rheumatoid arthritis. *J Appl Physiol* 75: 1691-1695, 1993.
10. Baum M, Geitner T, Liesen H. The role of the spleen in the leucocytosis of exercise: Consequences for physiology and pathophysiology. *Int J Sports Med* 17: 604-607, 1996.
11. Beiter T, Zimmerman M, Fragasso A, Armeanu S, Klauer UM, Bitzer M, Su H, Young WL, Niess AM, Simon P. Establishing a novel single-copy primer-internal intron-spanning PCR (spiPCR) procedure for the direct detection of gene doping. *Ex Immunol Rev* 14: 71-85, 2008.
12. Berenson RJ, Bensinger WI, Kalamaaz D. Positive selection of viable cell populations using avidin-biotin immunoabsorption. *J Immunol Methods* 91: 11-19, 1986.
13. Berg R, Ringertz O, Espmark Å. Australian antigen in hepatitis among Swedish track finders. *Acta Pathologica Microbiol Immunol Scand B* 79: 423-427, 1971.
14. Berk LS, Tan SA, Nieman DC, Eby WC. The suppressive effect of stress from acute exhaustion on T-lymphocyte helper/suppressor cell ratio in athletes and non-athletes. *Med Sci Sports Exerc* 17: 492, 1985.
15. Bieger WP, Weiss M, Michel G, Weicker H. Exercise-induced monocytosis and modulation of monocyte function. *Int J Sports Med* 1: 30-36, 1980.
16. Blalock JE. A molecular basis for bidirectional communication between the immune and neuroendocrine systems. *Physiol Rev* 69: 1-32, 1989.
17. Blank SE, Tiidus PM, Hoffman-Goetz L. Neutrophil response to prolonged exercise in immune-competent and RAG2/gamma c null mice. *Can J Physiol Pharmacol* 79: 490-495, 2001.
18. Bosenberg AT, Brock-Utne JG, Gaffin SL, Well MTB, Blake TW. Strenuous exercise causes systemic endotoxemia. *J Appl Physiol* 65: 106-108, 1988.
19. Bowman JF. Infectious hepatitis in a college football player. *Am J Sports Med* 4: 101-106, 1976.
20. Brenner IK, Severs YD, Rhind SG, Shephard RJ, Shek PN. Immune function and incidence of infection during basic infantry training. *Milit Med* 165: 878-883, 2000.
21. Brenner IK, Shek PN, Shephard RJ. Infection in athletes. *Sports Med* 17: 86-107, 1994.
22. Brooks SV, Vasilaki A, Larkin LM, McArdle A, Jackson MJ. Repeated bouts of aerobic exercise lead to reductions in skeletal muscle free radical generation and nuclear factor kappaB activation. *J Physiol* 587: 927-928, 2009.
23. Bruunsgard H, Hartkopp A, Mohr T, Konradsen H, Heron I, Mordhorst CH, Pedersen BK. In vivo cell-mediated immunity and vaccination response following prolonged intense exercise. *Med Sci Sports Exerc* 29: 1176-1181, 1997.
24. Buchwald D, Cheney PR, Peterson B, Henry SB, Wormsley A, Geiger DV, Ablashi SV, Salahuddin SZ, Saxinger C, Biddle R. A chronic illness characterized by fatigue, neurologic and immunologic disorders. *Ann Int Med* 115: 103-113, 1992.
25. Burch GF. Viral diseases of the heart. *Acta Cardiol* 34: 5-9, 1979.
26. Camus G, Deby-Dupont G, DuChateau J, Deby A, Pincemail J, Lamy M. Are similar inflammatory factors involved in strenuous exercise and sepsis? *Intens Care Med* 24: 602-610, 1994.
27. Cannon JG. Exercise and resistance to infection. *J Appl Physiol* 74: 973-981, 1993.
28. Cannon JG, Evans WJ, Hughes VA, Meredith CN, Dinarello CA. Physiological mechanisms contributing to increased interleukin-1 secretion. *J Appl Physiol* 61: 1869-1874, 1986.
29. Cannon JG, Kluger MJ. Exercise enhances survival in mice infected with *Salmonella typhimurium*. *Proc Soc Exp Biol* 175: 518-521, 1984.
30. Casper H. Veranderung des weissen Blutbildes nach dem 100m Lauf bei Frau [Changes in white cell concentration in women after running 100 meters]. *Z Ges Phys Therap* 30: 5-14, 1925.
31. Castell LM, Newsholme EA. Glutamine and the effects of exhaustive exercise upon the immune response. *Can J Physiol Pharmacol* 76: 524-532, 1998.

32. Cavarec L, Quillet-Mary A, Fradelizi D, Conjeaud H. An improved double fluorescence flow cytometry method for the quantification of killer cell/target cell conjugate formation. *J Immunol Methods* 130: 251-261, 1990.
33. Chalmers TC, Eckhardt RD, Reynolds WE, Cigarroa JG, Deane N, Reifenstein RW, Smith CW, Davidson CS. The treatment of acute infectious hepatitis; Controlled studies of the effects of diet, rest, and physical reconditioning on the acute course of the disease and on the incidence of relapses and residual abnormalities. *J Clin Invest* 34: 1163-1235, 1955.
34. Christenson B, Böttiger M, Grillner L. The prevalence of hepatitis B in Sweden: A statistical serosurvey of 3381 Swedish inhabitants. *Epidemiol Infect* 119: 221-225, 1997.
35. Clow A, Hucklebridge F. The impact of psychological stress on immune function in the athletic population. *Ex Immunol Rev* 7: 5-17, 2001.
36. Colbert LH, Westerlind KC, Perkins SN, Haines DC, Berrigan D, Donehower LA, Fuchs-Young R, Hursting SD. Exercise effects on tumorigenesis in a p-53-deficient mouse model of breast cancer. *Med Sci Sports Exerc* 41: 1597-1605, 2009.
37. Concordet JP, Ferry A. Physiological programmed cell death in thymocytes is induced by physical stress (exercise). *Am J Physiol* 265: C626-C629, 1993.
38. Cosentino LM, Cathcart MK. A multi-step isolation scheme for obtaining human natural killer cells. *J Immunol Methods* 103: 195-204, 1987.
39. Coulter WH. *Means for Counting Particles Suspended in a Fluid.* U.S. Patent Office, 1953.
40. Cowles WN. Fatigue as a contributory cause of pneumonia. *Boston Med Surg J* 179: 555-556, 1918.
41. Cupps TR, Fauci AS. Corticosteroid-mediated immunoregulation in man. *Immunol Rev* 65: 133-155, 1982.
42. Daniels WL, Sharp DS, Wright JE, Vogel JA, Friman G, Beisel WR, Knapik JJ. Effects of virus infection on physical performance in man. *Milit Med* 150: 8-14, 1985.
43. Davis JM, Gallin JI. The neutrophil. In: Oppenheim JJ, Rosentreich DL, Potter M, eds. *Cellular Functions in Immunity and Inflammation.* New York: Elsevier-North Holland, 1981, 77.
44. De Lanne R, Barnes JR, Brouha L. Hematological changes during muscular activity and recovery. *J Appl Physiol* 15: 31-36, 1960.
45. Deuster PA, Curiale AM, Cowan ML, Finkelman FD. Exercise-induced changes in populations of peripheral blood mononuclear cells. *Med Sci Sports Exerc* 20: 276-280, 1988.
46. Dittrich W, Gohde W. Impulsfluoremetrie bei Einzelzellen in Suspensionen [Impulse fluorometry of single cells in suspension]. *Z Naturforsch B* 24: 360-361, 1969.
47. Donehower LA, Godley LA, Aldaz CM, Pyle R, Shi YP, Pinkel D, Gray J, Bradley DA, Medina D, Varmus HE. Deficiency of p53 accelerates mammary tumorigenesis in Wnt-1 transgenic mice and promotes chromosomal instability. *Genes Dev* 9: 882-895, 1995.
48. Douglas DJ, Hansen PG. Upper respiratory infections in the conditioned athlete. *Med Sci Sports Exerc* 10: 55(Abstr.), 1978.
49. Douglas JH. The effects of physical training on the immunological response to diphtheria toxoid immunization. *J Sports Med* 14: 48-54, 1974.
50. Dùner H, Pernow B. Histamine and leukocytes in blood during muscular work in man. *Scand J Lab Clin Invest* 10: 394-396, 1958.
51. Eberhardt A. Effect of motor activity on some immunological mechanisms of non-specific resistance of the body. II Effect of strenuous effort. *Acta Physiol Pol* 22: 185-194, 1971.
52. Edlund A. The effect of defined physical exercise in the early convalescence of viral hepatitis. *Scand J Infect Dis* 3: 189-196, 1971.
53. Eichner R. Infectious mononucleosis: Recognition and management in athletes. *Phys Sports Med* 15: 125-135, 1987.
54. Eskola J, Ruuskanen O, Soppi E, Viljanen MK, Järvinen M, Toivonen H, Kouvailainen K. Effect of sport stress on lymphocyte transformation and antibody formation. *Clin Exp Immunol* 32: 339-345, 1978.
55. Fehrenbach E, Veith R, Schmid M, Dickhuth HH, Northoff H, Niess AM. Inverse response of leukocyte heat shock proteins and DNA damage to exercise and heat. *Free Rad Res* 37: 975-982, 2003.
56. Fiatarone MA, Morley JE, Bloom ET, Benton D, Makinodan T, Solomon GF. Endogenous opioids and the exercise-induced augmentation of natural killer cell activity. *J Lab Clin Med* 112: 544-552, 1988.
57. Fiatarone MA, Morley JE, Bloom ET, Benton D, Solomon GF, Makinodan T. The effect of exercise on natural killer cell activity in young and old subjects. *J Gerontol* 44: M37-M45, 1989.
58. Frick M, Pachinger O, Põlzl G. Sudden cardiac death in athletes. *Herz* 34: 299-304, 2009.
59. Friman G, Wright J, Ilbäck N-G, Beisel WR, White JD, Sharp DS, Stephen EL, Daniels WL, Vogel JA. Does fever or myalgia indicate reduced physical performance capacity in viral infections? *Acta Med Scand* 217: 353-361, 1985.
60. Fulwyler MJ. Electronic separation of biologicl cells by volume. *Science* 150: 910-911, 1965.
61. Gabriel H, Kindermann W. Flow cytometry: Principles and applications in exercise immunology. *Sports Med* 20: 302-320, 1995.
62. Gabriel H, Urhausen A, Brechtel L, Muller HJ, Kindermann W. Alterations of regular and mature monocytes are distinct, and dependent on intensity and duration of exercise. *Eur J Appl Physiol* 69: 179-181, 1994.
63. Gabriel HH, Urhausen A, Valet G, Heidelbach U, Kindermann W. Overtraining and immune system: A prospective longitudinal study in endurance athletes. *Med Sci Sports Exerc* 30: 1151-1157, 1998.

64. Garagiola U, Buzzetti M, Cardella E, Confalonieri F, Giani E, Polini V, Ferrante P, Marcuso R, Montanari M, Grossi E, Pecori A. Immunological patterns during regular intensive training in athletes: Quantification and evaluation of a preventive pharmacological approach. *J Int Med Res* 23: 85-95, 1995.
65. Gedge VL, Mackinnon LT, Hooper SL. Effects of 4 wk intensified training on natural killer (NK) cells in competitive swimmers. *Med Sci Sports Exerc* 29: S158, 1997.
66. Gemsa D, Leser EG, Deimann W, Resch K. Suppression of T lymphocyte proliferation during lymphoma growth in mice: Role of PGEE-2-producing macrophages. *Immunobiol* 161: 385-392, 1982.
67. Gimenez M, Mohan-Kumar T, Humbert JC, de Talance N, Buisine J. Leukocyte, lymphocyte and platelet response to dynamic exercise. *Eur J Appl Physiol* 55: 465-470, 1986.
68. Gleeson M, Pyne DB. Exercise effects on mucosal immunity. *Immunol Cell Biol* 78: 536-544, 2000.
69. Good RA, Fernandes G. Enhancement of immunological and resistance to tumor growth in Balb C mice by exercise. *Fed Proc* 40: 1040, 1981.
70. Gran JT, Hjetland R, Andreassen AHSJR. Pneumonia, myocarditis and reactive arthritis due to *Chlamydia pneumoniae*. *Scand J Rheumatol* 22: 43-44, 1996.
71. Green RL, Kaplan SS, Rabin BS, Stanitski CL, Zdziarski U. Immune function in marathon runners. *Ann Allergy* 47: 73-75, 1981.
72. Grimby G. Exercise in man during pyrogen-induced fever. *Scand J Lab Clin Invest* 14(Suppl. 67): 1962.
73. Hack V, Weiss C, Friedmann B, Suttner S, Schykowski M, Erbe N, Benner A, Bartsch P, Droge W. Decreased plasma glutamine level and CD4+ T cell number in response to 8 wk of anaerobic training. *Am J Physiol* 272: E788-E795, 1997.
74. Haralambie G. Serum glycoproteins and physical exercise. *Clin Chem Acta* 26: 287-291, 1969.
75. Hargreaves ER. Poliomyelitis. Effect of exertion during the paralytic stage. *Br Med J* 2: 1021-1022, 1948.
76. Hay JB, Andrade WH. Lymphocyte recirculation, exercise and immune responses. *Can J Physiol Pharmacol* 76: 490-496, 1998.
77. Heath GW, Ford ES, Craven TE, Macera CA, Jackson KL, Pate RR. Exercise and the incidence of upper respiratory tract infections. *Med Sci Sports Exerc* 23: 152-157, 1991.
78. Hedfors E, Biberfeld P, Wahren J. Mobilization to the blood of human non-T and K lymphocytes during physical exercise. *J Clin Lab Invest* 1: 159-162, 1978.
79. Hedfors E, Holm G, Ivansen M, Wahren J. Physiological variation of blood lymphocyte reactivity: T-cell subsets, immunoglobulin production, and mixed-lymphocyte reactivity. *Clin Immunol Immunopathol* 27: 9-14, 1983.
80. Heir T, Aanestad G, Carlsen KH, Larsen S. Respiratory tract infection and bronchial responsiveness in elite athletes and sedentary control subjects. *Scand J Med Sci Sports* 5: 94-99, 1995.
81. Hemmingsson P, Hammarstrom L. Nasal administration of immunoglobulin as effective prophylaxis against infections in elite cross-country skiers. *Scand J Infect Dis* 25: 783-785, 1993.
82. Hernando V, Soler P, Pedro R, García L, Castilla J, García MA, Quiñones C, García V, Gallardo V, Echevarria JM, Jardi R, Bleda MJ, De Mateo S. Seroprevalence study of hepatitis B among orienteers. *Med Clin (Barc)* 132: 649-657, 2009.
83. Hoffman-Goetz L, Keir R, Thorne R, Houston ME, Young C. Chronic exercise stress in mice depresses splenic T lymphocyte mitogenesis in vitro. *Clin Exp Immunol* 66: 551-557, 1986.
84. Hoffman-Goetz L, Kubes P, Shephard RJ. Immune response to inflammation and trauma: A physical training model. *Can J Physiol Pharmacol* 76: 469-597, 1998.
85. Hoffman-Goetz L, Spagnuolo PA. Freewheel exercise training modifies pro- and anti-apoptotic protein expression in mouse splenic lymphocytes. *Int J Sports Med* 28: 787-791, 2007.
86. Horstmann DM. Acute poliomyelitis. Relation of physical activity to the time of onset and to the course of the disease. *J Am Med Assoc* 142: 236-239, 1950.
87. Hulett HR, Bonner WA, Sweet RG, Herzenberg LA. Development and application of a rapid cell counter. *Clin Chem* 19: 813-816, 1973.
88. Jiménez-Jiménez R, Cuevas MJ, Almar M, Lima E, García-López D, De Paz JA. Eccentric training impairs NF-kappaB activation and over-expression of inflammation-related genes induced by acute eccentric exercise in the elderly. *Mech Ageing Develop* 129: 313-321, 2008.
89. Jokl E. The immunological status of athletes. *J Sports Med Phys Fitness* 14: 165-167, 1974.
90. Jonsdottir IH. Neuropeptides and their interaction with exercise and immune function. *Immunol Cell Biol* 78: 562-570, 2000.
91. Kappel M, Tvede N, Galbo H, Haahr PM, Kjaer M, Linstow M, Klarlund K, Pedersen BK. Evidence that the effect of physical exercise on NK cell activity is mediated by epinephrine. *J Appl Physiol* 70: 2530-2534, 1991.
92. Karpovich PV. The effect of basketball, wrestling and swimming upon white blood corpuscles. *Res Q* 6: 42-48, 1935.
93. Kiecolt-Glaser JK, Glaser R. Psychoneuroimmunology: Can psychological interventions modulate immunity? *J Consult Clin Psychol* 60: 569-575, 1992.
94. Kingston SG, Hoffman-Goetz L. Effect of environmental enrichment and housing density on immune system reactivity to acute exercise stress. *Physiol Behav* 60: 145-150, 1996.

95. Koivisto VA, Soman VR, Felig P. Effects of exercise on insulin-binding to monocytes in obesity. *Metabolism* 29: 168-171, 1980.
96. Krinkler DM, Zilberg B. Activity and hepatitis. *Lancet* 2: 1046-1047, 1966.
97. Krzyszton CP, Sparkman NL, Grant RW, Buchanan JB, Broussard SR, Woods J, Johnson RW. Exacerbated fatigue and motor deficits in interleukin-10-deficient mice after peripheral immune stimulation. *Am J Physiol* 295: R1109-R1114, 2008.
98. Krüger K, Frost S, Most E, Völker K, Pallauf J, Mooren FC. Exercise affects tissue lymphocyte apoptosis via redox-sensitive and Fas-dependent signaling pathways. *Am J Physiol* 296: R1518-R1527, 2009.
99. König D, Berg A, Baumstark MW, Grathwohl D, Halle M, Keul J. Correlation between plasma fatty acids and the immunological response to exercise. *Med Sci Sports Exerc* 28(Suppl.): S30, 1996.
100. Landmann RMA, Müller FB, Perini C, Wesp M, Erne P, Bühler FR. Changes of immunoregulatory cells induced by psychological and physical stress: Relationship to plasma catecholamines. *Clin Exp Immunol* 58: 127-135, 1984.
101. Lannergård O, Fohlman J, Wesslen L, Rolf C, Friman G. Immune function in Swedish elite orienteers. *Scand J Med Sci Sports* 11: 274-279, 2001.
102. LaPerriere A, Fletcher MA, Antoni MH, Klimas NG, Ironson G, Schneiderman N. Aerobic exercise training in AIDS risk group. *Int J Sports Med* 12: S53-S57, 1991.
103. LaPerriere A, Ironson G, Antoni MH, Schneiderman N, Klimas N, Fletcher MA. Exercise and psychoneuroimmunology. *Med Sci Sports Exerc* 26: 182-190, 1994.
104. Larrabee RC. Leukocytosis after violent exercise. *J Med Res* 7: 76-82, 1902.
105. Lehman M, Mann H, Gastmann U, Keul J, Vette D, Steinacker JM, Haussinger D. Unaccustomed high mileage vs. intensity training-related changes in performance and serum amino acid levels. *Int J Sports Med* 17: 187-192, 1996.
106. Levinson SO, Milzer A, Lewin P. Effect of fatigue, chilling, and mechanical trauma on resistance to experimental poliomyelitis. *Am J Hygiene* 42: 204-213, 1945.
107. Lewicki R, Tchorzewski H, Denys A, Kowalska M, Golinska A. Effect of physical exercise on some parameters of immunity in conditioned sportsmen. *Int J Sports Med* 8: 309-314, 1987.
108. Lewicki R, Tchorzewski H, Majewska E, Nozak Z, Baj Z. Effect of maximal physical exercise on T lymphocyte subpopulations and on interleukin 1 (IL-1) and interleukin 2 (IL-2) production in vitro. *Int J Sports Med* 9: 114-117, 1988.
109. Li Y, Hively WP, Varmus HE. Use of MMTV-Wnt-1 transgenic mice for studying the genetic basis of cancer. *Oncogene* 19: 1002-1009, 2000.
110. Liesen H, Uhlenbruck G. Sports immunology. *Sport Sci Rev* 1: 94-116, 1992.
111. Lindberg K, Berglund B. Effect of treatment with nasal IgA on the incidence of infectious disease in world-class canoeists. *Int J Sports Med* 17: 235-238, 1996.
112. Linde F. Running and upper respiratory infections. *Scand J Sports Sci* 9: 21-23, 1987.
113. Mackinnon LT. Current challenges and future expectations in exercise immunology: Back to the future. *Med Sci Sports Exerc* 26: 191-194, 1994.
114. Mackinnon LT. Immunoglobulin, antibody and exercise. *Ex Immunol Rev* 2: 1-34, 1996.
115. Mackinnon LT. Overtraining effects on immunity and performance in athletes. *Immunol Cell Biol* 78: 502-509, 2000.
116. Mackinnon LT, Hooper SL. Plasma glutamine and upper respiratory tract infection during intensified training in swimmers. *Med Sci Sports Exerc* 28: 285-290, 1996.
117. Mackinnon LT, Hooper SL, Jones S, Gordon RD, Bachman AW. Humoral, immunological and hematological responses to intensified training in swimmers. *Med Sci Sports Exerc* 29: 1637-1645, 1997.
118. Macneil B, Hoffman-Goetz L. Effect of exercise on natural cytotoxicity and pulmonary tumor metastases in mice. *Med Sci Sports Exerc* 25: 922-928, 1993.
119. Madden KS, Felten DL. Experimental basis for neural-immune interactions. *Physiol Rev* 75: 77-106, 1995.
120. Maron BJ, Shirani J, Poliac L, Mathenge R, Toberts W, Mueller F. Sudden death in young competitive athletes. *J Am Med Assoc* 256: 2696-2699, 1996.
121. Mast EE, Goodman RA, Bond WW, Favero MS, Drotman DP. Transmission of blood-borne pathogens during sports: Risk and prevention. *Ann Int Med* 122: 283-285, 1995.
122. Maurer HR. Potential pitfalls of 3H thymidine techniques to measure cell proliferation. *Cell Tissue Kinet* 14: 111-120, 1981.
123. McKenzie DC. Markers of excessive exercise. *Can J Appl Physiol* 24: 66-73, 1999.
124. Melamed MR, Mullaney PF, Shapiro HM. An historical review of the development of flow cytometers and sorters. In: Melamed MR, Lindmo T, Mendelsohn ML, eds. *Flow Cytometry and Sorting.* 2nd ed. Hoboken, NJ: Wiley, 1990.
125. Moldovan A. Photoelectric technique for the counting of microscopical cells. *Science* 80: 188-189, 1934.
126. Moldoveanu AI, Shephard RJ, Shek PN. Exercise elevates plasma levels but not gene expression of IL-1b, IL-6, and TNF-a in blood mononuclear cells. *J Appl Physiol* 89: 1499-1504, 1995.
127. Mooren FC, Blöming D, Lechterman A, Lerch MM, Völker K. Lymphocyte apoptosis after exhaustive and moderate exercise. *J Appl Physiol* 93: 147-153, 2002.

128. Mooren FC, Lechterman A, Völker K. Exercise-induced apoptosis of lymphocytes depends on training status. *Med Sci Sports Exerc* 36: 1476-1483, 2004.
129. Morse LJ, Bryan JA, Hurley JP, Murphy JF, O'Brien TF, Wacker WEC. The Holy Cross college football team hepatitis outbreak. *J Am Med Assoc* 219: 706-708, 1972.
130. Natelson BH, Ye N, Moul DE, Jenkins FJ, Oren DA, Tapp WN, Cheng YC. High titers of anti-Epstein Barr virus DNA polymerase are found in patients with severe fatiguing illness. *J Med Virol* 42: 42-46, 1994.
131. Nefzger MD, Chalmers TC. The treatment of acute infectious hepatitis: Treatment and ten-year follow-up study of the effects of diet and rest. *Am J Med* 35: 299-309, 1963.
132. Nehlsen-Cannarella SL, Fagoaga OR, Nieman DC, Henson DA, Butterworth DE, Schmitt RL, Bailey EM, Warren BL, Davis JM. Carbohydrate and the cytokine response to 2.5 hours of running. *J Appl Physiol* 82: 1662-1667, 1997.
133. Nelson M, Popp H, Sharpe K, Ashenden M. Proof of homologous blood transfusion through quantification of blood group antigens. *Haematologica* 88: 1284-1295, 2003.
134. Neubauer O, Reichold S, Nerseyan A, König D, Wagner K-H. Exercise-induced DNA damage: Is there a relationship with inflammatory responses? *Ex Immunol Rev* 14: 51-72, 2008.
135. Newsholme EA, Crabtree B, Ardawi M. Glutamine metabolism in lymphocytes: Its biochemical, physiological and clinical importance. *Q J Exp Physiol* 70: 473-489, 1985.
136. Nichol KE, Poon WW, Parachikova AI, Cribbs CG, Cotman CW. Exercise alters the immune profile in Tg2576 Alzheimer mice toward a response coincident with improved cognitive performance and decreased amyloid. *J Neuroinflamm* 5: 13, 2008.
137. Nichols EE, Spaeth RA. The relation between fatigue and suceptibility of guinea pigs to injections of type I pneumococcus. *Am J Hygiene* 2. 527-535, 1922.
138. Nieman DC. Exercise immunology: Practical applications. *Int J Sports Med* 18(Suppl. 1): S91-S100, 1997.
139. Nieman DC. Exercise, infection, and immunity. *Int J Sports Med* 15(Suppl. 3): S131-S141, 1994.
140. Nieman DC, Henson DA, Butterworth DE, Warren BJ, Davis JM, Fagoaga OR, Nehlsen-Cannarella SL. Vitamin C supplementation does not alter the immune response to 2.5 hours of running. *Int J Sports Nutr* 7: 173-184, 1997.
141. Nieman DC, Henson DA, Gusewitch G, Warren BJ, Dotson RC, Butterworth DE, Nehlsen-Cannarella SL. Physical activity and immune function in elderly women. *Med Sci Sports Exerc* 25: 823-831, 1993.
142. Nieman DC, Johannsen LM, Lee JW, Arabatzis K. Infectious episodes in runners before and after the Los Angeles marathon (LAM). *Med Sci Sports Exerc* 30: 316-328, 1988.
143. Nieman DC, Johanssen LM, Lee JW. Infectious episodes in runners before and after a road race. *J Sports Med Phys Fitness* 29: 289-296, 1989.
144. Nieman DC, Nehlsen-Cannarella S. Exercise and infection. In: Watson RR, Eisinger M, eds. *Exercise and Disease*. Boca Raton, FL: CRC Press, 1992, 121-148.
145. Nieman DC, Nehlsen-Cannarella SL, Markoff PA, Balk-Lamberton AJ, Yang H, Chritton DB, Lee JW, Arabatzis K. The effects of moderate exercise training on natural killer cells and acute upper respiratory tract infections. *Int J Sports Med* 11: 467-473, 1990.
146. Northoff H, Berg A. Immunological mediators as parameters of the reaction to strenuous exercise. *Int J Sports Med* 12(Suppl. 1): S9-S15, 1991.
147. Northoff H, Enkel S, Weinstock C. Exercise, injury and immune function. *Ex Immunol Rev* 1: 26-48, 1995.
148. Ortaldo JR. Cytotoxicity by natural killer cells: Analysis of large granular lymphocytes. *Meth Emzymol* 132: 445-457, 1986.
149. Pedersen B, Hoffman-Goetz L. Exercise and the immune system: Regulation, integration and adaptation. *Physiol Rev* 80: 1053-1081, 2000.
150. Pedersen BK, Steensberg A, Fischer C, Keller C, Keller P, Plomgaard P, Wolsk-Petersen E, Febbraio M. The metabolic role of IL-6 produced during exercise: Is IL-6 an exercise factor? *Proc Nutr Soc* 63: 263-267, 2004.
151. Pedersen BK, Toft AD. Effects of exercise on lymphocytes and cytokines. *Br J Sports Med* 34: 246-251, 2000.
152. Pedersen BK, Tvede N, Karlund K, Christensen LD, Hansen FR, Galbo H, Kharazami A, Halkjaer-Kristensen J. Indomethacin in vitro and in vivo abolishes post-exercise suppression of natural killer cell activity in peripheral blood. *Int J Sports Med* 11: 127-131, 1990.
153. Peters EM, Bateman ED. Ultramarathon running and upper respiratory tract infections. An epidemiological survey. *S Afr Med J* 64: 582-584, 1983.
154. Peters EM, Goetzsche JM, Grobbelaar B, Noakes TD. Vitamin C supplementation reduces the incidence of postrace symptoms of upper-respiratory-tract infection in ultramarathon runners. *Am J Clin Nutr* 57: 170-174, 1993.
155. Petrova IV, Kuzmin SN, Kurshakova TS, Suzdalnitsky RS, Pershin BB. Fagotsitarnaia aktivnost' neitrofilov i gumoral'nye faktory obshchego i mestnogo immuniteta pri intensivnykh fizicheskikh nagruzkakh [Neutrophil phagocytic acticity and the humoral factors of general and local immunity under intensive physical loading]. *Z Mikrobiol Epidemiol Immunobiol* 12: 53-57, 1983.
156. Plomgaard P, Fischer CP, Ibfelt T, Pedersen BK, van Hall G. Tumor necrosis factor-alpha modulates human in vivo lipolysis. *J Clin Endocrinol Metab* 93: 543-549, 2008.

157. Poortmans JR. Serum protein determination during short exhaustive physical activity. *J Appl Physiol* 30: 190-192, 1971.
158. Pross HF, Baines MG, Rubin P, Shragge P, Patterson MS. Spontaneous human lymphocyte-mediated cytotoxicity against tumor target cells IX. The quantitation of natural killer cell activity. *J Clin Immunol* 1: 51-63, 1981.
159. Pyne DB, Baker PA, Frickere WA, McDonald WA, Telford RD, Weidemann MJ. Effects of an intensive 12-week training program by elite swimmers on neutrophil oxidative activity. *Med Sci Sports Exerc* 27: 536-542, 1995.
160. Pyne DB, Gleeson M. Effects of intensive exercise training on immunity in athletes. *Int J Sports Med* 19(Suppl. 3): S191-S194, 1998.
161. Rabin BS, Moyna NM, Acker GR, Robertson RJ. Exercised induced alterations of cytokines by Th1 and Th2 lymphocytes. *Med Sci Sports Exerc* 27: S177, 1995.
162. Rall RC, Roubenoff R, Cannon JG, Abad LW, Dinarello CA, Meydani SN. Effects of progressive resistance training on immune response in aging and chronic inflammation. *Med Sci Sports Exerc* 28: 1356-1365, 1996.
163. Rasmussen AM, Smeland EB, Erikstein BK, Caignault L, Funderud S. A new method for detachment of dynabeads from positively selected B lymphocytes. *J Immunol Methods* 146: 195-202, 1992.
164. Redziniak DE, Diduch DR, Turman K, Hart J, Grindstaff TL, Macknight JM, Mistry DJ. Methicillin-resistant *Staphylococcus aureus* (MRSA) in the athlete. *Int J Sports Med* 30: 557-562, 2009.
165. Rhind SG, Shek PN, Shephard RJ. The impact of exercise on cytokines and receptor expression. *Ex Immunol Rev* 1: 97-148, 1995.
166. Ringertz O, Zetterberg BNEJM. Serum hepatitis among Swedish track-finders. An epidemiologic study. *N Engl J Med* 276: 540-546, 1967.
167. Roberts JA, Wilson JA, Clements GB. Virus infections and sports performance—A prospective study. *Br J Sports Med* 22: 161-162, 1988.
168. Rogers CJ, Zaharoff DA, Hance KW, Perkins SN, Hursting SD, Schlom J, Greiner JW. Exercise enhances vaccine-induced antigen-specific T cell responses. *Vaccine* 26: 5407-5415, 2009.
169. Rohde T, Ullum H, Rasmussen JP, Kristensen JH, Newsholme E, Pedersen BK. Effects of glutamine on the immune system: Influence of muscular exercise and HIV infection. *J Appl Physiol* 79: 146-150, 1995.
170. Rother K. Leukocyte mobilizing factor: A new biological activity derived from the third component of complement. *Eur J Immunol* 2: 550-558, 1972.
171. Russack V. Image cytometry: Current applications and future trends. *Clin Lab Sci* 31: 1-34, 1994.
172. Russell RW. Paralytic poliomyelitis. The early symptoms and the effect of physical activity on the course of the disease. *Br Med J* 1: 465-471, 1949.
173. Russell RW. Poliomyelitis. The pre-paralytic stage, and the effects of physical activity on the severity of the paralysis. *Br Med J* 2: 1023-1028, 1947.
174. Schlenzig C, Jäger H, Rieder H. Einfluss von Sporttherapie auf die zelluläre immunabwehr und die Psyche HIV-infizierter Männer [Influence of sport therapy upon cellular immune function and the pysche of HIV infected men]. *Dtsche Z Sportmed* 41: 156-160, 1990.
175. Schubert C, Schüssler G. Psychoneuroimmouology: An update. *Z Psychom Med Psychother* 55: 1-2, 2009.
176. Shek PN, Sabiston BH, Paucod JC, Vidal D. Strenuous exercise and immune changes. In: Buguet A, Radomski MW, eds. *Accord Franco-Canadien, Vol. 3. Physical exercise, hyperthermia, immune system and recovery sleep in man.* La Tronche, France: Centre de Recherches du Service de Santé des Armées, 1994, 121-137.
177. Shek PN, Shephard RJ. Physical exercise as a human model of limited inflammatory response. *Can J Physiol Pharmacol* 76: 589-597, 1998.
178. Shephard RJ. *Aging, Physical Activity and Training.* Champaign, IL: Human Kinetics, 1997.
179. Shephard RJ. Development of the discipline of exercise immunology. *Ex Immunol Rev* 16: 194-222, 2010.
180. Shephard RJ. *Ischaemic Heart Disease and Exercise.* London: Croom Helm, 1981.
181. Shephard RJ. Medical surveillance of endurance sport. In: Shephard RJ, Åstrand P-O, eds. *Endurance in Sport.* Oxford, UK: Blackwell Scientific, 2000, 653-666.
182. Shephard RJ. *Physical Activity, Training and the Immune Response.* Carmel, IN: Cooper, 1997.
183. Shephard RJ. Effects of exercise on the immune system: Overview of the epidemiology of exercise immunology. *Immunol Cell Biol* 78: 485-495, 2000.
184. Shephard RJ, Futcher R. Physical activity and cancer: How may protection be maximized? *Crit Rev Oncogen* 8: 219-272, 1997.
185. Shephard RJ, Gannon G, Hay JB, Shek PN. Adhesion molecule expression in acute and chronic exercise. *Crit Rev Immunol* 20: 245-266, 2000.
186. Shephard RJ, Kavanagh T, Mertens DJ. Personal health benefits of Masters athletic competition. *Br J Sports Med* 29: 35-40, 1995.
187. Shephard RJ, Shek PN. Associations between physical activity and susceptibility to cancer: Possible mechanisms. *Sports Med* 26: 293-315, 1998.
188. Shephard RJ, Shek PN. Heavy exercise, nutrition and immune function. Is there a connection? *Int J Sports Med* 16: 491-497, 1995.
189. Shinkai S, Konishi M, Shephard RJ. Aging and immune response to exercise. *Can J Physiol Pharmacol* 76: 562-572, 1998.
190. Shinkai S, Kurokawa Y, Hino S, Hirose M, Torii J, Watanabe S, Shiraishi S, Oka K, Wartanabe T. Triathlon competition induced a transient immunosup-

pressive change in the peripheral blood of athletes. *J Sports Med Phys Fitness* 33: 70-78, 1993.

191. Shinkai S, Shore S, Shek PN, Shephard RJ. Acute exercise and immune function: Relationship between lymphocyte activity and changes in subset counts. *Int J Sports Med* 13: 452-461, 1992.
192. Simon HB. The immunology of exercise: A brief review. *J Am Med Assoc* 252: 2735-2738, 1984.
193. Stock C, Schaller K, Baum M, Liesen H, Weiss M. Catecholamines, lymphocyte subsets, and cyclic adenosine monophosphate production in mononuclear cells and CD4+ cells in response to submaximal resistance exercise. *Eur J Appl Physiol* 71: 166-172, 1995.
194. Swift WE, Gardner HT, Moore D, Streitfeld FH, Havens WP. Clinical course of viral hepatitis and the effect of exercise during convalescence. *Am J Med* 8: 614, 1950.
195. Tharp GD, Barnes MW. Basketball exercise and secretory immunoglobulin A. *Eur J Appl Physiol* 63: 312-314, 1991.
196. Tharp GD, Barnes MW. Reduction of saliva immunoglobulin levels by swim training. *Eur J Appl Physiol* 60: 61-64, 1990.
197. Thomas NE, Williams DRR. Inflammatory factors, physical activity, and physical fitness in young people. *Scand J Med Sci Sports* 18: 543-556, 2008.
198. Thompson HJ, Ronan AM, Ritacco KA, Tagliaferro AR. Effect of type and amount of dietary fat on the enhancement of mammary tumorigenesis by exercise. *Cancer Res* 49: 1904-1908, 1989.
199. Thompson HJ, Westerlind KC, Snedden J, Briggs S, Singh M. Exercise intensity dependent inhibition of 1-methyl-1-nitrosourea induced mammary carcinogenesis in female F-344 rats. *Carcinogenesis* 16: 1783-1786, 1995.
200. Tisi PV, Hulse M, Chulakadabba A, Gosling P, Shearman CP. Exercise training for intermittent claudication: Does it adversely affect biochemical markers of the exercise-induced inflammatory response? *Eur J Vasc Endovasc Surg* 14: 344-350, 1997.
201. Tomasi TB, Trudeau FB, Czerwinski D, Erredge S. Immune parameters in athletes before and after strenuous exercise. *J Clin Immunol* 2: 173-178, 1982.
202. Uhlenbruck G, Order U. Perspectiven, Probleme und Prioritäten: Sport-immunologie—Die nächsten 75 Jahre [Perspectives, problems and priorities: Sports immunology—The next 75 years]. *Dtsche Z Sportmed* 38: 40-47, 1987.
203. Verde T, Thomas S, Shephard RJ. Potential markers of heavy training in highly trained distance runners. *Br J Sports Med* 26: 167-175, 1992.
204. Weicker H, Werle E. Interaction between hormones and the immune system. *Int J Sports Med* 12(Suppl. 1): S30-S37, 1991.
205. Weiss M. Anamnestische, klinische und laborchemische Daten von 1 300 Sporttauglichkeitsuntersuchungen im Hinblick auf Infekte und deren Prophylaxe bei Leistungssportlern [Anamnestic, clinical and lab-chemical data in 1 300 studies of sport participants in relation to infections and their prophylaxis in competitive athletes]. *Dtsche Z Sportmed* 33: 399-402, 1993a.
206. Weiss M. Infektprophylaxe mit polyvalent Immunoglobulinen: Diskussion anlässlich der Anwendung vor der Olympiade 1992 bei der deutschen Box-Nationalstaffel [Prophylaxis of infections with polyvalent immunoglobulin in top athletes: A discusssion on the occasion of its application in the German boxing team before the Olympic Games of 1992]. *Dtsche Z Sportmed* 33: 466-471, 1993b.
207. Weiss M, Fuhrmansky J, Lulay R, Weicker H. Haufigkeit und Ursache von Immunoglobulin-mangeln bei Sportlern [Frequency and cause of immunoglobulin lack in athletes]. *Dtsche Z Sportmed* 36: 146-153, 1985.
208. Wentworth P, Jentz LA, Croal AE. Analysis of sudden unexpected death in Southern Ontario with emphasis on myocarditis. *Can Med Assoc J* 120: 676-680, 1979.
209. Wesslen L, Pahlson C, Lindquistt O, Hjelm E, Gnarpe J, Larssont E, Baandrup U, Eriksson L, Fohlman J, Engstrand L, Linglof T, Nystrom-Rosander C, Gnarpe H, Magnius L, Rolf C, Friman G. An increase in sudden unexpected cardiac deaths among young Swedish orienteers during 1979-1992. *Eur Heart J* 17: 902-910, 1996.
210. Westerlind KC. Physical activity and cancer prevention—Mechanisms. *Med Sci Sports Exerc* 35: 1834-1850, 2003.
211. Woods JA, Ceddia MA, Wolters BW, Evan JK, Lu Q, McAuley E. Effects of 6 months of moderate aerobic exercise training on immune function in the elderly. *Mech Ageing Dev* 109: 1-19, 1999.

CHAPTER 22

The Skeletal System

Sarah L. Manske, PhD

Grant C. Goulet, PhD

Ronald F. Zernicke, PhD, DSc

Physical activity and exercise are among the most potent nonpharmacological means for developing and maintaining a healthy skeleton. The mechanical forces imparted on the skeleton by weight bearing and muscle contractions are strong effectors of bone homeostasis. Consequently, defining and promoting effective exercise strategies to achieve optimal bone mass during adolescence and maintain bone mass across the life span may have a significant effect on public health by minimizing cost, morbidity and mortality, and maximizing quality of life.

The end of the 19th century was notable for significant progress in the mechanical and morphological understanding of bone structure, function, and growth. Of particular importance was the emergence of the concept that bone form follows function. Over the past 100 yr, many investigators have attempted to elucidate the mechanisms underlying the ability of bone to adapt its external shape and internal microarchitecture in response to its mechanical environment (table 22.1). Aided by technological advancements in noninvasive imaging, our knowledge of the effects of exercise and disuse on bone has increased dramatically in the past century, which has allowed for enhanced assessment of bone structure and composition.

Here we describe the interplay between physical activity and bone and highlight the significant discoveries that contributed to our current understanding of how exercise influences the skeleton. First, we briefly review skeletal biology and highlight mechanisms involved in the adaptation of the skeleton to its functional requirements. We then explore how components of mechanical forces influence the ability of a loading regimen to elicit an osteogenic response, and how bone cells perceive their local mechanical environment and transduce signals into an appropriate cellular response.

Bone Functions and Structure

The skeletal system comprises individual bones and associated tissues—cartilage, tendons, and ligaments. Bone, the primary constituent of this system, serves important biomechanical and metabolic functions. The rigidity and hardness of bone tissue enable the skeleton to support the body; protect vital organs and the cranial, thoracic, and pelvic cavities from mechanical injury; supply the framework for hematopoiesis; and transmit the force of muscular contractions. Metabolically, bone is the body's principal reservoir of calcium and phosphorus. These minerals are essential for the proper function of a variety of systems in the body and are necessary for enzyme reactions, blood clotting, and transmission of nerve impulses. Although typically perceived as an inactive structure, the skeleton is dynamic and is continuously evolving to meet the functional requirements of its physical environment.

Bone tissue is a complex biomaterial consisting of cells and extracellular matrix. Bone matrix is a composite material consisting of 65% mineral and 35% organic phase. The organic phase contains 90% type I collagen and 10% amorphous ground substance (primarily glycoproteins and glycosaminoglycans). The inorganic mineral comprises calcium crystals primarily in the form of hydroxyapatite. The high collagen content of the organic phase gives bone its tensile mechanical properties, and the mineral phase provides bone with superior resistance to compressive stress.

On a gross structural level, bone can be divided into cortical and cancellous bone. Cortical bone constitutes approximately 80% of skeletal mass in the adult human and is largely responsible for the supportive and protective functions of the skeleton. It is a dense, solid mass with only microscopic channels. The remaining 20% of bone mass is cancellous bone—a lattice of plates and rods called trabeculae.

Table 22.1 Discoveries (1910-2010) Linking Exercise and Bone

Year	Investigator(s)	Milestone
1917	Thompson (158)	Suggested relation between bone architecture and tissue deformation (strain) rather than load (stress).
1920	Jores (60)	Proposed that bone cells act as sensors for structural alterations.
1931	Greig (44)	Suggested that microscopic damage may stimulate bone remodeling.
1971	Hert et al. (50)	Demonstrated influence of dynamic loading on bone remodeling.
1984	Drinkwater et al. (25)	Bone deficiencies with the female athlete triad.
1987	Frost (35, 39)	Proposed the mechanostat theory to explain how bone mass adapts to mechanical loads.
1995	Turner et al. (161)	Demonstrated that bone formation response was highly influenced by strain rate.
1995	Kannus et al. (63)	Reported large side-to-side differences in bone mineral content in playing arm versus nonplaying arm of tennis players.
1996-1998	Multiple research groups (10, 13, 111)	First exercise interventions in children in order to promote bone health.
2000	Robling (134)	Experimentally demonstrated the utility of incorporating rest periods between loading bouts in order to augment osteogenic response.

Macroscopically, long bones of the appendicular skeleton (e.g., humerus, ulna, femur, and tibia) serve as the typical example for the structure of bone. Generally, an adult long bone consists of a central cylindrical cortical shaft (called the diaphysis) filled with yellow marrow and two wider and rounded ends, where the cortical shell thins and cancellous bone is prevalent (called the metaphysis), filled with red marrow. A growth plate separates metaphyses from epiphyses, which are adjacent to the joint space.

Cell Populations

Bone tissue is highly populated with living cells. The principal cellular components include osteoclasts, osteoblasts, and osteocytes. Osteoclasts are giant, multinucleated (with up to 50 nuclei) cells that range in diameter from 20 μm to more than 100 μm. They are derived from cells of the hematopoietic marrow and are responsible for resorbing bone through solubilization of both the mineral and the organic component of the bone.

Osteoblasts are mononuclear cells that are approximately 15 to 30 μm in diameter. They are derived from fibroblast-like mesenchymal precursors, which are located near bone surfaces (i.e., periosteum, endosteum, and marrow stroma). Osteoblasts are principally bone-forming cells that synthesize and secrete unmineralized bone matrix (osteoid). They also participate in the calcification of bone, regulation of calcium and phosphorous flux, and regulation of osteoclast function via secreted hormones. When osteoblasts are not actively forming bone, they are flattened, elongated cells that cover bone surfaces and are referred to as bone-lining cells.

During bone formation some osteoblasts are incorporated into the newly deposited matrix. These osteoblasts differentiate into osteocytes by losing much of their organelles and acquiring long, slender processes. Osteocytes, the most abundant cell type in mature human bone, account for more than 90% of all bone cells. The osteocyte body and its processes are contained in voids called lacunae and channels called canaliculi, respectively. Osteocytes play a key role in homeostatic, morphogenetic, and restructuring processes of bone mass.

Functional Adaptation of Bone

The architecture of the skeleton arises from genetic constraints and epigenetic conditions. The local mechanical environment also plays a significant role in the morphology and density of bone tissue. The ability of the skeleton to adapt to its mechanical milieu is dependent on specific parameters of the mechanical signal and can be influenced by a variety of systemic factors, including genetic, hormonal, and dietary considerations.

Early Foundations

The concept that bone can sense its mechanical environment and respond accordingly has received significant attention for more than a century. In 1892, Wolff published his treatise *Das Gesetz der Transformation der Knochen* (*The Law of Bone Transformation*, translation of the German edition, published in 1896; 173), which described his observations that bone adapts its external shape and internal architecture to meet the demands of its mechanical environment (figure 22.1). Wolff's law is commonly cited in reference to the ability of bone's form to follow function. Wolff's contemporary, Roux, was actually the originator of that concept (136), but his theory was not validated until more recently (37). Although Wolff did not offer any specific principles for predicting bone adaptations in response to mechanical loading, he did postulate the existence of mathematical laws governing bone form and function and suggested that they related to principal compressive and tensile stresses. In the 1960s, Frost (38) and Pauwels (125) contributed significantly to the development of explicit mathematical models relating stress to alterations in bone geometry.

Figure 22.1 Julius Wolff (1836-1902). After decades of research, Wolff published his observations of the functional adaptation of bone in *The Law of Transformation of Bone*.
Courtesy of the Julius Wolff Institute, Charité- University Medicine Berlin, Germany.

Modeling and Remodeling

Bone adaptation through architectural modification relating to shape, size, content, and distribution of bony tissue is achieved through modeling and remodeling processes. Modeling is the independent activity of osteoblasts or osteoclasts that determines a bone's shape and outside and inside cortical diameters (figure 22.2*a*). Modeling is affected by two cellular mechanisms called drifts. Osteoblastic formation drifts add new circumferential lamellae over the bone's surfaces after high-intensity or unusual patterns of mechanical loading, whereas osteoclastic resorption drifts remove bone. Bone modeling is most active in infants and tends to decrease after skeletal maturity. Remodeling is the turnover of bone in discrete packets; new bone replaces affected bone without changing the geometry (figure 22.2*b*). In 1963, Frost referred to the complex mediator mechanism responsible for effecting bone remodeling as the basic multicellular unit (34) (figures 22.2*c* and 22.3). In cortical bone, basic multicellular units proceed by osteonal tunneling, during which osteoclasts excavate a canal that is subsequently refilled by osteoblasts. The resulting secondary osteons are typically 100 to 200 μm in diameter and may extend for as long as 10 mm. Unlike bone modeling that normally subsides after skeletal maturity, basic multicellular unit-based remodeling has been observed in the human fetus and continues throughout life.

Mechanical Parameters

Experimental investigations, particularly over the past 50 yr, have discovered several mechanical parameters that influence the ability of a given loading regimen to elicit an adaptive response in bone.

Dynamic Loading

In the early 1920s the importance of dynamic—as opposed to continuous (i.e., static)—mechanical loading in initiating an osteogenic response was suggested by Jores (60) and Loeschke and Weinnoldt (92). The first systematic series of experiments that investigated the effect of static versus dynamic loading was that conducted by Hert and colleagues in the late 1960s and early 1970s (50, 51, 91). They concluded that whereas all animals (rabbits) reacted to dynamic loading by periosteal and endocortical appositions on the stressed sides of the cortex, static loads caused no such effect. Shortly thereafter, Chamay and Tschantz (19) arrived at a similar conclusion from experiments using canine ulnae, which demonstrated that adaptive hypertrophy of long bones was elicited only by dynamic loads. In 1984, Lanyon and Rubin's (87, 141) studies on func-

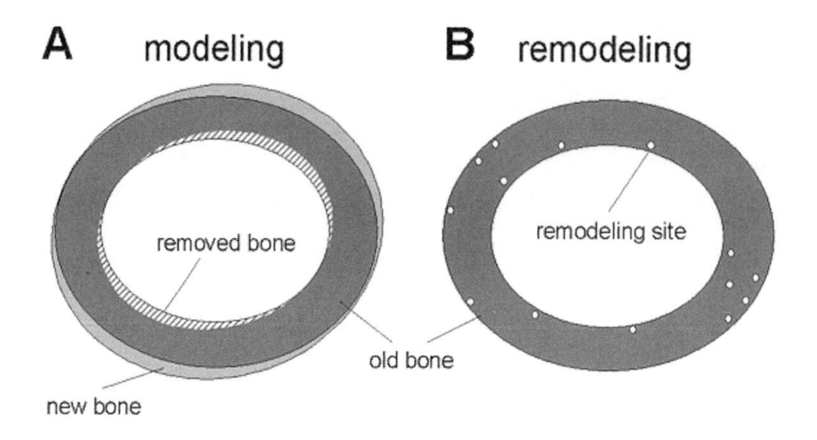

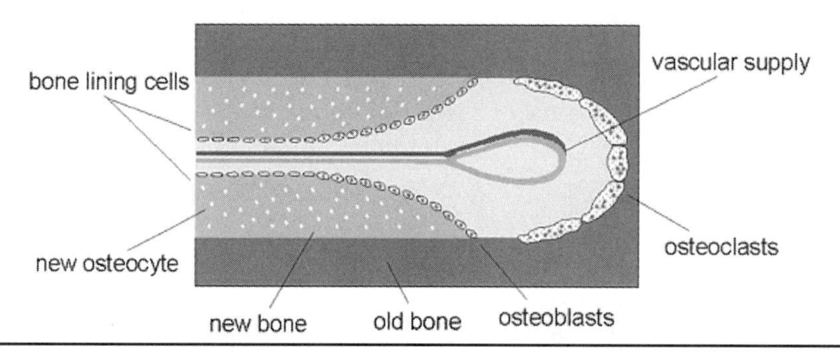

Figure 22.2 Schematic representation of modeling and remodeling in a bone cross-section. (*a*) During modeling, osteoblast formation drifts add new bone to the periosteal (shown as light gray) and endocortical (shown as hatched regions) surfaces. (*b*) During remodeling, discrete packets of pre-existing bone are replaced with secondary bone. (*c*) In cortical remodeling, osteoclasts tunnel through bone and osteoblasts follow, laying down new, unmineralized matrix. A vascular supply is required to supply the cells with adequate nutrition and waste removal.

Reprinted, by permission, from R.F. Zernicke, G.R. Wohl, and J.M. LaMothe, 2006, The skeletal-articular system. In *ACSM's advanced exercise physiology*, edited by C.M. Tipton (Philadelphia: Lippincott, Williams & Wilkins), 98.

tionally isolated avian bone identified the nonosteogenic potential of static loading as well as its deleterious effect on bone. Both statically loaded and disuse groups demonstrated an increase in endocortical diameter and intracortical porosity, which produced a significant decrease in cross-sectional area (figure 22.4). When contemplating the natural requirements placed on the skeleton throughout evolution, it was suggested that the lack of an osteogenic response to static load bearing was not surprising because there was no need for the skeleton to adapt to such an environment (140).

Strain Magnitude

Measurement has shown that limb bone dimensions are scaled geometrically with body size. However, Rubin and Lanyon (139) first argued that peak stresses are comparable regardless of animal size. They suggested that similar factors of safety (i.e., strength proportional to expected loads) are maintained not by allometrically scaling bone dimensions but rather by scaling the magnitude of peak strains. Using the functionally isolated turkey ulna, Rubin and Lanyon (142) demonstrated that peak longitudinal strains less than 1,000 microstrain (με) were associated with increased intracortical remodeling and porosity and with endocortical resorption. In contrast, peak strains greater than 1,000 με were associated with little intracortical remodeling activity and substantial periosteal and endocortical bone apposition. In 1983, Frost (36) proposed a basic bone property called the minimum effective strain (MES), which predicted when and where lamellar bone architectural adaptations would occur in response to mechanical loading. According to Frost's model—called the mechanostat—only strain magnitudes larger than the MES would elicit bone architectural adaptations. Consequently, in MES terms, an adapted skeleton is one in which the architecture remains stable (i.e., strain rarely or never surpasses the MES).

In 1996, Burr and colleagues (17) (figure 22.5) provided the first measurement of in vivo human tibial strains during vigorous activity. They demonstrated that principal compressive and shear strains reached nearly 2,000 με during uphill and downhill zigzag running—nearly three times greater than strains engendered during walking.

Figure 22.3 Harold M. Frost (1921-2004) is considered to be one of the most influential researchers in the field of skeletal biology and medicine in the past century. Among his many contributions were the development of bone histomorphometry, the concept of the basic multicellular unit, and the mechanostat model. Frost pioneered much of the research relating muscle to bone and participated in many studies examining the effects of exercise on bone.

Reprinted, by permission, from R.R. Recker et al., 2001, "Harold M. Frost. William F. Neuman awardee 2001," *Journal of Musculoskeletal Neuronal Interact* 2: 117-119.

Figure 22.4 Clinton T. Rubin (left) and Lance E. Lanyon (right), who are notable for the creative development of experimental methods for studying how mechanical stimuli influence bone growth, healing, and homeostasis.

Photos courtesy of Clinton T. Rubin and Lance E. Lanyon.

Strain Rate

The necessity for a dynamic loading regimen to elicit an osteogenic response led to the conclusion that the rate at which a mechanical load is applied to bone is an important factor in the adaptation process. In 1995, Turner and colleagues (161) (figure 22.5) were the first to demonstrate experimentally that bone formation response was highly influenced by strain rate rather than by absolute magnitude alone. By maintaining a constant peak strain magnitude, they examined bone formation in relation to strain rate by applying bending loads to the tibiae of adult rats at monotonically increasing strain rates. They concluded that mechanical loading increased bone formation and mineral apposition rates in proportion to the applied rate of strain. Similarly, Mosley and Lanyon (112) demonstrated that a high strain rate increased periosteal new bone formation relative to low or moderate strain rates. High-impact jumping exercise in roosters (62), noninvasive cantilever bending of the tibiae of mice (84), and endogenous loading of the radii and ulnae of sheep (122) further supported the role of strain rate as a major determinant of bone's adaptive response to mechanical loading.

Strain Gradient

Using functionally isolated turkey radii and an exogenous loading regimen coupled with animal-specific computational models, Gross and colleagues (45) first examined in 1997 the correlation of peak strain gradients with periosteal bone formation. They concluded that circumferential strain gradients were best correlated with experimentally observed bone formation. At the same time, Judex and colleagues (61) demonstrated

Figure 22.5 Charles H. Turner (left, 1961-2010) and David B. Burr (right), who contributed significantly to understanding the response of bone to mechanical stimuli at the cell and tissue levels.

Photos courtesy of Charles H. Turner and David B. Burr.

a similarly strong correlation between the amount of periosteal bone formation and peak circumferential strain gradients induced during high-speed running in roosters.

Loading Frequency

In addition to the large-amplitude strains typically associated with functional activity, low-magnitude strains arise through muscular activity in the frequency band of 10 to 50 Hz. In 2000, Fritton and colleagues (33) published a seminal study that quantified the daily in vivo strain events in dog, sheep, and turkey bones. They showed that large strains (>1,000 $\mu\varepsilon$) occurred relatively few times per day, whereas very small strains (<10 $\mu\varepsilon$) occurred thousands of times per day. Previously, in 1994, Rubin and colleagues (143) departed from the premise that only the large-amplitude signals were anabolic to bone and showed that low-level, high-frequency strain signals had a significant influence on bone morphology. Hsieh and Turner (53) also demonstrated the anabolic effects of a high loading frequency in the ulnar diaphysis of adult rats. They observed that periosteal bone formation increased in a dose-response manner with loading frequency.

In the past 10 yr, numerous studies have further investigated the anabolic potential of low-level vibrations (138). Recently, Rubin and colleagues performed clinical trials to assess the efficacy of low-magnitude, high-frequency mechanical stimulus in preventing bone loss in humans. Brief periods (<10 min/d) of high-frequency (30 Hz) vibration stimuli effectively inhibited bone loss in the trochanter region of the femur and the lumbar spine of postmenopausal women (137) and increased bone mass in the proximal tibia of adolescent children with musculoskeletal disorders (e.g., cerebral palsy and muscular dystrophy) (168).

Temporal Distribution

Experimental evidence suggests that bone cells exhibit a refractory period (14) such that after stimulation they do not respond to further stimulation for some time. Consequently, loading cycles that occur toward the end of a loading regimen are not as potent as those that occur toward the beginning, a phenomenon described by Turner in 1998 as "diminishing returns" (160). Investigators have noted the significance of a recovery period in enhancing mechanically induced bone formation. In 1993, Chambers and colleagues (20) and Chow and colleagues (21) reported that the formation rate for trabecular bone in loaded rat vertebrae was approximately seven and four times higher, respectively, in animals subjected to a loading regimen distributed over several days than in animals that had been administered a single continuous loading bout. Subsequently, Robling and colleagues (135) showed that only 8 h of recovery between loading sessions was necessary to restore mechanosensitivity to adapted bone cells. The same study also concluded that inserting short-term recovery periods between each loading cycle resulted in significantly higher bone-formation rates. Srinivasan and colleagues (153) concluded that 10 s of rest inserted between each load cycle of a low-magnitude loading protocol significantly enhanced the osteogenic potential of a murine loading regimen.

Mechanotransduction

Although several mechanical parameters have been identified over the years as having an influence on the osteogenic potential of an exogenous loading regimen, the mechanism by which the biophysical force is converted into a cellular response—a process called mechanotransduction—remains incompletely understood. Duncan and Turner (28) described mechanotransduction in bone as a four-step process involving the following: (1) mechanocoupling (the transduction of mechanical force into a local mechanical signal perceived by a sensor cell), (2) biochemical coupling (the transduction of a local mechanical signal into a biochemical signal), (3) transmission of the signal from the sensor cell to the effector cell, and (4) effector cell response (the appropriate tissue-level response; i.e., altered bone resorption or formation). For bone to adapt to its functional influences, feedback derived from stimuli produced in the tissue must be perceived by the bone cells.

Osteocytes as Mechanosensors

Jores in 1920 was the first to propose that bone cells act as sensors for structural alterations (60). Subsequent research confirmed his hypothesis. In the 1960s, it was shown that a fraction of active osteoblasts become surrounded by and then incorporated into the newly synthesized bone matrix and subsequently transform into osteocytes (8, 26). During the transformation from bone-forming osteoblast to embedded osteocyte, the cell changes its shape and generates long processes. Doty (24) was instrumental in providing evidence that the osteocytic processes from neighboring cells connect at gap junctions. A gap junction allows ions and compounds of low molecular weight to pass between the connected cells without passing through the extracellular space. Aarden and colleagues (2) and Knothe Tate (77) were pivotal in describing the interconnected cells as a functional syncytium (figure 22.6). The placement and distribution of this syncytial network is architec-

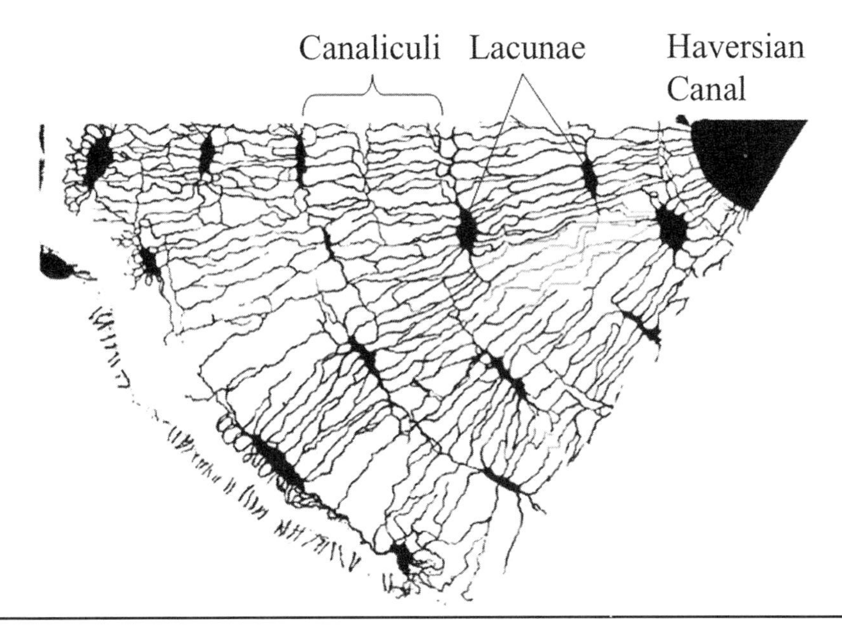

Figure 22.6 A pie-shaped section of the lacunocanalicular network. Osteocytes, housed in lacunae, are in contact with neighboring cells via gap junctions in the canalicular microchannels. Osteocytes are believed to be the cells responsible for sensing and responding to mechanical stimuli.

turally well suited to sense local mechanical signals in the mineralized tissue.

Transduction Mechanisms

The precise local signal, or combination thereof, that results in stimulatory deformation of the osteocytic cytoskeleton remains unclear. In general, the potential stimuli are categorized as direct or indirect.

Direct stimulation refers to cellular deformation resulting from strain of the extracellular matrix. In the late 1970s and early 1980s, investigators began reporting on the effects of mechanical strain on bone cells in vitro (e.g., 47, 174). Mechanical stimulation using devices such as flexible cell-culture plates increased proliferation, deoxyribonucleic acid biosynthesis, metabolic activity, and the production of growth factors. It was generally assumed that the level of strain magnitude responsible for bone adaptation was that measured by strain gauges in vivo (i.e., 50-2,500 $\mu\varepsilon$ in humans). However, to elicit a cellular response by direct mechanical deformation, strain levels one to two orders of magnitude greater than strains normally experienced by the whole bone in vivo were required. These large strains necessary for osteocytic stimulation could not be derived directly from matrix deformations because they would result in bone fracture. Consequently, indirect mechanisms of cellular stimulation have been the focus of numerous studies over the past 20 yr.

Dynamic loading elicits changes in intralacunar fluid pressure, fluid flow through the lacunocanalicular spaces, and electrical potentials from the flow of charged particles. Current hypotheses focus on these strain-derived, fluid-related stimuli rather than direct cellular deformation as primary means of mechanotransduction.

Bone Fluid as a Coupling Medium

The nature of the fluid–solid interaction in bone resembles that of a stiff, water-saturated sponge. In 1977, Piekarski and Munro (126) first investigated theoretically the flow of fluid in bone. They suggested that when bone was deformed, the resulting compressive and tensile strain caused the fluid pressure to increase and decrease, respectively, which in turn caused fluid movement from regions of high pressure to regions of low pressure (figure 22.7). The velocity at which fluid moved through the solid was dependent on the rate at which the deformation was applied and the permeability of the bone. Dynamic loads were, therefore, associated with pressure-driven fluid flow whereas prolonged static loading was characterized by an absence of interstitial fluid movement. The correlation among dynamic loading regimens, osteogenic response, and fluid flow suggested that bone fluid served as the coupling medium through which external mechanical forces were transduced into effects at the cellular level (77). Pericellular bone fluid may participate in osteocytic stimulation via a number of mechanisms, in-

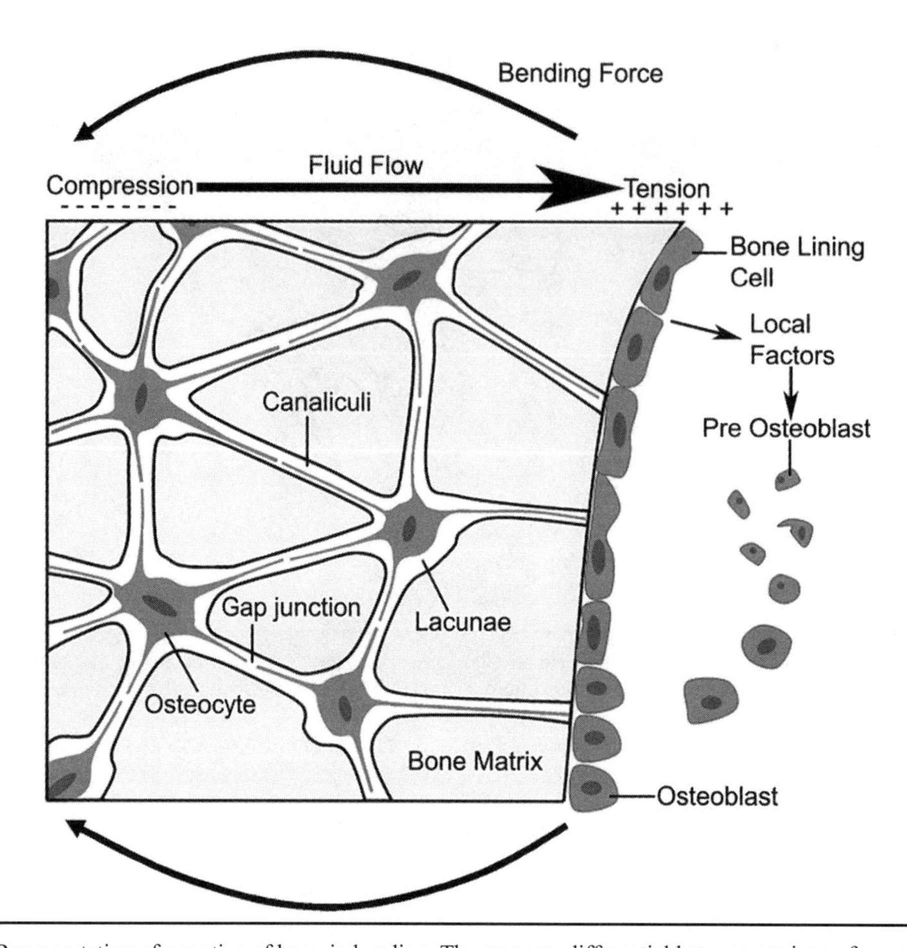

Figure 22.7 Representation of a section of bone in bending. The pressure differential between regions of compressive stress and tensile stress forces fluid through the lacunocanalicular network. The flow of fluid across the cell surface and its processes generates shear stress, which may be a primary means of mechanotransduction. Mechanotransductive signals are transmitted throughout the lacunocanalicular network via osteocytic cell processes connected by gap junctions. In response to those signals, bone-lining cells can release factors that stimulate differentiation of preosteoblasts into active bone-forming osteoblasts.

cluding fluid shear stress, fluid pressure, chemotransport, and streaming potentials.

Fluid Shear Stress

Fluid shear stress is generated by the motion of fluids across surfaces. Osteocytes are subjected to shear stress as a result of the flow of interstitial fluid through the lacunocanalicular network. The mineralized nature of bone and the small scale of the pore spaces have precluded direct measurement of in situ fluid forces experienced by bone cells. Consequently, shear stress magnitudes used in experiments have been derived primarily from influential theoretical studies by Weinbaum and colleagues (e.g., 170, 176). In the late 1990s, Owan and colleagues (123) and Smalt and colleagues (152) compared the sensitivity of bone cells to direct mechanical strain with that of fluid shear stress. They demonstrated that cells were more responsive to physiologically relevant shear stress than to direct deformation with respect to production of nitric oxide and prostaglandin E_2, regulators of bone metabolism.

In addition to theoretical studies, over the past 20 yr several investigators have examined the response of bone cells to fluid shear stress in the absence of mechanical strain. The studies of Jacobs (58) and Klein-Nulend and colleagues (72) confirmed the mechanosensitivity of bone cells to shearing regimens in vitro. In 2003, Qin and colleagues (130) described the regulatory role of fluid flow in a novel in vivo experiment in which flow was generated in the absence of matrix strain. The ulnae of adult turkeys were functionally isolated and the adaptive response to disuse or disuse plus 10 min/d of fluid flow was quantified. Fluid flow was induced by low-magnitude (8 kPa), high-frequency (20

Hz) oscillations of intramedullary pressure. Disuse alone resulted in bone loss via endocortical resorption and increased intracortical porosity, whereas disuse in conjunction with fluid flow increased bone mass via periosteal and endocortical bone formation. Furthermore, the sites of greatest osteogenesis correlated with the greatest transcortical fluid pressure gradient, suggesting that in vivo pressure-driven fluid flow may directly influence bone's adaptive response to mechanical loading.

Fluid Pressure

Bone cells are continuously exposed to fluid pressure as the pore fluid spaces compress or dilate in response to compressive or tensile stresses, respectively. It was established by Zimmerman in 1970 (180) that a variety of biological processes (e.g., protein synthesis and protoplasmic streaming) were affected by changes in hydrostatic pressure in the range of 0.1 to 70 MPa. To investigate the potential effect of hydrostatic pressure in relation to bone, Klein-Nulend and colleagues (74) first examined the influence of intermittent hydrostatic compression (IHC) on calcification of growth plate cartilage. Hypertrophic chondrocytes responded to IHC and continuous hydrostatic compression (CHC) by an increased deposition of calcium-phosphate mineral in the matrix. However, intermittent mechanical stimulation elicited a higher cellular response than did continuous stimulation.

Consistent with previous studies on static loading, Ozawa and colleagues (56, 124) demonstrated a negative effect of CHC on osteoblast-like cells. Collagen synthesis, calcification, and alkaline phosphatase activity—an early marker for bone formation—were decreased whereas production of prostaglandin E_2 was enhanced. Results suggested that in addition to inhibiting osteoblast differentiation, CHC stimulated bone resorption by generating new osteoclasts through a mechanism involving prostaglandin E_2 production.

The influence of IHC on osteoblastic and osteoclastic activity was further examined by several investigators. In the late 1980s and early 1990s, Klein-Nulend and colleagues (75, 76) reported that exposure to IHC resulted in increased bone formation as well as decreased bone resorption and showed that IHC stimulated the release of an autocrine growth factor and a paracrine factor that inhibited the differentiation of osteoclast precursor cells (73).

Chemotransport

Encased in the calcified matrix, osteocytes are dependent on patent metabolic pathways defined by the lacunocanalicular system for the supply of adequate molecular transport. Diffusion is a major contributing mechanism for molecular transport in the extravascular spaces of organs and soft tissues. However, given the relative impermeability of bone tissue, Piekarski and Munro suggested in 1977 (126) that diffusive transport mechanisms alone were likely unable to ensure a requisite supply of nutrients to maintain anabolic activity or to maintain pathways for waste removal in the case of catabolic activities. Rather, physiological deformations of bone may promote molecular transport through convection from capillaries to osteocytes via canaliculi. Therefore, in addition to its direct effects on cells (i.e., shear stress and pressure), mechanical load-induced fluid flow may indirectly elicit an osteogenic response through enhancement of chemotransport. Although it was suggested 20 yr earlier by Piekarski and Munro (126), Knothe Tate and colleagues were the first to demonstrate experimentally the existence of load-induced convective transport phenomena in cortical bone. A series of elegant ex vivo (78), in vitro (79), and in vivo (80, 81) methods were developed for the study of transport processes under controlled mechanical-loading conditions. Mechanical loading increased convective transport of both small and large molecular tracers through the lacunocanalicular network. Donahue and colleagues (23) later suggested that chemotransport alone would not elicit a cellular response but rather modulated the mechanosensitivity of bone cells to fluid shear stress.

Streaming Potentials

The conversion of mechanical loads to bioelectrical signals in bone has been suggested to regulate cellular activity. These signals in bone are attributed to electrokinetic behavior where kinetic energy is converted into electrical energy due to the motion of an ion-carrying extracellular fluid across the oppositely charged channel wall of the bone matrix. In 1962, Bassett and Becker (7) were the first to experimentally quantify electric potentials in response to mechanical loads using feline fibulae subjected to bending loads. The amplitude of the generated potential was dependent on the rate and magnitude of the applied deformation, whereas the polarity was determined by the stress state (the compressive region developed negative potentials with respect to other areas). Starkebaum and colleagues (154) later refined macroscopic studies by implementing a microelectrode that enabled the study of streaming potentials to a spatial resolution of 5 µm. Electric fields one to three orders of magnitude larger than those measured by macroscopic methods were discovered; maxima occurred at the osteonal canal boundary in human and bovine cortical bone. Based on the observation of large electric potentials surrounding the canals, it was suggested that streaming potential effects

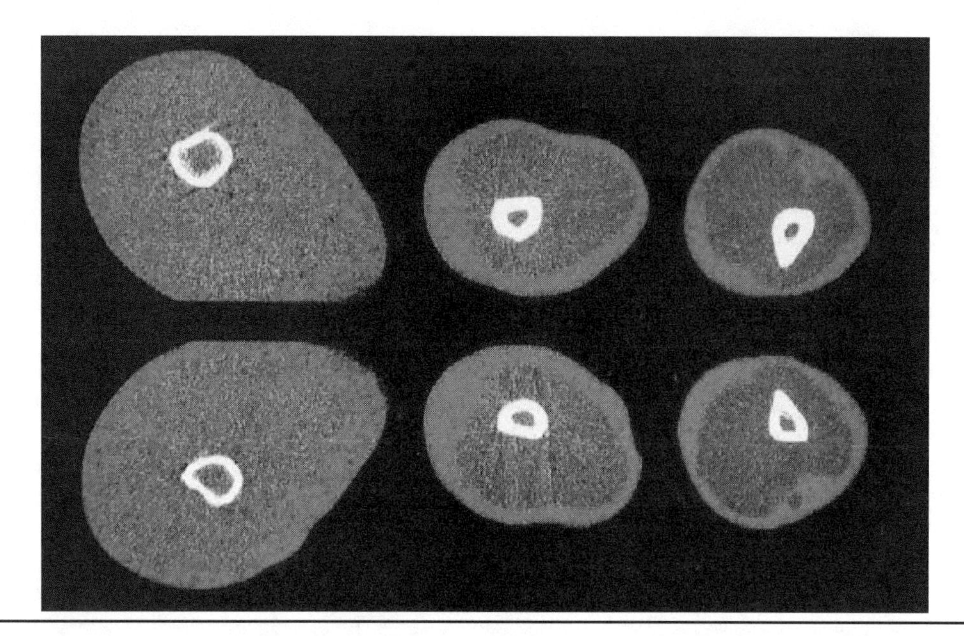

Figure 22.8 Humeral peripheral quantitative computed tomography scan of a 26-yr-old male tennis player showing greater bone cross-sectional area in the dominant arm (upper panel) than in the nondominant arm (lower panel). Slices from left to right are from the proximal humerus, humeral shaft, and distal humerus.

may initiate or influence cell function by acting first at the canal lining. However, because fluid shear stress gained acceptance as the primary means of mechanotransduction, relatively few studies now focus on streaming potentials in relation to functional adaptation of bone.

Adaptations to Exercise

Positive Effects of Exercise

Human studies of exercise and bone have corroborated many of the findings from animal and cell studies and have applied this knowledge to develop physical (exercise) interventions for accumulating and maintaining bone mass. In recent years, the sensitivity and ability to describe the nature of changes induced by physical activity in humans have improved with the development of new imaging technologies (see "Imaging Modalities"). Here, we describe the effects of mechanical factors and exercise on bone health, which is collectively determined by bone mass, structure, quality, and rate of turnover. Many of the outcomes measured include bone mineral content (BMC), areal bone mineral density (aBMD; see "X-ray and DXA"), and, more recently, structural measures such as cortical area.

Observational studies of elite athletes and nonathletic controls have provided insight into the types of exercise that have positive effects on bone mass and structure. For example, Finnish researchers (46, 63) reported large side-to-side differences in BMC and aBMD between the dominant and nondominant humerus and radius in high-caliber tennis players. For instance, they found 17% greater BMC in the dominant humeral shaft in experienced tennis players (figure 22.8). There were similar trends between dominant and nondominant arms in subjects who did not play tennis, although the reported differences were smaller in magnitude. They were one of the first groups to examine the effects of exercise with peripheral quantitative computed tomography (pQCT) (see "Quantitative Computed Tomography"). As such, they were able to demonstrate that the between-arm difference in BMC was attributable to increased bone size and not volumetric bone mineral density.

Similarly, a 3 yr longitudinal study that compared female gymnasts with normally active controls showed that the gymnasts had consistently greater total-body BMC from prepubertal to postpubertal stages (121). The large increases in BMC and aBMD observed in gymnasts were attributed to the substantial strain magnitudes and rates experienced by these athletes during training (157). Likewise, nonelite ballet dancers had significantly greater prepubertal BMC than did controls (105).

Interventional Studies

After discovering the potential to use exercise to improve bone mass, researchers conducted clinical trials and interventions. Due to the propensity of osteoporosis and osteoporotic fractures, the majority of intervention studies have focused on improving bone health in postmenopausal women, although osteoporosis in men has gained increasing attention (147). By the end of the 20th century, a critical mass of interventional studies had been conducted for meta-analyses to be performed. The most thorough meta-analysis of controlled clinical trials with postmenopausal women, using individual patient data, indicated a trend toward increased aBMD (a 0.51% increase in aBMD in the exercise subjects and 0.13% in the control subjects) in the femoral neck in response to exercise (67). During the era of dual-energy X-ray absorptiometry (DXA) measurements, study outcomes were often measured at the femoral neck because of the frequency and severity of fracture in that region and the strong relationship between measurements at that site and risk of hip fracture (22). The lack of a significant effect found in this meta-analysis may have been due to the different types of exercise used in these interventions, including walking (29), resistance training (129), other forms of weight-bearing exercise (18, 93), and cycling (11) as well as the large range in intensity, frequency, and duration. The work in animal studies suggests that higher-impact activities would be more beneficial to bone; however, lower-impact activities have commonly been used due to practical issues in implementation. Several groups have used higher-impact weight-bearing activities such as jumping (e.g., 64, 163). For example, Uusi-Rasi and colleagues (163) found that a 3 d/wk progressive exercise program that incorporated multidirectional jumps and calisthenics conducted for 12 mo increased section modulus (an estimate of bone strength) by 3.6% in the distal tibia compared with controls. Another study that applied the concept that strain rate, in addition to strain magnitude, is an important osteogenic stimulus found that powerlifting that included fast and explosive concentric exercises was more effective at improving lumbar spine and hip aBMD in postmenopausal women than regular strength training (156).

At the end of the 20th century, researchers began to intervene in children with the hope of increasing peak bone mass and ultimately aiming to optimize and maintain bone mass through adulthood (10, 13, 111). Bailey and colleagues (4) found that approximately 26% of adult bone mass is accrued in the 2 yr during which peak gains in height are made. Earlier research also suggested that young, growing animals appeared to be more sensitive to mechanical loading than adult animals (155, 162). Further, the Finnish studies on tennis and squash players found that the age at which exercise is started has a significant effect on bone mass and quality. Girls who began playing before menarche had nearly double the benefit on BMC than those who began playing after menarche (63).

Some of the interventional trials with children have incorporated resistance training (10), whereas other interventions focused on activities with high strain rates such as jumping (41, 48, 101). For example, a program in which children were active for 12 min 3 d/wk that incorporated diverse weight-bearing activities into regular physical education classes resulted in a 4% to 5% greater increase in lumbar spine and femoral neck BMC in prepubertal and early pubertal girls compared with controls after 2 yr (101). Similar interventions were employed by other groups with reasonable success, particularly for prepubertal and early pubertal children (41, 48, 52, 102, 111) compared with postpubertal children (10).

With advances in imaging, researchers have been able to better understand how bone changes in response to exercise rather than examining only the magnitude of the response. Macdonald and colleagues (99) found that a daily school-based jumping program increased estimated anterior–posterior tibia bending strength and tended to add bone in the anterior, medial, and posterior (but not lateral) tibial compartments.

Non-Weight-Bearing Activities

Frost's mechanostat theory suggested that bone has a lazy zone. As such, most evidence suggests that repetitive non-weight-bearing activities, such as cycling and swimming, have minimal benefit to bone because they may not elicit strain effects large enough to promote adaptive osteogenesis (27, 119, 157). However, a recent pQCT study compared several types of sports and found that the polar section modulus (a geometric factor associated with bone strength) of the humeral midshaft was similar for swimmers and participants of impact-loading sports such as volleyball and racket sports (120). This suggested that in addition to large ground-reaction forces produced from weight-bearing activities, muscular contraction can produce large-magnitude forces that can significantly alter bone geometry and quality.

Negative Effects of Exercise

Overuse and Stress Fractures

Although repetitive loading can induce increased bone formation, injury may occur when the magnitude and frequency of loading exceed the bone's ability to adapt. Microcracks and microfractures can develop in bone.

Such microcracks can lead to an increased activation of remodeling (9, 110). It is believed that this remodeling removes damaged bone and prevents the continuation to a fracture (103). A continuum of overuse responses —a pain-free stress reaction, a stress reaction characterized by bony tenderness, or a stress fracture—can occur in bone (70). The term *stress reaction* describes bone with evidence of heightened remodeling but with an absence of radiological evidence of fracture (171). Such stress reactions are common and are detectable using a combination of radiographs, bone scans, or magnetic resonance imaging scans.

Stress fractures were first reported in military personnel in the 1800s and in the past several decades have become more prevalent among participants in sport and exercise (15, 175). In 1855, Briethaupt noted the painful, swollen feet of several Prussian military recruits who had engaged in long, forced marches, and Stechow in 1897 reported the first X-rays of stress fractures (march fractures) (57).

Stress fractures occur less frequently than pure mechanical loading (i.e., material fatigue failure) alone would predict, suggesting that the progression to stress reaction and subsequent stress fracture involves physiological processes of bone adaptation to mechanical loading. Further, in normal conditions, muscle forces may protect the bone from damage because strain and strain rates increase with muscle fatigue (109). This is not to discount the role of mechanical fatigue, however, because microfractures have been detected at remodeling sites. If sufficient microdamage accumulates, macrocracks propagate and stress fracture lines develop (146).

Stress fractures can happen in essentially any bone in the body but are most frequently found in the tibia, which accounts for up to 50% of all stress fractures (175). The mechanical loads of specific movements play a prominent role in determining the fracture site. Runners—the most common victims of tibial stress fracture—exhibit fractures focused between the middle and distal thirds of the tibia, athletes in jumping sports tend to have proximal tibial stress fractures, and dancers sustain more midshaft tibial stress fractures (171).

Female Athlete Triad

The female athlete triad is characterized by low energy availability (due to inadequate dietary intake or excessive physical training), menstrual irregularities, and subsequent loss of bone mass (117). Energy availability can be defined as "dietary energy intake minus exercise energy expenditure" (95) and is considered the primary nutritional concern for the female athlete. Low energy availability has been highlighted in the most recent position statement of the American College of Sports Medicine as the most important factor of the triad that impairs skeletal health (117). Short-term (96-98) and long-term (16, 172) studies have reported that low energy availability, especially in the form of fat (95), adversely affects reproductive function by disrupting luteinizing hormone pulsatility and negatively affecting ovarian function and estrogen production.

Low estrogen levels, as found with amenorrhea (low estrogen level and an abnormally low number of menstrual cycles), had been linked to low bone mineral density, but as late as 1984 (25) the interrelation between exercise and amenorrhea remained equivocal. A pioneering study by Drinkwater and colleagues was the first to reveal that amenorrheic women ran significantly longer distances than did eumenorrheic women and, more importantly, that the average bone mineral density in the lumbar vertebrae of amenorrheic athletes was nearly 14% lower than that of matched eumenorrheic athletes and was equivalent to that of women 51.2 yr of age (figure 22.9). The bone density deficit in amenorrheic athletes might not be fully recoverable even upon resumption of normal menses (66). This link between amenorrhea and low bone density was fundamental to defining the female athlete triad.

Low energy availability and disrupted menstrual status impair bone turnover, essentially uncoupling the formation and resorption processes, in favor of resorption (95). Bone density is commonly low in athletes experiencing amenorrhea or oligomenorrhea (95). In a randomized clinical trial, bone resorption markers increased and bone formation markers decreased in as

Figure 22.9 Barbara L. Drinkwater is a pioneer in the field of exercise physiology. Her special areas of interest are the female athlete and her physical performance under environmental stressors, the effect of exercise-associated amenorrhea on bone health, and the role of exercise and calcium in preventing osteoporosis.

Photo courtesy of Barbara L. Drinkwater.

little as 5 d after energy availability was reduced below 30 $kcal \cdot kg^{-1} \cdot d^{-1}$ of fat-free mass (healthy, adequately nourished, sedentary young adult women achieve energy balance at an average of 45 $kcal \cdot kg^{-1} \cdot d^{-1}$ of fat-free mass) (55). By comparison, whereas eumenorrheic athletes may restrict their energy availability by approximately 30% (>30 $kcal \cdot kg^{-1} \cdot d^{-1}$ of fat-free mass), amenorrheic athletes may restrict theirs by 44% to 67% (15-25 $kcal \cdot kg^{-1} \cdot d^{-1}$ of fat-free mass) (159). Maintaining healthy bone mass is critical because low bone density is an etiological factor in stress fractures in athletes (115), and low bone mass is the best predictor of postmenopausal fractures (54).

Effect of Exercise on Stature

Although exercise has many positive effects on bone quality, there has long been concern that intensive exercise may attenuate stature or linear growth. Shorter limb lengths are typically observed in currently active gymnasts and dancers (127, 149). Self-selection into sports based on body shape and size, however, likely contributed greatly to cross-sectional studies that found shorter limb lengths in gymnasts and dancers.

A thorough longitudinal study by Bass and colleagues (5) found that the reduced leg length in gymnasts may be a result of selection bias because their leg lengths were shorter than controls at baseline and increased at the same rate as controls. And although growth velocity was reduced in active gymnasts, catch-up growth occurred in retired gymnasts: There was an increase and maintenance of growth velocity above that expected for their age. A later study following maturation in ballet dancers came to a similar conclusion: There was no difference between ballet dancers and controls in the age at peak height velocity, nor were there differences in absolute height, sitting height, or leg length (106). Further, in a long-term follow-up of gymnasts, swimmers, and tennis players, no significant differences existed in adult heights (31).

Because it is difficult to control for genetic factors, this issue remains controversial (70). However, recent research with proper controls suggests that intense training does not appear to result in reduced stature in adulthood and should not be discouraged.

Interactions With Exercise

Many factors influence the effect of exercise on the skeleton. These include genetics, diet, and hormones. Genetics can influence up to 80% of peak bone mineral density (178). Micronutrients, including calcium and vitamin D (82, 172), as well as macronutrients, such as fat and sucrose (90, 145, 177), also have important influences on bone health. Whether there are synergistic effects of exercise and calcium supplementation, however, is unclear (6, 32, 164, 169). A detailed review of how genetics, diet, and hormones interact and affect bone growth and development with exercise is beyond the scope of this chapter.

Adaptations to Disuse

Microgravity, prolonged bed rest, and immobilization can dramatically impair bone-formation rates and increase resorption with a net loss in bone mineral density. In space flight, bone loss has been reported to approach 1.6%/mo in regions such as the spine, femoral neck, and long bone metaphyses (166). The implications of significant increases in fracture risk during long missions in space are apparent, especially upon re-exposure to Earth's gravitational field (68, 86).

Mack and colleagues (100), using noninvasive bone density measurements based on carefully calibrated X-ray images, were the first to report significant bone loss in the calcaneus, talus, and fingers in crew members on the Gemini missions. In contrast, Vogel (167) later used single-photon absorptiometry and reported that astronauts in the Skylab missions did not show decrements in bone at the wrist, but significant losses were found in the normally weight-bearing calcaneus. Rambaut and Johnston (131) indicated that reduced bone mineral density, on average, became more pronounced with increasing duration of the spaceflight (e.g., 28, 56, and 84 d). Long-term recovery measurements have been made only rarely. However, a model using DXA data available from both the Russian and North American space agencies suggested that only 50% of bone loss over a 4 to 6 mo mission would be recovered 9 mo postflight (150).

Substantial research has been and is being conducted to develop effective countermeasures, including exercise and loading stimuli (e.g., 42, 43), to prevent bone loss during spaceflight. Nonetheless, the U.S. National Aeronautics and Space Administration and a distinguished team of bone scientists recently summarized the effects that a crew member experienced after a long-duration (6 mo) spaceflight (1).

- Bone resorption—uncoupled from bone formation—could be pronounced, particularly in normally weight-bearing bone, and decrements in bone mineral density could range from 3% to 9%.
- A decline in volumetric bone mineral density of the proximal femur may markedly reduce its strength.

- Bone mineral density may return postflight, but the period of recovery would exceed microgravity exposure, and whether the bone strength is completely restored remains an open question.
- Negative calcium balance and downregulated calcium-relating hormones could be compounded by potential nutritional deficiencies and lack of ultraviolet sunlight for vitamin D conversion.

From an exercise, bone loss, and spaceflight perspective, data to date suggest that in-flight exercise cannot prevent substantial bone loss, comparable to expected lifetime bone losses in some individuals, during spaceflight (68, 85). This bone loss occurs despite the up to 2 h/d allotted for exercise and the variation in exercise strategies used by space travelers (e.g., resistive exercise, treadmill, or cycle ergometer) (68, 85). Problems include lack of standardized exercise prescription and the failure to have reliable in-flight exercise equipment, protocols, and data collection.

Bed rest has been used for more than three decades as an analog for bone loss in spaceflight and, more recently, to test potential countermeasures (89). One of the first studies on bone loss with bed rest was reported by LeBlanc and colleagues (88). Their results and others (89) have generally found that bed rest-induced changes in calcium excretion and balance, bone mineral density, and bone markers were qualitatively similar to those reported during spaceflight, but the changes induced with bed rest were smaller in magnitude than those reported during spaceflight. In a series of bed rest studies, reductions in aBMD and BMC were significant. Regional differences occurred in the magnitude of the decrements, and the greatest losses occurred in weight-bearing bones (88, 133, 148). As with spaceflight, bed rest increases bone resorption (179), supporting the concept that musculoskeletal disuse uncouples the bone-remodeling cycle (89). Recovery of bone after prolonged bed rest does occur, but the return to baseline bone mineral density happens at a rate two to three times lower than the rate of original bone loss (88).

Technical Advancements

Histological Techniques

Dynamic Histomorphometry

Major advances in knowledge of bone adaptation to exercise have been enabled by technical developments that increased our ability to visualize and quantify changes in bone. Fluorochrome labeling was first used by Milch (108) and pioneered by Frost (40). The fluorochrome, such as tetracycline or calcein green, binds to hydroxyapatite as it is being deposited on mineralizing bone surfaces. Thus, injection of a fluorochrome at key intervals allows visualization of dynamic bone formation in humans and animals. Using bone biopsies, it is possible to assess dynamic bone formation in humans. However, due to the invasive nature of this technique, much of the research in the interaction of exercise and bone has applied this method only in animal models (e.g., 116).

Imaging Modalities

X-ray and DXA

X-ray technologies have profoundly affected bone research. Plain film X-ray had long been used to assess bone status (e.g., fracture), but photon absorptiometry allowed measurement of bone thickness using a low-level radionuclide source such as ^{125}I (107). Later, DXA brought a cost-effective means for quantifying changes in bone mineral density in response to exercise, beginning in the 1990s with low-energy X-rays rather than a radionuclide source (104). Using X-ray beams with two energy levels, DXA can allow the researcher to examine aBMD, or mass per unit area (g/cm^2), and BMC from two-dimensional images. aBMD represents a combination of properties, including bone mass and material properties, because it includes mineralization and the area scanned. As a result, aBMD measures are highly size dependent (70). In addition, because of the two-dimensional nature of the images, consistent positioning of the participant's limbs is required to ensure reliability. Both clinicians and researchers appreciate DXA because the wide availability of instruments fostered research questions that could be answered by permitting noninvasive means of assessing bone.

Quantitative Computed Tomography

Although the technology for computed tomography was developed earlier than DXA, only recently has it become common to demonstrate structural changes in bone in response to loading via quantitative computed tomography (QCT). Reasons for this upsurge in usage include the development of special-purpose devices that image only peripheral skeletal sites to minimize radiation dose (113, 144) and advancements in computed tomography technology such as complete and multiple rings of detectors and spiral rotation of the X-ray tube (151). In 1999, Jarvinen and colleagues (59) highlighted that DXA may lead to an underestimation of the effects of mechanical loading on bone because changes

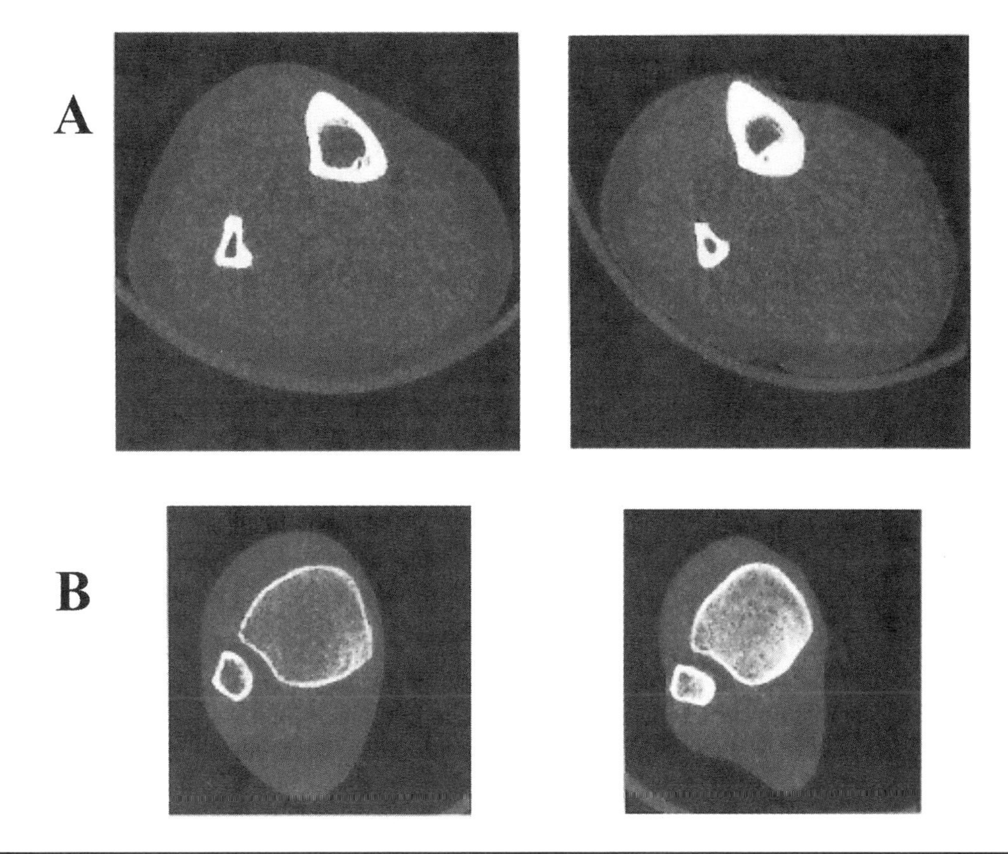

Figure 22.10 pQCT scan of the (*a*) tibial midshaft and (*b*) distal tibia of a female control subject (left panel) and a female triple jumper (right panel). The triple jumper had 30% greater tibial midshaft cortical area and 67% greater trabecular density than did the control subject.

in bone shape may lead to large changes in bone strength that cannot be accounted for using DXA outcomes such as BMC and aBMD. Unlike DXA, QCT can provide three-dimensional images through tomographic reconstruction applied to either two-dimensional regions of interest from single or multiple slices or three-dimensional volumes of interest from contiguous stacks of images (3, 30). In addition, QCT images are calibrated using a phantom to convert Hounsfield units to hydroxyapatite (HA) densities (mg/cm^3 of HA), which is interpreted as bone mineral density.

Dedicated scanners for assessing peripheral sites such as the humerus, radius, and tibia have been commonly used to quantify the effectiveness of exercise. pQCT scanners have moderate resolution (200-500 μm in plane) and are used to assess bone structure and average volumetric bone mineral density. Early pQCT studies demonstrated differences in bone structure between elite athletes and healthy controls. For example, Heinonen and colleagues reported 20% to 50% greater tibial wall thicknesses in triple jumpers compared with controls (49) (figure 22.10).

Microcomputed Tomography

At the end of the 20th century, microcomputed tomography (MCT) and high-resolution pQCT scanners were developed. MCT had sufficient resolution to resolve individual trabeculae, but because of limitations in the size of the field of view MCT was used to image animals only. MCT has quickly become a standard tool for evaluating exercise interventions in animals (figure 22.11). In contrast, high-resolution pQCT can be used in humans, but it is limited to a resolution equivalent to the thickness of the average human trabecula (83). In the future, this technology should provide insights into microarchitectural changes (i.e., changes in trabecular structure and cortical porosity) that result from exercise. Guidelines for conducting and reporting studies using

MCT were recently published to help standardize studies in this growing field (12).

Finite Element Analysis

Finite element (FE) analysis has been applied extensively in evaluating mechanical properties of skeletal structures (e.g., 65, 132). Models can be generated directly from many types of images (e.g., 68, 69, 114, 118, 165) and may thus be patient specific. For example, digital image voxels (three-dimensional pixels) from computed tomography can be converted directly to finite elements and used to generate models for estimating bone strength (69, 94) as well as tissue-level strains and stresses (165). To date, with respect to mechanical loading, FE models have been used mainly to predict bone strength after loading simulations (128). However, with the ability to convert images directly to FE models using automated methods, strength predictions from FE models will likely become one of the new standard outcomes for assessing the effectiveness of exercise interventions.

Future Directions

Over the past 100 yr, considerable advancements have been made in understanding how exercise can affect bone health. Progress has been in large part due to improvements in the ability to assess bone outcomes, particularly with noninvasive imaging technologies. In like manner, new advancements and refinements in noninvasive imaging and biological markers will lead the way for quantifying tissue-specific effects of exercise in the future. In the past 10 yr we have begun to realize that simply quantifying changes in bone mass is insufficient for understanding how exercise comprehensively affects bone. As higher-resolution images can be acquired at more relevant sites, such as the proximal femur, we will more easily be able to assess the potential effect of an exercise intervention on future fracture risk. With the technologies now readily in use in humans (QCT, pQCT, and high-resolution pQCT), we are able to assess cortical bone geometry and trabecular bone microarchitecture, which will improve our ability to predict the effect of exercise on bone. Further, other imaging modalities such as ultrasound and magnetic resonance imaging may offer alternative means for assessing bone outcomes without the need for ionizing radiation. As well, analytical techniques such as finite element modeling will aid our ability to interpret images by predicting bone strength.

To date, bone researchers have been quick to implement findings from theoretical modeling and animal studies to human interventions, particularly in terms of the types of exercise that can be effective in promoting bone gains. As mechanisms underlying bone adaptation (e.g., fluid shear stress and chemotransport) become more clearly elucidated, we will have a better understanding of how bone adapts to targeted mechanical loading.

A key focus for future research is to improve the understanding of exercise as a viable intervention for preventing fractures associated with low bone strength. Trials using fracture as an outcome necessitate large sample sizes and long durations and thus remain a challenge to complete. Extensive longitudinal studies are essential. Another key question is whether advances in bone mass and structure acquired in childhood can be maintained through adulthood. There is evidence that

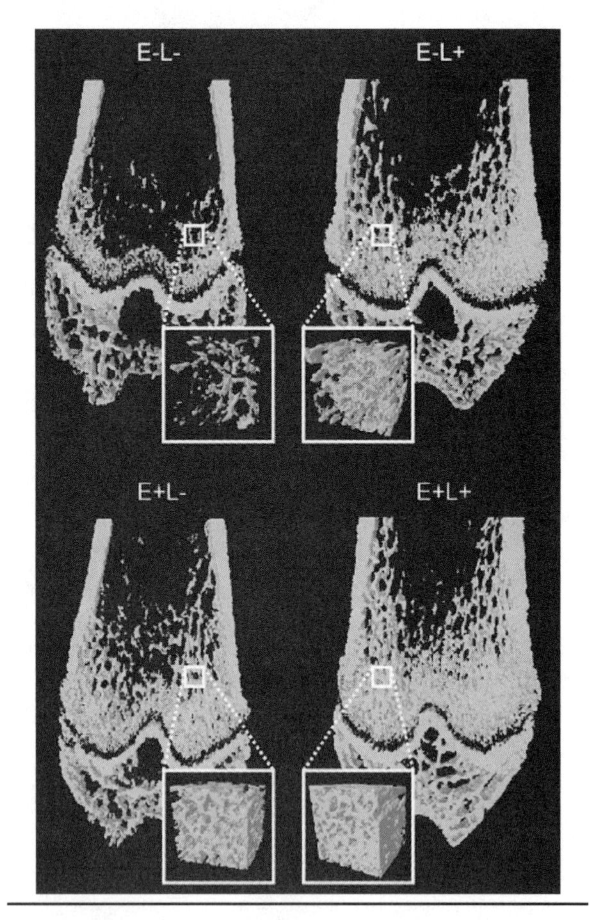

Figure 22.11 Three-dimensional reconstruction of microcomputed tomography scans of rat distal femur at 14.5 μm voxel size. Animals were assigned to ovariectomy (E^-) or sham (E^+) surgery and normal loading (L^+) or cast immobilization (L^-). Loading had a positive effect on bone volume fraction, trabecular thickness, and trabecular number.

retired professional elite ballet dancers had greater BMD in the lower limb compared with controls, but self-selection bias in such studies limits conclusions that can be drawn (71). Retrospective studies on this topic have been informative but not definitive in answering this important question. Until we have such studies, we will continue to question whether exercise-induced bone gains achieved in childhood can be maintained until adulthood.

The past three decades have brought considerable new knowledge about how exercise influences bone. Will progress continue to be as rapid or potentially accelerate? Health problems such as stress fractures and osteoporosis, particularly due to the aging population, will likely drive future progress. Even with the anticipated technical advancements and mechanistic discoveries, however, the Herculean challenge to develop and promulgate exercise regimens that capitalize on newly emerging knowledge and that are actually used widely by society will remain.

References

1. NASA. *NASA Human Research Program: Human Health Countermeasures Element--Evidence Book: Risk of Accelerated Osteoporosis (HRP-47072).* Houston, TX: NASA, 2008.
2. Aarden EM, Burger EH, Nijweide PJ. Function of osteocytes in bone. *J Cell Biochem* 55: 287-299, 1994.
3. Adams JE. Quantitative computed tomography. *Eur J Radiol* 71: 415-424, 2009.
4. Bailey DA, McKay HA, Mirwald RL, Crocker PR, Faulkner RA. A six-year longitudinal study of the relationship of physical activity to bone mineral accrual in growing children: The university of Saskatchewan bone mineral accrual study. *J Bone Miner Res* 14: 1672-1679, 1999.
5. Bass S, Pearce G, Bradney M, Hendrich E, Delmas PD, Harding A, Seeman E. Exercise before puberty may confer residual benefits in bone density in adulthood: Studies in active prepubertal and retired female gymnasts. *J Bone Miner Res* 13: 500-507, 1998.
6. Bass SL, Naughton G, Saxon L, Iuliano-Burns S, Daly R, Briganti EM, Hume C, Nowson C. Exercise and calcium combined results in a greater osteogenic effect than either factor alone: A blinded randomized placebo-controlled trial in boys. *J Bone Miner Res* 22: 458-464, 2007.
7. Bassett CA, Becker RO. Generation of electric potentials by bone in response to mechanical stress. *Science* 137: 1063-1064, 1962.
8. Baud CA. Submicroscopic structure and functional aspects of the osteocyte. *Clin Orthop Relat Res* 56: 227-236, 1968.
9. Bentolila V, Boyce TM, Fyhrie DP, Drumb R, Skerry TM, Schaffler MB. Intracortical remodeling in adult rat long bones after fatigue loading. *Bone* 23: 275-281, 1998.
10. Blimkie CJ, Rice S, Webber CE, Martin J, Levy D, Gordon CL. Effects of resistance training on bone mineral content and density in adolescent females. *Can J Physiol Pharmacol* 74: 1025-1033, 1996.
11. Bloomfield SA, Williams NI, Lamb DR, Jackson RD. Non-weightbearing exercise may increase lumbar spine bone mineral density in healthy postmenopausal women. *Am J Phys Med Rehab* 72: 204-209, 1993.
12. Bouxsein ML, Boyd SK, Christiansen BA, Guldberg RE, Jepsen KJ, Müller R. Guidelines for assessment of bone microstructure in rodents using micro-computed tomography. *J Bone Miner Res* 25: 1468-1486, 2010.
13. Bradney M, Pearce G, Naughton G, Sullivan C, Bass S, Beck T, Carlson J, Seeman E. Moderate exercise during growth in prepubertal boys: Changes in bone mass, size, volumetric density, and bone strength: A controlled prospective study. *J Bone Miner Res* 13: 1814-1821, 1998.
14. Brand RA, Stanford CM. How connective tissues temporally process mechanical stimuli. *Med Hypotheses* 42: 99-104, 1994.
15. Bruckner P, Bennell K, Matheson G. *Stress Fractures.* Melbourne: Blackwell Scientific, 1999.
16. Bullen BA, Skrinar GS, Beitins IZ, von Mering G, Turnbull BA, McArthur JW. Induction of menstrual disorders by strenuous exercise in untrained women. *N Engl J Med* 312: 1349-1353, 1985.
17. Burr DB, Milgrom C, Fyhrie D, Forwood M, Nyska M, Finestone A, Hoshaw S, Saiag E, Simkin A. In vivo measurement of human tibial strains during vigorous activity. *Bone* 18: 405-410, 1996.
18. Caplan GA, Ward JA, Lord SR. The benefits of exercise in postmenopausal women. *Aus J Public Health* 17: 23-26, 1993.
19. Chamay A, Tschantz P. Mechanical influences in bone remodeling. Experimental research on Wolff's law. *J Biomech* 5: 173-180, 1972.
20. Chambers TJ, Evans M, Gardner TN, Turner-Smith A, Chow JW. Induction of bone formation in rat tail vertebrae by mechanical loading. *Bone Miner* 20: 167-178, 1993.
21. Chow JW, Jagger CJ, Chambers TJ. Characterization of osteogenic response to mechanical stimulation in cancellous bone of rat caudal vertebrae. *Am J Physiol* 265: E340-E347, 1993.
22. Cummings SR, Black DM, Nevitt MC, Browner W, Cauley J, Ensrud K, Genant HK, Palermo L, Scott J, Vogt TM. Bone density at various sites for pre-

diction of hip fractures. The Study of Osteoporotic Fractures Research Group. *Lancet* 341: 72-75, 1993.

23. Donahue TL, Haut TR, Yellowley CE, Donahue HJ, Jacobs CR. Mechanosensitivity of bone cells to oscillating fluid flow induced shear stress may be modulated by chemotransport. *J Biomech* 36: 1363-1371, 2003.
24. Doty SB. Morphological evidence of gap junctions between bone cells. *Calcif Tissue Int* 33: 509-512, 1981.
25. Drinkwater BL, Nilson K, Chesnut CH III, Bremner WJ, Shainholtz S, Southworth MB. Bone mineral content of amenorrheic and eumenorrheic athletes. *N Engl J Med* 311: 277-281, 1984.
26. Dudley HR, Spiro D. The fine structure of bone cells. *J Biophys Biochem Cytol* 11: 627-649, 1961.
27. Duncan CS, Blimkie CJ, Cowell CT, Burke ST, Briody JN, Howman-Giles R. Bone mineral density in adolescent female athletes: Relationship to exercise type and muscle strength. *Med Sci Sports Exerc* 34: 286-294, 2002.
28. Duncan RL, Turner CH. Mechanotransduction and the functional response of bone to mechanical strain. *Calcif Tissue Int* 57: 344-358, 1995.
29. Ebrahim S, Thompson PW, Baskaran V, Evans K. Randomized placebo-controlled trial of brisk walking in the prevention of postmenopausal osteoporosis. *Age Ageing* 26: 253-260, 1997.
30. Engelke K, Adams JE, Armbrecht G, Augat P, Bogado CE, Bouxsein ML, Felsenberg D, Ito M, Prevrhal S, Hans DB, Lewiecki EM. Clinical use of quantitative computed tomography and peripheral quantitative computed tomography in the management of osteoporosis in adults: The 2007 ISCD Official Positions. *J Clin Densitom* 11: 123-162, 2008.
31. Erlandson MC, Sherar LB, Mirwald RL, Maffulli N, Baxter-Jones AD. Growth and maturation of adolescent female gymnasts, swimmers, and tennis players. *Med Sci Sports Exerc* 40: 34-42, 2008.
32. Friedlander AL, Genant HK, Sadowsky S, Byl NN, Glüer CC. A two-year program of aerobics and weight training enhances bone mineral density of young women. *J Bone Miner Res* 10: 574-585, 1995.
33. Fritton SP, McLeod KJ, Rubin CT. Quantifying the strain history of bone: Spatial uniformity and self-similarity of low-magnitude strains. *J Biomech* 33: 317-325, 2000.
34. Frost HM. *Bone Remodeling Dynamics*. Springfield, IL: Charles C. Thomas, 1963.
35. Frost HM. Bone "mass" and the "mechanostat": A proposal. *Anat Record* 219: 1-9, 1987.
36. Frost HM. A determinant of bone architecture. The minimum effective strain. *Clin Orthop Relat Res* 286-292, 1983.
37. Frost HM. *Intermediary Organization of the Skeleton*. Boca Raton, FL: CRC Press, 1986.
38. Frost HM. *The Laws of Bone Structure*. Springfield, IL: Charles C. Thomas, 1964.
39. Frost HM. The mechanostat: A proposed pathogenic mechanism of osteoporoses and the bone mass effects of mechanical and nonmechanical agents. *Bone Miner* 2: 73-85, 1987.
40. Frost HM, Villaneuva AR, Roth H. Measurement of bone formation in a 57 year old man by means of tetracyclines. *Henry Ford Hospital Med Bull* 8: 239-254, 1960.
41. Fuchs RK, Bauer JJ, Snow CM. Jumping improves hip and lumbar spine bone mass in prepubescent children: A randomized controlled trial. *J Bone Miner Res* 16: 148-156, 2001.
42. Genc KO, Humphreys BT, Cavanagh PR. Enhanced daily load stimulus to bone in spaceflight and on earth. *Aviat Space Environ Med* 80: 919-926, 2009.
43. Goodship AE, Cunningham JL, Oganov V, Darling J, Miles AW, Owen GW. Bone loss during long term space flight is prevented by the application of a short term impulsive mechanical stimulus. *Acta Astronaut* 43: 65-75, 1998.
44. Greig DM. *Clinical Observations on the Surgical Pathology of Bone*. Edinburgh: Olive, 1931.
45. Gross TS, Edwards JL, McLeod KJ, Rubin CT. Strain gradients correlate with sites of periosteal bone formation. *J Bone Miner Res* 12: 982-988, 1997.
46. Haapasalo H, Kannus P, Sievänen H, Pasanen M, Uusi-Rasi K, Heinonen A, Oja P, Vuori I. Effect of long-term unilateral activity on bone mineral density of female junior tennis players. *J Bone Miner Res* 13: 310-319, 1998.
47. Harell A, Dekel S, Binderman I. Biochemical effect of mechanical stress on cultured bone cells. *Calcif Tissue Res* 22(Suppl): 202-207, 1977.
48. Heinonen A, Sievänen H, Kannus P, Oja P, Pasanen M, Vuori I. High-impact exercise and bones of growing girls: A 9-month controlled trial. *Osteoporos Int* 11: 1010-1017, 2000.
49. Heinonen A, Sievänen H, Kyrolainen H, Perttunen J, Kannus P. Mineral mass, size, and estimated mechanical strength of triple jumpers' lower limb. *Bone* 29: 279-285, 2001.
50. Hert J, Liskova M, Landa J. Reaction of bone to mechanical stimuli. 1. Continuous and intermittent loading of tibia in rabbit. *Folia Morphol (Praha)* 19: 290-300, 1971.
51. Hert J, Liskova M, Landrgot B. Influence of the long-term, continuous bending on the bone. An experimental study on the tibia of the rabbit. *Folia Morphol (Praha)* 17: 389-399, 1969.

52. Hind K, Burrows M. Weight-bearing exercise and bone mineral accrual in children and adolescents: A review of controlled trials. *Bone* 40: 14-27, 2007.
53. Hsieh YF, Turner CH. Effects of loading frequency on mechanically induced bone formation. *J Bone Miner Res* 16: 918-924, 2001.
54. Hui SL, Slemenda CW, Johnston CC Jr. Baseline measurement of bone mass predicts fracture in white women. *Ann Intern Med* 111: 355-361, 1989.
55. Ihle R, Loucks AB. Dose-response relationships between energy availability and bone turnover in young exercising women. *J Bone Miner Res* 19: 1231-1240, 2004.
56. Imamura K, Ozawa H, Hiraide T, Takahashi N, Shibasaki Y, Fukuhara T, Suda T. Continuously applied compressive pressure induces bone resorption by a mechanism involving prostaglandin E2 synthesis. *J Cell Physiol* 144: 222-228, 1990.
57. Ireland J. Hip, thigh, pelvis. In: Sherry E, Bokor D, eds. *Sports Medicine: Problems and Practical Management*. London: Oxford University Press, 1997.
58. Jacobs CR, Yellowley CE, Davis BR, Zhou Z, Cimbala JM, Donahue HJ. Differential effect of steady versus oscillating flow on bone cells. *J Biomech* 31: 969-976, 1998.
59. Järvinen TL, Kannus P, Sievänen H. Have the DXA-based exercise studies seriously underestimated the effects of mechanical loading on bone? *J Bone Miner Res* 14: 1634-1635, 1999.
60. Jores L. Experimentelle untersuchungen uber die einwirkung mechanischen druckes auf den knochen. *Beit Path Anat Allge Path* 66: 433-469, 1920.
61. Judex S, Gross TS, Zernicke RF. Strain gradients correlate with sites of exercise-induced bone-forming surfaces in the adult skeleton. *J Bone Miner Res* 12: 1737-1745, 1997.
62. Judex S, Zernicke RF. High-impact exercise and growing bone: Relation between high strain rates and enhanced bone formation. *J Appl Physiol* 88: 2183-2191, 2000.
63. Kannus P, Haapasalo H, Sankelo M, Sievänen H, Pasanen M, Heinonen A, Oja P, Vuori I. Effect of starting age of physical activity on bone mass in the dominant arm of tennis and squash players. *Ann Intern Med* 123: 27-31, 1995.
64. Karinkanta S, Heinonen A, Sievänen H, Uusi-Rasi K, Pasanen M, Ojala K, Fogelholm M, Kannus P. A multi-component exercise regimen to prevent functional decline and bone fragility in home-dwelling elderly women: Randomized, controlled trial. *Osteoporos Int* 18: 453-462, 2007.
65. Keaveny TM, Donley DW, Hoffmann PF, Mitlak BH, Glass EV, San Martin JA. Effects of teriparatide and alendronate on vertebral strength as assessed by finite element modeling of QCT scans in women with osteoporosis. *J Bone Miner Res* 22: 149-157, 2007.
66. Keen AD, Drinkwater BL. Irreversible bone loss in former amenorrheic athletes. *Osteoporos Int* 7: 311-315, 1997.
67. Kelley GA, Kelley KS. Exercise and bone mineral density at the femoral neck in postmenopausal women: A meta-analysis of controlled clinical trials with individual patient data. *Am J Obstet Gynecol* 194: 760-767, 2006.
68. Keyak JH, Koyama AK, LeBlanc A, Lu Y, Lang TF. Reduction in proximal femoral strength due to long-duration spaceflight. *Bone* 44: 449-453, 2009.
69. Keyak JH, Meagher JM, Skinner HB, Mote JCD. Automated three-dimensional finite element modelling of bone: A new method. *J Biomed Eng* 12: 389-397, 1990.
70. Khan K, McKay H, Kannus P, Bailey D, Wark J, Bennell K. *Physical Activity and Bone Health*. Champaign, IL: Human Kinetics, 2001.
71. Khan KM, Green RM, Saul A, Bennell KL, Crichton KJ, Hopper JL, Wark JD. Retired elite female ballet dancers and nonathletic controls have similar bone mineral density at weightbearing sites. *J Bone Miner Res* 11: 1566-1574, 1996.
72. Klein-Nulend J, Semeins CM, Ajubi NE, Nijweide PJ, Burger EH. Pulsating fluid flow increases nitric oxide (NO) synthesis by osteocytes but not periosteal fibroblasts—Correlation with prostaglandin upregulation. *Biochem Biophys Res Commun* 217: 640-648, 1995.
73. Klein-Nulend J, Semeins CM, Veldhuijzen JP, Burger EH. Effect of mechanical stimulation on the production of soluble bone factors in cultured fetal mouse calvariae. *Cell Tissue Res* 271: 513-517, 1993.
74. Klein-Nulend J, Veldhuijzen JP, Burger EH. Increased calcification of growth plate cartilage as a result of compressive force in vitro. *Arthritis Rheum* 29: 1002-1009, 1986.
75. Klein-Nulend J, Veldhuijzen JP, de Jong M, Burger EH. Increased bone formation and decreased bone resorption in fetal mouse calvaria as a result of intermittent compressive force in vitro. *Bone Miner* 2: 441-448, 1987.
76. Klein-Nulend J, Veldhuijzen JP, van Strien ME, de Jong M, Burger EH. Inhibition of osteoclastic bone resorption by mechanical stimulation in vitro. *Arthritis Rheum* 33: 66-72, 1990.
77. Knothe Tate ML. "Whither flows the fluid in bone?" An osteocyte's perspective. *J Biomech* 36: 1409-1424, 2003.
78. Knothe Tate ML, Knothe U. An ex vivo model to study transport processes and fluid flow in loaded bone. *J Biomech* 33: 247-254, 2000.
79. Knothe Tate ML, Knothe U, Niederer P. Experimental elucidation of mechanical load-induced fluid flow and its potential role in bone metabolism and

functional adaptation. *Am J Med Sci* 316: 189-195, 1998.

80. Knothe Tate ML, Niederer P, Knothe U. In vivo tracer transport through the lacunocanalicular system of rat bone in an environment devoid of mechanical loading. *Bone* 22: 107-117, 1998.
81. Knothe Tate ML, Steck R, Forwood MR, Niederer P. In vivo demonstration of load-induced fluid flow in the rat tibia and its potential implications for processes associated with functional adaptation. *J Exp Biol* 203: 2737-2745, 2000.
82. Kunstel K. Calcium requirements for the athlete. *Curr Sports Med Rep* 4: 203-206, 2005.
83. Laib A, Häuselmann HJ, Rügsegger P. In vivo high resolution 3D-QCT of the human forearm. *Technol Health Care* 6: 329-337, 1998.
84. LaMothe JM, Hamilton NH, Zernicke RF. Strain rate influences periosteal adaptation in mature bone. *Med Eng Phys* 27: 277-284, 2005.
85. Lang T, LeBlanc A, Evans H, Lu Y, Genant H, Yu A. Cortical and trabecular bone mineral loss from the spine and hip in long-duration spaceflight. *J Bone Miner Res* 19: 1006-1012, 2004.
86. Lang TF. What do we know about fracture risk in long-duration spaceflight? *J Musculoskel Neuron Interact* 6: 319-321, 2006.
87. Lanyon LE, Rubin CT. Static vs dynamic loads as an influence on bone remodeling. *J Biomech* 17: 897-905, 1984.
88. Leblanc AD, Schneider VS, Evans HJ, Engelbretson DA, Krebs JM. Bone mineral loss and recovery after 17 weeks of bed rest. *J Bone Miner Res* 5: 843-850, 1990.
89. LeBlanc AD, Spector ER, Evans HJ, Sibonga JD. Skeletal responses to space flight and the bed rest analog: A review. *J Musculoskelet Neuronal Interact* 7: 33-47, 2007.
90. Li KC, Zernicke RF, Barnard RJ, Li AF. Effects of a high fat-sucrose diet on cortical bone morphology and biomechanics. *Calcif Tissue Int* 47: 308-313, 1990.
91. Liskova M, Hert J. Reaction of bone to mechanical stimuli. 2. Periosteal and endosteal reaction of tibial diaphysis in rabbit to intermittent loading. *Folia Morphol (Praha)* 19: 301-317, 1971.
92. Loeschke H, Weinnoldt H. Uber den einflus von druck und entspannung auf das knochenwachstum des schadels. *Beit Path Anat Allge Path* 70: 406-439, 1922.
93. Lord SR, Ward JA, Williams P, Zivanovic E. The effects of a community exercise program on fracture risk factors in older women. *Osteoporos Int* 6: 361-367, 1996.
94. Lotz JC, Hayes WC. The use of quantitative computed tomography to estimate risk of fracture of the hip from falls. *J Bone Joint Surg* 72: 689-700, 1990.
95. Loucks AB. The response of luteinizing hormone pulsatility to 5 days of low energy availability disappears by 14 years of gynecological age. *J Clin Endocrinol Metab* 91: 3158-3164, 2006.
96. Loucks AB, Heath EM. Dietary restriction reduces luteinizing hormone (LH) pulse frequency during waking hours and increases LH pulse amplitude during sleep in young menstruating women. *J Clin Endocrinol Metab* 78: 910-915, 1994.
97. Loucks AB, Stachenfeld NS, DiPietro L. The female athlete triad: Do female athletes need to take special care to avoid low energy availability? *Med Sci Sports Exerc* 38: 1694-1700, 2006.
98. Loucks AB, Verdun M, Heath EM. Low energy availability, not stress of exercise, alters LH pulsatility in exercising women. *J Appl Physiol* 84: 37-46, 1998.
99. Macdonald HM, Cooper DM, McKay HA. Anterior-posterior bending strength at the tibial shaft increases with physical activity in boys: Evidence for non-uniform geometric adaptation. *Osteoporos Int* 20: 61-70, 2009.
100. Mack PB, LaChance PA, Vose GP, Vogt FB. Bone demineralization of foot and hand of gemini-titan IV, V and VII astronauts during orbital flight. *Am J Roentgenol Radium Ther Nucl Med* 100: 503-511, 1967.
101. MacKelvie KJ, Khan KM, Petit MA, Janssen PA, McKay HA. A school-based exercise intervention elicits substantial bone health benefits: A 2-year randomized controlled trial in girls. *Pediatrics* 112: e447, 2003.
102. MacKelvie KJ, Petit MA, Khan KM, Beck TJ, McKay HA. Bone mass and structure are enhanced following a 2-year randomized controlled trial of exercise in prepubertal boys. *Bone* 34: 755-764, 2004.
103. Martin RB. Toward a unifying theory of bone remodeling. *Bone* 26: 1-6, 2000.
104. Martin RB, Burr DB, Sharkey NA. *Skeletal Tissue Mechanics*. New York: Springer-Verlag, 1998.
105. Matthews BL, Bennell KL, McKay HA, Khan KM, Baxter-Jones AD, Mirwald RL, Wark JD. Dancing for bone health: A 3-year longitudinal study of bone mineral accrual across puberty in female non-elite dancers and controls. *Osteoporos Int* 17: 1043-1054, 2006.
106. Matthews BL, Bennell KL, McKay HA, Khan KM, Baxter-Jones AD, Mirwald RL, Wark JD. The influence of dance training on growth and maturation of young females: A mixed longitudinal study. *Ann Human Biol* 33: 342-356, 2006.
107. Mazess RB, Cameron JR. Direct readout of bone mineral content using radionuclide absorptiometry. *Int J Applied Radiation Isotopes* 23: 471-479, 1972.
108. Milch RA, Rall DP, Tobie JE. Fluorescence of tetracycline antibiotics in bone. *J Bone Joint Surg* 40-A: 897-910, 1958.

109. Milgrom C, Radeva-Petrova DR, Finestone A, Nyska M, Mendelson S, Benjuya N, Simkin A, Burr D. The effect of muscle fatigue on in vivo tibial strains. *J Biomech* 40: 845-850, 2007.
110. Mori S, Burr DB. Increased intracortical remodeling following fatigue damage. *Bone* 14: 103-109, 1993.
111. Morris FL, Naughton GA, Gibbs JL, Carlson JS, Wark JD. Prospective ten-month exercise intervention in premenarcheal girls: Positive effects on bone and lean mass. *J Bone Miner Res* 12: 1453-1462, 1997.
112. Mosley JR, Lanyon LE. Strain rate as a controlling influence on adaptive modeling in response to dynamic loading of the ulna in growing male rats. *Bone* 23: 313-318, 1998.
113. Müller A, Rüegsegger E, Rüegsegger P. Peripheral QCT: A low-risk procedure to identify women predisposed to osteoporosis. *Phys Med Biol* 34: 741-749, 1989.
114. Müller R, Rügsegger P. Three-dimensional finite element modelling of non-invasively assessed trabecular bone structures. *Med Eng Phys* 17: 126-133, 1995.
115. Myburgh KH, Hutchins J, Fataar AB, Hough SF, Noakes TD. Low bone density is an etiologic factor for stress fractures in athletes. *Ann Intern Med* 113: 754-759, 1990.
116. Myburgh KH, Noakes TD, Roodt M, Hough FS. Effect of exercise on the development of osteoporosis in adult rats. *J Appl Physiol* 66: 14-19, 1989.
117. Nattiv A, Loucks AB, Manore MM, Sanborn CF, Sundgot-Borgen J, Warren MP. American College of Sports Medicine position stand. The female athlete triad. *Med Sci Sports Exerc* 39: 1867-1882, 2007.
118. Newitt DC, van Rietbergen B, Majumdar S. Processing and analysis of in vivo high-resolution MR images of trabecular bone for longitudinal studies: Reproducibility of structural measures and micro-finite element analysis derived mechanical properties. *Osteoporos Int* 13: 278-287, 2002.
119. Nikander R, Sievänen H, Heinonen A, Kannus P. Femoral neck structure in adult female athletes subjected to different loading modalities. *J Bone Miner Res* 20: 520-528, 2005.
120. Nikander R, Sievänen H, Uusi-Rasi K, Heinonen A, Kannus P. Loading modalities and bone structures at nonweight-bearing upper extremity and weight-bearing lower extremity: A pQCT study of adult female athletes. *Bone* 39: 886-894, 2006.
121. Nurmi-Lawton JA, Baxter-Jones AD, Mirwald RL, Bishop JA, Taylor P, Cooper C, New SA. Evidence of sustained skeletal benefits from impact-loading exercise in young females: A 3-year longitudinal study. *J Bone Miner Res* 19: 314-322, 2004.
122. O'Connor JA, Lanyon LE, MacFie H. The influence of strain rate on adaptive bone remodeling. *J Biomech* 15: 767-781, 1982.
123. Owan I, Burr DB, Turner CH, Qiu J, Tu Y, Onyia JE, Duncan RL. Mechanotransduction in bone: Osteoblasts are more responsive to fluid forces than mechanical strain. *Am J Physiol* 273: C810-C815, 1997.
124. Ozawa H, Imamura K, Abe E, Takahashi N, Hiraide T, Shibasaki Y, Fukuhara T, Suda T. Effect of a continuously applied compressive pressure on mouse osteoblast-like cells (MC3T3-E1) in vitro. *J Cell Physiol* 142: 177-185, 1990.
125. Pauwels F. *Gesammelte Abhandlungen zur Funktionellen Anatomie des Bewegungsapparates*. Berlin: Springer, 1965.
126. Piekarski K, Munro M. Transport mechanism operating between blood supply and osteocytes in long bones. *Nature* 269: 80-82, 1977.
127. Pigeon P, Oliver I, Charlet JP, Rochiccioli P. Intensive dance practice. Repercussions on growth and puberty. *Am J Sports Med* 25: 243-247, 1997.
128. Pistoia W, van Rietbergen B, Rügsegger P. Mechanical consequences of different scenarios for simulated bone atrophy and recovery in the distal radius. *Bone* 33: 937-945, 2003.
129. Pruitt LA, Jackson RD, Bartels RL, Lehnhard HJ. Weight-training effects on bone mineral density in early postmenopausal women. *J Bone Miner Res* 7: 179-185, 1992.
130. Qin YX, Kaplan T, Saldanha A, Rubin C. Fluid pressure gradients, arising from oscillations in intramedullary pressure, is correlated with the formation of bone and inhibition of intracortical porosity. *J Biomech* 36: 1427-1437, 2003.
131. Rambaut PC, Johnston RS. Prolonged weightlessness and calcium loss in man. *Acta Astronaut* 6: 1113-1122, 1979.
132. Reddy JN. *An Introduction to the Finite Element Method*. 2nd ed. New York: McGraw-Hill, 1993.
133. Rittweger J, Beller G, Armbrecht G, Mulder E, Buehring B, Gast U, Dimeo F, Schubert H, de Haan A, Stegeman DF, Schiessl H, Felsenberg D. Prevention of bone loss during 56 days of strict bed rest by side-alternating resistive vibration exercise. *Bone* 46: 137-147, 2010.
134. Robling AG, Burr DB, Turner CH. Partitioning a daily mechanical stimulus into discrete loading bouts improves the osteogenic response to loading. *J Bone Miner Res* 15: 1596-1602, 2000.
135. Robling AG, Burr DB, Turner CH. Recovery periods restore mechanosensitivity to dynamically loaded bone. *J Exp Biol* 204: 3389-3399, 2001.
136. Roux W. *Der zuchtende Kampf der Theile, oder die 'Theilauslese' im Organismus [The struggle of the components within organisms]*. Leipzig: Wilhelm Englemann, 1881.

137. Rubin C, Recker R, Cullen D, Ryaby J, McCabe J, McLeod K. Prevention of postmenopausal bone loss by a low-magnitude, high-frequency mechanical stimuli: A clinical trial assessing compliance, efficacy, and safety. *J Bone Miner Res* 19: 343-351, 2004.
138. Rubin C, Turner AS, Bain S, Mallinckrodt C, McLeod K. Anabolism. Low mechanical signals strengthen long bones. *Nature* 412: 603-604, 2001.
139. Rubin CT, Lanyon LE. Dynamic strain similarity in vertebrates: An alternative to allometric limb bone scaling. *J Theor Biol* 107: 321-327, 1984.
140. Rubin CT, Lanyon LE. Kappa Delta Award paper. Osteoregulatory nature of mechanical stimuli: Function as a determinant for adaptive remodeling in bone. *J Orthop Res* 5: 300-310, 1987.
141. Rubin CT, Lanyon LE. Regulation of bone formation by applied dynamic loads. *J Bone Joint Surg Am* 66: 397-402, 1984.
142. Rubin CT, Lanyon LE. Regulation of bone mass by mechanical strain magnitude. *Calcif Tissue Int* 37: 411-417, 1985.
143. Rubin CT, McLeod KJ. Promotion of bony ingrowth by frequency-specific, low-amplitude mechanical strain. *Clin Orthop Relat Res* 165-174, 1994.
144. Rüegsegger P, Stebler B, Dambacher M. Quantitative computed tomography of bone. *Mayo Clinic Proc* 57(Suppl): 96-103, 1982.
145. Salem GJ, Zernicke RF, Barnard RJ. Diet-related changes in mechanical properties of rat vertebrae. *Am J Physiol* 262: R318-R321, 1992.
146. Schaffler MB, Radin EL, Burr DB. Long-term fatigue behavior of compact bone at low strain magnitude and rate. *Bone* 11: 321-326, 1990.
147. Seeman E, Bianchi G, Khosla S, Kanis JA, Orwoll E. Bone fragility in men—Where are we? *Osteoporos Int* 17: 1577-1583, 2006.
148. Shackelford LC, LeBlanc AD, Driscoll TB, Evans HJ, Rianon NJ, Smith SM, Spector E, Feeback DL, Lai D. Resistance exercise as a countermeasure to disuse-induced bone loss. *J Appl Physiol* 97: 119-129, 2004.
149. Siatris T, Skaperda M, Mameletzki D. Anthropometric characteristics and delayed growth in young artistic gymnasts. *Med Probl Perform Art* 24: 91-96, 2009.
150. Sibonga JD, Evans HJ, Sung HG, Spector ER, Lang TF, Oganov VS, Bakulin AV, Shackelford LC, LeBlanc AD. Recovery of spaceflight-induced bone loss: Bone mineral density after long-duration missions as fitted with an exponential function. *Bone* 41: 973-978, 2007.
151. Silverman PM, Kalender WA, Hazle JD. Common terminology for single and multislice helical CT. *AJR Am J Roentgenol* 176: 1135-1136, 2001.
152. Smalt R, Mitchell FT, Howard RL, Chambers TJ. Mechanotransduction in bone cells: Induction of nitric oxide and prostaglandin synthesis by fluid shear stress, but not by mechanical strain. *Adv Exp Med Biol* 433: 311-314, 1997.
153. Srinivasan S, Weimer DA, Agans SC, Bain SD, Gross TS. Low-magnitude mechanical loading becomes osteogenic when rest is inserted between each load cycle. *J Bone Miner Res* 17: 1613-1620, 2002.
154. Starkebaum W, Pollack SR, Korostoff E. Microelectrode studies of stress-generated potentials in four-point bending of bone. *J Biomed Mater Res* 13: 729-751, 1979.
155. Steinberg ME, Trueta J. Effects of activity on bone growth and development in the rat. *Clin Orthop Relat Res* 52-60, 1981.
156. Stengel SV, Kemmler W, Pintag R, Beeskow C, Weineck J, Lauber D, Kalender WA, Engelke K. Power training is more effective than strength training for maintaining bone mineral density in postmenopausal women. *J Appl Physiol* 99: 181-188, 2005.
157. Taaffe DR, Robinson TL, Snow CM, Marcus R. High-impact exercise promotes bone gain in well-trained female athletes. *J Bone Miner Res* 12: 255-260, 1997.
158. Thompson DW. *On Growth and Form*. London: Cambridge University Press, 1917.
159. Thong FS, McLean C, Graham TE. Plasma leptin in female athletes: Relationship with body fat, reproductive, nutritional, and endocrine factors. *J Appl Physiol* 88: 2037-2044, 2000.
160. Turner CH. Three rules for bone adaptation to mechanical stimuli. *Bone* 23: 399-407, 1998.
161. Turner CH, Owan I, Takano Y. Mechanotransduction in bone: Role of strain rate. *Am J Physiol* 269: E438-E442, 1995.
162. Turner CH, Takano Y, Owan I. Aging changes mechanical loading thresholds for bone formation in rats. *J Bone Miner Res* 10: 1544-1549, 1995.
163. Uusi-Rasi K, Kannus P, Cheng S, Sievänen H, Pasanen M, Heinonen A, Nenonen A, Halleen J, Fuerst T, Genant H, Vuori I. Effect of alendronate and exercise on bone and physical performance of postmenopausal women: A randomized controlled trial. *Bone* 33: 132-143, 2003.
164. Uusi-Rasi K, Sievänen H, Pasanen M, Beck TJ, Kannus P. Influence of calcium intake and physical activity on proximal femur bone mass and structure among pre- and postmenopausal women. A 10-year prospective study. *Calcif Tissue Int* 82: 171-181, 2008.
165. Van Rietbergen B, Weinans H, Huiskes R, Odgaard A. A new method to determine trabecular bone elastic properties and loading using micromechanical finite-element models. *J Biomech* 28: 69-81, 1995.

166. Vico L, Collet P, Guignandon A, Lafage-Proust MH, Thomas T, Rehaillia M, Alexandre C. Effects of long-term microgravity exposure on cancellous and cortical weight-bearing bones of cosmonauts. *Lancet* 355: 1607-1611, 2000.

167. Vogel JM, Whittle MW. Bone mineral changes: The second manned Skylab mission. *Aviat Space Environ Med* 47: 396-400, 1976.

168. Ward K, Alsop C, Caulton J, Rubin C, Adams J, Mughal Z. Low magnitude mechanical loading is osteogenic in children with disabling conditions. *J Bone Miner Res* 19: 360-369, 2004.

169. Ward KA, Roberts SA, Adams JE, Lanham-New S, Mughal MZ. Calcium supplementation and weight bearing physical activity—Do they have a combined effect on the bone density of pre-pubertal children? *Bone* 41: 496-504, 2007.

170. Weinbaum S, Cowin SC, Zeng Y. A model for the excitation of osteocytes by mechanical loading-induced bone fluid shear stresses. *J Biomech* 27: 339-360, 1994.

171. Whiting WC, Zernicke RF. *Biomechanics of Musculoskeletal Injury*. Champaign, IL: Human Kinetics, 2008.

172. Williams MH. Dietary supplements and sports performance: Minerals. *J Int Soc Sports Nutr* 2: 43-49, 2005.

173. Wolff J. *Das Gesetz der Transformation der Knochen*. Berlin: Hirschwald, 1892. Translated by Manquet P, Furlong R as *The Law of Bone Remodeling*. Springer: Berlin, 1986.

174. Yeh CK, Rodan GA. Tensile forces enhance prostaglandin E synthesis in osteoblastic cells grown on collagen ribbons. *Calcif Tissue Int* 36(Suppl 1): S67-S71, 1984.

175. Young AJ, McAllister DR. Evaluation and treatment of tibial stress fractures. *Clin Sports Med* 25: 117-128, 2006.

176. Zeng Y, Cowin SC, Weinbaum S. A fiber matrix model for fluid flow and streaming potentials in the canaliculi of an osteon. *Ann Biomed Eng* 22: 280-292, 1994.

177. Zernicke RF, Salem GJ, Barnard RJ, Schramm E. Long-term, high-fat-sucrose diet alters rat femoral neck and vertebral morphology, bone mineral content, and mechanical properties. *Bone* 16: 25-31, 1995.

178. Zernicke RF, Wohl GR, LaMothe JM. The skeletal-articular system. In: Tipton CM, ed. *ACSM's Advanced Exercise Physiology*. Philadelphia: Lippincott Williams & Wilkins, 2006, 95-111.

179. Zerwekh JE, Ruml LA, Gottschalk F, Pak CY. The effects of twelve weeks of bed rest on bone histology, biochemical markers of bone turnover, and calcium homeostasis in eleven normal subjects. *J Bone Miner Res* 13: 1594-1601, 1998.

180. Zimmerman AM. *High Pressure Effects on Cellular Processes*. New York: Academic Press, 1970, 324.

Index

Page numbers followed by an *f* or a *t* indicate a figure or table, respectively.

About the Editor

Courtesy of Charles Tipton.

Charles M. Tipton, PhD, is an active emeritus professor of physiology at the University of Arizona. He received a PhD in physiology from the University of Illinois in 1962. He retired after 35 years of directing exercise physiology laboratories that investigated physiological mechanisms associated with the effects of acute and chronic exercise. He is recognized as a leading authority of exercise physiology.

Professor Tipton taught physiology and exercise physiology courses to undergraduate, graduate, medical, and professional students at the University of Iowa and the University of Arizona and mentored 21 PhD students at these locations. He has written, coauthored, or edited 6 books, 33 chapters and proceedings, and approximately 180 articles. In addition, he served as editor in chief of Medicine and Science in Sports and Exercise and was an associate editor of the Journal of Applied Physiology for nearly a decade. He has been both member and chair of select National Institutes of Health (NIH) study sections and of several American College of Sports Medicine (ACSM) research committees. A past president of ACSM, Professor Tipton has been appointed to microgravity advisory committees that include the NASA Review Panel on Space Medicine and Countermeasures, the External Advisory Committee for the National Space Biomedical Research Institute (NSBRI), and the Congress-directed National Research Council Steering Committee on Recapturing a Future for Space Exploration. For his research and professional endeavors, he received Honor Awards from ACSM and from the Environmental and Exercise Physiology Section of the American Physiological Society. Recently, Fellow Tipton received the Clark W. Hetherington Award from the National Kinesiology Academy.